Energy and Empire

A biographical study of Lord Kelvin

Lord Kelvin caricatured by *Vanity Fair Album* in 1897. Classing him as one of its 'Statesmen', *Vanity Fair* published the accompanying eulogy of the seventy-two-year-old peer.

LORD KELVIN

His father was Professor of Mathematics at Glasgow. Himself was born in Belfast seventy-two years ago, and educated at Glasgow University and at St Peter's, Cambridge; of which College, after making himself Second Wrangler and Smith's Prizeman, he was made a Fellow. Unlike a Scotchman, he presently returned to Glasgow – as Professor of Natural Philosophy; and since then he has invented so much and, despite his mathematical knowledge, has done so much good, that his name – which is William Thomson – is known not only throughout the civilised world but also on every sea. For when he was a mere knight he invented Sir William Thomson's mariner's compass as well as a navigational sounding machine, that is, unhappily, less well known. He has also done much electrical service at sea: as engineer for various Atlantic cables, as inventor of the mirror-galvanometer and siphon recorder, and much else that is not only scientific but useful. He is so good a man, indeed, that four years ago he was ennobled as Baron Kelvin of Largs; yet he is still full of wisdom, for his Peerage has not spoiled him.

He has been President of the Royal Society once, and of the Royal Society of Edinburgh three times. He has been honoured by nine universities – from Oxford to Bologna; he is the modest wearer of German, Belgian, French, and Italian Orders; and he has been twice married. He knows all there is to know about Heat, all that is yet known about Magnetism, and all that he can find out about Electricity. He is a very great, honest, and humble Scientist who has written much and done more.

With all his greatness, he remains a very modest man of very charming manner.

[*Vanity Fair Album* (1897)]

Energy and Empire

A biographical study of Lord Kelvin

Crosbie Smith
Lecturer,
History of Science Unit,
University of Kent at Canterbury

and

M. Norton Wise
Professor, Department of History,
University of California,
Los Angeles

Any one may commence the writing of a book,
but what may be its extent, or the time of its completion,
no one can tell . . .

Dr James to William Thomson, 9th April, 1843.

CAMBRIDGE UNIVERSITY PRESS
Cambridge
New York Port Chester
Melbourne Sydney

Published by the Press Syndicate of the University of Cambridge
The Pitt Building, Trumpington Street, Cambridge CB2 1RP
40 West 20th Street, New York, NY 10011, USA
10 Stamford Road, Oakleigh, Melbourne 3166, Australia

First published 1989

Printed in Great Britain at the University Press, Cambridge

British Library cataloguing in publication data

Smith, Crosbie W.
Energy and empire
1. Physics. Kelvin, William Thomson,
Baron, 1824–1907. Biographies
I. Title II. Wise, M. Norton
530′.092′4

Library of Congress cataloguing in publication data

Smith, Crosbie.
Energy and empire.
Bibliography: p.
Includes index.
1. Kelvin, William Thomson, Baron, 1824–1907.
2. Physics–History. 3. Physicists–Great Britain–
Biography. I. Wise, M. Norton, II. Title.
QC16.K3S65 1989 530′.092′4 [B] 88-25685

ISBN 0 521 26173 2

Printed and Bound in Great Britain by
Redwood Burn Limited, Trowbridge, Wiltshire

SE

Contents

Illustrations

Preface

Addressing the British Association for the Advancement of Science from the presidential chair in 1871, Sir William Thomson expressed his enthusiasm for intellectual capitalism:

> Scientific wealth tends to accumulation according to the law of compound interest. Every addition to knowledge of properties of matter supplies the naturalist with new instrumental means for discovering and interpreting phenomena of nature, which in their turn afford foundations for fresh generalizations, bringing gains of permanent value into the great storehouse of [natural] philosophy.[1]

Literally as well as metaphorically, scientific knowledge meant wealth for Thomson. In his pioneering Glasgow physical laboratory he had united the pursuit of scientific and industrial wealth through precision measurements on the properties of matter, measurements which he judged equally fundamental to science-based industry and to science itself.

Twentieth-century commentators have not been generally sympathetic to William Thomson's industrial vision, which is so closely associated with his elevation to Lord Kelvin. In his Royal Society obituary notice, the Lucasian professor of mathematics at Cambridge, Sir Joseph Larmor, regarded Kelvin's commercial interests as distractions from the proper intellectual concerns of a natural philosopher, most notably the constitution of matter and ether.[2] Larmor's remarks reflect a traditional Oxbridge disdain for economic man, a disdain also manifest in the character of much twentieth-century British physics, as dominated by the Cavendish Laboratory in Cambridge.

An even less sympathetic, but better known, commentary on the Kelvin style is that of Pierre Duhem. Duhem found the evil of the industrial spirit pervading textbooks of British physics: 'It has penetrated everywhere,' he said, 'propagated by the hatreds and prejudices of the multitude of people who confuse science with industry, who, seeing the dusty, smoky, and smelly automobile, regard it as the triumphal chariot of the human spirit.' Sir William Thomson's Baltimore Lectures supplied Duhem with numerous illustrations of the factory mentality of British physics, epitomized, as he saw it, in the use of mechanical models.[3]

[1] PL, 2, 175–6. [2] JL, iii, liv.

[3] Pierre Duhem, *The aim and structure of physical theory*, P.P. Wiener (ed.) (New York, 1962), pp. 71, 93. Originally published in 1906 from articles of 1904–5.

J.G. Crowther also saw in Kelvin's commercial interests the contamination of his science, which he severely criticized in comparison to the purer pursuits of James Clerk Maxwell:

> He [Kelvin] was the leading symbol of the scientific ideology of the British nineteenth-century governing class . . . He was an intellectual colossus who saw one-half of the aspects of material nature with unsurpassed clarity and power, but was blind to the other half. His great personality was an expression, in the realm of ideas, of the power and of the blindness of capitalism.[4]

Crowther maintained that physical nature is there to be discovered, finally and objectively, but that Kelvin's capitalist blinkers prevented him from grasping its deeper truths, despite his great scientific powers.

These three commentators approached Lord Kelvin's work from very different cultural and social perspectives: Larmor as elite Cambridge mathematician, heir to Isaac Newton, and staunch upholder of the political unity of Britain and Empire; Duhem as conservative French Catholic and apologetic defender of orthodox Christianity; Crowther as early twentieth-century Marxist, hostile to nineteenth-century capitalism but committed to knowledge as power. Nevertheless, in their common recognition of Kelvin's industrial orientation, his critics offered perceptive insights into the character of his activities.

To recognize that William Thomson related his physics to a vision of industry and empire, however, does not engage the question of how his science expressed that vision. In this biographical study our primary object has been not simply to recount the ideology and activities of a Victorian scientific entrepreneur, but rather to analyze the practices through which that ideology was realized in Thomson's best work, in his instruments and patents certainly, but also in his mathematical physics. By contrast to Larmor, who regarded Thomson's ventures as distractions from his science, we show that the scientific and industrial pursuits were essential to one another. It was therefore not merely a case of science applied to industry, but of industry applied to science, for Thomson's industrial vision thoroughly permeated his understanding of the natural world and the theoretical and experimental research which he pursued.

Again, Thomson's approach was not merely a matter of metaphysical predisposition, but of the making and doing of mechanics, thermodynamics, and electromagnetism, of problem solutions, models, measurements, and absolute values. We do not argue, however, that Thomson's social context determined the content of his science. Rather, we show that he drew extensively on conceptual and material resources available in his industrial culture and, with motivations structured by that culture, arrived at rational explanations of physical phenomena and at means of controlling those phenomena. Thus we

[4] J.G. Crowther, *British scientists of the nineteenth century* (London, 1935), p. 256.

seek to show, concretely and in detail, how the science that Thomson produced was inseparably integrated with the industrial culture that he represented.

Of overriding importance to his intellectual achievement were particular objects of the material culture. The steam-engine, electric telegraph, and vortex turbine, which he drew upon for model solutions of the most fundamental problems of natural philosophy, were constitutive of his thermodynamics, electrodynamics, and cosmology. And just as such artefacts served Thomson as generators of scientific theory, they also served him as guarantors of the reality, or validity, of theoretical entities. Machines, in short, represented the industrial culture to natural philosophy and natural philosophy to the industrial culture. By embodying the idea of work, for example, the steam-engine translated between the fundamental concept of labour value in political economy and the fundamental concept of energy in natural philosophy.

In analyzing Thomson's life and work we have employed a set of labels for interrelated aspects of his political, religious, and scientific ideology. The most general term is 'whig'. In the political sense it refers not to the Whig party as such, but to a liberal, progressive, reforming stance imagined to be above party affiliation. Its religious correlate is 'latitudinarian', connoting an anti-sectarian (if also anti-Catholic and anti-fundamentalist) point of view, and opposed to nothing so much as to perceived authority and dogma in theological matters.

In scientific methodology, the labels are 'non-hypothetical' and 'anti-metaphysical'. As in the political and religious spheres, these words must not be understood to mean that Thomson succeeded in developing a science without hypotheses and without metaphysics. Metaphysics, for him, carried the odour of idealism and *a priori* knowledge, while hypotheses lacked what he regarded as direct empirical evidence. His proscription of both sources of error expressed his view that truth in natural philosophy is based on direct sensory perception rather than mental construction.

Finally, and most ubiquitously, we use the term 'practical'. Thomson himself employed it from the beginning to the end of his career. Though covering a wide spectrum of meanings, it suggests the all-pervasive spirit of engineering and industry that we elaborate in his life and work.

All of these terms together – whig, latitudinarian, non-hypothetical, anti-metaphysical, and practical – label Thomson's commitments rather than any fully articulated or particularly consistent philosophy. They evoke the motivations which operated, sometimes implicitly and sometimes explicitly, in his everyday life. As such we use them, leaving their definitions somewhat elastic, subject to expansion and contraction in the specific contextual instances of their application. In Chapter 13, however, under 'The methodology of look and see' and 'Mephistopheles' we have attempted to articulate more fully the differences between metaphysical ideality, which he despised, and practical reality, which he worshipped.

A problem of constant concern to us has been the fact that, although the major Kelvin archives in Cambridge and Glasgow contain many thousands of letters, very few of those letters reveal William Thomson's innermost thoughts or his emotional responses to major crises, such as the death of his father. His relation to his first wife, who was ill virtually from their marriage to her death over seventeen years later, remains obscure. Thus we have had to choose between psychological speculation on meagre evidence and restricting Thomson's personality largely to its professional and public side. We have chosen the public side, first because there is so much of it and because of its historical importance, and secondly because we suspect that many sources of a personal nature have been destroyed, leaving a distorted sample. From William's Cambridge diary of 1843, for example, small and large pieces have been excised, while his brother James's surviving papers in Belfast contain no family correspondence. In dramatic contrast, a few letters from William in a separate regional archive offer a glimpse of an intense but unrequited love, prior to his marriage. These letters are unlike anything in the 'official' archives, suggesting that the latter conceal what they do not reveal. The truly private Thomson must remain a veiled figure.

Thomson's career covers a wide variety of subjects, each with a coherence of its own and each carrying an appeal to a different group of readers. We have therefore arranged our four-part study thematically rather than chronologically. In Part I we locate the young William and his close-knit family within their local context in the 1830s and 1840s. Pivotal is the dynamic city of Glasgow, destined to become, with its growing reputation for reliable and economic steamships, the shipyard of the Empire and its Second City during his maturer years. We explore Thomson's political, religious, and economic context in relation to the first major phase of his career, which saw him advance from Glasgow University student and Cambridge undergraduate to Cambridge Fellow and ultimately to Glasgow professor at the early age of twenty-two.

Part II concerns the development of Thomson's mathematical physics. Here we aim at portraying, through the activities of a single individual, one of the major 'revolutions' in the history of science. From 1840 to 1870 Thomson played a starring role in the birth-dramas of electromagnetic field theory and thermodynamics, as well as in the transformation of mechanics, such that classical physics could no longer bear the epithet 'Newtonian'. Energy and extremum principles replaced force as the foundation of dynamical explanation; temporal development became as essential to physical systems as to biological ones; and for the basis of material reality, British natural philosophers looked to structures of motion in a universal plenum, rather than to hard atoms. Thomson sometimes imagined himself the modern Newton. We attempt to show why, giving as much conceptual content as possible. A modicum of mathematics is required. Later parts of our study can nevertheless be read without extensive reference to this one.

Thomson's construction of a system of the world is the subject of Part III. Concerned here with his cosmological and geological theories, we discuss them less in terms of their general scientific and popular impact through the question of the earth's age – the focus of previous studies – and more in relation to debates with his colleagues in natural philosophy, most notably G.G. Stokes and Hermann von Helmholtz. Drawing heavily on engineering systems, Thomson constructed his economy of nature in accordance with the laws of energy. These laws ruled the material world. But he also devoted much thought to the nature of mind, to free-will, and to the relation of mind to matter, as well as to the origin of life on earth. An analysis of the latter problems concludes Part III.

Part IV takes us from a study of Thomson's involvement in the grandest of all Victorian engineering projects, the Atlantic telegraph cable which earned him his knighthood, through his profitable business and industrial activities, to his elevation to the British peerage as Baron Kelvin of Largs. Part I located him within a developing Scottish city of heavy industry. Now we see the mature professor and entrepreneur reaping the social prestige and economic wealth of a physics perceived to serve the needs of all Britain. It was a physics which offered qualitative improvements to the new telegraphic and maritime communications so fundamental to the physical unity and political identity of the Empire on which the sun never set.

This book originated in a manuscript written by Crosbie Smith some ten years ago. At his invitation Norton Wise joined the project in 1980. In the drafting and redrafting that has gone on since then our individual contributions have become highly mixed. Although primary responsibility for Parts I, III, and IV (Thomson's early years, his system of the world, and his entrepreneurial activities) has rested with Crosbie Smith, and for Part II (Thomson's mathematical physics) with Norton Wise, many chapters (4, 5, 9, 10, 18, 19), and numerous sections of other chapters, are fully coauthored. Above all, the book's unifying themes are the product of frequent and lively discussion between us, leading to a full integration of our separate ideas and original expertise.

For permission to use unpublished material, we are especially indebted to Cambridge University Library, where the main part of the Kelvin archive is deposited. The staff of the Manuscripts Room have been always helpful over a long period of research. Our special thanks go also to Dr Nigel Thorp and his colleagues in the Special Collections Department of Glasgow University Library, which holds substantial amounts of Kelvin material. We are further indebted to Rex Whitehead and to the late J.T. Lloyd in the Department of Natural Philosophy at Glasgow for their personal interest and for such invaluable services as directing us to Thomson's annotated copy of Maxwell's

Treatise on electricity and magnetism. The staff of Queen's University Library, Belfast, have been most helpful in providing access to the James Thomson papers. We thank the Strathclyde Regional Archives for the use of Archibald Smith's family papers, St Andrew's University Library for J.D. Forbes's correspondence, and the Salisbury Collections, Hatfield, as well as the Devonshire Collections, Chatsworth, for important correspondence cited in our final chapter.

Among very many colleagues who have supported us in our task, we are especially grateful to David Gooding and Geoffrey Cantor for critically reviewing the entire manuscript. We are similarly indebted to Nancy Cartwright and Robert Westman for their comments on large sections, and to Alex Dolby and Tim Lenoir for critical insights in the course of informal discussions. Norton Wise recognises the continuing value of concrete methodological lessons learned long ago from Tom Kuhn. David Wilson has relieved our research task immeasurably by giving us access to his fully edited typescript of the immense Stokes–Kelvin correspondence. To other colleagues we owe specific intellectual debts for interpretative insights: Simon Schaffer on J.P. Nichol's science of progress (ch. 4); Joan Richards on British algebra (ch. 6); David Gooding on Faraday's field theory (chs. 7 and 8); Menachem Fisch on Whewell's mechanics (ch. 11); Sam Schweber on political economy and optimization conditions in relation to biology and physics (ch. 11); Jed Buchwald and Bruce Hunt on Maxwellian electromagnetic theory (ch. 13); Jack Morrell on John Phillips (ch. 16); Ted Porter on Maxwell's statistical physics (ch. 18); and David Cannadine on the nineteenth-century Devonshires (ch. 23). On the ideas of all of these scholars our own work is built.

Financial support has been forthcoming for Crosbie Smith from the Royal Society of London, the Nuffield Foundation, and the Unit for the History of Science at the University of Kent, Canterbury. Without the generous hospitality of David Cannadine he would have found researches in the Kelvin archive in Cambridge immensely more difficult. Several agencies have supplied Norton Wise with travel funds and research assistance: the National Science Foundation, the American Council of Learned Societies, and the Academic Senate of the University of California at Los Angeles. Time, the most elusive commodity, came from fellowships at the Center for Interdisciplinary Research of the University of Bielefeld (special thanks to Lorenz Krüger), the Physics Department at the University of Pavia (Fabio Bevilacqua), the Edelstein Center at the Hebrew University in Jerusalem (Tim Lenoir), and the Institute for Advanced Study in Berlin.

The efficient typing of this book, as well as transcriptions of a large quantity of archival material, is due to the dedicated and good-humoured efforts of Veronica Craig-Mair, Yvonne Procter, and Veronica Ansley. Our thanks must also go to William Ginn for invaluable help with proofs and index; to the University of Kent Photographic Unit for preparation of many of the illustrations; to Irene Pizzie for efficient sub-editing; and not least to Simon Capelin for his patience, enthusiasm, and advice.

Crosbie Smith acknowledges a debt for the enduring support of his family, and in particular for that of his late father who introduced him at a very early age to the beautiful Firth of Clyde around Largs and to the fascinating city of Glasgow while it was still the Second City of the Empire. Without such youthful inspiration, he would scarcely have undertaken this project.

Norton Wise acknowledges his deepest debt and appreciation to Elaine Wise for continual discussion of theses, arguments, style, and methodology over many years, as well as for unswerving psychological support. (To her simultaneous exercise of word processing and culinary skills he dedicates his royalties.) For their patience and for the time they have given up, he thanks Erin and Licia Wise.

Crosbie Smith
Canterbury

M. Norton Wise
Berlin

Footnote abbreviations

BAAS Report *Report of the British Association for the Advancement of Science.*

BL William Thomson, *Notes of lectures on molecular dynamics and the wave theory of light* (Baltimore, 1884).

DNB *Dictionary of national biography*

DSB *Dictionary of scientific biography*, C.C. Gillispie (ed.) (16 vols., New York, 1970–80).

E&M William Thomson, *Reprint of papers on electrostatics and magnetism* (London, 1872).

James Thomson, *Papers* James Thomson, *Collected papers in physics and engineering*, Sir Joseph Larmor and James Thomson (eds.) (Cambridge, 1912).

JL Sir Joseph Larmor, 'William Thomson, Baron Kelvin of Largs. 1824–1907' (obituary notice), *Proc. Royal Soc.*, [series A], 81 (1908), i–lxxvi.

MPP William Thomson, *Mathematical and physical papers* (6 vols., Cambridge, 1882–1911).

MSP W.J.M. Rankine, *Miscellaneous scientific papers*, W.J. Millar (ed.), (London, 1881).

PL William Thomson, *Popular lectures and addresses* (3 vols., London, 1889–94).

QUB Thomson papers, Queen's University Library, Belfast.

SPT S.P. Thompson, *The Life of William Thomson, Baron Kelvin of Largs* (2 vols., London, 1910).

ULC Kelvin Collection, Add. MS 7342, University Library, Cambridge. Kelvin letters in the Stokes Collection, Add. MS 7656, University Library, Cambridge are listed as Stokes correspondence, ULC.

ULG Kelvin Collection, University Library, Glasgow.

I

The making of the natural philosopher

1

From the ashes of revolution

After a walk of about a mile and a half, a considerable part of which lay in the grounds of Lord Moira, we entered the camp of that body of men who were to sever Ireland from the dominion of Britain and to give her a separate existence, and a name among the nations – who were to give liberty and equality to their countrymen – to abolish tithes and taxes – in a word, to make Ireland, at least, as happy as the United States and the French Republic were considered, in the ardent conceptions of the republicans of the day. *James Thomson, 'The Battle of Ballynahinch [1798]'*[1]

Lord Kelvin, revered and respected statesman of science in the golden age of late nineteenth-century British Imperialism, began life not in Great Britain, but in Ireland. Addressing a Birmingham audience in 1883, he spoke humorously of the Irishman's seventh sense as common sense, believing 'that the possession of that virtue by my countrymen – I speak as an Irishman – . . . will do more to alleviate the woes of Ireland, than even the removal of the "melancholy ocean" which surrounds its shores'.[2] Less than a decade later, Sir William Thomson's elevation to the peerage as Baron Kelvin of Largs symbolized the social summit of a remarkable life lived in the context of Victorian Britain. Yet that ultimate acclaim did not flow from scientific and technical achievement alone, but also from his direct involvement in the political cause of Liberal Unionism during the 1880s. That involvement derived from the Irish context into which he had been born, a context of cultural and social liberalism upon which his enduring personal values were founded. It is thus that in beginning our story we turn to an age and a country radically different from that of late Victorian Britain, to an age which shaped the resolute and resourceful character of Lord Kelvin's father, James Thomson (1786–1849).

No definitive study of Lord Kelvin can ignore the profound significance of James Thomson for his son's social values and cultural beliefs. James Thomson, by the power of his convictions and strength of his character, played the major role in the making of the young natural philosopher. This chapter is therefore

[1] [James Thomson], 'Recollections of the Battle of Ballynahinch', *Belfast Mag.*, **1** (1825), p. 57. Eighteenth-century writers refer to Ballinahinch or Ballinehinch; nineteenth- and twentieth-century spelling holds to Ballynahinch. [2] PL, **1**, 254–5.

devoted to an analysis of the sources and development of James Thomson's beliefs and values from the momentous historical events of the 1798 rebellion (which, in the North of Ireland, explicitly embodied liberal presbyterian values) to his removal to the Glasgow College chair of mathematics as a widower with young family in the cholera year of 1832. By so doing, we provide a study of the deepest roots of Lord Kelvin's own convictions which were to permeate his scientific, religious, and public life alike.

Throughout the nineteenth century and beyond, Ireland and her social, cultural, and economic problems seemed forever set to threaten the stability of Britain and the Empire. No other single problem, by the magnitude and range of its implications, dominated for so long British political life in the late nineteenth and early twentieth centuries. The Home Rule battles – which ended with the partition of Ireland in 1920 – represented only the latest of a succession of political crises extending from the failure of revolution in 1798 through the tragic famines of the 1840s. In this period, then, liberal presbyterians no longer aimed to protect their values of civil and religious liberty by the separation of Ireland from the dominion of Britain as in 1798, but by the development of a United Kingdom of Great Britain and Ireland as a guarantee of those values against the nationalistic, Catholic values of many of their fellow Irishmen.

Ireland in rebellion: 1798

In the time of the French Revolution, and for many years before, Ballynahinch had little more distinction than any other Irish market town. Lying land-locked at the very centre of the County of Down, it was the meeting place for roads leading from market towns in the west, and from Belfast in the north, to ancient Downpatrick, and from there to the eastern coasts of Down. For the most part, these coasts, like those of neighbouring Antrim, faced the south-west shores of Scotland across twenty or so miles of turbulent sea. The waters of the North Channel and the Irish Sea, with their tidal streams flowing swiftly to and from the Atlantic Ocean and with their frequent winter gales, formed a natural barrier, which, however, was never enough to sever the strong ties of trade, culture, and religion for so long existing between the North of Ireland, or Ulster, and Scotland.

Ballynahinch was physically far removed from the centres of European civilization and of early British industrialization. Eighteenth-century travellers complained of the broken and narrow causeways which were the roads leading to it through rough and untamed countryside, and it obtained on this account the name of Magheredroll, or the field of difficulties. Yet in 1744 one observant visitor wrote that 'the Vallies and Sides of the Hills produce Oats and Flax in plenty, and the morass Grounds seldom fail of yielding a full Crop of Rye . . .

The staple Commodity . . . is linen yarn, which is sold at little Fairs here . . .'.[3] And so, unlike much of Ireland, it possessed economic prosperity of a modest kind. A temperate climate, combined with reasonable soils, had been exploited to the full by the seventeenth-century settlers, who for the most part had come from south-western Scotland. One such settler near Ballynahinch around 1641 was John Thomson.

Apart from the harsh slopes of Slieve Croob to the south-west of the town, the landscape comprised innumerable small hills, low whale-backed ridges of glacial boulder clay with their long axis parallel to the direction of ice movement. The southern slopes of these drumlins made them attractive for pasture and cultivation, and so they became the well-drained sites of relatively prosperous farmhouses. In 1786, at a farmhouse on the slopes of a hill known as Annaghmore, about two miles to the south of Ballynahinch, John Thomson's great-great-grandson James Thomson was born. This area had rather more distinction than Ballynahinch itself, being in possession, as our eighteenth-century traveller expressed it, of an 'excellent chalybeo–sulphureous Spa' to which many troubled pilgrims came to benefit from 'a very clear Water, and Withal very cold, of very disagreeable Taste and Smell . . . The Quantity of this Water commonly taken is from three Pints to three Quarts; some it vomits, others it purges . . .'.[4] Yet, despite unpleasant waters, the Spa was a fashionable meeting place for worthy citizens of the North of Ireland and being at the foot of Annaghmore may indeed have provided James Thomson with an early knowledge of the world beyond the confines of the farm. While we may marvel at the way in which a farm labourer – the youngest by some ten years of three sons – taught himself mathematics and eventually became a distinguished professor at Glasgow College, we need more especially to understand his attitudes to education, politics and religion, views which remained as major guiding principles throughout his life and the lives of all his children.

The complex problems which beset Ireland in the eighteenth century may be summarized in terms of three distinct religious denominations, each representing powerful social interests. The Irish Church, the state-established equivalent to the Church of England, was frequently associated with the Anglo–Irish and English landowners of Ireland. In theory, its members alone had full political rights and privileges. By contrast, the Irish Roman Catholics consisted of the great mass of native Irish, the vast numbers of peasants surviving as tenants in a subsistence economy throughout the island. In numbers only had the Catholics power and superiority: most other rights had been removed by the penal laws of

[3] [Walter Harris], *The antient and present state of the County of Down* (Dublin, 1744), pp. 76–8. Harris's classic account was written 'to counter misrepresentations' that 'the People are uncivilized, rude, barbarous . . .', that 'they count it no Infamy to commit Robberies, and that Violence and Murder are, in their Opinions, no way displeasing to God . . .' and that 'Wolves still abound too much in this country . . .'. [4] Harris, p. 176.

the late seventeenth century. The third group was that of the protestant dissenters who, unlike their English counterparts, were not thinly distributed over the land, but were a powerful and cohesive political force, organized through the well-defined structure of presbyterianism. Concentrated in the North of Ireland, the presbyterians looked to Scotland as the source of their culture and church. During troubled times, fear of Catholicism made them the natural allies of the established Irish Church, yet in more normal times the Irish Church felt threatened by the power of these dissenters and refused to grant them legal toleration. For their part, the presbyterians were probably powerful enough in practice not to need the toleration. Though essentially tenants, Ulster's yeomen farmers differed from the Catholic peasantry in so far as by custom (rather than by law) they were deemed to possess a saleable interest in their holdings, a feature which provided them with comparative prosperity and security.[5]

Emigration of dissenters to North America had begun on a small scale as early as 1717. In the mid-1770s, however, some 30 000 protestants left the North of Ireland in the space of two or three years. Many of them claimed that they emigrated to escape persecution by the established Church. But economic conditions – restrictions on Irish trade, high rents, and poor harvests – almost certainly prompted the first large-scale emigrations from Ulster. Thereafter the image of a land free from religious intolerance must have been a strong attraction for the dissenters, even if intolerance in Ireland and toleration in the New World were not as great in reality as in the minds of the presbyterian emigrants.[6]

The American War of Independence of 1776 helped to focus the issues still further. The Ulster presbyterians were identified, in the Irish government's eyes, with the rebellious element in America. Some presbyterians, moreover, were known to possess dangerously republican principles. By 1778, the presbyterians strongly supported volunteer companies, ostensibly against the threat of invasion. Demand for full toleration now had the backing of force, although the primary interest was free trade for Ireland. Religious toleration, so long existing in practice, became a legal right by 1780, and alongside it went the claims of a largely presbyterian mercantile class which found expression through the volunteer movement. The danger to the established interests of church and state was far from over by the 1780s, however, for, added to the inspiration of the American War of Independence, came the French Revolution in 1789. The movement for reform in Ireland began thereafter to threaten a political and social revolution.[7]

[5] J.C. Beckett, *Protestant dissent in Ireland, 1687–1780* (London, 1948), esp. pp. 13–19; *The making of modern Ireland, 1603–1923* (London, 1966), pp. 179–81.

[6] *Ibid.*, pp. 88–90; see also R.J. Dickson, *Ulster emigration to Colonial America, 1718–1775* (London, 1966).

[7] Beckett, *Modern Ireland*, esp. pp. 246–67, provides an excellent account of this period in Irish history.

Enthusiasm for the French Revolution was strong in Ulster. In 1792, many Ulstermen subscribed to assist the French in repelling invasion, and Belfast Volunteers celebrated the end of the monarchy and the establishment of a French Republic. The volunteer movement progressively slipped from the control of the landed gentry and became more and more radical. In 1791 the Society of United Irishmen had been founded to bring together Irishmen of all creeds, to establish complete religious equality, and to demand radical reform of parliament. Among those in the forefront of such activities were the famous Theobald Wolfe Tone, who was to lead attempted French invasions of Ireland in 1796 and 1798, and William Drennan, a medical doctor and the author of the Society's test – with strong echoes of French revolutionary principles – who later became an influential friend of James Thomson. Although Drennan ended his connection with the United Irishmen in 1794 after being tried and acquitted for 'wicked and seditious libel', he retained to the end of his life a firm belief in those principles which the United Irishmen failed to achieve for their country when they took up arms against the Irish government in 1798.[8]

James Thomson was twelve years old when the 'rebellion' or 'revolution' broke out in the North of Ireland. He was witness to the most traumatic event ever to take place in his town – the so-called Battle of Ballynahinch. In his vivid account, written in 1825 for the *Belfast Magazine*, he recollected the arrival of the rebel army on a hill to one side of the town and the visit of his family to supply provisions for the insurgents' needs. The visitors were shown the primitive weapons – mostly pikes – and the symbols of liberty and freedom, before fleeing for safety to an adjoining hill. During the evening two bodies of the King's forces – one from Downpatrick and the other from Belfast – approached, burning farmhouses indiscriminately to induce terror. Plagued by desertions, internal divisions, and inadequate training and weapons, the insurgents were easily routed by the following morning, but not before the King's forces had set fire to Ballynahinch and 'in a short time, a great proportion of the best houses in it were enveloped in flames, and hastening to inevitable destruction'.[9] Of a hundred houses in the town before the battle, sixty-three had been gutted by fire at the end, and most of the remainder were wrecked and looted. A prosperous market town had been transformed overnight into one of ruin, poverty and decay.[10]

Throughout Ireland in 1798 the defeat of the United Irishmen followed a similar pattern, and, in the aftermath, the Act of Union between Great Britain

[8] See William Drennan, *The Drennan letters being a selection from the correspondence which passed between William Drennan M.D. and his brother-in-law and sister Samuel and Martha McTier during the years 1776–1819*, D.A. Chart (ed.), (Belfast, 1931), p. viii. For a popular account of the United Irishmen, see Thomas Pakenham, *The year of liberty. The story of the great Irish rebellion of 1798* (London, 1969). For an interesting parallel to James Thomson, see T.L. Hankins, *Sir William Rowan Hamilton* (Baltimore and London, 1980), pp. 3–19.

[9] James Thomson, 'Recollections', pp. 56–64, esp. p. 62. See also Pakenham, pp. 246–64.

[10] S. McCullough, *Ballynahinch: centre of Down* (Belfast, 1968), p. 83.

and Ireland was drawn up, to take effect from 1st January, 1801. The scene was thereby set for Anglo–Irish relations throughout the entire nineteenth century. Never again would the presbyterians of Ulster take up arms on behalf of a United Ireland. Immense social and economic changes occurred during the time – up to 1920 – that the political scene was concentrated in London. The North of Ireland became, in the nineteenth century, almost a twin sister of Scotland's Clydeside – one of the great workshops of the British Empire[11] – while the remainder of the island continued in a state of commercial and industrial backwardness.

James Thomson's account of the battle emphasized the futility of the rebellion, while at the same time expressing an implicit sympathy with the ideals of the rebels – the ideals of liberty, equality, and freedom from sectarian dogmatism – and an explicit hatred of the repression and atrocities perpetrated by the establishment forces. Again and again we find these themes appearing in his life, themes which undoubtedly had their origins in the European and American movements for 'enlightenment' in the eighteenth century, and which were focussed upon his very doorstep by the Battle of Ballynahinch. A minor skirmish though the battle was, it was an expression of many of the tensions which had for long been developing in Ireland.

James Thomson's formal education began around the year 1800, at which time the Rev Samuel Edgar purchased a house and farm at nearby Ballykine for the purpose of preparing young men for the ministry and other professions. Edgar (later professor of theology at the Belfast Academical Institution) was minister of the secession church in Ballynahinch, a breakaway from the main presbyterian church in Ulster. At Edgar's school James Thomson studied mathematics and classics and eventually began to assist with the teaching.[12] Around the age of twenty-one 'I was teaching eight hours a day at Dr Edgar's, and during the extra hours – often fagged and comparatively listless – I was reading Greek and Latin to prepare me for entering College, which I did not do till nearly two years after'.[13]

Thomson, having entered Glasgow College in 1810, continued to act as assistant teacher to Edgar during the six month summer vacation, earning enough to see him through each college session from November to May, until he graduated a Master of Arts in 1812. Two more sessions of study at Glasgow College followed in which he took both medical and theological courses with the intention of becoming a presbyterian minister.[14] For an intelligent younger son, unlikely to inherit any property, such a profession was an obvious choice.

[11] See, for example, Beckett, *Modern Ireland*, pp. 268–83; J.C. Beckett and R.E. Glasscock (eds.), *Belfast. The origin and growth of an industrial city* (London, 1967), esp. pp. 67–131; D.J. Owen, *History of Belfast* (Belfast, 1921), pp. 296–305.

[12] SPT, **1**, 3; Elizabeth King, *Lord Kelvin's early home* (London, 1909), pp. 7–11; James Thomson, *Papers*, p. xv; McCullough, p. 83.

[13] Dr James Thomson to William Thomson, 1st July, 1845, T315, ULC.

[14] King, *Early home*, p. 10; James Thomson, *Papers* p. xv.

However, his original aims were redirected by the emergence of a remarkable new educational establishment in the dynamic town of Belfast – the Belfast Academical Institution, in which Thomson taught mathematics for eighteen years.

The Belfast Academical Institution: 1814–32

Between 1780 and 1800, Belfast had been transformed from a market centre for a domestic linen industry into a major commercial town through the introduction of a cotton industry on the factory system. A tough realism in business went hand in hand with the political ideals of tolerance and enlightenment, for Belfast was perhaps the main centre of Irish radicalism in the late eighteenth century. The town not only celebrated in 1791 the anniversary of the Fall of the Bastille and produced an openly radical newspaper, the *Northern Star* (until the militia destroyed the printing press in 1797), but also expressed its desire for 'enlightenment' and 'improvement' through the more restrained channels of an active philosophical society.[15] What Belfast lacked, however, was an educational establishment which could give expression to its liberalism, not to say its radicalism. By 1806, a public meeting sought to create such an institution.[16]

By the following year, the promoters had launched a scheme for a combined school and college, and the very considerable sum of £16 000 was raised by public subscription. Government support was weak, as presbyterians (and Belfast was still dominated by presbyterians of one kind or another) continued to be associated with unsound and often republican principles. Had the new Institution been founded as a sectarian one, the government might have had less to fear, but the Institution was radically non-sectarian, as the Secretary, Joseph Stevenson, stated: 'the Institution is not connected with any religious persuasion ... The subscribers to the Institution are composed of all religious persuasions'.[17] One of the most active and enthusiastic supporters was the former United Irishman, William Drennan. Drennan belonged to the radical tradition, although he had left the United Irishmen before the 1798 rebellion. An ardent supporter of Catholic emancipation and parliamentary reform, he was convinced of the need for union among Irishmen and the abandonment of sectarianism. As he once wrote in characteristic vein:

> His Lordship [Lord Redesdale] complains of the Catholics' want of Christian charity and their excluding the Protestants from Heaven. In return, however, the Protestants exclude the Catholics from the good things of earth ... Indeed, every established church becomes necessarily intolerant and exclusive in doctrine as well as in practice . . .[18]

[15] Beckett and Glasscock (eds.), *Belfast*, pp. 55–87.

[16] John Jamieson, *The history of the Royal Belfast Academical Institution, 1810–1960* (Belfast, 1959), pp. 1–2; T.W. Moody and J.C. Beckett, *Queen's Belfast 1845–1949* (2 vols., London, 1959), **1**, pp. xli–liii.

[17] Jamieson, pp. 11–12.

[18] Drennan, *Letters*, p. 335.

In his opening address in 1814, Drennan defined the aims of the Institution: to diffuse the benefits of education 'both useful and liberal' in Ulster, to provide for the education of children at home rather than Scotland, and to remove religious tests.[19] A full range of teaching was available, for the Institution had appointed specialist masters for the school departments, among them James Thomson, who had in 1812 proposed himself 'as a candidate for the department of teaching of Mathematicks or Natural Philosophy, or both, if they should be taught by the same person'.[20]

James Thomson shared the educational and religious goals of the new Institution. Anti-tory and anti-establishment in politics, Thomson sought the reform and advancement of the human condition through the diffusion of education to the common man. Latitudinarian in religion, he opposed all religious tests. As a professional mathematician, he aimed to advance his subject in a manner free from denominational or dogmatic constraints. And to achieve that end, he emphasized the practical, rather than the 'theoretical' or 'metaphysical', dimensions of his profession. Thus he later wrote in the preface to his *Algebra* of 1844:

> Throughout the work, the Author has carefully kept clear of every thing of a metaphysical or disputed character. With regard to all the practically useful applications and interpretations of algebra, there is no difference of opinion among men of sound science and judgement; and it is only with such matters that the mere learner should have any concern.[21]

A 'practical' approach led to fruit and advancement, while a 'metaphysical' one led to barren controversy.

On the 8th January, 1814, James Thomson accepted the post in the mathematical and arithmetical department in the Belfast Academical Institution.[22] He soon began a characteristic expansion and advancement of his modest empire. He wrote to the Assistant Secretary stating that 'additional desks & forms are immediately necessary in my schoolroom. I also very much need globes & maps [for the teaching of geography]. The globes with which I am at present furnished are unfit for performing except a very few problems'. Early in the following year, 1815, he wrote again to the Board to say that his classroom had become too small. At times he had almost a hundred pupils in a room built for less than half that number. He stressed that he had 'the largest classes & the smallest room', with the consequence that the heat and general discomfort were rapidly becoming intolerable. The quiet, unassuming farmer's son from

[19] Moody and Beckett, *Queen's Belfast*, **1**, pp. xlvi–xlvii.

[20] James Thomson to Joseph Stevenson, SCH 524/3C/1, Public Record Office of Northern Ireland.

[21] James Thomson, *An elementary treatise on algebra, theoretical and practical* (London, 1844), p. vi.

[22] James Thomson to Joseph Stevenson, SCH 524/7B/8/1, Public Record Office of Northern Ireland. He also asked in a postscript to the same letter whether the teaching of geography belonged to his department.

Ballynahinch was proving to be a firm and determined strategist intent on expanding his empire, as well as a competent, thorough, and successful teacher: '. . . It may be asked what remedy I would propose. I know of none but to take seven or eight feet in length from Mr Spence's room [the master of the writing school] . . .'.[23]

Appointed professor of mathematics in the collegiate department – essentially a Scottish university on a small scale – of the Belfast Academical Institution in 1815, Thomson consolidated his position, increased his property both materially and intellectually, and retained both posts until his move to Glasgow in 1832. His college classes in mathematics were frequently the largest of any subject in the Belfast Institution, numbering in 1829 a hundred pupils in the junior mathematics and twenty-six in the senior.[24] He insisted too on full recognition of the equality of his school department with the English and classics schools, and he protested strongly against an attempt by the English master to encroach on the territory of arithmetic and geography.[25]

While in the first quarter of the nineteenth century the North of Ireland did not suffer from the kind of warfare which it had endured in 1798, it was not surprising that the Institution, given its roots, should often have been the subject of political tensions. The initial government grant of £1500 per annum had enabled some professorial salaries to be as high as £150 – to which could be added student fees. However, largely because several masters attended a St Patrick's day banquet in 1816 during which radical toasts were allegedly drunk, the Irish Chief Secretary, Sir Robert Peel, required that the Institution must surrender its independence or lose the grant. The Institution sacrificed the grant, and struggled to remain solvent. So great was his loyalty to the independence and principles of the Institution that Thomson gave up his professorial salary of £50 during a severe financial crisis in 1823, and until affairs improved. The grant was, however, restored by Peel in 1828, and such was the change in relations that the word 'Royal' was added to the title of the Institution in 1831.[26]

The other major tension was with the powerful evangelical wing of the Irish Presbyterian Church, dominated during the 1820s by the fiery Rev Henry Cooke. The dispute was ultimately one of a sectarian versus a non-sectarian education, and illustrates why the Thomsons, during the 1840s, were such enthusiastic supporters of the abolition of Scottish university tests. Irish Presbyterianism had fragmented in the middle and late eighteenth century into an

[23] James Thomson to Robert Simms, 19th August, 1814, SCH 524/7B/8/20; 7th February, 1815, SCH/7B/9/15, Public Record Office of Northern Ireland. Robert Simms had been a founding member of the Society of United Irishmen in 1791, and was Assistant Secretary of the Academical Institution from 1812 to 1843. See Jamieson, p. 210.

[24] Return of students from the Faculty, December 1829, SCH 524/7B/23/46, Public Record Office of Northern Ireland. [25] Jamieson, pp. 15–16, 78.

[26] Moody and Beckett, *Queen's Belfast*, **1**, pp. xlviii–xlix; Jamieson, pp. 16, 23–47. The addition of the word 'Royal' may have been a shrewd move by Peel to ensure the permanent loyalty of an institution whose allegiances had hitherto been suspect.

almost bewildering variety of sects and synods ranging from evangelicals to unitarians. Cooke, a loyal Tory, launched a series of attacks on the Institution as an establishment infected by Arians who, in denying the Trinity, were liable to corrupt that large proportion of college students studying for the ministry, even though the students' final theological education was conducted by professors from the respective presbyterian synods.[27]

Whatever the radical religious and political views of the individual professors, the non-sectarian principles of the Institution were ultimately upheld. As James Thomson stated: 'we rather make a point, lest there should appear to be any distinction or anything unpleasant to the pupils, not to know, unless casually, what the religious principles of the pupils may be'.[28] Nevertheless, sectarianism was regaining ground in the North of Ireland through the first half of the century. This clear drift away from the old liberalism contrasted with Scotland, where the passage of the century tended to lessen, rather than exacerbate, sectarian differences, and made Scotland a more attractive home for the latitudinarian Professor William Thomson.[29]

The remarkable personality of William Drennan was significant not merely as a United Irishman and key founder of the Institution but as a highly esteemed personal friend of James Thomson. How they became acquainted is not certain, though from Drennan's surviving letters it is clear that during 1808 he stayed for a time at the Ballynahinch Spa where he attended different denominational services, on alternate Sundays, including the seceders of which Samuel Edgar, Thomson's teacher, was minister.[30] Drennan's characteristic latitudinarianism here not only heralds James Thomson's distinctive position, but is identical with William Thomson's subsequent religious practice of attending Church of Scotland, Free Church of Scotland, and Scottish Episcopal (or Anglican) services without regard for denominational differences.

James Thomson's friendship with Dr and Mrs Drennan led to his first meeting, at the Drennan's home near Belfast, with his future wife. Professor William Cairns, a Scottish minister, had been appointed to the chair of logic and belles-lettres at the Institution. In 1816, his cousin, Margaret Gardner (*c.*1790–1830), from a mercantile family, travelled from her home in Glasgow to visit the Cairns, and in July they, along with James Thomson, were invited to spend a day with the Drennans. It is said that Dr Drennan had arranged for Professor Thomson to be present as company for the attractive young lady, and indeed Margaret Gardner herself recorded it thus:

[27] Moody and Beckett, *Queen's Belfast*, **1**, pp. xliv–liii; Jamieson, pp. 36–58.

[28] *Ibid.*, p. 45.

[29] See, for example, Beckett, *Modern Ireland*, pp. 299–434; Moody and Beckett, *Queen's Belfast*, **1**, pp. 1–39.

[30] Drennan, *Letters*, p. 380. Edgar also contributed to *The Belfast monthly magazine* which Drennan edited (1808–14).

I had the honour of the Mathematician as my walking companion. His first appearance is about as awkward as can be; he looks as if he were thinking of a problem and so modest he can scarcely speak, but when tête-à-tête he improves amazingly in the way of speaking. On our forenoon walk we had a most edifying and feeling discussion on sea-sickness and the best mode of preventing it. But in the evening we were much more sublime. I suppose the moon rising in great beauty, and Jupiter shining with uncommon lustre, called forth the Professor's energies, and I got a very instructive and amusing lecture upon astronomy.[31]

Mrs Drennan's concern that the 'Mathematician', as he was known in the Belfast academic world, would be too quiet and studious to be good company for Miss Gardner proved unfounded as the engagement followed shortly, with marriage in the next year, 1817. The Thomsons settled in a town house in College Square East, facing the Institution, and it was there that the children were born – Elizabeth (1818), Anna (1820), James (1822), William (1824), John (1826), Margaret (1827), and Robert (1829).

On a visit to Ireland in September, 1827, the famous Scottish preacher and theologian Thomas Chalmers (1780–1847) resided with the Thomsons in their Belfast home, and his private diary affords a glimpse of life in the lively intellectual and social environment:

Landed at Mr Thomson's, whose wife is the cousin of the Grahams and Patisons of Glasgow, and really a very domestic and kindly person. The house was quite thronged with callers. Dined at Professor Thomson's after having reposed and written at some length in the easy and comfortable bedroom which has been assigned to me. Several at dinner, among others Professor Cairns of the logic, whom I think a very interesting person . . . on the whole, I spent one of the most agreeable days I have had since leaving home . . . [Mr and Mrs Thomson] have treated me with the utmost affection, and I love both them and their children.[32]

During his visit Chalmers opened the new Fisherwick Place Presbyterian Church in Belfast, not far from the Institution, to meet the spiritual needs of a growing town. The Thomson family were members of this Church which belonged to the orthodox (that is, trinitarian rather than unitarian) side of the presbyterian spectrum, though it did not at all share in the heated evangelical fervour of Cooke's oratory. It is important to appreciate that the Thomsons, in private especially, were deeply religious people, as Elizabeth's account shows.[33] Thoroughly versed in the scriptures and regular attenders at church, their beliefs were all-pervasive in their lives. However, as with Drennan, they sought to avoid association with openly sectarian organizations and establishments. On the whole it is not possible to categorize the Thomsons as evangelicals or

[31] Margaret to Agnes Gardner, July 1816, published in James Thomson, *Papers*, pp. xv–xvi. Margaret Gardner's visit to Belfast is also described in King, *Early home*, pp. 14–19.

[32] William Hanna, *Memoirs of the life and writings of Thomas Chalmers*, (4 vols., Edinburgh, 1849–52), **3**, pp. 177–9.

[33] King, *Early home* esp. pp. 47–9.

moderates (to use the Scottish terms). James Thomson could oppose the views of an Irish evangelical such as Henry Cooke, while finding agreeable those of a Scottish evangelical such as Thomas Chalmers. By contrast to Cooke's narrow vision, Chalmers' breadth of knowledge in areas from theology and political economy to natural science and mathematics appealed to Thomson who remained a life-long friend. As Chalmers wrote in 1847, the year of his own death:

> I was much interested by the comments which I received of your son the Professor. I have ever had a pre-eminent respect for high mathematical talent; and I am aware of the transcendental and lofty walk on which he has entered . . . may your son be guided by the wisdom from above to the clear and influential discernment of this precious truth – that a *sound faith and a sound [natural] philosophy are at one.*[34]

The population of Belfast had grown from around 20 000 in 1800 to 30 000 in 1815, and by 1840 it would be some 70 000 persons. No other town in Ireland could rival the prosperity and expansion of its commerce and industry. The port, still small, began to develop, and by 1819 what was probably the world's first cross-channel steamer service operated between Belfast and the Clyde. A revolution in sea transport was in sight.[35] At the same time, the town had benefited from the intellectual circle centred on the new Institution which, though heavily dependent upon students for the ministry, had been constituted with the needs of commerce in mind. James Thomson's classes in arithmetic and geography were understandably popular, and he developed their scope over the years.

James Thomson's mathematics and geography was of a practical, commercially useful kind, supplying the demands of a new mercantile order. During 1815, for example, his list of books employed in the mathematical school included a text devoted entirely to navigation.[36] Four years later James Thomson published the first edition of his own textbook, *A treatise on arithmetic, in theory and practice*, which went through seventy-two editions by 1880. The

[34] Thomas Chalmers to James Thomson, 14th February, 1847, C80, ULC. Our emphasis.

[35] Beckett and Glasscock (eds.), *Belfast*, pp. 78–108. See also E.R.R. Green, *The Lagan valley. 1800–1850. A local history of the Industrial Revolution* (London, 1949).

[36] James Thomson to Robert Simms, 1815, SCH 524/7B/9/83, Public Record Office of Northern Ireland. The full titles of the texts used were: John Playfair, *Elements of geometry; containing the first six books of Euclid* (Edinburgh, 1795); John Bonnycastle, *An introduction to algebra* (London, 1782); 10th edn. (London, 1815); Andrew MacKay, *The complete navigator; or an easy and familiar guide to the theory and practice of navigation* (London, 1804); 2nd edn. (London, 1810); Samuel Vince, *The principles of fluxions designed for the use of students* (Cambridge, 1790–99), being vol. 2 of the four-volume work by James Wood and S. Vince, *The principles of mathematics and natural philosophy*; 4th edn., of Vince's *Fluxions* (Cambridge, 1812); Bewick Bridge, *A compendious and practical treatise on the construction, properties, and analogies of the three conic sections; in eight lectures*, Part 3 of *Mathematical lectures* (2 vols., London, 1811); John Sharman, *An introduction to astronomy, geography and the use of the globes*, 2nd edn. (Dublin, 1794); 3rd edn. (Dublin, 1801); John Gough, *Practical arithmetick* (Belfast, 1797); and John Bonnycastle, *The scholar's guide to arithmetic; or a complete exercise-book for the use of schools*, 2nd edn. (London, 1780); 10th edn. (London, 1813). Reith on the globes and the first editions of Sharman and Bonnycastle's arithmetic cannot be traced.

treatise opened with the statement that its object was 'to present a full and regular course of whatever is useful in Arithmetic', the emphasis being very much upon what the author referred to as 'Mercantile Arithmetic' exemplified in the following question taken from a chapter on commission and insurance in the 1825 edition:

EX. 17. INVOICE of 100 firkins butter, shipped by Pollock and Blackwell in the Henry of Belfast, James McMillan, master, for Jamaica, to address to Messrs Mackay and Dunn, Kingston, by order and for account of John Maxwell & Co. Belfast. 100 firkins, neat 55c. 1q. 17 lbs. at £5 17 per cwt. – CHARGES. – Duty 4 per cent (on the prime cost:) Cooperage, Iron Hoops, Nails, Cartage, Shipping, Inspection, 8c. £16 5: Commission at $2\frac{1}{2}$ per cent. on £353 6 3. (Signed) Pollock and Blackwell. Cork, June 28, 1824. Answ. £362 2 11.[37]

Throughout the work, James Thomson's examples manifested a similarly striking degree of commercial realism contrasting markedly with the slightly earlier 1823 edition of Walker's *A new system of arithmetic*, published in London, which set questions of the form: 'A gentleman owed to different tradesmen the following sums: . . . what was the amount of the whole debts?'[38] Where Thomson taught arithmetic in relation to land, commodities, manufactures, and navigation, with the intention of diffusing practical and useful knowledge to all, Walker largely confined his text to serve the needs of gentlemen for whom trade was a distasteful fact of modern life.

The *Elements of plane and spherical trigonometry* (1820) followed the first edition of his *Arithmetic*, with *An introduction to the differential and integral calculus* and his edition of *The first six and the eleventh and twelfth books of Euclid's elements* appearing in 1831 and 1834, respectively. The original aim of these works was to provide textbooks for the use of students in the Belfast Institution, and the first edition of the *Trigonometry*, for example, 'was therefore written, not as a regular and complete treatise on Trigonometry, but as an outline to be filled up, and illustrated orally in the Lectures'. A favourable reception led him to expand and improve the work in the second edition (1830), and, as with his other texts, the book became widely used as one of the few suitable, readily available, and elementary British mathematical texts of the time.[39]

Published in 1827, Thomson's textbook on geography gave expression to his liberal views on social and political questions, views which emphasized the intimate links between education, commercial prosperity, and civil and religious liberty. These links are especially well illustrated by his accounts of Ireland,

[37] James Thomson, *A treatise on arithmetic in theory and practice*, 2nd edn. (Belfast, 1825), pp. 167, iii–iv. See also SPT, **1**, 6*n*–7*n*.

[38] J. Walker, *A new system of arithmetic . . . A new edition, with an appendix by W. Russell* (London, 1823), p. 49. On *English* schools and their relation to Cambridge University in this period, see Sheldon Rothblatt, *The revolution of the dons. Cambridge and society in Victorian England* (Cambridge, 1968), pp. 29–47.

[39] James Thomson, 'Advertisement', *Elements of plane and spherical trigonometry, with the first principles of analytical geometry*, 2nd edn., (Belfast, 1830).

England, Spain, and Norway. Of Ireland he explained that 'education has been greatly neglected, particularly among the lower classes':

The lower Irish are considered a lively, shrewd people, and warm in their attachments and antipathies. In many instances however, particularly in the south and west of the kingdom, they commit acts of turbulence and cruelty, arising from bad education and bad habits. These, it is to be hoped, will gradually yield to the active means, now employed by different parties, for the improvement of the population.[40]

By contrast, the English, 'accustomed to habits of good order and industry' and 'tenacious, in a high degree, of their civil rights', had 'perhaps higher degrees of comfort, in their mode of life, than the inhabitants of any other country'. The agriculture and manufactures of Spain, on the other hand, for all that country's great natural advantages, 'are far from being in a prosperous state; and its commerce is very limited'. Though he did not make the connection between commercial prosperity and religious liberty explicit, he explained further the fact that 'the established, and only tolerated, religion in Spain, is Roman Catholic', adding in a footnote that 'The Inquisition has been long established in Spain; and in no country has its power been exerted so cruelly and so effectually against any deviation from the doctrine and discipline of the Romish church'. He spoke also of the Norwegian people's superior habits and characters in consequence of their having long enjoyed civil liberty, and elsewhere he made clear his dislike of the African slave trade: 'this inhuman traffic has been productive of misery to millions of individuals, thus exiled from their homes and friends; and has been the cause of cruelty, oppression, and wars, by which thousands perish . . .'. His views matched those of the citizens of Belfast, who in 1806 had rejected the slave trade, largely through the efforts of a friend of Dr Drennan, Thomas McCabe. He had written in the proposal book: 'May God eternally damn the soul of the man who subscribes the first guinea'![41]

James Thomson's three-fold classification of the subject-matter of geography (defined as 'a description of the earth') anticipates some of his son's later divisions of the subject-matter of physics. First, *mathematical geography* 'treats of the figure and magnitude of the earth, of the latitudes and longitudes of places; and of globes, maps, and other artificial contrivances and instruments for illustration'.[42] His representation of the phenomena by geometrical descriptions and mechanisms foreshadows William's geometrical, visualizable and practical approach to physical phenomena.

Second, *physical geography* 'treats of the materials of which the earth is composed; of the forms of the various parts of its surface; of the atmosphere; of climate; of the various productions, animal and vegetable, found on its surface; and of other particulars respecting its natural condition'. This branch of geogra-

[40] James Thomson, *An introduction to modern geography: with an appendix containing an outline of astronomy, and the use of the globes* (Belfast, 1827), pp. 34–5.

[41] *Ibid.*, pp. 22, 49, 113, 179; Drennan, *Letters*, p. 380.

[42] Thomson, *Geography*, pp. 1*n*, 256–66.

phy comes close to William's major concerns with the properties of matter, terrestrial and celestial, under varying conditions, a concern which finds expression in a large programme of laboratory and field researches, ranging from terrestrial magnetism and atmospheric electricity to the effects of pressure on the melting points of various substances. And third, *political geography* 'treats of laws, modes of government, religion, learning, customs, and other subjects arising from the agency of man considered as a moral and political being'.[43] The distinctive role ascribed to 'the agency of man', as opposed to material agency, will have an important function in William Thomson's thinking on man's directing power (ch. 18).

During 1822, Thomson had travelled to London and spent over ten pounds in purchasing fifteen mathematical and physical books from Johnston Neilson for the Institution's library. About half were Continental works, and the list included Lagrange's *Leçons sur le calcul des fonctions* (1806), Biot's *Astronomie physique* (1811), Euler's *Introductio in analysis infinitorum* (1797), Thomas Young's *Lectures on natural philosophy* (1807), and William Nicholson's *An introduction to natural philosophy* (1782–96), as well as older works by John Barrow and Roger Cotes.[44] For a provincial educational establishment, with no previous mathematical tradition, it was an impressive list of quite advanced mathematical and physical volumes, not least because Young's *Lectures* contained a massive catalogue of most European writings on natural philosophy up to 1800. Indeed, as Kelvin's successor at Glasgow, Andrew Gray, remarked, since James Thomson had been bound by no examination traditions, he freely adopted the methods set out in the memoirs and treatises of the continental analysts. Such methods had only just begun to be appreciated in Cambridge by men such as Charles Babbage (1792–1871), John Herschel (1792–1871), George Peacock (1791–1858) and William Whewell (1794–1866).[45]

James Thomson was almost certainly the author of anonymous reviews of mathematics and the state of science in Scotland and Ireland published in the short-lived *Belfast Magazine* (1825). The reviewer, after the manner of John Playfair in the whig *Edinburgh Review* (1808), severely criticized the methods retained by British mathematicians:

> Since the days of Newton . . . the British mathematicians have been far surpassed in several branches of science by their neighbours on the continent. This has been particularly the case in the higher and more difficult parts of pure mathematics, and in physical

[43] *Ibid.*, p. 1*n*.

[44] The receipt is preserved as SCH 524/7B/16/34 in the Public Record Office of Northern Ireland. The full titles of the major works on the list were: J.L. Lagrange, *Leçon sur le calcul des fonctions* (Paris, 1804); 2nd edn. (Paris, 1806); J.B. Biot, *Traité elementaire d'astronomie physique* (Paris, 1805); 2nd edn. (3 vols., Paris, 1810–11); Leonhard Euler, *Introductio in analysis infinitorum* (Lausanne, 1748); 2nd edn. (2 vols., Lugduni, 1797); Thomas Young, *A course of lectures on Natural Philosophy and the mechanical arts* (2 vols., London, 1807); William Nicholson, *An introduction to Natural Philosophy* (2 vols., London, 1782–96).

[45] Andrew Gray, *Lord Kelvin. An account of his scientific life and work* (London, 1908), pp. 2–4.

astronomy; in which the Bernoullis, Clairault [sic], D'Alembert, Euler, Lagrange, and Laplace, have made such discoveries as will form monuments of glory, not only to themselves, but to Mankind. These great men pursued the path pointed out by Newton, and explored the mechanism of the universe, with such masterly power, and such distinguished success, as must ever be considered among the most glorious triumphs of mankind . . . While . . . the men of science in Britain were wasting their time and talents, some in restoring the ancient geometry of Greece, and some in following servilely and implicitly the *manner* in which Newton presented his investigations, without being actuated by the *spirit* by which he was directed in his researches.[46]

The reviewer unashamedly advocated continental analysis, remarking that 'to follow, at the present day, the modes of investigation employed by the original discoverers . . . to the exclusion of the new aids of science, is as absurd as it would be to reject the use of the steam-engine as a prime mover for machinery'.[47] He here employed one of the most potent images of the reform movement, analysis as an engine of progress in mathematics analogous to the engine of industry.

The expanding commercial town of Belfast could not have found a more progressive teacher of mathematics than Thomson, who numbered among his pupils Thomas Andrews (1813–85), the chemist, and William Bottomley, Belfast businessman and his future son-in-law. Rising to work on his textbooks at four o'clock in the morning, lecturing from eight to nine and from eleven to twelve, and holding the school classes in the afternoon before visiting the News Room and Commercial Buildings to obtain materials for his books 'from the latest and most authentic sources', James Thomson maintained a strikingly high standard of efficiency and thoroughness in every facet of his personal and professional life. The evenings he devoted to the education and well-being of his children, while Sunday involved the study of the Old Testament in the mornings and the New Testament in the evenings. Just as in Glasgow College, academic sessions lasted from 1st November to 30th April and then the Thomsons were free to rent a summer residence along the sea coast or on the shores of Belfast Lough.[48]

There is every reason to suppose that this prosperous and almost idyllic, if hard-working, life would have continued unchanged had not Mrs Thomson died in May, 1830, not long after the birth of Robert. James Thomson, then aged forty-four, was left with seven children, the eldest of whom was only eleven.[49] A vacancy in the mathematical chair at Glasgow College offered the combined

46 [James Thomson], 'State of science in Scotland', *Belfast Mag.*, **1**, (1825), 269–78, esp. pp. 270–1; [James Thomson], 'State of science in Ireland', *Belfast Mag.*, **1**, (1825), 459–69. As Scottish-educated mathematics professor at the Belfast Institution and as contributor of the article on the 'Battle of Ballynahinch' to the same volume, the *prima facie* case for Thomson's authorship is plausible. But the text of the article on Scottish science is almost identical in parts to that of his introductory lecture presented at Glasgow College from 1832 and discussed in our next chapter. John Playfair's famous critique of British science was published in the *Edinburgh Rev.*, **22** (1808), 249–84. 47 Thomson, 'State of science in Scotland', p. 273.

48 Thomson, *Geography*, preface; King, *Early home*, pp. 49–51, 62–5.

49 *Ibid.*, pp. 69–72.

possibilities of a new life, academic advancement and closer ties with Scottish relations. He wrote to the Belfast Institution's Board of Management in December, 1830, requesting a testimonial from the Board 'such as they would think he deserved'. It was almost a year before he secured the Glasgow appointment and could offer his resignation to the Institution where, as he put it, he spent 'a large portion of the best, and, I have no doubt, the happiest days of my life'.[50]

Thus he ended with regret his long association with the radical Institution through some of its most difficult and uncertain years. The path from the Ballynahinch farm to the Glasgow chair of mathematics had been arduous but immensely rewarding. Only the personal loss of his wife marred the meteoric rise. By comparison, his eldest brother, Robert Thomson, left with the charge of the Ballynahinch farm, remained in all the obscurity of a peaceful rural existence. John and Anna Thomson (William's younger brother and elder sister) recorded a visit to the Annaghmore farm during 1845, by which time its fortunes may have been declining sharply:

> We came upon Uncle and Aunt Thomson quite unexpectedly . . . They looked poor; and I am afraid they are not so well off as they used to be; but perhaps it is only owing to them becoming like most old people careless about their dress. I think Uncle Thomson is a very nice old man. He is very like papa; although his appearance quite wants the energy and activity of papa's. He seemed very glad to see us and to talk of papa and bygone days. He and Aunt Thomson say that they talk ten or a dozen times a day of papa . . .[51]

The Ballynahinch connection was finally severed in 1847 during the depression and misery of the Irish famine when the Thomson family sold the farm, although the house still stands as a reminder of the remarkable family which emerged from the ashes of revolution into a new century of industrial and scientific progress.

[50] Belfast Academical Institution Minute Book of the Joint Board of Managers and Visitors, 2nd December 1828–5th July 1836, pp. 102, 183–4; SCH 524/3A/4, Public Record Office of Northern Ireland. (Minutes of 21st December, 1830 and 3rd July, 1832); Jamieson, p. 16.

[51] John to William Thomson, 2nd November, 1845, T522, ULC.

2

Clydeside

> The noble winding river, with its fine bridges, and dazzling lights, is enchantingly beautiful. The city itself seems a great starry constellation, occasional luminaries disappearing from the view like so many meteors from the higher heavens. On nearer approach, those broad features become more minutely defined – numerous spires and lofty chimneys being now distinctly visible . . . What a mortal struggle for bread! – What delusive schemes of ambition afloat! – What vice and virtue! – What life, and withal, what death, fancy conjures up in disclosing to view the secret mysteries of the mighty city! *'Shadow' on Glasgow, 1858*[1]

Writing to his friend G.G. Stokes (1819–1903) in 1859 to urge him to consider applying for the vacant chair of practical astronomy at the University of Glasgow, William Thomson remarked upon the importance to science of getting Stokes out of London and Cambridge, 'those great Juggernauts under which so much potential energy for original investigation is crushed'.[2] Thomson's own subsequent refusal to occupy the Cavendish chair of experimental physics on no less than three separate occasions further underlines a lifelong commitment to his adopted city of Glasgow. With the Thomson family's removal to Glasgow College in 1832, indissoluble bonds came to be forged between the future Lord Kelvin and Glasgow's city, river, and college, bonds which depended not upon sentiment but upon the academic and industrial milieu developing on Clydeside in the first half of the nineteenth century.

To begin with, the radical transformation of Glasgow from a prosperous commercial town with a population of 77 000 in 1801 into a thriving industrial city of over 200 000 people by 1831, a city whose wealth depended more and more upon iron, steam, and above all ships and shipbuilding, sets the specific scene for much of William Thomson's subsequent life and work. In particular, the Glasgow Philosophical Society (founded 1802) provided an important meeting place for industrialists such as Walter Crum (1796–1867) and J.R. Napier (1821–80) whose respective cotton and shipbuilding firms represented the economic advancement of Clydeside, with academics such as J.P. Nichol (1804–59) and William Thomson whose scientific approaches symbolized the

[1] J.F. McCaffrey (ed.), *Glasgow, 1858. Shadow's midnight scenes and social photographs* (Glasgow, 1976), pp. 81–2. Reprint of Shadow [Alexander Brown], *Midnight scenes and social photographs: being sketches of life in the streets, wynds, and dens of the city* (Glasgow, 1858).

[2] William Thomson to G.G. Stokes, 6th October, 1859, K106, Stokes correspondence, ULC.

triumph of reform within the ancient Glasgow College.

When Dr James Thomson arrived in Glasgow during the year of the Reform Bill, the fourteen College professors had few links with the rapidly expanding city of heavy industry. During the next fifteen years, culminating in the appointment of his son to the chair of natural philosophy, Dr Thomson played a leading role in challenging and subverting the conservatism of this ruling college oligarchy. While at first concerned with the reform and advancement of his own profession, his universal reforming zeal as an 'ultra-whig' received immense support after the appointment of the new and radical professor of practical astronomy, John Pringle Nichol, in 1836, as well as that of medicine, Dr William Thomson, in 1841. In a small, intimate college, the advancing age of the tory establishment ensured a remarkable turnover of professors in the decade or so after 1836 such that, by 1848, only the principal himself and the professor of logic, Robert Buchanan, remained of the appointments made prior to 1831. The activities of the whig professors ensured that most of the new appointments would serve their cause.

Dr Thomson's growing power-base paved the way for the appointment of Glasgow College professors on academic, professional criteria rather than for their party, family, or religious loyalties. Nevertheless, Dr Thomson's professed 'independence' and 'objectivity' must not obscure his considerable commitment to whig politics embracing a latitudinarian stance in religious matters and a whole series of reforming crusades. The professor of law, Allan Maconochie, for example, believed Dr Thomson 'to have turned a most pestilent Whig, ready to do anything for Johnny Russell'.[3]

We shall examine how Dr Thomson's whig and latitudinarian convictions directed the early careers of his three eldest sons. Dr Thomson's belief that the most effective means of professional advancement lay with practical application rather than theoretical or metaphysical dispute found expression in his sons' aspirations to study engineering, mathematics and natural philosophy, and medicine, respectively. Similarly, he believed that a latitudinarian attitude guaranteed social advancement, as he advised William in his first term in Cambridge: 'recollect my invaluable maxim never to quarrel with a man (but to waive the subject) about religion or politics'.[4] Thus the cultural and social values which had developed in the Ulster context in an age of revolution came to bear new fruit in Clydeside during an age of reform and improvement.

Glasgow and the Clyde

The name of Clydeside recalls the Second City of the British Empire: Cunard liners, heavy engineering, industrial depression, and some of the grimmest

[3] Allan Maconochie to Dr James Thomson, 22nd June, 1846, in SPT, **1**, 182–3. As James Thomson received in 1829 an honorary degree of doctor of laws from Glasgow, we shall refer to him as Dr Thomson. See SPT, **1**, 6.

[4] Dr Thomson to William Thomson, 6th December, 1841, T186, ULC; SPT, **1**, 32.

tenements in Europe.[5] This glory and misery were products of Victorian times, for if we look back to 1800 we find an industrial city only in its merest childhood. The Glasgow of the eighteenth century had built its prosperity on the foundation of a market town centred around a bridge over the River Clyde. The town's university dated back to 1451 as one of Scotland's ancient academic institutions, the other universities being at St Andrews, Edinburgh and Aberdeen. The town's port was established in 1668 at Port Glasgow, some twenty miles down river, from where the goods were transshipped in shallow-draught barges to be sailed, poled or rowed upstream on the flood tide until horses could take over along the narrower reaches. The inconvenience, however, of a passage which could take three days did not prevent Glasgow becoming a successful trading town, dealing in tobacco from North America and sugar from the West Indies, while exporting fish, coal, salt and linen. Archibald Smith's family, for example, which figures prominently in our study of William Thomson, had in 1800 purchased the Jordanhill estate, about four miles west of Glasgow, by means of wealth generated by the West Indian trade. Similarly, the very prosperous eighteenth-century 'Tobacco Lords' accumulated considerable capital for the town.[6]

From 1770 the Town Council narrowed and deepened the River Clyde, and five years later vessels with a draught of six feet could reach the Broomielaw near the town centre. The confinement of the river by the construction of stone jetties, and the eventual joining of the ends of the jetties to make a new river bank, had set up a natural scouring action which, along with artificial dredging, turned the old shallows into what was effectively a ship canal without locks. By 1830, ships drawing fifteen feet could reach the Broomielaw, and the depth progressively increased during the nineteenth century and beyond as Glasgow became the world's leading shipbuilding centre in the years up to 1914. There was much truth in the local saying that 'Glasgow made the Clyde, and the Clyde made Glasgow'.[7]

Largely as a result of a decline in the tobacco trade after the American War of Independence, cotton had become, by the early nineteenth century, Glasgow's major industry and staple trade, increasingly concentrated in and around the town as water power gave way to steam and coal. Often, however, the industry remained in villages around the Clyde valley. One such village was Thornliebank, 'the whole of which, with the exception of two or three small

[5] See, for example, David Daiches, *Glasgow* (London, 1977), esp. pp. 95–179; C.A. Oakley, *The second city* (London and Glasgow, 1946); Geoffrey Best, 'The Scottish Victorian city', *Victorian Stud.*, **11**, (1967–8), 329–58.

[6] See L.J. Saunders, *Scottish democracy: 1800–1840* (Edinburgh, 1950), pp. 97–117; John Shields, *Clyde built* (Glasgow, 1949), pp. 12–23; T.M. Devine, *The tobacco lords. A study of the tobacco merchants of Glasgow and their trading activities c.1740–1790* (Edinburgh, 1975). An invaluable nineteenth-century source is *The new statistical account of Scotland, 6: Lanark* (London and Edinburgh, 1845), pp. 101–246, on the city of Glasgow, its population, trade and industry.

[7] Saunders, p. 99; Shields, pp. 15–17.

houses, belongs to Messrs Crum, and is almost wholly occupied by persons in their employment'. Men, women and children working twelve hours a day, Mondays to Fridays, and nine hours on Saturdays, at cotton-spinning, power- and hand-loom weaving, bleaching and finishing, and calico-printing, made up the Crums' labour force of some 850 persons engaged in cotton manufacturing and finishing.[8] The Crums, politically committed to whig causes, were an intimate part of the Thomson circle in Glasgow, and it was the young Margaret Crum (*c.* 1827–70) whom William Thomson would marry in 1852. As a sign of social and economic change, their commercial and industrial prosperity in the nineteenth century contrasted with the Smiths of Jordanhill, who by the 1830s were hard-pressed to maintain their economic status against the new generations of Clydeside industrialists.[9]

The second phase of industrialization in the Clyde valley occurred in the 1830s and 1840s with the emergence of heavy industry based on coal and iron. The local invention by J.B. Neilson of the hot-blast process for iron production in 1828 aided a seven-fold increase in Scottish iron output between 1835 and 1845. New machinery demanded by the cotton industry and the beginnings of railway construction, as well as the local abundance of iron and coal deposits, meant the transformation of forges into foundries.[10] In this period, Robert Napier, the father of Clyde shipbuilding and marine engineering, built up foundries at nearby Camlachie, Vulcan, and Lancefield.[11] In 1841, Napier began iron shipbuilding at Govan; while in 1842 the Glasgow to Edinburgh railway line was opened, and a locomotive works was established.

The old commercial town, with its Clyde crossing, its distant port at the mouth of the river, its trading links with North America and the West Indies, and its remoteness from any equivalent English town, bore little resemblance to the Glasgow which was emerging in early Victorian times. The 1820s saw the growth of regular steamer links between the Clyde and Belfast and between the Clyde and Liverpool, as well as the success of internal steamer routes within the Clyde River and Firth. The uncertainty of travel under sail to and from Glasgow College at the very birth of the steamship age is captured in Lord Kelvin's description of the passages undertaken by his father in 1812–14:

> The passage took three or four days; in the course of which the little vessel [an old lime smack], becalmed, was carried three times round Ailsa Craig [a conspicuous conical island in the approaches to the Firth], by flow and ebb of the tide. At the beginning of his fourth and last University session, 1813–14, my father and a party of fellow-students after landing at Greenock, walked thence to Glasgow. On their way they saw a prodigy – a black chimney moving rapidly beyond a field on the left side of the road. They jumped

[8] *The new statistical account of Scotland*, 7: *Renfrew-Argyle* (London and Edinburgh, 1845), p. 41.

[9] For example, Archibald to Susan Smith, 17th October, 1854, TD1/887, Smith papers, Strathclyde Regional Archives.

[10] Saunders, pp. 104–6; *Lanark*, pp. 138–65.

[11] James Napier, *Life of Robert Napier of West Shandon* (Edinburgh, 1904), pp. 29–38, 48–99, 148–65.

the fence, ran across the field, and saw to their astonishment Henry Bell's *Comet* (then not a year old) travelling on the river Clyde between Glasgow and Greenock.[12]

The earliest steamboats, tested on the rivers and canals of France, America, or Britain, had been little more than technical novelties, quite unsuitable for sea-going employment. Henry Bell's *Comet*, first on the River Clyde and later on Scotland's West Coast, inspired the cousins David and Robert Napier in their development of viable marine steam-engines. These founding fathers of marine steam engineering and iron shipbuilding on the Clyde had a fundamental concern with the improvement of the performance (economy and reliability) both of their marine engines and their hulls.

Sir William Thomson's association with maritime interests in developing and marketing his navigational patents (chs. 21 and 22) derives from his friendship with Robert Napier's son, James Robert, particularly through the forum provided by the Glasgow Philosophical Society. This favoured meeting place for Clydeside academics and industrialists had originated in 1802, with David Napier and William Meikleham (1771–1846) (professor of natural philosophy) as founder members, and it was at the meetings of the Society in the period 1850 to 1870 that much discussion of ship performance and safety took place among members such as J.R. Napier, W.J.M. Rankine (1820–72) and Sir William Thomson.

J.R. Napier, having attended Glasgow College at much the same time as James and William Thomson, attained a high place in Dr Thomson's mathematical class. The naval architect in his father's firm, William Denny, provided Napier with his practical training in the subject, and when Denny left to become a famous Clyde shipbuilder in his own right, Napier succeeded him. With his friend Rankine, J.R. Napier produced in 1866 a treatise on shipbuilding which became a standard work. His concerns were primarily with ship performance rather than the immediate business of shipbuilding, which he left to his brother, John, who carried on his father's works from about 1856.[13] J.R. Napier belonged to a tough entrepreneurial breed of Scots and, according to his gentlemanly friend Archibald Smith (1813–72), was 'not very careful about personal appearance & I was amused by seeing him get from one side of . . . [a listing] vessel to the other by sitting down on a tarred deck & sliding down'.[14]

The importance of the Glasgow Philosophical Society, significantly not a literary and philosophical society of the kind which appealed to gentlemen of more taste and cultivation than Napier, lies primarily in its role as a strong link between town and gown even from its foundation. While William Thomson did not become a member until his return to Glasgow as professor in 1846, his

[12] MPP, **6**, 371–2; SPT, **1**, 3–4.

[13] Napier, *Life of Robert Napier*, pp. 19, 199–201; J.R. Napier and W.J.M. Rankine, *Shipbuilding. Theoretical and practical* (London, 1866).

[14] Archibald to Susan Smith, 13th October, 1854, TD1/887, Smith papers, Strathclyde Regional Archives.

father had been elected in 1839, and his brother, James, in 1841. Other academics from the University included Thomas Thomson (elected 1834 and subsequently president), J.P. Nichol (1836), William Ramsay (1841), Lewis Gordon (1842), and W.J.M. Rankine (1852). All but Dr Thomson and Ramsay were regius professors in the University, appointed by the Crown and largely excluded from exercising much power within the College oligarchy. Yet their strong practical interests, whig politics, and above all their conception of the professional man possessed of marketable knowledge, aligned them with the many industrialists – Walter Crum, Charles Randolph, John Elder, for example – who participated in the Society meetings. With these city–college links in mind, we may now turn to an analysis of Glasgow College itself.[15]

The University of Glasgow

The ancient foundation of Glasgow College had been a rather passive witness to the social and economic transformations of the town in the early nineteenth century. Within its pleasant precincts, distinguished scholars such as Adam Smith, Thomas Reid, John Robison, William Cullen and Joseph Black had flourished at various times in the middle or second half of the eighteenth century. Their intellectual vigour had been worthy of a dynamic commercial centre. The spacious college gardens, the streets of elegant merchants' houses, and the 'walks on the Green and banks of the Clyde, where air was sweet and trees were fresh and the water was pure and clear in the river' must have made for an agreeable social and intellectual atmosphere, especially when the houses of the poor were much less obtrusive (and certainly less numerous) than they were to become.[16] The early nineteenth century, on the other hand, was hardly a period of great intellectual achievement for the Glasgow College professors. By and large they were an aging and inactive body, concerned to preserve their privileges against outside threats and unwelcome changes.

Tory (often ultra-tory) in politics and ardent supporters of the established (presbyterian) Church of Scotland, the majority of College professors had little in common with the rising industrial city founded on whig values. Growing pressures from within the College and from outside during an age of reform brought about a virtual revolution in the character of the College during the years that Dr James Thomson occupied the chair of mathematics. To understand the changes we begin with an outline of the institutional structure of Glasgow University.

[15] For membership of the Society, see *Proc. Glasgow Phil. Soc.*, **1** (1841–4), and subsequent volumes.

[16] See for example, the vivid account by H.G. Graham, 'Glasgow University life in olden times', *The book of the Jubilee. 1451–1901* (Glasgow, 1901), pp. 12–25. The classic history of the University is given in James Coutts, *A history of the University of Glasgow from its foundation in 1451 to 1901* (Glasgow, 1909). A more recent account is J.D. Mackie, *The University of Glasgow. 1451–1951* (Glasgow, 1954), esp. pp. 243–68 on the period 1801–58. See also Saunders, pp. 307–71.

Glasgow High Street running north and Glasgow College with tower as they would have looked during William Thomson's early years. [From a drawing by his niece, A.G. King, in Elizabeth King, *Lord Kelvin's early home*.]

Three bodies comprised the University. The Senate, a mainly ceremonial body, oversaw election and admission of the Chancellor and Dean of Faculty, admission of the Vice-Chancellor and Vice-Rector, election of a representative to the General Assembly of the Church of Scotland, conferring of degrees, and management of the libraries. It was made up of the Rector (who presided), the Dean, and all the professors. The second body, the Comitia, added the College Principal and the matriculated students to its membership for the purpose of electing and admitting the new Rector, for hearing the inaugural discourses of the principal and professors, prior to their admission to their respective offices, and other ritual activities. The real power of the institution, however, rested with the principal and thirteen professors making up the third body, the Faculty or College.[17]

The College had full control of its own not inconsiderable revenue and property, unlike the municipally controlled University of Edinburgh. The richest of the Scottish colleges, its annual income between 1825 and 1848 remained at almost £8000. With such an income, the occupants of College chairs averaged a salary of around £300 per year, to which must be added the fees from the students. For very good reasons, the College professors were keen to exclude the new regius professors – who often received a mere £50 per year from the Crown – from a share in the College revenue, and so men like the chemist Thomas Thomson and the engineer Lewis Gordon (unlike the five older regius professors of astronomy, church history, civil law, medicine, and anatomy) were without administrative and academic power in the Glasgow institution. The Principal of the College, Duncan Macfarlan, expressed his conviction that reform – particularly of the status of the regius chairs – was 'calculated only to excite and foster a spirit of jealousy and contention, injurious alike to the interests of the University and of its individual members'.[18] Needless to say, lack of reform had precisely that effect.

Added to the administrative power of the College was its control of eight professorships vested in the College. In the election of new professors, the Rector and Dean also had a vote, but the new regius professors had no say whatever, even though there were, by 1835, some half dozen such men.[19] Sir William Hamilton delivered a scathing attack on the abuse of patronage in the Scottish Universities in the whig *Edinburgh Review* in 1834, an attack which included a direct onslaught on the Glasgow system:

In the first place, by conjoining in the same persons the right of appointment and the right of possession, it tends to confound patronage with property . . . In the second place, as it disposes the patron to forget that he is a trustee, so it also primes him with every incentive to act as a proprietor. Natural affection to children and kindred; personal friendship and enmity; party, (and was there ever a University without this curse?); jealousy of superior intelligence and learning, operating the stronger the lower the University is degraded . . . The standard of professional competence must be kept down – it seldom needed to be lowered – to the average level of their relatives and partisans.[20]

[17] Coutts, pp. 337–79; *Lanark*, pp. 173–4. See also the anonymous article 'Notice of the University of Glasgow – its professors and students', *Edinburgh Mag. Literary Miscellany*, **16** (1825), 513–23. The most valuable source for this early period is undoubtedly the 'Report on the University of Glasgow', *Parliamentary Papers*, **12** (1831), 211–302.

[18] For a valuable study of the politics of Glasgow University, notably during the 1820s and 1830s, see J.B. Morrell, 'Thomas Thomson: Professor of Chemistry and University reformer', *Brit. J. Hist. Sci.*, **4** (1969), 245–65, on p. 252. On College revenues, see Coutts, p. 362.

[19] *Lanark*, p. 175*n*.

[20] [Sir William Hamilton], 'Patronage of universities', *Edinburgh rev.*, **59** (1834), 196–227, on pp. 221–2.

Adding as a footnote a quotation from Thomas Chalmers condemning the hereditary successions in colleges which brought about 'the decaying lustre of chairs once occupied by men of highest celebrity and talent', Hamilton concluded that, of all the forms of university patronage, this one 'tends to make celebrity the exception, obscurity the rule'.[21]

The only reform within the College with regard to appointments of new professors occurring before 1831 had been the election to the Greek chair in 1821 of Oxford-educated Daniel K. Sandford. The Oxbridge trend continued with the election of Cambridge-educated William Ramsay to the chair of humanity in 1831. With the notable exceptions of Dr Thomson and J.P. Nichol, as well as the election of Edmund Lushington (another Cambridge product) to replace Sandford in 1838, the College remained solidly in the hands of the tory oligarchy until the early 1840s. The oligarchy, indeed, temporarily strengthened its hold by the replacement of the elderly but radical professor of moral philosophy, James Mylne, with the ultra-tory William Fleming in 1839, whose own chair of oriental languages went to another tory, George Gray. Alexander Hill's election as divinity professor in 1840 in succession to the whig Stevenson MacGill (son of a Port Glasgow shipbuilder) and in preference to the distinguished Chalmers heralded yet another consolidation of the tory establishment.[22]

By 1840, then, four of the eight chairs vested in the College were occupied by tory professors (Buchanan, Fleming, Hill, and Gray). The Principal, of course, led the tory faction. Of the other four, Dr Thomson was the ultra-whig, Ramsay and Lushington were ready to support reform campaigns, and William Meikleham had been removed from the scene in 1839 through ill-health. Of the five other College professors (all Crown appointments) in 1840, only Nichol represented a new generation, the others being of advanced age or out of commission. The professor of anatomy, James Jeffray, appointed in 1790, supported the Principal, but the others seem to have ceased political activity. Thus the replacement of these four professors during the 1840s, as well as the election of a new natural philosophy professor, would mark a decisive shift in the balance of power *within* the College, and a break-up of the old oligarchy. The Crown appointment of the brothers William and Allen Thomson (no relation to Dr Thomson) as professors of medicine and anatomy in 1841 and 1848, respectively, by Whig governments (in which Lord John Russell was first Home Secretary and then Prime Minister) brought into the College two more very ardent whig reformers. Again, the appointment of J.S. Reid to the chair of church history in 1841 introduced another former teacher from the progressive Belfast Academical Institution, while Allan Maconochie's appointment by Peel's government in 1842 provided a professor of civil law who, though initially associated with the Principal's party in the College, became more and

Glasgow College: the inner quadrangle looking west. The archway below the tower led to the outer quadrangle and thence to the High Street. The windows to the right of the pentagonal turret belonged: (ground floor) to the Blackstone examination room which Thomson later annexed for his growing laboratory; (first floor) to the natural philosophy apparatus room; and (attic) to the engineering class-room. This building dated from 1656. [From David Murray, *Lord Kelvin as professor in the Old College of Glasgow*.]

more estranged from it, being as he was sympathetic to moderate reform and well disposed towards Dr Thomson.[23]

Professors appointed to regius chairs founded after 1807 remained outside the College elite. All six chairs founded between 1807 and 1831 – natural history (1807), surgery (1815), midwifery (1815), chemistry (1818), botany (1818), and materia medica (1831) – related to scientific (and especially medical) subjects; all except natural history belonged to the University's faculty of medicine. As a result of their practical scientific character they gave a new, progressive image to the University, contrasting with the faculties of art and theology, and providing strong bonds between city and university both through the marketing of much-needed medical expertise to students and through the Philosophical Society. Pressures from the whig and radical citizens of Glasgow for rejuvenation of their famous University found a sympathetic ear at court, and after the 1832 Reform Act when the Scottish electorate increased from under 4500 to over 60 000,[24] conservative reformers as well as whigs began to take note of Glasgow's needs. The elections of Lord Stanley, Sir Robert Peel, and Sir James Graham as successive Rectors of the University in the 1830s provided national leaders (albeit in opposition at the time) with a direct grasp of affairs in the city and College. Peel took great pride in winning the support of Glasgow students, whom he recognized as 'the descendants of the inveterate Whigs of Bothwell Brigg and the Ayrshire Covenanters, or the sons of the reforming merchants of Glasgow, who were so deeply imbued with democratic principles in 1832'.[25]

Probably in a bid to regain support, the Whig government (with Lord John Russell as Home Secretary) established regius chairs of forensic medicine and theory of physic in 1839, and generously endowed a regius chair of civil engineering in 1840 with an annual grant of £275. Lewis Gordon (1815–76), himself a whig, occupied the first engineering chair in a British university – a new chair which symbolized the importance of engineering for Clydeside and which coincided with the Glasgow meeting of the British Association.[26]

The failures of these regius professors to infiltrate the College elite, however, is vividly illustrated by the case of Thomas Thomson (1773–1852). Appointed to the chair of chemistry in 1818, he obtained in the same year at the College's

[23] *Ibid.*, pp. 390–5, 520–5. On Maconochie's defection to the whigs, see Dr James Thomson to William Thomson, 8th March, 1846, T330, ULC.

[24] Quoted in Derek Beales, *From Castlereagh to Gladstone* (London, 1969), p. 83.

[25] Sir Archibald Alison to Sir Robert Peel, 21st November, 1836, in C.S. Parker (ed.), *Sir Robert Peel from his private correspondence* (3 vols., London, 1899), **2**, pp. 327–8. Stanley and Graham had deserted the Whigs in 1834 in support of Anglican Church property. See Beales, pp. 97–8. Neither they nor Peel (the son of a Lancashire calico-printing industrialist) would have had much in common with the ultra-tories in the College. Rather, they would have been more concerned to win over former Scottish whigs to their new conservatism. See Norman Gash, *Sir Robert Peel. The life of Sir Robert Peel after 1830* (London, 1972), pp. 151–7; C.S. Parker, *Life and letters of Sir James Graham, 1792–1861* (2 vols., London, 1907), **1**, pp. 272–6.

[26] Thomas Constable, *Memoir of Lewis D.B. Gordon* (Edinburgh, 1877); C.A. Oakley, *A history of a faculty. Engineering at Glasgow University* (Glasgow, 1973), pp. 4–8; Coutts, pp. 537–41.

Glasgow College: the inner quadrangle looking north-west. The natural philosophy class-room occupied the space between the circular turrets on the north side (first floor), while moral philosophy occupied that part of the ground floor extending from the shadowy figure to the right beyond the picture. This building dated from 1632. [From Andrew Gray. *Lord Kelvin. An account of his scientific life and work.*]

expense a laboratory whose aims were to teach chemistry and 'to raise up a race of practical chemists', many of whom in fact became industrial chemists. The number of students annually attending his classes by the late 1820s exceeded 200. And in 1831 the College financed new chemistry buildings at a cost of £5000. Nevertheless, for all his academic successes and for all his reforming zeal,

Thomson failed to make any significant improvement to the status of regius professors. The college remained unredeemed, and Thomson slipped into retirement over the period 1841–6 at the very time James Thomson's crusade from within the College reached its maximum intensity.[27]The rapid infiltration of the establishment occurred after 1840, coinciding with the deaths of elderly regius professors within the College and with the campaigns of Dr Thomson and J.P. Nichol. For much of the early 1830s, however, Dr Thomson spoke with almost a lone dissenting voice within the College.

The new mathematics professor

By all accounts, James Millar's endeavours as professor of mathematics at Glasgow College were disastrous. The Scottish University Commissioners reported in 1830 that Millar had held his chair since 1789 and, though not yet very old, had 'no power of exerting authority in his class, or enforcing discipline among his students'. They also suggested that he owed his position less to his own abilities than to the influence and reputation of his father, 'the late eminent professor of law'. One-time teacher of James Thomson and of Archibald Smith, Millar's long reign was marred by frequent disorder and student frivolities, including a ducking of the professor in the deepest part of the College's Molendinar burn during a lesson on practical surveying. The Commissioners urged that 'the interest of science pleads for his release from the labours of . . . [his] charge'; at length the College persuaded the incompetent professor to retire and resign his College house while permitting him to retain his salary for life.[28]

On the 16th December, 1831, the College elected Dr James Thomson to the mathematics chair, with Dr Thomson apparently under the misapprehension that he would be entitled not only to the College house but also the usual professorial salary. On arrival in Glasgow Dr Thomson found that he had to pay out more than he received on account of the smaller-than-expected size of the class and the consequent reduction in income from fees. It would seem that, for this reason, Dr Thomson could not afford to send his two eldest sons to school in Glasgow. To solve the financial problems, the new professor opened afternoon lecture courses on popular geography and astronomy for ladies, courses which went on successfully for two or three years. At the same time he managed to subdue the previously chaotic mathematics classes, and build up the number of students once more, so that gradually he overcame the problems of maintaining and educating a large family on a non-existent salary.[29]

Dr Thomson was both a professional and a progressive mathematician. As a professional, marketing mathematical knowledge and expertise largely in the

[27] Morrell, 'Thomas Thomson', pp. 253–62.

[28] *Parliamentary Papers*, **12** (1831), 246–7; Elizabeth King, *Lord Kelvin's early home*, (London, 1909), pp. 94–7; Coutts, pp. 352–83.

[29] King, *Early home*, pp. 94–7.

university context, he proved an outstanding success. Ten years after his arrival, he recorded his largest-ever classes – fifty in the senior and 110 in the junior – providing through fees a substantial supplement to the restored professorial salary.[30] At the same time, he advanced his profession and supplemented his income by the writing of mathematical textbooks. He edited a version of Euclid's *Elements of geometry* in 1834, and published *An elementary treatise on algebra, theoretical and practical* in 1844, as well as revising his previous works on arithmetic, trigonometry, and calculus. He stressed in his *Albegra* the immense value of that science:

> In itself, indeed, and its application and extension in the differential and integral calculus . . . it is a most powerful, and indispensable, instrument for prosecuting investigations in mechanics, astronomy, and other subjects in physical science; and, without its aid, it is impossible to understand, or duly to appreciate, the discoveries of Newton, Laplace, and the other great men who have done such wonders in extending the boundaries of modern science.[31]

As he expressed the point in his *Introduction to the differential and integral calculus*, he saw physical science, natural philosophy, as the sequel to the study of pure mathematics,[32] which there functioned primarily in an instrumental role. An appreciation of this instrumental role of analysis (and other abstract and symbolic mathematical or logical tools) is crucially important for our understanding of Scottish natural philosophy (particularly as practised by J.P. Nichol and William Thomson) where analysis, symbols, and any metaphysical system had to be rejected as an end-in-itself in favour of a visualizable, geometrical, and empirical conceptualization.

Dr Thomson's *Algebra* went through three editions in less than two years, around 3000 copies going to the Irish Education Board, which had already taken 5000 copies of his *Arithmetic*, and the same number of his Euclid for introduction into the recently established National Schools in Ireland. These schools, promoted by the liberal catholic Thomas Wyse, and initiated by Lord Stanley in 1831 before he left the Whig party, aimed to provide an elementary, non-denominational education for all the people, and so conformed to Dr Thomson's commitment to the diffusion of knowledge as a means of advancing and freeing the human condition. This trend towards assimilation of the Irish people to a 'utilitarian' British culture aroused the wrath of middle-class, town-dwelling nationalists who idealized the rural, peasant life (ch. 23). Furthermore, as Nichol polemically observed, 'we have to deal in Ireland with a very peculiar, as well as a very powerful, Church, which, as a principle, rejects our fundamental tenet of

[30] Anna to William Thomson, 3rd December, 1842, B185, ULC; Dr to William Thomson, 7th December, 1842, T228, ULC.

[31] James Thomson, *An elementary treatise on algebra, theoretical and practical* (London, 1844), p. vi.

[32] James Thomson, *An introduction to the differential and integral calculus* (Belfast, 1831), p. iii; 2nd edn. (London, 1848), p. iv (modified slightly).

the FREEDOM OF EDUCATION.' Nevertheless, Dr Thomson's net share of the profits on all his text-books amounted, in 1845 alone, to £378 14s 8d.[33]

In his approach to mathematics, Dr Thomson was as progressive as in his approach to politics and religion. He favoured the full introduction of the most modern mathematical practices even in elementary texts. The *Belfast Magazine* article of 1825 had already subjected to censure Scottish mathematics, or 'pure science' as it was often called to distinguish it from 'physical science'. In particular, the review criticized works by two Scottish professors, John Leslie's *Geometrical analysis, and geometry of curved lines, being Vol.* II *of a course of mathematics, and designed as an introduction to the study of Natural Philosophy* (1821), and Thomas Duncan's *Supplement to Playfair's Geometry and Wood's Algebra, completing a course of mathematics, in theory and practice* (1824). Leslie was professor of natural philosophy at Edinburgh University from 1820 to 1832, but had previously been mathematics professor there from 1805 to 1819, and was a staunch supporter of geometrical rather than analytical methods.[34] Duncan was professor of mathematics at St Andrews. The reviewer noted that these two works demonstrated the negligible improvement so far effected in Scotland compared to the advances made in adopting continental methods at Cambridge and Dublin: neither work 'affords any reason to believe that the works of Euler, Lagrange, Lacroix, or Laplace, are properly known or appreciated in either [Scottish] seminary'. 'We have understood, also,' he continued, 'that the new science has not been introduced in either of the universities of Aberdeen; and we know that in the university of Glasgow, mathematical science, if it can be said even to exist, is a century behind.'

The reviewer condemned Leslie's book especially, not only for its exclusion of the new methods, but as being in parts 'very ill adapted for the purposes of teaching, as the higher properties which it develops may be derived far more easily and satisfactorily by means of the modern analysis'. Analysis would enable the student 'to prosecute the study of the higher and more difficult branches' of science. He dismissed Duncan's book too as trivial by the standards set by French mathematical texts: 'The entire work, indeed, is of a very slight and superficial nature. Too much is grasped at, and nothing is done well. Such is too frequently the character of our English courses of mathematics; and few of them exceed the

[33] The sales figures were recorded in letters from Dr James Thomson to William Thomson, 23rd October, 1842, T225, and 8th February, 1846, T328, ULC. The profits figure was reported in a letter from Dr Thomson to William Thomson, 19th May, 1846, T347, ULC. On the National Schools in Ireland, see T.W. Moody and J.C. Beckett, *Queen's Belfast 1845–1949* (2 vols., London, 1959), **1**, liii–lviii. Nichol's remarks appear in his 'Preliminary dissertation' for J. Willm, *The education of the people: a practical treatise on the means of extending its sphere and improving its character* (Glasgow, 1847), p. i.

[34] [James Thomson], 'State of science in Scotland', *Belfast Mag.*, **1** (1825), 269–78. See, for example, R.G. Olson, *Scottish philosophy and British physics 1750–1880. A study in the foundations of the Victorian scientific style* (Princeton, 1975), pp. 192–3. Leslie was apparently acquainted with analytical methods, but he gave his support to the geometrical approach. See also R.G. Olson, 'Scottish philosophy and mathematics', *J. Hist. Ideas*, **32** (1971), 29–44.

work before us, in this respect. How different are the courses of Garnier and Lacroix, in French!'

From such remarks, one might wrongly infer that Dr Thomson had little regard for the intuitive appeal of geometry compared to the power of abstract analysis. All of his textbooks, however, minimize abstraction. As a progressive reformer he believed that the most powerful methods could be and should be introduced at an elementary level, thus 'diffusing' them to the widest possible audience. In his *Introduction to the differential and integral calculus* of 1831, for example, he sought 'to render the investigations as simple and easy as possible', and included a large number of examples, many of them geometrical. If he adopted for his first edition Lagrange's algebraic foundations rather than Leibniz's much more efficient infinitesimals or the method of limits (not to speak of Newton's cumbersome fluxions), he did so only because of the conceptual obscurity of these rival approaches: 'the Author long hesitated before he ventured to make this [Lagrange's] method the basis of the present work, from a fear that the elementary investigations, derived by means of it, would be too tedious and difficult for beginners'. When further development of the limit approach by Cournot, Duhamel, Moigno, and a variety of British authors, including de Morgan, had removed many objections to that simplifying method, Dr Thomson welcomed it for his second (1848) edition with evident relief.[35]

While Dr Thomson was severe on the traditional mathematical methods employed in Scottish universities, he wrote in praise of the Scottish institutions in general:

> Scotland, with its universities and its system of parish schools, possesses, perhaps, the noblest foundation for national instruction that any country has ever enjoyed; and the efficiency of the means is fully proved by the production of a more general diffusion of education and intelligence, than an equal population in any other country can exhibit. Its means for the cultivation of the exact sciences are also considerable; as in its five universities there are five professorships of pure mathematics, five of natural philosophy, and two of astronomy . . .[36]

Clearly, the Scottish system of education conformed quite closely in principle, though not always in practice, to Dr Thomson's own commitment to reform and advancement. The availability of a university education to a very wide range of students, from rich to poor, was perhaps the most attractive feature to someone who had himself advanced far from a comparatively poor back-

[35] James Thomson, *Differential and integral calculus*, pp. iii, 241–2; 2nd edn., p. 308. Significantly, he made no mention of Cauchy in his second edition nor of the rigour of Cauchy's new epsilon-delta procedures. His basic commitment lay with methods sufficiently rigorous to guarantee validity, but more especially with methods powerful enough to open up the most advanced *applications* of mathematics, notably in mechanics and physical astronomy. William followed his father's lead in his attitude towards the value and best methods of mathematics (ch. 6).

[36] [James Thomson], 'State of science in Scotland', p. 270.

ground. The academic framework, too, he felt, was a beneficial one, the existence of mathematics, natural philosophy, and astronomy chairs side by side being particularly characteristic of Scottish universities. Such a situation differed from the specialist mathematical emphasis at Cambridge, for example, where there was little natural philosophy at all. During his years at Glasgow, therefore, Dr Thomson aimed to revolutionize the teaching of mathematics while at the same time to remove the deficiencies in what he saw as an otherwise sound system of education.

Dr Thomson's 'Introductory lecture' at Glasgow College on the 6th November, 1832, took the same line as the 1825 review article, praising the successes of the continental men of science compared to those of Newton's British disciples. Dr Thomson admitted that the Gregorys, Maclaurin, Simson, Robison and Playfair in Scotland had all achieved 'well merited and lasting fame' but he urged a reawakening of Scottish genius after the laurels of science had been allowed 'to fall to Laplace & his great associates'. Two years later he felt able to add to his lecture the remark that a significant change had commenced in British mathematics: unlike a quarter century earlier, Laplace and Lagrange were now intelligible to and read by hundreds of mathematical students, many of whom 'if they have time & talents, are thus placed in such circumstances as may enable them to contribute, in a greater or less degree, to the extension of science by new discoveries'.[37]

His increased optimism may well have been encouraged by the appointment of J.D. Forbes (1809–68) to the Edinburgh natural philosophy chair in succession to Leslie in 1833. Forbes became in due course a good friend of both Dr Thomson and especially of his son, once William Thomson had become the Glasgow natural philosophy professor. Although not educated at Cambridge, Forbes, a tory and an episcopalian, had been an admirer of the analytical methods introduced into Cambridge mathematics, and he had put to good use the advice given to him by William Whewell regarding suitable texts on physical science – Poisson's *Mécanique*, Airy's *Tracts* and Fourier's *Théorie analytique de la chaleur* among others – which employed analytical methods.[38] During the 1830s Forbes firmly established his standards at Edinburgh. In the session of 1836–7, he had as pupils the first two professors of engineering at Glasgow, Lewis Gordon and Macquorn Rankine; subsequent pupils were P.G. Tait (1831–1901) and James Clerk Maxwell (1831–79).[39] With emphasis on quality and merit Forbes's classes were generally small, but his career at

[37] James Thomson, 'Introductory lecture at Glasgow College, 6th November 1832', QUB. For a very similar perspective, see [J.F.W. Herschel], 'Mechanism of the heavens', *Quart. Rev.*, **47** (1832), 537–59.

[38] See Crosbie Smith, '"Mechanical Philosophy" and the emergence of physics in Britain: 1800–1850', *Ann. Sci.*, **33** (1976), 3–29 on pp. 25–6. The full titles of the principal recommended texts were: S.D. Poisson, *Traité de mécanique* (Paris, 1811); G.B. Airy, *Mathematical tracts* (Cambridge, 1826); Joseph Fourier, *Théorie analytique de la chaleur* (Paris, 1822).

[39] Smith, 'Emergence of physics', p. 27.

Edinburgh marked another crucial phase in the reform of science and mathematics at the Scottish universities.

Forbes's election, however, in preference to David Brewster (1781–1868), represented an anglicizing tendency in Edinburgh, parallel to the Oxbridge trend (Sandford and Ramsay) evident in Glasgow, and further confirmed by the election of the Cambridge mathematician Philip Kelland (1808–79) rather than D.F. Gregory (1813–44) to the Edinburgh mathematics chair in 1838.[40] Nevertheless, the strong tide of reform did not flow entirely from the south, for in 1836 John Pringle Nichol was appointed to the Glasgow chair of practical astronomy.

John Pringle Nichol: the education of the people

The death of Professor James Couper at the age of about eighty-three left no astronomy classes untaught, for the aged professor had not lectured since 1808. The chair was in the gift of the Crown but, unlike the newer regius chairs, carried with it a salary of £270 per annum. The College, furthermore, had £900 ready for the purchase of instruments, with a new observatory and house in prospect also. Not surprisingly, Thomas Thomson made a bid for the chair as a way of improving his worldly comforts ahead of old age.[41] A much younger and more promising candidate, Archibald Smith, about to become senior wrangler in the Cambridge Senate House Examination of 1836, was urged to apply to the Whig Home Secretary, Lord John Russell, by the Glasgow professors of Greek, Sandford, and of humanity, Ramsay, but exhibited a curious indecisiveness. Since Sandford and Ramsay represented a new Oxbridge tendency in the College, we may understand their enthusiastic support for Smith of Trinity. Both seemed convinced that with the support of William Whewell and especially G.B. Airy (1801–92), the Astronomer Royal, Smith would carry the day. Ramsay, indeed, as a former pupil of Whewell, spoke reverently of the twin names of Whewell and Airy as 'a tower of strength'.[42]

By contrast, the ultimately successful candidate, J.P. Nichol, lacked the high mathematical reputation of a senior wranglership and the status of Newton's Cambridge college. Educated at King's College, Aberdeen, and originally destined for the Church, he had distinguished himself at Aberdeen in mathematics and physics. Subsequently headmaster in three schools, he applied unsuccessfully for the chair of political economy at the Collège de France, having been recommended by James Mill. He began a lengthy correspondence with John Stuart Mill in 1833 concerning political economy. This correspondence demonstrated Nichol's alignment with the 'philosophic radicals', represented now by

[40] G.E. Davie, *The democratic intellect. Scotland and her universities in the nineteenth century* (Edinburgh, 1961), pp. 116–26. [41] Morrell, 'Thomas Thomson', pp. 255–6.

[42] D.K. Sandford to Archibald Smith, 15th January, 1836; William Ramsay to Archibald Smith, *c.*17th January, 1836; Archibald to James Smith, 15th January, 1836, TD1/745/1–3, Smith papers, Strathclyde Regional Archives; Coutts, p. 346.

the younger Mill, radicals who sought to diffuse knowledge to the common man and to replace the confused tangles of establishment patronage and rule with scientifically based, professional administration. As Mill put the matter to Nichol in 1834: 'At all events, whoever is in place, the march of Reform is wonderfully accelerated. How nobly and with what wisdom the people have acted.' Nichol's son, later professor of English literature at Glasgow University, confirmed his father's radical views in a biographical reference to his 'horrifying the landed interests of Fife by [publishing] revolutionary articles in the *Herald*'.[43]

Nichol's great gift was his power of rhetoric. One observer noted that 'his lectures were works of art, calculated at once to exercise and enlighten the intellect, quicken the imagination, and move the heart'.[44] William Thomson himself (as Lord Kelvin) praised his great teacher thus in 1903:

> You can imagine with how much gratitude I look upon John Pringle Nichol and upon his friendship with my father. His appointment as professor of astronomy conferred benefit, not only upon the University of Glasgow, but also upon the city and upon Edinburgh, and the far wider regions of the world, where his lectures were given and his books read. The benefit we had from coming under his inspiring influence, that creative influence, that creative imagination, that power which makes structures of splendour and beauty out of the material of bare dry knowledge, cannot be overestimated.[45]

Given the choice between J.P. Nichol and Archibald Smith for the astronomy chair, the key criterion applied was that of popular teacher over the quiet, donnish, and very indecisive qualities of the senior wrangler. Nichol thus represented a reversal of the Oxbridge trend in favour, not of the Glasgow establishment, but of a much more radical and democratic reform. Here, Nichol's friendship with the Mills must have helped his cause considerably. As J.S. Mill wrote to him in 1834: 'My father thinks that a professorship in a Scotch university would suit you; and it may be in his power to be of some aid to you in obtaining one, if it were vacant'.[46] Crucially, the astronomy chair was the gift of the Crown and not the College electors, and so Dr Thomson found himself with a new, dynamic ally within the College, an ally ready to break the power of the oligarchy.

Nichol's appointment is thus of considerable significance in our story. First, the failure of the Cambridge lobby has a direct bearing on the campaign for William Thomson's election to the Glasgow chair of natural philosophy exactly a decade later (ch. 4). Second, Nichol's success as a lecturer in attracting an

[43] W. Knight, *Memoir of John Nichol* (Glasgow, 1896), pp. xi–xiii, 6–7; 'Unpublished letters from John Stuart Mill to Professor Nichol', *Fortnightly Rev.*, **61** (1897), 660–78, on p. 671. For a useful summary of the role of the 'philosophic radicals' in the age of reform, see Beales, pp. 144–5. We are especially grateful to Simon Schaffer for making available the draft of his article 'The nebular hypothesis and the science of progress', in J.R. Moore (ed.), *The humanity of evolution* (Cambridge, forthcoming).

[44] Quoted in Knight, *Memoir of John Nichol*, p. xii.

[45] 'Lord Kelvin and his first teacher in natural philosophy', *Nature*, **68** (1903), 623–4. Quoted in D.B. Wilson, 'Kelvin's scientific realism: the theological context', *Phil. J.*, **11** (1974), 41–60, on p. 45.

[46] *Fortnightly Rev.*, **61** (1897), 669.

audience where none existed before reflected a major criterion for the election to a Glasgow chair – a criterion inapplicable to an Oxbridge chair, for example. Third, Nichol himself proved not only a well-known source of intellectual inspiration for William Thomson, but became an important political ally of the Thomsons in their reforming crusades. And, finally, Smith's lack of resolution – expressed at the time in a letter to his father in the words 'I am tired of Mathematics and look upon a legal life with more pleasure than I used to do'[47] – in deciding not to apply and subsequently applying, will provide a recurring contrast with the decisive, but often impulsive, character of William Thomson, not least in his marriage proposals to Smith's sister, Sabina (ch. 5).

Nichol's views on education, politics, religion, and science are best expressed in his 'Preliminary dissertation' written for the English translation of Willm's *The education of the people* published in Glasgow in 1847. Nichol sought a national, non-denominational system of education universally available. As with Dr Thomson, the words 'reform', 'usefulness' and 'practical' resound through the text. In particular, he approved of Willm's theory of education because it was no '*mere* theory or an abstraction without aim, [but] will be found to have bearings, on every side, on practical questions of the gravest import, and, in fact, virtually to control every arrangement admissible among our positive operations'. Theory and practice had to be in harmony. The point emerged even more forcefully when Nichol outlined the aim of education: 'to draw out man into freedom, and to establish between him and the universe, a solid and practical harmony'.[48]

Nichol warned of the danger of confounding the teacher's 'mere means or instruments necessary to his work, with the work itself'. Reading and writing, for example, were not education, but merely instrumental. Similarly, the 'logical function of the understanding' was not an end, whether through languages (Oxford classics) or abstract science [Cambridge mathematics], but 'the exercise of, nothing more than an *instrument*'. Failure to distinguish such instruments from ends produced dogma, mere opinion. In particular:

During a critical study of the physical sciences in their existing attitude, one is often required to marvel at the extent to which expert analysts become deceived as to the use and substance of their existing phalanxes of symbols; for it seems most unlikely, *a priori*, that any acute mind could so mistake the value and position of a mere instrument – as, in the pride of his power to wield it – to feel at liberty to withdraw from that careful and sagacious scrutiny of elementary processes, which are the essential foundation of every just philosophy of Nature. Now, this very error prevails, and to an extent perhaps not often imagined, through the whole range of moral and intellectual inquiry . . . Many and many a Ptolemaic system – an ingenious product of mere logic – exists still around us! Many a potent mind still wastes its energies in coordinating and adjusting phantasms: for, after all, men contend the most bitterly regarding the houses they construct of shadows![49]

[47] Archibald to James Smith, 15th January, 1836, TD1/745/1–3, Smith papers, Strathclyde Regional Archives. [48] Nichol, 'Preliminary dissertation', pp. xiv–xv.
[49] *Ibid.*, pp. xxvi–xxix.

Metaphysics, dogmatism, symbols, abstractions: these terms William Thomson would despise throughout his life. In this passage we see the very essence of a Scottish approach to intellectual inquiry, confirming Dr Thomson's methods, and above all Nichol's anxiety to stress the distinction between instruments and ends, between the tools of the knowledge-maker and the knowledge itself.

In a footnote to the above passage, Nichol denounced abstract mathematics, whether geometry or modern analysis, as 'a very imperfect or partial intellectual discipline' since, 'being almost wholly a science of *deduction*' it seldom required 'one to refer, in the way of strict scrutiny, to first principles'. Hence also in applied mathematics 'we so often find a fatal inattention to the necessary ground-work of fact – the substitution of *Logic* for *Philosophy*'. Let it never be forgotten, he concluded, 'that it is far more important to cultivate *powers of thought* than mere *cleverness*'. Here Nichol quite explicitly advanced Scottish educational values over English (notably Oxbridge) ones, for, as a recent historian of James Clerk Maxwell has summarized the difference, the Scots aimed to make a general education (organized around philosophy) the foundation for particular technical studies, whereas the English aimed to make particular technical studies (with an emphasis on intellectual discipline) the basis for a general education.[50] Correlatively, the Scots saw their education system as fundamentally democratic, the English as elitist.

In the later parts of his 'Preliminary dissertation', Nichol examined not only the means of realizing a universal system of education, but also the major hindrance to its implementation on account of sectarian differences. Here, Nichol's latitudinarian approach strongly coincided with that of Dr Thomson. The question, he argued, was 'not about separating the training of the intellect from training in religion – but *how far* are we precluded, by respect due *to the discrepancies of sects*, from accompanying the training of the intellect, in a common school, with all the aids and illustration it might receive from its connection with man's religious nature?' In other words, he believed that a religious spirit 'shall pervade all teaching' but he denounced the system of tests whereby in many educational institutions the teacher had to subscribe to the special articles of a church. Both the existence of design in the universe and the view which represented God as a Providence 'are sufficiently remote from relationship with the matters concerning which our churches are divided'. Thus 'to secure that the teacher be a religiously disposed man, it is unnecessary to descend among these disputed details . . . *such subscriptions are wholly unfitted to realize that object*'.[51]

Nichol linked the issue with his previous criticism of a false metaphysics in physical science:

> that anomaly is easily explained which presents us so frequently with high and severe Churchmen – stern and rigid supporters of systems of Articles, and other dogmatic forms

[50] C.W.F. Everitt, 'Maxwell's scientific creativity', in Rutherford Aris, H.T. Davis and R.H. Stuewer (eds.), *Springs of scientific creativity* (Minneapolis, 1983), pp. 71–141, on p. 84.

[51] Nichol, 'Preliminary dissertation', pp. liv–lxiii.

> . . . [as] arising in the activity of the logical faculties, and the comparative inertness of the powers of contemplation; and it has an exact counterpart in a phenomenon already referred to, connected with the cultivation of physical science. Men . . . are far from uncommon, who, while enjoying the greatest pleasure in the analytic representation and development of assumed Physical Laws, have yet but imperfect powers to sift thoroughly the physical facts on which alone laws can be founded; and, in the same manner, it is quite possible that a mind have much interest in the processes and investigations of systematic, or, rather, of dogmatic theology, without a corresponding power to descend into the far profounder region of INTUITIONS.[52]

Elsewhere Nichol contrasted 'the universal method of Christ' with 'the varied interpretation of St Paul's views', 'the principal source of sectarian discordances'. He stressed the importance of discussing the subject of a common or united education 'not as Sectarians, but as Christian men'.[53] By correlating dogmatism in theology with hypothetical systems in physical science, 'ingenious product[s] of mere logic', Nichol expressed a theme central to William Thomson's deepest scientific commitments.

Dr James Thomson: 'latitudinarian' and university reformer

If the 1830s saw Dr Thomson professionally occupied in building the reputation and restoring the respectability of the mathematics classes, then the 1840s saw him deeply involved in a series of major crusades for reform in Glasgow College. The programme centred on two major issues: the status of the regius professors and the abolition of the religious tests. An outright victory for the whigs on either of these issues would have overthrown the tory oligarchy irreversibly, for most of the regius professors outside the College were reformers, identified with science, and the abolition of tests would have opened up College chairs to Anglicans and dissenters as well as presbyterians, thereby undermining the whole basis of the tory establishment.

As the campaign developed, it became clear that the whig party within the College was moving ever-closer to toppling the old, high-tory establishment. From a minority position in 1832, the reformers, led by Dr Thomson, J.P. Nichol, and Dr William Thomson from 1841, began to command a majority in Senate which, consisting of all professors, elected the Dean of Faculty, who in turn could, along with the Rector elected by Senate and matriculated students, vote in the election of College professors.

By mid-1842, death had removed three of the four remaining regius professors within the College. The balance of power now shifted. The fifteen votes available for decision making within the College, and especially the fourteen for the likely election of a new natural philosophy professor, were made up as follows: five ultra-tories (Fleming, Jeffray, Hill, Gray, and Buchanan), three ultra-whigs (Dr Thomson, Nichol, and Dr William Thomson), three professors

52 *Ibid.*, p. lxiv. 53 *Ibid.*, pp. lxvi–lxvii*n*.

Dr James Thomson at the age of about sixty-two. To friends he was a 'stern disciplinarian' who 'did not relax his discipline when he applied it to his children, and yet the aim of his life was their advancement'. To his children he seemed both father and mother. [From a pencil drawing by his daughter Elizabeth, National Portrait Gallery, London.]

likely to support reform (Ramsay, Reid and Lushington), and one professor of moderate conservative views (Maconochie). The Rector, Dean, and absent Meikleham made up the remainder. In such a finely balanced power structure the election of Rector and Dean would be a matter of great moment, since six tory and six anti-tory votes from the professors could be turned into an overall tory victory if either the Rector or Dean were tory, the Principal having a casting vote. A whig Rector *and* Dean, on the other hand, would almost guarantee a whig success. The struggle for control began with the tests.

University tests, by which a new professor had to sign both the Confession of Faith demanded by the established (Presbyterian) Church of Scotland and the Formula of the Church of Scotland declaring an intention to conform to its worship and discipline as established by law, naturally aroused Dr Thomson's opposition. That opposition takes on a special significance if we recall that his presbyterianism made him a member of the Established Church in Scotland, but a dissenter in Ireland.

At Edinburgh, the university tests had been allowed to lapse – Forbes and Kelland after all were Episcopalians – but Glasgow quite rigidly adhered to them. Conservatives in theology as in politics, Principal Macfarlan's party opposed any change in the status of the university professors as established presbyterians, and saw the emergence of dissenting evangelicals and other such groups as threatening to break up the entire structure of Scottish society. The failure of the College to appoint the academically distinguished evangelical Thomas Chalmers to the divinity chair on his academic merits probably gave an impulse to the issue of university tests from 1840, particularly as his successful rival, Hill, was a firm supporter of their retention. Dr Thomson dissented from his appointment.[54]

The bitterness aroused by Hill's appointment erupted again in May, 1841, over the election by Senate of a new Dean of Faculties. The regius professors and reformers seized an opportunity to exert their as-yet limited power. Thomas Thomson proposed, and Dr Thomson seconded, the whig Lord Dunfermline, former Speaker of the House of Commons in the late 1830s and MP for Edinburgh. The Principal supported the tory Sir Archibald Campbell. As a result of the (possibly unconstitutional) vote of the retiring (tory) Dean, ten electors voted for each candidate, and the Rector, the Marquis of Breadalbane (president of the Glasgow meeting of the British Association and so sympathetic towards the science professors) gave his casting vote for Dunfermline. Meanwhile, members of the Principal's party began a virtual witch-hunt. They accused Nichol and John Couper (materia medica) of not having given evidence

[54] For a general sketch of the debate over university tests at Glasgow College, see Coutts, pp. 419–21. On Chalmers, see William Hanna, *Memoirs of the life and writings of Thomas Chalmers* (4 vols., Edinburgh, 1849–52), **4**, pp. 212–13. William Thomson himself provided a good analysis of the nature of the tests at Glasgow and Edinburgh in a letter to H.W. Cookson, 28th November, 1847, C151 (copy), ULC.

of signing the Confession of Faith; they protested against John Burns (surgery) and W.J. Hooker (botany) as Episcopalians; and they objected to Lewis Gordon's signing the Confession before the Presbytery of Edinburgh.[55] All this wrath was thus directed against the regius professors.

The defeated candidate went much further, however, and sent a revealing letter (which ended up in the hands of Dr Thomson) to the most recently elected College professor, Lushington, the Cambridge-educated professor of Greek. Campbell reminded Lushington that there had been three or four candidates of equal merit for the Greek chair, and that if the electors had not been '*misled by you in regard to your political opinions*, you would not have been placed in the situation you now occupy'. Sir Archibald went on to express the central tenets of the Principal's party in the College:

Indeed I consider it absolutely essential and hold that nobody ought to fill the Greek Chair who is not a first rate Greek scholar: but I consider it also to be of great importance that those who fill the Chairs of the University should be staunch upholders of the British Constitution in Church & State, and this more particularly applies to the person who is to teach the *Greek* class, because he has the superintendence of pupils so young and inexperienced that their political opinions neither are nor ought to be formed . . . One object of your letter seems to be to persuade me that in voting for a Whig Dean of Faculties, you did not depart from the political sentiments you professed yourself to hold in one interview in my house in Edinbro'. I must be allowed to say that *you have departed from them* by voting for a Noble Lord certainly of first rate abilities, but who has been all his life an Ultra Whig, & who when in the House of Commons carried those principles farther than most of his party.[56]

Campbell also noted shrewdly that the real aim in the appointment was 'to get a majority of Whigs in the *Senate* of the University, and by that means to obtain a majority of them also *in the* Faculty who elect a considerable number of the Professors to effect which no stone has been left unturned for some years, *with the assistance of the Government*, who have established several new Regius Professors appointed by the Crown'. Lushington, Campbell concluded, had betrayed the sound basis upon which he had been elected.

The election of Dunfermline set the scene for a long series of battles between the crumbling, aging Principal's party and the new generation. In 1842, Dr Thomson entered in the Faculty Records his opposition to a paper agreed upon by the majority of the faculty to be sent to the government rejecting the claims of the regius professors for equality with the others. Thomson's arguments were given a very hostile reception, and the Principal and Dr Fleming produced a document filling eight folio pages of closely written manuscript to be inserted in the Minutes of Faculty. It contained, wrote Thomson, 'from the brains of its two learned parents so many hard hits and home thrusts, that . . . I am still in a very

[55] Coutts, pp. 419–20.
[56] Sir Archibald Campbell to Edmund Lushington, 13th May, 1841, C10, ULC.

precarious and doubtful state'.[57] Nevertheless, Thomson had made his point, and in doing so had become the first member of the sacred inner circle of College professors to campaign vigorously for the rights of their regius colleagues.

1843 marked the crucial point in the debate over the tests, as it was the year of the Disruption in the Church of Scotland – when Thomas Chalmers led his many supporters out of the established Kirk, and founded the Free Church of Scotland. The Disruption occurred over the question of patronage, that is, Chalmers formed the Free Church in support of the principle that each congregation had the sole right to choose its minister, whereas the Established Church had agreed to allow a measure of indirect State (through a wealthy or influential patron, for example) intervention in the appointment of a minister to a parish.[58] As far as Glasgow College was concerned, the immediate effect of the Disruption would have been the exclusion from university chairs of Free Church members, who could not sign the tests.

In October the whig Rector, Fox Maule, and the whig Dean, Sir Thomas Brisbane, brought forward at a Senate meeting a motion petitioning for the opening of the University to all academics whatever their denomination. An extraordinary scene ensued. The Principal found that 'business of a public nature rendered it impossible for him to attend', and he stated further that he doubted whether Maule had a right to act as Rector, having joined the Free Church. Dr Fleming, the eccentric professor of moral philosophy, 'found a country walk necessary for his health' just before the meeting and Dr Hill was to be seen hastening after him, while a fourth, but less committed, member of the Principal's party, Maconochie, took to bed. Open hostilities broke out when the reformers decided to call another meeting of Senate the next day, and if that did not take place, then one was to be arranged for each successive day thereafter, until the Principal's party found it convenient to attend.[59]

On 7th November, 1843, the Senate carried a motion for petitioning parliament by eleven votes to seven. It was, as Anna Thomson noted in a letter to her brother William at Cambridge, a 'very great triumph [of] the Whig party in the College'. The Rector and Dean joined with many of the 'whig' professors to dine at the Thomson's after the Senate success and celebrated the defeat of the moderate – synonymous by now with the tory – party. Anna added with tongue-in-cheek:

> The poor Principal must be sadly mortified at such a signal defeat and I must say I cannot help feeling sorry for him; it must be with a heavy heart that he sees the power he has enjoyed so long ebbing gradually from him and falling into the hands of a set of agitators and demagogues . . . What would your Cambridge friends say to all the movements in

[57] Dr James Thomson to William Thomson, 21st February , 1842, T197; 6th April, 1842, T207, ULC.

[58] The question of church patronage was discussed in a letter from Rev David King to William Thomson, 15th August, 1847, K86, ULC.

[59] Anna to William Thomson, *c.*10th October, 1843, B199, ULC.

the Scotch Universities? I am sure they must think it very radical and if they heard of papa taking such a lead you would be expelled from the College for being a Chartist, so I suppose the less you say about it the better. Only do not let yourself become imbued with toryism as I have begun to think it a narrow selfish creed and should not like any one I take so much interest in as I do to become a tory.[60]

Glasgow College now looked very much less like a stronghold of conservatism. The Principal's party, however, fought back, alleging that the petition was contrary to a basic condition of the Act of Union between England and Scotland, inconsistent with the constitution of the British Empire and with the oath taken by Queen Victoria at her accession to maintain the Church of Scotland, and injurious to the University. They also attempted to relieve Fox Maule of his rectorship on the grounds of his being a staunch Free Church man. At St Andrews too the Presbytery proposed on the same grounds that Sir David Brewster be expelled as Principal.[61]

The warfare at Glasgow erupted in public during the spring of 1844 in response to a report of a speech delivered by Principal Macfarlan to the Synod of Glasgow and Ayr. According to the report, the Principal had argued for the maintenance of the tests as a means of transmitting high principles from teachers to classes so that if these classes would show 'by their profession and the purity of their lives, their attachment to . . . Christian truths, their examples might have the happiest effect over all parts of the people, in raising a religious and intelligent race of men, to promote, from generation to generation, the interests of this great country'. Failure to retain the tests could in the end mean having 'the sciences taught by a Roman Catholic, Hebrew by a Jewish Rabbi, the Eastern Languages by a Mahometan Mollah, and, above all, they might have a Jesuit to teach Ecclesiastical History'.[62]

Dr Thomson published a carefully worded but caustic reply in the *Glasgow Argus*. He argued that attendance at the College Chapel had been reduced to one or two professors and not many more than a dozen students. Only four of the non-theological professors said a short prayer once a day. Was this all, he asked of the Principal, 'that is produced by the vaunted operation of the university tests, any relaxation of which we are wished to believe would injure religion?'. He then struck hard at what he saw as the real motive for the Principal's stand: 'Were the Rev Principal not led away by party spirit, and by a wish to continue to a sect a privilege, which, if it were not already secured by law, no legislature of the present day would ever grant, he would not speak in the face of the facts . . .'. Even the Principal had admitted, he went on, that the extreme consequences of abolishing the tests, in the form of various notorious infidels taking over the hallowed University chairs, was 'perhaps not probable'. Arguing that 'in

[60] Dr James Thomson to William Thomson, 24th October, 1843, T250, ULC; Anna to William Thomson, early November, 1843, B200, ULC. [61] *Ibid.* See also Coutts, pp. 420–1.

[62] Quoted in the *Glasgow Argus* off-print of an article by Dr James Thomson, enclosed in a letter from John Thomson to William Thomson, 27th April, 1844, T504, ULC.

universities, and all other seminaries, respect should constantly be shown to religion', Thomson at the same time set out the case for the abolition of exclusiveness and narrowness from the institution:

If the Professors in our universities wish, as they naturally and properly do, to have their classes attended by dissenting students, ought they, in common justice, to endeavour to have those dissenters excluded from academic chairs in their maturer years? They have themselves trained them. They have received fees from them . . . and those students have often formed a large proportion of their classes . . . Ought they to lose their services to the University and the public, merely because they belong to a religious society not connected with the Establishment, but, in the great majority of instances, differing from it in no essential point in reference to doctrine? Can they feel comfortable in the idea that many a man might say, with justice, 'I see by the newspapers that such a one, an old class-fellow of mine, has been appointed to such a Professorship; at college I was much his superior, but *he is a churchman and I am a dissenter*. I might be a Member of Parliament, or I might hold any responsible office under her Majesty, but, by the rigid enforcement of a small remnant of former intolerance, I am excluded from a university chair'.

Early in 1845, the issue of the tests reached parliament. A deputation which included Sir David Brewster and the Lord Provost of Edinburgh went to London to wait upon the government. Brewster carried with him a petition from the majority of the professors at St Andrews, and he kept Dr Thomson informed on the progress of the talks with Prime Minister, Sir Robert Peel, and Home Secretary, Sir James Graham. Thomson noted with satisfaction in a letter to William that Peel 'is clearly a more ultra whig than Lord John Russell'. By May, 1845, the new whig Glasgow College Rector, Andrew Rutherfurd, MP, and former Lord Advocate, had, with the government's consent, got leave to bring in a private bill for the abolition of the Scottish university tests. Surrounded by a mood of optimism among his supporters, Rutherfurd introduced the Bill into parliament in late May, 1845, but it was rejected by eight votes at the second reading. The professor of humanity, Ramsay, Dr Thomson and a former opponent of the abolition of the tests, Maconochie, summoned the University Senate in March, 1846, which then agreed by a large majority to petition parliament again. After much delay a Bill was put through in 1853, substituting for the old tests the requirement that all future professors outside the divinity faculty should instead declare that they would not use their office to subvert the Church of Scotland, the Westminster Confession, or the divine authority of the Scriptures.[63]

William Thomson staunchly supported his father's latitudinarian outlook. On his appointment in 1846 to the Glasgow chair of natural philosophy, William was not required, 'through accidental circumstances', or a deliberate oversight, to subscribe to the Confession and Formula. He subsequently stated

[63] Dr James Thomson to William Thomson, 8th March, 1846, T330, ULC. See also Coutts, p. 421.

that 'I have long made up my mind, if asked, to refuse to sign, now, or any time in future'.[64] Meanwhile, William had adopted 'the habit of regularly conforming to the Episcopal Church [the Anglican Church in Scotland], & not appearing more than once or twice or three times in the course of a session at an Established Church [the Church of Scotland]'.[65] This confession led one of his Cambridge friends to congratulate him on acquiring the title of 'latitudinarian' at Glasgow – 'there I suppose it is equivalent to being a member of ye Episcopal Church'.[66]

More seriously for Glasgow College, the continuation of the tests until 1853 was sufficient to deter the Anglican G.G. Stokes from becoming a candidate for the Glasgow chair of mathematics made vacant by the death of Dr Thomson early in 1849. William Thomson tried his utmost to persuade Stokes to become a candidate, the tests notwithstanding. Stokes, however, stated that as he did not intend becoming a presbyterian, he could not, without acting hypocritically, sign the tests in a lax sense *unless* the University professors as a body were agreed that it could be signed in such a non-rigorous manner. Since the electors were clearly divided over the force of the tests, Stokes felt that, if elected, the end result for him would be wholly undesirable: 'I should be more disposed to skulk away like a dog with his tail between his legs than to defend my conduct unless driven into a corner & compelled to fight'.[67]

Although heated arguments over issues such as the Tests within Glasgow College were rare after 1850, William Thomson throughout the remainder of his life was always ready to take a firm stand for non-sectarian principles. Nowhere was this point more clearly illustrated than during a public meeting held in 1871 to promote Scottish 'united non-sectarian-compulsory education on the same model as the Irish national education which Roman Catholics, Presbyterians, and Church-people would not give fair play to in Ireland'.[68]

William chaired the meeting and spoke of the need to avoid the 'utter and destructive denominationalism' characteristic of Ireland. Rather, he argued, 'Scotland, of all the countries in Christendom, was the one more prepared, most ready to accept a united non-sectarian national system of education', which would be 'a thoroughly-religious system' and not a 'godless' one. The Bible, he insisted, should not be excluded from the schools, for it was 'truly and avowedly national', and not distinctive of any particular denomination. What had to be excluded from the teaching was a catechism or formulary belonging only to one or other of the several religious denominations within Scotland.[69]

Subsequent chapters will ramify these non-sectarian principles for William

[64] William Thomson *op. cit.* (note 54).

[65] William Thomson to G.G. Stokes, 14th January, 1849, K29, ULC.

[66] W.W. Herringham to William Thomson, March 1847, H97, ULC.

[67] William Thomson, *op. cit.* (note 65); William Thomson to G.G. Stokes, 20th February, 1849, K30; 24th February, 1849, K31; G.G. Stokes to William Thomson, 12th February, 1849, S340; 16th February, 1849, S341; 17th February, 1849, S342, ULC.

[68] William Thomson to Mrs P.G. Tait, *c.*23rd April, 1871, in SPT, **2**, 590.

[69] SPT, **2**, 590–1. Extract from the report of Sir William Thomson's speech in the *Glasgow herald*.

Thomson, displaying the intimate relation of his views on religion and science. Throughout his scientific career a desire to avoid metaphysics and dogma dominated his methodology, making his latitudinarianism manifest in his academic work as in his religious practice.

Dr Thomson's family

J.P. Nichol's son wrote of Dr James Thomson that 'he was a stern disciplinarian, and did not relax his discipline when he applied it to his children, and *yet the aim of his life was their advancement*'. Certainly his cultural beliefs shaped the views of his children. The lives of three of his sons, James, William, and John, express directly their father's ambitions and aspirations. Such willing conformity to a father's teaching may seem anomalous in the light of present-day assumptions about the centrality of oedipal conflict, until one recalls that Mrs Thomson, who might have been the essential object of such conflict, died when William was only six. Dr Thomson then devoted his entire energy to the emotional well-being of the family, playing, according to Elizabeth, the roles of both father and mother.[70]

Dr Thomson's children received much of their early education at home. Once he had initiated them in the wonders of knowledge, their education became an almost self-regulating, self-motivating process, with an older child teaching a younger, and the particular interests of one providing inspiration for the advancement of the others. Apart from a brief spell in the school department of the Belfast Institution, and attendance as spectators at their father's mathematical class in the 1832–3 session at Glasgow, William and his elder brother James did not receive any formal education until they matriculated at Glasgow College in 1834 – William being then aged ten – and enrolled in Professor Ramsay's humanity class. In the following session of 1835–6, William and James attended natural history and Greek, while in 1836–7 they both entered the junior mathematical class conducted by their father.[71]

The session 1837–8 saw them in the senior mathematical class, with attendance also at Professor Buchanan's logic classes, but not until 1838–9 did they begin to study either natural philosophy under Meikleham (the junior class) or chemistry under Thomas Thomson. James and William were then entitled to the Bachelor of Arts degree, although only James took his BA, which for him was a prelude to an MA in mathematics and natural philosophy in 1839–40. In his final sessions at Glasgow College, William followed the moral philosophy (Professor Fleming) and the senior natural philosophy (largely Professor Nichol) courses in 1839–40 and the senior humanity class (Professor Ramsay) in 1840–1. In 1838, Dr Thomson's third son, John, who had previously attended

70 Knight, *Memoir of John Nichol*, pp. 19–20. Our emphasis. King, *Early home*, p. 87.
71 King, *Early home*, pp. 87, 107–9, 115; SPT, **1**, 7–9.

Glasgow High School, matriculated at Glasgow College to study for his BA degree.[72]

The long summer vacations, extending fully from May until October, provided a unique opportunity for escape from the grim city environment and the darkness of a Scottish winter to 'the verdant banks of the muse-inspiring Clyde'.[73] The Isle of Arran (to which the Nichol family also often migrated) and Knock Castle, near Largs, were favourite places for the summer sessions, and to these William Thomson returned again and again in later life. With its rich geological and biological storehouse, Arran almost certainly inspired his remark in 1885 that 'the laboratory of the geologist and the naturalist is the face of this beautiful world. The geologist's laboratory is the mountain, the ravine and the seashore'.[74] The boating and fishing and the mountain-walking formed the very special attractions of the remote Arran, which also provided such a splendid landscape for Elizabeth's artistic sketches. Equally, Arran in 1834 or 1835 was the setting for an early expression of William Thomson's own failure to respond seriously to religious enthusiasm, as his sister Elizabeth recorded:

On Sunday mornings the family went, in an open cart of the country, seated with rough benches and straw, across from Brodick to Lamlash, where was the parish church. It happened one Sunday that there was a revival service, and the congregation grew much excited, uttering loud exclamations and groans; and at last some of the old women began to give vent to their feelings by tossing their Bibles in the air. This tickled Willie's sense of humour, and he shook with smothered laughter, which started all the other boys laughing too. Our pew was close under the pulpit in full view of the preacher, who, looking down, administered a grave rebuke. The smothered laughter then exploded, and the minister, pointing his finger at the ringleader, exclaimed, 'Ye'll no lach when ye're in hell!' This was too much: and Willie rolled clean over on to the floor. For some reason or other our father was not with us that day, and I was in charge of the party. Crimson with shame, I bustled them all out of the church as quickly as I could.[75]

Dr Thomson had become prosperous enough to afford two major departures from the usual family vacations in 1839 and 1840. In 1839 the family forsook the preferred 'land of mountain and the flood' for a visit to London and eventually to Paris. Dr Thomson characteristically conducted members of his family around not only the sights of London, but also took them to see those parts of the great metropolis which were seldom on a tourist itinerary. The family saw the poorer districts as well as the fashionable squares and crescents. The following year, they travelled by London, Rotterdam and the Rhine to Frankfurt, having been joined by the Nichol family at Bonn. While most of the party remained in Frankfurt, Dr Nichol embarked on a walking tour of the Black Forest with James and William.[76]

[72] SPT, **1**, 9–14; James Thomson, *Papers*, p. xix; NBs 9, 10, 21, ULC; King, *Early home*, pp. 118–30. [73] Robert to William Thomson, 23rd July, 1843, T547, ULC.

[74] PL, **2**, 476. [75] King, *Early home*, pp. 120–1.

[76] SPT, **1**, 15–18, 20–2; King, *Early home*, pp. 146–90; William Thomson, Journal of trips to London, France, and Germany, NB7 and 8, ULC.

The Firth of Clyde: looking from the foreshore near Largs (later the Kelvin country seat) across the Cumbrae Isles to the rugged granite peaks of Arran, the Thomsons' favourite summer retreat after the rigours of a Glasgow winter. Professors Nichol and Meikleham also frequented Arran. [From a water-colour drawing by William's niece, E.T. King, in Elizabeth King, *Lord Kelvin's early home*.]

James became the first member of the family to embark upon a profession – that of engineering. Knowing that his eldest son had developed a passionate interest in all things mechanical, Dr Thomson consulted a professional engineer, John McNeil, in London during the 1839 visit, and once the German trip was over James entered McNeil's Dublin office along with Professor Meikleham's son, Edward. The firm was engaged in civil engineering projects, notably the development of the Irish railway system. James, however, had to return to Glasgow after less than three months as a result of a knee injury sustained while walking in the Black Forest, and the misfortune proved to be only one of many setbacks through ill-health and bad luck which were to dog him throughout the early part of his career.[77] James returned to Glasgow College for the 1841–2 session, when he was able to study engineering under the new professor, Lewis Gordon, to develop his practical interests through the Glasgow Philosophical Society, and to widen his activities with a course on botany. In April, 1841, for example, he wrote to William regarding a meeting of the junior Philosophical Society at which the professor of medicine's son, John Thomson, read a paper on the steam-engine 'illustrated by my model which did its duty capitally' and in February, 1842, of the senior Philosophical Society at which James read a paper on a new kind of river boat. Other Society meetings of special interest to James

[77] King, *Early home*, pp. 190–1; James Thomson, *Papers*, p. xx.

James Thomson at the age of sixteen. Eldest son of Dr Thomson, James's engineering enthusiasms from an early age provided enduring inspiration for his brother's physics. [From a pencil drawing by his sister Elizabeth in 1838, National Portrait Gallery, London.]

concerned losses of power in hydraulic engineering – a major issue in the development of the brothers' thermodynamics (ch. 9).[78]

One year prior to William's entry into St Peter's College, and only a couple of months after the German tour, an event of considerable significance, not least for the two eldest Thomson brothers, took place in Glasgow. In September, 1840, the British Association for the Advancement of Science gathered in Glasgow on the occasion of its tenth annual meeting since its foundation at York in 1831. As

[78] James to William Thomson, 1st April, 1841, 12th February, 1842, and early 1840s, T378–80, ULC.

shown in a recent detailed study of the early years of the BAAS, the ideology of the 'gentlemen of science' or managers of the organization tended to encompass Whig or Peelite Conservative politics, liberal Anglican or Broad Church, latitudinarian religious attitudes, and a dominant concern with university life and education.[79] The ideology of these gentlemen of science thus coincided to some extent with the ideology of Dr Thomson (although his commitment to democracy and diffusion scarcely matched their elitism).

The idea of a Glasgow meeting originated in the Philosophical Society, and subsequently received enthusiastic support from all quarters of the city. Four specialist committees dealt with finance, accommodation, an exhibition of models and manufactures, and a museum of minerals found in the West of Scotland. Dr Thomson's eldest son, James, became secretary of the models and manufactures committee at the centre of a major correspondence network spread over more than forty towns.[80] Then only eighteen, James approached the task of organization with all the zeal and efficiency characteristic of his father. His sixteen-year-old brother William also gave secretarial assistance as required. In printed circulars to members of the model committee James urged:

> the immediate necessity of their strenuous individual exertions in obtaining MODELS, WORKS OF ART, SPECIMENS OF LOCAL MANUFACTURE, &c. &c. for the proposed Exhibition . . . I may add, that without such personal exertions on the part of the Model Committee, it will be impossible to mature [*sic*] an Exhibition in any degree worthy of our City, or of the British Association . . . All Models, &c . . . will be conveyed by any of the Glasgow Steam Packet Companies, free of expense, from Liverpool, Bristol, Dublin, and Belfast, &c; and by the Forth and Clyde Canal Shipping Company from Edinburgh. Communications addressed to the Secretary will receive immediate attention.[81]

Professor Nichol, who organized invitations to foreign visitors, agreed to give four popular evening lectures on recent discoveries in science in connection with a preliminary exhibition of models held in April, 1840, well ahead of the September meeting itself, and aimed at mechanics and operatives 'at a very reduced rate [i.e. charge] of admission'.[82]

The Thomson brothers' role here highlights several features of our study. First, the exhibition itself gave pride of place to the steam-engine model, which had inspired James Watt's discoveries during his association with the University and to the engine of Bell's *Comet*, two major symbols of Glasgow's industrial progress. The family's optimistic view of human advancement could find no clearer expression than in the economic and technical progress taking place on Clydeside. Second, even at this early stage in their lives, the Thomson brothers were no mere passive spectators, but were already moving to the centre stage of

[79] Jack Morrell and Arnold Thackray, *Gentlemen of science. Early years of the British Association for the Advancement of Science* (Oxford, 1981), pp. 25–6. [80] *Ibid.*, pp. 212–13.

[81] James Thomson, Circulars to members of the model committee, TD 68/6/83–TD 68/6/86, Strathclyde Regional Archives.

[82] James Thomson to Secretary of the Glasgow Mechanics' Institution, 20th April, 1840, University of Strathclyde Archives; *Gentlemen of science*, p. 213.

William Thomson at the age of sixteen. The year 1840 brought his first reading of Joseph Fourier's *Théorie analytique de la chaleur* as well as active participation in the Glasgow meeting of the British Association for the Advancement of Science by helping his brother James organize an exhibition of mechanical models and manufactures. [From a pencil drawing by his sister Elizabeth, National Portrait Gallery, London.]

British science, aided not only by their father but by the active participation of Nichol. And, third, the interest in machine models not only emphasizes the practical, engineering pursuits of James, but gives concrete origins to William's life-long enthusiasm for the practical in general and for mechanical models in particular.

The aspirations and early career of Dr Thomson's third son, John, provides another illustration of the combination of professional advancement with practical activities which characterized these three children. John gained his BA at Glasgow College in 1843, and took second place in the written examination in natural philosophy under Professor Nichol. As with his brothers, natural philosophy seems to have been a favourite subject, particularly with Nichol to teach it:

> we are going on still at railway speed in the Natural Philosophy. Dr Nichol has been giving a full account of the nebular theories, w^h I like very much. His lectures have been most delightful lately. He has given me some of (your rival!!) Gauss's observations regarding magnetism to translate for him to w^h I am just going as soon as I close this.[83]

Experience at the large Candleriggs Street warehouse of J. and W. Campbell in 1843-4 disillusioned him in the ways of merchandising, and, after taking his MA in 1844, he turned to 'all the fatigues and anxieties of a medical man'. The new medical student was awarded a bursary of £50 a year for four years, and aimed eventually to follow in Dr William Thomson's footsteps by studying in Paris.[84]

If Glasgow were perfectly suited to the advancement of a medical vocation, the same could not yet be said of physical science. With William's mathematical precocity, any decision to advance further would almost inevitably have involved turning to a Cambridge education, for no other British institution had acquired the same reputation for a specialist and intensive training in advanced mathematics. Men such as Forbes and Brewster, though conceptually and experimentally second to none, were at a serious disadvantage where the new French mathematical physics was concerned, expressed, for example, in the treatises of Fourier and Laplace. Dr Thomson knew that a Scottish education was no longer of sufficient depth for a person intent on profound scientific researches which involved mathematics. Of this fact, Forbes had been painfully aware, and according to the famous mathematical coach, William Hopkins (1793–1866), 'men from Glasgow and Edinburgh require a great deal of drilling'.[85] To read, and be drilled in, mathematics at Cambridge was thus a logical step for William Thomson.

[83] John to William Thomson, 18th February, 1843, T499, ULC; Robert to William Thomson, 11th March, 1843, T543, ULC; Anna to William Thomson, 29th April, 1843, B190, ULC.

[84] Agnes Gall to William Thomson, 13th May, 1843, G1, ULC; John to William Thomson, 27th April and 10th May, 1844, T504–5, ULC; *Lanark*, pp. 69–70; John to William Thomson, 10th November, 1844 and 18th April, 1845, T512 and T518, ULC; Dr Thomson to William, 15th February, 1845, T297, ULC.

[85] William to Dr Thomson, 17th March, 1844, T254, ULC.

3

A Cambridge undergraduate

> . . . you are unravelling the mazy intricacies of a mathematical problem. There you sit all absorbed, deep thought broods upon your brow, now you pause and look with earnest eyes at the vacant wall, and now with sudden inspiration you turn to commit the happy thought to paper which lies before you. Your fire is almost out, your candle is unsnuffed, your once hot tea stands untasted on your table. As for the bread and butter, your neighbour Stow, his own store having run short, came in, and unobserved carried them off for his own supper. *Elizabeth to William Thomson, 1842.*[1]

'Three years of Cambridge drilling is quite enough for anybody.'[2] William Thomson uttered this indictment of Cambridge mathematics, with its system of deadening discipline, in a letter of March, 1844, to his father, almost a year before sitting the Senate House Examination which would mark the end of three years of intensive study. A decade earlier, another bright Glasgow student, Archibald Smith, had expressed a still more passionate disillusionment with Cambridge mathematics, declaring to his sister Christina in October, 1835, that he was 'getting heartily sick of mathematics – and the pleasure I anticipate from being again at home is much increased by the thought that I shall by that time have done for ever with the drudgery of mathematics and be able to apply myself to more pleasant and more profitable studies'. Even after becoming senior wrangler, Smith emphasized that he was 'quite tired of, I might almost say disgusted with, mathematics'.[3]

William's remark concerned the entry into Cambridge of one of Dr Thomson's most promising Glasgow College students, W.A. Porter (1825–90) whose brother James (1827–1900) later became master of Peterhouse and whose father, a friend of Dr Thomson, was an Irish clergyman of such limited means that Porter entered Cambridge as a sizar, the poorest financial and social category of undergraduate. In an illuminating appraisal of Porter's prospects of an eventual fellowship at a Cambridge college, William explained

[1] Elizabeth to William Thomson, *c.*31st October, 1842, K61, ULC.

[2] William to Dr Thomson, 6th March, 1844, T253, ULC.

[3] Archibald to Christina Smith, 4th October, 1835, TD1/693/10; Archibald Smith to William Ramsay, 22nd January, 1836, TD1/745/5, Strathclyde Regional Archives.

to his father the complexities, exclusions and restrictions which filled the Cambridge landscape in the mid-1840s:

If he [Porter] would not care to hold his fellowship more than ten years, Hopkins says that Trinity would answer very well, but if he would wish to hold it longer, without going into [holy] orders Trinity would not answer. The Peterhouse fellowships are rather better than the Trinity ones, at least at first, but then there is always some degree of uncertainty as to vacancies at a small college. If however he takes a very high degree, Hopkins says he would probably be tolerably safe to get a fellowship here . . . I think for a sizar Peterhouse would be a much pleasanter college than Trinity . . . as the sizars here mix entirely with all the other undergraduates, and are on quite an equal footing, while at a large college, they are necessarily thrown a good deal by themselves. Caius would also be a good college for Porter as the fellowships are tolerably unrestricted, but I believe they are very poor . . . All the other colleges, except Pembroke, either require their fellows to take orders almost directly, or restrict them to particular counties.[4]

This passage encapsulates much of what Cambridge represented: systematic social division at undergraduate level into sizars, pensioners and poll men, corresponding, respectively, to students without financial support, fee-paying students, and idle aristocrats who would not take a degree; fellowships richly endowed for the benefit of the Anglican establishment; few fellowships combining freedom from religious restriction with adequate endowment; and the advantages and disadvantages of small colleges such as Peterhouse over large ones such as Trinity.[5]

Nevertheless, the Mathematical Tripos offered the best mathematical foundations for the advancement of physical science available in Britain, and the decision to enter William in the college with which Hopkins was most closely associated reflects Dr Thomson's awareness of this fact. As Lord Kelvin wrote to George Darwin (1845–1912) in 1906 concerning changes in the Tripos:

I think it would be a damage to the University of Cambridge and to the advancement of Science throughout the world, if any regulations should be adopted by which the efficiency of the Mathematical Tripos, and the number of undergraduates taking it, might be seriously diminished. The unique strength of Cambridge, as a place of experimental research, and as a leader in the advancement of Science generally, has depended greatly on the mathematical foundations given to a large proportion of all the undergraduates by the Mathematical Tripos, within the last hundred and fifty years.[6]

[4] William to Dr Thomson, 6th March, 1844, T253, ULC; Dr to William Thomson, 21st February, 1842, T197, ULC. The Porter brothers entered Peterhouse. William became third wrangler in 1849 and James seventh wrangler in 1851. P.G. Tait married one of their sisters in 1857. See C.G. Knott, *Life and scientific work of Peter Guthrie Tait* (Cambridge, 1911), p. 14.

[5] For a useful summary of Cambridge values, see G. Kitson Clark, *The making of Victorian England* (Edinburgh, 1962), pp. 263–4. For a general view, see M.M. Garland, *Cambridge before Darwin. The ideal of a liberal education 1800–1860* (Cambridge, 1980). A study of Cambridge 'physics' education is provided by D.B. Wilson, 'The educational matrix: physics education at early-Victorian Cambridge, Edinburgh and Glasgow Universities', in P.M. Harman (ed.), *Wranglers and physicists. Studies on Cambridge mathematical physics in the nineteenth century* (Manchester, 1985), pp. 12–48, esp. pp. 14–19.

[6] Lord Kelvin to G.H. Darwin, 12th November, 1906, D43, ULC; SPT, **2**, 1132–3.

Lord Kelvin added the suggestion, however, based on his own experience of what Cambridge had rather conspicuously lacked in the 1840s, that 'it might be a good thing, tending to fertilize the mathematical studies of undergraduates taking the Mathematical Tripos, if it were made a condition for going in for its examination that every candidate should have attended lecture courses and gone through practical work, during at least two of his three years, in either the Cavendish Laboratory or the Chemical Department'. Cambridge drilling in the 1840s, then, had its limitations as well as its benefits for a relatively undisciplined Clydeside undergraduate.

Of a liberal education

One of the most striking features of Cambridge admissions over the period 1800 to 1850 is the very small percentage (6%) of the undergraduates from business families compared to those of sons of the Anglican clergy (32%), professional men (21%), and the landed classes (31%). By contrast, matriculations at Glasgow College between 1790 and 1839 show that almost *one-half* of students had fathers employed in industry and commerce. The next largest group, the sons of tenant farmers, was only 17%, while sons of the clergy on the one hand and sons of the nobility and landed classes on the other, averaged around 10% each.[7] Thus although William Thomson's father belonged to the category of professional men, the young undergraduate's strong ties with the commercial world of Clydeside through family, friends, and interests, distanced him from the prevailing Cambridge distaste for a commercial or industrial spirit and for the diffusion of useful knowledge.

As early as his first term at Cambridge, William, sensing something of the particular character and ideals of a Cambridge approach to the education of her youth, reacted scornfully:

> Today we had the first classical lecture, or rather the introduction to the classical lectures. The lecture (by Freeman, who is transplanted from Trinity) was rather a curious one. He told us a good deal about the university's ideas of education, as opposed to the modern diffusion-of-useful-knowledge-Societies' ideas; that her idea of education was not as a collection of useful facts, so much as of a training and strengthening of the mind. That now we have in a manner given our consent to her dogma (not expressed in any decree or book, but indicated in her system) and so we should give it a fair trial by being good boys and attending lectures punctually. That the ultimate object of the lectures was not altogether for the examination, and not merely for annoyance (though perhaps that might be its immediate object) but to carry out the university's idea of education, and to ensure at least one hour a day being spent properly and according to her ideas &c &c.[8]

[7] Sheldon Rothblatt, *The revolution of the dons. Cambridge and society in Victorian England* (Cambridge, 1981), p. 87; W.M. Mathew, 'The origins and occupations of Glasgow students, 1740–1839', *Past and present*, no. 33 (1966), 74–94.

[8] William to Dr Thomson, 30th October, 1841, T181, ULC.

Apart from professorial lectures, there were at this time no university lecture courses. The so-called college lectures, much criticized for their superficiality, provided the formal teaching in mathematics and classics but failed to satisfy the needs of an honours candidate. The principal indebtedness of William Thomson to his Cambridge years relates much less to the University or College as such, than to his private mathematical coach, William Hopkins.

William Thomson's remarks, however, reflect not only an early dissatisfaction with the formal teaching, but more especially his scepticism towards the University's ideals of a liberal education. Although both the Belfast Academical Institution and Glasgow College offered a liberal education, they placed far more emphasis upon a 'useful' and commercially applicable dimension. The mathematics classes of Dr Thomson in Belfast catered to the demands of an expanding commercial centre, for the founders of the Belfast Institution had stated that one of its primary aims was to diffuse the benefits of education 'both useful and liberal'. Cambridge, on the other hand, seemed increasingly to emphasize and define its undergraduate education as wholly liberal, marginally practical, and not at all commercial.

Many of the leading features of Cambridge – particularly Trinity College – ideals found expression in Adam Sedgwick's 1833 *A discourse on the studies of the University*, originally delivered as a sermon in Trinity College Chapel on the occasion of the annual Commemoration of Benefactors in 1832. Its publication owed much to William Whewell's petitioning of the preacher, and by 1850 had reached its fifth, greatly expanded, edition. Like Whewell, Sedgwick (1785–1873) belonged to the first generation of Cambridge reformers who, from the 1810s onwards, moved the University from the intellectual stagnation of previous decades into a long period of academic pre-eminence.

In his *Discourse*, Sedgwick divided the studies of the University into three branches – first, 'the study of the laws of nature, comprehending all parts of inductive philosophy'; second, the study of ancient literature; and third, 'the study of ourselves, considered as individuals and as social beings'. Although the second and third branches occupied an important part in Cambridge studies – the second aided by the establishment of a Classics Tripos in 1822 and the third through a sustained Sedgwick and Cantabrigian attack on Benthamite utilitarianism – the first branch, pursued in the spirit of Isaac Newton's *Mathematical principles of natural philosophy*, took pride of place in the University of Newton. Sedgwick thus asserted that 'studies of this kind not merely contain their own intellectual reward, but give the mind a habit of abstraction, most difficult to acquire by ordinary means, and a power of concentration of inestimable value in the business of life'.[9]

These 'severe studies', then, provided the key not to success in commercial life, but to success in spheres of intellectual conflict, notably the Church, law,

[9] Adam Sedgwick, *A discourse on the studies of the University*, 2nd edn. (Cambridge, 1834), pp. 10–12.

and politics. Disciplining the mind by a study of 'that symbolical language [mathematics] by which alone these laws [of nature] can be fully decyphered'[10] served not merely to train a professional man of mathematical science – those few men of high mathematical honours, such as Airy, Herschel, Whewell, and Hopkins, who would make the advancement of science a central part of their lives – but to provide a fundamental education for those professional careers which employed analogous skills of disciplined argument and rigorous method. Archibald Smith opted for the Bar, while William Thomson reflected uneasily in his 1843 diary that 'If something else fail, I think I could reconcile myself to the Bar, though it would be a great shock to my feelings at present to have to make up my mind to cut Mathematics, which I am afraid I should have to do if I wished to get on at the Bar'.[11]

Apart from these academic and intellectual functions, Sedgwick's *Discourse* placed much weight on the benefits to man's moral and spiritual character which flowed from a study of the laws of nature: 'Simplicity of character, humility, and love of truth, ought therefore to be (and I believe generally have been) among the attributes of minds well trained in philosophy'. But, above all, a study of the Newtonian philosophy 'teaches us to see the finger of God in all things' and so make us ready 'for the reception of that higher illumination, which brings the rebellious faculties into obedience to the divine will'.[12] Mathematical natural philosophy, then, seen as an inductive science, led directly to natural theology. Properly studied, it prepared man for revelation, for discipline of man's spiritual being in accordance with divine volition.

The significance of interpreting Sedgwick's remarks within their religious context cannot be overemphasized. The religious ethos of Cambridge was thoroughly Anglican, but tended, among the Trinity reformers at least, to be liberal and latitudinarian in character, contrasting with the growing Tractarianism of Oxford University and the popular appeal of biblical literalism. The maintenance of close bonds between the ideals of mathematical science (represented by physical astronomy) and natural theology at Cambridge not only vindicated a study of the sciences from all charges of atheism or materialism but gave positive encouragement to these studies of the natural world, as exemplified in Sedgwick's own geology.[13] At the same time, Sedgwick took care to stress the limitations of natural theology as but a preparation for revelation through the scriptures.

The non-denominational nature of natural theology thus viewed went hand-in-hand with Sedgwick's enthusiasm for the abolition of university tests in 1834.

[10] *Ibid.*, p. 11.

[11] William Thomson, 24th March, 1843, Diary kept at Cambridge, 13th February to 23rd October, 1843, NB29, ULC; SPT, **1**, 51.

[12] Sedgwick, *Discourse*, pp. 12–14.

[13] For example, Jack Morrell and Arnold Thackray, *Gentlemen of science. Early years of the British Association for the Advancement of Science* (Oxford, 1981), pp. 236–45; Crosbie Smith, 'Geologists and mathematicians: the rise of physical geology', in Harman (ed.), *Wranglers and physicists*, pp. 49–83.

The undergraduate Archibald Smith shared the enthusiasm of the Trinity reformers, but from his own account it is clear that most Cantabrigians did not:

> There are about 1100 resident Bachelors and Undergraduates of whom 809 signed the petition against [the admission of dissenters] – I was one of a small minority who signed the other [petition in favour of admission]. Among the members of the Senate the proportion was less than 2 to 1 against admitting dissenters while among the undergraduates it is 8 to 1 . . . However it is consoling to think that their opinion and petition will have very little weight . . . as it will show the need of some infusion of liberality into the Church-of-Englandism . . . of the universities.[14]

Given that Cambridge degrees were conferred only upon candidates prepared to conform to the Established Church, such voting figures are scarcely surprising. But it is evident that without the latitudinarian position of many leading Cambridge academics, William Thomson would never have entered St Peter's College in 1841 at all. Thus the significant religious and political differences between the presbyterian Dr Thomson and the Anglican Rev Sedgwick were often minimal: they shared whig political and latitudinarian religious views.

Little or nothing of the content of Sedgwick's *Discourse* would have provoked disagreement with the Thomsons. Rather, it was what Sedgwick omitted which marked one real divergence between the northern whigs and the Cambridge reformers. For Sedgwick included little or nothing in his *Discourse* concerning the material benefits of a study of mathematical science, nothing relating to the relief of man's estate through the fruits of the advancement of learning or on the diffusion of useful knowledge to the common people, and certainly nothing with reference to the new commercial and industrial orders. A gulf indeed existed in the 1830s between the Cambridge disdain for economic man and the interests of the industrial entrepreneurs of Clydeside.

Cambridge ideals of education received their most vigorous exposition in Whewell's 1845 *Of a liberal education*. A higher or liberal education meant the education of the 'upper classes', the elites of Church and State, 'those who must direct the course of the community' and influence the conduct of the general body. Whewell directed attention not to moral or religious education, however, but to intellectual education only, with specific reference to his university. He divided intellectual education into two kinds of study. Permanent studies comprised subjects 'which have long taken their permanent shape; – ancient languages with their literature, and long-established demonstrated sciences', while progressive studies incorporated 'the results of the mental activity of our own times; the literature of our own age, and the sciences in which men are making progress from day to day'.[15]

Permanent studies aimed to draw forth and unfold, to educe, 'two principal

[14] Archibald to Christina Smith, 28th April, 1834, TD1/692/2, Strathclyde Regional Archives. See also Garland, pp. 7–8, 14–27, 70–89.

[15] William Whewell, *Of a liberal education* (Cambridge, 1845), pp. 1–6; Garland, pp. 39–49.

faculties of man, considered as an intellectual being; namely, Language and Reason'. The classical literature of Greece and Rome exhibited in their most complete form the powers and properties of the faculty of Language, while mathematics exemplified Reason in its most complete form. Thus in mathematical works 'truths respecting measurable quantities are demonstrated by chains of the most rigorous reasoning, proceeding from Principles self-evident, or at least certain'. Whewell included the first six books of Euclid, solid geometry, conic sections, mechanics, hydrostatics, optics, and astronomy, in his proposals for the permanent part of mathematical studies at the University. Mechanics and hydrostatics especially extended solid and evident reasoning from space and number to motion and force, taught truths which 'are perpetually exemplified in the external world, and [which] serve to explain the practical properties of machines, structures, and fluids', and provide 'the key to the understanding of all the great discoveries of modern times with regard to the constitution of the universe, and especially the discoveries of Newton'.[16]

Permanent studies 'provided the essential foundation to be mastered before the others are entered on, in order to secure such an intellectual culture as we aim at'. Intellectual discipline thus preceded intellectual progress, and so 'the Progressive Sciences are to be begun towards the end of a Liberal Education'. With regard to Cambridge University and its emphasis on mathematical studies, geometry, consisting 'entirely of manifest examples of perfect reasoning', and not algebra, which offered 'not so much examples of Reasoning, as of Applications of Rules' or reasoning denoted by symbols and rules of symbolic combination, was best suited to the intellectual discipline fundamental to a liberal education. Algebraic mathematics, or analysis, though 'in the highest degree ingenious and beautiful', could be 'an intellectual discipline to those only who fully master the higher steps of generalization and abstraction, by a firm and connected mental progress from the lower of such steps upward; and this requires rather a professional mathematical education, than such a study of mathematics as must properly form part of a liberal education'.[17]

In geometry:

> ... the student is rendered familiar with the most perfect examples of strict inference; he is compelled habitually to fix his attention on those conditions on which the cogency of the demonstration depends; ... He is accustomed to a chain of deduction in which each link hangs from the preceding, yet without any insecurity in the whole ... Hence he learns continuity of attention, coherency of thought, and confidence in the power of human Reason to arrive at the truth. These great advantages, resulting from the study of Geometry, have justly made it a part of every good system of Liberal Education, from the time of the Greeks to our own.[18]

[16] Whewell, *Of a liberal education*, pp. 8–15, 32–4.

[17] *Ibid.*, pp. 28–30, 38–9. Whewell explains that he is adopting the common, though not very exact, meaning of 'analytical' to refer to processes 'conducted by means of algebraical symbols'. On Whewell's views, see also Garland, pp. 34–6; D.S.L. Cardwell, *The organization of science in England* (revised edn., London, 1972), pp. 53–7.

[18] Whewell, *Of a liberal education*, pp. 30–1.

Similarly, propositions contained in the *Principia* of Newton are beautiful examples of mathematical combination and invention, following the course of ancient geometry; and for the purpose of general education, a portion might be selected from this work without difficulty. By contrast, Lagrange's *Mécanique analytique* or Laplace's *Mécanique céleste* formed 'no part of the standard portion of our educational course'.[19]

Less esoteric progressive studies, however, did form part of Whewell's idea of a liberal education. Thus it was requisite that 'a person should so far become acquainted with some portion of this body of accumulated and imperishable knowledge, as to know of what nature it is, what is the evidence of its reality, by what means additions to it are made from time to time, and what are the prospects which it opens to the present generation of mankind'. Among the progressive sciences, he included the ancient sciences of astronomy, optics, and harmonics, the classificatory sciences (botany, for example), and the palaetiological sciences (geology).

While Whewell banned analytical substitutes for geometrical forms of trigonometry, statics, dynamics, and other mathematical parts of permanent studies, he accepted the new approaches as part of progressive studies, remarking that almost all the 'modern portions of Mathematics are of the analytical kind':

> When, by the pursuit of Permanent Mathematical Studies, the reasoning powers have been educed and confirmed, the Student's powers of symbolical calculation, and his pleasure in symbolical symmetry and generality, need no longer be repressed or limited ... When the Student is once well disciplined in geometrical mathematics, he may pursue analysis safely and surely to any extent.[20]

The select candidates 'for our highest mathematical honours' ought to study 'capital works' (Newton's *Principia* or Laplace's *Mécanique céleste*, for example) – 'the great original works are the proper study of a man who would pursue mathematics for the highest purpose of intellectual culture' – rather than systematic or elementary treatises derived from the capital works. He further denounced the use by even the brightest students of original memoirs in the learned transactions lest it bewilder and overwhelm them, though he made clear that he did not wish to prevent undergraduates from 'pursuing such a line of reading, as far as their taste and time will allow'. But in general 'such a course of reading is fit only for those who make mathematics a main business of life'.[21]

On the other hand, Whewell wanted to encourage 'the study of new investigations with regard to practical problems' rather than 'in recondite and speculative subjects':

> Problems of Engineering and Practical Mechanics naturally receive new solutions, as, in the progress of Art, they take new forms; and it is desirable that our mathematical education should bring our students acquainted with the best and most recent solutions of such problems; because, by this means their mathematical knowledge has solidity and

[19] *Ibid.*, pp. 35–7. [20] *Ibid.*, pp. 63–5. [21] *Ibid.*, pp. 66–70.

permanence given it, by its verification in facts, and its coincidence with the experience of practical men.[22]

To this end, Whewell advised that 'we should introduce among the books of which we encourage the study among our best mathematical students, three excellent works of recent date': first, Poncelet's *Mécanique industrielle* 'in which he has given modes of calculating the amount and expenditure of Labouring Force'; second, Willis's *Principles of mechanism* 'in which he has classified all modes of communicating motion by machinery, and investigated their properties'; and third, Count de Pambour's *Theory of the steam engine* 'in which a sound mathematical theory is confirmed by judicious experiments'.

Each of these issues will have specific relevance to William Thomson's own scientific work. Thus the concept of labouring force or mechanical effect (discussed at greater length in Whewell's 1841 *Mechanics of engineering*) is fundamental to Thomson's approach to physical phenomena. Again, the 'kinematics' of Whewell and his Cambridge engineering colleague, Professor Robert Willis (1800–75), provides Thomson with the foundation for his dynamical theories. And the theory of the steam-engine, the basis of his thermodynamics, appeared as a question in the Smith's prize examination of January, 1845, when Whewell was one of the examiners and Thomson one of the candidates:

In the common mode of calculating the effect of a steam-engine, the effect depends on the pressure of steam in the boiler, and the space described by the piston. Shew that this method is erroneous. In what cases is it most erroneous? Give a better method. What evidence is there of its being better?[23]

A more general significance of Whewell's remarks lies in their illustration of a new Cambridge interest in engineering, an area so central to our study. There is indeed much in Whewell's concluding remarks to which Thomson would have subsequently assented: 'the study of these works would put our students in possession of the largest and most philosophical doctrines which apply to Engineering; and would thus give a tangible reality and practical value to their mathematical acquirements; while, at the same time they would find . . . excellent examples of mathematical rigour, ingenuity, and beauty'.[24]

Whewell's move to make engineering part of a liberal education in Cambridge during the early 1840s follows the establishment of Lewis Gordon's chair in Glasgow (Gordon having been a pupil of Whewell's old friend Forbes), the work of Charles Babbage *On the economy of machinery and manufactures* (1832), and the successful growth of the British Association (in which Whewell played a leading role) in an industrial as well as academic climate. Whewell related engineering and a liberal education to one another through geometry, but in these early years the new subject hardly cohered with the traditional aims and social function of the University. Although Willis had become a well-

22 *Ibid.*, p. 70. 23 *University of Cambridge examination papers, 1845* (Cambridge, 1845).
24 Whewell, *Of a liberal education*, pp. 70–1.

established professor, only after the middle of the century, with the Cavendish Laboratory for experimental physics in prospect, did engineering find a more natural, less marginal place in the ancient institution.

Although Whewell participated in the 'analytic revolution', which brought modern mathematics to Cambridge,[25] he may be characterized as a conservative reformer. His reforming zeal diminished with increasing age and status, and so, to the younger reformers, his conservatism appeared predominant. A political supporter of Peel – to whom he was indebted for the mastership of Trinity in 1841 – Whewell was no die-hard tory in the mould of the Glasgow College oligarchy. He was indeed not much liked by Glasgow academics in general, being as he was too much of a reformer for the high Scottish tories and too conservative for the Scottish whigs. Whewell's division of a liberal education into permanent and progressive studies exemplified perfectly his conservative and reforming outlook. William Thomson himself neatly referred to the Master of Trinity as 'the despotic Whewell' in 1845.[26]

We have outlined Whewell's position at some length for two principal reasons. First, his moderate conservatism will serve as a foil to the founders of the *Cambridge Mathematical Journal* with which William Thomson quickly identified himself. Thus Cambridge wranglers such as Archibald Smith, D.F. Gregory, R.L. Ellis (1817–59), and William Thomson were all 'whigs in mathematics as in politics', aiming to advance Cambridge mathematics and in a sense to encourage that 'school of eminent mathematicians' with which Whewell's liberal education was not concerned. These young turks emphasized not the training of the mind through geometrical drudgery – which is how they viewed the consequences of Whewell's programme – but rather the professionalization of their subject (ch. 6).

Second, Whewell's recognition of the advantages of geometry cannot be seen merely as an attempt to return Cambridge to the eighteenth century. He, like the younger reformers, saw the significance of geometry as the foundation of the physical sciences – of natural philosophy or physics in particular – and so warned against the loss of physical reality amid an unnecessary profusion of algebraic symbols. His conception of the inductive sciences in relation to geometry provides an important focus for our subsequent understanding of William Thomson's scientific style and not least for our analysis of its practical cast. In this respect, then, Whewell's views, emphasizing tangible, visualizable, geometrical exemplars, come tantalizingly close to Thomson's, while his concept of a liberal education, reserving research for the elect and regarding undergraduate originality as leading towards unsafe and uncertain speculations, contrasts vividly with Thomson and his circle of whig mathematicians.

[25] Harvey Becher, 'William Whewell and Cambridge mathematics', *Hist. Stud. Phys. Sci.*, **11** (1980), 1–48.

[26] William to Dr Thomson, 10th October, 1845, T320, ULC; James to William Thomson, 12th May, 1846, T418, ULC.

Although of the same generation as Adam Sedgwick and William Whewell, William Hopkins came to occupy a quite different role in the University. Son of a gentleman farmer and himself a political tory, he graduated as seventh wrangler in 1827 from St Peter's College comparatively late in life. On account of marriage, he was ineligible for a college fellowship; instead, he became one of Cambridge's most celebrated mathematical coaches or private tutors with the title of 'the senior wrangler maker'. By 1849 he could claim nearly 200 wranglers among his pupils of whom seventeen were senior wranglers and over forty in one of the top three places, among them Stokes and Thomson.[27]

Hopkins knew precisely how to derive maximum benefit from the Cambridge system of drilling, both for himself and for his distinguished pupils. He admitted that a successful private tutor could earn a comfortable £700–800 per annum which almost equalled the £900–1000 estimate of an upper middle class income in this period. Each pupil indeed paid Hopkins upwards of £100 per annum in fees.[28] But finance apart, Hopkins's personality clearly helped not a little in a system which must have seemed to most undergraduates cold, unworldly, and monastic. As Francis Galton described his coach's character: 'Hopkins to use a Cantab expression is a regular brick; tells funny stories connected with different problems and is no way Donnish; he rattles us on at a splendid pace and makes mathematics anything but a dry subject by entering thoroughly into its metaphysics. I never enjoyed anything so much before'.[29] Thus the educational and emotional needs of the brightest Cambridge undergraduates, not served by the low-level college lectures or remote college fellows, found satisfaction in William Hopkins.

In his views on the University's goals and on reform of the Tripos, Hopkins aimed to maintain and preserve that which he saw as best in the system. He was particularly concerned lest standards be lowered to favour the average undergraduate at the expense of candidates for the highest mathematical honours. While Whewell sought to define and shape the University's ideals of a liberal education for the undergraduate programme as a whole, Hopkins emphasized the importance of maintaining 'the standard of acquirement in our higher class of students'. In 1841, for example, he published his remarks opposing the exclusion under proposed Tripos changes of physical astronomy and physical optics, partial differential equations, the higher parts of geometry of three dimensions, and most of hydrodynamics from the course of undergraduate study, subjects 'which contribute most to elevate and enoble the character of our mathematical studies' by making the student appreciate fully 'the real importance and value of pure mathematical science, as the only instrument of investigation by which man could possibly have attained to a knowledge of so

[27] William Hopkins, *DNB*.

[28] Rothblatt, pp. 68*n*, 201. Hopkins's nomination of the Tory parliamentary candidate is noted in William to Dr Thomson, 24th March, 1843, T233, ULC. William expressed his own regrets that the Tory won.

[29] Rothblatt, pp. 191–2*n*.

much of what is perfect and beautiful in the structure of the material universe, and in the laws which govern it'. By this means the student 'can form an adequate conception of the genius which has been developed in the framing of those theories, and can feel himself under those salutary influences which must ever be exercized on the mind of youth by the contemplation of the workings of lofty genius'. Most significantly:

> . . . the importance of that small class of our students is not to be estimated by their number . . . their immediate influence on the tone and character of our undergraduate studies is of the first importance . . . We may also remark, that comparatively small as this higher class of students may be, it includes perhaps the majority of those who are selected by the University on account of their mathematical acquirements as fit objects for its rewards. They are the men who . . . remain among us, and afterwards form the tutorial body of the University, occupy its important offices, and give to it its prevailing tone and character.[30]

Hopkins's concerns had a strongly conservative and moral dimension, emphasizing the preservation of the University's high moral tone by the selection of an intellectual elite. Given the peculiar nature of Cambridge teaching in which undergraduates were taught, not by professors appointed 'for the matured acquirements of their manhood' but by men rewarded 'for the proficience of their youth', Hopkins believed a lowering of examination standards to be injurious to the character of the University in the absence of that career-motive present in most analogous continental institutions. Significantly, far from advocating a radical reform of the system to promote active exertion in pursuit of career advancement, Hopkins argued that the existing high degree of mathematical attainment remained the best guarantee of the University's 'high relative scientific position which she has hitherto maintained', enabling her not only 'to further the progress of original research, but also, through the influence of some of her most distinguished members, to diffuse much of a healthy tone of sentiment and feeling on the most serious subjects among the body of scientific men in this country'. In keeping with this gentlemanly view of the University, Hopkins denounced 'a merely utilitarian view of the subject' arguing against 'merely professional' aims and duties 'as if the philosophical portion of our studies possessed no value in themselves, or were incapable of presenting to us objects of contemplation and thought by which the mind of the educated man may be elevated above that of the uneducated one'.[31]

For William Thomson and many other candidates for high mathematical honours, then, Hopkins stands at the very centre of their Cambridge studies. Though *prima facie* outside the college and university system, it is one of the ironies of that idiosyncratic institution that an undergraduate's prospects – a college fellowship in particular – should have depended on the right choice of mathematical coach. Hopkins, however, for his part could also command the

[30] William Hopkins, *Remarks on certain proposed regulations respecting the studies of the University* (Cambridge, 1841), on pp. 10–12. [31] *Ibid.*, pp. 12–16.

right students on the basis of a distinguished record, and so to his reputation goes the credit of coaching men such as Stokes, Thomson, Tait, Maxwell, Galton and Todhunter.

Father and son

With no financial assistance available for a Glasgow student intent on a Cambridge mathematical education, Dr Thomson was to be burdened with another serious drain on his finances for several years once his son had been entered as a pensioner (that is, an undergraduate who paid his own way) at St Peter's College, Cambridge.[32] *Economy*, therefore, was a subject of much of the extensive correspondence between father and son during these Cambridge undergraduate years, Dr Thomson continually exhorting his son to minimize waste of time, effort, and money. These concerns serve also to underline the differences between William's middle-class, professional values, concerned primarily with the marketing of expertise, and those of the aristocratic, 'idle rich' poll men.

An undergraduate required no special entrance qualifications to go into residence at Cambridge and, as he had not taken either his BA or MA at Glasgow College, William Thomson could enter as an undergraduate. Thus the tutor of St Peter's College, H.W. Cookson (1810–76), wrote to Dr Thomson early in 1841:

> In reply to your obliging note . . . I beg to inform you that I shall be happy to enter your son upon the boards of this College as soon as you might wish it . . . If you are anxious that your son should have quiet and comfortable rooms, I should recommend him to be admitted without delay. A certificate is usually required to show that the person's education has not been wholly neglected, but as this is implied with respect to your son, in your application for his admission, I should not ask for any thing more. If the caution money, fifteen pounds, be paid to my bankers . . . I should conclude that you authorised me to admit your son without delay.[33]

Dr Thomson duly paid the caution money – effectively a security against any future debts – and William was entered on 6th April, 1841. Cookson informed Dr Thomson of the subjects of lectures in the first term, commencing in October of that year. The mathematical texts would be Simson's *Euclid*, Hind's *Algebra*, and Snowball's or Hymers' *Trigonometry*. Also on the course would be Paley's *Evidences of Christianity*, one of the Gospels in Greek, and the *Prometheus vinctus*.

[32] The valuable Snell exhibitions applied only to bright students sent from Glasgow to Oxford. See *The new statistical account of Scotland, 6: Lanark* (London and Edinburgh, 1845), pp. 175–6. 'In the year 1688, Mr John Snell, with a view to support Episcopacy in Scotland, devised to trustees a considerable estate near Leamington, in Warwickshire, for educating Scotch students at Baliol [*sic*] College, Oxford.' By 1835, the fund afforded about £130 per annum to each of ten exhibitioners, tenable for ten years. Dr Thomson thus lost the prospect of considerable financial support.

[33] H.W. Cookson to Dr Thomson, 26th March, 1841, C132, ULC. See also SPT, **1**, 23–4.

Other requirements were an academical dress, and, of a more domestic nature, 'I should recommend him to bring from home sheets & towels and a sufficient quantity of spoons, knives & table linen for his own breakfast & tea table'.[34]

On their arrival at Cambridge in late October, 1841, Dr Thomson and William had gone to see Hopkins. Hopkins advised William not to have a tutor (i.e. coach) for the first term, and although he 'will not teach me himself until my second year, he will examine me now and then, and supply me with a tutor when I need one'. Father and son also called on James Challis (1803–82), the Cambridge professor of astronomy, and D.F. Gregory, the founder editor of the *Cambridge Mathematical Journal*. Gregory, who graduated in 1838 and became a fellow of Trinity in 1840, was a personal friend of Dr Thomson, while the Cambridge-educated Professor Ramsay of Glasgow College, who also knew Hopkins well, had provided an introduction to Challis.[35] College lectures in mathematics and classics occupied William's first term. Cookson guided him in the elementary mathematical reading, and examined him from time to time by written problems. His tutor for mathematics in the second (Lent) term was Frederick Fuller (1819–1909), who had only just graduated and who subsequently became professor of mathematics at Aberdeen. Fuller set him three papers a week, including some on three dimensional analytical geometry.[36]

These formal requirements of the Cambridge system, however, pale into relative insignificance beside William's unheard-of precocity in contributing not only profound but numerous original papers to the *Cambridge Mathematical Journal*. His contributions were at the very forefront of research, and far exceeded what most of his competitors for high mathematical honours in the Senate House Examinations could have understood, let alone produced. With a reputation such as this, established even before his entry into Peterhouse, we can understand why, from the beginning, his contemporaries expected him to become senior wrangler of his year, and why he soon acquainted himself with such academically distinguished fellows of Trinity as D.F. Gregory and Archibald Smith. William's philosophy here, quite at variance with Whewell's, was that an undergraduate 'wisely spends some of his spare time in reading high physical subjects w^{h} will not occur in his regular reading. I have lent him [Porter] Fourier's theory of heat'.[37]

34 H.W. Cookson to Dr Thomson, 6th April and 17th September, 1841, C133–4, ULC. The full titles of the books listed were: Robert Simson, *The elements of Euclid* (Edinburgh, 1771); John Hind, *The elements of algebra: designed for the use of students in the University* (Cambridge, 1829); J.C. Snowball, *The elements of plane and spherical trigonometry, with the construction and use of mathematical tables* (Cambridge, 1834); John Hymers, *A treatise on plane trigonometry, and on trigonometrical tables and logarithms* (Cambridge, 1837); John Hymers, *A treatise on spherical trigonometry, together with a selection of problems* (Cambridge, 1841); William Paley, *A view of the evidences of Christianity*, 2nd edn., (2 vols., London, 1794); Aeschylus (Euripides), *Promethius vinctus*, C.J. Blomfield (ed.) (Cambridge, 1810). 35 Elizabeth King, *Lord Kelvin's early home* (London, 1909), p. 199.

36 William to Anna Thomson, 23rd October, 1841, B182, ULC; William to Dr Thomson, 5th November, 1841, T183, ULC; 15th January, 1842, T192, ULC; 6th February, 1842, T195, ULC. See also SPT, **1**, 31–5. 37 William to Dr Thomson, 1st November, 1845, T326, ULC.

Archibald Smith, senior wrangler and first Smith's prizeman in 1836 and whose Glasgow family appeared in Chapter 2, had called at Glasgow College on Dr Thomson's return there in late October. According to Dr Thomson, Smith expressed a great interest in William's success, Gregory having 'told him that you were the author of the paper "P. Q. R." [in the *Cambridge Mathematical Journal*]; and he expressed himself greatly pleased with it. He asked your age, and was surprised you could have written it. He expressed a strong wish to know you, and gave the enclosed card, wishing you to call on him, should you happen to be in London. Take care of it, and see him if there at any time.'[38] Shortly afterwards, both Smith and Gregory honoured William by a visit themselves:

It was certainly a great honour for a freshman of St Peter's to have two fellows of Trinity calling on him. They stayed, I suppose, nearly three quarters of an hour in my room, looking into my books and talking on various subjects connected with them. Smith looked over the senate house problems of his year, and showed one which he says he did in the examination, but could never do since. They also entered into a long disquisition about a circle rolling along a parabola, or other curve, what it would do when it came to a point at which the radius of curvature is smaller than that of the circle, and got hold of my paper and drew no end of figures regarding it.[39]

Although he almost became a rival candidate for the Glasgow natural philosophy chair in 1846, Archibald Smith remained a central figure among Thomson's friends.

After the break with the family, William inevitably felt his rooms, though comfortable, a little more lonely than the parlour in Glasgow. His sister Elizabeth assured him that she, at least, missed him very much, and admitted to having poured tea for him, fully expecting him 'to make your appearance to drink it, [it] being by no means uncommon for you to be somewhat late at mealtimes'. William's comforting reply was that the tea would be rather cold before he returned to drink it. Nonetheless, he rapidly settled into the new life-style, such that his other sister, Anna, speculated that he was to be seen studying with his feet on the fender, having strewn his books over the table, floor and sofa. He confessed that the truth of this imaginative picture was only modified by the fact that he had not yet more than half a dozen books to strew around.[40] A year later, Elizabeth, writing from Knock, painted the even more vivid portrait of William's life-style quoted at the beginning of this chapter.

Although every member of the Thomson family communicated regularly with William, Dr Thomson's substantive correspondence with his son is most revealing of the Thomsons' social and cultural values, especially in relation to the

[38] Dr to William Thomson, 28th October, 1841, T180, ULC; William to Elizabeth Thomson, *c.* October–November, 1841, in King, *Early home*, p. 204.

[39] William to Dr Thomson, 21st November, 1841, T185, ULC.

[40] Elizabeth to William Thomson, 27th February, 1842, K56, ULC; William to Elizabeth Thomson, 6th March, 1842, in King, *Early home*, pp. 211–12; William to Dr Thomson, 30th October, 1841, T181, ULC.

family economy. Given the widespread reputation of Cambridge undergraduates (especially the idle aristocrats) for luxurious habits, extravagant entertainments, and riotous living, it is not surprising to find Dr Thomson constantly urging William to exercise 'a strict and proper economy' not only in financial matters, but in 'health and habits' also, making 'moral correctness and propriety your aim above all things'. At the same time, however, he insisted that 'you must keep up a gentlemanly appearance, and live like others, keeping, however, rather behind than in advance'.[41]

Following this advice, William resisted, for a time at least, a strong temptation to join the college boat club, though his opinion of the Peterhouse boat was hardly dispassionate: 'our boat goes out every day, and will be head of the river in the next races, now that I have come here, though it was far from it before'.[42] He had already, it seems, ventured forth on the River Cam himself in a narrow rowing boat called a 'funny', designed to carry one or two people. His father wished him to avoid the boat club and regretted to hear that his son had been boating 'not on account of the thing itself, as I think there can be no danger; but that you be brought into loose society, a thing that would ruin you for ever'.[43] In his reply, William reassured Dr Thomson both with regard to wine parties and boating:

I have given no wine parties . . . Most of our freshmen have given parties of all kinds already, but I fear they are considered by the second and third year men as an extraordinarily 'fast' set of freshmen, and that it is not at all necessary for a freshman to give any parties in his first term. The separation of the freshmen of this college into the two classes of 'rowing men' (pronounced rouing, and meaning men who are fond of rows and 'rowing' parties) and 'reading men' has very soon become distinct. All my *friends* are among the latter class, and I am gradually dropping acquaintance with the former, as much as possible. I find that even to know them is a very troublesome thing if we want to read, as they are always going about troubling people in their rooms . . . With regard to boating you need not be in the least afraid. As I do not belong to the boatclub, I always row by myself in a funny . . . or at least go in a two oared boat, with some friend with whom I should otherwise be walking . . .[44]

In order to underline his opposition to what he termed 'the dissipated men', William included in his letter a detailed proposal, not 'so impracticable as at first sight appears', for the division of his time and its best use to advance his studies. The proposal involved rising at 5 a.m., lighting his own fire before the arrival of the college servant known as the gyp – 'which appears rather impracticable to most men' – reading until 8.15 a.m., attending the daily lecture and reading until 1 p.m., exercising until 4 p.m., attending chapel until 7 p.m., reading until 8.30

[41] Dr to William Thomson, 28th October, 1841, T180, ULC; SPT, **1**, 29. See also Rothblatt, pp. 65–75, 186–7; Garland, pp. 2–12.

[42] William to Elizabeth Thomson, *c.* October–November, 1841, in King, *Early home*, p. 204.

[43] Dr to William Thomson, 6th December, 1841, T186, ULC.

[44] William to Dr Thomson, 12th December, 1841, T187, ULC; SPT, **1**, 32–3.

p.m., and retiring to bed at 9 p.m. While it may be doubted that William actually adhered to such a disciplined allocation of time, its importance lies in his intention to minimize waste of time, in contrast to the idle set of dissipated, rowing men. Economy of time, as well as economy of capital, played a vital role in the Thomson life-style, as Admiral of the Fleet, Lord Fisher, recorded of Lord Kelvin many years later:

> One great exercise of Faith is 'Redeeming the Time', as Paul says. (I'm told the literal meaning of the original Greek is 'buying up opportunities') . . . Lord Kelvin often used to tell me of his continuous desire of 'redeeming the time'. Even in dressing himself he sought every opportunity of saving time (so he told me) in thinking of the next operation. However his busy brain sometimes got away from the business in hand, as he once put his necktie in his pocket and his handkerchief round his neck . . . And yet I am told he was an extraordinarily acute business man. Every sailor owes him undying gratitude for his 'buying up opportunities' in the way he utilised a broken thigh, which compelled him to go in a yacht, to invent his marvellous compass and sounding machine.[45]

During the course of his second term, William announced suddenly to his father that he had bought, with another student, a secondhand boat for seven pounds 'built of oak, and as good as new'. To mitigate the crime of not having consulted his financial and moral guide, he explained that he had had to act quickly, lest such a bargain be lost to another customer. Furthermore 'the expense will be just about the same as belonging to the boat club, and, as the temptation to join the club, (which I have been very much solicited to do of late there being a great want of rowers) has thus been completely removed, I felt confident that you would not be displeased'.[46] His father was most displeased at not having been consulted and reprimanded his son for allowing himself to be cajoled and probably cheated by persons who had represented the boat as being of some wonderful value:

> seven pounds for a tub that will hold only one person!!! My first impulse was to order you to give up at once what you term 'a wonderful bargain'. I shall wait, however, till I next write to Mr Cookson . . . and if he approve of the matter, I shall then, *for this time*, send you money to pay for your purchase. You talk of getting rid of 'the temptation to join the Club'. I had hoped, that after what you said yourself regarding that association, you would have been proof against all attempts to induce you to join it.[47]

Having issued his rebuke, Dr Thomson added that he hoped that *if* William retained the 'bargain', it would be a lesson for the future, and *if* 'I thought that such persons as Messrs Cookson, Hopkins, and Gregory, would approve of what you have done, I would then only have to complain about your not having consulted me'. Seeing the stern disciplinarian thus weakening, William replied that Hopkins 'approves very much of rowing, and always recommends it' while

[45] [J.A. Fisher] *Records by Admiral of the Fleet Lord Fisher* (London 1919), pp. 61–2.
[46] William to Dr Thomson, 19th February, 1842, T196, ULC. Compare SPT, **1**, 36–8.
[47] Dr to William Thomson, 21st February, 1842, T197, ULC.

'I have not the least doubt but Mr Cookson will also perfectly approve of it'. If Cookson did not approve, then William would sell his share in the boat which, nevertheless, would save a considerable sum of money in the long run, as he would no longer have to pay to hire a boat.[48] Elizabeth subsequently informed her cunning and high-spirited brother that, in a letter to their father, Cookson had spoken:

very favourably of your abilities and diligence. He says that you have got good companions who will not be likely to lead you astray. Though papa was at first displeased with your purchase he is now I think quite reconciled to it and will not disapprove of your rowing for exercise. You must always be most cautious. Remember how many accidents have happened, even when young men thought there was not the slightest danger.[49]

The affair of the boat illustrates more clearly than any other single episode the father–son relationship which existed throughout William Thomson's early years. The son was high-spirited and enthusiastic, if at times rather the most impetuous member of the family. The father was most concerned to train and guide him to be more cautious and prudent, and especially to avoid the acquisition of any kind of reputation for idleness and dissipation – a reputation which would not be appreciated in Calvinist Glasgow. Nevertheless, Dr Thomson's stern words were not those of an autocratic and authoritarian father-figure who rules his subject-children with an iron rod, and William, with the shrewdness characteristic of his father, recognized the real voice behind the rebukes, and responded accordingly with reason and not rebellion.

William's new boat settled down to a successful career. The 'sculling', as rowing in a single-seater craft was known, proceeded 'with great vigour, and is keeping me in excellent preservation'. The owner also reported that everyone 'says that I am looking much better now than I did some time ago, and I find that I can read with much greater vigour than I could when I had no exercise but walking, in the inexpressibly dull country around Cambridge'. At the same time, his interest in the College boat had not been entirely forgotten, the Peterhouse eight having indeed moved up, as William had predicted, from third place to head of the river. By the end of his first year at Cambridge, William had bought out the other student's share in the boat with money he had earned tutoring his younger brothers, and within a year his inhibitions about involvement in the College boat had been overcome.[50] His father was for his part quite satisfied that rowing, if moderately used, was a very good exercise, and he admitted candidly:

I frequently write to you in the admonitory style – I trust unnecessarily. It can do no harm, however; and it will show you my anxiety (which is participated in by all your friends) that you should do in every respect what is right, taking care not to allow yourself

[48] William to Dr Thomson, 25th February, 1842, T198, ULC.

[49] Elizabeth to William Thomson, 27th February, 1842, K56, ULC.

[50] William to Dr Thomson, 3rd March, 1842, T199, ULC; 14th April, 1842, T209, ULC; 6th May, 1842, T213, ULC; SPT, **1**, 39.

Peterhouse—The Main Court, looking East

St Peter's College (Peterhouse), Cambridge. The oldest of the colleges, St Peter's provided a less hierarchical ambience for Scottish undergraduates than did the larger colleges such as Trinity. Its links with the successful mathematical coach, William Hopkins, provided another strong incentive to gain admission.

to be led away. A few years hence there will be nothing to fear; but at your period of life, and placed as you are among so many persons of different characters and habits, you require to be most circumspect . . .[51]

Apart from the heavy drain on Dr Thomson's income which a Cambridge education entailed – the first academic year cost over £230 for maintenance at college alone – he was very concerned that William should acquire 'accurate business habits'. As a particularly vivid example, he cited early in 1842 the case of Dr Nichol who, 'for want of such habits, is a ruined man'. On a basic salary of £270 a year, Nichol had managed to incur debts of at least £5000, largely, it seems, because of his extravagance in the purchase of new instruments for the Glasgow observatory. Thus bankrupt, he was in danger of depriving himself and his family of bread. According to Dr Thomson, Nichol was talking of publishing, 'but how miserable a thing it is when a man, instead of merely *adding* to some more certain income, must live on the precarious produce of his pen'. Elizabeth also wrote to William to say that Dr Nichol's elegant hall was to be deprived of 'all his beautiful books that he was so proud of'. The lesson for

[51] Dr to William Thomson, 17th March, 1842, T203, ULC.

William was clear, as his father warned: 'to prevent such irregularities in your own case . . . get the habit of taking, at the moment, memorandums of anything of importance which there is the slightest chance of your forgetting; and, in particular, take a note of your expenditure the very day you pay away the money'. As a result, many of William's letters contain a detailed accounting of his expenditure to the halfpenny. Dr Thomson for his part continued to urge his son to 'use all economy consistent with respectability' and to 'contract no debts' except through his tutor.[52]

Dr Nichol's financial crash was perhaps an extreme case, and a stern moral for William, but it was by no means unique. Professor Buchanan, of the logic chair, lost all his possessions, except for a small house at Dunoon, through the failure of the Renfrewshire Bank in 1842. And as one perceptive commentator has remarked of the early Victorian age in general, 'in a period when hectic booms alternated with financial panics and there was no such thing as limited liability, the business magnate and the public investor were haunted by spectres of bankruptcy and the debtor's jail'. The dreaded loss of social status, the fear of losing all one's money and possessions, the possible disgrace of bankruptcy, the consequent anxiety, and the great physical and mental strain, were all characteristics of that age, and professional men, as public investors, were no more exempt from such worries than anyone else. The age was a turbulent one, marked by Chartist demonstrations, economic collapses and other manifestations of social chaos,[53] and it was with no degree of self-satisfaction that Anna Thomson, also speaking of their unfortunate friend, Dr Nichol, said:

> I am sure he must have many a heavy thought when he remembers the ruin he has involved himself & his family in. I really never heard of anything more sad than this whole affair has been . . . I do feel very sorry both for him & Mrs Nichol and for his children. It will make a sad change for he will not be able to give them the advantages they might have had. How very differently papa has done for all of us![54]

Nichol did recover from his loss by publishing successful popular books on astronomy and by delivering popular lectures. His lectures in Glasgow City Hall in 1845, for example, attracted an estimated two or three thousand people.[55] However, the strain had evidently told, for he died in 1859, while editing his well-known *Cyclopaedia of the physical sciences*, at the comparatively early age of fifty-five. Certainly his financial troubles of 1841–2 would remain as a painful reminder to his friends and family for many years afterwards that domestic

[52] Dr to William Thomson, 12th January and 21st February, 1842, T191 and T197, ULC; Elizabeth to William Thomson, 27th February, 1842, K56, ULC; SPT, **1**, 37.

[53] Dr to William Thomson, 6th April, 1842, T207, ULC. See also W.E. Houghton, *The Victorian frame of mind. 1830–1870* (New Haven and London, 1957), pp. 54–61; David Daiches, *Glasgow* (London, 1977), pp. 130–5.

[54] Anna to William Thomson, 8th November, 1842, B184, ULC.

[55] John to William Thomson, 5th January, 1845, T515, ULC. J.P. Nichol's *Views of the architecture of the heavens. In a series of letters to a lady* (Edinburgh and London, 1837), was in its ninth edition by 1851.

security in the 1840s was often a very precious and precarious state which could not be taken for granted as a right.

Dr Nichol's plight illuminates the reasons for Dr Thomson's frequent exhortations to William on all matters pertaining to economy. William, lest he be accused of only writing to ask for a further allowance, would sometimes make the communication more acceptable to his father by accompanying the request with a mathematical theorem which, he hoped, would be of some use in the Glasgow College examinations! Towards the end of his second term, William had nevertheless to justify his college expenditure:

On Saturday night I received my college bills. They come to rather much . . . but you may believe that I have not been in the least extravagant, as, on comparing accounts with the most economical I know among our freshmen, I find that my expenditure has been if anything less than his . . . One of our '*fast*' freshmen has spent about £200 last term (though of course not half of that was sent in to the tutor) and another has spent fully more.[56]

William's expenditure for three academic years at Cambridge, excluding the final term in the autumn of 1844 which led up to the Senate House examination but including two years of coaching under Hopkins, amounted to over £770, a very considerable sum indeed for those years, and indicative of the sacrifices that Dr Thomson was prepared to make for his family, better off than he had been in the 1830s though he now was.[57]

Towards graduation

When William returned to Cambridge in October, 1842, to enter his second year, he began reading for the first time with Hopkins in a relatively small class of about five selected students. As William explained to his father:

What we have had already approximates very much to the plan w[h] you pursue with your class. He asked us all questions on various points in the diff[l] calculus, in the order of his manuscript, w[h] he has given us to transcribe, and gave us exercises on the different subjects discussed, w[h] we are to bring with us tomorrow. He says he never can be quite satisfied that a man has got correct ideas on any math[l] subject till he has questioned him viva voce. I can judge very little yet of any of the other men whom I meet with him, but I hope they are not extremely formidable.[58]

The final sentence of this extract underlines the very competitive nature of the Cambridge Senate House Examinations. More especially, William himself was not only highly conscious of that competition, but his own cultural values,

[56] William to Dr Thomson, *c*. February–March, 1842, T201, ULC.

[57] Dr to William Thomson, 12th October, 1844, T274, ULC. See also SPT, **1**, 41, 87; Rothblatt, pp. 66–7. An example of William's technique of making his requests for money more acceptable by enclosing a theorem is given in William to John Thomson, 13th April, 1844, T503, ULC.

[58] William to Dr Thomson, October, 1842, T223, ULC.

commercial, whig, and Scottish, with no inherited wealth or property, fired in him an intense desire to win every race he entered, whether mathematical, scientific, or rowing.

The much-criticized 'previous' or 'littlego' examination came at the end of the Lent term of 1843. A non-mathematical examination consisting of Greek passages from St Luke's gospel for translation into English prose, questions on Paley's *Evidences* – such as 'how does Paley deal with Hume's theorem on the incredulity of miracles by the process of experiment?' – and other Latin and Greek passages for translation, the 'littlego' was a rather unsatisfactory attempt, elementary in nature, to prevent idleness among undergraduates between the time of entry to the University and the Senate House Examination at the end of their sojourn.[59] Not surprisingly, therefore, William's interests were not entirely directed towards his studies.

By and large the cause of William's diversions was the active social life into which he had entered. By March, 1843, his resolve not to join the college boat club seriously weakened. At first, as his personal diary shows, he kept to his regular sculling outings in his own craft. The sight of the boat races then enticed him to *think* of joining the boat club, but 'both on account of reading, & of what my father would think of me joining, I shall delay. I mean however to join the boat club sometime before I take my degree, perhaps this time next year'.

On 24th March, however, his friend Gisborne persuaded him to join a 'cannibal' (miscellaneous) crew. So, 'off I went, and I am to pull again tomorrow. I should like exceedingly to pull regularly, but, though I may join the club, I *shan't* pull in the races next term'. He admitted his offence to his father, stressing that to pull in the races next term, however, would make him too sleepy in the evenings. Nevertheless, he continued with the 'cannibals', and, finding that he was getting on surprisingly well, the exercise being not at all heavy and sleep-inducing, he reflected that 'if the cannibal crew get going on in the races, I *think* I shall pull . . .'. He also confessed shortly after that 'I have been thinking on the boat more than anything else all day'. Inevitably, he conceded to join the first Peterhouse boat in the last term of his second year. Even Cookson seemed half-convinced by William's arguments: 'after chapel I took tea with Cookson, and after exhausting mathematics we started boating. After a long discussion, I nearly managed to persuade him that I may try the Cannibals next term without committing suicide, mathematically, physically, & morally'.[60]

After a period of intensive training, Thomson rowed 'bow', or number one, in most of the Easter Term races, six in all, spread over two or three weeks. In between races he 'sculled quietly' but 'could do no reading'. In the second race:

[59] *University of Cambridge examination papers, 1843* (Cambridge, 1843). See also D.A. Winstanley, *Early Victorian Cambridge* (Cambridge, 1955), pp. 68, 167; Garland, pp. 115–6; Whewell, *Of a liberal education*, pp. 210–11.

[60] Thomson, diary, *c.*15th March; 24th March–10th April, 1843; William to Dr Thomson, 24th March, 1843, T233, ULC.

We rowed easy, just to keep a few inches ahead of the Caius boat, for about three-quarters of the course, till we got to 'the Willows'; then we laid out. We won all the rest of the races, though the betting at first was ten to one that Caius would bump . . . During those three weeks nothing occurring on the whole earth seemed of the slightest importance: we could talk and think of nothing else. It was three weeks clean cut out of my time for working at Cambridge; so I determined to do no more rowing . . . six months afterwards I won the silver sculls.[61]

Even Dr Thomson, who visited his son in May, 1843, entered into the competitive excitement, and wished to be kept posted about the events on the river. 'You see', he wrote, 'that though I consider it necessary you should give them up for the future, yet I feel an interest in them so far as you are concerned'. When the time came in October, 1843, for William to justify once more his time devoted to boating, he provoked Anna into penning a revealing comment:

I got your letter today containing all your reasons for having joined the boat races, which has one good effect at least – that of convincing us all that you are a most excellent logician, and that, in common with all the rest of your family you possess the excellent talent of being able to defend yourself most eloquently when anything you do is in the least blamed . . .[62]

Time-consuming though it was, boating on the Cam was insufficient to consume William's immense energy. His 1843 diary shows him rising at six or seven on a February or March morning, and retiring to bed seldom before midnight. Music – practising on his cornopean and helping to establish the University Music Society in the spring of 1844 – walking, skating, swimming and, of course, reading, filled his days, along with entering into the heated discussions which took place within his large circle of friends. Topics of conversation ranged from mathematics, through science in general, to politics, literature, and, not least, women. Indeed, many of the subsequent 'excisions' from the diary may have been carried out by the author in order to avoid embarrassment to his future marriage partners.[63]

Training for the Senate House examinations was fully under way throughout the academic year 1843–4. Rehearsals constantly took place according to the methods of Hopkins who, in the tradition of Cambridge 'reading parties', took his pupils to the seaside for two months of the long vacation in 1844, on account of the health of his own children. William Thomson was 'exceedingly glad of it, as it will be much pleasanter than Cambridge in summer', although it would add to an already-expensive year.[64] However, his father agreed, and he spent a pleasant time reading with Hopkins at Cromer, on the Norfolk coast. His friends Hugh Blackburn (1823–1909), who later succeeded Dr Thomson in the

[61] SPT, **1**, 60–1 (as told by Lord Kelvin to S.P. Thompson in the 1900s).

[62] Dr to William Thomson, 23rd May, 1843, T243, ULC; SPT, **1**, 62; Anna to William Thomson, *c.* October, 1843, B197–8, ULC; SPT, **1**, 65.

[63] Thomson, diary, especially entry for 21st March, 1843.

[64] William to Dr Thomson, 4th May, 1844, T259, ULC.

Glasgow mathematics chair, and W.F.L. Fischer (1814–90), subsequently professor of natural philosophy at St Andrews, also joined the reading party. The texts they studied included Airy's and O'Brien's mathematical tracts as part of the intensive study leading up to the examinations in January, 1845.[65] Towards the end of the Cromer session, Hopkins reported optimistically to Dr Thomson on William's progress:

I am happy to say that he has given us entire satisfaction. His *style* is very much improved, and though still perhaps somewhat too *redundant* for examinations in which the time allowed is strictly limited, it is very excellent as exhibiting the copiousness of his knowledge as well as its accuracy. I consider his place as quite certain at the *tip-top*, but I am anxious that he should recollect that he has his own reputation to contend against.[66]

The much-dreaded Senate House examination duly began on 1st January, 1845, and ended on 7th January, twelve mathematical papers later. Morning sessions were of two-and-a-half, and afternoon sessions of three, hours' duration. All the questions were issued to the candidate in printed form, the old practice of examiners dictating book-work questions having been abandoned in 1827. The number of questions set for each paper varied from eight to twenty-four and, of course, the final result would depend on quantity, as well as quality, of answers. The first two papers carried the rubric that the differential calculus was not to be employed, but thereafter the use of the calculus was freely allowed. Of the twelve papers, two were explicitly headed 'problems'; the rest contained a high proportion of book-work and related questions.[67]

The content of the examination covered a wide spectrum of pure and mixed mathematics. Euclidean geometry, arithmetical mensuration, algebraic equation-solving, trigonometry, spherical trigonometry and geometry, conic sections, statical problems – such as equilibrium of forces and centre of gravity – dynamical questions – such as the laws of motion, elastic impact, and pulleys – hydrostatics, geometrical optics, and practical astronomy all appeared in the first two papers of 1845. Newton, once prominent among the Senate House questions, was limited to one question only: 'enunciate and prove Newton's tenth Lemma'. The third and sixth papers, the so-called problem papers, reflected a similar range of questions such as:

A particle is placed on the surface of an ellipsoid, in the center of which is resident an attractive force. Determine the direction in which the particle will begin to move.

Account for Halos and mock suns. Find the greatest altitude of the sun at which the latter can be seen.

[65] William to Dr Thomson, 2nd June, 1844, T264; ULC; 13th June, 1844, T265, ULC; 10th July, 1844, T267, ULC; SPT, **1**, 78–83. The texts were G.B. Airy, *Mathematical tracts* (Cambridge, 1826) and Matthew O'Brien, *Mathematical tracts* (Cambridge, 1840).

[66] William Hopkins to Dr Thomson, 7th August, 1844, H122, ULC.

[67] *University of Cambridge examination papers, 1845* (Cambridge, 1845).

Find the position of a small rectilinear magnet, the center of which is fixed, when the action upon a distant particle of free magnetism is in a given direction.[68]

Paper four was devoted to mixed mathematics; paper five to pure mathematics, including the differential calculus; while papers seven to twelve offered both pure and mixed mathematical questions, at least two of which related specifically to the wave theory of light, one to the integration of partial differential equations, and one to the application of D'Alembert's principle to establish the equations of motion of a rigid body and the fundamental equation of hydrodynamics.

Contrary to everyone's predictions Thomson emerged as second wrangler. First place went to Stephen Parkinson of St John's College.[69] The Thomson family as a whole, though naturally disappointed at the result, were nevertheless ready to be content with the verdict. John Thomson's comments typified the family's attitude:

Although we hoped to have heard that you would have been *first*, we are all very glad that your place is so high & that you have been beaten by no one, but a Johnian. Mr Lushington [Cambridge-educated professor of Greek at Glasgow] in speaking to me about it, remarked that these Johnians are so very '*crafty*' that they constantly manage to slip in at examinations contrary to the expectations of all . . .[70]

One of the Tripos examiners, Ellis, who had succeeded Gregory as editor of the *Cambridge Mathematical Journal*, commented thus on the result to his fellow examiner, Harvey Goodwin: 'You and I are just about fit to mend his [Thomson's] pens', while Whewell noted to Forbes: 'Thomson of Glasgow is much the greatest mathematical genius: the Senior Wrangler was better drilled'.[71]

In the succeeding Smith's prize examination, held towards the end of January, 1845, Thomson emerged first, and Parkinson second. There were four papers in all, each with a high proportion of mixed mathematical problems relating to statics, dynamics, elasticity, the wave theory of light, astronomy, and the steam-engine – all of which Thomson found more congenial than the intense book-work demands of the Senate House examination papers.[72] Meanwhile, William Hopkins had sent a very careful and revealing analysis of his pupil's performance to Dr Thomson, an analysis which ended any possible doubts as to William's ability:

To one as intimately acquainted as myself with the difficult orders of intellect brought into competition in our University examinations, and with the requisites for success in

[68] *Ibid.*

[69] An interesting account of the 1845 Senate House examination is given in A.R. Bristed, *Five years in an English University* (2 vols., New York, 1852), **1**, pp. 325–7. See also SPT, **1**, 90–105.

[70] John to William Thomson, 21st January, 1845, T517, ULC.

[71] SPT, **1**, 97–8, 103.

[72] *University of Cambridge examination papers, 1845.*

St Peter's College graduates in 1852. P.G. Tait (first from left) became co-author with Thomson of the *Treatise on natural philosophy* (ch. 11), and W.J. Steele (third from left) had been one of Thomson's most promising Glasgow students. Tait was senior, and Steele second, wrangler in 1852. [From C.G. Knott, *Life of Peter Guthrie Tait.*]

those examinations, the fact of your son being *second* is perfectly explicable without lessening the conviction that in the high philosophic character both of mind and knowledge, he is decidedly *first*. I have before told you that the prevailing defect of his mind consisted in the want of *arrangement* and *method*, and though the defect has been conquered in a very great degree it has doubtless affected more or less his power of preparing himself for an examination so determinate in its character as our mathematical examinations. While others are simply answering a question, he will often be writing a dissertation upon it, admirable in itself, and indicative of the fullness of his knowledge, but still not exactly what is asked for.[73]

Hopkins went on to say that Parkinson's quality of mind was effectively the reverse of Thomson's. Parkinson seemed to possess 'the most extraordinary power of acquiring knowledge and of bringing it to bear in the extempori process of examinations but apparently with little of that expansiveness of view, which is so eminently characteristic of your son, and affords so sure an indication

[73] William Hopkins to Dr Thomson, *c.*16th January, 1845, H124, ULC.

of future distinction'. One of the examiners told Hopkins that while Thomson would be 'hereafter building up for himself a *European* reputation, his opponent might be scarcely known beyond the bounds of the University', his senior wranglership having been his culminating point, beyond which he had no capability of rising by any power of original research.

Hopkins's analysis summed up all the weaknesses inherent in the Cambridge system of the day. While it was a mathematics course more specialized and profound than that of any other British institution, and thus of great value as a training, not just of the mind, but for a person intent on pursuing mathematical physics, the examination system itself was not designed to produce original minds. On the other hand, William Thomson's years under Hopkins were most certainly of inestimable value to his future work in physical science, for Hopkins not only had the capacity for disciplining minds but the enthusiasm and ability for original work himself. At the same time, Thomson's failure to become senior wrangler did not prevent him from being elected a fellow of Peterhouse in June, 1845, when he had only just reached the age of twenty-one – an age at which, as Dr Thomson reminded him, his father had been teaching eight hours a day at Dr Edgar's school, and had not yet even entered Glasgow College.[74]

Nevertheless, there was still the question of William's longer term future, and the feelings of many members of his circle could have been summed up in two particular letters from college friends in 1845. David Foggo expressed the serious side of William's potential: 'I do not know at all what may be your views or intuitions, but I must say I hope to see you setting up, and coming out strong, as a philosopher and mathematician; and that neither Law, nor Theology, nor Politics, nor much coaching are destined to lead you away from science, wherein may the greatest success crown your efforts'. On the other hand, H.L. Parry, writing in humorous vein to congratulate him on his fellowship, concluded 'by wishing the fellows above you *well disposed of*, & you quietly settled with Eliza [Ker] in the Master's Lodge . . .'[75] But if William Thomson's future glory was indeed to lie with his scientific achievements, his professional and personal life was to be centred not on Cambridge, but on Glasgow.

[74] Dr to William Thomson, 1st July, 1845, T315, ULC (ch. 1).

[75] David Foggo to William Thomson, 27th January, 1845, F152, ULC; H.L. Parry to William Thomson, *c.* June, 1845, P19, ULC. Our emphasis. Eliza Ker, of Dalmuir on the Clyde, was perhaps the favourite among William's girlfriends of the 1840s.

4

The changing tradition of natural philosophy

> There is no candidate [for the Glasgow chair of natural philosophy] besides yourself from Cambridge, none I suppose from Dublin, and Oxford has no one fit. And there can be no one else, except some pet of the principal's, a minister who could not preach and must not starve, of whom I used to hear, who was ready for any chair from Natural Philosophy down. Such men as these get into sectarian colleges . . . but I believe it will be impossible hence forward for such men to get chairs in any national college like that of Glasgow. *W.A. Porter to William Thomson, 1846*[1]

Appointed to the Glasgow College chair of natural philosophy in 1803, and described by Elizabeth Thomson as 'a good-natured, fat, little hunchback with a very red face', Professor William Meikleham had none of the fierce conservatism of Principal Macfarlan or Dr Fleming. A founder member of the Glasgow Philosophical Society in 1802, his activities distinguished him from some of his incompetent or inert colleagues such as James Millar (mathematics) and James Couper (astronomy). In particular, he introduced his students to Lagrange's *Mécanique analytique*, and Laplace's *Mécanique céleste* within an enduring pedagogical and methodological tradition of Scottish natural philosophy.[2]

By the time William Thomson succeeded Meikleham in 1846, the intellectual assumptions of natural philosophy had changed dramatically from those in operation when Dr Thomson had been Meikleham's student in 1811–12. Yet the new professor inherited the pedagogical aims of his predecessor. Meikleham's course had been directed at large numbers of students (122 in 1823–4 and eighty-seven in 1826–7, for example, paying fees of four or five guineas each).[3] Such a popularly based course could scarcely do justice to the modern French analysts. The first challenge facing William Thomson, in his bid for the chair, would be to demonstrate his command of advanced mathematical and

[1] W.A. Porter to William Thomson, 5th September, 1846, P116, ULC.

[2] Elizabeth King, *Lord Kelvin's early home* (London, 1909), p. 121. See the entry for William Meikleham in W. Innes Addison, *The matriculation album of the University of Glasgow from 1728 to 1858* (Glasgow, 1913); SPT, **1**, 12; D.B. Wilson, 'The educational matrix: physics education at early-Victorian Cambridge, Edinburgh and Glasgow Universities', in P.M. Harman (ed.), *Wranglers and physicists. Studies on Cambridge physics in the nineteenth century* (Manchester, 1985), pp. 12–48, esp. pp. 16–33.

[3] 'Report of the Universities Commission 1826–1830', *Parliamentary papers*, **36** (1837), 115–17.

experimental practices in a new intellectual context; the second, to demonstrate his ability to communicate this sophisticated material to large, ill-prepared classes.

The laws of nature

In his introductory lecture to students in the senior natural philosophy class at Glasgow College, Meikleham put forward three general reasons for the study of natural philosophy. First, it 'extends our power over nature by unfolding the principles of the most useful arts'. Second, 'it gratifies the mind by the certainty of its conclusions, by its great extent, & by explaining phenomena'. And third, 'above all it leads us to view the Creator as the Great First Cause, and as maintaining the energies of nature'. Thomson's own 1846 introductory lecture advocated a study of natural philosophy for much the same reasons: practical benefits, intellectual satisfaction, and, most importantly, the approach to God by obtaining knowledge of the laws of nature established by Him 'for maintaining the harmony and permanence of his Works'.[4] The intimate connections suggested here between God, man, and nature characterize the role of natural philosophy in Glasgow College.

Writing in 1847, J.P. Nichol recognized two (and only two) points of connection between religion and a survey of the material world, both points being 'remote from relationship with the matters concerning which our churches are divided'. The first, the existence of design in the universe, is 'altogether apart from religious controversy'. The second, 'the view which represents GOD as a PROVIDENCE; which discerns the energies of Nature as his ministers; nay which, as its culmination, recognises in the Material World no energy or activity save HIS – the omnipresence of a SPIRIT whose distinguishing characteristic is LIFE', is also 'entirely removed from the matter of sectarian disputes'.[5] These two issues, design and providence, will appear as twin features of William Thomson's natural philosophy.

Dr Thomson set out his own early views on providence in an essay in his Glasgow College notebook (1811–12). He emphasized that all operations of nature were ultimately effects of divine power and that God manifested this power through the constant action of laws established by Him for the government of the world. While the aim of natural philosophy was to discover these laws, and perhaps reduce them to more general principles, there existed a definite limit to man's ability to trace such secondary causes to their source: 'an insuperable barrier will at length present itself' to terminate further knowledge. Natural philosophy was concerned with matter, motion, and the laws of nature,

4 William Thomson, 'Notebook of Natural Philosophy class, 1839–40', NB9, ULC; 'Introductory lecture [1846]', in SPT, **1**, 246–50.

5 J.P. Nichol, 'Preliminary dissertation', to the English translation of J. Willm, *The education of the people* (Glasgow, 1847), pp. lx–lxii.

but its range did not include an explanation of the means by which God created and sustained such laws. It was at this point that theology and natural philosophy came into contact.[6]

The brief sketch in Dr Thomson's essay might at first sight seem consistent with a 'deistic' theology of nature in which nature's laws were immanent in the fabric of the universe and so independent from God. Dr Thomson's phrase 'laws established by Him' appears to suggest just such a separation between the Creator and nature. On the other hand, the emphasis on secondary causes, and the claim that all the operations of nature were ultimately the effects of divine power, is also consistent with 'voluntarist' or 'Common Sense' theological traditions. Although Dr Thomson almost certainly did not intend to appear deistic, the very possibility of such an interpretation – and the need to avoid it – led to a fundamental reappraisal of current theologies of nature by Thomas Chalmers and J.P. Nichol, a reappraisal which would provide the theological framework for William Thomson's cosmos. We must briefly analyze, however, 'voluntarist' and 'Common Sense' theologies of nature.

Deriving from a thirteenth-century reaction against the Aristotelian view of St Thomas Aquinas that laws were immanent, necessary and eternal, early 'voluntarist' theologians such as Duns Scotus (*c.*1270–1308) gave primacy to the divine *will* over the divine *intellect* in order to counter the fear that Aquinas's view endangered the traditional conception of God's freedom and omnipotence over His creation. William of Ockham, for example, grounded all natural laws and all ethical values on the will of God, God being the supreme law-giver. The voluntarist believed in one God, the Father Almighty, maker of heaven and earth, and of all things seen and unseen. The whole of His creation had to be wholly contingent upon the undetermined decisions of the divine will. God's absolute power, *potentia absoluta*, had created the universe. His ordained power, *potentia ordinata*, maintained the framework of absolute and constant laws which were direct manifestations of divine will. Only by a further exercise of absolute power – as in miracles – could these immutable laws be suspended or abrogated.[7]

Given the existence of an enduring Christian tradition of 'voluntarism', which included seventeenth-century natural philosophers such as Robert Boyle, René Descartes, and Isaac Newton, it is not surprising to find eighteenth- and nineteenth-century men of science and theology – John Robison (1739–1805), Thomas Chalmers, William Whewell, James Joule (1818–89), and the Thomsons, for example – sharing very similar voluntarist theologies of nature. In the 1830s Chalmers wrote of 'God's creative energy originating all, and of His sustaining providence upholding all' – an unambiguous reference to the absolute and ordaining powers of God. Again, Whewell referred in 1833 to the 'universal

[6] James Thomson, 'Exercises in Natural Philosophy, 1811–12', QUB.

[7] See especially Francis Oakley, 'Christian theology and the Newtonian science', *Church Hist.*, **30** (1961), 433–57, for an account of the origins and nature of the 'voluntarist' tradition.

Creator and Preserver'. Joule in 1847 asserted that the whole universe was 'governed by the sovereign will of God'. And William Thomson in 1851 used the phrase 'supreme ruler', while his brother James stated in a letter of 1846 that he could 'view what we call the Laws of Nature in no other light than merely as expressions of the *will* of an Omnipresent and Ever-acting Creator'.[8]

Eighteenth-century debates in Scotland concerning the status of laws of nature – particularly the controversy provoked by David Hume – had stimulated an important new variant of voluntarist theology of nature which, while stressing divine will and power, distinguished laws of nature from efficient causes. For Scottish moral philosophers – most notably Thomas Reid (1710–96), who occupied the Glasgow chair of moral philosophy from 1763 until his death – the laws of nature were 'the rules according to which the effects are produced; but there must be a cause which operates according to these rules'. Agreeing with Hume on the nature of laws as mere rules, therefore, he nevertheless insisted that 'the rules of navigation never navigated a ship; the rules of architecture never built a house'. Laws of nature were discoverable by observation and experiment. They were the definite object of natural philosophy. Efficient causes, on the other hand, being related ultimately to divine agency and will, were not the object of science: 'upon the theatre of nature we see innumerable effects, which require an agent endowed with active power; but the agent is behind the scene'.[9]

In both the voluntaristic and Common Sense views, the ultimate and primary cause was God, who actively governed the universe by the exercise of His will. Laws of nature, as secondary causes, had no power or agency in themselves. Voluntarists, however, had generally held the laws of nature to be *direct* manifestations of divine will, whereas the Common Sense philosophers distinguished God's will from God's action through efficient causes. They thus interposed efficient causes between laws of nature and God's will. Reid in particular allowed that the efficient causes could be either God acting directly or 'by instruments under his direction'. But, either way, the efficient action was

[8] Thomas Chalmers, *The works of Thomas Chalmers* (25 vols., Glasgow, 1836–42), **4**, pp. 387–8; William Whewell, *Astronomy and general physics considered with reference to natural theology* (London, 1833), p. 201; J.P. Joule, 'On matter, living force, and heat' [1847], *The scientific papers of J.P. Joule* (2 vols., London, 1887), **1**, pp. 265–76, on p. 273; William Thomson, Preliminary draft for the 'Dynamical theory of heat', PA 128, ULC, p. 5; James Thomson to Robert Douglas, 15th August, 1846, in James Thomson, *Papers*, xxxvii.

[9] Thomas Reid, 'Essays on the active powers of man' [1788], in Sir William Hamilton (ed.), *The works of Thomas Reid, D.D.* (2 vols., Edinburgh, 1846–63), **2**, pp. 509–679, esp. p. 527. See L.L. Laudan, 'Thomas Reid and the Newtonian turn of British methodological thought', in R.E. Butts and J.W. Davis (eds.), *The methodological heritage of Newton* (Oxford, 1970), pp. 103–31, for an account of Reid's Common Sense philosophy. P.M. Heimann, 'Voluntarism and immanence: conceptions of nature in eighteenth century thought', *Journal hist. Ideas*, **39** (1978), 271–83, esp. pp. 278–9, outlines Reid's theology of nature. Heimann's statement that Reid was seeking 'to validate the notion of divine sustenance without stressing divine volition as the cause of phenomena' seems to us unclear. Divine volition – intimately bound up with divine power – was fundamental to Reid's theology of nature.

beyond human knowledge. What was important for Reid was that the laws of nature, as mere rules, *required* an efficient cause, an agent endowed with active power, ultimately dependent on but not identical with God's will.[10] Reid concerned himself, however, not so much with material nature as with moral nature, and the primary moral agent was man.

In Sir William Hamilton's famous dictum, Reid's philosophy postulated that 'On earth there is nothing great but man; in man there is nothing great but mind'. Man, made in the image of God, had been endowed with certain active powers which, though limited compared to God's, distinguished man from all other beings on earth and gave to man the independence to act morally, for good or evil. As Reid argued:

> It is evidently the intention of our Maker that man should be an active and not merely a speculative being. For this purpose, certain active powers have been given to him, limited indeed in many respects, but suited to his rank and place in the creation. Knowledge derives its value from this, that it enlarges our power, and directs us in application of it. For, in the right employment of our active power consists all the honour, dignity, and worth of a man, and in the abuse and perversion of it, all vice, corruption, and depravity.[11]

While Reid understood the active powers of mind primarily in a moral sense, his views were sometimes read in the more practical sense of man's capacity to direct the operations of nature towards useful ends. This theme received development in the work of the distinguished Edinburgh natural philosopher and near contemporary of Reid, John Robison. His interpretation of 'mind', human or divine, illuminates the three-fold character of natural philosophy in Scotland: worship of God through natural theology, intellectual satisfaction through science, and practical application through the useful arts. Here efficient cause, in the case of man, acquired the connotation of utility.

Distinguishing mind and matter, Robison comprehended mind not merely as immaterial substance, but in terms of 'art'. While matter was simply the substance immediately cognizable by our senses, mind alone exhibited that phenomenon of art, 'the employment of means to gain ends'. From this view of mind, the relation of the Supreme Mind, God, to the material world could be understood as the relation of the artist, the employer of skills and instruments, to his work of art:

> the world appears a WORK OF ART, a system of means employed for gaining proposed ends, and it carries the thoughts forward to an ARTIST; and we infer a degree of skill, power, and good intention in this Artist, proportioned to the ingenuity, extent, and happy effect which *we are able to discern* in his works. Such a contemplation of nature, therefore,

[10] Archdeacon William Paley's conception of law was virtually indistinguishable from Reid's: 'A law presupposes an agent; for it is only the mode, according to which an agent proceeds: it implies a power; for it is the order, according to which that power acts. Without this agent . . . the *law* does nothing; is nothing.' See *The works of William Paley, D.D.* (5 vols., London, 1823), **4**, pp. 13–14.

[11] Reid, *Works*, **2**, p. 511.

terminates in NATURAL THEOLOGY, or the discovery of the existence and attributes of GOD.[12]

The human mind, made in the image of the Supreme Mind, had been endowed with similar, if limited, qualities of art. As a result, Robison comprehended 'sciences' and 'arts' as mutually dependent:

[Natural philosophy] stands in no need of panegyric: its ultimate connection with the arts gives it a sufficient recommendation to the attention of every person. It is the foundation of many arts, and it gives liberal assistance to all. Indebted to them for its origin and birth, it has ever retained its filial attachment, and repaid all their favours with the most partial affection . . . [For example] It is only from the most refined mechanics that we can hope for sure principles to direct us in the construction and management of a ship, the boast of human art, and the great means of union and communication between the different quarters of the globe.[13]

Such a view of the intimate connection of natural philosophy with practical arts contrasts with the values of a Cambridge liberal education. To underline this Scottish approach, we may note that of over forty articles contributed by Robison to the third edition of the *Encyclopaedia Britannica* (1797–1801), more than one-third treated practical arts such as projectiles, pumps, steam-engines, waterworks, arches, carpentry, seamanship, rope-making, and machinery.[14] William Thomson would follow directly in this Scottish tradition of natural philosophy with its very strong emphasis on the unity of science and art, conceived in terms of the relation of mind to the material world.

As a major interpreter of Bacon, as well as a disciple of Newton, Robison's influential articles 'Physics' and 'Philosophy' in the *Encyclopaedia Britannica* had mapped out a grand classification of the sciences in the manner he believed accorded most closely with our experience, treating laws of nature as inductive generalizations from experience, and condemning speculative hypotheses. He attributed his methodology to Bacon who began with classification and description of phenomena (the method of natural history) and proceeded cautiously to generalizations and laws (the method of natural philosophy)[15].

For Robison, natural history treated of pure descriptions and arrangements.

[12] [John Robison], 'Physics', *Encyclopaedia Britannica*, 3rd edn. (Edinburgh, 1797–1801), **16**, pp. 637–59, on p. 641. On Robison's voluntarism, see J.B. Morrell, 'Professors Robison and Playfair, and the *theophobia gallica*: natural philosophy, religion and politics in Edinburgh, 1789–1815', *Notes and Records Royal Soc. London*, **26** (1971), 43–63.

[13] Robison, 'Physics', pp. 645–57.

[14] Thomas Young, 'Life of Robison', in George Peacock (ed.), *Miscellaneous works of the late Thomas Young* (3 vols., London, 1855), **2**, pp. 505–17; *A course of lectures on Natural Philosophy and the mechanical arts* (2 vols., London, 1807), **2**, pp. 130–8.

[15] Robison, 'Physics', p. 647; 'Philosophy', *Encyclopaedia Britannica*, 3rd edn. (Edinburgh, 1797–1801), **14**, pp. 573–600, esp. pp. 581–600. For a further discussion of Robison's conceptual framework see Crosbie Smith. '"Mechanical Philosophy" and the emergence of physics in Britain: 1800–1850', *Ann. Sci.*, **33**, (1976), 3–29, esp. pp. 7–11. Thomas Thomson outlined a similar framework in his *History of the Royal Society from its institution to the end of the eighteenth century* (London, 1812), p. 311.

As the science of *contemporaneous* nature, natural history aimed to classify the resemblances which existed among the objects – animate and inanimate – of the material universe. Natural history therefore concerned the structure, collocation, arrangement, or disposition of objects, whether of plants or planets. Natural philosophy, on the other hand, was the science of *successive* nature. It aimed to classify the resemblances which took place among the events of the material universe by treating of the incessant, but regular and law-like, changes observed in nature through, for example, the formulation of laws of matter and motion. Natural philosophy further divided into natural philosophy, 'strictly and indeed usually so called', and chemistry. The former division dealt with the laws of *sensible*, visible motions; and the latter with the phenomena, or visible effects, of *insensible* motions. Hypotheses concerning the nature of the insensible motions themselves he regarded as premature and undesirable.

Sensible motions in Robison's classification provided the subject-matter of natural or mechanical philosophy, centred on dynamics. Robison thus defined mechanical philosophy as 'the study of the sensible motions of the bodies of the universe, and of their actions producing sensible motions, with the view to discover their causes, to explain subordinate phenomena, and to improve art'. Scottish natural philosophy adhered to this definition until the mid-nineteenth century. The phenomena of insensible motions provided the subject-matter for Scottish professors of chemistry.[16] These divisions of natural knowledge also had important theological implications, relating on the one hand to order and structure in the cosmos (natural history and natural theology), and on the other to the laws of motion by which God governed the universe (natural philosophy and providence).

Professor Meikleham's natural philosophy course continued the system established by John Robison. The distinction between natural history and natural philosophy, the progressive narrowing of the subject-matter of natural philosophy until it had been defined in terms of sensible motion and force, and the consequent emphasis on the science of force or dynamics, are all characteristics of Robison's *Encyclopaedia Britannica* articles, characteristics which conspicuously reappear in the lecture notes of Meikleham's pupils.[17]

Politics and progression

Like most mechanical philosophy in the nineteenth century, the tradition handed down from Robison to Meikleham to Dr Thomson, and including Robison's successor at Edinburgh, John Playfair (1748–1819), took Laplace's *Mécanique céleste* to epitomize perfection and harmony in the operations of nature. Its outstanding feature was the proof that the solar system would remain

[16] Robison, 'Physics', pp. 643, 647; 'Philosophy', pp. 582–5.

[17] James Thomson, 'Exercises in Natural Philosophy, 1811–12', QUB; William Thomson, 'Notebook of Natural Philosophy class, 1839–40', NB9, ULC.

eternally stable, in a state of dynamic equilibrium, despite a profusion of disturbing forces and attendant irregularities in its motion. Because of this remarkable order in the midst of apparent chaos, the Laplacian system of the world served in a much broader context as a symbol of enlightened liberal ideology, of the faith in natural laws to govern the most perfect system of the world whether material or social.

Playfair's 1808 essay on Laplace's great work, which helped to crystallize the movement for reform in British natural philosophy and mathematics, reflects the symbolic significance of natural order in its political metaphors: 'all the inequalities in our [solar] system are periodical . . . our system is thus secured against natural decay; order and regularity preserved in the midst of so many disturbing causes; and anarchy and misrule eternally proscribed'. Adam Smith, whose 'History of astronomy' (1795) Playfair cited here, had recognized the analogical value in Newton's gravitational explanation of the solar system long before its perfection by Laplace, and had constructed his political economy as another such system of dynamical equilibrium.[18] For the Scottish whigs, therefore, like reformers elsewhere who sought to replace the imperfections of authoritarian rule with the harmonies of natural law, a great deal depended on the stability of the systems of nature. Equilibrium dynamics appeared in every sphere.

Playfair is best known for his *Illustrations of the Huttonian theory* (1802), in which he employed both geological and astronomical arguments to support his commitment to a cyclical universe where one could discern no marks of origins or endings:

> [In geology] we neither see the beginning nor the end . . . In the continuation of the different species of animals and vegetables that inhabit the earth, we discern neither a beginning nor an end; in the planetary motions . . . we discover no mark either of the commencement or the termination of the present order . . . The Author of nature has not given laws to the universe, which, like the institutions of men, carry in themselves the elements of their own destruction. He has not permitted in His works any symptoms of infancy, or of old age, or any sign by which we may estimate either their future or their past duration.[19]

Implicitly, then, if the institutions of men were reformed to reflect natural laws, social life too could approach the stable order of God's works, with 'anarchy and misrule eternally proscribed'.

In 1831 John Herschel would use much the same words in reviewing Mary Somerville's weighty 'popularization' of the *Mécanique céleste* (ch. 6). Shortly

[18] [John Playfair], 'La Place, Traité de méchanique [*sic*] céleste', *Edinburgh Rev.*, **22** (1808), 249–84, on pp. 277–9. See also Adam Smith, 'The history of astronomy', in W.P.D. Wightman and J.C. Bryce (eds.), *Adam Smith. Essays on philosophical subjects* (Oxford, 1980), pp. 33–105. (First published London, 1795).

[19] John Playfair, *Illustrations of the Huttonian theory* (Edinburgh, 1802), pp. 119–21. Quoted *verbatim* by William Thomson in 1868 (ch. 15). See PL, **2**, 11–12.

afterwards Dr Thomson also echoed Playfair's cosmological creed in his introductory lecture at Glasgow College (1832):

They [the European astronomers and mathematicians] have also arrived at one of the most interesting conclusions to which science has ever led: they have found that contrary to some appearances, the eternal stability of the solar system is fully provided for in its own mechanism: that it contains no seed of its own dissolution, no principle leading to old age or decay; but that every phenomenon appearing to give such indication is periodical, and is regulated by a corrective power arising from the principle of universal gravitation, which prevents it from exceeding a fixed amount.[20]

In sharp contrast to Laplace himself, however, each of these reformers insisted that the incomparable order of nature could not have arisen from a chance arrangement, but gave indisputable evidence of God's handiwork. Dr Thomson, for example, added that the 'grand conclusion' of stability afforded 'a striking manifestation of design in the formation of the system, as it is proved that had the arrangements & mechanism of its parts been much different from what they are found to be, there w^d have been no such permanence . . . [and] the disturbing actions of the bodies on one another w^d in time have brought them into destructive collision'. As the system was constructed, however, the planets 'must continue to move forever with pristine vigour and with undeviating regularity, unless arrested in their career by that Almighty power by w^h they were first called into existence, and launched along their respective paths'. The planetary system was thus eternal, except by exercise of God's absolute power.

Within a year of Dr Thomson's remarks the symbol to which he and so many others had attached their natural theological and political optimism suddenly lost its potency. Conservation of motion in the solar system seemed to be vitiated by observations on Encke's comet, which indicated the action of a resisting medium in interplanetary space. If confirmed for planetary motion, the retarding action would inevitably, given sufficient time, cause the entire system to collapse into the sun.

Both William Whewell in his *Bridgewater treatise* (1833) and John Herschel in his *Treatise on astronomy* (1833) discussed this sobering implication, with Whewell confidently asserting that 'facts have been observed which show, in the opinion of some of the best mathematicians of Europe, that such a very rare medium does really occupy the spaces in which the planets move':

Since there is such a retarding force perpetually acting, however slight it be, it must in the end destroy all the celestial motions . . . still the day will come (if the same Providence which formed the system, should permit it to continue so long) when this cause will entirely change the length of our year and the course of our seasons, and finally stop the

[20] Dr James Thomson, 'Introductory lecture at Glasgow College, 6th November, 1832', QUB; [J.F.W. Herschel], 'Mechanism of the heavens', *Quart. Rev.* **47** (1832), 537–59.

earth's motion round the sun altogether. The smallness of the resistance . . . does not allow us to escape this certainty.[21]

Dr Thomson deleted from his lecture the now dubious claim of eternal stability but added a footnote indicating his reluctance:

It may be proper to mention that a supposed change in the movements of one of the comets had led some to believe that there is a highly rarefied medium which offers some slight resistance to its motions from the lightness of its parts. This however requires confirmation and nothing of the kind has been observed in regard to the planets.

Ignoring all such quibbles, Whewell pressed home the newly refreshed lessons for natural theology: 'no one who has dwelt on the thought of a universal Creator and Preserver, will be surprised to find the conviction forced upon the mind of every new train of speculation, that viewed in reference to Him, our space is a point, our time a moment, our millions a handful, our permanence a quick decay'.[22] This 'universal law of creation' implied not only an end of the material world but also a transient destiny for man and his institutions.

In place of the Laplacian image of eternal conservation and harmony Whewell thus substituted an image of instability and decay, of 'perpetual change, perpetual progression', in which 'motion is perpetually destroyed, except it be repaired by some living power':

To maintain either the past or future eternity of the world, *does not appear consistent with physical principles, as it certainly does not fall in with the convictions of the religious man*, in whatever way obtained. We conceive that this state of things has had a beginning; we conceive that it will have an end. But in the meantime we find it fitted, by a number of remarkable *arrangements*, to be the habitation of living creatures.[23]

Whewell associated design with beneficial *arrangements* in the creation which would endure for a time, but which tended ultimately to dissolution. And although he employed the familiar analogy of a divine watch-maker to illustrate the designer–creation link, he was much more concerned to use the analogy to emphasize, with Newton, the limitations of mechanical enquiry. Even if we could trace the present arrangement of the solar system to other pre-existing mechanical systems, we must ultimately go outside mechanical causes to creation *ex nihilo*, by 'a First Cause whch is not mechanical'.[24]

[21] William Whewell, *Astronomy and general physics considered with reference to natural theology* (London, 1833), pp. 191–209; J.F.W. Herschel, *A treatise on astronomy* (London, 1833), pp. 308–9. Whewell's treatise was one of Meikleham's recommended texts in 1839. See William Thomson, 'Note-book of natural philosophy class 1839–40', NB9, ULC. Lecturing to his own natural philosophy class in 1862, William gave a definitive statement about Encke's comet: it 'manifested irregularities in motion which prove a resisting medium'. See William Thomson, 6th November, 1862, lecture 3, in David Murray, 'Lecture notes in *classe physica*, bench II, November, 1862', MS Murray 325, ULG.

[22] Whewell, *Astronomy and general physics*, p. 201.

[23] *Ibid.*, pp. 203–4. Our emphasis.

[24] *Ibid.*, pp. 206–7. Whewell was quoting Newton here. See Isaac Newton, *Opticks* (4th edn., London, 1730), p. 369. For a very similar perspective to that of Whewell in this period, see Humphrey Lloyd, 'On the rise and progress of Mechanical Philosophy' [1834], *Miscellaneous papers connected with physical science* (London, 1877), pp. 414–36, esp. pp. 433–4.

Whewell's triumphant message highlights a major turning point in the scientific culture of Britain and in the intellectual position of the Thomson family. In previous disputes over the adequacy of natural laws to guarantee stable systems in a perfect, or perfectible, world, all sides had taken conservation and decay as antitheses. Could the ideals of rational order and progressive politics be recovered in a world where decay was the 'universal law of creation'? Two friends of the Thomson family, Thomas Chalmers and J.P. Nichol, took positions which illustrate the dilemma. Both sought optimistic answers, but from divergent perspectives, the former through revealed religion and the latter through a reinterpretation of enlightenment ideals.

Speaking as Scotland's most noted theologian and evangelical preacher, Chalmers wrote in the introduction to his 1833 *Bridgewater treatise* of the 'fine generalisation by the late Professor Robison, of Edinburgh, which ranges all philosophy into two sciences', natural history and natural philosophy. But Chalmers recognized that natural philosophy, the science of successive nature, was perfectly consistent with 'infidel' beliefs in immutable laws, immanent in the fabric of nature, rigidly deterministic, and leaving no place for miracles or prayer. Natural philosophy, therefore, could not provide a basis for natural theology, for arguments to the existence and attributes of God. Natural philosophy had to be interpreted through revelation, which implied, for Chalmers, a voluntarist theology of nature. To provide a basis for natural theology Chalmers turned to natural history, to the arrangements and dispositions of matter throughout the cosmos.[25]

Given a clear distinction between natural history – concerned with arrangements of objects – and natural philosophy – concerned with the laws of nature – Chalmers developed the notion that the laws of natural philosophy were by themselves inadequate to give rise to the *arrangements* of our existing natural history. Furthermore, he believed that if these arrangements were destroyed, no powers of nature, no laws of natural philosophy, would be able to replace or renew them. He argued that this claim was most important 'for the cause of natural theology' since, although the laws of nature might 'account for the evolution of things or substances collocated in a certain way . . . they did not originate the collocations':

The laws of nature may keep up the working of the machinery – but they did not and could not set up the machine . . . For the continuance of the system and of all its operations, we might imagine a sufficiency in the laws of nature; but it is the first

[25] Thomas Chalmers, *The adaptation of external nature to the moral and intellectual constitution of man* (London, 1834), pp. 25–7; *The works of Thomas Chalmers* (25 vols., Glasgow, 1836–42), **1**, pp. 222–5; **7**, pp. 234–62, esp. pp. 234–6. See also D.F. Rice, 'Natural theology and the Scottish philosophy in the thought of Thomas Chalmers', *Scott. J. Theol.*, **24** (1971), 23–46, for a discussion of Chalmers's evangelical theology in relation to Scottish Common Sense philosophy, and Crosbie Smith, 'From design to dissolution: Thomas Chalmers's debt to John Robison', *Brit. J. Hist. Sci.*, **12** (1979), 59–70, for a further analysis of Chalmers's theology of nature in relation to his natural theology. For a modern biographical study, see S.J. Brown, *Thomas Chalmers and the Godly Commonwealth in Scotland* (Oxford, 1982).

construction of the system which so palpably calls for the intervention of an artificer, or demonstrates so powerfully the fiat and finger of a God.[26]

Similarly, living organisms required the interposition of a living and purposing agent to bring together their parts and mould them into existence. Chalmers therefore argued in his *Bridgewater treatise* that recognizing divine order and plan depended far more on natural history than on natural philosophy.

Chalmers first met Whewell at the 1833 meeting of the British Association in Cambridge. Perhaps as a result of their discussions and of reading Whewell's *Bridgewater treatise*, which he called 'truly admirable', Chalmers began to stress the theme of decay in his own sermons.[27] He recognized that a progressionist cosmology offered much greater support for natural theology than did a steady-state cosmology, for it would vitiate both deistic and naturalistic interpretations of laws of nature. David Hume had skilfully articulated the sceptical view in his *Dialogues concerning natural religion* (1779). To avoid going beyond the present world, Hume argued, it would be better to suppose the principle of order within nature than without: 'an ideal system [the divine mind], arranged of itself, is not a whit more explicable than a material one which attains its order in like manner'.[28] Chalmers now answered:

Let there only be evidence, whether in nature or in history, by which to get quit of the hypothesis that this world with all its present laws and harmonies must be eternal – and then, on the stepping stone of a world so beauteously ordered and so bountifully filled [in the absence of eternal laws], might we rise to the second hypothesis of an Eternal Mind from whom this universe is an emanation. This would give full introduction to the reasonings *a posteriori* – carrying us at once from the indications of design to a primary designer. All that is needed is satisfactory evidence that these indications are not from Eternity.[29]

With respect to the solar system, that evidence existed in the effects of a resisting medium.

Chalmers and Whewell both carried their critique of equilibrium dynamics in the material world into the political arena, with writings on political economy. As in the economy of nature, so in the social economy natural force alone would lead not to harmonious equilibrium but to progression, and very likely in the direction of dissolution. Both authors, however, believed that the moral force of Christianity could arrest the process and even lead to improvement in the material condition of man, allowing peasants and workers to escape the Malthusian condemnation to a subsistence existence. Whewell naturally did not share Chalmers's anti-hierarchical, presbyterian perspective, because he

[26] Chalmers, *Works* **1**, p. 225.

[27] Smith, 'From design to dissolution', p. 65.

[28] David Hume [1779], reprinted in H.D. Aitken (ed.), *Dialogues concerning natural religion* (New York, 1948), pp. 20, 34–6.

[29] Thomas Chalmers, 'On the non-eternity of the present order of things', *The works of Thomas Chalmers* (25 vols., Glasgow, 1836–42), **1**, pp. 161–87, on p. 176.

vested moral power in the established Anglican Church rather than in personal salvation. Nevertheless, while Malthus regarded sexual passion and hunger as antagonistic forces of nature which inevitably would keep the lower classes in a state of miserable equilibrium, both Chalmers and Whewell believed that, through religious guidance, peasants and workers could control their overpopulating urges.

Whewell challenged the political economy of Malthus and Ricardo by attacking equilibrium theory directly, arguing that the tendencies of supply and demand, as of population and food supply, could give only inadequate expressions of the full dynamics of natural law, even if one assumed that they represented human nature, an assumption which he also disputed.[30] In contrast, Chalmers accepted Malthus's assessment of human nature, but advocated a socially radical solution based on transcendence of this depraved nature. The working classes, he argued, could take control of their own destiny by controlling the supply, and thus the cost, of labour. They could therefore market labour in the same way a capitalist would market commodities: 'It is at the bidding of their collective will, what the remuneration of labour shall be; for they have entire and absolute command over the supply of labour'. Such control, however, required education of the collective will through 'general instruction, or by the spread of common, and more especially of sound Christian education over the country'.[31]

A progressive view of man, then, for the conservative Whewell and the evangelical radical Chalmers, required abandoning the faith of the enlightenment in natural law. The alternative available to liberals and radicals in the natural law tradition was to reconstruct their ideology and its relation to natural philosophy. That alternative is the one followed by J.P. Nichol.

In 1836, Nichol launched a powerful attack in John Stuart Mill's *London and Westminster Review* on Chalmers's distinction between the dispositions of matter 'fixed at the original setting up of the machine' and the laws of nature 'ordained for the conducting of the machine'. Nichol argued:

> The truth is, a 'collocation' *per se* excites nothing but wonder . . . and this feeling . . . operates as an incitement . . . to the intellect to seek out the physical cause or the origin of the new and not comprehended scheme . . . Persuade weak minds that the existence of the Deity is best shown by 'collocations' as distinct from law . . . and we have at once the

[30] William Whewell, 'Mathematical exposition of some doctrines of political economy', *Trans. Cam. Phil. Soc.*, **3** (1830), 191–230; 'Mathematical exposition of some of the leading doctrines in Mr Ricardo's "Principles of political economy and taxation"', *Trans. Cam. Phil. Soc.*, **4** (1833), 155–98. For further discussion, see M. Norton Wise (in collaboration with Crosbie Smith), 'Work and waste: political economy and natural philosophy in nineteenth century Britain', *Hist. Sci.* (forthcoming).

[31] Thomas Chalmers, *On political economy, in connexion with the moral state and moral prospects of society* (Glasgow, 1832), pp. 25–6. See especially R.M. Young, 'Malthus and the evolutionists: the common context of biological and social theory', *Past and present*, no. 43 (1969), 109–45, esp. pp. 120–5.

reluctance of the intolerant mob of Greece to permit the extension of physical law – which is, the extension of knowledge . . . Had not the argument for the being of a God been often placed, by a similar mistaken dogmatism, in apparent collision with progressing and legitimate inquiry, we should have had fewer Atheists and less intolerance.[32]

Nichol's radicalism obviously had different foundations from Chalmers's. He made his critique at the end of a remarkable discussion of the nebular hypothesis, incorporating an essentially progressive and evolutionary view of the heavens, which he developed at length in his 1838 *Views of the architecture of the heavens.*

At once popular and profound, poetic and persuasive, Nichol's book traced the whole range of explanatory powers of a nebular hypothesis, placing special emphasis on the origins, development and future of the solar system, a grand *progression* from beginnings into new forms and arrangements. Nichol disagreed fundamentally with Chalmers's inference that original collocations alone placed natural theology *qua* design on a secure foundation, and insisted that development in nature followed fixed natural laws, laws which expressed both beneficent design and providence. He carefully stressed that the laws manifested divine power and not self-actuating processes:

If uneasy feelings are suggested – and I have heard of such – by the idea of a process which may appear to substitute *progress* for *creation*, and place *law* in the room of *providence*, their origin lies in the misconception of a name. LAW of itself is no substantive or independent power; no causal influence sprung of blind necessity, which carries on events of its own will and energizes without command.[33]

Nichol's view of laws required a new conception, for in traditional natural philosophy the laws were not temporal, were not in themselves laws of progression. Although the successive development of the states of a system could be derived from them, that development could as well proceed in the opposite direction. Unity in a system, furthermore, depended on the motions governed by the laws being eternally repeated. In contrast, progressing systems exhibited no such constancy in the spatial relations of their parts. They did not form 'clusters of perfect regularity', in Nichol's phrase.

In correspondence with J.S. Mill (and from his familiarity with German idealism), Nichol developed the view that a spatial arrangement at any given

32 J.P. N[ichol], 'State of discovery and speculation concerning the nebulae', *London and Westminster Rev.*, **3** (1836), 390–409, on pp. 406–9.

33 J.P. Nichol, *Views of the architecture of the heavens. In a series of letters of a lady* (Edinburgh and London, 1837), p. 183. James Thomson, *op. cit.* (note 8), criticized Robert Chambers's controversial *Vestiges of the natural history of creation* (London, 1844) in the same terms: 'it appears to me that the author substitutes for the Creator, what he calls *Law*; and that, if he gives assent to the existence of God as a First Cause, he, at least, supposes Him to be now infinitely *removed* from all the Works of Nature, and that everything goes on now *of itself* just as a clock after its weights have been wound up. Now I am strongly impressed with the idea that a law is in itself nothing and has no power; and I can view what we call the Laws of Nature in no other light than merely as expressions of the *will* of an Omnipresent and Ever Acting Creator'. Ironically, Chambers had drawn extensively on Nichol's astronomical publications.

time ought to be understood as one state in a succession of states: 'Parting from the notion of a cluster of perfect regularity, the idea thus rises before us, of a *series* ... upon which MUTABILITY is stamped'. The unity of the parts would now reside in their derivation from a unitary created origin. If one part of a system – considered at a given moment – seemed isolated from the rest, the problem lay in not considering temporal relations: 'its character must be determined through its relations, not with Space but Time; and its very isolation among surrounding things ... constrains us to regard it as only one part of a term of a Series'.[34] Mill and Nichol regarded this doctrine as the cornerstone of a new 'science of progress', social progress in the first instance, but founded on an analogy with progression in nature. Thus Mill asserted that 'The mutual correlation between the different elements of each state of society is ... a derivative law, resulting from the laws which regulate the succession between one state of society and another'.[35]

Like Mill, Nichol believed that the natural progression of man would be upward. A sound education would promote this advancement if it followed nature's model. 'The education of Man is the most visible among the purposes of the existing scheme of things ... in seeking to advance it by Education, we therefore act in harmony with manifold resistless agencies'.[36] It should therefore be democratic, universal, non-sectarian, useful, positive, and improving (ch. 2).

In advancing this doctrine, Nichol had to contend with a most problematic relation between the nebular hypothesis and the supposed upward course of man, for the solar system was decaying:

> the system, though strong, is not framed to be EVERLASTING; and our Hypothesis also develops the mode of the certain decay and final dissolution of its arrangements. Remember the effects of the Solar Ether! Although no mark of age has yet been recognized in the planetary paths, as sure as that filmy comet is drawing in its orbit, must they too approach the sun, and at the destined term of their separate existence, be resumed into his mass. The first indefinite germs of this great organization, provision for its long existence, and finally its shroud, are thus all involved in that master conception from which we can now survey the mechanisms amid which we are![37]

Inverting the usual pessimistic assessment of this decay, Nichol argued that it derived, not (as Newton had thought) 'by accident, derangement, or disease, but through the midst of harmony': 'the inheritance of this same Nebulous parent-

[34] Nichol, *Architecture of the heavens*, p. 197; *Thoughts on some important points relating to the system of the world*, 2nd edn. (Edinburgh, 1848), p. 198. See also 'Unpublished letters from John Stuart Mill to Professor Nichol', *Fortnightly Rev.*, **59** (1897), 660–78. In Chapter 14 we discuss the importance of these ideas for Thomson's ether.

[35] We owe this interpretation of Nichol's and Mill's views of law and of the nebular hypothesis to Simon Schaffer, 'The nebular hypothesis and the science of progress', in J.R. Moore (ed.), *The humanity of evolution* (Cambridge, forthcoming). Mill's remarks are quoted in Schaffer. See J.S. Mill, *A system of logic*, 9th edn. (2 vols., London, 875), **2**, p. 510.

[36] Nichol, 'Preliminary dissertation', pp. xv–xix.

[37] Nichol, *Architecture of the heavens*, pp. 188–9.

age [of the planets], viz., the existence of an ether, leads gently to their decline'. Furthermore, while 'The idea of the ultimate dissolution of the solar system has usually been felt as painful, and forcibly resisted by philosophers', Nichol argued against any such view on the grounds that 'Absolute permanence is visible nowhere around us, and the fact of change merely intimates that in the exhaustless womb of the future, unevolved wonders are in store'.[38]

The optimism of the enlightenment was to be recovered by building the upward progress of man on an underlying course of decay in the processes of nature:

> Nay, what though *all* should pass? What though the close of this epoch in the history of the solar orb should be accompanied . . . by the dissolution and disappearing of all these shining spheres? Then would our Universe not have failed in its functions, but only been gathered up and rolled away these functions being complete. That gorgeous material framework, wherewith the Eternal hath adorned and varied the abysses of space, is only an instrument by which the myriads of spirits borne upon its orbs, may be told of their origin, and educated for more exalted being.[39]

Gone was the eternal harmony of equilibrium states in the economy of nature, gone were their unchanging collocations; the law of nature was now eternal change, leading to completion of a built-in end, and to ultimate dissolution of the instrument of this glorious process. Nichol, however, proposed no law within natural philsophy which would govern the progressive development of the solar system and thus could serve as a model for the laws of society. He proposed only a 'nebulous' example.

Nichol's perspective on the relation between decay in nature and progress in society could hardly have been further from that of Chalmers. While both men saw progression in nature, Nichol made it progress and Chalmers decay. Both envisaged the improvement of man's condition, and both advocated universal education of the people as the means to that end, but Nichol would found progress on natural law itself while Chalmers would transcend natural law through the moral power of Christianity. That both men were close friends of the Thomson family brings into sharp focus that dilemma in which the Thomsons found themselves as proponents of both social rationalism and providential theology. Prior to the 1830s, rationalism had been associated with conservation of the states and powers of nature, while providential theology might assume either conservation or decay. Now one had to assume progression of some kind, but natural philosophy offered as yet only conservation. To follow Chalmers would mean (*a*) giving up the former basis of rationalism within natural philosophy, and (*b*) emphasizing the directing power of man in society; to follow Nichol meant (*a*′) extending the province of natural law to encompass development, and (*b*′) seeing man's progress as similarly governed. We shall find William deeply committed to (*a*′) the strict rule of law in nature

[38] *Ibid.*, pp. 154–5. [39] *Ibid.*, p. 191.

and to (*b*) the necessity of man's directing power in social affairs. He could follow *in toto* neither Chalmers nor Nichol and would struggle mightily to arrive at a resolution and a new synthesis (ch. 10).

At the beginning of his career, then, Thomson found himself in an intellectual situation entirely different from that of the previous generation. The conflict over conservation and progression had become the issue that marked a new epoch. New answers were required in natural philosophy, just as in geology and biology, with their famous debates over the steady-state earth and the evolution of species (chs. 6 and 14–18). All of these problems, as well as their concomitants in natural theology and social theory, would impinge directly on Thomson's intellectual development during the 1840s, both while at Cambridge until 1846 and afterwards as he attempted to bring the problems and prospective answers before his classes in natural philosophy. Before he fully entered the fray, however, he and his father embarked on a campaign to secure his professional status.

Politics and pedagogy: the campaign for succession

'The democratic intellect' is an epithet well applied to Scotland and her universities in the nineteenth century. With an absence of academic entrance requirements, with prizes awarded not by the authority of the professor but by the vote of the class, and with a vigorous pedagogical tradition of daily examination meetings (or class discussions) and class essays, the Scottish universities aimed to provide four years of general education, both useful and liberal.[40] In its liberal, as well as its useful function, a Scottish university education thus differed markedly from an Oxbridge liberal education through an egalitarian, non-elitist training of undergraduate minds.

Above all, a university education in Scotland cost comparatively little, providing professorial lectures for students from a wide spectrum of social backgrounds, in contrast to the expensive, hierarchical structure of Cambridge. As Dr James Thomson, himself a notable beneficiary of the Scottish system, made the point in 1825, 'the efficiency of the means is fully proved by the production of a more general diffusion of education and intelligence than an equal population in any other country can exhibit'.[41] From a small country steeped in such presbyterian values as opposition to hierarchical authority (especially episcopacy and papal supremacy), election of the elders of the kirk, and the right of congregations to call the ministers of their choosing, there had emerged a distinctive university system at once democratic and presbyterian, a system which appealed, in different ways, to Scottish liberals and tories alike.

By the 1820s and 1830s, however, criticisms of the Scottish universities on

[40] G.E. Davie, *The democratic intellect. Scotland and her universities in the nineteenth century* (Edinburgh, 1961), pp. 15–20.

[41] [James Thomson], 'State of science in Scotland', *Belfast Mag.*, **1** (1825), 270.

both academic and religious grounds had become commonplace. These criticisms manifested particular religious and political forms at Glasgow College, dividing the institution along whig and tory political lines. The whigs, led by professors such as Thomas and James Thomson, sought to free the College from its remaining sectarian ties and to make professional merit the principal criterion for election to an academic chair, while the tories, led by Principal Macfarlan and other professors with close ties to the established Church of Scotland, aimed to preserve the traditional religious affiliations of the College (ch. 2).

The Glasgow College tories could unite, therefore, both in opposition to any weakening of the role of the Church and to any change in the popular form of the pedagogical system. Professor Robert Buchanan, for example, appointed to the chair of logic in 1827, followed his celebrated predecessor and teacher, George Jardine, in upholding the democratic approach.[42] Buchanan, an ordained minister, became a loyal member of what Dr Thomson called 'the Principal's party'. By contrast, the reformers in the College had a more difficult task. While the fragmentation of the Church in the early 1840s gave new impetus to the campaign for the abolition of the tests, reformers divided over the wider issue of educational reform.

On the one hand, the Oxbridge educated reformers favoured the anglicization of Scottish universities, while the more radical professors promoted Scottish democracy over Oxbridge elitism. Sir Daniel Sandford, the Oxford-educated professor of Greek and a conservative reformer, sought higher academic standards along Oxford lines, while James Mylne, the elderly Scottish-educated professor of moral philosophy and much more radical political reformer, wanted to preserve the democratic traditions of Scotland.[43] Again Sandford and Ramsay supported the senior wrangler Archibald Smith, himself a whig and once addressed by his sister as 'my dearest Radical Archy',[44] for the vacant chair of practical astronomy in 1836, but lost to the Scottish-educated radical J.P. Nichol, whose great asset lay in an ability to communicate, popularize, and bring alive his subject (ch. 2). Furthermore, opposition to the 1826 Royal Commission's recommendations, aristocratic and anglicizing in tone, came from Scottish whigs and tories alike.[45]

Paradoxically, then, William's Cambridge successes threatened to weigh as a liability in his bid for the Glasgow chair. Dr Thomson's dilemma may be illustrated by an issue which arose when in 1845 the Marquis of Breadalbane provided Glasgow University with a prize worth £100 per annum. The Senate divided over whether the prize should be given as a three-year exhibition to Cambridge (a proposal supported by Dr Thomson, Dr William Thomson, and Nichol) or as a fellowship to a Glasgow MA '*instead of sending our best men away to acquire unScottish ideas in England* and to lead them to the English bar or

[42] Davie, *The democratic intellect*, pp. 28–30. [43] *Ibid.*, pp. 38–9.

[44] Joanna Smith to Archibald Smith, 8th January, 1835, TD1/715, Smith papers, Strathclyde regional archives. [45] Davie, *The democratic intellect*, pp. 26–40.

church'.[46] Having lost the support of not only some of the College professors, including Ramsay, and all the regius professors, Dr Thomson admitted defeat, acknowledging that '*perhaps* their plan might be more beneficial to this University, but that, considering the superior capabilities of Cambridge, I thought my plan would be more beneficial to science'.

Given the character of the existing Scottish pedagogical tradition, the principal remedy for William's potential weaknesses lay in promoting his experimental strengths. Experimental demonstrations formed a crucial part of a popular natural philosophy course, while at a deeper level Scottish natural philosophy viewed experimental apparatus or illustrative mechanisms as the very embodiment and practical realization (and not merely the application or exemplification) of theory. Experiment is thereby integral to (and not merely illustrative of) the meaning and derivation of a concept. Furthermore, a Scottish dislike of speculative hypothesis provided an agreeable complement to the stress on experiment and observation.[47]

By the beginning of the 1841–2 session at Glasgow College, William's first year at Cambridge, Professor Meikleham had reached the age of seventy. He had not returned to the teaching of the natural philosophy classes after his collapse during the 1839–40 session, and to his colleagues it was becoming clear that he was unlikely ever to do so again. Towards the end of 1841 Dr Thomson began a careful appraisal of potential candidates for the natural philosophy chair. In so doing, he had to satisfy several difficult criteria.

First, the university tests, if rigorously applied, restricted the chair to a candidate prepared to conform to the established Church of Scotland. As a result, many distinguished Cambridge men, who were for the most part Anglican, would be excluded unless their latitudinarianism permitted a stretching of their religious principles. Second, since Cambridge alone produced the high mathematical attainments which Dr Thomson prized, yet discouraged dissenters, the possibility of a candidate combining presbyterian loyalties and mathematical accomplishments seemed rare indeed. And third, while a Cambridge man would satisfy both Dr Thomson's mathematical criterion *and* Glasgow reformers aiming to promote more rigorous professional standards (Ramsay and Lushington in particular), Scottish whigs and tories alike wished to preserve popular values. Thus a candidate had above all to be a professional teacher, with the ability to market his expertise to large numbers of fee-paying students, in the manner of J.P. Nichol or Dr Thomson himself.

At first sight, D.F. Gregory seemed the most obvious candidate. Archibald Smith and J.D. Forbes were also possibilities, though Smith's lack of commitment to the 1836 contest and Forbes's Episcopalian connections weighed against

46 Dr to William Thomson, 22nd March, 1845, T302, ULC. Dr Thomson was quoting the views of his opponents here.

47 See, for example, G.N. Cantor, 'Henry Brougham and the Scottish methodological tradition', *Stud. Hist. Phil. Sci.*, **2** (1971), 69–89.

them. Gregory had been defeated by Kelland in the 1838 contest for the Edinburgh chair of mathematics, Forbes having apparently written testimonials for both candidates with the deliberate aim of supporting the senior wrangler Kelland against the fifth wrangler Gregory, Kelland representing continental analysis and Gregory representing a new creative style that owed much to his Scottish roots (ch. 6).

With the aftermath of the 1838 election evidently in mind, Dr Thomson asked William in November, 1841, to take soundings about Gregory:

> Have you any means of forming an opinion regarding his powers of communicating? Of his attainments there can be no doubt, but has he popular talents? Quietly look after this matter, and let me know what you gather on this subject either from your own observation or otherwise. Would he or Forbes or some other person still best suit for our Natural Philosophy chair?[48]

Not until early 1842 did William send an estimate of Gregory's qualities to Dr Thomson. By that time he had seen a good deal of both Gregory and Archibald Smith, the two mathematicians who 'have written nearly all the good articles in the *[Cambridge Mathematical] Journal*'. William favoured Gregory who, for reasons of the prevailing religious tests at Cambridge, could not continue as a fellow of Trinity College:

> Gregory especially has introduced a great deal of new & original matter [into the *Journal*], in his articles on the separation of symbols, the application of algebra to geometry, impossible logarithms, &c., &c. I think he would make a splendid professor for you, and would probably be very glad to get the situation as he will not enter the English Church, and must therefore give up his fellowship in seven years. He is a considerable Whig (both in mathematics and politics) but that need not be known. I think he is undoubtedly the best and most original math^{n} in Cam[bridge], and it is also said that he has a great knowledge of experimental and physical subjects. He certainly wishes to encourage such subjects here, as he set one question from Willis's Practical Mechanics . . . and he told me afterwards that he wishes to encourage the men to read such subjects.[49]

As a Scot, and a presbyterian, Gregory was well-qualified to pass the Scottish tests, and as a brilliant mathematical scholar and whig 'both in mathematics and politics' he would have made an excellent colleague and ally for Dr Thomson. Dr Thomson, however, continued to have doubts, wishing that 'Gregory had given proofs of his qualifications as a teacher and of his ability to popularise'.[50]

Towards the end of 1842, Dr Thomson considered the possibility of supporting Forbes, and was of the opinion that the '*independent* party in the College will be able to carry the election' against any sectarian candidate put forward by the Principal's party. As Forbes had held the Edinburgh chair without sectarian conflict since 1833, the delicate problem of the tests at Glasgow would presum-

[48] Dr to William Thomson, 15th November, 1841, T184, ULC; Davie, *The democratic intellect*, pp. 105–6. [49] William to Dr Thomson, February or March, 1842, T201, ULC.

[50] Dr to William Thomson, 7th December, 1842, T228, ULC.

ably not be insoluble. In other words, Glasgow College would look particularly foolish if, on the grounds of the tests, it refused a post to a man who had held a prestigious Edinburgh chair for ten years. However, Forbes was not the ideal candidate: 'so far as "Physique" [i.e. experimental physics] is concerned, he is perhaps the best we could get; but as to the math[l] parts of Nat[l] Phil[l], I doubt his preeminence and I am not sure about his power in teaching according to *our* system'.[51]

Dr Thomson's doubts concerning Forbes's 'power in teaching according to *our* system' are especially revealing. Though a reformer, Forbes's aristocratic and tory alignments placed him much nearer to the anglicizing aims of the 1826 Royal Commission and to Oxbridge goals than to the Glasgow College whigs whose reforms aimed to maintain the democratic, popular ideals of the Scottish system. In his fight against the whig David Brewster to secure the Edinburgh chair of natural philosophy in 1833, through his correspondence with William Whewell critical of the low level of Scottish teaching, by his ruthless support of Kelland against Gregory, and by his encouragement of a small, elite senior class similar to Hopkins's classes, Forbes had shown himself unsympathetic to the traditional Scottish aims of the diffusion of knowledge both liberal and useful to large, poorly prepared classes of students.[52] Thus, although Forbes had a reputation for experimental acquirements and as a good teacher, Dr Thomson could scarcely view Forbes as a particularly suitable candidate representing Scottish democratic values against Whewellian elitism.

Meanwhile, the problem of finding a suitable candidate for the 'independent party' became more pressing by the precarious state of Dr Meikleham's health in late 1842. He had had a second attack of his 'distemper', as his colleague in the mathematics chair expressed it, and could not be expected to recover significantly. Dr Thomson decided to sound out the Anglican William Hopkins, while also keeping an open mind with regard to Archibald Smith. The prestige of the chair would be a strong attraction, and the emoluments of £600–800 a year with a free residence were certainly considerable, but as both men were busily engaged in establishing reputations in their respective fields of mathematical coaching and law, Dr Thomson believed that neither Hopkins nor Smith would be likely to throw themselves into the election campaign with single-minded ambition.[53]

Dr Thomson had also his son William's future to reflect upon, for the Cambridge undergraduate days would not last for ever. He had consulted his old colleague in the Belfast Academical Institution's natural philosophy chair, John Stevelly, about the prospects of a fellowship for William at Trinity College, Dublin, and Stevelly had replied early in 1842 that he had no doubts that William could obtain a fellowship in Dublin at a very early age, but that the

[51] *Ibid.* [52] Davie, *The democratic intellect*, pp. 116–19, 169–89,
[53] Dr Thomson, *op. cit.* (note 50).

prospects at Cambridge seemed more favourable, both in terms of number and quality of posts as well as the greater opportunity of forging influential connections.[54] The Thomsons followed the wisdom of this advice and, as we have seen, the immediate aim of William's Cambridge education was to secure the position of senior wrangler and to obtain a college fellowship. However, with the increasing probability of a sudden vacancy in the Glasgow chair, Dr Thomson naturally felt obliged to consider the qualifications of William as a candidate, and by early 1843 he had openly discussed the possibility with his son and with his much-valued Glasgow colleague, Dr William Thomson, professor of medicine and fellow reformer. In fact, the earliest support on William's behalf seems to have come, not from his father, but from Professor Nichol, as a letter of April, 1843, from Dr Thomson to his son would suggest. The same letter illustrates the important premium placed on experimentation as a qualification for the Glasgow chair:

I felt that . . . I ought to mention to him [Dr William Thomson] my views regarding you. In doing this I asked him whether Dr Nichol had ever conversed with him about the chair, and finding that he had not I told him about Dr N's views regarding you. He was naturally struck with the idea of your youth, etc.; but he received the proposition as favourable as could be expected. He asked about your *experimental* acquirements, particularly in Chemistry; and he mentioned Forbes as being *in this respect* . . . of 'European reputation'. He seemed also to wish Gregory to be found to be a good experimentalist, as well as what he is acknowledged to be, a good mathematician, and he said that a mere mathematician would not be able to keep up the class.[55]

As early as 1843, therefore, when William was eighteen, Dr Thomson began orchestrating a campaign aimed at securing the election of his son to the natural philosophy chair. In fact, three years were to pass before Dr Meikleham's death, but that period was sufficient to permit not only the completion of William's course at Cambridge but also to allow time for a strengthening of support from within and without Glasgow College, and most of all to allow William to acquire for the first time truly experimental skills which would enable him to compete with someone as eminent as Forbes.

Since there was no doubt of William's mathematical ability, even before the Senate House examinations, the main concern lay with his experimental training. It was clear that experimentation was not a part of the conventional Cambridge course in mathematics, and such shortcomings in the Cambridge system would certainly have favoured a non-Cambridge man such as Forbes should he have made a bid for the Glasgow chair. Advised by his close colleagues, Dr William Thomson and Dr Nichol, Dr Thomson encouraged his son to take up practical experimental work whenever possible, even to the

54 John Stevelly to Dr Thomson, 5th January, 1842, S302, ULC.
55 Dr to William Thomson, 29th April, 1843, T236, ULC; SPT, **1**, 53–5.

extent of employing simple apparatus in his Cambridge rooms, and to 'get a proper introduction to Cumming [Cambridge professor of chemistry] – you might tell him you wished to practise in some small degree in performing experiments (keeping, of course, your main object concealed from him and all others); and he, if you could get no means in his laboratory, would probably direct you regarding some simple apparatus and some suitable books; and a certificate from him or any such person on this subject might be of great consequence'.[56] William had already attended Thomas Thomson's Glasgow chemistry class in 1838–9 and 'also attended the Laboratory, where he was a practical student for some time'. William apparently returned to the laboratory, however, during part of the 1843–4 session, by which time Thomas Thomson had left his nephew, R.D. Thomson, in charge.[57]

William attended Professor Challis's lectures on experimental natural philosophy in 1843, and those on practical astronomy and astronomical instruments in 1844. Indeed, he attended the experimental natural philosophy lectures for a second time in 1844 in order to have a further opportunity of seeing and handling the apparatus. In this way, William exploited the limited scope for experimental work at Cambridge, and in so doing followed closely the interests of his Glasgow teacher, Nichol, whose enthusiasm for astronomy, for the wave theory of light, and for astronomical instruments paralleled that of Challis.[58]

Experimentation formed no part of conventional Cambridge mathematical studies. While Cambridge mathematical texts were abundant, experimental texts in the mould of French writers such as Biot, Lamé, and Pouillet were unknown. William, however, had already got hold of Lamé's *Cours de physique* – 'an entirely experimental work' – when a new directive arrived from Glasgow. Dr William Thomson (who had studied medicine in Paris) and Nichol (who had applied for the chair of political economy at the Collège de France) advised William to travel to Paris in order to further his knowledge of experimental techniques: 'He [Dr William Thomson] still speaks emphatically about the necessity of your giving very great attention to the experimental part as soon as you can; as he says no one will have any doubt as to your mathematical attainments, but that some may even think them to be such as to make you neglect the popular parts of Natural Philosophy'. Nichol was more blunt, informing Dr Thomson that William's 'attaining the object you have in view would be "wormwood and gall" to certain parties; but that your going to Paris,

56 *Ibid.*

57 Printed copy of William Thomson's testimonials for the Glasgow chair of natural philosophy, pp. 15–16, PA34, ULC; SPT, **1**, 167–83, esp. p. 174 (testimonials from Thomas and R.D. Thomson); J.B. Morrell, 'Thomas Thomson: professor of chemistry and university reformer', *Brit. J. Hist. Sci.*, **4** (1969), 245–65, esp. p. 262.

58 William to Dr Thomson, 8th May, 1843, T242, ULC; 22nd April, 1844, T257, ULC; SPT, **1**, 68–9; James to William Thomson, 12th May, 1846, T418, ULC. This last letter suggests that Challis and Nichol were well acquainted.

besides improving you, would strengthen the hands of your friends and weaken the objections of others'.[59]

William travelled to Paris as soon as he had taken his degree early in 1845. Professor Forbes provided some of the key introductions for William to the French *savants* (notably Cauchy). Forbes also warned Dr Thomson that the state of science in France was not all that might be desired:

> In these times it is somewhat a delicate thing either to give or take letters to the French savants what with the jealousies of England & those of one another, no one can be secure against a reception different from what they & their recommender expect . . . it is useful to remember that it is dangerous to praise indiscriminately the colleagues of these gentlemen . . . I wish your son a happy journey; but I do not predict that he will fall in love with scientific character as seen in Paris.[60]

With the requisite diplomacy, William Thomson was nevertheless well received in Paris. His notebook of February, 1845, shows that he attended the *Leçons de physique* by Pouillet and the *Leçons de chimie* by Dumas at the Sorbonne. The former course provided lectures on statical electricity (with elementary experiments on the electrophorous, proof plane, condensers, and non-conductors), galvanism, thermo-electricity, and cells.[61]

William's father, ever conscious of the need for a return on capital, passed on Dr William Thomson's latest advice on both economy and *la physique expérimentale*:

> Dr Wm. Thomson says you should spend your 20 guineas [prize money from St Peter's College] in purchasing books in Paris, where you will be able to get much for your money, and he says you should spend much of it in purchasing books on *la physique expérimentale*. To the lectures on this he says you ought to pay the greatest attention, and is glad Blackburn is with you, and he says you and he will be able to repeat, as it were, the lectures to one another in the evening, and, above all, he says you should be writing discourses or lectures in the plainest and most attractive terms in your power, and improving your elocution by constant, free and open practice . . . He says people may think you too *deep* to have *popular* talent. Do all in your power to obviate this impression. Use all economy consistent with comfort and respectability.[62]

As a result of this advice, William purchased Pouillet's *Traité de physique*, and evidently took careful note, not so much of the content of Pouillet's and Dumas's lectures, but of their presentation and use of illustrations. He found that Dumas's were 'exceedingly well illustrated by experiments. All the things which are required are prepared with great care beforehand, so that he has

[59] Dr to William Thomson, 22nd September, 1844, T271, ULC; SPT, **1**, 85–6; Dr to William Thomson, 12th October, 1844, T274, ULC. See also Gabriel Lamé, *Cours de physique* (3 vols., Paris, 1836), and for Lamé's position in French science, see Robert Fox, *The caloric theory of gases from Lavoisier to Regnault* (Oxford, 1971), pp. 263, 268–70, 316.

[60] J.D. Forbes to Dr Thomson, 22nd January, 1845, F160, ULC.

[61] William Thomson, 'Notes on lectures and reading in Paris in 1845', NB30, ULC; SPT, **1**, 114–32.

[62] Dr to William Thomson, 4th February, 1845, T294, ULC; SPT, **1**, 115–16.

always a great many experiments to show'. And of Pouillet, Thomson marked down 'the particular experiments he makes, and how they succeed, and what seems to be more appreciated by the audience, which is very numerous, and "popular"'. The aim was thus to compensate for the belief, as expressed by Dr William Thomson, 'that a Cambridge education did not always give the power of easy expression or of commanding the attention of an audience'.[63]

Thomas Chalmers and Sir David Brewster provided William with introductions to J.B. Biot, who informed him that Victor Regnault (1810–78) 'is the best physicien here'. According to S.P. Thompson, 'it was the aged Biot who, taking him literally by the hand, introduced him to Regnault'. William himself described his introductory visit to Regnault's *laboratoire de physique* at the Collège de France:

> On Monday Biot introduced me to Regnault (the professor of Natural Philosophy at the Collège de France), and told me to go to M Regnault at the end of his lecture any day, and that he would show me his *cabinet de physique* (i.e. apparatus room). I went yesterday, and . . . he sent his assistant to show me all the apparatus. I was greatly interested in it, and saw a great many pretty things, of which they have a great abundance here, as the Government gives them a great deal of money for apparatus for popular experiments and historical illustrations in the lectures.[64]

William soon offered to assist Regnault in his laboratory. According to Thomson, Regnault 'seemed to be quite willing to let me come as often as I choose, and I suppose I may now and then have a job in the way of holding a tube for him when he is sealing it, or working an air-pump, as I had the privilege of doing yesterday'. Very soon, William was 'occupied the whole day in Regnault's physical laboratory at the Collège de France', arriving at eight in the morning and seldom leaving before five or six in the evening.[65]

On learning of William's activities in the laboratory, his father wrote back enthusiastically: 'I think that were it only to hold a tube or work an air-pump, you should by all means go on in Regnault's cabinet. You will see what instruments he has, and you should take lists of them as far as you can. Besides, *certificates* from him, Pouillet, Dumas, etc., with reference to *practical* matters might serve you much'. Dr William Thomson in particular hoped for a good testimonial from Regnault and others 'regarding your knowledge of physique, and showing that you are not merely an expert x plus y man. He [Dr W. T.] still says that you should try to get practice in the mere *manipulations*, so as to acquire expertise in the mechanical operations'.[66] Here is one of several different senses

[63] William to Dr Thomson, 10th February, 1845, T295, ULC; SPT, **1**, 116–18.

[64] James to William Thomson, 23rd January, 1845, T407, ULC; William to Robert Thomson, 5th March, 1845, T557, ULC; SPT, **1**, 115–23.

[65] William to Dr Thomson, 23rd February and 16th March, 1845, T298 and T301, ULC; SPT, **1**, 120–6.

[66] Dr to William Thomson, 22nd March, 1845, T302, ULC; SPT, **1**, 127; Dr to William Thomson, 8th April, 1845, T305, ULC; SPT, **1**, 129–30.

of the term 'practical' that figure prominently in Thomson's career, referring to the ability to perform demonstration experiments in a popular lecture: the experimentalist was a practical man whereas the mere mathematician was not.

William Thomson's involvement in the laboratory and his indebtedness to Regnault went much further than this popular sense of the practical. As he said later, his principal debt to Regnault was 'a faultless technique, a love of precision in all things, and the highest virtue of the experimenter – patience'.[67] Thomson's developing commitment to precision experimentation, which probably originated with his early studies under Nichol, owed much to his acquaintance with Regnault. In addition, the content of Regnault's investigation will prove important for Thomson's electrical and thermodynamic researches (chs. 8 and 9). As William himself summed up almost casually the extent of his involvement: 'I always get plenty to do, and Regnault speaks a great deal to me about what he is doing, and has of late employed me in working, along with him, some of the formulas necessary for the reduction of the experiments'.[68]

Meanwhile, the range of candidates for the Glasgow chair had narrowed dramatically. In the autumn of 1843, Gregory became so seriously ill that he was unable to go up to Cambridge to deliver his lectures. R.L. Ellis took over his duties, including the editing of the *Cambridge Mathematical Journal*. Dr Thomson discovered that 'some of the most eminent medical men in Edinburgh think Gregory is in a very dangerous state, which is beyond the power of medicine'. The diagnosis proved all too correct, for Gregory died early in 1844, much to the regret of all who knew of his mathematical potential.[69] Forbes, too, became poorly, being ordered by his medical advisers to spend the winter of 1843–4 in Italy. Forbes indeed had already made himself unpopular with his erstwhile Glasgow supporters by attempting to lay down his own terms for accepting the chair. The Thomsons judged that Forbes, although anxious for the position, had thought himself so sure of it that he had requested a reduction of lecture hours – a request quite unacceptable to the electors.[70] Nevertheless, Forbes did remain a *possible* candidate, if an unlikely one, until 1846, when he finally stated his intention not to stand.

David Thomson (1817–80), a former pupil of Dr Thomson's at Glasgow College, graduate of Trinity College and a cousin of Faraday, had conducted the natural philosophy classes for a number of sessions from 1840–1 until 1844–5. His testimonial to Dr Thomson in support of William in 1846 explained that 'I should certainly have sunk under the labours of my first session [1840–1], had not your son kindly undertaken to assist me in the preparation of the experi-

[67] SPT, **2**, 1154.

[68] William to Dr Thomson, 30th March, 1845, T303, ULC; SPT, **1**, 128–9.

[69] William to Dr Thomson, 22nd October, 1843, T249, ULC; Dr to William Thomson, 24th October, 1843, T250, ULC; William to Dr Thomson, 6th March, 1844, ULC.

[70] Anna to William Thomson, *c.* October, 1843, B202, ULC; William to Dr Thomson, 8th May, 1843, T242, ULC.

ments by which my lectures were illustrated'.[71] But, in 1844, David Thomson himself seemed a very likely candidate for the chair, and must have been carefully observed by Dr Thomson. However, a vacancy suddenly occurred elsewhere in December, 1844: 'Dr Knight, professor of Natural Philosophy in Marischal College, Aberdeen, died a few days ago, in consequence of a cold caught, while he was observing the late lunar eclipse. David Thomson is applying to succeed him'. Doubtless encouraged by Dr Thomson, David Thomson obtained the Aberdeen chair just before the 1845–6 session. Up to that time, Dr Thomson had suspected that he was still looking with a longing eye to the Glasgow professorship, but, once he had secured the Aberdeen post, David Thomson did not attempt to return to Glasgow. Another temporary replacement was found for Meikleham, William Thomson himself being advised not to give up his Cambridge commitments for such an uncertain position.[72]

Dr James Thomson sounded out the strength of support within the College in May, 1845, by speaking with some of the professors before the summer separation. Ramsay 'had looked forward to the thing for some time and had regarded you as far the most likely person for the situation', though he had had Archibald Smith in view and 'in one of his easy fits he had promised him his vote'. While Ramsay felt that Smith, recalling perhaps the 1836 contest, would not come forward, Dr Thomson believed that he would still like the chair: 'if so, you as well as I should consider what is best to be done in a fair way to forward your views'. Of Lushington's support, Dr Thomson had 'good hopes'. And, though opposed to Dr Thomson politically, the professor of law, Maconochie, no longer supported the Principal's party, thereby making possible an independent assessment of William's merits.[73]

To Buchanan, Hill, Fleming, and the Principal he could not talk, as William's 'appointment would be regarded as a strengthening of the opposite party; which, however, is quite too strong for them already'. Dr William Thomson and Dr Nichol, were already favourable to the cause, and the remaining professors as well as Rutherfurd, the Rector, could in general be relied upon to select the new professor on relevant merits alone. Indeed, by the autumn of 1845, even Dr Fleming had spoken to Dr James Thomson about William, 'saying that he had heard exceedingly gratifying accounts' of his success, and leading Dr Thomson to speculate that Fleming was going to make a virtue out of necessity by not, in the end, opposing William's election.[74] Dr Thomson, however, was

[71] W.L. Low, *David Thomson, M.A. Professor of Natural Philosophy in the University of Aberdeen. A sketch of his life and character* (Aberdeen, 1894); James Coutts, *A history of the University of Glasgow from its foundations in 1451 to 1909* (Glasgow, 1909), p. 384; SPT, **1**, 19–20; Testimonials, *op. cit.* (note 57), p. 26. David Thomson made an unsuccessful bid to succeed Dr Thomson in the mathematics chair. See his printed testimonials of 1849 among the Kelvin pamphlets, ULG.

[72] John to William Thomson, 8th December, 1844, and 18th April, 1845, T513 and T518, ULC; William to Dr Thomson, 19th October, 1845, T323, ULC; Dr to William Thomson, 21st October, 1845, T324, ULC; 29th October, 1845, T325, ULC.

[73] Dr to William Thomson, 4th May, 1845, T309, ULC.

[74] John to William Thomson, 2nd November, 1845, T522, ULC.

Archibald Smith, senior wrangler in 1836 and one of the founders of the *Cambridge Mathematical Journal*, had strong Glasgow connections. Although he never declared himself a candidate for the vacant natural philosophy chair, the possibility of his doing so posed the greatest threat to William Thomson's campaign. As a friend from his own undergraduate days, Thomson regarded Smith as a man of rare 'mathematical tact and practical ability'. [From Smith papers, Strathclyde Regional Archives.]

not the kind of man to base his judgements on flattery, and he therefore continued his thorough preparation of the ground.

Another development during the summer of 1845 drove the question of the Glasgow chair temporarily aside. Dr Thomson at first seemed certain to be appointed Principal of the new Northern Irish College, subsequently the Queen's University of Belfast. In Ireland there had been a clear need for educational reform. By 1840 the National Schools had been established, but as yet there was effectively only one university (Trinity College) for over eight million people. When the Conservative Party took office in 1841, with Peel as Prime Minister and Sir James Graham as Home Secretary, O'Connell's agitation for repeal of the Act of Union dominated Irish politics. To counter this challenge, Peel adopted a policy of conciliation towards Ireland from 1843, including the establishment of three new university colleges placed strategically throughout the country. The models were to be the Scottish universities and University College, London, incorporating particularly the non-sectarian ideals of the latter but, in order to maintain control, having appointments made by the Crown.[75]

The men of the Belfast Academical Institution naturally welcomed the prospect of a new, non-sectarian college, as did most liberal presbyterians, Churchmen, and Catholics. Dr Thomson's old colleagues and friends in Belfast, including Samuel Edgar, Thomas Andrews and Simms, editor of the liberal Belfast newspaper, *The Northern Whig*, were most anxious that he accept the post of principal were it offered to him, and they prepared to present a public memorial to the Peel government in his favour.[76] Dr Thomson was at first in a great dilemma, for on the one hand he was much tempted by the prospect of returning to his old town with all the honour of being principal of a new non-sectarian college, while on the other hand he had already set his hopes upon consolidating his position at Glasgow College, around which he had also built a solid circle of friends and colleagues. For his part, William expressed the hope that his father would not leave Glasgow, although 'I can quite conceive that he may be greatly induced to go back to Belfast by old associations'. If William could obtain the natural philosophy chair, then 'we could easily manage together, to get the business of both Classes done, even when he may not wish to work so hard as he has done hitherto'. Besides, he added somewhat facetiously, 'I do not know how we could get on, or how things in the College could get on without him in Glasgow'.[77] William's loyalties lay very much with Glasgow, and indeed remained so for the rest of his life.

Dr Thomson, by August, 1845, inclined towards Belfast, and was confident in his hope of having William appointed as his successor in Glasgow, there being no one 'of any weight who could oppose you: while for [the] Natural Philos-

[75] T.W. Moody and J.C. Beckett, *Queen's Belfast 1845–1949* (2 vols., London, 1959), **1**, pp. liii–lxvii, 1–9. [76] Agnes Gall to William Thomson, *c.*23rd July, 1845, G20, ULC.

[77] William Thomson to Agnes Gall, 28th July, 1845, G21, ULC.

ophy Chair there might be Professor Forbes whose standing at least is far before yours'. On the other hand, James (William's brother) was of the opinion that William would not readily choose the mathematics chair – an indication of how much William was committed to natural philosophy rather than to mathematics. The family therefore eagerly awaited William's return from Cambridge so that they might discuss thoroughly the issues.[78]

To Dr Thomson's circle it seemed certain that he would be appointed principal of the Belfast College by the government, and that William would be his successor at Glasgow. However, Peel's government, having settled on Belfast, had tacitly agreed that the Northern College must have an orthodox presbyterian minister as president or principal in order to forestall the almost inevitable conflict with Cooke's party – for long the enemy of the Belfast Insitution and a staunch tory – which would be brought about by the establishment of a non-sectarian college. The Presbyterian Church was anxious for a college under its own control and authority just as the Catholics had Maynooth College for the training of priests, and it viewed with disfavour, not to say aversion, the proposal for a non-sectarian Northern College. Cooke therefore warned Peel that if any unitarian or Catholic were appointed to a chair in any obligatory subject of the arts course, the General Assembly would debar its students from the College. Equally, the Catholic Archbishop MacHale took a hostile line towards the new colleges, and demanded a separate education for his flock. In his fears MacHale was not altogether unjustified, for, indeed, Graham had remarked privately that he saw the colleges as a means of liberating the new generations from 'priestly domination'.[79]

So it was that Peel's government denied the £1000 per annum position to both Dr Thomson and Cooke himself by appointing a polite, moderate clergyman, the Rev Dr Pooley Shuldham Henry, who was more of an administrator than a scholar. Dr Thomson, as a consolation prize, was offered the vice-presidency, at a salary of £500, but all his family urged him to refuse what they regarded as an insult to his reputation and ability.[80] Anna, William's sister, for one, received the news of Dr Henry's appointment with deep anger and intense emotion as she wrote to her aunt:

> if this really be the end of it, it will be one of the most scandalous jobs ever perpetrated and any one who has had a helping hand in it deserves to be whipped out of the country as a traitor to the real interests of Ireland. I am very sorry papa ever thought of it and I feel more disappointed now than ever since you say you were disappointed and that papa feels it so much . . .[81]

James stated that the appointment was well known to be 'owing to Dr Henry being a private personal favourite of the Lord Lieutenant'. Agnes Gall, Wil-

78 Agnes Gall to William Thomson, 8th August, 1845, G22, ULC.
79 Moody and Beckett, *Queen's Belfast*, **1**, pp. 19–32.
80 *Ibid.*, pp. 33–8; James to William Thomson, 1st December, 1845, T413, ULC.
81 Anna Bottomley to Agnes Gall, 10th November, 1845, B221, ULC.

liam's aunt, was similarly aroused, hoping that her brother-in-law would have nothing further to do with these Irish colleges 'which seem intended as merely instruments of party'.[82] By December, 1845, the immediate issue had come to an end, and Thomas Andrews had been appointed vice-president. But, in the longer term, the triumphant institutionalization of sectarianism in Ireland would reinforce William Thomson's own staunch latitudinarianism, leading him to campaign actively for the policies of liberal unionism in the 1880s and to assume the most public political mantle of his entire career in the years prior to his elevation to the peerage (ch. 23).

With these false hopes for Irish colleges at an end, 1846 opened with a return to the question of the Glasgow chair. Professor Meikleham had maintained his delicate balance between life and death through periodic attacks. As long ago as April, 1844, William had expressed his ambivalent concern. He was 'sorry to hear about Dr Meikleham's precarious state. I have now got so near the end of my Cambridge course that even on my own account I should be very sorry not to get completing it. For the project we have it is certainly much to be wished that he should live till after the commencement of next session'.[83] Meikleham did live through the 1844–5 session, and through most of the 1845–6 session as well. He died on 6th May, 1846: 'How sudden it was in the end. He had been just as usual the whole day till 10.45 at night when Dr [William] Thomson was sent for & in a quarter hour it was hopeless'. From that moment the contest for the Glasgow chair was on. Agnes Gall warned William that 'there will be opposition to your obtaining the situation and you must not set your heart too much upon it . . .'. Dr Thomson had recently been winning his reforming campaign against Principal Macfarlan and Dr Fleming, and he was well aware that they would do everything in their power to prevent the election of a third ultra-whig Thomson to the inner circle of College professors.[84]

Dr Thomson and his son took immediate steps to deluge the Glasgow College electors with printed testimonials from a variety of distinguished persons. William and his old tutor, Cookson, were inclined to favour a few carefully chosen references, but Dr Thomson knew Glasgow College rather better: that 'the Hills, Grays, Flemings &c. would be influenced . . . by number as by weight'.[85] In due course, testimonials were collected from almost thirty scientific men as well as the Master and Resident Fellows of St Peter's College. The list, which reads like a roll of honour of mid-nineteenth-century scientific worthies, included Augustus de Morgan, Arthur Cayley, Sir William Rowan Hamilton, George Boole, G.G. Stokes, Victor Regnault, J.D. Forbes, and Thomas Thom-

[82] James to William Thomson, 1st December, 1845, T413, ULC; Agnes Gall to William Thomson, 14th November, 1845, G25, ULC.

[83] William to Dr Thomson, 22nd April, 1844, T257, ULC.

[84] Robert to William Thomson, 9th May, 1846, T560, ULC; Agnes Gall to William Thomson, 7th May, 1846, G28, ULC.

[85] James (writing on behalf of Dr Thomson) to William Thomson, 12th May, 1846, T418, ULC.

son in addition to William's Cambridge examiners. Archibald Smith remained conspicuously absent from the list.

While Dr Thomson and William were friends of Archibald Smith, they regarded his father as wily, cunning, and plausible, if not unscrupulous, and in danger of extracting promises from such persons as Dr Hill, Dr Gray, and above all Dr Fleming who would 'as papa says, rather see Satan in the professorship than you'. Dr Thomson took the precaution of reminding Fleming that 'from the Electors I wish no pledge or promise in his [William's] favour; and I have no doubt of their keeping themselves equally unpledged in reference to any other candidates that may come under their notice . . .'. Smith seemed to be the only dangerous rival in the last part of the long campaign. By June, 1846, however, William explained to his father that Smith was unlikely to leave the bar to resume mathematics. This analysis proved correct, as the deadline for submission of testimonials came and went without Smith applying, and he subsequently apologized to William for any uneasiness caused by his indecision. There were no further rival candidates of any academic weight and significance. Dr Thomson, meanwhile, believed that Smith's erstwhile sponsor, Ramsay, should receive private letters on William's behalf from Challis and Hopkins: 'At the present moment *his position is such* that I think he could turn the scales'.[86]

Nonetheless, the Thomsons did not relax the intensity of their campaign. The need for William to appear to the electors as a promising and popular teacher was made all the more urgent by rumours which had been circulating in Cambridge that he, as college lecturer in mathematics, did not bring down his instruction to the capacity of ordinary students. These rumours had originated just before Meikleham's death from one Thompson of Trinity, and Dr Thomson warned his son that 'such a report may seriously injure you as that is the only doubtful point in reference to the N.P. chair here; and you *must* take care to cure the evil if it exists; and if not, to teach so simply, clearly, & slowly, that you may be able to get decidedly good testimonials on that point'. A few days later, Dr Thomson requested William to obtain at once Kelland's new edition of Thomas Young's *Lectures on natural philosophy and the mechanical arts* as a work of 'a popular character'. In the same letter he reminded his son of the dissatisfaction felt against Oxbridge men in Scotland's universities, providing much leverage for the tory faction at Glasgow College: keep 'in mind, what I find to be more and more the feeling, that Oxford and Cambridge men (Lushington, Kelland, Hitchens, etc.) have not given satisfaction here, and that you will have to contend against the feeling thus produced, and against the *handle* it will afford to the Dr Flemings *et hoc genus omne*'. William hotly denied the allegations against

[86] J.D. Forbes to Dr Thomson, 11th May, 1846, in SPT, **1**, 162; Archibald Smith to William Thomson, 14th May, 1846, S154, ULC; John to William Thomson, 20th June, 1846, T527, ULC; Dr Thomson to William Fleming, 20th June, 1846, F22, ULC; Dr to William Thomson, 21st June, 1846, T355, ULC; William to Dr Thomson, 19th June, 1846, T354, ULC; Archibald Smith to William Thomson, 18th September, 1846, S155, ULC; Archibald Smith to Isabella Smith, 10th, 14th, and 16th May, 1846, TD1/676/4–6, Smith papers, Strathclyde regional archives.

William Thomson, aged twenty-two, at the time of his election to the Glasgow College chair of natural philosophy, a position which he held until his retirement in 1899. [From Andrew Gray, *Lord Kelvin. An account of his scientific life and work.*]

his own teaching, but still the rumours persisted.[87] Special caution had therefore to be exercised in the preparation of testimonials and, where possible, referees were encouraged to write favourably of William's teaching abilities.

Cookson, in his testimonial, spoke of the fact that William 'may be quite depended on for adapting his instructions to his class'. Fuller stated that 'he combines the greatest clearness and precision with the most extended views in science, and he has always been as much distinguished for the simplicity and accuracy of his demonstrations of the more elementary propositions of natural philosophy, as for his talent in treating the abstrusest problems'. Fuller added that, above all, the energy of his character and his great enthusiasm for science could not fail to be communicated in some degree at least to his pupils, while Hopkins wrote that 'the amiableness of his character, and the simplicity of his manners can hardly fail to render him as *popular* in Glasgow, as he has been with all classes of his acquaintance in Cambridge'.[88]

On 11th September, 1846, William Thomson was unanimously elected to the Glasgow chair of natural philosophy to the unrestrained delight of Dr Thomson and the family. Elizabeth felt that the event would give new health and strength to her father and tend to prolong his days. William, she thought, did not look in the slightest degree elated but was perfectly composed. Anna was for her part overjoyed, exclaiming, amid her congratulations sent from Belfast, that she felt 'almost as if I could take even Dr Fleming into my heart today'.[89]

It was indeed a remarkable victory for the Glasgow College whigs. Out of the eight votes required for a majority (a seven-vote tie would have required the casting vote of the die-hard Macfarlan), the whigs had at first been certain of only three votes (Nichol and the two Thomsons). Not only had they secured the support of the whig Rector (Rutherfurd) and the three professors sympathetic to reform and high academic standards (Ramsay, Lushington and Reid) but they had won over the conservative professor of law (Maconochie) and his father (the Dean, Lord Meadowbank) and had encountered no opposition from the five tory professors. With the unanimous election of a twenty-two-year-old professor, the University of Glasgow ceased to be identified with an introverted, tory-dominated oligarchy and became an institution ready to advance the reputation and wealth of the Second City of the Empire.

[87] Dr to William Thomson, 2nd, 10th, and 16th May, 1846, T333, T337, and T341, ULC; William to Dr Thomson, 8th May, 1846, T335, ULC.

[88] Testimonials, *op. cit.* (note 57), pp. 6–10.

[89] Elizabeth King to Agnes Gall, in King, *Early home*, p. 233; Anna Bottomley to William Thomson, 15th September, 1846, B226, ULC.

5
Professor William Thomson

Graduates and undergraduates . . . You have been creating property. You have not been making money, nor adding field to field, nor building houses. But you have been creating a property more precious than gold or silver, or broad acres, or houses that may be burned or ruined. The property you have created is your own for ever, indestructible, imperishable, inalienable. The splendid university organisation, with its material resources, and the living influence of its teachers and students, has helped you. But every one of you has, by himself and for himself, by the power of God working in him, made the property which he brings away with him. May it to every one of you be a joy and a blessing for ever. *Lord Kelvin, Glasgow University graduation ceremony, 1898*[1]

On the 13th October, 1846, William Thomson was admitted to the office of professor of natural philosophy in Glasgow College. His rapid ascent at an extraordinarily early age marked his arrival at the top of a broad plateau in his academic career. From 1846 until retirement in 1899 he rose no higher in institutional terms, and even the offer of the Cavendish chair of experimental physics at Cambridge no less than three times in the 1870s and 1880s failed to tempt him away from the position which he and his supporters had conquered with such intensity of effort. Yet his spectacular ascent continued in other directions. During the first decade as Glasgow College professor he had effectively revolutionized the practice of natural philosophy in that local context, and had contributed in no small measure to a major transformation of British science in terms both of incipient professionalization and of the emergence of laboratory science. Writing to him in 1857, Daniel Halloran, retired under-keeper of the College's Hunterian Museum, captured in simple words the essence of William Thomson's achievement:

Continue on in your career, till you be what Liebig is to Chemistry and none will more rejoice than your humble serv[ant], who told your father when he saw you, a white headed boy, solving problems before the class that you would be a professor in the College yet. My prediction is more than verified . . . and if Dr Meikleham was to rise from

[1] SPT, **2**, 1005–6.

the grave, he would not know his class room nor half the language that is used in modern [natural] philosophy . . . He would have to [at]tend the class himself.[2]

In this chapter, a prelude to our analysis of William Thomson's mathematical physics, we shall be concerned first with his professional practice of natural philosophy and second with his personal life in the period 1846–70. The first decade of this period embraces a shift from the old paternal household and professoriate of Dr Thomson to the newly established household and physical laboratory of William. Three years after Dr Thomson's death in 1849, William's marriage to Margaret Crum strengthened further his links with industrial Glasgow and set his personal life on a course of social advancement hindered only by Margaret's protracted illness and eventual death after almost eighteen years of unsuccessful treatment.

An examination of Thomson's first decade as natural philosophy professor raises the wider issue of professionalization of science in nineteenth-century Britain. We may understand the result of professionalization in very broad sociological terms to be 'the development of knowledge-based occupational groups conferring status on their members *qua* members'.[3] In the context of nineteenth-century social history, the process of professionalization of science centred on social transformations accompanying industrialization. In particular, the 'professions' (traditionally law, medicine and theology) provided ambitious and intelligent persons lacking inherited status with the opportunity for advancement. 'Whig' values, then, of personal progress, rather than the old 'tory' values of inherited property, were intimately linked with the moves to extend the range of the professions far beyond the traditional subjects.

While in a general sense every professor at a university was a professional man similar in status and role to members of the traditional professions of law and medicine, a more specific sense of the 'professional man' attaches to the industrial society of nineteenth-century Britain. Professional men in this sense marketed their book learning for fees, fees which rose and fell in accordance with marketing skills and demand. This sense applied in the Scottish university context, where the occupants of chairs were primarily teachers whose professional expertise was marketed to a large number of students. A substantial proportion of a professor's income, as we have seen, derived from class fees, and the mark of professional reputation and accomplishment could be measured in part by pedagogic success in expanding the size of classes. While Glasgow

2 Daniel Halloran to William Thomson, 4th June, 1857, H3, ULG.

3 Henrika Kuklick, 'Professionalization', in W.F. Bynum, E.J. Browne and Roy Porter (eds.), *Dictionary of the history of science* (London, 1981), pp. 341–2. For discussion of some of the defining features of a profession, see Joseph Ben-David, 'The profession of science and its powers', *Minerva*, **10** (1972), 363–83. For a general (though dated) European perspective, see Everett Mendelsohn, 'The emergence of science as a profession in nineteenth-century Europe', in Karl Hill (ed.), *The management of scientists* (Boston, 1964), pp. 3–48. See also W.J. Reader, *Professional men. The rise of the professional classes in nineteenth-century England* (London, 1966); J.B. Morrell, 'Individualism and the structure of British Science in 1830', *Hist. Stud. Phys. Sci.*, **3** (1971), 183–204.

College professors derived a very adequate salary from the College itself, the new regius professors had to derive most of their income from fees, a policy which reflected the strong medical emphasis of the regius chairs. Similarly, Edinburgh University professors depended wholly for their income on class sizes.[4] By contrast, Cambridge professors, largely independent of fees, were professional men in the general sense which makes no distinction between entrepreneurial and service professionals.

In Glasgow, the newer regius professors' position, bridging town and gown but without a share in College power and property, opened the way for a more extensive marketing of knowledge to the public (medicine) and to industry (chemistry and engineering). At the same time, from a combination of economic need and a commitment to the diffusion of knowledge, both Dr Thomson, professor of mathematics, and Dr Nichol, professor of practical astronomy, met with quite spectacular results in their respective marketing of mathematics texts and astronomical books. Here indeed was the creation of property through sound investment in education.

The contrast with the development of professional ideals in the reformed Cambridge of the nineteenth century could scarcely be more striking. With an aristocratic, Anglican heritage, Cambridge dons sought to maintain and even sharpen the distinction between service (the professional man) and profit (the business man). The leading features of the professional man were not merely occupational, but moral, involving notions of self-renunciation, responsibility and obligations to society or clients, and high standards of ethical behaviour. The Cambridge image of the professional man's reputation, reflecting discretion, tact, and expertise, thus differed markedly from that of the commercial man, whose values and reputation rested on financial success, competition, and all the aggressive, *ungentlemanly* characteristics of the market place.[5] In short, the conflict between service and profit by which Cambridge preserved its moral superiority and high moral tone in the face of industrial expansion did not exist in anything like the same form in the University of Glasgow.

Professor William Thomson, as seen in our opening quotation, regarded scientific knowledge as intellectual property, differing from other kinds of property in that it *had* to be created by each individual, not inherited, and possessing more-nearly spiritual values of indestructibility compared to the perishable character of mere material property. Science was for him a form of wealth, a superior form, as he made clear in a commitment to an unambiguous form of intellectual capitalism for the benefit of the 1871 meeting of the British Association for the Advancement of Science:

[4] J.B. Morrell, 'Practical chemistry in the University of Edinburgh, 1799–1843', *Ambix*, **16** (1969), 66–80; 'Thomas Thomson: professor of chemistry and university reformer', *Brit. J. Hist. Sci.*, **4** (1969), 245–65.

[5] Sheldon Rothblatt, *The revolution of the dons. Cambridge and society in Victorian England* (Cambridge, 1981), pp. 90–3. See also Martin Wiener, *English culture and the decline of the industrial spirit, 1850–1980* (Cambridge, 1981), pp. 22–4, 90.

Scientific wealth tends to accumulation according to the law of compound interest. Every addition to knowledge of properties of matter supplies the naturalist with new instrumental means for discovering and interpreting phenomena of nature, which in their turn afford foundations for fresh generalisations, bringing gains of permanent value into the great storehouse of [natural] philosophy.[6]

This progressive accumulation of scientific capital at compound interest captures the essence of not only Thomson's whig outlook but also his commitment to an emergent professionalization of science involving the still more specific notion of the research imperative. In addition, therefore, to the marketing of knowledge, and indeed as a prerequisite to it, the advancement of knowledge through the research laboratory was to become the fundamental responsibility of the professional man of science.

William Thomson, following the lead of Thomas Thomson, Liebig and others with regard to the goals of laboratory research, developed this imperative into true intellectual capitalism. He began from the existing marketing system of Glasgow College professors, transformed the popular experimental course inherited from his predecessor, and established a related research laboratory of international renown. On the one hand he thus provided the means by which intellectual capital could accumulate at compound interest, while on the other hand he also created the conditions whereby that captial could be marketed to industry partly for personal profit and partly to fund further research through reinvestment in intellectual capital.

The practice of natural philosophy

On 4th November, at the commencement of the 1846–7 session, William Thomson delivered his introductory lecture, which, as he wrote the next day to his Cambridge friend, G.G. Stokes, 'was rather a failure as I had it all written, and I read it very fast'. This lecture, presented in part to the class at the beginning of almost every subsequent session, provides valuable insight into Thomson's programme for the natural philosophy class and for his developing physical laboratory during his long reign.[7]

In his opening remarks, he made clear that 'attempts to give sharp and complete definitions, especially to define branches of science, have generally proved failures'. His aim was not 'to lay down with logical precision any definite and sharp line round our province' but to emphasize that definition and subdivision 'become practically valuable' by giving method and by promoting 'order and regularity in the prosecution of a study'.[8] Thomson's conception of

[6] PL, **2**, 175–6.

[7] William Thomson to G.G. Stokes, 5th November, 1846, K15, Stokes correspondence, ULC. See SPT, **1**, 239–51, where Thomson's 'Introductory lecture' is printed in its entirety. We must stress, however, the need for a cautious approach to this version as the original manuscript cannot be traced.

[8] SPT, **1**, 239.

Glasgow College: the natural philosophy class-room (first floor, and above, between circular turrets). To the east, the more modern Hamilton building (1811) contained the Common Hall. [From David Murray, *Lord Kelvin as professor in the Old College of Glasgow*.]

natural philosophy was thus not that of a static, logical structure but rather of a progressive, practical study comprising the key skills and *instruments* of experiment and mathematics.

As with his predecessors in Scottish natural philosophy chairs, Thomson divided mind and matter into distinct provinces for investigation, corresponding to mental science and natural science. Fundamental to investigative procedures in both sciences, but especially 'in the progressive study of natural phenomena', were two successive stages: first, 'to observe and classify facts' (the natural history stage), and second, 'the process of inductive generalisation . . . in which the laws of nature are the objects of research' (the natural philosophy stage):

in the study of external nature, the first stage is the description and classification of facts observed with reference to the various kinds of matter of which the properties are to be investigated; and this is the legitimate work of Natural History. The establishment of general laws in any province of the material world, by induction from the facts collected in natural history, may with like propriety be called Natural Philosophy.[9]

Thomson went on to explain, however, that the ordinary use of the terms natural history and natural philosophy did not correspond to this logical distinction. Natural history 'is commonly restricted to the description and classification of the various natural products in the mineral, vegetable, and animal kingdoms of the earth', that is, to non-quantitative and non-experimental subjects. Consequently, natural philosophy under ordinary usage took on broad quantitative experimental investigations of the properties of matter and the phenomena of nature (including such subjects as meteorology and descriptive astronomy). These concerns would *properly* be included in natural history but the 'systematic observations and experiments which have for their object the establishment of laws and the formation of theories' made them the concerns of natural philosophy, forming the 'experimental' part of the study. The proper sense of natural philosophy, the establishment of mathematical laws by induction and mathematical deduction from laws to phenomena, now constituted the 'theoretical' part of the course.

Corresponding to the experimental and theoretical parts of natural philosophy were the two instruments of investigation, experiment and mathematics. The aim of experiment was the discovery and perfection of the laws of nature, inductively, through an investigation of the properties of matter by measurement. The aim of mathematics in the theoretical part was to subsume particular phenomena or properties of matter under general laws, deductively, through mathematical analysis. In neither case, we may note, did Thomson regard instruments as vague or qualitative tools but as precise instruments of analysis which served to break down the complex systems of nature into controllable and predictable relations.[10]

While divisions of natural science such as geology, chemistry, and vegetable and animal physiology went beyond a study of the characteristic properties of minerals, plants, and animals (natural history in its restricted sense) to the establishment of general laws, these sciences had been formally excluded from natural philosophy in its ordinary sense. Natural philosophy, then, embraced the 'great province that remains' which had as its fundamental subject 'mechanics, or the science of force'. Force was the primary concept:

[9] *Ibid.*, pp. 239–40; PL, **2**, 455.

[10] William Smith, Notes of the Glasgow College natural philosophy class taken during the 1849–50 session, Ms Gen. 142, ULG. Here the divisions of the course according to the instruments of mathematics and experiment are quite explicit. For a longer historical perspective, compare T.S. Kuhn, 'Mathematical versus experimental traditions in the development of physical science', *The essential tension. Selected studies in scientific tradition and change* (Chicago, 1977), pp. 31–65, esp. pp. 60–5. Kuhn suggests that the emergence of physics in the nineteenth century brought a lowering of the historical barriers between classical (mathematical) and Baconian (experimental) sciences. William Thomson's physics exemplifies the harmony of the mathematical and experimental components.

Every phenomenon in nature is a manifestation of *force*. There is no phenomenon in nature which takes place independently of force, or which cannot be influenced in some way by its action; hence mechanics has application in all the natural sciences; and before any considerable progress can be made in a philosophical study of nature a thorough knowledge of mechanical principles is absolutely necessary. It is on this account that mechanics is placed by universal consent at the head of the physical sciences. It deserves this position, no less for its completeness as a science, than for its general importance.[11]

Mechanics (replaced in 1862 by the term 'dynamics') thus occupied the central role in natural philosophy that it had for Robison and Meikleham (ch. 4).

Furthermore, Thomson explained that from 'a few simple, almost axiomatic principles, founded on our common experience of the effects of force, the general laws which regulate all the phenomena, presented in any conceivable mechanical action, are established; and it is thus put within our power by a strict process of deductive reasoning to go back from these general laws to the actual results in particular cases of the operation of force; the instrument . . . by which this deductive process is conducted being mathematical analysis'. Here Thomson reiterated Nichol's (and Dr Thomson's) view of mathematical analysis as a means to an end and not an end in itself. Mechanics, Thomson concluded, belonged to mixed or applied mathematics.[12] Natural philosophy (for Thomson as for Meikleham and Robison before him) in effect meant *mechanical philosophy*.

A set of remarkable lecture notes by one of Thomson's pupils, William Smith, taken in the junior class during the 1849–50 session, allows us to trace more accurately the subsequent programme of William Thomson's lectures at Glasgow College.[13] These notes, rewritten in copperplate form, serve to counter the commonly held view that Thomson emerged as an unsystematic, discursive lecturer whose enthusiasm for his latest researches far outweighed his ability to communicate effectively with his lowly audience. One of his students, David Murray, said later:

Lord Kelvin possessed the gift of lucid exposition in ordinary language remarkably free from technicalities. Occasionally he got outwith the range of the majority of his class; but there was no obscurity in his statement, it was simply beyond their grasp . . . He had no syllabus of lectures and used no notes in lecturing. He had his subject clearly before him and dealt with it in logical order. He was not dictating a manual of Natural Philosophy to

[11] SPT, **1**, 241–2. The term 'dynamics' replaces 'mechanics' in 1862. See David Murray, 'Lecture notes in *classe physica*, bench II, November, 1862', MS Murray 325, ULG; MPP, **3**, 317–18*n*. SPT, **1**, 192, however, states that in 1878 Thomson 'substituted "Dynamics" for "Mechanics" wherever the word occurred' in this introductory lecture. We have reverted here to the original term 'mechanics' because this is the term employed in William Smith's notes of 1849–50. Both words, however, were used to refer to the 'science of force' in the context of Thomson's natural philosophy. A similar view of dynamics, the science of force and motion, is briefly expressed in J.F.W. Herschel, *Preliminary discourse on the study of natural philosophy* (London, 1830), p. 96.

[12] SPT, **1**, 242.

[13] William Thomson, March, 1850, lecture 73, in Smith, *op. cit.* (note 10), indicates that these notes refer to the junior natural philosophy class: 'All that remains [in hydrostatics] is to give you a differential equation of equilibrium when there is any force but this must be reserved for my senior course'.

his students . . . He considered that it was unnecessary for him to teach what could be got in an ordinary text-book and that his province was to supplement this.[14]

The two great divisions of mechanical philosophy, Thomson stated in his first lecture, are statics and dynamics: 'the former is the relation of force to bodies at rest; the latter is the relation of force to bodies in motion'. Mere motion, 'cinematics', without reference to force, 'does not belong to dynamics' but 'may be wrought by pure geometry'.[15] These important distinctions will appear repeatedly in Thomson's mathematical physics, most notably in the birth of the famous Thomson and Tait *Treatise on natural philosophy* (ch. 11).

Thomson also drew attention to the 'subdivisions of statics and dynamics depending on the name of the bodies which are the object of force'. Thus 'Pneumatics is that part of Hydrostatics in which is [treated] the equilibrium of the air' while acoustics, a subdivision of dynamics, is 'the vibrations of the air to produce sounds'. A second major division of natural philosophy was optics, or the properties of light, while the remaining subjects consisted of 'heat, electricity, and magnetism', our knowledge of these being 'not so definite as to reduce them [entirely] to laws'. The division of the course in 1849-50 appeared as follows:

NATURAL PHILOSOPHY

I Mechanics 1 Statics
Hydrostatics
Pneumatics
2 Dynamics Cinematics
Astronomy
Hydrodynamics
Acoustics

II Optics
III Heat
IV Electricity
V Magnetism

Experimental course
Heat
Hydrostatics

[14] David Murray, *Lord Kelvin as professor in the Old College of Glasgow* (Glasgow, 1924), p. 3. For the more usual view, see, for example, SPT, **1**, 444–6; **2**, 651–3. The sheer quantity and intensity of Thomson's scientific and business activities from the 1860s probably brought about this reputation for discursiveness, but in his first decade or so able students such as William Smith or John Nichol (son of J.P. Nichol) had no difficulty compiling systematic notes. Nichol, for example, wrote of the 1848–9 session: 'The lectures I heard were on electricity and magnetism. I took careful notes, read, thought, and made experiments on subjects which interested me intensely.' See W. Knight, *Memoir of John Nichol* (Glasgow, 1896), p. 93; SPT, **1**, 209–10.

[15] William Thomson, 1st November, 1849, lecture 1, in Smith, *op. cit.* (note 10); SPT, **1**, 242.

Pneumatics
Acoustics
Magnetism
Electricity

As Thomson explained elsewhere, only in the subdivisions of mechanics (I) and optics (II) had our knowledge advanced sufficiently far 'to enable us to reduce all the various phenomena to a few simple laws from which, as in mechanics, by means of mathematical reasoning every particular result may be obtained'. The remaining branches – heat, electricity, and magnetism – had not advanced so far, making observation and experiment the 'principal means by which our knowledge . . . can be enlarged'. These branches thus also belonged to the experimental or physical course, while 'the more perfect sciences of mechanics and optics, being really mathematical subjects, form a distinct division of the studies prescribed by the University for the complete course of Natural Philosophy'.[16] During his second lecture of 1849–50, Thomson explained his actual timetable: 'Monday, Wednesday and Friday will be devoted to the theoretical part. Tuesday, Thursday and Saturday to the experimental. Saturday is for examination not lecture.'[17]

In the theoretical part of the course, following a discussion of the two entities, matter and force, fundamental to mechanical philosophy, Thomson moved directly to a consideration of some of the general qualities attributed to matter. The essential or primary qualities were extension, impenetrability and inertia in virtue of which three qualities we perceive matter. Another was temperature, not taken into account in general mechanics. The third lecture added important secondary qualities of matter (divisibility, porosity, compressibility, plasticity and elasticity) which occupied the class until the tenth lecture on 19th November. Then began statics proper, a branch of mechanics which extended as far as the seventy-second lecture on 6th March, when dynamics took over, ending with the ninety-third lecture on 12th April, 1850. Thomson had left himself no time for optics.

The large amount of time devoted to statics suggests that in its structure the theoretical or mathematical part of the course ('natural philosophy' proper) was intended not as a general, popular survey in the manner of Thomson's predecessors but rather as a detailed treatment of one or two branches of the subject in order to make his students practitioners of the art. His treatment of statics provided not a mere survey of principles (nor a drilling for subsequent reproduction in a written examination as at Cambridge) but offered an intensive exposition of the practices which embodied the principles, combined with various techniques of simplification and visualization. For example, in his discussion of the subject 'centre of gravity', he defined the 'base' of a body

[16] SPT, **1**, 245.
[17] William Thomson, 2nd November, 1849, lecture 2, in Smith, *op. cit.* (note 10).

supported on points as the 'figure which a cord will assume if drawn tightly round the points of support'. Then the 'conditions of equilibrium are extremely simple: "The centre of gravity must be over some part of the base"', a rule which he illustrated with a diagram of two tables, one with short and the other with long legs, standing on an inclined plane.[18] Such practical techniques coupled with rapid calculating devices were intended to make natural philosophy students practitioners of the art of statics in the same way as medical students would learn the art of dissection, not through a survey of principles but by practice.

Thomson's lectures in the mathematical part of the course also regarded mathematics as valuable in natural philosophy only to the extent of its practical usefulness through its applicability to concrete cases. In his presentation Thomson employed numerous familiar examples, readily visualizable objects, which in many cases related directly to the concerns of an industrial city of machinery, steamships and railway locomotives. An important section of his discussion of 'centre of gravity', for example, featured 'a general practical method for finding by calculation the centre of gravity as in the case of a ship', the 'largest body that we ever require to find the centre of gravity':

> Imagine the ship divided into sections . . . The weight of each part is known before it is put into the ship. The centre of gravity of each section may be found by the method of coordinates . . . We will not go far wrong in omitting the first [stern] section altogether but we must take into account all the sections onward to the bow of the ship where the sections get narrower. If we make the sections at the bow sufficiently small we will get the result . . .[19]

Further application of the method yielded the height of the centre of gravity above the keel. Such a comprehensive, simple treatment of a concrete statical problem, the bread-and-butter of a Clyde shipbuilder's art, highlights the very practical character of Thomson's 'theoretical' part of the natural philosophy course.

Viewed from a symmetrical perspective, Thomson's lectures in the experimental part of the course represent a major shift from popular illustrations to a detailed discussion of precision measurement in a research laboratory. Here indeed we may discern clearly the emergence of Thomson the professional man of science, offering his students not a survey of the field but a thorough treatment of selected subjects, notably heat, which formed the core of the professor's current experimental research in 1849-50. For example, he announced to his class on 17th January, 1850, his very recent experimental verification 'with a very delicate thermometer' of his brother's prediction of the depression of the freezing point of water under pressure (ch. 9).[20]

[18] William Thomson, 21st January, 1850, lecture 47, in Smith, *op. cit.* (note 10).

[19] William Thomson, 9th–16th January, 1850, lectures 38–44, in Smith, *op. cit.* (note 10). Locomotive stability was discussed in lecture 47 on 21st January.

[20] William Thomson, 17th January, 1850, lecture 45, in Smith, *op. cit.* (note 10). His experimental course in previous years appears to have been concerned largely with electricity and magnetism. See

The experimental course had almost become a seminar in which Thomson discussed with his students the state of the art of precision experimental techniques, the results of his latest investigations, and the implications of those results. Years later he explained his convictions in this respect:

> The object of a university is teaching, not testing . . . the object of examination is to promote the teaching. The examination should be, in the first place, daily. No professor should meet his class without talking to them. He should talk to them and they to him. The French call a lecture a *conférence*, and I admire the idea involved in that name. Every lecture should be a conference of teacher and students. It is the true ideal of a professorial lecture . . . Written examinations are very important, as training the student to express with clearness and accuracy the knowledge he has gained, and to work out problems, or numerical results, but they should be once a week to be beneficial.[21]

In this sense, Thomson had committed himself to upholding a distinctive Scottish pedagogical tradition. Yet while his brother-in-law, David King, could remark at the end of the first session that William had become '*the most popular professor* in the College', the new professor's approach signified a radical new professionalism marked by a clear research orientation. As David Murray noted:

> Everything which was observed was explained, discussed and commented on, and this commentary might be continued on the next day or even longer as fresh suggestions occurred. The Professor was always on the hunt for information; his students . . . became partners in the quest.[22]

In the experimental course, Thomson often employed the noun 'appreciation' as a synonym for 'measurement'. This usage indicates that measurement carried for Thomson an aesthetic quality which he expressed in the oft-repeated adjectives 'delicate', 'sensitive', and 'precise' applied to measurements and measuring instruments. Introducing the subject of heat on the first day of the course, he remarked: 'The principal object of a theory of heat is to examine and appreciate the nature of temperature', an appreciation obtained with a 'delicate thermometer'.[23]

This aesthetic value of precision measurement mirrored Thomson's love of the beauty of properly constructed mathematical theory, as in his 1862 reference to Fourier's theory of heat conduction as a 'great mathematical poem'.[24] Precision and sensitivity alone, however, did not constitute the aesthetics of

especially Nichol, *op. cit.* (note 14), who in his first session of 1848–9 attended the experimental course; William Thomson to J.D. Forbes, 22nd November, 1846, Forbes papers, St Andrews University Library, where he explained that 'In the experimental course . . . we have only got through part of the subject of magnetism. At this rate we might get through magnetism & common electricity in one session, galvanism & electromagnetism in another, and get something of heat &c. in a third. I shall have therefore to push on . . . '. He also explained that the experimental course might be taken without attending the whole course. [21] PL, **2**, 498–9.

[22] Anna Bottomley to William Thomson, 22nd April, 1847, B233, ULC, quoting her brother-in-law's verdict; Murray, *Lord Kelvin*, p. 4.

[23] William Thomson, 6th November, 1849, lecture 4, in Smith, *op. cit.* (note 10). Other references to these terms may be found in lectures 6, 9, 11, 13, 16, 27, 43 and 45.

[24] MPP, **3**, 296.

temperature and quantities of heat. Appreciation also required reference standards to universalize and objectify the beauty of measurement: 'The object of the thermometer is to get a perfectly determinate standard.'[25] Precision measurements reduced to absolute units thus represented the exemplar of the experimenter's art, an art which Thomson taught his students to practise just as he taught them the practice of mathematical theory, albeit in an elementary way. If Fourier stood as the heroic master of mathematical analysis in the theory of heat, Regnault stood as an equal master of experimental measurements.

From what we have seen so far, particularly in regard to the experimental part of the course, William Thomson had plainly transformed the teaching of natural philosophy at Glasgow College while maintaining the basic framework of subject-matter comprehended in the Scottish tradition. Gone were the old popular surveys with illustrative experiments, and in their place were the mighty investigative instruments of mathematics and precision measurement. Little wonder that William Thomson and P.G. Tait in 1861 could feel highly critical of all existing natural philosophy texts,[26] for Thomson's methods were dramatically new. His approach had an even more radical component, however: the development of a research laboratory aimed at making precision measurement a reality for professor and students alike.

Early years of the physical laboratory

Apparatus of 'appliances for lecture-demonstrations' had been crucial to the teaching of natural philosophy at Glasgow College under Thomson's predecessors. The new professor found the existing apparatus of 'a very old-fashioned kind': 'much of it was more than a hundred years old, little of it less than fifty years old, and most of it was of worm-eaten mahogany'. Yet he acknowledged that 'with such appliances year after year students of natural philosophy had been brought together and taught' and that 'the principles of dynamics and electricity had been well illustrated and well taught: as well taught as lectures and so imperfect apparatus – but apparatus merely of the lecture-illustration kind – could teach'. With a Faculty grant of £100 in the first year, William Thomson initiated an extensive reform programme. During his Paris visit in the summer of 1847, he spent some time ordering apparatus, notably acoustical apparatus, for his teaching, and 'felt inclined to order everything I saw, as indeed I did very nearly'. After his first session he also visited Faraday at the Royal Institution where he could view the apparatus relating particularly to electricity and magnetism.[27]

25 William Thomson, 13th December, 1849, lecture 27, in Smith, *op. cit.* (note 10).

26 P.G. Tait to Thomas Andrews, 18th December, 1861, in C.G. Knott, (ed.), *Life and scientific work of Peter Guthrie Tait* (Cambridge, 1911), p. 177.

27 PL, **2**, 483–4; SPT, **1**, 193–4; William Thomson to J.D. Forbes, 30th July, 1847, Forbes papers, St Andrews University Library; William Thomson to Michael Faraday, 11th June, 1847, in SPT, **1**,

Allan Maconochie and William Ramsay presented a report in November, 1847, on behalf of the committee appointed by the Faculty to oversee the expenditure on apparatus. The report noted that only £80 had been expended 'showing the caution and ceremony with which the purchase and selection of the valuable instruments was carried into effect', and therefore recommended that another £100 be placed at the committee's disposal, with an additional £50 grant towards storage boxes for preservation of the instruments. In fact, over the five years following Thomson's appointment, the Faculty sanctioned £550 for new apparatus.[28]

These grants allowed Thomson gradually to establish a modest physical laboratory in Glasgow College. As he later wrote to Stokes, explaining its precise purpose and organization, the 'primary and essential' work involved 'the preparation of illustrations for my lectures during the winter six months'.[29] Thus the laboratory served *prima facie* as an integral part of his role as a university professor, marketing knowledge to students. At the same time, we have seen how his experimental course focussed not on the old 'appliances' illustrative of the principles of natural philosophy but on the use of delicate instruments for accurate measurements.

This 'primary and essential' work, however, already existed alongside a completely new dimension added to Glasgow natural philosophy, a dimension with far-reaching consequences for Thomson's professional activities, for he had come very early to see the laboratory as an agent of industrial, as well as of scientific, progress. By 1846, there existed several important precedents for the connection of research laboratories with industry. Of these precedents, William Thomson had been most recently and directly concerned with Regnault's precision experiments on high pressure steam aimed at perfecting the great motive power of nineteenth-century industry. In Glasgow itself, however, Thomas Thomson's celebrated chemical laboratory had a strong industrial dimension familiar to William through his own time there as a student and through his future father-in-law, Walter Crum, whose primary interests in industrial chemistry had also brought him into the circles of Justus von Liebig and Michael Faraday (1791–1867). Liebig's chemical laboratory, opened at the University of Giessen in 1824, subsequently became famous for its concern with agricultural chemistry, and was renowned for the number of accomplished chemists trained there. William's own efforts to establish a physical laboratory

203–4; Michael Faraday to William Thomson, 14th June, 1847, F32, ULC. Thomson dealt principally with the French instrument makers Pixii and Marloye. Among the Kelvin pamphlets in ULC are two catalogues, Marloye's *Catalogue des principaux appareils d'acoustique et autres objets qui se fabriquent chez Marloye*, 2nd edn. (Paris, 1845), and Pixii's *Catalogue des principaux instruments . . . à l'usage des sciences* (Paris, 1845–9).

[28] SPT, **1**, 194–6; James Coutts, *A history of the University of Glasgow from its foundations in 1451 to 1909* (Glasgow, 1909), p. 385.

[29] William Thomson to G.G. Stokes, 7th February, 1860, K111, Stokes correspondence, ULC.

are thus an instance of a general relation between the development of research laboratories and industry.[30]

Even as early as his first session at Glasgow College, two critical experimental investigations had emerged for William Thomson. First, he employed the concept of 'work' in a study of the force acting on an inductively magnetized object (ch. 8); and, second, he wanted to know the nature of 'waste' or losses of power through conduction and stirring by consideration of Stirling's air engine.[31] Both 'work' and 'waste' developed in parallel with his brother James' engineering concerns. While serving his apprenticeship in William Fairbairn's Thames shipbuilding yard, James developed an improved dynamometer for the precise measurement of 'work', being at the same time profoundly involved with the reduction of 'waste' and the improvement of economy in marine steam-engines (ch. 9).

William's orientation towards an experimental research laboratory which focussed on work and waste had thus taken shape by 1846–7. This orientation, with its special attention to precision measurements of the properties of matter, formed the foundation for his entrepreneurial activity as well as his physics. By 1854 he had, following his brother's lead with a vortex-turbine patent, entered on the entrepreneurial activity which would win him his knighthood twelve years later and establish him as a leading figure in the development of science-based industry (ch. 19). The accumulation of scientific capital at compound interest would indeed yield social and economic dividends for the intellectual capitalist.

The subsequent story of Thomson's research laboratory is not one of the development of a private institution for personal research in which interested individuals from all over Europe might come to work with the master of physical science. Rather, the laboratory research, though related intimately to Thomson's own very wide concerns, rapidly became central to his professional activities as university professor and teacher. He later explained the development of the experimental researches himself:

Soon after I entered my present chair . . . I had occasion to undertake some investigations of certain electrodynamic qualities of matter, to answer questions which had been suggested by the results of mathematical theory, questions which could only be answered by direct experiment. The labour of observing proved too heavy, much of it could scarcely be carried on without two or more persons working together. I therefore invited

[30] See especially J.B. Morrell, 'The chemist breeders: the research schools of Liebig and Thomas Thomson', *Ambix*, **19** (1972), 1–46, esp. pp. 21–3. Crum worked under both chemistry professors. For a general survey of the development of physical laboratories in Britain, see Romualdas Sviedrys, 'The rise of physics laboratories in Britain', *Hist. Stud. Phys. Sci.*, **7** (1976), 405–36. Walter Crum was a student of practical chemistry under Thomas Thomson in 1818–19, and he subsequently published work on industrial chemistry in Thomson's *Annals of philosophy*. See F.H. Thomson, 'Opening address by the president', *Proc. Glasgow Phil. Soc.*, **6** (1865–8), 233–6.

[31] William Thomson to J.D. Forbes, 1st March, 1847, Forbes papers, St Andrews University Library.

students to aid in the work. They willingly accepted the invitation . . . Soon after, other students, hearing that some of their class-fellows had got experimental work to do, came to me and volunteered to assist in the investigation. I could not give them all work in the particular investigation with which I had commenced . . . but I did all in my power to find work for them on allied subjects (Electrodynamic Properties of Metals, Moduluses of Elasticity of Metals, Elastic Fatigue, Atmospheric Electricity, &c).[32]

Members of Thomson's one-hundred-strong junior class had often to support themselves by working after class hours as teachers or as city-missionaries, for example. But about twenty-five, three-quarters of whom would become theology students, 'found time to come to me for experimental work several hours every day'. All these pupils were 'volunteers', paying no laboratory fee, and among them Thomson generally found several 'efficient for original investigation to whom I give work specified in my application to the Royal Society Grant Committee'. These carefully selected pupils, then, pursued the part of Thomson's laboratory work which he referred to as the investigation of 'new truth'. To this end he kept his laboratory open during nearly all of the summer six months. His permanent assistant, whom he engaged during the whole year at his own expense, 'is in the summer months solely occupied in assisting me in such investigations, and in carrying them on according to my directions when I am not on the spot'. One of his laboratory volunteers was also frequently employed to act during the summer as a second assistant, receiving out of the grants a small payment of twenty pounds.[33]

If we are to understand more fully the nature and significance of the group of twenty or so volunteers, we must interpret their place within the Scottish educational system. The natural philosophy class generally marked the end of the broad philosophical curriculum at Glasgow University which preceded the professional training in theology, for example. The traditional pedagogical thrust, as we have emphasized, had been a democratic and popular one, contrasting with Anglican, Oxbridge values. William Thomson's institution of 'a system of experimental exercise for laboratory pupils' marked a considerable extension of those democratic values in which the participation of the student in class discussion and examination now reached much further – into the realm of handling and employing the apparatus itself to produce definite measurements. Thus 'we put the junior students at once into investigations, and let them measure and weigh whatever requires measurement and weighing in the course of the investigation'.[34] This strategy, of course, echoes precisely Thomson's own experiences in Paris where he derived the greatest benefit not from mere attendance at lectures but from seeing and handling the apparatus of Regnault.

[32] PL, **2**, 484–5.

[33] *Ibid.*, pp. 485–6; Thomson to Stokes, *op. cit.* (note 29). The Royal Society grants funded original experimental research into particular topics such as the effects of magnetization on the thermo-electric qualities and on the electric conductivity of iron. See, for example, MPP, **2**, 178.

[34] PL, **2**, 492.

At the same time, Thomson's philosophy here (as in the famous *Treatise*) is centred on a remarkable belief that the humblest person, crofter or artisan, is capable of achieving the deepest insights into natural philosophy, a belief stemming from both the career of his father and the democratic culture of Scotland. Helmholtz's account of his visit to Thomson's laboratory in 1863 provides further insight into the character of the professor and his work:

In the intervals I have seen a quantity of new and most ingenious apparatus and experiments of W. Thomson . . . He thinks so rapidly, however, that one has to get at the necessary information about the make of the instruments, etc., by a long string of questions, which he shies at. How his students understand him, without keeping him as strictly to the subject as I venture to do, is a puzzle to me; still, there were numbers of students in the laboratory hard at work, and apparently quite understanding what they were about.[35]

In order to accommodate his 'new volunteer laboratory corps' beyond the existing lecture room and adjoining apparatus room, the Faculty allotted Professor Thomson an unused wine-cellar around 1850, part of an old professor's house. The subsequent fitting out of this new property by 1852 generated enough noise to disturb the neighbouring classes of his less amiable colleague, Professor Fleming, who was never slow in lodging a complaint.[36] With the abolition of the College's Blackstone examination by the University Commissioners around 1858, Thomson annexed the examination room adjoining the wine-cellar, and so extended still further his laboratory empire. The advancement and expansion of his professional property by these means bore a striking resemblance to his late father's expansive moves at the Belfast Academical Institution. As William explained in 1885:

The examination room was left unprotected, its talisman, the old 'Blackstone Chair', removed. I instantly annexed it (it was very convenient, adjoining the old wine-cellar and below the apparatus room); and, as soon as it could conveniently be done, obtained the sanction of the Faculty for the annexation. The Blackstone room and the old wine-cellar served well for physical laboratory till 1870, when the University removed from its old site imbedded in the densest part of the city, to the airy hill-top on which it now stands. In the new University buildings ample and commodious provision was made for experimental work.[37]

The generally smooth development of Thomson's professional activities within Glasgow College had been greatly facilitated by the growing progressive character of the institution. Following William's appointment, Allen Thomson's replacement of James Jeffray in the anatomy chair in 1848 ensured a

[35] SPT, **1**, 430.

[36] Coutts, pp. 385–6. The year 1852 seems to mark the opening of the physical laboratory proper, for in 1861 Thomson explained that the system of laboratory instruction had 'gradually grown up during the last nine years'. See William Thomson to the Scottish University Commissioners (draft), 23rd December, 1861, S514, ULC; J.T. Bottomley, 'Physical science in Glasgow University', *Nature*, **6** (1872), 29–32.

[37] PL, **2**, 487–8.

Glasgow College: the physical laboratory. The laboratory windows are on the ground floor of the north side of the inner quadrangle (to the middle right of the illustration) below the natural philosophy classroom. The Blackstone room on the ground floor of the west side (middle left) provided Thomson with additional space, with the old apparatus room above. From there the professor entered his classroom. [From David Murray, *Lord Kelvin as professor in the Old College of Glasgow*.]

strong whig group consisting now of Nichol and four Thomsons (Dr James, William, Allen, and Dr William, brother of Allen), supported on most occasions by Reid, Ramsay, Lushington, and Maconochie. The Principal, left only with Fleming, Hill, and Buchanan after 1850, lingered until 1857. But even within the hitherto largely tory Divinity Faculty, changes came about with the appointments of the very liberal Thomas Barclay as Principal in 1857, John Caird as professor of divinity in 1862, and Duncan Weir as professor of oriental languages in 1850. Edward Caird, John's brother, took over moral philosophy in 1866. Both Cairds had been old class-mates of the Thomsons and shared a similarly broad theological outlook.[38] Meanwhile, the Universities Act of 1858 gave the nine regius professors equal rights and functions as professors of the University and College of Glasgow and also abolished finally the religious tests.[39]

Under new financial arrangements, the former College professors continued to derive their salaries from university revenues, supplemented by student fees and parliamentary vote or land revenues. Thus the professor of natural philosophy received almost £270 from university revenues, a parliamentary vote of £21, an average sum of £20 from the rents derived from the west coast island of Shuna, and estimated student fees of £300, totalling just over £600. In addition, he now received £100 for the salary of a teaching assistant and £100 allowances for class expenses.[40]

The university reformers of the mid-1840s had invested much effort in a bid to have the College removed to a new site at Woodlands to the west of the city. A deteriorating economic climate affecting the railway company which would have redeveloped the old College site in High Street and funded the new site and buildings led to an abandonment of the early scheme, but throughout the 1850s Allen Thomson spearheaded a fresh campaign for removal from the increasingly overcrowded College.[41]

The City of Glasgow Union Railway Company eventually agreed in 1864 to a purchase price of £100 000. With additional funds of over £20 000 from public subscriptions and a similar amount as a parliamentary grant, the University purchased the lands of Gilmorehill and Donaldshill to the west of the city and on the north of the river Kelvin, a tributary of the Clyde, for around £63 000. The first stone was laid in April, 1867, and the new buildings opened only just in time for the 1870–1 session. Costs rose to an estimated £363 000, of which the government of Lord Derby, supported by the Chancellor of the Exchequer, provided a grant of £120 000, a sum matched by public enthusiasm

[38] Coutts, pp. 379–475; James Thomson, *Papers*, pp. lxix–lxx; Thomas Barclay and John and Edward Caird, *DNB*.

[39] *General report of the Commissioners under the Universities (Scotland) Act, 1858, with an appendix containing ordinances, minutes, reports on special subjects, and other documents* (Edinburgh, 1863), p. xiv.

[40] *Ibid.*, pp. 39–45. In various representations to the Commissioners between December, 1861, and March, 1862, Thomson requested an additional £100 to fund a laboratory assistant 'with the view of enabling him to continue the system of laboratory instruction introduced by him in his class' (pp. 155–62).

[41] Coutts, pp. 415–9.

on the part of Glasgow industrialists and other benefactors, notably the Marquis of Bute, a coal magnate, and Charles Randolph, partner of John Elder the shipbuilder.[42]

The result was a grand cathedral of learning, an enduring monument to Victorian confidence in the unified progress of human knowledge and the wealth of nations. The greatest upheaval in the history of the ancient university provided William Thomson with a new physical laboratory vastly superior in scale and design to the old accommodation. With the laboratory available from 1870,[43] Thomson was poised to undertake the enormous extension of his entrepreneurial activities in telegraphy, electrical instruments, and navigational aids. Yet the firm foundations for such activities had already been laid in the last two decades of the old College where so much of William Thomson's career had taken shape.

Human fragility

The six year period following William's appointment as professor brought immense and unforeseen changes to his personal life. The spectres of disease and death came to haunt the rural communities and industrial cities of western and northern Britain following the onset of the fearful Irish famine in 1845 and left immune neither peasant nor professional. Then, from 1850, social conditions improved once more, and the tide of economic and technical progress seemed set fair to sweep industrial Britain ever more rapidly forward into a new Victorian age of prosperity and progress, confirming the strength of whig values which had almost foundered amid the darkness of the hungry forties.

The failure of the Irish potato crop from 1845 left millions of people on or beyond the edge of starvation, defenceless against the ravages of typhus fever and other potentially lethal diseases. As Anna (who had married William Bottomley) wrote from Belfast to William in 1847:

> on all sides we hear of nothing but bad business, sickness and distress . . . I hope that none of us will ever see such another winter for really I had not heard of the unparalleled misery that has been suffered . . . the Belfast streets are quite infested with beggars and whatever objection one may feel to begging yet the hunger depicted on the countenance of most of them makes it seem hard to refuse.[44]

Agnes Gall, visiting Belfast at that time, added that on one day alone there were forty persons at the Bottomleys' door asking for relief, and 'I do not like to name

[42] *Ibid.*, pp. 442–7; Allen Thomson, 'Prefatory notice of the new College buildings', *Introductory addresses delivered at the opening of the University of Glasgow session 1870–71* (Glasgow, 1870), pp. vi–xxvi.

[43] Andrew Gray, 'Lord Kelvin's laboratory in the University of Glasgow', *Nature*, **55** (1896–7), 486–92.

[44] Anna Bottomley to William Thomson, 3rd February, 1847, B228, ULC. On the Irish famine, see Cecil Woodham-Smith, *The great hunger. Ireland 1845–1849* (London, 1962); J.C. Beckett, *The making of modern Ireland 1603–1923* (London, 1966), pp. 336–50.

the thousands in Belfast at this moment depending upon charity lest I should exaggerate'. A long time must pass, she concluded, before the effects of these last two years' famine cease to be felt. All this distress, too, was observed in by far and away the most prosperous part of Ireland.[45]

The immediate effects of the famine were the fever outbreaks and the desperate exodus of human beings from the stricken island, more often than not taking the infection with them. Glasgow's long-standing social and sanitary problems could only be made worse by this additional overcrowding. The housing suffered from faulty drainage, bad ventilation, polluted water, narrow thoroughfares and almost every other domestic defect which could be conceived. A visitor back in 1818 had to: 'pick his steps among every species of disgusting filth, through a long valley from four to five feet wide, flanked by houses five floors high, with here and there an opening for a pool of water from which there is no drain and in which all the nuisances of the neighbourhood are deposited in endless succession to float and putrefy and waste away in noxious gases'.[46] Twenty years later, with its little improved social conditions, Glasgow was still the ideal environment for the rapid spread of fevers of all kinds, and the residents of the College knew only too well the personal grief brought about by such afflictions. In 1838, the professor of Greek, Sir Daniel Sandford, had died from typhus fever at the age of forty. In 1843, just before the famine, Dr William Thomson's nine-year-old daughter had died of the 'fever' which raged 'fearfully in town' and was 'particularly among the poor cutting off immense numbers'.[47]

By 1845–6, John Thomson was in the second session of his medical course, attending the Infirmary three hours a day, dissecting two hours a day, and taking the classes in surgery, materia medica and the practice of physic. During the summer of 1846, there was little or no escape to the peace and relaxation of the Firth. The winter of 1846–7 proved for the medical men of Glasgow to be the severest of all their ordeals, with the particularly harsh weather and the virtually uncontrollable outbreaks of fever. Early in 1847, therefore, his resistance weakened by the unending duties, John Thomson contracted what was probably typhus, and died at the Infirmary a few days later on 7th February, one of over 4000 fever-deaths recorded for Glasgow in 1847.[48]

Anna, writing to William from Belfast, gave open expression to the family's grief and to the sources of consolation:

There would have been no death among our number for which I would have been so totally unprepared . . . he seemed so strong both in body and mind, so able to bear up and

[45] Agnes Gall to William Thomson, 23rd February, 1847, G29, ULC.

[46] Robert Graham, *Practical observations on continued fever, especially that form at present existing as an epidemic* (Glasgow, 1818), p. 56. Quoted in L.J. Saunders, *Scottish democracy: 1800–1840* (Edinburgh, 1950), p. 181.

[47] Coutts, p. 381; Dr Thomson to William Thomson, 9th April, 1843, T234, ULC; Anna to William Thomson, *c.* October 1843, B197–8, ULC.

[48] John to William Thomson, 9th October, 1845, T521, ULC; Agnes Gall to William Thomson, 24th March, 1846, G38, ULC; John to William Thomson, 18th May, 1846, T524, ULC; J. Cunnison and J.B.S. Gilfillan (eds.), *The third statistical account of Scotland: Glasgow* (Glasgow, 1958), p. 476.

support others in their trials and difficulties of life that anything else than a long life of usefulness for him had never occurred to me. But we are continuously made to see that God's ways are not our ways, and of this we must be certain that there is some great good end to be served by such a stroke as this coming in the midst of so much prosperity as we were all enjoying so lately. With what a voice does it not cry to us 'All flesh is grass & all the glory of man is as the flower of grass'. I am sure we were all allowing ourselves to rest too well content with our positions here and forgetting that this is not the world of rest but that every thing here must be changing and uncertain.[49]

Anna consoled herself with the assurance that it was 'the object of such heavy afflictions as we have had to *force* us from earth to heaven and to teach us that this is not our home, and that our rest is not here . . .'.[50]

Her bleaker forebodings were realized less than two years after John's death, when her father, aged sixty-two, died from cholera on 12th January, 1849. The dreaded Asiatic disease had returned to Glasgow, killing over 3500 persons in the winter of 1848–9 and removing from the Thomson family its central figure. Elizabeth's health had also broken down, and she and her husband (Rev David King) were in Jamaica for the winter when the bitter news arrived. H.W. Cookson, in his letter of sympathy to William, summed up Dr Thomson's qualities in a tribute to the man who had done so much to shape his family's destiny:

Though ripe in years and honours you would not have expected him to pass away so soon, with his faculties in full vigour and apparently equal to many more years of useful and meritorious exertion. You will have much to console you . . . in the recollection of the great esteem to which your father was held, the great respect paid to his talents, the great services which he has rendered to his pupils and the valuable additions he has made to knowledge by his published works and which are more particularly to be regarded for their usefulness in removing the sources of inaccuracy & error of all kinds. I know that you will feel your loss severely and that all your family must grieve for your departed Head.[51]

After John's death and William's academic successes and public acclaim, James had been the member of the family closest to Dr Thomson. Forced early in 1845 to abandon the beginning of an engineering career in Millwall and Manchester through a 'quickness of the pulse', he had returned to Glasgow College where he had been subjected to various forms of uncertain medical treatment, including 'an infusion of digitalis which has a great effect in bringing down the pulse', a blister over his heart which 'kept me in bed for a fortnight', a piece of silk cord put through his skin 'with the ends left out so as to cause a permanent running' which 'brought on a number of feverish attacks' and which

[49] Anna Bottomley to William Thomson, 15th February, 1847, B229, ULC. See also Elizabeth King, *Lord Kelvin's early home* (London, 1909), pp. 234–5, for Elizabeth's reaction to her brother's death.

[50] Anna Bottomley to William Thomson, 20th March, 1847, B230, ULC; Elizabeth King to William Thomson, 29th March, 1847, K82, ULC; Agnes Gall to William Thomson, 23rd February, 1847, G29, ULC; Anna Bottomley to William Thomson, 22nd April, 1847, B233, ULC.

[51] H.W. Cookson to William Thomson, 15th January, 1849, C152, ULC.

'weakened me a good deal', and a very restricted diet with neither exercise nor excitement.[52] The slowness of his recovery kept him within the familiar Clydeside setting, and it was there that he endured the loss of his brother and father.

On his father's passing, James wrote to his old friend, J.R. McClean:

> The loss of our much loved father is the cause of great sorrow to all of us. On me, who was so dependent on him, this calamity falls very heavily. He was the chief object of my love, and my main adviser and supporter in all my difficulties. He participated and sympathized in all my joys and sorrows, and was the dearest friend I had.[53]

To his father alone does he appear to have confided a change of religious beliefs. He had abandoned the orthodox doctrines of Christianity and had become a unitarian, a faith well-enough known in Belfast intellectual circles. Only when Dr Thomson died did James reveal his secret to his friends, and such were his feelings of anguish that he wrote to his family stating that he did not plan to return, feeling perhaps that its members' latitudinarianism would not extend as far as abandonment of the Trinity. Anna felt his distress keenly, and it was to her household that James was soon to return after working in London with Lewis Gordon. She cautioned William to pity rather than to rebuke their brother – William being more accustomed to debating issues with James – as this crisis was an especially severe personal trial for him. She considered his unhappy religious opinions as the greatest of his misfortunes, and determined to help him in every way she could other than by argument. She hoped, therefore, that 'some judicious reading upon the subject might, with God's blessing, correct him where he is wrong'.[54]

Anna's concern contributed much to James's return to Belfast in 1851. With improved health, he entered fully the intellectual and social life of the town. With friends such as Thomas Andrews and William Bottomley he was able to pursue with new vigour his engineering activities during the 1850s and beyond. Elected a member of the Belfast Literary Society in 1853, he was also a frequent participant, as his brother was, in the meetings of the British Association, and he read papers before a variety of provincial scientific societies in both Great Britain and Ireland. He secured a temporary appointment to the chair of civil engineering in the Queen's College, Belfast, in 1854, and three years later obtained the chair on a permanent basis until he moved to the Glasgow chair of engineering as W.J.M. Rankine's successor in 1873. P.G. Tait's appointment to Queen's College mathematics chair in 1854 brought another notable personality into the Thomsons' academic and social orbit. Unfortunately James's spectacular recov-

[52] James to William Thomson, 15th February, 1845, T297, ULC; Agnes Gall to William Thomson, 26th March, 1845, G16, ULC; Anna Bottomley to William Thomson, 2nd April, 1845, B218, ULC.

[53] James Thomson to J.R. McClean, 26th January, 1849, in James Thomson, *Papers*, p. xxxviii.

[54] Anna Bottomley to William Thomson, 7th March, 1847, B238, ULC; James Thomson, *Papers*, pp. xxxix–xl.

ery from the misfortunes of the 1840s and his rapid rise to academic eminence was marred by the death in 1857 of his devoted sister, Anna, at the early age of thirty-seven.[55]

James married Elizabeth Hancock in late 1863. Her brother Neilson was professor of political economy at Queen's College, Belfast, and her father had been a Poor Law Commissioner until his death from typhus following the famine, having contracted the disease during his personal efforts to minimize hardships among the numerous poor. Elizabeth Hancock's liberal (even radical) views were very much in harmony with her husband's and both husband and wife contributed much to reform and social improvement in Belfast. Appointed Engineer to the Belfast Water Commissioners, James advised on the much-needed supply of clean water to towns. As a member of the new Belfast Social Enquiry Society formed in 1851, with his brothers-in-law Hancock and Bottomley as Secretary and Treasurer, James presented plans for financing public parks in manufacturing towns, particularly Belfast. Both projects reflected his involvement with problems of political economy. Elizabeth encouraged movements for the higher education of women and the Married Women's Property Act.[56]

William Thomson's youngest brother, Robert, was the only one of Dr Thomson's four sons not to aim for a profession. Early years of poor health discouraged him from academic studies, and, though he matriculated at Glasgow College in 1842, he left without taking a degree, and thus gave up thoughts of any learned profession. In May, 1846, he commenced work in the famous Scottish Amicable Insurance Company's offices at a salary of £20 per year, working from nine to five o'clock each weekday, and on Saturdays until two o'clock. He exhorted his academic brother to 'insure your life in the Scottish Amicable when you become Professor', and he continued in the office with improving health until 1850 or 1851. The prevalence of disease and misery not only in Glasgow but throughout Great Britain and Ireland, the loss of his father following on the loss of his brother, and his memories of a childhood of fevers and poor health, doubtless made him decide, however, to leave for all time the city and land of death. In July, 1850, Agnes Gall wrote to William to see if he could prevent a further breakup of the Glasgow College circle, but it was too late. Shortly afterwards, Robert Thomson emigrated to Australia, where he subsequently joined the Colonial Mutual Life Assurance Society in Melbourne. He never returned to Europe, and died in 1905, aged seventy-five.[57]

[55] James Thomson, *Papers*, pp. xl–xlix; J.P. Joule to William Thomson, 22nd February, 1857, J243, ULC. [56] *Ibid.*, pp. xlvii–xlix, 464–72.

[57] Robert to William Thomson, *c.* June, 1843, T546, ULC; July 1843, T548, ULC; Dr Thomson to William Thomson, 6th January, 1845, T283, ULC; Robert to William Thomson, 28th February, 1845, T300, ULC; 14th December, 1845, T559, ULC; 9th May, 1846, T560, ULC; 21st May, 1846, T561, ULC; Agnes Gall to William Thomson, 15th July, 1850, G34, ULC; Frederick Fuller to William Thomson, 30th March, 1851, F306, ULC; King *Early home*, p. 113; SPT, **1**, 5; Robert to William Thomson, 13th April, 1885, T565, ULC.

The unwelcome changes at Glasgow College which had taken place with great rapidity between 1847 and 1851 had left William and his aunt as the remaining residents at No. 2, the College, Glasgow. While the surviving members of the Thomson family undertook frequent visits to each other, the old society which had lasted from 1832 until the late 1840s had been shaken to its foundations. Nevertheless, the Thomsons' acquaintances, notably the Kers of Dalmuir, the Smiths of Jordanhill, and the Crums of Thornliebank, were still very much in evidence around the College. It was to these neighbouring families that William could look for friendship and, ultimately, marriage.

William's sisters had long teased him about his girlfriends. Back in 1842, for instance, Elizabeth wrote to him announcing that two of his girlfriends had become engaged in his absence. 'I almost fear,' she said sympathetically, 'to mention them lest you should be cast into despair.' Again in 1845, his aunt maliciously described to him a party in the College when 'Miss Crum and Miss Bottomley were the *belles* of the evening, every young man of spirit . . . running after them'.[58] Margaret, daughter of Walter Crum, was often referred to as 'the poet' by the Thomsons. A selection of her poems was privately printed in 1866. She was clearly a most eligible young lady and, according to Elizabeth, 'very clever'. Having attended a school in London, she spent some time with Professor Liebig in Giessen.[59] Meanwhile, William also had his eyes on other young ladies, his sister Elizabeth encouraging him in such pursuits as she wrote to the professor in 1847:

> Eliza Ker is looking extremely pretty. I think when I have helped you to furnish the less important parts of your house she would be an excellent assistant for the drawing room and then she would look very nice sitting in it. I would come very often to see you for she is a very kind and gentle hostess. Indeed I think you would have plenty of visitors with her to entertain them.[60]

William's Cambridge friends, especially Frederick Fuller, also eagerly teased him on the slightest hint of his flirtations in Glasgow. According to Fuller in 1846 'there is a scandalous report afloat in Cambridge about you – viz. – that you are engaged to be married'. Fuller also warned the new professor to be very careful not to 'get married before you are aware of it – you are in a *very dangerous* position now – all the prudent mammas in Glasgow will be asking you to tea &c., but take care!!'[61]

After his father's death, William embarked on a romance which would scarcely have gained Dr Thomson's imprimatur. James Smith of Jordanhill

[58] Elizabeth to William Thomson, 26th October, 1842, K60, ULC; Agnes Gall to William Thomson, 10th March, 1845, G15, ULC.

[59] Elizabeth to William Thomson, 27th February, 1842, K56, ULC; John to William Thomson, 18th February, 1843, T499, ULC; Agnes Gall to William Thomson, 30th May, 1845, G17, ULC; SPT, **1**, 533.

[60] Elizabeth King to William Thomson, 13th September, 1847, K88, ULC.

[61] Frederick Fuller to William Thomson, 14th September, 1846, F295, ULC.

(1782–1866), father of Archibald, had seven daughters, one of whom, Sabina, was just a year younger than William. Dr Thomson's verdict in 1846 on the elder Smith had been that he 'is a cunning, wily man who in an interview would be as likely to draw out from you matters which he would turn, if possible, to your disadvantage. Of the son I have a much better opinion, though not an *unqualified* one; and Mr Buchanan does not seem to admire him much as to manner and disposition'.[62] The context, of course, was the threat of Archibald becoming a candidate for the Glasgow chair. Perpetually of two minds, Archibald's indecision haunted him even when the mathematical chair itself became vacant in 1849.[63] By 1850, however, and in marked contrast to Archibald's excessive caution in all things, William had fallen head-over-heels in love with Sabina.

Sabina's first rejection of William's marriage proposal prompted a remarkable response from William on 1st January, 1851:

Dearest Sabina

I cannot express to you how dreadfully I have been grieved by your letter and how ill prepared to receive it I was after an almost sleepless night. Ever since you left me yesterday I have been most distressingly conscious of how wrong I was to imagine that you could be prepared already to give me the inestimable gift I asked. Not a word from you ought to have been necessary to tell me what my own feelings should have told me, what they did really tell me, although I allowed myself to be blinded by my sanguine hopes. Do not I implore you judge me too hardly for this rashness but try to forget it and do continue to give me your friendship as frankly as you would have done if it had not taken place.

I cannot rest now till I have opened my whole heart to you. I *must* have a long long conversation with you. I wish it could be today. I am determined not to expect any answer from you but to be patient and leave to time the alteration of your mind on which all my happiness in this world rest.

With the most earnest prayers for your happiness dearest Sabina I remain

Ever your's

William Thomson[64]

Other Smith family letters ensued, the most notable of which Archibald wrote on 3rd January to Sabina expressing himself 'well pleased by the result because I really do not think you would be suited to each other' and advising Sabina never to yield to feelings of pity.[65] William's passion, meanwhile, cooled only slightly until, in April, 1851, he made a second bid for Sabina's hand.

Once again Sabina refused. Writing to another sister, Christina, on the day

[62] Dr to William Thomson, 21st June, 1846, T355, ULC.

[63] Archibald to Isabella Smith, 27th January, 1849, TD1/676, Smith papers, Strathclyde regional archives.

[64] William Thomson to Sabina Smith, 1st January, 1851, TD1/485, Smith papers, Strathclyde regional archives.

[65] Archibald to Sabina Smith, 3rd January, 1851, TD1/485, Smith papers, Strathclyde regional archives.

after, William confessed experiencing 'bitter bitter grief far more than I felt prepared for when I went away last night . . . My consolation at present is the constant prayer that God will move her heart yet. I try to take comfort in the assurance that all things work together for the good of those who love God. It seems so very good for both of us that I cannot help thinking that it may come in time that we shall be united'.[66] Archibald, for his part, wrote to Sabina to assure her that 'There is no fear of my showing any coldness to P[rofessor] T[homson]. I have a great regard and liking for him . . . I have never thought his succeeding with you a probable enough matter to disturb myself about it'. In May, 1851, however, Archibald took steps to discourage William's hopes, announcing to Sabina that he intended to exclude all domestic news from correspondence with William during the Professor's continental visit that summer.[67]

By April, 1852, William prepared the ground for a third and final offer by sending Sabina a copy of John Ruskin's *The stones of Venice*.[68] Sabina's rejection yet again sometime in the spring or early summer caused her far-reaching and life-long regrets. Twenty-five years later she explained that, before the third proposal, she 'knew quite well that my happiness depended on it, & yet . . . I "instinctively" (with a sort of unreasoning impulse) put it away from me – I think it must have been because I had just before received a letter from Archy saying it w^{d} be very foolish for me to accept him'. Feeling confident of her influence over William 'I did not disturb myself, thinking that the *next time* I w^{d} say yes – when in a very short time I was informed he was engaged to someone else. It was *then* I became conscious of my "mistake" & that the feeling of its being "too late" & by my own fault took possession of me'.[69]

Although Archibald had strongly disapproved of the prospective marriage, and although Sabina in later years increasingly made him the scapegoat for her misfortune, she freely admitted that the fault was her own. She wrote to her sister Isabella almost forty years after the affair:

I suppose you must know that I have regretted all my life having refused him? . . . The first two times I had no doubt what to say & even if I had, Archy's opinion w^{d} have influenced my decision . . . But by the third time I was fully prepared to accept him, when Archy's letter arrived w^{h} put a stop to it. Even then I thought it was only for a time, & always felt sure it must come to pass in the end. Of course it was my own fault being so swayed by another, & it was the extremity of folly to think I c^{d} go on refusing a man, & yet have him at my disposal whenever I choose! . . . certainly *he* was not to blame. But it w^{d} be difficult

[66] William Thomson to Christina Smith, 20th April, 1851, TD1/485, Smith papers, Strathclyde regional archives.

[67] Archibald to Sabina Smith, *c.*24th April and 12th May, 1851, TD1/485, Smith papers, Strathclyde regional archives.

[68] William Thomson to Sabina Smith, 22nd April, 1852, TD1/485, Smith papers, Strathclyde regional archives.

[69] Sabina Smith to Rev Dr Robert Paisley, 10th February, 1877, TD1/485, Smith papers, Strathclyde regional archives.

Sabina Smith, daughter of James Smith of Jordanhill, near Glasgow, and sister of Archibald. Refusing William's proposal of marriage on three separate occasions, Sabina lived to regret her decision. [From Smith papers, Strathclyde Regional Archives.]

to make any one understand the feeling I had for Archy & my reverence for his opinion & wishes . . .[70]

Sabina's long letter, written at the time of William's elevation to the peerage, went on to explain how, after receiving the news of William's engagement in July, 1852, to Margaret Crum, she had travelled to the September meeting of the British Association in Belfast 'in the vague hope that I might yet win him back' such was her 'belief in my own power over him'. But she failed, and William's marriage took place on 15th September.

Sabina remained unmarried until 1878 when the Rev Dr Robert Paisley, a widower, married her more out of pity than from love. Dr Paisley, however, died in 1879. To him, as 'a regular Father Confessor' Sabina had poured forth her regrets that William who had promised that he 'w^d love me all his life' had been prevented from doing so on account of Sabina's successive rebuffs. She thought she could understand 'in some faint degree what it w^d be to have reached the end of one's life, & thrown away by one's own madness, the hope of salvation . . .'. As though in fulfilment of an earthly torment, indeed, Sabina not only outlived all her sisters, but her erstwhile suitor also. She died at Helensburgh in 1915, aged ninety.[71]

If Sabina, like Archibald, tended too often to draw back from the threshold of decisive action, her other qualities of intelligence, good looks, and simplicity of character had evidently made her the subject of Professor Thomson's heartfelt devotion. Sharing with him a love of European travel and a taste for philosophical and religious simplicity, she would have brought him into the socially advantageous circle of Glasgow's old commercial aristocracy, while he in turn might have brought to Jordanhill both the scientific distinction and financial security for which Archibald longed. When one reads in her later correspondence that 'the word metaphysics is one that makes me shudder as I like things plain & easy of comprehension w^h are much more suited to my powers',[72] it is difficult not to conclude that Archibald's verdict that the professor and Sabina were not suited to one another was profoundly mistaken.

Disappointed in love, William fell victim to a powerful rebound. In late July, 1852, he announced to G.G. Stokes his forthcoming marriage, stating that 'sometime, probably early in September, I am going to be married to a Miss Crum. I cannot describe her exactly to you, but I am sure that is unnecessary to ensure your good wishes . . .'. Equally unenthusiastically, he wrote to Forbes to

70 Sabina to Isabella Gore Booth (nee Smith), 2nd December, 1891, TD1/485, Smith papers, Strathclyde regional archives.

71 Sabina Smith to Robert Paisley, *op. cit.* (note 69); 'Obituary of Sabina Douglas Clavering Paisley (nee Smith)', *Helensburgh and Gareloch Times*, 29th December, 1915. On Robert Paisley's death (apparently from suicide following mental breakdown), see Sabina's correspondence with Archibald's wife, Susan, in 1879, TD1/928, Smith papers, Strathclyde regional archives.

72 Sabina Smith to Rev Dr Robert Paisley, 7th February, 1877, TD1/485, Smith papers, Strathclyde regional archives.

Professor William Thomson aged twenty-eight. Failing in his final bid to win the hand of Sabina in 1852, William married the accomplished Margaret Crum in September of that year. Compared to the faded prosperity of the old Glasgow commercial aristocracy typified by the Smiths, the cotton manufacturing Crums of Thornliebank represented the new industrial order. [From SPT.]

reveal his plans of marriage to 'a Miss Crum, whom I must have first met sometime about November 1832. Perhaps from this you will think she must be 30 or 40 years old, but she is in reality just three years younger than myself'. To his sister, Elizabeth, he also announced the news 'which I think will please you as much as it will surprise you'. According to Sabina, Margaret Crum had been 'an old love, who had refused him before I knew him, & who I suppose had seen *her* mistake in time to rectify it'.[73] During the wedding tour in Wales, William reported to Elizabeth with just a hint that Margaret lacked the passionate enthusiasm he might have wished for in a Professor's young wife:

[we] are now sitting in our parlour writing letters. The day is somewhat dark and cold, and some people might say dreary, but it does not seem so at all to me. I scarcely think it does so to Margaret either, although she has just been saying that it is, and what is more, laying particular emphasis on the most dismal parts. Perhaps she is only joking, but whether or not, she looks cheerful, and has quite got rid of her cold. In fact, I do not think either of us are going to apply to Dr Brown [the minister] to undo what he did on Tuesday.[74]

More revealingly, none of William's letters to his relations and scientific friends speak in terms of a fervent relationship with Margaret, in sharp contrast to his evident feelings for Sabina. Whether as the result of an unsatisfactory marriage or of other causes, Margaret's health declined rapidly within the first year. Soon after their return from a tour to Gibraltar, Malta, and Sicily in May, 1853, she became seriously ill, apparently from the physical over-exertions of travel and walking.[75]

All attempts at treatment were largely unsuccessful, including several visits to the German spa of Kreuznach (where Thomson first met Helmholtz in 1855). There is no existing personal record of those years from 1852 to 1870 except for the occasional glimpses provided by William Thomson's letters to his scientific friends which give some indication of the devotion he gave to his accomplished but sadly stricken wife. A letter to his brother James in 1855 was a typical expression of these unhappy circumstances:

I still have tolerably good accounts of Margaret's progress. She suffers much after the driving & walking and is quite unable to sit up without much pain in her own room, but Dr Johnson is very attentive & watches the effect of the exercise. He says it will do good notwithstanding the pain & fatigue, to a limited degree, but he says she is not in a fit state for almost any exercise. He trusts that the system of treatment he follows will make her gradually able for more.[76]

After prolonged suffering and many set-backs, Margaret died at Largs on 17th June, 1870.

[73] William Thomson to G.G. Stokes, 31st July, 1852, K60, Stokes correspondence, ULC; William Thomson to J.D. Forbes, 19th July, 1852, Forbes papers, St Andrews University Library; William Thomson to Elizabeth King, 13th July, 1852, in SPT, **1**, 232–3; Sabina Smith to Robert Paisley, *op. cit.* (note 72).

[74] William Thomson to Elizabeth King, 19th September, 1852, in SPT, **1** 233–4.

[75] SPT, **1**, 238, 308, 530–4.

[76] William to James Thomson, 17th March, 1855, (copy), T445, ULC.

II

The transformation of classical physics

6

The language of mathematical physics

Mathematical analysis is as extensive as nature itself; it defines all perceptible relations, measures times, spaces, forces, temperatures; this difficult science is formed slowly, but it preserves every principle which it has once acquired; it grows and strengthens itself incessantly in the midst of the many variations and errors of the human mind. Its chief attribute is clearness; it has no marks to express confused notions. It brings together phenomena the most diverse, and discovers the hidden analogies which unite them . . . It makes them present and measurable, and seems to be a faculty of the human mind destined to supplement the shortness of life and the imperfections of the senses; and what is still more remarkable, it follows the same course in the study of all phenomena; it interprets them by the same language, as if to attest the unity and simplicity of the plan of the universe, and to make still more evident that unchangeable order which presides over all natural causes. *Joseph Fourier, 1822*[1]

'Fourier made Thomson.' Peter Guthrie Tait's reported judgement has merit. From a mathematical love affair, beginning in 1840 with the 'splendour and poetry of Fourier', emerged the style of mathematics that Thomson represented throughout his career. Indeed, he devoted the first ten years of his professional life to defending, nurturing, and propagating the language of the 'mathematical poem' that was the *Théorie analytique de la chaleur*.[2] Thomson, however, was not alone in his captivation by the French master. Fourier may well be taken to symbolize the distinctive style of Victorian mathematical physics that emerged in the 1840s.

Considerable attention has been devoted to the transformation of British mathematics attending the spread of continental (largely French) methods of analysis. But the literature has not explored the question of what slant the process of reception gave to continental analysis: which authors were selected, what methods, and with what goals; how did the reception process produce a

[1] Joseph Fourier, *The analytical theory of heat*, 1st edn. (Paris, 1822), trans. A. Freeman (Cambridge, 1878), pp. 7–8.

[2] C.G. Knott, *Life and scientific work of Peter Guthrie Tait* (Cambridge, 1911), p. 191. SPT, **1**, 14; **2**, 1139; MPP, **3**, 296. 'It has been called "an exquisite mathematical poem", not once but many times, independently, by mathematicians of different schools' was the verdict of the reviewer of Freeman's translation, *Nature*, **18** (1878), 192.

language of mathematical physics that was essentially British?[3] Differentiating two generations of British reformers, active before and after about 1835, we argue below that Fourier ultimately emerged as a model for the first generation as part of a general perception of bankruptcy in the Laplacian programme for science. Of great importance to this perception were apparent failures of equilibrium systems controlled by natural law – Laplace's *System of the world* being the prime exemplar – to describe adequately anything from the solar system to the political economy (ch. 4). Of equal importance, and providing our focus here, was a reassertion of native traditions in mathematics and natural philosophy: non-hypothetical theory, geometrical analysis, and utility, which quickly put their stamp on the transmitted body of mathematics.

Mathematicians of the second generation, among whom William Thomson attained mathematical maturity, organized themselves around the *Cambridge Mathematical Journal*. Self-consciously British in criticizing their French forebears, they strove for international recognition and for fundamentally new mathematical methods.

The younger generation may be characterized by the mathematics it produced, the calculus of operations. Thomson's famous use of mathematical analogies in physical theory should be seen as a corollary to this calculus, which united diverse areas of mathematics by formal analogy. Like him, the developers of the new formalism drew inspiration from Fourier's hymn to mathematical analysis as a faculty of the human mind, a faculty that revealed the unity and simplicity of the natural order by allowing man to speak the universal language of nature. This vision united mathematics with an optimistic assessment of man's ability rationally to control his own destiny, to advance and to improve, both spiritually and materially. Thus the theme of universal analogy involves also the theme of social and political reform, which supplied a constant background to Thomson's career. Whigs in politics as in mathematics, the young men around the *Cambridge Mathematical Journal* directed their progressive social views into the channel of progressive mathematics.

Whig mathematics leads to two related issues, 'progression' and 'kinematics'. 'Progression' refers specifically to the vexed question of whether or not the geological history of the earth exhibited a secular cooling from an initial hot, fluid mass to its present state. Fourier supplied the main resource for progressionists, and Thomson applied Fourier to establish mathematical conditions for the 'age' of any given temperature distribution. His method was 'kinematical', referring to the view that a sharp distinction ought to be drawn between the geometry of motion, or *kinematics*, and the causes of motion, *dynamics*. That view emerged at Cambridge in the late 1830s.

[3] Maurice Crosland and Crosbie Smith, 'The transmission of physics from France to Britain: 1800–1840', *Hist. Stud. Phys. Sci.*, **9** (1978), 1–61, provides an extensive survey of French texts, translations, reviews, and progeny in Britain, as well as secondary literature.

Reformers of the first generation

When in 1811 John Herschel, Charles Babbage, and George Peacock, along with a handful of other Cambridge undergraduates, constituted themselves as the Analytical Society, they formally set in motion reforms that would revitalize British mathematics and natural philosophy. Joined by William Whewell, George Biddell Airy, and Augustus de Morgan (1806–71), they represented by the mid-1820s a major centre of power in the much larger movement to reform British science, playing major roles in the founding of the Cambridge Philosophical Society (1819), the Astronomical Society of London (1820), and the British Association (1831).[4] Politically, the group extended from the radical Babbage to the conservative Whewell, but it centred around moderate whigs such as Herschel, Peacock and Airy. Dr Thomson led his own component of this reform movement in Belfast and Glasgow, though with a much more practical and democratic intent than Whewell, Airy, or even Herschel. His zeal for diffusion and his antipathy for doctrine mirror more nearly the position that De Morgan represented when in 1828 he acquired his first position as professor of mathematics at the newly founded London University. Of unitarian persuasion, De Morgan left Cambridge in 1826 without his MA degree, having rejected the religious tests, and took up his London post with high hopes for the success of an institution dedicated to non-discrimination against dissenters and Jews. In twenty-five years of writing for the *Penny Cyclopaedia* of the Society for the Diffusion of Useful Knowledge, De Morgan contributed at least 800 articles. His *Differential and integral calculus*, which set a new standard for British texts – including the second edition of Dr Thomson's *Calculus* – appeared originally in serialized form, beginning in 1836, in the society's Library of Useful Knowledge.[5]

The reformers originally saw it as their mission to bring the most powerful techniques of mathematical analysis, the techniques of Lagrange and Laplace, to the moribund centres of mathematical non-learning in Britain. Of greatest immediate importance was replacing the cumbersome system of dots in the Newtonian fluxional notation with the d's of Leibnizian differentials. Symbols nearly as political as they were mathematical, the d's represented youth and progress in the modern age. Babbage later claimed that in 1812, as the *Memoirs of the Analytical Society* neared publication, he had suggested a more apt title: 'The Principle of pure D-ism in opposition to the Dot-age of the University'. His pun

[4] On the Analytical Society and Herschel see S.S. Schweber, 'Prefatory Essay', in S.S. Schweber (ed.), *Aspects of the life and thought of Sir John Frederick Herschel* (2 vols., New York, 1981), **1**, pp. 54–67. On inbreeding among the new scientific societies see Jack Morrell and Arnold Thackray, *Gentlemen of science: early years of the British Association for the Advancement of Science* (Oxford, 1981), pp. 26–9.

[5] Sophie de Morgan, *Memoir of Augustus de Morgan* (London, 1882), pp. 108, 22–8, 50–3, 86; Augustus de Morgan, *The differential and integral calculus* (London, 1842).

on deism suggests the ideology of natural law that he espoused and how it could be embedded in the symbols of mathematics. Dr Thomson, while no deist in the religious sense, agreed in his *Calculus* of 1831 that the 'inferiority' of dot-age had been 'a principle cause of the small progress made in later times by British mathematicians'. He exemplified its 'clumsiness' in such comparisons as $\ddot{\ddot{y}}$ for d^4y and $(a^x)\cdot$ for da^x.[6]

Of several inadequacies of Newton's dots, most basic was that they suggested no consistent and easily applicable set of rules for differentiation in complex cases, rules that one could apply without rethinking at each application the whole process of geometrical reasoning that the dots symbolized. The dots confused the untrained mind of a student, demanding the sophisticated intuition of a master, but robbed the master of the yet deeper and more powerful command of his subject obtainable through a more efficient language. The call of the Analytical Society was the call of Bacon and of the Enlightenment: reform the language of science by stripping it of its esoteric trappings, thus rendering knowledge both accessible and powerful. Industrialization suggested another ideal; the language of analysis would do for mathematics what machinery did for industry, freeing skilled labour from exhausting manual tasks. Babbage and Herschel remarked: 'It is the spirit of this symbolic language, by that mechanical tact, (so much in unison with all our faculties) which carries the eye at one glance through the most intricate modifications of quantity, to condense pages into lines, and volumes into pages; shortening the road to discovery, and preserving the mind unfatigued by continued efforts of attention to the minor parts, that it may exert its whole vigor on those which are more important.' Inevitably, mathematical analysis called to mind the steam-engine, and, on that analogy, wishing that 'we could calculate by steam', Babbage developed his famous calculating engines, the difference engine and the analytical engine. J.P. Nichol, in the same mode, would refer to mathematical analysis as 'the engine of deduction', while J.J. Sylvester (1814–97) would call it an 'Engine of Development'.[7]

In the hands of Laplace, according to Herschel, this engine had shown, 'that a brief and simple sentence [Newton's law of gravitation], intelligible to a child of ten years of age, accompanied with a few determinate numbers, capable of being written down on half a sheet of paper, comprehends within its meaning the history of all the complicated movements of our globe, and the mighty system to which it belongs'. That one sentence, developed through the language of

[6] Charles Babbage, *Passages from the life of a philosopher* (London, 1864), p. 30; Dr James Thomson, *An introduction to the differential and integral calculus* (Belfast, 1831), pp. 240–1.

[7] Charles Babbage and J.F.W. Herschel, 'Preface', *Memoirs of the Analytical Society* (Cambridge, 1813), pp. i–ii; reprint in Schweber, *Herschel.* Charles Babbage, *History of the invention of the calculating engines*, 10, Buxton papers, Museum for the history of science, Oxford; quoted in Anthony Hyman, *Charles Babbage: pioneer of the computer* (Princeton, 1982), p. 49. William Thomson, 'Notebook of Natural Philosophy class', 1839–40, NB9, ULC; J.J. Sylvester to William Thomson, 13th March, 1855, S600, ULC.

analysis, reduced 'the mazy and mystic dance of the planets and their satellites' to 'a single feature in creation, independent of the lapse of time, and registered only in the unprogressive annals of eternity'.[8]

Analytic mathematics thus revealed to man that he could have complete confidence in the everlasting order of the solar system, for it constituted a self-stabilizing system, despite its many irregularities. By implication, all other systems in the economy of nature would exhibit a similar stability if allowed to function freely under the rule of natural law alone. For early British mathematical reformers, French mathematical physics, if not the French Revolution, symbolized rationality, stability, and progress, and Laplace's *Mécanique céleste* epitomized the goal of natural order. We noted (ch. 4) how John Playfair's 1808 review, which helped to spur modernization of British mathematics, captured the transition from mathematics to politics through his use of metaphors for order contrasted with anarchy: 'all the inequalities in our [solar] system are periodical . . . our system is thus secured against natural decay; order and regularity preserved in the midst of so many disturbing causes; – and anarchy and misrule eternally proscribed'. Dr Thomson echoed the same optimism in his Introductory Lecture at Glasgow in 1832, speaking of 'eternal stability', 'every phenomenon . . . periodical', and 'regulated by a corrective power arising from the principle of universal gravitation'.[9] By analogy, the natural laws of human action, if free of interference, would produce an equally orderly society, in which the irregular behaviours of individuals would cancel each other.[10]

In 1831 Mary Somerville introduced the stabilizing virtues of Laplace's system to the larger reading public with her *Mechanism of the heavens*, originally written for the Diffusion Society and inscribed to its prime mover, Lord Brougham. Her 'popular' volume of over 600 pages schooled the reader in the wide variety of self-correcting mechanisms in the solar system and in the mathematical methods required for their analysis. John Herschel, who had himself stressed these mechanisms in his 1823 article on 'Physical astronomy' for the *Encyclopaedia Metropolitana*, gave Somerville's survey a warm (if condescending) reception in the *Quarterly Review*. Dwelling on the wonders of simplicity and harmony that emerged from the intricacies of the planetary orbits, he too gave a transparently political interpretation: 'yet this intricacy has its laws, which distinguish it from confusion, and its limits, which preserve it from degenerating into anarchy. It is in this conservation of the principle of order in the midst of perplexity – in this ultimate compensation, brought about

[8] [J.F.W. Herschel], 'Mechanism of the heavens', *Quarterly review*, **47** (1832), 537–59; *Essays from the Edinburgh and Quarterly reviews, with essays and other pieces* (London, 1857), pp. 21–62, on pp. 21–2.

[9] [John Playfair], 'La Place, traité de méchanique [*sic*] céleste', *Edinburgh Rev.*, **22** (1808), 249–84, on p. 277; Dr James Thomson, 'Introductory lecture at Glasgow College, 6th November, 1832', QUB.

[10] For a more extended development of this model see M.N. Wise (with the collaboration of Crosbie Smith), 'Work and waste: political economy and natural philosophy in 19th century Britain', *Hist. Sci.* (forthcoming).

by the continued action of causes, which appear at first sight pregnant only with subversion and decay – that we trace the Master-workman with whom the darkness is even as the light.' Behind the apparent disarray lay the beneficent rule of the Creator, who, acting through natural laws, would prohibit ultimate chaos.[11]

Acutely embarrassed that the land of Newton had played no role in developing the mathematics that demonstrated this eternal order in nature, the first generation of reformers had levelled a devastating critique at home-grown adherence to Newton's view that 'algebraic processes' formed 'mere auxiliaries to geometrical construction and demonstration'. To Herschel, 'the last twenty years of the eighteenth century were not more remarkable for the triumphs of both the pure and applied mathematics abroad, than for their decline, and, indeed, all but extinction at home'. The Analytical Society had set out, therefore, 'to re-import the exotic, with nearly a century of foreign improvement, and to render it once more indigenous among us'.[12] In opting for the power and generality of analysis, they originally absorbed to a considerable degree Lagrange's disparagement of any appeal to geometrical intuition, as lacking in rigour, and his alternative algebraic foundations for the differential calculus. 'Algebraic' here refers to Lagrange's abstract method of defining successive derivatives of a function as successive coefficients in its Taylor series expansion. 'Geometric' foundations, by contrast, refer to such procedures as defining the derivative as the tangent to a curve and defining the limit process as the approach of a secant to a tangent. Newton's fluxional calculus treated the curve as generated by the motion of a point and the tangent as the direction of its velocity.[13] Dr Thomson too, in rejecting such geometrical foundations, adopted Lagrange's algebraic methods for the first edition of his *Calculus* (ch. 2).

In physical, rather than purely mathematical, analysis the reformers initially adopted Laplace's notion of rigour, which required the reduction of gravitational attractions, and all other physical phenomena, to actions at a distance between point masses or atoms, whether ponderable or imponderable. To this essentially static model one applied mathematical analysis to deduce observable consequences, usually requiring power-series solutions of the relevant partial differential equations. Algebraic mathematics and reductionist physics went together in Laplace's programme for a rigorous *physique mathématique*. Its

[11] Mary Somerville, *Mechanism of the heavens* (London, 1831), pp. v–lxx. Herschel, 'Mechanism', p. 28; 'Physical astronomy' (1823), *Enc. Met.*, **4** (London, 1830), pp. 341–586. See Schweber, *Herschel*, **1**, pp. 71, 77. On design see Playfair, 'La Place', pp. 278–9.

[12] Herschel, 'Mechanism', pp. 19–31. Babbage and Herschel, 'Preface', p. iv. On Cambridge mathematics see also Playfair, 'La Place', pp. 283–4.

[13] Judith V. Grabiner, 'Changing attitudes toward mathematical rigor: Lagrange and analysis in the eighteenth and nineteenth centuries', and Ivor Grattan-Guinness, 'Mathematical physics, 1800–1835: genesis in France, and development in Germany', both in H.N. Jahnke and M. Otte (eds.), *Epistemological and social problems of the sciences in the early nineteenth century* (Dordrecht, 1981), pp. 311–30; 349–70.

reductions to hypothetical point atoms of ponderable matter and imponderable fluids – electric, magnetic, and caloric – were justified only by the agreement of mathematical deductions with observations. The physical reductions could be no more valid than the mathematical deductions, which therefore required rigorous methods, i.e. algebraic. The Laplacian schema thus contained three intertwined aspects: point-atom reductions, hypothetico-deductive methodology, and algebraic mathematics.

Although importers of French physics originally adopted all three components of this foreign technology, they rapidly turned critical once they had attained 'the final *domestication* of the peculiar notation of the differential calculus'. They then reasserted native traditions to seek an accommodation between analysis and geometry. In 1832 Herschel detected 'the gradual formation of what, at length, begins to merit the appellation of a British School of Geometry'.[14] Indeed, the reformers had never been interested in mathematics primarily for its own sake, but always for its power of revealing the secrets of nature, both mental and physical. The emergence of the British school in the late 1820s and early 1830s can be described internally as, on the one hand, an increasing preference for the new works of Fresnel and Fourier over those of Laplace and his disciples, notably Poisson, and, on the other, a rejection of the newly rigorized epsilon–delta foundations for the calculus, developed by Cauchy, as well as his atomistic reduction of ether mechanics. While Fresnel and Fourier offered theories of light and heat that were in the first instance non-reductionist, non-hypothetical, and geometrical, Cauchy offered abstract mathematics and speculative physics. A factor of central importance to the former alternative was practical utility. Fresnel and Fourier represented the engineering orientation of Napoleon's *Ecole polytechnique* as well as its mathematical physics. They provided adequate mathematical descriptions of phenomena rather than ultimate mechanical theories.

Opposing styles and a skewed reception

Laplace has often been described as a Newtonian, largely on the strength of his programme of atomistic reduction of all physical phenomena. But the differences between Newton's and Laplace's physical views are nearly as striking as those between Newton's and Lagrange's views of the calculus.[15] Newton justified his programme through the 'analogy of nature', arguing that whatever

[14] Herschel, 'Mechanism', p. 37.

[15] Roger Hahn, *Laplace as a Newtonian scientist* (Los Angeles, 1967), pp. 17–19, emphasizes the differences, while the similarities are stressed by Robert Fox, 'The rise and fall of Laplacian physics', *Hist. Stud. Phys. Sci.*, **4** (1974), 89–136, esp. pp. 92–101. See also 'Pierre-Simon, Marquis de Laplace' *DSB*, **15**, 277–403, esp. pp. 358–62; and for the Laplacian programme, D.H. Arnold, 'The *mécanique physique* of Siméon Denis Poisson: the evolution and isolation in France of his approach to physical theory (1800–1840). II. The Laplacian programme, *Arch. Hist. Exact Sci.*, **28** (1983), 267–87.

held true for all observable objects had also to hold for their unobservable parts, for 'nature is wont to be simple, and always consonant to itself'. Since all observable objects possessed the qualities of extension, hardness, impenetrability, mobility, inertia, and gravitation, so also did its parts; they were infinitely hard atoms of finite size that attracted one another with a force varying as the inverse square of the distance between their centres, like perfect planets or marbles. Newton's analogy of nature thus gave priority to the observable properties of objects, and to generalization by induction, in establishing the properties of atoms.

By constrast, Laplace gave priority to mathematical form. Although 'analogy' appeared often in his writings, he did not use it to establish the properties of atoms but only to universalize the properties of potentially observable objects in different parts of the universe. As mathematical points, rather than marbles, his atoms could obey the inverse square law throughout its range of mathematical validity, even down to infinitesimal distances, whereas Newton's extended atoms would collide when their centres were separated by two radii. Clearly one could not justify point atoms by any physical analogy of nature, nor by induction. Similarly, Laplace recognized that no induction could fully justify regarding the gravitational force as exactly an inverse square force, rather than a slightly different power law like 2.0001. Both errors in measurement and perturbing forces would set a limit to the accuracy of any observations. He therefore appealed to the aesthetic criterion of mathematical simplicity and to the hypothetico-deductive justification noted above.[16]

Laplace probably adapted his point atoms from those of Roger Boscovich, whose conception contained the advantage over Newton's that the atoms could never collide, firstly because they were mere points, but secondly because they repelled one another with increasing force at increasingly short distances. Since Laplace's atoms could not collide they could never lose *vis viva* (kinetic energy), while Newton's infinitely hard marbles, possessed of no elasticity, lost motion in every collision. Laplace's mathematical universe, consequently, had no need for Newton's God, who continually acted to replenish motion in a world that otherwise would run down, nor for teleology of any kind.[17]

The abstraction and rationalism evident in the Laplacian scheme did not always appeal either to his peers or to younger mathematicians, particularly those who supported the practical utilitarian thrust of educational reform following the revolution. In particular, Lazare Carnot, military engineer and

[16] Alexander Koyré and I.B. Cohen (eds.), *Isaac Newton's Philosophiae naturalis principia mathematica: the third edition (1726) with variant readings* (2 vols., Cambridge, 1972), **2**, pp. 552–5; P.S. de Laplace, *Exposition du système du monde*, 4th edn. (2 vols., Paris, 1813), **2**, pp. 10–11, 204–8, 217–95 ('De l'attraction moléculaire' on atomistic reduction).

[17] Roger Boscovich, *A theory of natural philosophy*, trans. J.M. Child, from the 2nd edn. (1763), (London, 1922), p. 141, note (m). See T.L. Hankins, 'Eighteenth-century attempts to resolve the *vis viva* controversy', *Isis*, **56** (1965), 281–97, esp. pp. 291–7. Laplace, *Système du monde*, **1**, pp. 248, 280–2.

statesman, and Gaspard Monge, first director and driving force of the *Ecole polytechnique*, maintained close contact with the needs of engineering, emphasizing visualizable synthetic geometry and the mechanics of machines in their books and teaching. Their younger followers in this tradition, including Dupin, Poncelet, Chaslés, Hachette, Navier, and Cariolis, attempted to ground some of the most abstract of mathematical concepts, such as imaginary numbers, in projective geometry, which would generate concrete conceptions of unfamiliar entities by showing how they could be evolved continuously from the familiar. Simultaneously they transformed engineering mechanics into a new science of work and spent much of their energy on attempts at social reform through the education of craftsmen and practical engineers.[18] Within the elitist ranks of the *Académie des sciences* during the Restoration, when analytic mathematics and reductionist physics ruled, this practical, progressive orientation carried little weight. Carnot and Monge had been removed as unregenerate Republicans, and few of their followers gained entry.[19]

In the space between Laplacian physics and theoretical engineering, however, new varieties of physical theory emerged, including those of Fourier, Fresnel, Arago, Petit, Dulong, and Ampère.[20] Fourier admired greatly the power of mathematical analysis applied to physical problems, but he disapproved of mathematical abstraction and hypothetical models. From that perspective he interpreted the analogy of nature in a way entirely different from Newton's, as an analogy of macroscopic mathematical forms, independent of underlying micro-structure. He could therefore treat heat conduction as though it were a phenomenon of continuous flow, without regard to its true physical nature, whether imponderable fluid or mode of motion. The technique brought the power of analysis directly to bear on empirical laws of thermal behaviour without appeal to microscopic models or to their hypothetico-deductive justification. It may usefully be viewed as continuing the Laplacian priority on mathematical over physical concepts, but at a practical rather than hypothetical level.

Mathematically, Fourier's reputation rests on his expansion of an arbitrary function in a 'Fourier series', consisting of a sum of periodic sine and cosine functions. Any sound, for example, may be decomposed into such periodic components, but the general decomposition has wide application to the solution of differential equations that describe physical processes. It greatly simplified Fourier's analysis of problems of heat conduction. Nevertheless, in the decade

[18] This fascinating complex of issues has been opened up by Lorraine J. Daston, 'The physicalist tradition in early nineteenth century French geometry', *Stud. Hist. Phil. Sci.*, **17** (1986), 269–95, and Ivor Grattan-Guinness, 'Work for the workers: advances in engineering mechanics and instruction in France, 1800–1830', *Ann. Sci.*, **41** (1984), 1–33.

[19] Daston, p. 294.

[20] See Fox, 'Rise and fall', pp. 109–32, for valuable discussion and references. Fox, however, does not treat the engineering tradition, nor does Robert H. Silliman in the accompanying article, 'Fresnel and the emergence of physics as a discipline', *Hist. Stud. Phys. Sci.*, **4** (1974), 137–62.

following submission of his original essay of 1807 to the *Institut de France*, neither Lagrange (his main opponent, who died in 1813) nor Laplace nor Poisson considered his solutions in trigonometric series, rather than power series, to be rigorously founded, even though he won the prize competition of the *Institut* in 1811.[21]

Fourier's method of obtaining the partial differential equation of heat conduction excited equal controversy, for he very nearly wrote it down as a direct expression of experimental results. To the prevailing orthodoxy, rigour in *physique mathématique* meant rigorous deduction of observable consequences from a detailed microscopic model of atoms or molecules and their interactions. Conceptually, the net action of a system at any point in space derived from a sum (or less rigorously an integral) over the independent actions of all the discrete sources in the system. Rigour therefore required one to begin with an integral conception, then carefully to convert it to the more manageable form of a partial differential equation, and finally to find a power-series solution. Fourier violated every step in the process, beginning with his appeal directly to experimental laws.

Two kinds of experiment are relevant. One showed that a warm body placed in flowing air at a constant lower temperature loses heat at a rate very nearly proportional to the temperature difference. The other showed that radiant heat will not penetrate even very thin foil. Together the results justified Fourier's 'principle of the communication of heat': 'If two molecules of the same body are extremely near, and are at unequal temperatures, that which is most heated communicates directly to the other during one instant a certain quantity of heat; which quantity is proportional to the extremely small difference of the temperature'. He immediately translated the temperature difference between molecules into a temperature differential across an imaginary surface in a solid substance, arguing that 'layers in contact are the only ones which communicate their heat directly'.[22] In this form, the principle expressed directly, albeit in terms of differentials, the experimental observations. It thus remained a *macroscopic* relation, independent of microscopic structure. Conduction occurred as a continuous flux of heat F across an arbitrary geometrical surface, in response to a temperature gradient across that surface. In amount, finally, the flux depended on a macroscopic property of the medium, its relative ability to conduct heat, or its conductivity K, yielding,

$$\vec{F} = -K \textit{ gradient } V,$$

where V is temperature.[23]

[21] Ivor Grattan-Guinness in collaboration with J.R. Ravetz, *Joseph Fourier, 1768–1830: a survey of his life and work, based on a critical edition of his monograph on the propagation of heat, presented to the Institut de France in 1807* (Cambridge, MA and London, 1972), pp. 441–90, summarizes Fourier's mathematical controversies and accomplishments, suggesting that Laplace regarded Fourier in a much more favourable light than is normally supposed.

[22] Fourier, *Heat*, pp. 41–5, 456–7, 460.

[23] *Ibid.*, pp. 45–52.

Fourier's argument, while simple, involved a subtle ambiguity in his use of 'molecule'. In his introductory discussion he used the term as Laplace did, referring to discrete physical objects between which attractive and repulsive forces acted and which radiated heat as point sources. This usage made 'a point m' and 'a molecule m' equivalent. But gradually he began to use 'molecule' in the sense of a differential element of volume $dxdydz$ located at the point (x,y,z). This change in usage attended the change in the principle of communication of heat from microscopic to macroscopic form, with all formal mathematical arguments depending only on the latter. The shift became immensely important to British theory. Just as Fourier's 'molecules' were not molecular, we shall find that their 'particles' (sometimes a translation for *molécules*) were often not particulate.[24]

Assuming that heat is conserved, whatever its nature, Fourier set the amount of heat leaving – diverging from – any 'molecule' $dxdydz$ per unit time equal to the rate at which its heat content was decreasing, expressed as its heat capacity (again a macroscopic property) times the rate of decrease of its temperature. The resulting 'equation of continuity' was

$$\text{divergence } \vec{F} = -CD\partial V/\partial t,$$

where C is heat capacity per unit mass, or specific heat, D is mass per unit volume, or density, and t is time. In essence, by substituting the first equation in the second, and assuming K to be constant, he obtained his general equation for the motion of heat in a solid body:

$$K \text{ divergence(gradient } V) = CD\partial V/\partial t,$$

or, fully written out,

$$K\,(\partial^2 V/\partial x^2 + \partial^2 V/\partial y^2 + \partial^2 V/\partial z^2) = CD\partial V/\partial t.$$

Thus Fourier established the most basic equation of his *Théorie analytique* in three lines.[25] In an equally elegant manner he established the equation of the bounding surface of the solid when radiating into flowing air at a constant temperature. The two equations constituted his general theory of the motion of heat. He applied the theory to a variety of geometries to obtain their particular equations and found the forms of the solutions under various conditions, again and again exhibiting the simplicity and power of his techniques.

Notwithstanding the 'splendour and poetry' that William Thomson would see in Fourier's treatise, his Laplacian contemporaries did not see its beauty. After more than twenty-five years of controversy, Poisson continued to elaborate in his own *Théorie mathématique de la chaleur* the same kinds of objections that Laplace originally raised in 1808. All of their complaints involve the charge that

[24] *Ibid.*, compare pp. 39–41 progressively with pp. 47, 74, 80, 84, and many later passages, such as pp. 98, 112. This ambiguity in the terms 'molecule' and 'particle' was quite typical in hydrodynamics, e.g. Laplace, *Système du monde*, **1**, p. 301, for 'molecule' and P.S. de Laplace, *The system of the world*, trans. J. Pond (2 vols., London, 1809), **1**, p. 355, for 'particle'.

[25] Fourier, *Heat*, pp. 112–14. Fourier did not use the phrase 'equation of continuity', but Laplace and others used it to express conservation in fluid flow, e.g. P.S. de Laplace, *Traité de mécanique céleste* (5 vols., Paris, 1798–1827), **1**, p. 95.

Fourier's principle for the communication of heat obscured the physical process of diffusion by incorporating *from the outset* two approximations: radiation extends only to infinitesimal distances in a solid, rather than to small but finite distances; and radiation rate is strictly proportional to temperature difference, rather than nearly so. Behind these complaints stood the Laplacian creed, as expressed by Laplace himself in responding to Fourier's original memoir in 1808: 'I have wanted to establish that the phenomena of nature reduce in the final analysis to action *ad distans* from molecule to molecule, and that the consideration of these actions ought to serve as the basis of the mathematical theory of these phenomena'.[26]

A proper derivation of the equations of heat conduction, as Poisson presented it, began with an explicit model of the relation between ponderable molecules and caloric fluid in a solid, incorporating both the free caloric radiated from the molecules and responsible for temperature, and the bound or latent caloric involved in changes of phase. His model attributed to the radiating molecules the full complexity of observable objects, including radiation to finite distances, radiation rates between molecules proportional to finite temperature differences, and a correction factor depending on absolute temperature to account for possible non-linearity. From this speculative picture, he extracted by rigorous, if tedious, expansions and approximations the general equation for the motion of heat. It involved a conductivity factor derived from the model that included variations of conductivity with temperature and with inhomogeneities in the medium.[27] For homogeneous media he obtained:

$$CD\partial V/\partial t = K\left(\frac{\partial^2 V}{\partial x^2}+\frac{\partial^2 V}{\partial y^2}+\frac{\partial^2 V}{\partial z^2}\right)+\frac{\mathrm{d}K}{\mathrm{d}V}\left[\left(\frac{\partial V}{\partial x}\right)^2+\left(\frac{\partial V}{\partial y}\right)^2+\left(\frac{\partial V}{\partial z}\right)^2\right].$$

This equation differs from Fourier's by the second term on the right, which is non-zero if conductivity is a function of temperature. Poisson's derivation implied that this term must be non-zero if radiation occurs between molecules separated by finite distances, and thus by finite temperature differences, for then the conductivity would indeed depend on absolute temperature. He regarded his much more complex procedure, therefore, as not only more rigorous but also as leading beyond a mere restatement of experimental results.[28] It allowed

[26] P.S. de Laplace, 'Mémoire sur les mouvements de la lumière dans les milieux diaphanes', *Mémoire de l'Institut*, **10** (1809), 300–42; *Oeuvres complètes de Laplace* (14 vols., Paris, 1878–1912), **12**, p. 295.

[27] S.D. Poisson, *Théorie mathématique de la chaleur* (Paris, 1835), pp. 6, 92. Most of the results of this treatise had appeared already in papers of 1815 and 1823. For Poisson's mathematical and physical objections to Fourier, see Arnold, 'Poisson. IV. Disquiet with respect to Fourier's treatment of heat', and 'IX. Poisson's closing synthesis: Traité de physique mathématique', *Arch. Hist. Exact Sci.*, **28** (1983), 299–320, and **29** (1983), 73–94, esp. pp. 78–85; M.N. Wise, 'The flow analogy to electricity and magnetism – Part I: William Thomson's reformulation of action at a distance', *Arch. Hist. Exact Sci.*, **25** (1981), 19–70, esp. pp. 23–9.

[28] Poisson, *Chaleur*, pp. 117–18.

theory to lead experiment by suggesting what features to test, much as one might claim today for the hypothetico-deductive method. For his own part, Fourier had in essence already defended his phenomenological procedure on the ground that it too would lead to Poisson's equation if conductivity varied with temperature, since in that case the substitution of the flux equation into the continuity equation would yield Poisson's extra term. Until experiments actually showed such a temperature dependence, however, the mathematical theory should not extend beyond observations, particularly when serious doubts existed concerning the validity of the caloric theory of heat.[29]

To summarize, both Poisson and Fourier expected mathematical analysis to open up the secrets of nature, but where Poisson put his faith in mathematically constructed models subjected to experimental testing, Fourier relied on macroscopic mathematical forms themselves as direct expressions of observable reality. By analogy of mathematical form, heat conduction was heat flow. That identification was all one needed for intellectual satisfaction and practical application. As Fourier put it, 'the theory of heat will always attract the attention of mathematicians, by the rigorous exactness of its elements and the analytical difficulties peculiar to it, and above all by the extent and usefulness of its applications; for all its consequences concern at the same time general physics, the operations of the arts, domestic uses and civil economy'.[30]

Fourier's theory merits the following labels: non-hypothetical, macroscopic, geometrical, and practical. But those terms miss two features of his mathematical style that require additional notice. The first concerns his treatment of heat flow as continuous motion rather than the net effect of radiating points; heat flux was simply flow across a geometrical surface. In the interior of a solid the continuous flux could be supposed to be produced by a continuously varying temperature across such surfaces, but at the boundaries of the solid the procedure implied a discontinuous change from the interior temperature to the temperature of the surrounding space. Poisson objected that the discontinuity was non-physical, like the rest of the theory, because it replaced what was in fact a finite layer near the surface, in which the temperature varied continuously, with an artificial geometrical surface across which it jumped. Ever concerned that the transition from finite differences to infinitesimals and from physical surfaces to mathematical boundaries be rigorously argued, he insisted on following through a series of approximations from his model to obtain Fourier's equation of the bounding surface.[31] The differences are profound. They are much the same as the differences between the theoretical mechanics of Laplace and Poisson, on the one side, and the engineering mechanics of Poncelet and Navier, on the other. Against Poisson's view from *inside* the machine, so-to-speak, concerned with its ultimate parts, Fourier set the engineer's view from the

[29] Fourier, *Heat*, pp. 458, 464–6. [30] *Ibid.*, p. 26.

[31] Poisson, *Chaleur*, pp. 119–24. Arnold, 'Poisson. VIII. Applications of the mécanique physique', *Arch. Hist. Exact Sci.*, **29** (1983), 53–72, esp. pp. 59–64.

outside, concerned with what went in and what came out and what motions were involved, with the gross features of the machine and what work it could do.[32] The bounding surface from this perspective became merely a 'boundary condition' to be imposed on the general solution of the differential equation. Similarly, whatever the sources or sinks might be that, for example, maintained a boundary at constant temperature, they appeared in the theory as the geometrical surface of a heat reservoir rather than as distinct physical entities.

In the second place, Fourier's view from the outside highlighted the role of mathematical analysis as an instrument for probing the interior of a body when one knew already the condition of its surface. An adequate knowledge of the boundary conditions at any one time could be used to determine the state of the interior for all past and future times. This boundary value technique for solving differential equations was the stock in trade of Laplace and Poisson as well, but within Fourier's scheme it acquired a new importance, because he used it, not to test any particular hypothetical model of the interior, but to learn what the general conditions of the interior necessarily had to be. This issue reappears below with respect to Fourier's theory of secular cooling of the earth.

We have considered at some length the differences in style between Fourier and the Laplacians in order to emphasize that the French suppliers of mathematical physics offered distinct options to British importers. Once analysis had become familiar in Britain, the methods epitomized by Fourier fared much better than those of latter-day followers of Laplace and Lagrange, such as Cauchy. Partly the change reflected an anti-Laplacian mood in French physics itself,[33] but only in Britain did that mood lead to a coherent new style. Between 1825 and 1835 atomistic reduction virtually disappeared from Britain, as did the purely algebraic approach to analysis. Fourier may be considered as setting the capstone to the reception process. Although no comprehensive study of this process exists, its results in specific areas are quite clear.

Most extensively studied has been the reception of the wave theory of light, where Fresnel played a role similar to Fourier's for heat. The wave theory presupposed an etherial medium to propagate the undulations of light. Because only undulations transverse to the direction of propagation could explain polarization, this luminiferous ether had to behave like an elastic solid rather than a rarefied fluid, and, in order to propagate the undulations at the speed of light, it had to be incredibly rigid. However, solid objects like planets apparently moved through it unresisted. No known substances behaved in this unresisting

[32] See Grattan-Guinness, 'Work for the workers', *passim*, and Chapter 8 below. Compare Arnold, 'Poisson. VII. Mécanique physique', *Arch. Hist. Exact Sci.*, **29** (1983), 37–52. Arnold notes several difficulties for a stylistic alignment of Fourier with Navier (also with Lagrange; cf. Fox, 'Rise and fall', pp. 118–19). Similar remarks apply to Fresnel. We ignore these subtleties since British interpreters often did not respect them. See, for example, Airy's distinction of 'mechanical' and 'geometrical' below.

[33] Fox, 'Rise and fall', pp. 127–32.

but rigid manner, yet the wave theory beautifully and simply explained phenomena like interference and diffraction, which embarrassed particle emission theories. In defence of the wave theory, therefore, and in lieu of an adequate ether theory, Fresnel often adopted a phenomenological approach.[34]

This approach is the only one many British readers ever encountered, for they typically read the theory in the descriptive form that Fresnel himself gave it in an elementary survey published in English in 1827 or in a more extended survey in the same year by John Herschel for the *Encyclopaedia Metropolitana*.[35] Whatever the physical nature of luminiferous ether, the wave equations describing the macroscopic behaviour of light were valid because they represented the phenomena in the only natural manner available. While streams of particles passing through two slits in a screen could hardly be imagined to annihilate each other to produce the familiar black lines of an interference pattern, such patterns emerged naturally from the action of waves. Fresnel himself tended to justify the wave theory in a hypothetico-deductive manner, but his British interpreters often presented it as a mathematical description of the phenomena, which represented their true form independent of their true cause. The principle of interference, Herschel remarked, 'if not founded in nature, is certainly one of the happiest fictions that the genius of man has yet invented to group together natural phenomena'.[36]

Like Fresnel, Herschel would have preferred a full mechanical model of the ether from which the phenomena could be deduced, presumably a theory of atoms and forces, but unlike Fresnel he devoted little effort to generating such a theory. Instead, he accepted the mathematical wave description as an intermediate stage of development. The same can be said, only more so, of George Airy, who in 1831 added a section on the undulatory theory to his revised *Mathematical Tracts*. He there distinguished the 'geometrical part', which concerned only those properties of light associated with the mathematical form of the wave equation, and which he considered 'certainly true', from the 'mechanical part', depending on suppositions of etherial structure. In general, the fact that the mere mathematical form of the theory reproduced major phenomena of light appears to have been quite important to its overall reception in Britain. Adherents of the

[34] J.Z. Buchwald, 'Optics and the theory of the punctiform ether', *Arch. Hist. Exact Sci.*, **21** (1980), 245–78, esp. pp. 247–50, discusses Fresnel's dependence on molecular ether dynamics in his more sophisticated works.

[35] Augustin Fresnel, 'Elementary view of the undulatory theory of light', *Quart. J. Sci.*, **23** (1827), 127–40, 441–54; **24** (1827), 113–35, 431–48; **25** (1828), 198–215; **26** (1829), 168–91, 389–407; **27** (1829), 159–65. J.F.W. Herschel. 'Light' (1827), *Enc. Met.*, **4** (London, 1830), pp. 341–586. For discussion of Herschel's communications with Fresnel and his move from molecular to wave optics, see Crosland and Smith, 'Transmission', pp. 41–3. The best general discussions of the British response to the wave theory are G.N. Cantor 'The reception of the wave theory of light in Britain: a case study illustrating the role of methodology in scientific debate', *Hist. Stud. Phys. Sci.*, **6** (1975), 109–32; *Optics after Newton: theories of light in Britain and Ireland, 1704–1840* (Manchester, 1983), pp. 159–72.

[36] Herschel, 'Light', p. 456.

wave theory were predominantly mathematicians while particle people were predominantly experimentalists.[37]

Not all mathematicians in Britain limited themselves to the 'geometrical part' of the wave theory. William Rowan Hamilton, Baden Powell, Matthew O'Brien, Samuel Earnshaw, Philip Kelland, and John Tovey all sought microscopic theories of the ether based on point atoms. In this they followed Fresnel himself, Laplace, Poisson, and others. Most successful and mathematically powerful of the French theorists was Cauchy, whose papers in the 1830s set the standard for British attempts. It is symptomatic of the general rejection of Cauchy's abstract and reductionist methods, however, that by the early 1840s nearly all British theorists had either turned to macroscopic theory or abandoned the subject, Hamilton being the only notable exception.[38]

James MacCullagh's phenomenological paper of 1837 illustrated the point. He sought a mathematical description of the process of propagation of waves across a reflecting and refracting boundary independent of any model of that process. MacCullagh (1809–47) won a medal for his efforts from the Royal Irish Academy, presented by his colleague in Dublin and president of the Academy W.R. Hamilton. In his remarks, Hamilton distinguished between MacCullagh's method of 'mathematical induction' (like Kepler's) and his own preferred method of 'dynamic deduction' from a postulated system of attracting and repelling molecules (like Newton's). But he applauded 'the preparatory but important task of discovering, from the phenomena themselves, the mathematical laws which connect and represent those phenomena, and are in a manner intermediate between facts and principles, between appearances and causes'. Herschel, in a letter to Hamilton, strongly supported the idea of 'abandoning for a while the *a priori* or deductive path, and searching among the phenomena for laws simple in their geometrical enunciation' without for the moment worrying about a dynamical theory. Their distinction mirrors that of Airy between 'geometrical' and 'mechanical' parts of the theory. When MacCullagh in 1839 did develop what he considered a 'dynamical theory', however, it provided nothing to which Airy, Herschel, and Hamilton would have attached that name, but only a set of macroscopic physical properties of differential volume elements and corresponding generalized dynamical equations that behaved like no known mechanical system.[39]

[37] G.B. Airy, *Mathematical tracts on the lunar and planetary theories, the figure of the earth, precession and nutation, the calculus of variations and undulatory theory of optics; designed for the use of students in the University*, 2nd edn. (Cambridge, 1831), p. iv. See Cantor, *Optics after Newton*, pp. 173–87; 'Reception', pp. 109–32.

[38] Buchwald, 'Punctiform ether', pp. 257–60; 'The quantitative ether in the first half of the nineteenth century', in G.N. Cantor and M.J.S. Hodge (eds.), *Conceptions of ether: studies in the history of ether theories, 1740–1900* (Cambridge, 1981), pp. 215–37.

[39] James MacCullagh, 'On the laws of crystalline reflection and refraction', *Trans. Royal Irish Acad.*, **18** (1837), 31–74; J.H. Jellet and S. Haughton (eds.), *The collected works of James MacCullagh* (Dublin, 1880), pp. 87–137, and 'An essay towards a dynamical theory of crystalline reflexion and refraction', *Trans. Royal Irish Acad.*, **21** (1848), 17–50; *Works*, pp. 145–84. W.R. Hamilton, 'Address by the President' (1837), *Proc. Royal Irish Acad.*, **1** (1841), 212–21, on pp. 215–16. See D.F. Moyer, 'James MacCullagh', *DSB*, **8**, 591–3, and Buchwald, 'Punctiform ether', pp. 260–3.

More typical of the developing British style were papers of the late 1830s by George Green (1793–1841), which provided MacCullagh's point of departure. Green sought a theory of the luminiferous ether, regarded as an elastic solid, that would rely only on macroscopically defined parameters of elasticity, and would encompass quantitatively such phenomena as polarization by reflection, double refraction, and the intensities of reflected and refracted rays at a geometrical boundary between two media. His theory rivalled Cauchy's best efforts but without appeal to any particular interactions of whatever molecules constituted matter:

> We are so perfectly ignorant of the mode of action of the elements of the luminiferous ether on each other, that it would seem a safer method to take some general physical principle as the basis of our reasoning, rather than assume certain modes of action, which, after all, may be widely different from the mechanism employed by nature; more especially if this include in itself, as a particular case, those before used by M Cauchy and others, and also lead to a much more simple process of calculation.

He considered the behaviour of 'elements', supposed to contain 'a very great number of molecules' acting by any kind of forces but only to very small distances, and applied Lagrange's generalized form of D'Alembert's principle (ch. 11) to obtain equations of motion of the elements, just as MacCullagh would do.[40] The action thus became action between contiguous elements in a continuous medium. In style, Green's analysis compares with Cauchy's much as Fourier's compares with Poisson's. The analogy is not coincidental.

In 1828 Green had published privately *An essay on the application of mathematical analysis to the theories of electricity and magnetism*, a work that would be of great importance to William Thomson. Acknowledging Fourier in his preface, Green simplified and greatly extended the pathbreaking works of Poisson on electricity (1811) and magnetism (1822) in the same way as Fourier's analysis of heat conduction bypassed Poisson's detailed modelling. For example, where Poisson went to some trouble to analyze conditions at the surface of an electrified conductor due to the finite thickness of his supposed electrical fluids, both positive and negative, Green simply treated the surface as a geometrical boundary with positive or negative electrical density and did not concern himself with the physical nature of electricity. Similarly, where Poisson based his theory of magnetism on a model of magnetic molecules containing equal quantities of northern and southern magnetic fluids, Green substituted a macroscopic density of magnetization, or polarization per unit volume.[41]

Neither Poisson's theories of electricity and magnetism nor Fourier's style, however, were transmitted through Green's *Essay*, since it remained unknown.

[40] George Green, 'On the laws of the reflexion and refraction of light at the common surface of two non-crystallized media' (1837) and 'On the propagation of light in crystallized media' (1839), *Trans. Cam. Phil. Soc.*, **7** (1842), 113–20; 121–40; N.M. Ferrers (ed.), *Mathematical papers of the late George Green* (London, 1871), pp. 243–69, 281–311, on pp. 245, 248.

[41] George Green. *An essay on the application of mathematical analysis to the theories of electricity and magnetism* (Nottingham, 1828); Ferrers (ed.), *Mathematical papers*, pp. 1–115, for example pp. 10, 92.

Instead, William Whewell served as the primary agent for both. In 1827 he wrote the article 'Theory of electricity' for the *Encyclopaedia Metropolitana*. Aiming only to communicate Poisson's theory as a new domain of mixed mathematics, he presented it in the form Poisson had left it, as a theory of action at a distance between point masses of electrical fluid. Soon Whewell adopted a more critical stance. In his 1835 'Report on the recent progress and present condition of the mathematical theories of electricity, magnetism, and heat' for the British Association he largely repeated his earlier discussion of electricity, but now recognized that Fourier offered an alternative to the Laplacian programme. (The alternative was also an alternative to Laplace's steady-state cosmology and the support it gave Lyell's geology, as discussed below.) To Laplace's verdict that 'Fourier's equations are right, but the true bases of them are to be found in the doctrine of the action of molecules *ad distans*', he strenuously objected:

Fourier maintained that the quantity of heat transferred from one slice to the next in unit of time was a finite quantity, independently of molecular reasoning . . . Fourier's reasoning no more requires the introduction of molecular action, than do the reasonings by which the common formulae of Hydrostatics (formulae much resembling those of Fourier) are established in Mechanical Treatises.

Whewell also remarked on Fourier's method of solving differential equations in trigonometrical series, noting that it had led to 'some remarkable disquisitions on points of pure analysis'. He concluded that 'the skill and resource shown by Fourier in this investigation, and the interesting and instructive nature of the results, make the series of his labours one of the most important of the physico-mathematical researches of the present century'. Following such a report, Fourier could no longer go unread.[42]

He might, however, be read badly. Philip Kelland published his elementary *Theory of heat* in 1837. He adhered to an extreme form of the Laplacian programme expecting the reduction of all physical phenomena to inverse square laws alone. He once wrote: 'The law of the inverse square of the distance has always appeared to me a *necessary* law, necessary, I mean, as regards the actual state of the constitution of the Universe: and although I could allow that the particles of matter might have been impressed with any law at their creation, I cannot, in consistence with the *simplicity* of all *known* actions, conceive any other than Newton's law'. His *Heat* represented a strange hybrid which mixed this commitment to 'mechanical theory' with Fourier's mathematics, for he found Poisson's procedures 'devoid of simplicity' and thus unsuitable for an elementary work.[43] But neither was he happy with Fourier's mathematics. His

[42] William Whewell, 'Theory of electricity', *Enc. Met.*, **4** (London, 1830), pp. 140–70; 'Report on the recent progress and present condition of the mathematical theories of electricity, magnetism, and heat', *BAAS Report*, **5** (1835), 1–34, on pp. 20–2, 24–7.

[43] Philip Kelland, *Theory of heat* (Cambridge, 1837), p. x. For the inverse square law see pp. 7, 143–82, and Kelland's 'On the dispersion of light, as explained by the hypothesis of finite intervals', *Trans. Cam. Phil. Soc.*, **6** (1838), 153–84, on p. 184.

incautious criticisms soon aroused the ire of the sixteen-year-old prodigy from Glasgow, William Thomson.

In 1839–40 J.P. Nichol had taught natural philosophy to young William in the Scottish manner (ch. 4), emphasizing sensible motions, non-hypothetical theory, and practical application. The reductionist programme had never been compatible with this tradition, although Scots like Robison, Playfair, and Meikleham had been quicker than others in Britain to advocate study of the modern texts and methods of analysis. From Meikleham in the previous year, and on his own in the summer, Thomson had learned enough of the methods of Lagrange, Legendre, and Laplace (not to mention Airy and J.H. Pratt) to write a sophisticated essay on the 'Figure of the earth', which won a University Medal for 1839–40. But Nichol, with his charismatic personality and popular lecture style, made Fresnel and Fourier the heroes of truth and beauty in the analytic art. No doubt he also used Fourier, along with the nebular hypothesis, to support his ideology of progression in the affairs of man and nature (ch. 4). Nichol's presentation of Fourier's accomplishments (despite his never having actually read the mathematical parts) made such an impression on his avid student that, immediately after the session ended in May, Thomson began to devour Fourier on his own. 'In a fortnight', he recalled, 'I had mastered it–gone right through it'.[44]

Soon afterwards he read Kelland's *Heat*, only to find his new idol attacked. On a family educational vacation to Frankfurt-am-Main, where the Nichols and Thomsons took houses for two months in the summer, he carried with him his Fourier. Apparently he spent much of his language-learning time proving Kelland an incompetent mathematician, thus irritating Dr Thomson who had gone to considerable trouble and expense to arrange the trip. The 'waste' nevertheless led to William's first publication. By February, 1841, he had readied his counterattack, which began by quoting Kelland's blunder:

It is remarked by Mr Kelland, in his 'Theory of Heat', p. 64, in reference to the expansion of discontinuous functions in trigonometrical series, that 'there can be little doubt to any one who carefully examines the subject, that nearly all M Fourier's series on this branch of the subject are erroneous'. It appears to me, after a careful examination of the subject, that this remark is incorrect. The two series, given by Fourier, to which Mr Kelland refers as differing from his own series, are really expansions of different functions.

The Glasgow student proceeded to give the Edinburgh professor an incisive lecture on the expansion of functions in Fourier series.[45]

With ironic justice Dr Thomson communicated the paper to his young acquaintance D.F. Gregory, editor of the new *Cambridge Mathematical Journal*, whom Kelland had recently defeated for the chair at Edinburgh.[46] Gregory

[44] SPT, **1**, 9–14. William Thomson, NB9, ULC, and 'Essay on the figure of the earth' (1840), NB11, ULC.

[45] SPT, **1**, 15–19. William Thomson to D.F. Gregory (draft), February, 1841, G179, ULC.

[46] For the election to the chair, see G.E. Davie, *The democratic intellect: Scotland and her universities in the nineteenth century* (Edinburgh, 1961), pp. 116–26, 158–68; see also Chapter 2 above.

agreed that 'the flippant manner in which Mr Kelland speaks of Fourier would deserve pretty strong terms of reprobation', but thought that, as the article was 'of the nature of an attack', Kelland 'should know the name of his opponent'. Dr Thomson took the precaution of sending the still anonymous paper to Kelland and received a reply 'evidently written under a feeling of irritation', although Kelland did admit that his criticism of Fourier had been too strong. The astute veteran of many Glasgow College political campaigns returned William's revised version, emphasizing that the author's 'sole object is to establish what is true', and expressed his confidence that 'if any injustice has been done to Fourier thro' misconceptions or otherwise, you will be glad to have the mistake rectified'. Outmanoeuvred and outclassed, Kelland reversed his ground, praised the author's investigation, and generously conceded that 'if he works it up well into a paper, it will be most interesting'.[47]

So ended a last skirmish in the war between the styles of Laplace and Fourier. When Laplace had insisted on 'action *ad distans*' in 1808, no one could have imagined that his grand programme would end up defended by Kelland's flabby dogma of 1838. This parting shot, however, launched the career of an avid disciple of Fourier, charged with the excitement of competition and success. Subsequent papers, also based on Fourier, would sustain his intellectual energy through the drilling of Cambridge, when he sometimes longed for the days of May, 1840: 'I then commenced reading Fourier & had the prospect of our tour in Germany before me'.[48]

Whig mathematicians of the second generation

When Thomson went up to Cambridge in the autumn of 1841 he immediately joined a self-consciously progressive group of recent wranglers. Not since the Analytical Society had such a coherent and purposeful group of reformers assembled at Cambridge. Like its predecessors, who had attempted to found a journal, the *Memoirs of the Analytical Society*, the new generation organized itself around the *Cambridge Mathematical Journal*. But the interests and the mathematics had changed. No longer primarily importers of French methods, and much less concerned with new societies (Gregory's role in founding the Chemical Society in Cambridge is an exception), the group pursued a distinctive mathematical programme looking towards a near future in which it would compete on an equal and professional basis with their continental counterparts. To appreciate the shift we shall look briefly at the later stages of mathematics itself under the older generation.

[47] D.F. Gregory to Dr Thomson, 28th February and 6th March, 1841, G180 and G182, 1841, ULC; Dr Thomson to Gregory, 9th March, 1841, G183, ULC; Philip Kelland to Dr Thomson, 4th March and *c*.8th March, 1841, K4 and K6, ULC; Dr Thomson to Kelland, 6th March, 1841, K5, ULC.

[48] William Thomson, 14th March, 1843, Diary kept at Cambridge, 13th February to 23rd October, 1843, NB29, ULC.

As in natural philosophy, much changed in the reception of mathematics after 1830. The importers abandoned, for example, Lagrange's algebraic foundation for the differential calculus, adopting instead various formulations of the limit concept. To some extent, they were still importing French works, but newer ones that returned to an intuitive conception of limits rather than following Lagrange, or even Cauchy's newly formalized definition of limits in terms of the epsilon–delta procedure familiar today. Lagrange had attempted to avoid the idea of infinitely small elements by defining derivatives in terms of finite algebraic quantities, the coefficients in a Taylor series expansion. Detractors argued that the infinitesimals were present anyway, but hidden, and questioned his proof that one could always expand a function in positive integral powers of the variable. Critics in Britain objected especially to the abstractness of his method. The limit approach, in contrast, brought infinitesimals into the open. For a function $f(x)$ one defined the derivative $df(x)/dx$ as the limit of the ratio $[f(x+h)-f(x)]/h$ as the quantity h approached zero. Justifying that procedure required assurance that the ratio remained finite as h became evanescent. Some kind of explicit physical, geometrical, or logical reasoning was required, but Cauchy's abstract formalization would not do.

William Whewell had been a conservative reformer from the 1820s, and long advocated geometrical demonstration as the proper foundation for mathematical teaching, so far as 'permanent studies' were concerned (ch. 3). In his *Doctrine of limits* of 1838 he employed geometrical techniques almost exclusively, relying on intuition as the arbiter of truth in the limit process. The *Doctrine* formed one component in the conservative programme that Whewell supported at Cambridge. Most reactionary was the campaign to eliminate the modern subjects of electricity, magnetism and heat from the Tripos, a project which succeeded by 1850, over the opposition of the younger fellows of Thomson's circle.[49] 'The despotic Whewell' thus served as a foil against which they defined themselves. We observed how his arguments grounded a politically conservative and elitist view of liberal education, calculated to continue in the old universities the classical education begun in the public schools (including classical geometry) and to prepare students for high office in the Establishment of church and state. It must be stressed, however, that he did not oppose modern methods for students in the 'progressive studies', who had already internalized right reasoning and correct moral views.

Others of the older generation modified their analytic procedures in apparently similar, but actually very different, ways. Augustus de Morgan regarded two features of his *Calculus* of 1836 (which he had apparently been teaching for some time at London University) as distinctive: 'In the first place, it has been endeavoured to make the theory of *limits*, or *ultimate ratios* . . . the sole foundation

[49] Harvey Becher, 'William Whewell and Cambridge mathematics', *Hist. Stud. Phys. Sci.*, **11** (1980), 1–48, esp. pp. 27–9, 39–41.

of the science, without any aid whatsoever from the theory of series, or algebraical expansions . . . Secondly, I have introduced applications to mechanics as well as geometry'. While this connection of limits with mechanics and geometry agrees with some of Whewell's concerns, De Morgan's goal pushed far beyond the intuitive and visualizable. He sought to retain the concreteness of geometry and mechanics, while avoiding their limitations. He focussed attention, therefore, on the *process of reasoning* involved in forming the concept of the limit, independent of the particular medium of that reasoning. For students brought up on geometry, he argued, difficulties arise 'from the student depending somewhat too much on ocular demonstration, and not entirely on reasoning, in his preceding course, and can only be overcome by close attention to the reasoning'.[50] On the other hand, one could not dispense with concrete interpretation of the reasoning, the precondition for creative thinking in his view.

De Morgan's *Calculus* was not an abstract calculus but a general one. It generalized processes of reasoning that were exemplified in geometry and mechanics. If we recall his liberal political and religious position, and the fact that he wrote his *Calculus* for the Diffusion Society and London University, then the larger significance of his approach versus Whewell's becomes apparent. A staunch opponent of Establishment elites, De Morgan devoted his life to realizing his belief that the highest attainments of mathematics should be available to the widest possible audience, and that many minds from the not-so-high strata of society could put such learning to good use.[51] His *Calculus* would provide, first, a course of instruction appropriate to a general education, which, like traditional geometry, would train all students in right reasoning; and, second, it would carry talented but unsophisticated students from elementary analysis to the forefronts of research by the shortest path.

De Morgan's aims were simultaneously democratic and professional. Again, they mirror closely those of Dr Thomson, who would revise his own *Calculus* in 1848 to incorporate the limit approach. For beginners Dr Thomson wanted 'investigations as simple and easy as possible', but he wished also to prepare ambitious students for 'the most interesting and valuable applications of the Differential and Integral Calculus in the modern works on Mechanics, Physical Astronomy, and other branches of Natural Philosophy'. By 1848 he could cite three major French authors, Cournot, Duhamel, and Moigno, and six British ones, including De Morgan, among the 'latest and best writers . . . who have made the Method of Limits, or, what is virtually equivalent, the Infinitesimal Method, the basis of their treatises'. His opinion owed much to advice from

[50] De Morgan, *Calculus*, pp. 3, 19. For 'reasoning' as opposed to 'symbol drumming' see De Morgan to William Whewell, 20th January, 1861, in Sophie de Morgan, *Memoir*, pp. 304–5. For De Morgan's conceptual emphasis, its relation to his progressiveness, and the contrast with Cauchy, as well as Whewell, see J.L. Richards, 'Augustus de Morgan, the history of mathematics and the foundations of algebra', *Isis*, **78** (1987), 7–30 esp. pp. 18–24.

[51] Sophie de Morgan, *Memoir*, pp. 79, 222–8, 278.

William, who conveyed Gregory's praise of Moigno and De Morgan.[52] In the matter of foundations for the calculus, then, the most progressive elements of the older generation merged with the younger.

The same is true in the domain of symbolic algebra, where Peacock and De Morgan might be said to have founded a new school of British mathematics. At the very least they broke new mathematical ground and set the stage for their juniors. Recent studies by Joan Richards of the views on algebra held by Peacock, Whewell, Herschel, and De Morgan show that all of them, although they developed algebra as a formal logic of operations, or rules of combination of symbols, nevertheless required that the operations be grounded in particular subject matter, in interpretation, rather than in internal consistency of the rules of combination alone. Their algebra was not abstract algebra in the modern sense and never could be, which is a point of considerable importance to the British context. Two approaches need to be distinguished.[53]

George Peacock published works seminal to the movement in 1830 and 1833, while a tutor at Trinity College. He based symbolic algebra on a 'principle of equivalent forms', which claimed that the relations discoverable in the generalized algebra of arithmetic quantity, or universal arithmetic, would continue to hold as 'equivalent forms' in symbolic algebra, independent of reference. The equivalent forms, however, could not be regarded as generalizations by induction from arithmetic, for they encompassed such entities as negative and imaginary numbers, which had no interpretation in arithmetic; they were not arithmetic quantities. For example, the form $a(b-c)=ab-ac$ would require in arithmetic, but not in symbolic algebra, the restriction $b>c$. Peacock did not think that symbolic forms were necessarily interpretable at all. Their validity rested simply on their correspondence with the *forms* of universal arithmetic, or on the principle of equivalent forms.[54]

Universal arithmetic stood in relation to symbolic algebra, for Peacock, as a 'science of suggestion' of forms. William Whewell took that insight over into his epistemology to buttress the doctrine of 'fundamental ideas'. These innate

[52] Dr James Thomson, *An introduction to the differential and integral calculus*, 2nd edn. (London, 1848), pp. iii, 308. William wrote to Dr Thomson, 15th January, 1842, T192, ULC: De Morgan's book 'is very queer but contains a great many good ideas'; and 20th June, 1847, T366, ULC: 'I still think you should most decidedly take the method of limits absolutely and entirely as the foundation'. See Thomson's retrospective appreciation of De Morgan's 'great book' in his 'Presidential Address', *BAAS Report*, **41** (1871), lxxxiv–cv; PL, **2**, 132–205, on p. 140.

[53] J.L. Richards, 'The art and science of British algebra: a study in the perception of mathematical truth', *Hist. Math.* **7** (1980), 343–65; 'De Morgan', pp. 7–30. We follow her presentation here. See also Helena Pycior, 'George Peacock and the British origins of symbolical algebra', *Hist. Math.*, **8** (1981), 23–45; 'Early criticisms of the symbolical approach to algebra', *Hist. Math.* **9** (1982), 413–40; and the detailed study by Elaine Koppelman, 'The calculus of operations and the rise of abstract algebra', *Arch. Hist. Exact Sci.*, **8** (1971), 155–242. As Richards observes, Koppelman's excellent analysis is marred by its treatment of the British works as steps toward abstract algebra in an unacceptably modern sense (see, for example, pp. 157, 188, 215–16).

[54] George Peacock, *A treatise on algebra* (Cambridge, 1830); 'Report on the recent progress and present state of certain branches of analysis, *BAAS Report*, **3** (1833), 185–352, esp. pp. 194–9.

ideas served as ordering concepts prerequisite to any induction: 'In each inductive process, there is some general idea introduced, which is given, not by the phenomena, but by the mind'. Phenomena called attention to the pre-existing ideas. In inductive sciences, like mechanics, one had to add truly empirical discoveries about motion to obtain complete laws of phenomena (ch. 11), but mathematics required only reasoning upon fundamental ideas. It did not involve discoveries of essentially new concepts. The ideas of number and space thus yielded arithmetic and geometry as *a priori* sciences, although not abstract ones. Since the forms of arithmetic and geometry expressed relations of space and of number, they were necessarily interpretable in this definite subject matter. Whewell made the step to symbolic algebra through Peacock's principle of equivalent forms, which he regarded as another fundamental idea. But Whewell, unlike Peacock, read the principle as an *a priori* guarantee of interpretability: '*The absolute universality of the interpretation of symbols* is the fundamental principle of their use. This has been shown very ably by Dr Peacock in his *Algebra*'.[55]

To Whewell's and Peacock's intuitive approach, De Morgan and Herschel opposed more empiricist views. Although De Morgan believed that one could choose to pursue algebra as a purely abstract system of internally consistent propositions, without external reference, he denied that such an exercise had any significance or utility: '[it] can be of no use to any one in the business of life'. He distinguished this vacuous 'art' of manipulating symbols according to rules from the rich 'science' of algebra, a discovering science which started by expressing symbolically relations pertaining to a concrete subject-matter, and which progressed by seeking ever wider interpretations of the symbolic relations, thus giving rise to ever more general subject-matter. He conceived the science of algebra as a search for a universal logic of meaning, 'the science which investigates the method of giving meaning to the primary symbols, and of interpreting all subsequent symbolic results'. Richards has stressed the essentially historical, contingent, and progressive nature of De Morgan's interpretive enterprise, in contrast to Whewell's.[56] Deduction from fundamental ideas would not serve any better than induction from arithmetic. One had to proceed empirically to discover more general meanings for familiar symbols so as to encompass problematic entities: e.g., interpret $\surd a$ so as to include not only $\surd 4$ but also $\surd(-1)$.

De Morgan's friend Herschel went further. He admitted no propositions '*other than as truths inductively collected from observation*, even in geometry itself'. In

[55] Richards, 'De Morgan', pp. 13–15; 'British algebra', pp. 351–3. Peacock, *Report*, p. 199. William Whewell, *The mechanical Euclid* (Cambridge, 1837), p. 173; *The Philosophy of the inductive sciences* (2 vols., London, 1840), **1**, pp. 139–44 ('Of the foundations of the higher mathematics'), on p. 149.

[56] Richards, 'British algebra', pp. 354–5; 'De Morgan', pp. 20–3. Augustus de Morgan, 'Preface', *Elements of algebra* (London, 1837), pp. 1–6; 'On the foundations of algebra', *Trans. Cam. Phil. Soc.*, **7** (1842), 173–87, on p. 173.

his sharp critique in 1841 of Whewell's *Philosophy of the inductive sciences* he put the matter in its strongest form, denying that internal consistency in a set of axiomatic propositions could be tested at all independent of reference:

> The test of truth by its application to particulars being laid aside, nothing remains but its self-consistency to guide us in its recognition. But this in axiomatic propositions amounts to no test at all. *It is the essence of such propositions to stand aloof and insulated from each other* . . . Axioms, rigorously such, can admit of no meaning in common. *Their mutual compatibility, as fundamental elements of the same body of truth, can only be shown by experience* – by the observed fact of their co-existence as *literal truths* in a particular case produced.[57]

Herschel's view that '*literal truth*' provided the only guarantee of the truth of mathematical structures corresponds quite closely to the view William Thomson would soon adopt for the mathematical structures of physics. For *pure* mathematics, Thomson would not quibble with Whewell's insistence that 'the results of systematic symbolical reasoning must always express general truths, by their nature, and do not, for their justification, require each of the steps of the process to represent some definite operation upon quantity'.[58] But with De Morgan, he regarded these truths as empty, useless. The most famous utterance of Thomson's entire career, from the Baltimore lectures of 1884, had its origin in such considerations: 'I never satisfy myself until I can make a mechanical model of a [mathematical] thing. If I can make a mechanical model I can understand it. As long as I cannot make a mechanical model all the way through I cannot understand'.[59] Thomson's views on interpretation matured through his association with the *Cambridge Mathematical Journal*, and its extension of symbolic algebra to the differential and integral calculus, as a calculus of operations.

Writing to Herschel in 1845 with advice for a forthcoming Presidential Address to the British Association meeting at Cambridge, De Morgan said: 'You should not forget the Cambridge "Mathematical Journal". It is done by the younger men . . . It is full of very original contributions. It is, as is natural in the doings of young mathematicians, very full of symbols. The late Dr [sic] F. Gregory, whom you must notice most honourably . . . gave his extensions of the Calculus of Operations . . . in it. He was the first editor. He was the most rising man among the juniors'.[60] Herschel did include this appreciation of Gregory

[57] [J.F.W. Herschel], 'Whewell on inductive sciences', *Quart. Rev.*, **68** (1841), 177–238; *Essays*, pp. 142–256, on pp. 216, 222. [58] Whewell, *Philosophy*, **1**, p. 143.

[59] William Thomson, *Notes of lectures on molecular dynamics and the wave theory of light* (papyrographed stenographic notes by A.S. Hathaway) (Baltimore, 1884), pp. 270–1. cf. 'Report of the Commissioners appointed to inquire into the state, discipline, studies and revenues of the University and Colleges of Cambridge', *Parliamentary papers, 1852–53*, **14**, 113: 'the candidate for mathematical distinction who uses symbolic processes should be called on to interpret his equation at each point'; quoted in J.L. Richards, 'Projective geometry and mathematical progress in mid-Victorian Britain', *Stud. Hist. Phil. Sci.*, **17** (1986), 297–325, on p. 307.

[60] Augustus de Morgan to J.F.W. Herschel, 28th May, 1845, in Sophie de Morgan, *Memoir*, pp. 150–1.

and the *Journal*, using them as examples of how the University produced 'not merely expert algebraists, but sound and original *thinkers*'. His remarks are noteworthy because he recognized in symbolic mathematics a significance for physical theory that William Thomson had already begun to exploit. The search for ever-wider interpretations of familiar symbols, Herschel noted, led to new knowledge of nature. Fourier's remarks on how mathematical analysis discovers 'hidden analogies' come naturally to mind. Herschel anticipated that 'our notions of Light, Heat, Electricity, and other agents of this class' would soon include the propagation of forces, 'the transference of physical causation from point to point in space – nay, even the generation or development of attractive, repulsive, or directive forces at their points of arrival'.[61] Whether or not Herschel had read Thomson's several papers in the *Journal* discussing the analogy between Fourier's heat conduction and electrical action at a distance, or even had prior knowledge of Thomson's classic paper on the subject for this very meeting, he envisaged a new potential in the Cambridge context of symbolic algebra and the calculus of operations.

That context was now sharply focussed on the small minority of members of the University who supported the twin goals of diffusion and professionalization. They were the progressive young whigs who founded and edited the *Journal*: Gregory – 'a considerable Whig (both in mathematics and politics)'; Archibald Smith – 'my dearest Radical Archy'; Robert Leslie Ellis – 'he was not a very earnest politician, but always professed himself a Whig, a profession which was probably strengthened by his intimacy with Sir William Napier, to whom he always expressed himself as much attached'; and Thomson – son of the 'pestilent Whig' from Glasgow College.[62] Ellis's intimacy with Sir William Napier meant identification with a popular and radical whig controversialist. Being whigs, of course, did not distinguish this generation of reformers from the previous one; but representing Scottish values, and/or commercial and industrial interests, did. We have already observed these identifying characteristics for Gregory and Smith. Ellis, an intimate friend of Gregory, was the only son of a wealthy Bath merchant. His emphasis on the practical utility of knowledge is apparent in his analysis of Bacon's philosophy.[63] Of the previous generation, only Babbage took a direct personal interest in commerce and

[61] J.F.W. Herschel, 'Presidential address', *BAAS Report*, **15** (1845), xxvii–xliv; *Essays*, pp. 634–82, on pp. 638–9. The use of 'thinkers' here corresponds to De Morgan's 'reasoning', above, and contrasts with the 'abstract' of abstract algebra.

[62] See Chapter 3. William to Dr Thomson, February or March, 1842, T201, ULC. Joanna Smith to Archibald Smith, 8th January, 1835, TDl/715, Smith papers, Strathclyde regional archives. Harvey Goodwin, 'Biographical memoir', in William Walton (ed.), *The mathematical and other writings of Robert Leslie Ellis* (Cambridge, 1863), pp. 417–427, xvii. Goodwin was second wrangler in 1840 when Ellis was first. Walton, eighth wrangler in 1836, was another important member of their circle.

[63] R.L. Ellis, 'General preface to Bacon's philosophical works', in J. Spedding, R.L. Ellis, and D.D. Heath (eds.), *The works of Francis Bacon* (15 vols., Boston, 1861–), **1**, pp. 61–127, for example on p. 92: 'Bacon connected the doctrine of Forms with practical operations, because this doctrine . . . had altogether a practical significance'.

manufacturing. The otherwise quite radical De Morgan exhibited that disdain for tradesmen which typified the gentrified middle classes. Having resigned his position at London University in 1831 to protest at the dismissal of a colleague, he wished to be assured, before reassuming his chair in 1836, that new regulations would guarantee the 'respectability which a gentleman (meaning only by education and sentiments . . .) requires'. Of the liberal businessmen who had financed the institution and formerly made up its Board of Proprietors, he said: 'a body of commercial Englishmen got together upon a point of trade (and with these gentlemen . . . the honour and character of a Professor was avowedly, and almost *ipsissimis verbis*, made a question of trade) knows neither right from wrong, nor reason from anything else'.[64] Thus even De Morgan was inclined to affect the high moral tone and gentlemanly airs that characterized the London scientific societies, or clubs, such as the Astronomical Society, where he, Herschel, and Airy regularly communed.

The 'juniors' pursued their mathematics much less in the spirit of these 'Gentlemen of Science' than as professional researchers training themselves to advance the frontiers of their discipline, often with a view to the utility of their study and to the possibility of making it a livelihood. Archibald Smith did exhibit a marked tendency to gentrification, but even he never eradicated his commercial and Scottish roots. As Thomson presented Smith's mathematical style in an obituary notice, it derived from his father James Smith, FRS, 'who had literary and scientific tastes with a strongly practical turn, fostered no doubt by his education in the University of Glasgow and his family connexion with some of the chief founders of the great commercial community which has grown up by its side'. It was 'much to be regretted', Thomson remarked, that Archibald had been unable to secure a regular position in mathematical and physical science, thus making it 'the *professional* work of his life', rather than law, to which he turned to support his family.[65] No doubt these observations say as much about Thomson as Smith, but they reflect also the practical and professionalizing interests of the group.

It was Smith who in 1836, after taking his degree as senior wrangler and first Smith's prizeman, suggested to Gregory that they establish 'an English periodical for the publication of short papers on mathematical subjects'. Gregory happily agreed to serve as editor once he had completed the chore of his Senate House examination in 1837: 'But all this must be done after the degree; for "business before pleasure," as Richard said when he went to kill the king before he murdered the babes'. The 'babes' in this case were to be the creative productions of budding Cambridge mathematicians, and those creations not killed would be published. Papers were to be short because they would present

[64] Augustus de Morgan to Sir Harris Nicolas (his solicitor), 10th October, 1836, in Sophie de Morgan, *Memoir*, pp. 70–3.

[65] William Thomson, 'Archibald Smith, and the magnetism of ships', *Proc. Royal Soc.*, **22** (1874), i–xxiv; MPP, **6**, 306–34, on pp. 306, 309. Our emphasis.

only the latest results, and not extended treatises on established subjects. As Smith assessed the early years of the *Journal* in 1845, when Thomson proposed further professionalization and expansion, 'The C.M.J. has hitherto been useful and successful and [,] in the particular of stimulating men reading for and who have recently taken their degrees to put into shape and presence any thing pretty or ingenious which they hit on [,] it is probably more useful than a journal of a more general character would be'.[66]

In short, the *Journal* had been established to convert students engaged in Whewell's 'progressive studies' into publishing mathematicians. Whewell's complaint to Herschel in 1845 of 'the most active students being encouraged to study rather the last improvements, contained in memoirs, journals, and pamphlets, than the standard works of mathematical literature' surely referred in part to the *Journal*.[67] Gregory had announced its 'primary object' in the first issue as supplying 'a means of publication for original papers' and secondarily as 'publishing abstracts of important and interesting papers that have appeared in the Memoirs of foreign Academies, and in works not easily accessible to the generality of students. We hope in this way to keep our readers . . . on a level with the progressive state of Mathematical science'. Thomson judged in 1874 that the new publication had 'inaugurated a most fruitful revival of mathematics in England, of which Herschel, Peacock, Babbage, and Green had been the prophets and precursors'.[68]

Our claim, then, is that the northern whigs and their commercially oriented peers did much to transform the pursuits of their gentlemanly 'precursors' into a regular profession of mathematics. The most striking symbol of that transformation was their encouragement of publication by undergraduates, a practice strongly discouraged by their seniors (ch. 3). Ellis and Arthur Cayley both began publishing in the *Journal* prior to their degrees, and Thomson followed their example. Many youthful contributors published under pseudonymous initials, again like Thomson, who signed himself 'P.Q.R.'. Smith employed no less than nine sets of initials in the first volume. Partly this practice simply made publication less threatening to insecure authors like Smith, who regarded many of his minor articles as 'the mere sweepings of my undergraduate M.S.S. to which I was ashamed even to put my initials'.[69] More importantly, it protected gentlemanly propriety and thus future careers at the bar and in the pulpit.

In 1845, following Gregory's death in 1844 and a year of stand-in editing by

[66] *Ibid.*, p. 309. Archibald Smith to William Thomson, 16th July, 1845, S145, ULC.

[67] William Whewell to J.F.W. Herschel, 20th August, 1845, Herschel papers, HS18.172, Royal Society (quoted in Becher, 'Whewell', p. 33). Becher also notes Airy's complete agreement with Whewell. De Morgan, however, in his testimonial, strongly supported Thomson's having been 'employed in original research during his undergraduateship'. See 'Printed copy of William Thomson's testimonials for the Glasgow chair of natural philosophy', PA34, ULC, p. 18.

[68] D.F. Gregory, 'Preface', *Cam. Math. J.*, **1** (1837), 1–2, on p. 1; MPP, **6**, 309.

[69] Archibald Smith to William Thomson, 5th August and 15th December, 1845, S146 and S151, ULC.

Ellis, Thomson was to take over the *Journal* with plans for expanding it into a truly 'national' publication. His first two acts expressed increasing professionalization. First, he discontinued publishing anonymous articles and supplied an index to earlier volumes that included the name of authors he could identify. Second, he adopted the title *Cambridge and Dublin Mathematical Journal* to win the support of Irish mathematicians (including those outside Trinity College, Dublin), thereby forging between two major scientific centres a kind of 'liberal union' in political terms, which would transcend cultural and religious differences.[70] The colourful and brilliant Jewish mathematician, J.J. Sylvester, who suffered considerably from religious discrimination, wrote to 'hail . . . the auspicious conjunction of Cambridge and Dublin Mathematics', which he judged 'likely to tend in a most material degree to introduce a Catholic [i.e. universal] Spirit among your readers'. Thomson responded with his own progressive vision, welcoming Sylvester's offer to contribute a series on his new 'Theory of combinatorial aggregation' and emphasizing that in his view 'the principal object of any scientific journal *should* be the publication of original investigations and discoveries'. The combined ideal of national unity and professional research naturally involved international competition. As Richard Townsend, a young disciple of MacCullagh and one of the new Irish contributors, remarked 'the time has come when we of the British Islands must endeavor to hold our own, and not put forward any thing which could be seized upon to our disadvantage, when there are those abroad so ready and so willing to take advantage of any slip we may chance to make'.[71]

The wider national and professional goals that Thomson pursued were represented in the early years by such contributors as George Boole of Lincoln (1815–64), since famed for Boolean logic, but more appropriately for his attempt to ground *The laws of thought* in a mathematical (thus non-metaphysical) notion of unity and duality. The son of a small tradesman closely associated with the diffusion movement of the Mechanics Institutes, Boole was largely self-educated in mathematics and pursued his highly creative researches while teaching in a small school that he founded in Lincoln. Radically anti-sectarian in religion and a self-help socialist in politics, he epitomized the goals of Christian reform, rational progress, and universal truth. In 1849, with the help of

[70] Archibald Smith to William Thomson, 16th July, 1845, S145, ULC, and 5th August, 1845, S146, ULC. 'General index to the first series', *Cam. and Dublin Math. J.*, **1** (1846), i–viii. R.L. Ellis to William Thomson, 17th July, 1845, E57, ULC, and 24th July, 1845, E59, ULC. William Thomson to and from George Boole, 17th July–6th August, 1845, B10–12, ULG and B143–6, ULC. Thomson chose originally simply 'The mathematical journal', hoping 'that the Dublin men will join and not allow any [national, *del.*] provincial feelings to keep them from cooperating with a thing which would be much more useful when quite general than it could be if restricted in any manner' (B12, 27th July). The latitudinarian sense of 'useful' and 'general' here mirrors the mathematical sense in the text below.

[71] J.J. Sylvester to William Thomson, 18th November, 1845, S594, ULC, and Thomson to Sylvester, 19th November, 1845, S595, ULC; Richard Townsend to William Thomson, 30th January, 1847, T584, ULC.

his many mathematical friends, including Thomson, he would obtain a professorship at the newly founded Queen's College in Cork, allowing him for the first time to make his living as a mathematician.[72]

Boole discussed 'the general question of the use of symbolical language in mathematics' in 1847 in the first presentation of his logic, which was in part a defence of De Morgan in a bitter priority attack launched by Sir William Hamilton. The metaphysically inclined Hamilton had also attacked the study of mathematics (symbolic reasoning) by comparison with traditional Aristotelian logic in a liberal education. Boole considered the issue from two perspectives: 'First . . . with reference to the progress of scientific discovery, and secondly, with reference to its bearing upon the discipline of the intellect'. For progressiveness, symbolism had all the advantages, which no one contested. For training the moral mind, Hamilton, like Whewell, considered it dangerous, as producing an uncritical acceptance of whatever premises were proposed and a mechanical, unthinking view of demonstration (compare Herschel on Gregory, above). In response Boole argued that all logical reasoning could be represented symbolically and that Hamilton's favoured Aristotelian logic merely constituted an inferior symbolism. To Hamilton's claim that philosophy, as logical metaphysics, discovered real existences and true causes, while mathematics merely described and measured, Boole retorted that 'according to [my] view of the nature of philosophy [metaphysics], *Logic forms no part of it* . . . we ought no longer to associate Logic and Metaphysics, but Logic and Mathematics'. De Morgan agreed: 'I would not dissuade a student from metaphysical inquiry . . . but I would warn him, when he tries to look down his own throat with a candle in his hand, to take care that he does not set his head on fire'.[73]

'Progressive' and 'anti-metaphysical' nearly define the 'science' of Thomson and his friends, for whom opposition to metaphysics encompassed latitudinarian religion, politics above party (ostensibly), and non-hypothetical, practical knowledge.[74] Science in this sense entailed concomitant emphases in the mathematical style of the new professionals. They sought generality of

[72] Ivor Gratten-Guinness, 'Psychology in the foundations of logic and mathematics: the cases of Boole, Cantor and Brouwer', *Hist. Phil. Logic*, **3** (1982), 33–53; John Richards, 'Boole and Mill: differing perspectives on logical psychologism', *Hist. Phil. Logic*, **1** (1980), 19–36; L.M. Laita, 'Boolean logic and its extra-logical sources: the testimony of Mary Everest Boole', *Hist. Phil. Logic*, **1** (1980), 37–60. On diffusion and mechanics institutes see Steven Shapin and Barry Barnes, 'Head and hand: rhetorical resources in British pedagogical writing, 1770–1850', *Ox. Rev. Ed.*, **2** (1976), 231–54; M.D. Stephens and G.W. Roderick, 'Science, the working class and mechanics' institutes', *Ann. Sci.*, **29** (1972), 353–9. On Boole's appointment, see George Boole to William Thomson, 17th August, 1846, B151, ULC, and 26th August, 1846, B152, ULC, as well as succeeding letters through to December, 1847, B153–B166, ULC, and B15–21, ULG.

[73] George Boole, *The mathematical analysis of logic: being an essay towards a calculus of deductive reasoning* (Cambridge, 1847), pp. 9, 13. De Morgan presented his view of the dispute in Augustus de Morgan, *Formal logic, or, the calculus of inference, necessary and probable* (London, 1847), pp. 297–323; remark on metaphysics on p. 27*n*. See also L.M. Laita, 'Influences on Boole's logic: the controversy between William Hamilton and Augustus de Morgan', *Ann. Sci.*, **36** (1979), 45–65.

[74] See especially William to Dr Thomson, February or March, 1842, T201, ULC, quoted in Chapter 4, note 49, for William's description of Gregory. Ellis's view of *a priori* metaphysics appears

expression, simplicity of technique, and utility of results, or *power* and *efficiency*. We shall examine these features in two of their preoccupations: symmetrical method and the calculus of operations.

To employ symmetrical method meant to express the equations of analytic geometry in such a way that figures would be described as forms in space, independent of particular axes of reference. For example, rather than expressing the equation of a straight line as $y = kx + b$, one would choose $mx + ny = c$, so that the coordinates appear symmetrically and the equation has the same form for all orientations of the axes. Archibald Smith applied such symmetrical method in his first published paper, read to the Cambridge Philosophical Society in 1835, thereby greatly simplifying a derivation that Ampère had given of Fresnel's wave surface in the undulatory theory of light. Smith wrote Fresnel's equations in symmetrical form and extracted the wave surface in two pages of simple analysis. Ampère's derivation had occupied thirty-two pages, presenting, in Thomson's phrase, 'so repulsive an aspect that few mathematicians would be pleased to face the task of going through it', while Smith's required but 'a few short lines of beautifully symmetrical algebraic geometry'. Those few short lines set a new standard: 'It was one of the first applications in England, and it remains to this day a model example, of the symmetrical method of treating analytical geometry, which soon after (chiefly through the influence of the *Cambridge Mathematical Journal*) grew up in Cambridge, and prevailed over the unsymmetrical and frequently cumbrous methods previously in use.'[75] Smith and others applied the method to simplify a wide variety of problems: relations of straight lines and planes (S.S. Greatheed and Boole), inflection points (William Walton), parabolas (unknown), lines of curvature on an ellipsoid (Ellis), transformations of coordinates (Thomson), and more sophisticated problems. Boole expressed their common attitude: 'the most general method of resolving the problem is [often] also the most simple'.[76]

Thomson's publishing initials 'P.Q.R.' actually symbolized symmetrical

in his interpretation of Bacon, *op. cit.* (note 63), pp. 95–102. His associated rejection of Bacon's doctrine of forms, (pp. 78–93), challenged Whewell's doctrine of fundamental ideas in inductive science, for Ellis insisted that 'the progress of science continually requires the formation of new conceptions' (p. 86), not merely the unfolding of innate ones. Richards, 'Projective geometry' has shown that Cayley, Sylvester, and others of the younger generation applied this progressive, inductive view even to mathematical conceptions, as Ellis apparently also wished to do here (pp. 98–9).

[75] Archibald Smith, 'Investigation of the equation of Fresnel's wave surface', *Trans. Cam. Phil. Soc.*, **6** (1838), 85–9; 'Notes on the undulatory theory of light', *Cam. Math. J.*, **1** (1839), 3–9; MPP, **6**, 308.

[76] These papers appear in the *Cambridge Mathematical Journal* as: S.S. Greatheed, 'Application of the symmetrical equations of a straight line [and plane] to various problems in analytical geometry of three dimensions', **1** (1839), 37–42; 135–43; George Boole, 'Symmetrical solutions of problems respecting the straight line and plane', **2** (1841), 179–88, on p. 187; William Walton, 'Symmetrical investigation of points of inflection', **4** (1845), 13–17; G. (?), 'On a symmetrical form of the equation of the parabola', **2** (1841), 14–17; R.L. Ellis, 'On the lines of curvature of an ellipsoid', **2** (1841), 133–8; William Thomson, 'On the relations between the direction cosines of a line referred to two systems of rectangular coordinates', **3** (1843), 247–8; and 'On the equations of motion of heat referred to curvilinear co-ordinates', **4** (1845), 33–42.

method for they denoted three symmetric coefficients in standard expressions of analytic geometry, for example, the differential coefficients of a function $f(x,y,z)=0$, representing a figure in three dimensions,

$$df(x,y,z) = Pdx + Qdy + Rdz = 0.$$

In his Cambridge research notebook he regularly employed this notation and simultaneously emphasized symmetric expression. One example comes from his review of analytic geometry, including Boole's papers, for the Senate House examinations. Concerned with symmetric transformations, he expressed himself much as Boole had on generality and simplicity, referring to one solution as worked out in 'the simplest manner so as to give the transformation in the most general case'.[77]

Symmetrical method provides the most obvious illustration of the younger generation's programme for obtaining mathematical power and efficiency. Through their emphasis on *generality* they would release the power of symbolic representation from the cumbrous constrictions of particular interpretations. By demanding *simplicity* along with generality they would make the essential features of derivations stand out, revealing the forms common to problems normally considered different. Finally, simplicity made generality useful in solving particular problems, so that *utility* ever attended the other two emphases, as in Smith's remark, 'It is also of great use . . . if all the expressions be put in symmetrical form . . . their symmetry greatly facilitates the practical application of the method'.[78]

These same features characterize the calculus of operations, as developed from French beginnings by several early writers in the *Journal*. The new calculus epitomized the vision of a British engine of mathematical progress, as Thomson reflected in 1871: 'With the French writers . . . this was rather a short method of writing formulae than the *analytical engine* which it became in the hands of . . . Sylvester and Gregory . . . and Boole and Cayley. This method was greatly advanced by Gregory, who first gave to its *working-power* a secure and philosophical foundation and so prepared the way for the marvellous extension it has received from Boole, Sylvester, and Cayley'.[79] By symbolizing algebraically the operations of differentiation and integration they wished to extend to the differential and integral calculus the ideas that Peacock and De Morgan had developed for symbolic algebra. The status of these ideas in the Cambridge context is well illustrated in a letter from William to Dr Thomson after only

77 William Thomson, 20th June to 3rd August, 1844, 'Journal and research notebook, 1843–45', NB33, ULC, pp. 41–51, on p. 44, referring to his own 'Elementary demonstration of Dupin's theorem', *Cam. Math. J.*, **4** (1845), 62–4; MPP, **1**, 36–8, and to George Boole, 'On the transformation of multiple integrals', *Cam. Math. J.*, **4** (1845), 20–8, and 'Symmetrical solutions', pp. 179–88. For the meaning of 'P.Q.R.' compare Thomson's usage in his notebook (pp. 42, 84–90) with, for example, that of M.N.N. (?) 'On certain cases of consecutive surfaces', *Cam. Math. J.*, **1** (1839), 187–92.

78 Archibald Smith, 'Elimination by means of cross multiplication', *Cam. Math. J.*, **1** (1839), 46.

79 William Thomson, 'Presidential Address', British Association, PL, **2**, 138–9. Our emphasis.

three months in residence, when Gregory had already become his deeply admired model and his letters were filled with 'Gregory says', 'Gregory told me', and 'Gregory thinks':

On Thursday I got an examination paper in algebra from Hopkins. Almost every question in it was on the principles of algebra, and seemed to inculcate most of Peacock's views, with which I am beginning to agree. Mr Cookson had lent me Peacock, and so I was able to answer most of the questions, I hope in the way he wished. I have been calling two or three times on Gregory, and he seems to take quite the same view of the principles of algebra. He has been doing a great deal in finding the values of definite integrals, in a very curious way, by the separation of symbols. He has given one specimen of the method, which I believe is his own discovery in the Math. Journal, and others in his 'Examples'. He is not however quite clear about the principles of it yet.[80]

A dominant approach had clearly developed, and Thomson's immersion – via coach, tutor, and adviser – was total.

The 'separation of symbols' refers to the idea of separating the symbols of operation from those of quantity. Gregory rewrote such linear differential equations as

$$\mathrm{d}^2y/\mathrm{d}x^2 + \mathrm{d}y/\mathrm{d}x + ay = g(x)$$

in the form

$$(\mathrm{d}^2/\mathrm{d}x^2 + \mathrm{d}/\mathrm{d}x + a)y = g(x).$$

Considering the bracketed expression as a single operation on y, symbolized by $f(\mathrm{d}/\mathrm{d}x)$ and treated as a linear function of the elemental operation of differentiation $\mathrm{d}/\mathrm{d}x$, he wrote

$$f(\mathrm{d}/\mathrm{d}x)y = g(x)$$

and the inverse equation,

$$y = f^{-1}g(x).$$

The latter expression would in principle solve the differential equation for y, or integrate it, if f^{-1} represented a meaningful inverse operation in the calculus just as it represented an inverse quantity in ordinary algebra; that is, if f^{-1} could actually be taken to represent a linear function of the operation of integration in the same way as f represented a linear function of differentiation.[81]

Such symbolic representations had earlier been employed by Lagrange, Fourier, and Herschel, among others, but only for obtaining non-rigorous

[80] William to Dr Thomson, 27th December, 1841, T188, ULC. William to Dr Thomson, 3rd March, 1842, T199, ULC, states that he has been using Gregory's *Examples*: 'I like it exceedingly'. The problem of handling definite integrals was regarded by many as the single most important issue of contemporary mathematical physics, as in heat conduction and the wave theory of light, where discontinuities at boundaries figured so prominently. Although most of the major authors in the Journal contributed to the subject, including Thomson, we omit it here. See R.L. Ellis, 'Memoir of the late D.F. Gregory', *Cam. Math. J.*, **4** (1845), 145–52, esp. p. 150; *Writings*, pp. 193–201, esp. pp. 198–9; also Richard Townsend to William Thomson, 5th April, 1847, T586, ULC: 'I am convinced that some powerful method of managing definite integrals is what alone will enable us to make any advance in modern physics'.

[81] D.F. Gregory, 'On the solution of linear differential equations with constant coefficients', *Cam. Math. J.*, **1** (1839), 22–32.

results by analogy to algebra, and not for demonstration, because the status of the analogies remained uncertain. One typically supposed that symbols of operation differed in some essential way from ordinary algebraic symbols of quantity. Gregory too remained unclear about the principles of his method, as Thomson reported, but he made the bold assertion that the difference between the calculus and algebra was not a difference of operations versus quantities, for symbols of quantity like a and x could with equal propriety be regarded as operations on unity, $a(1)$ and $x(1)$. Similarly $a(x)$ would be the operation a performed on x and $a^n(x)$ the operation a performed n times in succession on x. From this perspective, the theorems of algebra were true only by virtue of the rules of combination of the *operations* being true. Therefore, Gregory argued, 'whatever is proved of the latter [algebraic] symbols, from their known laws of combination, must be equally true of all other symbols which are subject to the same laws of combination'. The rules of algebraic operation,

$$a^m a^n(x) = a^{m+n}(x) \qquad \text{(index)}$$
$$a[b(x)] = b[a(x)] \qquad \text{(commutative)}$$
$$a(x) + a(y) = a(x+y) \qquad \text{(distributive)},$$

were the same as those of differentiation,

$$\mathrm{d}^m \mathrm{d}^n(x) = \mathrm{d}^{m+n}(x)$$
$$\mathrm{d}/\mathrm{d}x[\mathrm{d}/\mathrm{d}y(z)] = \mathrm{d}/\mathrm{d}y[\mathrm{d}/\mathrm{d}x(z)]$$
$$\mathrm{d}(x) + \mathrm{d}(y) = \mathrm{d}(x+y).$$

Consequently, all theorems derivable from the first set of rules in algebra would hold also for differentiation. Having once proved the binomial theorem in algebra, for example, no separate proof was wanted for differentiation. Gregory therefore employed the theorem to factor the operating function $f(\mathrm{d}/\mathrm{d}x)$ and thereby to reduce the solution to a series of much simpler inverse operations, or integrations.[82]

Quickly taking up Gregory's algebraic procedure, Boole applied it to decompose the entire inverse operator into a sum of partial fractions, obtaining 'the simplest and most symmetrical form into which the solution of the equation can be brought'. He re-emphasized Gregory's justification for the algebraic analogy, its 'being founded only on the common laws of the combination of the symbols'.[83]

Gregory and Boole employed symbolic algebra to solve equations of finite differences in the same way as they solved differential equations, revealing a 'close analogy' between the two kinds of equations. Boole commented, 'The analogy . . . is very remarkable, and unless we employed a method of solution common to both problems, it would not be easy to see the reason for so close a resemblance in the solution of two different kinds of equations'.[84] His paper

[82] *Ibid.*, pp. 30–1.

[83] George Boole, 'On the integration of linear differential equations with constant coefficients', *Cam. Math. J.*, **2** (1841), 114–19, on p. 119.

[84] D.F. Gregory, 'On the solution of linear equations of finite and mixed differences', *Cam. Math. J.*, **1** (1839), 54–61, on p. 56; Boole, 'Linear differential equations', p. 119.

appeared in May, 1840, when William Thomson had just begun to read Fourier and over a year before he would submit his third paper to the *Journal*, the one displaying the mathematical analogy between heat conduction and electrostatic action (ch. 7). He would argue, in essence, that if a problem in one area of physics has the same mathematical form as a problem in another area, then any theorem that can be established from physical considerations in the one must hold also for the other. The argument relates different areas of physics by mathematical analogy in the same way as the operational calculus relates different areas of mathematics by what might be called meta-mathematical analogy. If he were reading the *Journal*, as seems likely, he would have found considerable precedent for such meta-mathematical analogy, in papers by Greatheed and Ellis as well as by Boole and Gregory.[85] Thomson's scheme differs from their's, however, in that symbolic representation of *physical processes* replaces symbolic representation of *mathematical operations*.

Gregory applied his analogical methods at every opportunity to simplify and extend Fourier's results.[86] In 1838 he responded to the beauty of Fourier and 'the great resources of his analysis' much as Thomson would in 1840, remarking that 'Indeed there is a freshness and originality in the writings of Fourier which make them in no ordinary degree arrest the attention of the reader'. And Ellis remembered him saying, as he paged through Fourier, 'All of these things seem to me to be a kind of mathematical paradise'. Among other exotic fruits, Gregory found that Fourier had 'very frequently' separated symbols of operation from those of quantity in order to express the series solutions of partial differential equations, although only for convenience of expression and not as 'the proper solutions of the equations'.[87] Evidently, then, Fourier offered modes of analysis that were highly suggestive for both the analogies of physical process that Thomson would exploit and those of mathematical operations that Gregory developed. It is worth attempting to specify the nature of the connection between those analogies.

Fourier's analysis of heat conduction had been macroscopic and geometric, focussing on heat flow as seen from the outside, and employing such concepts as flux across a surface and the net quantity of heat entering or leaving a volume element. These were general concepts, independent of the physical interpretation of heat, but which nevertheless described the *forms* of the physical processes. They were also concepts whose expression was essentially mathematical. Flux resulted directly in Fourier's picture from a temperature gradient across a surface, expressed mathematically as the operation of taking the

[85] S.S. Greatheed, 'On general differentiation', *Cam. Math. J*, **1** (1839), 11–21; 109–17; R.L. Ellis, 'On the integration of certain differential equations', *Cam. Math. J.*, **2** (1841), 169–77; 193–201 ('analogy' on p. 195).

[86] D.F. Gregory, 'Solution of linear equations', pp. 59–61; 'Notes on Fourier's heat', and 'On the solution of certain partial differential equations', *Cam. Math. J.*, **1** (1839), 104–7; 123–31.

[87] Gregory, 'Fourier's heat', p. 104; Ellis, 'Memoir of the late D.F. Gregory', p. 151; *Writings*, p. 199; Gregory, 'Partial differential equations', p. 123.

derivative of temperature across the surface. Similarly, the net heat leaving a volume element derived directly from the change in flux between two surfaces, again a derivative. The transition from physical process to mathematical operation was nearly immediate in Fourier's analysis (in contrast to Poisson's). It is this immediacy which seems crucial for correlating the readings of Thomson and Gregory. While Thomson saw the equations as expressing physical processes, Gregory saw them as expressing mathematical operations.

This relation between process and operation relates closely to the Newtonian tradition of geometrical fluxional analysis, with lines generated by moving points, derivatives as velocities, etc. Of all his applications of the calculus of operations, Gregory considered that 'Geometry is the most important'. He wished to understand why it is that we can pursue geometry algebraically. Why, at the simplest level of the relation between plane geometry and arithmetic, can we represent the area of a parallelogram as a product of the symbols for two lines? 'This question', he said, 'has always appeared to me to be one of great difficulty in the application of Algebra to Geometry . . . It is not sufficient to say, as is usually done, that if we divide each of the lines into a certain number of units, the number of superficial units in the parallelogram will be equal to the product of the number of units in the two lines; it is also necessary to show how a superficial unit can be represented by the product of two linear units, and this I think cannot be done except on the principle which has here been used.' His principle interpreted the symbol a as an operation on a point (.), so that $a(.)$ was '*transference* in *one direction*' of the point through the space a. The combination of symbols $ba(.)$ represented transference of the point in one direction by a followed by transference of the resulting line in another direction by b, producing a parallelogram with sides a and b. Since the operation of transference obeyed commutative and distributive laws,

$$ab = ba$$
$$a(b + c) = ab + ac,$$

all the results of arithmetic depending only on these rules of combination were immediately available in geometry.[88]

In this example an operation literally represents a process of transference; the operation generates the figure. The idea of generation was central to the extensions of projective geometry, which Ellis, in his Royal Commission testimony of 1852, labelled the 'new geometry': 'What may be called the new geometry seems to be little studied in the University; yet the method of which it makes so much use, namely, the generation and transformation of figures by ideal motion, is more natural and philosophical than the (so to speak) rigid geometry to which our attention has been confined. It has been well said that the differential calculus is the symbolical expression of the law of continuity'.[89] This

[88] D.F. Gregory, 'On the elementary principles of the application of algebraical symbols to geometry', *Cam. Math. J.*, **2** (1841), 1–9, on pp. 1–5.

[89] R.L. Ellis, 'The course of mathematical studies', *Writings*, 423*n*. The observation that Ellis was referring in particular to projective geometry we owe to Richards, 'Projective geometry', p. 314.

new analytic geometry, in which relations in space were conceived as generated continuously in time by the operations of differentiation, will figure importantly in our discussion of Thomson and Tait's *Treatise* (ch. 11). Gregory, Boole, Cayley, and others devoted considerable effort to it in the *Journal*.[90]

In the same vein, Thomson would represent the process of heat conduction given by Fourier's flux equation, $\vec{F} = -K$ gradient V, as lines of motion of heat directed perpendicularly across surfaces of constant temperature, or isothermal surfaces, the lines being generated by the operation of differentiation. To write the equation in vector form as we have done here, however, is to take the operational calculus one step too far for Thomson, since it suggests that he should have been happy later with the idea of vector operators like gradient, divergence, and curl, when in fact he regarded them as merely the cartesian component expressions 'translated into gibberish'.[91] But the operational calculus did provide the context out of which vector analysis emerged. W.R. Hamilton would publish a considerable portion of his analysis of quaternions, or 'symbolic geometry' (in analogy to symbolic algebra), in the *Journal* in 1846 under Thomson's editorship.[92] We have arrived, then, at a point where Thomson's path separated from that of many of his peers. A look at the problem of interpretation will show how.

Among writers on symbolic algebra Gregory was one who paid particular attention to illustrating the meaning of results by analogies to geometry and arithmetic, but by no means did he believe interpretation always possible. With respect to the analogy above of arithmetical algebra to geometry, he wrote: 'Whatever, therefore, may have been proved in Arithmetic, in dependence solely on these laws [of combination], is equally true in Geometry, provided always that we can interpret the result; for there is no reason why we should always be able to interpret a symbolical result either geometrically or arithmetically. And indeed, in geometry the interpretability is soon presented to us in the combination of more than three symbols of transference [figures of more than three dimensions]'. Similarly, logarithms of negative numbers, involving the symbol $\sqrt{(-1)}$, he regarded as 'impossible logarithms' rather than 'imaginary', for the operations were not imaginary but only '*uninterpretable in arithmetic*'.[93] In principle, therefore, the 'General Theory of the Science of

The hallmark of the subject was the 'principle of continuity', and its relations to analytic geometry were developed especially by Cayley and Sylvester. See also Daston, 'Physicalist tradition'.

90 D.F. Gregory, 'On the existence of branches of curves in several planes', and 'On the theory of maxima and minima of functions of two variables', *Cam. Math. J.*, **1** (1839), 259–66; **2** (1841), 138–40.

91 Annotation signed 'W.T. Aug. 13/88' in Sir William's copy of James Clerk Maxwell, *A treatise on electricity and magnetism* (2 vols., Oxford, 1873), **2**, p. 222, in the Natural Philosophy Department, University of Glasgow.

92 W.R. Hamilton to William Thomson, 29th September, 1845, H7, ULC. W.R. Hamilton, 'On symbolical geometry', *Cam. and Dublin Math. J.*, **1** (1846), 45–57; 137–54; 256–63; **2** (1847), 47–52; 130–3; 204–9; **3** (1848), 68–84; 220–5; **4** (1849), 84–9; 105–18.

93 D.F. Gregory, 'Elementary principles', p. 4; 'On the impossible logarithms of quantities', *Cam. Math. J.*, **1** (1839), 226–34, on p. 232.

Symbols' could have been carried on as a purely abstract science, making no reference even to such justifying ideas as Peacock's principle of equivalent forms. In practice, Gregory excluded any laws of combination for which no meaningful interpretation could be found in arithmetical algebra: 'It is solely from the previous knowledge which we have of the combinations of arithmetical symbols, that we are enabled to facilitate our researches by the application of Algebra to Geometry, or to any science whatever'.[94] Thus interpretability, or utility, for Gregory as for De Morgan, restricted the acceptable forms of symbolic algebra. Nevertheless, a considerable distance had opened between forms and interpretation, especially physical interpretation.

Much the same remarks apply to Boole's famous extension of the calculus of operations into 'Laws of thought'. Other laws of combination of mental operations than the ones he employed might be imagined to exist, but if they did 'the entire mechanism of reasoning, nay the very laws and constitution of the human intellect, would be vitally changed. A Logic might indeed exist, but it would no longer be the Logic we possess'. Boole struggled to free symbolic logic from restriction to particular interpretations of the symbols employed, but at the same time he restricted their laws of combination to those which seemed to conform to human mental operations and which maintained relatively close analogy to arithmetic (although he excluded the commutative law). Like other exact sciences, logic did not discover necessary truth, in the sense of the ultimate causes of things being as they are; its business was with 'laws and phenomena'. Thus logic was an empirical science, albeit of internal mental processes. It was not, however, an inductive science, such as Boole considered other exact sciences to be, for one did not discover the laws of thought by generalization from many instances, but rather by direct perception in any single instance. The laws of thought, therefore, while contingent in themselves, were nevertheless necessary in relation to any possible external subject-matter, for they did not depend on physical interpretability. To obtain interpretations in any particular science, or even an area of mathematics, might well require more restrictive laws than those of thought in general. For instance, in Gregory's rendering, the calculus of operations required the commutative law for application to arithmetic, geometry, and the differential calculus, while the operation 'log' required special rules of its own.[95]

As the distance of the operational calculus from physical interpretation grew larger, William Thomson's interest in it grew smaller. He would agree with Boole's remarks in his *Differential equations* of 1859 that 'the mere processes of symbolical reasoning are independent of the conditions of their interpretation'

[94] D.F. Gregory, 'Branches of curves', p. 265; 'Elementary principles', pp. 3, 5.

[95] Boole, *Mathematical analysis of logic*, pp. 6, 12–13; *An investigation of the laws of thought, on which are founded the mathematical theories of logic and probabilities* (London, 1854), pp. 399–424 ('On the nature of science and the constitution of the intellect'), especially p. 404. Gregory, 'Linear differential equations', p. 31; 'Impossible logarithms', p. 230.

and that they should be regarded as 'in some sort the visible manifestation of truths relating to the intimate and vital connection of language with thought'; but he would emphasize Boole's further remark: 'As discussions about words can never remove the difficulties that exist in things, so no skill in the use of those aids to thought which language furnishes can relieve us from the necessity of a prior and more direct study of the things which are the subjects of our reasonings. And the more exact, and the more complete, that study of things has been, the more likely shall we be to employ with advantage all instrumental aids and appliances'. Gregory put the distinction of thought and things even more succinctly: 'Algebra takes cognizance only of the laws of combination of the symbols, and not of their meaning – in the eye of that science the symbol and the operation are identical. When we turn to the interpretation of our results, we must of course consider the meanings of the symbols – but such interpretation is out of the province of Algebra, and belongs to the science, the operations of which are symbolized'.[96]

Symbolic algebra epitomized pure mathematics, 'the only true metaphysics' in Thomson's later phrase, which agrees with the viewpoint of his *Journal* friends.[97] But the topics of natural philosophy, as sciences 'the operations of which are symbolized', required interpretation. One began with geometrical interpretation, for natural philosophy described objects in space and time; but even geometrical forms required additional interpretation, specific to the phenomena of heat, or electricity, etc., to obtain definite physical meaning. Thomson construed this demand for physical interpretation in a most literal manner (cf. Herschel above), requiring direct empirical reference for every physical entity symbolized mathematically and for every equation obtained. That requirement severely limited his use of mathematical analogy as a heuristic device for constructing physical explanations. He would reject, for example, Maxwell's transfer to electrostatics of Fourier's complete geometrical form for heat conduction because it required the invention by analogy of an electrical entity with no empirical referent (ch. 7). In a sense, he simply applied to mathematical analogies between physical processes the same restrictions on interpretation that Gregory and Boole applied to meta-mathematical analogies, but he inverted their priorities, putting physical meaning first and mathematical form second. In the same way, because quaternions and vectors replaced direct physical reference with mathematical forms taken as having a general physical meaning, Thomson rejected them.

These empiricist strictures on symbolism and analogy rapidly separated Thomson's concerns from those of friends like Boole and Arthur Cayley (1821–95) once he left the mathematical world of Cambridge for his Glasgow laboratory of natural philosophy and industrial application. He corresponded

[96] George Boole, *A treatise on differential equations* (Cambridge, 1859), pp. 389, vii–viii. Gregory, 'Elementary principles', p. 2.

[97] SPT, **2**, 1124.

regularly with them for a year or two but only occasionally later. By 1847 he already contemplated giving up the *Journal* and became increasingly disenchanted with it afterwards, largely because of the dearth of papers on physical subjects. Townsend, who wrote only on geometry, sent encouragement: 'you say the present number is very *analytical*, perhaps *I* might not like it as well as either geometry or physics but there are others who will like it far better in consequence – for instance, *I* do not feel as much interest in Mr Cayley's papers as in yours, but there is no writer of the present day that one of our first Analysts, if not our very first, Sir William Hamilton, praises like Mr Cayley'.[98] Townsend could hardly have chosen a more suspect example than Hamilton, for the combination of metaphysics and symbolism that Hamilton produced in his calculus of quaternions struck Thomson as at least physically unhelpful if not an outright mystification of phenomena.

He apparently expressed his misgivings to Boole as early as 1845, when attempting to win Hamilton and John Graves (another Dublin mathematician) as allies. Boole reassured him that 'They will soon . . . get through their quaternions and triplets and interesting as the subject is I must confess that I should be glad to see them turning their attention to the Int. Calc. and to physical science'. Hamilton continued for four years to publish his quaternion geometry in the *Journal*, always hoping to make Thomson see that it opened new vistas in 'analytical physics', but in vain. 'I do not flatter myself with the thought that . . . [you have had time] to examine into that theory', he wrote in 1847, and so sent a second copy of an abstract on quaternions, anticipating a 'possible future' when Thomson might have the inclination to study it. He never did. His recognition in 1871 that 'some of the most thoughtful mathematical naturalists of the day' regarded quaternion analysis as 'destined to become an *engine* of perhaps hitherto unimagined power for investigating and expressing results in Natural Philosophy', did nothing to sway him. Writing in 1892 to R.B. Hayward, who had authored an *Algebra of coplanar vectors and trigonometry*, he opined that 'it would lose nothing by omitting the word "vector" throughout. It adds nothing to the clearness or simplicity of the geometry, whether of two dimensions or three dimensions. Quaternions came from Hamilton after his really good work had been done; and, though beautifully ingenious, have been an unmixed evil to those who have touched them in any way, including Clerk Maxwell'.[99]

Boole himself, and Cayley, originally maintained a lively interest in physical mathematics, and Thomson valued their contributions highly. 'Exceedingly good', described Cayley's work on the differential equations of dynamics, and 'a

[98] Richard Townsend to William Thomson, 30th January and 1st March, 1847, T584 and T585, ULC, quotation from the latter.

[99] George Boole to William Thomson, 6th August, 1845, B146, ULC. W.R. Hamilton to William Thomson, 17th May and 22nd October, 1847, H22 and H24, ULC. William Thomson, 'Presidential Address', British Association, PL, **2**, 140. Our emphasis. William Thomson to R.B. Hayward, December, 1892, in SPT, **2**, 1138.

subject in which I am much interested' applied to one of Boole's papers on potential theory that produced practical solutions by symbolical means. But Thomson pointedly did not comment on Boole's accompanying paper, 'On a certain symbolical equation', which Cayley appreciated as 'a very pretty (tho' as he [Boole] says somewhat unpractical) paper'.[100] Upon receiving Boole's *Mathematical analysis of logic*, Thomson wrote back: 'I hope to be able to read it carefully. The advocates of "symbolical algebra" must be delighted with such an unlooked for extension of the class of subjects, for laws of symbolical operations'. Obviously he was not one of the advocates, though he wished Boole every success and published a summary of parts of the work in the *Journal*. Similarly, despite his great respect for Cayley's mathematical genius and their personal friendship, he grew impatient at Cayley's preoccupation with pure mathematics. Reflecting on G.R. Kirchhoff's perceptive work on elasticity in 1864, he exclaimed to Hermann Helmholtz (1821–94): 'Oh! that the CAYLEYS would devote what skill they have to such things instead of to pieces of algebra which possibly interest four people in the world, certainly not more, and possibly also only the one person who works. It is really too bad that they don't take their part in the advancement of the world'.[101] 'Advancement' refers to material progress through the Baconian unity of knowledge and power. It would come, not from egocentric mental gymnastics, but from mathematical analysis of the material world.

As Thomson's mathematical interests narrowed so too did his competence. He wrote to Boole in March 1847: 'I enclose another paper . . . & as it is in the symbolical way, I shall be glad to have your opinion . . . I have some compunction about the way in which I save myself trouble & put it on you and Cayley; but . . . I could not come to sufficiently satisfactory opinions on some of these papers not being *au courant* of the subjects'.[102] With his own research programme now at full speed he could ill afford time for pure mathematics, and, since few contributions of a physical kind appeared, his friends bore much of the editorial burden. Although he carried on for several years, his disenchantment grew. In 1851 he expressed his frustration to George Stokes, his newer and much more physically oriented Cambridge friend: 'I shall be very glad to get publishing your paper in the Journal, as I am very desirous of getting such papers on physical subjects sometimes in place of the endless algebra & combinations wh.

[100] Thomson to Boole, 21st February, 1847, B19, ULG, referring to Arthur Cayley, 'On the differential equations which occur in dynamical problems', *Cam. and Dublin Math. J.*, (1847), 210–19. See also, Arthur Cayley to William Thomson, 19th January, 1847, C44, ULC. William Thomson to George Boole, 2nd September, 1846, B15, ULG, acknowledges receipt of Boole's 'On the attraction of a solid of revolution on an external point' and 'On a certain symbolical equation', *Cam. and Dublin Math. J.*, **2** (1847), 1–7 and 7–12, respectively.

[101] William Thomson to George Boole, 5th December, 1847, B21, ULG. Thomson to Boole, 18th April, 1848, B22, ULG, concerns the summary, 'The calculus of logic', *Cam. and Dublin Math. J.*, **3** (1848), 183–198. William Thomson to Hermann Helmholtz, 31st July, 1864, in SPT, **1**, 433.

[102] William Thomson to George Boole, 27th March, 1847, B20, ULG.

so abound'. That problem, together with the *Journal*'s constant financial difficulties and his own distance from Cambridge, led Thomson to try to persuade Stokes, Lucasian professor of mathematics since 1849, to replace him as editor. Stokes declined, but by June, 1852, N.M. Ferrers, a more appropriate Cambridge mathematician, had been drafted, soon to be joined by Sylvester in a successor journal, *The Quarterly Journal of Mathematics*.[103]

The change in editors effectively formalized the division that had been emerging for some time between mathematicians and mathematical physicists, as a glance at the table of contents of Ferrers's volumes of 1853 and 1854 makes plain; for virtually no physics papers appear. Professionalization had indeed produced the specialization of interests that had so disturbed William Whewell. The division is equally plain in Thomson's view of Cayley in 1859. The chair of astronomy at Glasgow had fallen vacant on Nichol's death, and Thomson was attempting to recruit Stokes: 'Cayley is thinking on being a candidate, but it will probably be considered & I believe justly that he is not physical enough . . . I think Cayley ought to be provided for by the country – mathematician laureate would be his right post – but there is no doubt but that popular or physical lines of science are not in his way . . . Now I wish you most seriously to consider whether you will not apply for, and therefore I trust take, the Glasgow Professorship, & resign your Cambridge Prof^{p} in favour of Cayley?'[104] But Stokes remained in Cambridge.

The difference in roles between pure and physical mathematicians that seemed so obvious to Thomson in the 1850s had clear internal reference to the calculus of operations, with its separation of the symbols of operation from their interpretation. The separation was not so much one of abstract versus concrete as of general versus particular or theoretical versus applied. To consider a given operational equation from symbolic and from interpreted perspectives, however, involved a critical shift in emphasis from a mathematical operation to a physical process. The difference, while sometimes subtle, as in the case of the new analytic geometry, had been important to Thomson from his Cambridge days. Gregory's and Thomson's separate treatments of a particular problem from Fourier will illustrate the divide.

In 1838 Gregory discussed the solutions of Fourier's equation for the linear propagation of heat, that is, along the x-axis in an infinite solid, with the temperature varying along that axis but uniform over any plane perpendicular to it. The equation,

$$\partial V/\partial t = a\partial^2 V/\partial x^2,$$

has two types of solutions, depending on whether one integrates with respect to t or to x. For the single integration with respect to t the calculus of operations gave

[103] William Thomson to G.G. Stokes, 25th February, 1851, K46, Stokes correspondence, ULC (On financial difficulties, etc.). N.M. Ferrers, 30th June, 1852 and 27th March, 1855, F60A and F62A, ULC. See also Richard Townsend to William Thomson, 12th January, 1849, T589, ULC (on difficulties).

[104] William Thomson to G.G. Stokes, 6th October, 1859, K106, Stokes correspondence, ULC.

by inversion a Taylor series solution involving a single arbitrary function of x, the initial distribution of temperature $V(x,0)$ along the axis at the time $t=0$. Integrating twice with respect to x, on the other hand, gave a solution consisting of two series and involving two arbitrary functions of time, namely the temperature $V(0,t)$ and the gradient of temperature $\partial V(0,t)/\partial x$ at the plane $x=0$. Why, Gregory asked, should the same equation 'have two solutions so different in character'? He answered in terms of the general relation of the form of the equation to the differential coefficients in its power-series solutions. That answer could be called purely mathematical.[105]

Considering the same equation and the same problem, Thomson, in 1842, began from the assumption that if a determinate physical problem has two different forms of solution they must necessarily be physically connected: 'I propose to deduce the latter of these solutions from the former, and to shew, so far as possible, the relation which they bear to one another, with regard to the physical problem'. He proceeded to demonstrate that the arbitrary initial temperature distribution $V(x,0)$, given along the entire x-axis in the integral with respect to time, could be replaced by two conditions: the half of this initial distribution *to the right* of the zero plane ($x>0$); plus either of the two arbitrary functions in the integral with respect to x, that is, either the temperature at the zero plane $V(0,t)$ or the temperature gradient at the zero plane $\partial V(0,t)/\partial x$ given over all time. Thus the initial distribution *to the left* of the zero plane could be regarded as physically equivalent to, or as producing over time, either of the latter two distributions. He had thereby shown how the solutions for infinitely extended (mathematical) bodies were related to those for bounded (physical) bodies, e.g. the body to the right of the zero plane. For bounded bodies, data on the temperature distribution in the interior at some initial time, plus the time dependence of the temperature at the surface, would completely determine the interior temperature for all future time. Alternatively, the same determination would follow from the initial interior distribution plus the rate of heat loss at the surface (proportional to temperature gradient).[106]

Thomson's conception of the relation of the solutions, then, was essentially physical and practical. It concerned such questions as what one could learn mathematically about the unobservable interior state of a system from observations on the state of its surface. This example is the first of many we shall encounter that exemplify what A.E.H. Love, in his eulogy of Lord Kelvin before the London Mathematical Society in 1908, would call the 'ideal of mathematical physics': 'According to the standard that Kelvin had set up it is not sufficient to obtain an analytical result, and to reduce it to numerical computation, every step in the process must be associated with some intuition, the whole argument must be capable of being conducted in concrete physical terms'. In

[105] Gregory, 'Partial differential equations', pp. 126–8.

[106] William Thomson, 'On the linear motion of heat. Part I' and 'Part II', *Cam. Math. J.*, **3** (1843), 170–4; 206–11; MPP, **1**, 10–21, on p. 10.

Thomson's own phrase, 'mathematics . . . is merely the etherealization of common sense'.[107]

Progression and kinematics

The problem of relating external observations to interior conditions introduces two subjects to which Thomson would devote much effort: the cooling of the earth, or the problem of progression (chs. 10 and 16); and the distinction between kinematics and dynamics (chs. 7, 8, and 11). These issues concern a more general Cambridge context than that of the *Journal*.

At the conclusion of his first alternative, above, for specifying the temperature distribution of a bounded body over time, Thomson considered the question of extending the solution backwards in time rather than forwards. He observed that in general the initial distribution 'is not of such a form as to be any stage, except the first, in a system of varying possible temperatures, or is not producible by any previous possible distribution'.[108] If so, and if one applied the analysis to the earth, regarded as a cooling body, then it might be possible to fix a time for the beginning of the cooling process, or to set a maximum limit on the earth's age. Almost certainly he had that goal in view from this first paper of 1842, for both Fourier and Poisson had discussed closely related issues, and Charles Lyell in the 1830s had made the question a burning one for Cambridge academics such as William Whewell, Adam Sedgwick, and William Hopkins. Their concerns, coupled with Nichol's nebular hypothesis, turned Fourier's mathematical poem into a major resource for politico-theological debate.[109]

In the 'Preliminary discourse' of his *Théorie analytique* Fourier set the problem of the earth's heat in the grand context of 'the system of the world', on a par with gravity. He stressed terrestrial phenomena of climates, winds, currents, and diurnal and annual temperature variations, but also drew attention to three possible sources of the earth's heat: primitive heat in the earth's mass, the sun's heat, and the natural heat of the heavens.

Only the effects of the sun received mathematical development in the 1822 treatise. In a paper of 1820, however, Fourier had analyzed the 'secular cooling' resulting from a primitive store of heat in the earth. He followed this mathematical discussion with a qualitative one in 1824 on the temperature of both the earth and planetary spaces. William Whewell praised Fourier's approach in his 1835 'Report', but Poisson and Lyell both criticized the notion of a primitive central heat. Poisson objected that the observed increase of temperature with depth might arise simply from the solar system having moved into a cosmical region of lower temperature, to which the earth's interior temperature was slowly adjusting. Lyell had much more pressing reasons for opposing the central heat

[107] SPT, **2**, 1144, 1139. [108] Thomson, 'Linear motion', p. 15.

[109] Simon Schaffer stresses this point in 'The nebular hypothesis and the science of progress', in J.R. Moore (ed.), *The humanity of evolution* (Cambridge, forthcoming), notes 17–18.

view. It challenged the entire theory of steady-state averages, or uniformitarianism, that he presented in his famous *Principles of geology* (1830–3). If internal temperatures were decreasing then the volcanic forces producing elevation at the surface of the earth could not in the long run balance the forces of erosion.[110]

The Cambridge geologist Sedgwick and his colleagues Whewell and Hopkins strongly opposed Lyell's views; they saw the uniformitarian theory as an arbitrary hypothesis that gratuitously eliminated progression and teleology from the history of the habitation of man. Whewell formulated his notion of 'palaetiological' sciences with the historical changes of geological conditions in mind: 'the object is to ascend from the present state of things to a more ancient condition, from which the present is derived by intelligible causes'.[111] Nichol's nebular hypothesis, though not yet adequately supported, fell under palaetiological sciences, as did the secular cooling of the earth.

In Cambridge, then, Fourier was known as much for his views on the central heat as for his mathematics. Both attributes supported generative, or genetic, explanations of phenomena. The flux equation generates heat flow from temperature differentials, and the central heat doctrine applies this analysis to show how previous thermal conditions generate later ones. Thomson saw it that way in the 1840s. His 1842 paper constitutes his first attempt to reconcile simultaneous rationalist and progressionist beliefs about nature, the one requiring the operation of fixed natural laws and the other a directional, never repeating, series of states (ch. 4).

Dissatisfied with the complexity and inadequacy of his first attempt to analyze the mathematical conditions for the existence of temperature distributions antecedent to any given one, Thomson published a new version in February, 1844. Clearly intended for a more popular audience, this discussion reduced mathematical details to a minimum and gave a primarily verbal analysis of the crucial issue: 'the arbitrary initial distribution may be of such a nature that it cannot be the natural result of any previous possible distribution, or that it cannot be any stage except the first in a system of varying temperatures'. No distributions with jumps, cusps, edges, or angular points, he thought, could result from prior distributions, 'for though we may suppose such a distribution to be arbitrarily made, it could obviously not be produced by the spontaneous motion of heat from any other preceding distribution, and all such abrupt transitions or angles which may exist in an initial distribution, will disappear instantaneously

[110] Joseph Fourier, 'Extrait d'un mémoire sur le refroidissement séculare du globe terrestre', *Ann. Chim. Phys.*, **13** (1820), 418–38; 'Remarques générales sur les températures du globe terrestre et des espaces planétaires', *Ann. Chim. Phys.*, **27** (1824), 136–67. See John Herivel, *Joseph Fourier: the man and the physicist* (Oxford, 1975), pp. 197–202. Whewell, 'Report', pp. 28, 304; *History of the inductive sciences from the earliest to the present times* (3 vols., London, 1837), **3** pp. 559–63. Poisson, *Chaleur*, 'Supplement'. Charles Lyell, *Principles of geology, being an attempt to explain the former changes of the earth's surface by reference to causes now in operation* (3 vols., London, 1830–3), **1**, pp. 141–3; much expanded discussion in 3rd edn. (3 vols., London, 1835), **1**, pp. 203–8; **2**, pp. 273–322.

[111] Whewell, *Philosophy*, **2**, p. 95.

after the motion has commenced'.[112] Now the conclusion that abrupt transitions cannot result from prior distributions is not at all 'obvious' and certainly does not follow from the final remark on instantaneous disappearance. Nature presents many examples of sharp breaks resulting from superposition of waves, as anyone can observe on the sea, and there is nothing obvious about Fourier analysis in general, which Thomson used here, to suggest that waves spreading out from an initial sharp break could not run in reverse.[113] Evidently he simply assumed that heat flow is a process of progressive diffusion, proceeding in one direction only, and that it always produces a levelling of temperature rather than peaks. Evidently also, he believed that the earth had a beginning, and he was using Fourier analysis to confirm his belief, although no mention of the earth occurs in the paper.

Thomson's mathematical analysis epitomizes the intuitive physical style that he polished at Cambridge. By looking at the 'simplest case' which nevertheless presented the problem in its 'most general form' the correct answer would vividly present itself. Beginning from the general equation for conduction along a homogeneous rod, independent of sources and internal conditions, he wrote down the general solution in a Fourier series and examined the conditions on the coefficients such that the series would converge for negative values of time. Diverging series he took, following Gregory's terminology, to represent 'impossible' distributions, impossible in the processes of physical nature though perfectly representable as mathematical operations. With convergence as a criterion for the existence of antecedent distributions, he distinguished three types of initial states, according to their 'age': those which could not be 'any stage but the first in a system of varying temperatures' (divergence for all past time); those of 'finite *age*' (convergence for finite past time); and those for which 'no limit can be assigned to the *age*' (convergence for all past time).[114] 'Age' thereby took on the aspect of a perfectly natural, non-hypothetical and non-dogmatic parameter in the generation of phenomena according to natural law, a parameter that would require only adequate data on existing temperatures for its determination. Neutral rationality, in Thomson's view, could now reign in place of the speculative contentions of sects over the year, day, or even hour of the Genesis account of creation.

He developed these 'age' arguments for his 1846 inaugural dissertation at Glasgow, 'On the distribution of heat through the body of the earth'.[115] It would be hard to imagine a more fitting conclusion to the highly political campaign that his father had been waging to win William's election to the chair

112 William Thomson, 'Note on some points in the theory of heat', *Cam. Math. J.*, **4** (1845), 67–72; MPP, **1** 39–47, on p. 40.

113 Within a year Thomson had recognized the hole in his reasoning, *ibid.*, pp. 45–7 ('Note added 26th June, 1881').

114 *Ibid.*, pp. 40, 42–3.

115 Only one draft page survives of the Latin text of the dissertation, but it shows that Thomson applied the same three categories of 'age' as in his 1844 paper. Text in PA103, ULC; SPT, **1**, 187; translation by Dr J.C. McKeown. See also Kelland, *Heat*, pp. 125–42 ('Terrestrial heat').

than a dissertation which epitomized the ideology of 'apolitical' politics and latitudinarian religion that Dr Thomson represented and which placed that ideology in the service of progressionist cosmology. The tory party in the College could find no fault with the evidence of God's design and of beginnings and endings that the analysis suggested, while the whigs could find in the young professor's style of analysis all the evidence of progressive scientific rationality they might wish. This same style would characterize his physical mathematics throughout the 1840s and beyond, as when he turned his talents to clarifying the ideas on electricity and magnetism of Michael Faraday in 1845, and when he deployed those same talents as a weapon in the great geological debates. Both applications owe much to the merger of Thomson's Glasgow background and sophisticated *Journal* mathematics with the particular methodology of science taught by Hopkins and a variety of others at Cambridge.

In 1835 Hopkins had read to the Cambridge Philosophical Society the first major memoir of his career, which introduced a new subject, 'Researches in physical geology'. 'Physical geology' would bring the power and prestige of mathematical analysis to bear on geological phenomena in much the same way as Newton and Laplace had brought astronomy under the rule of universal gravitation and the laws of motion, thereby creating physical astronomy. In both cases the resulting physical theories were dynamical theories, dealing with observable motions and the forces that caused them. As a science of observable motions, physical geology excluded chemical phenomena just as natural philosophy in general excluded the insensible motions of chemistry.[116]

Hopkins carefully divided the method of physical geology, and his memoir, into three parts or stages, analogous to the mythologized stages of physical astronomy. One first obtained a geometrical description of the relevant motions (by analogy to Kepler's laws), then postulated a very general force that would produce the motions (like gravity), and finally derived the motions from the force according to dynamical principles (the laws of motion). The key to this scheme for geology was of course the first step, discovering 'geometrical laws' of faults, fissures, mineral veins and other phenomena of fracture. Hopkins argued that 'notwithstanding the appearances of irregularity and confusion in the formation of the crust of our globe . . . geologists have been able in numerous instances to detect, in the arrangement and position of its stratified masses, distinct approximations to geometrical laws'. Most important was the 'law of approximate parallelism' according to which fissures within a given geological region tend to occur in two sets, the fissures within each set running parallel to each other but perpendicular to the fissures of the other set, thereby forming a

[116] William Hopkins, 'Researches in physical geology', *Trans. Cam. Phil. Soc.*, **6** (1835), 1–84. For Hopkins's geological and academic context, see Crosbie Smith, 'Geologists and mathematicians: the rise of physical geology', in P.M. Harman (ed.), *Wranglers and physicists. Studies on Cambridge mathematical physics in the nineteenth century* (Manchester, 1985), pp. 49–83. See also Chapter 16 below.

cross-hatched pattern. This pattern represented the geometry of the motions whose dynamics were required.[117]

The same pattern seemed to be repeated over the whole surface of the globe and to characterize equally anticlinal lines, faults, and mineral veins, thus affording '*a priori* a strong probability that they are all assignable to the same general cause'. The geometrical form of the pattern, furthermore, suggested by analogy a force pushing the earth's crust up from below with a fairly uniform pressure over a given uplifted region. Hopkins thus postulated, as his second step, a corresponding 'elevatory force': 'I suppose this elevatory force, whatever may be its origin, to act upon the lower surface of the uplifted mass through the medium of some fluid, which may be conceived to be an elastic vapour, or in other cases a mass of matter in a state of fusion from heat. Every geologist, I conceive, who admits the action of elevatory forces at all, will be disposed to admit the legitimacy of these assumptions'.

Although this 'distinct mechanical cause' constituted a hypothesis, it amounted to little more than a descriptive name for the observed phenomena. He had avoided 'any speculations respecting the interior constitution of our globe', as he noted in a summary of the paper for the *Philosophical Magazine*.[118] In the summary, however, Hopkins cautiously relaxed his anti-hypothetical position to indicate that he favoured the primitive central heat view. If one assumed the earth had once been a hot fluid mass that gradually solidified as it cooled, then it seemed probable that cavities would have formed and that hot vapours or fluid forced into those cavities from below would produce the elevatory pressure in local regions. He judged this hypothesis the simplest possible and to bear the closest analogy to other cases in nature. In a series of papers for the Royal Society between 1838 and 1842 he sought to justify it by demonstrating its adequacy and the relative inadequacy of alternatives (ch. 16).

Having arrived at geometrical laws and a general mechanical cause, it remained only 'to institute an investigation, founded on mechanical and physical principles, of the necessary relations which may exist between our observed phenomena and the general cause to which we attribute them'. Hopkins therefore devoted the third, and by far the largest, part of his memoir to deriving the motions from the forces by means of the general laws of dynamics, much as Newton derived planetary motions from the action of gravitational force by applying the same laws. Aware that he might be charged with a premature application of mathematical analysis to a subject which did not admit of such precision, he insisted on the value of his procedure: 'the phenomena do distinctly approximate to obvious geometrical laws, and there is a simple cause to which they may be referred, the effects of which it has been my object . . . to investigate

117 Hopkins, 'Researches in physical geology', pp. 1, 2–9.

118 *Ibid.*, pp. 9–11. William Hopkins, 'An abstract of a memoir on physical geology, with a further exposition of certain points connected with the subject', *Phil. Mag.* [series 3], **8** (1836), 227–36, 272–81, 357–66, on p. 230.

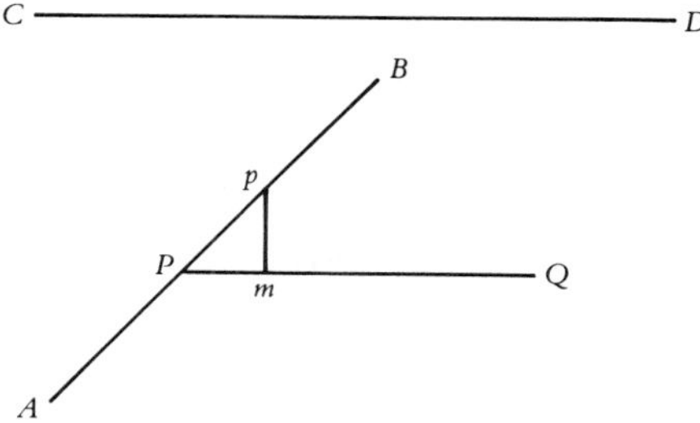

Figure 6.1. Hopkins analyzed the direction of fissure in a geological stratum as the direction of tearing in a thin sheet under uniform tension.

on mechanical principles, in order that we may compare the laws obtained from these results with those to which the observed phenomena are found to approximate'. He challenged potential critics to supply a better method.[119]

Hopkins's actual derivations provide considerable insight into the style of mathematical physics which his students would pursue, including Thomson. In his simplest case he considered a thin lamina, the plane of the paper in figure 6.1, and sought the orientation of the fissure that would necessarily result from a uniform tension parallel to *CD*. He put the problem as follows: to find 'the tendency of the forces of tension to separate the *particles* which are *contiguous*, but on opposite sides of the geometrical line *AB*, by causing them to move parallel to *CD*'. Taking *Pp* as any infinitesimal element of *AB*, *pm* as its projection perpendicular to *CD*, and *F* as the total tension per unit length across *pm*, he obtained for the required tendency, $pm \times F$. Since this quantity is maximum when *AB* is perpendicular to *CD*, the fissure will occur along a line perpendicular to the tension, as it should.[120] From this form of solution for the lamina, Hopkins gradually worked up to complex systems of tension in thick strata to derive the law of parallelism from elevatory forces.

We note immediately that the analysis is geometric and macroscopic. More specifically, the term 'particle' refers to a macroscopic element of the lamina, not to a microscopic molecule. The tension *F*, similarly, is a macroscopic force acting between 'contiguous' elements. These are the same features we observed in Fourier's analysis of heat conduction, in Airy's 'geometrical' part of the wave theory of light, in Green's and MacCullagh's theories of the luminiferous ether, and in Green's theories of electricity and magnetism. This fact implies that the correlation of Hopkins's physical geology with the schemes of Newton and Laplace, who depended on a reduction to atoms, is misleading. It should be a correlation with the Cambridge reformulation of their approach. Airy's geometrical wave theory depended on no model of the ether, but only on the

[119] Hopkins, 'Researches in physical geology', pp. 9–10.
[120] *Ibid.*, pp. 13–14. Our emphasis.

assumption of a linear restoring force for small transverse displacements of its elements. As he put it, 'we can hardly conceive any law of constitution of a medium in which undulations are propagated, where this does not hold'. In the same vein, but with respect to the phenomena of geometrical geology, Hopkins considered it 'impossible not to be struck with the idea of their being referable to the action of some powerful elevatory force', whatever the underlying structure of the earth might be.[121]

One final feature of Hopkins's dynamical theory correlates closely with heat conduction and wave propagation. He derived not only statical relations between tensions and fissures but properly dynamical results for the generation of a fissure from its point of inception. For example, he showed that a fissure, once started, would continue to open in the direction perpendicular to the line of greatest tension and that the velocity of propagation would be very large. More importantly, he pointed out explicitly a feature which characterizes propagating effects as effects of contiguous action: 'whatever may be the direction first given to the fissure by any local cause, its subsequent direction will soon become independent of that cause'.[122] This perception is one Thomson would employ in developing mathematically Faraday's idea of electric or magnetic fields as sets of lines of force propagated by contiguous action. Such propagation of forces in a field depends only on local conditions and not on the original sources.

Hopkins presented in his physical geology, then, a well-defined scheme for forming a dynamical theory. His method was not new in applying the Newtonian staging of geometry – cause – dynamical theory, nor in employing a Fourier-like macroscopic description, but in combining both elements to produce the theory directly as a dynamical theory of a continuous medium, as continuum dynamics. The essential motivation for such theorizing consisted in the elimination of speculative modelling of internal, unobservable causes. Hopkins wished originally to proceed without an interior model of the earth, and he did proceed without an interior model of a 'particle'. Because the procedure eliminated that reduction to hypostatized atomic parts characteristic of Newtonian dynamics it eliminated also the need to conceive of the problem in terms of sums (or integrals) over actions between such parts. Dynamics in the macroscopic theory rested on contiguous action, whose mathematical expression was a differential between adjoining elements. The action therefore propagated from element to element, and differential relations for any given element most naturally represented the entire system.

Although macroscopic dynamics was certainly not unique with Hopkins – Green's and MacCullagh's ether theories were immeasurably more sophisticated versions – no one else seems to have presented its complete methodology with such simplicity and clarity. The matter of presentation must be accorded considerable weight, for it was from the great 'wrangler maker' and not from

121 *Ibid.*, p. 9. Airy, *Tracts*, pp. 257–9.
122 Hopkins, 'Researches in physical geology', p. 24.

Green or MacCullagh, that Cambridge students learned their mathematical physics. On the other hand, by the late 1830s the methodology was the possession of the Cambridge mathematical network rather than of any individual. Hopkins held a crucial position in that network because his function required him to know precisely what kinds of questions would appear on the examinations and to transmit his insider's knowledge to his pupils. He formed a central locus through which the prevailing views of past wranglers – the moderators, examiners, and textbook writers of the present – passed on to wranglers of the future. In Hopkins's geological dynamics we meet with ideas which were rapidly acquiring canonical form.

'Geometrical laws' of motion constituted one such canonical idea. By 1840 William Whewell and Robert Willis had adopted the term 'kinematics', from the French *cinématique* of Ampère, to describe the science of pure motion, independent of force. Significantly, the term arose within the practical tradition of engineering mechanics and projective geometry, going back to Lazare Carnot and Gaspard Monge, and not within either Laplacian molecular mechanics or Lagrangian abstract mathematics.[123] It was pre-eminently a descriptive science, focussing on geometrical forms produced by constrained motions, as by a rod attached to a crank. Distinguishing geometrical relations from causal relations, Whewell and Willis wrote complementary textbooks on machines, with Willis contributing the descriptive kinematics in his *Principles of mechanism* and Whewell the action of forces in his *Mechanics of engineering*. Reading this distinction of the kinematics and dynamics of machines back into mechanics in general, the geometrical laws of planetary and geological motions are kinematical. Whewell employed the term in this general sense, although not with respect to geology, in his *Philosophy of the inductive sciences* in 1840.[124]

By the mid-1840s 'kinematics' had wide currency among Cambridge mathematicians. Arthur Cayley in 1846 found no need to explain himself when he stated the following standard result as a 'kinematical theorem': 'Any motion of a solid body round a fixed point may be represented as the rolling motion of a conical surface upon a second conic surface fixed in space'. Such theorems on the geometry of motion made up the subject that Thomson insisted on spelling *cinematics* (referring to Ampère and not Whewell) from his earliest Glasgow lectures until the 1860s. Ellis concurred in using 'cinematics in the large sense in which it is equivalent to the doctrine of motion'.[125]

123 André-Marie Ampère, *Essai sur la philosophie des sciences, ou exposition analytique d'une classification naturelle de toutes les connaissances humaines* (Paris, 1834), pp. 50–3. See Daston, 'Physicalist tradition', p. 283, for Carnot's call for the new science in relation to projective geometry.

124 Robert Willis, *Principles of mechanism, designed for the use of students in the universities, and engineering students generally* (London, 1841), pp. vi–ix. William Whewell, *The mechanics of engineering, intended for use in universities, and in colleges of engineers* (Cambridge, 1841), pp. iii, 1; *Philosophy*, **1**; pp. 144–7.

125 Arthur Cayley, 'On the geometrical representation of the motion of a solid body', *Cam. and Dublin Math. J.*, **1** (1846), 164–7, on p, 166. R.L. Ellis, 'Some remarks on the theory of matter', *Trans. Cam. Phil. Soc.*, **8** (1849), 600–5; *Writings*, pp. 38–48, on p. 43.

Clearly the central idea in kinematics was that of geometry generated by motion, as in projective geometry generally, or what Ellis called the 'new geometry'. The motions generating a figure were represented analytically by the operations of differentiation in the equation of the figure. Beginning at any one point, the operations would map out the entire figure by continuous extension from differential element to differential element. Connecting kinematics with differential operations in this way immediately suggests the idea of propagation in a continuous field. We begin to see already how important to Thomson's mathematical formulation of fields was the interrelation of several ideas: macroscopic description, propagation, the new geometry, and the analogy of mathematical forms, all of which we must recognize in the single term 'kinematics' as that term emerged in the Cambridge context.

By far the most important of those who developed the kinematics–dynamics distinction for the differential equations of continuous systems was one who did not use the terminology, George Stokes. Although Stokes studied with Hopkins and was senior wrangler in 1841, he was not among the whig mathematicians of the *Journal*, and Thomson seems to have come to know him only after completing his degree in 1845.[126] In that year Stokes published his masterpiece 'On the theories of the internal friction of fluids in motion, and of the equilibrium and motion of elastic solids', through which he hoped to treat the luminiferous ether as a medium intermediate between a solid and a fluid. In the common Cambridge manner, he used Poisson as a foil. Where Poisson spoke of molecules and their interactions, Stokes spoke of 'elements' (also 'particles') and the geometrical description of their motions: 'It will be necessary now to examine the nature of the most general instantaneous motion of an element of a fluid. The proposition in this article is however purely geometrical, and may be thus enunciated: "supposing space, or any portion of space, to be filled with an infinite number of points which move in any continuous manner, retaining their identity, to examine the nature of the instantaneous motion of any elementary portion of these points"'.[127] The solution led to four motions: pure translation and pure rotation of the element as a whole, 'uniform dilation' (change of volume), and 'two motions of shifting' (change of shape, or shear).

On this 'purely geometrical' foundation Stokes constructed his dynamics for homogeneous and isotropic fluids and solids, starting from the 'undoubted result of observation' that molecular forces are sensible only at 'insensible distances'. That assumption, like Fourier's equivalent statement for the commu-

[126] In his testimonial for Thomson, dated 6th July, 1846, Stokes called himself a 'personal acquaintance'. See 'Testimonials', PA34, ULC, p. 24. A mention of Stokes, in relation to Thomson's friend F.W.L. Fischer, appears on 12th February, 1845, in William Thomson, 'Notes on lectures and reading in Paris in 1845', NB30, ULC.

[127] G.G. Stokes, 'On the theories of the internal friction of fluids in motion, and of the equilibrium and motion of elastic solids', *Trans. Cam. Phil. Soc.*, **8** (1849), 287–319; *Mathematical and physical papers* (5 vols., Cambridge, 1880–1905), **1**, pp. 75–129, on p. 80.

nication of heat between layers in contact, immediately allowed him to eliminate molecular considerations altogether, reducing all action within the fluid to contiguous action between elements. He described all internal forces in terms of normal pressures across, and tangential actions along, three perpendicular planes defining a tetrahedral element of fluid. These forces he supposed proportional to differential changes of volume and of shape of any element. Having thus obtained the forces, the equations of motion of the fluid followed immediately from general dynamics (with Lagrange's formulation of D'Alembert's principle substituting for Newton's laws of motion).

Stokes's procedure for obtaining the partial differential equations of fluid motion reproduced at a sophisticated level Hopkins's three steps in geological dynamics: kinematics, forces, dynamical theory. Again, the critically important aspect was his direct application of dynamical principles to macroscopic elements, which yielded equations of an effectively continuous medium, being independent of internal structure. The constants of proportionality, which connected the forces acting between elements with the associated motions of the elements (stress with strain), were in this scheme simply macroscopic measures of elasticity, similar to Fourier's conductivity, and so they required no derivation from a molecular model, but only experiments to determine their values and their constancy for different materials. These constants of elasticity, furthermore, obtained their definition from the kinematics of the problem. A given element could change independently in volume and in shape. Thus two constants were required: 'There are two different kinds of elasticity: one, that by which a body which is uniformly compressed tends to assume its original volume, the other, that by which a body which is constrained in a manner independent of compression tends to assume its original form'.[128]

The utter simplicity of this statement masks its profundity. It ascribes physical properties to matter – kinds of elasticity – on the basis of geometrical considerations alone, insisting that there must in general be at least two such properties (and many more for inhomogeneous and anisotropic media). The contrast with the theory of Poisson and several other French mathematicians is striking. From an argument purporting to show that the elasticity of matter could be understood in terms of forces acting along the lines between point atoms, Poisson claimed that a fixed ratio had to exist between the constants for dilation and shear. Stokes could only register his amazement that such notable theorists as Lamé and Clapeyron should so implicitly trust Poisson's molecular theory that they would claim it applicable to indiarubber, for which the theory would apparently yield a compressibility comparable to that of a gas. In the hands of Thomson and Maxwell, the absurdity of ignoring the obvious geometrical generality of two constants in favour of a pure speculation on molecular

[128] Stokes, *Mathematical and physical papers*, pp. 125–6.

structure would serve as a convenient stick for beating (French) 'mathematicians', an activity in which they took some glee.[129]

With Stokes's theory we have arrived at the first fully articulated example of what came for a time to be understood as a proper dynamical theory in Britain. Classic examples are his 'Dynamical theory of diffraction' (1849), Thomson's 'Dynamical theory of heat' (1851) and Maxwell's 'Dynamical theory of the electromagnetic field' (1867). The six chapters which follow analyze Thomson's intensive engagement with this style of theorizing and the products he generated in electric and magnetic theory, thermodynamics, dynamics, and ether theory.

[129] *Ibid.*, p. 121. See the final section of Chapter 10 below for Thomson's ridicule of Poisson. See also J.C. Maxwell, 'On the equilibrium of elastic solids', *Trans. Royal Soc. Edinburgh*, **20** (1853), 87–120; W.D. Niven (ed.), *The scientific papers of James Clerk Maxwell* (2 vols., Cambridge, 1890), **1**, pp. 30–73, on p. 30 (for 'mathematicians' versus 'experimental philosophers').

7

The kinematics of field theory and the nature of electricity

> I have been sitting half asleep before the fire, for a long time thinking whether gravity and electrical attraction might not be the effect of the action of contiguous particles, communicated from one surface of =m [equilibrium] to another. In Cavendish's experiment, will the attraction of the balls depend at all on the intervening medium? *William Thomson, Cambridge Diary, 1843*[1]

In this dreamy speculation the second-year undergraduate recorded an insight central to much of his work in the 1840s. That insight connected the geometry of action at a distance with the physical alternative of action propagated between contiguous particles. It suggested how Fourier's techniques might be used to develop a mathematical alternative to Laplace and Poisson in the areas of their greatest strength, gravitation and electrostatics. No project could have been more appropriate within the context of Cambridge mathematics. Still, it is doubtful that the mathematical transformation itself would have had great impact if it had not matched in detail the *Experimental researches* of Michael Faraday. Thomson's speculation probably reflects rudimentary acquaintance with Faraday's ideas, but from 1845 Thomson's analysis would proceed step for step with Faraday's investigations. The result was field theory.

This interaction of youthful mathematical theorist with venerable sage of experiment sets the theme of the present chapter and the succeeding one. To emphasize the distinction of geometrical description from causal explanation in Thomson's maturing methodology, and to highlight the way in which 'work' (energy) first entered his considerations, we have divided the chapters as the *kinematics* and the *dynamics* of field theory, respectively. This division also largely follows the chronological order in which Thomson focussed his attention, first on electricity and then on magnetism.

Electrostatics: the analogy to heat flow

In September, 1841, at the age of seventeen, and just before going up to Cambridge, William sent to D.F. Gregory his third paper for the *Mathematical*

[1] William Thomson, 24th February, 1843, Diary kept at Cambridge, 13th February to 23rd October, 1843, NB29, ULC.

Journal, 'On the uniform motion of heat in homogeneous solid bodies, and its connexion with the mathematical theory of electricity'.[2] It was the work of a prodigy. He suddenly found himself in the sophisticated world of Philarète Chaslés and even C.F. Gauss, both of whom had recently published proofs of his results. Their papers appeared in 'far the best journal of the time', as he reported Gregory's opinion, 'and indeed better than any wh. has appeared yet, for pure mathematics'. The *Journal de Mathématique*, established by Joseph Liouville (1809–82) in 1836 but difficult to obtain in Britain, proved essential reading for our aspiring professional, 'anxious to get the latest intelligence on various points'.[3] By 1843 he had become a contributor and soon after would travel to Paris, still the centre of scientific activity, to study with the great men themselves.

William's letters to his father in these early years display the excitement, and anxiety, of high-powered competition. Upon beginning his second year at Cambridge, after a summer spent in extending his earlier results, he wrote home: 'Since I came here I have been able fortunately to see the August No. of Liouville, containing the second part of Gauss's Mémoire, and, as I feared, he has proved exactly the same theorems as I had done in my previous paper, and in nearly the same way. This does not at all interfere however with the paper I was writing at Knock [a family vacation spot near Largs], and so, lest Gauss, or Chaslés, may be doing anything more I have sent it off to Gregory, explaining the circumstances, and asking whether he has room for it, in the November No.'.[4] Only in 1845 would he discover that in an almost unknown work of 1828 George Green had proved all of the theorems, and many more, using virtually the same mathematical techniques as William had himself developed.

[2] William Thomson, 'On the uniform motion of heat in homogeneous solid bodies, and its connexion with the mathematical theory of electricity', *Cam. Math. J.*, **3** (1843), 71–84; E&M, 1–14. The best general discussion of the significance of Thomson's analogical method and its place in British physics is Ole Knudsen, 'Mathematics and physical reality in William Thomson's electromagnetic theory', in P.M. Harman (ed.), *Wranglers and physicists. Studies on Cambridge physics in the nineteenth century* (Manchester, 1985), pp. 149–79, esp. pp. 150–7, for the heat analogy. For a detailed examination of the heat analogy, see J.Z. Buchwald, 'William Thomson and the mathematization of Faraday's electrostatics', *Hist. Stud. Phys. Sci.*, **8** (1977), 101–36. We disagree, however, with essential aspects of Buchwald's interpretation. cf., M.N. Wise, 'The flow analogy to electricity and magnetism – Part I: William Thomson's reformulation of action at a distance', *Arch. Hist. Exact Sci.*, **25** (1981), 19–70.

[3] See William to Dr Thomson, 30th October, 1841, and 1st October, 1842, T181 and T220, ULC, for Thomson's discovery of Chaslés'; and Gauss's priorities, respectively. For his remarks on Liouville, see William to Dr Thomson, 26th October and 1st November, 1842, T226 and T182, ULC. The latter, which refers also to Gauss, is misdated 1841 in the ULC catalogue.

[4] William to Dr Thomson, 1st October, 1842, T220, ULC. Thomson's paper was 'Propositions in the theory of attraction', *Cam. Math. J.*, **3** (1843), 189–96, 201–6; E&M, 126–38. cf. C.F. Gauss, 'Allgemeine Lehrsätze in Beziehung auf die im verkehrten Verhältnis des Quadrats der Entfernung wirkenden Anziehungs– und Abstossungs– Kräfte', in C.F. Gauss and Wilhelm Weber (eds.), *Resultate aus den Beobachtungen des magnetischen Vereins im Jahre 1839* (Leipzig, 1840), pp. 1–51; translated for Joseph Liouville's *Journal de Mathématique*, **7** (1842), 273–324, and Richard Taylor's *Scientific Memoirs*, **3** (1843), 153–96 (hereafter 'General propositions').

Thomson's early paper on heat and electricity is noteworthy because, while Chaslés' methods were 'nearly entirely geometrical' and Gauss's 'analytical', William relied on 'physical considerations'.[5] That is, he did not prove the results mathematically at all, but employed the physical process of heat conduction to justify mathematical conclusions which he transferred to electricity by mathematical analogy. The physical reasoning made certain theorems in electricity obvious, theorems which, considered analytically, were anything but palpable. He began his argument by stating for heat flow the main theorem that he wished to establish for electricity:

> It is obvious that the temperature of any point without [outside] a given isothermal surface, depends merely on the form and temperature of the surface, being independent of the actual sources of heat by which this temperature is produced, provided there are no sources without the surface. The temperature of an external point is consequently the same as if all the sources were distributed over this surface in such a manner as to produce the given constant temperature.[6]

The theorem is important because it allows one to replace by a boundary value problem (which, for simple geometries, yields immediate solutions) the problem of relating the actual sources to the fluxes they produce.

If we ask why the theorem was obvious for heat flow, an answer follows immediately from Fourier's theory (although not from Poisson's). Assuming heat to be propagated by contiguous action, its motion after leaving its sources would be independent of those sources. The continuing motion through any region would depend only on the local temperature gradient and conductivity. One could therefore determine the distribution of flux, in the space outside a certain bounded region containing the sources, as easily from knowledge of the flux of heat through the boundary as from knowledge of the actual sources. Supposing conservation of heat, furthermore, it would necessarily be possible to arrange the actual sources on the surface so as to produce the given fluxes there. If the bounding surface were an isothermal surface, heat would flow perpendicular to it – the gradient along the surface being zero – and the flux across any infinitesimal area would be equal to the source strength on that area. This physical reasoning showed how to replace actual sources by sources on an isothermal bounding surface, which became the source for succeeding surfaces. The flux, therefore, could be regarded as 'communicated from one surface of equilibrium to another', as Thomson was soon to remark in his 1843 diary, having adopted the term 'surface of equilibrium' from Gauss's mathematical theory of forces acting at a distance and having united it with Fourier's analysis of heat flow and with Faraday's conception of lines of force.[7]

Considering the problem from the side of electrical action at a distance – meaning, for Thomson in 1841, Poisson's theory – electrical force did not travel

[5] 'Uniform motion', E&M, 2: 'Propositions', E&M, 132; William to Dr Thomson, 24th April, 1843, T238, ULC. [6] 'Uniform motion', E&M, 2. [7] 'Propositions', E&M, 126.

to its point of action along a route depending on local conditions. One could not argue on physical grounds, therefore, that the original sources could be replaced by the forces over a surface enclosing them. But, regarded mathematically, heat and electricity behaved quite similarly. For Poisson's electrical forces, like Laplace's gravitational ones, could be derived from an unnamed function V, 'the sum of the molecules . . . divided by their respective distances to the point attracted':

$$V = \iiint (\rho/r)\mathrm{d}\tau, \tag{1}$$

where ρ is the density of attracting matter (electrical or gravitational) in the volume element $\mathrm{d}\tau$, and r its distance from the 'point attracted'.[8] The force at the point was then,

$$\vec{F} = -\text{gradient } V, \tag{2}$$

while V obeyed the equation,

$$\text{divergence (gradient } V) = -4\pi\rho. \tag{3}$$

Similarly, with $\vec{F}$ and V symbolizing flux of heat and temperature Fourier's equations were:

$$\vec{F} = -K \text{ gradient } V, \tag{4}$$

and

$$\text{divergence (gradient } V) = (CD/K)\partial V/\partial t. \tag{5}$$

Thus, outside electrified matter ($\rho = 0$), Poisson's theory looks the same as Fourier's would for steady-state heat flow ($(\partial V/\partial t) = 0$), except for the conductivity K in equation (4). For an infinite homogeneous medium, K could be set arbitrarily equal to unity, yielding full mathematical equivalence between the two systems.

This identity justified Thomson's analogy between electrostatic action and heat conduction. Recognition of the analogy, however, was neither new nor necessarily significant. In a section of the *Mécanique céleste*, which may well have provided Thomson's inspiration, Laplace had noted the same equivalence and had observed that his methods for gravitating spheroids could be applied to heat conduction.[9] Poisson too had recognized an analogy between the equation he derived for magnetization in homogeneous materials and the continuity equation for fluid flow, but declined to pursue it: 'we will not stop to develop this

[8] P.S. de Laplace, *Traité de mécanique céleste* (5 vols., Paris, 1799–1825), **1**, book 2, chap. 1; **2**, book 3 ('de la figure des corps célestes'). Thomson had read the latter, at least, in preparation during 1839 for his essay 'On the figure of the earth', which won a Glasgow University Medal. See SPT, **1**, 9–10, 15. His knowledge of Poisson's theory probably began in his chemistry class in 1838–9 with Thomas Thomson, who gave a qualitative discussion in his published lectures, *An outline of the sciences of heat and electricity* (London and Edinburgh, 1830), chap. VI. He apparently did not read Poisson's papers until his visit to Paris in 1845, but probably read one of the derivatives: William Whewell, 'Theory of electricity', *Enc. Met.*, **4** (London, 1830), pp. 140–70; or Robert Murphy, *Elementary principles of the theories of electricity, heat, and molecular actions: Part I, On electricity* (Cambridge, 1833) (other parts never published). Poisson's original papers are: 'Mémoire sur la distribution de l'électricité à la surface des corps conducteurs' and 'Second mémoire sur la distribution de l'électricité à la surface des corps conducteurs', *Mémoire de l'Institut* (1811), 1–92; 163–274.

[9] Laplace, *Mécanique céleste*, vol. **5**, pp. 72–85, discusses the earth's cooling. Laplace has added this section to a discussion of the figure of the earth.

analogy, which would be of no utility for the solution of the problem that occupies us, and which might lead into error on the nature of magnetism'.[10] Neither Laplace nor Poisson, it seems, saw any use in applying the ideas of flow phenomena to solve problems in attractions. Committed to microscopic description, action at a distance, and rigorous mathematical analysis as requisites of proper physical theory, they could not have approved, much less pursued, young Thomson's analogical reasoning, which carried Fourier's failings to a higher order.

Situated in a very different context, where macroscopic description, geometric relations, and simplicity held priority, the precocious youth saw possibilities where the mathematical giants had failed to look. While they had regarded their function V as a tool of analysis without physical significance, Thomson began in his 1841 paper to view it through Fourier's spectacles as a physical state like temperature, whereby its gradient not only *yielded* force but *was* force. An electrified body would tend to move along a potential gradient just as heat flowed along a temperature gradient, from one isothermal surface (surface of equilibrium) to another.

In order to establish that analogy Thomson had to add a factor absent from the works of Laplace, Poisson, and even Fourier. Nowhere had they developed for flow systems a relation like equation (1). Sources of heat, for example, did not contribute their independent actions to the temperature at a distant point in the way molecules of attracting matter contributed to the function V. Fourier actually eliminated radiating sources in order to obtain his macroscopic picture of flux emanating from heat reservoirs across isothermal boundaries. Thomson restored them to show that every isothermal surface could be regarded as a surface covered with sources. Perhaps because the step was so simple, commentators have not remarked that his entire analogy depended on it, and that it was completely new. It required noticing only that a point source in an infinite homogeneous medium must produce a flux which decreases as the inverse square of the distance, and therefore, from equation (4), a temperature which decreases as $1/r$.[11] Assuming linear superposition, the temperature produced by a distribution of sources of 'intensity', or density, ρ, was, just as for attracting matter,

$$V=\iiint(\rho/r)\mathrm{d}\tau.$$

With that result established Thomson could move back and forth at will between electrical and heat flow arguments. Most important to the main theorem on replacement surfaces was his analogy between an isothermal surface, maintained by surface sources alone, and the surface of an electrified conductor

[10] S.D. Poisson, 'Mémoire sur la théorie du magnétisme' and 'Second mémoire sur la théorie du magnétisme', *Mémoire Acad. Sci.*, **5** (1821–2), 247–338 and 488–533, on p. 302.

[11] Poisson had in fact begun his analysis from the inverse square nature of radiation in the void and had pointed out that the flux in a medium, when integrated over a closed surface, would have to equal the heat lost inside, but without obtaining equation (1). See 'Propositions', E&M, 135, for Thomson's reference to Poisson's *Théorie mathématique de la chaleur* (Paris, 1835), p. 177.

at equilibrium. From conservation of flux he showed that the temperature everywhere inside the isothermal surface had to be constant and equal to the surface temperature; the internal flux therefore vanished. Substituting force for flux yielded the empirically observed condition of zero electrical force inside conductors. For heat, the result followed simply from the fact that the surface was isothermal, or equivalently, that the flux was perpendicular to the surface. 'Hence', Thomson argued by analogy, 'the sole condition of equilibrium of electricity, distributed over the surface of a body, is, that it must be so distributed that the attraction on a point at the surface, oppositely electrified, may be perpendicular to the surface'.[12] An electrified conducting surface behaved exactly like the isothermal surface maintained by surface sources. In general, any closed surface of constant V, with its actual internal sources of electricity imagined spread over it in equilibrium, would produce exactly the same external forces as the actual sources and could be taken to replace them. From the perpendicularity condition, Thomson showed that the surface sources would have an intensity given by the force (or flux) F_n perpendicular to the surface, or the gradient of V in the direction of the normal n,

$$4\pi\rho = F_n = -\partial V/\partial n.$$

This result completed the replacement surface theorem. The reasoning provided a clear and simple image of electrical force as something conducted, like heat, from one surface of equilibrium to another (figure 7.1):

> Since at any of the isothermal surfaces, V is a constant, it follows that $-\partial V/\partial n$, where n is the length of a curve which cuts all the surfaces perpendicularly, measured from a fixed point to the point attracted, is the total attraction on the latter point; and that this attraction is in a tangent to the curve n, or in a normal to the isothermal surface passing through the point. For the same reason also, if ρ represent a flux of heat, and not an electrical intensity, $-\partial V/\partial n$ will be the total flux of heat at the variable extremity of n, and the direction of this flux will be along n, or perpendicular to the isothermal surface.[13]

(For later reference, we note that Thomson's lines of flux originate in sources ρ but their conception does not require that they terminate in negative sources, $-\rho$.)

This geometrical picture of a family of equilibrium surfaces and perpendicular force lines, representing the entire space of action of an electrostatic system, is symbolized in the term 'potential' for Thomson's function V. Within a year he had adopted Gauss's use of the term in this sense, and in 1845 he discovered Green's identical usage. 'Potential' signifies not merely an analytical function but a new way of thinking about force distributions in terms of potential gradients and equipotential surfaces.

For understanding Thomson's conception of this 'physical mathematics' the term 'electrical intensity' is important. It appears variously as source strength, flux across a surface, and the origin of a line of flux. It referred in general, then, to

12 'Uniform motion', E&M, 4. 13 *Ibid.*, 4–5.

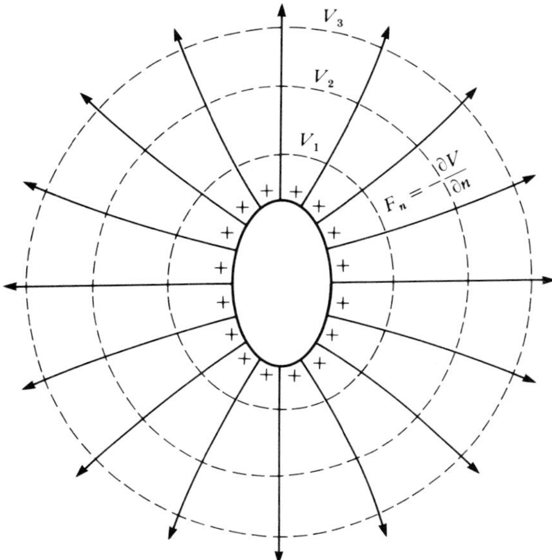

Figure 7.1. Lines of force (flux), originating in sources, flow always perpendicular to equipotential (isothermal) surfaces. Their termination is unspecified.

a state of electrification as measured by the concentration of force (flux) produced. He used electrical 'intensity' interchangeably with 'density' of electrical matter, but meant by the latter only source strength, and not necessarily the density of an electrical fluid, although electrical matter behaved in many ways as though it were a fluid. In this way he avoided any hypothesis on the nature of electricity, such as Poisson's assumption of positive and negative electrical fluids. He also avoided the difficulty Poisson had encountered with the condition of equilibrium of a finitely thick electrical layer at the surface of a conductor. Such a layer could not correspond rigorously to an equipotential surface, and perpendicularity of force could not supply its 'sole condition of equilibrium'.[14] Thomson's electrical analysis thus differed from Poisson's in the same way that Fourier's treatment of thermal boundaries, as geometrical surfaces across which temperature changed discontinuously, differed from Poisson's modelling of physical layers in which temperature changed continuously. The difference, to Thomson, defined proper 'mathematical theory'. In Chapter 9 we shall find that

[14] Jed Buchwald has emphasized this important aspect of Poisson's surface conditions, although with a view of Thomson's response different from ours, in Buchwald, 'William Thomson', pp. 107–19. See also D.H. Arnold, 'The mécanique physique of Siméon Denis Poisson: The evolution and isolation in France of his approach to physical theory (1800–1840). VIII. Applications of the mécanique physique', *Arch. Hist. Exact Sci.*, **29** (1983), 53–72, esp. pp. 59–64; and Chapter 6 above.

the idea of heat as a state of a body, similar to electricity as a state of intensity, was crucial to his early conception of thermodynamics as well.

Reference to Fourier, however, does not provide sufficient motivation for treating sources as states of intensity. More appropriate is young William's Scottish context of natural philosophy (ch. 4), which discouraged appeal to imponderable fluids of all kinds. His notes from Meikleham's lectures assert that, 'Light heat electricity magnetism, &c are termed imponderable. This is incorrect, as we know them, not as substances, but as states of bodies'. For the cause of such states Meikleham chose the neutral term 'agent': e.g., '[Electricity is] Another agent that seems now of universal agency'. Again, 'by electricity or galvanism all substances may be put into such a state as to be affected by the magnet. And hence mag. & elect. are modifications of the same [substance, *del.*] agent'. Meikleham's remarks should be juxtaposed with those of Nichol, from the same notebook, on the priority of geometry: 'the quality of form is the simplest of all the qualities of matter, and hence geometry, which treats of it, stands at the head of Natural Philosophy'.[15] Together, the two statements suggest why macroscopic states and geometrical relations figure so prominently in Thomson's physical mathematics. They were essential components of the tradition of natural philosophy at Glasgow, with its emphasis on non-hypothetical theory.

This observation leads us to characterize Thomson's developing method of mathematical analogy more carefully, especially since his first analogy exhibits features that endured throughout his life. Here he transferred an obvious theorem from heat conduction to electrostatics by demonstrating mathematical equivalence between action at a distance and contiguous action under certain general conditions. The technique therefore implied that the truth of the theorem for electrostatics depended only on the *form* of the mathematical theory and not on its physical interpretation. Given the basic equations relating ρ and V the same theorems would be true whether V were temperature or a physically uninterpreted function, for the operation $\partial/\partial n$ on V generated a line passing from one surface of constant V to another whether that operation represented conduction or not. The mathematical theory was true, therefore, not by virtue of expressing the correct physical model, but by virtue of generating the correct geometry. That much seems relatively unproblematic; but 'generating' and 'correct geometry' require comment.

Thomson always understood mathematical operations as more or less perspicuous generalizations of physical processes. The operation $\partial/\partial n$ generated $\vec{F}$ in the same way as temperature gradient generated flux. For that reason physical processes could be taken to generate theorems – meaning literally to prove them – with as much validity as mathematical analysis itself. In a notebook entry of 1844 Thomson recalled that this insight had been his starting point for establish-

[15] William Thomson, 'Notebook of Natural Philosophy class, 1839–40', NB9, ULC. On the Scottish methodological context see especially G.N. Cantor, 'Henry Brougham and the Scottish methodological tradition', *Stud. Hist. Phil. Sci.*, **2** (1971), 69–89.

ing the replacement surface theorem: 'This [theorem], being a physical axiom [for heat], enables us *physically to prove* the theorem, with ref^{ce} to attr^{n}, as a purely analytical theorem. This was what first occurred to me in the winter of 1840–41'.[16] Thomson apparently regarded his physical method of proof as a natural extension of the methods employed in projective geometry. This interpretation correlates closely with Richards's analysis of the place of projective geometry among Thomson's friends on the *Cambridge Mathematical Journal*, who stressed the generation of geometrical forms by ideal motions, motions represented by operations of differentiation. Indeed, it should now be clear in addition that Thomson's method of mathematical analogy for transferring results from one area of physics to another functioned in much the same way as Gregory's operational calculus did for transferring results between different areas of mathematics. One had only to know that two physical processes generated results according to the same laws of mathematical operation, i.e. generated the same geometrical relations, to know that the same theorems held for those processes. This insight is the core of what the Cambridge circle called kinematics – or, to preserve its French heritage, cinematics – the geometry of motion.[17] The point is underscored by the fact that Thomson's analogy appeared as one of the questions on the Smith's Prize Examination in 1845 (with Thomson himself competing):

The equilibrium of temperature, the attraction of bodies, and under a certain hypothesis the *motion* of a fluid, are all represented by a differential equation of the same form [divergence $V=0$], of which an integral is $V=\sum(c/r)$ . . . What physical properties are embodied in this integral in each case?[18]

In correspondence with his conception of the relation between physics and mathematics Thomson required that mathematical theory express only observable relations between observable physical properties of bodies. The entire theory should operate at a non-hypothetical, macroscopic level, providing the mathematical structure, or better, the geometrical structure of *true* physical laws. Many passages in his writings illustrate this demand, but none more clearly than the following, from his 1845 extension of the heat flow analogy, to be discussed below:

Now the laws of motion for heat which Fourier lays down in his *Théorie analytique de la chaleur*, are of that simple elementary kind which constitute a mathematical theory properly so called; and therefore, when we find corresponding laws to be true for the phenomena presented by electrified bodies, we may make them the foundation of the mathematical theory of electricity: and this may be done if we consider them merely as

[16] William Thomson, 14th August, 1844, 'Journal and research notebook, 1843–1845', NB33, ULC, p. 51.

[17] J.L. Richards, 'Projective geometry and mathematical progress in mid-Victorian Britain', *Stud. Hist. Phil. Sci.*, **17** (1986), 297–325; L.J. Daston, 'The physicalist tradition in early nineteenth century French geometry', *Stud. Hist. Phil. Sci.*, **17** (1986), 269–95; and Chapter 6 above.

[18] Smith's Prize Examination, 23rd January, 1845, question 20, *University of Cambridge examination papers, 1845* (Cambridge, 1845). Our emphasis.

actual truths, without adopting any physical hypothesis, although the idea they naturally suggest is that of the propagation of some effect by means of the mutual action of contiguous particles; just as Coulomb, although his laws naturally suggest the idea of material particles attracting or repelling one another at a distance, most carefully avoids making this a *physical hypothesis*, and confines himself to the consideration of the mechanical effects which he observes and their necessary consequences.[19]

Thomson's limitation of mathematical theory to macroscopic geometrical structure involved a second restriction, implicit but nonetheless severe, without which he would not accept the geometry as correct. We have seen that, while he described his analogy as one between flux and force, mathematically it correlated temperature gradient with potential gradient. The difference vanishes for conduction in an infinite homogeneous medium, for which he set $K=1$ in equation (4), because then flux equals temperature gradient just as force equals potential gradient. But if there were any inhomogeneities in the system, or interfaces between media of different conductivity, then the physical difference between flux and temperature gradient would show up mathematically. While heat, treated as a conserved quantity, flows continuously across such interfaces, preserving the normal component flux, the temperature gradient normal to the surface changes discontinuously. Thus, two entities, with very different mathematical properties, necessarily appear in the physical description of conduction, while only one, force = potential gradient, appears in electrical action at a distance. Although of no importance in the 1841 analogy, we shall see that all of Thomson's later analogies depend on mathematical identity in the subjects related. He regularly restricted one side of an analogy physically so as to produce mathematical identity, but never appealed to analogy to enlarge the physical scope of a theory. That is, he never employed mathematical analogy to justify introducing new physical terms. James Clerk Maxwell, by contrast, would develop speculative analogies with great effect, thereby expanding Thomson's limited analogies into his own quite different field theory. Thomson's stricture indicates once again the intensity of his commitment to (what he regarded as) non-hypothetical theory. Recognizing that analogy alone could lead to uninterpretable phantasies, he required full empirical reference on both sides of an analogy. Without it, the analogy could not support a mathematical theory 'properly so called'. We shall return below to a deeper analysis of this commitment.

Mathematical theorist meets experimental theorist: Thomson and Faraday

Exactly when Thomson learned of Faraday's idea of electrostatic 'induction in curved lines' is uncertain. From 1840 to 1845, David Thomson, a cousin of

[19] William Thomson, 'On the mathematical theory of electricity in equilibrium: I. On the elementary laws of statical electricity', *Cam. and Dublin Math. J.*, **1** (1846), 75–95; E&M, 15–37, on p. 29.

Faraday's, replaced the ailing Meikleham in the junior course of natural philosophy at Glasgow. He reported in his testimonial for William's application for the chair that he would never have survived the first session if young Thomson had not assisted him. No doubt, then, William learned something of Faraday's ideas already in 1840–1, when he was formulating his heat analogy. He reported to S.P. Thompson that he had been 'innoculated with Faraday fire' by David Thomson; but the fire can have ignited only somewhat later, for his letters and diary show no mention of Faraday before 1843, and then show only condescension.[20]

The epigraph of this chapter is Thomson's first mention of an 'action of contiguous particles' and of a possible dependence of the communication of force 'on the intervening medium'. Looking for just such an effect, Faraday had in 1837 discovered 'specific inductive capacity'. But Thomson ridiculed Faraday's notions. On 16th March, 1843, he visited Gregory, noting in his diary: 'I asked him about where I could see anything on Elec^{y}, and we had then a long conversation, in wh. Faraday & Daniell got (abused)^{n} [abused to the nth power]'. The following day he recorded: 'I have been reading Daniell's book & the account he gives of Faraday's researches. I have been very much disgusted with his [Faraday's] way of *speaking* of the phenomena, for his theory can be called nothing else'.[21]

It is certainly true, now as then, that to appreciate Faraday one must first enter into his private language. But, more than that, Faraday regarded his demonstrations of the curvature of lines of electrical action as a telling objection to action at a distance, which he supposed required action in straight lines. Thomson recognized that claim for what it was: total ignorance of the mathematical theory, in which superposition of actions in straight lines leads to a resultant that changes direction from point to point. Having just spent two years laying an analytic foundation for the insights of his heat analogy, the proud mathematician looked with disgust on Faraday's incomprehension.

Thomson's analysis, entitled 'Propositions in the theory of attractions', established a number of theorems basic to potential theory, among them several theorems of Gauss and what is called Green's theorem.[22] So original was this work that he contemplated an entire treatise on electricity based on potential theory. In September, 1844, he united that plan with his ambitious project for expanding and professionalizing the *Cambridge Mathematical Journal*, (ch. 6). The treatise would be a series of articles for the new journal to be written as soon as he completed his degree. For the moment he wrote out ' in a disjointed manner' the basic ideas for a theory that would depend on no particular conception of the

[20] Printed copy of William Thomson's testimonials for the Glasgow chair of natural philosophy, PA34, ULC, p. 26. SPT, **1**. 179.

[21] William Thomson, 16th–17th March, 1843, Diary, NB29, ULC.

[22] 'Propositions', E&M, 126–38.

nature of electricity but only on the fact of electrical intensity and the results of observation concerning it.[23]

The object of the series 'On the mathematical theory of statical electricity' would be: 'to determine the manner in which electrical intensity is distributed over a conducting body bounded by a surface of any form . . . and acted upon by any quantity of electricity, distributed according to a given law, on nonconducting matter. The only results of observation which are required for the elements of the mathematical theory are the following'. He outlined four sets of 'facts':

(I) '*Absolute charge of electricity*. The whole quantity of electricity in connection with an isolated conducting body remains unchanged by any external action . . . '

(II) '*Law of electrical force*.' Between any two small charged bodies the force varies as the inverse square of the distance between them.

(III) '*Distribution of electrical intensity*.' The distribution of electrical intensity (1) inside a conductor 'at points situated at any finite distance, however small from the surface, is always zero' and (2) on any 'finite area of its surface, however small' cannot be zero for every point of the area if it is non-zero for any point of the surface.

(IV) '*Direction of electrical action at the surface of conductors*. The force upon any electrical element, at the surface of a conductor . . . is in a direction perpendicular to the surface'.

These facts [principles, *del.*], which are well supported by all strict experiments which have yet been made, are in perfect accordance with all the phenomena observed in operating upon statical electricity, and have become part of our fundamental notions of the subject, so that, whatever hypothesis may be made relative to the absolute nature of electricity, they must be considered as the facts to be accounted for.[24]

Unfortunately, point (IV), the perpendicularity condition, had not quite the status of the others. While they could be regarded as 'established directly and rigorously by experiment', this fact was 'not capable of direct verification. Its truth as an assumption must therefore depend on the complete agreement of the results deduced from it, along with the other three, by [mathematical reasoning, *del.*] mechanical principles, with the phenomena'.[25] Both Thomson's heat flow analogy and his analytic proofs assumed a strictly superficial distribution of electrical intensity, having no thickness, in order to obtain perpendicularity as 'the sole condition of equilibrium'. If the 'finite distance, however small' of (III) (1) did not actually reduce to an infinitesimal distance, then neither the analogy nor the analysis was rigorous. Thus he had to rest his ostensibly non-hypothetical theory on the consistency of experiment with deduced results, i.e. on a hypothetico-deductive justification.

[23] William Thomson, 10th September, 1844, NB33, ULC, p. 67.
[24] *Ibid.*, pp. 69–77. [25] *Ibid.*, p. 75.

This uncomfortable situation did not long continue. The Tripos examination in January, 1845, opened a new door. 'One of Goodwin's problems', the second wrangler wrote in his notebook even before learning the results, 'suggested some consideration about the equilibrium of particles acted on by forces varying inversely as the square of the distance'.[26] He quickly drafted a paper showing that, if a 'mechanical theory' were adopted, in which electricity actually consisted of such particles, then all of the electricity on a conductor would have to be in contact with the surface and perpendicularity of force at the surface would supply 'the sole condition of equilibrium'. His proof rested on Earnshaw's theorem of 1839, which states that a material point cannot remain in equilibrium under inverse square forces alone. He concluded:

Since every particle is on the surface, the whole *medium* (if it can be properly so called), will be an indefinitely thin stratum, the thickness being in fact the ultimate breadth of an atom or material point. If we suppose these atoms to be merely centres of force, the thickness will therefore be absolutely nothing, and thus the *fluid* will be absolutely compressible and inelastic. Any thickness which the stratum can have must depend on a force of elasticity, or on a force generated by the contact of material points, and in either case will therefore require an ultimate law of repulsion more intense than that of the inverse square . . . As, however, all experiments yet made serve to confirm . . . that the stratum has absolutely no thickness, we conclude that there is no elasticity in the assumed electric fluid, and thus the law of force, deduced independently by direct experiments, is confirmed.[27]

Since the inverse square law now implied the perpendicularity condition, Thomson eliminated that assumption from his list of 'fundamental facts' in later works. He had reduced Poisson's electrical fluids to the condition of conservation of net charge (point (I), above). When in 1848 he finally wrote the introductory instalment of his projected treatise he remarked: 'In this, and in all the papers which will follow, instead of the expression "the thickness of the stratum", Coulomb's far more philosophical term, *Electrical Density*, will be employed . . . without involving even the idea of a hypothesis regarding the nature of electricity'.[28]

By that time the original plan for the treatise had evaporated, for on 27th January, 1845, eleven days after clarifying its principles and on the very day he was to leave for his debut in Paris, Hopkins had presented him with two copies of

[26] William Thomson, 11th January, 1845, NB33, ULC, p. 110. Harvey Goodwin, Fellow of Caius College, was Junior Moderator for the exam. His question stated: 'Eight centers of force, resident in the corners of a cube, attract, according to the same laws and with the same absolute intensity, a particle placed very near the center of the cube: shew that their resultant action passes through the center of the cube, unless the law of force be that of the inverse square of the distance.' See *University of Cambridge examination papers, 1845* (Cambridge, 1845).

[27] William Thomson, draft of 'Demonstration of a fundamental proposition in the mechanical theory of electricity', 11th January, 1845, NB33, ULC, pp, 110–15; extended in *Cam. Math. J.*, **4** (1845), 223–6; E&M, 100–3, on pp. 102–3.

[28] William Thomson, 'On the mathematical theory of electricity in equilibrium: II. A statement of the principles on which the mathematical theory of electricity is founded', *Cam. and Dublin Math. J.*, **7** (1848), 131–40; E&M, 42–51, on p. 48.

Green's *Essay*. 'I have just met with Green's memoir', Thomson added to his sketch, 'which renders a separate treatise on electricity less necessary'. After brief study the full extent of the damage became apparent: 'I see that Green has given *all* the general theorems in attraction (perhaps there is one excep^{n}) and that I have, most unwittingly, trodden almost exactly in his steps as far as regards electricity. He has not touched upon heat'.[29] Shortly, however, the situation would change again as Thomson began to grapple with the authentic Faraday, with the result that his own analysis deepened considerably and his speculations on contiguous action threatened to swamp his antihypothetical methodology.

Upon arriving in Paris he presented himself to Liouville. They struck an immediate friendship and met often thereafter, sometimes spending the time working through mathematical papers. Thomson had brought with him papers from Arthur Cayley and Augustus de Morgan, among others, as well as copies of the *Mathematical Journal*, partly by way of introduction and partly to promote his project for British mathematics. He also gave to Liouville a copy of Green's *Essay*, which 'creates a great sensation here, Chaslés and Sturm find their own results and demonstrations in it.'[30] The mathematical theory of electricity held a prominent place in their discussions. But news of Faraday had also reached Paris: 'Arago, it seems, has recently heard of Faraday's objections, and the uncertainty thus thrown on the theory prevented, as Liouville told me, its being made the subject for the mathematical prize of the Institute this year . . . However, as Poisson before he died wished Liouville to do anything he could for it, I think it will very likely be proposed again'. This exchange supplied the motivation for Thomson's reconsideration of Faraday: 'I told Liouville what I had always thought on the subject of those objections (i.e. that they are simple verifications), and . . . he asked me to write a paper on it'.[31]

The well-known result, 'On the elementary laws of statical electricity', went through four versions: a draft, a paper in French for Liouville, an abstract for the British Association, and a much-extended English version.[32] In each revision Thomson's respect for Faraday increased, especially after meeting the famous experimenter at the British Association in June and beginning a continuing correspondence. His initial attitude towards all qualitative investigators like Faraday, but especially another British experimentalist named William Snow Harris (1791–1867), whose significance we shall discuss in the next chapter, was

[29] William Thomson, 27th January and 4th February, 1845, NB33, ULC, pp. 68, 117. George Green, *An essay on the application of mathematical analysis to the theories of electricity and magnetism* (Nottingham, 1828); in N.M. Ferrers (ed.), *Mathematical papers of the late George Green* (London, 1871), pp. 1–115.

[30] William Thomson, 27th January, 1845, NB33, ULC, p. 68.

[31] William to Dr Thomson, 30th March, 1845, T303, ULC. See also William Thomson, 27th March, 1845, NB33, ULC, p. 156.

[32] William Thomson, 'On the [fundamental, *del.*] laws of the theory of statical electricity' (draft), NB32, ULC (the date 12th April, 1845 appears on p. 11); 'Note: sur les lois élémentaires de l'électricité statique', *J. de Math.*, **10** (1845), 209–21; 'On the elementary laws of statical electricity' (abstract, read 23rd June, 1845), *BAAS Report*, **15** (1845), 11–12; 'On the elementary laws' (dated 22nd November, 1845), E&M, 15–37.

thinly disguised contempt. The deletion from the following passage of the draft is typical. 'On an attentive consideration of their investigations however [it must be obvious to every person conversant with the, *del.*] it seems that all the experiments that they have made, relatively to the distribution of electricity on conductors, are results of the laws of Coulomb, and instead of objections are verifications of his theory'.[33] Snow Harris, the draft suggests, was not only incapable of comprehending mathematics, but incompetent to employ a proof plane correctly let alone to carry out a precise experiment on the forces between two electrified spheres that would offer a quantitative comparison with theory.

That sarcastic tone continued with respect to Faraday's theory, labelled earlier nothing but a 'way of *speaking*'. Faraday 'commences his investigation with experiments on *electrical induction in curved lines* [according to his own mode of expression which according to the views of Mr Harris cannot exist, *del.*]. This expression he uses to describe the electrical action which exists at any point exterior to a charged body, or group of charged bodies'. Immediately Thomson translated the expression into his own vocabulary: 'a line of inductive action or more generally a line of electrical force with reference to any electrified body, may be defined to be a line such that the resultant force upon any point in it, is in the direction of the tangent, or, which amounts to the same, as an orthogonal trajectory of the surfaces of equilibrium relative to the system'.[34] This translation of Faraday's 'way of speaking' into the geometrical picture of equipotential surfaces and perpendicular lines of force, as presented in the heat analogy, constituted an initial reconciliation of Faraday with mathematics.

But the details of the reconciliation show that the confident Cambridge wrangler had not yet grasped essential features of Faraday's view. Describing the closed surfaces of equilibrium surrounding a charged conductor, surfaces which would become spherical at infinity, he emphasized: '*Thus, if we commence with any point of the conducting surface, we may draw a line of electrical action, commencing at right angles to the surface, never meeting it again, and becoming ultimately a radius of the system of spheres*'.[35] Faraday would immediately have pointed out that lines of inductive action never trail off into space as though emanating from an isolated source of electricity (cf. figure 7.1). Such ideas might be satisfactory in an idealized mathematical theory, but physically no positive electricity ever appears without equal negative electricity. Thomson would have answered that the negative electricity could be regarded as at infinity, but the answer is irrevelant; he had missed the experimentalist's point that, since neither electricity could be produced alone, both should be explicitly present in every theoretical description. That position was characteristic of Faraday's view that the two electricities were nothing more than terminations of lines of inductive action, the manifestation of tension in the intervening space.

In other cases, more nearly like Faraday's, involving equipotential surfaces

[33] William Thomson, NB32, ULC, p. 2. [34] *Ibid.*, pp. 18–19. [35] *Ibid.*, p. 20.

between two conductors with opposite charges, Thomson allowed for inequality of the charges and thus once again for lines of electrical action with 'infinite branches'. These conceptual errors, on Faraday's theory (and an outright mistake or two on any theory), allowed him to assert that 'the complete theory of "induction in curved lines" . . . is a fundamental part of the mathematical theory'. He had as yet failed to appreciate that Faraday offered, not a verification of, but a comprehensive alternative to, the mathematics of action at a distance. By the end of the draft, however, he had surely begun to recognize the importance of the two electricities always appearing together, for he quoted from Faraday's *Experimental researches* (para. 1173): '"The conclusion I have come to is that non-conducting [bodies], as well as conductors, have never yet had an absolute & independent [charge] . . . comm[unicate]^{d} to them and that to all app[earance]^{s} such a state of matter is imposs."'.[36]

Subsequent versions of the paper contain no isolated charged bodies with lines of force vanishing at infinity. Lines typically begin on a charged conductor *A* and end on a grounded one *B*. In a somewhat disingenuous manner Thomson acted as though the entire idea had existed in his original heat analogy. Still, Faraday had acquired new status: 'All the views which Faraday has brought forward, and illustrated or demonstrated by experiment, lead to this [analogical] method of establishing the mathematical theory, and, as far as the analysis is concerned, it would, in most *general* propositions, be even more simple, if possible, than that of Coulomb . . . It is thus that Faraday arrives at a knowledge of some of the most important of the general theorems, which, from their nature, seemed destined never to be perceived except as mathematical truths'. By reading into the idea of lines of conserved flux an explicit concern with terminations *at both ends*, Thomson had assimilated Faraday's thoughts to his own. He continued (see figure 7.2):

> Thus, in his [Faraday's] theory, the following proposition is an elementary principle: Let any portion α of the surface of *A* be projected on *B*, by means of lines (which will be in general curved) possessing the property that the resultant electrical force at any point of each of them is in the direction of the tangent: the quantity of electricity produced by induction on this projection is equal to the quantity of the opposite kind of electricity on α. The lines thus defined are what Faraday calls the 'curved lines of inductive action'.[37]

This projection theorem had marked the culmination of the Eleventh Series (1837) of the *Experimental researches*, which Thomson had read even for his original draft. Faraday wrote:

> Thus *induction* appears to be essentially an action of contiguous particles, through the intermediation of which the electric force, originating or appearing at a certain place, is propagated to or sustained at a distance, appearing there as a force of the same kind exactly equal in amount, but opposite in its direction and tendencies.[38]

36 *Ibid.*, pp. 24, 28. 37 'On the elementary laws', E&M, 29–30.

38 Michael Faraday, *Experimental researches in electricity* (3 vols., London, 1839, 1844, 1855), **1**, para. 1295.

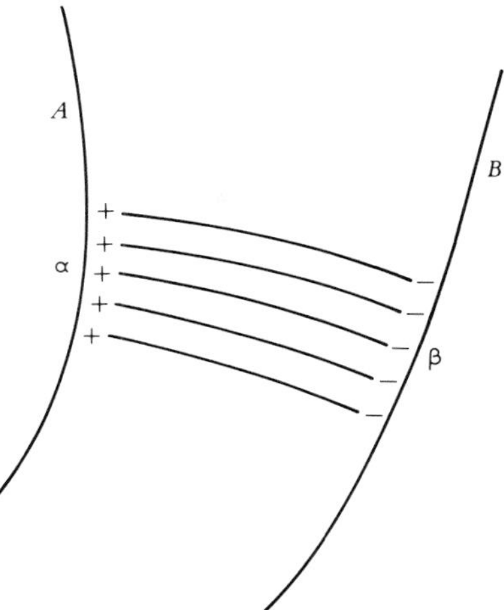

Figure 7.2. After assimilating Faraday's ideas, Thomson conceived lines of force (flux) as lines of projection between a given quantity of positive electricity and an equal quantity of negative electricity.

Inductive capacity and the quantity–intensity distinction

So far we have discussed only what might be called Faraday's ideological objection to the doctrine of action at a distance. But his alternative doctrine of contiguous action had led him to search for an effect of the medium through which propagation occurred. He discovered what the mathematical theory had never contemplated, a 'specific inductive capacity' in solid non-conductors, or 'dielectrics'. His confidence in contiguous action, evident in the last quotation, depended largely on that experimental measurement. Thomson originally understood the measurement as imperfectly as he had Faraday's lines of induction, and for much the same reason. He stated the result in his draft as follows:

If any insulator such as glass, sulpher or lac be interposed between a charged body, and an uninsulated [grounded] body, the *quantity of electricity* called out by influence, on the latter is greater than if air or any gas be the only intervening insulating medium. The quality of the medium which produces this effect which he calls its 'dielectric power', he finds to be different for different non conductors and he gives numerical determinations for sulpher, lac, and glass which, whatever may be their *real meaning*, are at least in order of magnitude of the effects represented.[39]

Aside from a negative attitude towards Faraday's language and towards the accuracy of his measurements, this passage displays a view of the inductive effect which would be incoherent on Faraday's principles. For a given charged body A, the 'quantity of electricity called out by influence' on the grounded conductor B must always be the same, according to his projection theorem. Faraday had

[39] William Thomson, NB32, ULC, pp. 25–6. Our emphasis.

in fact not measured the quantity of electricity on the grounded conductor *B* at all, but the potential (in Thomson's terms) on *A*, and had shown that this potential decreased by a definite amount when a dielectric medium was interposed between them. Thomson recognized that fact by the time he made his own calculations of the effect (with *k* the specific inductive capacity): 'Hence the potential in the interior of *A* will be V/k, or the fraction $1/k$ of the potential, with the same charge on *A*, and with a gaseous dialectric'.[40] He had also learned that Faraday's measurements indicated not merely the 'order of magnitude of the effects represented', but the effects precisely. Only their 'real meaning' remained in question.

The mathematical theorist was learning rapidly to appreciate that the experimental philosopher had important lessons to teach him. He never would learn, however, to accept the full range of Faraday's 'way of speaking'. For Faraday loaded the concept of force with a duality of aspects that Thomson always regarded as gratuitous, both in Faraday's theory and in Maxwell's, where the electromagnetic theory of light would depend on it. The duality was one of *quantity* and *intensity*, and it exercised Thomson for over half a century, ultimately symbolizing the failure of his methodology in competition with Maxwell's. To see the initial problem clearly we must consider further Faraday's measurement of inductive capacity.

The measurement reflects procedures that Faraday had developed earlier for electrochemical experiments, which in turn reflect his early attempts to unify statical electricity with electricity of currents. The former involved the high tension or 'intensity' evident in the sparks produced by an electrical machine, while the latter involved the large 'quantity' of electricity stored in a voltaic cell.[41] Both statical and current electricity, Faraday showed, would produce chemical decomposition, which provided one ground for treating intensity and quantity as two aspects of the same basic power. The language of lines of force allowed him to represent intensity as tension along a line and quantity as the number of lines crossing unit area. The quantity–intensity distinction, therefore, provided essential meaning to Faraday's lines.

While he regarded the lines themselves as, in the first instance, a descriptive device, he also regarded them as representing, in a non-hypothetical manner, the physical state in an electrochemical cell. Each line represented a chain of polarized particles, such as he supposed must precede decomposition with its release of chemical constituents at the electrodes and concomitant release of positive and negative electricities. Conduction in such a cell consisted of a successive establishment and release of tension; with the two electricities propagated in opposite directions along the lines of tension. Faraday thought of the

[40] 'On the elementary laws', E&M, 33.

[41] Faraday, *Experimental researches*, **1**, paras. 265–379. For further discussion of issues relevant here, see David Gooding, 'Metaphysics versus measurement: the conversion and conservation of force in Faraday's physics', *Ann. Sci.*, **37** (1980), 1–29.

electricities as 'powers' or 'forces'. Polarity involved separation of contrary powers; and a line of force, whether of current or only of tension preceding a current, was '*an axis of power having contrary forces, exactly equal in amount, in contrary directions*'.[42]

This complex of ideas constituted Faraday's model of the lines of inductive action between electrified conductors. Just as during conduction the total current through a cross-section remained constant for an entire circuit, so also induction conserved the total electrical force across the lines: 'The idea is, that any section taken through the dielectric across the lines of inductive force, and including *all of them*, would be equal, in the sum of the forces, to the sum of the forces in any other section'.[43] Here is the projection theorem again. Electrical forces propagated through the dielectric appear at the conducting surfaces in amounts equal to the total quantity of force anywhere along the lines.

Faraday's conception of quantity and intensity as measures of electrical force across and along the lines, respectively, mirrors his experimental arrangements. Often the electrochemical experiments involved combinations of decomposition cells and voltaic cells connected in parallel (adding quantities) and in series (adding intensities), displaying attendant changes in quantity of decomposition and in intensity required to initiate decomposition.[44] Such observations also involved his general idea of conservation of all powers and of conversion of power from one form to another. In a simple circuit containing a voltaic cell and a decomposition cell, for example, the power set free in the first had its equivalent in the power expended in the second. In such cells Faraday supposed quantity and intensity to be proportional, the proportion depending on the nature of the substances: 'The intensity of an electric current traversing conductors alike in their nature, quality, and length, is probably as [proportional to] the quantity of electricity passing through a given sectional area perpendicular to the current, divided by the time'.[45]

All of these ideas appeared in the researches on induction as well. Faraday's basic induction apparatus, by analogy to an electrochemical cell, consisted of two conductors separated by a dielectric. Assuming conservation of total quantity in a charged system, he changed the plate area, the distance between the plates, and the dielectric medium in order to observe resulting changes in intensity. By changing the medium he established the effect of dielectrics on induction.

To measure precisely the new effect, Faraday constructed two identical sets of apparatus, each a Leyden jar (capacitor) consisting of two concentric spherical

[42] Faraday, *Experimental researches*, **1**, para. 517; cf. para. 1168.

[43] *Ibid.*, **1**, para. 1369, from Series Twelve, which followed one month after Series Eleven and in which Faraday showed parallels between induction and conduction. See also para. 504 on currents.

[44] *Ibid.*, **1**, para. 908 on quantities and cf. para. 726; on intensities see paras. 985, 990, 1025.

[45] *Ibid.*, **1**, para. 1369, where Faraday goes on to show how quantities and intensities can be measured by placing a standard water decomposition cell in series and in parallel, respectively, with the current to be tested.

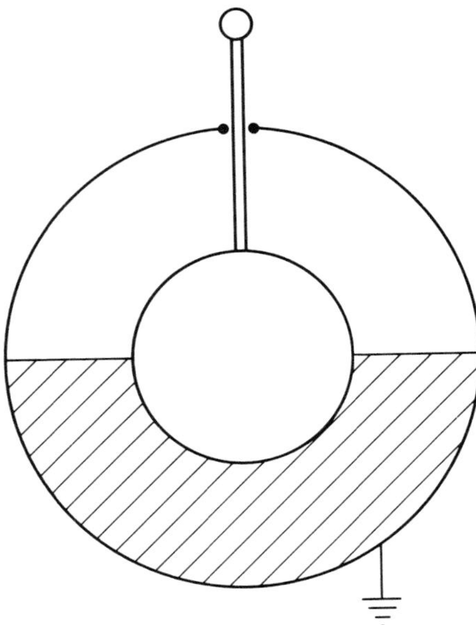

Figure 7.3. Faraday employed twin spherical Leyden jars for measuring specific inductive capacity. The one represented here is half-filled with a dielectric substance and has a protruding rod for measuring the 'degree of charge' of the apparatus, the outer sphere being grounded.

conductors, as in figure 7.3. The outer shell he grounded and the inner one he made accessible through a small hole. An insulated wire protruding through the hole connected the inner shell to a small metal ball from which the inner shell could be charged and its 'degree of charge' (intensity) tested. To touch the ball of one charged apparatus to the other uncharged one was to connect them in parallel and to 'divide the charge' (quantity) between them. This division Faraday regarded as a 'conversion' from power in the one to power in the other until their 'degrees of charge' were equal.[46]

Testing 'degrees of charge' with Coulomb's torsion electrometer, he measured the effect on the division of charge of inserting various dielectric media between the shells of one of the Leyden arrangements. Apparently setting the loss of quantity in one equal to the gain in the other, and the change in each proportional both to its change of 'degree of charge' and to its inductive capacity, Faraday obtained a quantitative result which corresponds to the equation $k_1V_1 = k_2V_2$. Using air as a standard in the first apparatus ($k_1 = 1$), charged to degree V_0, with the second originally uncharged, and after division

[46] *Ibid.*, e.g. **1**, paras. 1234, 1260.

labelling the common degree of charge V, the specific inductive capacity of the medium in the second apparatus is

$$k = k_2/k_1 = (V_0 - V)/V.$$

This result is the one Faraday obtained, and it agrees with Thomson's, stated above, for the reduction of potential between conductors A and B upon introduction of a dielectric of capacity k between them.[47]

The entire account is very simple, but only after we have translated it from the language of 'division of charge' and 'degree of charge' first into Faraday's quantities and intensities and then into the quantities of electricity and potentials of the mathematical theory. Thomson faced the same difficulties, and surmounted them through his heat analogy. Flux and temperature gradient bear close analogy to quantity and intensity of lines of force, including their relation to sources; and Faraday's description produces the same geometrical picture as the heat analogy. Nevertheless, quantity and intensity as dual aspects of force remained as foreign to the mathematical theory of electricity as the distinction of flux and temperature gradient. We shall find that Thomson collapsed the new concepts into a mathematical identity in the same way as he had those of the earlier analogy.

In spite of his own speculations on contiguous action, Thomson recognized that Faraday's discovery of inductive capacity did not *demand* a propagation theory. Poisson had developed long before a theory of magnetic polarization based on action at a distance. Thomson studied those papers, along with Green's elegant reformulation of them, before beginning his new reading of Faraday. Noting in his draft that, 'when any medium is found to make the inductive effect [at constant potential] different from that which is produced across a gas, it is always increase [of charge] that is produced' ($k > 1$), he inferred an action like magnetic polarization, with no effect in 'non-magnetic' media like air:

> From this fact we are irresistably led to the conclusion that, if the insulating medium be a gas, the inductive action [on a conductor] is solely due to the attraction of the charged body, but that if any of the ordinary solid non-conductors be interposed there is an *additional* action, due to some effect of the attraction of the charged body on the particles of the 'dielectric'. The experimental data we have at present are not sufficient to enable us to pronounce with certainty upon the nature of this action but as far as they go, they seem to indicate an internal 'polarisation' of the same nature as that produced in a vessel [?] of soft iron by a magnet.[48]

[47] *Ibid.*, **1**, para. 1259. Because the dielectric in Faraday's spherical induction apparatus occupied only a hemisphere, the measured value k_m was related to k by the relation $k = 2k_m - 1$, which follows from treating the two hemispherical halves as two capacitors in parallel. cf. Buchwald, 'William Thomson', pp. 125–30, for a different interpretation of conservation and conversion.

[48] William Thomson, NB32, ULC, pp. 26–7. As he quoted in his final version (p. 32*n*), Faraday had proposed a similar model in *Experimental researches*, **1**, para. 1679: 'If the space round a charged globe were filled with a mixture of an insulating dielectric . . . and small globular conductors, as shot . . . then these [latter] in their condition and action exactly resemble what I consider to be the condition and action of the particles of the insulating dielectric itself. If the globe were charged, these little conductors would be all polar'.

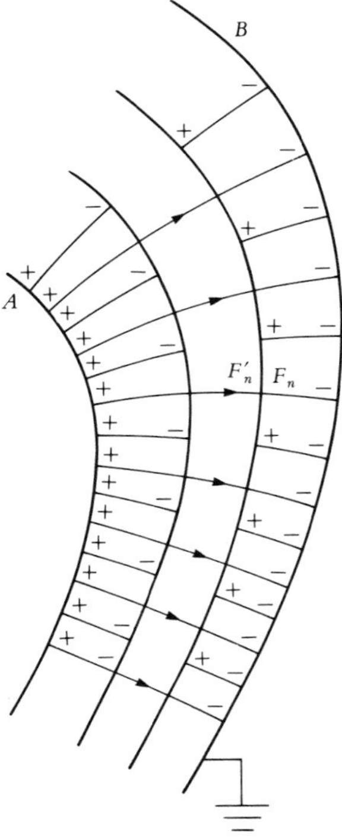

Figure 7.4. A dielectric placed between two charged conducting sheets, with the surfaces of the dielectric matching equipotential surfaces of the original charged system, becomes polarized in such a way that its effects can be represented by electricity on its surface. This electricity decreases the force (potential gradient) inside the dielectric and therefore decreases the total potential between the charged sheets.

He then stated immediately the basic theorem that he would employ in his final version to show complete equivalence between the effect on potential of this polarization by action at a distance and the effect Faraday had observed. The theorem was essentially the one on replacement surfaces that he had originally established via the heat analogy.

Complete proof, however, required him to show that the dielectric could be replaced by imaginary sources ρ on its surface, as in figure 7.4, such that the normal component of force across the surface changed by $4\pi\rho$,

$$F_n - F_n' = 4\pi\rho,$$

where F_n is the normal force outside and F_n' is that inside the polarized medium. He also showed that the ratio of normal components could be expressed by a constant k, 'depending on the capacity [of the medium] for dielectric induction',

$$F_n'/F_n = 1/k.$$

These two expressions, together with the replacement surface theorem, yielded Faraday's result.[49] Here then is the simple but impressive defence of the mathematical theory that Liouville had commissioned. The question remains whether it constituted a reconciliation between action at a distance and contiguous action via the heat analogy.

Already in the draft Thomson had claimed: 'that the phenomena of electricity are modified by the action of dielectrics in exactly the same manner as those of the uniform motion of heat would be if, in the spaces corresponding to the dielectrics, have [sic] a conducting power which varies with the temperature, while this quantity is constant for the rest of the body'. He pursued the idea of variable conductivity (though with respect to change in material, not temperature) as he worked out the conditions of polarization, recording success on 12th July after his return to Cambridge: 'I have just made out definitely that the action of soft iron or (hypoth^{y}) of sulpher &c in electricity as in the math^{l} theory, the same as in heat of the presence of a body of the same shape but of greater conducting power'. One must therefore take seriously a footnote in the final version attached to the boundary condition, $F_n'/F_n = 1/k$: 'From this it follows that, in the case of heat, C [the dielectric] must be replaced by a body whose conducting power is k times as great as that of the matter occupying the remainder of the space between A and B'.[50]

We recall that Thomson correlated force with flux in his heat analogy. That correlation provided his intuitive grasp of Faraday's projection theorem, for conservation of heat required continuity of flux along lines of flow. But here we find him expressing a *discontinuous* change of force across an interface where, on the heat analogy, flux would remain constant, although conductivity and temperature gradient would change in reciprocal proportion, as in his equation. In his notebook, in fact, the entire calculation is carried out in terms of potential gradients, yielding,

$$(\partial V'/\partial n)/(\partial V/\partial n) = 1/k.$$

Evidently, then, he actually considered temperature gradient, not flux, to be the proper mathematical analogue of force, i.e. of potential gradient. Unfortunately, that identification destroys the projection theorem in Faraday's sense.

The problem is that no neutral mathematical theory can encompass completely both Faraday's propagation theory and Poisson's action at a distance theory, for the former relates sources to flux – quantity – and the latter relates them to potential gradient – intensity. Only in homogeneous media do flux and potential gradient behave similarly. One can hardly doubt that Thomson

[49] 'On the elementary laws', E&M, 33–5. Thomson's original derivation, 12th July, 1845, NB34, ULC, pp. 8–10, is adapted from Green, *Essay*, pp. 92–3. Analysis is given in Buchwald, 'William Thomson', pp. 131–3; Wise, 'Flow analogy', pp. 48–9.

[50] William Thomson, NB32, ULC, p. 28; 12th July, 1845, NB34, ULC, p. 8; 'On the elementary laws', E&M, 33. For discussion of this boundary condition, see Knudsen, 'Mathematics and physical reality', pp. 155–6; Wise, 'Flow analogy', pp. 49–50.

recognized the distinctions of quantity–intensity and flux–gradient in the works of his heroes of contiguous action, Faraday and Fourier, yet he continued to insist on the strict analogy of force lines and flux lines. In 1854 he added a second note to the boundary condition above, referring now to a body magnetized by induction: 'the lines of magnetic force will be altered by [the body] precisely as the lines of motion of heat in corresponding thermal circumstances would be altered by introducing a body of greater or less conducting power for heat. Hence we see how strict is the foundation for an analogy on which the *conducting power of a magnetic medium for lines of force* may be spoken of'.[51]

Obviously Thomson believed he had achieved the sought-after reconciliation. Study of his unpublished papers, however, shows that the reconciliation was not the one he allowed his readers to see and that commentators have commonly ascribed to him. He did not in fact believe that Faraday's conception of electricity, as nothing but terminations of lines of force, was justified experimentally. Rather he treated electrical intensity as though it were a real density of a conserved substance. He did believe that force was propagated, but only as a tension produced by concentrations of electricity. Because he treated electricity as a conserved substance and force as a propagated tension, his own propagation theory (as opposed to Faraday's) was equivalent mathematically to Poisson's theory, but he neglected to specify what the differences were between his and Faraday's views.

That differences existed in 1845 is apparent from the inconsistency, noted above, in the strict analogy of force to flux. Only in notes taken by a student in his natural philosophy class in 1852–3, however, do we find positive evidence of what his conception must have been.[52] Discussing the various theories of electricity, single fluid, double fluid, and Faraday's, he remarked: 'The one hypothesis of one electric fluid seems far the most probable. Faraday says that it [electricity] is the air which is put into a state of tension – bodies are pulled asunder by air', and slightly later, 'Hypothesis of one electric fluid may be perhaps reconciled to probability by Faraday's theory of tension'. Again, 'An electrified body puts the air around it in tension & transmits this tension all round towards all points. Most probably the atmosphere is excited'. For Thomson the question was not whether electrical force was propagated as a tension but whether that tension was electricity. He thought not. Faraday's own finding, that induction through gases did not depend on their density, suggested that tension in air did not involve anything like polarization in a solid dielectric, so that electrical density at the surface of a conductor, for example, could not be regarded as simply the manifestation of tension in the surrounding air: 'The 2

[51] 'On the elementary laws', E&M, 33*n*. This note of March, 1854, is one of several added to the reprint in *Phil. Mag.* [series 4], **8** (1854), 42–62.

[52] William Thomson, in William Jack, 'Notes of the Glasgow College natural philosophy class taken during the 1852–53 session', MS Gen. 130, ULG. The notes have no page numbers and few dates or lecture numbers.

electricities for all we know, cannot be separated in air. It is possible that air may have, for very minute portions, such a separation. In rare & dense air the same property is held to the same degree, and in all gases the same . . . There may be or may not be then inductive electrification of air. Probably there is not'.

These remarks refer back to Faraday's measurements of specific inductive capacity, on which Thomson had based his reconciliation in 1845. He here repeated the basic idea, drawing it from Faraday (see note 48) rather than Poisson:

> General explanation (Experimental researches, [para.] 1679 or 1639). A plate of sulphur acts like a solid resister of electricity filled with conductors throughout. He [Faraday] supposes that there is some such polar action in air. It appears that there is probably no such polar electrification in air.

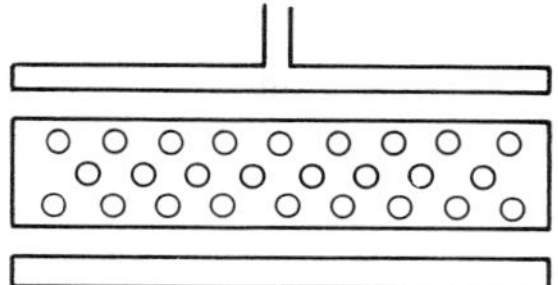

An important corollary to Thomson's view is that he, in contrast to Faraday, believed that electrical force acted even inside conductors, but that the net force vanished because contributions from different portions of electricity on the surface cancelled one another: 'All the electrical forces give a resultant equal to 0 on any interior point. Faraday & Snow Harris say that elect^{y} dont operate through conductors. Faraday considered that it was a tension of the air on the outside entirely. But electrical force acts everywhere, in every direction'. Since this view is precisely one that Thomson expressed in the draft of his 1845 paper, it seems safe to assume that he also entertained at that time the accompanying conceptions of electricity and tension. We shall find considerable support for this view in the next chapter, when we take up his work of 1843 on the question of how the surrounding medium (air or aether) exerts ponderomotive force on an electrified body. Still, his ambivalence towards any public commitment to a physical theory of propagation is manifest. For example, responding in 1848 to a request from Whewell (apparently) for a statement on the relation of Faraday's views to Coulomb's, Thomson responded: 'It appears to me that Faraday's hypothesis [is a step farther, *del.*] of propagation may be considered [the foundation, *del.*] very probably [the] *physical explanation* of the positive mathematical theory founded on Coulomb's laws'.[53]

If the preceding picture represents correctly the situation in 1845, then the heat analogy should be read as follows. The continuity equation,

$$\text{divergence } k\vec{F} = 4\pi\rho,$$

can be taken to represent electrostatic action through dielectrics, where $\vec{F} = -\text{gradient } V$ is force and $k\vec{F}$ is the analogue of flux produced by free electricity ρ. In reality, however, the 'conductivity' k is only a parameter which represents the effect of induced electricity located on the surface of the dielectric (or wherever inhomogeneities occur). This induced electricity results

[53] William Thomson to [William Whewell?], 30th November, 1848, K12, ULC.

from a real polarization, in which electrical substance moves within small conducting elements inside the dielectric. Taking these induced sources ρ' into account, the continuity equation involves no variable conductivity and becomes:

$$\text{divergence } \vec{F} = 4\pi(\rho + \rho').$$

$\vec{F}$ is now analogous to flux, as Thomson always claimed, but it is *not* continuous across dielectric boundaries, as a true flux would be across interfaces between media of different conductivity.

In this restricted form the heat analogy constitutes a literal translation of Poisson's action at a distance theory into Thomson's conception of propagated electrical action. A phenomenological electrical intensity replaces density of electrical fluid and a propagated electrical tension replaces force acting at a distance. Otherwise the theories are identical. Thomson therefore recognized that, logically, Faraday's discovery of dielectric action gave no more support to the doctrine of propagation than to the ordinary doctrine. He concluded in a characteristically neutral vein:

> It is, no doubt, possible that such [electrical] forces at a distance may be discovered to be produced entirely by the action of contiguous particles of some intervening medium, and we have an analogy for this in the case of heat, where certain effects which follow the same laws are undoubtedly propagated from particle to particle. It might also be found that magnetic forces are propagated by means of a second medium, and the force of gravitation by means of a third. We know nothing, however, of the molecular action by which such effects could be produced, and in the present state of physical science it is necessary to admit the known facts in each theory as the foundation of the ultimate laws of action at a distance.[54]

The latitudinarian theorist and the practical imperative

Having established by 1845 a basic view of electricity, Thomson maintained essentially the same perspective until his death. At no time did he accept the idea of a 'quantity' such as polarization accompanying electrical 'intensity' (tension) through space. That is, he rejected both Faraday's and Maxwell's concepts of electrification and, consequently, in Maxwell's case, would reject the flux called 'displacement current' which made possible the beautifully symmetric mathematical structure of 'Maxwell's Equations'.[55] No more direct experimental evidence existed for Maxwell's displacements in space than for Faraday's polarizations. Damning the irresponsibility of Maxwellian theorists in 1896, he wrote: 'It is not the equations I object to. It is the being satisfied with them, and with the pseudo-symmetry (pseudo, I mean, in respect to the physical subject) between electricity and magnetism. I also object to the damagingly misleading

[54] 'On the elementary laws', E&M, 37.

[55] On Maxwell's concept of electrification, see J.Z. Buchwald, *From Maxwell to microphysics* (Chicago, 1985), chap. 3.

way in which the word "flux" is often used, as if it were a physical reality for electric and magnetic force, instead of merely an analogue in an utterly different physical subject for which the same equations apply, see [my heat flow analogy of 1841]'.[56]

That Thomson adhered to his non-hypothetical perspective with such consistency throughout his life points to deep commitment. Of course he had been educated in a tradition of natural philosophy that stressed descriptive empirical law and avoided speculation, but his style of empiricism was far more specific, more personal, than national or even university traditions suggest. It bears the stamp of the Thomson family. We have stressed earlier Dr James Thomson's position – on religion, politics, and mathematics alike – as at once *latitudinarian* and *practical*, avoiding doctrinal disputes of every kind while seeking the fruits of shared truth. Recall the preface to his 1844 *Algebra*: 'With regard to all the practically useful applications and interpretations of algebra, there is no difference of opinion among men of sound science and judgement'.[57] William expressed this same combination of values in his rapidly maturing physical mathematics of 1845.

The latitudinarian commitment is most apparent in the heat analogy itself, employed as it was to reconcile supposed incompatibilities between the mathematical theory and Faraday's discoveries, and to elicit the truth that necessarily underlay them both:

As it is impossible that the phenomena observed by Faraday can be incompatible with the results of experiment which constitute Coulomb's theory, it is to be expected that the difference of his ideas from those of Coulomb must arise solely from a different method of stating, and interpreting physically, the same laws: and farther, it may, I think, be shown that either method of viewing the subject, when carried sufficiently far, may be made the foundation of a mathematical theory which would lead to the elementary principles of the other as consequences. This theory would accordingly be the expression of the ultimate law of the phenomena, independently of any physical hypothesis we might, from other circumstances, be led to adopt.[58]

If we simply call this viewpoint non-hypothetical we miss its ideological content, the 'non-sectarian', 'non-dogmatic', 'above-party-and-conflict' meanings that permeated the lessons Dr Thomson taught his family. Having fully absorbed those values into his own life and work, William could hardly have espoused publicly one or another doctrine of electricity without exceedingly strong reason or a personal crisis, despite his preference for propagation. The crisis would in fact come soon enough, but with respect to heat rather than electricity, and coinciding with Dr Thomson's death. The point to be made here is that William's non-hypothetical methodology should be understood to involve his deepest emotional commitments and family ties, along with his

[56] William Thomson to G.F. Fitzgerald, 29th April, 1896, in SPT, **2**, 1071.

[57] James Thomson, *An elementary treatise on algebra, theoretical and practical* (London, 1844), p. vi. See Chapter 1 above.

[58] 'On the elementary laws', E&M, 26.

social and political commitments, and not merely a rational view of what constitutes good science. All of these aspects appear repeatedly in his life and his science.

But how was the theorist of latitudinarian persuasion to locate the essential truths of physical phenomena, above doctrine and independent of a hypothetical model? Through mathematics, answered the son of the mathematics professor, who in his *Algebra* had 'carefully kept clear of every thing of a metaphysical or disputed character'. William would teach his natural philosophy class that mathematics constituted 'the only true metaphysics'.[59] Not every mathematical theory would serve, of course. Only those founded on laws that expressed *directly* the results of experiment, such as Coulomb had discovered in the case of electricity, would qualify: 'the elementary laws which regulate the distribution of electricity on conducting bodies have been determined by means of direct experiments, by Coulomb, and in the form he has given them, which is independent of any hypothesis, they have long been considered as rigorously established'.[60]

A proper mathematical theory would seek initially to describe the macroscopic geometry which the elementary laws implied, as Fourier had done for heat and as Thomson himself had done for electrical forces. These theories, furthermore, gave the geometry of heat and of force as generated by their 'motion', their propagation through space. They were therefore 'kinematical' theories, in the sense of geometry of motion. The term is appropriate here also, however, because it designates a theory of motion which is independent of the true nature of the thing moved and independent of beliefs about causal mechanisms, that is, prior to dynamics. It was in that sense that Thomson regarded kinematics as 'a part of metaphysics', meaning the purely mathematical part of theory.[61] His reconciliation of action at a distance with contiguous action had uncovered the true metaphysics of electricity, the neutral geometry that expressed the essential truths on which the divisive 'sects' of action at a distance and contiguous action would have to agree. In retrospect, then, the entire heat analogy can best be labelled kinematics, the kinematics of forces disbributing themselves through space, or the kinematics of field theory.

The second essential aspect of Thomson's non-hypothetical commitment, practicality, also concerns his emphasis on geometry. Just as elementary laws were to express directly the results of experiment, so geometrical development of those laws would provide a directly visualizable, intuitive grasp of the mathematical theory, a grasp immediately applicable to the solution of practical problems. Like the mechanical models of ether for which he would later become famous, geometrical visualization and geometrical methods of solution were to

[59] As reported by SPT, **2**, 1124.

[60] 'On the elementary laws', E&M, 15.

[61] William Thomson, 5th November, 1862, in David Murray, 'Lecture notes in *classe physica*, bench II, November, 1862', MS Murray 325, ULG.

Thomson great 'brain savers'.[62] Again, the heat analogy first displayed this characteristic of his method. It asked, in general, what is the relation between the equation governing physical processes in a given region of space and the conditions at the boundary of that region? Representing the space by a series of isothermal surfaces and flow lines gave a particularly clear view of the problem and the replacement surface theorem allowed simple solutions for particular cases. In fact, this representation gave him in his 1841 paper the distribution of temperature (or potential) everywhere inside or outside an ellipsoidal isothermal surface, as a family of confocal isothermal ellipsoids.[63] It also gave him the distribution of sources of heat on the original ellipsoid and the distribution of flux throughout space, all with a minimum of analysis. Such power epitomized practicality. The method of physical geometry had allowed a seventeen-year-old boy to compete with Gauss and Chaslés.

The ellipsoid presents the next simplest geometry to a sphere. To solve problems with less symmetry in a similarly fluent manner, Thomson sought new methods. While in Paris he invented two, the first eminently practical – the method of images – and the second of great theoretical significance also – relating forces to mechanical effect, or work done. Both concerned the heat analogy and both provided techniques for actually calculating the distributions of electricity and force for two mutually influencing spheres, a case to which Poisson had applied his most powerful analytic techniques and which Snow Harris had investigated experimentally. We shall treat here the purely geometrical method of images, reserving mechanical effect for the following chapter.

In describing the 'principle of electrical images' to the British Association in 1847, Thomson remarked that it 'is suggested by Green's elementary propositions'. Indeed, on 4th February, 1845, while drafting a paper inspired by Green, he broke off to enter in his notebook the earliest existing record of the principle.[64] Reading Green from his own perspective, he seems to have recognized that the replacement surface theorem could be inverted, to ask for imaginary sources inside the surface of a conductor that would replace actual electricity on the surface. One could thus 'reflect' a point source in a conducting surface to find an image point inside the surface that would produce the same force outside as the induced electrification of the surface. By extension, several conducting surfaces would reflect each others' image points, and the net effect of induction in all the surfaces would be the sum of the effects of the 'successive'

[62] William Thomson, *Notes of lectures on molecular dynamics and the wave theory of light*, papyrographed stenographic notes by A.S. Hathaway (Baltimore, 1884), pp. 37, 80, 171, give mechanical models that 'save brain'. Other brain savers are symmetrical method, p. 176, and chalk (writing out in full), p. 32. [63] 'Uniform motion', E&M, 7–14.

[64] William Thomson, 'On electrical images', *BAAS Report*, **17** (1847), 6–7; 4th February, 1845, NB33, ULC, p. 133. The draft is of 'Démonstration d'un théorèm d'analyse', *J. de Math.*, **10** (1845), 137–47, which derived from George Green, 'On the determination of the exterior and interior attractions of ellipsoids of variable densities', *Trans. Cam. Phil. Soc.*, **5** (1835); *Mathematical papers*, pp. 185–222, esp. p. 222.

images. Thomson's original note expressed this insight for a point source (of light) between two parallel reflecting planes, the point being reflected back and forth an infinite number of times.

By far the most important early application of the method of images supplied a solution of Poisson's two-spheres problem. On 15th March he recorded: 'I am occupied the whole day in Regnault's physical laboratory at the Collège de France. At spare times I have been reading Poisson's memoirs on Electricity, w^h I find among the memoirs of the Institute, in Regnault's cabinet. I have applied my ideas on induction in spheres and the principle of successive influences, & get a very simple solution, in the form Poisson gives it, for two spheres. I think I can work it out for i spheres, & express the $dist^n$ by means of $(i-1)^{ple}$ int^{ls}. I can also by means of i^{ple} int^{ls} find very simply the $attr^n$ on any one, of the others'.[65] He then sketched the method for later development.

The problem for a single sphere and a point is contained almost explicitly in Green's *Essay*, which exhibits generally a fascination with reciprocity relations. With respect to grounded spheres, Green considered the effect on an exterior (or interior) point p of electricity induced on the surface by a unit charge located at an interior (or exterior) point P. The resultant effect of P and the surface must always vanish at p. By letting the point p move until it contacted the surface Green was able very simply to show that the electrical distribution induced on the surface by a point charge, has the same form whether the charge is inside or outside, and varies as the inverse cube of the distance f from the influencing point P to the surface.[66] Both cases are represented in figure 7.5, where P and P' are influencing points inside and outside, respectively, f, f' and b, b' are their distances to the surface at p and to the centre O of the sphere, and a is its radius. The surface densities per unit inducing charge are:

$$\rho = (b^2 - a^2)/4\pi a f^3$$
$$\rho' = -(b'^2 - a^2)/4\pi a f'^3.$$

One can imagine Thomson looking at those results with his replacement surface theorem in mind and asking himself the following question. If the distribution induced on the surface by P inside has the same general form as that which would be induced by P' outside, can a point P occupied by a charge Q not be found that will replace the actual surface distribution induced by a unit charge at P', producing the same effects everywhere outside? He would soon have found the answer in what he later called 'a well-known geometrical theorem' (given in Dr Thomson's *Euclid*).[67] If b, a, and b' form a continued proportion, $b/a = a/b'$, then f and f' are proportional for all points p on the surface,

[65] William Thomson, 15th March, 1845, NB33, ULC, p. 152.

[66] Green, *Essay*, pp. 50–4. See also pp. 36–9 for a similar reciprocity of points inside a sphere.

[67] William Thomson, 'On the mathematical theory of electricity in equilibrium: III. Geometrical investigations with reference to the distribution of electricity on spherical conductors', *Cam. and Dublin Math. J.*, **3** (1848), 141–8, 266–74; **4** (1849), 276–84; **5** (1850), 1–9; E&M, 52–85, on pp. 64 and 58*n*. The same analysis for spheres appears already on 5th July, 1845, NB34, ULC, pp. 5–6, which also contains the reference to Dr Thomson's *Euclid*, VI. Prop. G.

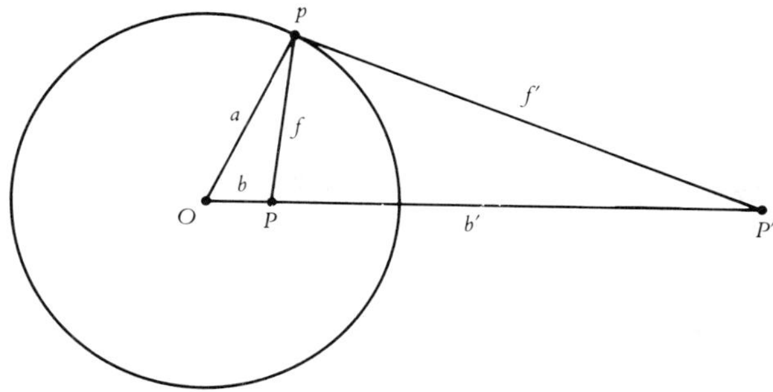

Figure 7.5. Green analyzed the reciprocal relation between charge distributions induced on a spherical conductor by a point charge placed either at *P* (inside) or *P′* (outside).

$$f'/f = b'/a = a/b = \text{constant}.$$

Thus Thomson began his sketch: 'The *image* of an exterior point in a conducting sphere, is a p^t in the interior, with opposite electr^y. $b = a^2/b'$, $Q = -a/b' = -b/a$'.[68] The latter result follows from requiring equality of the densities ρ and ρ', above. He moved immediately to the problem of two spheres, one charged and insulated and the other grounded, by regarding all of the charge on the first as at its centre (equivalent to an initial uniform distribution on the surface). Reflecting the images of this initial point charge back and forth between the spheres, using the relations above for locations and magnitudes, yielded an infinite series of image charges, each derived from the preceding by a simple recursion formula. To find the actual distribution of electricity induced on either sphere required only summing contributions corresponding to the images within it. Similarly, the total force acting between the spheres could be regarded as the sum of the forces between the image charges. The problem therefore reduced to calculating the required sums, or to approximating them by the first few terms of the rapidly converging series.[69]

While extremely simple conceptually, and while yielding approximate solutions quite readily, the method of images nevertheless gave for the exact force between two spheres a messy double sum over the images in both spheres. Thomson realized that the double sum could be replaced by a double integral, thus simplifying the expression somewhat.[70] Still the double integral defied

[68] William Thomson, 15th March, 1845, NB33, ULC, p. 153. We have altered the notation.

[69] Only in 1853 did Thomson publish a full development of this sketch, 'On the mutual attraction or repulsion between two electrified spherical conductors', *Phil. Mag.*, [series 4], **5** (1853), 287–97; **6** (1853), 114–15; E&M, 86–97, on pp. 86–92. He there remarks (p. 87) that the 'principle of successive influences' was suggested by Murphy, *Electricity*, p. 93. Since no reference to Murphy appears in his notebooks, however, it is doubtful that Thomson had recognized Murphy's suggestion earlier.

[70] William Thomson, 28th March, 1845, NB33, ULC, pp. 152, 159. In 1849 Thomson finally found the direct reduction, for which he had 'searched so long', of the double series to a simple series, 12th July, 1849, NB34 ULC, pp. 124–31.

exact evaluation. Early in April he would discover an entirely different approach, discussed in the next chapter, to alleviate this problem.

Despite the difficulty with exact solutions, the idea of images remained intuitively clear and, above all, practical, being perfectly suited for conceptualizing complex problems and for obtaining numerical solutions. Thomson soon applied it to geometries which pure analysis could not touch, such as the electrical analogue of David Brewster's kaleidoscope, a set of any number of planes or spheres intersecting at angles equal to submultiples of π and subject to induction from a point charge. He also showed how to obtain simplifying mathematical transformations from one geometry to another, as a plane to a sphere and a disc to a portion of a sphere. The geometrically inclined among his friends, such as Liouville, Ellis, and Boole, naturally saw the new technique as stunning.[71]

To their creator, the images were so many stones for his sling, allowing him to play young David to Poisson's Goliath. Triumphantly he proclaimed to the British Association in 1847 the victory of direct perception over analytic sophistication:

> There is no branch of natural philosophy of which the elementary laws are more simple than those which regulate the distribution of electricity in equilibrium, upon conducting bodies. Yet its impracticability has always been the reproach of the Mathematical Theory of Electricity. Very few of the varied and interesting problems which it presents have been made subjects for investigation, on account of the apparently extreme complexity of the conditions to be satisfied; and even when results have been forced from it by the analytical skill and energy of a Poisson, the physical interest has been almost lost in the struggle with mathematical difficulties, and the complexity of the solution has eluded that full interpretation without which the mind cannot be satisfied in any analytical operations having for their object the investigation or expression of truth in natural philosophy.[72]

'Truth' refers, as above, to the latitudinarian essence constituting 'the legitimate foundation of every perfect mathematical structure that is to be made from the materials furnished in the experimental laws by Coulomb'. Here Thomson ascribed the unveiling of those truths to 'our countryman Green', whose British virtues had yielded up full explanations of 'the beautiful qualitative experiments'

[71] Thomson's original work was done 4th July–9th September, 1845, NB34, ULC, pp. 1–22. For responses see the letters between Thomson and Liouville published in *J. de Math.*, **10** (1845), 364–7; **12** (1847), 256–90; E&M, 144–77, esp. p. 177; William Thomson to George Boole, 2nd September, 1845, B13, ULG, and Boole to Thomson, 10th September, 1845, B147, ULC; and Ellis's remarks to the Parliamentary commission investigating Cambridge, 'On the course of mathematical studies', in William Walton (ed.), *The mathematical and other writings of Robert Leslie Ellis*, (Cambridge, 1863), pp. 417–27, esp. p. 424. In Frederick Fuller to William Thomson, 16th February, 1845, R290, ULC, Fuller jokes about Sir David Brewster's discovery 'that all the phenomena of electricity are only optical delusions. This will be clearly shewn and explained in an Appendix to the new edition of his "Natural magic"; and the same thing is beautifully illustrated by a recent improvement in his kaleidoscope, for which he is about to take out a patent'.

[72] William Thomson, 'On electrical images', p. 6.

of 'practical electricians' and simultaneously suggested to mathematicians 'the simplest and most powerful methods of dealing with problems, which if attacked by the mere force of analysis must have remained forever unsolved'. Green's methods had suggested electrical images, which would carry the banner forward.

If his audience had not already recognized that he saw himself presenting an ideal for all British science, his euphoric ending can have left no doubt:

In conclusion I would call attention to the circumstance that the method which I have . . . explained leads to the full solution of . . . all problems concerning spherical conductors which have hitherto been comtemplated by purely synthetical [geometrical] processes, without the aid of any mathematical analysis whatever. This powerful instrument is left for problems presenting difficulties of a higher order, now within our reach, which will be solved and as in all the branches of progressive knowledge will in their turn become elementary leaving still a boundless field for exercising the ingenuity and gratifying the curiosity which lead the human mind on in its search after the truths of nature.

Simplicity and utility yielded power, progress, and truth: the method of images truly carried an inspirational moral for Victorian science.

An even more direct homage to British practicality and geometrical insight, as against continental analysis, appears in Thomson's response in 1854 to a request from J.D. Forbes at Edinburgh for, 'A few jottings . . . of the Historical progress of the subject [electricity] with your opinion of how far each author has contributed an important share'. The resulting history exhibits an intensified patriotism, with a rediscovered Henry Cavendish now replacing Coulomb as the hero of experimental science, just as the rediscovered Green had already replaced Poisson in mathematical theory. Three times Thomson labels the fertile and intuitive approach of Cavendish and Green 'practical', to distinguish it from Poisson's repulsive sterility. 'Practical view', 'practical sense', and 'practical interest' contrast both with Poisson's lack of any feel for the physical magnitudes essential to experiment and with his limited and contorted analysis that have brought to the mathematical theory 'a character for barrenness that it deserves less than any other branch of applied mathematics'.[73]

Thomson's heat analogy figures prominently in his story of practical victories, as do his images: 'I have shown that all electrical problems regarding the distribution of electricity on one sphere or two, or any number . . . may be treated in a very simple synthetical manner . . . and that all that has been done by Laplace's coefficients [Poisson's method] and a great deal more, may be so obtained, without any properly so called math[l] analysis'. Only Poisson's theory of magnetism, 'which is much more profound & less barren', escapes the broadsword, but then magnetic problems 'do not admit of a corresponding

[73] J.D. Forbes to William Thomson, 19th December, 1853, F212, ULC; Thomson to Forbes, 4th January, 1854, F213, ULC. Poisson's aversion for laboratory work is noted by Arnold, 'Poisson. II. The Laplacian program', *Arch. Hist. Exact Sci.*, **28** (1983), 267–87, on p. 286.

simplification'. Poisson's theory, as we shall see, also provided the basis for solving the pressing navigational problem of magnetism in iron ships, a problem which was playing havoc with the naval and merchant fleets of the Empire.

In the method of images we have obtained our first clear view of that constant interplay between practicality, geometrical visualization, and mathematical theory that was to typify Thomson's ideal for natural philosophy. Mathematical operations were to translate directly into manipulations of objects; the machine and its principle should picture each other. Engineering practice and theoretical structures would interpenetrate at every level of teaching and research. In previous chapters we have set the larger context for this practical imperative in the political–religious and industrial–economic terms of the university and the city of Glasgow. Here we have narrowed the imperative to the technical domain of electrical theory, but still only in its rather passive, descriptive signification, its 'kinematics'. Already that dimension has involved several interrelated aspects: laws derived directly from experiment; conceptually simple, visualizable mathematics; a feeling for relevant magnitudes; and techniques directly applicable to problem solution. The practical imperative in its more dynamic, progressive, and aggressive dimension will involve two further aspects: precision measurement, which Thomson regarded as the primary motor of discovery in science; and mechanical effect, or work, the expression of all force and power in the material world.

8

The dynamics of field theory: work, ponderomotive force, and extremum conditions

> Today . . . I got the idea, which gives the mechanical effect necessary to produce any given amount of free electricity, on a conducting or nonconducting body . . . This enables us to find the attraction or repulsion of two influencing spheres . . . [and] This has confirmed my resolution to commence experimental researches . . . with an investigation of the absolute force, of statical electricity. *William Thomson, 1845*[1]

Prior to 1845 Thomson exercised his latitudinarian style of mathematical theorizing primarily on a pure description of the distribution of forces in space, 'kinematics'. Increasingly, however, he sought to expose in an equally non-hypothetical manner the 'dynamics' of these force distributions, that is, the causal factors controlling them and, concomitantly, determining their action on moveable bodies, their ponderomotive force. The terminology of kinematics and dynamics acquired much of its modern significance through Thomson's efforts in the 1840s to create a mathematical structure for the dynamics of force 'fields'.

Between 1843 and 1847 two conceptual elements appeared in Thomson's work that had not been present in any previous mathematical theory of force distributions. He began, firstly, to consider the distribution as one of 'mechanical effect' or work – later energy – spread throughout a field of force. Secondly, he began to treat the change in mechanical effect that the entire system would experience if any one of its parts moved as a measure of the total ponderomotive force exerted on that part. These two ideas are constitutive of field theory and as such provide our primary focus here.

Secondarily, however, Thomson's reformulation of moving forces involved as an essential feature the notion that any given state of a system should be understood genetically, in terms of the requirements for its formation and its

[1] William Thomson, 8th April, 1845, 'Journal and research notebook, 1843–45', NB33, ULC, pp. 177–9.

change, and thus as one state in a succession of states. Apparently the new scheme for dynamics corresponds to the notion of generation in kinematics. It provides our second major example of Thomson's attempts to reconcile progression and natural law, the first having been the secular cooling of the earth. Here, however, progression will be seen to involve the *progressiveness* of human society, symbolized in the steam-engine, with its multiple connotations of power.

What makes an electrified body move?

In May of 1843 Thomson published in the *Cambridge Mathematical Journal* a paper of a mere two pages which marks his earliest consideration of ponderomotive forces on electrified bodies. 'On the attractions of conducting and non-conducting electrified bodies' showed that, for a given distribution of electricity on the surface of a body A (figure 8.1), the total moving force exerted on A by an arbitrary electrical mass M is the same whether A be a conductor or non-conductor. Almost certainly he had become acquainted with the problem in his chemistry class at Glasgow under Thomas Thomson, who gave in his lectures a qualitative overview of theories of electricity and heat. Concerning the nature of heat, the famous chemist concluded 'that it will be safest for us, in the present state of our knowledge, to confess that we are incompetent to decide whether it be a substance or a quality'. If we compare Thomas Thomson's language here – concerning the present state of our knowledge and confessing our ignorance – with, for example, William's concluding remarks for his reconciliation paper (ch. 7), we are struck by the familiarity of the theme within the Scottish context. On the nature of electricity, however, Thomas Thomson regarded the two-fluid theory adopted by Poisson as probably correct, because of its comprehensiveness and predictive power: 'Even, therefore, if we suppose it a mere mathematical hypothesis, its importance and utility to all who wish to understand the principles of this most important science, must be admitted to be very great'.[2]

A prominent issue in fluid theories had long been the question of what prevents a self-repulsive fluid from escaping from a charged conductor through which it is perfectly free to move. With Poisson, Thomas Thomson concluded that 'nothing but the pressure of the ambient atmosphere prevents it from making its escape'. But that theory made the cause of ponderomotive forces on conductors different from those on non-conductors. For electrified non-conductors one simply summed up the forces acting directly on their electrical fluid

[2] William Thomson, 'On the attractions of conducting and non-conducting electrified bodies', *Cam. Math. J.*, **3** (1843), 275–9; E&M, 98–9. See Thomas Thomson's published lectures, *An outline of the sciences of heat and electricity* (London & Edinburgh, 1830), pp. 335, 425. William's participation in the course is witnessed in Thomas Thomson's testimonial in the printed copy of William Thomson's testimonials for the Glasgow chair of natural philosophy, PA34, ULC, p. 15.

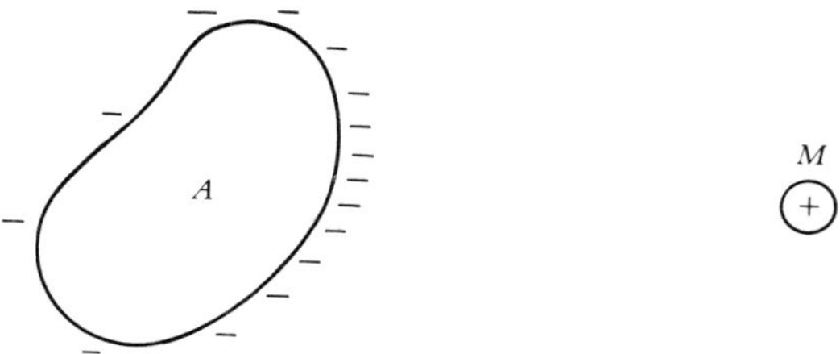

Figure 8.1. The ponderomotive force exerted by M on A is the same, for the same distribution of electricity, whether A is a conducting or a non-conducting body.

from any external source. The fluid then pulled along the non-conducting matter, to which it was attached by a friction-like force. For conductors, however, one supposed that the electricity, in 'pressing upon the ambient air', decreased the pressure which the air exerted against the surface of the conductor, this being a non-uniform decrease proportional to the thickness of the electrical stratum at different parts of the surface. Thus, for conductors, the ponderomotive force was owing, 'partly at least, to the unequal pressure of the air on the body, and of the electricity on the air'.[3]

William Thomson's short paper showing mathematical equivalence of the moving forces on conductors and non-conductors employed with almost schoolboy simplicity the expressions of his former teacher, including a 'thickness' for the electrical layer, as though he had forgotten temporarily his own more sophisticated treatment of electricity as an unspecified state of intensity, independent of any physical properties of fluid, like thickness, but conceived by mathematical analogy to a fluid. Indeed, the paper reads like an exercise resurrected from his Glasgow student days and published at an opportune moment. In any case, if we recall that only two to three months earlier he had entered in his diary his speculation about electrical attraction being 'the effect of the action of contiguous particles', then his return to the older language may be seen to mask a deeper result.[4] The actual calculations concerned not ponderomotive forces on two different bodies, but the force on one body A considered from two different perspectives: as a direct action at a distance of the electrical mass M on its surface electricity; and as an action of the surrounding (contiguous) air resulting from non-uniform pressures exerted against its surface. One suspects, then, that Thomson was more interested in this equivalence between action at a distance and contiguous action than in the physical difference between conducting and non-conducting bodies in relation to electricity.

[3] Thomas Thomson, *Heat and electricity*, pp. 424, 430.

[4] William Thomson, 24th February, 1843, Diary kept at Cambridge, 13th February to 23rd October, 1843, NB29, ULC.

In the contiguous action picture the effect of electricity on the air would involve the mechanism of propagation in the air (or ether). This idea is one he proposed in his lectures in 1852–3 to support a single-fluid theory of electricity in association with a theory of propagation by tension in air/ether: 'Hypothesis of one electric fluid may be perhaps reconciled to probability by Faraday's theory of tension [in air]. It may be that there is an elastic fluid pervading all space, and causing light and heat. This is very highly probable – we get both light and heat from electricity . . . Repulsion may be when the tension on one side is greater than the other'.[5] Though expressed here with more confidence than the state of Thomson's theorizing in 1843 would have allowed, the idea of moving forces produced by non-uniform tensions agrees with the particular manner in which he formulated his heat analogy for Faraday's specific inductive capacity in 1845. It agrees even more thoroughly with his analysis of Snow Harris's experiments at the same time.

Mechanical effect, mathematically and practically: the steam-engine as metaphor

In Paris, in the spring of 1845, the recent Cambridge graduate took on the task of reconciling the mathematical theory with Snow Harris's measurements of the force acting between two mutually influencing conducting spheres, for Harris claimed to have shown a contradiction between theory and experiment. From his recently discovered method of images, Thomson already knew how to obtain numerical results for this case from the mathematical theory. Harris's researches, however, raised the issue of air pressure in relation to electrical forces. While considering this issue, Thomson discovered a new solution of the two-spheres problem and a new conception generally of ponderomotive forces. The conception depends on mechanical effect, the context for which we must first develop.

In 1842, after establishing analytically the basic theorems of his earlier heat analogy in 'Propositions in the theory of attraction', Thomson discovered that Gauss had proved several of the theorems using a different method.[6] Considering a surface with arbitrarily distributed electrical density ρ and corresponding potential V, Gauss showed that, when the integral $\iiint \rho V d\tau$ was minimum, the potential would be constant over the surface, no part of the surface could be vacant, and the distribution would be unique. Thus the equipotential condition of a conducting surface at equilibrium required that electricity distribute itself so as to minimize the integral $\iiint \rho V d\tau$. Though simple in concept, the proof led to mathematical obscurities, so that Gauss himself acknowledged, 'we are com-

[5] William Thomson, in William Jack, 'Notes of the Glasgow College natural philosophy class taken during the 1852–53 session', MS Gen. 130, ULG.

[6] E&M, 126–38, esp. 136–7. C.F. Gauss, 'General propositions relating to the attractive and repulsive forces acting in the inverse ratio of the square of the distance', in Richard Taylor (ed.), *Scientific Memoirs*, **3** (1843), 153–96, esp. pp. 186–92. See also Chapter 7, note 4.

pelled to seek for the rigorous demonstration of this, the most important proposition of our whole investigation, by a somewhat more artificial path'.[7] Ever concerned with physical interpretation, Thomson wondered what the integral signified. Two years later an answer struck him.

> The original idea of Gauss's method of proving the general theorems about attraction has long appeared to me very mysterious. The following w^h has occurred to me to day seems to through [*sic*] some light upon it.
>
> Let m_1 m_2 &c be any no. of material points constrained to rest upon a surface S, and repelling one another acc^g [to] the law of grav^n, and let M be any fixed mass, repelling or attracting them. Then if u_i be the pot^l of M on a p^t coinciding with m_i when in its pos^n of equilibrium, the cond^n of equilibrium of the particles will be that
>
> $$\sum\sum m_i m_j/r + \sum m_i u_i \tag{1}$$
>
> shall be a minimum or maximum [where r is the distance between m_i and m_j]. The equilibrium will be stable in the former, unstable in the latter case. This is very readily shown, by the equation of virtual vel^s, or by the eq^ns of equilibrium of a point resting on a surface.[8]

He went on to argue that, since the expression was subject to a minimum, but never an absolute maximum, only stable equilibrium could obtain, and that for an infinite number of points, or continuous electrical matter, only one distribution could satisfy the minimum condition. (We should note that the *ponderomotive* force between this equilibrium distribution and the mass M had been the focus of the 1843 paper described above.)

Most importantly, Thomson here recognized that the minimized expression obtained a simple interpretation through the principle of virtual velocities, which Lagrange had made the foundation of all of rational mechanics, followed by Poisson who set the example for Cambridge texts. A virtual 'velocity' is actually the displacement that any point in a mechanical system would experience if the system were to undergo a possible motion. The principle requires, as the condition of equilibrium of the system, that the following sum should vanish, $\sum \vec{F_i} \cdot \vec{\delta r_i} = 0$, where $\vec{\delta r_i}$ is the virtual displacement of the point on which the force $\vec{F_i}$ acts. For conservative forces, expressible as gradients of potentials, the condition is, $\sum m_i \delta V_i = 0$, or $\delta \sum m_i V_i = 0$, where δ represents an arbitrary variation of the system. The latter condition is the condition of a maximum or minimum. If we write $V_i = \sum m_j/r + u_i$, it becomes Thomson's minimum condition for the equilibrium of electricity on a conducting surface under the action of a charged mass, which for continuous distributions is Gauss's minimized integral, $\iiint \rho V \mathrm{d}\tau$.

For conservative forces, the principle of virtual velocities was normally converted into the principle of *vis viva*, expressed for actual motions.

$$\mathrm{d}\sum m_i v_i^2 = 2\sum \vec{F_i} \cdot \vec{\mathrm{d}r_i} = 2\sum m_i \mathrm{d}V_i,$$

[7] *Ibid.*, p. 190.

[8] William Thomson, 14th August, 1844, 'Journal and research notebook, 1843–45', NB33, ULC, p. 53 (notation altered).

which says that the *vis viva* $\sum m_i v_i^2$ depends on functions V_i of position alone. When the system returns to the same configuration in space it will always have the same *vis viva*, or *vis viva* is conserved. Typically, in textbooks on rational mechanics that Thomson used at Cambridge, the product of a force times the distance through which it acted $\vec{F_i} \cdot \vec{dr_i}$ was given no distinct name (nor was V_i labelled potential). Samuel Earnshaw, for example, expressed the above equilibrium condition as follows:

> If the system pass through a position of equilibrium, then by the principle of *virtual velocities*,
>
> $$\sum \vec{F} \cdot \vec{dr} = 0,$$
> $$\text{and} \therefore \; d\sum(mv^2) = 0,$$
>
> that is, $\sum mv^2$ is a maximum or minimum . . . If the position be one of *stable* equilibrium, the forces tend to prevent the system from passing out of that position; that is, they have a tendency to reduce the system to a state of rest, and consequently to *diminish* the *vis viva*, which therefore has its maximum value in that position.[9]

Meaningful terms were *vis viva* and force. Only in engineering textbooks did one find the term 'work', or one of its many equivalents, for the product of a force and a distance.[10] For engineers, the concept of work was central, measuring both consumption and production with respect to every machine. It is of considerable interest therefore that Thomson developed his ideas on electrical equilibrium in relation to the engineering concerns of his brother James. Because their interaction will appear at length in the next chapter, we consider here only essential features of it.

Thomson's notebook entry above is dated 14th August, 1844, shortly after he returned to Scotland from Hopkins's 'reading party' in preparation for the Tripos in January. He had reviewed 'all of the first and second year subjects with Hopkins, and Hydrostatics, Optics, & 1/3 of Astronomy . . . of the third year subjects'. This list includes all of mechanics, which he had studied in detail from March to October, 1843.[11] With the principles of mechanics fresh in his mind he

[9] Samuel Earnshaw, *Dynamics, or a treatise on motion: to which is added a short treatise on attractions*, 3rd edn. (Cambridge, 1844), p. 180 (notation altered). We discuss mechanics textbooks in Chapter 11.

[10] For example, William Whewell, *The mechanics of engineering* (Cambridge, 1841), Whewell acknowledged that he had borrowed heavily from French engineering texts by Navier, Poncelet, and others, transforming their term 'travail' into 'labouring force'. See also the 'Addition' to the second edition of S.D. Poisson, *Traité de mécanique* (Paris, 1833); *A treatise of mechanics*, translated by H.H. Harte (2 vols., Dublin, 1842), **2**, sects. 679–96, entitled 'Relative to the application of living forces in the calculation of machines in motion'. Only here did Poisson discuss the significance for machines of the concept of work. The importance of the engineering concept of work has been stressed particularly by T.S. Kuhn, 'Energy conservation as an example of simultaneous discovery', in M. Clagett (ed.), *Critical problems in the history of science* (Madison, 1959), pp. 321–56, esp. pp. 330–4. See also Chapter 3, notes 22–4; Chapter 9, notes 9–11.

[11] William Thomson, 14th August, 1844. See PA11–17, ULC, which are Thomson's transcriptions of Hopkins's manuscripts on mechanics. See Chapter 3, note 62. Hopkins used these

received from James on 4th August a letter which very likely crystallized in a new form his thinking on equilibrium conditions generally. James's letter concerned steam-engines, specifically their efficiency and the source of the useful work derived from them. Such work he designated 'mechanical effect', following the usage of the professor of engineering at Glasgow, Lewis Gordon. The brothers had discussed the issues before, when William mentioned to James the paper of Emile Clapeyron on the now-classic theory of Sadi Carnot. James had not yet read their works, but in his letter he described a waterwheel analogy which they also had employed, whereby a heat engine derives its mechanical effect from the fall of heat from high to low 'intensity' in the same way as a waterwheel extracts mechanical effect from the fall of water from high to low elevation. Here a state of high intensity was a state of concentration, as well as of high temperature, and James opposed such a state to one of 'diffusion'. It connoted heat collected in a boiler, by analogy to water collected in a mountain reservoir, and as opposed to water spread out, or diffused, in the sea.[12]

Seen from James's perspective, which we now presume William to have seized, the heat flow analogy for electrical 'intensities' and forces extended the significance of mechanical effect from heat engines and waterfalls to electrostatics, as in figure 8.2. Just as heat and water attained equilibrium at their lowest levels of available mechanical effect, so too did electricity. Gauss's minimum condition minimized the mechanical effect contained in an electrostatic system.

Only one feature of this interpretation is missing from Thomson's notebook entry: the term 'mechanical effect' with its 'practical' engineering connotations. The deficiency takes us to Paris six months later, where Thomson had begun working in Victor Regnault's laboratory. We recall that the Paris visit was intended to remedy a defect in his Cambridge education so far as candidacy for the Glasgow chair of natural philosophy was concerned. He would counter the potential complaint that he was 'merely an expert x plus y man' by acquiring the 'practical' credentials expected of one who would present natural philosophy to unsophisticated students, largely through popular demonstration experiments.[13] This sense of the 'practical' correlates closely with the one we have loosely labelled 'kinematic' with respect to the method of images, emphasizing: experimentally derived laws; descriptive, visualizable methods; an intuition for magnitudes; and practical techniques for problem solution. Scottish traditions in natural philosophy went on from these descriptive and educational aspects of practical experimentation, however, to stress practicality in the more 'dynamic' sense of controlling nature for the material progress of man. Regnault's laboratory, with its current aim of carrying out precision measurements on high-pressure steam (measurements paid for by the French government and designed

manuscripts as an outline textbook, referring his students for details to Earnshaw (specially), Poisson, Pratt, and Whewell.

[12] James to William Thomson, 4th August, 1844, T402, ULC; see especially Chapter 9, notes 15–18. [13] See chapter 4, note 66.

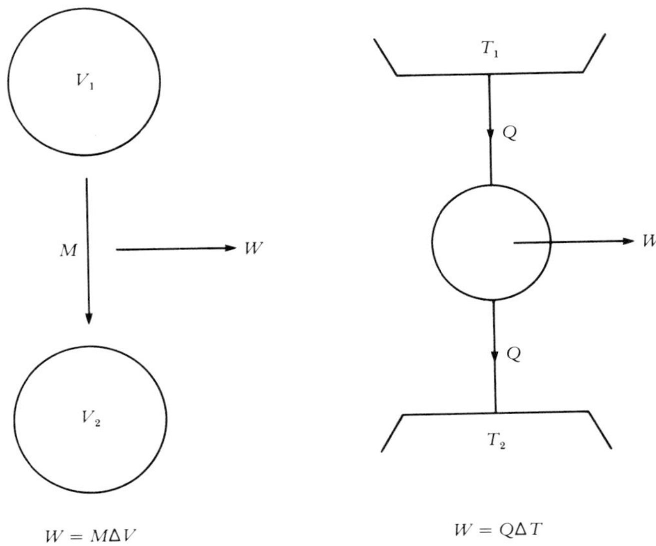

Figure 8.2. On the left are two charged conducting spheres at potentials V_1 and V_2. The electricity M tends to fall to its lowest level, which involves a capacity for doing work $W = M\Delta V = M(V_1 - V_2)$, just as on the right, for the fall of heat (water) Q to its lowest temperature (elevation) T, the amount of work available is $W = Q\Delta T = Q(T_1 - T_2)$. Similarly, electricity within any container, like heat and water, reaches equilibrium when it is spread out, or diffused, into the state of minimum mechanical effect.

to improve the efficiency of steam-engines for producing motive power, or mechanical effect) epitomized this engineering ideal for Thomson.[14]

He reported these experiments for James's benefit on 14th April, and announced his own goal of electrical experiments similarly conceived: 'I have been planning some experimental investigations in electricity, which, however, I could not commence till I have accurate apparatus at my command, as it is precise measurements I wish to make'.[15] In this practical context of precision measurements and mechanical effect, or more specifically of measurements of pressures that produced motive power, he stated explicitly the relation between mechanical effect and Gauss's minimum condition. But he also extended that relation, attaining: (1) a new conception of the ponderomotive force between two electrified spheres, (2) a new conception of absolute measurements, and (3) a new understanding of the significance of air pressure in relation to mechanical effect. This massive reorganization of thought appears in a notebook entry of 8th April, 1845, which we quote in full:

[14] Robert Fox, *The caloric theory of gases from Lavoisier to Regnault* (Oxford, 1971), esp. pp. 281–302. [15] William to Dr Thomson, 14th April, 1845. T306, ULC.

To day, in the laboratory (of Physique at the Coll. de France, M. Regnault, prof.) I got the idea, which gives the mechanical effect necessary to produce any given amount of free electricity, on a conducting or non conducting body. If m is any electrical element, V the potential of the whole system upon it, the mechanical effect necessary to produce the distribution m is $\sum mV$. [V may be due to other mass in add^{n} to the system it is required to produce, *del.*] If the body be conducting this exp^{n} becomes VM. This enables us to find the attraction or repulsion of two influencing spheres, without double integrals. Also the theorem of Gauss that $\sum mV$ is a min^{m} when V is const., shows how the double int^{l} which occurs when we wish to express the action directly, may be transformed into the diff. co. of a simple int^{l}, taken with reference to the dist^{ce} betw. the two spheres.

Let c be this dist^{ce}, V the pot^{l} in the interior of one sphere, and M the quantity of electricity, V'&M' the corresp^{g} magnitudes, for the other, a & a' the radii. We have, if Γ be the attr^{n}

$$\Gamma = -\frac{\partial}{\partial c}(MV + M'V' - M^2/a - M'^2/a')$$

$$= -\frac{\partial}{\partial c}(MV + M'V') = -\left(M\frac{\partial V}{\partial c} + M'\frac{\partial V'}{\partial c}\right).$$

This has confirmed my resolution to commence experimental researches, if I ever make any, with an investigation of the absolute force, of statical electricity. As yet each experimenter has only compared [quantities, *del.*] intensities by the dev^{ns} of their electrometers. They must be [compared with, *del.*] measured by pounds on the square inch, or by 'atmospheres'. Also the standard must be the greatest intensity which can be retained by air of given density.[16]

We shall consider this remarkable collage under the three headings listed above.

(1) *Ponderomotive force.* Thomson's new insight allowed him to solve in a perfectly transparent way a concrete problem that had defeated Poisson's most sophisticated analysis. The solution, however, depended on an entirely new mode of conceptualizing the problem. At the heart of Thomson's creation was his sharply focussed recognition that the expression $\sum mV$, or $\iiint \rho V d\tau$, represents the amount of work that would have to be done to assemble an electrical distribution from elements dispersed at infinity, or the total mechanical effect contained in the system after assembly, and that this quantity is a minimum at equilibrium. The genesis of the system, in terms of the work required to produce it from a state of infinite diffusion, therefore defined its state uniquely. With that focus he saw that the ponderomotive force acting within the assembled system to displace its parts, when multiplied by the displacement, would equal the total work expended during the motion, or the change in total mechanical effect required to move to the new state. Thus the relation of parts in the existing state of the system could be defined in terms of a succession of states, just as Nichol and

[16] William Thomson, 8th April, 1845, NB33, ULC, pp. 177–9.

Mill had proposed in their ideology of progress (ch. 4). Nichol admonished that the character of any part of a system 'must be determined through its relations, not with Space but Time; and its very isolation among surrounding things . . . constrains us to regard it as only one part of a term of a Series'.[17] By contrast, Poisson had envisaged the two-spheres problem as one of static spatial relations of parts, requiring a double integral over the forces acting between elements of electricity.

With Thomson's perspective on successive states, the attraction Γ between two electrified conducting spheres became the 'diff[erential] co[efficient] of a simple int[egral] taken with reference to the distance between the two spheres' c, or the gradient of total mechanical effect,

$$\Gamma = -\frac{\partial}{\partial c}\Sigma m V = -\frac{\partial}{\partial c}\iiint \rho V \mathrm{d}\tau.$$

Resolving this expression into the self-interactions of the two spheres considered as isolated, M^2/a and M'^2/a', plus their mutual interactions, MV and $M'V'$, yields Thomson's equation above. Since the self-interactions are independent of c, his final result follows.[18]

Thomson's immediate interest in this result derived from the need to compare measurements made by Snow Harris with the theory of inverse square attractions. Harris employed identical spheres, one grounded and one held by a battery at constant potential V, as in figure 8.3. In this case the quantities of electricity vary, rather than the potentials, but the same analysis in terms of mechanical effect applies. On 12th April, in the draft of his reconciliation paper for Liouville, Thomson wrote down the ponderomotive force, $\Gamma = -V\partial M/\partial c$, where M is the charge on the sphere connected to the battery, and hailed the formula as 'a case of a general theorem which we shall have occasion to announce subsequently'. The result allowed him to make qualitative and quantitative comparisons with Harris's experiments, although he judged them to be so incompetently performed as to be 'unavailable for the accurate *quantitative* verification of any law'.[19]

Thomson put off the 'subsequent' publication of his general theorem until July, 1849, when he developed it in a letter to Liouville that went astray but was intended for the *Journal de Mathématique*. The delay was occasioned by the fact that only then had he 'made out the demonstration for which I have searched so long'. Ironically, the long-sought demonstration was not a full development of

[17] J.P. Nichol, *Thoughts on some important points relating to the system of the world*, 2nd edn. (Edinburgh, 1848), p. 198.

[18] Thomson's summation counts the interaction of any two electrical elements twice, a fact he had noted earlier about Gauss's integral, 14th August, 1844, NB33, ULC, p. 54. He would insert a factor of $\frac{1}{2}$ in later works. Gauss, 'General propositions', p. 176, recognized the doubling as well as the fact that $MV = M'V'$ (in Thomson's notation), since 'either is no other than the aggregate of all the combinations mm'/r'.

[19] William Thomson, 'On the elementary laws', draft, *c.*12th April, 1845, NB32, ULC, p. 13; E&M, 22, 25. See Chapter 7, note 32. For Thomson's original calculations see 27th October, 1845, 'Journal and research notebook, 1845–56', NB34, ULC, pp. 15–20.

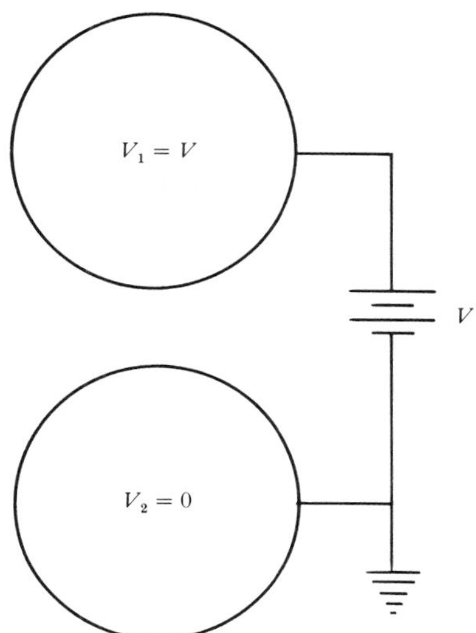

Figure 8.3. The battery does work $W = MV$ to raise the quantity of electricity M to potential V.

the method of mechanical effect but of the method of images, which he apparently regarded as more perspicuous, it being 'synthetic'.[20] That implicit judgement should remind us that work, mechanical effect, and energy were not yet fundamental concepts in physical theory. We are viewing the mathematical beginnings of the great energy revolution in the writings of one of its main actors, who recognized as yet only a powerful new technique for conceptualizing and solving problems. Nevertheless, his accomplishment was real.

He had shown that mechanical effect behaved like a total potential, from which ponderomotive force could be derived as a gradient. We shall henceforth symbolize this total potential by Φ to emphasize the clarity of his insight. In July, 1845, he explicitly referred to Φ as a '*force function*', underlining each usage. Only around 1849 would he begin to use 'total potential' and 'mechanical value'. Similarly, we have symbolized ponderomotive force on an extended body by Γ to distinguish it from point forces F.[21] The general relations are then,

[20] William Thomson, 12th July, 1849, NB34, ULC, pp. 124–31, contains a draft of the letter (crossed out) and a summary of the result ultimately published in William Thomson, 'On the mutual attraction or repulsion between two electrified spherical conductors', *BAAS Report*, **22** (1852), 17–18; *Phil. Mag.*, [series 4], **5** (1853), 287–97; **6** (1853), 114–15; E&M, 86–97.

[21] William Thomson, 5th July, 1845, NB34, ULC, pp. 6–7. Thomson typically differentiated the electrical force at a point (field intensity) from ponderomotive force as 'resultant force at a point' from 'attraction'. The term 'force function', though rare in his writings, occurs, with a somewhat different meaning, in William Thomson, 'On the theory of electromagnetic induction', *BAAS Report*, **18** (1848), 9–10; MPP, **1**, 91–3, which refers to the electromagnetic potential discovered by Franz Neumann. Although Gauss had occasionally employed the symbol Ω and Poisson had used ϕ, neither derived ponderomotive forces from the function.

$$\text{total potential} = \Phi = \iiint \rho V \mathrm{d}\tau = \text{total mechanical effect},$$

and,

$$\text{ponderomotive force} = \vec{\Gamma} = -\text{gradient } \Phi,$$

where this force is to be understood as the tendency of the system to attain its lowest state of mechanical effect or greatest diffusion. *The ponderomotive force and the extremum condition for equilibrium states are the same thing*. Thomson called attention to the origins of this idea in his published version. 'This proposition occurred to me in thinking over the demonstration which Gauss gave of the theorem *that a given quantity of matter may be distributed in one and only one way over a given surface so as to produce a given potential at every point of the surface*, and considering the mechanical signification of the function [Φ] on the rendering of which [function] a minimum that demonstration is founded'.[22]

(2) *Absolute measurement.* In his notebook entry Thomson moved directly from attractions determined by changes in mechanical effect to the absolute measurement of electrical intensities. This step, more directly than any other in the passage, foreshadows a fundamental role for work in physical theory; absolute units had been tied to that concept for the first time. The idea looks ridiculously simple in retrospect. Rewriting the equation of ponderomotive force, we have, for Harris's arrangement,

$$-\Gamma \mathrm{d}c = \mathrm{d}\Phi = \mathrm{d}(MV) = V\mathrm{d}M.$$

A measurement of the work done by the force Γ during a small displacement $\mathrm{d}c$ of two attracting conductors, therefore, would yield directly a measure of the change in quantity of electricity with respect to a standard potential V.

The radical implications of this idea for Thomson's physics can hardly be overestimated. In terms of the concept of mechanical effect, his idea was simply that the work done during the 'fall' of an element $\mathrm{d}M$ of electricity (or water or heat) through a potential (or height or temperature) V, as in figure 8.2 above, should measure the quantity $\mathrm{d}M$. Thereby the principle of waterwheels and steam-engines, the practical engineering principle that powered the ships and mills of Glasgow, became in a single stroke the principle that would reduce the theoretical entities of mathematical physics to practical experimental measure and control. And since Thomson was quickly coming to regard measurement as the source of all progress in natural philosophy, it is correct to say that the motor of progress in society was for him becoming one with the motor of progress in natural philosophy. Or, to reverse the connection, the concept of the steam-engine translated the *successive* states of nature into the *progressive* states of society, now more literally than Mill and Nichol had ever imagined. Such statements may seem grandiose here, out of all proportion to Thomson's simple notebook entry; we have seen already in Chapter 5, however, that they represent quite accurately the direction of his rapidly developing perspective on science in industrial Britain. Within that perspective the steam-engine became a metaphor for power and wealth. How appropriate that it should also become a

22 William Thomson, 'On the mutual attraction', E&M, 92*n*.

metaphor for absolute measurement of the properties of matter, measurements which in Thomson's view of 'scientific wealth' tended 'to accumulation according to the law of compound interest'.[23]

Upon assuming the Glasgow chair a year after returning from Paris, the twenty-two-year-old professor would immediately embark on his project for accumulating scientific wealth by establishing a laboratory for measurements of the properties of matter, precision measurements in absolute units. With him he brought the basic insight recorded in Paris. Mechanical effect contained in a system represented a total potential for doing work. Any and all systems capable of producing effects, therefore, could be treated as *engines*, analyzable in terms of the mechanical effect entering and leaving them. By that means physical agents like electricity and heat could be reduced to dynamical measure without any hypothesis regarding their physical nature.[24]

By the end of 1847 Thomson had proposed in a letter to Lewis Gordon the original version of the absolute temperature scale that makes the name Kelvin a popular possession. In the published analysis he expressed the idea of an engine in essentially the same terms as James had used in his letter of August, 1844, (but here following Clapeyron's account of Carnot's theory):

> Now Carnot . . . demonstrates that it is by the *letting down* of heat from a hot body to a cold body, through the medium of an engine (a steam-engine, or an air-engine for instance), that mechanical effect is to be obtained; and conversely, he proves that the same amount of heat may, by the expenditure of an equal amount of labouring force, be *raised* from the cold to the hot body (the engine being in this case *worked backwards*); just as mechanical effect may be obtained by the descent of water let down by a water-wheel, and by spending labouring force in turning the wheel backwards . . . water may be elevated to a higher level.

Writing $\Phi = MV$ for mechanical effect, whether as 'labouring force' (Whewell's term for work expended) to raise heat M to temperature V or as the total potential available for doing work, it is apparent that a measurement of Φ units of mechanical effect for M units of heat 'let down' would define V units of temperature. As Thomson put it: 'The characteristic of the scale which I now propose is, that all degrees have the same value; that is, that a unit of heat descending from a body A at the temperature $V°$ of this scale to a body B at the temperature $(V-1)°$, would give out the same mechanical effect, whatever may be the number V. This may justly be termed an absolute scale, since its characteristic is quite independent of the physical properties of any specific substance'.[25]

[23] PL, **2**, 175–6. See Chapter 5, note 6.

[24] cf., in Chapter 6, Nichol's ideal of mathematics as an 'engine of deduction'; Thomson's ideal of the calculus of operations as an 'analytical engine' (a kinematical engine), and the similar images of Babbage, and Sylvester.

[25] William Thomson, 'On an absolute thermometric scale, founded on Carnot's theory of the motive power of heat, and calculated from the results of Regnault's experiments on the pressure and latent heat of steam', *Phil. Mag.*, [series 3], **33** (1848), 313–17; MPP, **1**, 100–6, on pp. 103–4. See Chapter 9, notes 38–41.

This latter sense of 'absolute' represents one of two connotations of the term in Thomson's usage. The other sense, more important to his natural philosophy, required reduction to dynamical measure, as in 'the absolute amount of mechanical effect'. We shall return to this subject in Chapters 11 and 20, on dynamics and measurement, respectively. Here we need only note that, through the metaphor of the steam-engine, mechanical effect quickly became the measure of all things in his 'practical' conception of theory. The calculation in 1845 of the ponderomotive force between spherical conductors, for example, would provide the basis for his subsequent design of various electrometers reading in absolute measure.[26]

(3) *Air pressure and fields of force.* Decades later Thomson still recognized (backhandedly) that his electrometers originated in his critical appraisal of Snow Harris's experiments. A student in his natural philosophy class in 1881–2 recorded the following history of 'Electrostatic measurement':

Gold leaf electroscope very rough. Coulomb's torsion balance with Faraday's glass thread . . . very fair. Snow Harris constructed several [electrometers] but not properly arranged & really all electrometers got invented by Sir William Thomson. Snow Harris doubted mm'/r^2. [His] best instrument a sort of weighing instrument. A common balance.

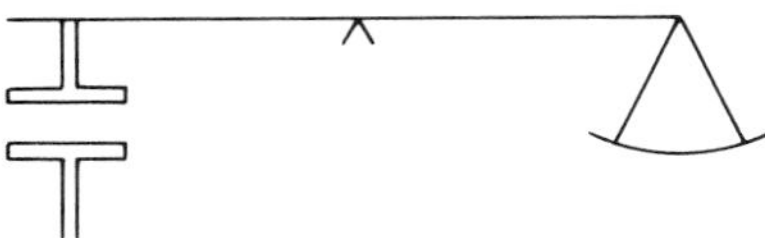

Experiments led him quite wrong[;] didn't understand sufficiently. Sir W. Thomson when undergraduate at Cambridge published papers that showed from Snow Harris's own figures that law almost correct.[27]

The diagram shows one of Harris's original arrangements, with two equally and oppositely charged plates on the left whose attraction is balanced by weights (not shown) in the pan at the right. It clarifies Thomson's remark in his Paris notebook that electrical intensity 'must be measured by pounds on the square inch, or by "atmospheres"'. Assuming large plate area relative to distance of separation, the electrical intensity would be approximately uniform over the opposing surfaces of the plates and negligible elsewhere, as would the electrical force – read either pressure or flux – at the surface, to which the intensity was

[26] William Thomson, 'Applications of the principle of mechanical effect to the measurement of electro-motive forces, and of galvanic resistances, in absolute units', *Phil. Mag.*, [series 4], **2** (1851), 551–62; MPP, **1**, 490–502, gives his first general discussion of absolute measurement, but the 'principle of mechanical effect' here has the much extended meaning of 'conservation of energy'. He first applied the 1845 theory directly to the design of an electrometer when he published that theory in 1853. 'On the mutual attraction', E&M, 96–7. See Chapter 20.

[27] William Thomson, in J.A. MacCallum, 'Notes of lectures in William Thomson's natural philosophy class, 1881–82', MS Gen. 481, ULG, pp. 44–5.

William Thomson's attracted disc electrometer (1888) based on the principle of weighing the mutual attraction of two parallel conducting discs. This 'absolute electrometer' yielded readings in absolute electrostatic units of potential in the range *c*.200–*c*.5000 volts and was used to standardize other electrometers. [From The Science Museum Library; further details of construction in George Green and J.T. Lloyd, *Kelvin's instruments and the Kelvin museum*, pp. 20–1, and *BAAS Report*, 37 (1867), 497–501.]

proportional. Thus the ponderomotive force measured by the balance, divided by the plate area S, would give directly the electrical force, as 'pressure', and thereby the electrical intensity, $\Gamma/A = \rho F = \rho(4\pi\rho) = 4\pi\rho^2$. Further, since F, as flux, would 'flow' directly across the gap it would be uniform across it and equal to the potential difference between the plates divided by their separation, $F = V/c$. In terms of those two basic conditions for plane parallel conductors Thomson originally conceptualized Harris's experiments: 'Now in the experiments of Mr Harris, the bodies A and B are generally of such a nature that a considerable portion of their surfaces are opposed to another, and separated by an interval which is small with respect to the radii of curvature at the opposed parts. Hence in these cases the intensity of electricity at any point of the opposed surface of A will, if the distance c be changed, vary nearly as $1/c$ and therefore the attractions between the two bodies will be $1/c^2$. This must of course be taken as a very restricted result, but its generality is probably quite equal to that of the measurements of Mr Harris'.[28]

Subsequent analysis proceeds 'by contemplating the equivalent problem in heat', to explain Harris's results on sparking distances. 'If A and B be equal spheres, the flux of heat at the nearest points will be very nearly the same, and proportional to the difference of their temperatures divided by their distance even when this distance is considerable with reference to the radius of the spheres. This accounts for the law with respect to the length of the spark', i.e. 'the quantity of electricity . . . necessary to produce a spark varies directly as the distance between the bodies'.[29]

By far the most intriguing aspect of this discussion is the analogy implicit throughout it between electrical force conceived as heat flux and as an effect of air pressure. The notebook entry (8th April) clearly makes the identification with air pressure, requiring the measure 'pounds on the square inch' or 'atmospheres'. The draft, similarly, states: 'The most important part of Mr Harris's researches is that in which he investigates the insulating power of air of different densities. The result which he arrives at is that the intensity necessary to produce a spark depends solely on the density of the air, and *not at all on the pressure*'. In the published versions Thomson added, 'He thus shows that the conducting power of flame, of heated bodies, and of a vacuum, are due solely to the rarefaction of the air in each case'.[30] Faraday had shown with respect to induction, on the other hand, that inductive capacity in air and other gases was independent of both pressure and density and equal to that of vacuum (ether). Concerning pressure, therefore, a potential conflict seemed to exist between Thomson's description of electrical intensity as producing a pressure which acted against the pressure of the atmosphere and the empirical non-dependence of both sparking and induction on pressure. Furthermore, if electrical force were

[28] William Thomson, NB32, ULC, pp. 13f. [29] *Ibid.*, pp. 14–15, 11.
[30] *Ibid.*, p. 16. 'On the elementary laws', E&M. 23. Our emphasis.

truly analogous to a flux, how was it possible that inductive capacity did not depend on density?

These problems informed Thomson's handling of the heat analogy in its final form in November, 1845. We saw in Chapter 6 that he did not, in fact, regard force as analogous to a true flux, but only to a tension, which resolves the problem of induction in gases. Gases, unlike liquids and solids, in Thomson's view, experienced no electrical 'displacement', no polarization in response to tension, for they contained no free electricity. On his model, therefore, inductive capacity would naturally be independent of pressure and density.

Similarly, on this model, however an increase in density affected sparking, it could not depend on any change of the specifically electrical response to tension in the gas, for no such effects existed. Perhaps the electrical intensity on the conductor simply behaved like a fluid which exerted a mechanical pressure against the air/ether, creating a mechanical stress within its structure, and a tendency to fracture, but the capacity of the air/ether to resist fracturing increased as a result of its structure involving more air molecules. Such a mechanism would explain Harris's results (and it corresponds to Thomson's later ideas, ch. 12). Knowing 'nothing, however, of the molecular action by which such effects could be produced', he would never, in 1845, have speculated publicly about them, much less have developed a theory that depended on them.[31] But he had pondered the problem. In his first letter to Faraday, after their meeting at the British Association in June, he included an 'important question': 'whether the air in the neighbourhood of an electrified body, if acted upon by a force of attraction or repulsion, shows any signs of such forces by a change of density, which, however, appears to me highly improbable'. Faraday gave no answer.[32] Any such action would have cast grave doubts on the theory Thomson entertained, which supposed that electrical tension (stress) in air/ether involved no corresponding displacement (strain).

The relation of electrical 'pressure' at the surface of conductors to electrical 'flux' in the surrounding space involved also less speculative and thus (in Thomson's vocabulary) more practical aspects. The issue returns us to mechanical effect, but now to the question of where the mechanical effect contained in a system was supposed to be located. The 'Propositions' of 1842 had already elaborated theorems given by Gauss which showed that the minimized function we have labelled Φ could be expessed as an integral over all space of the force squared,

$$\Phi=\iiint\rho V\mathrm{d}\tau=(1/4\pi)\iiint F^2\mathrm{d}\tau.$$

In the notebook entry in August, 1844, where he first interpreted Φ through the principle of virtual velocities, Thomson also noted Gauss's use of the expression

31 'On the elementary laws', E&M, 37. See Chapter 7, note 54.

32 William Thomson to Michael Faraday, 6th August, 1845; Faraday to Thomson, 8th August, 1845, in SPT, **1**, 146–9, on p. 148.

on the right in proving the uniqueness of the equilibrium state.[33] Having once thoroughly identified Φ with the mechanical effect contained in a system of sources at their respective potentials, therefore, an extension of that identification to the spatial integral $\iiint F^2 d\tau$ was, simply from the mathematical perspective, quite natural.

Three other factors reinforced such a conceptual extension. The heat analogy, first, related the steady-state output of sources to the distribution of flux throughout space, and thus to the motion of heat. 'The most general form of the [uniqueness] theorem, applied to the uniform motion of heat', Thomson remarked immediately following his note on Gauss's integral, 'is that it is always possible, by means of sources of heat distributed over any given surface, to retain each point of the surface at a given temperature, the motion of heat being supposed to have become uniform. This, being a physical axiom, enables us physically to prove the theorem, with ref^{ce} to attr^{n}, or as a purely analytical theorem. This was what first occurred to me in the winter of 1840–41'. Secondly, and in a somewhat similar fashion, the principle of virtual velocities itself, converted as usual to the principle of *vis viva*, transformed static mechanical effect into its equivalent in motions $\sum mv^2$. For fluid motions distributed in space that expression is, $\rho\iiint v^2 d\tau$. Finally, George Green, in proving his own uniqueness theorem for electrostatic potentials, minimized the spatial integral, $\iiint F^2 d\tau$.[34]

All of these factors were present in Thomson's reconciliation of Snow Harris and Faraday with Gauss and Green. Taken together with a preference for propagation over action at a distance, they formed a suggestive context for questioning whether the mechanical effect contained in an electrostatic system ought to be regarded as distributed in the space of propagation of the forces rather than localized in the electrical concentrations. On the former view, the ponderomotive force on an electrified body would be seen as the tendency of the entire distribution of lines of propagation to arrange itself so as to minimize its mechanical effect, while on the latter the tendency was in the electricity. No evidence exists that Thomson actually made these reinterpretations of the mathematical theory in 1845, but they became the focus of his research in electricity and magnetism within the next two years. Only a thorough ground-

33 William Thomson, 'Propositions', E&M, 137. Gauss, 'General propositions', pp. 183, 191. William Thomson, 14th August, 1845, NB33, ULC, p. 53.

34 George Green, 'On the determination of the exterior and interior attractions of ellipsoids of variable densities', *Trans. Cam. Phil. Soc.*, **5** (1835), 395–429; N.M. Ferrers (ed.), *Mathematical papers of the late George Green* (London, 1871), pp. 185–222, on pp. 192–4. See Chapter 7, note 64, for this paper in relation to Thomson's method of images. Once again, a possible source of mathematical methods was Robert Murphy, whose works Thomson had read early in his Cambridge career. Their relevance is difficult to judge, however, because Thomson almost never refers to them. For their mathematical significance see J.J. Cross, 'Integral theorems in Cambridge mathematical physics, 1830–55', in P.M. Harman (ed.), *Wranglers and physicists: studies on Cambridge physics in the nineteenth century* (Manchester, 1985), pp. 112–48, esp. pp. 121–9.

ing of the later ideas in the undigested set of connections established in this earlier work makes the germination of those ideas fully comprehensible.

The preceding section has necessarily been complex. Between August, 1844, and April, 1845, the potent solvent labelled mechanical effect washed through Thomson's physical mathematics and began to restructure it, transforming old meanings and creating new ones. We may briefly summarize the essential features.

Mechanical effect as total potential. Mechanics textbooks had long discussed stability conditions on mechanical systems through the principle of *vis viva*, in terms of maximum and minimum conditions on the total *vis viva* and on the expression $\sum mv$, representing the aggregated action of all forces acting on all points of the system. The function V gave the point forces as $\vec{F} = -\text{gradient } V$ and was therefore called a potential by Gauss and Green. In no standard usage, however, were ponderomotive forces acting between extended parts of systems derived from the composite function $\sum mv$. Drawing on an engineering interpretation of the capacity of steam-engines and waterwheels to produce mechanical effect from the successive states of whole systems, Thomson recognized that this function could be regarded as a total potential Φ for ponderomotive forces, $\vec{\Gamma} = -\text{gradient } \Phi$.

Mechanical effect in a field. Through Gauss's relation,

$$\Phi = \sum mV = (1/4\pi)\iiint F^2 \mathrm{d}\tau,$$

and through the heat analogy, Thomson began to see that the mechanical effect contained in a system might be regarded as spread throughout the space, or *field*, of its forces. That recognition would provide the central mathematical expression for his development of field theory. First, it would give physical existence to a field of force as a field of mechanical effect. Second, the minimization condition on this field would appear as a condition which controlled the distribution of lines of propagation in space. Whereas the heat analogy had described any particular distribution *kinematically*, as a geometry generated by the equations of conduction under given conditions, the minimization condition would explain the distribution *dynamically*, giving a causal account of its present state in terms of its tendency towards a later state if conditions changed. Third, ponderomotive force would be the tendency of the lines of propagation to arrange themselves so as to minimize the mechanical effect contained in their distribution. In short, through his insights of 1845, Thomson began to transform the traditional analysis of point forces and ponderomotive forces into the kinematics and dynamics of mechanical effect in fields.

Mechanical effect and absolute value. Interpreting physical systems in their dynamical aspect as engines, possessing the capacity to produce mechanical effect, or alternatively, as produced by the expenditure of mechanical effect, Thomson began in 1845 radically to transform the notion of measuring a physical entity. Whatever agents like electricity and heat might be in themselves they could be measured by their role in the production or expenditure of mechanical effect. This idea forms the deepest, most pervasive theme in Thomson's entire natural philosophy. We have stressed its origins in practicality, of both the experimental and engineering kind. Those two senses express the progressive, dynamic aspect of his science, as of his social–political perspective, deeply rooted as both were in an industrial vision of the city of Glasgow. The fundamental measure of value in economic terms, work, would now measure the absolute value of physical entities. Such measurements, conversely, would provide new economic wealth in industry and the wealth of new knowledge in science. In a word, mechanical effect attained its potency in Thomson's natural philosophy through the steam-engine, deployed as a metaphor for work, wealth, and progress.[35]

Field theory proper

On 1st July, 1846, following twelve months as a Fellow of Peterhouse, Thomson wrote in his notebook a brief summary of the year's activities, remarking: 'I have made many abortive attempts to commence my *treatise* on electricity', and adding between the lines: 'I have also, since the beginning of the Lent term, been often trying to connect the theory of *propagation* of elec^{y} & magnetism with solid transmission of force'.[36] Direct motivation for the new theory of propagation through a solid came from Faraday's most recent experimental masterpiece, the discovery that the plane of polarization of light was rotated when transmitted along lines of magnetic force through a special glass. Faraday had long sought such an effect, but he renewed the search after the meeting of the British Association in June 1845, where Thomson had called for research on 'whether a transparent dielectric in a highly *polarized* state affects light transmitted in the same manner as a uniaxial crystal', and after Thomson repeated the call in the letter that began their continued communication in August.[37]

Magneto-optic rotation turned Thomson's attention from electricity to magnetism. After taking up the Glasgow chair in the autumn of 1846, he continued to seek a coherent representation of the forces 'by the straining of an elastic solid constituted in a peculiar way'. On the evening of 31st October, in

[35] For development of this theme see M.N. Wise and Crosbie Smith, 'Measurement, work, and industry in Lord Kelvin's Britain', *Hist. Stud. Phys. Sci.*, **17** (1986), 147–73.

[36] William Thomson, 1st July, 1846, NB34, ULC, p. 23.

[37] William Thomson, 'On the elementary laws of statical electricity', *BAAS Report*, **15** (1845), 11–12. William Thomson to Michael Faraday, 6th August, 1845, and Faraday to Thomson, 8th August, 1845, in SPT, **1**, 146–9. Faraday had completed his experiments by November. Thomson learned of them in Archibald Smith to William Thomson, 19th November, 1845, S148, ULC. See also L. Pearce Williams, *Michael Faraday* (London, 1965), pp. 383–91.

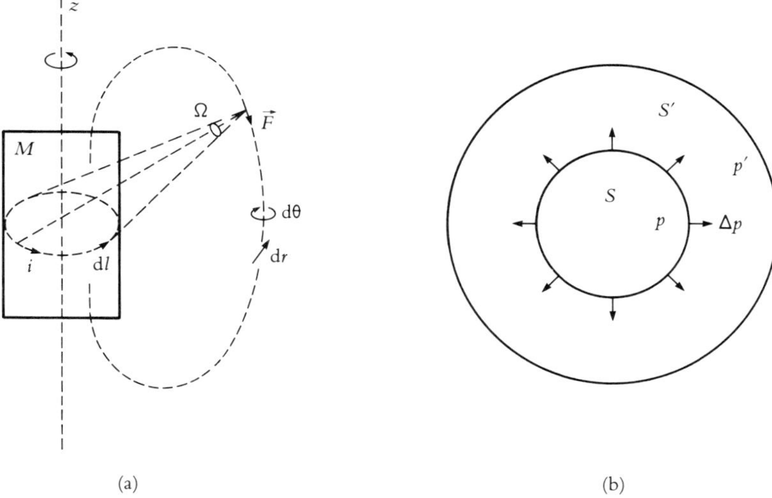

Figure 8.4.(a) Thomson sought to relate the representation of a bar magnet *M* given by current loops *i* around its surface to another representation given by a rotational deformation of the ether about the axis *z* of the magnet. Gauss had shown that the potential at any point due to a current loop is given by the solid angle Ω subtended by the current loop at that point. This solid angle, Thomson hoped, could be further shown to correspond to the ether displacement at the point arising from the axial rotation, the rotation itself being actually produced by the current dragging the ether round. Here $\vec{F}$ is the force at any point along a line of force (dashed), d*r* is the displacement at any point along the line, and dθ indicates the effective rotation about the line due to decreasing displacements at larger distances. (b) An electrified sphere, similarly, would displace the ether outwards as a result of increased pressure *p* in the interior space *S*, compared with the pressure *p′* in the exterior space *S′*.

the middle of preparing the introductory lecture for his first session, he announced, 'I *think* the following must be true' (see figure 8.4):

If particles along a closed curve of any form [in a solid medium] be displaced equal ∞ly [infinitely] small distances [d*l*] along the curve, the displacement produced [d*r*], at any pt of the medium can be represented in some way by means of diffl coeffts of the *solid angle* [Ω] whose vertex is the pt & base the closed curve. This solid ∀ [angle] is the potential due to the action of a voltc current [*i*], circulating in the closed curve, as is known [from Gauss].

Thus a bar magnet [*M*] would be represented by an axis [*z*] turned round in the elastic solid so as to drag points of the solid round along with it, & the force [$\vec{F}$ = gradient Ω] produced by the magnet at any pt would have some reln to the displacements [d*r*], in the case of the medium. An electrified sphere would be represented by a hollow spherical space [*S*] filled with the air of a higher pressure than the air without [$\Delta p = p - p'$], if the exterior boundary of the *solid* (the room in wh the electrifd body is placed) be a concentric sphere [*S′*]. *Possibly* the same might be true as I once thought, when the exterior &

interior boundaries are of any form. I am not at all sure of anything I have written just now, but I want to get it out of my head, as I shall have no time to spare, during the session.[38]

This passage shows the rudimentary physical reasoning from which the famous 'Mechanical representation of electric, magnetic, and galvanic forces' derived. Its essential feature is the attempt to represent a magnet by a current loop around its axis, and the effect of the current by a rotational displacement (shearing strain) throughout the surrounding medium (ether). The forces would then appear as some function of the varying displacement in this overall rotation. Since Thomson designed the model to account for Faraday's rotation of light waves propagated along magnetic lines, it was reasonable for him to try to represent the forces directly as differential rotations, or rotational strains circling about the lines of force. He would soon succeed.

Without break, Thomson's description jumps from magnetism to electric intensity at the surface of a charged conductor, which corresponds to a simple pressure Δp pushing against the surrounding medium, or 'air'. Electric force is presumably the displacement produced in the air by this pressure. The scheme differs little from the heat flux representation. It is intriguing because it shows distinctly how Thomson could move from an elastic *solid* medium (ether), to *air*, and back to the solid without stopping to question the transitions. Obviously his air was not simply an aeriform fluid, but an air with the same properties as the luminiferous ether: behaving like an elastic solid, with respect to the rapid vibrations of light and the actions of magnetism and electricity, even though it behaved like a non-resisting subtle fluid with respect to the motion of gross bodies within it.

Such an ether had been the focus of George Stokes's classic paper on the macroscopic mechanics of fluids and solids, read to the Cambridge Philosophical Society in April, 1845. Having got to know Stokes during his fellowship year, Thomson turned to the as-yet unpublished portions of the paper for the equations of the hypothetical medium, no doubt with particular attention to Stokes's 'Reflections on the constitution, and equations of motion of the luminiferous ether in vacuum'.[39] On 28th November, at 10.15 p.m., he found the elusive answer: 'I have at last succeeded in working out the *mechanico-cinematical*(!) representation of electric, magnetic, & galvanic f[ces]. I yesterday evening wrote to Cayley, the two first, but I have only this moment got out the last case'. By 2.45 a.m. he had 'just completed a paper for the *Journal* on the

38 William Thomson, 31st October, 1846, NB34, ULC, pp. 41–3. Our emphasis on 'solid'. Thomson knew Gauss's solid angle representation of potential from his 'General theory of terrestrial magnetism', in Richard Taylor (ed.), *Scientific memoirs*, **2** (1841), 184–251, on p. 230. See below, note 53.

39 G.G. Stokes, 'On the theories of the internal friction of fluids in motion, and of the equilibrium and motion of elastic solids' (1845), *Trans. Phil. Soc.*, **8** (1849), 287–319; *Mathematical and physical papers* (5 vols., Cambridge, 1880–1905), **1**, pp. 75–129. The ether section appears on pp. 124–9. See Chapter 6 for discussion of Stokes's macroscopic style.

subject'.[40] Assuming an infinite, homogeneous, isotropic, and *incompressible* medium, he found three states of strain for the three cases: a radial displacement varying as the inverse square of the distance from a point electrical charge, and two states with rotational components corresponding to the forces of an infinitesimal magnetic dipole and of an infinitesimal element cut from an electric current loop. While differing in detail from his original concept, the latter two states preserved the basic idea of rotation of the medium about the axis of a magnet, with each line of force represented by the relative rotation about it,

$$\vec{F} = \text{curl}\ \vec{A}.$$

This is the origin in British electromagnetic theory of the 'vector potential' $\vec{A}$, as Maxwell named it, for magnetic force.

Throughout his life Thomson would refer back to his 'mechanical representation' as a true picture of magnetism, so far as real rotations were concerned, although he soon would look for those rotations in the absolute motion of a fluid ether rather than in the relative displacements of an elastic solid. With respect to electric force, however, he never expressed such confidence, claiming fifty years later: 'I could not in Nov. 1846, nor have I, ever since that time, been able to regard "displacement" as anything better than a mere "mechanical representation of *electric* force"'.[41] 'Displacement' here carries the connotations of Maxwell's 'electric displacement' in the ether as the reality of electrification. Maxwell subscribed to Faraday's view of positive and negative electric charges as nothing other than the opposite terminations of lines of inductive action, but (unlike Faraday) interpreted that action as a continuous polarization or flux in the ether. As we have argued previously, Thomson never accepted either Faraday's or Maxwell's version of the idea, reasoning that no evidence existed to suggest a change in the structure of either gases or ether due to electric action. His representation of air/ether as incompressible (divergence $\vec{F} = 0$, where $\vec{F}$ is electric force) is thoroughly consistent with that position. It means, for example, that an electrified plane conductor, exerting its pressure against the air/ether, would produce no other effect than to push or pull the medium as a block, displacing it, but without any *internal* change, no volume compression or polarization, and thus no electrical induction within it.

Under magnetic action, by contrast, the ether would undergo rotational strain as a true elastic deformation, or shear, not merely a rotation of the entire medium as a block. Assuming light waves to be propagated as transverse undulations, to account for their polarization, magnetism might thus be capable of causing a rotation of the waves about lines of force. Electric force could produce no action on light waves simply because it did not affect the medium. Indeed, Faraday had observed no action despite his best efforts, and so Thomson's etherial medium represented the established facts. It had the addi-

[40] William Thomson, 28th November, 1846, NB34, ULC, pp. 45–7. 'On a mechanical representation of electric, magnetic, and galvanic forces', *Cam. and Dublin Math. J.*, **2** (1847), 61–4; MPP, **1**, 76–80. [41] Lord Kelvin to G.F. Fitzgerald, 29th April, 1896, in SPT, **2**, 1069.

tional advantage that it eliminated from the wave theory of light the troubling longitudinal compression waves that should have accompanied transverse waves through any compressible elastic solid, but which were never observed. In an incompressible medium such waves would travel with infinite velocity (ch. 13).

To label his accomplishment Thomson chose the phrase '*mechanico-cinematical*' representation. This usage of 'kinematic' is the one we have adopted previously for his purely descriptive treatments of the 'motion' of forces in space, their distribution or propagation, independent of the causes of the motion. He might have called the heat flow analogy 'thermo-cinematical' in the same sense.[42] The use of the term here probably derived from the fact that the idea for the new analogy arose while he worked out his introductory lecture, which contained the division of mechanics into statics, cinematics, and dynamics (ch. 5). The point is that the representation, whatever its heuristic value, did not give anything like a causal physical theory of the propagation of force in the ether. It served only to generate the distribution of forces by analogy to a realistic form of motion. Thomson described his attitude, and the limits of the analogy, to Faraday:

> I enclose the paper which I mentioned to you as giving an analogy for the electric and magnetic forces by means of the *strain*, propagated through an elastic solid. What I have written is merely a sketch of the mathematical analogy. I did not venture even to hint at the possibility of making it the foundation of a physical theory of the propagation of electric and magnetic forces, which, if established at all, would express as a necessary result the connection between electrical and magnetic forces, and would show how the purely *statical* phenomena of magnetism may originate either from electricity in motion, or from an inert mass such as a magnet. If such a theory could be discovered, it would also, when taken in connection with the undulatory theory of light, in all probability explain the effect of magnetism on polarized light.[43]

In these lines lay the research programme of a lifetime, the source of the chronic disease that Thomson later called 'ether dipsomania', attacks of which he suffered until his death.[44] In 1847 he would restrict his efforts to working out the mathematical relations that governed the distribution of magnetic forces in the ether and the associated ponderomotive forces on inductively magnetized substances, that is, to the dynamics of a magnetic field.

When Thomson had queried Faraday in August, 1845, about the effects of polarized dielectrics on light, he had also asked about attraction or repulsion of such dielectrics, as suggested by the analogy to magnetic materials. Although Faraday found neither effect in dielectrics, he found both in what he soon called

42 Thomson called attention to the similarity in method of the two analogies in 'Note on the integration of the equations of equilibrium of an elastic solid', *Cam. and Dublin Math. J.*, **3** (1848), 87–9; MPP, **1**, 97–9.

43 William Thomson to Michael Faraday, 11th June, 1847, in SPT, **1**, 203–4.

44 Lord Kelvin to G.F. Fitzgerald, 9th April, 1896, in SPT, **2**, 1065.

diamagnetics.[45] The same glass that rotated light was also weakly repelled from the poles of a powerful magnet, in contrast to all previously known magnetic substances, which were attracted. Subsequently he discovered that many other materials behaved similarly, bismuth and copper being notable. He believed that he had finally demonstrated a quite general dependence of magnetic action on the medium through which it passed.

Repulsion itself seemed quite inexplicable in terms of the usual theory, but Faraday found something even more puzzling. 'Repulsion' did not properly characterize the effect, for the moving force acted not simply in the reverse direction along lines of force, but sometimes perpendicularly across them. Having never believed in attraction and repulsion at a distance in any case, Faraday described the general behaviour as a tendency 'to move from stronger to weaker places of magnetic force'. To capture the sense of a space filled with forces in which objects moved between places of differential strength, Faraday introduced the term 'field' in his November, 1845, *Researches*. This terminology and the effects it connoted provided the specific incentive for Thomson to extend his own derivation, from April, 1845, of the ponderomotive force between electrified conductors. From the confluence of their two modes of analysis came the basic tenets of classical field theory.

Immediately following the 'Mechanical representation' in his notebook, but one month later, 30th December, 1846, Thomson entered the following sketch:

Let μ be the magnetic moment of a small sphere of any magnetic or diamagnetic substance, if magnetized by a unit of force. If this sphere be put at any p^{t} (x,y,z) in the neighbourhood of a magnet, it will be magnetized so that its magnetic moment will be μF [where] $(\vec{F}^2 = [(\text{gradient } V)^2] = X^2 + Y^2 + Z^2)$ & the comp^{t} of the force with w^{h} it will be urged, in the dir^{n} parallel to OX, will be

$$\mu(X\frac{\partial X}{\partial x} + Y\frac{\partial Y}{\partial y} + Z\frac{\partial Z}{\partial z})$$

$$\text{or } \frac{1}{2}\mu\frac{\partial}{\partial x}[(\text{gradient } V)^2]$$

$$\text{or } \mu\frac{1}{2}\frac{\partial}{\partial x}(F^2).$$

Hence the sphere, if of a magnetic substance, will be urged by a [ponderomotive] f^{ce} in the dir^{n} along w^{h} the absolute result^{t} [magnetic] f^{ce} increases; if diamagnetic, in the contrary direc^{n}. (from 'weaker to stronger lines of force' or the reverse. Faraday – § 2418 also § 2269).

45 Thomson to Faraday, in SPT, **1**, 146–9. Michael Faraday, *Experimental researches in electricity* (3 vols., London, 1834, 1844, 1855), **3**, series XX and XXI. See especially David Gooding, 'Final steps to the field theory: Faraday's investigation of magnetic phenomena: 1845–1850, *Hist. Stud. Phys. Sci.*, **11** (1981) 231–75, and 'Faraday, Thomson, and the magnetic field', *Brit. J. Hist. Sci.*, **13** (1980), 91–120.

He went on to describe the conditions of equilibrium for magnetic and diamagnetic spheres located at positions of maximum and minimum F, giving examples of how such conditions could be realized experimentally.[46]

In the published version Thomson replaced the induced magnetic moment $\mu\vec{\mathrm{F}}$ by a continuously variable 'intensity of magnetism' $\vec{\mathrm{I}}$ (magnetic moment per unit volume, or magnetization) resulting from an inductive capacity k_0,

$$\vec{\mathrm{I}} = k_0 \vec{\mathrm{F}}.$$

For a small volume σ he therefore obtained

$$\vec{\Gamma} = \text{gradient } (\tfrac{1}{2}k_0\sigma F^2),$$

with the interpretation: 'the components of the force in different directions [are] expressible by the differential coefficients of the function $\frac{1}{2}k_0\sigma F^2$'. The function behaved, then, as a total potential for the ponderomotive force and bears close comparison to the relation $\vec{\Gamma} = -\text{gradient } \Phi$, where Φ is mechanical effect.[47] Throughout his notebook in 1847 Thomson interpreted the function in this way. He gave the interpretation publicly in a paper of 1851 that extended the results to crystalline media. Simplified to the non-crystalline case, his general principle was:

> the entire action which [an infinitely small sphere] experiences, whether directive tendency or tendency to move from one part of the field to another, is defined by the following proposition: The quantity of mechanical work which is required to bring the body from a position where the intensity of the force is F . . . to a position where the intensity of the force is F' . . . is equal to . . .
>
> $$\tfrac{1}{2}k_0\sigma(F'^2 - F^2),$$
>
> and the proposition is equivalent to the mathematical expression of Faraday's law regarding the tendency to places of stronger or weaker force.[48]

Two aspects of this 1851 statement are notable. The formal proposition relating ponderomotive forces to work, firstly, has the connotations of the fully generalized 'principle of mechanical effect', or the principle of energy conservation (ch. 9). In essentially the form given here, Thomson and Peter Guthrie Tait, in their classic *Treatise on natural philosophy* of 1867, would regard the proposition as 'comprehending the whole of abstract dynamics', for it transformed all questions of force into questions of energy.

Secondly, Thomson has adopted Faraday's term 'field' for the system of

46 William Thomson, 30th December, 1846, NB34, ULC, pp. 48–9.

47 William Thomson, 'On the forces experienced by small spheres under magnetic influence; and on some of the phenomena presented by diamagnetic substances', *Cam. and Dublin Math. J.*, **2** (1847), 230–5; E&M, 493–9. Thomson, following Green, actually used $3i/4\pi$ rather than our k_0, $\frac{3}{4}\pi$ being the magnetic moment of a unit sphere which freely conducts magnetic fluids. He changed to the simpler form in later papers. See his 'Remarks on the forces experienced by inductively magnetised ferromagnetic or diamagnetic non-crystalline substances', *Phil. Mag.*, [series 3], **37** (1850), 241–53; E&M, 500–13.

48 William Thomson, 'On the theory of magnetic induction in crystalline and non-crystalline substances', *Phil. Mag.*, [series 4], **1** (1851), 177–86; E&M, 465–81. The abstract appeared in *BAAS Report*, **20** (1850), 23.

forces, consisting of 'places' of stronger and weaker force. He now formally defined a 'magnetic field' as 'any space at every point of which there is a finite magnetic force', and a 'line of force' as 'a line drawn through a magnetic field in the direction of the force at each point through which it passes'. These definitions may easily be misinterpreted to involve no more than the concept of action at a distance would suggest; point forces acting in a definite direction at each point in space. Thomson conceived the action of the field on bodies within it, however, not in terms of point forces, but in terms of differences in field 'intensity' between different 'places', an action which might or might not be along the lines of force. In each publication he called attention to this remarkable feature of Faraday's discovery, noting in 1847: 'Thus in some cases [a magnetized body] may actually be urged across the direction of the magnetizing force'.[49] The same result would follow from summing over the point forces acting on the body, but as a conceptual matter the field treatment of ponderomotive forces was radically new.

In 1847 Thomson not only developed the essentials of these two ideas separately but began explicitly to unite them in the idea of the field as a field constituted of mechanical effect. The work done to establish a magnetic system became mechanical effect distributed throughout the field, rather than localized in discrete sources. Concomitantly, work expended or extracted in moving a body from one place to another became work done to increase or decrease the mechanical effect constituting the field, rather than mechanical effect localized in the position of the body. Through mechanical effect the field became a real physical entity rather than a description of a distribution of abstract forces. Its reality, conversely, was limited to the reality of mechanical effect as a physical entity, or as energy. The concept 'field' and the concept 'energy' thus depended on each other and emerged in tandem.

The hydrodynamic analogy

In the period 1847–50 Thomson transformed his kinematics of fields, as embodied in the analogies to heat and to elastic solids, into field dynamics. Once again an analogy led him through the physical reasoning while he searched for analytic proofs. Water, or hydrodynamics, the special province of his new friend Stokes, this time provided the requisite medium of thought. And once again a compact notebook entry, of 29th March, 1847, signalled the shift in focus. We give it in full before analyzing its multiple facets.

Let F_n be the attraction in the direction of the normal, produced by a distribution (ρ) of matter over a surface S.

I. Is it possible to determine ρ, F_n being given (arbitrarily) with the limitation $\iint F_n d\sigma = 0$?

[49] 'On the forces', E&M, 496.

II. Is it possible to distribute closed wires over S so as to produce the same attraction as a given distribution (F_n) of matter?
III. Is it possible to magnetize a solid within S so as to produce the same force on an external point as a given dist^{n} (F_n) over the surface? [(This is at once shown to be ident^{l} with question I.), *del.*]
IV. Is it possible to find a function ϕ which [gives] by coefficients the [velocity, *del.*] motion of water in the interior of S, the motion across an element ω of the surface being $F_n\omega$? (Of course we may add to any solution obtained an arbitrary part which would represent motion which could take place *within* S.)

It is possible to find a state of motion of water through S such that the motion across the surface will be represented by a(n arbitrarily) given function F_n (but perhaps not one in which $\vec{v} \cdot \vec{dr} = d\phi$ [where $\vec{v}$ = velocity]). Hence III. is answered in the affirmative!

It has occurred to me this evening that any motion of water in the interior of S gives a distribution of magnetism in the interior of a piece of steel bounded by a surface similar to S, such that no force will be exerted on an external point. For instance, a closed tube, with water flowing round through it, is comparable with a bar magnet ($I\omega$ constant) bent with its ends together.

$$\left.\begin{matrix}\text{velocity}\\ \\ \text{direction}\end{matrix}\right\}\text{ of motion,}\qquad \left.\begin{matrix}\text{intensity}\\ \\ \text{direction}\end{matrix}\right\}\begin{matrix}\text{of magnetic}\\ \text{polarisation}\end{matrix}$$

Wrote to Stokes tonight asking him IV., &c.[50]

Taken as a whole, this passage indicates both the definite goal and the ambiguity of his initial hydrodynamic analogy. The ambiguity has two roots. Most obviously, he took the motion of water as analogous to both intensity of magnetic force $\vec{F}$ and intensity $\vec{I}$ (magnetic moment per unit volume, or magnetization). This seems to be the same ambiguity as that in the correlation of electric force with temperature gradient and flux of heat simultaneously. Recognizing, however, that Thomson regarded magnetic force, in contrast to electric force, to be propagated as a true displacement in the interior of the ether, as in the elastic solid model, we shall have no difficulty in identifying his magnetic force with magnetic polarization, and thus as a flux. A deeper ambiguity derives from the fact that the sources of magnetic force are both magnetic matter, in (I), and electric currents, in (II). He expressed his worry about this double origin of magnetic force in the letter to Faraday in June (quoted above) in which he observed that the 'mechanical representation' would have to show 'how the purely *statical* phenomena of magnetism may originate either from electricity in motion, or from an inert mass such as a magnet'. We consider the simpler issue first.

In the analysis of ponderomotive forces on induced magnets, discussed above,

[50] William Thomson, 29th March, 1847, NB34, ULC, pp. 52–3. William Thomson to G.G. Stokes, 30th March, 1847, K18, ULC, contains questions IV and the remarks that follow, which initiated extensive and important discussions in April.

magnetization $\vec{I}$ is proportional to magnetizing force $\vec{F}$, $\vec{I} = k_0\vec{F}$, where k_0 is positive for magnetic bodies, negative for diamagnetics, and zero for free space (ether). With this convention the two kinds of materials have opposite polarizations and space is non-magnetic. In his original publication on diamagnetism, however, Faraday had recognized that all substances could be taken to have the same kind of inductive capacity, diamagnetics to a lesser degree than space and magnetics to a greater:

> Whichever view we take of solid and liquid substances, whether as forming two lists, or one great magnetic class . . . it will not, as far as I can perceive, affect the question. They are all subject to the influence of the magnetic lines of force passing through them, and the virtual difference in property and character between any two substances taken from different places in the list . . . will be the same; for it is the differential relation of the two which governs their mutual effects.[51]

In support of that view, Faraday had done experiments on a vial containing a weak solution of an iron salt, showing that it behaved like a normally magnetic substance in air but like a diamagnetic when immersed in a stronger iron solution.

Thomson too recognized the conventionality of measuring inductive capacities relative to zero or to unity for air/ether. In his notebook on 10th March he wrote $(k-1)$ for the k_0 above and on 30th September, under 'Possibility of problem of magnetic induction', he specified: 'Let k be the *inductive capacity* [conducting power in analogous prob^{m} of heat, *inserted*] at any p^{t} of a mass of matter susceptible of induc^{n}; & let us consider k as a fu^{n} w^{h} is equal to 1 for all p^{ts} of space unoccupied by this matter'.[52] The convention $k=1$ for air/ether made the hydrodynamic analogy applicable to all space, with the motion of water analogous to a continuous flux of magnetic force, as polarization, through any region outside the sources.

From March to November Thomson drove this hydrodynamic analogy towards solutions of the four 'possibility' questions above, often consulting Stokes on his 'favorite subject' and receiving lengthy replies. On 7th April Thomson wrote: 'I have been for a long time thinking on subjects such as those you write about, & helping myself to understand them by illustrations from the theories of heat, electricity, magnetism, & especially galvanism; sometimes also water. I can strongly recommend heat for clearing the head on *all* such considerations, but I suppose you prefer cold water'. He thought Stokes would be 'delighted if you read a few pages or even a few words of a paper of Gauss . . . by which you will be supplied by a flood of illustrations for water'. The paper was Gauss's 'General theory of terrestrial magnetism', which contained illustrations and maps of magnetic lines over the earth's surface. Thomson referred specifically to a section outlining how magnetic forces produced by galvanic

[51] Faraday, *Experimental researches*, **3**, paras. 2443ff.

[52] William Thomson, 10th March and 30th September, 1847, NB34, ULC, pp. 50, 83.

currents could be represented by distributions of imaginary magnetic matter (northern and southern fluids), remarking, 'I am very anxious to find whether Gauss has found the opportunity he waited for, of writing more'.[53]

This concern changes the issue from induced magnetization to the two kinds of permanent magnetic sources, questions (I) and (II), their relation to each other and to the hydrodynamic analogy. Water supplies useful illustrations for magnetism because it behaves (nearly) like an incompressible fluid, in which lines of motion must always form closed curves. For every line leaving a closed surface S containing magnetic sources, another must enter, like the lines of force leaving and entering a magnet at its north and south poles. The integral of the flux over the entire surface must therefore vanish, $\iint F_n d\sigma = 0$. Beginning from this fact (though not from the analogy) Gauss had sought the most general representation possible of the observed distribution of magnetic forces over the surface of the earth, and the relation of that representation to the possible sources of terrestrial magnetism, whether magnetic fluids in the interior materials or electric currents at the surface.

He called attention, in particular, to Ampère's theorem, 'that in place of each linear current bounding an arbitrary surface, we may substitute a [uniform] distribution of the magnetic fluids on both sides of this surface', as in figure 8.5. Thomson would soon name the surface a 'magnetic shell'. Gauss deferred a full development of the theory to 'another opportunity' but stated two essential theorems. The solid angle Ω subtended at any point by a current loop, multiplied by the strength of the current i, gives the magnetic potential at the point; and, a corollary, when the current loop is replaced by a magnetic shell, the potential must change in crossing the surface by 4π (the entire solid angle of a closed surface surrounding the point) times the current strength.[54]

On the basis of those two results, Thomson sought new versions of his replacement surface theorem of 1841 that would show how distributions of fluid (I) and current (II) could be found over any closed surface (not necessarily an equipotential surface) which would reproduce an arbitrarily given distribution of magnetic force over the surface, and therefore throughout the external space. While heat flow had generated the earlier theorem, water flow would generate the new ones. Having completed his first session at Glasgow he spent May and much of June, 1847, back at Cambridge: 'getting out various pieces of work, along with Stokes, connected with some problems in electricity, fluid motion, etc., that I have been thinking on for years, and I am now seeing my way better than I could ever have done by myself, or with any other person than Stokes'.[55]

53 William Thomson to G.G. Stokes, 30th March and 7th April, 1847, K18 and K19, Stokes correspondence, ULC. Thomson was referring to Gauss, 'General theory', p. 230.

54 Gauss, 'General theory', pp. 230–1. Thomson used the term 'magnetic shell' by 19th August, 1848, NB34, ULC, p. 101. It renders his understanding of Gauss rather too explicitly here. Thomson and Stokes only worked out the definite meaning of Gauss's theorems in April, during their lengthy exchange regarding the hydrodynamical interpretation.

55 William to Dr Thomson, 20 June, 1847, T366, ULC.

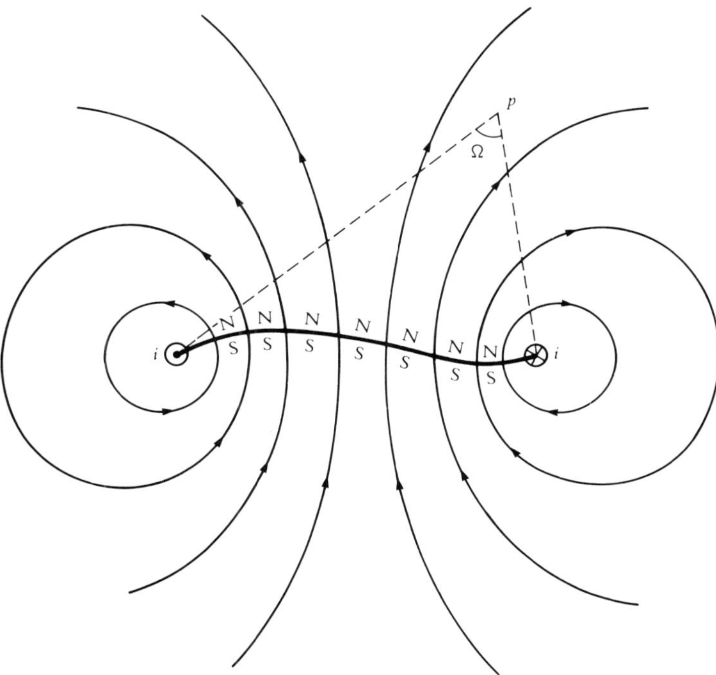

Figure 8.5. The two small circles represent cross-sections through a current loop of strength *i*, directed into the paper on the right and outwards on the left. The loop bounds an arbitrary surface which has uniform distributions of northern N and southern S magnetic matter on its two sides and subtends a solid angle Ω at an arbitrary point *p*.

He announced one such piece of work, the replacement surface theorems for terrestrial magnetism, at the meeting of the British Association in June. Green had already answered question (I) affirmatively, he recognized, so that 'a certain [superficial] distribution of imaginary magnetic matter may be found, which would produce all the phænomena of terrestrial magnetism observed at the surface of the earth or above it', $\rho = F_n/4\pi$.

The trick was to obtain the superficial currents suggested by Gauss (and Ampère) from Green's theorem, which Thomson said he had done, 'by an analogous theorem of which a physical demonstration may be given by considerations connected with fluid motion'. He gave only the result. From ρ find a potential U for the *internal* space such that $\partial U/\partial n = \rho$ and divergence $U = 0$. (This would be a potential for the motion of water in the space, as guaranteed by Stokes's positive answer to question (IV).) Then: 'Construct on the surface a "map of the values of U". If wires be laid along the lines round the surface corresponding to sufficiently close equidifferent values of U, as indicated by this

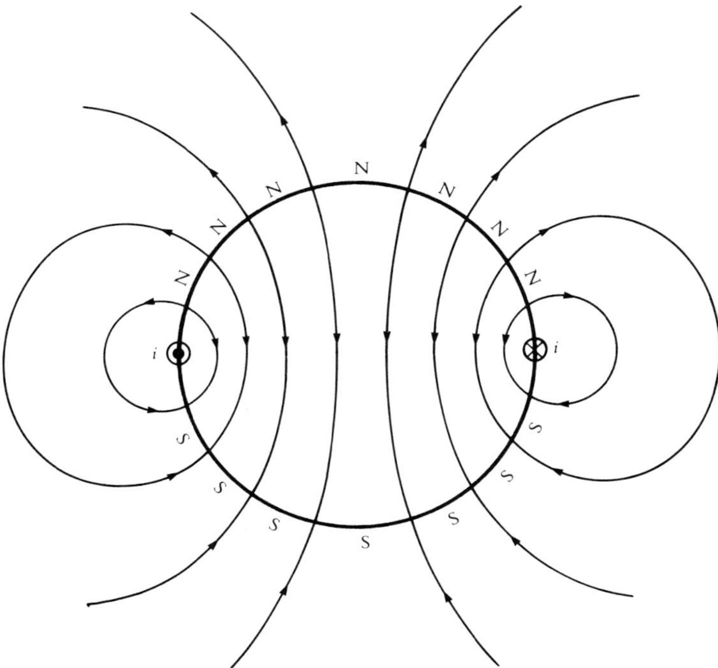

Figure 8.6. The two sides of the single surface in figure 8.5 have here been separated to open an interior space, with the lines of force inside directed oppositely to those outside.

map, and if currents of equal intensity be made to circulate through them (each being a closed curve), the electro-magnetic force that will result, upon external points, will be the same as the force of terrestrial magnetism'.[56]

Thomson seems to have reasoned backwards from Gauss's analysis as follows. For any supposed linear current i ringing the earth's surface, imagine the two faces of the equivalent 'magnetic shell' to be separated, opening up an internal space having the shape of the earth (figure 8.6). Then the imaginary magnetic matter at the surface will produce the same external forces as the current loop, and the internal potential U at the surface will change across the line of the current by i. Repeating the process for every such linear current, and summing the resultant densities of magnetic matter and the potentials at each point, will generate the appropriate distributions ρ and U over the surface. Conversely, from a map of equipotential lines of U, one may find the appropriate distribution of currents, as Thomson stated.[57]

56 William Thomson, 'On the electric currents by which the phænomena of terrestrial magnetism may be produced', *BAAS Report*, **17** (1847), 38–9; E&M, 462–5. Draft of 22nd June, 1847, in NB34, ULC, pp. 66–72. Stokes gave his definite answer to (IV) in G.G. Stokes to William Thomson, 10th April, 1847, S327, ULC.

57 This reconstruction follows from the analysis and diagrams given by G.G. Stokes to William

While apparently very simple, the solution contains a profound ambiguity with respect to internal conditions. On the magnetic matter representation the internal lines of force run in the opposite direction from the external ones, while on the current representation they form continuous loops, like lines of motion in a space filled with water. In one sense the problem is purely mathematical. If one attempts to define a potential ϕ_2 for the re-entering motions (ϕ_1 being the potential for non-re-entrant motions), its value must change by $4\pi i$ on every return to the same point. Should the potential be regarded as multiple valued, or as discontinuous? Stokes initially remarked, 'that ϕ_2 must *necessarily be discontinuous*. If it were not, the permanent temperature of a solid of the same form as S might vary from point to point without there being any flux of heat across the bounding surface; or a fluid at rest within a closed rigid surface might set itself in motion'. But Thomson responded: 'There is no reason for saying that ϕ_2 is "discontinuous", but rather we should say it has an infinite number of values at each point. For . . . we may go round the ring continually in the direction of motion and find a continually increasing value of ϕ_2 . . . but if we obtained a corresponding expres^n for p ([pressure] in a problem of hydrostatics...) of for V ([temperature] in a problem of heat) then when, in going round the ring, we reach the surface of equilibrium or isoth^l surf. from w^h we started, we must place an impermeable barrier, to keep up the diff^ce betw. the values of p, or betw. those of V'.[58]

In physical terms, therefore, the problem would not go away. Depending on whether one supposed the sources of terrestrial magnetism to act like magnetic fluids or like electric currents, the lines of force would be discontinuous or continuous, respectively, across the surface of the earth (or other permanent magnet). The old problem in the heat analogy to electrostatic inductive action, of an intensity/quantity or gradient/flux duality, therefore re-emerged in the hydrodynamic analogy to magnetism as the problem of sources. On the magnetic matter representation, magnetic forces appeared as discontinuous gradients of a proper, single-valued potential, like gradients of pressure in hydrostatics. On the current representation, they appeared as properly continuous fluxes.

Magnetism, however, presented phenomena very different from electrostatics: magnetic lines always re-entered the magnet from which they emerged; diamagnetism existed, with an inductive capacity less than space; and a true displacement of matter seemed to be required to explain magneto-optic rotation. All of these factors favoured a flux representation of magnetic lines. While clearly preferring this view, Thomson still had no adequate theory of magnetism that would explain it. He confidently rejected the imaginary magnetic

Thomson, 16th April, 1847, S329, ULC, and from a remark Thomson appended to a slightly different derivation on 16th August, 1848, in NB34, ULC: 'When I wrote the paper (June 22 [1847]) I only contemplated the method by imagining a body *potentially* magnetized, & cutting it into slices by equipot^l surfaces, &c'.

[58] G.G Stokes to William Thomson, 1st April, 1847, S324, ULC, and Thomson to Stokes, 7th April, 1847, K19, Stokes correspondence, ULC.

matter representation of the earth's magnetism, asserting, 'This proposition . . . cannot be entertained as expressing a physical fact'. On the other hand: 'we have now many reasons for believing that the existence of terrestrial electric currents, producing wholly or in part the magnetic phaenomena, is a physical fact'. Still he did not believe in Ampère's theory of electrodynamic molecules, which assumed all magnetism to be caused by electric currents surrounding the molecules of magnetic materials; 'Ampère's theory . . must be regarded as merely theoretical; for it is absolutely impossible to conceive of the currents which he describes round the molecules of matter as having a physical existence'.[59] There can be no question, then, of Thomson having already in mind his famous vortex theory of atoms. His electric currents on the earth were surface currents, and in general he imagined intermolecular, not intramolecular, currents. He had not yet discovered a connection, as he told Faraday, between the statical and the dynamical aspects of magnetism.

Nevertheless, by the autumn of 1847, the hydrodynamical analogy had taken on aspects of a quite general physical theory. Between August and October Thomson worked out two theorems, and two papers, which transformed the analogy into a theory of mechanical effect and ponderomotive force in a field of magnetization.

The first theorem, published (in its converse form) as , 'On the vis-viva of a liquid in motion', answered questions (I), (III), and (IV) affirmatively, by proving that the solutions existed when the integral $\iiint F^2 \mathrm{d}\tau$, interpreted as *vis viva* in the hydrodynamical analogy, was minimum. He wrote to Stokes:

> The following occurred to me when I was trying to get an analytical proof of the possibility of distributing matter over a surface, S, so that the force $[F_n]$, along the normal, on p^ts^ infinitely near the surface may have an arbitrary expression . . . Let [divergence $\vec{F}$] $=0$ . . . F_n being arb^y^ with the sole rest^n^ $\iint F_n{}^2 \mathrm{d}\sigma = 0$. There exists a solution which renders $\iiint F^2 \mathrm{d}\tau$ less than any other . . . You will easily see that this analysis, when applied to hydrodynamics, will give a proof, by means of the principle of least action that $\vec{F}\cdot\vec{\mathrm{d}r}$ is a complete diff^l^ $[\mathrm{d}\phi]$ in the case of a mass of fluid, bounded by a surface, (changing of course in form) S, and set in motion by unequal pressure at the surface, thus if t be the time, from rest, we must have $\int \mathrm{d}t \iiint F^2 \mathrm{d}\tau$ a minimum, and the result is easily found, as above. I sh^d^ be glad to hear from you whether this is one of the proofs w^h^ have been given.

Stokes answered: 'The proof you give that $\vec{F}\cdot\vec{\mathrm{d}r}$ is an exact diff^l^ when the motion begins from rest is perfectly new to me, and appears quite satisfactory. Of course you will publish it'.[60]

He did, but without the principle of least action. He relied instead on the idea of motion generated suddenly, by impulsive forces at the boundary, which avoids a problem Stokes raised (below) about the effects of fluid friction when forces act over finite times. Nevertheless, in the principle of least action, read

[59] William Thomson, 'Terrestrial magnetism', E&M, 462–5.

[60] William Thomson, 'Notes on hydrodynamics. V. On the vis-viva of a liquid in motion', *Cam. and Dublin Math. J.*, **4** (1849), 90–4; MPP, **1**, 107–12. William Thomson to G.G. Stokes, 20th

loosely as a principle of least work done to establish a flow system, Thomson had already recognized a more general solution of the problem of distributions of magnetic forces, applicable to spaces of variable inductive capacity. On 27th September, 1847, he was considering the work done in pushing fluid through a space enclosed by a surface S held fixed. On 30th September he began working on the 'Possibility of [the] problem of magnetic induction', by minimizing the integral $\iiint k(\text{gradient } V)^2 d\tau$ over all space, subject to the continuity equation, divergence(gradient V) $= -4\pi\rho$. This integral simply extends to the entire magnetic system the potential $\Phi = -k_0\sigma F^2$ for the ponderomotive force on a sphere of volume σ and inductive capacity k_0. The procedure is flawed, however, because it assumes the same potential V and magnetizing force $\vec{F} = -$gradient V as would exist in free space. That is, it does not take into account any redistribution of the lines of force (conceived now as lines of polarization, analogous to lines of fluid motion) resulting from inserting the arbitrary distribution of capacities k into a previously homogeneous free space where $k=1$.

After several attempts to perfect the proof, he arrived on 7th October at the analysis published in 'Theorems with reference to the solution of certain partial differential equations', which established existence and uniqueness of solutions to the equation

$$\text{divergence}(k \text{ gradient } V) = -4\pi\rho.$$

This is a continuity equation for the polarization ($-k$ gradient V) which 'flows' continuously through all space outside the sources. The proof depends on the existence of a minimum in a function which, suitably interpreted, is the work done to redistribute the lines of magnetization throughout space when an arbitrary function k replaces $k=1$.

This incredibly cryptic paper of a mere three pages makes no reference to work, *vis viva*, mechanical effect or least action, and supplies no aids to understanding other than the concluding remark: 'The analysis given above, especially when interpreted in various cases of abrupt variations in the value of k, and of infinite or evanescent values, through finite spaces, possesses very important applications in the theories of heat, electricity, magnetism, and hydrodynamics, which may form the subject of future communications'. Yet Thomson wrote in his notebook: '*The possibility of all the problems I have yet considered in the allied physical theories* is I believe established by this theorem'. He returned to his 'possibility' questions of 29th March to write 'yes' by each of them and to note: 'See Oct. 7, 1847 for proof of a theorem on which the answers are founded'.[61]

October, 1847, K21, Stokes correspondence, ULC; Stokes to Thomson, (received) 30th October, 1847, S330, ULC. Thomson had written out the proof and its *vis-viva* interpretation on 13th–14th August, 1847, in NB34, ULC, pp. 74–9.

[61] William Thomson, 'Theorems with reference to the solution of certain partial differential equations', *Cam. and Dublin Math. J.*, **3** (1848), 84–7, and (with added note) *J. de Math.*, **12** (1847), 493–6; E&M, 139–43; MPP, **1**, 93–6. In the French version, a curious change of sign occurs, which would make $k=0$ in free space. See William Thomson, 29th March, and 7th October, 1847, in

Perhaps Thomson published his theorem in its occult form just because of its immense significance, which he had not yet had time to digest. Stokes had responded to his announcement of the theorem on minimum *vis viva* with a suggestion: 'To prevent misconception it would be well to state *explicitly* that it only holds on the hypothesis of the absence of tangential action [friction], in wh case alone the principle of least action can be applied. You have not it seems considered the case of elastic fluids'.[62]

Stokes's remark is striking because both men had heard James Joule, at the British Association in June, present a paper on the conversion of work to heat by the friction of fluids. Thomson wrote retrospectively in 1872: 'I did not then know that motion is the very essence of what has been hitherto called matter. At the 1847 meeting of the British Association in Oxford, I learned from Joule the dynamical theory of heat, and was forced to abandon at once many, and gradually from year to year all other, statical preconceptions regarding the ultimate causes of apparently statical phenomena'.[63] The degree to which he had begun that process in 1847 is uncertain, but the problems of heat would indeed soon force him to pursue a dynamical theory of magnetism based on molecular electric currents, and a dynamical theory of elasticity based on the same currents.

In the very letter to which Stokes was responding, in fact, Thomson reported:

> I perceived a fine instance of elasticity in an incompressible liquid, in a very simple observation made at Paris, on a cup of thick 'chocolat au lait'. When I made the liquid revolve in the cup, by stirring it, and then took out the spoon [the velocity of rotation diminished very rapidly, *del.*] the twisting motion (in eddies, and in the general variation of angular vel. on acct of the action of the spoon overcoming the inertia of the liquid, and the fricn at the sides) in becoming effaced, always gave rise to several oscillations so that before the liquid began to move as a rigid body, it performed oscillations like an elastic (incompressible) solid.

This passage suggests already a close relation between the elastic-solid analogy of the 'mechanical representation' and the hydrodynamical analogy for magnetism. Since Thomson did not develop that idea until after 1849, we shall take it up only after discussing his interaction with Joule and thermodynamics (chs. 9 and 10) and after showing how he recovered the principle of least action in dynamics (ch. 11).

The theorem of minimum *vis viva* provided Thomson with the dynamical principle of hydrodynamics, the principle controlling the states of motion in a fluid. It apparently motivated the series of 'Notes on hydrodynamics' that he announced in the same letter, noting that he had already written the first Note, 'On the equation of continuity', and soliciting contributions from Stokes. As

NB34, ULC, pp. 52, 87. For additional interpretation of this theorem and its relation to Thomson's mathematical theory of magnetism see M.N. Wise, 'The flow analogy to electricity and magnetism, Part I: William Thomson's reformulation of action at a distance', *Arch. Hist. Exact Sci.*, **25** (1981), 19–70, esp. pp. 56–61; for a mathematical description see Cross, 'Integral theorems', pp. 140–1.

62 Stokes to Thomson, (received) 30th October, 1847, S330, ULC. 63 E&M, 419*n*.

Stokes could not immediately comply, Thomson wrote Note II as well, 'On the equation of the bounding surface', to precede Stokes's Note III, 'On the dynamical equations'. The Notes present an explicit formulation of the distinction between kinematics and dynamics. In requesting Note III, Thomson suggested a preface, 'to the effect that the *cinematical* equations which depend on the characteristics peculiar to fluid bodies, have been investigated in the preceding articles, & that you now proceed to apply the ordinary [mechanical, *del.*] dynamical principles, connecting actual motions, with the forces by w^h they are produced or sustained'. Stokes responded with a fuller version: 'In reducing to calculation the motion of a system of rigid bodies, or of material points, there are two sorts of equations with which we are concerned; the one expressing the geometrical connexions of the bodies or particles with one another, or with curves or surfaces external to the system, the other expressing the relations between the changes of motion which take place in the system and the forces producing such changes. The equations belonging to these two classes may be called, respectively, the geometrical, and the dynamical, equations. Precisely the same remarks apply to the motion of fluids'.[64]

Thomson's *kinematical* analogies, therefore, comprised only geometrical descriptions of motions, in particular those governed by the continuity equation. The analogy between magnetism and the motions of water became a hydro*dynamical* analogy precisely when the causes of the motions entered. Stokes's Note III, and also IV, dealt with the 'ordinary dynamical principles' of point forces and pressures.[65] Thomson's Note V, however, on the minimum in *vis viva*, which he intended originally to express the principle of least action, imposed a dynamical principle on the fluid as a whole. It made the behaviour of any element depend on the requirement that the entire field of motion contain the minimum possible mechanical effect, or, alternatively, that the mechanical effect expended to establish the field be minimum.

In the latter form it is particularly obvious that the dynamical principle expressed an identity between the ponderomotive forces on objects in a field and the tendency of the field to arrange itself so as to minimize its total mechanical effect. That identification constitutes the essential significance of Thomson's analysis of his extended continuity equation. His proof of existence and uniqueness of solutions simply says that every region of space, and every object, that participates in the field will acquire the particular magnetization that minimizes the total mechanical effect, and, concomitantly, that every object experiences a force determined in direction and magnitude by the maximum rate of decrease

[64] Thomson to Stokes, 20th October, 1847, K21, and 1st February, 1848, K23, Stokes correspondence, ULC; Stokes to Thomson, 25th November, 1847, S331 and S332 (copy), ULC. 'Notes on hydrodynamics. I., II., and III.', *Cam. and Dublin Math. J.*, **2** (1847), 282–6; **3** (1848), 89–93, 121–7; II. in MPP, **1** 83–7; III. in Stokes, *Papers*, **2**, pp. 1–7.

[65] G.G. Stokes, 'Notes on hydrodynamics. IV. Demonstration of a fundamental theorem', *Cam. and Dublin Math. J.*, **3** (1848), 209–19; *Papers*, **2**, pp. 36–47. Stokes finished the series with 'VI. On waves', *Cam. and Dublin. Math. J.*, **4** (1849), 219–40; *Papers*, **2**, pp. 221–42.

of the total mechanical effect. The equilibrium state of the system is thus to be understood in essentially temporal terms, the terms of genesis and further development in a natural succession of states.

These are the same results he had established in April, 1845, for electrostatic systems, but he had not then distributed the mechanical effect in a field (at least not explicitly), having located it in the sources at their various potentials through the integral $\iiint \rho V \mathrm{d}\tau$. Although he had been perfectly aware of the mathematical equivalence between this integral and the field integral $\iiint F^2 \mathrm{d}\tau$, it was his new conviction that an effect (polarization) actually took place in the field which legitimized treating mechanical effect as spread through the field. Conviction came only with Faraday's new experiments on magneto-optic rotation and diamagnetic action. The experiments thus provided a crucial motivation in Thomson's relocalization of mechanical effect through the flow analogy. We may now symbolize his magnetic field theory in three equations, for magnetization, mechanical effect, and ponderomotive force, respectively:

$$\text{divergence } (k \text{ gradient } V) = -4\pi\rho,$$

$$\Phi = \iiint k \,(\text{gradient } V)^2 \mathrm{d}\tau = \iiint \rho V \mathrm{d}\tau,$$

and

$$\vec{\Gamma} = -\text{gradient } \Phi.$$

Although he continued to shield the field interpretation of these equations from public view, he was prepared to apply it when necessary. In a well-known letter to Faraday he explained why Faraday's description of magnetic and diamagnetic action in terms of 'conducting power' would require both magnetic and diamagnetic bars to align themselves with the lines of force in a magnetic field:

> Let the diagram represent a field of force naturally uniform, but influenced by the presence of a ball of diamagnetic substance. It is clear that in the localities *A* and *B* the lines of force will be less densely arranged, and in the localities *D* and *C* they will be more densely arranged than in the undisturbed field. Hence a second ball placed at *A* or at *B* would meet and disturb fewer lines than if the first ball were removed; but a second ball placed at *D* or *C* would meet and disturb more lines of force than if the first ball were removed. It follows that two equal balls of diamagnetic substance would produce more disturbance on the lines of force of the field if the line joining their centres is perpendicular to the lines of force than if it is parallel to them. But the disturbance produced by a diamagnetic substance is an effect of worse 'conducting power', and the less of such an effect the better. Hence two balls of diamagnetic substance, fixed to one another by an unmagnetic framework, would, if placed obliquely and allowed to turn freely round an axis, set with the line joining their centres *along* the lines of force.

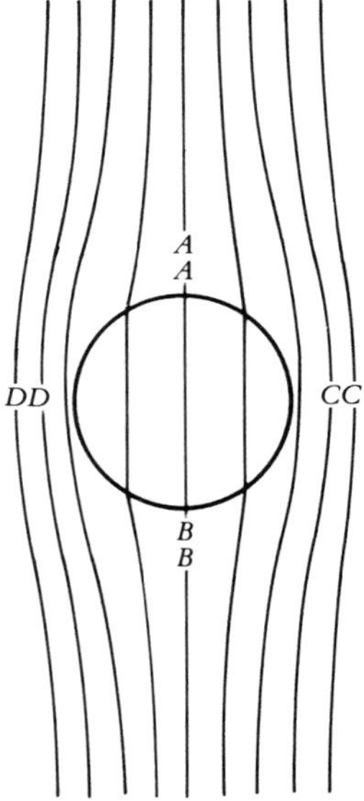

The same argument, for a contrary reason, shows that two balls of soft iron similarly arranged would set with the line joining their centres also *along* the lines of force. For this position more disturbance is produced on the lines of force than in any other, but now the *more* disturbance (being of better 'conduction') the better. Hence the conclusion. Of course similar conclusions follow for bars, or elongated masses, of the substances.[66]

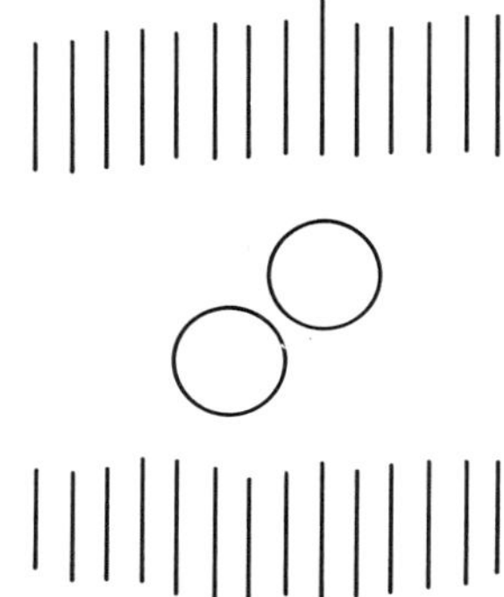

This interpretation follows immediately from the hydrodynamic analogy for magnetization and the minimum condition on mechanical effect Φ. It unites completely Faraday's intuitive and familiar rendering of fields, in terms of lines of force and differential conducting powers, with the mathematical theory that Thomson had spent seven years developing.

Small wonder that Faraday was to acquire the status in scientific mythology of implicit mathematical genius. That image is usually associated with Maxwell's many expressions of it. But Thomson too promoted the myth; for example, in his Glasgow lectures in 1881–2: 'Faraday not mathematician but had highly mathematical mind. Math feeling that always led him right when math question came up. Gave a name done much for science, field of force'.[67] We may pass by the question of whom this myth was intended to glorify, Faraday or those who rendered him mathematical. Having started out to show that Faraday's electrostatic experiments were merely confirmations of the mathematical theory – meaning originally a theory that elaborated the inverse square law for point forces – Thomson had quickly begun to translate, via analogy, the action at a distance interpretation of that theory of contiguous action. When he wrote of the proof of the extended continuity equation for magnetization that established, '*The possibility of all the problems I have yet considered in the allied physical theories*', he had completed the translation. In the process Faraday's theory had acquired a power far beyond that of the descriptive lines of force, but the mathematical theory too had become a much more powerful, and quite different tool: it had become field theory.

[66] William Thomson to Michael Faraday, Saturday, 19th June, in SPT, **1**, 215–16. S.P. Thompson dates the letter 1849, but it may have been 1848 or, conceivably, 1847, for 19th June was Saturday in 1847. See David Gooding, 'A convergence of opinion on the divergence of lines: Faraday and Thomson's discussion of diamagnetism', *Notes and records Royal Soc. London*, **36** (1982), 243–59. The diagram with two balls (referring to Faraday, para. 2264) appeared in Thomson's notebook on 24th March, 1847, in NB34, ULC, p. 51, albeit interpreted in terms of 'mutual action' between the two magnetized spheres and analyzed in terms of the rotational couple on an elongated ellipsoid.

[67] William Thomson, in MacCallum, 'Notes', p. 40.

Practical considerations and 'positive' theory

At the 1847 meeting of the British Association Thomson had presented his work on the method of images, emphasizing its practical dimension. At the same meeting he read the paper on terrestrial magnetism, and we have seen that the associated theory of magnetic fields derived in essential ways from an engineering interpretation of potentials, making mechanical effect the central concept for ponderomotive forces, absolute measurements, and the field itself. Both images and fields thus express the practical imperative in the young Glasgow professor's theorizing. Lest this most pervasive theme be lost in the theoretical complexities, we conclude with a reminder.

Thomson's analysis of terrestrial magnetism extended that of Gauss, who began from the problem of mapping the magnetic field over the surface of the earth. Such maps served primarily the interests of navigation: mercantile, military, and exploratory. Gauss acknowledged that 'science delights to render such useful services', while protesting that the mapping had been 'prompted by the pure love of science'. Whatever Gauss's personal motives, his data derived from state supported projects, particularly British projects in the 1830s. He relied especially on magnetic observations made or collected by Colonel Edward Sabine (1788–1883; Sir Edward after 1869 and President of the Royal Society from 1861–71) and published in the *BAAS Report* for 1837. Sabine had by then become the prime mover of the so-called Magnetic Crusade, the largest scientific project ever undertaken in Britain. Its proponents, well named the 'magnetic lobby' in a recent analysis, included such notables as John Herschel, Humphrey Lloyd and William Whewell. Sponsored jointly by the British Association and the Royal Society, and paid for by the Admiralty, the War Office, and the East India Company, the Magnetic Crusade illustrates as well as any other single project the intertwining of interests between the institutions of science and those of the Empire, an association that Lord Kelvin would epitomize. With few exceptions the lobby and its supporters constituted a list of scientific Whigs, with a Peelite Conservative or two like Whewell and the lone Tory, J.D. Forbes. Herschel's close ties with the Whig government of Lord Melbourne played a crucial role in the success of the crusade.[68]

It was through his whig friend Archibald Smith that Thomson, while still a student at Cambridge, first became acquainted with Sabine. After a visit to Sabine's Woolwich home, William wrote to Dr Thomson:

> I was very much pleased with Col and Mrs Sabine . . . Col Sabine is at present employed by government in reducing the magnetic observations made on the late Antarctic expedition [of Captain James Ross, another major supporter of the Crusade]. He has just been setting up in his garden the magnetic observatory wh. they used, and he says that

[68] John Cawood, 'The Magnetic Crusade: science and politics in early Victorian Britain', *Isis*, **70** (1979), 493–518; Jack Morrell and Arnold Thackray, *Gentlemen of science: early years of the British Association for the Advancement of Science* (Oxford, 1981), pp. 353–70.

> after this he will always have a variety of magnetic instruments in it, and invited me to come down any time I can spare time, when I may be in London, and amuse myself with them. Mrs Sabine . . . translated all the very valuable papers of Gauss on Terrestrial Magnetism, in Taylor's Scientific Memoirs, and Smith looked over the mathematical parts for her. She also, to complete the series, translated Gauss' Memoir on General theorems in Attraction, wh. has been so interesting to me, though she says that when she translated it she was very much afraid it would never be of use to any body.[69]

No doubt Sabine represented to Thomson the possibility of support in high places for the forthcoming campaign for the Glasgow chair, but, for his part, Sabine recognized in Thomson a source of badly needed theoretical expertise. Gauss had no counterpart in Britain, and the Astronomer Royal, G.B. Airy, who had most knowledge of the mathematics of magnetism, opposed the project after 1840. Sabine distrusted Gauss's theory but could hardly have understood it theoretically, while Thomson knew it in every detail. Their association was to bear fruit for both parties when Sabine, as President of the Royal Society, communicated to that body in 1849 Thomson's massive 'Mathematical theory of magnetism', and when Thomson was elected a Fellow in 1851.

Already by 1845 Thomson had acquired a certain status in this larger sphere. When Sabine received from the Italian Giovanni Plana his '*Mémoire sur la distribution de l'électricité*', he sent it to Thomson through Smith, with a message: 'As you are about the only person in the country who knows much about the subject he thought the best use to make of it was to send it to you'. Three weeks later Smith passed on the intelligence, received in conversation with Charles Babbage, that 'within the last few days Faraday has discovered that every body whatever is magnetic [diamagnetic] & if properly suspended places itself at right angles to the poles of a magnet'.[70] Thomson thus began to find access to the highest circles of national science, and the politics of Empire, at the same time as he began work on the general theory of magnetic induction and magnetic fields. His paper on terrestrial magnetism may be regarded as a first contribution to the ongoing Magnetic Crusade.

Somewhat later we find him occupied, at Sabine's request, with the inadequacy of Gauss's theory to explain diurnal variations detected in magnetic observations, and with the possibility of explaining them on the basis of an idea of Faraday's that the atmosphere of the earth becomes magnetized by induction, its inductive capacity varying with temperature. Thomson told Smith: 'I have been puzzling myself horribly with this and attempts to get out something like the diurnal variations known in Col Sabine's three books'. Nothing seems to have come of the attempts except reinforcement of his earlier concern with the relation of pressure and density in air/ether to polarization, whether electric or

[69] William to Dr Thomson, 4th April, 1844, T255, ULC.
[70] Archibald Smith to William Thomson, 19th November, and 11th December, 1845, S148 and S150, ULC.

magnetic. 'I have been thinking every now and then since the summer of 1846', he claimed, 'on writing a paper on this subject and on what would correspond in common electricity if Faraday's hypothesis regarding the electropolarization of the air were true'.[71] The practical problem of predicting precisely the intensity of the earth's magnetic field for all points of space and time thereby returned to the deepest problems of natural philosophy.

A more successful application of the theory of magnetic induction arose in the same context of Smith, Sabine, and the Admiralty. As the shipbuilders of the Clyde, the Thames, and other centres increasingly replaced wood with iron, the major difficulty for navigating with a magnetic compass became that of correcting the compass readings for deviations caused by the magnetism induced in the ship's iron, rather than that of constructing an accurate map of the earth's field. The existing technique of correction, developed by Airy from a limited version of Poisson's theory of magnetic induction, was not adequate. By July, 1848, Thomson had produced results within his theory of magnetic induction which suggested a new solution to the problem. Soon it would lead to extensive work with Smith on ships' magnetism and ultimately to the considerable wealth which he derived from his patented magnetic compass (ch. 22). Thus the engineering concern with sources of power and wealth, which had informed Thomson's earliest ideas in field theory, would remain an integral part of his work on magnetism throughout his life. In fact, the analysis of magnetic induction in terms of mechanical effect produced by an engine provided the key to his solution of the compass problem already in 1848. Without straining the metaphor, we may assert that in Thomson's mind the power of the British fleet, from the engines that drove it to the compasses that steered it, rested on the single principle of mechanical effect.

The universal dynamics of mechanical effect represents the inseparable unity of theoretical and practical issues in Thomson's life. It also brings us, however, to a great turning point in his natural philosophy (as in natural philosophy generally). The nature of this turning point will only emerge when we examine the subject of heat in the next two chapters. We cannot leave magnetism, however, without considering a strange dinosaur in the evolution of Thomson's thought, a creature of the earlier era that would be rendered obsolete by the upheaval of the dynamical theory of heat.

On 16th August, 1848, he began drafting 'Memoranda for 1st memoir on magnetism; recollected from last summer' (when the letter to Liouville was lost). He had in mind an entire series constituting a formal treatise. But on 28th October, 1848, Joule's name appeared in his notebook for the first time, with the observation, 'For the last two days I have been writing to Regnault & Joule'.[72]

[71] William Thomson to Archibald Smith, 5th December, and 26th November, 1850, respectively, TD1/757/2 and TD/757/1, Strathclyde Regional Archives, Glasgow. In a notebook entry of 22nd September, 1846, NB34, ULC, p. 38, Thomson concluded that magnetization of a gaseous fluid would not cause any change in its density.

[72] William Thomson, 16th August and 28th October, 1848, NB34, ULC, pp. 99, 106.

Over the next two years he would spend many more days writing to Joule than he would spend writing his 'Magnetism'.

The published portions of the 'Mathematical theory of magnetism' recall the purely descriptive 'Mathematical theory of electricity', the kinematics of force distributions. The unpublished parts in Thomson's notebook, however, deal overwhelmingly with the dynamics of mechanical effect. As a whole, the projected series recapitulates the development of his thought from 1840 to 1850, and offers a sharp preview of the transition from the physics of point forces to the physics of work (or mechanical energy).

In both his abstract and introduction of June, 1849, the prospective candidate for the Royal Society stressed that a proper theory must be independent of any hypothesis, especially that of magnetic fluids. He would present a purely '*positive*' theory, endeavouring to show 'that a *complete* mathematical theory of magnetism may be established upon the sole foundation of facts generally known'. This 'positive' and 'complete' theory would rest on a purely macroscopic and empirical concept of 'intensity of magnetization', defined as magnetic moment per unit volume, and on Coulomb's equally empirical inverse square law for the force between magnetic poles.[73] (Compare 'electrical intensity' and Coulomb's electrical inverse square law.) Clearly this theory was to be a practical theory in the latitudinarian sense of the previous chapter.

An arbitrary distribution of magnetization constituted a magnet, and the immediate task was to calculate the resultant force of a magnet at every point of space. Only source magnets were to be considered initially – magnetic induction would come in later instalments – so that all space had an inductive capacity of either unity or zero. To simplify the mathematics, source magnetization could be represented by 'imaginary magnetic matter' (fictional magnetic fluids), net densities of which existed only at discontinuities of magnetization, as at the ends of a bar magnet. Here, however, the positive theory ran into an ambiguity. How was the force *inside* a magnet defined? Experimentally, one might cut a cavity in the magnet to measure the force inside. Unfortunately, on the magnetic matter representation, the force would depend on the shape of the cavity. Wherever a line of magnetization was cut, a source would appear that did not exist in the uncut magnet. A long thin cavity tangential to the lines of magnetization would cut no lines and produce no effects, while a disc-shaped cavity perpendicular to the lines would have opposite sources on its two faces, and the force inside F_p would be in the opposite direction from that in the tengential cavity F_t (see figure 8.7). On the other hand, at the ends of the magnet, F_t would change direction discontinuously across the boundary while F_p would form continuous closed loops. Concomitantly, F_t could be derived from a single-valued potential while F_p could not.

Because he had not yet found a completely general representation of magne-

[73] William Thomson, 'A mathematical theory of magnetism', *Proc. Royal Soc.*, **5** (1849), 845–6 (abstract), and *Phil. Trans.*, **141** (1849), 243–85 (chap. I–V); E&M, 340–405, on pp. 340–1, 347, 351–2. Our emphasis on 'positive' and 'complete'. See Wise, 'Flow analogy', pp. 61–7.

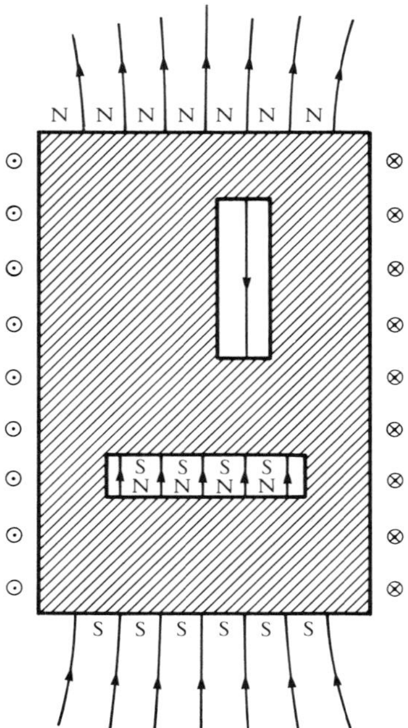

Figure 8.7 Tangential and perpendicular cavities cut in a cylindrical magnet exhibit (on the magnetic matter representation) lines of force in opposite directions. The cavities are supposed infinitely thin so that the tangential cavity has infinitesimal magnetic matter at its ends and the lines inside the perpendicular cavity are not distorted.

ization in terms of electric currents, Thomson intended in the first instalment of his series to refer all forces to the magnetic matter representation. By November, 1849, however, he had found the desired alternative. The question for the positive theory, then, was whether to use F_t, as he had naturally done for the initial magnetic matter description or F_p, which suited an electric current picture. But the positive theory could not provide an answer to what was in fact a physical question: what *were* the sources of magnetism? Unwilling to violate his latitudinarian commitment, Thomson refused to make a choice. He developed both perspectives in parallel, along with the transformation between them.[74]

Very little of this parallel development was published, being limited to a geometrically elegant analysis of 'lamellar' and 'solenoidal' geometries of mag-

[74] For F_t see E&M, 362*n*. He found the galvanic representation on 7th November, 1849, and the perpendicular cavity on 16th November, 1849. See William Thomson, NB34, ULC, pp. 134–46; E&M, 378–405.

netization, the first corresponding to a magnet cut up into 'magnetic shells' perpendicular to the lines of magnetization and the second to a magnet cut into tubes along the lines.[75] The lamellar division produced the mathematics of electric currents, and the solenoidal that of magnetic matter. This double partitioning of magnetized space was soon to be of great importance to James Clerk Maxwell's first attempt at an electromagnetic field theory. In 1855 Maxwell would resynthesize Thomson's double representation of magnetic force into a geometrical version of Faraday's duality of quantity and intensity. Thomson, of course, had no interest in such dualities, which violated his empiricist methodology. Even more important, he was quickly losing interest in purely geometrical description.

Ironically, 'young Clerk Maxwell' enters Thomson's notebook for the first time on the same page as Rudolf Clausius's first appearance, on 15th August, 1850.[76] Afterwards, electricity and magnetism appear only in relation to thermodynamics and interconversions of mechanical effect. Thomson had become convinced that electricity in motion was the most likely source of both heat and magnetism, and that measurements of mechanical effect offered the best probe of their unified structure. Looking back over the entries on magnetism of the preceding two years, one recognizes two characteristics: they nearly all concern the electric current representation, and questions of mechanical effect appear continually. Can the law of electromagnetic induction be derived from considerations of the mechanical effect of the induced current? Is the mechanical effect of a lamellar distribution less than that of any other producing the same external effect? How much work is required to assemble a magnet assuming it to be made up of pre-existing current loops? None of these questions made their way into the 'Mathematical theory of magnetism'.[77] They provided the driving force, however, for subsequent physical theories.

[75] Thomson first used 'lamellar' and 'magnetic shell' on 19th August, 1848; then 'filamentary' on 7th November, 1849, which changed to 'solenoidal' by 11th June, 1850, in NB34, ULC, pp. 100, 137, 151. In the course of this work Thomson discovered what has become known as 'Stokes theorem'; William Thomson to G.G. Stokes, 2nd July, 1850, K39, Stokes correspondence, ULC.

[76] William Thomson, 15th August, 1850, NB34, ULC, p. 154.

[77] *Ibid.*, pp. 98, 105, 143. See, however, paras. 502–3 added to the 'Mathematical theory' in June, 1850, on the mechanical value of a distribution of magnetization, E&M, 375–8. 'Chapter VI. – On electromagnets', was published for the first time in E&M, 405–25.

9

Thermodynamics: the years of uncertainty

> My brother James sent in a paper to the Society of Arts in Edinburgh stating that water freezes under a pressure of 760 mm at .0075 Cent[igrade] below 0°. This has never been verified by any experimenter till I did it the other day by myself, with a very delicate thermometer. This thermometer is assuredly the most delicate that ever was made, there being 71 divisions in a single degree of Fahr[enhei]t. It is filled with ether: mercury would not do . . . A pressure of 18 atmo[sphere]s was applied & the temperature was lowered by 17½ of the divisions. 0.246 was the lowering acc[ording] to experiment; 0.2295 acc[ording] to theory. Thus there is an evident agreement. I shall give you the result of another experiment. Under a pressure of 8.1 atmos. the temp[erature] was lowered 7½ divisions. 0.106 is the lowering by experiment; 0.109 by theory. This is within 1/3000 of a Fahr[enheit] degree. *Professor William Thomson to his class, 1850*[1]

Spoken in the context of a lecture to his experimental natural philosophy class on 17th January, 1850, during his fourth session as a Glasgow College professor, these remarks communicated the experimental verification of James Thomson's famous theoretical prediction of the lowering of the freezing point of ice under pressure. Crucial to the confirmation of Sadi Carnot's theory of the motive power of heat in the face of recent criticism from James Prescott Joule, James's prediction had been fulfilled through the precise measurements of one wholly captivated by the laboratory of Regnault five years earlier. Yet such words, uttered a full year before William's satisfactory enunciation of two laws of thermodynamics, only came at the end of a decade of debate and discussion concerning the principles underlying the motive power of heat. This decade constitutes the years of uncertainty.

Sir Joseph Larmor referred to James as 'the philosopher, who plagued his pragmatical brother', a reference to his insistence on comprehensive understanding of any problem, his combination of theoretical concepts with engineering tasks, and his introduction of new and unambiguous terms into scientific language. James exhibited also a determined single-mindedness, manifested as much in his compulsive involvement with engineering problems as in his

[1] William Thomson, 17th January, 1850, lecture 45, in William Smith, 'Notes of the Glasgow College natural philosophy class taken during the 1849–50 session, MS Gen. 142, ULG.

devotion to a unitarian faith; a single-mindedness which contrasted vividly with William's impulsive attacks on a variety of loosely related problems and his notable lack of commitment to any particular religious denomination. As Helmholtz put it in 1863, James 'is a level-headed fellow, full of good ideas, but cares for nothing except engineering, and talks about it ceaselessly all day and all night, so that nothing else can be got in when he is present. It is really comic to see how the two brothers talk at one another, and neither listens, and each holds forth about quite different matters. But the engineer is the most stubborn, and generally gets through with his subject'.[2]

The role that James Thomson played in the accomplishments of his younger brother has not been adequately appreciated. William Thomson's development of thermodynamics, for example, often concerned the requirements of marine engineering, at a time when Clyde and Thames shipbuilders were making an economic reality of ocean steam navigation. James's direct involvement with problems of marine engineering, through his friendship with Lewis Gordon, his membership of the Glasgow Philosophical Society, and his apprenticeship at William Fairbairn's Thames shipbuilding yard, provided William with immediate access to contemporary issues of steam engineering. Thus William's ideas on heat developed through close interaction with James, much as his ideas on electricity in 1845 had incorporated James's engineering perspective on engines (ch. 8). In fact, William's ideas on both heat and electricity during the 1840s expressed the practical reality of the steam-engine. If electrified bodies and hot bodies could produce work, then that capacity would supply the absolute measures and the basic concepts of electricity and heat, which behaved analogously in the production of work.

This perspective yields a second revisionist theme. Thomson had considerable difficulty accepting James Joule's argument for interconversion of heat and mechanical work when he heard it in 1847. His attachment to Sadi Carnot's view, in which work derived from the 'fall' of heat, has been regarded as a concomitant of his supposed belief either in a general principle of conservation of heat or in a more specific caloric theory in which heat was a material substance.[3] But Thomson treated heat in the same way as he dealt with electricity, not as the quantity of a substance, but as a *state of a body*. The distinction will allow an explanation of how he could consistently accept Joule's experiments on the conversion of work into heat without being willing to accept the reverse conversion of heat into work. In other words, it will shed light on Thomson's dilemma in reconciling Joule's conservation of mechanical effect (energy) with Carnot's theory of the heat engine.

[2] D'Arcy W. Thompson, 'Obituary of Joseph Larmor', *Year book of the Royal Society of Edinburgh*, 1941–2, p. 12; Hermann Helmholtz to Frau Helmholtz, *c.* April, 1863, in SPT, **1**, 429–30.

[3] H.I. Sharlin, *Lord Kelvin. The dynamic Victorian* (Pennsylvania, 1979), p. 97; D.S.L. Cardwell, *From Watt to Clausius. The rise of thermodynamics in the early industrial age* (London, 1971), p. 244; SPT, **1**, 252–95.

James Thomson, seen here in later years, probably when he succeeded W.J.M. Rankine as Glasgow University professor of engineering in 1873. Though afflicted by ill-health prior to a successful academic career, James played a central role in the creation of his brother's thermodynamics, energy physics, cosmology, and matter theory. [From James Thomson, *Papers*.]

Prelude to 'the dynamical theory of heat': the Thomson brothers

James and William Thomson made their scientific debut in 1840 through the models and manufactures committee for the Glasgow meeting of the British Association (ch. 2). The resulting exhibition, displaying James Watt's steam-engine model and the engine of Bell's *Comet*, provided one of the earliest expressions of the Thomson brothers' commitment to the engineering and industrial values of Clydeside. At the same meeting, according to his recollection in 1862, James had explained to Professor Kelland some ideas concerning 'the possibility of having a tidal mill which could perpetually grind corn or overcome friction, and that this must involve an expenditure of the power stored in the motions of the earth or moon or of one of them'. Kelland (with whom William had also wrangled over Fourier) 'insisted that I was wrong and that the water could effect no deduction from the power or work stored in the motions of the earth and moon jointly'.[4]

Although these recollections are by no means infallible, James apparently recorded the problem of tidal retardation of the earth in his private notebook. He told William in 1862 that he had found 'under the date September 1841 the following: "Some years ago it occurred to me that the tides caused on the earth by the motion of the moon round it must produce some retardation in the motion of the moon or acceleration in the motion of the earth round its axis or some other similar change". I give you the memorandum just as it stands. It is stupidly blundered in the expression . . . but assuredly I had the main idea correctly in my mind long before'.[5] Tidal retardation, then, may be the earliest instance of the Thomson brothers' concern with 'losses' of power, both in nature and in machines. No doubt J.P. Nichol's nebular cosmology and the problem of Encke's comet informed these early concerns (ch. 4).

Elected to the senior Glasgow Philosophical Society in 1841, James described in a letter to William a paper read to the Society 'on a curious method of raising water to the tops of mills from the cisterns on the tops of engine houses'. Observing that there was 'evidently a great loss of power', he drew attention to Mr Gordon's remarks 'that a similar apparatus had been used for raising water for irrigation'.[6] James had consolidated his acquaintance with the recently appointed professor of engineering by attending Gordon's engineering class during the 1841–2 session, and both professor and pupil found a forum for their mutual interests in the active Philosophical Society.

James soon embarked on his practical training in heavy industry, first at Walsall in 1842, then with the Horseley Iron Company at Tipton, and finally with Fairbairn at his shipbuilding yard on the Thames and briefly at his

[4] James to William Thomson, 9th July, 1862, T117, ULG. The context of the long letter was John Tyndall's priority claim for J.R. Mayer both with regard to tidal retardation and to 'Joule's' mechanical equivalent of heat. [5] *Ibid.*

[6] James to William Thomson, *c.*1841–2, T380, ULC.

Manchester engine-building works. A long series of letters exchanged between the brothers from 1841 shows James's deep interest in engineering problems of all kinds: steam-engines, shipbuilding, marine engineering, water power, and civil engineering. According to James's own account, the brothers' interest in the question of 'loss' of power continued during his stay at Walsall. He wrote to William in 1862:

> Even you and I at Walsall (in 1842 I think) when watching the consumption of power in the flow of water into a canal lock were speculating on what became of the power as we could not suppose the water *worn* and therefore altered like as solids might be supposed to be when power is consumed in their friction.[7]

In the famous English canal network centred in the Black Country (including Walsall), water which might have been used to raise a barge from one contour to another often simply cascaded over the lock gates. The power to lift boats or turn a waterwheel could thus be 'consumed' even if not 'absolutely lost'. The analogous problem in heat engines, when heat is conducted from a high to a low temperature without performing useful work in an engine, was subsequently to haunt William.

While the Thomsons did not know of a satisfactory answer to the question of power consumption, the query itself may have been prompted by a reading of Whewell's *Mechanics of engineering*, published during the preceding year when William Thomson had entered St Peter's College. Above all, their reference to solids being worn suggests that they had Whewell's new text in mind. The importance here of Whewell's book lies primarily in its discussion of 'labouring force', adapted from the French *travail* (work) employed in texts on engineering mechanics by Navier, Poncelet, and others. Apart from being the first major British textbook to employ the notion of 'work' as a central feature in the theoretical structure of mechanics, Whewell's use of the term 'labouring force' expressed his parallel interest in political economy, especially in the labour theory of value:

> *Labouring Force is the labour that we pay for.* In many cases the work to be done may be performed by various agencies; by men, by horses, by water, by wind, by steam. In these cases, that is the cheapest mode of doing the work which gives us the requisite labouring force at the smallest expense: and the price men are willing to pay, and customarily do pay, is proportional to the quantity of labouring force which they purchase. Labouring force enters into the prices of articles produced by man; as the wages of labour, so far as the labouring force is executed directly by man; as the reward for capital, so far as the labouring force arises from accumulated capital. But wages of labour are paid, not only for man's labour, but for labouring force when arising from machinery.[8]

[7] James to William Thomson, 13th August, 1863, T119, ULG.

[8] William Whewell, *The mechanics of engineering* (Cambridge, 1841), pp. 148–9. On the role of the engineering concept 'work', see T.S. Kuhn, 'Energy conservation as an example of simultaneous discovery' in M. Clagett (ed.), *Critical problems in the history of science* (Madison, 1959), pp. 321–56, esp. pp. 330–4.

These remarks could largely have been quoted from Adam Smith or virtually any other political economist. But Whewell went further, assimilating labour (for which wages are paid) to work done by machines, and accumulated labour or capital (for which profits are paid) to work accumulated in such storehouses as reservoirs of water and the flywheels of machines.

Following the approach of the French engineering texts, Whewell explained that we measure labouring force 'by the product of the resistance overcome, and the space through which it is overcome', though more specifically it may be measured 'by a given weight raised through a given vertical space'. He also gave a clear statement that the labouring force was proportional to the *vis viva* (mass times velocity squared) which the forces of the machine could have generated, acting through the same spaces, and he found that 'if we take the same measures on the two sides of the equation, *the labouring force is half the vis viva*'. He then proceeded to consider how the labouring force was consumed in doing work. The work exerted a so-called *useful resistance* which had to be overcome by the machine. *Impeding resistances* hindered the work; they included air, friction, and 'the forces producing waste and change of form in parts which we wish to have durable and invariable', and which exist continually in machines. Taken together, the useful and impeding resistances consumed the total labouring force.

Given the context of Whewell's writings on political economy and his definition of labouring force, this distinction of useful and impeding resistances should be understood in parallel with the much-discussed distinction of productive and unproductive labour. Productive labour referred to labour that produced a commodity, something that could be sold or stored up. Unproductive labour referred to labour consumed in the doing, such as the labour of personal servants. Many political economists considered the latter form of labour to contribute nothing to the wealth of a nation, and in that sense to be wasted.[9] Such waste, by mere consumption, was to figure prominently in Thomson's thermodynamics. As part of an ever-present economic context, it carried the same significance as the term 'dissipation', which had appeared regularly in correspondence with his father during his undergraduate years. The opposite term, 'economy', in Thomson's frequent usage, will be seen to refer not only to the efficiency of transforming mechanical effect from one form to another, as in steam-engines, but also to minimizing waste of economic value.

In his *Mechanics of engineering*, Whewell emphasized that only in cases 'where the force [is] employed in raising a weight, moving a mass, or bending a spring', can the force be stored up and brought again into play. In all other cases, the labouring force was consumed, whether productively or unproductively, in overcoming useful or impeding resistances. It is 'lost', he said, 'and cannot be

[9] Whewell, *Mechanics of engineering*, pp. 146, 153–6. See M. Norton Wise (with the collaboration of Crosbie Smith), 'Work and waste: political economy and natural philosophy in nineteenth-century Britain', *Hist. Sci.* (forthcoming).

recovered after being used'.[10] For the Thomsons it was precisely this 'loss' which demanded explanation.

With respect to water flowing into a canal lock, the brothers rejected the account Whewell had given for solids: 'work consists in shaping or moving certain portions of matter. Thus, we have to grind bodies, to polish them, to divide them into parts . . . In these cases we have to overcome the cohesion of matter, inertia, elasticity, weight'.[11] The Thomsons could not see how such an alteration might account for fluid friction. We do not know what positive ideas they had in 1842, if any, but their concern with 'what became of the power' indicates that they did not accept Whewell's implication, that power or work could be annihilated. Absolute loss of labouring force seems not to have been an option.

Early in 1843 James moved to the Horseley Iron Works at Tipton, Staffordshire. The Horseley Iron Works claimed the distinction of having constructed in 1821 the first iron steamer to put to sea, a strange feat for a works so far from the coast. In the letters to William, James showed himself highly critical of the firm's approach, and in particular of its assignment of the task of estimating the cost of an iron bridge contract, worth twenty or thirty thousand pounds, to young apprentices. He also criticized the way in which allowances were made by the 'eye', the true 'rule of thumb' technique, rather than by more precise methods. The firm lost the bridge contract, ran short of orders, and found itself in the hands of its creditors.[12]

In the autumn of 1843, Dr Thomson paid £100, under the premium apprenticeship system, for his son to become an apprentice to the shipbuilding subsidiary, at Millwall on the Thames, of William Fairbairn's famous Manchester engineering firm. One of the great Victorian entrepreneurs, and a pioneer of the heavy machinery industry, Fairbairn had branched out into iron shipbuilding and marine engineering as early as 1835. Laying out the yard at Millwall, on the Isle of Dogs, had involved a large amount of capital, upward of £50 000, provided entirely by borrowed money. Fairbairn's project on the Thames suggested that London's river rather than the Clyde in his native Scotland would become the leading shipbuilding centre in the new age of steam navigation. David Napier's removal to the Thames in 1836, and John Scott Russell's abandonment of an association with Greenock shipbuilders and engineers to find fame and fortune in London from 1844, seemed to confirm the trend. In the forefront of technical progress, and with a workforce of around 2000 men, Fairbairn's yard had in its early years plentiful orders, many of them placed on the strength of his reputation. However, local competition kept profits down, the large capital invested in plant and machinery required interest payments, and management of the firm, divided between London and Manchester, proved

[10] Whewell, *Mechanics of engineering*, pp. 155–8. [11] *Ibid.*, p. 145.

[12] James Thomson, *Papers*, pp. xxi–xxii; William Pole (ed.), *The life of Sir William Fairbairn, Bart.* (London, 1877), p. 335; James to William Thomson, *c.* late 1842–May, 1843, T382–8, ULC.

especially difficult. The shipbuilding 'factory' soon became far more of a liability than an asset.[13]

Yet James Thomson's work at Millwall provided the setting for a most important discussion with William. For the first time the ideas of Clapeyron and Carnot on the motive power of heat entered the brothers' dialogue. In a letter to William in August, 1844, James inquired who it was that had proved there was a definite quantity of mechanical effect (work) given out during the passage of heat from one body to another.[14] He stated his intention of writing an article for the *Artisan* about the theoretical possibility of working steam-engines without fuel by using over again the heat which was thrown out in the hot water from the condenser, noting that 'I shall have to enter on the subject of the paper you mentioned to me'. This paper was almost certainly the 1837 translation for Taylor's *Scientific memoirs* of Clapeyron's 'Memoir on the motive power of heat'.[15]

The discussions in James's letter of 1844 are the earliest record of implicit references to Clapeyron and Carnot by the Thomsons. They show a considerable understanding of the basic principles involved in Carnot's theory. As James interpreted the problem:

during the passage of heat from a given state of intensity to a given state of diffusion a certain quantity of mec[hanical] eff[ect] is given out whatever gaseous substances are acted on, and that no more can be given out when it acts on solids or liquids.[16]

This statement, he said, was all he could prove, because he did not know whether in solids or liquids the fall of a certain quantity of heat would produce a certain quantity of mechanical effect, and that the same mechanical effect '*will give back as much heat*'. That is, 'I don't know that the heat and mec[hanical] eff[ect] are interchangeable in solids and liquids, though we know they are so in gases'. It must be emphasized that James did not have in mind here an *interconversion* of heat and work but only the possibility of a universal relation between the fall (or rise) in intensity of heat and the mechanical effect produced (or consumed) in an ideal, reversible engine. This view made the problem of mechanical effect, not its production, but its loss in non-ideal engines.

James Thomson exhibited his familiarity with the waterfall analogy employed by Carnot. In the analogy, the quantity of water M was conserved during its fall, just as in Carnot's theory the quantity of heat Q remained constant, and in each case a definite quantity of mechanical effect W was obtained, either by the mass M falling from height h_1 to h_2, or the heat Q falling

[13] James Thomson, *Papers*, p. xxii; W. Pole (ed.), *Fairbairn*, pp. 335–42; Philip Banbury, *Shipbuilders of the Thames and Medway* (Newton Abbot, 1971), pp. 171–4; G.S. Emmerson, *John Scott Russell. A great Victorian engineer and naval architect* (London, 1977), pp. 26–7.

[14] James to William Thomson, 4th August, 1844, T402, ULC. See SPT, **1**, 275*n* for a very brief reference to his letter as 'containing a curious piece of primitive thermodynamics'.

[15] Emile Clapeyron, 'Memoir on the motive power of heat', in Richard Taylor (ed.), *Scientific Memoirs*, **1** (1837), 347–76.

[16] James Thomson, *op. cit.* (note 14).

from temperature T_1 to T_2. That is, $W \propto M(h_2 - h_1)$ in the one case and $W \propto Q(T_2 - T_1)$ in the other:

The whole subject you will see bears a remarkable resemblance to the action of a fall of water. Thus we get mec[hanical] eff[ect] when we can let water fall from one level to another or when we can let heat fall from one degree of intensity to another. In each case a definite quantity is given out but we may get more or less according to the nature of the machines we use to receive it. Thus a water mill wastes part by letting water spill from the buckets before it has arrived at the lowest level and a steam engine wastes part by throwing out the water before it has come to be of the same temperature as the sea. Then again, in a water wheel, much depends on our not allowing the water to fall through the air before it commences acting on the wheel and in a steam engine the greatest loss of all is, that we do allow the heat to fall perhaps from 1000° to 220°, or so, before it commences doing any work. We have not materials by means of which we are able to catch the heat at a high level. At the same time, if we did generate the steam at 1000° a great part of the heat would pass unused up the chimney and with the products of combustion. If we had a water wheel sufficiently high to receive the water of a stream almost at its source we would waste all the tributary streams which run in at a lower level.[17]

These reflections contain considerable insight into problems of less-than-ideal engines and waste generally. Not only did James Thomson grasp the fundamental principles of Carnot and Clapeyron, but he also recognized that, in reality, by extending the analogy with waterwheels, heat was wasted – produced no useful work – on passage from a state of intensity to one of diffusion, and that as a result real engines fell short of ideal ones. The main issue was that of economizing heat in the production of work, but James also introduced the notion of the sea as a sink of lost work, both for water running to the sea in rivers and for heat diffusing itself to the temperature of the sea, a notion crucial to William's later formulation of the second law of thermodynamics.

This important letter of August, 1844, was written during James's stay at the Fairbairn shipyard. His concern with marine engines occasioned the phrase 'a steam engine wastes part by throwing out the water before it has come to be of the same temperature as the sea'. Other letters of the period also show his desire to improve the 'economy' (efficiency and profitability) of steam-engines for marine use. Concern with losses in marine engines far exceeded any similar concerns with land engines since steamships on ocean routes not only had to take into account capital and running costs but above all *availability* of fuel during the voyage itself and at the outward destination.[18]

The Thomsons's use of the key term 'mechanical effect' rather than 'work' or 'labouring force' requires comment. 'Effect' had been the term of their Glasgow

[17] *Ibid.*

[18] For example, James to William Thomson, 19th June, 1844, T401. ULC. See, for example, W.J.M. Rankine, *A memoir of John Elder* (Edinburgh, 1872) for a study of some of the problems which marine engineers had to overcome before ocean steam navigation became a viable alternative to sail.

forebear, James Watt, whose steam-engine model they exhibited at the British Association in 1840. Occasionally they followed his usage. The more definite term 'mechanical effect', not generally used in Britain, appears to have entered their vocabulary through Lewis Gordon. Gordon, having left Edinburgh University about 1833, was employed by Marc Isambard Brunel from 1835 to 1837 on the building of the first Thames Tunnel, and then completed his engineering studies at the Royal Mining Academy of Freiburg and at the *Ecole Polytechnique* in Paris about the year 1838. While at the Freiburg Academy he studied widely in mineralogy, geology, physics, chemistry, metallurgy, mining operations, and mathematics. In mathematics he was a pupil of Julius Weisbach who, along with other German writers, spoke of *mechanische Wirkung*, which Gordon translated as 'mechanical effect'.[19] He read a paper to the Glasgow Philosophical Society, 'On dynamometrical apparatus; or, the measurement of the mechanical effect of moving powers', on 23rd February, 1842, shortly after James Thomson had been elected a member of the Society. Gordon gave a detailed account of the dynamometrical apparatus of M. Morin which he had seen during his visit to Metz in 1839. Claiming that British dynamometers were unreliable for the correct measurement of mechanical effect developed by machines or 'moving powers', Gordon emphasized that it was not the product of the *effort* and the *duration*, but the product of the *effort* and the *distance through which it is exerted* which should be obtained directly from a dynamometer.[20]

In his *Synopsis of lectures* to be delivered in the 1847–8 session at Glasgow College, Gordon defined the measure of mechanical effect in the same way, and summarized his preference for the term 'mechanical effect' over the variety of equivalent terms from French and British writers:

> Different names have been given to *Mechanical effect*, as we have now defined it. Smeaton used the term '*Mechanical power*', Carnot the term '*Moment d'activité*', Monge and Hachette '*Effet dynamique*', Coulomb and Navier '*Quantité d'action*', Coriolis and Poncelet '*Quantité de travail*', and '*Travail mécanique*', Dr Whewell proposes '*Labouring force*', Mr Moseley uses the term '*Work*', Weisbach, and German writers generally, speak of '*Mechanische Wirkung*', or Mechanical Effect. This latter term seems to be the least ambiguous in its application.[21]

Gordon's awareness here of British, French, and German terminology derived from his continental training in Freiburg and Paris. By comparison, Whewell's discussion of terminology in *The mechanics of engineering* did not include German sources.[22]

[19] *Memoir of Lewis D.B. Gordon F.R.S.E.* (Edinburgh, 1877); Julius Weisbach, *Principles of the mechanics of machinery and engineering* (2 vols., London, 1848), **2**, p. 66. Gordon was the translator of this second volume.

[20] Lewis Gordon, 'On dynamometrical apparatus; or, the measurement of the mechanical effect of moving powers', *Proc. Phil. Soc. Glasgow*, **1** (1841–4), 41–2.

[21] Lewis Gordon, *A synopsis of lectures to be delivered. Session 1847–8* (Glasgow, 1847), p. 5.

[22] Whewell, *Mechanics of engineering*, p. 149.

James Thomson's friendship with Gordon, through the Philosophical Society in particular, explains the Thomsons' preference for mechanical effect. James too was intensely conscious of the need for accurate measurement of mechanical effect when he began to prepare drawings at Fairbairn's during May, 1844, for a modified Morin dynamometer. He suggested to Fairbairn's son, for instance, that the improved instrument might be employed 'for measuring the exact quantity of mechanical effect' used by anyone who purchased power from the main engine at the works. The dynamometer would thus serve, not merely for testing purposes, but directly to meter economic costs.[23] When William became professor, he included a discussion of Morin's dynamometer in his lectures and in the 1867 *Treatise*. It supplied a precise and practical measure of the quantity (mechanical effect) which, in his formulation, provided the foundation for dynamics itself and for all absolute measurements (chs. 11, 20).

James Thomson had arrived at Millwall just as the financial crisis there deepened. Within a year, William Fairbairn decided to cut his losses and dispose of the yard. While James had been able to learn much about the latest heavy engineering techniques, to see marine engines and iron ships at close hand, and to make drawings for some of the engine contracts, he was unsettled yet again in his professional career by circumstances of a kind which had never interfered with William's intellectual life at the unworldly University of Cambridge. 'So you see', he wrote to his brother, 'I am not getting settled in any works. Just when I begin to take root I have to leave the place'.[24] The summer of 1844 saw the gradual run-down at Millwall. In the autumn of that year James moved to Fairbairn's Manchester Works where he commenced work in the fitting shop. Steam-engine building was the main source of his practical experience during this period.

Writing to his brother before the Cambridge Senate House examinations, he wistfully remarked: 'I wish my apprenticeship was as nearly done as yours – but even when it is done, I fear I shall have no such comfortable berth to step into as that which is probably waiting for you'.[25] The words were prophetic, as James afterwards developed that 'quickness of the pulse', which detained him at Glasgow from Christmas, 1844, and prevented any hope of a return to Manchester (ch. 5). For the immediate future he had to abandon all hope of an engineering career.

James's enforced inactivity, however, yielded particular fruit in 1846–7 when he began to develop his vortex turbine. Although he had long been interested in waterwheels for motive power, especially horizontal wheels or turbines, his slightly improved health and the return of William as natural philosophy

[23] James to William Thomson, *c.*19th May, 1844, T399, ULC. William very occasionally employed Whewell's term 'labouring force'. See MPP, **1**, 103.

[24] James to William Thomson, July, 1843–May, 1844, T391–8, ULC.

[25] James to William Thomson, October–December, 1844, T403–4, ULC; James Thomson, *Papers*, p. xxiii.

professor seem to have stimulated him to aim for a patent. This form of intellectual property cohered well with the Thomson's commercial values. William would soon begin his own phenomenal exploitation of such property in patents for electrical and navigational instruments (chs. 20–2). But the present significance of James's work, which finally led to a patent in July, 1850, lies in its relation to the brothers' continuing concern with waste and loss of power in nature and in machines.

Following the making of his own small model of a horizontal waterwheel towards the end of 1846, James had a large model constructed at Walter Crum's Thornliebank works in the spring of 1847, a model subsequently tested by Prony's friction dynamometer. Even with some defects in workmanship, the model (working to one-tenth horsepower) produced around 70% 'of the total work due to the water expended' compared to about 75% in the best overshot waterwheels.[26]

These encouraging results, along with much guidance from William and his father, led to many more tests. James delegated to William the task of exploring the possibility of a patent in France during William's visit to Paris in July, 1847. As a result, William found the opportunity to discuss in detail the latest French designs with the elderly Poncelet himself.[27] Thus the brothers' intense interest in waterwheels coincided both with William's work on hydrodynamics with Stokes and with his first encounter with Joule's researches on fluid friction. Meanwhile, James's vortex turbine continued to evolve. A Glasgow Philosophical Society committee headed by W.J.M. Rankine later offered the following description of its eventual design:

> Turbines, or horizontal wheels, acted upon by a vortex or whirlpool of water, have long been used in a rude and imperfect form; but the bringing of them into an efficient state has been the work of recent inventors . . . In Fourneyron's Turbine, a vortex moving spirally outwards, drives a vane-wheel surrounding the case from which the water is supplied . . . In Professor James Thomson's Turbine, or Vortex-Wheel, the vortex moves spirally inwards, and drives a vane-wheel, surrounded by the casing that supplies it with water. This wheel possesses over the others the advantage of being easily regulated, and of requiring a less speed for the production of its maximum efficiency. Its invention was the result of an investigation, by Mr Thomson, of the theory of the motion of fluids in whirlpools.[28]

Although James became intensely occupied with the waterwheel in 1846–7, the brothers' parallel interest in the motive power of heat had been developing steadily since August, 1844. We saw that, while in Paris early in 1845, William

[26] James Thomson to J.R. McClean, 25th December, 1846, and 12th April, 1847, in James Thomson, *Papers*, pp. xxvii–xxviii.

[27] William to James Thomson, 22nd July, 1847; James to William Thomson, 29th July, 1847, in James Thomson, *Papers*, pp. xxviii–xxxiii.

[28] J.R. Napier, Walter Neilson, and W.J.M. Rankine, 'Report on the progress and state of applied mechanics', *Proc. Phil. Soc. Glasgow*, **4** (1855–60), 207–30.

had apparently read Clapeyron in its French version for the first time and searched in vain for Carnot's original text. On his return he discussed the preliminary part of Clapeyron's paper with James, who soon located it in the *Journal de l'Ecole Polytechnique* for himself. Thus in February, 1846, he remarked that the first sections constituted 'a very beautiful piece of reasoning, and of course perfectly satisfactory', remarks echoed by William three years later when he said of Carnot's approach that 'nothing in the whole range of Natural Philosophy is more remarkable than the establishment of general laws by such a process of reasoning'.[29]

Then, during his first busy session as professor of natural philosophy, William made a discovery of a piece of machinery, a Stirling air engine, which focussed his mind once more on Carnot's theory. On 1st March, 1847, he announced his find to J.D. Forbes:

> I have found a Stirling's air engine in our Augean stables, and got it taken to pieces, as it was clogged with dust & oil, and I expect to have it going as soon as I have time. I think this [the following] consideration will make it clear that there is really a loss of effect, in conduction of heat through a solid. There is neither expenditure nor gain of mechanical effect in melting ice in an atmosphere at 32°, as is easily proved. Now we may have a fire or source of heat in the interior of a hollow conducting shell, spending all its effect in melting ice at 32°. It seems very mysterious how power can be lost in such a way [by conduction of heat from hot to cold], but perhaps not more so than that power should be lost in the friction of fluids (a plumb line with the weight in water for instance) by which there does not seem to be any heat generated, nor any physical change effected.[30]

This letter was written just a few months *before* Thomson's famous meeting in June, 1847, with Joule. It also came just before Thomson began his intensive interaction with Stokes on hydrodynamics and its relation to electricity, magnetism, and heat (ch. 8).

Judging from a remark of Joule's in June – 'I have felt very gratified in meeting with two at least, Mr Stokes and yourself, who enter into my views of this subject' (the conversion of work into heat) – Stokes and Thomson agreed that mechanical effect could not be annihilated by fluid friction.[31] When Thomson observed in his letter to Forbes, therefore, that there did not *seem* to be any heat generated, he was acknowledging that heat *might* indeed have been generated. He was thus well prepared for Joule's forthcoming announcement of just that result. More especially, however, Thomson had here already associated loss of mechanical effect in fluid friction with the new claim that mechanical effect was likewise 'lost' when heat was merely conducted from high to low temperature without operating a heat engine. In other words, he had already pointed to the problem that would continue to plague him in reconciling Joule's discovery with Carnot's theory. Both frictional and conduction losses, indeed,

[29] James to William Thomson, 22nd February, 1846, T415, ULC; MPP, **1**, 143; SPT, **1**, 132–3.
[30] William Thomson to J.D. Forbes, 1st March, 1847, Forbes papers, St Andrews University Library. [31] J.P. Joule to William Thomson, 29th June, 1847, J59, ULC.

Stirling hot-air engine – a working model of the kind employed by William Thomson in his Glasgow laboratory. [From The Science Museum Library, London.]

were analogous to the loss of useful work when water fell from a high to a low level *through a resisting medium*, without turning a waterwheel.

In extending the mystery of loss of power or mechanical effect to the case of using a heat source to melt ice rather than to produce mechanical effect, Thomson had not as yet explained the precise connection of the melting or freezing of ice with the Stirling air engine, but he soon illuminated that connection in a paper on the theory of the engine, deduced from Carnot's principles. As his first paper read to the Glasgow Philosophical Society (to which he had just been elected a member in December, 1846), this work of April, 1847, began a long and fruitful relationship with the Society.

Thomson's account began by stating that, at a previous meeting of the Society, Professor Gordon had given an explanation of Carnot's theory and that, in accordance with this theory, the mechanical effect to be obtained from an air engine by the transmission of a given quantity of heat depended 'on the difference between the temperature of the air in the cold space above and the heated space below the plunger'. Since this temperature difference was considerably greater than that in the best condensing steam-engines, it was argued that, given the removal of the practical difficulties of constructing an efficient air engine, 'a much greater amount of mechanical effect would be obtained by the consumption of a given quantity of fuel' in the case of the air engine.[32] Figure 9.1 gives Thomson's description of the engine's operation.[33]

His notice to the Philosophical Society continued with 'some illustrations, afforded by the Air Engine, of general physical principles', and it was here that the discussion of Stirling's engine involved questions of ice and mechanical effect. If the engine were turned *forwards*, William argued, and if no heat were applied, the reservoir below the plunger would become colder than the surrounding atmosphere and the space above hotter. That is to say, given the usual way the engine operated, with a hot reservoir below and a cold reservoir above the plunger, the engine would turn forwards of itself, cooling the lower and warming the upper reservoirs until their temperatures were equalized. But by continuing to crank it forwards the engine would act further to cool the lower reservoir, reversing the temperature relation. Once this new relation of temperatures was established, 'contrary to that which is necessary to *cause* the engine to turn forwards [of itself]', expenditure of work would be necessary to turn it.

[32] William Thomson, 'Notice of Stirling's air engine', *Proc. Phil. Soc. Glasgow*, **2** (1847), 169–70. For a full account of the Stirling hot air engine, see E.E. Daub, 'The regenerator principle in the Stirling and Ericsson hot air engines', *Brit. J. Sci.*, **7** (1974), 259–77. The original hot air engine by Robert Stirling dated from 1816, and was improved in 1827 by the development of the so-called regenerator or economizer principle. The regenerator was essentially a heat exchanger employed within the air engine cycle in the belief that the large loss of heat incurred in the condenser of the steam-engine could be eliminated in the air engine. Suggested by John Ericsson in the 1830s, the term 'regenerator' implied that the lost heat could be reused to produce mechanical work. The 'real' value of the regenerator was appreciated by W.J.M. Rankine. See also 'Stirling's Air-Engine', *Mech. Mag.*, **45** (1846), 559–66.

[33] William Thomson, 15th January, 1850, lecture 43, in Smith *op. cit.* (note 1).

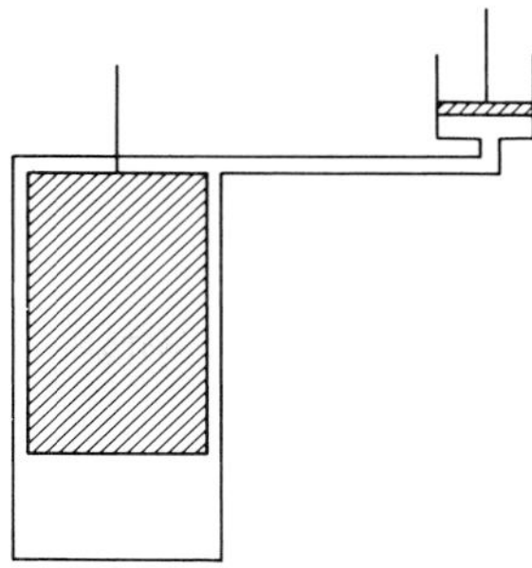

Figure 9.1. William Thomson gave this account of Stirling's air engine to his natural philosophy class in January, 1850:

On the principle of the motive power of heat Stirling's Air Engine is constructed. It is very simple. One mass of air alone is necessary to drive it. Here we have a large cylinder with a plunger in it. Suppose it to be at the top. There is a considerable quantity of air below. If we apply a spirit lamp below & heat that air it expands and rushes up along the sides of the plunger, along the tube and forces up the piston in the other small cylinder. There is a wheel placed between the two cylinders (which is not represented). There is a crank attached to each end of the axle of the wheel. When the small piston rises it turns round the wheel which brings the plunger down and this drives out most of the heated air. The air in coming in contact with the cool metal at the top contracts and draws down the piston which raises the plunger and again the air is heated & so on. In order to condense the air better it is expedient to have a stream of water rushing over the upper part thus carrying away the heat.

If, however, the temperature in the upper reservoir were prevented from rising, and in the lower from falling, 'the engine may be turned without the expenditure of any work (except what is necessary in an actual machine for overcoming friction, &c.)'. Apart from the obvious way of achieving this result by immersing the machine in a stream of water, Thomson advocated finding a solid body which melted at the temperature at which it was required to retain the engine. Thus, he suggested letting a stream of water at 32° Fahrenheit run across the upper part of the engine, while the lower part was held in a basin of water at the same temperature. When the engine was turned forwards, heat would be taken from below the plunger and deposited in the space above, this heat being supplied by water in the basin, all of which would be converted gradually into ice at 32°, *without expenditure of work.*

In Thomson's view, then, the Stirling engine could be used to illustrate what he had termed here a 'physical principle', that the making of ice, consistent with

Carnot's theory, involved no expenditure or production of work, for it involved no temperature difference. Furthermore, if work depended only on temperature differences, it would follow, as he had stated in the letter to Forbes, that heat generated at a high temperature, and then spent in the melting of ice, truly involved a loss of work which might have been produced in operating an engine.

This conclusion, however, was not entirely adequate. As James soon pointed out, since water expands on freezing, the freezing of water could involve the production of work. In May, 1848, he carried out his calculation of the lowering of the freezing point by pressure, and in October of that year he commented in his notebook: 'William and I have examined the investigation on the last page. The principles and the numerical result are extremely nearly true'.[34] William read an account of these researches to the Royal Society of Edinburgh in January, 1849, and the paper was published the same year. In it James explained that his brother's conclusion 'at first appeared to me to involve an impossibility, because water expands while freezing; and therefore it seemed to follow, that if a quantity of it were merely enclosed in a vessel with a movable piston and frozen, the motion of the piston, consequent on the expansion, being resisted by pressure, mechanical work would be given out without any corresponding expenditure; or, in other words, a perpetual source of mechanical work, commonly called a perpetual motion, would be possible'. In order to avoid such a perpetual motion, James explained, 'it occurred to me that it is necessary farther to conclude, that the freezing point becomes lower as the pressure to which the water is subjected is increased'.[35]

James's argument, modelled on the operation of the Stirling engine, proceeded in two steps. The first step confirmed William's result for a free expansion of freezing water. No work was required of an engine for removing heat from a quantity of water at 0° centigrade and atmospheric pressure in order

[34] James Thomson, 'Motive power of heat: air engine', Notebook A14(A), QUB. This notebook also contains an account of a meeting between the Rev Dr Robert Stirling, minister of the Parish Church of Galston, an Ayrshire town some twenty miles south of Glasgow, and James, in the presence of Dr Thomson, William and Mansell (William's assistant at Glasgow College). According to James, Stirling did not understand the principle of his own engine.

[35] James Thomson, 'Theoretical considerations on the effects of pressure in lowering the freezing point of water', [Read 2nd January, 1849], *Trans. Royal Soc. Edinburgh*, **16** (1849), 575–80; *Papers*, pp. 196–203; MPP, **1**, 156–64. In a subsequent letter, William Thomson to J.D. Forbes, 7th December, 1848, F199 (copy), ULC, William took credit for this conclusion concerning the lowering of the freezing point. As he there expressed it: 'In conversing with my brother James about this proposition, it struck me that we may prove as a consequence that the freezing point of water under heavy pressure must be lower than 32°'. Lacking independent evidence, we would suggest three alternative explanations: (i) that William allowed James the credit in the published paper as a form of recompense for his elder brother's other disappointments in health and career and as an acknowledgement of his invaluable guidance in the subject; or (ii) that William had simply given an erroneous account to Forbes; or (iii) that the idea was so much the product of both minds that a separation of their respective contributions was impossible. We do know that James carried out the actual calculation of the theoretical lowering of the freezing point, which William then verified experimentally.

to freeze it while merely transferring the heat to an infinite reservoir at the same temperature, because the temperature of the working substance in the engine – air in James's example – always remained constant at 0°C. With temperature fixed, pressure and volume were uniquely related. Whatever changes in pressure and volume attended compression, therefore, had to be followed in reverse on expansion such that the work done to turn the engine in a complete cycle would necessarily be zero.

If, on the other hand, the freezing water did not expand freely but were resisted, it would produce external work. This second step went beyond William's result. Such external work could only derive from work done to turn the engine, implying a difference between the changes of pressure and volume in the engine on compression and on expansion, which in turn implied that on expansion (when in contact with the freezing water) the temperature of the working substance, and consequently that of the water, was below 0°C. In complete conformity with Carnot's theory, then, an engine requiring work to turn it had to be operating between two different temperatures, moving heat from a lower to a higher temperature.

Basing his detailed argument on the Carnot cycle, James derived an approximate relation for the depression of the freezing point with increased pressure. Around January, 1850, William verified that relation experimentally, thereby making of it an important confirmation of the Carnot–Clapeyron theory. More precisely, he had confirmed the principal result of the theory, its prediction of the intimate relation between work and temperature differences.[36] Stokes noted the importance of the experiment soon after:

> I congratulate you and your brother on the success of the heat experiment, which is a very remarkable one. I will tell Hopkins about it. It goes rather contrary to his notions, or rather conjectures, relative to the possible solidity of the centre of the Earth (depending on pressure) while the surrounding parts are in a state of fusion. I don't know that however; for some substances contract in freezing (like Mercury) and in such cases I presume pressure would favour solidity.[37]

Hopkins's researches in physical geology concern an important subtext for Thomson's thermodynamics, the doctrine of progression in the earth's geological history (chs. 6, 16).

The principal result of Carnot's theory, confirmed by William in his measurement of the depression of the freezing point in 1850, had already become the basis of his famous absolute scale of temperature. At the end of 1847, he had sent some notes on Clapeyron's memoir to Gordon, adding in an accompanying letter:

[36] William Thomson, 'The effect of pressure in lowering the freezing point of water, experimentally demonstrated', [Read 21st January, 1850], *Proc. Royal Soc. Edinburgh*, **2** (1951), 267–71; MPP, **1**, 165–9.

[37] G.G. Stokes to William Thomson, *c.*22nd January, 1850, S558, ULC.

A good deal has been said in various treatises about fixing an absolute standard of temperature. The air thermometer is chosen merely for [convenience] of comparison. Now would it not be a good *absolute* definition of equal degrees, to say that they are such that the same *quantity* (determined in an absolute way by the melting of ice) descending a degree will always produce the same mech^{l} effect?[38]

Here was the Kelvin scale of absolute temperature in its earliest version. Presentations to the Glasgow and Cambridge Philosophical Societies followed in April and June, respectively, of 1848, with publication by October. Thomson observed that the experimental determination of temperature had been the subject of some of the very elaborate and refined researches of Regnault, such that there now existed as complete a practical solution as could be desired. All these practical methods, however, made reference to a specific body as the standard thermometric substance; as yet no absolute scale, independent of a particular substance, was available. Thus, 'we can only regard, in strictness, the scale actually adopted as an arbitrary series of numbered points of reference sufficiently close for the requirements of practical thermometry'.[39] But the concept of an engine producing mechanical effect provided a solution:

The relation between motive power and heat, as established by Carnot, is such that *quantities of heat* and *intervals of temperature* are involved as the sole elements in the expression for the amount of mechanical effect to be obtained through the agency of heat; and since we have, independently, a definite system for the measurement of quantities of heat, we are thus furnished with a measure for intervals according to which absolute differences of temperature may be estimated.[40]

The characteristic property of the new scale was that each of its degrees had the same relation to work available from a heat engine; that is, 'a unit of heat descending from a body A at the temperature $T°$ of this scale, to a body B at the temperature $(T-1)°$ would give out the same mechanical effect, whatever be the number T. This may justly be termed an absolute scale, since its characteristic is quite independent of the physical properties of any specific substance'. With the aid of his pupil, William Steele, Thomson undertook an approximate comparison of the calculated values from the absolute scale with experimental values from the air thermometer, and recorded good agreement.[41] The agreement confirmed once again the validity of Carnot's relation between work and the fall of heat.

Repeated references to Regnault throughout this classic paper, and the dependence of the calculations on Regnault's measurements, should remind us that Thomson's earliest attempt to employ the relation of mechanical effect to 'fall' in intensity as an absolute measure of a physical quantity took place in 1845

[38] William Thomson to Lewis Gordon, 20th December, 1847, G124 (copy), ULC.

[39] William Thomson, 'On an absolute thermometric scale, founded on Carnot's theory of the motive power of heat, and calculated from the results of Regnault's experiments on the pressure and latent heat of steam', *Phil. Mag.*, [series 3], **33** (1848), 313–7; MPP, **1**, 100–6.

[40] Thomson, 'On an absolute thermometric scale', pp. 314–15.

[41] *Ibid.*, pp. 315–17.

when he was working in Regnault's laboratory. He had then conceived the absolute quantity of electricity on a conductor at any given potential in terms of the work required to raise the electricity to that potential. This view of absolute measure apparently derived in part from his prior familiarity with the waterwheel analogy for a steam-engine (through James, Clapeyron and Regnault). His new idea for measuring temperature simply applied to heat the same scheme as he had applied to electricity, but for measuring temperature (analogous to potential) rather than quantity of heat (analogous to electricity). We shall see repeatedly that Thomson's concern with the mechanical effect of heat in the years 1847–50 paralleled closely his work in the same period on mechanical effect in electric and magnetic fields.

Meanwhile in April, 1848, Forbes had urged him to make the wonders of the Carnot theory more widely known: 'as you have taken so much trouble about this Theory of Carnot's I think it would be reasonable to expect you to print a little notice of it for the benefit of people in general'.[42] Forbes pressed him further in November, 1848: 'I write to remind you of your promise to give us an abstract of the Motive Power of Heat for the R[oyal] S[ociety] [of Edinburgh]. When can we have it?'[43] William read his abstract on 2nd January, 1849. And having at long last received a copy of Carnot's original memoir from Gordon in late 1848, he prepared for the Society's *Transactions* his detailed 'Account of Carnot's theory'.[44]

Shortly after these communications, the death of Dr Thomson temporarily unsettled the remarkable thermodynamic partnership of William and James. James joined Lewis Gordon in London with the intention of collaborating in a patent for the waterwheel. William waited in the spring of 1849 for Gordon's comments on his 'Account' before its full publication, and Gordon wrote to say he was 'much delighted and enlightened' by it. He found that William's 'analytical expression of the results of the theory are more distinct than those of Clapeyron and your conclusions as to the parity of the amounts of mechanical effect derivable from all different vapours and all different gases at the same temperature is more clearly stated and seen than in either Carnot or Clapeyron'.[45] With the publication of the 'Account' in 1849, then, the Carnot theme in the work of the Thomson brothers had reached a critical juncture. In both the depression of the freezing point and the absolute temperature scale they had confirmed theoretically and experimentally the power of Carnot's theory. How, then, could it be fundamentally in error? How, indeed, could James

[42] J.D. Forbes to William Thomson, 20th April, 1848, F194, ULC.

[43] J.D. Forbes to William Thomson, 27th November, 1848, F198, ULC.

[44] William Thomson, 'An account of Carnot's theory of the motive power of heat, with numerical results deduced from Regnault's experiments on steam', *Trans. Royal Soc. Edinburgh*, **16** (1849), 541–74; MPP, **1**, 133–55. See also SPT, **1**, 133, 269–70.

[45] Lewis Gordon to William Thomson, 24th May, 1849, G128, ULC; James to William Thomson, 18th June, 1849, T438, ULC.

Joule's claim that mechanical effect derived not from the fall in intensity of heat, but from its conversion, be correct?

The Joule–Thomson debates: 1847–50

The famous encounter between Joule and William Thomson took place at the Oxford meeting of the British Association in June, 1847. After the close of this, his first meeting since becoming professor, William wrote to his father from Cambridge, where he had been residing since the end of the Glasgow College session:

I have just returned (viz. last night) from Oxford, where I enjoyed a week very much. The meeting of the Association was quite delightful, from the opportunity it afforded of seeing so many people engaged in various interesting researches. I need not give you any details, as you will of course see the Athenaeum containing the report (tell James to look for the account of Joule's paper on the dynamical equivalent of heat. I am going to write to James about it and enclose him a set of papers I received from Joule, whose acquaintance I made, as soon as I have time. Joule is I am sure wrong in many of his ideas, but he seems to have discovered some facts of extreme importance, as for instance that heat is developed by the fric^{n} of fluids in motion). I met Dr Nichol, Forbes, Faraday, Sir W. Hamilton, Snow Harris.[46]

We have seen the concern of both James and William with the loss of *vis viva* through fluid friction, most particularly in William's letter of March, 1847, to Forbes. William naturally received Joule's new 'fact', that heat is developed by the friction of fluids in motion, with great interest, having considered the possibility himself. We need, then, to examine why William would also have regarded Joule as 'wrong in many of his ideas'.

Something dramatic, indeed, seems to have captured Thomson's attention. As he recalled the event in 1882:

I heard his [Joule's] paper read at the section, and felt strongly impelled to rise and say that it must be wrong, because the true mechanical value of heat given, suppose to warm water, must for small differences of temperature be proportional to the square of its quantity. I knew from Carnot's law that this must be true (and it *is* true; only now I call it 'motivity' in order not to clash with Joule's 'mechanical value'). But as I listened on and on I saw that (though Carnot had vitally important truth not to be abandoned) Joule had certainly a great truth and a great discovery, and a most important measurement to bring forward. So instead of rising with my objection at the meeting, I waited till it was over, and said my say to Joule himself at the end of the meeting.[47]

Thomson's immediate reaction, that the mechanical value of heat must be proportional to the *square* of the quantity of heat, no doubt derived directly from

[46] William to Dr Thomson, 1st July, 1847, T367, ULC. See also SPT, **1**, 263–5.

[47] William Thomson, *Nature*, **26** (1882), 618; SPT, **1**, 264. For the mathematical significance of this passage see M. Norton Wise, 'William Thomson's mathematical route to energy conservation: a case study of the role of mathematics in concept formation', *Hist. Stud. Phys. Sci.*, **10** (1979), 49–83, esp. pp. 78–80.

James Prescott Joule in 1882 aged about sixty-four. Thomson had first met Joule at the 1847 Oxford meeting of the British Association, and Joule's accurate measures of the mechanical value of heat became the centrepiece of Thomson's new energy physics, with work or mechanical effect the absolute measure of all physical agencies. [From *Nature*, 26 (1882), facing p. 617.]

the waterfall analogy. On that analogy the work done when Q falls through a temperature T is $W \propto QT$. But adding heat to a body in order to raise its temperature is like rebuilding a column of water (or a stack of bricks) which has fallen to the ground. After reaching a height h with mass m, the increment of work needed to lift another increment of mass dm is hdm. The total work for the column of total mass M is:

$$W \propto \int_0^M h\,dm \quad \propto \int_0^M m\,dm \quad \propto \quad M^2.$$

In the same way, therefore, the work done in raising the temperature of a body would be proportional to the square of the heat added:

$$W \propto \int_0^Q T\mathrm{d}q \propto \int_0^Q q\mathrm{d}q \propto Q^2.$$

The implication was immediate. Although Thomson had previously, in his letter to Forbes, been willing to consider the possibility of the conversion of work to heat in fluid friction, he now realized that, if Joule's result ($W \propto Q$) were correct, he could never reconcile such a conversion with Carnot's account of the motive power of heat ($W \propto Q^2$).

A second, complementary, reason for Thomson's response derived from his prior work on mechanical effect in electric and magnetic fields, which showed mechanical effect always proportional to force squared. The result formed part of the flow analogy. If, then, heat flux were actually analogous to force, the work expended in heating a body should again have been proportional to the *square* of the heat added. Since Thomson was engaged with Stokes in working out this hydrodynamic analogy for magnetism, electricity, and heat – with a focus on mechanical effect – at the very time he encountered Joule, it would have been surprising indeed if he had not reacted strongly to Joule's claim.

According to Joule, it was William Thomson's interest in his paper presented to the Oxford meeting which effectively rescued him from obscurity.[48] Although such obscurity may have derived in part from Joule's general relationship to the scientific communities in Manchester and London, his hypothesis that heat was the result of some form of motion – subsequently the dynamical theory of heat – would not have impressed anyone committed to a non-speculative methodology. Most strikingly, in his 1850 paper 'On the mechanical equivalent of heat', Joule had to eliminate the major conclusion that 'friction consisted in the conversion of mechanical power into heat' before the memoir was published in the *Philosophical Transactions*. Only the stark, experimentally established conclusion that the quantity of heat produced was *proportional* to the quantity of work done, according to an exact numerical equivalent, was acceptable.[49]

As early as 1841 Joule had spoken of heat as the vibration of the atmospheres of electricity and magnetism surrounding atoms. Again, in his 1843 paper 'On the calorific effects of magneto-electricity, and on the mechanical value of heat', he had observed that 'when we consider heat not as a *substance*, but as a *state of vibration*, there appears to be no reason why it should not be induced by an action of a simply mechanical character'. In other words, the mutual conversion of work and heat would follow from the general hypothesis that heat was dynamical in character. And, in 1845, Joule wrote of the *vis viva* of the particles of water and of the *vis viva* of the atmospheres of elastic fluids, concluding that

[48] J.P. Joule, *The scientific papers of J.P. Joule* (2 vols., London, 1887), **2**, p. 215.

[49] See John Forrester, 'Chemistry and the conservation of energy: the work of James Prescott Joule', *Stud. Hist. Phil. Sci.*, **6** (1975), 273–313, for a discussion of Joule's own conceptual and social affiliations, particularly prior to his meeting with Thomson. See Crosbie Smith, 'Faraday as referee of Joule's Royal Society paper "On the mechanical equivalent of heat"', *Isis*, **67** (1976), 444–9, for an account of the modifications to Joule's 1850 paper.

'an enormous quantity of *vis viva* exists in matter'.[50] The general part of this hypothesis, that heat was not a material substance but a state of motion, measured by *vis viva* (later kinetic energy), would become known in Joule's and Thomson's terminology as the dynamical theory of heat.

Joule read his paper not to the mathematics and physics section of the 1847 British Association meeting, but to the chemistry section. That Thomson should have been present at all in that section probably indicates his special interest in Joule's subject-matter. Although the chairman of the section confined Joule to a short verbal description of his experiments, this restriction did not prevent Thomson from initiating the long debate with Joule that followed. Before the close of the Oxford meeting, Joule had left copies of two important papers for Thomson, expressing relief and gratitude at having found a potential supporter:

I was greatly disappointed in not having an opportunity of calling upon you yesterday before leaving Oxford. I called however at Pembroke College, and left two papers with the porter, one 'On the changes of temperature produced by condensation &c. of air' [1844], the other, which was the first published by me on the subject, 'On the calorific effects of magneto-electricity & the mechanical value of heat' [1843]. I beg your acceptance of the above and hope they will interest you. I have felt very gratified in meeting with two at least, Mr Stokes and yourself, who enter into my views on this subject and hope to be able to cultivate an acquaintance which I find so delightful.[51]

As noted earlier, it seems that the enthusiasm of both Thomson and Stokes derived from their prior conviction of the impossibility of annihilating mechanical effect. They received Joule's results within their own dialogue on fluid friction and mechanical effect.

On 12th July, 1847, William sent Joule's two papers to James, remarking that they would astonish him and that 'I think at present that some great flaws must be found. Look especially to the rarefaction and condensation of air, where something is decidedly neglected, in estimating the total change effected, in some of the cases'.[52] James replied:

There is one blunder certainly. He [Joule] encloses some compressed air in one vessel, connects that with another which is vacuous, and allows the air of the former to rush into the latter till the pressure is the same in both. Both vessels were immersed in water, and after the operation the temperature of the water remains the same as before. Joule says that no mechanical effect has been developed outside of the vessels during the operation,

[50] Joule, *Papers*, **1**, pp. 53, 134, 204–5.

[51] J.P. Joule to William Thomson, 29th June, 1847, J59, ULC. Nearly all the letters from Joule to Thomson are preserved in ULC and ULG. For an account of Joule's life, see Osborne Reynolds, 'Memoir of James Prescott Joule', *Proc. Manchester Lit. Phil. Soc.*, **6** (1892). The 1843 and 1844 papers referred to were published as J.P. Joule, 'On the calorific effects of magneto-electricity, and on the mechanical value of heat', *Phil. Mag.*, [series 3], **23** (1843), 263–76, 347–55, 435–43; Joule, *Papers*, **1**, pp. 123–59; and J.P. Joule, 'On the changes of temperature produced by the rarefaction and condensation of air', *Phil. Mag.*, [series 3], **26** (1845), 369–83; Joule, *Papers*, **1**, pp. 172–89.

[52] William to James Thomson, 12th July, 1847, T429, ULC; SPT, **1**, 266; James Thomson, *Papers*, pp. xxviii.

and that therefore the heat remains unchanged. But in reality mechanical effect *was* developed outside, as the two vessels became of different temperature.[53]

Whatever the source of the (slight) temperature difference, the attitudes of Joule and of James Thomson towards it reveal their quite different concepts of the motive power of heat. Joule thought in terms of work converted to heat, and Thomson in terms of work done to raise heat. And so, for Joule, no net change in the overall temperature of the water bath outside meant no work done and no heat generated. He regarded the temperature difference of the two vessels as experimental error. For Thomson, work had necessarily been done to produce the temperature difference.

Joule's 1844 paper contained his well-known criticism of the Carnot–Clapeyron view of the steam-engine whereby mechanical power arose simply from the passage of heat from a hot to a cold body, no heat being lost during the transfer. By analogy with the fall of water through a height, the caloric would, if unresisted (which never occurred for caloric), acquire *vis viva* through the temperature fall from boiler to condenser. If resisted by a perfect engine the potential *vis viva* would be converted into available work, while if merely conducted through a resisting medium it would be 'lost'. Joule, however, was already firmly committed to a principle of conservation or indestructibility of *vis viva* throughout nature, as he stated in the 1844 paper: 'I conceive that this [Clapeyron's] theory, however ingenious, is opposed to the recognized principles of philosophy, because it leads to the conclusion that *vis viva* may be destroyed by an improper disposition of the apparatus'. His own views, he claimed, avoided just such a difficulty. The steam 'expanding in the cylinder loses heat in quantity exactly proportional to the mechanical force which it communicates by means of the piston, and that on condensation of the steam the heat thus converted into power is *not* given back'.[54]

Joule supported his claim for the indestructibility of *vis viva* with a theological statement: 'believing that the power to destroy belongs to the Creator alone, I entirely coincide with Roget and Faraday in the opinion that any theory which, when carried out, demands the annihilation of force, is necessarily erroneous'. William subsequently admitted Joule's objection to the Carnot–Clapeyron theory 'in its full force, agreeing as I do with you when you say you coincide with Faraday and Roget'.[55] In other words, Thomson vigorously supported Joule in his claim that the power to annihilate mechanical effect was the privilege of God alone. Their statements belong to the voluntarist tradition (ch. 4). An omnipotent God created and governed the universe. The laws of nature, as instruments of divine providence, had no independent existence nor could they be altered or destroyed by any agent except the divine will.

[53] James to William Thomson, 24th July, 1847, T433, ULC; James Thomson, *Papers*, pp. xxx–xxxi. [54] J.P. Joule, *Papers*, **1**, pp. 188–9.

[55] William Thomson to J.P. Joule, 27th October, 1848, J62 (copy), ULC.

More especially here, the basic entities of the universe (matter and 'energy') were subject to conservation laws, and could neither be created nor destroyed by man. The sources for William's assent to this interpretation – Nichol, Chalmers and others – all converge towards the omnipotent and providential rule of divine will. Even his lecture notes on Fleming's moral philosophy in 1839–40 contain the voluntarist assumptions: 'all created things must be sustained' or 'the continual presence of God is necessary to preserve as to create'. Furthermore, 'there is a difference between saying that man makes who merely gives different arrangements. God makes in another sense'. When man created, merely the different arrangements of things were involved, but when God created He created in the fundamental sense of making things out of nothing, by the exercise of His absolute power, and thereafter sustained them in being by His ordained power, except in the event of a miracle.[56] William would frequently place energy conservation within this theological perspective.

Joule's criticism of Clapeyron caused James Thomson to admit to William in July, 1847, that, even given the 'blunder', some of Joule's views 'have a slight tendency to unsettle one's mind as to the accuracy of Clapeyron's principles. If some of the heat can absolutely be turned into mechanical effect, Clapeyron may be wrong'. He saw the solution of the difficulty as requiring a more accurate definition of a '*certain quantity of heat* as applied to two bodies at different temperatures'. He then made a bold attempt to reconcile the Clapeyron and Joule views:

> Perhaps Joule would say that if a hot pound of water lose a degree of heat to a cold one, the cold one may receive a greater absolute amount of heat than that lost by the hot one; the increase being due to the mechanical effect which might have been produced during the fall of heat from the high temperature to the low one.[57]

James attempted here to reinterpret Joule within his own framework, in effect by turning him on his head. Thus, on the one hand, he retained the Clapeyron principle of the production of mechanical effect by the fall of heat from a hot to a cold body in the ideal case. In reality, however, mechanical effect was often not produced when heat passed from the high to the low temperature, as in conduction. James therefore suggested that in such cases this potential, but unrealized, mechanical effect would appear as additional heat, which

[56] William Thomson, 'Notebook, *c*.1839–40', NB21, ULC. A transcript of 'essay xvi', constituting part of this notebook, is published in D.B. Wilson, 'Kelvin's scientific realism: the theological context', *Phil. J.*, **11** (1974), 55–8. See also William Fleming, *A manual of moral philosophy* (London, 1867), esp. Book III, 'Theistic ethics or natural theology'. For a more detailed analysis of these lecture notes, see Crosbie Smith, 'Natural philosophy and thermodynamics: William Thomson and "The Dynamical Theory of Heat"', *Brit. J. Hist. Sci.*, **9** (1976), 298–304.

[57] James to William Thomson, *op. cit.* (note 53). For an attempt by James to work out an alternative dynamical theory of heat which conserves quantity of heat by analogy to conservation of quantity of motion and which treats loss of mechanical effect in conduction by analogy to loss of momentum in inelastic collisions, see Crosbie Smith, 'William Thomson and the creation of thermodynamics: 1840–1855', *Arch. Hist. Exact Sci.*, **16** (1976), 231–88, esp. pp. 278–9.

supplemented the heat that merely passed from the hot to the cold body. This speculation involved, not heat converted into external work by an engine, as in Joule's standard example, but mechanical effect converted internally into heat during conduction. The conflicting frameworks of Joule and James Thomson are represented schematically in figure 9.2.

Joule, committed to a dynamical theory of heat, heat as *vis viva*, focussed criticism on the production of work without loss of heat. James Thomson, bound to no theory of the nature of heat, but committed to the idea of work done by a fall in intensity, focussed criticism on a fall without compensating work, as in conduction. Both views conserved mechanical effect. Indeed, the Thomson brothers welcomed the conversion of work into heat during processes involving internal resistance; such conversion solved, for example, the problem of loss of *vis viva* by fluid friction. By making fall in intensity the fundamental issue, however, their view directed attention to a quite different problem, namely recoverability, or the question of whether or not the additional heat could be converted again into available mechanical effect, or into heat at a higher temperature. *It made the question of dissipation during conduction into the same question as whether or not heat could be converted back into work.* We shall indeed see the importance of this move in relation to William's continuing unwillingness to accept fully Joule's view of the *mutual* conversion of heat and work.[58] Like James, Lewis Gordon also believed at this stage that Joule's 'denial of Carnot's beautiful idea will not necessarily overthrow so fertile a theory', while he relied upon William to 're-enlighten' him on the subject.[59]

The Joule–Thomson correspondence lapsed until the autumn of 1848, but in the interval both William and James were busy assessing the various implications of Joule's view. Towards the end of 1847, William embarked upon a series of experiments on fluid friction. He wrote to Forbes of what he was trying to achieve:

[By the end of the week] I may have succeeded in boiling water by friction. This is not very probable however as the machine I have made already is not strong enough. In a first experiment made with it the temperature rose from 45° or 46° to 57°, at about the rate of 1° every five minutes during a continued turning of the instrument (a very flat disc, with narrow vanes, turning in a thin tin box . . . of wh[ich] the bottom & lid were furnished with fixed vanes. The bearings of the paddles were entirely without the box which was full of water). In a second experiment, we began with water about 98° and the temperature rose at a much slower rate to about 99°, when part of the machine gave way, before the experiment could be considered as quite decisive [My] assistant was preparing to make an experiment yesterday, to commence with water at 80° or 90°, & to go on

[58] Cf. Cardwell, *From Watt to Clausius*, pp. 241–2, who regards Thomson's acceptance of Joule's experiments on the frictional heating of liquids as quite inconsistent with his refusal to accept the converse process, viz. the conversion of heat into work. We show that, while such a position would have been inconsistent for Joule, it was not so for the Thomsons.

[59] Lewis Gordon to William Thomson, 2nd July, 1847, G120, ULC.

Joule's view:

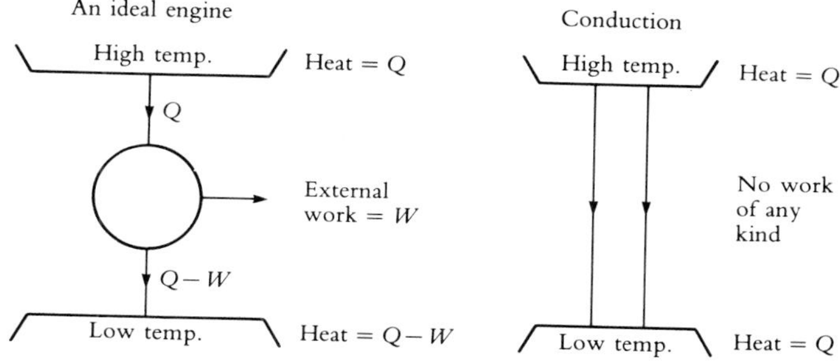

James Thomson's view:

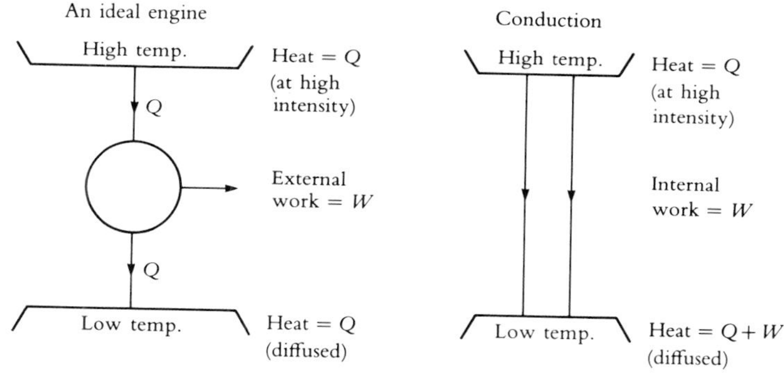

Figure 9.2. The conceptual differences between Joule and James Thomson are apparent in differing relations of work W and heat Q during the operation of an ideal engine and during conduction.

grinding (along with another, for relief) for about 4 hours, but I have not heard yet whether it went on.[60].

Such was the zealous nature of research in the Glasgow laboratory. With his usual enthusiasm, Thomson wanted not simply to raise the temperature of water, but to boil it!

Thomson's demonstration of the generation of heat by expenditure of work sufficiently states his belief in the principle that mechanical effect could not be annihilated. Not until August, 1848, however, did he explicitly apply this

[60] William Thomson to J.D. Forbes, 5th December, 1847, F190 (copy), ULC.

conservation principle to an apparently non-conservative system. In a paper 'On the theory of electromagnetic induction' for the British Association meeting in Swansea, he gave an elegantly simple derivation of Franz Neumann's recent theorem that the electric current induced in a closed wire by a magnet in relative motion could be expressed as the time rate of change of a potential function for the force between the magnet and the current. 'A very simple *a priori* demonstration of this theorem', he observed, 'may be founded on the axiom that the amount of work expended in producing the relative motion must be equivalent to the mechanical effect lost by the current induced in the wire'.

The structure of this beautiful application of the conservation principle differs little from his earlier derivation of electric and magnetic ponderomotive forces as gradients of total potentials, where total potential represented the work done to assemble a system. Here he extended that idea to the rate of doing work. Of present interest, however, is his reliance on the proposition that whatever the *losses* occurring during conduction of a current through a resisting wire, they had to equal the work done in producing the current: 'The amount of the mechanical effect continually *lost* or spent in some physical agency (according to Joule the generation of heat) during the existence of a galvanic current . . . [is] proportional to the square of the intensity of the current' (i.e. proportional to the work spent in producing the current).[61]

Thomson's cryptic mode of expression, taken in the context of his other work, indicates his readiness to accept publicly Joule's view that lost mechanical effect could appear as heat. The axiom, however, did not entail the reverse conversion of heat into mechanical effect, as Thomson made clear in his paper on the absolute thermometric scale (begun at the end of 1847 and published in October, 1848). In a well-known footnote he commented on Joule's

> very remarkable discoveries which he has made with reference to the *generation* of heat by the friction of fluids in motion, and some known experiments with magneto-electric machines, seeming to indicate an actual conversion of mechanical effect into caloric. No experiment however is adduced in which the converse operation is exhibited; but it must be confessed that as yet much is involved in mystery with reference to these fundamental questions of Natural Philosophy.[62]

So long as William did not admit the conversion of heat into mechanical effect, he could maintain both that motive power originated only in a fall of intensity, or diffusion, and that motive power could be converted into heat. James's previous attempt to reconcile these two doctrines, however, together with the brothers' continual perplexity about waste, suggests that William's reluctance to admit the reverse conversion had a deeper root, namely a prior commitment to directionality, or progression, in the processes of nature. His own work in 1844 on the 'age' of temperature distributions supports that

[61] William Thomson, 'On the theory of electromagnetic induction', *BAAS Report*, **18** (1848), 9–10; MPP, **1**, 88–90. [62] MPP, **1**, 102*n*. See also SPT, **1**, 268–9.

reading, especially in the context of progressionist geology and Nichol's nebular hypothesis (ch. 6). Indeed, he seems to have regarded conversion of mechanical effect to heat as an irreversible process of *dissipation*, like conduction of heat, in which available work was lost to man. Hence, conversion of work to heat did not imply reconversion to work, or *mutual* convertibility, even when one believed firmly that mechanical effect could never be destroyed.

Joule, however, had no access to Thomson's reasoning and a considerable commitment to his own. He quickly responded to Thomson's 'Account' in a letter dated 6th October, 1848, which marks the resumption of their correspondence after a lapse of over a year. Joule took 'the present opportunity of communicating some of my notions on heat &c. and asking your opinion thereon'. He noted that Thomson still adhered to Carnot's theory of the motive power of heat and he stated his conviction that Thomson's views 'will lose none of their interest or value even if Carnot's theory be ultimately found incorrect'.[63]

What concerned Joule most, however, was Thomson's refusal in the above footnote to admit the conversion of heat into mechanical effect. Joule confessed that some points in the experiment with the electromagnetic engine were not demonstrated with regard to the proof of the conversion of heat into mechanical effect. In his 1843 apparatus he had employed a battery first to generate heat in the coil of an electromagnetic engine (electric motor) held stationary, and second to generate heat *and* work with the coil revolving. By comparison he found that the heating effect of the battery current was reduced by the work done. Joule attempted to persuade Thomson that in such an engine the chemical force between atoms which in ordinary cases would be converted into heat was in this case turned into mechanical effect. Such a consistency argument, however, did not demonstrate conversion of heat into mechanical effect, and so Thomson simply replied politely to Joule stating that he would have to defer saying anything about the magneto-electric experiments, for 'indeed I have not yet sufficiently considered the subject to see it in its bearings to our views on the Heat question'.[64]

Joule's letter of 6th October proceeded to a discussion of his rarefaction and condensation experiments in which 'I thought I had proved the convertibility of heat into power; for I found that on letting the compressed air escape into the atmosphere, a degree of cold was produced *equivalent* to the mechanical effect estimated by the column of atmosphere displaced'. Furthermore:

It appears to me that a theory of the steam engine which does not admit of the conversion of heat into power leads to an absurd conclusion. For instance, suppose that a quantity of fuel *A* will raise 1000 lbs. of water 1°. Then according to a theory which does not admit the convertibility of heat into power the same quantity *A* of fuel working a steam engine will produce a certain mechanical effect, and besides that will be found to have raised 1000

[63] J.P. Joule to William Thomson, 6th October, 1848, J61, ULC.
[64] Thomson, *op. cit.* (note 55).

lbs. of water 1°. But the mechanical effect of the engine might have been employed in agitating water and thereby raising 100 lbs. of water 1°, which added to the other makes 1100 lbs. of water heated 1° in the case of the engine. But in the other case, namely without the engine the same amount of fuel only heats 1000 lbs. of water. The conclusion from this would be that a steam engine is a *manufacturer* of heat, which seems to me contrary to all analogy and reason.[65]

Thus Joule, using the kind of reasoning employed by Carnot and Clapeyron, pointed out the absurdity of Thomson's partial acquiescence in his views, namely in accepting the conversion of power into heat, as in the friction of fluids, while remaining sceptical of the converse process. Joule's powerful arguments forced Thomson to re-examine the foundations both of the Carnot theory and of his own reticence to accept fully Joule's position.

In his reply of 27th October, 1848, Thomson began by saying that he despaired of stating everything in one letter 'especially as I must think and work upon the subject a good deal longer before I can collect my ideas . . . and I now merely write a few remarks which will I hope lead towards an ultimate reconciliation of our views'.[66] The two main themes of the letter were the friction of fluids and the conceptual questions of heat engines. On the friction of fluids, Thomson described, as he had done to Forbes at the end of 1847, apparatus which he had constructed for investigating the heat developed by a rotating disc of tin plate with radial vanes on each side, and he suggested that by this means 'we may be able to boil water by friction alone'. He thus dramatized his agreement with Joule on the conversion of work into heat, a dramatization which Joule himself quickly seized in order to convince the sceptics in Manchester and which inspired his definitive Royal Society paper of 1850 'On the mechanical equivalent of heat'.[67]

Thomson then emphasized the serious difficulty that remained, namely that Joule had not provided any resolution of what happens to work lost during conduction. Until some answer could be found, he could not totally accept Joule's views and reject the Carnot theory. Thomson confessed that he had never seen any way of explaining the difficulty, 'although I have tried to do so since I read Clapeyron's paper; but I do not see any modification of the general hypothesis which Carnot adopted in common with many others, which will clear up the difficulty. That there really is a difficulty in nature to be explained with reference to this point (just as there is with reference to the loss of mechanical effect in fluid friction) the consideration of the following case will I think convince you'. He outlined a thought experiment dealing with two different processes which yielded the same change in the state of a body, the one a reversible compression and expansion producing work and the other an irreversible conduction of heat producing no work. The work not produced in the

[65] Joule, *op. cit.* (note 63). [66] Thomson, *op. cit.* (note 55).
[67] J.P. Joule to William Thomson, 6th November, 1848, J63, ULC.

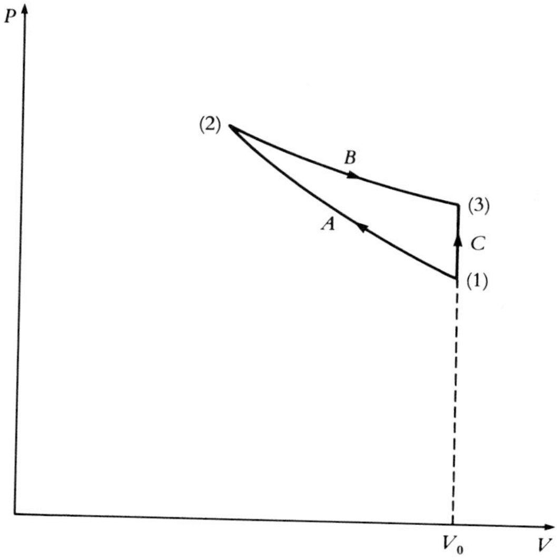

Figure 9.3. On a pressure–volume diagram the change of state of a cylinder of air from (1) to (3) may proceed either along the path of adiabatic compression *A*, followed by the isothermal expansion *B*, or along the path of constant volume *C*. Thomson believed the heat absorbed in the two processes would have to be the same, because he regarded heat as a state function.

latter case was thus 'lost'. Figure 9.3 illustrates Thomson's reasoning, where the work done along any path is always the area under it, $W = \int P\mathrm{d}V$.

The first process began with a mass of air filling a perfectly insulated cylinder of volume V_0, the mouth of the cylinder being closed by a piston and 'the temperature of the air being lower than that of the sea' (point 1 on the figure). Then, 'let the piston be pushed down till the temperature of the air becomes that of the sea' (along the adiabatic path *A*), and let the whole setup be plunged below the sea (point 2). Then:

let the bottom & sides of the cylinder become perfectly permeable to heat. The piston may now be allowed to rise gradually [along the isothermal path *B*], doing work, the temperature of the air expanding in the cylinder remaining the same as that of the sea. When the piston has arrived at its original position [point (3)] we shall have the mass of air at its original volume [V_0] but raised in temperature to that of the sea. Now the work spent in compressing the air [along *A*] when its temperature was being raised is clearly less than the work obtained by allowing it to expand [along *B*] retaining the higher temperature. Hence there is an amount of work gained.[68]

But consider the second process. 'The original mass of air, with the piston held fixed, might at once have been plunged into the sea & been allowed to have its

[68] Thomson, *op. cit.* (note 55).

temperature inc[reased] by conduction gradually' [path C] and so 'the effect might have been produced without getting any work. What has become of the work that might have been gained in arriving at the same result after having gone through the process first described?' Thomson as yet saw no way of explaining this difficulty, but insisted there *must* be an answer: 'I do not see how it can be explained by saying that a greater quantity of heat is taken from the sea in the first process than in the second; and although perhaps an experimental test of the truth is necessary, I believe it would be universally admitted that the quantity of heat absorbed in the two cases would [be] precisely the same'.

Thomson had here finally isolated the most fundamental basis for his reticence in adopting Joule's view on work extracted from 'condensation and rarefaction' of a gas. The heat contained in a body was for him purely a function of the body's physical condition. The heat contained was therefore what would now be called a 'state function'; any gain or loss of heat depended only on initial and final states and not on the path between them.

This concept of heat mirrors his concept of electricity as a state rather than a substance. Electricity, as a state of a body, was related to force by the continuity equation. The net 'flux' of force into or out of a body measured its state of electrification. Thomson had elaborated that concept through the analogy to heat conduction. If the analogy held in reverse, the net flow of heat into or out of a body also determined the thermal state of the body. Thomson's lectures to his natural philosophy class in November, 1849, represent this view precisely. Defining matter as that which 'may be perceived either directly or indirectly by means of the muscular sense of touch or the sensation of resistance', he noted that 'By this [definition] electricity and heat are excluded and it is probable that neither of them are matter'. At the beginning of his lectures on heat measurement and thermometry he amplified that view: 'Temperature may be defined as the state of a body as to heat or cold. We cannot explain it any more than that it is a state, a state which we are enabled to judge of by the [non-muscular] sense of touch'.[69]

Thomson took as axiomatic, then, the principle that the amount of heat flowing into or out of a body depended only on its change of state. If there were no change of state, as much heat left the body as entered it. That principle was for him at least as profound as the indestructibility of mechanical effect, and so, until he could either reconcile the two principles, or discard one of them, he could not proceed. For his part, Joule faced no such obstacles. Heat was for him *vis viva*, and could thus enter and leave a body either as the *vis viva*, of heat, or as work. He responded to Thomson accordingly:

In reference to the hypothetical experiment you mention, that of filling a cylinder with air at a lower temperature than the sea and then having pushed down the piston so as to

[69] William Thomson, 2nd and 6th November, 1849, lectures 2 and 4, in Smith, *op. cit.* (note 1). Thomson differentiated here two senses of touch: 'resistance and temperature are two very distinct feelings'.

raise the temperature equal to that of the sea, to immerse it in the sea, and allow the air to keep the temperature of the sea on expanding, I certainly should say that more heat will in that case be abstracted from the sea than if the temperature of the air had been raised by it without any motion of the piston developing force.[70]

Joule remarked further that he would like to see the experiment tried in practice, even though it would present difficulties. If its result could in any way be decisive proof against Carnot's theory, then the experiment would be especially worthwhile. He proposed a bath of mercury to represent the sea, and suggested that Thomson carry out the experiment because of Joule's own bias.

Still lacking such experimental tests, however, Thomson further formalized his position early in 1849, in his 'Account of Carnot's theory of the motive power of heat', by quoting from Carnot's 1824 treatise:

In our demonstration, we tacitly assume that after a body has experienced a certain number of transformations, if it be brought identically to its primitive physical state as to density, temperature, and molecular constitution, it must contain the same quantity of heat as that which it initially possessed; or, in other words, we suppose that the quantities of heat lost by the body under one set of operations, are precisely compensated by those which are absorbed in the others. This fact has never been doubted . . . To deny it would be to overturn the whole theory of heat, in which it is a fundamental principle.[71]

For that reason, 'there cannot, at the end of a complete cycle . . . have been any absolute absorption of heat, and consequently no conversion of heat, or caloric, into mechanical effect'. Motive power, it seemed to follow necessarily, derived only from 'the transference of heat from one body to another at a lower temperature'.[72] If that were the case, however, then mechanical effect was lost whenever heat was merely conducted from a high to a low temperature without driving an engine. In a well known footnote of 1849, Thomson stated both his full commitment to conservation of mechanical effect and his precise dilemma:

When 'thermal agency' is thus spent in conducting heat through a solid, what becomes of the mechanical effect which it might produce? Nothing can be lost in the operations of nature – no energy can be destroyed. What effect then is produced in place of the mechanical effect which is lost?

In his lectures during the spring of 1850, he stated the same dilemma with respect to loss of *vis viva* by fluid friction: 'there is absolute loss of *vis viva* in the interior of the liquid . . . What becomes of the *vis viva*? Now there is no such thing as loss in nature. There must be some physical effect = to it'.[73]

Thomson obviously did not think of heat as a substance, for he freely accepted

[70] Joule, *op. cit.* (note 67).

[71] William Thomson, 'An account of Carnot's theory of the motive power of heat . . .,' *Proc. Royal Soc. Edinburgh*, **2** (1851), 198–9. Although Carnot was not quoted *verbatim* in the full version of the 'Account' published in the *Transactions*, the same cyclic argument is stated there. See MPP, **1**, 115.

[72] MPP, **1**, 117–18.

[73] *Ibid.*, p. 118*n*; William Thomson, *c.*April, 1850, lecture 90, in Smith, *op. cit.* (note 1).

the conversion of work to heat in fluid friction. Instead, he treated heat as a state function, implying that no heat could be converted into work in a cyclic process. From his insistence on this argument in the 1849 'Account', however, it is equally clear that he could not yet have been thinking of heat as dynamical, as *vis viva*, for if he had his logic would have been severely flawed.[74] Heat entering a body as *vis viva* would not have to leave as the *vis viva* of heat, but could equally well leave in the form of mechanical work.

We have seen, then, that Thomson's dilemma rested not on any unwillingness to accept that work might be converted into heat. Rather, he could not accept the reverse conversion of heat into work, and therefore could not accept Joule's principle of interconvertibility. Joule, in thinking of heat as *vis viva* – convertible into work in a cycle of expansions and contractions – denied Thomson's argument from the cycle of states, but could offer no help with the problem of where the mechanical effect went that was lost in conduction. As late as the spring of 1850, Joule still looked to the Thomsons for a resolution:

> as your brother [James] said, there ought to be some connecting link between the results I have arrived at and those deduced from Carnot's theory. Perhaps you will succeed before long in discovering it. For my own part it quite baffles me.[75]

Although such a 'connecting link' did not come immediately, there is evidence that the Thomsons still hoped to retain both Carnot's theory of the heat engine and the conversion of work to heat by their emphasis on heat as a state function. In a slightly revised version of his 1849 paper on the depression of the freezing point, printed in November, 1850, James substituted the cyclic argument for the conservation of heat principle, stating that they were the same '*if* Carnot's principles are admitted'.[76] The cyclic argument, however, did not help with the fundamental questions of conduction: where did the lost mechanical effect go; could it be recovered as useful work?

74 See Cardwell, *Watt to Clausius*, pp. 241–2 and note 58 above.

75 J.P. Joule to William Thomson, 26th March, 1850, J68, ULC.

76 MPP, **1**, 161; James Thomson, *Papers*, p. 201. M.J. Klein, 'Closing the Carnot Cycle', in *Sadi Carnot et l'essor de la thermodynamique* (Paris, 1974), pp. 213–19, argues that Sadi Carnot himself, being uncertain about the truth of the caloric theory, had set up his cycle in such a way – beginning with an arbitrary initial state – that the cycle makes no reference to conservation of caloric. Clapeyron, on the other hand, had fewer scruples. William's 'Account' and James's 1849 version of his freezing point paper followed Clapeyron quite closely, though with much emphasis on Carnot's concern about the caloric theory and conservation of heat. By 1850, however, James had set up the cycle in his own way to avoid explicit reference to conservation of heat. See also Larmor's comments in James Thomson, *Papers*, pp. xxxiv–v, and William's own 1881 notes about his brother's move in MPP, **1**, 122–3, 127–8.

10

Thermodynamics: the years of resolution

> Everything in the material world is progressive. The material world could not come back to any previous state without a violation of the laws which have been manifested to man, that is without a creative act or an act possessing similar power . . . 'The earth shall wax old &c'. The permanence of the present forms & circumstances is limited. Mechanical effect escapes not only from agencies immediately controlled by man, but from all parts of the material world, in the shape of heat, & escapes *irrecoverably*, though without *loss of vis viva*. *William Thomson, 1851*[1]

Thomson's dramatic shift from a broad-minded uncertainty about inter-conversion in 1849 to this uncompromising assertion of irreversible dissipation in 1851 signals a transformation in both the style and content of his science. From now on he would engage in one of the grandest schemes of molecular and cosmological model building that mathematical natural philosophy had yet seen. This freedom rested on a new unification of two of the deepest commitments of his life: to the universal rule of natural law within a material world created and governed by divine power, and to the progressive development of that world towards an inevitable end.

He had been faced with the problem of reconciling those two beliefs during his early Glasgow days, from the very different perspectives of Thomas Chalmers and J.P. Nichol (ch. 4). His first attempt at reconciliation came in 1844, when he buttressed the progressionist geological cause with his theoretical analysis of the earth's 'age' using Fourier's theory of heat conduction, an analysis which he presented as his inaugural lecture at Glasgow University. In addition, by developing the behaviour of force systems through an analogy to this directional heat flow, he had between 1845 and 1847 generated a new analysis of ponderomotive forces in conservative mechanical systems, showing that any given state of such a system was analyzable in terms of a natural succession of states, given by a tendency to move in the direction in which the mechanical effect contained in the system most rapidly approached an extremum (ch. 8).

[1] William Thomson, preliminary draft for the 'Dynamical theory of heat', PA128, ULC, pp. 6, 9. (Hereafter 'Draft'.) For a transcript of this draft, see Crosbie Smith, 'William Thomson and the creation of thermodynamics: 1840–1855', *Arch. Hist. Exact Sci.*, **16** (1976), 280–8.

Although in conservative mechanical systems any development was in principle reversible, the development was nevertheless controlled by this directional tendency, yielding, for example, an oscillation about an equilibrium state. Thus directionality had become a hallmark of Thomson's analysis of the operations of nature long before he encountered Joule's disturbing arguments. The problem of waste of mechanical effect, furthermore, during such processes as heat conduction and fluid friction, had long puzzled the Thomsons in their attempt to understand directionality by analogy to the 'fall' of gravitational, electrical, and magnetic 'masses', for waste was apparently involved in the directional tendency becoming actual. Joule's arguments for interconversion of heat and work were so disturbing just because they vitiated this attempt at understanding. Only a new reconciliation of natural law with progression could relieve the tension. That reconciliation came with the adoption of a complementary principle, the Second Law of Thermodynamics.

The beginnings of thermodynamics: Rankine, Clausius, and Thomson

The engineering concerns of Victor Regnault, Lewis Gordon, and James Thomson provided a crucial context for William Thomson's concern with the motive power of heat during the 1840s. Now, in 1850, another Scotsman with strong engineering enthusiasms joined the network surrounding the Glasgow professor.

W.J.M. Rankine was educated at Ayr Academy (1828–9), Glasgow High School (1830–6), and Edinburgh University (1836–8) where he studied natural philosophy under J.D. Forbes. At Edinburgh he won the Gold Medal prize for 'An essay on the undulatory theory of light' – a subject closely related to Forbes's own researches on radiant heat during the 1830s – and later a prize for 'An essay on methods in physical investigation'. These essays grounded two of Rankine's subsequent interests: a series of papers on molecular–mechanical hypotheses which included radiation within an ingenious vortex model; and a strong interest in the aims and limitations of scientific method. After Edinburgh, Rankine's career became, for a decade or more, intimately bound up with the rapid expansion of the Scottish and Irish railways during the 1840s – a practical engineering experience which he subsequently widened to include most branches of nineteenth-century civil and marine engineering but with special emphasis on heat engines and naval architecture. From 1855, when he succeeded Lewis Gordon, until his premature death in 1872, Rankine held the Glasgow chair of engineering, a position which facilitated interaction not only with his natural philosophy colleague, but also with Clydeside industrialists, most notably shipbuilders and marine engineers, during a period of rapid technological development in ocean navigation. In particular, his active role in the Glasgow Philosophical Society greatly enhanced those interactions. He was succeeded by James Thomson, whose interests had so clearly paralleled his own.[2]

2 P.G. Tait, 'Memoir [of W.J.M. Rankine]', MSP, xix–xxxvi.

W.J.M. Rankine, Lewis Gordon's successor as Glasgow University professor of engineering in 1855, whose interests in the economy of motive power so closely paralleled those of the Thomson brothers. Rankine's colourful personality (he published, for example, an amusing collection of *Songs and ballads*) contrasted with the relative sobriety of some of Thomson's mathematical friends such as Stokes and Helmholtz. [From MSP.]

On 4th February, 1850, Rankine read his paper 'On the mechanical action of heat, especially in gases and vapours' to the Royal Society of Edinburgh. The introduction contained a summary 'of the principles of the hypothesis of molecular vortices, and its application to the theory of temperature, elasticity, and real specific heat'. The author claimed to have commenced his researches in

1842, but, because of a lack of accurate data, to have laid them aside until 1849 when Regnault's experiments on gases and vapours became available. Rankine's hypothesis of molecular vortices had many conceptual affinities with prevailing views, and some important differences. He mentioned Franklin, Aepinus, Mossotti, Davy, and Joule as having put forward suppositions similar to his own hypothesis on the nature of matter. Each atom of matter consisted of a nucleus surrounded by an elastic atmosphere, self-repulsive but retained in position by attraction to the nucleus. For Rankine, as for Davy and Joule, quantity of heat was the *vis viva* of revolutions or oscillations among the particles of the atmospheres, which Rankine supposed to constitute vortices about the nuclei. His claim to originality, apart from developing the mathematical consequences of the vortex hypothesis, was that the luminiferous medium transmitting light and radiant heat consisted of the nuclei of the atoms, which vibrated independently, or nearly so, of their atmospheres.[3]

The importance for Thomson of Rankine's 1850 paper was two-fold. To begin with Thomson wrote a report on the paper before it was published in the *Transactions of the Royal Society of Edinburgh*, a report which established their interaction.[4] And, on the conceptual side, Rankine's concrete model of heat as *vis viva* helped Thomson to accept the mutual convertibility of heat and work, abandoning the state-function view of heat which required no net loss or gain during the production of mechanical effect in cyclic processes. Even though Thomson did not accept Rankine's *specific* mechanical hypothesis of the nature of heat, he was soon prepared to accept a *general* dynamical theory of heat, namely that heat was *vis viva* of some kind.

In August, 1850, Rankine wrote to Thomson to thank him for 'calling my attention to the paper by Clausius, in Poggendorff's *Annalen*, on the Mechanical Theory of Heat. I approve of your suggestion to send a copy of my paper either to Clausius or Poggendorff'. Thomson was therefore aware of Rudolph Clausius's (1822–88) first paper, published by April, 1850, 'On the motive power of heat' in which Clausius first enunciated and established his version of the second law of thermodynamics. This fact, however, does not imply that Thomson had assimilated its contents, and so the critical period between August, 1850, and March, 1851, when Thomson published his own theory, requires careful interpretation.[5]

[3] W.J.M. Rankine, 'On the mechanical action of heat, especially in gases and vapours', *Trans. Royal Soc. Edinburgh*, **20** (1852), 147–90; MSP, 234–84. The first volume of Regnault's results appeared as *Relation des expériences entreprises par ordre de monsieur le ministre des travaux publics pour déterminer les lois et les données physiques nécessaires au calcul* (Paris, 1847). Subsequent volumes appeared in 1862 and 1870. For a general description of Rankine's early work on thermodynamics and molecular vortices, see Keith Hutchison, 'W.J.M. Rankine and the rise of thermodynamics', *Brit. J. Hist. Sci.*, **14** (1981), 1–26.

[4] A rough draft of the report on Rankine's paper is preserved as PA119, ULC. The Rankine–Thomson correspondence dates from the time of this report.

[5] W.J.M. Rankine to William Thomson, 19th August, 1850, R18, ULC. See also William Thomson, 15th August, 1850, NB34, ULC, p. 154: 'I have just written to Rankine telling him of Clausius' paper in Poggendorff (incompl. in the Number [last April] I have seen) on the Motive

In his 1850 paper, Rankine took as his problem the calculation of the variations in heat in a body resulting from variation of volume and temperature, with the object of determining the effect of cyclic processes in a heat engine *à la* Carnot. Arguing from the vortex hypothesis, he showed that the variation of heat would arise from three factors: mere change of volume (external work), change of molecular distribution dependent on change of volume, and change of molecular distribution dependent on change of temperature. The latter two 'internal' changes were represented as changes in his well-known function *U*,[6] which could be determined in any given case from the principle of conservation of *vis viva*. Thus Rankine provided quite a detailed illustration of how a dynamical theory of heat could explain conversion of heat to work. In summary:

> According to the theory of this essay . . . and to every conceivable theory which regards heat as a modification of motion, no mechanical power can be given out in the shape of expansion, unless the quantity of heat emitted by the body in returning to its primitive temperature and volume is *less* than the quantity of heat originally received.[7]

He thus emphasized that, in his view, the complete interconvertibility of heat and work – over which issue William Thomson had been hesitating – was a consequence of any dynamical theory of heat.

In his letter of August, 1850, to Thomson, Rankine explained that his first attempt to apply mathematical reasoning to the subject arose from his seeing the translation of Clapeyron's paper 'on the opposite theory', opposite, that is, to a mechanical theory of heat:

> The mechanical convertibility of heat has always (since I was first able to reason on the subject) appeared to me as approaching the nature of a necessary truth. I do not of course believe that it is really so; but I speak merely of the feeling, from whatsoever cause arising, which it has produced in my own mind. I have consequently always felt a confident anticipation of its being proved by experiment.[8]

The reference to a 'necessary truth', illustrating the strength of Rankine's prior convictions, occurred again in a letter of November, 1853, to the *Philosophical Magazine*. There he claimed that 'the law of the mutual convertibility [of physical powers] has long been a subject of abstract speculation, and may appear to some minds in the light of a necessary truth. As we cannot, however, expect it to be generally received as such, its practical demonstration must be considered as having been effected by the experiments of Mr Joule'.[9]

Power of heat & c'. For an account of Clausius's 1850 paper, see D.S.L. Cardwell, *From Watt to Clausius. The rise of thermodynamics in the early industrial age* (London, 1971), pp. 244–9.

[6] MSP, 250. See also Keith Hutchison, 'Der Ursprung der Entropiefunktion bei Rankine und Clausius', *Ann. Sci.*, **30** (1973), 341–64. [7] MSP, 253.

[8] Rankine to Thomson, *op. cit.* (note 5).

[9] W.J.M. Rankine to the editors, *Phil. Mag.* [series 4], **7** (1854), 1–3. Rankine had earlier believed the true mechanical equivalent to be considerably less than Joule's values (MSP, 244–5), but by the end of 1850 he had accepted fully Joule's figure. See J.P. Joule to William Thomson, 30th August, 1850, J69, ULC; W.J.M. Rankine to William Thomson, 28th November, 1850, R20, ULC.

An important proposition in Rankine's 1850 paper took on special relevance for Thomson and Joule with regard to the old theme of fluid friction. Rankine's proposition was simply that 'if vapour at saturation is allowed to expand, and at the same time is maintained at the [decreasing] temperature of saturation, the heat which disappears in producing the expansion is greater than that set free by the fall of temperature, and the deficiency of heat must be supplied from without, *otherwise a portion of the vapour will be liquefied in order to supply the heat necessary for the expansion of the rest*'.[10] In the latter case, steam escaping from a high-pressure boiler would be wet, and would 'scald' any passersby, which in practice it did not. Thomson apparently brought the proposition to Joule's notice, for in August Joule wrote:

The point you mention about dry steam from a high pressure boiler is very interesting, and I hope you will not delay to publish your remarks upon the subject in the Phil[osophical] magazine . . . The friction against the orifice will undoubtedly liberate heat [preventing liquefaction]. This circumstance also will account for the good duty performed by some steam engines, although the narrowness of the passages to the cylinder appears such as must seriously obstruct the flow of steam from the boiler.[11]

The 'non-scalding' property of dry steam from a high-pressure boiler was, like the boiling of water by friction, a dramatic empirical 'fact' for everyone to see for himself. Thomson fully discussed the implications in a letter to Joule in October, 1850, which Joule published in the *Philosophical Magazine*.[12] That liquefaction did not take place was shown by the dry nature of the steam and therefore, Thomson claimed, Rankine's conclusions could be reconciled with the facts only by Joule's discovery that heat is evolved by the friction of fluids in motion, that is, heat is acquired by the steam as it issues through the orifice.

So far, Thomson's discussion of Rankine and Joule merely provided further, if striking, evidence that work could be converted into heat. He went on, however, to an even more significant result, a probable conversion of heat into work:

. . . if your fundamental principle regarding the convertibility of heat and mechanical effect, adopted also by Mr Rankine, be true, a quantity of water raised from the freezing point to any higher temperature, converted into saturated vapour at that temperature and then allowed to expand through a small orifice wasting all its 'work' in friction, will, in its expanded state, possess the 'total heat' which has been given to it; but, on the contrary, if it be allowed to expand, pushing out a piston against a resisting force, it will in the expanded state possess less than that total heat by the amount corresponding to the mechanical effect developed.[13]

[10] MSP, 260–1. [11] Joule to Thomson, *op. cit.* (note 9).

[12] William Thomson to J.P. Joule, 15th October, 1850, *Phil. Mag.*, [series 3], **37** (1850), 387–9; MPP, **1**, 170–3.

[13] MPP, **1**, 171–2. Rankine himself approved of Thomson's interpretation as he wrote to Thomson in the letter of November, 1850: 'I am glad that you have published the suggestion you mentioned to me last summer that the dryness of high-pressure steam rushing from an orifice to a

At last Thomson had recognized possible experimental conditions for the conversion of heat into work. The amount so converted, he noted, would bear 'a very considerable ratio to the total heat, instead of being, as I believe all experimenters except yourself have hitherto considered it to be, quite inappreciable'. Were the result to be established, he would no longer be able to withhold assent to Joule's principle of complete interconversion of work and heat.

The remainder of the letter examined the hypothetical foundation of Rankine's proposition, concluding that, if one were content to accept available experimental data, 'we may demonstrate Mr Rankine's remarkable theorem without any other hypothesis than the convertibility of heat and mechanical effect', that is, without Rankine's specific molecular vortex hypothesis. Thomson had made a similar criticism previously in his report on Rankine's paper, to which Rankine responded:

> As to the hypothetical part of my investigations, although it undoubtedly rests on a much less firm basis than that which is founded on the general law of the mechanical convertibility of heat, and although I believe you did my paper essential service by inducing me to make it less prominent & less detailed than it was originally, still I conceive it may lead to some useful results.[14]

Indeed, the vortex model had allowed Thomson to see the viability of the more general principle; having seen it, he attacked the specific model, 'forgetting' its critical role.

Although Thomson only developed his full argument on the interconversion of heat and work in the letter to Joule of October, 1850, there is evidence that he had at least tentatively adopted Joule's and Rankine's position on interconvertibility in the preceding spring. In March, 1851, he claimed to have employed 'more than a year ago' what we shall henceforth call 'Carnot's criterion' and to have begun rewriting the theory of the motive power of heat based on this theorem, but apparently without the supposition that heat was a state function. In order to clarify his position (and problems) during 1850, we state explicitly the logic of Carnot's theory as abstracted from Thomson's 1849 'Account'.

(A) The heat in a body is a state function (Thomson's version of Carnot's 'fundamental axiom', 'fundamental principle' or simply 'hypothesis').

boiler is owing to the reproduction by friction of part of the heat consumed by expansion'. About 1864, Thomson stated that the non-scalding property of dry steam was *false*, and that it had been suggested originally by James. See William Thomson, Research notebook on thermodynamics, (1864?), NB52, ULC.

[14] Rankine to Thomson, *op. cit.* (note 5). For an account of the philosophical sources for Rankine's later views, see Richard Olsen, *Scottish philosophy and British physics. 1750–1880* (Princeton, 1975), pp. 271–86. It seems to us most likely that Rankine reinterpreted the role of hypotheses in response to Thomson. Hypotheses for Rankine became valuable for their suggestiveness rather than for their truth. Such a view would then, as Olsen argues, be consistent with Scottish Common Sense methodology, which emphasized laws containing only observables.

(B) Any work obtained from a cyclic change of state therefore derives from the only change that can occur in such a cycle: namely, transfer of heat (without loss) from high to low temperature.

(C) Applying (B) to a reversible process, and denying perpetual motion, yields: no engine is more efficient than a reversible engine. This statement Thomson called Carnot's 'criterion' for a perfect engine.

(D) From (C) it follows that the maximum efficiency obtainable from any engine operating between heat reservoirs at temperatures T_1 and T_2 is a function of those temperatures:

$$\text{efficiency} = \frac{\text{output}}{\text{input}} = \frac{W}{Q} = f(T_1, T_2) = \int_{T_1}^{T_2} \mu \mathrm{d}T$$

where μ is 'Carnot's coefficient' (replaced by 'Carnot's function' in an alternative formula for $f(T_1, T_2)$ in the dynamical theory).[15]

Adopting Joule's interconversion principle, then, meant rejecting (A) and (B), leaving Thomson with no proof of (C) and no derivation of the Carnot coefficient on which he based all his calculations. As he stated his conviction of (C) in an early draft of his 1851 published paper 'On the dynamical theory of heat':

> The same conclusion has been arrived at by Clausius, to whom the merit of having first enunciated and demonstrated it is due. It is with no wish to claim priority that the author of the present paper states that *more than a year ago* he had gone through all the fundamental investigations depending on it which are at present laid before the Royal Society, at that time considering the conclusion as highly probable even should Carnot's hypothesis be replaced by the contrary axiom of the dynamical theory; and that more recently he succeeded in convincing himself demonstratively of its truth, without any knowledge of its having been either enunciated or demonstrated previously, except by Carnot.[16]

If this recollection is correct, at the beginning of 1850 Thomson was tentatively investigating the consequences of adopting conjointly Joule's hypothesis and Carnot's criterion.

There is in fact some evidence for Thomson's claim in William Smith's notes from the natural philosophy lectures. As early as November, 1849, Thomson told his class: 'We cannot lay down anything for absolute temperature', indicating that he no longer trusted his own absolute temperature scale, based on the Carnot theory of motive power, which had guaranteed the universal relation

15 William Thomson, 'An account of Carnot's theory of the motive power of heat; with numerical results deduced from Regnault's experiments on steam', *Trans. Royal Soc. Edinburgh*, **16** (1849), 541–74; MPP, **1**, 113–64, esp. pp. 115, 117, 132–3, 134.

16 William Thomson, Early draft of the 'Dynamical theory of heat', PA132, p. 10. Our emphasis. This draft is separate from the crucial preliminary draft (note 1) discussed below. For the published version of this claim, see MPP, **1**, 181.

between mechanical effect and the fall of heat from higher to lower intensity. Lecturing very briefly on the motive power of heat in mid-January, 1850, he avoided a sharp distinction between Carnot's and Joule's hypotheses, noting only that for Carnot the 'disengagement' (also 'generation') of heat by work done in compressing a gas depended on the temperature of the gas, while for Joule it did not. Although he basically followed Carnot, he could also cheerfully report during the following week's lectures on Joule's (and his own) proofs of generation of heat by fluid friction.[17]

Apart from his denial of the absolute temperature scale, the most telling evidence that he doubted the Carnot theory derives from the immediately succeeding lecture on heat conduction where he stated: 'The general law is "heat is conducted from a hotter to a colder body["]. The laws on which this conduction depend have not been fully investigated'.[18] Since Thomson had no doubts about Fourier's equation of conduction, it would appear that he no longer considered a difference in temperature (intensity) a sufficient causal explanation of the motion of heat. We have every reason to suppose then that he had begun searching for an alternative explanation of why heat moves from hot to cold and never the reverse. The earlier gravitational analogy of fall of heat to fall of water (or electricity or magnetism), upon which Thomson had for so long relied, had now broken down.

Later in the 1849–50 session, while discussing hydrodynamics, Thomson returned to the problem of fluid friction and 'absolute loss of *vis viva* in the interior of the liquid'. 'What becomes of the *vis viva*?' he asked. 'Now there is no such thing as loss in nature. There must be some physical effect = to it'. To this familiar dilemma he now added Sir Humphry Davy's demonstration of melting ice by rubbing it and concluded: 'Thus there is a hypothesis that heat is just motion between the particles of a body'.[19] Tentatively, ever so cautiously, Thomson had begun to toy with the dynamical theory of heat. How far he actually progressed in reconciling it with Carnot's criterion (C) before reading Clausius, we can only surmise. His completely independent statement of the law and problem of conduction, that 'heat is conducted from a hotter to a colder body', while remarkably close to Clausius' formulation, suggests, however, a separate route to a 'Second Law of Thermodynamics'.

Towards the end of 1850, Rankine's response to Clausius's first paper on the motive power of heat (published earlier that year in Poggendorff's *Annalen*) reinforced Thomson's awareness of the need for a new 'proof' of Carnot's

[17] William Thomson, 17th November, 1849, lecture 16; 15th January, 1850, lecture 43; 22nd January, 1850, lecture 48, in William Smith, 'Notes of the Glasgow College natural philosophy class taken during the 1849–50 session', Ms Gen. 142, ULG.

[18] *Ibid.*, January, 1850, lecture 50. James's lecture notes taken during his attendance at William's 1847–8 session refer to Fourier's law of heat conduction as 'probably rigorously true'. Motive power of heat was not discussed explicitly, and there was nothing in the twenty-five page heat section to indicate difficulties. See James Thomson, Notebook A7, Thomson papers, QUB.

[19] *Ibid.*, April, 1850, lecture 90.

criterion (C) and perhaps also for a new interpretation of the Carnot principle (D) derived from it. In a postscript to a letter of September, 1850, Rankine wrote to Thomson that he had looked over 'the Second part of the paper of Clausius "Ueber die bewegende Kraft der Wärme . . .". The First Part, consisting entirely of deductions from the law of convertibility of Heat and Power, agrees, as far as it goes, with my own investigations'.[20] Rankine and Clausius fully concurred in adopting a principle of the equivalence of heat and work based on a dynamical theory of heat. And they both recognized that the principle did not depend on the specific kind of motion assumed to take place within bodies. For Rankine 'those phenomena, according to the hypothesis [of molecular vortices] now under consideration as well as every hypothesis which ascribes heat to motion, are simply the transformation of mechanical power from one shape to another'.[21]

The case of the second part of Clausius's paper was more complex. Rankine continued his letter to Thomson:

The Second Part consists of deductions from the same law [of heat–work equivalence], taken in conjunction with a *portion* of the principle of Carnot, viz., [(D) above] that the power produced by transmitting a given quantity of heat *through* any substance, is equal to the quantity of heat transmitted multiplied by a function of the temperature only: in other words, [the modified portion] that the ratio of the q. [quantity] of heat converted into expansive power to the q. [of heat] not so converted, is a function of the temp[erature] only. Clausius gives a sort of *a priori* proof of this second law, which so far as I have yet been able to consider the subject, seems to me very unsatisfactory.[22]

Later, in a letter of March, 1851, Rankine made his own position, retrospectively at least, a little clearer: 'I have always thought the principle of Clausius to which you refer had an appearance of probability; but I was not satisfied with his mode of proving it'.[23] The '*a priori* proof' was Clausius's argument that if the Carnot criterion (C), from which the principle derived, were false, then 'it would be possible, without any expenditure of force or any other change, to transfer as much heat as we please from a cold to a hot body, and this is not in accord with the other relations of heat, since it always shows a tendency to equalize temperature differences and therefore to pass from hotter to colder bodies'.[24] Clausius therefore founded the modified Carnot principle (D) on what he saw as a widely based assumption, namely that the transfer of heat from a cold to a hot body was impossible without compensation.

20 W.J.M. Rankine to William Thomson, 9th September, 1850, R19, ULC; Rudolf Clausius, 'On the motive power of heat, and on the laws which can be deduced from it for the theory of heat', *Ann. der Phys. Chem.*, **79** (1850), 368–97, 500–24. Reprinted in E. Mendoza (ed.), *Reflections on the motive power of fire by Sadi Carnot and other papers on the second law of thermodynamics by E. Clapeyron and R. Clausius* (New York, 1960), pp. 109–152, to which version references below are made.

21 MSP, 246.

22 Rankine to Thomson, *op. cit.* (note 20).

23 W.J.M. Rankine to William Thomson 17th March, 1851, R23, ULC.

24 Clausius, 'On the motive power of heat', pp. 132–4. See also Cardwell, *From Watt to Clausius*, pp. 247–9, 253–4.

Rankine had described Clausius's modification of Carnot's principle to constitute what Rankine termed a 'second law' without giving this *a priori* principle, which shortly became known as Clausius's statement of the Second Law of Thermodynamics. Nor had Rankine noted Clausius's intermediate step, namely his use of the Carnot criterion (C). Rankine's account, therefore, did not encourage Thomson to pursue Clausius's arguments further. He wrote at the end of his October letter to Joule: 'I have not yet been able to make myself fully acquainted with this [Clausius's] paper'.[25] Even after October, 1850, then, we must suppose that Thomson's route to a Second Law ran largely separate from Clausius's, although his knowledge that Clausius had attempted to solve the same problem can hardly have been irrelevant. In whatever way one judges Thomson's independence from Clausius, his interaction with Rankine during 1850 was crucial. First, it provided a detailed model for interconversion of heat and work, and led Thomson to see for himself probable experimental conditions under which heat could be converted into work. And, second, it highlighted the need for a new interpretation and a new derivation of Carnot's coefficient in the equation of efficiency (D).

The establishment of classical thermodynamics

By early February, 1851, William Thomson had reconciled Carnot and Joule to his own satisfaction, as a letter from Joule shows:

> I have read with very great pleasure your letter received this morning by which you seem to have completely solved the difficulty which before seemed to render the results of Carnot's theory and what we must consider the true theory irreconcilable. The subject is so important that I must beg of you to lose no time in following it out to all its legitimate deductions and send it to the R[oyal] S[ociety] [of] E[dinburgh]. . . .[26]

From the reconciliation came Thomson's famous series of papers 'On the dynamical theory of heat' published between 1851 and 1855.[27]

In an early draft (February to March 1851) Thomson made clear that he aimed only to communicate the new theory, just as he had done in his 'Account of Carnot's theory' in 1849, rather than to lay down a work of original discovery. He would show what general and numerical conclusions in the 'Account' still held under the dynamical theory.[28] The draft constitutes, nevertheless, the most creative phase in the development of Thomson's thermodynamics and indeed of his whole cosmological perspective.

Written in fragments over a period of about three weeks in a free-flowing style with a variety of deletions, repetitions, digressions, and side-notes which for the most part disappeared from the published version, the draft provides a revealing look at William Thomson's unprotected, more or less private, ideas

[25] MPP, **1**, 173. [26] J.P. Joule to William Thomson, 6th February, 1851, J76, ULC.
[27] MPP, **1**, 174–332. [28] Draft, pp. 1, 3.

and motivations. Since the preliminary and the published versions illuminate each other, we shall discuss them together, but we can best analyze the significance of the whole theory by dividing it into two components: the macroscopic, phenomenological theory consisting essentially of the two familiar laws of thermodynamics; and the dynamical theory of heat, or heat as molecular *vis viva*. In this section we shall treat only the macroscopic theory, focussing on Thomson's two published propositions, on which 'the whole theory of the motive power of heat is founded', and shall treat the dynamical theory itself in the next section.

In his first proposition, Thomson adopted Joule's principle of mutual convertibility of heat and mechanical effect, replacing (B) in his former argument:

PROP. 1. (Joule). – When equal quantities of mechanical effect are produced by any means whatever from purely thermal sources, or lost in purely thermal effects, equal quantities of heat are put out of existence or are generated.[29]

What should we regard as the foundation of this new commitment? From the published paper one would suppose it was experiment. A late part of the draft, however, denied the force of experiment and promoted Joule's arguments: 'the author considers that as yet no experiment can be quoted which directly demonstrates the disappearance of heat when mechanical effect is evolved; but he considers it certain that the fact has only to be tried to be established experimentally, having been convinced of the mutual convertibility of the agencies by Mr Joule's able arguments'.[30] Yet it was precisely the lack of such experimental demonstration which Thomson had previously cited against Joule's arguments, and no new experiments could be adduced.

There can be no doubt that Joule's arguments weighed heavily on Thomson, as did experimental evidence, the probability of new evidence, and Rankine's and Clausius's attacks on the problem. All of these pressures may have sharpened Thomson's dilemma, but can have done little to resolve it. They did not touch the long-standing perplexity of both William and James over losses – waste or dissipation – as the crucial phenomenon requiring explanation. To accept mutual convertibility, whether based on a dynamical theory or not, was merely to push the problem further into obscurity; for, if heat could be converted into work, then work lost by conduction could be recovered as work again. In principle, no losses would occur. But of the fact that losses did occur irrevocably, the Thomsons had long been certain. Thus William's susceptibility to Joule's powers of persuasion, and equally to the experimental evidence, hinged on his resolution of dissipation.

Thomson in his draft isolated this critical hinge in the long saga of thermodynamics:

[29] William Thomson, 'On the dynamical theory of heat, with numerical results deduced from Mr Joule's equivalent of the thermal unit, and M. Regnault's observations on steam', *Proc. Royal Soc. Edinburgh*, **3** (1851), 48–52; *Trans. Royal Soc. Edinburgh*, **20** (1853), 261–88; MPP, **1**, 178.

[30] Draft, p. 21.

The difficulty which weighed principally with me in not accepting the theory so ably supported by Mr Joule was that the mechanical effect stated in Carnot's Theory to be *absolutely lost* by conduction, is not accounted for in the dynamical theory otherwise than by asserting that *it is not lost*; and it is not known that it is available to mankind. The fact is, it may I believe be demonstrated that the work is *lost to man* irrecoverably; but not lost in the material world. Although no destruction of energy can take place in the material world without an act of power possessed only by the supreme ruler, yet transformations take place which remove irrecoverably from the control of man sources of power which, if the opportunity of turning them to his own account had been made use of, might have been rendered available.[31]

We must thus regard the foundation of Thomson's belief in his Proposition 1 to have been both his new view of dissipation and his use of that view to establish Carnot's criterion (C) for an ideal engine as a second proposition:

PROP. 2. (Carnot and Clausius). – If an engine be such that, when it is worked backwards, the physical and mechanical agencies in every part of its motions are all reversed, it produces as much mechanical effect as can be produced by any thermodynamic engine, with the same temperatures of source and refrigerator, from a given quantity of heat.[32]

Thomson founded his demonstration of this proposition on an axiom which appears to have emerged first in the 1851 draft:

Is it possible to continually get work by abstracting heat from a body till all its heat is removed? Is it possible to get work by cooling a body below the temperature of the medium in which it exists: I believe we may consider a negative answer as axiomatic. Then we deduce the proposition that μ [Carnot's coefficient] is the same for all substances at a given temperature.[33]

In part one of the 'Dynamical theory' these remarks emerged in the form since become famous as Kelvin's statement of the Second Law of Thermodynamics: 'it is impossible, by means of inanimate material agency, to derive mechanical effect from any portion of matter by cooling it below the temperature of the coldest of the surrounding objects'.[34]

Closely linked to this statement was a footnote pointing out that denial of such an axiom would entail that 'a self-acting machine might be set to work and produce mechanical effect by cooling the sea or earth, with no limit but the total loss of heat from the earth and sea, or, in reality, from the whole material world'.[35] This footnote reveals his conviction that dissipation of energy was a universal feature of nature, to be accepted as axiomatic and not to be explained further. A self-acting machine which could convert all the heat of the sea, land, and whole material world into useful mechanical effect would clearly recover energy which Thomson believed irrecoverable to, and by, man. In other words, a marine engine could utilize the heat in the sea and not only run without the need for coal or other fuel, but eventually remove *all* the heat from the oceans! Since such possibilities ran counter to the cosmological framework which

[31] *Ibid.*, pp. 5, 6. [32] MPP, **1**, 178. [33] Draft, p. 10. [34] MPP, **1**, 179. [35] *Ibid.*, 179*n*.

Thomson had accepted by 1851, he used them as axiomatic impossibilities by which to 'demonstrate' Carnot's theorem, and so provide a 'proof' quite different from that of Clausius in its formulation.

William Thomson's 'demonstration' of the second proposition began by supposing two thermodynamic engines *A* and *B*, *B* reversible, and *A*, if possible, more efficient than *B* (figure 10.1). Both operated between the same two heat reservoirs such that *B* ran backwards and restored at the higher temperature whatever heat *A* extracted. Now, since *A* was supposed more efficient than *B*, it would deliver less heat at the lower temperature than *B* extracted and besides would deliver more work than was required to operate *B*. A complex engine might thus be constructed with *A* powering *B* and at the same time delivering work:

> We should thus have a self-acting machine, capable of drawing heat constantly from a body surrounded by others at a higher temperature, and converting it into mechanical effect. But this is contrary to the axiom, and therefore we conclude that the hypothesis that *A* derives more mechanical effect from the same quantity of heat drawn from the source than *B*, is false. Hence no engine whatever, with source and refrigerator at the same temperatures, can get more work from a given quantity of heat introduced than any engine which satisfies the condition of reversibility, which was to be proved.[36]

Appreciation of Thomson's demonstration requires recognition of the depth of his belief that irremediable losses must occur. In the draft he expressed that belief with religious conviction: 'Everything in the material world is progressive. The material world could not come back to any previous state without a violation of the laws which have been manifested to man, that is, without a creative act or an act possessing similar power'. He then recited his basic creed in several forms:

> I believe the tendency in the material world is for motion to become diffused, and that as a whole the reverse of concentration is gradually going on – I believe that no physical action can ever restore the heat emitted from the sun, and that this source is not inexhaustible; also that the motions of the earth & other planets are losing vis viva which is converted into heat. . . .[37]

The old problem of loss of useful work which had begun as an engineering concern – the loss of power in canal locks and the need to minimize losses in marine steam-engines – and which had become identified with 'loss' in conduction and fluid friction, Thomson now expressed as a cosmological, and indeed theological, principle, the historical foundations of which we examined in Chapter 4.

Thomson's new formulation required acceptance of temporal directionality as a fundamental feature of the creation. This directional property of time soon became familiar in the scientific literature through the terms 'dissipation' and 'irreversibility'. A failure on man's part to harness the inevitable diffusion of energy from sources of concentration to provide power for his uses (whether by

36 *Ibid.*, 179–80. 37 Draft, pp. 6, 8.

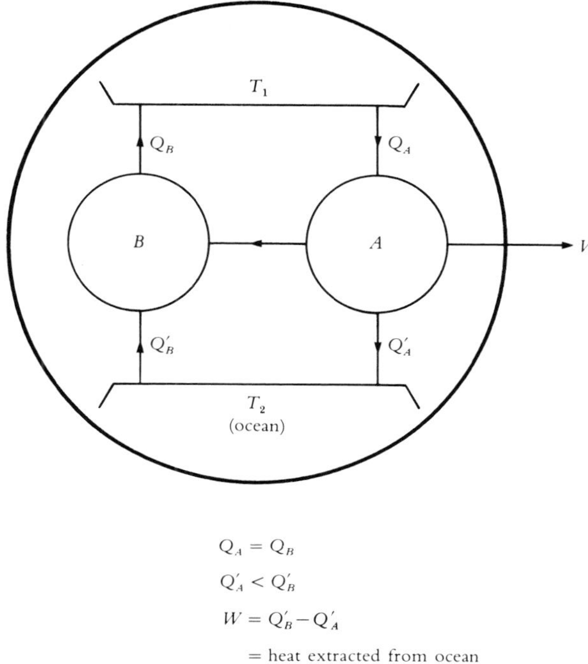

$$Q_A = Q_B$$
$$Q'_A < Q'_B$$
$$W = Q'_B - Q'_A$$
$$= \text{heat extracted from ocean}$$

Figure 10.1. Thomson proved Carnot's theorem by reference to a compound engine, in which engine *A* (supposed more efficient than a reversible engine) drives refrigerator *B* (reversible) and the combination produces net work *W*, with no other effect than extraction of heat from the ocean, contrary to his second axiom.

an inefficient machine or by not deploying a machine at all) was to 'lose' or 'waste' the opportunity of benefiting from a fundamental feature of the creation. Thomson summed up, therefore, with the passage quoted at the opening of the present chapter, which made implicit reference to the 102nd Psalm:

25. Of old hast thou laid the foundation of the earth: and the heavens are the work of thy hands.
26. They shall perish, but thou shalt endure: yea, *all of them shall wax old like a garment*; as a vesture shalt thou change them, and they shall be changed.
27. But thou art the same, and thy years shall have no end.[38]

[38] Psalm 102:25–7. Our emphasis. An alternative verse in Isaiah 51:6 reads: 'for the heavens shall vanish away like smoke, and the earth shall wax old like a garment'. See also Thomas Chalmers, 'The transitory nature of visible things', *Works*, **7**, 263–79, esp. pp. 266–7. In this published sermon, Chalmers chose as his text 2 Corinthians 4:18: 'the things which are seen are temporal; but the things which are not seen are eternal'. He interpreted the text to mean that visible things 'had a beginning and they will have an end', and he reinforced his verdict with a direct reference to the above Psalm. Given the Thomsons' friendship with Chalmers and attendance at his services, the coincidence of interpretations is not surprising.

The new emphasis brought theological and cosmological arguments to the very core of the draft, arguments which illuminated the problems of loss of useful work in machines by relating them to a necessary diffusion of energy which only God Himself could restore to its original concentration. Dissipation had finally become an issue independent of conservation. God alone could create or annihilate energy, *and* God alone could restore the original distribution or arrangement of energy in the created universe. By accepting such a creed, grounded on theological conceptions of an omnipotent God whose action alone made man's use of power possible, William Thomson synthesized a worldview which incorporated within itself a problem not amenable to purely dynamical understanding.

Having laid down two propositions in his 'Dynamical theory' of 1851, and having given 'demonstrations' of them, Thomson had attained a commanding position from which to draw together many of the detailed themes discussed during the years 1847 to 1851, to incorporate them into a new framework, and to formulate his theory of the motive power of heat in mathematical language. He thus stated the aim of thermodynamics in part one of the 'Dynamical theory': 'a complete theory of the motive power of heat would consist of the application of the two propositions demonstrated above, to every possible method of producing mechanical effect from thermal agency'.[39] He added a reference in a footnote to his 1849 'Account of Carnot's theory' indicating that there were at present only two distinct ways in which this could be done – by alterations of volume which bodies experienced through the action of heat, or through the medium of electric agency.

Dealing with cases of electric agency first, Thomson observed that here the second proposition, with its criterion of a perfect engine, had not as yet been applied. However, the application of the first proposition had been thoroughly investigated through Joule's work, especially in his 1843 paper. Joule's achievement there, as Thomson saw it, was to express the heat generated as proportional to the whole work spent, in a process by which mechanical work through a magneto-electric machine produced galvanism, and ultimately heat, and to conclude that heat may be created by working such a machine. Provided all the current was used to produce heat, the total quantity of heat produced was exactly proportional to the quantity of work spent.

Thomson also integrated Joule's views on the Peltier effect (that 'cold is produced by an electrical current passing from bismuth to antimony'), quoting from Joule's letter of July, 1847, as well as from his 1843 paper. In the letter to Thomson, Joule had argued that 'in Peltier's experiment on cold produced at the bismuth and antimony solder we have an instance of the conversion [of heat] into the mechanical force of the current, for an *increase* of electromotive force arises from the solder which is cooled'. These discussions of Peltier and of electric agency in general formed the starting point for Thomson's extensive work on

[39] MPP, **1**, 181.

thermo-electricity from 1851, and a lengthy paper read in May, 1854, subsequently part six of the 'Dynamical theory', dealt with that subject in relation to both fundamental propositions.[40]

Turning to the other method of producing mechanical effect from thermal agency (*via* expansive engines) Thomson stated that here both of the fundamental propositions could be applied in a perfectly rigorous manner, an application he admitted as having been already achieved by Rankine and Clausius with respect to the first proposition, and by Clausius with respect to the second by employing 'Carnot's unmodified investigation of the relation between the mechanical effect produced and the thermal circumstances from which it originates, in the case of an expansive engine working within an infinitely small range of temperatures'. Thomson then proceeded, quite directly, to derive analytical expressions for each proposition.

The first equation expressed, 'in a perfectly comprehensive manner, the application of the first fundamental proposition to the thermal and mechanical circumstances of any substance whatever, under uniform pressure in all directions, when subjected to any possible variations of temperature, volume and pressure'. The second equation gave values of μ, the Carnot coefficient, the same for all substances at the same temperature. Hence, 'all Carnot's conclusions, and all conclusions derived by others from his theory, which depend merely . . . [on this equation] . . . require no modification when the dynamical theory is adopted'. Furthermore it followed:

> that Carnot's expression for the mechanical effect derivable from a given quantity of heat by means of a perfect engine in which the range of temperatures is infinitely small, expresses truly the greatest effect which can possibly be obtained in the circumstances; although it is in reality only an infinitely small fraction of the whole mechanical equivalent of the heat supplied; the remainder being irrevocably lost to man and therefore 'wasted', although not *annihilated*.[41]

If, on the other hand, Thomson argued, the quantities of mechanical effect obtained were finite, a finite quantity of heat must be converted, in falling through a finite range of temperature, to give results which 'will differ most materially from those of Carnot'. The investigation of this aspect of the theory was contained in part two of the 'Dynamical theory', which Thomson, as we saw earlier, had asserted was worked out prior to his writing of the first part.

The dynamical theory of heat

In founding his second proposition on the axiom of dissipation, William Thomson opened the way to cosmological speculations on a grand scale.

[40] *Ibid.*, pp. 182–3, 232–55. A detailed study of the history of thermo-electricity has been made by B.S. Finn, *Developments in thermo-electricity: 1850–1920* (University of Wisconsin PhD dissertation, 1963), esp. pp. 26–132 on Thomson. See also B.S. Finn, 'Thomson's dilemma', *Phys. Today*, **20** (1967), 54–9. [41] MPP, **1**, 183*n*, 185–9.

Similarly, but in the opposite direction, the foundation he adopted for his first proposition would lead him to speculations on the molecular structure of matter. To establish mutual convertibility of heat and work he adopted the hypothesis that heat is 'a dynamical form of mechanical effect', consisting in molecular motions. The proposition followed automatically.[42] In thus committing himself to a dynamical theory of heat, Thomson violated the fundamental methodological doctrine of his prior work, the stricture against physical hypotheses, which limited 'positive' mathematical theory to the development of experimentally established laws. As in his discussion of the dissipation axiom, he spoke of the dynamical theory in terms of belief, 'feeling convinced that the theory . . . is true'. And his conviction allowed him to argue, in spite of the lack of definitive evidence for conversion of heat to work, that 'the fact has only to be tried to be established experimentally'.[43] The dynamical theory of heat, indeed, marked a critical watershed in Thomson's career, a watershed whose character we must investigate. But first, why did he abandon the non-committal view, that heat is a state of a body, in favour of the dynamical theory?

To note the obvious, if conversion of heat to work were to explain the work derived from heat engines, then the heat in a body could not be a state function, for in a closed cycle of states the body would have to absorb more heat along one portion of the cycle than it emitted along the return portion. Since the change in heat would not be independent of the path, the heat could not be a state function. The same argument would apply to a refrigerator (a reversed engine) converting work into heat. In order to maintain mutual convertibility, Thomson required a new view of heat, and the dynamical theory satisfied his needs. Its adequacy, however, did not make it necessary. More profound issues lay behind Thomson's adoption of the new theory.

The additional argument that he actually gave will require brief reflection on his fundamental beliefs about natural philosophy. Those beliefs, as we know from Chapter 4, derived in large part from Scottish tradition. Beginning from a dualistic distinction between moral and material realms, one located natural philosophy solely in the latter realm and identified it with mechanical philosophy. In his draft Thomson put the traditional point in extreme form, albeit with respect to the new mechanical philosophy of irreversible processes:

> The problem of Natural Philosophy wh[ich] includes all physical science is this. Given at any instant the position & motion of each atom of matter. Required the position & motion of each at any time past or future. This is a problem the conception of which is possible to man's intelligence; although of course the solution can never be effected. I believe those data [position and motion] are sufficient to imply the solution; & that this is the great distinction . . . between the ways of God in the physical & the moral world; that with distinct and exceptional cases wh[ich] we are justified in calling miracles man can foresee the future with certainty in the material world; that he cannot & that in this world

[42] *Ibid.*, pp. 175, 178–9. [43] Draft, pp. 3, 21.

he never will be able to foresee even the simplest fact with certainty in the operations of the mind.[44]

Although Thomson's concern here with miracles referred to violation of the dissipation axiom, his general point about natural philosophy implied that mechanics provided the only conceivable basis for understanding mutual convertibility.

'Traditional' captures some sense of Thomson's belief. In detail, however, his mechanical philosophy pushed considerably beyond that of his predecessors. The Scottish tradition had limited natural philosophy to the sensible motions of bodies, leaving insensible molecular motions to chemistry. Even chemistry, in fact, dealt only with the sensible effects of supposed insensible motions. This anti-hypothetical tradition had motivated Thomson's own view of heat and electricity as states of matter rather than material fluids. His extension of natural philosophy to the molecular domain, therefore, signalled a major departure. That departure stands out starkly in Thomson's own lectures. In 1847, beginning the experimental course with heat, he remarked, 'the effects of heat in producing chemical changes do not come under our consideration in this class. The physical effects of heat will be considered'. The latter he limited to changes of state and expansion.[45] By 1852, however, he included chemistry (of the constitution of bodies) as one of four areas of the mathematical course of natural philosophy, just after mechanics (which included heat and light) and before electricity and magnetism. Noting that chemistry had usually been classed under natural history, he insisted, 'Chemistry pertains almost wholly to Nat. Phil^{y} not Nat. Hist^{y}'.[46] With Joule and Faraday as his guides, Thomson had begun co-opting chemistry for natural philosophy. For him, as for them, interconversions transcended the disciplinary boundary. He could thus maintain in an 1885 speech that 'there is no philosophical division whatever between chemistry and physics. The distinction is that different properties are investigated by different sets of apparatus. The distinction between chemistry and physics must be merely a distinction of detail and of division of labour'.[47]

Reference to Faraday in the present context, however, leads in the direction opposite from Thomson's, for Faraday would never have identified natural philosophy with mechanical philosophy. Magnificent though his interconversion experiments were, he rejected all attempts at mechanical reduction, and spoke instead of the mutual conversion of different forms of 'power' or 'force'. 'Force', in the general sense of a natural power, served as a primitive category in Faraday's philosophy. It played the same role in a wide variety of German idealist constructions from Kant and Hegel to J.R. Mayer (Joule's main

44 *Ibid.*, pp. 6, 7. 45 James Thomson, Notebook A7, Thomson papers, QUB.

46 William Thomson, 2nd November, 1852, lecture 1, in William Jack, 'Notes of the Glasgow College [senior] natural philosophy class taken during the 1852–3 session', Ms Gen. 130, ULG.

47 PL, **2**, 484.

rival as 'discoverer' of energy conservation) and to Wilhelm Ostwald (founder of physical chemistry and energeticism). Their examples should remind us of how little necessity lay behind Thomson's belief in mechanical philosophy. They should remind us also how very different was his conception of dynamical theory – reducing heat, light, magnetism and other agencies, to matter in motion – from theirs, which made matter itself a state of dynamic equilibrium between opposing 'forces'. To Thomson, all reliance on such disembodied powers contributed to the delusive metaphysics epitomized by Hegel's speculative *Naturphilosophie*.[48] 'The Germans', he warned his class in 1852, 'apply the term Nat. Phil^{y} to metaphysical science'. *Naturphilosophie* conflated the sciences of mind and of matter, while in English, natural philosophy referred to matter alone and obtained its results strictly by 'inductive inference and generalization'.[49] From that position, Thomson repeatedly denigrated the accomplishments of J.R. Mayer, for example, in favour of Joule.[50]

The same view lay behind the arguments that Thomson gave for the dynamical theory of heat. He opened his famous series of papers with the following theme:

> Sir Humphry Davy, by his experiment of melting two pieces of ice by rubbing them together, established the following proposition: 'The phenomena of repulsion are not dependent on a peculiar elastic fluid for their existence, or caloric does not exist'. And he concludes that heat consists of a motion excited among the particles of bodies . . . The recent discoveries made by Mayer and Joule, of the generation of heat through the friction of fluids in motion, and by the magneto-electric excitation of galvanic currents, would either of them be sufficient to demonstrate the immateriality of heat; and would so afford, if required, a perfect confirmation of Sir Humphry Davy's views.[51]

If heat were not material it had to be dynamical. No alternative, certainly not heat as a natural power, existed. Thomson did not even mention his own subtle view of heat as a state. The myth thus emerged that Davy had established the dynamical theory over fifty years earlier by rubbing pieces of ice together (an argument Thomson had not accepted himself one year earlier) and that Joule had put the argument beyond attack.

In the draft Thomson expressed his new-found faith in Davy in syllogistic – and theological – form:

[48] See M.N. Wise, 'German concepts of force, energy and the electromagnetic ether: 1845–1880', in G.N. Cantor and M.J.S. Hodge (eds.), *Conceptions of ether: studies in the history of ether theories, 1740–1900* (Cambridge, 1981), pp. 271–6. Thomson adhered to this view to the end. He wrote to Larmor in 1906: 'Young persons who have grown up in scientific work within the last fifteen or twenty years seem to have forgotten that energy is not an absolute existence. Even the Germans laugh at the "energetikers". I do not know if even Ostwald knows that energy is a capacity for doing work; and that work done implies mutual force between different parts of one body relatively movable, or between two bodies or two pieces of matter, or between two atoms of matter . . .'. See Lord Kelvin to Joseph Larmor, 9th October, 1906, L37, ULC.

[49] William Thomson, 2nd November, 1852, lecture 1, in Jack, 'Notes'.

[50] For example, MPP, **1**, 175. [51] *Ibid.*, p. 174.

Man cannot create matter.
Heat may be created by man.
∴ Heat is not matter.
or Matter cannot be created by operations under human control. Heat may be created by operations under human control.
∴ Heat is not matter.[52]

Having thus buttressed Joule's experimental evidence with a logic derived from his own commitment to mechanical philosophy, Thomson contented himself that the dynamical theory required no further proofs. From now on heat was unequivocally a state of molecular motion.

There remains, nevertheless, a further puzzle. A state of motion in a system, according to Thomson's mechanics, generally contained two forms of mechanical effect: *vis viva* and total potential (the work done against forces between the parts to assemble the system with no motion from infinitely separated parts). What part of this mechanical effect constituted heat? The question is of considerable importance because it raises once again the problem of force, not disembodied forces now but molecular forces, supposed in the usual Newtonian and Laplacian mechanics to be actions at a distance between molecules. We have seen that since 1846 at the latest Thomson had been searching for ways to replace electric and magnetic action at a distance by contiguous action in an ether field, either by analogy to elastic deformation in a solid or to fluid motion. One must ask what relation he now supposed to exist between molecular force and the state of motion constituting heat.

Naturally he stood in no position to solve such a problem, and his remarks are accordingly ambiguous. The draft, however, shows a strong inclination to identify heat exclusively with actual *vis viva*, thus separating off the problem of molecular forces and static potentials, or internal work, from that of heat. With respect to conduction, he said:

According to the Dynamical Theory of Heat, the quantity of heat in a body is measured by the vis viva of the molecular motions which exist within it and the evolution of mechanical effect from thermal agency consists in the diminution of such motions by resistance. [The conduction of heat through matter from one part of a body to another *del.*] The propagation of heat consists in the communication of vis viva from molecules in motion to *contiguous* molecules; and unless any portion of the vis viva be lost in producing changes in the dimension or arrangements of bodies against resistance, or some be gained by the reverse the quantity of heat remains constant. Thus the ordinary method of estimating quantities of heat is consistent with the dynamical theory.[53]

[52] Draft, note facing p. 19. Repeated in Jack, 'Notes': 'Heat was made by Davy, but he can't make matter – ergo – what can [the] rubbing do? Only inertia or matter set in motion'. (No date, but shortly after lecture 10.)

[53] Draft, pp. 12, 13. In the transcript published in Smith, 'William Thomson and the creation of thermodynamics', p. 284, the pages printed as thirteen and fourteen should be reversed and the word 'produces' inserted at the end of the corrected p. 13.

In these remarks, conservation of *vis viva* during conduction replaced the older doctrine of conservation of caloric, except when work was done, whether internal work of molecular rearrangement or external work of volume change. Under such circumstances, heat as caloric would merely have changed from the sensible to the latent state, but now in Thomson's description heat as *vis viva* was lost or gained. Thus the heat in a body was strictly sensible heat. It did not include latent heat. Interconversions between the latent and sensible states were special cases of the interconvertibility between work and heat generally.

In an earlier portion of the draft Thomson presented the same view for the mixing of bodies of different temperatures, a case often employed to exemplify conservation of caloric. Noting that *vis viva* would remain constant during mixing and that 'the measurement of "quantity" in heat implies that in those circumstances [mixing] the quantity of heat is unchanged', he concluded: 'qu[antit]y of heat corresponds to vis viva'. Again he added: 'Here it ought to be observed that if any of the bodies expand or contract so as to become of different (resilience) innere arbeit [internal work, from Clausius] the vis viva in the system will be changed & the quy of heat will be changed'.[54]

Thomson's attempt in the draft to define the heat in a body strictly as *vis viva* was not a happy one, for it suggested that heat could be added to a body, held in equilibrium internally by molecular forces, without altering its total internal potential. By the time of the published version, he had ceased to speak of the heat *in* a body at all, referring only to heat absorbed or emitted with respect to the surroundings, and of 'thermal motions' inside. Thus, when a body emitted heat, 'the work which any external forces do upon it, the work done by its own molecular forces, and the amount by which the half *vis viva* of the thermal motions of all its parts is diminished, must together be equal to the mechanical effect produced from it; and consequently to the mechanical equivalent of the heat which it emits'.[55] In short, heat emitted now equalled external work plus internal work plus decrease in internal *vis viva*. 'Heat' had become a purely macroscopic and external concept, while 'thermal motion' encompassed both static and dynamic (potential and kinetic) forms of mechanical effect.

It is difficult to say why Thomson originally insisted that heat was *vis viva* alone. One might argue that he simply adopted the definition of the dynamical theory as it had come to him from Joule, Rankine, and Clausius. One could also observe that the distinction of heat as *vis viva* from work as potential immediately justified his first proposition, the mutual convertibility of heat and work. We shall pursue here yet a third interpretation, more speculative, but more consonant with his prior and later work, emphasizing that the basic thrust of his theorizing in electricity and magnetism had been to replace static forces by dynamic effects propagating in a field of contiguous actions. On this interpretation, force, at the molecular level, had become something to explain rather than

[54] Draft, p. 4. [55] MPP, **1**, 179.

to explain with. The primitive categories available in macroscopic dynamics remained matter, motion, and force, but in molecular dynamics Thomson recognized only matter and motion. Thus his first theoretical impulse made heat dynamical in the sense of *vis viva* and not molecular potential, or internal work. If he included internal work in his published account of thermal motion in order to make the concept more coherent formally, that inclusion only signified that the underlying problem had as yet defied explanation. The true mechanical philosophy remained elusive.

The best evidence for this interpretation of Thomson's concept of dynamical theory consists in his repeated references to Sir Humphry Davy. Davy, of course, supplied the argument that heat is not matter, but his idea of 'repulsive motion' as the alternative interested Thomson even more. In both the draft and the published paper he quoted Davy's proposition: 'The phenomena of repulsion are not dependent on a peculiar elastic fluid for their existence, or caloric does not exist'. While admitting in the draft that Davy's logical demonstration of this proposition was hardly convincing, Thomson claimed that the single experiment of rubbing ice established it. In both versions he continued, 'D[avy] concludes that heat consists of a motion excited among the corpuscles of bodies; and he says "to distinguish this motion from others, and to signify the cause of our sensation of heat" and of the expansion, or expansive pressure, produced in matter by heat "the name *repulsive motion* has been adopted"'.[56]

Now it is all too easy to ignore the phrase 'repulsive motion', because it has no essential function in the published paper, where repulsion arises from molecular *forces*. Thomson, however, remembered the phrase long afterwards in his 1884 lecture on 'Steps towards a kinetic theory of matter' as providing a crucial insight, and his students in the winter of 1852 heard the notion several times.[57] Elasticity was the problem. He introduced the class to it through his archetype of misguided natural philosophers, Poisson. Poisson's microscopic theory (following the infamous Boscovich in Thomson's story) assumed that bodies are made up of atoms exerting central forces of attraction and repulsion on each other. From this hypothesis of central forces Poisson and others had concluded that the rigidity and the compressibility of any 'perfect solid' were necessarily proportional, for they depended on the same central forces. Unfortunately, theory and fact bore no comparison. Stokes had rescued natural philosophy by arguing that rigidity and compressibility were as independent as shape and volume and required independent coefficients (ch. 6).

This much of the story Thomson had been relating since his second session, when he told the class that 'Poisson's Theory of Elastic Solids is totally erroneous' and that 'Stokes . . . was the first to take a correct view of this matter'.[58] Now in his 1852–3 lectures he shifted the ground to heat. Poisson had gone

[56] Draft, pp. 17–19; MPP, **1**, 174. [57] PL, **1**, 228; Jack, 'Notes'.
[58] James Thomson, Notebook A7, QUB.

wrong, he said, by founding his theory 'on erroneous considerations of heat'. Indeed, the repulsion in Poisson's scheme derived from self-repulsive atmospheres of caloric attracted to the atoms. Dynamics required that they act effectively as *central* repulsions between atoms, and therein lay the necessity of proportional rigidity and compressibility. Rankine had pointed the way to an improved theory early in 1851, based on 'the true nature of heat', with his vortex hypothesis, for the molecular vortices did not have to act centrally. In fact, Rankine provided not only independent coefficients of rigidity and compressibility in non-crystalline solids but different rigidities in three dimensions for crystals.[59] Rankine's hypothesis, however, did not completely satisfy Thomson. It assumed cohesion between atomic nuclei and self-repulsive atmospheres, like Poisson's perfect solid, adding only the *vis viva* of heat to those atmospheres: 'A solid without heat is a perfect solid', Thomson remarked. Here Davy's repulsive motion suggested a more pleasing picture. Thomson had also discovered Herapath's kinetic theory:

> Suppose a mass of air destitute entirely of thermal motions *may* collapse till it go to nothing. Rankine has assumed that there are nuclei or solid atoms of air & elastic atmospheres possessed of repulsive properties. These atmospheres have an expansive power, & the nuclei have a power influencing each other. If the atmospheres exist, there must be cohesion between particles [nuclei]. Davy considered all expansion due to thermal motions, the repulsive motions of heat. This would be true, if Herapath's theory that atoms act on each other only by impact [is true] . . . Rankine thinks that rotating motions are given, in heating, to the atoms [creating] a centrifugal force in the elastic atmospheres. There is work done by expanding air then, against the cohesion of the atoms, besides the external effect of heat.[60]

Granting the professor's lecture slightly more coherence than the student's notes, Thomson here suggested Davy's repulsive motion as an alternative to Rankine's rotating motions in the elastic atmospheres, the difference being that Davy's motions required no internal work done against static attractions at a distance, nor any mysterious repulsive properties. The properties of elasticity and expansion on heating apparent in macroscopic bodies would derive directly from matter in motion, without regressing to a property of elasticity in a new and unobservable kind of matter. As usual in areas so speculative, Thomson did not state his preference directly, but merely implied it by pointing out how such-and-such would work if thus-and-so were true. It seems apparent, nevertheless, that when he began his 'Dynamical theory of heat' by citing Davy's repulsive motion, he meant to imply that an ultimate dynamical theory would employ only two primitive terms, matter and motion. 'Force' would appear only macroscopically or where gaps remained in the molecular picture. While he excluded his concern with 'contiguous action' from the published paper in

[59] W.J.M. Rankine, 'Laws of the elasticity of solids', *Cam. and Dublin Math. J.*, **10** (1851), 47–80.

[60] Jack, 'Notes'; MSP, 67–100. On Herapath, see S.G. Brush (ed.), *Mathematical physics . . . and selected papers by John Herapath* (New York, 1972), esp. pp. vii–xxxiii.

1851 he retrieved it in 1854 in part six of the 'Dynamical theory', redefining thermodynamics as consisting of two subjects: '*the relation of heat to the forces acting between contiguous parts of bodies*, and *the relation of heat to electrical agency*'.[61] With those subjects he opened the path to his famous speculations on the nature of matter and force.

The watershed: molecular energy physics

The year 1851 marked a distinct reorientation in William Thomson's public scientific face. From 1847 to 1851 he had struggled to maintain his purely macroscopic ideal of mathematical theory against increasingly pressing demands for a physical conception of molecular reality. His new commitment to the dynamical theory of heat fixed his gaze unswervingly on the goal of a unified physics and suggested the form of the new structure. We shall attempt here to capture the main outlines of this change of countenance.

In 1851 Thomson abandoned the project that had epitomized his former goals, 'The mathematical theory of magnetism'. Formal, didactic, and rigorous, the mathematical theory avoided assuming anything about the nature of magnetism, resting instead on a purely 'positive' definition of the measure of magnetization, or magnetic moment per unit volume, obtained as the infinitesimal limit for a small macroscopic element $dxdydz$ and treated as a continuous density. The 'theory' consisted of an extended mathematical analysis of theorems applicable to this type of distribution, essentially the mathematics of the continuity equation or the potential theory of a modern textbook. It realized the insights he had won from the analogies of heat flow and hydrodynamics, fully justifying his enthusiasm when, having finally attained his existence and uniqueness proof for solutions of the extended continuity equation, he wrote in his notebook in 1847: '*The possibility of all the problems I have yet considered in the allied physical theories* is I believe established by this theorem'.[62]

Thomson's 'Magnetism' showed the power inherent in his method of mathematical analogy, but it also showed the weakness of that method, which set 'the allied physical theories' in parallel without relating them physically. Immensely important for its mathematical style – Maxwell found many of the theorems of his field theory in it – Thomson's analysis failed at just those points where the new physics demanded answers: wherever interrelationships between different forces entered. Unable to choose, for example, between magnetic matter and electric current as the source of magnetic forces, he could not settle on a definition of magnetic force, much less develop a theory of electromagnetism or interrelate light with magnetism, as required by Faraday's magneto-optic experiments. The theory extended no further than providing alternative repre-

[61] MPP, **1**, 232.

[62] William Thomson, 7th October, 1847, Journal and research notebook, 1845–56, NB34, ULC, p. 87.

sentations of the continuity equation, with alternative generalized solutions, assuming one or the other type of magnetic source and either potential gradient or flux as the representative of force. History has shown two routes out of this impasse: a less restrictive use of mathematical analogy (Maxwell's choice in 1855 for 'Faraday's lines of force'), and molecular theory, which Thomson adopted.

If his 'Magnetism' of 1851 represented the last gasp of Thomson's old style, he had long taken his first hesitant step towards a unified physical theory with his 'Mechanical representation of electric, magnetic, and galvanic forces' of 1847, motivated indeed by Faraday's magneto-optic effect. Although presented as a mathematical analogy, it nevertheless treated the three forces on a common basis, as potentially coexistent states of strain in an elastic solid. It thus began to suggest how the forces would be interrelated if the macroscopic analogy to an elastic solid were read literally, as a theory of the real structure of the underlying ether. No doubt Thomson already believed in 1847 that some such theory would prove correct. His struggles with heat, however, show how agonizing his path to molecular theory and ether theory actually was. His positive methodology lay in the balance. Lacking unassailable evidence for a dynamical theory of heat, as for an etherial basis of forces, he could hardly adopt a physical theory without compromising the basic values that his science embodied.

Not only positive science, however, was at stake. Thomson took his anti-hypothetical stance supported by the much more general latitudinarian religious commitment shared with his father. Mathematical theory, like true religion, expressed only essential truths, the common doctrines upon which all men of sound judgement could agree. Only full conviction of the truth of the dynamical theory, therefore, could move him from the secure ground where electricity, magnetism, and heat were macroscopic states of bodies, otherwise unexplained, to molecular theory. In assessing William's new commitment one should keep in mind that in 1849 Dr Thomson, the moral rock of the family, suddenly died, and that James Thomson in the same year made his wrenching decision openly to disavow the Trinity. The direct relevance of this breakdown in the former stability of William's family life to the restructuring of his scientific position is difficult to assess, since he rarely betrayed his personal feelings, even in private correspondence. That his own reorientation involved a soul-searching examination of the grounds of his scientific beliefs, however, is readily apparent.

The evidence of struggle lies all around. What we have been calling the 'draft' consists in fact of at least three separate attempts to begin the dynamical theory. In the first attempt, Thomson tried to explain why for so long he 'carefully avoided committing myself by any decisive expression of my own opinion on the subject'.[63] In the second, he merely stated the content of the theory. Only in the third did he come to the argument from Davy's ice-rubbing experiment

[63] Draft, p. 3. New attempts begin on pp. 12 and 17.

against any material theory, which he thereafter took as a definitive argument for the dynamical theory.

The stages of this transition are significant. The first stage retained self-consciousness of the struggle and sought self-justification. There the ground of 'belief' in an 'hypothesis' was the main issue. As such, Thomson resolved it in explicitly religious terms, citing the nature of the Supreme Ruler and His creation to support both conservation and dissipation. The second stage distanced both the man and his religion from the new theory, while the third, by creating the myth of the crucial experiment, eliminated the need for any hypothesis and eradicated the four-year personal struggle, replacing it with a purely rational choice facing the scientific community. No matter that the crucial experiment was only seen to be so after fifty years, for it now served its purpose as rationalizer of scientific progress.

In the end, the symptoms of Thomson's passage would remain only in such artefacts as his testy attitude towards Clausius's priority in enunciating the basic propositions of the theory. Having wrestled for so long himself to establish that Carnot's coefficient must be independent of the working substance in a heat engine, and having ultimately derived that result from his religiously grounded dissipation axiom, Thomson felt some irritation that Clausius should get the credit, for in Thomson's eyes Clausius had based his reasoning on a mere hypothesis, and a false one at that. 'The memoir of Clausius', Thomson wrote in his third draft, 'contains a most satisfactory & nearly complete working out of the theory of the motive power of heat by Carnot's peculiar method of reasoning but hypothesis is so mixed with sound theory that the general effect is lost'. Both Clausius and Rankine had assumed 'certain hypotheses equivalent to this [essentially Mayer's hypothesis] that the "innere arbeit" of air is not altered by compression & that the heat evolved is the equiv[alent] of the work. Regnault's experimental data make any *hypoth[esis]* of the kind unnecessary; and, as far as they can be depended upon even betw[een] the limits 0 & 100° Cent[igrade], they show that in reality that hypothesis is very appreciably at fault. In the present paper the practical part is worked out without any hypothesis'.[64]

Thomson did recognize that Clausius had additionally enunciated an axiom (subsequently the Second Law of Thermodynamics) equivalent to his own, from which the result followed rigorously, but still in 1864, when he sent a set of notes on the history of thermodynamics to Tait for the latter's textbook, and when Clausius's hypothesis had been fully justified, he stated concerning Clausius's 1850 paper: 'It is *this* [paper] that gives Clausius his greatest claim. In it he takes Carnot's theory & *adapts it to true thermodynamics*. But he does so mixedly with a hypothesis [Mayer's] justly assumed as *probably APPROXIMATE* for gases. It is only by separating out what depends on this hypoth., from what does not

[64] *Ibid.*, pp. 22, 23.

that I was able to see that Clausius in his gaseous part does really use the Carnot pure & simple'.[65]

We thus see how important for Thomson's move towards molecular theory was his conviction that he had established the dynamical theory of heat on non-hypothetical grounds, however we may judge the grounds themselves. No sooner had he arrived at this all-important resolution of his conflicting commitments than he embarked on his life-long pursuit of a theory of matter that would unify all physical forces. The notes of his lectures of 1852–3 provide a sensitive barometer of the shift. One meets there, in sharp contrast to his lectures of 1849–50, speculations of the following kind: 'Probably in the friction of two solids, there is change of shape, and the heat produced is very probably on account of electrical currents'. Thomson now saw in the 'principle of mechanical effect' (soon the First Law of Thermodynamics or Conservation of Energy) a powerful probe of physical reality. Just as the principle connected measurable external work with the molecular motions of heat, so it would relate work with any other internal property of matter. The preceding remark on friction concluded a brief discourse on mechanical effect, which began: 'The principle of mechanical effect is axiomatic. All the effects produced by a working force are equivalent to the work [done] by the force. We require to measure the effects, and test their mech^{l} value'. Such statements occur throughout the notes. The measure of mechanical effect – now suggestively labelled 'mechanical value' – would be the ultimate measure of the value to man of every physical agency.[66]

In his second lecture of 1852–3 Thomson set the theme for the course, which as usual began with properties of matter as a preliminary to mechanics: 'Are magnetism, elect. heat and light matter? We must consider general principles of work & mechanical effects'. From several repetitions of this rhetorical formula, the students learned to generalize the lesson so recently internalized by their professor: 'Heat was made by Davy, but he can't make matter – Ergo – What can [the] rubbing do? Only inertia overcome or matter set in motion'. The task of natural philosphy was to discover the states of molecular motion which constituted the various physical forces. The new philosopher would begin with careful measurements on the balance sheet of mechanical effect in observable interchanges. That scheme constituted not a solution but a programme for research. In 1862 Thomson was still teaching the lesson of 1851 under 'Friction & Imperfect Elasticity': 'Theory of friction one of the most important in the whole of Natural Science. Shows what becomes of Energy lost. Force of Friction most probably due to Electricity. Friction may be deprived of a portion of its

[65] William Thomson, Research notebook on thermodynamics, 1864(?), NB52, ULC. The illegitimacy of Mayer's hypothesis had already been the basis for Thomson's and Tait's polemic against Tyndall in 1862–3, when Tyndall pressed Mayer's priority over Joule in the discovery of energy conservation (see Chapter 11, note 11). Thomson's intense personal struggle to enunciate a non-hypothetical basis for thermodynamics helps to explain his role in this dispute.

[66] Jack, 'Notes'.

equivalent heat, if we separate the two electricities & charge a Leyden Jar. Heat of discharge of Jar is a portion of heat of Friction'.[67]

Clearly one had to comprehend all the forces at once. With tremendous energy, and little order, the professor sent his students from one subject to the next and back again in search of the interrelations that would yield a unique order. Nowhere is Thomson's new style more evident than in his jumbled lectures of 1852. No neat analogy between one well-defined area and another, controlled by a fixed mathematical form, would yield the single mechanical system that would explain all the forces of nature. In his compulsive rush to find that all-encompassing system Thomson seems to have forgotten the careful pedagogy that marked his 1849 course and to have acquired the reputation for disordered exuberance that he thereafter maintained, for he skipped with no apparent motivation between mechanics, heat, electricity, elasticity, galvanic batteries, and the power of animated creatures to produce mechanical effect. One can only observe that he saw everywhere the manifestations of mechanical effect and saw them as a single problem.

We can now state succinctly the essence of Thomson's reorientation. Until 1847 his entire programme had been unified by the continuity equation, which subsumed heat conduction, distributions of electric and magnetic forces and hydrodynamics under a single mathematical analogy, the geometry of flow. But each area constituted a *separate* physical interpretation of the basic mathematical form. The unity was a unity of language rather than of things. The analogy nevertheless taught Thomson a great deal about the formal identity between statical and dynamical phenomena. In particular it taught him that the total potential of a static system of inverse square forces could always be translated into the form of *vis viva* in a flow system, and that the forces themselves could be derived as a gradient of the total potential. The analogy thus suggested that mechanical effect could be taken as the common measure of all of the mathematically similar phenomena. By 1847 Thomson had begun to exploit extensively this common measure and to seek its physical basis, though always through his macroscopic analogies and never publicly entertaining speculations on molecular reality. The dynamical theory of heat altered his stance. It necessarily turned the common macroscopic measure into a physical probe and a mathematical unity into a physical unity. The chief agent of that transformation was the principle of mechanical effect, for it required – in Thomson's mechanical philosophy – that any mechanical effect lost from the states of gross bodies should re-emerge in *mechanical* states of molecules and ether. The dissipation axiom contributed significantly to this demand because it required – again in Thomson's formulation – a molecular mechanics to explain the movement of mechanical effect from states of concentration to states of diffusion, or from states of gross bodies to states of molecules and ether.

[67] William Thomson, in David Murray, 'Lecture notes in *classe physica*, bench II, November, 1862', MS Murray 325, ULG.

One may picture Thomson's change of direction as the turn of a hinge, but a hinge with two joints is required: the first his extensive work on the mathematics of mechanical effect concentrated in 1847, and the second his adoption of the dynamical theory of heat in 1851. Such an image leads us to qualify the preceding emphasis on the differences between Thomson's pre-1847 and post-1851 styles, and to recognize that his turn from macroscopic physics to molecular physics involved a great deal of continuity as well as discontinuity. Indeed, his subsequent molecular theorizing was to constitute more nearly an attempt at describing the necessary macro-structure of molecules than a claim about their ultimate nature. Molecular physics did not imply speculative hypotheses. The dynamical theory of heat, for example, claimed only that heat consists in a state of molecular motion. Thomson specified no particular motion, and he reprimanded Rankine for tying the theory to a vortex hypothesis. We have seen in the previous section an important aspect of this continuing anti-hypothetical commitment. Believing that an ultimate mechanical theory would not involve static force as a primitive term, but only matter and motion, Thomson nevertheless included both static potential (internal work) and *vis viva* in the mechanical effect of thermal motions. To the degree that he continued to employ the descriptive distinction without further explanation, his molecules remained macroscopic objects with undefined internal structure and undefined interactions. Late in 1851 he began to formalize this macro-molecular terminology when he introduced in part five of 'On the dynamical theory of heat' the concept of 'mechanical energy' to describe the total mechanical effect internal to a body:

> The total mechanical energy of a body might be defined as the mechanical value of all the effect it would produce in heat emitted and in resistances overcome, if it were cooled to the utmost, and allowed to contract indefinitely or to expand indefinitely according as the forces between its particles are attractive or repulsive, when the thermal motions within it are all stopped; but in our present state of ignorance regarding perfect cold, and the nature of molecular forces, we cannot determine this 'total mechanical energy' for any portion of matter, nor even can we be sure that it is not infinitely great for a finite portion of matter. Hence it is convenient to choose a certain state as standard for the body under consideration, and to use the unqualified term, *mechanical energy*, with reference to this standard state; so that the 'mechanical energy of a body in a given state' will denote the mechanical value of the effects the body would produce in passing from the state in which it is given, to the standard state, or the mechanical value of the whole agency that would be required to bring the body from the standard state to the state in which it is given.[68]

Having thus defined mechanical energy as a state function (the familiar 'internal energy' of macroscopic thermodynamics), Thomson soon went on formally to differentiate 'statical' and 'dynamical' stores of mechanical energy, again in a macro-molecular fashion. Electrostatic and chemical energies, analo-

[68] MPP, **1**, 222–3.

gous to the gravitational potential of raised weights, he termed statical, presumably because electricity and the chemical elements, like gravitational matter, could not be created by man, whatever their underlying 'forces' might consist in. Light and heat, analogous to creatable motions of matter, or *vis viva*, he termed dynamical stores of energy. The principle of mechanical effect thereby became explicitly a principle of conservation of mechanical energy, static plus dynamic. Thomson afterwards proudly believed this to be 'the first division of Energy into two kinds'. Since the division was hardly unique among conservation advocates, we must emphasize 'Energy', and recognize that his pride lay in his having enunciated a new term for a newly generalized concept, independent of any particular hypothesis of atoms, forces, or other underlying reality. His doctrine of mechanical energy simply asserted that the entire, conserved store of mechanical effect in the world could be divided up into the spatial relations (forces) of permanent things labelled matter – statical energy – and the motions of matter – dynamical energy. By late 1852 he had adopted Rankine's suggested terms, 'potential' and 'actual' energy, which he would change to the now-standard 'potential' and 'kinetic' in 1862.[69]

Under Thomson's scheme of macro-molecular energy physics, the task for physical research was to study mathematically and experimentally the relations and transformations of the two forms of mechanical energy so as to specify with increasing precision what the structure of molecular matter had to be. Particular mechanical models were important heuristically, but also to guarantee the credibility of a theory as a mechanically possible one. Thomson would always require that they be, first, conceivable, in the sense of observable mechanical systems in the real world, and, second, empirically grounded, possessing no elements corresponding to purely hypothetical physical quantities. His original empiricist methodology of mathematical theory, which allowed only positive, macroscopically defined, physical entities, thus carried over into his molecular 'speculations' as a requirement that a mathematical theory be representable by a mechanical model and that the model be firmly grounded in empirical reality. We shall develop these ideas in detail in Chapters 12 and 13. First, however, we shall examine the theoretical foundation for the new energy physics that Thomson laid down in collaboration with Tait in their epoch-making *Treatise on natural philosophy*.

[69] *Ibid.*, pp. 505–10, 523, 541; NB52, ULC, 36; Jack, 'Notes'. In a personal annotation of 29th March, 1896, to MPP, **1**, 541, Kelvin noted: 'I had not by this time [June, 1853] replaced by "kinetic" Maquorn [*sic*] Rankine's "actual" (which I temporarily adopted from him for what I had previously called "dynamical energy")'.

11

T & T′ or Treatise on natural philosophy

I am getting quite sick of the great Book . . . if you send only scraps and these at rare intervals, what can I do? *You have not given me even a hint as to what you want done in our present chapter about statics of liquids and gases*! I have kinetics of a particle almost ready, nearly the whole of the next chapter, but I don't see the fun of paying 30/- [shillings] for sending the MSS to you [in Germany] for revision, when in all probability you won't look at it till some indefinite period when you are in Arran, where it would be *certain* of reaching you – and for 8d. Now all this is very pitiable: I declare you did twice as much during the winter as you are doing now. *P.G. Tait to William Thomson, 1864.*[1]

The classic *Treatise on natural philosophy* known as T & T′ began its tortured gestation late in 1861 when William Thomson (T) offered to join Peter Guthrie Tait (T′) in writing a textbook suitable for their respective natural philosophy classes at Glasgow and Edinburgh. A year earlier Tait had left his mathematics professorship at Queen's College in Belfast to succeed J.D. Forbes in the Edinburgh chair of natural philosophy. He brought ambitious plans for expanding Forbes's domain, and his own income, by attracting more students to a popular experimental course, but felt sorely the lack of a textbook. After six years of collaboration with Thomson a book did emerge, but, far from an elementary textbook, it consituted a highly creative treatise on the mathematical methods of physics. If few students in the experimental course can ever have read it, it nevertheless provided an entire generation of advanced students and Cambridge wranglers with a model for a new energy-based dynamics. The model reflected the Victorian mathematical style that Thomson shared with Stokes, Maxwell, and the many others who emphasized non-hypothetical, geometrical theory; but it also embodied specifically Thomsonian stylistic elements and idiosyncrasies. We shall highlight both aspects, while focussing on the role of energy in shaping the new dynamics.[2]

1 P.G. Tait to William Thomson, 20th June, 1864, T6X, ULC.

2 Since good selections of letters between Thomson and Tait regarding the writing of the *Treatise* have been published in SPT, **1**, chap. 10, and in C.G. Knott, *Life and scientific work of Peter Guthrie Tait* (Cambridge, 1911), chap. 5, we concentrate here on analyzing its content. Note, however, that Knott falsely ascribes most of the *Treatise* to Tait (see discussion below, note 25).

No existing text adequately represented the conditions for pursuing natural philosophy in the new context of energy conservation and the dynamical theory of heat. To supply the want in a non-speculative manner was the authors' primary goal. But some idea was required of a physical basis for the different forms of energy, and of the relation of energy to force, in order to structure the mathematical theory. Thomson thought that he had found the beginnings of such a physical foundation in a continuum theory of matter and forces, which formed an ever-present background to the writing. Another such background feature was the engineering orientation that we have seen so ubiquitously in earlier works and in the Scottish tradition of natural philosophy. Closely coupled with democratic ideals of Scottish education and with the diffusion of useful knowledge, this practical dimension arguably dominated Thomson's and Tait's conceptualization. And that conceptualization would establish a new set of canonical concepts for classical mechanics: kinematics separate from dynamics; dynamics inclusive of static as well as moving systems; energy as the fundamental entity in nature, rather than force; and extremum principles as the foundation of dynamics.

The collaboration

Although Thomson and Tait seem not to have been acquainted with one another much before 1860, James Thomson and Tait had been colleagues at Queen's since 1854. Together with Thomas Andrews (professor of chemistry and former pupil of Dr Thomson), they supported the reform movement associated in part with the College (heir to the collegiate wing of the Academical Institution) and with the Belfast Literary Society.

Tait had originally planned an elementary text in two volumes for a popular and largely experimental course, in line with his teaching needs and democratic ideals, as well as with the necessity of attracting fee-paying students. 'Experimental' referred, not so much to a laboratory course, as to a lecture course grounded on experimental demonstration, as opposed to mathematical deduction. The collaboration made possible an ambitious mathematical extension of the text, and Tait sent off a revised plan to the publisher, Macmillan: 'I proposed 3 volumes . . . and I expressly stipulated that *one* of these (or $\frac{1}{3}$ of the work, if the Exp[l] & Math[l] parts were to be *mixed*) should be devoted to Math[l] Physics – including all sorts of Potentials, Dynamical Theory of Heat &c &c. I said also that as we were both . . . at present *in training* we could turn out a volume by August next, another in Jan[y] or Feb[y]/63 & the third in the ensuing summer. The professors here, to whom I have mentioned the affair, are particularly enthusiastic, & I have little doubt of our increasing our students by 30 per cent (at least) if we do the business well'. Thomson had himself been contemplating a mathematical text for some time, originally as an inexpensive republication of some of his own papers in electricity, magnetism, and heat. But Charles Millar,

Peter Guthrie Tait, professor of natural philosophy in the University of Edinburgh and co-author with Thomson of T & T′. In his obituary, Thomson wrote of Tait: 'We never agreed to differ, always fought it out. But it was almost as great a pleasure to fight with Tait as to agree with him'. [From C.G. Knott, *Life of Peter Guthrie Tait*.]

who had been involved in the publication of J.P. Nichol's *Cyclopaedia*, advised him that he could not 'depend on a sufficient sale to risk the book as a *cheap* book. In the "*Nichol*" style, it would never do . . . Your volume is not intended so much for students who in fact only buy the books which they are forced to buy as Class Books and grumble even at that – but for highly cultivated men who do not grudge a price – but this class is small'. Millar proposed instead 'either a

"Systematic Work on Electricity & Galvanism &c" or a "Handbook of Physics". Either of these, I think, would sell & be profitable as they would interest a large class of readers'.[3] Such considerations apparently stimulated Thomson and Tait's projected third volume.

In addition, personal appeals came from a variety of friends for assistance in mastering the new energy concepts. Ludwig Fischer (1814–90), Thomson's Cambridge friend and now professor of natural philosophy at St Andrews University, wished to know in 1855 of 'any elementary work on Mechanics starting with the idea of "mechanical energy" or "work"'. He was reading Thomson's papers and Helmholtz's classic '*Ueber die Erhaltung der Kraft*' but required instruction in 'how I should read them or, rather, how I could get hold of the theory of mechanical energy, first as applied in ordinary mechanics and then to other Physical problems. I have commenced several times, but not having proceeded far enough I have derived no benefit from this desultory kind of study, and yet I feel the want of a clear insight in these doctrines most severely'.[4] Fischer's case is particularly revealing of the difficulty of adjusting to the new conceptual structure, for he had extensive knowledge of continental mathematics and physics. A native German, he had come to Cambridge after over three years of study in Berlin with Encke, Dirichlet, Mitscherlich, Erman, Dove, and others, and after three years in Paris with Poisson, Liouville, Dulong, Pouillet, Dumas, Savart, and company.[5] Yet he could not easily find his way about in Thomson's and Helmholtz's papers.

Tait thus anticipated overnight fame: 'I am convinced that if we get the *essence* of our lectures into two volumes & do a third on Math. Physics (the *unique* one as I call it in my note to Macmillan, and which I could never have ventured on alone) we shall make a great hit – besides being translated into Russian & perhaps Hindi & Chinese'. Concerning this '*unique*' volume he wrote to Andrews: 'I know of no such work in any language . . . Such a book is one I would willingly have paid almost any price for during the last ten years – but it does not yet exist. And I think Thomson and I *can* do it'.[6] Since the first two volumes were to contain 'simply the essence of the Glasgow and Edinburgh Experimental Lectures blended into (I hope) an harmonious whole', Tait imagined, with characteristic exuberance, that 'An average of three or four (or less) hours a day would give us the [first] volume in six weeks in such a state as to require little correction in the present state of the science'. Evidently he assumed he could control the timetable of these volumes, since he was to take on the primary

[3] P.G. Tait to William Thomson, 10th December, 1861, T2, ULG. Charles Millar to William Thomson, 5th December, 1859, M37, ULG.

[4] Ludwig Fischer to William Thomson, 20th October and 6th November, 1855, F101 and F102, ULC. See also the request for guidance in J.C. Maxwell to William Thomson, 13th November, 1854, M89, ULC, and J.J. Sylvester to William Thomson, 8th May, 1856, S601, ULC.

[5] Ludwig Fischer to William Thomson, 15th February, 1847, F83, ULC.

[6] P.G. Tait to William Thomson, 13th December, 1861, T3, ULG. P.G. Tait to Thomas Andrews, 18th December, 1861, in Knott, *Life of Tait*, p. 177.

responsibility for drafting them. 'The major part of the writing will be done by me', he wrote to Andrews on 20th January, 1862, 'as Thomson feels a repugnance to it which is not common'.[7] By that date, however, he should already have realized that Thomson's expansive imagination suffered no one's control. On 25th December Tait had sent him a list of publication 'postulates' which explicitly separated the experimental from the mathematical volumes. Thomson responded with two postulates of his own (no longer extant) which clearly compromised the entire scheme for popular, inexpensive, and rapid publication by mixing mathematical analysis into the experimental volumes. Tait's misgivings were prophetic:

> The only objection I see to this, but it is a grave one, is the *expense to the students*, especially the Scotch ones. We may mulct & bleed Oxford & Cambridge & Rugby &c &c to any extent, but how about our own classes? What we want *AT ONCE* is not the fame of authorship, but the supply of a want in elementary teaching. Now I have a vague feeling, which may soon become stronger, that our best course will be to issue a smaller, but thoroughly trustworthy volume or volumes *first* – suited for general educational purposes – and *THEN* come out with a really great work in a good many volumes on your enlarged idea of our first plan. In fact my impression is of this kind, do the *extensively useful*, but thoroughly accurate, popular book first – and then astonish the world with, at all events an attempt at, what it has not yet seen, a complete course of Natural Philosophy, Exp[l] & Mathematical.[8]

Ostensibly the two authors compromised, agreeing to insert only brief mathematical notes in the descriptive text. Tait reported to Andrews: 'No mathematics will be admitted (except in notes, and these will be more or less copious throughout the volume, being printed in the text but in smaller type). But we shall give very little in that way as my great object in joining Thomson in this work is to have *him joined to me* in the great work which is to follow, on the Mathematics of Nat. Phil., which I do not believe any living man could attempt alone, not even Helmholtz'.[9] Thus arose the distinction of 'large print' and 'small print' portions of the ultimate product. They corresponded to the popular experimental and the more advanced mathematical parts, respectively, of Thomson's natural philosophy course at Glasgow, on which he lectured at different hours. In the succeeding five years of writing, however, the small print would gradually engulf the large as the 'great work' swallowed the popular one. Experimental physics would barely see print of any size, and in over 700 pages T & T′ would not complete one-half of even their intended first volume. Nevertheless, their '*Principia mathematica*', as Tait predicted, was to 'go over Europe like a statical charge', replacing Newton's *Principia* of force with a new *Principia* of energy and extrema.

7 *Ibid.* P.G. Tait to William Thomson, 25th December, 1861, T6C, ULC. P.G. Tait to Thomas Andrews, 20th January, 1862, in Knott, *Life of Tait*, pp. 178–9.

8 P.G. Tait to William Thomson, 28th December, 1861, T6D, ULC.

9 Tait to Andrews, *op. cit.* (note 7).

Tait sent to Andrews the following table of contents for Volume I, in which we indicate in square brackets the chapters eventually published:

Section I.
Chap. I. Introductory. [Thomson's introductory lecture – omitted]
II. Matter, Motion, Mass, etc. [II]
III. Measures and Instruments of Precision. [IV]
IV. Energy, Vis viva, Work. [II]
V. Kinematics. [I]
VI. Experience (Experiment and Observation). [III]
Section II. Abstract Mechanics (*Perfect* solids, fluids, etc.).
Chap. I. Introductory (I have written this and will let you see it soon). [V]
II. Statics. [VI and VII]
III. Dynamics (Laws of Motion, Newton. Did you ever read his Latin? Do.). [II]
IV. Hydrostatics [VII] and [Hydro]Dynamics. [Never written]
Section III. Properties of Matter, Elasticity, Capillarity, Cohesion, Gravity, Inertia, etc. etc. (This is to be mine.) [Outlined by Tait; portions expanded by Thomson; never completed]
Section IV. Sound. [Never written]
Section V. Light. [Never written]

This will give you as good an idea as I yet possess as to the contents of our first volume. All the other physical forces [heat, magnetism, electricity, electrodynamics] will be included in Vol. II, which will finish up with a great section on the *one* law of the Universe, the Conservation of Energy.[10]

From beginning to end, in fact, T & T′ were to organize their book around energy, with other topics either supporting or deriving from that unifying concept. Their popular 1862 article 'Energy' in *Good Words* reflected the all-encompassing role they assigned to 'the ONE GREAT LAW of Physical Science, known as the *Conservation of Energy*'.[11] Even Newton's laws of motion, as Tait signalled in Section II, Chapter III of his outline, were to be reread from the Latin to bolster the doctrine of energy. We shall focus on that rereading below.

Of the outlined sections of Volume I, only the first reached completion. Statics and hydrostatics in Section II grew under Thomson's hand to cover over half of the published work, while dynamics, which should have formed the most extensive part, found expression only as general laws and principles. In essence the published 'fragment' consisted of four parts in two divisions:

[10] *Ibid.* This outline reflects Thomson's comments on a less detailed scheme in Tait to Thomson, *op. cit.* (note 7).

[11] William Thomson and P.G. Tait, 'Energy', *Good Words*, **3** (1862), 601–7. The article is written largely in Tait's polemical style and attacks both Tyndall's popularizations and his support of Mayer's priority over Joule in the discovery of energy conservation. The ensuing battle can be followed through the Tait–Thomson correspondence in 1863 and through a series of bitter exchanges in *Phil. Mag.*, [series 4], **25** (1863), 220–4, 263–6, 368–87, 429–31, and **26**, (1863), 65–7, 144–7. See Chapter 10, note 65, and Chapter 15.

Division I – 'preliminary notions'
  Kinematics (160 pages)
  Dynamical laws and principles (144 pages)
Division II – 'abstract dynamics'
  Statics of a particle (potential theory, 71 pages)
  Statics of solids and fluids (300 pages).

Tait had already outlined the never-completed section on 'Properties of matter' by mid-December of 1861 and sent it to Thomson along with the proof sheets for an article on 'Force'. He suspected trouble in these areas, writing to Andrews that 'A little difficulty arises at the outset, Thomson is dead against the existence of atoms; I though not a violent partisan yet find them useful in explanation – but I suppose we can mix these views well enough'. But of course they could not. Here we see in fact the second major feature, additional to energy, which shaped the *Treatise*. Beneath the surface lurked always Thomson's belief in a continuous substance as a substratum for atoms, molecules, and forces, or for matter and energy. His comments on the 'Force' article apparently showed Tait for the first time the extent of this commitment, for Tait responded: 'It is amusing to see how definitely you go into the ease of conception and treatment of the continuous uniform medium in which atoms (or at all events matter) are supposed to float. I am quite willing to adopt your views, but I should like you to send me as soon as you have leisure a little sketch of your proposed mathematical treatment of such a fluid or solid – or refer me to the works, Stokes' or others', in which it is found, – if already in print'.[12]

No such work existed, except as a visionary sketch in Thomson's private research notebook for 1858 and as suggestive hints in a paper of 1856 on the implications of Faraday's discovery of the magnetic rotation of light (ch. 12). During the writing of the great book, at first under the pressure of Tait's scepticism and then with his full collaboration, Thomson's speculations became increasingly comprehensive, exciting, and unwieldy. And as the goal of a unified physical theory of ether and matter loomed larger, the *Treatise on natural philosophy* became an attempt to lay its dynamical foundation. This imperative helps to explain the unique, often puzzling, character of the book. It also explains why the section on 'Properties of matter' was never completed. Lacking an adequate foundation in the dynamics of continuous media, T & T′ could hardly erect the structure of interconnected properties that the dynamics was intended to support. The *Treatise* remained, therefore, a programme, but one which merged with such complementary 'programmes' as Maxwell's *Treatise on electricity and magnetism* (1873) to provide a generation of British mathematical physicists with a coherent, if largely intuitive, goal for natural philosophy.[13]

[12] Tait to Andrews, *op. cit.* (note 6). Tait to Thomson, *op. cit.* (note 7). See note 25.

[13] For an excellent account of the 'Maxwellian' programme see J.Z. Buchwald, *From Maxwell to microphysics* (Chicago, 1985), Parts II and III. Cf. Thomson's programme in Chapter 13 below.

Clues to this dynamical theory of nature will emerge in the kinematics and dynamics of the *Treatise*. Its heuristic nature, characteristic of Thomson's style generally, has been well-expressed by one of the closest readers of the book, Joseph Larmor:

It was but rarely that his expositions were calculated to satisfy a reader whose interests were mainly logical; though they were always adapted to stimulate the scientific discontent and the further inquiry of students trained towards fresh outlook on the complex problem of reality, rather than to logical refinement and precision in knowledge already ascertained. Each step gained was thus a stimulus to further effort. This fluent character, and want of definite focus, has been a great obstacle to the appreciation of 'Thomson and Tait', as it is still to Maxwell's 'Electricity', for such readers as ask for demonstration, but find only suggestion and exploration.[14]

One reader, dismayed by this lack of logical rigour, was Pierre Duhem, whose critique of the broad but shallow mind of the British physicist in contrast to the narrow but deep French mind has become a classic. Thomson provided his best examples. Notwithstanding his polemical purpose, however, Duhem recognized the heuristic power of the British methods and the fact that they were grounded in practicality, which he associated with the factory mentality of the nation of shopkeepers. He could well have applied the same description to the *Treatise*, for a third major aspect of it is a thorough interpenetration of theoretical exposition and practical conception. In writing their textbook for Scottish students, neither author forgot his commitment to the diffusion of useful knowledge at an affordable price, nor his belief, however unrealistic, that even the most sophisticated subjects could be made accessible to willing students and 'practical men'. As Tait said to Andrews of his contribution to kinematics, 'It is all about Motion, Actual and Relative, and such matters as Rotations, Displacements, &c., and I hope to make the large type part of it intelligible even to savages and gorillas'.[15]

Progressive, reforming, and practical, Thomson and Tait aimed at an education suited to their latitudinarian, whig ideology, in which the esoteric and sectarian would bow to the democratic and meritorious, all in the service of national honour, wealth, and power. As far removed from theoretical dynamics as these issues may seem, they established preconditions even for such questions as the most suitable measures of energy and force, issues which penetrated deeply into the structure of the *Treatise*. A single example will capture the mood.

In an 1862 draft on 'Expenditure of work and generation of kinetic energy thereby', which Thomson wrote in his usual fashion as part manuscript and part letter to Tait, he began carefully with large and small print portions to establish the work–energy theorem. For a body moving along a line from position *a* to

[14] JL, lv.

[15] Pierre Duhem, 'Abstract theories and mechanical models', in *The aim and structure of physical theory*, translated by P.P. Wiener (New York, 1962), pp. 55–104 (originally published in 1906 from articles of 1904–5). P.G. Tait to Thomas Andrews, 9th September, 1862, in Knott, *Life of Tait*, p. 179.

position s, with velocity changing from u to v under a constant acceleration f, he showed that the distance travelled multiplied by the acceleration determines the change in the square of the velocity:

$$f(s-a)=\tfrac{1}{2}(v^2-u^2).$$

Remarking that he had been presenting this analysis in his small-print hour and that he had given 'a lot wh I wish I had got taken down, about $P=Wf$' (Newton's second law, with $P=$ force and $W=$ mass) he wrote out the relation between work done and change in kinetic energy,

$$P(s-a)=\tfrac{1}{2}Wv^2-\tfrac{1}{2}Wu^2,$$

adding the notation, 'Interpret &c'.

This orderly presentation introduced a number of scrawled pages interpreting the work–energy relation as an absolute kinetic measure for work, and thereby for force. In contrast to the common unit of force based on weight, which varied with gravitational force, the kinetic unit was universal, independent of geographical location (not to say local prejudice). Thomson set out his view of universality while waiting his turn to speak in a meeting 'to promote the rifling among the students' (the volunteer civil defence corps in which he held the rank of captain). Touched himself with a little of the 'rifle fever',[16] he wrote at furious speed with total abandon:

> Consider [the] unit in which $P=Wf$ is measured. It is the absolute unit. It is a universal unit *of force* wherever British or other national standards of mass & length can be carried & the earth's angular motion observed . . . There is a great advantage in using it in scientific expressions. The same as men of letters formerly found in using Latin as their language, practically a universal language. It has not the same advantage now, when men of letters know modern languages & is practically retained by only one branch of the old stock [the Catholic church] & for quite a different reason (preventing people from acquiring their learning &c &c. A cut & ten [below] roman catholic latin wh was understood tho' no names named). The expression of force & work in terms of the absolute kinetic unit has now & must always continue to have the same advantage (or greater) that said use of Latin formerly had. But it has also a corresponding disadvantage. It is not the vernacular. We keep it & shall keep it forever because it alone is or can ever be a universal language: but with truly X'ian' benevolence (how difft from the papistic spirit above alluded to) we translate every scientific result out of a language which not one of the 1000 intelligent mechanics of Glasgow (each with as good a head as any mathematician) would understand into terms perfectly appreciated by every inhabitant of this great prosperous & influential city. We do so by dividing by 32.2. . . .
>
> The above is my recollection of some twaddle interspersed with the $\int(u+ft)dt$, $P=Wf$ &c &c of today. But being the small print hour, the nimble fingers let it slip – which you will perceive must have been a great bon [boon] to mankind & the PROTESTANT RELIGION in particular.[17]

[16] PL, **1**, 448. SPT, **1**, 405–6, describes Thomson's involvement in the Volunteer Rifle Corps movement from 1859. Rankine and Thomson both held the rank of captain.

[17] William Thomson, 'Expenditure of work and generation of kinetic energy thereby' (19th November, [1862]), PA146B, ULC.

'Twaddle', perhaps; unguarded opinion, certainly; but no more telling portrait could be created of the mixture of liberal, latitudinarian and practical commitments – mixed with anti-catholicism – which informed Thomson's personal and professional life. Dividing by 32.2 refers to dividing the absolute kinetic unit, based on motion produced, by the acceleration of gravity at Glasgow, to yield a gravitational unit equal to the weight of a British pound, the measure employed by all intelligent mechanics of the 'great, prosperous, and influential city', whose grandeur supposedly depended on the free exercise of reason by each of its productive citizens.

This wild passage formed the nucleus of a section of the *Treatise* advocating Gauss's absolute unit of force and explaining its relation to weight (paras. 220–6). The rhetoric disappeared from the *Treatise*, just as it did from notes on Thomson's small-print hour by David Murray, whose 'nimble fingers' extracted from the 'twaddle' only the fact that 'our vernacular unit of force shall be the weight [of] a pound weight at Glasgow'.[18] But Thomson's sentiments were not so fleeting. Two features deserve emphasis, for they appear also in Murray's notes from the general lectures.

First, Thomson's rifle-corps patriotism and anti-catholicism expressed his notion of universality in the same way. He sought national unity in the unionist sense, meaning the union of Ireland with the rest of Britain under one set of standards and laws (chs. 1, 23). Catholicism, within his ideology, represented at once the tyranny of sects, parties, superstition, and authority. With less venom, but similar ideas, he attacked all other particularist impediments to national unity. Cambridge 'mathematicians' came under the same barrage:

> by the Cambridge mathematicians [Whewell and company] the unit of mass is defined as the mass whose weight is unity divided by the force of gravity. Hence we have the equation
>
> $$m = W/g$$
>
> *W* being the weight in the locality. This is an absurd distinction. Gauss' system has been adopted in this university. Our unit of mass shall be a national standard unit & shall be the same for all latitudes.[19]

Second, one of the former Glasgow mechanics whom Thomson had particularly in mind was James Watt. His steam-engine stood behind the analysis here just as it had behind Thomson's vision of absolute units of electricity and

[18] William Thomson, in David Murray, 'Lecture notes in *classe physica*, bench II, 1862–63', MS Murray, 326, ULG, p. 27 (for the 'small-print' hour). Thomson's remarks exemplify the climate of discord between such pro-science Cambridge liberal Anglicans as Charles Kingsley (who saw in Baconian ideology the advancement of Britain's prosperity under God) and anti-science Oxford Tractarians such as Cardinal Newman (who looked increasingly to the ancient Church of Rome for escape from materialism in all its forms). See W.E. Houghton, *The Victorian frame of mind, 1830–1870* (New Haven and London, 1957), esp. pp. 43–4; Jack Morrell and Arnold Thackray, *Gentlemen of science: early years of the British Association for the Advancement of Science* (Oxford, 1981), pp. 229–45.

[19] William Thomson, in Murray, *op. cit.* (note 18), p. 89. See pp. 165–9 for the absolute unit of energy applied to bullets.

temperature in the 1840s (chs. 7–9). A latter-day Watt, in the person of Thomson's brother James, stood too behind the reinterpretation of dynamics that the *Treatise* was to offer. Those two connections should remind us at the outset that practical engineering always involved economic considerations, and that wherever the steam-engine entered Thomson's thinking, there also the concept of work as labour value entered. Here he reminded his general class that work was 'labouring force' and that 'when Watt had so far perfected the Steam Engine as to be prepared to recommend its use for raising weights from Mines &c' he invented the units of horsepower as a comparative measure of its effectiveness 'on a safe commercial principle'.[20] Thomson's national and economic concerns with the measure of work thus went hand in hand as concerns with the wealth of the nation. They were in fact concerns with political economy.

We have identified three preoccupations of T & T′ which characterize the style of mathematical physics they pursued: energy, continuum theory, and practicality. The style of the book, however, requires a further comment. It was written under the loosest of relationships between the two authors and with the printer. Tait was to do the actual drafting of the final manuscript, working from notebooks posted back and forth between Glasgow and Edinburgh, in which each author entered additions and corrections to the other's rough sketches of assigned portions. The scheme worked well enough in the beginning, but as neither of them paid particular attention to order it soon went awry. Tait wrote in March, 1862, 'I couldn't find your note-book till this morning . . . It is a facsimile of my wife's House-Keeping Expense Book so that on tossing over the piles on my table I never thought of looking into *it*. Accident opened it this morning, else you might have wanted it for weeks'.[21]

To Tait, nevertheless, must go the credit for publication. Thomson had no concept whatsoever of schedules or deadlines. In addition, he seemed always to be either travelling to various spas on the continent with his very ill wife, making plans for the Atlantic telegraph, designing electrical instruments, running his Glasgow laboratory, or participating in geological controversies. He failed to respond to Tait's letters, sent bits and pieces of analysis, and failed to return proofs. Sometimes he rewrote entire sections on the proof sheets, adding more than the originals contained. This extemporaneous aspect contributed to the lack of focus that Larmor noted. The interminable delays also produced an explosion of costs, which thoroughly undermined the original intention of producing a cheap book for student use. The following excerpts from Tait's missals are typical:

20 *Ibid.*, p. 123; see also pp. 113–15 (for Watt). Generally, see M.N. Wise (with the collaboration of Crosbie Smith), 'Work and waste: political economy and natural philosophy in 19th century Britain', *Hist. Sci.* (forthcoming).

21 P.G. Tait to William Thomson, 27th March, 1862, T14, ULG.

[January 1862] I have received two very strange fragments from you lately (one today). I suppose the latter was written by you on an envelope addressed to some third party . . . and by him returned.

[March, 1863] What in the name of goodness are you doing with the proof sheets? Gray [the printer] is growling, and I am getting bewildered lest some of the scraps you are always sending should go astray till the last moment and then turn a sheet upside down when imprimatur is just about to be thankfully given. Another bother is that for the next two or three days I shall be very busy – but I must do double duty I suppose.

SEND THE PROOFS BACK THE SECOND DAY AFTER RECEIPT AT LATEST.

GOLDEN RULE.

[May, 1864] *Do* look alive with my M.SS. It should have been all in type this week. I fear that you have *not* gone, after all, to the Cavendish Hotel – the name of which is so ominous of all that you pretend to detest [elitism]. This, of course, in such a case won't find you either – so that I am at a dead lock; and must wait till you send some other indication.

Your recent comments on the Attr^{n}. of an Ellipsoid are monstrous – as you had the M.SS. in your hands for a week, and I made all the alterations you indicated *THEN* . . .

Send me my M.SS. & then go & see McMillan & account to him for deficits & extensions &c &c and promise him on *your* part (and then you need have no fears about adding *mine*) that Vol. I will be ready in the end of July. The small Vol. *MUST* appear in the end of October, else we shall have to republish the L.M.[22]

While nurturing their final product, T & T′ published a *Sketch of elementary dynamics* (the 'L.M.' above), with the notice that 'The Authors give in this Pamphlet a rough sketch of the fundamental notions of Dynamics, for the use of their Classes in the Session 1863–64'. It provided a cursory discussion of 'Kinematics' and 'Dynamical Laws and Principles', which would largely constitute Division I of the *Treatise*. A somewhat more elaborate publication of a portion of Thomson's Glasgow lectures on statics, entitled *Elements of dynamics*, supplemented the *Sketch* as an interim textbook (for Thomson at least) and provided the starting point for Division II of the *Treatise*. By 1864, however, they recognized that the 'great book' itself would never do as a textbook and therefore planned to publish separately a 'small Vol.' (referred to above) consisting mainly of the 'large print' of the *Treatise*. Thus originated the *Elements of natural philosophy*, replacing the earlier *Sketch* and *Elements*.[23]

In discussing the relation of these elementary works to the *Treatise*, C.G. Knott claimed that the *Sketch* 'was almost entirely the work of Tait', because he

[22] P.G. Tait to William Thomson, 15th January, 1862, T6G, ULC; 17th March, 1863, T37, ULG; 13th May, 1864, T6W, ULC.

[23] William Thomson and P.G. Tait, *Sketch of elementary dynamics* (Edinburgh, 1863); *Elements of dynamics* (edited by John Ferguson from notes of lectures by William Thomson) (Glasgow, 1863); William Thomson and P.G. Tait, *Elements of natural philosophy* (Oxford, 1873), hereafter '*Elements*' (available to their students already in 1867 but consisting only of the 'large print' of Division I).

wrote the manuscript. And, since the *Sketch* contains many of the central ideas of the larger work, Knott regarded Tait as their author. Subsequent writers have sometimes followed his lead.[24] Since archival materials suggest more nearly the opposite conclusion, we shall not discuss Tait's perspective in any detail. With respect to both kinematics and dynamics, Thomson had been teaching many of the leading ideas of the *Treatise* since the late 1840s. Of particular significance here is the relation of force and energy. Tait presented his ideas on the subject in his inaugural lecture at Edinburgh in 1860, giving no hint of the viewpoint of T & T′. Instead he followed Helmholtz's rendering of the conservation doctrine, based on point atoms and central forces, the basis that Thomson aimed to supersede with his ethereal continuum.[25] We proceed then to Thomson's views on kinematics and dynamics and their representation in the *Treatise*.

Kinematics

'We adopt the suggestion of Ampère', said T & T′, 'and use the term *Kinematics* for the purely geometrical science of motion in the abstract, independent of its causes'. From his earliest lectures at Glasgow, Thomson had introduced the divisions of natural philosophy with such remarks as, 'Mere motion does not belong to dynamics. The name given to this by the French is Cinematics from κινημα, the science of mere motion without reference to force. For instance in the steam engine, given the height of the piston and the whole stroke, to tell how far the flywheel will have turned round. *Thus Cinematics may be wrought by pure geometry*'.[26] Two concerns appear here in immediate interrelation: motion as pure geometry and motion as the geometry of machines. Importantly, in Thomson's conception the two subjects were one; he did not differentiate them as theory and application but unified them as theory and practice.

The idea of studying motion as pure geometry originated long before the nineteenth century. Descartes' entire theory of nature, for example, is today often labelled 'kinematical' because it was supposed to derive from *necessary* laws

[24] Knott, *Life of Tait*, p. 199.

[25] P.G. Tait, *The position and prospects of physical science: a public inaugural lecture* (Edinburgh, 1860), esp. pp. 29, 32. Compare his post-*Treatise* lecture at the Glasgow meeting of the British Association on 'Force', *Nature*, **14** (1876), 459–63; *Scientific papers* (2 vols., Cambridge, 1898), **1**, pp. 256–69.

[26] William Thomson and P.G. Tait, *Treatise on natural philosophy* (Oxford, 1867), p. vi; William Thomson, in William Smith, 'Notes of the Glasgow College natural philosophy class taken during the 1849–50 session', MS Gen. 142, ULG, p. 3. See also William Thomson, in James Thomson, 'Lecture notes from William Thomson's natural philosophy class, 1847–48', A7, QUB, p. 1, where 'cinematics' is 'the science of motion of masses connected in any mechanical way, as toothed wheels, cranks, connecting rods &c without reference to the agency which produces the motion'. Early in 1862, Thomson and Tait carried on a running war of words over 'cinematics' versus 'kinematics'. For example, P.G. Tait to William Thomson, 28th January, 1862, T6K, ULC: 'Well, be it Cinematics; (do you propose to say Sinematics?) I wonder where we got Kine, & Kyloes & Kangaroos, besides Kirk and Kirns &c &c – But I can't help looking on Cinématique & Conductibilité as being equally French and equally erroneous'.

of motion, developed as pure geometry. They required conservation of the quantity and direction of inertial motion of individual bodies and conservation of the total quantity of motion of two or more bodies during impact.[27] But the cartesian sense of the geometry of motion, as compared with Thomson's usage, confused kinematics and dynamics, for it conflated motion with the causes of motion, geometry with inertia and force, and the necessary with the contingent. Meikleham's lectures at Glasgow when Thomson was a student exemplify the distinctions that he sought to protect. Defining natural philosophy as 'the science of the pressure or motion of bodies, the science of the forces by which press. & motion are produced', Meikleham remarked: 'Everyone has had ideas of comparison of space passed over in conjunction with time . . . Motion [is] purely geometrical. Hence the connection between mathematics and natural philosophy. The amount of the force [is] measured by motion'. But he continued, 'Motion [of bodies is] discovered by trial and this is the distinction [between mathematics and natural philosophy]. The foundation of mathematical science is our own ideas, of Nat. Phil. it is experiment'.[28] Like Meikleham, Thomson insisted on distinguishing *a priori* reasoning from true natural philosophy: or metaphysics from physics, Descartes from Newton, and his own view of inductive science from that of William Whewell.

To see the special role kinematics played in Thomson's anti-metaphysical natural philosophy, it is helpful to delimit his kinematics as a uniquely acceptable form of metaphysics standing between two other forms which he rejected, the one simply false, the other misdirected. Taking metaphysics in general to connote the endeavour to establish universal truths from reasoning upon one's own ideas, the empirically minded Thomson recognized only one *true* metaphysics, namely pure mathematics. *Misdirected* metaphysics was the pursuit of mathematics for its own sake, or as abstract symbolic algebra (ch. 6). It was misdirected because it lacked utility, and thus wasted mental powers, but not because it gave false results. *False* metaphysics was speculative metaphysics, the pretension to establishing *a priori* truths about the external world. For Thomson, Hegel's *Naturphilosophie* epitomized its ill effects. S.P. Thompson tells of his reading aloud to his class the attack of the 'arrant impostor' on Newtonian mechanics, and exclaiming: 'If, gentlemen, these be his physics, think what his metaphysics must be!'[29]

A more immediate challenge from the speculative camp came from the idealistically inclined Whewell. Although Whewell had been among the first in Britain to introduce the term kinematics for the pure geometry of motion, his

[27] See R.S. Westfall, *Force in Newton's physics: the science of dynamics in the seventeenth century* (London and New York, 1971), chap. 2, for an account of Descartes' mechanical philosophy.

[28] William Thomson, 'Notebook of Natural Philosophy class, 1839–40', NB9, ULC. Meikleham's lectures (apparently the mathematical part of the course, assumed by Nichol upon Meikleham's illness) begin from the front, while Nichol's own lectures (the experimental part) begin from the back.

[29] SPT, **2**, 1123–4.

philosophy did not allow the sharp separation between *a priori* geometry and inductive natural philosophy that Thomson demanded. In his 'On the nature of the truth of the laws of motion' of 1834, as well as in later editions of the *Elementary treatise on mechanics* and in the *Philosophy of the inductive sciences*, Whewell derived the 'terms' of the laws of motion, 'the form which is necessarily true', from metaphysical axioms of causality, referring the reader 'to his own thoughts for the axiomatic evidence which belongs to them'. Only 'the meaning of the terms' depended on experience. 'The laws may be considered as a formula derived from *a priori* reasonings, where experience assigns the value of the terms which enter into the formula'. The law of inertia, for example, was supposed to follow from the axiom that no change can take place without a cause, supplemented by the contingent empirical fact that *time* alone is not a cause of change of velocity, but only *force*, defined with reference to *space*. While anxious to insist, on the basis of this empirical element, that the law of inertia had resulted from a long and difficult inductive process, Whewell stressed that the law depended on the 'fundamental ideas' of cause, space, and time, which organize our experience and are prior to it.[30] Like inertia, the other laws of motion also depended on antecedent metaphysical principles.

Avoiding such an infection of natural philosophy by false metaphysics, Thomson placed inertia strictly in the inductive domain of dynamics, isolated from kinematics. 'Kinematics', he emphatically stated in his 1862 introductory lecture, 'has nothing to do with Inertia'. As 'a branch of pure mathematics', kinematics was 'a part of [true] metaphysics'. Dynamics, by contrast, was 'a Science of Observation', conveniently called 'Mixed Mathematics – it is mathematics applied to the calculation of certain [empirical] properties of matter', among them inertia. Thomson's words echo those of his Glasgow professor J.P. Nichol who regarded the laws of motion as 'general facts'. Of the law of inertia Nichol had asserted, 'this law is not deduced from metaphysics, but from extensive observation'.[31]

In their *Treatise* T & T′ also took pains to counter such views as those of Whewell. Introducing Newton's laws as *Axiomata*, they reminded their readers:

30 William Whewell, 'On the nature of the truth of the laws of motion', *Trans. Cam. Phil. Soc.*, **5** (1835), 149–72 on pp. 166, 172, 153–5; appended as 'Essay I' to his *The philosophy of the inductive sciences, founded upon their history*, 2nd edn. (2 vols., London, 1847), **2**, pp. 573–94. A copy of Whewell's paper exists among the bound pamphlets originally in the possession of William Thomson and now in ULG. See also Whewell, *An elementary treatise on mechanics: designed for the use of students in the university* (5th edn., Cambridge, 1836), pp. 138–9, 160–1; *Philosophy*, **1**, book 3, chaps. 4, 7, quotation on p. 177; cf. his critiques, 'On the principles of dynamics, particularly as stated by French writers', *Edinburgh J. Sci.*, **8** (1828), 27–38, and 'Observations on some passages of Dr. Lardner's Treatise on Mechanics', *Edinburgh J. Sci.*, [new series] **3** (1830), 148–55. For a sympathetic reading of Whewell's view of induction and for the interpretation of his laws of motion given here see Menachem Fisch, 'Necessary and contingent truth in William Whewell's antithetical theory of knowledge', *Stud. Hist. Phil. Sci.*, **16** (1985), 275–314, esp. pp. 294–8. We thank Fisch for helpful discussion and references.

31 William Thomson, 4th and 5th November, 1862, in Murray, *op. cit.* (note 18), pp. 7, 9. William Thomson, 5th November [1839], NB9, ULC (from the back).

'As the properties of matter *might* have been such as to render a totally different set of laws axiomatic, these laws must be considered as resting on convictions drawn from observation and experiment, *not* on intuitive perception'. They regarded themselves as 'Restorers' of Newton, and so aimed to recover the true *Mathematical principles of natural philosophy*.[32] In this goal lay their uncompromising separation – their Newtonian separation – between the geometry of space and the motion of matter. We do not find in their kinematics, therefore, any reference to mass, momentum, or energy. Although modern texts on classical mechanics sometimes regard those quantities as kinematical, because they obey conservation laws which limit the possible motions of a system independently of the particular forces acting, they were to Thomson and Tait strictly dynamical, for they involved inertia and the character of force in general.

If we approach kinematics from the side of merely misdirected rather than false metaphysics, we recognize another essential characteristic of the *Treatise*. Thomson had long insisted that mathematical description should represent directly observable relations or processes. As he once remarked, mathematics 'is merely the etherealization of common sense', meaning that it represents what we see, hear, and otherwise perceive in everyday life, from the rise and fall of cotton prices on the Liverpool market to the complex sounds of a combined orchestra and chorus.[33] This view he shared fully with Tait and also with Maxwell, but his compatriots possessed a taste for speculation on the relation between the laws of mind and the laws of matter that Thomson found objectionable, for they believed that mathematical forms could serve as powerful extensions to physical theory, while he demanded that extension from physical systems through mathematical symbolism should proceed not one step further than physical intuition could reach. Tait's favourite mathematical subject, quaternions, went one step too far for Thomson (ch. 6).

Tait had already published several papers on quaternions before the collaboration began. He wrote to Thomson early in January, 1862, that he was 'grinding hard' at a more popular treatment 'to make it easily intelligible'. He naturally expected that the subject would find a significant place in their

[32] *Treatise*, pp. 178, vi. The strong statement against intuition emerged from a negotiation between Thomson and Tait. While writing the original section on laws of motion, Thomson had been immediately concerned with the *conventional* aspects of inertia and force, particularly with the conventionality of equal units of time and of rest, which made the laws unprovable, although they were empirically adeqvate. By adopting other conventions, contradictory laws would emerge. He therefore regarded the laws as self-evident results of experience once the conventions had been adopted. 'Alas alas', he concluded, 'Behind all is the convention "what is force". It too is merely relative; but in our 1st Ed^{n} perhaps we had better prop up the tottering & doomed systems which make it absolute'. Tait apparently agreed, for he wished to stress the empirical foundation of the laws and to downplay their conventionality. See the series of notes in Thomson's draft, recasting Newton's second chapter, 'Axiomata sive Leges Motûs', NB47, ULC, which Tait received on 16th January, 1862, according to letter T6H, ULC. Compare *Treatise*, pp. 178–80, which follows closely, often verbatim, the original draft.

[33] William Thomson, 'The six gateways of knowledge' (3rd October, 1883), PL, **1**, 253–99, on p. 273.

kinematics as the basis for treating displacement, velocity, and acceleration as vectors, and that it would be useful in the small print throughout the volume. After receiving a letter later in the month from Sir William Rowan Hamilton himself (inventor of quaternion analysis) asking, 'whether I would think it a bore if he would arrange his ideas on Quaternion Differentials by writing me a series of long letters on the subject', Tait became increasingly committed to it.[34] He wrote to Thomson in March, 1863, for example, of a quaternion transformation which 'is really of the greatest value (as witness all your, Maxwell's, Neumann's, etc., attempts at potential function expressions for distributions of magnetism, electricity, static or kinetic, etc., etc.)'. Setting down a one line quaternion relation, he exclaimed, it 'contains all about potential and will go into the Book in splendid style'. One month later he wrote: 'I find that in your "Mechanical Representation of Electric etc. Forces" you need never have taken the *rotations* of the solid – you can always get *displacement* suiting any of the forms of force, and they are in fact simpler than yours as given in the Mathe. Journal. I will send you a proof to-morrow so that you may advise on it before it goes to press'. He now expected that quaternions, along with Thomson's spherical harmonics, would go into the *Treatise* as an appendix.[35]

Thomson's response can be surmised. By 1863 no single aspect of the underlying dynamics of the ether seemed more certain to him than that its rotations constituted magnetism. He had won that result through long and hard physical–mathematical reasoning (ch. 12). No set of symbolic abstractions, however seductive, could shake either his physical intuition or his belief that symbolic reduction destroyed physical insight. Tait had apparently fallen prey to the evils of symbolism, for in its published form his paper simply gave Thomson's own analysis in quaternion notation, including the rotational form for magnetism.[36] Writing much later of their collaboration, Thomson recalled:

> We had a thirty-eight years' war over quaternions. He had been captivated by the originality and extraordinary beauty of Hamilton's genius in this respect; and had accepted, I believe, definitely from Hamilton to take charge of them after his death, which he has most loyally executed. Times without number I offered to let quaternions into Thomson and Tait if he could only show that in any case our work would be helped by their use. You will see that from beginning to end they were never introduced.[37]

'Unfortunately', one may remark, in agreement with Maxwell who regretted in his review of Thomson's and Tait's *Elements of natural philosophy* (1873) 'that the authors, one of whom at least is an ardent disciple of Hamilton, have not at once

34 P.G. Tait to William Thomson, *op. cit.* (note 26); 11th January, 1862, T6F, ULC.

35 P.G. Tait to William Thomson, 30th March and 24th April, 1863, in Knott, *Life of Tait*, pp. 182–3. On 28th July, 1863, T42, ULG, Tait planned to send the appendix on quarternions within a week.

36 P.G. Tait, 'Note on a quaternion transformation', *Proc. Royal Soc. Edinburgh*, **5** (1863), 115–19; *Scientific papers*, **1**, 43–6.

37 William Thomson to George Chrystal, 13th July, 1901, in SPT, **1**, 452–3; Knott, *Life of Tait*, p. 185.

pointed out that every displacement is a vector, and taken the opportunity of explaining the addition of vectors as a process, which applied primarily to displacements, is equally applicable to velocities, or the rates of change of velocities [i.e. to the whole of kinematics]'.[38]

If Thomson and Tait differed on how far mathematical form could extend physical understanding, their disagreement only enhanced their common goal of making mathematical forms express physical forms. Kinematics, the geometry of their subject, would describe lines, surfaces, and solids not as given objects but as processes by which lines, surfaces, and solids were generated. In reviewing the second edition of T & T′, Maxwell remarked: 'The guiding idea . . . which, though it has long exerted its influence on the best geometers, is now for the first time boldly and explicitly put forward, is that geometry itself is a part of the science of motion, and that it treats, not of the relations between figures already existing in space, but of the process by which these figures are generated by the motion of a point or a line'.[39] We have noted previously the importance of kinematics in the emergence of the Cambridge style of mathematical physics (ch. 6). A more extended survey of the heritage of kinematics will further illuminate the new style by displaying its deep interrelation with practical mechanics.

In calling originally (1834) for a separate science of motion, considered independently of force, Ampère had borrowed his idea from French engineers such as Lazare Carnot writing on '*le mouvement considéré géométriquement*' and Lanz and Bétancourt whose *Essai sur la composition des machines* was adopted for the *Ecole polytechnique* in 1808. In fact, Gaspard Monge had already in 1794 called for two months of study in the first year at *Polytechnique* devoted to the elements of machines: 'By these *elements* are to be understood the means by which the directions of motion are changed'. Thus Ampère required a new definition of a machine; rather than 'an instrument by the aid of which one may change the direction and the intensity of a given *force*' it should be 'an instrument by the aid of which one may change the direction and the velocity of a given *motion*'. With respect to machines, then, one would study the modes of interconnection of parts and the motions they allowed. He generalized to any material system: 'kinematics must above all concern itself with the ratios which exist between the velocities of the different points of a machine, and in general, of any system of material points, for all the motions of which this machine or this system is capable; in a word with the determination of what one calls *virtual velocities*'. The study of possible motions, independent of forces, would prepare the student for

[38] J.C. Maxwell, *Nature*, **7** (1873), 399–400; *The scientific papers of James Clerk Maxwell*, W.D. Niven (ed.) (2 vols., Cambridge, 1890), **2**, pp. 324–8, on p. 327.

[39] J.C. Maxwell, *Nature*, **20** (1879), 213–6; *Scientific papers*, **2**, pp. 776–85, on p. 777. For the mathematical context in which Maxwell's remarks should be seen, see J.L. Richards, 'Projective geometry and mathematical progress in mid-Victorian Britain', *Stud. Hist. Phil. Sci.*, **17** (1986), 297–325.

the principle of virtual velocities as the foundation of statics, the study of forces in equilibrium, or of 'force independent of motion'.[40]

Ampère's suggestions for the general science of motion gradually found expression in textbooks on theoretical mechanics at the *Ecole polytechnique*, though in limited form. Thus J.M.C. Duhamel in 1845, in a book of 700 pages, included twenty-six pages of preliminary material: the projection of lines on coordinate axes, change of coordinates, rigid body rotations, and combined translational and rotational motion. He expanded this section in 1853 into a short second book, not explicitly labelled kinematics, but described as 'movement considered independently of its causes' and inserted between Statics and Dynamics. Finally, in the third edition of 1862 (coincident with Thomson and Tait's beginnings), he moved the section to the front as an eighty-six page introduction to the entire subject of *Mécanique*. Here it began with linear velocities and accelerations, their compositions and transformations, and went on to a much expanded analytic treatment of rigid-body motion, including such topics as the instantaneous axis of rotation of a body moving arbitrarily.[41] These are problems typical of what Cambridge mathematicians such as Cayley, Ellis, and Thomson had called kinematics in the 1840s. Indeed, Thomson had been referring his students to Duhamel's text since his earliest lectures, perhaps having become acquainted with it in Paris in 1845. But Duhamel lacked the emphasis on machines that Ampère had given to *cinématique* and that Thomson always associated with it. Thomson defined 'cinematics' in 1847 as 'the science of motion of masses connected in any mechanical way, as toothed wheels, cranks, connecting rods &c [without reference to the agency which produces the motion, *added*]'.[42]

A second French text, by C.E. Delaunay, also went into its third edition in 1862, advertising that it contained all the theoretical material for the course on mechanics and machines at the *Ecole polytechnique*. It too began with an introductory book on the geometry of motion (ninety-nine pages), explicitly labelled '*cinématique*', and covering much the same material as Duhamel. Delaunay, however, reflecting his interest in machines, presented a much more descriptive

[40] André-Marie Ampère, *Essai sur la philosophie des sciences, ou exposition analytique d'une classification naturelle de toutes les connaissances humaines* (Paris, 1834), pp. 50–3. For the mathematical context in France see L.J. Daston, 'The physicalist tradition in early nineteenth century French geometry', *Stud. Hist. Phil. Sci.*, **17** (1986), 269–95, esp. p. 283 (for the new science of motion).

[41] J.M.C. Duhamel, *Cours de mécanique de l'école polytechnique*, 1st edn. (Paris, 1845); 2nd edn. (2 vols., Paris, 1853–4), on p. iii; 3rd edn. (2 vols., Paris, 1862). These volumes reproduce Duhamel's lectures at the *Polytechnique* and depend heavily on the development of mechanics, especially the theory of couples, by Louis Poinsot, whose book, *Théorie générale de l'équilibre et du mouvement des système*, Thomson had apparently purchased in 1840. See the books and prices listed at the rear of Thomson's notes from Nichol's natural philosophy class (beginning with 21st January, 1840), NB10, ULC.

[42] William Thomson, in James Thomson, 'Notes', on pp. 1, 35. William Thomson, in William Jack, 'Notes of the Glasgow College natural philosophy class taken during the 1852–53 session', MS Gen. 130, ULG, lecture 2.

and geometrical treatment. Where Duhamel gave analytic geometry and no diagrams, Delaunay gave diagrams on nearly every page and very little differential calculus. He concerned himself especially with how various motions might be envisaged and with their resulting geometrical description.[43] Thus Delaunay developed *cinématique* more nearly in the way Ampère had envisaged, as a study founded on the motions of machines, but even he did not generalize the subject as a science of mechanism.

For machines, Ampère found his expositor not among his countrymen but in Robert Willis. Willis's book on *Principles of mechanism, designed for the use of students in the universities, and for engineering students generally*, was very probably the source of both Thomson's and Tait's first acquaintance with kinematics as a developed subject.[44] Tait attended Willis's course at Cambridge in 1849. Thomson was also on familiar terms with Willis, but no evidence exists that he attended Willis's lectures. No doubt Hopkins's friendship with Willis helps to explain the rapid dissemination of his views among Cambridge graduates in the 1840s and 1850s.[45] Maxwell, for example, wrote to Thomson from Marischal College, Aberdeen, on 30th January, 1859, 'This year I introduced practical kinetics [read kinematics] in the form of toothed wheels, cranks, Hooke's joint &c., which we studied with respect to their *motion* only, bringing in the forces in a different set of lectures. I devoted a little time to more theoretical matters such as the motion of the nail of a wheel tracing a cycloid, an ellipse traced by epicycloidal motion &c'.[46]

Willis had set as his object 'to form a system that would embrace all the elementary combinations of mechanism and at the same time admit of a mathematical investigation of the laws by which their modifications of motion are governed'. Focussing his theory on the '*relations of motion* between two pieces', he gave a complete taxonomy of mechanical connections.[47] The importance of Willis's theory of mechanism for the kinematics of T & T′, however, lay not so much in the theory of mechanical connections *per se* (which they called either constraints or kinematical conditions) as in the general idea of

[43] C.E. Delaunay, *Traité de mécanique rationelle*, 1st edn. (Paris, 1856); 2nd edn. (Paris, 1857); 3rd edn. (Paris, 1862). See also his *Cours élémentaire de mécanique théorique et appliquée*, 4th edn. (Paris, 1857), which treats machines entirely but does not take up kinematics as such. Thomson refers to various papers of Delaunay in his notebooks (e.g., 6th November, 1844, NB33, ULC, p. 102) but not to the textbooks, although it seems very likely he was familiar with them.

[44] See Chapter 6 above; T.J.N. Hilken, *Engineering at Cambridge University, 1783–1965* (Cambridge, 1967), pp. 50–7.

[45] *Treatise*, 2nd edn. (2 vols., Cambridge, 1879), **1**, p. 152*n*. Robert Willis to William Thomson, 14th November, 1849, W129, ULC; William Thomson to G.G. Stokes, 19th November, 1849, K37, Stokes correspondence, ULC. Stokes attended Willis's lectures in 1837 and maintained contact with him. See PA25 and PA26 in the Stokes Papers, ULC, as well as Stokes's correspondence with Willis. For Hopkins and Willis see William Hopkins to William Fairbairn, 25th April, 1851, in William Pole (ed.), *The Life of Sir William Fairbairn, Bart.* (London, 1877), p. 290.

[46] J.C. Maxwell to William Thomson, 30th January, 1858, M11, ULG.

[47] Robert Willis, *Principles of mechanism, designed for the use of students in the universities, and for engineering students generally* (London, 1841), pp. xii–xiv.

treating geometry mechanically. As Willis pointed out, kinematics in his sense would include 'the description of all the mechanical curves, as epicycloids and conchoids [Maxwell's 'more theoretical matters'] . . . a great mass of matter that has hitherto been classed with geometry'.[48] The kinematics of the *Treatise* consisted of geometry as produced by free and constrained mechanical motions.

From this perspective we can understand why, on page four, the unsuspecting reader encounters, not an elementary discussion of the resolution of velocity into cartesian components, but an unfamiliar term for a complex subject, the 'tortuosity' of curves. The topic extended to three dimensions an already sophisticated analytic treatment on page three of curvature in a plane. Ordinarily, three-dimensional curves were labelled curves of double curvature, to indicate that the curvature could be described by projections on two perpendicular planes. 'The term "curve of double curvature" ', scoffed Thomson and Tait, 'is a very bad one . . . The fact is, that there are not two curvatures, but only a curvature . . . of which the plane is continuously changing, or twisting, round the tangent line; thus exhibiting a torsion. The course of such a curve is, in common language, well called "tortuous"; and the measure of the corresponding property is conveniently called *Tortuosity*'. In their *Elements* they substituted an illustration for two small-print pages of analytic geometry in the *Treatise*: 'The simplest illustration of a tortuous curve is the thread of a screw'.[49]

Since a screw is also one of the simple machines, it is clear that the student should already have learned two lessons: mechanism was the place to begin learning the geometry of motion; and an appropriate (i.e. physical) choice of descriptive parameters in the simple case would yield, via analysis, deep insight and a powerful grasp of complex cases. The problem of cords presents a particularly significant example. Moving quickly through the total curvature and tortuosity of such figures as the epicycloid, T & T′ remarked: 'The use of a cord in mechanism presents us with many practical applications of this theory . . . We shall say nothing here about such cases as knots, knitting, weaving, etc., as being excessively difficult in their general development, and too simple in the ordinary case to require explanation'. One might suppose that this statement referred transparently to the mechanisms for handling threads employed in Walter Crum's Thornliebank textile mills. Partly it did; but at a deeper level it referred to the passion the two authors had developed for complex vortex filaments in a continuous fluid, the famous vortex atoms. This underlying meaning, while inaccessible to uninitiated readers, certainly enlivened the book for Thomson's and Tait's own students. In the second edition they enlightened their larger audience slightly by adding: 'we intend to return to the subject, under vortex-motion in Hydrokinetics'.[50]

Having reached page ten, the attentive reader would already have acquired a

[48] *Ibid.*, p. xii. [49] *Treatise*, pp. 3–4; *Elements*, p. 3.
[50] *Treatise*, p. 9; 2nd edn., pp. 1, 10.

thorough initiation into the idiosyncratic kinematics of Thomson and Tait. He would then have encountered more traditional discussions of velocity and acceleration, their resolution and composition, and expression in relative co-ordinates, albeit not without meeting such unusual topics as the 'Hodograph' of Hamilton (the path of a point in velocity space) and 'Curves of Pursuit', produced by 'the old rule of steering a privateer always directly for the vessel pursued'.[51] Both topics reminded the reader that curves were produced by motions and that the character of a curve could be expressed in terms of the process, or strategy, for producing it.

Succeeding sections signalled that Thomson and Tait were writing a treatise on dynamical theory as the presently realizable foundation of *all* natural philosophy, and not, as in previous texts, on a branch called mechanics bearing little relation to the other branches. They aimed at a thorough dynamical theory of heat, light, electricity, magnetism, and electrodynamics. Kinematics was to include the mathematical techniques basic to those 'motions' as well as to the standard mechanical ones. For Thomson, especially, that requirement pointed to Fourier's theorem of harmonic analysis, on which he contributed a twenty page development.

Beginning with the idea of a point moving uniformly in a circle, and defining simple harmonic motion as the projection of its motion on any diameter, he illustrated the definition with circular planetary motion and with 'those common kinds of mechanisms, for producing rectilineal from circular motion, or *vice versa*, in which a crank moving in a circle works in a straight slot belonging to a body which can only move in a straight line'. He proceeded with gradually increasing complexity through compositions of simple harmonic motions in a line – with different amplitudes, phases, and periods – to similar compositions in two dimensions, supplying graphical illustrations and the promise of describing afterwards 'mechanical methods of obtaining such combinations'. Finally, referring to its importance as 'an indispensable instrument in the treatment of nearly every recondite question in modern physics', from rotation of the plane of polarized light by crystals and magnetic fields to the propagation of telegraph signals and the conduction of heat by the earth's crust, Thomson gave the general Fourier theorem: 'A complex harmonic function, with a constant term added, is the *proper* expression, in mathematical language, for any arbitrary periodic function; and consequently can express any function whatever between definite values of the variable'.[52]

We emphasize the evaluation '*proper*' because its use here captures the tone of the entire kinematics. It signifies Thomson's lifelong attachment to Fourier's mathematical style and methods. Fourier showed how to reduce the most arbitrarily complex of periodic functions to a series of terms which any thinking

[51] *Treatise*, pp. 24–8, on p. 27. Dr Thomson had discussed the curve of pursuit in his *Introduction to the differential and integral calculus* (Belfast, 1831), pp. 159–61.

[52] *Treatise*, pp. 35–56, on pp. 37, 47, 50, 51. Our emphasis.

person could easily visualize; that is, which anyone could reproduce for himself, either mentally or graphically, but especially mechanically. He thereby made many of the most recondite problems of physics not only simple but, in present-day terminology, 'computable', mechanically calculable and thereby practically meaningful. Thomson placed great store in just this sense of proper analysis, for example in his evaluation of Archibald Smith's application of 'Fourier's grand and fertile theory' to the problem of ships' magnetism and its effect on the compass, a problem that had defeated both Poisson and the Astronomer Royal, G.B. Airy (ch. 22). 'To facilitate the practical working out of this analysis', Thomson remarked, 'he gave . . . tabular forms and simple practical rules for performing the requisite arithmetical operations'. As a result, the compass deviations of 'every ship in Her Majesty's Navy, in whatever part of the world' were subjected once a year to harmonic analysis, which had 'undoubtedly done more than anything else to promote the usefulness of the compass, and to render its use safe throughout the British Navy'. Not least of Smith's accomplishments, in Thomson's view, was his invention of the 'dygogram', a mode of representing important Fourier components in the resultant of the earth's and the ship's magnetic force for all orientations of the ship, a representation which could be produced practically by a simple form of epicycle and deferent mechanism, or by one circle rolling on another.[53]

The grounding of Thomson's thinking about kinematics in general, and Fourier analysis in particular, in the practical theory of mechanism is illustrated most clearly in his work on a calculating machine for predicting the tides, which he appended to the second edition of the *Treatise*. As the prime mover behind the 'Committee of the British Association appointed for the purpose of promoting the extension, improvement, and harmonic analysis of tidal observations', meeting from 1867 to 1876, he designed a 'kinematic machine' for predicting not merely 'the times and heights of high water, but the depths of water at any and every instant, showing them by a continuous curve, for a year, or for any number of years in advance'. The machine (figure 11.1) superposed numerous harmonic components of tidal motion, from the 'mean lunar semi-diurnal' to the 'luni-solar quarter-diurnal, shallow water tide'. Its use depended on a separate harmonic analyzer for determining the amplitudes of these various components from an actual plot of the tidal variations for any particular port, as registered by a tide gauge. Thomson had based the harmonic analyzer on 'a new kinematic principle', the so-called disc-globe-and-cylinder integrator, discovered by his brother. James Thomson's integrator provided the starting point as well for William's designs of other 'continuous calculating machines' (analogue computers) for solving simultaneous linear equations and integrating differential equations. He included the entire series in the new appendix to the *Treatise*.[54]

53 William Thomson, 'Obituary notice: Archibald Smith, and the magnetism of ships', *Proc. Royal Soc.*, **22** (1874), i–xxiv; MPP, **6**, 306–34, on p. 313.

54 *Treatise*, 2nd edn., **1**, pp. 479–508, on pp. 483, 479, 481, 488. See also William Thomson, 'The tide gauge, tidal harmonic analyser, and tide predicter', *Proc. Inst. Civil Eng.*, **65** (1881), 2–31, 58–64;

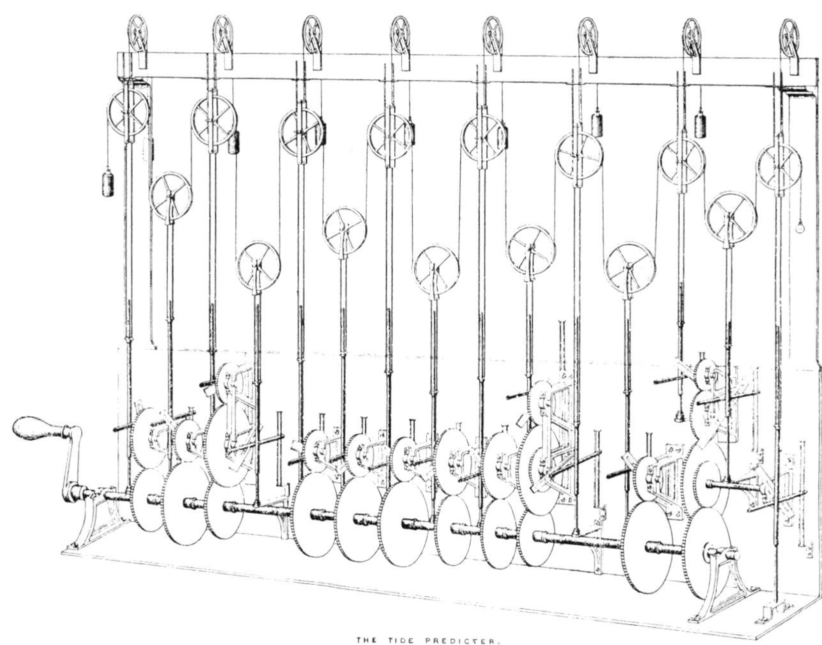

Figure 11.1. Thomson's most sophisticated tide predictor superposed fifteen simple harmonic motions, each produced by a gearing mechanism attached to the main shaft and each driving a 'tidal shaft' with an attached pulley, which affects the length of the continuous wire running over and under the whole series of pulleys. The weight at the left, which rises and falls as the tide would, is replaced by an ink bottle and pen to record the predictions.

Other kinematical subjects that naturally found extensive development in the *Treatise* were those that had originally occupied Thomson and Stokes in the 1840s, especially the continuity equation and the description of strain in solids and fluids. They would figure prominently in any formulation of the laws of electricity, magnetism, heat, and light, for they described the macroscopic motions of the ether. Thomson now added an extensive development of 'Laplace's coefficients', renamed 'spherical harmonics', which he had come to see as nearly as beautiful in their symmetry and visualizability as Fourier series, requiring for their full appreciation only a more intuitive presentation than Laplace had originally given them. For problems exhibiting spherical symmetry, the spherical harmonics provided the natural and 'proper' solution of the restricted continuity equation for inverse square forces and steady-state fluxes (Laplace's equation),

$$\text{divergence } \vec{F} = \text{divergence } (-\text{gradient } V) = 0.$$

MPP, **6**, 272–305, figure between p. 304 and 305. The reports of the BAAS committee appeared in the annual volumes of 1868, 1870, 1871, 1872, 1876 and 1878.

They gave power-series solutions in the radius *r* from the centre of symmetry, with coefficients varying harmonically in the angle variables for any sphere about the centre. Thus they behaved in their angular dependence much like Fourier series.

It is Thomson's attitude to the mathematical problem, however, that is of particular interest. Noting that an absolutely general analytical investigation of the equation could probably be found, he nevertheless asserted that 'with us the only value or interest which any such investigation can have, depends on the availability of its results for solutions fulfilling the conditions at bounding surfaces presented by physical problems'.[55] Even in the most esoteric aspects of kinematics, then, the practical and the physical exercised their hegemony over mathematical generality and rigour. Actual mechanical contrivances, whether of nature or art, lurked just behind the mathematical forms as the moving systems which realized their geometrical properties.

The background to dynamics

As the pure geometry of motion, kinematics corresponded to the natural history stage of natural philosophy (ch. 5). 'In the progressive knowledge of nature', Thomson regularly remarked in his introductory lecture, 'we classify facts and then we reason upon those facts. These two stages of science are designated by the name of Natural History and Natural Philosophy'. The transition to natural philosophy required a general theory of the causes of motion, of the operation of forces on matter. That science had usually been called mechanics, subdivided into statics, or forces producing rest, and dynamics, or forces producing motion. But what principles constituted the science of mechanics in the mid-nineteenth century?

It is all too easy to suppose that mechanics had long-since acquired an unproblematic foundation in Newton's three laws of motion, and that more sophisticated formulations, such as the principle of virtual velocities and d'Alembert's principle, were derived from the Newtonian basis. But French treatises and textbooks typically gave no set of axioms like Newton's. The British did give three axioms but usually not Newton's. Finally, the relation between statics and dynamics had become problematic in Britain but not in France. This chaotic state formed the backdrop for Thomson's and Tait's reinterpretation and reunification of mechanics. Appropriating the term 'dynamics' for the science of force, whether producing rest or motion, they would 'restore' Newton's laws to their proper place.

Newton's laws of motion, stripped of interpretation, may be labelled:

[55] *Treatise*, p. 141. For Thomson's authorship of the spherical harmonics see P.G. Tait to William Thomson, 25th March and 27th May, 1863, T39 and T40, ULG.

(I) inertia;
(II) $\vec{F} = m\vec{a}$ (moving force = mass × accelerating force);
(III) action = reaction (for moving forces and attendant changes of momenta).

Among the classic texts preceding T & T′, Lagrange's *Mécanique analytique* held first place. In it no list of laws quite like the above appeared. Lagrange gave a historical summary of various pieces of fundamental doctrine, contributed by authors from Galileo to himself, without attempting to distil the minimum number of axioms. For comparison with the Newtonian scheme, Lagrange's historical presentation of dynamical principles may be summarized as consisting of three parts:[56]

(I) inertia (Galileo);
(II) (*a*) composition of velocities follows the parallelogram rule (Galileo);
(*b*) measure of accelerating force = acceleration (largely Newton);
(*c*) moving force = mass × acceleration (Newton not mentioned);
(III) (*a*) d'Alembert's principle;
(*b*) principle of virtual velocities.

Although (II) involves three steps, it contains only what Newton included in his second law. The primary divergence appears in (III). Without mentioning Newton's principle of action and reaction, Lagrange replaced it with d'Alembert's more general statement: if one applies to several bodies m_i forces $\vec{F_i}$ which, acting alone, would produce motions which the bodies cannot in fact take because of their mutual connections and external constraints, then one may regard these fictitious motions as composed of those which the bodies actually take (measuring the 'effective' forces $\vec{F_i'}$) and those which are destroyed (measuring forces 'destroyed', or forces of constraint $\vec{f_i}$), yielding $\vec{F_i} = \vec{F_i'} + \vec{f_i}$. The destroyed forces, because they produce no movement, must be in equilibrium, $\sum \vec{f_i} = 0$. This gives,

$$\sum(\vec{F_i} - \vec{F_i'}) = 0,$$

implying that the reversed effective forces are always in equilibrium with the impressed forces. This conclusion constitutes d'Alembert's principle. Since the reversed effective forces are just the reaction forces, the principle generalizes the law of action and reaction for many-body systems. Replacing the effective forces with their measure in terms of motion produced, or mass times acceleration, one obtains,

$$\sum(\vec{F_i} - m_i\vec{a_i}) = 0,$$

which is the form in which the relation was usually applied. D'Alembert's principle, Lagrange showed, offered 'a direct and general method for resolving, or at least for putting into equations, all the problems of dynamics that one can imagine. This method reduces all the laws of motion of bodies to those of their equilibrium, and thus brings dynamics back to statics'.[57]

[56] Joseph-Louis Lagrange, *Mécanique analytique*, 4th edn. (2 vols., Paris, 1888–9), **1**, pp. 237–63. First published 1788. [57] *Ibid.*, **1**, p. 255.

The latter subsumption of dynamics under statics constituted Lagrange's unique contribution to the basic principles of mechanics. He showed that the equations of motion of any mechanical system followed from inserting d'Alembert's equilibrium principle into an old principle of statics, generalized by Jean Bernoulli early in the eighteenth century, called the principle of virtual velocities: 'if any system of arbitrarily many bodies or points, each acted upon by any force, is in [static] equilibrium; and if one gives to this system any small movement, as a result of which each point traverses an infinitely small space, called its virtual velocity; the sum of the forces, each multiplied by the space which its point of application traverses in the direction of this same force, will always be equal to zero; regarding as positive the small spaces traversed in the direction of the pressure and as negative the spaces traversed in the opposite sense'.[58] Virtual velocities are any motions consistent with the constraints on the system, but differ from actual motions in that they do not involve time and are not caused by the actual forces in the system. They are merely possible changes in its configuration. The principle is well illustrated for statics by a balance on which two weights exerting forces F_1 and F_2 rest at distances l_1 and l_2 from the fulcrum. The balance principle, $F_1l_1 = F_2l_2$, may be obtained by requiring that for any small displacement in which one weight would fall by δr_1 and the other rise by δr_2 (their virtual velocities), the sum of the forces multiplied by the virtual velocities is zero, $F_1\delta r_1 + F_2\delta r_2 = 0$. For the special case of actual motions $\mathrm{d}r_1$ and $\mathrm{d}r_2$, which would necessarily be proportional to l_1 and l_2, respectively, one has, $F_1l_1 - F_2l_2 = 0$.

Generally, for an arbitrary system of forces in static equilibrium,

$$\sum \vec{\mathrm{F}}_i \cdot \delta\vec{\mathrm{r}}_i = 0.$$

Applying the principle of virtual velocities to d'Alembert's equilibrium of action and reaction, Lagrange obtained his most general relation,

$$\sum (\vec{\mathrm{F}}_i - m_i\vec{\mathrm{a}}_i) \cdot \delta\vec{\mathrm{r}}_i = 0.$$

From this extended principle of virtual velocities the equations of motion of any system followed, as did several general conservation laws and the theorem of least action. For conservative forces (forces derivable from a potential function V_i), and considering actual rather than virtual motions, the principle gave immediately the conservation of *vis viva*,

$$2\sum m_i \mathrm{d}V_i + \sum \mathrm{d}(m_i v_i^2) = 0.$$

Lagrange went on to develop his famous 'Lagrangian' formulation of the equations of motion in generalized coordinates.[59] Not until Thomson and Tait revived it, however, would the latter scheme receive wide attention in textbooks.

Lagrange's *Mécanique analytique* set the standard for elegance in theoretical mechanics, but Poisson's thorough *Traité de mécanique* served as the storehouse of

[58] *Ibid.*, **1**, pp. 21–2.

[59] *Ibid.*, **1**, p. 267 (extended virtual velocities), pp. 273–324 (conservation laws), p. 307 (*vis viva*), pp. 334–5 (Lagrange's equations).

acquired French wisdom from which most British authors drew for their more limited textbooks. Even Poisson, however, did not include the Lagrangian equations of motion. He did attempt to clarify the ideas of inertia, accelerating force (measured by acceleration), and motive force (measured by mass times acceleration produced), but without extracting laws of motion. In general his analysis followed that of Lagrange, treating d'Alembert's principle, extended by virtual velocities, as *the* principle of dynamics, without attention to action and reaction.[60]

Cambridge authors such as Whewell, Earnshaw, and Pratt recovered three 'Newtonian' laws of motion as an axiomatic foundation of dynamics, but action–reaction was not one of them. They regarded that principle as a deduction from the other laws, based on an obvious truth about mutual pressures (Earnshaw and Pratt) or an *a priori* truth about causation applied to pressures (Whewell). Their laws, as given by Earnshaw, were:

(I) inertia

(II) 'If a particle of matter already in motion be acted on by any external force, the change of motion is in the direction of the [accelerating] force, and is in magnitude the same as if the particle had no previous motion' (i.e. acceleration = accelerating force);

(III) 'The pressure (*measured as in Statics by the weight which it can support*) which produces motion in any body is equal to the product of the mass of the body into the accelerating force'.[61]

Law (II) asserts that the change of motion which a body experiences is not affected by its existing motion, which allows for composition of inertial motion with acceleration, as in Lagrange's (II.*a*) and (II.*b*). Law (III) corresponds to Newton's Law (II) and to Lagrange's (II.*c*), but with an important difference. Newton and Lagrange understood the relation as the fundamental measure of force, whether the force actually produced motion or only tended to do so. The Cambridge authors took their law as a relation between two separate concepts of force, pressure (measured statically by the equivalent weight) and moving force (measured dynamically by mass times acceleration produced). As Earnshaw restated Law (III), 'The pressure which produces motion varies as the moving force'. The main object of the law, then, was to establish an empirical relation

[60] S.D. Poisson, *A treatise on mechanics*, translated by H.H. Harte (2 vols., London and Dublin, 1842), **1**, pp. 170–89 (on forces); **2**, pp. 1–8, 317–20 (d'Alembert's principle), p. 360 (action–reaction). First published 1811; translation from the second edition of 1833.

[61] Samuel Earnshaw, *Dynamics, or a treatise on motion; to which is added a short treatise on attractions*, 3rd. edn. (Cambridge, 1844), p. 17 (third law as deduction), pp. 5, 12 (quotations). J.H. Pratt, *The mathematical principles of mechanical philosophy and their application to the theory of universal gravitation* (Cambridge, 1836), pp. 176–204. Whewell, 'On the nature of the truth of the laws of motion, pp. 152, 158–63. *Elementary treatise*, pp. viii, 150, and 160–1; also, *Philosophy*, **1**, pp. 177–84, 215–44. Older Cambridge authors, such as James Wood (Dean of Ely and Master of St John's College), employed Newton's laws in their original form. See *The principles of mechanics; designed for the use of students in the University*, 8th. edn. (Cambridge, 1830), pp. 16–27.

between statics and dynamics, two quite distinct subjects. As a corollary, since the equality of action and reaction was true *a priori* for pressures, it would hold also for moving forces and momenta, yielding the more usual statement of the third law (here from Whewell): 'In the direct mutual action of bodies, the momentum gained and lost in any time are equal'.[62]

By contrast, Lagrange regarded pressure and moving force as merely two different words for the same quantity, depending on whether it referred to a body acting or being acted upon: 'the product of the mass and the accelerating force . . . if it is considered as the measure of the effect that the body can exert because of the velocity which it has assumed or which it tends to assume, constitutes what is called *pressure*; but if it is regarded as a measure of the force or power necessary to impart this same velocity, it is then what is called *motive force*'. He defined '*force*' in general as 'the cause . . . which impresses or tends to impress motion on bodies', so that even the static equilibrium of a body between two opposed forces was an equilibrium of impressed motions, or tendencies to motion, which 'destroyed' one another.[63] Thus all equilibrium, for Lagrange, Poisson, and later French authors, was dynamic equilibrium, and statics and dynamics were two aspects of a single subject. The principle of virtual velocities grounded this unity most deeply, for virtual velocities were those mutually destroyed under conditions of static equilibrium.

To emphasize the radical separation of statics from dynamics among Cambridge authors, we may quote the 'despotic' Whewell, who declared: 'I have always insisted earnestly upon the distinction of Statics and Dynamics, the doctrines of equilibrium and of motion. These two branches of the subject rest upon different fundamental principles; and the mixture of the two has been a fertile source of confused thought and vicious reasoning. It has given rise to many false or unphilosophical steps in mechanical treatises; as, for instance, the attempt to prove the law of the composition of pressures by the consideration of motion; the attempt to prove the third law of motion by *defining* momentum gained and lost to be action and reaction; and the like'. Whewell went so far as to denounce the principle of virtual velocities and to strike it out of his textbooks from 1824 onwards.[64] Though never so extreme as Whewell, William Hop-

62 Earnshaw, *Dynamics*, p. 13. For the relation of pressure to moving force, see especially Whewell's 'On the principles of dynamics', pp. 29–33, and 'On the nature of the truth of the laws of motion', pp. 159–63, 169, quotation on p. 163. To obtain his third law for momenta from the necessarily true form of the action–reaction axiom for static pressures and from the equally necessary conviction that pressure and motive force must have the same measure, Whewell required from experience a measure of moving force or 'action', which would make the axiom empirically true. He concluded that either momentum or *vis viva* would do. To obtain d'Alembert's more general principle for a connected system of bodies he required the additional empirical condition that neither the connections nor the actions already present changed the effect of an additional action. See Fisch, 'Necessary and contingent truth', pp. 296–7.

63 Lagrange, *Mécanique*, **1**, pp. 245–6, 1.

64 Whewell, *Elementary treatise*, p. vi. Again, older authors made no such radical distinction of statics from dynamics; see Wood, *Mechanics*, pp. 28–35, for the 'vicious' error of composition that

kins, in a coaching manuscript copied out by the young Thomson, also separated statical and dynamical measures of force and gave the laws of motion in the Cambridge form. In sharp contradistinction to Whewell, however, and like Earnshaw, he presented a hypothetico-deductive justification of the laws (making no mention of the *a priori* content upon which Whewell so adamantly insisted) and he used the principle of virtual velocities.[65]

The lectures of Thomson's professors in natural philosophy at Glasgow, Meikleham and Nichol, followed the French texts much more closely than did their Cambridge counterparts. Although Meikleham admitted that 'much assistance may be obtained from Whewell's mech. & dynam.', he left no question as to his sympathies when he issued the blanket endorsement, 'Newton, Laplace, and Lagrange are those who have chiefly enriched Natural Philosophy', and at some points he took examples and notation directly from Poisson. Measuring force only in terms of motion, as we saw previously in Chapter 4 and above under kinematics, he subdivided mechanics into 'two classes of questions, first to find the resultant or equivalent to the combined action of composing forces . . . 2nd . . . to find the force requisite to produce a given amount & direction of motion. Hence the two branches, statics & dynamics'.[66] In this formulation, statics simply analyzed the composition of balancing forces, or forces in dynamic equilibrium, each measured by the motion it would have produced if acting alone. Nichol treated statics and dynamics similarly, remarking that, in rational mechanics, 'forces are mere motions (or motions produced or generated in a unit of time)'. He also gave independent (and empirical) status to the action–reaction law: 'If one body strikes another body, one receives as much motion as the other gives. The celebrated theorem of D'Alembert is nothing else than a great generalization of this law, that action and reaction are equal'.[67]

Given this Glasgow background, with its opposition to Whewellian metaphysics and Whewellian mechanics alike, it is not surprising that Thomson suggested to Tait that for their own book they dispense altogether with the term 'mechanics' and substitute 'dynamics', as we learn from Tait's reply: 'As to the title of the whole, I think there are great advantages in using '*Dynamics*' instead of Mechanics – *first*, that in reality there is no such thing as Statics – *only dynamical equilibrium* – *Secondly*, & very happily, *Dynamics* really means the science of force

Whewell refers to. Menachem Fisch points out that the first edition (1819) of Whewell's *Elementary treatise* contains virtual velocities, as does the *Treatise on dynamics* of 1823, but the second edition (1824) of the *Elementary treatise* does not. See his 'A philosopher's coming of age – a study in erotetic intellectual history', in Menachem Fisch and Simon Schaffer (eds.), *William Whewell – a composite portrait – studies of his life, work, and influence* (Oxford, forthcoming), section 3.3; and *William Whewell, philosopher of science* (Oxford, forthcoming), section 2.2341.

[65] William Thomson, 'Dynamics, commenced [copying from Hopkins's manuscript] March 20th 1843', PA11, ULC. See also notebooks PA12–17, which continue Hopkins's manuscript notes. The notes refer regularly to Earnshaw and less frequently to Pratt.

[66] William Thomson, 9th and 5th November, 1839, NB9, ULC.

[67] *Ibid.*, 7th November, 1839 (from the back).

or Power – and is erroneously used as a contrast to Statics. I am perfectly willing to drop Mechanics entirely and make Dynamics the general title. What would you propose as a substitute for the phrases *mechanical equivalent of heat* &c? – *THIS QUESTION IS IMPORTANT AT THE OUTSET*'.[68]

Tait's remarks supplied the immediate impetus for the division of dynamics in the *Treatise* into statics and 'kinetics' rather than of mechanics into statics and dynamics. But 'kinetics' had not yet appeared. In February, beginning a long draft, 'towards Friction & Cohesion', Thomson broke off: 'N.B. "statical" and "dynamical" friction won't do now. To say force is not *dynamical* is nonsense . . . When it prevents sliding it is called statical friction. When it resists without preventing, it is called ? active friction? or what. This is not good'. Soon he found 'kinetic friction' to answer the query, giving in general 'kinetics' in place of 'dynamics'. Thomson then coined 'kinetic energy' to replace both his earlier 'dynamical energy' and Rankine's 'actual energy'.[69] For consistency, apparently, he finally accepted 'kinematics' over his long-favoured 'cinematics'. Familiar phrases such as 'kinetic theory of gases' seem also to derive from this transformation. Ironically, the basic form 'kinetics' never found acceptance beyond the *Treatise*, although 'dynamics' is now usually taken to encompass the whole of mechanics, including statics.

The extended usage of 'dynamics' proposed by T & T′ had important precedents. Both authors were probably aware of Delaunay's *Mécanique rationelle* (2nd edn., 1857) which subsumed equilibrium and motion under *dynamique*, in addition to using *cinématique* as they did. In March, however, Tait discovered a far superior source, which anchored their pedigree in Scottish tradition. He wrote to Thomson: 'Robison's Elements of Mechanical Philosophy [1804] ignores Statics altogether, and employs Dynamics as we purpose. He alludes to having given the contents of this book as lectures for 30 years – so the idea is 90 years old at least'. John Robison, professor of natural philosophy at Edinburgh from 1774 to 1805, argued quite generally that mechanical philosophy depended entirely on the study of motion, that its changes were 'the only marks and measures of the changing forces'. His successor John Playfair, professor from 1805 to 1819, closely followed Robison's views, arguing that, 'Dynamics is the most elementary branch of the doctrine of motion and the most general in its principles. The term signifies literally the doctrine of *power*; power or force being known to us only as the cause of motion, and being measured by the motion it produces'. Statics arose from the study of constrained motions.[70]

68 P.G. Tait to William Thomson, 20th January, 1862, T6I, ULC.

69 William Thomson, 'Towards friction and cohesion', NB48, ULC, p. 1. Tait thanked Thomson for this draft on 17th February, 1862, T6R, ULC, and announced his adherence to 'kinetics' on 19th March, 1862, T12, ULG. See also William Thomson to G.G. Stokes, 4th April, 1862, K135, Stokes correspondence, ULC, where Thomson discusses the appearance of 'kinetics' in his paper 'On the rigidity of the earth', *Phil. Trans.*, (1863), 573–82; MPP, **3**, 312–36, esp. p. 317*n*. For the various energy terms see Chapter 10, note 96.

70 P.G. Tait to William Thomson, 27th March, 1862, T14, ULG. On Robison and Playfair see

Probably Meikleham too had adopted much of his viewpoint from Robison and Playfair, and Nichol simply continued the tradition (ch. 4). Thus Thomson and Tait in 1862 were indeed returning their natural philosophy to its Scottish roots.

These affiliations suggest that Tait's remarks above contained no especially innovative perspective on dynamics within the Scottish context. Indeed, that was certainly the case for Tait, as we saw previously from his inaugural lecture of 1860. Thomson, however, had long harboured radical views. To recapitulate, Cambridge texts reified the statical conception of force by giving independence, and thus priority, to its statical measure. They required a special law to connect the statical measure, pressure, to the dynamical measure, motion produced. To restore Newton, therefore, meant in the first instance to restore the dynamical measure of force, force being then known only by motion produced or tending to be produced. To this degree Thomson and Tait completely agreed. Tait, however, had not yet entered fully into the much deeper restoration that Thomson envisaged, the restoration of the original mechanical philosophy of the seventeenth century, in which force not only was measured by, but *was*, matter in motion. Thomson's view may be regarded as a radical version of Nichol's dictum that in rational mechanics 'forces are mere motions'. It implied ultimately, along with the restoration of Newton's laws, the destruction of his physical theory of atoms and forces.

At the deepest level, then, Thomson's programme of dynamical theory implied replacing forces acting at a distance with matter in motion. He preferred any theory tending in that direction to one requiring Newton's (and Laplace's and Poisson's) 'statical' atoms and forces. Only during 1862, however, did Thomson declare publicly his opposition to atoms, and then in a most obscure fashion. Writing in January to James Joule, who published his letter, Thomson described an experiment which would give 'a definite limit for the sizes of atoms', but corrected himself, 'or rather as I do not believe in atoms, for the dimensions of molecular structures'.[71] We saw earlier that Tait had already become nervous about Thomson's antipathy to atoms in December, 1861. Twice more in January Tait objected that he could 'scarcely admit the *ultimate* compressibility' of the molecules of matter, and complained that Thomson had not yet referred him 'to the paper or treatise in which are to be found the mathematical conditions of an homogeneous (incompressible or nearly so) continuous gas or liquid' which would serve as a substrate for the compressible molecules of matter. Neither Stokes's papers on continuum mechanics nor

Crosbie Smith, '"Mechanical philosophy" and the emergence of physics in Britain: 1800–1850', *Ann. Sci.*, **33** (1976), 3–29, esp. pp. 6–13. John Robison, 'Physics', *Encyclopaedia Britannica*, 3rd edn. (Edinburgh, 1797–1801), **16**, pp. 637–59, on pp. 647–8. John Playfair, *Outlines of natural philosophy, being the heads of lectures delivered in the University of Edinburgh*, 3rd edn. (Edinburgh, 1819), p. 16.

71 William Thomson, 'New proof of contact electricity', *Proc. Manchester Lit. Phil. Soc.*, **2** (1862), 176–8; E&M, 317–18. See also his 'The size of atoms', *Proc. Royal Inst.*, **10** (1884), 185–213; PL. **1**, 147–217, esp. p. 177*n*.

Helmholtz's *Wirbelbewegungen* (vortex motions) seemed to him 'to realize the aforementioned idea'.[72]

Thomson seems not to have been aware of Helmholtz's paper, on which he would later base the vortex atoms. In his draft on friction and cohesion in February, however, he put forward the idea of a cellular net of continuous matter to explain both atomicity and cohesion. 'Cohesion requires no *attractive* law as supposed in the Boscovich–Laplace theory, other than the law of gravitation, when the fact that all matter is cellular (? "nets in space") hitherto overlooked, is taken into account. By cellular I mean intensely heterogeneous as to density, with continuity through all the densest. (– what a contrast to the common idea of atoms, or even of molecules!)'. A quick sketch of the calculation showed how adjacent nets of parallelepipeds, with arbitrarily dense walls, would attract each other with arbitrarily large gravitational force when brought within the cohesive distance. He wondered if Tait would be interested in communicating a note on the subject to the Royal Society of Edinburgh. Tait agreed but wished to be enlightened as to why the nets would not collapse and how their walls became transparent. Undeflected from his basic insight, Thomson stressed in his published 'Note on gravity and cohesion' that 'any sufficiently intense heterogeneousness of structure whatever' would produce the required result. 'All that is valid of the unfortunately so-called "atomic" theory of chemistry seems to be an assumption of such heterogeneousness in explaining the combination of substances'.[73]

These considerations suggest the degree to which the *Treatise* developed inseparably with the continuum theory of atoms and forces. As Thomson wrote to Tait concerning his cellular hypothesis, 'I have hinted at it ever so many years in my lectures, but of late, the compulsion to think on the book has made me feel it to be of more importance than it seemed to me before. I am persuaded that it has some positive truth in it'.[74]

Revisionist dynamics – work, energy, and Newton's laws

While the view that all physical phenomena are dynamical, including those apparently statical, lay implicit in many parts of the *Treatise*, 'the Book' did not teach this doctrine of *physical* dynamics. It taught, rather, the doctrine of energy applied to *abstract* dynamics; but in a form almost as radical as the full dynamical programme. All of abstract dynamics was to be regarded as contained in the 'law of energy':

[72] P.G. Tait to William Thomson, 15th and 23rd January, 1862, T6G and T6J, ULC. Hermann Helmholtz, 'Ueber Integrale der hydrodynamischen Gleichungen welche den Wirbelbewegungen entsprechen', *Journal für die reine und angewandte Mathematik*, **55** (1858), 25–55; *Wissenschaftliche Abhandlungen* (3 vols., Leipzig, 1882), **1**, 101–34.

[73] William Thomson, 'Friction and cohesion', pp. 34–40. P.G. Tait to William Thomson, 17th February, 1862, T6R, ULC. William Thomson, 'Note on gravity and cohesion', *Proc. Royal Soc. Edinburgh*, **4** (1857–62), 604–6; PL, **1**, 59–63, on pp. 62–3.

[74] William Thomson, 'Friction and cohesion', p. 40.

The whole work [W] done in any time, on any limited material system, by applied forces [$\vec{F_i}$], is equal to the whole effect in the forms of potential [V] and kinetic energy [T] produced in the system, together with the work lost in friction [W_f]. This principle may be regarded as comprehending the whole of abstract dynamics, because . . . the conditions of equilibrium and of motion, in every possible case, may be immediately derived from it.[75]

In symbols,

$$\mathrm{d}W = \sum \vec{F_i} \cdot \mathrm{d}\vec{r_i} = \mathrm{d}V + \mathrm{d}T + \mathrm{d}W_f.$$

'Abstract' meant in essence that, with the exception of work lost in sliding friction between solids, W_f, no processes involving dissipation would be considered, neither viscosity of fluids nor imperfect elasticity nor electric and magnetic dissipation. Work not waste, the first law and not the second, would rule. Even W_f would enter the dynamical laws as though it were produced by conservative forces. T & T′ did believe that the reversibility of abstract dynamics applied to the excluded phenomena, but that its application required a molecular theory to see why the macroscopic effects were irreversible: 'it is only when the inscrutably minute motions among small parts, possibly the ultimate molecules of matter . . . are taken into account . . . that we can recognise the universally conservative character of all natural dynamic action, and perceive the bearing of the principle of reversibility on the whole class of natural actions involving resistance, which seem to violate it'.[76]

'Abstract' thus meant that only gross parts of bodies would be considered and therefore that dynamics applied to systems composed of only a finite number of parts (particles), related by only a 'finite (and generally small) number of forces, instead of a practically infinite number'. In such circumstances, abstractions like 'rigid' and 'elastic' became reasonable approximations. It will be important later to recall this limitation of abstract dynamics to *finite* systems of particles. T & T′ considered it from two aspects: the limitations on human knowledge ('our utter ignorance as to the true and complete solution of any physical question by the only perfect method, that of the consideration of the circumstances which affect the motion of every portion, separately, of each body concerned') and the practical truth of the methods ('*the limitations introduced being themselves deduced from experience*, and being therefore Nature's own solution . . . of the infinite additional number of equations'). These aspects of the finite will have particular relevance for Thomson's views on continuum mechanics, statistics, and the relation of physical dynamics to free will (chs. 12, 18).

The preceding remarks describe the place of the 'law of energy' in abstract dynamics. Its significance will require elaboration. The basic problem in under-

[75] *Treatise*, p. 200. It is clear from P.G. Tait to William Thomson, 7th and 11th October, 1862, T26 and T27, ULG, that Thomson was to write the bulk of their Chapter II on dynamical laws and principles, including 'Force, Time, Energy, Work, Laws of Motion &c &c', also 'Least Action & Virtual Velocities'. Ultimately, he followed Tait's outline for the Hamiltonian theory of action (below). [76] *Ibid.*, p. 195.

standing the *Treatise* is to understand the relation of energy to the laws of motion, for Thomson and Tait claimed that their general work–energy principle was contained in Newton's third law, the action–reaction principle. They regarded Newton's first and second laws as providing only 'a *definition* and a *measure* of force'. Building on this definition, the third law expressed the transformation of energy necessary for the mutual interaction between two or more bodies, 'such as, for instance, attractions, or pressures, or transference of energy in any form'. As Thomson said in his lectures in the autumn of 1862, 'The action of a force . . . *is* [not 'involves'] . . . a transformation . . . of Energy'.[77] Energy was a substantial reality, on a par with mass, while force became a derivative term, however useful.

The earliest surviving expression of this view appears in Thomson's lectures for 1849–50, the year he first publicly proclaimed his belief in the 'principle of mechanical effect', or conservation of energy. There he redefined 'action': 'The action of a change means its rate of performing work . . . Newton seems clearly to have given the definition of the rate of performing work'. Thomson was referring to the Latin edition of the *Principia* where, in the scholium to the third law, Newton discussed '*actio agentis*' and '*actio resistentis*' with respect to machines. In the later reading of the *Treatise*: 'If the Action of an agent be measured by its amount and its velocity conjointly; and if, similarly, the Reaction of the resistance be measured by the velocities of its several parts and their several amounts conjointly, whether these arise from friction, cohesion, weight, or acceleration; – Action and Reaction, in all combinations of machines, will be equal and opposite'.[78]

Thomson stressed in 1862 that '*Newton looks upon resistance to acceleration as a reaction* . . . We are to consider the resistance against acceleration as a real reaction'.[79] An example would be the action of a force F on a mass m moving with velocity v in one dimension. The force produces acceleration a while doing work W. Setting the action of the force equal and opposite to the reaction of the (reversed) acceleration, one obtains:

$$Fv = -(-ma)v$$
$$F\mathrm{d}x/\mathrm{d}t = mv\,dv/\mathrm{d}t$$
$$\text{action} = \mathrm{d}W/\mathrm{d}t = \mathrm{d}/\mathrm{d}t(mv^2) = -\text{reaction}.$$

77 *Ibid.*, p. 184; William Thomson, in Murray, *op. cit.* (note 18), p. 177. Our emphasis. As usual, Whewell provides an important precedent and foil for Thomson. In his 'On the nature of the truth of the laws of motion', Whewell claimed not only that the necessarily true action – reaction axiom would become empirically true with 'action' measured either as mv or as mv^2, but that the two resulting propositions 'are necessarily connected, and one of them may be deduced from the other' (p. 163). He settled on mv by convention alone. This view is particularly puzzling for Whewell in 1834 since he had stressed the empirical failure of conservation of *vis viva* for the solar system in his *Bridgewater Treatise* of 1833. See Chapter 4 above and James Thomson's reflections below, note 89.

78 William Thomson, 20th March, 1850, lecture 80, in Smith, *op. cit.* (note 26). Isaac Newton, *Philosophiae naturalis principia mathematica* (London, 1687), p. 25; *Mathematical principles of natural philosophy*, translated by A. Motte (1729), F. Cajori (ed.) (Berkeley, 1934), p. 28. *Treatise*, **1**, p. 185.

79 William Thomson, in Murray, *op. cit.* (note 18), p. 101.

Thus the rate of working is equal to the rate of production of kinetic energy, which Thomson already in 1849 called 'Newton's law of dynamical effect' and 'the great principle of work spent in dynamics'. 'This', he taught, 'is the most important part of dynamics'. Characteristically, he included a hydrodynamic example. Considering a cylinder of water pressed from above by a piston and producing a jet from an orifice in the side, he showed that the '*actio agentis*' (work) of the piston was equal to the '*actio resistentis*' (kinetic energy) of the issuing stream. Much later Tait would claim for the replacement of force by energy: 'modern science shows us that force is merely a convenient term employed for the present (very usefully) to shorten what would otherwise be cumbrous expressions; but it is not to be regarded as a *thing*, any more than the bank *rate of interest* . . . is to be looked upon as a sum of money . . . *Force is the rate at which an agent does work per unit of length*'.[80]

If the law of energy – read as Newton's law of action and reaction – were to perform its universal function, it had to replace both d'Alembert's principle and the principle of virtual velocities, the combination of which Lagrange had made the universal foundation of dynamics and which substituted for the third law in his scheme. Lagrange had obtained a work–energy relation, $dW = dT$, from the general virtual velocity principle by restricting it to actual movements. He obtained conservation of *vis viva* by further specializing to the case of forces derivable from a potential function, as noted above. Thomson and Tait had now to be able to reverse that derivation, showing that in fact special cases exhausted all cases.

This view appeared already in Thomson's lectures in 1849–50. He there treated the principle of virtual velocities in statics as nothing more than an application of the work–energy relation: 'Every problem in nature of which we have forces in equilibrium may be brought under this [work–energy] equation. Newton lays it down in a very general way . . . I will give you a view that will bring it under the law of dynamical motion . . . The condition necessary & sufficient for equilibrium is that for every infinitely small displacement the mechanical effect produced by one set of forces is precisely = to the work spent in the others'.[81] Although the stated condition is unexceptionable, if read as applying to *virtual* velocities, its identification with the energy equation is not, for the latter equation applies to *actual* motions, 'natural motions' for T & T′, produced by the forces acting on the system, while the principle of virtual velocities refers in addition to any other *possible* motions that the system might be 'guided' to take by constraints acting perpendicular to the motion. The work–energy relation yields only a single equation and therefore cannot determine the motion of a system completely unless that system has only one possible motion. The principle of virtual velocities, on the other hand, yields as many

[80] William Thomson, 10th April, 1850, lecture 91; 8th March, lecture 74; 18th March, lecture 79, in Smith, *op. cit.* (note 26). Tait, 'Force', p. 266.

[81] William Thomson, [4th] March, 1850, lecture 70, in Smith, *op. cit.* (note 26).

independent equations of motion as there are degrees of freedom in the system.[82]

The point is so elementary and had been stated so often, by Lagrange and others,[83] that one must ask what Thomson could have been thinking. His view appears in an 1852 draft of the article 'On a universal tendency to the dissipation of mech^{l} energy in nature'. He began:

The principle of mech^{l} effect [conservation of energy] first stated in all its generality by Newton at the concl^{n} of the scholium to his 3rd Law of Motion, and enunciated more or less explicitly by subsequent writers in the two propositions commonly called 'the principle of Virtual Velocities' and 'the principle of the conservation of vis-viva', is first deduced from the fundamental axioms of mechanics as a theorem applicable to cases in which either working or resisting forces may be regarded as arbitrarily applied.

That is, the work–energy relation, in the form of work done on a conservative system, $\mathrm{d}W = \mathrm{d}T + \mathrm{d}V$, had previously been regarded as a restricted result, applicable only when the effect of all forces is independent of the path taken between any two states (i.e., 'arbitrarily applied') and potential functions are definable. It would not be true for the force of sliding friction, for example, when the term W_f in the general principle would enter and would depend on the path taken. Thomson continued:

But as soon as it is established in Natural History that all the working or resisting forces of inanimate matter, as well as all the mech^{l} actions of living Creatures either are due to the inertia of matter, or are mutual forces between material particles which, when overcome through any spaces are always ready to restore the work spent, by working backwards through the same spaces; [then] the postulate (that [a potential function exists for every force]) assumed in the theorem of the 'conservation of vis viva' becomes known as a Universal Truth.[84]

Once conservation of energy had been established empirically (by Joule and others), the existence of a potential function for *every* force in nature would be guaranteed. W and W_f in the general work–energy principle could be collapsed into V and the effect of every force would be independent of path. If so, then one could substitute any *possible* motion for any *actual* motion and thereby elevate the work–energy relation into a completely general variational equation, $\delta V = \delta T$, which would give immediately the principle of virtual velocities and the general equations of motion, or, more basically, Newton's second law.

The derivation is no more general than the claim that every component of work done can be expressed in terms of either a kinetic or a potential energy.

[82] For a standard modern derivation see Herbert Goldstein, *Classical mechanics* (Reading, MA and London, 1950), pp. 16–18. Note, however, that Goldstein refers to Lagrange's generalized principle of virtual velocities as d'Alembert's principle.

[83] For example, Lagrange, *Mécanique*, **1**, p. 307.

[84] William Thomson, 'On a universal tendency to the dissipation of mech^{l} energy in nature' (draft), PA137, ULC.

Any heat generated in a system, for example, must be reduced to conservative motions, otherwise Newton's second law cannot be obtained. Strangely, therefore, T & T′ deduced the principle of virtual velocities from their work–energy theorem, which included sliding friction (although not internal friction). Aware, apparently, that the derivation would seem problematic at best, they claimed further on that frictional forces could always be replaced by conservative forces for infinitesimal motions. They offered no argument.[85] It seems apparent that their confidence expressed their faith in the universality of energy conservation and their attendant belief that it guaranteed potential functions. After all, the laws of action of forces were to be derived from 'the *one* law of the Universe, the Conservation of Energy'. Their faith in this doctrine, however, calls for further examination of its connotations. Familiar themes immediately arise: field theory, practicality, economics, and continuum mechanics.

In the late 1840s, while developing his ideas on ponderomotive forces acting in electric and magnetic fields, Thomson had obtained a theorem very like the grand claim of the *Treatise*: the force acting to move an inductively magnetized body could in every case be derived from the expression for the work done during the motion. This claim constituted the essence of his theory of fields. It translated the idea of forces acting at a distance between points, or point forces, into ponderomotive forces acting to move macroscopic bodies as a result of their participation in a continuous field of energy (ch. 8).

These ideas suggest more particular aspects of energy physics. The idea of ponderomotive forces, for example, involved immediately two claims. The common abstraction in dynamics of a *material point* was not to be regarded as an idealization of an unobservable atom, as had been common on the continent, but as an idealization of an ordinary body that one could manipulate in experiments. And, similarly, forces acting on material points were not idealizations of atomic forces but of ponderomotive forces exerted on gross objects. The sections of the *Treatise* on 'Particle and point' and 'Place of application of a force' derive directly from Thomson's 1849–50 lectures, where he defined a material point as, ' "the smallest portion of matter with [on] which a force may act" sometimes I may use balls or rings'. We are reminded of a response given in Dr Thomson's mathematics class to the question, what is a point: 'It's just a dab sir'. In his own class Thomson instructed his charges to 'Let the fig. represent a material point on a horizontal plane', and he drew a block lying on a table (figure 11.2). To illustrate the meaning of the 'point' of application of a force, he pressed on the desk with a ruler and with his finger and called attention to the finite extent of the areas of contact. In the same way T & T′ would insist that, 'the place of application of a force is always either a surface or a space of three dimensions occupied by matter'.[86]

[85] *Treatise*, pp. 178, 203–5, 341.

[86] William Thomson, 19th November, 1849, lecture 10; [18th] February, 1850, lecture 61, in Smith, *op. cit.* (note 26). SPT, **1**, 7. *Treatise*, pp. 164–5.

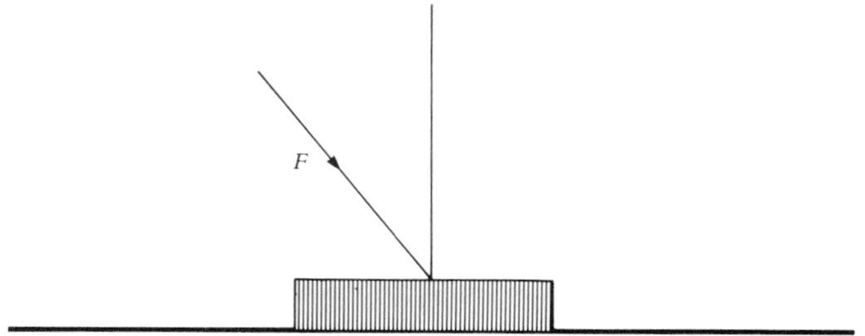

Figure 11.2. Thomson represented a 'material point' as a block lying on a table.

This macroscopic thrust carried Thomson's usual demand for practicality, as well as the connotations of 'abstract dynamics' discussed above. The *Treatise* would employ only concepts realizable in everyday objects and acts. In the same vein a force was to be measured by its rate of doing work, its 'action' in the newly discovered 'Newtonian' definition. When introducing his Glasgow students to the principle of virtual velocities as the basis for the equilibrium of forces, for example, Thomson remarked: 'The view which I shall bring before you is founded on the *practical* consideration of work done and of mechanical effect produced'. One often hears of the 'hands-on' approach. Thomson used the language literally: 'Suppose we apply forces that do not equilibrate. Suppose we apply *hands* to a body. Let us produce motion'. With that expression he introduced 'the great principle of work spent in dynamics'.[87]

These examples indicate the degree to which Thomson insisted on regarding 'force' as a term derivative from practical work done. He stood firmly on the side of 'the 1000 intelligent mechanics of Glasgow' as he sought to eradicate the hypothetical abstraction of point forces.

We have seen previously, for electricity, magnetism, and heat, the central role that the image of the steam-engine played in Thomson's thinking about force and work. The same image infused the *Treatise*. In a draft manuscript giving a hydrodynamic interpretation of Newton's third law, we find:

According to the estimate here laid down of 'action' in dynamics the 'action' of a 'labouring force' at any instant is its *rate of performing work* . . . The term thus defined has not the same signification as that which has unfortunately been attached to it by the immortal author of the 'Mécanique Analytique' [Lagrange] in his 'principle of least action', and which it has boren [borne] in the most valuable writings on dynamics since published; but the practical use of the term is such as to present no definite idea to the mind unless Newton's definition be adopted. Thus the *action of a steam engine* is a very intelligible phrase, and may be used without any vagueness to express the energy with which the engine is working, and is measured by the number of foot-pounds of work

[87] William Thomson, 24th February, 1850, lecture 65; 8th March, lecture 74, in Smith, *op. cit.* (note 26). Our emphasis.

produced in a given time, or according to the general practice of engineers, by the 'horse power' at which it is working.[88]

One recognizes in this statement the proximity of William's brother James and his intensive involvement with steam-engines and water turbines. In fact, the evidence suggests that James's engineering perspective may have been the original source for the rereading of Newton's third law as a law about work and as containing Newton's second law. In his notes from Lewis Gordon's lectures on civil engineering from 1841, following an elementary discussion of the relation of mechanical effect to *vis viva*, James inserted a 'proof that the velocity produced in a certain time is proportional to the force':

$$\mathrm{d}W = F\mathrm{d}s = \mathrm{d}(mv^2/2)$$
$$F\mathrm{d}s = mv\mathrm{d}v$$
$$\mathrm{d}s = v\mathrm{d}t$$
$$F\mathrm{d}t = \mathrm{d}(mv).$$

This derivation of Newton's second law is signed 'J.T. [at] Tipton', implying a date in 1843 when James was working at the Horseley Iron Works in Tipton.[89] We know that James had begun by this time to take special interest in the practical production and measurement of work and that William's conception of ponderomotive forces as derived from the work contained in a system emerged in part from his interaction with James (chs. 8, 9). It is not surprising, therefore, that James's critical role should emerge again here. Further evidence of that role is contained in several letters of Tait to Thomson dunning him for a draft of their chapter two, on dynamical laws and principles, and referring to its source as 'Digitalis': e.g., '*Do* let me have spharcs [spherical harmonics] and Digitalis at once; else I shall write to Mrs Thomson'. 'Digitalis' seems to refer to James, who had been taking the drug for his heart condition since the 1840s.[90]

This enhanced role for James in the central theorem of the *Treatise* should remind us of the importance of the Morin dynamometer in William's lectures and of its presence in the *Treatise* as one of the instruments included in the chapter on 'Measures and instruments'. If energy were to be the crucial concept in

[88] William Thomson, 'Dft. MSS for T and T″', PA177, ULC. This draft continues without break from PA113; see below, note 96.

[89] James Thomson, 'Civil engineering class notes, 1841', A4, QUB. We have substituted differential relations for James's integrated ones. In two notes added sometime later James worried that the proof assumed the second law, presumably in obtaining the work–*vis viva* relation. He compared his derivation with Whewell's claim that Newton's second law (the Cambridge third law) could not be proved *a priori*, concluding that 'If the proof given here do not contain a fallacy, it is still not an a-priori proof as it is founded on the other laws of motion, and involves the principle that all the mec. eff. given by a force to a body is absorbed in its inertia'. The latter remark suggests Lagrange's extended principle of virtual velocities, which William would soon claim was contained in Newton's interpretation of action and reaction.

[90] P.G. Tait to William Thomson, 27th May, 1863, T40, ULG: see also Tait to Thomson, 25th May, 1863, T39, ULC. For James's use of digitalis see Chapter 5, note 54. cf. Knott, *Life of Tait*, p. 182*n*.

dynamics, and work its measure, then a measuring device for work carried considerable symbolical as well as practical significance. James's improvements and extensive use of the dynamometer supplied the immediate context for the measure, as for the concept itself.

The attempt to redefine force through the relation $F{\cdot}dr = dT$ suggests that force ought to be defined as the rate of change of kinetic energy per unit distance, in conflict with Newton's form of the second law, which gives force as rate of change of *momentum* per unit *time*. That is unfortunate, said Thomson, for 'We w^d prefer to define momentum by the product of the mass into the square of the velocity' (kinetic energy), giving for the second law, $F = dT/dr$ in one dimension, or generally $\vec{F} =$ gradient T. His reasons are of special interest: 'The term moment is used very much in accordance with its use in ordinary life as = important . . . The moment of a body is the importance of its motion'. To Thomson, 'importance' meant work or energy, but usage opposed him. He compromised: 'There are therefore two ways of reckoning the importance of motion, the one depending on the time the other on the space. When we consider the Energy produced it is the latter we must take. The Dynamical value of motion is the quantity of Energy required to produce it'.[91]

'Dynamical value' here replaces 'mechanical value', with the same meaning: work content or labour value. The labour theory of value permeates Thomson's lectures in the autumn of 1862. *Vis viva*, or living force as opposed to *vis mortua*, or dead force, is to correlate with Whewell's 'labouring force', also 'working force', 'like every day work'. 'The work done by a living agent is comparable, as to work done, in every respect with the attraction of the earth'; '*Actio agentis* is made definite when applied to a labouring [force, *del.*] Agent. The *action of a labouring force is the rate at wch he is doing work*'.[92]

Dynamical value, therefore, is not simply the numerical value or measure of energy, but 'the value [to man] of the effect produced by a working force'. It responds to the question: 'How are we to value the effect produced?'[93] And it answers with economic value or labour value. The ubiquitous theme of political economy thus informs the most fundamental concept of the *Treatise*. T & T′ would have preferred to reformulate Newton's laws to correspond to the industrial economy of mid-nineteenth-century Britain. In ideal form they would have read:

(1) inertia,
(2) $\vec{F} =$ gradient T,
(3) $\delta W = \delta T$.

Energy, therefore, would 'include the whole of kinetics'. And just as political economy dealt with the accumulation and expenditure of labour value, so,

[91] William Thomson, in Murray, *op. cit.* (note 18), pp. 85–7, 151.
[92] *Ibid.*, pp. 109, 111, 117, 121–3. [93] *Ibid.*, p. 141.

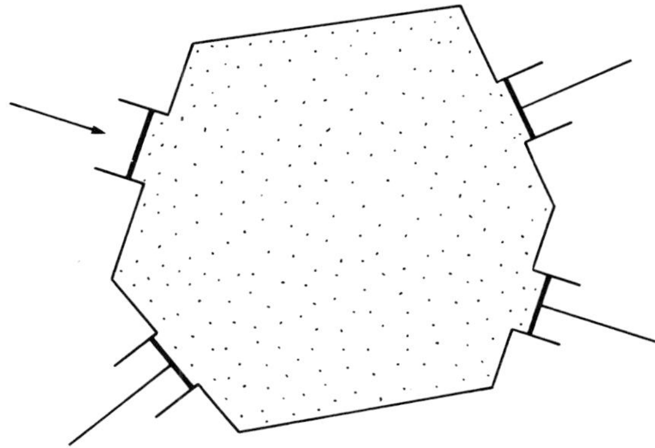

Figure 11.3. The principle of virtual velocities may be illustrated by a hydrostatic system, in which the possible 'action' (rate of working) of any one piston, or combination of pistons, is balanced by the 'reactions' of the others.

Thomson told his class, 'We shall view Kinetic problems as illustrating Energy, the accumulating & storing of Energy and the expenditure of Energy'.[94]

Looking to the practical dimensions of field theory for the significance of Thomson's commitments in the *Treatise* has led us from mathematical physics to engineering to economics. If we pursue instead the physical theory of fields, we are led back to the etherial continuum as the agent of ponderomotive forces, and to Thomson's fascination with hydrodynamic expressions of the force–energy relation. An early example appears in his Glasgow lectures for 1849–50, where he suggested a proof of the principle of virtual velocities in terms of a closed container of water with pistons pushing on it in various places (figure 11.3), a picture alternative to Lagrange's favoured image of a body suspended in equilibrium by a string-and-pulley network.[95] Equality of pressure throughout the former system would replace equality of tension throughout the single connecting string in the latter.

A second example attaches to the quotation above from the draft on 'action' and Newton's third law. Thomson there showed how a liquid enclosed in a bounding surface, subjected to arbitrary pressures at the boundary, and experiencing both gravitational and intermolecular forces, would obey both d'Alembert's principle and the general work–energy relation. The total pressure p at the boundary divided neatly into two parts, π and ω, of which π was the pressure that would exist if the liquid were at rest and ω the pressure that would produce the actual motion from rest, $p=\pi+\omega$. Now the 'action' of the pressures π in any time would have to be zero, according to 'Newton's' action–

94 *Ibid.*, p. 79.

95 William Thomson, 1st March, 1850, lecture 69, in Smith, *op. cit.* (note 26).

reaction theorem, because they produced equilibrium. Similarly the 'action' of ω would equal the 'reaction' against acceleration of the liquid. Thus the whole work done at the boundary in any time by ω would have to equal the kinetic energy produced in the liquid.[96]

These illustrations only begin to suggest the deep commitment to continuum dynamics that Thomson had developed by the 1860s (ch. 12). As examples of mechanical principles they established nothing particularly new; they merely showed how the usual theorems about forces on systems of material points could be translated into theorems about macroscopic pressures in continuous systems, thereby establishing a language for continuum dynamics. They did suggest how the quantities regarded as pressures at the boundary might, from another perspective, be regarded as symptoms of the distribution of energy inside. Nevertheless, converting the work–energy relation into a derivation of pressures from energy required more than suggestiveness, just as did the attempt to derive forces from energy. The problematic character of the examples leads to what posterity regards as the single most important aspect of the *Treatise*.

Extremum principles

The preceding discussion has indicated ways in which the force–energy relation of the *Treatise* was both logically and conceptually incomplete. Energy relations by themselves could not replace force as the foundation of dynamics. Only in variational form, as in the principle of virtual velocities, would the energy principle suffice, and it could not be derived in that form from the action–reaction law. If T & T′ did not worry unduly about the rigour of this part of their dynamics, it may be simply because they had at hand an alternative to their 'Newtonian' formulation, namely a variational formulation based on an extremum principle, Hamilton's extension of the principle of least action.

Already in 1845 Thomson's analysis of ponderomotive force within an equilibrium system, as the gradient of its contained mechanical effect, appeared as a consequence of the tendency of the entire system towards minimum mechanical effect. And when in 1847 he obtained the existence and uniqueness proof that subsumed all the problems he had yet tackled in electricity, magnetism, heat, and hydrodynamics, it was a variational proof that rested on a more general form of the same minimum condition.

More particularly, his 1847 theorem on the *vis viva* of a liquid showed that, with arbitrarily given normal velocity over a closed surface, a solution for the interior velocity existed when the total *vis viva* was minimum, and that this velocity – like a conservative force – was derivable from a potential. He originally regarded the solution, furthermore, as a consequence of the then little

[96] William Thomson, PA113, ULC, which concerns the '*principle of mechanical effect in dynamics*' and continues with PA177.

used 'principle of least action', according to which a conservative system will always move along the particular path which minimizes the time integral of its total *vis viva*. In other words, the natural motion uses less 'action' (as average *vis viva* in the older designation, rather than rate of change of *vis viva*) than any other possible motion, or the variation of the action must vanish,

$$\delta A = \delta \int (\sum mv^2) \mathrm{d}t = 0.$$

In this form, the extremum principle brings in explicitly the notion of explaining any given state of a system genetically, in terms of the process of forming it. Thus Thomson imagined starting the fluid from rest, by applying normal forces at the boundary for some time t, until the given normal velocities were produced. The extremum condition, applied to this entire process of production, resulted in the same solution for the motion as had been obtained previously by a direct application of d'Alembert's principle to an existing state. Thomson merely supplied an alternative viewpoint.

The principle of least action had received short shrift in British textbooks before the *Treatise*. At Cambridge, this circumstance was owing partly to the position of William Whewell, who by the 1830s had rejected the use of variational procedures altogether, especially virtual velocities. Others, such as Earnshaw, were more impressed by the fact that the least action principle presupposed the principle of *vis viva*, and therefore was less general than Newton's laws or the principle of virtual velocities, because it did not include frictional forces. For Whewell, who argued for the non-conservative character of planetary motion in his 1833 *Bridgewater treatise*, this deficiency meant that least action was no principle at all.[97]

Of more general import in the wider sphere is the fact that Lagrange, Laplace and other French authors were seen to have suppressed the principle because of the metaphysical interpretation supplied by its authors. Maupertuis and Euler, in developing the least action approach in the mid-eighteenth century and in arguing for its general validity, regarded it as a teleological principle, a principle which revealed the ends that God sought and thereby revealed the necessity behind the efficient causes, or forces, that acted from moment to moment. The direct action of forces expressed God's power, while the indirect principle of least action expressed His wisdom. Either approach, however, would suffice for determining the equations of motion of a mechanical system and for solving particular problems.[98]

Rationalist Enlighteners like Lagrange and Laplace would not allow such teleological metaphysics to enter the province of natural law. They recognized no direction in physical nature, no evolution, only eternally stable and ever-repeating equilibrium systems. But from his earliest works Thomson had been

[97] Fisch, *op. cit.* (note 64); Earnshaw, *Dynamics*, pp. 183–6. See also Chapter 4.

[98] For a rich account of the principle and its context, see Mary Terrall, 'Maupertuis and eighteenth-century scientific culture' (dissertation, University of California, Los Angeles, 1987).

concerned with beginnings and endings, with progression. After 1850, the Second Law of Thermodynamics made the temporal perspective not a choice but a necessity. In abstract mechanics, of course, where no dissipative effects were acknowledged, one could still choose to analyze any system directly, through the action of the forces, but already in conceptualizing those forces as energy exchanges the temporal perspective had implicitly entered. The role of least action in the *Treatise* should be seen as an extension of these earlier ideas and problem solutions from equilibrium states, where Thomson first employed them, to fully temporal dynamics.

He reported to his students in 1862–3: 'Maupertuis [,] taking a metaphysical view [,] thought that he established a principle of least action. Lagrange says that [it] is simply a deduction from the axioms of Science – it comes to man laying down laws for Creation'. But since T & T′ believed that natural philosophy concerned at least the evolution of the creation, if not the creation itself, they held a different view: 'We are strongly impressed with the conviction that a much more profound importance will be attached to it [least action], not only in abstract dynamics, but in the theory of the several branches of physical science now beginning to receive dynamic explanations'. Relinquishing Newton's 'action', they adopted the 'much less judiciously chosen word' as then universally used, which nevertheless referred explicitly to the genesis of a moving system: 'the average kinetic energy, which it has possessed for the time from any convenient epoch of reckoning, multiplied by the time'.[99]

The formal connection of action principles in the *Treatise* with Thomson's 1847 work is straightforward and proceeds through two intermediate steps, a generalization by Thomson and the introduction of Hamiltonian dynamics by Tait. Thomson had always sought to introduce extremum conditions on the work content, or total energy of a system, which he regarded as the natural expression of its state. Had 'action' been defined as he wished, as rate of working, total action integrated over time from rest would have given this total energy. But, as least energy, 'least action' was not generalizable. It sufficed for systems containing either purely potential or purely kinetic energy, but it could not give the relation between kinetic and potential energy in any mixed system, nor could it treat the temporal effects of friction.

To avoid these temporal problems, Thomson drew on one of his economic metaphors and extended a series of works by Euler, Lagrange, Delaunay, and Bertrand on the maximum–minimum properties of systems set in motion suddenly by impulsive forces. In fact the published version of this theorem on minimum *vis viva* of a fluid continued this genre of theorems on sudden generation of motion, in this case by impulsive forces at the boundary of the fluid. Early in 1863 Thomson found two generalizations, which he and Tait labelled 'Greed' and 'Laziness'. 'Greed' specified that: 'A material system of any kind, given at rest, and subjected to an impulse in any specified direction, and of

[99] William Thomson, in Murray, *op. cit.* (note 18), p. 177. *Treatise*, p. 231.

any given magnitude, moves off so as to take the greatest amount of kinetic energy which the specified impulse can give it'. This principle describes the 'natural' behaviour of a system free to respond in any way it 'chooses' to the given impulses. If, however, 'the system is guided to take, under the given impulse, any motion . . . different from the natural motion', the principle of 'Laziness' operates. Thus, if the system is required to take specified velocities, 'The motion actually taken by the system is that which has less kinetic energy than any other'.[100] Such 'Laziness' became apparent in systems in which the freedom of individual parts to move was constrained by connections. The parts given particular velocities then had to drag connected parts along with them, which moved as lazily as possible.

'Greed' and 'Laziness' contained an obvious moral, which Thomson summarized for his class: 'Impulses always do more work when Body free than when constrained. Even when constrained will take as much work as it can'. Apparently the natural behaviour of material systems mirrored that of the free enterprise economy. Indeed, when we recall the sense in which Thomson used 'mechanical value' as the work content of a system, and that he spoke of systems as being more or less 'valuable' according to their work content, we cannot help recognizing that he drew a conscious parallel between the principles of economy in mechanics and in political economy (ch. 9). Just as the work of machines substituted for labour in the labour theory of value, so the maximum–minimum conditions on mechanical work replaced the optimization conditions on the wealth of the nation.[101]

Although quite general, Thomson's extremum theorems did not yield the equations of motion of a system as it moved from state to state. For that purpose, T & T′ returned to the principle of least action and to its extensions by Hamilton. Interestingly, though Thomson's friends had been bringing Hamilton's and Jacobi's sophisticated treatments of mechanics to his attention since the 1840s, he had shown no interest in them. We are familiar with his general rejection of abstract formalism in physical theory (ch. 6). Hamilton's idealist metaphysics must have been at least as odious.[102] But Tait apparently made him see that the replacement of force by energy as the fundamental physical entity in nature

[100] P.G. Tait to William Thomson, 25th March, 1863, T39, ULG, and 4th April, 1863, in Knott, *Life of Tait*, p. 182. For an evaluation of the importance of these theorems see D.F. Moyer, 'Energy, dynamics, hidden machinery: Rankine, Thomson and Tait, Maxwell', *Stud. Hist. Phil. Sci.*, **8** (1977), 251–68, esp. 257–8. *Treatise*, pp. 216–25, on pp. 222–5. Original publication in William Thomson, 'On some kinematical and dynamical theorems' (read 6th April, 1863), *Proc. Royal Soc. Edinburgh*, **5** (1866), 113–15; MPP, **4**, 458–9.

[101] William Thomson, in Murray, *op. cit.* (note 18), p. 178. See also the corollary to theorem II ('Greed') in 'Kinematical and dynamical theorems', MPP, **4**, 459. We develop these ideas in Wise (and Smith) 'Work and waste'. Our interpretation originated in conversations with Silvan S. Schweber over Darwin. See his 'The wider British context of Darwin's theorizing', in David Kohn (ed.), *The Darwinian heritage* (Princeton, 1985), pp. 35–70.

[102] For a brief but lucid account of the metaphysics and mathematics of Hamilton's dynamics, see T.L. Hankins, *William Rowan Hamilton* (Baltimore, 1980), pp. 172–98.

would benefit from one of the powerful extremum approaches that Hamilton and Jacobi had developed.[103]

The Hamiltonian procedure first appears in Thomson's notebooks in March, 1863, in the form of a three-page summary of 'the whole Hamiltonian affair', written by Tait and pasted in. It consists of three parts: a derivation from the principle of least action of the Lagrangian equations of motion in generalized coordinates; a transformation of those second-order equations (n in number for n degrees of freedom in the system) into Hamilton's first-order 'canonical' equations of motion ($2n$ in number); and a derivation from Hamilton's principle of 'varying action' of a second set of $2n+1$ equations. The three parts formed the basis of the more elaborate presentation in the *Treatise*.[104]

Without entering here on a discussion of Hamiltonian dynamics, we can still summarize briefly its place in the *Treatise*. Most basic was the derivation of the principle of virtual velocities (and thereby both Lagrangian and Hamiltonian equations of motion) from two conditions: least action,

$$\delta A = 2\delta \int T \mathrm{d}t = 0,$$

and conservation of energy,

$$E = T + V.$$

In this way, with full rigour, energy conditions replaced force as the dynamical essence of conservative systems.[105] (T & T′ also derived the generalized Lagrangian equations for non-conservative systems, from the principle of virtual velocities. Their derivation, however, is no more general than their derivation of the principle, and, in any case they showed little interest in the non-conservative systems.[106]) The least action procedure involved fixing the end points and the energy of the motion of a system and varying its path, to show that the 'natural motion' has less action than 'any other motion, arbitrarily

[103] James Armitage to William Thomson, 27th October, 1845, A70, ULC, gives Liouville's remarks on one of Jacobi's papers, pointing out 'a principle as general as that of the Conservation of Areas or of Vis Viva which will at once in numberless mechanical problems give an independent equation at once'. See similar remarks in Arthur Cayley to William Thomson, 28th[?]January, 1847, C44, ULC, and R.L. Ellis to William Thomson, 18th February, 1847, E76, ULC. Cayley surveyed the entire field in his 'Report on the recent progress of theoretical dynamics', *BAAS Report*, **27** (1857), 1–42, from which Thomson claimed to have gained great benefit in his 'Presidential address', *BAAS Report*, **41** (1871), lxxxiv–cv; PL, **2**, 132–205, on p. 149.

[104] P.G. Tait, in William Thomson, 12th March, 1863, 'Notebook re Hamiltonian dynamics', NB50, ULC. On 26th December, 1862, T32, ULG, Tait approved of Thomson's initial draft with 'Least Action is very neat', but apparently it did not include the Hamiltonian formulation, for on 13th February, 1863, T34, ULG, he promised to 'send you my view of Hamilton's Varying Action in a day or two'.

[105] T & T′ regarded the variation of the action δA *as kinematical*. Dynamics entered when they required $\delta A = 0$ (an extremum) and simultaneously imposed conservation of energy for all varied paths, $\delta T = -\delta V$. *Treatise*, p. 233; 2nd edn., pp. 339, 341 (added). The latter variational relation, in their view, would yield the principle of virtual velocities (and thus the equations of motion) directly.

[106] *Treatise*, pp. 251–3.

guided and subject only to the law of energy'. It yielded differential equations specifying the *rate of change* of position and momentum anywhere along the path (Hamilton's canonical equations). Hamilton passed to a different principle by considering only natural motions which obeyed the equations of motion, but allowing the end points and energy to vary. He thereby derived a partial differential equation of the first order and second degree for what he now called the 'characteristic function' (the action expressed only in terms of end points and energy). If one could solve this 'characteristic equation', the momenta themselves p_i at the two end points were obtainable directly by differentiation with respect to the end coordinates q_i,

$$p_i = \partial A / \partial q_i,$$

and equally directly, the total time of travel was,

$$t = \partial A / \partial E.$$

Hamilton therefore called these equations 'integral' equations.[107]

For T & T′, the new formulation expressed something fundamental to their entire viewpoint. It showed how the state of a system as it developed over time could be expressed in terms of a single energy function: it is 'remarkable', they repeated over and over in various forms, 'that the single . . . equation . . . is sufficient to determine a function A, such that the equations [$p_i = \partial A / \partial q_i$] express the momenta in an actual motion of the system'.[108] Thereby, their evolutionary perspective on dynamical systems had been fully realized in the mathematical structure.

107 Hankins, *Hamilton*, pp. 193–4. *Treatise*, pp. 234–41, on pp. 234–5.
108 *Treatise*, p. 238; also pp. 241, 245, and 249.

12
The hydrodynamics of matter

> I have changed my mind greatly since my freshman's year when I thought it so much more satisfactory to have to do with electricity, than with hydrodynamics, which only first seemed at all attractive when I learned how you had fulfilled such solutions as Fourier's by your boxes of water. Now I think hydrodynamics is to be the root of all physical science, and is at present second to none in the beauty of its mathematics. *William Thomson to G.G. Stokes, 1857*[1]

In the preceding five chapters we have followed the development of Thomson's views on electromagnetism, heat, and mechanics in their general, typically macroscopic form, ending in each case with an underlying conception of dynamical theory as hydrodynamics. We turn now to his explicit attempts to realize this hydrodynamical view of the universe in a physical theory. While the earlier works produced major contributions to what is today known as 'classical' physics, the later ones did not. They have largely disappeared even from the mythology of physics. A variety of reasons for this divergence between earlier and later phases of Thomson's career may be cited: his theories became increasingly speculative and ill-founded, his methodology departed increasingly from that of the younger generation, and the industrial orientation of his work never became institutionalized in Britain. All of these factors will emerge in what follows. They constitute, however, a retrospective image, visible to his contemporaries only after about 1885. From the publication of his 'Vortex atoms' in 1867 to the Baltimore Lectures of 1884, Thomson rode the heights of an incomparably prestigious life of science. The vortex atoms belong to a rising image both of the man and of the prospects for a continuum theory of ether and matter. The present chapter treats this positive image. The one following describes its decline.

1 William Thomson to G.G. Stokes, 20th December, 1857, K101, Stokes correspondence, ULC. The boxes of water appear in G.G. Stokes, 'On some cases of fluid motion' (read 29th May, 1843), *Trans. Cam. Phil. Soc.*, **8** (1849), 105–37, 409–14; *Mathematical and physical papers* (5 vols., Cambridge, 1880–1905), **1**, pp. 17–68, esp. pp. 60–8.

General relations of ether and matter

As discussed previously, 1851 marks a great watershed in Thomson's career, after which the search for a consistent physical theory of ether and matter became his constant preoccupation (ch. 10). At the centre of that search sat the notion of air–ether, or simply 'aer', filling all space. As he remarked to Tait in 1862, the 'solar system is moving through space full of aer (ae-the-r) with a constantly changing motion'.[2] The term 'aer' implied a unity of ether and normal matter, a common foundation for both and a continuous transition from the one to the other.

Although the role of ether as the carrier of waves of light and radiant heat had become a commonplace subject of natural philosophy, its electrical, magnetic, and thermodynamic attributes had not, and especially not in relation to matter itself. Thus Thomson's proposal to treat matter and ether as structures of the same kind in an underlying continuous fluid found little immediate support. Tait's sceptical response to Thomson's interstellar 'air' is typical: '*That* is one of the great stumbling blocks between us as joint authors. I can't rightly appretiate [*sic*] your idea of unlimited atmosphere – I have seen *hints* of it in your papers, but *no reasoning*. Why not say matter?' Thomson merely shrugged. 'Matter or medium if you please – But air seems to me simpler'.[3]

Even George Stokes responded negatively when in 1854 Thomson first suggested to him the continuity between air and ether. Seeking at that time a mechanical explanation both of the source of the sun's heat and of its transmission to earth (ch. 14), Thomson revived his interest in a 'mechanical theory' of spectral lines, which Stokes had suggested earlier. Stokes's mechanical theory derived from the coincidence, first noted by Fraunhofer in 1823 and confirmed in accurate measurements by Stokes's acquaintance W.H. Miller in 1837, between the bright double D line of sodium and corresponding dark double lines in the solar spectrum. The theory actually consisted of an analogy between light and sound, especially between the emission and absorption of spectral lines by molecules and the analogous behaviour of piano wires.[4] Mechanical vibrations in the molecules of certain substances would excite mechanical vibrations in the ether and thus give rise to characteristic waves of light, or spectral lines, like the characteristic notes produced by piano wires vibrating in air. The same waves of light might be reabsorbed by resonant vibration if they impinged on the same substances elsewhere. Thus, absorption of the sun's radiation in its own

[2] *Treatise* draft, 'Axiomata sive Leges Motûs', NB47, ULC, facing p. 3 (received by Tait, 16th January, 1862). [3] *Ibid.*, facing p. 5.

[4] Compare the Thomson and Stokes letters between 20th February and 9th March, 1854, K62–6, Stokes correspondence, and S366–8, ULC, with those of 1st–7th July, 1871, K174, NB21.42, NB21.44, and K176, Stokes correspondence, ULC. The latter were written in preparation for Thomson's 'Presidential address', *BAAS Report*, **41** (1871), lxxxiv–cv; PL, **2**, 132–205, esp. pp. 169–72.

atmosphere or the earth's atmosphere would produce the familiar dark lines in its spectrum.

Stokes apparently thought of the piano analogy in 1851–2 at the time of his discovery of fluorescence, which he and Thomson discussed at length in relation to spectral lines.[5] Thomson jumped to the conclusion that Stokes's mechanical analogy offered not only a possible but a very nearly necessary explanation of the coincidence of bright and dark lines. He assumed that whatever dark lines existed in the solar spectrum necessarily corresponded to bright lines emitted by its constituent substances, thus promising a qualitative analysis of the sun's chemical composition. He began teaching that theory to his students in 1852–3 and exhorted Stokes repeatedly in 1854 to extend his experimental investigations. Ever afterwards he believed he had learned the theory from Stokes. His cautious friend remembered differently: 'I did not at the time think that all, or even I think most, of the dark solar lines were to be connected with bright lines . . . [but] were due to absorption by compound gases formed in the cooler parts of the Sun's atmosphere . . . When *you* jumped to the conclusion (since borne out) that to find what elements were in the sun and stars we must examine the bright lines in artificial flames *I* thought you were going too fast ahead'.[6]

Although chemical analysis of the sun was one of Thomson's immediate interests in 1854, he wished also to know whether Stokes could make any 'decided mathematical investigation of your mechanical theory'; and 'Can you investigate mechanically the undulatory theory of radiant heat?'[7] Presupposing the newly established dynamical theory of heat, Thomson sought the mechanical connection of bodily to radiant heat, or of molecular motions to ether motions. And thinking of that microscopic molecule–ether relation in analogy with the grand relation of sun and interplanetary space, he stated, perhaps for the first time, a version of the famous black-body problem: 'E.G. A hot black ball, in the centre of a hollow black sphere, each of given temperature. What is the wave length, or lengths, of the undulations. The wave lengths as experiment shows are less the [greater the] temperature of the hot body. It is a splendid subject for mathematical investigation'.[8] Although unable to produce a detailed model of molecular and ether vibrations, he nevertheless calculated macroscopic energy

[5] See the Stokes–Thomson correspondence from14th August to 15th November, 1851, K51–2, Stokes correspondence, and S362–5, ULC.

[6] Description summarized from letters of note 4. Quotation from Thomson to Stokes, 5th July, 1871, NB21.42, Stokes correspondence, ULC. Thomson wished to establish Stokes's priority over the famous works of Foucault, Bunsen, and Kirchhoff, at least in the 'dynamical theory', if not as 'practically established' by spectrum analysis. He first staked this claim publicly in a letter published by Kirchhoff in the 1866 English translation of his own classic paper (see Thomson's account in BL, 100–3) and then in his paper 'On vortex atoms' (read 18th February, 1867), *Proc. Royal Soc. Edinburgh*, **6** (1869), 94–105, on p. 96; MPP, **4**, 1–12, on p. 3. Stokes had in fact said in a letter of 7th March, 1854, S367, ULC: 'But we must not go on too fast. This explanation I have not seen, so far as I remember, in any book, nor do I know a single experiment to justify it'.

[7] William Thomson to G.G. Stokes, 2nd and 9th March, 1854, K64 and K66, Stokes correspondence, ULC (quotations from the latter).

[8] *Ibid.*, 9th March.

relations between sun and ether. The sun's total heat output per square foot per second, judging from the fraction intercepted at the earth's surface, was 7900 horsepower, comparable with the work produced by the most powerful marine engines. At the earth's distance from the sun, therefore, the ether contained at any time about 10^{-7} foot-pounds per cubic foot, or in Thomson's practical phrasing, 'The mechanical value of a cubic mile of sunlight is consequently 12050 foot-pounds, equivalent to the work of one-horse power for a third of a minute'.[9]

On the dynamical theory, all of this energy had to exist as kinetic and potential energy in a medium possessing mass, and 'forming a continuous material communication throughout space'. Its energy content allowed an estimate of its minimum mass density. On the reasonable assumption that the vibrating particles of ether had velocities no larger than one-hundredth the velocity of light, he obtained an ether density, at the earth's distance from the sun, of around 10^{-20} pounds per cubic foot. 'Whether or not this medium is (as appears to me most probable)', he speculated, 'a continuation of our own atmosphere, its existence is a fact that cannot be questioned'. For comparison he calculated the density the air would have if it continued into interplanetary space at a constant temperature, decreasing linearly with pressure according to Boyle's law. Incredibly, the ether had to be 10^{200} or 10^{300} times as dense as this continuation of the air. 'What is the lum^s medium then,' he asked Stokes.[10]

Stokes found nothing unreasonable in the density estimate for ether and had contemplated such a calculation himself. But he could not countenance Thomson's 'aer'. 'I am altogether sceptical about the existence of air in the planetary spaces', came Stokes's reply, 'but if it do exist I have no confidence in the truth of Boyle's law when pushed to such limits'.[11] Puzzled, Thomson asked, 'How can you think the air stops? Boyle's law need not hold of course. I have never seen or heard of any valid reason for supposing the air to stop.'[12] Stokes seems to have reasoned that if our air were an interplanetary medium its density ought to vary with the seasons as the earth's distance from the sun changed, just as the density of our planetary atmosphere decreases with altitude. But Thomson remained undeterred:

I still believe in the continuity of atmosphere through space, and I have no doubt but that the difficulty you show . . . will be explained by taking into acc^t the centrifugal force due to the revolution round the sun of a portion of it carried round with the earth, or else, by

[9] William Thomson, 'Note on the possible density of the luminiferous medium and on the mechanical value of a cubic mile of sunlight' (read 1st May, 1854), *Trans. Royal Soc. Edinburgh*, **21** (1857), 57–61; MPP, **2**, 28–33, on p. 29. This estimate and those below appeared first in the letters to Stokes.

[10] *Ibid.*, 28. William Thomson to G.G. Stokes, 2nd March, 1854, K64, Stokes correspondence, ULC. The larger estimate is the published one.

[11] G.G. Stokes to William Thomson, 7th March, 1854, S367, ULC.

[12] William Thomson to G.G. Stokes, 9th March, 1854, K66, Stokes correspondence, ULC.

considering, w^h is very probable, the whole interplanetary atmosphere to be revolving round the sun. I am much disposed to go back to the Vortices, differing only from DesCartes in being dragged round by the planets instead of drag[g]ing them round.[13]

The immediate reference for this latter speculation was Thomson's theory that the sun's heat derived from a vortex of meteors moving under solar gravitation and gradually collapsing into the sun after losing energy in friction with the ether (ch. 14). The meteoric vortex would thus drag with it an accompanying ether vortex, in which density and pressure were controlled not by gravity but by the hydrodynamics of the etherial fluid. This ether vortex expands dramatically the scope of Thomson's concern to relate matter and ether. The correspondence with Stokes establishes the following picture: between 1851 and 1854 Thomson developed a firm attachment to his air–ether as forming a 'continuous material communication' throughout space; that he proposed to analyze the dynamics and thermodynamics of ether just as he would any other material substance; and that he thought of the sun–ether relation as a cosmic analogue of the molecule–ether relation. As these points would suggest, the ether vortex did not stand in isolation. It belonged to Thomson's rapidly developing views on the molecular dynamics of heat and elasticity, that is, on molecular vortices.

He had long ago contemplated the effect of vortex-like motions in producing elasticity as he stirred his cup of *chocolat au lait* in Paris in 1847. 'The twisting motion (in eddies, and in the general variation of angular vel . . .) in becoming effaced, always gave rise to . . . oscillations like an elastic (incompressible) solid', he wrote to Stokes.[14] Not until he had read Davy's views on 'repulsive motion' in 1849–50, however, did he attempt to unite heat and elasticity as modes of molecular motion.[15] Rankine's molecular vortices certainly contributed to that effort (ch. 10). In fact, Rankine conceived his vortices about atomic centres much as Thomson would conceive the ether vortex round the sun, assuming that 'the changes of condition and elasticity due to heat arise from the centrifugal force of revolutions among the particles of the atmospheres'.[16] And Rankine had schematized a mechanical theory of the relation between radiant and bodily heat. He supposed that the bare atomic centres, possessing very small mass and exerting attractive forces on one another, constituted the elastic solid ether,

[13] William Thomson to G.G. Stokes, 21st March and 20th April, 1854, K68, Stokes correspondence, ULC (quotation from latter).

[14] William Thomson to G.G. Stokes, 20th October, 1847, K21, Stokes correspondence, ULC. Full quotation in ch. 8, following note 63.

[15] See ch. 10, notes 47, 80, 84, and 85, and Thomson's retrospective account in his presidential address to the British Association meeting at Montreal in 1884, 'Steps toward a kinetic theory of matter', *BAAS Report*, **54** (1884), 613–22; PL, **1**, 225–59, on pp. 229–31.

[16] W.J.M. Rankine, 'Laws of the elasticity of solids', *Cam. and Dublin Math. J.*, **10** (1851), 47–80, 178–81, 185–6, on p. 67. As editor, Thomson objected to one of the crucial proofs and enlisted Stokes to referee Rankine's new proof (pp. 178–81); see Thomson to Stokes, 25th February, 1851, Stokes correspondence, K46, ULC.

whose vibrations propagated waves of light and heat. Normal matter, with its atomic centres dressed in atmospheres of self-repulsive fluid, absorbed radiant energy by somehow converting the wave vibrations of its atomic centres into thermal rotations in the atmospheres. With respect to elasticity, such increase in thermal rotations increased volume compressibility at the expense of the extreme rigidity of the structure of atomic centres. Consequently, absorption of heat gradually converted elastic solids into liquids and finally gases, as it ought to do.

Discussing this scheme in his Glasgow lectures for 1852–3, Thomson already treated the atomic centres as 'nuclei or solid atoms of air', rather than ether, having remarked in November that 'An ether has been assumed – a luminiferous ether. No proof has been given of the sudden ceasing of air. It is more probable that it [the ether] is matter than that it is not'. The remark followed a discussion of the analogy between waves in air and ether, in which, despite vast differences in velocity of propagation, frequency, and manner of propagation, Thomson stressed similarities. Air propagated the slow longitudinal vibrations associated with a highly compressible fluid, while ether propagated the rapid transverse vibrations of a quite rigid solid, but he believed both media would exhibit both kinds of waves when observed under the proper circumstances. 'It is probable that vibrations like those of sound [longitudinal] are propagated in ether at the same time [as those of light]', he told his class, but since the ether was nearly incompressible the longitudinal waves travelled at effectively infinite velocity. He would insist on this conception to the end of the century (ch. 13). Reciprocally, he had queried Stokes in February whether 'the velocity of sound is affected by some solid elasticity of air [implying transverse vibrations] existing during the rapid vibrations of sound, but not existing at all or not appreciably in any *statical* circumstances of air'.[17]

Rankine's molecular vortices provided a suggestive model for the continuity between waves in ether and air, showing how to transform a rigid, nearly incompressible ether continuously into a perfectly fluid, but highly compressible air by adding rotational motion. His scheme suffered, however, from the

[17] William Jack, 'Notes of the Glasgow College natural philosophy class taken during the 1852–53 session', MS Gen. 130, ULG, prior to 25th November, 1852; William Thomson to G.G. Stokes, 16th February, 1852, K55, Stokes correspondence, ULC. Thomson's immediate concern in the letter was with Mayer's hypothesis, which required that isothermal compression of air involves no change in internal potential energy, but only heat expelled equivalent to work done. Initial experiments by Joule on the specific heat of air at constant pressure, coupled with Mayer's hypothesis, gave a ratio of specific heats at constant volume and at constant pressure markedly smaller than that determined from measurements on the velocity of sound. Thus either Mayer's hypothesis was false or the connection of sound velocity with specific heats required modification by including something like solid elasticity. Within a week Joule obtained experimental agreement and Thomson withdrew his suggestion of solid elasticity, in William Thomson to G.G. Stokes, 20th February, 1852, and 31st May, 1854, K56 and K71, Stokes correspondence, ULC. The original suggestion probably derived from remarks of Rankine's 'On the velocity of sound in liquid and solid bodies of limited dimensions, especially along prismatic masses of liquid', *Cam. and Dublin Math. J.*, **10** (1851), 238–67.

assumption of attraction between atomic centres and between the centres and their atmospheres, which violated Mayer's hypothesis and Joule's experiments, by implying that 'There is work done by expanding air then, against the cohesion of the atoms, besides the external effect of heat'.[18] It was precisely to settle this question that Thomson and Joule had planned their famous series of experiments on air forced through a porous plug in the spring of 1851. They intended, Thomson wrote to Stokes, 'to test the relation between the heat absorbed & the mechanical effect emitted by air expanding at a constant temperature, which we propose to do by making it waste all its work on fluid friction'. By late 1852 they had found some deviation from Mayer's hypothesis but nothing like that required by Rankine's model.[19] Thus Thomson had good reason to seek a molecular theory of gases in which energy added or subtracted would be nearly all kinetic, as in Herapath's theory of action by impact, but the new theory would have to preserve the continuous transition from elastic-solid ether to gaseous fluids of Rankine's molecular vortices.

Magnetic rotations

Elasticity and thermodynamics tell only half the story of Thomson's molecular theorizing; the other half derives from electricity and magnetism. Electricity he tended to conceive as a fluid, whose accumulations produced tensions in ether and air, and currents of which produced the rotational strains of magnetism. A late but telling indication of the relation of those views to the dynamical theory of heat appears in the reprint of his old 1845 paper on the impossibility of Poisson's inverse square electrical fluid having a thickness at the surface of conductors. He had concluded that 'Any thickness . . . must depend on a force of elasticity, or on a force generated by the contact of material points, and in either case will require an ultimate law of repulsion more intense than that of the inverse square'. He troubled to remark in 1869: 'This was written without knowledge of Davy's "repulsive motion," and without the slightest idea that elasticity of every kind is most probably a result of motion'.[20] Apparently he wished to stress not only that electricity could once again be regarded as a fluid, with forces of elasticity replaced by motions, but also that heat, Davy's repulsive motion, was electricity in motion.

[18] Jack, *op.cit.* (note 17), immediately preceding entry dated 26th January, 1853.

[19] Thomson suggested the experiments in 'On a method of discovering experimentally the relation between the mechanical work spent, and the heat produced by the compression of gaseous fluid' (read 21st April, 1851), *Trans. Royal Soc. Edinburgh*, **20** (1853), 289–98; MPP, **1**, 210–22, having queried Stokes on 10th April, 'Do you think, when air is compressed any appreciable portion of the mechanical effect produced by the work spent is statical?', K47, Stokes correspondence, ULC. Quotation, William Thomson to G.G. Stokes, 9th May, 1851, K50, Stokes correspondence, ULC. Thomson and Joule communicated initial results to the British Association in September 1852, published as 'On the thermal effects experienced by air in rushing through small apertures', *Phil. Mag.*, [series 4], **4** (1852), 481–91; MPP, **1**, 333–45.

[20] E&M, 103. See chapter 7, at note 27.

That these motions of electrical fluid, constituting heat and elasticity, were in fact vortex motions, stemmed from the connection of magnetism with elasticity. Thomson had shown in his 1847 'Mechanical representation' that magnetism, including electromagnetism, could be represented by rotational displacement about lines of magnetic force, which seemed also to be required by Faraday's magneto-optic rotation. Having adopted the dynamical theory of heat and elasticity by 1851, he began immediately to investigate the relation of heat to electricity and magnetism, or 'thermo-electricity' and 'thermo-magnetism'. His studies of thermo-electricity in linear conductors (e.g. wires) were important for establishing macroscopic energy relations between heat and electricity. Of greater importance for Thomson's molecular models, however, was the possibility of *rotatory* thermo-electric currents in solids, and even more intriguing, simple rotatory conduction of heat. The thermo-electric rotations would certainly produce magnetism, like any other electric current; but so might rotational heat conduction, if it involved any convection of electricity. Both effects would allow an investigation at the macroscopic level of connections between heat, electricity, and magnetism that could be expected to hold also for molecules.

This connection of macroscopic and molecular rotations is particularly evident in a remark that Thomson made to J.P. Joule in response to Joule's suggestion of an analogy between heat engines and electromagnetic engines (electric motors). 'I am inclined to think', wrote Thomson, 'that an electric current circulating in a closed conductor is *heat*, and becomes capable of producing thermometric effects by being frittered down into smaller local circuits or "molecular vortices"'.[21]

Discussions of rotational heat flow arose between Thomson and Stokes early in 1851, with respect to a paper Stokes was writing for the *Cambridge Mathematical Journal* on conduction of heat in crystals. Stokes argued from symmetry considerations that the nine arbitrary constants of conductivity, which in a crystal would connect the three rectangular components of flux of heat with the three corresponding temperature gradients, must necessarily reduce to six, independent of any hypothesis on the nature of heat conduction. Otherwise conduction could occur in a rotatory fashion, spiralling outwards from a point source. 'This *rotatory* sort of motion of heat' seemed at the least 'very strange'.[22] Coincidentally, Thomson had just completed a paper of his own on magnetic induction in crystals containing the analogous conclusion that the constants of

[21] William Thomson to J.P. Joule, 31st March, 1852, quoted in J.P. Joule, 'On the œconomical production of mechanical effect from chemical forces', *Phil. Mag.*, [series 4], **5** (1853), 1–9. In Joule to Thomson, 26th March, 1852, J111, ULC, Joule requests an opinion on the relation between heat and electricity and expresses his own suspicion that the fluid theory of electricity 'will be ultimately overturned, exactly as the theory of caloric'.

[22] G.G. Stokes, 'On the conduction of heat in crystals', *Cam. and Dublin Math. J.*, **10** (1851), 215–38, on p. 236. See G.G. Stokes to William Thomson, 6th January, 1851, S360, ULC, and Thomson to Stokes, 13th January and 25th February, 1851, K44 and K46, Stokes correspondence, ULC.

inductive capacity reduced to six. The two papers thus demonstrated once again the mathematical analogy of heat and magnetism. But Thomson's proof, although it too rested on the rotational character of a non-symmetric set of constants, derived from energy considerations. He had formulated the argument already in 1848 and applied it to the problem of induced magnetism in iron ships (ch. 22). It simply asserted the impossibility of any perpetual source of work, or, specifically, that 'a sphere of matter of any kind, placed in a uniform field of force, and made [constrained] to turn round an axis fixed perpendicular to the lines of force, cannot be an inexhaustible source of mechanical effect'.[23] Although Stokes's remark on the strangeness of rotatory conduction struck Thomson as 'corresponding to the proof I give', he apparently did not take it as definitive, perhaps because Stokes did not employ any general physical principles, such as those of energy, which had become a prerequisite of mathematical theory for Thomson.[24]

The central role of the laws of energy as heuristic principles guiding Thomson's research is apparent in a November letter where he reported being 'greatly engrossed with electrodynamics especially in connection with the principle of mechanical effect. I think I have got a good foundation for a theory of the mechanical effect of thermo-electric currents'. In the next paragraph he announced a rule for finding the 'mechanical value of a current of given strength' in terms of the work gained in assembling it, and immediately stated: 'I am quite convinced that diamagnetics are only bodies less magnetizable than space (i.e. I suppose the luminiferous medium)'. The only apparent unity in these loose associations, between heat, electric currents, magnetism, and ether, lies in mechanical effect. The transitions indicate the direction of Thomson's physical speculations, but also their turbulent state. For the moment he took thermo-electricity as his specific subject and worked out the purely macroscopic relations of mechanical effect required by the First and Second Laws of Thermodynamics. In addition he began a series of accurate experimental measurements of those relations, which continued until 1854 in his rapidly developing Glasgow laboratory.[25]

23 William Thomson, 'On the theory of magnetic induction in crystalline and non-crystalline substances', *Phil. Mag.*, [series 4], **1**, (1851), 177–86; E&M, 465–81, on p. 480.

24 Quotation, William Thomson to G.G. Stokes, 25th February, 1851, K46, Stokes correspondence, ULC. See Thomson's retrospective justification added in 1882 to 'On the dynamical theory of heat, part VI: thermo-electric currents' (read 1st May, 1854), *Trans. Royal Soc. Edinburgh*, **21** (1854), 123–71; MPP, **1**, 232–91, on pp. 280–1*n*. Discovery of the Hall effect for *electric currents* in a magnetic field had vindicated the principle of his original assumption about *heat conduction* in crystals.

25 See William Thomson to G.G. Stokes, 8th November, 1851 (quotation), 21st December, 1852, and 20th February, 1854, K52, K61, and K62, Stokes correspondence, ULC. William Thomson, 'On a mechanical theory of thermo-electric currents' (read 15th December, 1851), *Proc. Royal Soc. Edinburgh*, **3** (1851), 91–8; MPP, **1**, 316–23; 'Experimental researches in thermo-electricity' (dated 30th March, 1854), *Proc. Royal Soc.*, **7** (1854), 49–58; MPP, **1**, 460–8; 'Account of experimental researches in thermo-electricity', *BAAS Report*, **24** (1854), 13–14; MPP, **1**, 469–71. B.S. Finn, 'Development in thermoelectricity, 1850–1920', Dissertation, University of Wisconsin, 1963, pp. 21–53, gives a useful summary.

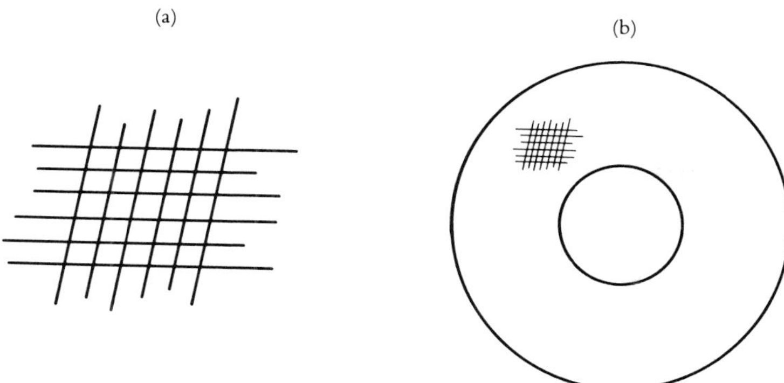

Figure 12.1. In a section of a cylindrical solid (b), having an oblique crystalline structure analogous to that of cloth (a) woven from brass wires in one direction and iron in the other, Thomson thought that a radial temperature gradient would produce circular thermo-electric currents.

Then began in earnest his fascination with rotations, from the solar to the molecular level. In the spring of 1854 he wrote to Stokes not only of his meteoric vortex theory of the sun's heat, his 'aer' density calculations, and the problem of spectral lines, but also of producing a bar magnet by means of thermo-electric rotations in suitably non-symmetric crystals and artificially constructed solids. For example:

> I have got a piece of wire cloth woven of brass and iron wire, in the two directions; & by pulling it oblique [figure 12.1a] I shall have a representative of thermo electric crystalline obliquity. If a solid possessing this property be cut into a cylindrical shape [figure 12.1b], and the axis hollowed & heated, while the outer surface is kept at a uniform temperature, circular currents will be produced, & it will for the time act like a bar magnet.

He now believed that he could 'prove, both for thermo-electricity, and for the conduction of heat, the possibility of making a structure' which would possess 'the wonderful [rotatory] conductive properties' which Stokes had thought would never occur.[26]

Stokes dissented, at least with respect to conduction. For three months the two argued back and forth over whether or not it would be possible 'to compose a structure possessing the obliquity as regards thermal conductivity, (or rotatory conductivity) out of materials which do not possess it', with Thomson proposing examples and Stokes finding flaws. Reluctantly Thomson finally gave up on this 'structural' rotational conduction, having convinced himself for the moment that his perpetual motion argument for induced magnetization applied also to heat conduction, when considered alone. But still he held on to thermo-

[26] William Thomson to G.G. Stokes, 21st March and 20th April, 1854, K68, Stokes correspondence, ULC. He had proposed this thought experiment already on 2nd March, 1854, K64, Stokes correspondence, ULC.

electric rotation, in the belief that what one agency alone could not do, 'the *two* agencies of electric conduction & thermo-electric force' could still accomplish. There the debate ceased, with Stokes unconvinced even of the latter effect and with Thomson quite sure that in general conduction of either heat or electricity would require nine independent coefficients.[27]

In his major paper on thermo-electricity, read to the Royal Society of Edinburgh on 1st May, 1854, Thomson devoted a large section to rotations, both in their mathematical form and in illustrative structures, without, however, having yet observed such rotations experimentally. His confidence seems to have derived largely from Faraday's magneto-optic rotation, which displayed the required type of asymmetry, called 'dipolarity' by Stokes and Thomson. Thus he expected to find that magnetized iron possessed a 'dipolar' axis of thermo-electric rotation parallel to the lines of magnetization, so that a disc cut perpendicular to the lines would behave like the oblique grate of iron and brass wires described above. If its centre were held at one constant temperature and its entire circular edge at another, the resulting thermo-electric current would exhibit at every point, in addition to the expected radial component, a tangential component, yielding a circular current around the disc. This relation of radial and circular motions would be the same as that for a light wave travelling along lines of magnetization, for the wave displacement transverse to the lines obtained a rotation about them. Thomson's experiments did not reveal the rotational currents. Nevertheless, he did not hesitate, in an 1882 note, to claim virtual priority for the 'Hall effect', at the expense of Stokes: 'Hall's recent great discovery shows that the hypothesis which 28 years ago I refused to admit [Stokes's hypothesis of symmetry], was incorrect, and proves the rotatory quality to exist for electrical conduction through metals in the magnetic field'.[28]

After an interlude marked by furious activity on telegraphy during the

[27] Quotations from Thomson's final letter, 7th June, 1854, and the preceding one, 31st May, K72 and K71. Stokes correspondence, ULC. Stokes sent his enduring reservations on 17th June, S373, ULC.

[28] William Thomson, 'On the dynamical theory of heat, Part VI', pp. 123–71; MPP, **1**, 232–91, esp. pp. 273–87; experimental results on magnetized iron in notes added July, 1854 and 13th September, 1854, on pp. 286–7; notes on Hall effect, pp. 280–1, 288; nine independent coefficients required, pp. 288, 291. See William Thomson, 'On the effects of mechanical strain on the thermo-electric qualities of metals' and 'On the electric qualities of magnetized iron', *BAAS Report*, **25** (1855), 17–18, 19–20; MPP, **2**, 173–4, 178–80. In 'Experiments on the electrodynamic properties of metals', PA140, ULC, (apparently a brief report to the Royal Society in 1854 or 1855 on the mechanical strain and magnetization experiments, which they funded) Thomson still concludes that either agency produces the 'thermo-electric properties of a crystal', with magnetization producing a single 'axis of thermo-electric symmetry' (the rotatory property). He described his continuing experiments at great length, but omitting rotations, in his Bakerian Lecture, 'On the electrodynamic qualities of metals' (read 28th February, 1856), *Phil. Trans. Royal Soc.*, (1856), 649–751; MPP, **2**, 189–327; esp. pp. 267–327. He proposed this and an alternative lecture 'On the origin and transformation of rotatory motion', produced by gravitation between bodies, in a letter to Edward Sabine, president of the Royal Society, on 17th October, 1855, S10A, ULC. On the discovery of the Hall effect, see J.Z. Buchwald, *From Maxwell to microphysics: aspects of electromagnetic theory in the last quarter of the nineteenth century* (Chicago, 1985), pp. 73–95.

autumn and winter (ch. 13), Thomson began anew on magnetic rotations, this time from the perspective of elasticity and its connection with heat. 'Can you tell me', he asked Stokes in March, 1855, 'whether Faraday's optical effect of magnetism has been reduced to elastic forces?' Once again adapting Stokes's analysis, he represented the required elasticity in terms of one set of constants for 'isotropic' strain, the same in all directions, and a second set for rotational strain about an axis of 'dipolarity'. So began a development of ideas which would culminate in 1856 in his seminal 'Dynamical illustrations of the magnetic and the helicoidal rotatory effects of transparent bodies on polarized light'.[29] Already in March, 1855, Thomson began a series 'On the thermo-elastic and thermo-magnetic properties of matter'. Part I, on thermo-elasticity, is all that appeared. In it he analyzed, from a thermodynamic viewpoint, the conditions on the six variables defining the reversible straining of a cube into a parallelepiped (the lengths of three edges and the angles between three planes, for dilation and shear, respectively) in relation to the six corresponding components of stress. Others had argued from various mechanical considerations that the thirty-six coefficients connecting the stress and strain components reduced to twenty-one. Thomson now showed that this result followed from the Second Law of Thermodynamics alone, as a perfectly general and macroscopic condition, independent of any hypothesis of molecular structure.[30]

He also asserted that any further reduction was in general invalid, for 'it is quite certain that an arrangement of actual pieces of matter may be made, constituting a homogeneous whole when considered on a large scale . . . which shall have an arbitrarily prescribed value for each one of these twenty-one coefficients'. Swiping at atomists (of the hard atoms sort) and 'mathematicians' who would impose arbitrary conditions 'on the infinitely inconceivable structure of the particles of a crystal . . . for the sake of shortening the equations', he pointed out that the principal-axis reduction in particular, which yielded three perpendicular axes of symmetry, was invalid. Such symmetry would have eliminated his dipolar axes, or axes of rotation.[31]

Thomson's analysis effectively showed, on the other hand, that no infinitely

[29] William Thomson to G.G. Stokes, 14th March, 1855, K79, Stokes correspondence, ULC; William Thomson, 'Dynamical illustrations of the magnetic and the helicoidal rotatory effects of transparent bodies on polarized light' (received 10th May, 1856; read 12th June, 1856), *Proc. Royal Soc.*, **8** (1856), 150–8; *Phil. Mag.*, [series 4]. **13** (1857), 138–204; *Baltimore Lectures on molecular dynamics and the wave theory of light* (London, 1904), pp. 569–77.

[30] William Thomson 'On the thermo-elastic and thermo-magnetic properties of matter, Part I' (dated 10th March, 1855), *Quart. J. Math.*, **1** (1857), 57–77; MPP, **1**, 293–313. See also, William Thomson, 'Elements of a mathematical theory of elasticity' (read 24th April, 1856), *Phil. Trans. Royal Soc.*, (1856), 481–98. On thermo-magnetism, Thomson published only a brief account on the change of magnetic susceptibilities with temperature and on heating and cooling effects that would attend motions of a magnetized crystal in a magnetic field. William Thomson, 'Thermomagnetism', in J.P. Nichol (ed.), *A cyclopaedia of the physical sciences*, 2nd edn. (London and Glasgow, 1860); extract in MPP, **1**, 313–15. In the notes on the history of thermodynamics that he sent to Tait in 1864(?), Thomson observed that 'No expt has ever been made in verification of Thermomagm.', NB52, ULC, pp. 39–40.

[31] MPP, **1**, 304–7.

homogeneous solid (homogeneous down to infinitesimal dimensions) could possess either helicoidal or dipolar asymmetry. On this fact he built his 'Dynamical illustrations', in the spring of 1856, arguing that both the natural, helicoidal rotations of the plane of polarized light (produced in, for example, quartz and syrup) and the magnetic, dipolar rotations discovered by Faraday necessarily involved a heterogeneousness of structure at dimensions not incomparably less than a wavelength, and thus a molecular heterogeneousness in normal matter. A liquid filled with spiral fibres, all left-handed or all right-handed, would illustrate the required heterogeneousness for helicoidal rotations, which exhibited always the same sense of rotation, independent of direction of transmission. Magnetic rotations, however, could not depend on such spiral structures, because the rotations changed sense from right-handed, for light transmitted along the lines of force, to left-handed for transmission in the opposite direction. Thomson concluded that magnetic rotations necessarily depended on 'the direction of motion of moving particles' about the lines of force, and claimed to be able to prove that no other dynamical explanation was possible.[32]

Faraday's discovery therefore demonstrated, in Thomson's view, that magnetism consisted in absolute rotations of matter, as in Ampère's electrodynamic molecules and Rankine's molecular vortices. If so, it provided 'a definition of magnetization in the dynamical theory of heat'. Magnetism would consist in a net alignment of axes of the rotational motions constituting heat, and the resultant angular momentum of any element would measure its magnetic moment. Confident of this basic rotational model, Thomson nevertheless veiled his more detailed ideas in cryptic qualifiers:

> The explanation of all phenomena of electro-magnetic attraction or repulsion, and of electro-magnetic induction, is to be looked for simply in the inertia and pressure of the matter of which the motions constitute heat. Whether this matter is or is not electricity, whether it is a continuous fluid interpermeating the spaces between molecular nuclei, or is itself molecularly grouped; or whether all matter is continuous, and molecular heterogeneousness consists in finite vortical or other relative motions of contiguous parts of a body; it is impossible to decide, and perhaps in vain to speculate, in the present state of science.[33]

In vain perhaps, but precisely the latter speculations on vortical motions in a continuous electrical fluid were his own.

During the remainder of 1856 and 1857 Thomson devoted much of his energy to the first (failed) Atlantic telegraph; but the technical activity did not dampen his speculations on ether and matter. In May, June, and December, 1857 he wrote to Stokes of his hopes for a theory of rotating 'motes' in a perfect fluid.

[32] William Thomson, 'Dynamical illustrations', pp. 150–8, on p. 152. For an excellent discussion of Thomson's argument see Ole Knudsen, 'The Faraday effect and physical theory, 1845–1873', *Arch. Hist. Exact Sci.*, **15** (1976), 235–81.

[33] William Thomson, 'Dynamical illustrations', p. 152.

He now expected not only that the rotations would represent heat, and that an alignment of their axes would constitute magnetism and produce the Faraday effect, but that the rotations would produce on the average an intense repulsion between motes, leading to a stable, stiff structure like that required for luminiferous vibrations. His confidence, from Stokes's perspective, was astonishing, for the mathematics remained completely uncertain. But Thomson's enthusiasm was born of the conviction that 'Faraday's property *cannot possibly be explained without* some such dynamical conditions being admitted . . . (['Dynamical illustrations . . .'] contains what I cannot see but as an unanswerable argument to this effect)'.[34] That conviction informs the epigram of the present chapter: 'hydrodynamics is to be the root of all physical science'.

Early in January, 1858, Thomson entered in his research notebook the most elaborate speculations he had yet dared to put in writing. 'Considering the probable truth of the doctrine of the Universal Plenum,' he began, 'I have been led to think of a fluid filling the interstices between detached solid particles or *molecules* not necessarily *atoms*, (indivisible) . . . and to endeavour to explain some of the known properties of sensible matter by investigating the motion of such a system on strict dynamical principles'. The problems seemed insurmountable: permanence of the motes or molecules, gravitation, universal proportionality between gravitation and inertia, etc. Multiple 'ifs' and 'mights' carried him forward. 'If it were possible to conceive the properties of one particular substance [the motes] to be owing to a particular form & order of motions or eddies in a fluid, and to remain as constant as they do in nature through all combinations and actions of all kinds to which they may be subjected; it might be possible to conceive that all the phenomena of matter might be explained by the consequences of contractility in a universal fluid constituting the material world . . . If this were true, inertia would be the cause of impenetrability, and all repulsion would be explained by Davy's "repulsive motion" '.[35] Thus contractility and inertia, taken as primitives, might function like gravity and inertia in the solar system, as centripetal and centrifugal principles, the former holding the parts of a moving system together and the latter preventing collapse, as though by repulsion. An analogy with Nichol's nebular hypothesis clearly pertained: 'Point atoms with inertia [and gravitation] could certainly never collapse to a point but could only constitute systems (solar systems & nebulae)'.[36] Similarly, but eschewing action at a distance through a vacuum, 'eddies' in the universal fluid might form stable motes, while systems of motes would make up the molecules of normal matter.

These permanent eddies and systems of eddies are the first explicit, if hesitant,

[34] William Thomson to G.G. Stokes, 23rd May, 17th June, 20th December, and 23rd December, 1857, K97, K98, K101, K102, Stokes correspondence, ULC (quotation from K98).

[35] William Thomson, 6th January, 1858. NB35, ULC, pp. 1–11, published, with an introduction and notes, by Ole Knudsen, 'From Lord Kelvin's notebook: ether speculations', *Centaurus*, **16** (1971), 41–53.

[36] Thomson *op. cit.* (note 35), p. 6; Knudsen, pp. 48–9.

form of Thomson's vortex atoms and molecules. Uncertain of their properties, however, he treated them provisionally as merely solid elastic particles. And he had 'not *yet* (!!!) *succeeded* in proving that the general character of the mutual action between two rotating motes, in virtue of fluid pressure, is repulsion (!)'. Nevertheless, the potential rewards were great:

> It seems certain that any motion of a large solid through such a liquid, of one of the motes through it, [or] of the liquid itself while the motes are held back by a strainer or sieve enclosing the space under consideration, will tend to generate, or will lose itself in, rotating motions of the motes in general. This seems an illustration, so far as it goes perfect, of the generation of heat in a liquid by stirring it, and of the generation of heat in a liquid or solid by a current of electricity through it.[37]

The latter speculation is particularly suggestive. If assemblies of motes were arranged in the stable structure of a solid, forming a sieve through which the fluid could flow, and if the fluid were electricity, then the resulting electric current would generate heat, consisting in rotations of the motes and accompanying fluid. Presumably also, a linear current would align the rotations about axes circling the current, like lines of magnetic force. And Thomson thought he could prove that there would be less repulsion along the axes of rotation than perpendicular to them, thus 'perfectly' illustrating magnetic attraction. But the capstone of the system was magneto-optic rotation: 'If the motes are in any way, by mutual repulsion . . . distributed and retained stably in or averaging near fixed positions with the kind of rigidity required for undulations of light . . . and if on the whole their rotations are round axes parallel to one line . . . then, polarized light propagated in this direction . . . must necessarily, as I can now fully demonstrate (see ['Dynamical illustrations . . .']) follow Faraday's magneto-optic law'. With that, he concluded. 'A complete dynamical [theory, *del.*] illustration of magnetism & electromagnetism seems not at all difficult or far off'.[38]

But first he required the abstract hydrodynamics of the motes and their interactions. He recorded his progress: 'Last month (December) I made out a complete foundation for the theory of the motion of a solid of any shape in a perfect liquid; & up to this time without success have made many attempts to find something simple & workable for the mutual action between two solids near one another moving through a perfect liquid'.[39] In December, in fact, he had written to Stokes a full outline of his analysis of translational and rotational motion of a single solid (or mote), pointing out how it would lead to right-

[37] Thomson, *op. cit.* (note 35), pp. 3–5; Knudsen, pp. 47–8. The symbol '(!!!)' is inserted and may have been added, with the emphasis, in February, 1871, the date of other notes. See also Thomson to Stokes, 23rd December, 1857, *op. cit.* (note 34), where Thomson hopes to include elasticity, chemical affinity, and thermo-electricity.

[38] Thomson, *op. cit.* (note 35), pp. 7–9; Knudsen, p. 49.

[39] Thomson, *op. cit.* (note 35), pp. 7; Knudsen, p. 49. Marginal note, probably from February 1871: 'All done long before Feb 1870', when he presented 'On the forces experienced by solids immersed in a moving liquid' (read 7th February, 1870), *Proc. Royal Soc. Edinburgh*, **7** (1872), 60–3; E&M, 567–71.

handed helicoidal rotations of light in a rigid structure of motes, if each were a fragment of a right-handed screw. Similarly, if the motes were rotating spheroids, 'magnetic' rotation would occur.[40] It is apparent here that Thomson is talking about both the elastic-solid lattice of ether and that of normal solids, but it is unclear how these differ. Probably he had in mind a difference in the scale of their latticeworks, since he had previously emphasized the necessity of heterogeneity comparable with a wavelength for magnetic rotation to occur. Two problems remained: to establish the stability of the fluid rotations constituting the motes, and to establish net repulsion between motes.

The first problem Thomson thought he had solved: 'instability, or a tendency to run to [dissipative] eddies, or any kind of dissipation of energy, is impossible in a perfect liquid (a liquid with neither viscosity nor compressibility)'. Stability, he argued, followed from his proof in 1847 that the continuity equation, for a mass of liquid originally at rest and subjected to any change in its boundary, possessed unique solutions. He had shown that the solutions existed when the *vis viva* was minimum. That all-important theorem, suitably generalized, had grounded all of his major results in field theory. He now believed that any instabilities in a liquid would have to derive from friction, or perhaps compressibility, because the minimization condition defined a state of stable equilibrium in a perfect fluid.[41]

Stokes had doubted the generality of this comfortable analysis all along. Not until February, 1858, however, did he spell out his objections to it and to other aspects of Thomson's scheme. The stability argument, he observed, depended on 'the *assumption* of continuity, and I have always been rather inclined to believe that surfaces of discontinuity would be formed in the fluid, i.e. surfaces in passing across which the velocity resolved in a direction tangential to the surface would alter abruptly'. On the latter assumption, he thought, a solid moving through the liquid would experience a resistance, which would dissipate the kinetic energy of the solid into kinetic energy of its wake. Among additional objections, Stokes noted that he had himself attempted long ago to explain Faraday's discovery and that he found Thomson's 'unanswerable' argument problematic. 'I naturally tried rotations of the luminiferous ether as suggested by Ampère's theory but found that the proper law for the rotation as a function of [wavelength] would not come out . . . I certainly am by no means clear that magnetic rotation must be due to motions going on independently of luminiferous vibrations'.[42]

On the following day Stokes added physical illustrations to support his view

[40] Thomson to Stokes, 23rd December, 1857, *op. cit.* (note 34). He had proposed the same idea to Stokes on 23rd May, 1857, *op. cit.* (note 34). A more detailed sketch of the theory, titled 'Abstract hydrodynamics', appears in Thomson's notebook, immediately following the speculation on the 'universal plenum' and with the same date, 6th January, 1858, NB35, ULC, pp. 11–15; portions later extracted for 'On the motion of free solids through a liquid' (read 20th February, 1871), *Proc. Royal Soc. Edinburgh*, **7** (1872), 384–90; MPP, **4**, 69–75.

[41] Thomson to Stokes, 23rd December, 1857, *op. cit.* (note 34).

[42] G.G. Stokes to William Thomson, 12th February, 1858, S391, ULC.

that the solutions obeying the continuity equation were not those realized in nature and that the idealized solutions were in fact *unstable*.[43] Thomson seems now to have found Stokes's arguments to be the unanswerable ones, for there is no evidence that he responded. He did not publish his mathematical results until much later, and did not attempt to develop the theory further until after 1862, by which time he had learned of Helmholtz's 1858 paper on *Wirbelbewegungen*. We return to that story after developing another aspect of the context for vortex atoms.

Engineering and economy

The preceding account of the first appearance of Thomson's vortex atoms has followed the high road of theoretical physics, treating only internal problems of heat, electricity, and magnetism. But, just as his theories in those areas depended profoundly on an engineering perspective, so too did the vortex atoms.

Thomson's view of the condition governing stability in fluid motion goes back to 1847 and the hydrodynamical analogy to magnetism (ch. 8). In July, while he was working out both the minimization condition on *vis viva* and the analogy, Thomson made a trip to Paris to purchase apparatus for his Glasgow laboratory. There he undertook a charge from James Thomson to find out whatever he could about French progress in the development of horizontal waterwheels, for James was perfecting a wheel which he hoped to patent (ch. 9). By good fortune, William met J.V. Poncelet, the doyen of French engineering and 'a most excellent old man'. Poncelet described his own horizontal wheels in some detail, giving data on size and efficiency, which William forwarded to James for comparison.[44] Although William seems initially to have had only a limited idea of the principles of James's wheel, following their exchange of letters and ensuing discussions in Glasgow, he knew them intimately.

James's 'vortex turbine', as he called the machine in his 1850 patent, injected water from outside the wheel in a spiral motion. Having transferred its energy to largely radial vanes as it travelled inwards, the water passed 'down the drain' at the centre (figure 12.2). This spiral injection against radial vanes was new. Poncelet's wheels employed circular injection along tangential vanes which curved gradually inwards. Both types sought to minimize losses of energy attendant on collisions between water and vanes, but while Poncelet achieved that result by gradually redirecting the water – which originally travelled much faster than the wheel – from tangential to radial, James injected the water *at the same speed* as the wheel, and at high pressure, relying on a gradual decrease in pressure towards the centre to carry the water inwards as it lost its motion against the vanes.[45]

The principle is elegantly simple. If the water were to travel continuously in a

[43] G.G. Stokes to William Thomson, 13th February, 1858, S392, ULC.

[44] William to James Thomson, 12th and 22nd July, 1847, T429 and T429X, ULC; James to William Thomson, 29th July, 1847, T434, ULC; James Thomson, *Papers*, xxvii–xxxiii.

[45] James Thomson, 'On the vortex water-wheel', *BAAS Report*, **22** (1852), 130; *Papers*, 2–16.

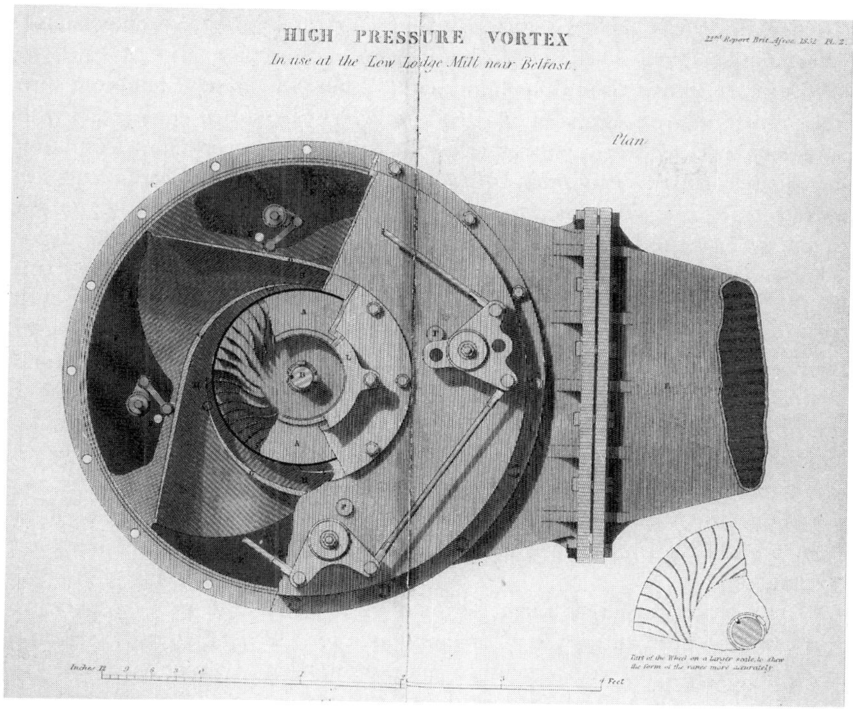

Figure 12.2. In James Thomson's vortex turbine (top, shown from above), water entering from the right is directed by guide blades G tangentially into the central turbine chamber A. As the water flows inwards, losing its tangential velocity, the turbine vanes (inset) curve from radial to nearly tangential, so that the water continues to press against the vanes until it flows quietly away at the centre.

circle at the velocity of a radial vane, a centripetal force would have to act to overcome the 'centrifugal force' of inertia. For a mass m travelling with velocity v at radius r, the centripetal force is mv^2/r. If this force were to push the mass to the centre as it lost its velocity it would do work $mv^2/2$, equal to the original kinetic energy. Therefore, if the water driving the vortex turbine derived its mechanical effect from a fall of height h, one-half the work of the fall ought to be expended in producing motion in the water and the other half in pushing the water inwards against centrifugal force. This principle aims at keeping any given portion of the water as close as possible to a state of equilibrium while it moves from high to low pressure, or equivalently, at extracting work by a reversible process. It is therefore the same general condition that Carnot applied to heat engines. James achieved efficiencies of 75%, equal to that of the best conventional overshot waterwheels. By contrast, Poncelet's scheme converted the entire work of the fall to kinetic energy of the water, relying on the curved vanes of the wheel for centripetal force, and yielded efficiencies of 60 to 68%.

William's hydrodynamic investigations of 1847 were obviously related to

James's concerns in the sense that both were working on the role of mechanical effect in hydrodynamics; but the relation has more structure. William's analysis of the minimization condition on mechanical effect was closely connected with the engineering conception of work, symbolized by the steam-engine, and with attendant questions of economy (chs. 8 and 9). The minimization condition represented nature's economic behaviour in establishing equilibrium states as the states requiring the least expenditure of work, or least action, as Thomson wrote to Stokes (ch. 8). It also represented a state of *conservative* systems, for which alone the equilibrium condition was valid. More generally, in the terms of 1847, conservation meant no dissipation. Reflecting this identification, both William and James understood equilibrium as simultaneously a condition of conservation of mechanical effect and a condition of economy in natural processes. For William equilibrium represented minimum waste, in the sense of minimum mechanical effect contained in a conservative system; for James equilibrium represented minimum waste in extracting mechanical effect. From the beginning of the two hydrodynamics projects, therefore, good reason exists for connecting them as two aspects of a common conceptual structure of work and waste, based, just as in their discussions of thermodynamics, largely on engineering.

This relation developed further around 1850, when William was separating conservation from dissipation in the First and Second Laws of Thermodynamics and when he began to regard heat as a 'repulsive motion'. Thought of in terms of Rankine's models, heat consisted in the energy of rotating vortex atmospheres, and the stability of the vortex derived from the balance of its centrifugal force with the centripetal attraction of the core. Now a steam-engine derives its work from a cyclic expansion and contraction of the steam, produced by adding heat at a high temperature and extracting it at a low temperature. In terms of vortex molecules, that process implies a cyclic expansion and contraction of the atmospheres, adding *vis viva* at the high temperature to produce rotations with larger radius and then collapsing them at the low temperature. The process differs from that of James's vortex turbine in that the work is extracted during the expansion of the vortex rather than during its collapse, and the one operates cyclicly while the other operates continuously, but both depend on adding *vis viva* to a vortex from outside sources and then extracting its mechanical effect by taking advantage of the relation of rotational *vis viva* to the size of a vortex. At the molecular level, then, the steam-engine was a microcosmic vortex machine, operating on the same principles as James's machine but as a reciprocating engine rather than a vortex turbine.

If that analogy seems far-fetched, one need only look to William's initial theory of solar heat for a closer one. The sun shone from a continual influx of meteors in a contracting vortex, for the meteors delivered up both the kinetic and potential energies of their rotational motion as they spiralled into the sun. Here, in 1854, William had produced literally a macrocosmic replica of James's vortex turbine (ch. 14).

The significance of James's engineering work for William's hydrodynamics of the physical world extended beyond the vortex turbine. In 1852, concerned with economizing power in centrifugal pumps, James introduced the idea of a 'Vortex or whirlpool of free mobility', as he had labelled it by 1858.[46] The usual centrifugal pumps wasted as much as 50% of the power expended to run them because they expelled water with a considerable rotational velocity from the circumference of their driving wheel. The kinetic energy of this rotational motion was consumed in friction in the discharge pipe. James proposed to avoid the waste by adding an exterior whirlpool chamber in which the revolving water would gradually lose its tangential velocity as it moved from the wheel outwards, acquiring in return either an increased height at discharge or an increased pressure capable of raising it to a greater height. This simple addition of an exterior whirlpool would increase the efficiency of centrifugal pumps from less than 50% to 70%. As usual, the increased efficiency depended on maintaining every portion of the vortex in equilibrium with neighbouring portions.

The vortex in James's external chamber, much like that in a draining sink, approximated a condition of 'free mobility', meaning that every particle, at whatever distance it happened to be from the central axis, possessed the same total energy as every other equal particle, and therefore was '*free to move* to any position within the whirlpool, without interfering with the general motion of the other particles, as each one . . . assumes of itself, subject simply to the laws of motion under a central force, the velocity due to its position in the whirlpool'.[47] Each particle moved with a velocity inversely proportional to its distance from the central axis. Assuming an infinite sea, originally level, and neglecting friction, the equilibrium surface of such a whirlpool would take the shape $y \propto 1/x^2$, where y is distance below the level sea and x is distance from the central axis (figure 12.3). Thus any point of the surface moves with the velocity it would acquire in falling freely from the level surface, or that it would lose in rising. A point in the interior, moving with the same velocity as the one directly above it in the surface, would gain gravitational potential energy if it rose to a higher level but would lose an equal energy with the attendant decrease in pressure. Similarly, in moving outwards, it would lose kinetic energy as the pressure increased. No work was required for either vertical or radial changes of position, ensuring equilibrium throughout the mass.

In his lectures at Belfast in 1858 James referred to this motion variously as a '*Whirlpool or Vortex of Equal Energies*, or *Of Free Mobility*, or *Of Maximum*

[46] James Thomson, 'On some properties of whirling fluids, with their application in improving the action of blowing fans, centrifugal pumps and certain kinds of turbines', *BAAS Report*, **22** (1852), 130; *Papers*, 1–2. 'On a centrifugal pump with exterior whirlpool, constructed for draining land' (read 27th October, 1858), *Trans. Inst. Eng. Scotland*, **2** (1858), 20–6; *Papers*, 16–24. In the same volume, see Rankine's 'Introductory address' as President of the Institution, which includes comments on James Thomson's work in relation to the economic importance of saving coal for use 'at sea or in locomotive engines', pp. 1–16, on p. 13.

[47] James Thomson, *Papers*, 19. Our emphasis.

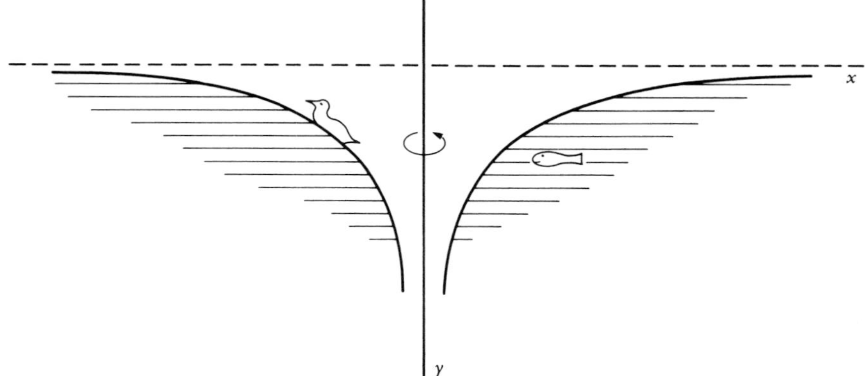

Figure 12.3. James Thomson's vortex of free mobility has the shape $y \propto 1/x^2$. A fish in the water and a duck on the surface, he remarked, could each swim freely from any one point to another without doing work (neglecting friction).

Velocities', and again as a '*Whirlpool of Permanent Motion*'.[48] He apparently understood the ability of his vortex to deliver power with minimum waste as depending on an interrelationship between equality, freedom, an optimum condition, and stability. The flavour of liberalism and political economy in this conception correlates well with William's understanding of extremum conditions (ch. 11).

Concerning the maximum condition and permanence, no particle in steady motion could be supposed to move faster than its given velocity without all particles at the same depth also moving faster, and tending to fly outwards. But then the particles in the layer above would move inwards to replace them, restoring the given conditions. The maximum condition on the velocities, as James saw it, was therefore a sufficient condition for permanence of the motion.[49] James's extensive work on these problems before and during 1858 makes it unlikely that William's speculations on 'eddies' in the universal plenum in January, 1858, proceeded independently. In particular, William's confidence in the constancy of such eddies, based on his 1847 extremum principle, parallels James's conception. Given William's usual insistence on grounding theoretical ideas in practical realities, one suspects that the dramatic improvement in efficiency effected by James's external whirlpool served as a major support for his hydrodynamic programme in the face of Stokes's objections.

When William began in September, 1868, to pursue with Tait experiments on the stability properties of various vortices, James wrote to remind him of his own much earlier knowledge of some of those properties and of discussing them with William. As James remembered describing the condition of free mobility,

[48] James Thomson, 'Civil engineering class, 1858', C1, QUB, pp. 48, 52.
[49] *Ibid.*, p. 52; James Thomson, *Papers*, 2.

a 'duck could swim from place to place without expenditure of energy in like manner as in still water: Also that a fish in the interior of any whirlpool, except the one of Free Mobility, would require to give out a real finite quantity of energy in order to move from place to place'. William, however, already had James's work in mind, for the 'vortex of free mobility' had appeared in his notebook in June.[50] The relations between extremum conditions and permanence would also figure prominently in his later papers on vortex motion, as they had in the *Treatise*, where they served as the new foundation for dynamics. This new foundation should be seen as a generalization of the work on hydrodynamics that William and James had both begun in 1847, from their different but intertwined perspectives.

Vortex atoms

No evidence exists that Thomson knew of Hermann Helmholtz's classic paper of 1858 on *Wirbelbewegungen* (vortex motion) prior to Tait's mention of it early in 1862 (ch. 11).[51] A year later, however, when Helmholtz visited Glasgow, he witnessed a typically enthusiastic Thomsonian demonstration of the explanatory power that his vortices represented, in this case for the rigidity of matter. The experience nearly cost Helmholtz his head: 'Thomson's experiments . . . did for my new hat. He had thrown a heavy metal disk into very rapid rotation; and it was revolving on a point. In order to show me how rigid it became in its rotation, he hit it with an iron hammer, but the disk resented this, and it flew off in one direction, and the iron foot on which it was revolving in another, carrying my hat away with it and ripping it up'.[52] Later experiments would demonstrate the kinetic rigidity of such objects as thin copper globes filled with water, when set in rotation, and their analogy with gyroscopes.[53]

[50] James to William Thomson, 30th September, 1868, T120, ULG. William Thomson, 4th–11th June, 1868, NB54 (1864–8), ULC.

[51] Hermann Helmholtz, 'Ueber Integrale der hydrodynamischen Gleichungen welche den Wirbelbewegungen entsprechen', *Journal für die reine und angewandte Mathematik*, **55** (1858), 25–55; translated by P.G. Tait, 'On integrals of the hydrodynamical equations, which express vortex-motion', *Phil. Mag.*, [series 4], **33** (1867), 485–512, with an addendum by Sir W. Thomson, pp. 511–12; MPP, **4**, 67–8. In a letter of 30th August, 1859, H65, ULC, Helmholtz reported to Thomson that he had been working on hydrodynamic equations including friction (see note 78), and asked about Stokes's classic paper, but did not mention *Wirbelbewegungen*.

[52] Hermann Helmholtz to Frau Helmholtz, April, 1863, in SPT, **1**, 430. The experiment on the rotating disc probably relates to an entry of 30th January, 1863 (see also 7th September, 1863) in NB35, ULC, concerning the oscillatory motion of a solid disc under an incompressible liquid, which continues Thomson's notebook calculations in January, 1858. Thomson asked Stokes to return his letter of that date, detailing the calculations, on 24th January, 1863. See William Thomson, 'On the motion of free solids', pp. 389–90.

[53] For example, William Thomson, 'On the precessional motion of a liquid', *BAAS Report*, **46** (1876), 33–5; MPP, **4**, 129–34. See also the experiments of Thomson's student, John Aitken, 'Experiments illustrating rigidity produced by centrifugal force' (read 20th December, 1875), *Proc. Royal Soc. Edinburgh*, **9** (1878), 73–8, and Thomson's discussion in 'Elasticity viewed as possibly a mode of motion' (read 4th March, 1881), *Proc. Royal Inst.*, **9** (1882), 520–1; PL, **1**, 142–6.

Following completion of the *Treatise* and the successful laying of the Atlantic cable of 1866 (ch. 19), the newly knighted Sir William Thomson turned his full attention to vortex motion. In January of 1867, he wrote to Helmholtz:

'Just now . . . *Wirbelbewegungen* have displaced everything else, since a few days ago Tait showed me in Edinburgh a magnificent way of producing them. Take one side (or the lid) off a box (any old packing-box will serve) and cut a large hole in the opposite side. Stop the open side *AB* loosely with a piece of cloth, and strike the middle of the cloth with your hand. If you leave anything smoking in the box, you will see a magnificent ring shot out by every blow . . . you will easily make rings of a foot in diameter and an inch or so in section, and be able to follow them and see the constituent rotary motion.[54]

Smoke-ring vortices differ from the whirlpool vortices of James Thomson as a tubular ring from a section of tube; or a smoke ring consists of a whirlpool of smoke, with its axis extended and wrapped back on itself.

Helmholtz had gone a considerable way towards demonstrating in 1858 what Sir William had been unable to convince Stokes of in the same year, namely, that *Wirbelbewegungen* would remain stable, if once established in a perfect fluid (frictionless and incompressible). Vortex motions, it seemed, would fulfil precisely the role Thomson had envisaged for his permanent eddies, or motes:

If there is a perfect fluid all through space, constituting the substance of all matter, a vortex-ring would be as permanent as the solid hard atoms assumed by Lucretius . . . to account for the permanent properties of bodies (as gold, lead, etc.) and the differences of their characters. Thus, if two vortex-rings were once

created in a perfect fluid, passing through one another like links of a chain, they never could come into collision, or break one another, they would form an indestructible atom; every variety of combinations might exist. Thus a long chain of vortex-rings, or three rings, each running through each of the others, would give each very characteristic reactions upon other such kinetic atoms.[55]

The last remark was largely visionary, for Thomson could not yet calculate the action of even two vortex rings on one another. And, in fact, Helmholtz had not proved the stability of a vortex ring, but only that its rotational motion was permanent. Sir William did not immediately see the difference. He broadcast in

[54] William Thomson to Hermann Helmholtz, 22nd January, 1867, in SPT, **1**, 513–14.
[55] *Ibid.*, pp. 514–15.

February, 1867, the news that vortex atoms would henceforth abolish 'the monstrous assumption of infinitely strong and infinitely rigid pieces of matter' as adopted by Lucretius, Newton, and 'some of the greatest chemists in their rashly-worded introductory statements'.[56]

If vortices were indestructible, they had the additional advantage that only God could create them: 'to generate or to destroy "wirbel-bewegung" in a perfect fluid can only be an act of creative power'.[57] And while the modern Lucretians could only explain properties of matter by attributing them to a variety of mysterious forces inhering in atoms, the economical Creator of vortices worked simply with primitive inertia in the universal fluid. Not even the 'contractility' of Thomson's 1858 speculations was necessary. The spectroscopic requirement for atoms that vibrated, for example, could now be filled by the 'kinetic elasticity of form' of 'vortex rings in a perfect liquid'. Tait's smoke rings suggested that 'the vortex atom has perfectly definite fundamental modes of vibration'. And preliminary calculations on longitudinal vibrations in a vortex tube, or columnar vortex, suggested that the sodium atom, with its prominent double D lines, might consist of two nearly identical vortex rings linked together, each vibrating in its fundamental mode.

This same kinetic elasticity, acting between atoms, would 'become the foundation of the proposed new kinetic theory of gases' of Clausius and Maxwell, for 'two smoke-rings were frequently seen to bound obliquely from one another, shaking violently from the effects of the shock'. No doubt the elasticity of liquids and solids and the thermodynamics of all matter would soon succumb to vortex explanation. The mathematical problems presented difficulties of 'an exciting character', which only enhanced Thomson's sense of power over the deep secrets of ether and matter.[58]

We should note immediately that the kinetic theory of gases came ready-made with an insistent problem: the actual size of atoms. In order to obtain empirically reasonable diffusion rates for the mixing of gases, Clausius had introduced in 1858 the concept of 'mean free path', the average distance between collisions, which depends on the effective size of the atoms. Maxwell too made the mean free path a basic parameter in his own theory of 1860, while in Vienna in 1865 Joseph Loschmidt used Maxwell's results to estimate the size of a

[56] William Thomson, 'Vortex atoms', p. 94. See also Thomson to Stokes, February, 1867, K372, Stokes correspondence, ULC, mentioning the 'very promising' vortex atoms and their vibrations.

[57] William Thomson, 'Vortex atoms', p. 94. The necessarily *created* character of vortices made an indelible impression on Thomson. See, for example, his letter to Helmholtz of 3rd October, 1868, H71, ULC, where he asked rhetorically for conditions under which vortex motion could arise, and answered: 'I think we may say without farther proof *through none*' (double underlining).

[58] William Thomson, 'Vortex atoms', pp. 96–8. Thomson published rigorous solutions for a columnar vortex only in 1880. 'Vibrations of a columnar vortex' (read 1st March, 1880), *Proc. Royal Soc. Edinburgh*, **10** (1880), 443–56; MPP, **4**, 152–65. He had worked out many of the solutions, however, much earlier; see William Thomson to G.G. Stokes, 1st–2nd January, 1873, K188, Stokes correspondence, ULC. See also Thomson's attempts in May, 1868 to give rules for the impact of two vortices, NB55, ULC.

molecule of air at about 10^{-7} centimetres on the assumption that in the liquefied state the molecules occupy the entire space. In 1870, unaware of Loschmidt's work but remarking that 'the Kinetic theory of gases has led me at last to come to terms as to the size of molecules', Thomson would juxtapose its results with three others – an energy argument from contact electricity that he had sketched in 1862, a new energy argument from capillary attraction, and an old argument of Cauchy on prismatic colours – to fix the size of molecules between 0.5×10^{-8} and 10^{-8} centimetres, or at least 1000 per wavelength. Cauchy's argument, which Thomson claimed to have been teaching since about the first year of his professorship, and which he had adapted for magnetic rotation in 1856, attributed refractive dispersion, or the fact that different wavelengths of light exhibit different velocities in normal matter, to its 'coarse grainedness' in comparison with ether.[59]

Throughout what follows, therefore, we are assuming that, though vortices may come in all sizes, those of ether must be much smaller than those of normal matter. This increasingly conscious differentiation does not yet compromise the integrity of 'aer' as the middle term between ether and matter, at least insofar as the propagation of light waves is concerned, because the difference is only one of different size grains in the medium of propagation. And though Thomson no longer spoke of 'aer' we find no essential change in his programme until the 1880s. Nevertheless, as the internal dynamics of ponderable molecules became an increasingly important location for the physical phenomena of gases and spectroscopy, 'aer' would lose its significance as a unifying concept. We return to this subject at the end of the present chapter.

In April, 1867, in a formal paper 'On vortex motion', Thomson began to publish mathematical foundations for 'the hypothesis, that space is continuously occupied by an incompressible frictionless liquid acted on by no force, and that material phenomena of every kind depend solely on motions created in this liquid'.[60] Helmholtz had defined a 'vortex line' as a line drawn so that it coincided with the instantaneous axis of rotation of each element of fluid through which it passed. A 'vortex filament' then denoted a tube of fluid of infinitesimal section bounded by vortex lines, something like the long thin funnel of a tornado. All such vortex filaments had either to form closed curves or to terminate in the bounding surface of the fluid. Helmholtz proved not only the

59 S.G. Brush, *Statistical physics and the atomic theory of matter, from Boyle and Newton to Landau and Onsager* (Princeton, 1983), pp. 50–8, describes the size problem; he gives detailed summaries of the relevant papers in *The kind of motion we call heat: a history of the kinetic theory of gases in the 19th century*, (2 vols., Amsterdam, 1976), **2**, pp. 335–56, 422–42. See Thomson to Stokes, notes 75 and 80 below, for Thomson's 1866 criticism of Maxwell's assumption of point atoms in his second paper on gas theory, and for Thomson's estimates, 'The size of atoms', *Nature*, **1** (1870), 551–3, and extracts of letters to Joule (as President of the Society), *Proc. Manchester Lit. Phil. Soc.*, **2** (1862), 176–8, and **9** (1870), 136–41; reprinted, respectively, in E & M, 317–18, and *Nature*, **2** (1870), 56–7, quotation on p. 56.

60 William Thomson, 'On vortex motion' (read 29th April, 1867, recast and augmented 28th August to 12th November, 1868), *Trans. Royal Soc. Edinburgh*, **25** (1869), 217–60; MPP, **4**, 13–66.

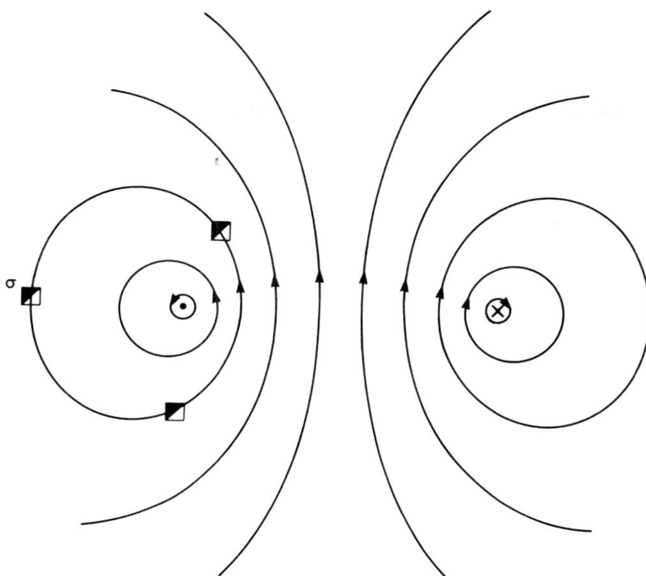

Figure 12.4. The 'flow' of magnetic lines of force through and around a current loop (of which the smallest circles are cross sections) is *irrotational*, meaning that an infinitesimal element σ does not change its orientation in moving around the line of flow.

permanence of vortex lines and filaments, but also that the same portions of fluid that belonged at any one time to a vortex line belonged forever afterwards to the same vortex line, however the line itself might move. That is, a vortex filament contained always the same fluid with the same volume. It would move through the fluid like a solid wire. Similarly, a vortex 'tube' of finite section would move like a more extended solid.

To illustrate the distinction of rotational from irrotational motion, Helmholtz developed further the analogy between fluid velocities and magnetic forces that Thomson had employed earlier (ch. 8).[61] Electromagnetism, with its never-ending lines of force (flow) circulating around electric currents, was particularly relevant (figure 12.4). Outside the currents the system of flow was irrotational; that is, it involved no 'molecular rotation', in the phrase adopted by Thomson and Stokes, meaning no rotation in space of any element, or 'molecule' $dxdydz$, in its motion along a flow line.[62] The currents, however, had to be represented by rotating vortex filaments or tubes, for the flow was directed

[61] See also J.C. Maxwell to William Thomson, 18th July, 1868, M102, ULC, where Maxwell develops a more general version of the analogy between vortices and electric currents, responding to a letter that Thomson sent to Tait.

[62] See, for example, Thomson's addendum to Helmholtz, 'Wirbelbewegungen', p. 511.

oppositely on opposite sides of the current. These rotational filaments served as 'cores' (Thomson's term) for the irrotational circulation around them. Since both rotational and irrotational motions appeared in the same system, Helmholtz included both motions under 'vortex motion', although strictly that term applied only to the rotational kind.[63] Finally, the unending lines circulating around electric currents could be represented by multi-valued potentials, and the total change in potential around one circuit measured the strength of the enclosed vortex rotation (assuming continuity between rotational and irrotational motions). Thomson, therefore, would label this all important parameter the 'circulation' or 'cyclic constant' of the vortex motion.

The electromagnetic analogy, in summary, illustrated the mathematical distinction of irrotational from rotational motion and the coexistence of both in a single flow system. From a more practical perspective, James Thomson's vortices illustrated the same principles for water, assuming no friction. His vortex turbine involved purely rotational motion, corresponding to the uniform angular velocity required by a turbine with radial vanes. The centrifugal pump, on the other hand, developed a rotational vortex within the space of its central vaned wheel surrounded by an irrotational vortex of free mobility. The central vortex would produce rotation in space of a cork carried round with the water, while the external vortex would not.

In his own paper of 1867, Sir William went over much the same ground as Helmholtz, while providing a more general analysis of multiply continuous (or multiply connected) spaces to facilitate the treatment of knotted and knitted vortex tubes and complex arrays of linked rings (figure 12.5).[64] As in 1847, he took pains to establish the uniqueness of any irrotational motion generated by impulses normal to the boundaries of the fluid or by action between vortex tubes, that is, any motion that could possibly occur in the history of the vortex universe once the original rotational cores had been created. This (apparent) uniqueness of the irrotational motions confirmed his basic assumption of determinism in mechanical philosophy: 'the actual motion is, of course (as the solution of every real problem is), unambiguous'.[65] We shall see below, however, that the assumption of complete determinacy in nature would require qualification with respect to the statistical results of the kinetic theory of gases.

Thomson drew attention also to the energy relations of irrotational motion,

[63] *Ibid.*, p. 491.

[64] 'Vortex motion', p. 244. See also Thomson's notebooks of 1867–9, containing numerous calculations and drawings of vortices in preparation for 'Vortex motion', NB54–7, ULC (NB55 precedes NB54).

[65] William Thomson, 'Vortex motion', p. 254. The uniqueness theorem is (p. 256): 'The motion of a liquid moving irrotationally within an $(n+1)$ply continuous space is determinate when the normal velocity at every point of the boundary, and the values of the circulations [cyclic constants] in the n circuits, are given'. 'Boundary' here includes imaginary membranes across the apertures of all closed vortex tubes, which instantaneously dissolve once their normal velocity is given.

Figure 12.5. The possibilities for permanently knotted, knitted, and linked vortex tubes are immense.

showing that the same fluid motion always required the same work to produce it, independent of the mode of generation. The analysis set the stage for what he dubbed 'kinetico-statics', the reduction of static forces and static potential energies to relations of kinetic energy.[66] This goal, which had motivated much of his research since 1847, especially the attempt in the *Treatise* to derive forces entirely from energy considerations, now seemed to be nearing reality. By 1870 Sir William was prepared to state that reality almost without qualification:

Kinetico-statics . . . is in reality a branch of physical statics simply. For we know of no case of true statics in which some if not all of the forces are not due to motion; whether as in the case of the hydrostatics of gases, thanks to Clausius and Maxwell, we perfectly understand the character of the motion, or, as in the statics of liquids and elastic solids, we only know that some kind of molecular motion is essentially concerned. The theorems which I now propose . . . are of some interest in physics as illustrating the great question of the 18th and 19th centuries: Is action at a distance a reality, or is gravitation to be explained, as we now believe magnetic and electric forces must be, by action of intervening matter?[67]

Only gravity left some doubt. A vision so universal naturally required new heroes of the eighteenth century. The *Treatise* had resurrected Maupertuis as the visionary of extremum principles. Now the name of Leonhard Euler, founder of mathematical hydrodynamics and author of a speculative kinetic theory of magnetic force much reviled by action at a distance advocates, began to appear

[66] *Ibid.*, pp. 258–60.

[67] William Thomson, 'On the forces experienced by solids', p. 60; E&M, 567–8.

in Thomson's magnetic papers, which themselves 'curiously illustrated' Euler's 'fanciful theory of magnetism'.[68]

From such a confident beginning, one would expect a demonstration of success to follow. Indeed, Thomson began by repeating the old kinematic analogy between an electromagnetic force distribution and the corresponding hydrokinetic velocity distribution. Passing to the dynamical problem, he could prove that the mutual action between two magnets would be identically reproduced in magnitude by two corresponding flow systems. Unhappily, however, these ponderomotive forces acted in the wrong direction. When the magnets attracted, the corresponding flow systems repelled. No remedy for this 'remarkable difference' was forthcoming; nevertheless, Sir William remained undeflected from his course.[69]

He clarified his attitude in a report to the Philosophical Society of Glasgow on the experiments of a competitor in the game of mediated action. Frederick Guthrie, of the Normal School of Science at South Kensington, had discovered that a vibrating tuning fork in the neighbourhood of a card suspended in air would attract the card, and he suspected that such an action might explain magnetic attraction.[70] Assimilating these experiments to his own equations, and displaying a variety of hydrokinetic analogies, Thomson showed that they all gave forces in the opposite direction from magnetic action. 'In concluding, the speaker [Thomson] remarked, that it would be very wrong if he were to say that these experiments on the hydro-kinetic analogue contained a direct opening up of the question of the mechanism of magnetic forces. They did not go any way towards explaining magnetic forces; but it was impossible to look upon them without feeling that they suggested the possibility of some very simple dynamical explanation'. A deep faith thus supported the assertion that 'the ultimate theory of magnetism is undoubtedly kinetic', a faith that would last into the 1890s, despite continued failure with forces between magnets, not to mention a similar failure with electrostatic forces (ch. 13).[71]

More secure results seemed to be emerging for the mechanical properties of

[68] William Thomson, 'On the forces experienced by solids', p. 61.

[69] *Ibid.*, p. 62. For a full treatment of the 'correspondence of opposites', or hydrokinetic analogy to magnetic action, both of permanent and induced magnetization, with mutual actions derived from derivatives of kinetic energy, see the 'General hydrokinetic analogy for induced magnetism' (February, 1872), E&M, 579–87, and the 'General problem of magnetic induction' (March 1872), E&M, 544–66. Also see the considerations in Thomson's notebook, NB58, 1870(?), ULC.

[70] Frederick Guthrie, 'On approach caused by vibration' (received 26th August, 1869), *Proc. Royal Soc.*, **18** (1870), 93–4; **19** (1871), 35–41; 'On approach caused by vibration', *Phil. Mag.*, [series 4], **39** (1870), 309; **40** (1870), 345–54. See also five letters from Thomson to Guthrie, 14th November, 1870 to 11th January, 1871, in *Phil. Mag.*, [series 4], **41** (1871), 423–9; extracts from letters of 14th and 23rd November, 1870, in E&M, 571–4.

[71] William Thomson, 'On the attractions and repulsions due to vibration observed by Guthrie and Schellbach' (read 14th December, 1870), *Proc. Glasgow Phil. Soc.*, **7** (1871), 401–4; E&M, 574–8, on pp. 578, 574.

matter, such as elasticity, when conceived as relations of vortices, or equivalent solids, swimming in the universal fluid. Picking up where he had left off in his notebook and letters to Stokes in January, 1858, Thomson worked out the equations of motion of such 'solids' moving both independently and in interaction, finding interpretable solutions for simple cases. A disc moving freely through quiescent fluid would oscillate about one of its diameters like a gyroscope. A sphere in irrotationally moving fluid (i.e. a fluid in which any number of fixed vortex filaments, arranged in rings, knots, etc., served as cores for irrotational motion around and through them) would behave exactly like a particle in a field of static force having the same potential as that of the fluid velocities.[72]

Although Thomson regularly related these calculations to magnetic analogies, they were not limited to magnetism. In an 1872 abstract of the projected continuation of his 'Vortex motion', he claimed that 'the difficulties of forming a complete theory of the elasticity of gases, liquids, and solids, with no other ultimate properties of matter than perfect fluidity and incompressibility are . . . in all probability, only dependent on the weakness of mathematics'.[73] He promised to include the following: gravitation, the kinetic theory of gases, extrema and dissipation of energy, and wave motion in elastic liquids and solids. We take up those four subjects in turn.

Gravity

Opposition to gravitational action at a distance reawakened the sleeping eighteenth-century genius of Le Sage, as magnetism had called up Euler.[74] Le Sage's hypothesis depended on two kinds of matter: a gravific fluid, consisting of innumerable ultra-mundane corpuscles flying in all possible directions in space, and mundane matter, consisting of arrays of atomic 'cages', empty cubes or octahedrons, with matter only in bars along their edges. (Compare Thomson's 1862 theory of cohesion, discussed in Chapter 11.) These edge bars in mundane bodies were so thin that the entire earth did not intercept more than one-ten-thousandth of the ultra-mundane corpuscles supposed to traverse it. Nevertheless, the shadow cast by any one body on another created a differential flow of

72 William Thomson, 'On the motion of free solids', esp. pp. 389–90; 'On the motion of rigid solids in a liquid circulating irrotationally through perforations in them or in any fixed solid' (read 4th March, 1872), *Proc. Royal Soc. Edinburgh*, **7** (1872), 668–82, esp. pp. 675–82; MPP, **4**, 101–14, esp. pp. 107–14.

73 William Thomson. 'On vortex motion' (abstract, read 18th December, 1871), *Proc. Royal Soc. Edinburgh*, **7** (1872), 576–7.

74 Thomson's knowledge of Le Sage probably derived from his business partner, Fleeming Jenkin, who, in 1867, playing the role of research assistant, sent brief reports on 'dynamicist' predecessors, including Bacon, Descartes, Leibniz, Malebranche, and, especially, Le Sage, 7th–26th June, 1867, J66–70, ULG. Jenkin did not initially approve of Le Sage's idea, though he recognized it to be very like Thomson's. See, however, Jenkin on Lucretius (ch. 18).

corpuscles, or differential momentum transfer, sufficient to cause a 'gravitational' force obeying the inverse square law.[75]

Modern thermodynamics required that some modifications be made in the scheme. Le Sage had assumed inelastic corpuscles, in order that they would rebound with decreased velocities, thereby creating the gravitational differentials; but conservation of energy would not allow ultimate inelasticity. Furthermore, if the corpuscles rebounded with diminished energy, depositing the remainder, bodies like Jupiter and the earth would vaporize in a fraction of a second. But Thomson's vortex atoms could solve both problems. Being perfectly elastic, they would conserve energy in collisions. They might also rebound with diminished velocity, while depositing almost no energy. One had only to suppose, with Le Sage, that the cage atoms possessed enormously greater mass than the corpuscles and, with Clausius in the kinetic theory of gases, that the corpuscles could vibrate and rotate, as vortex atoms could. The kinetic energy of the corpuscles, supposed entirely translational before impact, would then be preserved on collision with the cage atoms, but in a different form, part of it having been converted into vibrational and rotational kinetic energy.

Le Sage had recognized that the stock of ultra-mundane corpuscles with undiminished velocity would gradually decrease, and with it the gravitational force. This dissipation principle suited Thomson's philosophical and theological persuasion. He gave Le Sage's conclusion his own phrasing, characteristic of his works on the age of the sun and earth (chs. 14–17): 'at a not infinitely remote past time they [the corpuscles] were set in motion for the purpose of keeping gravitation throughout the world in action for a limited period of time; and . . . by their mutual collisions, and by collisions with mundane atoms, the whole stock of gravific energy is being gradually reduced, and therefore the intensity of gravity [is] gradually diminishing from age to age'.[76] Even gravitation would exhibit the beginning and ending of the created universe.

This satisfying theory, however, ran against a formidable obstacle in the disparity between the perfect isotropy of the gravitational force and the essential anisotropy (Thomson's aeolotropy) of crystalline substances. Crystals required a

[75] William Thomson, 'On the ultramundane corpuscles of Le Sage' (abstract, read 18th December, 1871), *Proc. Royal Soc. Edinburgh*, **7** (1872), 577–89; MPP, **5**, 64–76, where Thomson has translated a portion of Le Sage's 'Lucrèce Newtonien', *Nouveaux mémoires de l'Académie Royale des Sciences et Belles-lettres, année 1782* (Berlin, 1784), pp. 404–32. He comments also on Le Sage's posthumous *Traité de physique mécanique*, P. Prévost (ed.) (Geneva and Paris, 1818). The attractiveness of Le Sage's scheme to Thomson is apparent in, for example, his review of Maxwell's famous second paper on gas theory, where he criticizes the division of theories of matter into either the atomic one or 'the one that supposes all matter continuous *AND HOMOGENEOUS*. The only views that have ever appeared to me true or natural as to the constitution of matter are those that suppose all space to be full but the properties of known bodies to be due to or necessarily associated with molecular structure or of a sponge or other organic tissues or brick work, i.e. that there are vast variations of density from point to point within spaces of dimensions some small fraction of a wave length (though not inappreciably small)', Thomson to Stokes, 13th October, 1866, RR.6.179, Royal Society of London.

[76] William Thomson, 'Ultramundane corpuscles', pp. 585–6.

molecular structure that would exhibit different capacities in different directions with respect to light, heat, electricity, magnetism, elasticity, and other physical agencies. Failure to find a resolution compromised the entire project of generalizing the dynamical theory of heat (*Heat, a mode of motion*, in the title of Tyndall's popular book) into a kinetic theory of matter:

No finger-post pointing towards a way that can possibly lead to a surmounting of this difficulty, or a turning of its flank, has been discovered, or imagined as discoverable. Belief that no other theory of matter is possible is the only ground for anticipating that there is in store for the world another beautiful book to be called *Elasticity, a Mode of Motion*.[77]

Kinetic theory of gases

Thomson's speculative kinetic theory of gravitation depended on principles fundamental to the kinetic theory of gases. From his first paper on 'Vortex atoms' he had expected gas theory to yield most easily to the new hydrodynamics, for it seemed that any 'crowd' of freely moving elastic molecules would satisfy its basic requirements. One had only to work out how temperature would be related to the vortex motion of an atom. In a vortex filament the product of angular velocity and cross-sectional area remained constant all along the filament, as did the total volume of the filament. Therefore the angular velocity varied inversely as the cross-sectional area and directly as the length. Thomson always assumed in addition that no slippage occurred between portions of rotational fluid inside the filament and irrotational fluid outside.[78] Consequently, an increase of irrotational circulation through a ring atom would imply a larger ring. Equivalently, if the rotational and irrotational flow of a vortex ring were related as the electric current in a circular wire to its associated magnetic field, then the area of the ring aperture would measure the total kinetic energy of the irrotational flow. Adding kinetic energy to a 'crowd' of such atoms meant increasing their diameters. That idea promised a theory of temperature in a gas:

If, after any number of collisions or influences, a Helmholtz ring escapes to a great distance from others and is then free, or nearly free, from vibrations, its diameter will

[77] William Thomson, 'Elasticity . . . a mode of motion', pp. 145–6. Cf. 'On . . . Le Sage', *op. cit.* (note 75), 'Postscript, April 1872', 588–9. Thomson's praise of Tyndall's book must be balanced against his deep disapproval of Tyndall's materialism (ch. 18). Concerning gravity and the ether, Thomson in 1884 continued to believe that 'it is just as likely to be attracted to the sun as air is', but he had no gravitational theory, BL, 207.

[78] Thomson and Stokes had debated this issue with respect to stability in 1858 (above) and continued to discuss it in 1862, with respect to an experimental proof by Helmholtz and G. Piotrowski of slipping between a moving liquid and its containing vessel. See Stokes to Thomson, 22nd and 25th February, 1862, S83, ULG, and S398, ULC; also David Murray, 'Lecture notes in *classe physica*, bench II, 1862–63', MS Murray 325, ULG, 9th February, 1863. The debate continued into the 1890s without resolution. See Thomson's paper 'On the doctrine of discontinuity of fluid motion, in connection with the resistance against a solid moving through a fluid', *Nature*, **50** (1894), 524–5, 549, 573–5, 597–8; MPP, **4**, 215–230.

have been increased or diminished according as it has taken energy from, or given energy to, the others. A full theory of the swelling of vortex atoms by elevation of temperature is to be worked out from this principle.[79]

The full theory, however, would have required a complete determination of the relation of translational motion of the ring to its irrotational circulation, and of the relation between translational, vibrational, and rotational modes of motion. The latter problem became increasingly complex as the kinetic theory of gases developed. By 1866 at the latest Thomson was well aware of discussions by both Clausius and Maxwell of how the measured ratio of specific heats in a gas (specific heats at constant pressure and at constant volume) ought to be related to the partitioning of energy among the various modes of motion. From the measured ratio of specific heats Clausius in 1857 simply derived the proportion of translational to total energy without providing a theory of molecular motions that would justify the proportion. Maxwell insisted in 1860 on the so-called 'equipartition theorem', according to which kinetic energy must be allocated equally to the various degrees of freedom of the motion, but the theorem neither had a rigorous theoretical foundation nor did it yield the measured ratios of specific heats. Boltzmann, who entered the fray in 1868, also could find no satisfactory solution.[80]

With considerable excitement, Thomson wrote to Stokes in 1875 to announce that the gas theory of 'Maxwell & Boltzmann' (and Clausius), based on elastic-solid molecules colliding by actual contact, would necessarily fail, because, 'After an infinitely great number of collisions all the kinetic energy will have been converted into vibrations [and rotations, *del.*]'. That result, he convinced himself, 'must be right, and it is very important', the more so because it would leave the vortex theory without a rival. In 1884, after years of investigation, the vortices still held the high ground, 'because all I have been able to find out hitherto regarding the vibration of vortices . . . does not seem to imply the liability of translational or impulsive energies of the individual vortices becoming lost in energy of smaller and smaller vibrations'. Still, he had no proof.[81]

[79] William Thomson, 'Vortex atoms', p. 104. Thomson used 'crowd' for a collection of molecules from at least 1872; Thomson to Stokes, 19th December, 1872, K187, Stokes correspondence, ULC, postscript.

[80] Thomson never took seriously Maxwell's assumption in his second paper (1867) of point atoms. In his review he criticized Maxwell for treating molecules as centres of force (non-rotational), and thus for not taking account *physically* of the constant proportions of kinetic energy. He recommended a clarification: 'that the molecules are regarded *not* as centres of force but as really (according to their name) little heaps of matter, acting on one another with forces *not* in lines through their centres of inertia' (Thomson to Stokes, *op. cit.*, note 75). Maxwell himself clearly regarded his models more nearly as calculational devices than as realistic pictures. For references and general discussions of the problem of equipartition see Elizabeth Garber, 'Molecular science in late-nineteenth-century Britain', *Hist. Stud. Phys. Sci.*, **9** (1978), 265–97; Brush, *Statistical physics*, pp. 50–1, 65–6; Brush, *The kind of motion*, **2**, pp. 353–63.

[81] In 1892 Thomson would come to the opposite conclusion on vortex stability (ch. 13). William

Sir William here faced two interrelated requirements of his theory: stability of vortex atoms and dissipation of energy in all processes of physical nature. Consideration of the second will lead us back to the first. To satisfy the Second Law of Thermodynamics, any concentration of energy within a gas had to spread throughout the whole gas, yielding a randomized distribution in the velocity squared which followed the normal curve or Maxwell–Boltzmann distribution. Thomson regarded the dissipation principle as obtaining unquestionably in any kinetic theory of gases based on elastic collisions of molecules, and therefore also in the vortex atom theory. The elastic collision theory itself, when based on solid atoms, interested him only as an exercise in abstract dynamics. He had written to Tait already in 1864: 'Clausius has done a good deal & Maxwell too as to molecular impacts. I don't set much store by it all except as dynamical problems'. Maxwell agreed, especially after he saw the promise of the vortex theory.[82] The inadequate physical model did, however, allow one to make statistical arguments, which Thomson happily exploited in his published papers, following the line that Maxwell had developed with the help of his famous 'demon' (ch. 18).

As 'an intelligent being endowed with free-will and fine enough tactile and perceptive organisation to give him the faculty of observing and influencing individual molecules of matter', the demon could direct molecules with high and low velocities to opposite portions of a gas, thereby concentrating kinetic energy in a finite system of molecules, in violation of the Second Law.[83] His ability to act in this way illustrated that the same kind of violations would occur naturally, purely as a statistical matter, among a finite number of gas molecules. As their number increased in a given volume, however, the probability of violations would decrease, approaching zero as the number approached infinity. Conversely, beginning from a non-equilibrium distribution, the equilibrium distribution would rapidly ensue.

Thomson ignored the full problem of how equilibrium distributions were established, how an initially uniform distribution, for example, would attain the

Thomson to G.G. Stokes, 2nd November, 1875, K198, Stokes correspondence, ULC. William Thomson, 'Steps toward a kinetic theory of matter' (opening address to the mathematical and physical section of the BAAS, Montreal, 28th August, 1884), *BAAS Report*, **54** (1884), 613–22; PL, **1**, 218–52, on pp. 230, 251–2. For Thomson's continued belief that Maxwell–Boltzmann gases would dissipate all energy and his attempts to disprove the equipartition theorem see his papers 'On some test cases for the Maxwell–Boltzmann doctrine regarding distribution of energy', *Proc. Royal Soc.*, **50** (1892), 79–88; MPP, **4**, 484–94; 'On a decisive test-case disproving the Maxwell–Boltzmann doctrine regarding distribution of kinetic energy', *Proc. Royal Soc.*, **51** (1892), 397–9; MPP, **4**, 495–6; and 'Nineteenth century clouds over the dynamical theory of heat and light', *Phil. Mag.*, [series 6], **2** (1901), 1–40. Boltzmann registered his objections in Ludwig Boltzmann to William Thomson, 10th and 13th December, 1892 and 6th January, 1893, B137–9, ULC.

[82] Thomson, NB52, p. 40. Brush, *On the kind of motion*, **1**, 206–7; **2**, 354–6.

[83] William Thomson, 'The kinetic theory of the dissipation of energy' (read 16th February, 1874), *Proc. Royal Soc. Edinburgh*, **8** (1875), 325–34, on p. 326*n*; MPP, **5**, 11–20; see also 'The sorting demon of Maxwell' (read 28th February, 1879), *Proc. Royal Inst.*, **9** (1882), 113–14; MPP, **5**, 21–3; PL, **1**, 137–41.

Maxwell–Boltzmann distribution of velocities.[84] He discussed only the statistical equalization of velocities between two regions of a gas originally at different temperatures. Given his commitment to the vortex atom theory, and to the completely unambiguous nature of vortex motions, he certainly believed that this statistical result would follow from the laws of fluid motion, although just how is uncertain. Probably he thought that the interactions between the vortex atoms of a gas, and with the walls of their containing vessel, would produce a distribution of irrotational circulation among the molecules that tended to minimize the kinetic energy in each region, and thereby equalized the energy between regions. The Second Law would then have appeared as a necessary aspect of the hydrodynamic universe, even though it remained a merely statistical result within the mechanics of Lucretian atoms and forces.

A point of considerable importance arises here, with respect to Thomson's (and Maxwell's) view of the relation between ultimate physical nature, on the one hand, and man's possible knowledge of it, on the other. Man, as a finite being, possessed the capacity for only finite knowledge. But, in the infinitely extended continuum that Thomson envisaged, every vortex atom had to be regarded as dependent on an infinity of other atoms, and no object could be regarded as truly isolated. Only an infinite mind, therefore, could know completely the course of development of the system. Man could idealize the problem by treating a finite number of atoms statistically. In this way an apparent indeterminacy would arise through statistical irregularities, but the indeterminacy was purely a matter of man's limitations. To approach the true situation, he would have to take the limit of an infinite number of atoms. Then the result of his statistical calculation would be complete certainty, with no violations of the Second Law.

A second kind of infinity would arise even within a finite portion of the continuum if one sought to give a completely general account of the motion of every element $dxdydz$ within it, for every region, no matter how small, would contain an infinite number of such elements.[85] The usual 'abstract mechanics' of discrete particles could handle only finite numbers of particles and therefore did not apply to the continuum. Even hydrodynamics offered 'knowable' solutions only for perfectly regular, stable motions. Instabilities would lead immediately back to the problem of determining the motion of an infinite number of independent fluid elements. After 1872, Thomson apparently came to see this instability problem as the basis of the Second Law within the vortex theory of matter.

[84] Thomson never became a major figure in the debate over the adequacy of Boltzmann's probabilistic formulation of the second law, although Tait regularly wrote to him about it in the 1880s and 90s. P.G. Tait to William Thomson, T38–52, T87–90, T106–29, ULC.

[85] For the infinity of variables in continuum mechanics see Thomson and Tait, *Treatise on natural philosophy*, pp. 262–3; for a general discussion, see pp. 337–9.

Extrema and dissipation: the problem of stability

Evidence for this interpretation emerges from an intensive correspondence with Stokes in 1872–3, and a long series of papers up to 1880, on extremum conditions in relation to the steadiness and stability of vortex motion. 'Steady' meant to Thomson that, with constant energy, 'successive states of motion are precisely similar to one another', while stability required in addition that the same configuration of motion would continue even after a change of energy.[86] As candidates for atoms, vortices had to possess stability through a continuous range of energies. From general dynamical considerations, a vortex would be stable if its kinetic energy were either minimum or maximum for a given state of momentum (i.e., for given volume and given product of cross-sectional area and 'molecular' angular velocity within the vortex core; for given rotational impulse delivered to the core and to the fluid moving irrotationally through its apertures). If, as for most vortices, the kinetic energy were a maximum–minimum (maximum for one kind of transformation and minimum for another), the motion might in general be either stable or unstable. Analytical demonstrations of stability for specific maximum–minimum states, however, proved elusive.

In 1875, for example, Thomson presented an abstract on 'Vortex statics', examining conditions of stability for single vortex rings, with no knots.[87] The uniform circular Helmholtz ring seemed obviously stable, since real smoke rings existed. But one could pass continuously to a ring of larger kinetic energy, without changing the momentum state, by imagining the uniform ring to be thinned in some parts (smaller circular cross-section) and thickened in others. Similarly, in passing to a uniform ring of elongated, or oval cross-section (smaller cross-sectional area and larger circumference) the energy would decrease. Thus Helmholtz rings exhibited a maximum–minimum condition. Even for rings, however, and knowing the answer beforehand, Sir William had to admit defeat: 'Hitherto I have not indeed succeeded in rigorously demonstrating the stability of the Helmholtz ring in any case'.[88] He could only give examples of multiple solutions of the maximum–minimum problem for which symmetry – and a large portion of intuition – indicated stability, such as the series of figures obtained by winding a vortex filament on a torous through an integral number of turns (figure 12.6). The figures corresponded to standing waves in an elastic ring. Any of them, he believed, would be stable over a wide range of energies.

The question of stability with continuous change of energy, however, had been vexed. From 1872 to 1876 Thomson and Stokes debated the issue at the

[86] William Thomson to G.G. Stokes, 19th December, 1872, K187, Stokes correspondence, ULC.

[87] William Thomson, 'Vortex statics' (read 20th December, 1875), *Proc. Royal Soc. Edinburgh*, **9** (1878), 59–73; MPP, **4**, 115–28.

[88] William Thomson, 'Vortex statics', p. 68.

Figure 12.6. A vortex filament, structured as though wound on a torous through an integral number of turns, would probably be stable, Thomson believed.

most basic level of a two-dimensional vortex, or a section through a vortex tube. Typically they considered a rotational core surrounded by irrotationally circulating fluid, all enclosed by a circular boundary, supposed flexible but imperfectly elastic. Being flexible, the boundary allowed for disturbance of the vortex, and, being imperfectly elastic, it removed energy by its motion. The debate concerned the evolution of the vortex from an initially circular core at the centre, to the eccentric elongated cores of figures 12.7a and 12.7b, and thence to the spiral of figure 12.7c. Thomson argued for continuing stability, or quasi-stability, and Stokes for instability.

For some time Thomson had conceived the process of elongation as follows: 'suppose energy to be lost through inperfect [*sic*] rigidity and inperfect elasticity of the containing vessel; the vortically moving portion of liquid will spread itself to a greater length round the circumference, [*sic*] and if this degradation [of energy] be extremely gradual and the envelope become again rigid, I believe the altered motion would be found to be again stable and very approximately steady'. The transition to configurations such as the spiral of figure 12.7c had long puzzled him, but he had a new idea, based now on the belief that the elongated configurations would become *unstable* with continuing loss of kinetic energy:

> If we commence with a central circular vortex, draw off energy with extreme slowness till, through the head too nearly overtaking the tail, the motion becomes unstable, and then give perfect rigidity to the containing vessel, so that there shall be no more loss of energy, such a state of things as that represented in fig [12.7c] will supervene; the irrotational canal [unshaded] become longer and thinner and aquiring [*sic*] more and more turns, *ad infinitum*.

Thus the 'vortically moving fluid' would be drawn out into a 'streak' forming a 'labyrinthian' structure of 'convolutions', through which the irrotational fluid

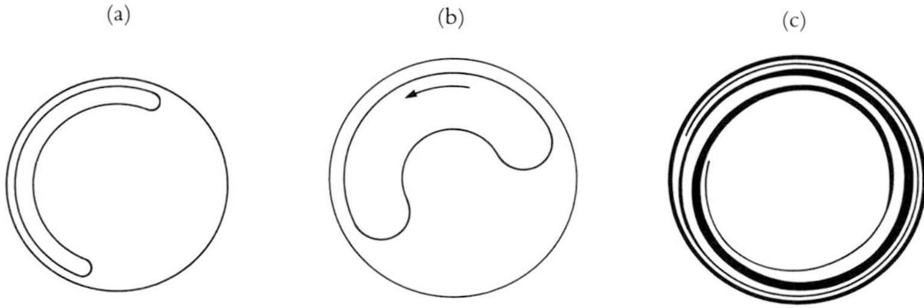

Figure 12.7. The vortex cores of (a) and (b) can be drawn out into an infinitely long spiral (c) by withdrawal of energy through the containing walls of the fluid.

would move to the centre as the rotational fluid moved to the boundary, until the absolute minimum of energy was reached.[89]

But Thomson had come to see that, in spite of the essential instability of the motion, if the energy degradation were stopped at any point, a layer of interstreaked rotational and irrotational fluid would form, 'infinitely nearly a shell vortex' with irrotational fluid both inside and outside. Although internally chaotic, the form as a whole would remain stable. Such quasi-stable states could form the continuous series of energy states required for vortex atoms. That position is the one he clung to in succeeding letters, over Stokes's continuing objections to the assumption of stability. He had also seen that the mixing at the boundary might occur, not by flow through the irrotational canal, but as a result of wave motions at the interface of the two portions of fluid.

Although anxious to publish these conjectures already in 1873, Thomson did so only in 1880, perhaps because, as he acknowledged to Stokes, 'This is an extremely difficult subject to write upon'.[90] For simplicity he then considered a very particular two-dimensional vortex, fulfilling, in fact, precisely the ideal conditions of James Thomson's centrifugal pump with external whirlpool of free mobility. The circular core rotated uniformly, like a solid, with velocity proportional to radius. Around it irrotational fluid circulated concentrically, subject to the condition of no finite 'slip' between rotational and irrotational portions, so that the velocity varied inversely with the radius, decreasing from the core to the circular boundary.[91]

James's vortex had maximum energy for given angular velocity of the core and given dimensions. It was therefore stable. Slight deformations of the boundary, Sir William argued, would cause vibrations in the core, and attendant

[89] William Thomson to G.G. Stokes, 19th December, 1872, K187, Stokes correspondence, ULC.

[90] *Ibid.*; publication plans in William Thomson to G.G. Stokes, 8th January, 1873, K190, Stokes correspondence, ULC.

[91] William Thomson, 'On maximum and minimum energy in vortex motion', *BAAS Report*, **50** (1880), 473–6; incorporated into 'On the stability of steady and of periodic fluid motion', *Phil. Mag.*, [series 5], **23** (1887), 459–64, 529–39; MPP, **4**, 166–85.

waves of various lengths travelling round it. These waves would in turn cause waves to travel round the boundary. If the boundary were supposed imperfectly elastic and were held fixed, its waves would consume both energy and angular momentum from the irrotational fluid (although not the 'molecular' angular momentum of the rotationally moving core). If the boundary were left free to rotate with the fluid, the waves would consume only energy. Thomson described what would happen if, after holding the boundary fixed for a brief period it were released:

The consumption of energy still goes on, and the way in which it goes on is this: the waves of shorter length are indefinitely multiplied and exalted till their crests run out into fine laminae of liquid [around the core], and those of greater length are abated. Thus a certain portion of the irrotationally revolving water [!] becomes mingled with the central vortex column. The process goes on until what may be called a vortex sponge is formed; a mixture homogeneous on a large scale, but consisting of portions of rotational and irrotational fluid, more and more finely mixed together as time advances. The mixture is . . . altogether analogous to the mixture of the substances of two eggs whipped together in the well-known culinary operation.[92]

Successive repetitions of the process, he reasoned, would cause the 'sponge' to thicken, until it occupied the entire space. After that, irrotational fluid would begin to separate out at the centre, and rotational fluid would do so at the boundary, until the system was inverted: 'The final condition towards which the whole tends is a belt constituted of the original vortex core now next the boundary; and the fluid which originally revolved irrotationally round it now placed at rest within it, being the condition . . . of absolute minimum energy'.[93] Interpreted thermodynamically, this condition would be the absolute zero of temperature.

Beginning from the stable vortex of maximum energy, successive operations on its boundary could carry it to the stable vortex of minimum energy through a continuous series of intermediate maximum–minimum states. These states of 'sponge', more clearly than the previous interstreaked states, appear to exhibit the quasi-stability necessary for states of atoms. Although Thomson did not make this point in his published papers, the letters to Stokes make his intentions relatively clear. The intermediate states were to correspond to the states of an atom as it dissipated its energy in interaction with other atoms, or macroscopically, as the atoms of a gas at high temperature (high kinetic energy) gave up their energy in interaction with a gas at a lower temperature.

Stokes, by contrast, believed that the maximum–minimum states would be neither stable nor steady, and therefore that they were not very promising candidates for states of real atoms: e.g., 'I see no reason whatsoever for supposing

92 William Thomson, 'Maximum and minimum energy', p. 475. Thomson apparently continued this discussion in 'On vortex sponge' (read 21st February, 1881), *Proc. Royal Soc. Edinburgh*, **11** (1882), 135; title only. On 'sponge', compare note 75.

93 William Thomson, 'Maximum and minimum energy', p. 476.

that the motion will be steady . . . This could only, I should say, be possible under very peculiar conditions'.[94] In his view, apparently, the labyrinthian interstreaking of rotational and irrotational portions of fluid would quickly lead to chaos in the entire system of motion.

Thomson's counter-argument in 1872 is of particular interest because it shows how important quasi-stability in vortex motion was to his view of dissipation. He made clear that the flexible and imperfectly elastic boundary of his model represented no true inelasticity in matter, but simply any boundary across which energy could pass out of one portion of matter and into another of lower temperature.

An imperfectly elastic solid is slow but sure poison to a vortex. The minutest portion of such matter would destroy all the atoms of any finite universe. I therefore do not admit that this [labyrinth] of structure, indefinitely increasing with the time does not seem very promising for the explanation of atoms. Any number of all true three-dimensional vortices in a finite frictionless liquid contained in a rigid vessel, however great the space, and however closely packed the vortices, can never become more than finitely interstreaked. I think indeed on the contrary that the more or less of labyrinthian which must supervene from a too crowded initial configuration of vortices is the true explanation of the condensation of a liquid on the sides of the containing vessel.[95]

Thomson discussed the issue of condensation again in 1876 in a paper 'On the vortex theory of gases' at the Royal Society of Edinburgh, but never published his speculations.[96] He seems to have believed that just as interstreaking occurred *within* single atoms of a gas, so some form of interstreaking *between* atoms would occur when enough energy had been dissipated to produce condensation, such that the atoms no longer moved independently.[97]

He continued his remarks to Stokes on the meaning of idealized boundaries by considering a rigid, elastic container, across which energy could not pass:

I of course must ignomineously [*sic*] use a rigid body or containing vessel for temporary illustrations, but the ladder is to be kicked away from under our feet when it has us where we wished. Of course all rigid bodies present a deceitful simpleness of rigidity; and real

[94] G.G. Stokes to William Thomson, 6th January, 1873, NB21.56, Stokes correspondence, ULC. This objection repeats one in an earlier letter, no longer extant.

[95] William Thomson to G.G. Stokes, 19th December, 1872, K187, Stokes correspondence, ULC.

[96] William Thomson, 'On the vortex theory of gases, of the condensation of gases on solids, and of the continuity between the gaseous and liquid state of matter' (read 3rd April, 1876), *Proc. Royal Soc. Edinburgh*, **9** (1878), 144; title only. No manuscript survives. In the period 1872–85 Thomson presented to the Royal Society of Edinburgh and to the British Association at least a dozen papers which were never published, and for which no manuscripts survive, on issues related to gas theory and the stability of vortex atoms.

[97] Thomson briefly mentioned the mixing of two vortices in 'Maximum and minimum energy', p. 474, using the same simile as for a single vortex, 'like two eggs "whipped" together'. He may have discussed the same process in 'On two-dimensional motion of mutually influencing vortex columns, and on two-dimensional approximately circular motion of a liquid' (read 3rd January, 1876), *Proc. Royal Soc. Edinburgh*, **9** (1878), 98 (title only).

elastic solids must when we understand them properly, be recognized as properly packed crowds of vortices, prevented from dissipating into space by vortices moving about more freely outside, and sometimes getting entangl[e]d in the group; sometimes swerving off, and giving on the average as much accession to as there is loss from the crowd.

A 'real' solid in empty space, on this view, or even in a region of lower temperature and pressure, would gradually evaporate. Since it could never be isolated from the surrounding vortices in the infinite continuum, it would necessarily participate in the tendency towards equalization of energies in the entire universe. Here the problem of treating atomic motions statistically might easily have entered, but Thomson aimed at atomic dynamics, rather than macroscopic statistics.[98]

He expressed his attitude to statistics much later in a letter to Larmor: 'As to statistics, I don't quite agree with you. It is not a method of ignorance but a matter of intelligent ignoring of details (something like the ignoration of coordinates in dynamics) and estimating & determining on true arithmetical principles great resultant effects'.[99] Interpreting this statement in the light of Thomson's 1875 paper on 'The kinetic theory of the dissipation of energy', he meant that the statistical behaviour of an idealized system of atoms could lead one to a true understanding – albeit at a merely descriptive level – of real physical phenomena. Simply by moving to the limit of the infinite number of atoms necessarily involved in any physical (non-isolated) system, one could see that no violations of the Second Law of Thermodynamics could ever occur. His analysis concerned the probability of a temperature disequilibrium arising either in a gas, consisting of a 'finite number of perfectly elastic molecules' enclosed by 'a perfectly rigid vessel', or in an isolated iron bar: 'Do away with this impossible ideal, and believe the number of molecules in the universe to be infinite; then we may say one-half of the bar will never become warmer than the other, except by the agency of external sources of heat or cold. This one instance suffices to explain the philosophy of the foundation on which the theory of the dissipation of energy rests'.[100] Of course the vortex continuum lay behind this statistical foundation and gave it its cogency.

Since 1872, equilibrium and dissipation had not been, to Thomson, essentially statistical matters, but questions of relative stability in hydrodynamical systems.

[98] Cf., Thomson's description in 'On the average pressure due to impulse of vortex-rings on a solid', *Nature*, **24** (1881), 47; MPP, **4**, 188: 'for every vortex-ring that gets entangled in the condensed layer of drawn-out vortex-rings another will get free, so that in the statistics of vortex-impacts the pressure exerted by a gas composed of vortex-atoms is exactly the same as is given by the ordinary kinetic theory, which regards the atoms as hard elastic particles'.

[99] William Thomson to Joseph Larmor, 2nd July, 1901, LB10.94, ULC. Thomson and Tait introduced the method of ignoration of coordinates in the second edition of their *Treatise*, pp. 320–7. The method allows one to write equations of motion without explicitly including an ever-repeating cyclic coordinate, such as the angle of rotation of a gyrostat. The method is closely related to cyclic constants in vortex motion, which express the permanence of atoms and the fact that, since the rotational motion will never change, one can ignore it for many dynamical purposes, as in gas theory.

[100] William Thomson, 'Kinetic theory', p. 330; MPP, **5**, 16.

Because the atoms were not perfectly stable, they would lose energy in interaction with other atoms whenever possible, or when the others possessed lower energy. Even a finite number of atoms would always equilibrate. Consequently, just as the gravitational force gradually lost intensity in the universe of ultramundane vortex atoms, so also the atoms of mundane matter would gradually dissipate their energy until they reached the state of minimum energy, the state of ultimate stability. In short, dissipation was the tendency of hydrodynamical systems to approach their most stable condition. In this state the atoms would not have lost their identities, which were determined by their permanent rotational motions, but they would have reached the lowest level of kinetic energy consistent with their rotational character and with conservation of energy in the universe.

This relation of permanence and dissipation receives a revealing interpretation in another objection of Sir William's to one of Stokes's ideas. Early in 1872, Stokes thought he had shown how to annihilate the rotational motion of a two-dimensional vortex with solid circular core and circular boundary. Thomson responded:

> You describe a method of throttling an irrotational vortex. If the core is brought slowly up till it is infinitely near the boundary, and then removed again, things will be as they were. If it is made absolutely to touch the boundary, the last spark of energy is destroyed, and the cyclic constant annihilated. But this is a discontinuity utterly to be excluded. The cyclic constant remains unchanged until the energy is infinitely small. The fact that the vortex has infinitely little energy when on the point of death is nothing against its stability when restored to pristine energy. A man who has been nearly drowned, but restored before the last spark of vital energy was extinguished, has never once lost his psyshic [*sic*] constant, and after restoration is as stable in his ways as ever.[101]

As the ultimate created entities in the physical universe, the vortices had to obey two principles of cosmogony. Nothing that God created could be annihilated; and the habitable world would have an end. Stokes apparently shared neither the first of these principles nor Sir William's faith in vortex atoms.

Another example of Thomson's views on the relation of stability, dissipation, and creation appears in an unpublished lecture which he wrote for delivery to the 'Royal Society' (of Edinburgh, probably) sometime during the 1870s. Its immediate purpose was to correct a 'tremendous error' in Thomson and Tait, 'which I am glad we found out ourselves before it has been detected by others'. They had argued that the dissipative effects of tidal action (but assuming no friction with the interplanetary medium) would bring the system of sun, earth, and moon to rotate forever about a common centre of inertia as though connected like a rigid body, for they had assumed that in such a state the energy

[101] William Thomson to G.G. Stokes, 22nd January, 1872, K182, Stokes correspondence, ULC. Stokes crystallized his own view a year later: 'motion which can be profoundly altered by communication of an infinitesimal amount of energy is unstable'; Stokes to Thomson, 20th January, 1873, S404, ULC, and NB21.58, Stokes correspondence, ULC.

would be an absolute minimum. But Thomson had since found that 'Any configuration of steady circular motion [of more than two bodies] fulfils the condition that, with given moment of momentum, kinetic energy is not maximum or minimum. It has a maximum–minimum quality'. In the absence of viscous influences it might be stable or unstable, but in their presence it would necessarily be unstable.[102]

Thomson regretted having insufficient time in his lecture to give applications to vortex motion, but noted as a cosmical example that there could be no permanent stability in Saturn's rings as collections of separate masses. Each ring would coalesce into a single satellite. Ultimately, as a result of friction with the medium of space, the entire solar system would collapse into the sun. These reflections took him back to his earlier cosmogony of 1854, with respect to the origins of the solar system and the source of the sun's heat. He now reiterated: these considerations show us the impossibility of accounting for motion in the history of the earth and the living beings upon the earth without the direct action of creative power and continual all-pervading influence of creative power.[103] In the beginning, Thomson's remarks suggest, God created a primordial fluid containing macroscopic solids, or mundane matter, at rest, but constructed of vortex atoms and subject to gravitational attraction, perhaps produced by ultra-mundane vortices with high velocities. Gravity, collisions and dissipation produced the solar system with its present seemingly stable motions. With continuing losses of kinetic energy it would collapse into the sun; and, as gravity itself succumbed to dissipation, the coalesced mass would evaporate as free atoms, leaving a gas of uniform density and temperature throughout space as the final state of the universe. Vortical motions from the atomic to the cosmic level would thereby solve the ultimate problem of permanence and change from the beginning to the end of the creation.

Waves in the vortex ether and a restructured methodology

If Sir William's fundamental problem throughout the 1870s had been the stability of individual vortex atoms, he nevertheless began also to examine the stability and vibrational properties of arrays of vortices, systems which might represent the luminiferous ether and normal solids, albeit with different

102 William Thomson, untitled manuscript, PA175, ULC, pp. 62, 69. We have found no record of this lecture having been delivered. It has obvious relation to his attempts to analyze the stability conditions of the earth, and to deduce the most likely state of its interior (ch. 16). See 'Turning the world upside down', *Nature*, **22** (1880), 493; 'On the precessional motion of a liquid', concerning the gyrostatic stability of a rotating oblate spheroid of fluid contained in a thin shell; also 'Presidential address' to the British Association on 7th September, 1876 in the same *Report*; PL, **2**, 238–72; and letters to Stokes of 16th and 23rd September, 1876, K363 and K207, Stokes correspondence, ULC. The latter give results for a prolate ellipsoid published in, 'On an experimental illustration of minimum energy in vortex motion', *BAAS Report*, **50** (1880), 491–2; MPP, **4**, 183–5.

103 William Thomson, PA175, ULC, p. 97. He subsequently deleted this passage having circled 'motion' with a question mark in the margin. See Chapter 18 for the problem of living beings.

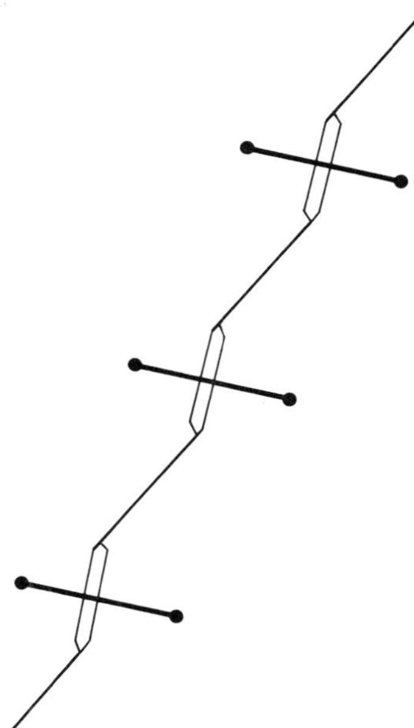

Figure 12.8. A chain of gyrostats, connected together by rods with universal joints at each end, would propagate circularly polarized transverse waves along its length.

grainedness. Almost none of this early work survives, except in the titles of unpublished papers delivered to the Royal Society of Edinburgh: for example, 'On the oscillation of a system of bodies with rotating portions. Part II. Vibrations of a stretched string of gyrostats (dynamics of Faraday's magneto-optic discovery) with experimental illustrations' (1875).[104] The theoretical part of the latter paper survives in a brief communication to the London Mathematical Society. Defining a gyrostat as 'a rapidly rotating fly-wheel, frictionlessly pivoted on a stiff moveable framework or containing case', Thomson considered a chain of gyrostats connected along their axes of symmetry by massless rods with universal flexure joints at each end (figure 12.8).[105] This line, he showed, would propagate circularly polarized, simple harmonic waves, suggesting that an elastic solid made up of a bundle of such lines would produce Faraday's magneto-optic rotation. Thomson presumably had in mind that each of his gyrostats represented, in the simplest case, a ring vortex rotating rapidly

[104] Read 19th April, 1875, *Proc. Royal Soc. Edinburgh*, **8** (1875), 521; part I is listed on p. 490.

[105] William Thomson, 'Vibrations and waves in a stretched uniform chain of symmetrical gyrostats' (read 8th April, 1875), *Proc. London Math. Soc.*, **6** (1875), 190–4; MPP, **4**, 533–8. Thomson added an extensive discussion of gyrostats, with illustrations, to the 1879 edition of the *Treatise*, **1**, para. 345, additions i–xxviii.

about its axis of symmetry. The scheme for an elastic solid, then, would depend on the possibility of constructing a stable, three-dimensional array of vortices.

Mathematical demonstration of the stability of vortex systems, pursued through analysis of maximum, minimum, and maximum–minimum conditions, proved as intractable as the demonstration for individual vortices. Various mechanical illustrations, however, and the analogy between vortices and magnets, offered, in effect, experimental confirmation of stability for simple cases. Sets of up to seven columnar vortices (or two-dimensional cross-sections) were amenable to testing by magnetic needles stuck through corks and floated vertically.[106] Triangular, square, and pentagonal arrays were stable, but not hexagonal or higher order polygons, which would collapse into denser configurations, such as a pentagon with a central vortex. The results indicated that stable elastic lattices could exist, if one could suppose the vortices themselves to be perfectly stable. They did not take into account the effects on stability of dissipation.

Acutely aware of the problems he faced, Thomson nevertheless pursued in the 1880s those aspects of the vortex solid that seemed amenable to solution. In the process he would enunciate a methodology calculated to justify highly speculative theorizing within what he still quaintly regarded as his non-hypothetical and anti-metaphysical approach to natural philosophy. That approach had originally rested on the claim that certainty resided in macroscopic empirical laws alone. Between 1850 and 1870, however, the complementary truths of the universal plenum and the molecular structure of matter had become as certain to Sir William as empirical law. His new faith required a new response to the aversion that had informed his macroscopic physics since the 1840s, namely the aversion to action at a distance. Formerly one could simply plead ignorance of unobservable causes and pursue continuum mechanics. But molecular physics had become a necessity. If action at a distance and the 'monstrous assumption' of hard atoms were not to be considered, then only continuum mechanics, the epitome of the old macroscopic physics, remained. It would now be applied from two directions at once, from the more properly 'macroscopic' properties of ether regarded as a homogeneous elastic solid (ignoring its constitutive lattice of vortices) and from the mechanical properties of molecular vortices, to squeeze down on molecular structure from complementary directions.

Two corresponding methodological emphases, previously only implicit, entered Thomson's physics in the 1880s: a differentiation of 'molar' from

[106] William Thomson, 'Floating magnets', *Nature*, **18** (1878), 13–14; MPP, **4**, 135–40. Closely related unpublished papers were: 'On vortex vibrations, and on instability of vortex motions' (read 15th April, 1878) and 'A mechanical illustration of the vibrations of a triad of columnar vortices' (read 20th May, 1878), *Proc. Royal Soc. Edinburgh*, **9** (1878), 613, 660. In 1887 Thomson came as close as he ever would to establishing the mathematical conditions for a vortex ether, but with respect to its stability he could only pronounce 'the Scottish verdict of *not proven*'; 'On the propagation of laminar motion through a turbulently moving inviscid liquid', *BAAS Report*, **57** (1887), 486–95; MPP, **4**, 308–20, on p. 320.

'molecular' explanation; and an association of 'measurable' with 'conceivable'. The molar–molecular distinction formally recognized the approach from two equally legitimate directions, while measurability supported the claim to practical, non-metaphysical theory in spite of molecular hypotheses.

'Measurable' became a term to conjure with in the opening lines of Thomson's popular lecture at the Royal Institution in 1883 on 'The size of atoms'. The formerly metaphysical entities were to be cleansed by measurement: 'atoms or molecules are not inconceivably, not immeasurably small . . . That which is measurable is not inconceivable, and therefore the two words put together constitute a tautology. We leave inconceivableness to metaphysicians'.[107] While 'metaphysical word-fencing' had sometimes led to the 'inconceivable absurdity' of a limit to divisibility of space and time, the question of such a limit for matter had become 'a very practical question', one of measurement. The equation,

$$\text{conceivable} = \text{measurable} = \text{practical},$$

not only made atoms into legitimate theoretical entities, but legitimated any theory about atoms so long as all the entities of the theory were subject to measurement. Two preconditions are implicit in this scheme. 'Conceivable' presupposes *mechanical* conceivability and 'measurable' means *directly* measurable. These two preconditions will prove critical for Thomson's rejection of Maxwell's electromagnetic theory of ether vibrations, as compared to his own elastic-solid theory, and we shall develop their significance further in that context (ch. 13).

Measuring atoms and molecules meant to Thomson establishing limits on how 'coarse-grained' and how 'fine-grained' ponderable matter could be. For the lecture he rejuvenated the four lines of argument of 1870 to establish that 'the mean distance between the centres of contiguous molecules is less than the 1/5 000 000th, and greater than the 1/1 000 000 000th of a centimetre'.[108] These figures bracket the presently accepted size of atoms of approximately 10^{-8} centimetres. It was in establishing this 'grainedness' for the behaviour of light in ponderable substances, that Thomson developed his two-sided method, differentiating molecular from molar theory, like bricks from a brick wall, or microscopic heterogeneity from macroscopic homogeneity, but with the more important differentiation of the action of molecules from the action of ether.[109]

The correspondence with Stokes reveals this development, beginning from the fact that prior to 1879 Thomson had never seriously studied the standard

[107] William Thomson, 'The size of atoms' (read 2nd February, 1883), *Proc. Royal Inst.*, **10** (1884), 185–213; PL, **1**, 147–217, on p. 147. [108] *Ibid.*, p. 217.

[109] Of course Thomson had long argued for building up the properties of bulk matter from parts that did not possess those properties, as in the debates with Stokes in the 1850s over rotational conduction. See also his 1866 remarks on Maxwell's gas theory (note 75, above) and the discussion in ch. 11 of his 'Note on gravity and cohesion', *Proc. Royal Soc. Edinburgh*, **4** (1857–62), 604–6; PL, **1**, 59–63.

mathematical problems of 'molar' wave optics, such as 'the dynamics of reflection and refraction of distortional [transverse] waves at a surface of separation of two homogeneous incompressible solids united so as to allow no *slip*'. Stokes informed him that Green, Cauchy, and MacCullagh had all worked on the problem and that Green had 'virtually solved' it for isotropic media.[110] Similarly, at the 'molecular' level, although Thomson had long believed, following Cauchy, that refractive dispersion (velocity dependent on wavelength) depended necessarily on the molecular structure of matter, he had never attempted to calculate that dependence but had assumed for perhaps thirty years that a transition from the finer grain of ether to the coarser grain of ponderable matter would suffice. Cauchy's theory shows that in a lattice of particles interacting by attractive forces the wave velocity decreases as the number of particles per wavelength increases. But Thomson's calculations on 7th January, 1883, with his lecture scheduled for 3rd February, proved distressing. The difference in velocity between red and violet light in normal substances allowed twenty molecules at most in the length of a wave, whereas his latest size estimates allowed no fewer than 200 and more likely 600. Mere heterogeneity clearly would not do. Stokes added only injury: 'I have never had much faith in that sort of thing', leaving Thomson with no clue to one of the simplest of all optical phenomena in matter, refraction, and with a gaping hole in his forthcoming lecture.[111]

On 21st January he was still planning to present a version of the deficient Cauchy theory illustrated by a 'wave machine', consisting of a series of wooden bars attached crosswise to a wire suspended from the ceiling. A periodic twist at one end would produce wavelike oscillations in the ends of the bars, regarded as a series of identical molecules (cf. figure 13.3, where this model is transferred to the *interior* structure of a single molecule). But on 26th January, working on the train from Glasgow to London, he invented a completely new solution, which did not depend on the particular size of the molecular grain but on the character of the molecules themselves, as he announced to Stokes: 'I have now a way of explaining refractive dispersion by rotating molecules in an infinitely fine-grained substance'. But he would not see his way through it until some 'seventeen hours' before the lecture.[112]

[110] William Thomson to G.G. Stokes, 4th May, 1879, K229, Stokes correspondence, ULC; Stokes to Thomson, 7th May, 1879, S410, ULC.

[111] William Thomson to G.G. Stokes, 7th January, 1883, K261, Stokes correspondence, ULC. G.G. Stokes to William Thomson, 9th January, 1883, S450, ULC. Thomson, 'The size of atoms' (1883), pp. 193, 217.

[112] William Thomson to G.G. Stokes, 21st January and 26th January, 1883, K263 and K266, Stokes correspondence, ULC. Thomson, 'The size of atoms (1883)', p. 195; model on p. 157. For Stokes's later insistence (and agreement with Thomson) on treating the molecules as active agents rather than mere modifiers of ether properties see G.G. Stokes to William Thomson, 23rd April, 20th May, and 17th November, 1884, S457, S459, and S461, ULC. In the first Stokes writes: 'in a crystal, where we have the vibrating medium existing in the interstices between the molecules of the ponderable matter . . . I have no faith in a priori conclusions derived from the results applicable to a

The linked gyrostats which he had employed in 1875 for the magnetic rotation of light, and which he had analyzed in detail for the 1879 edition of the *Treatise*, now seemed capable of explaining Cauchy's approximate law of dispersion,

$$\nu_\lambda = \nu_\infty\ (1 - C/\lambda^2),$$

with ν the velocity for wavelength λ, and C a parameter depending on the spin of the gyrostats and their elastic connection to the ether:

'I make this out for an infinitely fine-grained substance, (elastic solid) with rapidly rotating infinitesimal fly wheels, pivotted [*sic*] in vesicles of the solid, with their axes in all directions; and I find

$$C = S/n$$

where n denotes an average angular velocity of the fly wheels, and S a constant depending on the elasticity of the solid, and moment of inertia of fly wheel, and dimensions & shapes of cavities. Rotating liquid may be substituted for the solid flywheels.

Here is the final end of 'aer' and the systematic beginning of 'molar–molecular' methodology. Formerly 'aer' had stood as the middle term in a continuous series from the ether to normal objects, which assumed that light waves are propagated through ponderable elastic solids in the same way and according to the same equations as through the elastic solid ether. The two solids had differed in fineness of grain and in the internal structure of their molecules, but not in their constitutive relation to light waves. Now Thomson wished to separate off the behaviour of the ether as the basic carrier of light waves – a problem of molar continuum dynamics – from effects of the interaction of molecules and ether – a problem of molecular loading of the ether dependent on internal dynamics. This is an entirely different explanatory strategy. No longer would molecular structure simply modify the macroscopic parameters of the ether for wave transmission. The individual molecules would act on their own. His exuberance reached new heights: 'There is more & more behind it. I want to do away with elasticity of [ponderable] solid entirely [,] and with nothing but rotating liquid & perfectly rigid [ether] solid containing vessels & links, to get undulations proper for undulatory theory of light'.[113]

That research programme is the one Thomson would pursue in his famous

homogeneous elastic medium'; a marginal note reads: 'Nor have I. W.T. nor I. H. Helmholtz' (their hands, no date). For Helmholtz's similar mode of dealing with molecules and ether in 1875, as different systems, obeying different equations, but with a coupling term, see Buchwald, '*Maxwell to microphysics*', pp. 233–6. Thomson did not recognize the significance of Helmholtz's work, nor even of anomalous dispersion, which was Helmholtz's main focus, until early in 1884, when he wrote to Stokes for assistance (thinking it published 'two or three years ago . . . I forget the title', and unsure of its content) only to discover that Stokes himself had never studied the papers on the phenomenon. William Thomson to G.G. Stokes, 27th January, 1884, K269, Stokes correspondence, and G.G. Stokes to William Thomson, 30th January, 1884, S453, ULC.

[113] Thomson to Stokes, 26th January, 1883, *op. cit.* (note 112). For the interpretation given here see Thomson, 'The size of atoms' (1883), p. 195, and BL, 106–8, where it is clear that 'infinitely fine grained' refers to the grain of the ponderable molecules rather than the ether and where he gives the mathematics as a wave equation for the ether *plus* a molecular loading term.

Baltimore Lectures in the following year, with complex vibrational–rotational molecules embedded in a homogeneous elastic solid ether. Through his molar–molecular pincer strategy, he would attempt to squeeze down on the ether–matter interaction to crack its refractory shell and reveal the vortices on both sides. Meanwhile, however, the younger generation had colonized the ether anew and established their own order under the laws of electromagnetic fields laid down by Maxwell. Sir William would have to fight, it seems, not only to solve his problems, but to preserve his domain. We turn to the battle.

13

Telegraph signals and light waves: Thomson versus Maxwell

Regarding Lord K. I thought he got a proper grip of Maxwell 7 or 8 years ago. His paper on rotational ether showed that. I was rather amazed lately to see in *Nature* he was still bound to ideas founded on the *ordinary* incompressible elastic solid, involving instantaneous propagation of effects. *Oliver Heaviside to G.F. FitzGerald, 1896.*[1]

Sir William Thomson delivered his Baltimore Lectures *On molecular dynamics and the wave theory of light* in 1884, five years after the death of James Clerk Maxwell. Maxwell's electromagnetic theory of light, published in several versions from 1863 to its still incomplete form in 1873, already dominated research on both electromagnetism and light. But, in spite of its mathematical elegance, and in spite of its empirical adequacy, Maxwell's theory contained several elements which failed Thomson's tests of legitimacy for theoretical entities: conceivability and measurability. He now launched a public attack on 'the so-called Electromagnetic theory of light' while displaying the groundwork for his alternative elastic-solid theory.[2] The spectacle was grand, and highly moralistic. At the peak of his intellectual strength and scientific prestige, Sir William of Glasgow, armed with the lance of mechanical analogy and the shield of practicality, went out to slay the nihilistic infidels of 'mathematical', as opposed to 'physical', theory. But his weapons had lost their sheen. The knight of the industrial spirit would go on to become one of its Lords, while his crusade for virtue foundered and his scientific leadership became more formal than actual. Mathematical physics had taken a different road.

The single best indicator of Thomson's rise and decline is the electric telegraph. It embodied his electrical theory in 1854 and continued into the 1890s to act as heuristic guide and legitimizing exemplar while he developed ever more complex conceptions of the interrelation of light and electricity. The telegraph serves also to distinguish Thomson's mode of theorizing from that of the followers of Maxwell.

[1] Oliver Heaviside to G.F. FitzGerald, 9th June, 1896, FitzGerald collection, Royal Dublin Society, Dublin. We thank Bruce Hunt for this reference. [2] BL, 42.

Submarine telegraph theory

Inspiration for William Thomson's work on telegraphy came from a paper by Michael Faraday published in the *Philosophical Magazine* in March, 1854.[3] The introduction around 1848 of the highly effective insulating material, gutta-percha, had made possible underground and submarine telegraph lines. But over long distances a retardation of the signal was observed in the new lines, which limited the speed and clarity of signalling. Responding to this financial threat, the Electric Telegraph Company invited Faraday to its gutta-percha works at Lothbury for an experimental investigation.[4]

Working with submerged lines of one hundred miles, connected at one end through a battery and galvanometer to ground, Faraday found that the arrangement acted like a Leyden jar on a huge scale. The central copper wire, in becoming charged, acted by induction laterally through the gutta-percha, producing an opposite charge at the surface of the surrounding water. Given the immense surface area of the copper, with the surface of the water only 0.1 inch distant, the system possessed an enormous capacity for induction in comparison with normal wires suspended in air. Thus it acquired a prodigious 'quantity' of charge, even though at an 'intensity' equal only to the battery's intensity. 'When the wire is separated from the battery and the charge employed,' Faraday reported, 'it has all the powers of a considerable voltaic current, and gives results which the best ordinary electric machines and Leyden arrangements cannot as yet approach'.[5]

The large capacity of the water wire fits neatly into Faraday's general theory of electricity to provide an explanation of the retardation of signals. The speed at which a signal would propagate he took to be the same as the rate of conduction of electricity along the wire, proportional at any moment to the intensity or tension along it. This tension built up gradually by induction, which in Faraday's scheme always preceded conduction. To the degree that the initial induction directed itself laterally through the gutta-percha to the water, rather than longitudinally through the wire, the tension along the wire would build up more slowly, with a concomitant retardation of conduction: 'the induction consequent upon charge, instead of being exerted almost entirely at the moment within the wire, is to a very large extent determined externally; and so the discharge or conduction being caused by a lower tension, therefore requires a longer time. Hence the reason why, with 1500 miles of subterraneous wire, the wave was two seconds in passing from end to end; whilst with the same length of air wire, the time was almost inappreciable'.[6]

[3] Michael Faraday, 'On electric induction – associated cases of current and static effects', *Phil. Mag.*, [series 4], **7** (1854), 197–208; *Experimental researches in electricity and magnetism* (3 vols., London, 1839–55), **3**, pp. 508–20.

[4] L.P. Williams, *Michael Faraday* (London, 1965), pp. 483–7.

[5] Faraday, *Experimental researches*, **3**, pp. 510–11.

[6] *Ibid.*, p. 515. Here '1500 miles of subterraneous wire' refers to multiple wires between London and Manchester connected in series. They were covered with iron for protection.

Thomson responded to Faraday's analysis in June, 1854, by calculating theoretically the immense inductive capacity of submarine cables. He intended to append the calculations to a republication in the *Philosophical Magazine* of his earlier articles on the heat analogy (ch. 7).[7] Reiterating the correlation of heat flux with electrical force and of conductivity with inductive capacity, he invited the reader to consider:

> the conduction of heat that would take place across the gutta-percha, if the copper wire in its interior were kept continually at a temperature a little above that of the water which surrounds it. Here the quantity of heat flowing outwards from any length of the copper wire, the quantities flowing across different surfaces surrounding it in the gutta-percha, and the quantity flowing into the water from the same length of gutta-percha tube, in the same time, must be equal [flux conserved]. But the areas of the same length of different cylindrical surfaces are proportional to their radii, and therefore the flow of heat across equal areas of different cylindrical surfaces in the gutta-percha, coaxial with the wire, must be inversely as their radii.

The analogy yielded the now-standard formula for the inductive capacity c per unit length of a cylindrical capacitor having inner and outer radii R and R' and containing a dielectric of specific capacity K,

$$c = K/2\ln(R'/R).$$

Correspondingly, Thomson argued, the water wire used by Faraday had a total capacity at least as great as 'an ordinary Leyden battery of 8300 square feet of coated glass 1/23 of an inch thick'.[8]

Faraday's primary concern with this capacity had been its effect in retarding telegraph signals. Thomson took up that issue only in the autumn of 1854. 'I accidentally got on the theory of the propagation of electricity by submarine wires, one day in October . . .,' he wrote to James, 'which showed me at once what would be necessary to insure efficiency for great distances (300 miles or more)'.[9] The 'accident' was a letter of 16th October from Stokes, asking if he were correct in attributing 'the finiteness and even . . . considerable magnitude of the time concerned in the phenomena described by Faraday . . . to the following two causes'. Both of Stokes's causes were resistances to the free flow of electricity. He regarded the 'enormous Leyden jar' as forming a *closed circuit* with the charging battery during the process of induction. Although not closed by conductors, 'The circuit is in a sense closed: +^{ve} electricity flows into the wire, and as the water becomes −^{ve} at the inner surface (that in contact with the gutta-percha) by induction, +^{ve} elec^{y} flows into the water and earth regarded as the

[7] William Thomson, 'On the electro-statical capacity of a Leyden Phial and of a telegraph wire insulated in the axis of a cylindrical conducting sheath', *Phil. Mag.*, [series 4], **9** (1855), 531–5; E&M, 38–41. The paper is dated June, 1854, and is described in William Thomson to G.G. Stokes, 28th October, 1854, K73, Stokes correspondence, ULC.

[8] The symbols are those of his later telegraph papers.

[9] William to James Thomson, 13th January, 1855, T442, ULC; James Thomson, *Papers*, l. The date of the letter was erroneously written as 1854.

common reservoir which supplies the + ve elecy which has to flow in at the − ve end of the battery'.[10]

Stokes's first cause of retardation was effectively internal resistance in the battery, which would have to work for a considerable time to supply the required electricity. His second cause was resistance in the wire and in the water. This concern with resistance in the battery and the water, additional to that through the wire, derived from Stokes's closed-circuit picture of the induction process. To complete the circuit of resistances he suggested yet a third cause of retardation, a cause which had long been a part of Faraday's idea of contiguous action: time for induction. 'I have not mentioned a possible expenditure of time in producing possibly a change in the molecular state of the dielectric gutta-percha, because such a cause would be altogether hypothetical'.

Stokes here set aside the very speculation that would make Maxwell famous, a 'displacement current' closing the charging circuit through the dielectric. Concomitantly, Stokes made nothing of Faraday's repeated emphasis on the interrelation of induction and conduction as two aspects of a single phenomenon, while Maxwell, in his first paper on field theory one year later, would recognize this interrelation as a key ingredient of Faraday's view.

The contrast with Maxwell's reading of Faraday is even more striking for Thomson, who freely misread both Faraday and Stokes in his enthusiasm for his own new insights. On 28th October Thomson wrote: 'The two reasons you assign for the finite intervals occupied by the agencies concerned in the charging &c of the telegraph wire under water are inseparable & I believe contain the explanation of all the perceptible features of the phenomena as observed by Faraday. Your "Cause 1" appears to have been fully appreciated by Faraday himself, but your "Cause 2" I think I have not seen explicitly stated except in your letter & must be taken into account to get a complete investigation of the action'.[11] In stating his view Thomson read Stokes's 'Cause 1' as the inductive capacity of the wire, rather than resistance in the battery, and 'Cause 2' as resistance to conduction in the wire alone, rather than in wire and water conjointly. Faraday, of course, had fully appreciated both of these problems of induction and conduction and had stressed their intimate physical relation. Thomson's remarks thus require decoding, to read: I have already written a paper on the capacity of the wire (induction); and, I had not previously thought about its interaction with resistance (conduction).

Once challenged, Thomson stood in perfect position to reformulate the problem in his own style, for during September and October he had been reviewing his Fourier mathematics in his research notebook.[12] He apparently

10 G.G. Stokes to William Thomson, 16th October, 1854, S374, ULC.

11 William Thomson to G.G. Stokes, 28th October, 1854, K73, Stokes correspondence, ULC. The mathematical parts of this letter, and parts of several later letters, appeared in Thomson's 'On the theory of the electric telegraph', *Proc. Royal Soc.*, **7** (1854–5), 382–9; MPP, **2**, 61–76.

12 See the note introducing Thomson's 'Compendium of the Fourier mathematics for the conduction of heat in solids, and the mathematically allied physical subjects of diffusion of fluids and transmission of electric signals through submarine cables', MPP, **2**, 41.

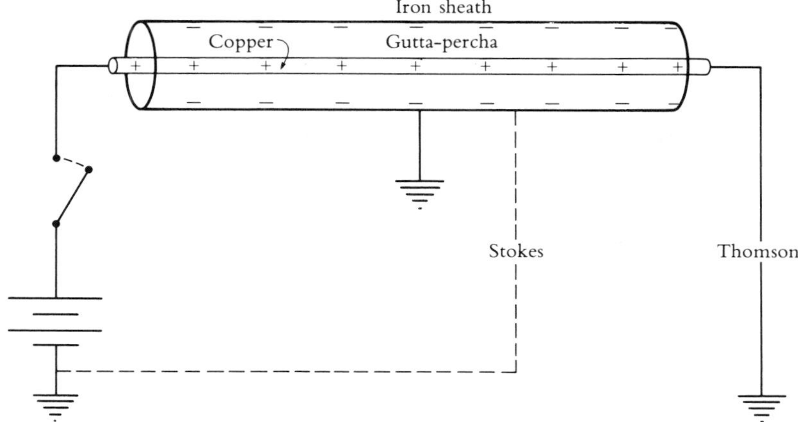

Figure 13.1. Stokes assumed the far end of a telegraph wire to be insulated, while Thomson assumed it grounded. Concomitantly, Thomson treated the basic problem of signalling to be linear conduction along the wire (solid lines), with lateral induction between wire and sheath as merely a parameter of capacity, while Stokes treated the lateral action as part of a closed circuit (dashed line).

recognized at once that propagation of electricity along the telegraph wire might be analyzed as though it were a simple case of linear conduction of heat along an insulated rod between high and low temperature reservoirs, rather than as a closed circuit. As Thomson pictured the problem, the process of lateral induction through the gutta-percha did not enter explicitly, except as a constant capacity for holding electricity on the central conducting wire, analogous to heat capacity. Nor did he include resistance in the battery or in the sea. Figure 13.1 depicts the basic difference of Stokes's (and Faraday's and Maxwell's) conception from Thomson's. Stokes consistently regarded the end of the conducting wire as insulated while Thomson grounded it. Thus Stokes's circuit was closed through the gutta-percha, by induction, while Thomson considered only linear conduction along the wire.

Implicitly, Thomson's circuit was closed through its two grounds. It also contained capacity, which might be represented as a set of small condensers attached along the wire with one side grounded. We omit these features in order to stress that Thomson's scheme isolated conceptually the longitudinal conduction process from the lateral induction process. Having previously employed a heat analogy for static induction to calculate the capacity of the wire, he now employed a different heat analogy for the conduction process. We have noted similar moves previously in Thomson's mathematical interpretations of Faraday (ch. 7). He regularly separated off, in analytical technique, intensity problems from flux problems, while nevertheless treating them as of a single kind through the flow analogy.

The difference from Maxwell is far-reaching. From early in 1854, when

Maxwell first began, in correspondence with Thomson, to develop his own flow analogy to fields, he took seriously Faraday's distinction of quantities and intensities, treating them as dual aspects of the mere form of any flow-like phenomenon involving resistance to conduction or its analogue in induction.[13] From this quantity–intensity distinction would come the dual electric and dual magnetic fields of 'Maxwell's equations': electric intensity E and 'displacement' D; magnetic intensity H and 'induction' B. Maxwell also recognized that Faraday's views required a unified treatment of induction and conduction, although he did not yet have any clear idea of their relation. In 1863, however, Maxwell proposed his 'displacement current', or current of induction, the rate of change of displacement $\partial D/\partial t$. Such displacement currents in non-conducting media, he assumed, would produce all the effects of normal conduction currents, including magnetic fields. This assumption entailed electromagnetic waves, because changing electric displacements and changing magnetic inductions mutually induced one another. The waves, Maxwell showed, would propagate through space at the velocity of light (see below).

With Faraday, Maxwell supposed that the opposite electricities appearing on the conductors of a Leyden arrangement during the charging process were manifestations of some underlying response in the medium – whether ether or ponderable matter – through which lines of electric force were propagated. This free electricity appeared at the terminations of the lines in quantities determined by the number of lines. While the lines were changing in quantity, therefore, a displacement current 'flowed' through the dielectric equal in quantity to the conduction current in the wires.[14] In general, the total or true current consisted of conduction current plus displacement current, and could be represented as the continuous flow of an incompressible fluid, even though electricity accumulated at interfaces between conductors and dielectrics.

Thomson had already rejected such notions by the early 1850s at the latest (chs. 7 and 8). Electricity for him had become the universal substratum whose motions constituted heat, whose aligned rotational motions constituted magnetism, and in which a structure of vortices produced the incompressible lattice of the elastic-solid ether. This fluid flowed freely through conducting wires to accumulate on the plates of a capacitor, but if only air or free ether separated the plates (air–ether or 'aer'), nothing happened in that space. A tension existed in the ether lattice, constituting the electric lines of force, but no response

[13] Joseph Larmor (ed.), 'The origins of Clerk Maxwell's electric ideas, as described in familiar letters to W. Thomson', *Proc. Cam. Phil. Soc.*, **32** (1936), 695–750; letters of February 20th and March 14th, 1854, pp. 697–701. Thomson's replies seem not to have survived. Maxwell's first paper on fields was 'On Faraday's lines of force' (read 10th December, 1855 and 11th February, 1856), *Trans. Cam. Phil. Soc.*, **10** (1856), 27–83; *The scientific papers of James Clerk Maxwell*, W.D. Niven (ed.), (2 vols., Cambridge, 1890), **1**, pp. 155–229.

[14] For detailed interpretations of Maxwell's views in 1863 and of later Maxwellian theory, respectively, see Daniel Siegel, 'The origin of the displacement current', *Hist. Stud. Phys. Sci.*, **17** (1986), 99–146, and J.Z. Buchwald, *From Maxwell to microphysics: aspects of electromagnetic theory in the last quarter of the nineteenth century* (Chicago, 1985), pp. 20–53.

occurred in this incompressible structure, no flow, polarization, or strain, no 'displacement'. The electricity on the plates derived from flow through the wires, not from tension in the ether. Concomitantly, the displacements involved in waves of light could not be electric displacements.

In true dielectrics, not including gases or ether on Thomson's view, a limited motion of electricity did occur, but only within small pores. Aer seemed to have no pores. In his 1852 lectures, he said: 'A plate of sulphur acts like a solid resister [*sic*] of electricity filled with [polarized] conductors throughout. [Faraday] supposes that there is some such polar action in air. It appears that there is no such polar electrification in air'. His point was empirical; no evidence of such action existed. One could only conclude that 'aer' transmitted electrical force as tension in an incompressible medium, instantaneously. In conductors the motion of electricity relieved this tension, or neutralized it by a back pressure of accumulated electricity. Similarly, in dielectrics, the limited motion partially neutralized the tension, but the electrical action existed everywhere, both inside and outside conductors and dielectrics. 'Faraday considered that it was a tension of the air on the outside entirely. But electrical force acts everywhere; in every direction'.[15]

Thomson's theory of electricity constituted a one-to-one translation of the terms of action at a distance into those of fields, replacing electrical force at a point by tension and instantaneous action at a distance by instantaneous propagation to a distance. His lectures dispel any doubts about whether he understood that Faraday's theory involved nothing like a real electrical density, or that it required a quantity–intensity duality for propagating forces. He understood both features and rejected them. His continued use of the term 'flux', furthermore, in the analogy between transmission of electrostatic force and heat conduction, had become purely metaphorical, for his flux was not continuous across interfaces between different media, as flux of heat would be for changes in conductivity. There could be no continuity of current, therefore, in the circuit that charged a Leyden jar. No electric displacement like Maxwell's could complete the circuit of electrical 'quantities'.

Concretely, then, Thomson did not unite his heat analogy for *conduction* along the telegraph wire with his analogy for *induction* through the gutta-percha to describe a closed circuit of fluxes, as Stokes had suggested. He treated inductive capacity only as a parameter in the conduction process. On the back of Stokes's initial letter he sketched the Fourier derivation for linear conduction in a few short lines, which he then elaborated. 'In taking up your letter this morning to answer it, I find that the whole may be worked out definitely as follows'.[16] With c as the capacity of the wire per unit length, r its resistance per unit length, V the instantaneous potential at any point x along the wire, and I the instantaneous

[15] William Thomson, in William Jack, 'Notes of the Glasgow College natural philosophy class taken during the 1852–3 session', Ms Gen. 130, ULG. The quotations precede an entry dated 25th November, 1852. [16] Thomson to Stokes, *op. cit.* (note 11).

current at that point, he set the rate of change of static electricity on any element dx, $cdx(\partial V/\partial t)$, equal to the rate at which the current deposited electricity at that element, $-(\partial I/\partial x)dx$, the two quantities being measured in electrostatic units (esu) and electromagnetic units (emu), respectively. Substituting $I = (1/r)\partial V/\partial x$ (Ohm's law), and with σ as the ratio of units (esu/emu), he obtained in electrostatic units the diffusion equation:

$$rc\partial V/\partial t = \sigma\partial^2 V/\partial x^2.$$

'This equation', Thomson observed, 'agrees with the well known equation of the linear motion of heat in a solid conductor, and various forms of solution which Fourier has given are perfectly adapted for answering practical questions regarding the use of the telegraph wire'. He gave details for the propagation of a pulse down an infinitely long wire (grounded at the distant end) and indicated the requisite modifications for a finite wire. To account for leakage through the gutta-percha ('most sensible in Faraday's experiments') he added a term $-hV$ to the above equation, corresponding in Fourier's scheme to radiation losses from the surface of a rod conducting heat along its length.

The simplicity of Thomson's approach testifies not only to his familiarity with Fourier's mathematics. At least as important was his intuition for relevant and irrelevant magnitudes, for only that sensibility to actual numbers allowed him to see at once the applicability of Fourier's formalism. He could disregard Stokes's worries about resistance in the water (it 'being incapable of preventing the inductive action from being completed instantaneously round each part of the wire') and resistance in the battery (it 'being small, say not more than that of a few yards of the wire').[17]

In letters of December, 1854, to February, 1855, Stokes continued to press his views and Thomson continued to reject them.[18] A great deal more than intellectual interest rested on the numbers. Writing to Stokes on 30th October, 1854, two days after his initial letter, Thomson announced the consequences of his theory. 'An application of the theory . . . which I omitted to mention in the haste of finishing my letter . . . shows how the question raised by Faraday as to the practicability of sending distinct signals along such a length as the 2000 or 3000 miles of wire that would be required for America, may be answered'.[19] The practicability of the Atlantic telegraph had been discussed at the 1854 meeting of the British Association,[20] but this statement is the first indication of Thomson's developing interest in the project. 'The general investigation', he claimed, 'will show exactly how much the sharpness of the signals will be worn

[17] *Ibid.*

[18] G.G. Stokes to William Thomson, 2nd December, 1854, S376, ULC. William Thomson to G.G. Stokes, 25th December, 1854, and 12th February, 1855, K76 and K78, Stokes correspondence, ULC.

[19] William Thomson to G.G. Stokes, 30th October, 1854, K74, Stokes correspondence, ULC.

[20] J.W. Brett, 'On the origin of the submarine telegraph and its extension to India and America', *BAAS Report*, **24**(1854), 7–8.

down, and will show what maximum strength of current through the apparatus in America, would be produced by a specified battery action on the end in England, with wire of given dimensions &c'. His general solution of the partial differential equation showed that 'the time required to reach a stated fraction of the maximum strength of current at the remote end (length l) will be proportional to rcl^2 [actually rcl^2/σ]. [Thus] we may be *sure* beforehand that the American telegraph will succeed, with a battery sufficient to give a sensible current at the remote end when kept long enough in action, but the time required for each deflection will be 16 times as long as would be with a wire a quarter the length, such for instance as in the French submarine telegraph to Sardinia & Africa'.[21]

The penalty of long cables, then, was that as the length increased, the time increased as the square of the length, or the effective velocity decreased in proportion to the length. As the most obvious resolution of a problem which cast doubt on the economic viability of a 2000 mile cable, Thomson suggested increasing both the diameter of the conducting wire to decrease its resistance and of the gutta-percha insulation to maintain constant capacity. Hence, 'when the French submarine telegraph is fairly tested, we may make sure of the same degree of success in an American telegraph by increasing all the dimensions of the wire in the ratio of the greatest distance to which it is to extend, to that for which the French one has been tried'. The signals would exhibit the same effective velocity and the same degree of degradation.

On 1st December Thomson made an unusual request of Stokes:

I should be much obliged if you would not mention to any one what I wrote to you regarding the remedy for the anticipated difficulty in telegraphic communication to America, at present, as Rankine has suggested that I should join with him in applying for a patent for a way of putting it in practice . . . In a very short time I believe there will be no necessity of keeping any part of our plan secret. In the mean time there can be no harm in publishing the theory, & if I can get time to write it out, I shall try to do so soon.[22]

Overburdened with multiple commitments and with anxiety over his ailing wife, Thomson wished simply to establish his priority by publishing excerpts from his letters. Conveniently, as Secretary of the Royal Society, Stokes could have the 'paper' published in the *Proceedings*. It appeared in May, 1855, by which time the patent had been secured. Thomson was 'not very hopeful of making anything of it, but it is possible that it may be profitable'.[23] Indeed, although unsuccessful in this first attempt, he would reap both profit and prestige from subsequent patents. The Atlantic telegraph would depend critically on his theory, his instruments, and his measurements (ch. 19), and would win him his knighthood.

[21] Thomson to Stokes, *op. cit.* (note 19).
[22] William Thomson to G.G. Stokes, 1st December, 1854, K75, Stokes correspondence, ULC.
[23] William to James Thomson, *op. cit.* (note 9).

Measurements, units, and velocity

Sir William's deep involvement in the Atlantic cable sheds considerable light on his lifelong rejection of Maxwellian electromagnetic theory. The laying and working of the cable required the complete unity of theory and practice that he had always preached. His lesson was both an economical and a moral-political one, for, in his Victorian capitalist vision, personal profit and public service were, ideally, one. Practicality would bring profits to investors, strength to the Empire, and material benefits to mankind. Speculation would bring uncertainty, divisiveness, and ruin. To the degree that theories lost direct contact with reality, therefore, they became not only methodologically wrong but morally wrong.

We have observed this ethic operating in Thomson's natural philosophy since his early days at Cambridge. It opposed the metaphysical to the practical. In proscribing mathematical analogies that extended beyond direct experimental observation, it eliminated the flux–force duality from his flow analogies to electricity and magnetism. Maxwell's theory, by contrast, introduced a physical entity which no one had ever observed, the displacement current. But, most specifically, Thomson opposed 'metaphysical' to 'measurable', and it is this aspect that the telegraph especially reinforced. The only aspects of Maxwell's theory that Thomson would ever applaud were those related to measurements.

The cable required precise measurements of the resistivity of copper, the specific inductive capacity of gutta-percha, and the ratio of units, esu/emu. Thomson noted in the published version of his letters to Stokes that 'It will be an economical problem, easily solved by the ordinary analytical problems of maxima and minima, to determine the dimensions of wire and covering which, with stated prices of copper, gutta-percha and iron, will give a stated rapidity of action with the smallest initial expense'.[24] But resistivities and capacities were only roughly known and never controlled in manufacture. Not even the requisite instruments existed, although that situation began rapidly to change when Thomson, in the early 1850s, established his Glasgow laboratory, devoted in its research function largely to electrical measurements and instruments. In 1857, upon testing four specimens of no. 22 gauge copper wire, manufactured specifically for submarine telegraphy by four different companies (A–D), he found that their resistivities varied nearly as two to one. He warned investors:

> It has only to be remarked that *a submarine telegraph constructed with copper wire of the quality of the manufactory* A *of only* $\frac{1}{21}$ *of an inch in diameter, covered with gutta-percha to a diameter of a quarter of an inch, would, with the same electrical power, and the same instruments, do more telegraphic work than one constructed with copper wire of the quality* D, *of* $\frac{1}{16}$ *of an inch diameter,*

[24] William Thomson, 'On the theory of the electric telegraph', *Proc. Royal Soc.*, **7** (1855), 382–99; MPP, **2**, 61–76, on p. 68.

covered with gutta-percha to a diameter of a third of an inch, to show how important it is to share-holders in submarine telegraph companies that only the best copper wire should be admitted for their use.[25]

Telegraph companies learned faster than natural philosophers about the value of measurements. In his oft-cited lecture on 'Electrical units of measurement', delivered in 1883, Sir William proudly observed that 'resistance coils and ohms, and standard condensers and microfarads had been for ten years familiar to the electricians of the submarine-cable factories and testing stations, before anything that could be called electric measurement had come to be regularly practised in almost any of the scientific laboratories of the world'. Such experience supported his conviction that 'The life and soul of science is its practical application', expressed particularly in direct measurements of the properties of matter. The lecture contains the famous lines: 'when you can measure what you are speaking about and express it in numbers you know something about it; but when you cannot measure it . . . you have scarcely, in your thoughts, advanced to the stage of *science*'.[26] Needless to say, Maxwell's displacement hardly met the test of 'science', for it had never been observed, let alone measured in the direct sense Thomson intended. The value of Maxwell's theory lay in calling attention to the probable physical relation existing between the ratio esu/emu and the velocity of light waves.[27] But of course Thomson himself had established the importance of this ratio for the effective velocity of telegraph signals in 1854. And on his theory rested the daily operation of a 3000 mile cable stretched across the Atlantic.

The ratio esu/emu thus carried special practical significance as the ratio of induction to conduction. Indeed, Faraday, in his 1854 analysis, had already stressed that 'The phenomena [of the submarine wire] altogether offer a beautiful case of the identity of static and dynamic electricity'. He noted further that the telegraph could be used to compare dynamic and static measures: 'The whole power of a considerable battery may in this way be worked off in separate portions, and measured out in units of static force'.[28] Thomson, in his first outline of the theory to Stokes, observed that the required ratio of electrostatic units to electromagnetic units 'may be determined by finding the velocity of propagation of a regular periodical effect . . . and I believe it may be actually estimated (roughly) from what Faraday has already done'.[29]

For accurate data he applied to the Astronomer Royal at Greenwich, G.B. Airy, who provided retardation times and dimensions for the cables from Greenwich to Edinburgh ($\frac{1}{17}$ sec. for a line nearly all above ground), Greenwich to Brussels ($\frac{1}{10}$ sec. for a 270 mile line two-thirds underwater), and London to

[25] William Thomson, 'On the electric conductivity of commercial copper of various kinds', *Proc. Royal Soc.*, **8** (1857), 550–5; MPP, **2**, 112–17, on p. 113.

[26] William Thomson, 'Electrical units of measurement' (delivered to the Institution of Civil Engineers, 1883), PL, **1**, 73–136, on pp. 79, 75–6, 73.

[27] *Ibid.*, pp. 83, 114–15.

[28] Faraday, 'On electric induction', p. 511.

[29] Thomson to Stokes, *op. cit.* (note 19).

Manchester (2 sec. for 1500 miles in series underground).[30] The information convinced Thomson:

that a kind of experiment . . . which I hope to be able before long to get in practice will be successful in giving a tolerably accurate comparison of the electrostatical & electrodynamic units and with a farther experiment to determine the specific inductive capacity of gutta percha which will present no difficulty *will enable me to give all the data required for estimating telegraph retardations without any data from telegraphic operations.* This experiment is simply to put two plane conducting discs in communication with the two poles of a Daniell's battery (or any other of known electromotive force), and to *weigh* the force between them.[31]

To Stokes, he reported that Airy's data showed 'there cannot be more than 419 000 000 nor less than 104 000 000 electrostatic units in the electrodynamic, & consequently that the attraction between the separated electricities flowing from the decomposition of a grain of water if concentrated in points 1 foot apart cannot be less than 10 tons nor more than 42: and that the attraction between two parallel discs each a squ. foot area, & $\frac{1}{100}$ foot apart, when connected with the two poles of a Daniell's battery of 100 cells (which I am now going to try & measure directly) cannot be less than 4 grains nor more than 71'.[32] This direct *weighing* experiment is the one Thomson had proposed already in 1845 for obtaining an absolute measure of static electricity, and forms part of a continuing story of absolute units (ch. 20).

Airy's data gave Thomson confidence in his 'law of the squares' for retardations. Joule supplied further good news in February, 1855. Comparing retardations in the Manchester–London–Manchester cable (about $\frac{1}{4}$ sec. over 380 miles) with those in the Greenwich–Brussels line, Joule reported, 'you will find your law of the square of the length completely borne out'.[33]

Secure in his basic equation, Thomson now gave it an analogical interpretation more consistent with an actual electrical fluid. He had in mind the motion of a fluid in an elastic tube when subjected to a pressure pulse at one end. This idea suggested new subtleties for telegraph cables carrying multiple wires, like parallel bores in an elastic cylinder. From the electrical perspective, if several wires ran parallel to one another, separated only by gutta-percha, electrostatic action between them would alter their equation of conduction. Any gradient in the state of charge and potential along one wire would produce corresponding gradients along the others by electrostatic induction, but these gradients would act back on the first wire to change its original gradient, and so on, reflecting back and forth *ad infinitum*. The mutually induced electrostatic gradients, furthermore, would produce electrical currents along the wires, presenting a strange case of *static* induction of *dynamic* effects. Avoiding the contradictory

[30] MPP, **2**, 74–5; G.B. Airy to Edward Sabine, 19th December, 1854, S8A, ULC.
[31] William Thomson to G.B. Airy, 2nd February, 1855, A22, ULC.
[32] Thomson to Stokes, 12th February, 1855, *op. cit.* (note 18).
[33] J.P. Joule to William Thomson, 22nd February, 1855, J192, ULC.

locution, 'electrostatic induction of electric current', Thomson invented the term '*peristaltic induction*' to distinguish this new form of electrodynamic action from the usual electromagnetic induction. The effects seemed completely analogous to those of pressure pulses in adjacent elastic tubes filled with fluid.

The new phenomena present a very perfect analogy with the mutual influences of a number of elastic tubes bound together laterally throughout their lengths, and surrounded and filled with a liquid which is forced through one or more of them, while the others are left with their ends open or closed. The hydrostatic pressure applied to force the liquid through any of the tubes will cause them to swell, and to press against the others, which will thus, by peristaltic action, compel the liquid contained in them to move in different parts of them in one direction or the other . . . the hydraulic motion will follow rigorously the same laws as the electrical conduction, and will be expressed by identical language in mathematics.[34]

The mathematics, while forbidding to look on, involved simply a system of simultaneous partial differential equations, which Thomson could write down by inspection from his analogy. Their general solutions followed forms given by Fourier; but for particular cases of two, three, four, and six wires Thomson deployed his 'method of images' to give immediate solutions in a form suitable for 'practical computation'.[35] The master of theory for the practical man had here given a virtuoso performance on his special instruments of analogy and images. Much impressed, Joule wrote: 'the analogy between the retardation of the current and that of fluids in pipes of narrow bore as pointed out by you is very instructive and must be appealed to whenever the real physical nature of the electric fluid is sought to be discovered'.[36] Thomson agreed. He would make the problem of fluids in pipes of narrow bore the key to the relation of electricity and ether, or telegraph signals and light waves.

In an article on 'Velocity of electricity' in 1860, three years before Maxwell's electromagnetic theory of light, Thomson correlated all of the basic properties of electricity in telegraph lines with properties of fluids. The three phenomena 'concerned in the transmission of an electric signal along an insulated conductor' were: electrostatic induction (characterized by capacity for electrical accumulation); electromagnetic induction (characterized by the tendency of a current to oppose changes in strength); and conduction (characterized by resistance). The electrical characteristics, all of which slowed telegraph signals, were 'in the present state of science not understood except as quite distinct from one another', but water in an elastic tube exhibited a similar set of properties: expansibility of the tube coupled with compressibility of the water (very small); inertia; and viscosity.

Thomson treated expansibility as the analogue of high capacity in submarine

[34] William Thomson, 'On peristaltic induction of electric currents in submarine telegraph wires', *BAAS Report*, **25** (1855), 22; MPP, **2**, 77–8. See also SPT, **1**, 310–11.
[35] MPP, **2**, 89.
[36] J.P. Joule to William Thomson, 19th March, 1855, J193, ULC.

cables insulated with gutta-percha and compressibility as the analogue of low capacity in air lines. Thus submarine cables acted like highly expansible rubber tubes while air lines behaved like rigid metal tubes. The inertial property of electromagnetic induction Thomson knew to be insignificant in long submarine cables; but in short cables, and in air lines, the frequency of signalling (rate of change of current) had only to be high enough for the 'slowness' due to electromagnetic induction (inertia) to dominate over the slowness due to electrostatic induction (expansibility). In this case the telegraph equation became a wave equation rather than a diffusion equation, as Kirchhoff had shown.[37] Thomson's analogy for telegraph signals changed accordingly, from non-inertial heat conduction to inertial oscillations. Electric signals in air lines propagated like longitudinal pressure pulses (sound waves) in a rigid tube of water, with a definite velocity depending only on compressibility and inertia. Viscosity, or resistance to conduction, would degrade the signal, but not affect this constant velocity.

Available data gave velocities for air wires from three-fifths to one and one-half times the velocity of light. Kirchhoff's theory supported the larger figure, but since electromagnetic inertia had never been accurately measured, Thomson refrained from speculations. He said little more on the subject until 1884, in his Baltimore Lectures, *On molecular dynamics and the wave theory of light.*

Longitudinal waves and light

In the Baltimore Lectures Thomson virtually claimed to have made Maxwell's great discovery for him:

> I am quite conscious . . . of what has been done in the so-called Electro-magnetic theory of light. I know the propagation of electric impulse along an insulated wire surrounded by gutta-percha, which I worked out myself, about the year 1854 and in which I found a velocity comparable with the velocity of light. We then did not know the relation between electrostatic and electro-magnetic units. If we had, that [velocity] might have been obtained[,] in the way Maxwell has brought out so beautifully[,] from the proper coefficients of capacity for the gutta percha. If we work that out for the case of air instead of gutta percha, we get practically the same v, I think, for the velocity of propagation of the impulse.[38]

These remarks are surprising because Thomson's calculations in 1854 concerned the effective diffusion velocity of an impulse rather than the definite propagation velocity of a wave. For submarine cable covered with gutta-percha the calculations gave an effective velocity on the order of one-hundredth of the velocity of light in the 3000 mile Atlantic cable. For a similar 300 mile cable, this effective velocity would have been one-tenth that of light. Supposing that

[37] William Thomson, 'Velocity of electricity', in J.P. Nichol (ed.), *Cyclopaedia of the physical sciences*, 2nd edn. (London and Glasgow, 1860); MPP, **2**, 131–7. [38] BL, 42.

Thomson had in mind his 1860 discussion of the definite velocity of waves in air wires, his claims still seem peculiar, for his theory was entirely different from Maxwell's. Thomson's velocity was the velocity of a longitudinal wave travelling within the central conductor; Maxwell's was the velocity of a transverse wave in the surrounding non-conductor, whether gutta-percha or air. This transverse wave consisted of oscillating electric and magnetic forces propagated by reciprocal induction between the field 'quantities', electric displacement and magnetic induction.

At least one young proponent of Maxwell's theory, but also a friend of Thomson's, expressed his horror at the apparent confusion between signal velocities in submarine cables and the velocity of light, but more generally at the idea that Thomson's analogy bore any relation to Maxwell's theory. Writing to *Nature*, George Francis FitzGerald (1851–1901) initially protested at what he thought must be a mistaken report of Thomson's Baltimore Lectures. Shortly afterwards, having examined a copy of them and having read the 1860 article, he extended his objections and sent the letter through Thomson for approval. Sir William complained of misrepresentation, for he had not asserted that submarine signals behaved like light waves nor that his theory was similar to Maxwell's, but only that for air wires his theory would similarly highlight the ratio emu/esu as very near the velocity of light and that he found Maxwell's theory unsatisfactory because it had no definite dynamical foundation. FitzGerald amended his wording and apologized for his excesses: 'I was so anxious to prevent what I find is a very common mistake namely its being supposed that the velocity of transmission of signals [in submarine lines] is the same thing as Maxwell's velocity of light that I wrote as if your whole paper had been upon this velocity of transmission of signals'. More colourfully: 'my letter to "Nature" is not intended to inform *you* of anything as I might as well teach my grandmother to suck eggs'.[39]

But FitzGerald certainly did intend to teach Thomson that one need not suppose that electricity behaved like a compressible fluid with inertia. 'According to my view of Maxwell's theory', he wrote, 'there is no known phenomenon in electricity exactly *analogous* to the compressibility of the water in the tube'. Furthermore, 'it is evident that the inertia of the water is a very bad analogue to electromagnetic induction, for this latter depends essentially upon the form of the circuit, and not only upon its section and length'.[40]

These two objections, if accepted, would demolish Thomson's telegraph theory, for his conception of wave propagation in air wires depended entirely on compressibility and inertia inside the tube. Maxwell's theory, FitzGerald ob-

[39] G.F. FitzGerald to William Thomson, 25th April, 1885, F118, ULC. See also G.F. FitzGerald, 'Molecular dynamics', *Nature*, **31** (1885), 503; 'Sir Wm. Thomson and Maxwell's electro-magnetic theory of light', *Nature*, **32** (1885), 4–5; William Thomson to G.F. FitzGerald, 17th April, 1885, in SPT, **2**, 1038.

[40] FitzGerald to Thomson, *op. cit.* (note 39); FitzGerald, 'Sir Wm. Thomson', p. 4.

served, located electromagnetic inertia, and the electromagnetic waves of light, outside the tube. To include anything at all analogous to Maxwell's theory, Thomson should have considered 'a tube bored in an indefinitely large lump of india-rubber', which would propagate waves away from the tube throughout all space. 'This is just the difference between Sir Wm. Thomson's and Maxwell's views. According to Maxwell's view there is a great deal more going on outside the conductor than inside it'.[41]

The terms of the controversy between Thomson and the Maxwellians had thus been clearly stated by 1885. Thomson took his telegraph theory, a theory limited to the motion of electricity in a conducting wire, as a perfectly clear, definite, and practical core for building outwards to a more general theory for non-conductors. This theory would eventually relate electric and magnetic forces to the motions of the elastic-solid ether, motions which for Thomson constituted light waves. Maxwellians regarded the telegraph, and electrical conduction generally, as a special problem to be solved within an already attained electromagnetic theory for non-conductors, which included light. The respective roles of inside and outside the telegraph wire therefore capture a critical difference in viewpoint. Thomson would move outwards through the gutta-percha to understand light while the Maxwellians would move inwards through gutta-percha to understand conduction.[42] From both perspectives, electrostatic induction in the gutta-percha provided the key puzzle. Thomson sought to treat it as modified conduction, because he thought he understood electricity, while the Maxwellians treated it as a modified action of the ether, whose behaviour Maxwell's equations satisfactorily described, albeit without providing a dynamical explanation.

Since Thomson could not accept the displacement current, he regarded Maxwell's equations as a highly misleading chimera, standing between a well-founded telegraph theory and an equally well-founded elastic-solid theory of light, while the Maxwellians, accepting neither the electrical fluid nor the elastic-solid ether, took Maxwell's phenomenological theory of displacement as the only sound basis for understanding either electricity or ether. FitzGerald put the point succinctly:

> As to Maxwell's notion of the ether not being sufficiently definite I think that is just one of its advantages . . . I think that a lot of theories as to the structure of the ether could be framed which would attribute to its elements properties obeying the laws of the symbols that Maxwell uses to represent its condition. As long as there are a lot of hypotheses possible I think it is unsafe to commit ourselves to any more than we know and that is the laws that its elements obey.[43]

On the surface, these remarks sound very like Thomson's anti-hypothetical

41 *Ibid.*

42 On the central role of the conduction problem in Maxwellian theory see Buchwald, *From Maxwell to microphysics*, pp. 20–53.

43 G.F. FitzGerald to William Thomson, 25th April, 1885, F118, ULC.

statements in his 1845 reconciliation between Faraday's theory and Poisson's action at a distance theory (ch. 7). But just as Thomson there rejected Faraday's notion of a continuous electrical flux, because no empirical evidence existed for it, he here rejected Maxwell's electric displacement. Similarly, FitzGerald's appeal to the 'laws of the symbols' should remind us of the operational calculus of Gregory and Boole and of the problem of interpretation associated with its symbolic forms (ch. 6). Symbols possessing neither direct empirical nor definite dynamical reference were dangerous wills-o'-the-wisp.

From FitzGerald's perspective, however, it was Thomson who threatened the scientific enterprise, with his 'reactionary' commitment to an elastic-solid ether. FitzGerald too sought dynamical models and had very recently devised one whose elements obeyed Maxwell's equations. Importantly, it showed how electric displacements might be regarded as 'changes in structure of the elements of the ether, and not actual displacements of the elements', as in elastic solid theories.[44] But FitzGerald insisted on its purely heuristic nature, for since 'it consists of wheels pumping liquid through diaphragms nobody could imagine it at all *like* the ether'.[45] He accused Thomson of dogmatism: 'I also think that Sir Wm. Thomson, notwithstanding his guarded statements on the subject, is lending his overwhelming authority to a view of the ether which is not justified by our present knowledge and which may lead to the same unfortunate results in delaying the progress of science as arose from Sir Isaac Newton's equally guarded advocacy of the corpuscular theory of optics'.[46]

This damning charge from Sir William's young friend suggests how radically his Baltimore Lectures had cut across the grain of Maxwellian theory, both methodologically and in content. Before considering the methodological issue more extensively, we must sharpen the relation of Thomson's telegraph theory to his ether theory.

Electricity and the telegraph actually appear only in isolated remarks in the Baltimore Lectures. They play an essential role, however, because they provide Thomson's ground for assuming the existence of longitudinal waves in the luminiferous ether, which justifies his treatment of the ether as an elastic solid rather than some other sort of substance which would transmit only transverse waves. FitzGerald concluded his initial letter to *Nature* with the crucial issue: 'Of course, as Sir Wm. Thomson asserts, something analogous to a longitudinal

[44] Bruce Hunt, '"How my model was right": G.F. FitzGerald and the reform of Maxwell's theory' (unpublished) analyzes FitzGerald's models and methodology and makes the point about structural theory. FitzGerald, 'Sir Wm. Thomson', p. 5. G.F. FitzGerald, 'On a model illustrating some properties of the ether', *Sci. Proc. Royal Dublin Soc.*, **4** (1885), 407–19; *Scientific writings of the late George Francis FitzGerald*, J. Larmor (ed.) (Dublin, 1902), pp. 142–56.

[45] FitzGerald to Thomson, *op. cit.* (note 43).

[46] FitzGerald, 'Sir Wm. Thomson', p. 5. For the continuing deep commitment of Maxwellians to mechanical models and a material ether see Bruce Hunt, 'Experimenting on the ether: Oliver J. Lodge and the great whirling machine', *Hist. Stud. Phys. Sci.*, **16** (1986), 111–34, who pays special attention to the role of experiment in stripping the ether of its physical properties towards the end of the century.

vibration may co-exist with [Maxwell's transverse waves], but Maxwell's theory shows that a medium which would transmit only transverse vibration [i.e. not an elastic solid] would explain electric and magnetic phenomena'.[47]

In his Lecture IV, Thomson considered a section of wire – essentially a telegraph wire – terminated by spherical conductors (figure 13.2) and surrounded by ether, or by air (consisting now of large molecules embedded in ether, ch. 12):

> Suppose that we have at any place in air, or in luminiferous ether (I cannot distinguish now between the two ideas) . . . two spherical conductors united by a fine wire, and that an alternating electromotive force is produced in that fine wire . . . It is absolutely certain that such an action as that going on would give rise to electrical waves [in air–ether]. Now it does seem to me probable that those electrical waves are condensational waves in luminiferous ether; and probably it would be that the propagation of these waves would be enormously faster than the propagation of ordinary light waves.[48]

Alternating compressions and rarefactions of electrical fluid in the spherical conductors would produce increasing and decreasing pressures or tensions in the surrounding – and containing – ether. Since no evidence existed that the ether responded with any finite displacement or polarization (no capacity for electrostatic induction), one ought to assume, on Thomson's view, that it behaved like a (nearly) incompressible medium, which would propagate longitudinal waves with (nearly) infinite velocity. Elastic-solid theories had always assumed that such waves existed in principle, but usually assumed as well that they had never been detected in association with the transverse waves of light because of their infinite velocity. The tension theory of electrical action gave independent support to both assumptions.

Thomson had been developing this reasoning ever since his discussion in 1843 of how air pressure might account for electrostatic forces acting on charged conductors. The 1847 'Mechanical representation' explicitly promoted a tension theory of electrostatic force in an incompressible elastic-solid ether, as did his 1852–3 lectures ('aer'), and the tension theory had always implied longitudinal waves: 'It is probable that vibrations like those of sound are propagated at the same time [as those of light]'.[49] But the unquestionable success of the telegraph theory gave an entirely new order of support to longitudinal waves. Telegraph signals in air wires now simply *were* longitudinal waves to Thomson, and their continuation in the space beyond the terminations of the wire seemed guaranteed: 'that is a case of excitation of a kind that we know; we know the a,b,c of it, and the laws of it, and feel certain that if this operation be performed but fast enough there will be [longitudinal] waves'.[50]

47 FitzGerald, 'Molecular dynamics'.

48 BL, 41–2. On the difference of size (grainedness) and mass of the molecules of air and of normal solids from those of ether see BL, 10, 106–7, 119, 199–208.

49 Jack, 'Notes', autumn of 1852.

50 BL, 42–3.

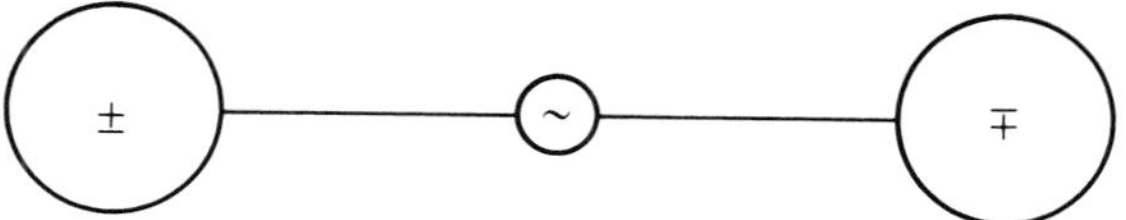

Figure 13.2. A telegraph wire, activated by an alternating electromotive force and terminated by two spheres, would, in Thomson's view, necessarily produce longitudinal waves in the surrounding and containing medium.

The 'a,b,c' of the telegraph thus stood implacably behind the claims in the Baltimore Lectures for an elastic-solid theory of ether:

If I knew what the magnetic theory of light is, I might be able to think of it in relation to the fundamental principles of the wave theory of light. But it seems to me that it is rather a backward step from an absolutely definite mechanical motion that is put before us by Fresnel and his followers to take up the so-called Electro-magnetic theory of light in the way it has been taken up by several writers of late. In passing, I may say that the one thing about it that seems intelligible to me, I scarcely think is admissable. What I mean is, that there should be an electric displacement perpendicular to the line of propagation and a magnetic disturbance perpendicular to both. It seems to me that when we have an electro-magnetic theory of light, we shall see electric displacement as in the direction of propagation . . . I merely say that in passing, as perhaps some apology is necessary for my insisting upon the plain matter of fact dynamics and the true elastic solid as giving what seems to me the only tenable foundation for the wave theory of light in the present state of our knowledge.[51]

Given his unshakable faith in the telegraph, Thomson apparently felt free to use his Baltimore Lectures as a forum for combating the leading young theorists of electromagnetism, who had strayed from the true path of latitudinarian practicality and into a false metaphysics. The forum was well chosen, for among his twenty-one hearers sat Lord Rayleigh (J.W. Strutt; 1842–1919), H.A. Rowland (1848–1901), A.A. Michelson (1852–1931), E.W. Morley (1838–1923), and other notables. He would attempt to show them, in a grand symphony of mechanical models, how a true dynamical theory might be constructed. But at the same time he would attempt to deliver a captivating sermon on proper empiricist methodology.

The methodology of 'look and see'

In characterizing Thomson's view of valid theoretical entities we have often used the term 'practical' to contrast with 'metaphysical', but more specifically we have used 'direct': direct observation, direct measurement, direct dynamical

[51] BL, 5–6. Among his references for the electromagnetic theory of light, Thomson listed FitzGerald, Gibbs, Glazebrook, Gordon, Lorenz, Maxwell, Rayleigh, Rowland, and Turmlirz (detailed references in Index at end of BL).

interpretation. These usages may have conveyed the idea of direct sensory perception. The Baltimore Lectures, however, offer a much more explicit and more integrated ground for that reading than his earlier works justify. Taken as a whole, the lectures argue that theoretical expressions are acceptable to the degree that they are accessible to 'seeing' and 'feeling'. The theorist, like the experimenter and the engineer, must 'look' and 'touch' in order to 'see' and 'feel', thereby to 'understand' and 'believe'. This methodology we label 'look and see'.

Thomson's methods have usually been understood more narrowly, in terms of the justification of theories by analogies and mechanical models. That view is symbolized by classic lines from the Baltimore Lectures: 'I never satisfy myself until I can make a mechanical model of a thing. If I can make a mechanical model I can understand it. As long as I cannot make a mechanical model all the way through I cannot understand; and that is why I cannot get the electro-magnetic theory'. Much less attention has been paid to such parallel proscriptions as, 'I have no satisfaction in formulas unless I *feel* their arithmetical magnitude'.[52] Getting to know theoretical structures by mechanical modelling plays much the same role as getting to know physical magnitudes by calculation, which in turn correlates with getting to know physical properties by experiment and by measurement. All provide a direct 'feeling' for how things vary under varying conditions. Thus Thomson proposed an 'arithmetical laboratory' for his audience of twenty-one 'coefficients' in Baltimore, just as he had developed an experimental laboratory for his students in Glasgow. He urged them, for example, to calculate wave velocities on different assumptions 'in order that you may *feel* for yourselves what these two or three symbols show us, but which we need to *look* at from a good many points of view before we can *make it our own* and *understand* it thoroughly'.[53] The work of the arithmetical laboratory figures as prominently in the lectures as do mechanical models, and it is their relation that we seek to elaborate here.

The idea of making-our-own through sensory perception, expressed in many different ways, implies that we ought in some sense to put ourselves in the place of the object under study, in order to feel as it feels. Objects get 'uneasy' under particular circumstances.[54] They are to be thought of almost as touching and feeling things. Through the lectures one gets to know two such lively objects: the elastic-solid ether and a ponderable molecule embedded in it, which Thomson pictures as a spherical cavity bounded by a massless shell, with vibrating internal structure and a massive core.

We shall try to *get into* the notion of this, that the molecule must be soft and that there must be an enormous mass in its interior. Its outer part *feels* and *touches* the luminiferous ether and the luminiferous ether *feels*, it may be, comparatively slight [light] to it. It is a very curious supposition to make, of a molecular cavity lined with a massless rigid spherical shell; but that something exists in the luminiferous ether and acts upon it in the

[52] BL, 270–1, 72. Our emphasis. [53] BL, 73, 113. Our emphasis. [54] BL, 151.

manner that is faultily illustrated by our mechanical model, I absolutely believe. I have no more doubt that something of the kind is true than I have of my own existence.[55]

To Descartes' 'I think therefore I am' Thomson might have answered, 'I feel therefore I am'. He based the reality of theoretical entities on our ability to sense their behaviour. A mechanical model of a possible molecule enabled one to manipulate it in all its variety of circumstances; to touch it, turn it, look at it, and thereby to know it intuitively as a potentially real thing. Knowing the relative magnitudes of things gave a similar sense of reality: e.g., 'calculate the energy ratios for each root. We shall then be able to put our formula into numbers; and I *feel* that I understand it much better when it is in numbers than when it is in a literal form'.[56]

To know *the thing* as object and magnitude thus contrasted with knowing symbols and words for the thing, or formulae and sentences about it. A 'word painting' or even a 'mathematical word painting', however pretty, could never substitute for direct perception. Repeatedly Thomson complained of the 'aphasia' of mathematics. Even the heroic Green had fallen victim to it in too loosely treating the coefficients of rigidity in the ether as like those in a gas: 'I spoke of the disease of aphasia. This is a manifestation of it. What does one know of the meaning of *A* and *B* who can only speak of the properties of matter by "*A*" and "*B*". If Green had thought of *the thing itself* and not of the letters he would have saved himself that reference to gases at all'. Again, in the middle of a long quotation from Rayleigh – who attended the lectures – Thomson came to a complicated passage on coupled vibrational motions and broke off. 'I cannot understand the meaning of that sentence, at all. There is a terrible difficulty with writers in abstruse subjects to make sentences that are intelligible. It is impossible to find out from *the words* what they mean; it is only from knowing *the thing* that you can do so'.[57]

To know the thing meant to have it in one's hands and eyes. Thomson demonstrated that kind of knowledge for his molecule in interaction with the ether by embedding a ball in a bowl of jelly, saying 'apply your *hand*' and 'produce vibrations in your jelly solid by *taking hold* of this ball and *shoving* it to and fro'. Again, to show coupled vibrations inside the molecule, resulting from an elastic internal structure, he constructed a 'wiggler', which consisted of a series of bars attached to a piano wire hanging vertically (figure 13.3). By learning to twist the lowest bar at appropriate frequencies, as though acting in the place of the driving ether, one could see and feel its characteristic modes of vibration. He

[55] BL, 145–6. Our emphasis.

[56] BL, 105. Our emphasis. The direct sensory basis of all our knowledge of external matter is the theme of Thomson's Presidential Address to the Birmingham and Midland Institute on 'The six gateways of knowledge' (3rd October, 1883), PL, **1**, 253–99.

[57] BL, 146, 85, 220, 76. Our emphasis. In its direct relation to sensory perception, Thomson regarded mathematics as 'the etherealisation of common sense', 'Six gateways', p. 273; cf. 'logic is etherealised grammar', p. 285.

(a)

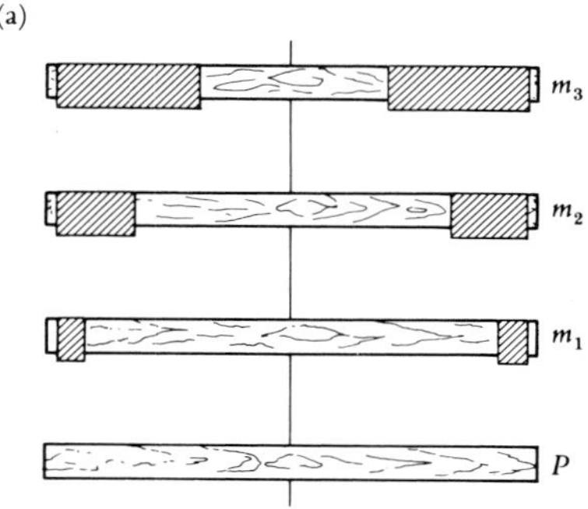

(b)

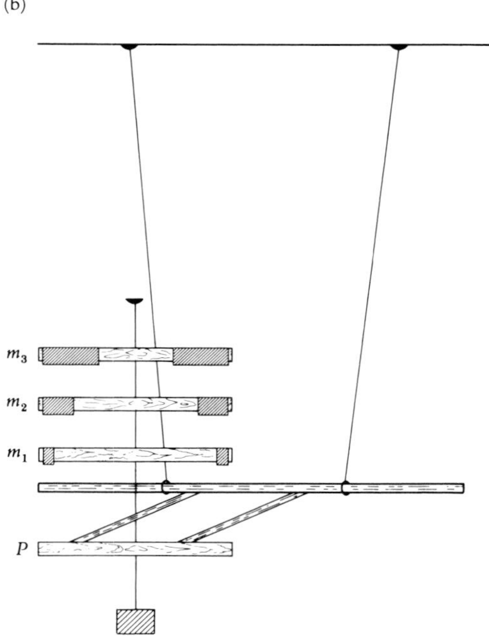

Figure 13.3. The 'wiggler' consists of (a) a set of weighted wooden bars attached to a piano wire and suspended from the ceiling. It can be driven by hand or (b) by a heavy metal rod, suspended bifilarly on rings, so that the driving period varies with the distance between the rings.

encouraged his audience 'to repeat these experiments, whether Professors or not. See how easily this model is made. Do the work at home *with your own hands*'.[58]

Similarly, Thomson wished to demonstrate the possibility of an elastic-solid ether, constructed merely from particles and springs, which possessed the maximum of twenty-one independent coefficients connecting stress and strain. Poisson's theory for such systems had supposedly proved that an invariable relation existed between compressibility and rigidity (chs. 6 and 10), thus precluding the *incompressible* elastic solid that Thomson needed. He had got the idea of running a cord twice around the edges of a simple box, yielding twenty-four connections, which reduced to twenty-one through three relations between them. But actually to run the cord posed problems of another order. 'I must confess that it is the most difficult thing in it, after I got the idea, to run a cord twice around the 12 edges of a parallelopipedon. Here you see the problem solved by these cords running around the edges of this parallelopipedon through a ring in each of the 8 corners . . . *just follow the cord and we will find how to do it*. In fact I am finding out how to do it again in a certain way, myself . . . We have got a cord thrice through each of these 8 points and the thing is done'.[59] By assuming the cord inextensible, the incompressible elastic solid resulted. Thus, hands and eyes following the cord did what Poisson's mathematics had said could not be done.

The general point seems to be that tactile and visual experience are of a different kind from, and more efficient and more trustworthy than, mental constructions. 'Brains' are so limited in their capacity that one ought to seek every possible aid from the senses. For example, Thomson wrote out on the board the full cartesian components of rotational stress and strain, even though from any one the others followed by symmetry, because 'The expenditure of chalk saves brain'. A reader of mathematics 'should have pencil and paper beside him to work *the thing* out', rather than thinking it out. Chalk served the same function, in this respect, as did the bowl of jelly: 'try it for yourselves. It allows you to *see* the vibrations we are speaking of. I wish I had it to show you just now, so that you might *see the thing* in force. It saves brain very much'.[60]

Thomson did not mean simply that sensory aids served a heuristic purpose. Much more, they provided a rigorous analysis when the brain failed. Green had said, 'I have no faith in speculations of this kind [hypothetical explanation] unless they can be reduced to regular analysis'. To supply a regular analysis of the coupled vibrations of his molecule, Thomson substituted an experimentally realizable model (figure 13.4) whose motions he described. He commented: 'From *looking* at *the thing*, and learning to understand it by making the experiment if you do not understand it by *brains* alone, you will see that everything that

58 BL, 80–2, 253. Our emphasis.
59 BL, 130. Our emphasis.
60 BL, 25, 32, 80. Our emphasis.

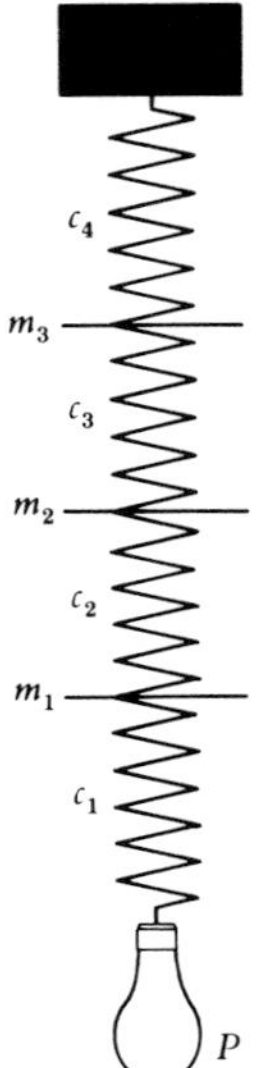

Figure 13.4. A simple demonstration of coupled vibrations can be obtained from a series of springs c connecting masses m, the whole being suspended from the ceiling and pulled with a periodic motion from below by the bell-pull P. The experimenter can both see and feel the various periods at which accumulation and loss of energy in the different masses occurs.

I am saying is obvious. It is not satisfactory to speak of these things in general terms unless we can submit them to a rigorous analysis'.[61]

Partly because a dynamical model provided a rigorous analysis, it operated as an agent of discovery, going beyond known phenomena to produce new ones. And if the model itself did not immediately display the phenomena, the equations describing the model, and interpreted by it, might reveal its possibilities. So it was with the properties of light waves interacting with molecules. 'All these properties, remarkable as they are, seem to come out as a matter of course from the dynamical consideration. So much so that any one not knowing these phenomena would have discovered them on working out these things dynamically. He would have discovered anomalous dispersion, fluorescence, phosphorescence . . . the dynamical treatment that discovers what is afterwards verified by experiment is a very competent piece of dynamics'. Thomson claimed to have 'discovered' in this way both anomalous dispersion and the colours produced by metallic reflection. 'It was known perfectly well, but the molecule first discovered it to me . . . There is no difficulty about explaining

[61] BL, 35, 37. Our emphasis.

these things; we can predict them from the consideration of the molecule without experimental knowledge'.[62]

If he here somewhat exaggerated the value of mechanical models,[63] Thomson nevertheless wanted to claim for them something of the status of experiment. Because they gave a similar, tangible, knowledge of phenomena, they offered convincing evidence of the reality of the objects they represented. He wished above all to convince his audience of the reality of the elastic-solid ether with heavy molecules vibrating within it. 'I do not want to part from you without letting you know all I can in the way of helping you think of the luminiferous ether as a reality, and that we are speaking of real bodies and that this is not a mystification of the mind'.[64]

'Mystification' versus reality offers the same contrast as that between 'brains' and hands, symbols and things, the ideal versus the real, but most especially the electromagnetic theory of light versus the elastic-solid theory. In fact the entire series of Baltimore Lectures should be read in its negative dimension as an attempt to denounce Maxwell's displacement current as mystification. Thomson introduced his assertion that the displacement current is 'scarcely admissible' by contrasting it with the real ether: 'we must not listen to any suggestion that we must look upon the luminiferous ether as an *ideal* way of putting the thing. A *real* matter between us and the remotest stars I believe there is and that light consists of real motions of that matter . . . motions in the way of transverse vibrations'.[65] Thereafter Maxwell's 'backward step' appears only rarely, its place being taken implicitly by the errors of Green, Helmholtz, and Rankine; or, most egregiously, Poisson.

Green introduced coefficients of 'extraneous force' for which Thomson could find no direct evidence. 'I say, therefore, it is a mistake to introduce the coefficients A,B,C if they correspond to nothing in nature'. Helmholtz, in his theory of anomalous dispersion, had correctly recognized that the phenomenon could be produced by a vibrator, which would absorb large amounts of energy at its resonant frequency. But, rather than accounting mechanically for the disappearance of energy, he simply assumed some unspecified 'viscous consumption of energy'. Such indefinite ideas might be 'beautiful' but they hardly sufficed for true science. 'In speaking of . . . Helmholtz's *beautiful* paper, which is quite a mathematical gem, I must still say that I think Helmholtz's modification is rather a retrograde step . . . in introducing it he is throwing up the sponge as it were, so far as the fight with the dynamical problem is concerned'.[66]

Of all his misguided friends, however, Rankine came in for the most ridicule. Thomson owed much to Rankine both for molecular vortices and now for the idea of different 'effective inertias' in different directions, with which Rankine

[62] BL, 120, 282–3.

[63] cf., BL, 171, where he makes algebra the better agent of discovery and the model only 'a corrective to brain sluggishness'.

[64] BL, 120.

[65] BL, 5. Our emphasis.

[66] BL, 191, 98. Our emphasis.

had wanted to explain double refraction in crystals. Applying that concept – aeolotropy of inertia – to his heavy molecules Thomson attempted to improve on Rankine's results. Indeed, this programme for double refraction is one of the most prominent features of the Baltimore Lectures. But of Rankine's paper on molecular vortices Thomson said, 'The title is of more importance than anything else in the work', and of his matter theory, 'there is no explanation of his kind of matter'. These remarks immediately precede the classic lines, 'I never satisfy myself until I can make a mechanical model of a thing . . . and that is why I cannot get the electro-magnetic theory'. Maxwell's '*beautiful* ideas of electro displacements', like Rankine's 'splendid failure', had no foundation in sensory reality.[67]

'Beautiful' is a term Thomson used regularly – though not exclusively – for mathematical simplicity and elegance. It connoted the poetic, the ideal, like Fourier's 'great mathematical poem'. Such products of the mind might or might not be realized in nature. Unfortunately for Rankine's 'beautiful idea', Stokes made a measurement on double refraction which destroyed even the potential reality of etherial aeolotropy of inertia. 'Stokes took away the poetry of it'.[68] He did so with rigorous calculations and precision measurements, hard numbers that dissolved the wishful fantasy.

At least six different sensory agents can be identified in Thomson's programme for 'feeling' and 'seeing' the truth about nature. Most prominent, of course, are (1) models, which allow us to know a thing by literally making it ourselves. Models supply the first-order test of a theoretical construction: can we produce it using only the materials and the interactions that we know as everyday working realities. Ultimately more important, however, are (2) experimental measurements, and (3) numerical calculations, which give a knowledge of the relative and varying magnitudes of things. Numbers not only justify models but guide the process of building them: 'what will be the disposition of the energy? How will it creep inwards among the masses? I think that our arithmetical work will help us to *see* our way to the answer to some of these questions; and through them we shall be able to form perfectly definite dynamical notions of fluorescence and phosphorescence and anomalous dispersion'. The 'great difficulty' with the wave theory of light was not to reproduce the mere phenomena, as the 'popular imagination' would have it, but 'to bring out the proper quantities in these effects'.[69]

Three other agents complement those above. We have mentioned (4) writing out equations in their full cartesian form, to which Thomson appended symmetrical expression (ch. 6): 'The symmetrical system is a great brain saving system'. Symbols and formulae were to be as transparently connected with things as possible, avoiding abstraction. The same applied to (5) words. Here Rankine had been of great service, clothing the formidable mathematics of deformable bodies

[67] BL, 270, 277. Our emphasis. [68] BL, 18. [69] BL, 123, 119. Our emphasis.

with the perspicuous language of 'stress' and 'strain', but also with a classicist's taxonomy of coefficients. 'Any one who will learn the meaning of all these words will obtain a large mass of knowledge with respect to an elastic solid . . . Hear the grand words "Thlipsinomic, Tasinomic, Platythliptic, Euthytatic . . . &c"'. If physicists today cannot quite see the force of those names, most would nevertheless agree with Thomson's emphasis on (6) graphs and diagrams. Over and over again he clubbed French mathematicians with mistakes traceable to their abstract, algebraic, method. 'Their mistake was due to the vicious habit in those days of not using examples and diagrams. In the *Mécanique céleste* you find no diagrams, nor in Lagrange, nor in Poisson's splendid memoir on waves'.[70]

These six 'brain saving' agents in Thomson's methodology of 'look and see' present nothing fundamentally new. But they define in its most articulated sense the distinction of metaphysical from practical that informed his life and his science. They provided for him the only ground for what he regularly called his 'faith' and 'belief'. 'But that there are such waves [longitudinal] I believe; and I believe that the velocity of propagation of electro-static force is the unknown condensational velocity that we are speaking of. I say "believe" here in a somewhat modified manner. I do not mean that I believe this as a matter of religious faith, but rather as a matter of strong scientific probability'.[71] The denial of religious reference gives itself away, so to speak, for Thomson held a direct sense of the Creator to be the only secure ground for religious belief (ch. 18). One could not acquire knowledge of the Creator by exercising brains. No more could one acquire knowledge of His creation through mental construction. The difference between religious and scientific belief lay in the modes of sensory awareness and the degree of certainty attainable in each.

Confrontation with Maxwell's ghost

Maxwell's system of equations was metaphysical to Thomson, a product of brains alone. It attached to neither of his practical realities: electricity as like a compressible fluid and ether as like an elastic solid. In the Baltimore Lectures he attempted systematically to incorporate massive, vibrating molecules within the ether and to make them just as real as it was, ignoring for the moment the role of electricity. As discussed in Chapter 12, these molecules of normal matter, which produced observable physical effects as a result of their individual internal behaviour, now constituted objects very different in character from ether, which produced its effects macroscopically. The lectures thus alternated between two quite different modes of analysis, the *molecular dynamics* of ponderable matter and the *molar dynamics* of ether, regarded for this purpose as a homogeneous elastic solid characterized by its macroscopic properties.

[70] BL, 176, 161, 129. [71] BL, 143.

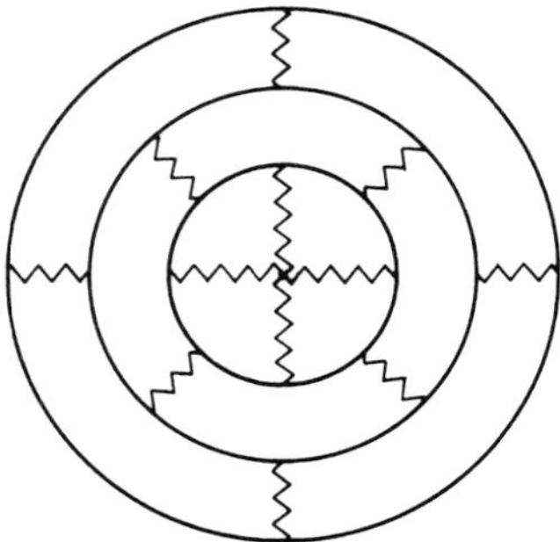

Figure 13.5. The outer (massless) shell in this series of concentric shells corresponds to the driving device *P* in the two preceding figures and to the boundary between the ether and a ponderable molecule suspended in it. The molecule is represented by the series of internal shells, with a massive core, connected to each other and to the ether by springs. At resonant frequencies the system will absorb energy from waves in the ether.

Thomson identified four apparent problem areas: etherial matter, dispersion, reflection and refraction, and double refraction. Of these, the first two presented no serious difficulties. Ether had to behave like a perfect fluid with respect to planets moving through it, but like an extremely rigid solid with respect to the motions of light. Such radical differences could be accounted for by time alone. The time scale on which the ether responded to planetary motion was almost infinitely long in comparison with that of light vibrations. Such differences existed in ordinary materials like 'Scotch shoemakers' wax', ice, jelly, and glycerine, as Stokes had been arguing since the 1840s.[72] Making such a solid both elastic and incompressible also seemed simple enough (above).

Phenomena related to dispersion offered a similar guarantee of reality to Thomson's molecule. One had only to construct a series of concentric rigid shells (figure 13.5), connected to one another by springs, with an extremely massive central core, and with some elastic connection to the ether, here represented by the outermost, ideally massless, shell.[73] Such a system would reproduce in detail discrete spectral radiation, anomalous dispersion, refractive dispersion, absorption and heating, phosphorescence, and fluorescence. The crude model, therefore, which one could actually 'have and look at and experiment upon', contained essential features of the true molecule. 'I cannot

[72] Among the array of demonstrations Thomson developed was one in which a slab of Scotch shoemakers' wax floated in a jar of water, with corks under it and bullets resting on its surface (BL, 7). Within the course of a year the bullets sank to the bottom and the corks rose to the surface, although the wax remained brittle and if shaped into a tuning fork would vibrate. He also set a glob of black pitch or tar moving like a glacier down a stepped slope. It is still flowing today in the museum of the Department of Natural Philosophy at Glasgow University.

[73] BL, 118.

but feel that there is a great reality in the theory of detached molecules', Thomson continually insisted, 'I cannot but believe that it is really true'.[74]

The details of reflection and double refraction, however, remained unexplained. Works of Stokes and Rayleigh on diffraction and polarization enabled Thomson to claim that 'we have absolutely proved' that the direction of vibration of light polarized by reflection is perpendicular to the plane of incidence and reflection.[75] That conclusion, together with an apparently convincing proof by Rayleigh that refraction and reflection at an interface required the media to be of equal rigidity and unequal density, fixed all of the disposable parameters in the theory of homogeneous elastic solids. But the numbers would not agree with experimental measurements on polarization by reflection. Thomson tried it with his molecules loading the ether but found no improvement. Reflection from metals posed similar obstacles.

The most intractable of all difficulties, however, seemed to be double refraction in crystals, which depended on different velocities of propagation for different directions of vibration. To produce that effect, Rankine had suggested different inertias in different directions. Thomson's molecule realized Rankine's suggestion, since if the massive core were vibrating in any given direction, the molecule as a whole would exhibit aeolotropy of effective inertia. But alas, in this respect, the aeolotropic molecule remained poor poetry.

Thomson sought a way out through his favourite device, rotation. Perhaps effective rigidities and inertias produced by rotationally stabilized molecules would ultimately crack the great mystery. First, however, he had to produce the correct functional form for the fundamental rotational phenomenon, Faraday's magneto-optic effect. He began by mounting a flywheel (gyrostat) inside the sheath of his molecule, disregarding the series of interior shells (figure 13.6). This is essentially the same model he had proposed for magnetic rotation in 1875 and that, with respect to refractive dispersion, had saved his 1883 lecture on the size of atoms (ch. 12).[76] Initially the rotation seemed incorrect, varying approximately as the inverse of the wavelength rather than as the inverse square. But Thomson's faith did not falter: 'I therefore lay it aside for the present, but with perfect faith that the principle of explanation of the thing is there'. Vindication came even before he left the United States. He had simply made a mistake; the thing was there after all.[77]

Even better, as Thomson found after returning to Glasgow, was a gyrostatic molecule with two gyrostats in line, connected by a ball joint (figure 13.7).[78] Its efficiency in rotating the plane of vibrations of a light wave depended essentially on translational rather than rotational motion of the sheath, which meant that the molecule could be reduced in size as much as one might wish without altering its rotative efficiency for polarized light. This property seemed particularly important for a theory that would ultimately reduce to motions in a

[74] BL, 198. [75] BL, 227. [76] BL, 242. [77] BL, 244–5. [78] BL, 321.

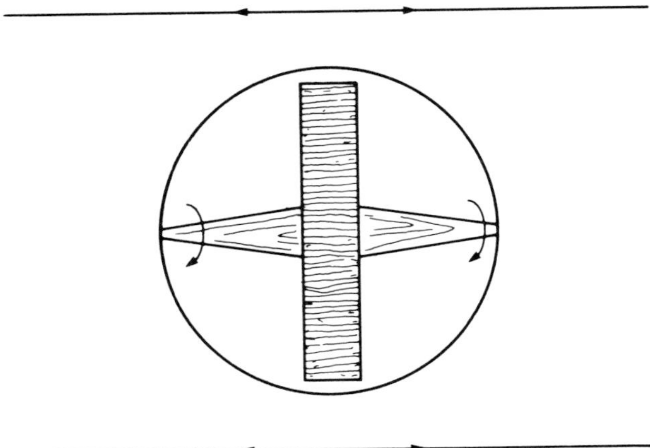

Figure 13.6. A gyrostat, mounted in the outer shell of figure 13.5, would mimic the rotational effect of ponderable molecules on light waves in Faraday's magneto-optic effect. The horizontal arrows indicate the direction of vibrational displacement of polarized waves travelling perpendicular to the axis of the gyrostat. Since the displacement is assumed to change significantly across the diameter of the molecule, there is a torque on its axis. It responds by turning out of the plane of the figure.

continuous substratum as envisaged by Thomson. For the moment, however, he could go no further.

The gyrostatic molecule suggests immediately why Thomson often spoke of a 'magnetic', rather than an 'electro-magnetic', theory of light. Rotations represented the magnetic character of the molecules. They behaved similarly to rotational molecules of the ether, except that the ether, being much finer grained and having molecules with a different rotational structure would produce neither refractive dispersion nor magneto-optic rotation.[79] The magnetic character alone, therefore, of the molecules of ether and normal matter, might well suffice for the transverse waves of light, independent of the relation of magnetism to electricity. Certainly there seemed no reason to insist on propagation by electromagnetic induction, when nothing like electric currents or displacements could be shown to exist in the ether. In 1884, therefore, no definitive evidence existed in favour of Maxwell's theory over Thomson's programme, except that Maxwell's equations constituted a consistent, and beautiful, mathematical structure.

In 1888 that situation changed dramatically. Word reached Britain of Hertz's experiments on the production, reception, reflection, polarization, and interference of electromagnetic waves: virtually everything required of light. The Maxwellian crowd immediately saw the triumph of their own commitments.

[79] BL, 10, 241. See remarks at end of Chapter 12, above.

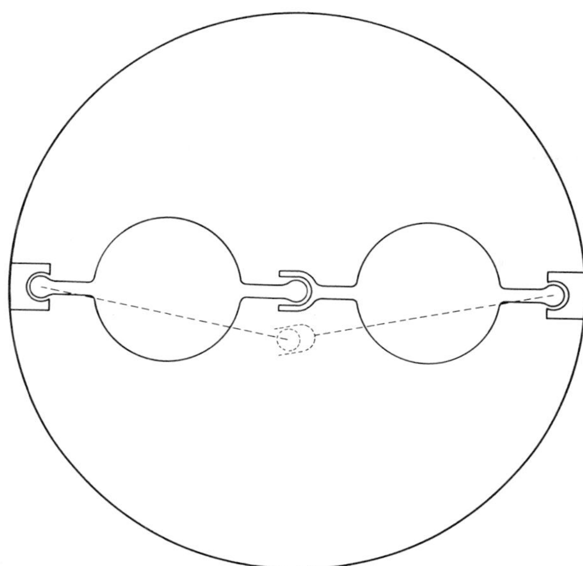

Figure 13.7. A molecule containing double gyrostats connected by a universal joint provides a better representation of the magneto-optic effect than the single-gyrostat model of figure 13.6.

Early in September, at the Bath meeting of the British Association, FitzGerald, their spokesman and President of Section A, would proclaim the grand implications of Hertz's researches: 'man has won the battle lost by the giants of old, has snatched the thunderbolt from Jove himself and enslaved the all-pervading ether'.[80] Already on the defensive, Sir William turned to the source of the heresy, Maxwell's *Treatise on electricity and magnetism* of 1873, to locate its deeper flaws. In 1885 he had told FitzGerald: 'I have never yet felt any satisfaction in Maxwell's paras. 783, 784, 790, 791, 792, 645, 646, 794, 797, 798, 824...829 [electromagnetic equations for light, energy and stress in the ether, and magneto-optic rotation]. I have never yet met any one who understood a definite dynamical foundation for para. 783 [general equations of electromagnetic disturbances]'.[81] But now deeper study was required. He intended to go to Bath armed for battle.

Thomson's copy of Maxwell's *Treatise* is signed and dated 19th October, 1878, but only in 1888 did he begin to make extensive annotations, dating every

[80] G.F. FitzGerald, 'Presidential Address to Section A', *BAAS Report*, **58** (1888), 557–62, on p. 562. For a good survey of some of the material discussed below see Ole Knudsen, 'Mathematics and physical reality in William Thomson's electromagnetic theory', in P.M. Harman (ed.), *Wranglers and physicists: studies on Cambridge physics in the nineteenth century* (Manchester, 1985), pp. 149–79, esp. pp. 171–9. [81] William Thomson to G.F. FitzGerald, 17th April, 1885, SPT, **2**, 1038.

Sir William Thomson aged sixty-four. The senior master of British mathematical physics is seen in this photograph by his niece, A.G. King, some four years after delivering his Baltimore Lectures in the United States. [From SPT.]

entry. His mood on 13th August is evident from an entry beside the equations for electromagnetic induction as written in quaternion form. These equations, he noted, merely 'translated in gibberish' the full cartesian component equations. Other such sympathetic remarks read: 'unintelligible, no definition' and 'notations mutually incongruous'.[82] By the time he arrived in Bath on 4th September he had been through Maxwell's analysis of displacement currents and the electromagnetic theory of light at least five times, finally on the train itself.

The angry knight aimed a sharp lance as he entered Bath, in the form of a 'Simple hypothesis for electro-magnetic induction of incomplete circuits, with consequent equations of electric motion in fixed homogeneous solid matter'. With the telegraph at his back, supporting electricity as a compressible, viscous substance, he charged into Maxwell's doctrine of closed circuits, which made electricity behave like an incompressible fluid. Tactically, he would adopt Maxwell's incompressibility assumption, derive a contradiction, and emerge with compressibility intact, having vanquished the evil displacement current, that 'curious and ingenious, but as seems to me not wholly tenable hypothesis'.[83]

Introducing displacement current in the *Treatise*, Maxwell had claimed that when electromotive intensity (electric field intensity $\vec{E}$) acted on any material body, *two effects* occurred, 'called by Faraday Induction and Conduction, the first being most conspicuous in dielectrics, and the second in conductors'. The first produced displacement, $\vec{D} = (K/4\pi)\vec{E}$, where K is dielectric capacity, and the second produced conduction current, $\vec{j} = C\vec{E}$, where C is conductivity. Then came 'one of the chief peculiarities of this treatise . . . the doctrine . . . that the true electric current $\vec{i}$, that on which the electromagnetic phenomena depend, is not the same thing as $\vec{j}$, the current of conduction, but that the time variation of $\vec{D}$, the electric displacement, must be taken into account in estimating the total movement of electricity, so that we must write',

$$\vec{i} = \vec{j} + \partial\vec{D}/\partial t, \tag{1}$$

or,

$$\vec{i} = [C + (K/4\pi)\partial/\partial t]\vec{E}. \tag{2}$$

Accumulations of 'free electricity' were given by the divergence of displacement,

$$\text{divergence } \vec{D} = \rho, \tag{3}$$

but Maxwell also claimed that the true current $\vec{i}$ 'is subject to the condition of

[82] William Thomson, annotations to James Clerk Maxwell, *Treatise on electricity and magnetism* (2 vols., Oxford, 1873), **2**, p. 222, para. 599, 13th August, 1888; **2**, p. 143, para. 498, 15th January, 1888; **2**, p. 209, para. 581, 27th August, 1888.

[83] William Thomson, 'Simple hypothesis for electro-magnetic induction of incomplete circuits, with consequent equations of electric motion in fixed homogeneous solid matter', *BAAS Report*, **58** (1888), 567–70; *Nature*, **38** (1888), 569–71; MPP, **4**, 539–44, on p. 543.

motion of an *incompressible fluid*, and that it must necessarily flow in closed circuits', so that,

$$\text{divergence } \vec{i} = 0. \tag{4}$$

This agreed with his equation for electromagnetism,

$$4\pi\vec{i} = \text{curl } \vec{H}, \tag{5}$$

where $\vec{H}$ is magnetic intensity.[84]

Introducing the magnetic induction, $\vec{B} = \mu\vec{H}$, where μ is magnetic permeability, and expressing $\vec{B}$ in terms of the vector potential $\vec{A}$,

$$\vec{B} = \text{curl } \vec{A}, \tag{6}$$

Maxwell obtained an electromagnetic equation relating the components of $\vec{i}$ to those of $\vec{A}$,

$$4\pi\mu i_i = \nabla^2 A_i \tag{7}$$

where $\nabla^2 A_i$ stands for, divergence (gradient A_i). Another equation between $\vec{i}$ and $\vec{A}$ derives from the equation of electromotive force $\vec{E}$, written as a sum of electromagnetically induced force (rate of change of vector potential $\vec{A}$, or 'electromagnetic momentum') and electrostatic force (gradient of potential ψ),

$$\vec{E} = -\partial\vec{A}/\partial t - \text{gradient } \psi. \tag{8}$$

Substituting equation (8) in equation (2) and comparing equation (2) with equation (7), Maxwell derived his 'general equations of electromagnetic disturbances',

$$\mu(4\pi C + K\partial/\partial t)\ (\partial\vec{A}/\partial t + \text{gradient } \psi) + \nabla^2 A = 0. \tag{9}$$

Within a homogeneous *non-conductor*, he argued, $C = 0$ and no volume density of electricity $\rho = (1/4\pi)\nabla^2\psi$ can develop, so that this equation becomes a wave equation in the vector potential,

$$\mu K\partial^2 A/\partial t^2 + \nabla^2 A = 0. \tag{10}$$

The waves propagate by electromagnetic induction at the velocity $c = (\mu K)^{-1/2}$, given by the famous ratio emu/esu.[85] Maxwell immediately went on to argue (paras. 785–98) that the waves are purely transverse waves of electric and magnetic quantity, with no longitudinal waves, in conformity with all known phenomena of light.

The few pages of the *Treatise* that lay out the above mathematical structure

[84] Maxwell, *Treatise*, **2**, on pp. 232, 233, 231, paras. 608, 610, 607. Emphasis added and notation altered.

[85] *Ibid.*, pp. 385–7, paras. 783–5. With Thomson, we omit terms in $\vec{J}$ = divergence $\vec{A}$, which means choosing the 'gauge' in which $\vec{J} = 0$: 'nothing is gained by $\vec{J} \neq 0$, or altered by $\vec{J} = 0$' (p. 385, para. 783, probably 31st January, 1890).

Thomson literally covered with annotations, in various colours of pencil and ink. Next to equation (2), he wrote in the right hand margin:

These give
divergence $\vec{\mathrm{i}} = [C + (K/4\pi)\partial/\partial t]$ divergence $\vec{\mathrm{E}}$
∴ *if* divergence $\vec{\mathrm{i}} = 0$, divergence $\vec{\mathrm{E}} = 0$, &
∴ this [divergence $\vec{\mathrm{D}} = \rho] = 0$!

In the left margin he repeated his emphasis:

∴ divergence $\vec{\mathrm{i}}$, *if it were* $= 0$, *would* give
$0 = [C + (K/4\pi)\partial/\partial t]\ (\partial\vec{\mathrm{J}}/\partial t - \text{“}\nabla\text{”}^2\psi)$ being [Maxwell's] para. 783 (8)
[with divergence $\vec{\mathrm{A}} = \vec{\mathrm{J}}$].[86]

From the latter equation, Maxwell had argued that in non-conductors, with $C = 0$, one could disregard both $\vec{\mathrm{J}}$ and ψ, since no free electricity, $\rho = (1/4\pi)\nabla^2\psi$, could accumulate.

Both of Thomson's notations, therefore, concern free electricity. They suggest that he understood Maxwell's assumption of closed currents to imply that no accumulations would exist within the interior of either conductors or non-conductors. If so, he may originally simply have blundered, since his derivations are valid only for homogeneous media. For inhomogeneous media, C and K cannot be taken outside of the divergence operator, so that, on Maxwell's theory, even if divergence $\vec{\mathrm{E}} = 0$ electricity will appear wherever there are variations in C or K.[87] In 1890 Thomson would add the qualification: 'within each homog^{s} part. Through all space we have,

$$\text{divergence } [C\vec{E} + (K/4\pi)\partial\vec{E}/\partial t] = 0\text{'}.$$

At the Bath joust Thomson's attack consisted in an analysis of heterogeneousness in conductors. He followed a simplified version of Maxwell's steps above, with $\mu = 1$ and $K = 1$ (no magnetization or induction), but he also assumed that the closed circuit assumption applied to conduction currents alone, divergence $\vec{\mathrm{j}} = 0$, or that the electrical 'fluid' alone, independent of displacement, was incompressible. He then derived a set of four equations in three unknowns which, although consistent for homogeneous media, were inconsistent for inhomogeneous media, implying that the incompressibility assumption was wrong and that accumulations of electricity would appear in inhomogeneous media. This conclusion he took to be news, and a sharp rebuke to the mathematical theorists:

An interesting and important practical conclusion is, that when currents are induced in any way, in a solid composed of parts having different electric conductivities (pieces of copper and lead, for example, fixed together in metallic contact), there must in general be changing electrification over every interface between these parts. This conclusion was

[86] William Thomson, annotations to Maxwell's *Treatise*, **2**, p. 233, paras. 610–11. Thomson's “∇”^{2} is the negative of Maxwell's. Thomson would not have approved our use of vector notation.
[87] For a full analysis see Buchwald, *Maxwell to microphysics*, pp. 23–34.

not at first obvious to me; but it ought to be so by anyone approaching the subject with mind undisturbed by mathematical formulas.[88]

Comparing Maxwell's equation (9 above) of para. 783, for the case of homogeneous conductors, Sir William drove home his lance: it 'cannot be right, I think, according to any conceivable hypothesis regarding electric conductivity, whether of metals, or stones, or gums . . . or gutta-percha . . . being, as seems to me, vitiated for complete circuits by the curious and ingenious, but as seems to me not wholly tenable hypothesis which he introduces . . . for incomplete circuits' (i.e., displacement current completing the circuit). He proposed as an alternative hypothesis the standard continuity equation for compressible fluids in incomplete circuits, but modified by a factor $4\pi c^2$ to correct for the difference of electromagnetic and electrostatic units,

$$(4\pi c^2) \text{ divergence } \vec{j} = \partial/\partial t(\nabla^2\psi) = -4\pi\partial\rho/\partial t.$$

In effect, Thomson attempted simply to excise the displacement current term $c^2\partial^2 A/\partial t^2$ from Maxwell's 'general equations' (9). Electromagnetic induction would thereby produce only diffusion of currents in conductors, rather than waves in non-conductors, demolishing the electromagnetic theory of light. For homogeneous conductors, he obtained essentially his telegraph equation for the variation of density of a compressible, viscous, electrical fluid.

$$(1/4\pi C)\partial/\partial t(\nabla^2\psi) = \partial^2\rho/\partial t^2 - c^2\nabla^2\rho. \tag{11}$$

Aside from the viscosity term on the left, this is a simple equation for compressional waves travelling at the velocity c.[89]

Alas, although c makes its appearance, the scheme provides no physical reason whatever why electricity should be related to the velocity of light. Of course its theoretical inadequacy in this respect carried little weight with Sir William in comparison with its practical adequacy. 'I find simple and natural solutions, with nothing vague or difficult to understand, or to believe when understood, by their application to practical problems, or to conceivable ideal problems, such as the transmission of ordinary or telephonic signals along submarine telegraph conductors and land lines, electric oscillations in a finite insulated conductor of any form, transference of electricity through an infinite solid, &c. &c'.[90] By contrast, Maxwell had remarked of his scheme: 'we have deduced everything from purely dynamical considerations, without any reference to quantitative experiments in electricity or magnetism. The only use we have made of experimental knowledge is to recognise, in the abstract quantities deduced from the theory, the concrete quantities discovered by experiment, and to denote them by names which indicate their physical relations rather than their math-

[88] William Thomson, 'Simple hypothesis', MPP, **4**, 542–3.

[89] William Thomson, 'On the transference of electricity within a homogeneous solid conductor', *BAAS Report*, **58** (1888), 570–1; *Nature*, **38** (1888), 571; MPP, **4**, 545–6. We use symbols consistent with those above.

[90] William Thomson, 'Simple hypothesis', MPP, **4**, 544.

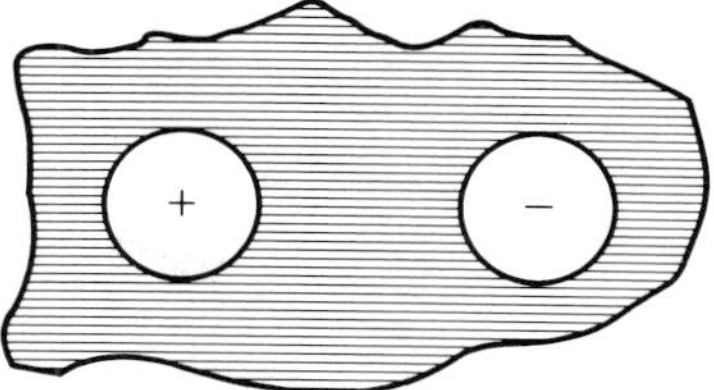

Figure 13.8. Two charged conducting spheres in a very slightly conductive medium, Thomson believed, would discharge along the lines of electrostatic force, thus as a direct longitudinal effect independent of any electromagnetic induction of displacement currents.

ematical generation'.[91] Abstract mathematics and no quantitative experiments! Thomson admitted that he could not prove his own hypothesis experimentally, but at least he was not one of those 'trying to evolve out of his inner consciousness a theory of the mutual force and induction between incomplete circuits'.[92]

Thomson's paper, according to the report of the meeting written by Oliver Lodge (1851–1940), excited 'a great deal of discussion, both in and out of the section' on the question of how electrostatic potential ψ would be propagated on Maxwell's theory. As Lodge understood it, Maxwell's transverse electromagnetic waves were not in question: 'about them there is no difficulty. Difficulties begin with the propagation of electrostatic displacements . . . Suddenly confer upon a conductor an electric charge: in all directions there is experienced a rise of potential equal to Q/r. How did that potential reach any given place?'[93] This issue is highly confusing in Maxwell's *Treatise*. It shows up the difficulty he had in trying to assimilate his electromagnetic equations to an elastic-solid ether. Lodge thought that Maxwell had regarded the ether as an incompressible solid, and the potential as transmitted instantaneously. This view sits badly, however, with Maxwell's emphasis on propagation of energy.[94] By 1888, not only FitzGerald, but also Oliver Heaviside (1850–1925), who did not attend the meeting, and Rowland, who did, had all come to regard the elastic solid as a poor likeness for ether. They consequently tried to give an account of electrostatic potential which showed that it propagated at the speed of light by electromagnetic action alone, and not by 'end thrust', or compression.

The discussion almost certainly concerned a problem that Thomson had discussed in a letter to Rayleigh on 22nd August:

Given two conducting globes oppositely charged (equal quantity) in blue sky [figure 13.8]. Let the blue suddenly become conductive; say as conductive as slate or as marble.

[91] Maxwell, *Treatise*, **2**, p. 229, para. 606.

[92] William Thomson, 'Simple hypothesis', MPP, **4**, 544.

[93] Oliver J. Lodge, 'Sketch of the electrical papers in Section A at the recent Bath meeting of the British Association', *Electrician*, **21** (1888), 622–5, on p. 624.

[94] Maxwell, *Treatise*, **2**, 435–8, paras. 861–6.

The discharge will be along the lines of the previous electrostatic force, and $\exp[-t4\pi Cc^2]$ will be the time law of subsidence, it being slow enough for no quasi-inertia [electromagnetic induction]. Where are the 'displacement' currents and where the possibility of expressing the result (with or without quasi-inertia) by any even pseudo circuits, or imagined analogues of incomplete circuits? Alas – alas, the whole thing breaks down the first time it is really put on trial.[95]

This arrangement of spheres is the same one Thomson had used in the Baltimore Lectures to argue for longitudinal electrical waves, only without the (telegraph) wire connecting them. The time decay derives from the viscous term in his equation (11) above.

FitzGerald and Rowland apparently expended great effort to develop the view, and to convince Thomson, that the growth or decay of electric potential ψ could always be understood on Maxwell's theory as an effect of changing magnetic potential $\vec{A}$. 'An electrostatic field is not developed *sui generis*', Lodge reported', but is always the consequence of a previously existing electromagnetic one, which, on subsiding, leaves it as its permanent record'. Thus, in terms of equation (8) for electromotive force, a potential ψ at any point would have to derive from a previously changing $\vec{A}$, which had produced a field $\vec{E}$ and a resultant motion of electricity, whether by conduction or displacement. The new distribution of electricity would control the potential ψ. 'If any one asks how soon will the pull of a suddenly electrified body be felt at a distance? one may answer, "As soon as the charging spark is seen." But if it be asked at what rate electrostatic potential travels, the answer is that it does not travel, but is generated *in situ* by the subsidence of a magnetic potential which travels with the velocity of light'. Lodge quite rightly suspected that Sir William might not agree with this rendering and might try to 'upset the whole thing once more'.[96]

Ether, electricity, and ponderable matter

Already at Bath, in fact, Sir William had laid plans to extend the viscous fluid model to relate electricity to his magnetic theory of light, without displacement currents. He quoted an analogy from Oliver Heaviside: 'Water in a round pipe is started from rest and set into a state of steady motion by the sudden and continued application of a steady longitudinal dragging or shearing force applied to *its boundary*'. Thomson planned to apply this analogy to a 'philosophical consideration of electricity, ether, and ponderable matter', by extending it 'to include the mutual induction between conductors separated by air or other insulators'.[97] As usual the telegraph anchored his analysis. Indeed, he delivered

95 William Thomson to Lord Rayleigh, 22nd August, 1888, in SPT, **2**, 1039.

96 Lodge, 'Sketch', pp. 624–5.

97 William Thomson, 'Five applications of Fourier's law of diffusion, illustrated by a diagram of curves with absolute numerical values', *BAAS Report*, **58** (1888), 571–4, on pp. 572–3; modified version in MPP, **3**, 428–35. Heaviside, a former telegraph engineer, had shown, on the basis of

the new synthesis as his Presidential Address to the former Society of Telegraph Engineers, now the Institution of Electrical Engineers. The Address epitomizes the mediating role he sought to play in uniting theory and practice. Whereas he had lectured the mathematical physicists on bringing their imagination down to concrete realities, he now lectured the engineers on the benefits of mathematical generalization and less intuitively obvious physical conceptions. Aside from this didactic function, however, the paper represents a magnificent attempt to reach back to his ether speculations of 1847 and 1856 to produce a unified physical worldview.

'Ether, electricity, and ponderable matter', rests on Thomson's longtime tenet, grounded in magneto-optic rotation, that 'Whatever the current of electricity may be, I believe *this* is a reality: *it does pull the ether round*'.[98] The fluid moving in his telegraph wire would do that if its viscosity created a drag on the ether lattice. It would produce a state of static rotational strain, with axes of rotation circling round the wire. To constitute lines of magnetic force, the ether displacements $\vec{A}$ would have to obey the relations, $\vec{B} = \text{curl}\,\vec{A}$ and $\vec{i} = \text{curl}\,\vec{B}$, as he had first shown in 1847. This magnetic rotational strain would propagate by transverse shear waves at a velocity depending only on the rigidity and density of the lattice, the velocity of light. Wherever the rotational strain was changing, furthermore, it would create a drag on any viscous fluid present, or an electromotive force, thus inducing currents in conductors. Assuming the drag to be given by the viscosity n times the rate of change of ether displacement, gives $\vec{E} = -n\partial\vec{A}/\partial t$, which shows, since induced force is independent of material, that the viscosity must be everywhere the same. Finally, if the fluid is free to move, a current will be induced proportional to its density ρ and to the moving force, $\vec{j} = \rho\vec{E} = C\vec{E}$, indicating that conductivity C is given by the *density* of free fluid, and does not depend on viscosity at all.[99] To put it differently, what looks like viscosity, so far as resistance to electric currents is concerned, is actually the limited amount of fluid per unit volume that is free to move through the ether lattice.

Maxwell's theory, that the inertial property of electromagnetic induction could be used to increase dramatically the rate of signalling in long distance cables. Thomson was here incorporating this crucial practical result into his own theory. Bruce Hunt argues in 'Imperial science: telegraphy and physics in Victorian Britain' (unpublished) that telegraph engineering was as important to the development of Maxwellian field theory as it was to Thomson. For the controversy between Maxwellians and practical telegraph engineers, see Bruce Hunt, '"Practice vs. theory": the British electrical debate, 1888–1891', *Isis*, 74 (1983), 341–55.

[98] William Thomson, 'Ether, electricity and ponderable matter', *Journal of the Institution of Electrical Engineers*, **18** (1890), 4–37, 128; **24** (1895), between p. 396 and p. 397; MPP, **3**, 484–515. It is not at first obvious that this paper presents a coherent set of models. We interpret it with extensive aid from the mathematical formulation Thomson wrote up for the third volume of MPP, 'Motion of a viscous liquid; equilibrium or motion of an elastic solid; equilibrium or motion of an ideal substance called for brevity *ether*; mechanical representation of magnetic force' (May, 1890), MPP, **3**, 436–65.

[99] Thomson, 'Ether, electricity and ponderable matter', MPP, **3**, 497–8; 'Motion of a viscous liquid', MPP, **3**, 440–1.

The key feature of this scheme is the viscous connection between ether strain and electric current; but the same viscosity must act also within the fluid itself, such that motion in one layer 'induces' motion in the next. The relations work out, Thomson showed, if static shear in the ether corresponds to rate of shearing in the fluid, or generally displacement in the ether to velocity in the fluid, and if rigidity in the ether corresponds to viscosity in the fluid. *But electric current cannot be given by the velocity of the viscous fluid itself.* The velocity of the fluid must be $\vec{A}$ and the current is $\vec{j} = \text{curl}\ (\text{curl}\ \vec{A}) = \nabla^2\vec{A}$. This means that in a wire carrying an electric current, supposed uniform over any cross-section, the velocity of the viscous fluid actually increases from zero at the walls to a maximum in the centre, in a parabolic curve. Its rate of shearing, here proportional to curl $\vec{A}$, would give $\vec{B}$, and the rate of change of shearing, proportional to curl $\vec{B}$, would give the uniform current. Thomson had actually employed this conception of the electric current at Bath.[100] Now, however, by relating velocity $\vec{A}$ in the viscous fluid to displacement $\vec{A}$ in the ether, he had connected his telegraph equation for electric motions inside conductors with a wave equation for magnetic shear in the ether. He had produced a model of how telegraph signals are related to light waves.

If only the relation of ether to conductors had been at issue, therefore, Thomson would have been 'perfectly satisfied with the problem of electromagnetic induction' by characterizing a conductor as a porous region of the elastic-solid ether, but possessing the same rigidity.[101] He could even envisage subsuming electrostatic induction by supposing that both conductors and dielectrics contained pores of invariable size filled with the compressible viscous fluid, but that in conductors the fluid could travel from one pore to another through passages that were at least partially closed in dielectrics by elastic partitions. 'Electrostatic stress depends on the curvature and extension of the partitions', he told the Edinburgh Royal Society in January, 1890, 'The law of capacity in the model is identical with that in conductors'. In such a model, electromagnetic induction (shearing forces) would produce polarization in the closed pores just as it produced currents through the open ones. Electrostatic action (with pressure gradients for potential gradients) would produce the same effects. 'A complete dynamical illustration of electro-dynamic action may be had' in this way, Thomson claimed.[102] Notice, however, that neither polarization nor currents – no displacement currents – exist in free ether, where there are no pores.

A difficulty with the elastic-solid ether arose for magnetic materials like iron,

[100] Compare Thomson, 'Simple hypothesis', MPP, **4**, 539–40, with 'Ether, electricity, and ponderable matter', MPP, **3**, 498–500, and with 'Motion of a viscous liquid', MPP, **3**, 441.

[101] Thomson, 'Ether, electricity, and ponderable matter', MPP, **3**, 502.

[102] William Thomson, 'On electrostatic stress' (abstract), *Nature*, **41** (1890), 358; MPP, **5**, 482. Throughout his 'Motion of a viscous liquid' Thomson included the effect of pressure gradients in his equations of motion.

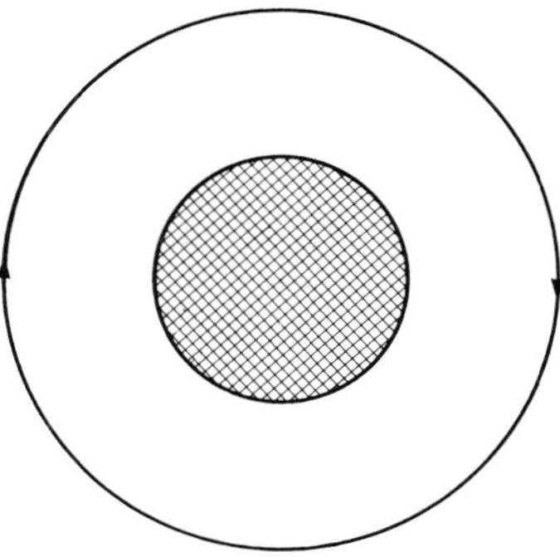

Figure 13.9. The concentric circles represent a cross section through a solenoid with an iron core and with a jelly ether both inside and out, which is supposed fixed at distant boundaries. If the steady current in the outer circle exerts a drag on the ether, the exterior jelly will experience a shearing rotation, while the interior, including the iron core, will experience only a uniform rotation.

in which the magnetic force, or rotational strain, had to increase over that in the surrounding ether. Such behaviour was not compatible with the idea of a jelly-like elastic solid, because the boundary conditions did not match. Supposing a steady current to move in a circle, as in a section of a long solenoid, with a central core of iron (figure 13.9), and supposing the surrounding space to contain a jelly, the current would ultimately pull the interior round in a uniform rotation, with no internal shear and thus no possibility for an increased rotation in the core, even if one assumed a lower rigidity there. Furthermore, the energy in such a jelly would reside in the space outside the current loop, where shear existed, rather than in the uniformly rotated interior.[103] Apparently a different sort of elastic solid was required, one in which magnetism and energy could be related to *absolute rotations*, independent of any attendant distortions.

Once again Thomson's great philosopher's stone, the gyrostat, came to the rescue. He displayed a two-dimensional ether lattice (figure 13.10) consisting of rigid squares (heavy lines) with corners connected by flexible, inextensible threads (light lines) running through eyes or pulleys at the corners. This 'molecular skeleton', considered by itself, represented an incompressible perfect fluid for small displacements. But in each square he placed a gyrostat (figure 13.11) consisting of a flywheel rotating rapidly on an axle in a rigid case. With the case mounted in bearings on a perpendicular axis, the skeleton became an elastic solid for rotations in its own plane. To maintain any given angle of rotation required a corresponding torque, with a concomitant rotation of the axles of the gyrostats out of the plane of their square frames. Mounting gyrostats

[103] Thomson, 'Ether, electricity, and ponderable matter', MPP, **3**, 504–5; 'Motion of a viscous liquid', MPP, **3**, 444–5, 449–50, 462–3.

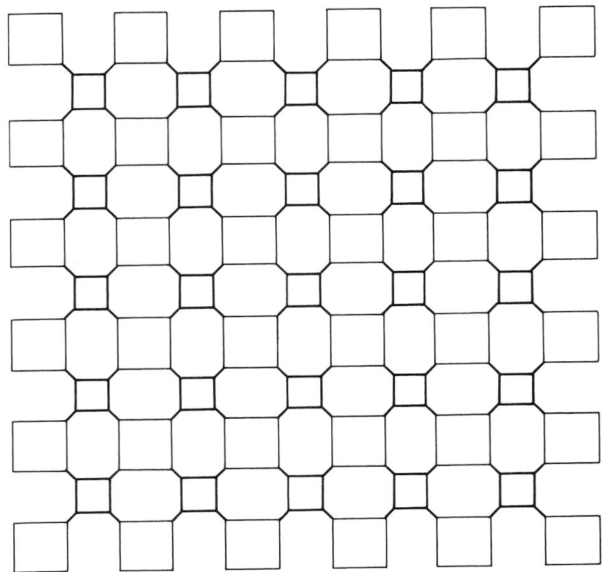

Figure 13.10. In this lattice structure the heavy lines are rigid squares representing molecules and the light lines are flexible but inextensible threads on which the molecular corners are free to slide.

of different angular momentum in the frames produced solids of different rigidity, so that the same torque would be balanced by different rotational strains. Equilibrium of torques across any interface required that rotation times rigidity be a constant. With rotations of the lattice representing magnetic force, $\vec{B} = \text{curl}\,\vec{A}$, and with rigidity the inverse of magnetic permeability μ, the ratio of the magnetic force just inside the core to that just outside had to be equal to the ratio of permeabilities, $\vec{B}/\vec{B}' = \mu/\mu'$. Since a given electric current would produce a given torque, the equation of currents became (implicitly), $\vec{j} = \text{curl}\,(\vec{B}/\mu)$.[104] Electromagnetism had been conquered.

Marvellous as it was, Thomson's ether had problems. Most seriously, it served only for infinitesimal rotations. Larger rotations produced compression and thus inevitably connected transverse light waves with longitudinal waves, even in the absence of electrical pressures. A better ether would uncouple the two sorts of waves. 'After many unavailing efforts' he found a 'gyrostatic adynamic

104 Thomson, 'Ether, electricity, and ponderable matter', MPP, **3**, 505–10; 'Motion of a viscous liquid', MPP, **3**, 463–4. Thomson uses his 'electromagnetic' definition of the force inside a magnet, which has a different connotation from Maxwell's magnetic induction $\vec{B}$, because Thomson did not accept Maxwell's dual fields (ch. 8). In Maxwell's *Treatise*, p. 240, para. 621, he wrote beside the expressions for the magnetic pair $\vec{B}$, $\vec{H}$ and the electrostatic pair $\vec{D}$, $\vec{E}$, 'not usefully a "pair", W.T. Aug. 12, 1890'.

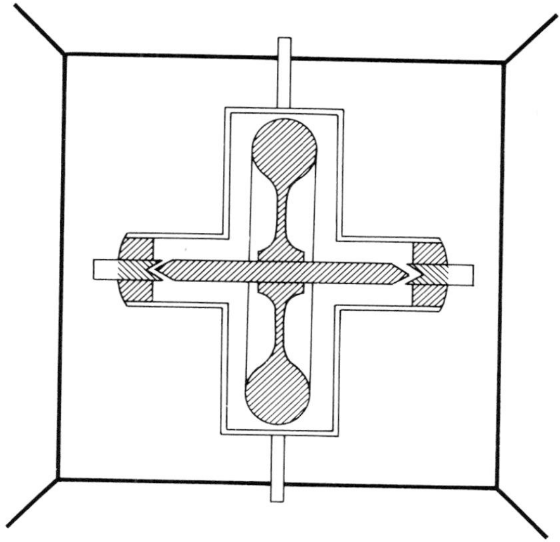

Figure 13.11. In each molecule of the ether lattice in figure 13.10, a gyrostat is spinning, producing rotational rigidity in the lattice.

constitution for ether' in March, 1890. Each corner point in the new three-dimensional lattice was effectively a 'ball-and-twelve-socket' joint connected to twelve nearest neighbours by freely extensible and elastically flexible rods consisting of bars sliding in tubes. Tetrahedronal cells in the resulting assembly had mounted within them rigid frames consisting of three bars meeting perpendicularly in their midpoints; and each bar carried four gyrostats, with two each mounted in the planes of the other two bars and spinning in opposite directions. The whole arrangement was to be shrunk to the size of ether molecules and the gyrostats were to have liquid flywheels to suggest their ultimate origin. This improbable adynamic assembly, or 'labile' ether as it became known, would 'transmit vibrations of light exactly as does the ether of nature; and it would be incapable of transmitting condensational–rarefactional waves, because it is absolutely devoid of resistance to condensation and rarefaction'.[105]

By this point the philosopher of 'look and see' had regained a great deal of his credibility. His ether lattice and his viscous fluid together could do mechanically most of what Maxwell's equations could do mathematically. In fact, FitzGerald, Heaviside, and soon Joseph Larmor (1857–1942), admired it greatly as an analogy which showed the possible reality of a vortex ether. Unfortunately, however, although Sir William's models described *fields* of magnetic force in

[105] William Thomson, 'On a gyrostatic adynamic constitution for "ether"', begun in *Comptes Rendus*, **109** (1889), 453–5; extended in *Proc. Royal Soc. Edinburgh*, **17** (1890), 127–32; MPP, **3**, 466–72, on p. 472.

wonderful detail, they gave not the slightest idea of how the *ponderomotive* force acting between two magnets might be produced. Nor did they describe the simplest ponderomotive effects of electrostatics, such as the attraction between a piece of paper and a comb. In this respect, the purely labile ether was the worst yet. It did not even provide a mechanism for the mere transmission of electrostatic fields, since it would not propagate pressures and it contained no displacement currents for electromagnetic induction. Thomson seems never to have taken it very seriously. 'Our first love was electrostatics', he lamented, 'That is absolutely left out in the cold; we do not touch it'.[106] To attain 'anything like a satisfactory material realisation of Maxwell's electro-magnetic theory of light' seemed to require a third category, additional to ether and electricity, to act as the locus of ponderomotive forces, namely, ponderable matter. Electricity would then somehow act as a '*tertium quid* . . . a fluid go between, serving to transmit force between ponderable matter and ether and to cause by its flow the molecular motions of ponderable matter which we call heat'. The goal of reducing the three substances to a single primordial one seemed to recede, but hope remained: 'I think we must feel at present that the triple alliance, ether, electricity, and ponderable matter is rather a result of our want of knowledge, and of capacity to imagine beyond the limited present horizon of physical science, than a reality of nature'.[107]

Mephistopheles

That the problem of ponderomotive forces should have presented the ultimate obstacle to Thomson's attempt to give Maxwell a material realization is particularly poignant, for it was from Thomson that Maxwell learned how to relate ponderomotive forces to fields, as gradients in field energy. That idea supplied the backbone of Maxwell's *Treatise*. Interpreting electrostatic and magnetic energy densities, $\vec{E}\cdot\vec{D}/2 = KE^2/8\pi$ and $\vec{H}\cdot\vec{B}/8\pi = \mu H^2/8\pi$, as potential and kinetic forms of energy, respectively, he supposed the corresponding forces to be obtainable from generalized dynamical principles (as given, for example, in T & T′). Maxwell treated the energies as though they derived from stresses in a medium, but without specifying the medium other than as elastic and possessing a certain density. 'Any further explanation of the state of stress', he said, 'must be regarded as a separate and independent part of the theory, which may stand or fall without affecting our present position'.[108]

Thomson's methodology of 'look and see' allowed no such luxury. He could not locate Maxwell's electrostatic energies in ether because Maxwell's stresses were not those of any known substance. As he told FitzGerald in 1896, 'mere

[106] Thomson, 'Ether, electricity and ponderable matter', MPP, **3**, 510.

[107] Thomson, 'Motion of a viscous liquid', MPP, **3**, 465.

[108] Maxwell, *Treatise*, **1**, 119–34, paras. 103–11, **2**, 246–58, paras. 630–46, on p. 257, para. 645. Buchwald, *Maxwell to microphysics*, pp. 34–7, 60–4.

displacement does not in an elastic solid or in any conceivable "ether" give rise to energy equal to $E^2/8\pi$ per unit volume of field'. Even if such large electrostatic energies could exist in transverse displacements of an elastic solid, furthermore, one would expect coexisting electrical compressions, thus returning to an old theme of the Baltimore Lectures: 'there would be a great deal of energy in condensational waves going about through space, and there would be a new force (to take an absurd mode of speaking of these things) that we know nothing of. There would be some tremendous action all through the universe'.[109] He seems, therefore, to have usually looked to the incompressible solid for the transmission of electrostatic pressure, although he sometimes entertained a very soft medium with velocities of longitudinal waves much lower than that of light.[110]

Sir William would never progress beyond his 1890 triad of ether, electricity, and ponderable matter, despite prodigious efforts. In 1892, furthermore, he became convinced on mathematical grounds that the periodic motions required for the vortex theory of ether and matter were essentially unstable. In the same year he was elevated to the peerage. Lord Kelvin thus had to face the prospect that Sir William's great quest had been in vain. At the Jubilee in 1896, celebrating his fifty years as professor of natural philosophy, Kelvin characterized his efforts to penetrate the secrets of the ether with one word, 'FAILURE. I know no more of electric and magnetic force, or of the relation between ether, electricity, and ponderable matter, or of chemical affinity, than I knew and tried to teach to my students of natural philosophy fifty years ago in my first session as Professor'.[111] He came to feel himself the victim of a deception, his own self-deception to be sure, in the expectation of reducing nature to vortices. Mephistopheles came to mind, as the great deceiver who played on Faust's scientific vanity. Kelvin himself now reluctantly played the devil's role as he withdrew the promise of knowledge. To Professor W.H. Julius in Holland, who, following Kelvin's hydrodynamic line, had thought to lift the veil with 'ether squirts', he had to write that 'in respect to all these Ether Theories, my own Vortex-Atom included, I must unhappily rank with Mephistopheles, "der Geist der stets verneint" . . . I cannot feel any happiness in any ether–theory which does not account for electro-static force and ordinary magnetic attraction . . .'[112]

Worse even than the vanity of vortices, was the promise of mathematical symbolism with which the younger generation deceived itself, for they had abandoned the principle of scientific virtue that had governed his own quest and in part redeemed it, the demand for practical, sensory knowledge. 'Nihilism', as

[109] Thomson, BL, 143. See also, pp. 40–1, 264.

[110] For example, Kelvin to G.F. FitzGerald, 23rd November, 1898, F17, ULG: 'I am becoming year by year after 1888 more and more coerced towards the belief that in pure ether the velocity of the condensational wave is small in comparison with that of the distortional wave'.

[111] Kelvin, reply to toast at his Jubilee, SPT, **2**, 984. On the essential instability of vortex motion see William Thomson to Hermann von Helmholtz, 20th June, 1892, in SPT, **2**, 921–2, and SPT, **2**, 1047.

[112] Kelvin to W.H. Julius, 3rd July, 1899, J206, ULG.

Kelvin labelled the mathematical conceit of the 1890s, now openly ruled. Heaviside epitomized that form of decadence in Britain, with his vector calculus and his attempt to extract the pure mathematical essence from Maxwell's *Treatise*. In Germany Hertz too, 'unwisely', had adopted 'Heaviside's nihilism', which resulted in his reduction of Maxwell's theory to the symmetrical set of field equations familiar today. Writing a preface for the English edition of Hertz's *Electric waves* (1893), Kelvin praised Hertz's experimental achievements while pointedly avoiding mention of his mathematical theory. The book soon made famous 'that unfortunate statement . . . that *Maxwell's Electromagnetic theory is Maxwell's formulas on the subject*', as Kelvin wrote to Pieter Zeeman in 1906. Equally egregious was Hertz's *Prinzipien der Mechanik*, which 'tended to turn students and even mathematicians from the right view of dynamical principles'.[113] For this same sin of undermining the 'right view', Kelvin had chastised FitzGerald in 1896: 'It is mere nihilism, having no part or lot in Natural Philosophy, to be contented with two formulas for energy, electromagnetic and electrostatic, and to be happy with a vector and delighted with a page of symmetrical formulas. "*Giebt nur ein Wort* (Mephistopheles) *Ein würdiges Pergament*! (Wagner)"'.[114] Words, symbols, not realities, constituted the Maxwellian incantations, as deceptive as Mephistopheles himself. The evil had spread in some degree to most of Thomson's friends. Rayleigh, by 1904, had 'long ago succumbed' to the 'nihilistic statement in the so-called electromagnetic theory of light as to "magnetic vector" and "electric vector"'.[115] Joseph Larmor, with whom Thomson began correspondence in the 1890s, adopted Thomson's labile gyrostatic ether for his own theory of electrons, treating each electron 'as a point-pole from which intrinsic rotational strain radiates'; but of its 'intimate structure' he did not wish to ask. 'I know this is not dynamics in your sense', Larmor wrote, 'the rotational aether is not like a material elastic solid'. He preferred '*generalized dynamics* [double emphasis by Kelvin, who spat 'equations!'] without a *definite model*, but with its possibilities *illustrated* by the potentialities of the vortex atom theory ['*zero*', double emphasis]'.[116] The entire younger generation, Kelvin rightly recognized, tended to reify the mathematical forms of energy physics. Not only radical energeticists, like those around Wilhelm Ostwald in Germany, had forgotten that energy was work, work done by and on matter:

Young persons who have grown up in scientific work within the last fifteen or twenty years, seem to have forgotten that energy is not an absolute existence. Even the Germans laugh at the 'Energetikers'. I do not know if even Ostwald knows that energy is a capacity

[113] Kelvin to G.F. FitzGerald, 29th April, 1896, in SPT, **2**, 1069–72, on p. 1070; Kelvin to Pieter Zeeman (copy), 22nd May, 1906, Z2, ULC.

[114] Kelvin to G.F. FitzGerald, 9th April, 1896, in SPT, **2**, 1064–9, on p. 1064.

[115] Kelvin to C.G. Knott, 12th September, 1904, K111, ULC.

[116] Joseph Larmor to Kelvin, 14th November, 1899 and 20th August, 1900, L23 and L25, ULC. All emphasis Kelvin's.

for doing work; and that work done implies mutual force between different parts of one body relatively movable, or between two bodies or two pieces of matter, or between two atoms of matter, or between an atom of matter and an electron, or, at the very least, between two electrons.[117]

There can be little doubt that for Thomson personally, the source of this heresy was Maxwell's *Treatise*. Greater attention to his own *Treatise* would have prevented it. But within the community of the 'new physicists', the 'so-called electro-magnetic theory' as opposed to his own telegraph theory – the metaphysical rather than the practical – dominated physical research. Over and over again Kelvin returned to Maxwell's *Treatise* to make annotations and to impress his name and date, as though opening a book of sorcery to locate its magic and to break its spell. Part of the magic surely resided in the sections on energy and stress in the electromagnetic field and in radiation, which Kelvin would dissolve with 'wholly unacceptable, K Oct. 16 1905' and 'no dynamical proof of this, K Oct. 16, 1905'. On 25th October, 1907, only two months before his own death, he was still seeking to banish Maxwell's ghost with his now shaking hand.[118] Mephistopheles, the spirit of nihilism, had won.

Kelvin's methodology died with him. It was a methodology that grounded mathematical physics in steam-engines, vortex turbines, and telegraph lines; a methodology of industrialization and Empire, suitable for Glasgow but not for Cambridge. Its decline should probably be seen as a symbol of what Martin Wiener has called 'the decline of the industrial spirit' in Britain.[119] But, on the positive side, Kelvin achieved great practical successes in tying his view of energy, as work done on matter, to the economic and political strength of the Empire. We shall close this series of chapters on his mathematical physics, therefore, by recalling his last fight to preserve the telegraph theory of electricity.

When in 1895 Conrad Röntgen discovered X-rays, he suggested that they might be the long-sought longitudinal waves of the elastic-solid theory of ether. Kelvin, who took great interest in the discovery, would soon become convinced that X-rays were merely transverse waves of high frequency, but meanwhile he began again to insist that the wave theory of light required the existence of longitudinal waves. For the purpose he rejuvenated his old thought experiment concerning two charged conducting globes, which either underwent continuing oscillations in charge through a connecting 'telegraph' wire, as in the Baltimore Lectures, or were discharged through the intervening space, as in his 1888 letter to Rayleigh. Such actions, he insisted, necessarily involved transmission of

[117] Kelvin to Joseph Larmor, 9th October, 1906, L37, ULC.

[118] Kelvin, annotation to Maxwell's *Treatise*, **2**, p. 391, para. 792 (referring back to para. 643 and to the last sentence of 792), p. 86, para. 448.

[119] M.J. Wiener, *English culture and the decline of the industrial spirit, 1850–1980* (Cambridge, 1981). See also Peter Alter, *The reluctant patron: science and the state in Britain, 1850–1920*, translated by A. Davis (Oxford, 1987), esp. pp. 98–137.

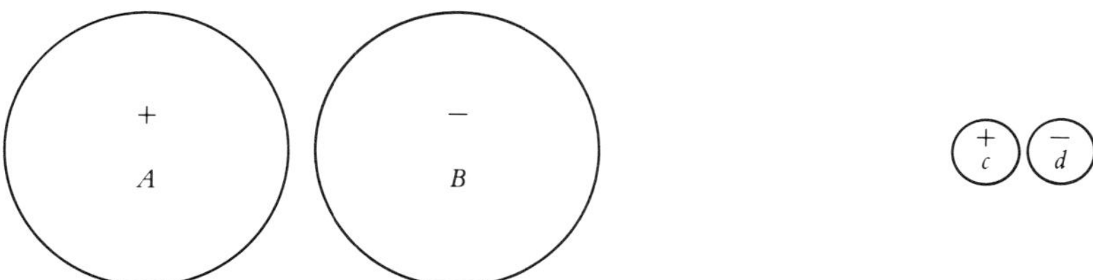

Figure 13.12. A spark discharge between the charged conducting spheres *A* and *B*, Thomson argued, would, even on Maxwell's theory, produce longitudinal waves, thereby causing a secondary discharge in the smaller spheres *c* and *d* held just below the sparking potential.

potential by pressure waves. The renewed thought experiment, appearing in *Nature* on 6th February, 1896, involved two large spheres between which a spark discharge would occur, producing a secondary discharge between two smaller spheres kept at a potential just below the sparking potential (figure 13.12). The question Kelvin raised was not whether the action would be propagated longitudinally, but only how fast. Maxwell's theory, he asserted, represented the elastic-solid theory restricted to incompressibility, thus instantaneous transmission, while the unrestricted theory would give 'the velocity of the condensational–rarefactional wave'. Acknowledging that the term 'theory' was misapplied when it did not explain 'the attraction between rubbed sealing-wax and a little fragment of paper', he nevertheless insisted that 'Elastic solid . . . we must have, or a definite mechanical analogue of it, for the undulatory theory of light and of magnetic waves and of electric waves'.[120]

FitzGerald jumped into the old argument on 12th February: 'Surely you are not right in your letter to "Nature" last week in stating that Maxwell's ether would give an instantaneous action in the case you mention. Whatever properties it may have in respect of gravitation &c &c so far as electric actions are concerned they are *all* represented by the equation

$$\nabla^2 U = K\mu \partial^2 U/\partial t^2$$

and so represent an action propagated with a velocity $(K\mu)^{-1/2}$.' He explained how Hertz had accounted for the kind of action Kelvin envisaged within the theory of transverse electromagnetic waves, but Kelvin stuck to his view. FitzGerald tried again two days later, with diagrams and more detail, insisting that 'Maxwell never suggested that his ether was an elastic solid'. He admitted that something analogous to compressibility might take place in the ether, but insisted that the effects would not be electromagnetic. Furthermore, and on this point he thought he had Kelvin trapped in his own methodology, the supposed effects could have 'any sort of properties we choose to give them until they are chained to some experimental result'.[121]

120 Kelvin, 'Velocity of propagation of electrostatic force', *Nature*, **53** (1896), 316.
121 G.F. FitzGerald to Kelvin, 12th and 14th February, 1896, F121 and F122, ULC.

Kelvin merely fired back on the 16th proofs of a new paper proposing an experimental test of the existence of X-rays in the region of a supposed longitudinal discharge, commenting: 'Read mark learn and inwardly digest Maxwell paras, 610 . . . 616 [displacement current and closed circuit hypothesis]. I think when you have done so you will see that the statements marked // on the proof are correct, and will reconsider the whole of your letter to me of 14th'. The marked statements no doubt included one asserting that 'either an instantaneous transmission of . . . electrostatic force, or a set of electric waves of almost purely longitudinal displacement, according as ether is incompressible or compressible', would occur.[122] Of course FitzGerald had long ago inwardly digested the Maxwell passages and pointed out to Kelvin the source of his anti-Maxwellian attitudes: '*I feel pretty confident that you are overlooking the electric effects* [displacement currents] *due to the magnetic force being generated while the spark [current, del.] is starting*'. Indeed Kelvin was ignoring displacement current, 'Maxwell's only real discovery' in FitzGerald's opinion, because it was to him no discovery at all but a devilish deception, merely a word.[123]

And so the friendly combat continued, with FitzGerald tirelessly expounding Maxwellian doctrine in eight long letters from February to May, 1896, and in five more during the month of November, 1898. 'Displacement' to Kelvin either meant displacement of matter or it meant nothing: 'Maxwell's expression, "electric displacement" is I believe, absolutely true so far as it indicates a true displacement of matter, as in the undulatory theory of light, but my difficulty is in respect to the electric quality concerned in this displacement'. For FitzGerald displacement was an indispensable structural element in the mathematical theory of electromagnetic waves, which Hertz had shown to exist: 'Your faith that Maxwell's "electric displacement" is "absolutely true so far as it indicates a true displacement of matter" goes far beyond anything I can feel about its certainty'.[124]

Among many ironies here, FitzGerald, like Larmor, believed that the reality of the ether would be found in some form of Kelvin's vortex theory of matter, which Kelvin himself had abandoned; but they employed the term 'matter' in such an expanded sense that Kelvin no longer recognized it. Matter was simply that which obeyed the mathematical laws corroborated by experiment. FitzGerald thus challenged directly Kelvin's most basic assumption, that matter had to have the properties of sensible matter: 'it seems to me very inconsistent to object to one set of symbols merely because they are only symbols and to use the others [of fluids and elastic solids] without objections'. To him, 'to work away

[122] Kelvin to G.F. FitzGerald, 16th February, 1896, F122, ULC (a copy of his note made on FitzGerald's letter of the 14th). Kelvin, 'On the generation of longitudinal waves in ether' (read 13th February), *Proc. Royal Soc.*, **59** (1896), 270–3; *Nature*, **53** (1896), 450–1, on p. 450.

[123] G.F. FitzGerald to Kelvin, 17th February, 1896, F123, ULC.

[124] Kelvin to G.F. FitzGerald, 29th April, 1896, in SPT, 1069—72, on p. 1069. G.F. FitzGerald to Kelvin, 11th May, 1896, F128, ULC.

upon the hypothesis' that the ether was an elastic solid was 'a pure waste of time'.[125]

To this condemnation of his practical realism and of his favourite analogy, Kelvin retorted: The analogy 'is certainly not an allegory on the banks of the Nile. It is more like an alligator. It certainly will swallow up all ideas for the undulatory theory of light, and dynamical theory of E&M not founded on force and inertia. I shall write more when I hear how you like this'. The answer came, 'I am not afraid of your alligator which swallows up theories not founded on force and inertia . . . I am quite open to conviction that the ether is *like* and not merely in some respects *analogous* to an elastic solid, but I will . . . wait till there is *some* experimental evidence thereof before I complicate my conceptions therewith'.[126] And there the debate ended. Larmor wrote the epitaph of the elastic-solid ether – and of the mechanical worldview as well – in his review of the 1904 publication of Kelvin's rewritten *Baltimore lectures*:

for better or worse most of us are now wedded to the electric theory of light, the creation of Lord Kelvin's most famous disciple, which forms a consistent scheme of the relations of electricity and radiation . . . Is it incumbent on us to treat the aether as strictly akin to the material bodies around us? or may we assign to it a constitution of its own, to be tested by its success in comprehending the complex of known relations of physical systems? . . . It would appear that Lord Kelvin cannot grant that such a constitution has been determined until . . . a precise mechanical model of it could be imagined; whereas, on the other side, it may be held to be the merit of the scheme that it evades such a hopeless task.[127]

Heaviside had put the point more humorously in 1896:

Lord K. has the defects of his qualities. He is powerful minded, and has devoted so much attention to the elastic solid, that it has crystallized his brain. Some years ago he forced his brain into another arrangement [the adynamic ether]. That he actually did it (all the harder because of his power) he proved by his papers. If he had kept on at it, it might have been a permanent change, but it seems he didn't; for his forced readjustment was *unstable*, and he has reverted automatically to old notions.[128]

[125] G.F. FitzGerald to Kelvin, 17th April, 1896, F127, ULC.

[126] Kelvin to G.F. FitzGerald, 28th November, 1898, F18, ULG. G.F. FitzGerald to Kelvin, 29th November, 1898, F137, ULC.

[127] Joseph Larmor, 'Lord Kelvin on optical and molecular dynamics', *Nature* (supplement), **70** (1904), iii–v, on pp. iv–v.

[128] Oliver Heaviside to G.F. FitzGerald, 12th June, 1896, FitzGerald collection, Royal Dublin Society, Dublin. We thank Bruce Hunt for this reference.

III

The economy of nature: the great storehouse of creation

14
The irreversible cosmos

Do you know of any experimental data from which the absolute mechanical value of as much of the sun's rays as fall on a unit of surface in a unit of time may be determined? . . . I am thinking on communicating a paper on the sources of the mechanical effect producible under the direction of man. I think that, with the exception of what might be got from tide mills, or the combustion of meteoric stones or other native metals, all is derived from the sun, and is merely a part of the mechanical value of the undulations which he has sent us from the epoch of the creation of plants. *William Thomson to G.G. Stokes, 1852*[1]

Thomson's commitment to a progressionist cosmology and geology had enabled him by 1851 to transform the problem of 'loss' of available energy from an engineering issue into a universal cosmological one with clearly defined theological support. This transformation marked the beginning of his quest for a new economy of nature – an irreversible cosmos – as a systematic and coherent vision founded upon the universal dissipation of energy. So strong, indeed, was Thomson's commitment that his reading of dissipation included phenomena which other physicists (notably Rankine) regarded as reversible. He viewed the conversion of the sun's energy into radiant heat, for example, as an irreversible process with an inevitable dissipation of energy.

The universal dissipation of energy

In his 1851 draft of 'The dynamical theory of heat', Thomson's remarkable attempt to comprehend the mysterious 'loss' of available energy took as its starting point the engineering problem. 'The fact is,' he had written, 'it may I believe be demonstrated that work is lost to man irrecoverably; but not lost in the material world'. Such a statement went some way to resolving the problem of the loss of work or mechanical effect in human artefacts – such as heat engines – by accepting that the *vis viva* remained undiminished 'in the material world', though irrecoverable as useful work. But he had not explained the 'irrecoverable' nature of that *vis viva*. He merely stated, as a fundamental belief, that

[1] William Thomson to G.G. Stokes, 13th January, 1852, K53, Stokes correspondence, ULC.

'everything in the material world is progressive. The material world could not come back to any previous state without a violation of the laws which have been manifested to man, that is, without a creative act or an act possessing similar power'.[2]

As early as 1849–50, Thomson told his class with reference to the sun's heat that 'Till 2 years ago it was thought that there was no heat but what was latent'. Joule (and he) had shown, however, that heat can derive from mechanical effect (the First Law). In his 1851 draft he drew the more extended implications arising from the new dissipation axiom (the Second Law). He believed 'that no physical action can ever restore the heat emitted from the sun, and that this source is not inexhaustible'. He also believed 'that the motions of the earth & other planets are losing *vis viva* which is converted into heat; and that although some *vis viva* may be restored for instance to the earth by heat received from the sun or by other means, that the loss cannot be *precisely* compensated & I think it probable that it is under-compensated'.[3] These two articles of faith were the starting point of his new cosmology. While they appeared in the draft, and had in fact grounded Thomson's 1851 paper 'On the dynamical theory of heat', they did not appear in the published memoir, which contained brief reference to the heat which, in practical machines, was 'irrecoverably lost to man, and therefore "wasted", although not annihilated'.[4] Public elaboration of the cosmological significance of the dissipation axiom had to await the appearance of 'On a universal tendency in nature to the dissipation of mechanical energy' (1852).

Having published nothing in 1851 on the general ideas underlying his dissipation axiom, Thomson now sought to obtain those ideas as deductions from the more restricted expression for heat engines. That is, he generalized from the economical loss of energy in machines to cosmological dissipation in the economy of nature:

> The present communication contains an examination of circumstances regarding a universal tendency to unreversible transformation in the condition of mechanical energy . . . as implied in the proposition originally stated by Carnot, and established on the modified principle of the dynamical theory of heat, that when heat is diffused by conduction there is an economical loss as regards its value as a source of mechanical effect.[5]

The nature of the link between 'loss' in machines and 'loss' in nature, however, had to be elucidated more precisely.

To begin with, Thomson divided stores of mechanical energy into two

[2] William Thomson, Preliminary draft for the 'Dynamical theory of heat', PA 128, ULC, pp. 5, 6. Hereafter 'Draft'.

[3] William Thomson, 22nd January, 1850, lecture 48, in William Smith, 'Notes of the Glasgow College natural philosophy class taken during the 1849–50 session', Ms Gen. 142, ULG; Draft, 8.

[4] MPP, **1**, 189.

[5] William Thomson, Draft of 'On a universal tendency in nature to the dissipation of mechanical energy', PA 137, ULC.

classes. Statical stores of energy – potential energy from late 1852 – included weights at a height, electrified bodies, and quantities of fuel. Dynamical stores of energy – kinetic energy from 1862 – included masses of matter in motion (flywheels, for example), undulations of light or radiant heat in a volume of space, and bodies having thermal motions among their particles. He then laid down four key propositions 'regarding the *dissipation* of mechanical energy from a given store, and the *restoration* of it to its primitive condition', propositions which, he asserted, were 'necessary consequences' of the axiom upon which Carnot's proposition was founded, namely, 'It is impossible, by means of inanimate material agency, to derive mechanical effect from any portion of matter by cooling it below the temperature of the coldest of the surrounding objects'. The first proposition referred to reversible processes – a perfect thermodynamic engine – while the other three treated of 'unreversible' processes – friction, conduction, and absorption of radiant heat or light:

I. When heat is created by a reversible process (so that the mechanical energy thus spent may be *restored* to its primitive condition) there is also a transference from a cold body to a hot body of a quantity of heat bearing to the quantity created a definite proportion depending on the temperature of the two bodies.
II. When heat is created by an unreversible process (such as friction), there is a *dissipation* of mechanical energy, and a full *restoration* of it to its primitive condition is impossible.
III. When heat is diffused by *conduction*, there is a *dissipation* of mechanical energy, and perfect *restoration* is impossible.
IV. When radiant heat or light is absorbed, otherwise than in vegetation, or in chemical action, there is a *dissipation* of mechanical energy, and perfect *restoration* is impossible.[6]

The paper ended with a rather terse announcement of three 'general conclusions' drawn from these four propositions and from 'known facts with reference to the mechanics of animal and vegetable bodies'. The first conclusion stated that 'There is at present in the material world a tendency to the dissipation of mechanical energy'. Since this 'conclusion' had been the very foundation of his axiom providing the new proof of Carnot's proposition, Thomson's paper reversed their historical order, giving priority to the axiom and deriving the 'belief' that had given rise to it.

The second 'general conclusion' stated that 'any *restoration* of mechanical energy, without more than an equivalent of dissipation, is impossible in inanimate material processes, and is probably never effected by means of organized matter, either endowed with vegetable life or subjected to the will of an animated creature'. It denied, therefore, the possibility of reversing the dissipation tendency in nature – at least in inanimate nature – thus complementing and elaborating the first general conclusion.

[6] William Thomson, 'On a universal tendency in nature to the dissipation of mechanical energy' *Phil. Mag.*, [series 4], **4** (1852), 304–6; MPP, **1**, 511–12.

The third 'general conclusion' concerned the limited time during which mankind could have inhabited, and would continue to inhabit, the earth unless 'operations have been, or are to be performed, which are impossible under the laws to which the known operations going on at present in the material world are subject'. In other words, given the first two conclusions regarding the universal dissipation of energy and the impossibility of its restoration, the third conclusion as to the finiteness of the earth as an abode for man would follow.[7]

Thomson did not suggest here a general heat death of the universe as a necessary consequence of his fundamental energy principles. A decade later, in his 1862 article 'On the age of the sun's heat', he reiterated his fundamental beliefs in the indestructibility and universal dissipation of energy, leading to diffusion of heat, cessation of motion, and exhaustion of potential energy through the material universe while clarifying his view of the consequences for the universe as a whole:

> The result would inevitably be a state of universal rest and death, if the universe were finite and left to obey existing laws. But it is impossible to conceive a limit to the extent of matter in the universe; and therefore science points rather to an endless progress, through an endless space, of action involving the transformation of potential energy into palpable motion and thence into heat, than to a single finite mechanism, running down like a clock, and stopping for ever.[8]

This explicit statement may well have emerged from discussions with two other major figures who shared Thomson's cosmological interests – Rankine and Helmholtz – or from their interpretations of his 1852 conclusions.

In a short paper aptly entitled 'On the reconcentration of the mechanical energy of the universe', Rankine said: 'there is . . . Professor Thomson concludes, so far as we understand the present condition of the universe, a tendency towards a state in which all physical energy will be in the state of heat, and that heat so diffused that all matter will be at the same temperature; so that there will be an end of all physical phenomena'.[9] Rankine thereby read considerably more into Thomson's 1852 paper than was actually claimed, for Thomson's third general conclusion had scarcely gone beyond a statement about the earth as a habitable place for living creatures and had said nothing about the 'end of all physical phenomena'. A little later, Helmholtz also widened Thomson's original claims and attributed to the Glasgow professor the idea of the heat death of the universe:

> we must admire the sagacity of Thomson, who, in the letters of a long-known little mathematical formula which speaks only of the heat, volume, and pressure of bodies, was

[7] *Ibid.*, p. 514.

[8] William Thomson, 'On the age of the sun's heat', *Macmillan's Mag.*, **5** (1862), 288–93; PL, **1**, 349–68.

[9] W.J.M. Rankine, 'On the reconcentration of the mechanical energy of the universe', *Phil. Mag.*, [series 4], **4** (1852), 358–60, on p. 352; MSP, 200–2.

able to discern consequences which threatened the universe, though certainly after an infinite period of time, with eternal death.[10]

Thomson, in 1862, thus effectively accepted Helmholtz's interpretation of his own paper.

Unlike Helmholtz, however, Rankine was not content to accept these pessimistic conclusions for the universe. In his 1852 paper, Rankine argued that, while such a vast speculation appeared soundly based on experimental data and 'to represent truly the present condition of the universe *so far as we know it*', yet a reconcentration of mechanical energy was possible. He speculated that radiant heat – 'the ultimate form to which all physical energy tends' – may be totally reflected at the boundaries of the very interstellar medium through which the radiation had been transmitted and diffused. Such total reflection would reconcentrate the energy into foci of intense heat – the renewed store of energy capable of resolving, for example, a mass of inert compounds into its constituent chemical elements and so reproducing a new source of chemical power. Rankine therefore concluded that 'the world, as now created, may possibly be provided within itself with the means of reconcentrating its physical energies, and renewing its activity and life'.[11]

The differences between Thomson, Helmholtz and Rankine on this cosmological issue are illuminating. Renewal of the universe – or even the earth or the solar system – by cyclical reconcentration of energy was a hypothesis which William Thomson never entertained. So powerful was his commitment to a principle of universal dissipation that such a violation was wholly unacceptable. Yet his commitment to dissipation was clearly not based on empirical evidence alone – convincing though the instances of 'loss' of available energy were to anyone familiar with heat engines. We have argued, therefore, that the full strength of that commitment derives from Thomson's theological view that God had not created a universe – or solar system – as an eternal entity, and that only He could restore the initial sources of energy. That is not to say, however, that the principle of dissipation itself *derives* from theological considerations, rather that Thomson's *commitment* to it as an inviolable axiom received its extraordinary strength from the axiom's place in his theology of nature. Helmholtz, on the other hand, was, in the light of man's limited knowledge, prepared to place Rankine's cyclical cosmology alongside a directional view. As he expressed the point in his 1854 lecture, 'we know not whether the medium, which transmits the undulations of light and heat, possesses an end where the rays must return, or whether they eternally pursue their way through infini-

[10] Hermann Helmholtz. 'On the interaction of natural forces', (lecture at Königsberg, 7th February, 1854), *Phil. Mag.*, [series 4], **11** (1856), 489–518, on p. 503. The lecture was translated by John Tyndall. Helmholtz, who had been at the 1853 meeting of the British Association at Hull – from which Thomson had been absent – did not in fact meet the Glasgow professor until 1855. See SPT, **1**, 308–10. These favourable remarks apparently date, however, from 1852. See SPT, **1**, 295.

[11] MSP, 201–2.

tude'. Rankine, for his part, had a very similar religious background to Thomson, but never *actively* used theological beliefs about the transitoriness of the creation to construct a new cosmology.

Following his enunciation in 1852 of universal dissipation of energy in nature, Thomson was to remain unwavering in his commitment to that principle for the rest of his life. Forty years on, in 1892, he wrote again of 'the irreversibility of actions' connected with all real engines – whether powered by heat, electricity, wind, or tide – forming 'only part of a physical law of irreversibility according to which there is a universal tendency to the dissipation of mechanical energy'.[12] During those forty years, however, he had employed the law to construct detailed theories of the nature of the earth and solar system.

Sources of energy: the nature of the sun's heat

In Thomson's cosmological picture, the sun's importance derived from his twofold belief that it provided the most striking instance of the cosmical dissipation of energy, and that it functioned as the principal source of those stores of energy available to man for the production of mechanical effect. In the 1851 draft he had stated his conviction 'that no physical action can ever restore the heat emitted from the Sun and that this source is not inexhaustible'. He did not necessarily mean that the sun had no external sources of energy by which it might be fuelled, but only that the heat given out by the sun could not be recovered or restored, and that, whatever the source of the sun's heat, that source must be finite. At the beginning of the following year, Thomson began to develop the significance of the sun as the principal source of energy available to man. In January, 1852, he asked Stokes if he knew of 'any experimental data from which the absolute mechanical value of as much of the Sun's rays as fall on a unit of surface in a unit of time may be determined'. At the same time, he told Stokes that he was thinking of communicating a paper on the sources of mechanical effect producible under the direction of man. In his view, 'with the exception of what might be got from tide mills, or the combustion of meteoric stones or other native metals, all is derived from the sun, and is merely a part of the mechanical value of the undulations which he sent us from the epoch of the creation of plants'.[13]

These ideas formed the basis of papers 'On the mechanical action of radiant heat or light' and 'On the sources available to man for the production of mechanical effect'. In the former, Thomson made reference to Pouillet's estimate of heat radiated from the sun in any time – an estimate most probably supplied by Stokes in answer to Thomson's request, and one which became the starting point for his subsequent calculations on the sun's heat. In the latter paper,

[12] William Thomson, 'On the dissipation of energy', *Fortnightly Rev.*, **51** (1892), 313–21; PL, **2**, 469.

[13] William Thomson, *op. cit.* (note 1). F.A.J.L. James, 'Thermodynamics and sources of solar heat, 1846–1862', *Brit. J. Hist. Sci.*, **15** (1982), 155–81 wrongly quotes 'planets' here.

he classified the sources available to man into mechanical effect derived from heat radiated from the sun (the principal source), from the motions of the earth, moon, and sun as well as from their mutual attractions, and from terrestrial (hot springs or the combustion of native inorganic substances, for example) or meteoric sources.[14]

Thomson went on to point out that the general notion of the sun as the principal source of motion on earth was not new. He drew special attention to Sir John Herschel's *Outlines of astronomy* (1849) where the following passage on solar influence occurred, a passage first published in Herschel's 1833 *Treatise*:

> The sun's rays are the ultimate source of almost every motion which takes place on the surface of the earth. By its heat are produced all winds, and those disturbances in the electric equilibrium of the atmosphere which gives rise to the phenomena [disturbances] of terrestrial magnetism. By their vivifying action vegetables are elaborated from inorganic matter, and become, in their turn, the support of animals and of man, and the sources of those great deposits of dynamical efficiency which are laid up for human use in our coal strata. By them the waters of the sea are made to circulate in vapour through the air, and irrigate the land, producing springs and rivers. By them are produced all disturbances of the chemical equilibrium of the elements of nature . . . Even the slow degradation of the solid constituents of the surface, in which its chief geological changes consist, and their diffusion among the waters of the ocean, are entirely due to the abrasion of wind and rain, and the alternate action of the seasons . . .[15]

Herschel's *Treatise on astronomy* may have been the inspiration for J.R. Mayer's 1848 paper 'Beiträge zur Dynamik des Himmels in populärer Darstellung' which made several explicit references to the elder and younger Herschel. This paper, eventually translated for the *Philosophical Magazine* in 1863, advanced a celestial dynamics very close to that developed by William Thomson in terms of a meteoric theory of the sun's heat, but remained unknown to him until about 1862.[16]

William Whewell, on the other hand, followed through Herschel's views in their economic, engineering context. In a chapter on the sources of labouring force published in his *Mechanics of engineering* (1841), Whewell emphasized the importance of solar heat. He believed that the various powers of water, wind, steam, and other gases, resolved themselves into the effects of heat, especially the sun's heat. Water provided not merely a reservoir of labouring force, but a source flowing from the springs or skies which in turn was constantly supplied by evaporation of moisture from the earth, and condensation or absorption of

[14] William Thomson, 'On the mechanical action of radiant heat or light'; 'On the power of animated creatures over matter'; 'On the sources available to man for the production of mechanical effect', *Phil. Mag.*, [series 4], **4** (1852), 256–60; MPP, **1**, 505–10. On Pouillet's estimate, see P.A. Kidwell, 'Prelude to solar energy: Pouillet, Herschel, Forbes and the solar constant', *Ann. Sci.*, **38** (1981), 457–76.

[15] J.F.W. Herschel, *Outlines of astronomy* (London, 1849), p. 237; *A treatise on astronomy* (London, 1833), p. 211.

[16] J.R. Mayer, 'On celestial dynamics', *Phil. Mag.*, [series 4], **25** (1863), 241–8, 387–409, 417–28.

the moisture thus evaporated: 'this evaporation is the result of heat. If clouds and springs were not constantly thus replenished the supply would fail; and water would cease to be, as it is, a perpetual and universal source of labouring force placed by nature at our disposal'. As with water, air too circulated by the heat of the sun through the production of differences of temperature in different parts of the atmosphere. The great cycles of circulation maintained the sources of labouring force: 'the natural powers of weather and wind, no less than the steam engine, depend on the continued operation of heat'.[17]

Thomson's own paper on sources of mechanical effect largely reformulated these ideas in the terms of the new thermodynamics. Having satisfied himself concerning sources on the globe, he took up the challenge explicitly laid down by Herschel in the section of his *Astronomy* immediately following the discussion of solar influence. This highly suggestive passage referred to friction, rather than chemical combustion, as a possible cause of the sun's heat:

> The great mystery [of the sun's heat] . . . is to conceive how so enormous a conflagration (if such it be) can be kept up. Every discovery in chemical science here leaves us completely at a loss, or rather, seems to remove farther the prospect of probable explanation. If conjecture might be hazarded, we should look rather to the known possibility of an indefinite generation of heat by friction, or to its excitement by the electrical discharge, than to any actual combustion of ponderable fuel, whether solid or gaseous, for the origin of solar radiation.[18]

Writing originally in 1833, long before the formulation of the energy principles, Herschel assumed that an indefinite generation of heat by friction was possible. Recognizing the fallacy of Herschel's assumption, Thomson sought elsewhere for a solution to the 'great mystery' of the nature of the sun's heat. By the spring of 1854 he had initiated an important debate with Stokes which led to his paper 'On the mechanical energies of the solar system' read to the Royal Society of Edinburgh on 17th April, 1854, and published with four important additions dated May to August, 1854.

In his opening paragraphs Thomson made clear that his interest in the source of the sun's heat derived from the principle of universal dissipation. Impressed with the vast development of energy by the sun compared to the energy of motion of all the planets, and by the 'almost inconceivably minute fraction of the Sun's heat and light reaching the earth [which] is the source of energy from which all the mechanical actions of organic life, and nearly every motion of inorganic nature at its surface, are derived', he stated:

> The energy, that of light and radiant heat, thus emitted [by the sun], is dissipated always more and more widely through endless space, and never has been, probably never can be,

[17] William Whewell, *The mechanics of engineering* (Cambridge, 1841), pp, 172–3. For a background to the 'cosmology of heat', see D.S.L. Cardwell, *From Watt to Clausius. The rise of thermodynamics in the early industrial age* (London, 1971), pp. 89–120.

[18] Herschel, *Outlines*, p. 238; *Treatise*, p. 212.

restored to the Sun, without acts as much beyond the scope of human intelligence as a creation or annihilation of energy, or of matter itself, would be. Hence the question arises, What is the source of mechanical energy, drawn upon by the Sun, in emitting heat, to be dissipated through space?[19]

He formulated the question in terms, first, of whether the source consisted of dynamical (kinetic) or potential energy, or both, and, second, of whether the source existed in the sun or in the surrounding matter – or as energy convertible into heat only by the mutual actions between the sun and the surrounding matter.

Thomson reasoned that if the source of the sun's heat were dynamical and *in* the sun, 'it can only be primitive heat'. In other words, 'the Sun is a heated body, losing heat'. If, however, the source were potential and likewise *in* the sun, 'it can only be energy of chemical forces ready to act'. In other words, 'the Sun is a great fire'. If, on the other hand, the source were external to the sun, that source could either be intrinsic to external matter, or be developed by mutual action between this matter and the sun itself. He had, then, three *general* hypotheses to consider: primitive heat, chemical activity, and meteors.[20] Much of the discussion with Stokes centred on an evaluation of the rival merits of these accounts, and on the appraisal of more specific versions of each in the light of available data regarding the sun's constitution and power.

In the first letter to Stokes on the issue, dated 2nd March, 1854, Thomson referred to the importance of investigating the solar spectrum because 'I think it will lead us to a qualitative analysis of the sun's atmosphere'. He urged Stokes to 'take up the whole subject of spectra, or solar & artif[l] lights, since you have already done so much on it', and he added in a characteristic way:

I am quite impatient to get another undoubted substance besides vapour of soda in the Sun's atmosphere. What you tell me looks very like as if there is potash too. I think copper . . . would be hopeful . . . Do try iron too. There *must* be a great deal of that about the Sun, seeing we have so many iron meteors falling in, & there must be immensely more such falling in to the Sun. I find the heat of combustion of a mass of iron w[d] be only about 1/34 000 of the heat derived from potential energy of gravitation, in approaching the Sun. Yet it w[d] take 2000 pounds of meteors per sq[uare] foot of the sun, falling annually, to account for his heat by gravitation alone.[21]

Immediately following this passage, Thomson dramatized the issue by comparing the emission of heat per square foot per second from the sun with the output from the largest man-made engine. On a rough estimate, based on his calculation from Pouillet's data that the heat emitted from a square foot of the

[19] William Thomson, 'On the mechanical energies of the solar system', (read 17th April, 1854), *Phil. Mag.*, [series 4], **8** (1854), 409–30 (with addenda); MPP, **2**, 1–25, on pp. 1–2.

[20] *Ibid.*, p. 3. See also J.D. Burchfield, *Lord Kelvin and the age of the earth* (London, 1975), pp. 23–5.

[21] William Thomson to G.G. Stokes, 2nd March, 1854, K64, Stokes correspondence, ULC. See also James, 'Thermodynamics and sources of solar heat', p. 162. The discussion of line spectra in relation to the solar spectrum had begun as early as 1851.

sun was 'equivalent in energy to about 7 or 8000 horse power', 'four sq[uare] feet of the Sun emit as much heat as the burning coals in the largest marine furnace – not altogether inconceivable, & quite in accordance with what we know of the excessive intensity of the Sun's heat'. Thus the marine furnace provided a concrete standard against which to consider the output of energy from that vast cosmic furnace, the sun.

As the passage quoted above shows, he was here contemplating two possible hypotheses: that the sun's heat derived from combustion of, for instance, iron, or from the transformation of gravitational potential energy into the energy of heat by the collision between material bodies and the sun. A week later, on 9th March, 1854, Thomson asked Stokes what he thought of the gravitation theory of solar heat, adding that he had been trying 'to make out what share of meteors the earth w^{d} take, if the sun gets enough to produce this heat, & I think it possibly reconcilable with what we have of falling stars &c'.[22] In his reply, delayed by illness, Stokes offered only a cautious evaluation of the theory:

As to the gravitation theory of the heat of the sun, I do not know of any objection to it . . . I have been in the habit of regarding the sun as an enormous body in a state of intense heat, emitting continually a portion of its original heat; as in fact 'growing dim with age', but at a rate not to be measured by the lives of us mortals. According to the gravitation theory the heat might be kept up, but there would be a progressive change of another kind, namely an increase of the sun's mass. This would of course produce a secular augmentation of the Earth's and retardation of the Moon's mean motion. I have not put the thing in numbers, but it would be requisite to attend to this point, to see if the augmentation of mass be sufficiently small.[23]

As Thomson subsequently made clear, he had 'got into the habit' of considering Stokes's 'not unfavourable' view of the gravitation theory as 'favourable' during the month following receipt of the above remarks. Thus 'it was during this time, principally during the last week of it when I was writing a paper w^{h} I had first intended merely to show some numerical relations as to heat of combustion, gravitation &c, that my own conviction became so strong as to the truth of the gravitation theory'. He had quite forgotten Stokes's mention of the primitive heat theory as the one he had been in the habit of holding. At the same time, Thomson also confessed that he had himself 'always inclined to the primitive heat theory till rather more than two years ago [late 1851 or early 1852] when I became convinced that neither as solid nor fluid the Sun (of his own finite dimensions) could have given out heat as he has done for 6000 years with no source but primitive heat to draw upon'.[24]

At the presentation of his paper, therefore, Thomson quickly dismissed the primitive heat hypothesis as 'quite untenable'. Primitive heat conducted out-

[22] William Thomson to G.G. Stokes, 9th March, 1854, K66, Stokes correspondence, ULC.

[23] G.G. Stokes to William Thomson, 28th March, 1854, S369, ULC.

[24] William Thomson to G.G. Stokes, 26th April, 1854, K69, Stokes correspondence, ULC. Compare above (note 3).

George Gabriel Stokes aged about fifty-six. 'I always consult my great authority, Stokes, whenever I get a chance,' declared Sir William in his 1884 Baltimore Lectures. Yet Stokes's cautious intellect and strong sense of establishment virtues stood in marked contrast to Thomson's impulsive, adventurous, and entrepreneurial spirit. [From *Nature*, **12** (1875), facing p. 201.]

wards to the sun's surface would allow the sun to become dark 'in two or three minutes, or days, or months, at his present rate of emission' – unless the sun 'be of matter inconceivably more conductive for heat, and less volatile, than any terrestrial meteoric matter we know'.[25]

Stokes reminded his Glasgow colleague on 24th April that his own view had been that the sun 'retained original heat, heat perhaps of condensation'. Furthermore, 'even if the numerical results of this hypothesis, made on the supposition of the sun's being solid, were not tenable . . . we may well suppose [instead] the sun to be in a state of fusion [i.e. in a fluid state], in which case if the surface cooled a little, and so became denser, the outer portions would descend, and be renewed from beneath'. Cautious as ever, Stokes concluded that he did not wish to argue in favour of this view; rather that 'I do not see it to be absurd on the face of it'.[26] Thomson, however, remained sceptical about the thermal capacity of such a model, adding on 4th May a footnote to his paper:

> it might be supposed that, as the Sun is no doubt a melted mass, the brightness of his surface is constantly refreshed by incandescent fluid rushing from below to take the place of matter falling upon the surface after becoming somewhat cooled and consequently denser – a process which might go on for many years without any sensible loss of brightness. If we consider, however, the whole annual emission at the present actual rate, we find, even if the Sun's thermal capacity were as great as that of an equal mass of water, that his mean temperature would be lowered by about 3° C in two years. We may, I think, safely conclude that primitive heat within the Sun is not a sufficient source for the emission which has continued without sensible (if any) abatement for 6000 years.[27]

Or, as he pointed out to Stokes in a letter of 26th April containing a full order-of-magnitude calculation, about 3° C per annum would mean '18000° C. in the 6000 years we know he has been giving out heat at something like the present rate'. This estimate made it 'certain (or as nearly certain as anything in w^{h} we consider possible properties of matter out of our reach) that the primitive heat theory cannot be true'.[28] With these remarks, Thomson upheld his verdict against the primitive heat theory in 1854 and turned to its alternatives.

The gravitation theory of the sun's heat

William Thomson's stated objects for his 1854 communication were 'to consider the relative capabilities of the second [the chemical] and third [the meteoric] hypothesis to account for the phenomena; to examine the relation of the gravitation theory to the meteoric theory in general; and to determine what form of the gravitation theory is required to explain solar heat consistently with other astronomical phenomena'. Lively enthusiasm for the gravitation theory

25 MPP, **2**, 3. 26 G.G. Stokes to William Thomson, 24th April, 1854, S370, ULC.

27 MPP, **2**, 3–4*n*.

28 William Thomson to G.G. Stokes, 26th April, 1854, K69, Stokes correspondence, ULC.

dominated his paper. That theory he attributed, in its first definite form, to J.J. Waterston's ideas put before the Hull meeting of the British Association in 1853. Waterston's theory – 'a remarkable speculation in cosmical dynamics' – proposed that 'solar heat is produced by the impact of meteors falling from extra-planetary space, and striking his [the sun's] surface with velocities which they have acquired by his attraction'. To this theory, Thomson gave the name 'gravitation theory of solar heat', 'itself included in the general meteoric theory'.[29]

With respect to the chemical hypothesis, Thomson discussed and rejected three of its possible forms. First, taking the estimate he had earlier offered to Stokes of the rate of emission of energy from each square foot of the sun's surface as equivalent to 7000 horse power, he concluded that 1500 pounds of coal per hour would be required to produce heat at the same rate. He asked, 'if all the fires of the whole Baltic fleet were heaped up and kept in full combustion, over one or two square yards of surface, and if the surface of a globe all round had every square yard so occupied, where could a sufficient supply of air come from to sustain the combustion?' On such a hypothesis, the products of combustion from the sun would either prevent a supply of fresh air from above, or a supply of new elements from below, and the fire would consequently be choked. He therefore affirmed that 'no such fire could be kept alight for more than a few minutes, by any conceivable adaptation of fuel and air'.[30]

Second, the sun could still be a burning mass if it were 'more analogous to burning gunpowder than to a fire burning in air'. Provided the solid mass contained within itself all the elements of combustion, and provided – unlike gunpowder – the products of combustion were permanently gaseous, this hypothesis might account for the sun's enormous heat. Estimating an upper limit for the heat of combustion of a pound of matter, and knowing the actual rate of emission of heat per square foot per second, Thomson calculated that about 0.7 pounds of matter would be lost per square foot per second from the sun – a quantity equivalent to a fifty-five mile thick layer over the sun's surface in a year. At the same rate, he concluded, 'a mass as large as the Sun is at present would burn away in 8000 years'. In order to burn at that rate in the past, the sun 'must have been of double diameter, of quadruple heating power, and of eight-fold mass, only 8000 years ago'. These dimensions, however, defied the evidence of history.

Finally, Thomson rejected a third version of the chemical hypothesis in which an influx of meteors provided the fuel. As in his letter of 2nd March, he argued that the quantity of heat thus obtained from a given mass would be insignificant

[29] MPP, **2**, 4.

[30] *Ibid.*, p. 10. The reference to the Baltic fleet is significant, for in 1854 a steam battlefleet went to war for the first time. This campaign also marked the opening of the Crimean War between Britain and Russia. See Basil Greenhill and Ann Giffard, *The British assault on Finland 1854–55: a forgotten naval war* (Teignmouth, 1988).

by comparison with the quantity obtained from the gravitational energy of the same mass.[31]

Justification for adopting the general meteoric theory not only flowed from a decisive rejection of the two rival hypotheses, but also derived from Thomson's claim that 'it is the only one of all conceivable causes which we know to exist from independent evidence'. The fact of meteors coming to earth 'proves the existence of such bodies moving in space', and even if these bodies move in elliptical or circular orbits about the sun, 'the effects of the resisting medium would gradually bring them in to strike his surface' with immensely greater relative velocities than those approaching the earth. The dynamical account of meteors or 'falling stars' with regard to the earth had been provided by Joule as early as 1847; and these arguments Thomson transferred to the sun, not by analogy alone, but by the key assumption that meteors abound in space:

> Now, Joule has shown what enormous quantities of heat must be generated from this relative motion in the case of meteors coming to the earth; and, by his explanation of 'falling stars' has made it all but certain that, in a vast majority of cases, this generation of heat is so intense as to raise the body in temperature gradually up to an intense white heat, and cause it ultimately to burst into sparks in the air (and burn if it be of metallic iron) before it reaches the surface. Such effects must be experienced to an enormously greater degree . . . by meteors falling to the Sun . . . Hence, it is certain that *some* heat and light radiating from the Sun is due to meteors.[32]

His methodology here conformed to Herschel's interpretation of Newton's well-known first rule of reasoning in philosophy. According to Herschel, *verae causae* were 'causes recognized as having a real existence in nature, and not being mere hypotheses or figments of the mind'. Thus, whenever 'any phenomenon presents itself for explanation, we naturally seek, in the first instance, to refer it to some one or other of those real causes which experience has shown to exist, and to be efficacious in producing similar phenomena'.[33]

Thomson argued further that, since the effects were so much greater for the sun than for the earth, and since 'the Sun, by his greater attraction, must draw in meteoric matter much more copiously with reference to equal areas of surface', the influence of meteors was far from being an insensible cause of solar heat. He modified, however, Waterston's view that a meteor 'either strikes the Sun or enters an atmosphere where the luminous and thermal excitation takes place, *without having previously experienced any sensible resistance*'. Waterston had supposed that the meteors fell from great distances into the sun in straight lines or in parabolic or hyperbolic paths – but to do so in sufficient quantities to generate all

[31] *Ibid.*, pp. 10–12; William Thomson to G.G. Stokes, 20th April, 1854, K68, Stokes correspondence, ULC.

[32] MPP, **2**, 4–5; J.P. Joule, 'On shooting-stars', *Phil. Mag.*, [series 3], **32** (1848), 349–51; *The Scientific papers of James Prescott Joule* (2 vols., London, 1887), **1**, pp. 286–8.

[33] J.F.W. Herschel, *A preliminary discourse on the study of natural philosophy* (London, 1830), pp. 144–9.

of the sun's heat would mean that the earth, in crossing their paths, would be 'at least struck much more copiously by meteors than we can believe it to be from what we observe'. Furthermore, 'the meteors we see appear to come generally in directions corresponding to motions which have been elliptic or circular'. Thus, the orbits of 'the meteors containing the stores of energy for future Sun light must be principally within the earth's orbit: and we actually see them there as the "Zodiacal Light", an illuminated shower or rather tornado of stones'.[34] Here he made specific reference to the discussion in Herschel's *Astronomy* of this phenomenon which Thomson now employed as clear and independent empirical evidence to justify not only the general meteoric theory but the specific gravitational version which he was developing.

Herschel himself had written in 1833 of the zodiacal light, observed on clear spring evenings after sunset as a cone or lens-shaped light, extending obliquely upwards from the position on the horizon where the sun disappeared, i.e. the sun's position in the zodiac. This phenomenon seemed 'to indicate the existence of some slight degree of nebulosity about the sun'. Although he admitted the phenomenon to be very faint and ill-defined in the northern latitudes – less so in tropical regions – Herschel concluded:

> It is manifestly in the nature of a thin lenticularly-formed atmosphere, surrounding the sun, and extending at least beyond the orbit of Mercury and even of Venus, and may be conjectured to be no other than the denser part of that medium, which, as we have [some] reason to believe resists the motion of comets; loaded, perhaps, with the actual materials of the tails of millions of those bodies, of which they have been stripped in their successive perihelion passages, and which may be slowly subsiding into the sun.[35]

In 1849, Herschel added several significant remarks to this account. To begin with, he claimed that the zodiacal light could not be a gaseous atmosphere of the sun. The existence of a gaseous envelope of such dimensions and ellipticity, propagating pressure from part to part and subject to mutual friction in its strata – and so rotating with the central body – was incompatible with dynamical laws, presumably because friction would stop the rotation of the sun. Supposing that the particles of the zodiacal light possessed inertia, each particle must instead be equivalent to a separate, if minute, planet of the sun. Furthermore, although the total mass was for Herschel 'almost nothing compared to that of the sun' – making mutual perturbation irrelevant – nevertheless mutual collision 'may operate in the course of indefinite ages to effect a subsidence of at least some portion of it [the zodiacal light] into the body of the sun or those of the planets'.[36]

Again in his *Outlines of astronomy*, Herschel related the zodiacal light to

[34] MPP, **2**, 5–7.

[35] Herschel, *Treatise*, pp. 407–8; *Outlines*, p. 616. The qualifying 'some' appears in the *Outlines* version. No doubt Thomson first learned of the zodiacal light from J.P. Nichol, for whom it was a remnant of the nebular origin of the solar system; *Views of the architecture of the heavens: in a series of letters to a lady* (Edinburgh, 1838), pp. 148–51.

[36] Herschel, *Outlines*, p. 616.

meteors. He argued that there was nothing to prevent at least some of the particles having a tangible size and being at great distances from one another. Thus, 'compared with planets visible in our most powerful telescopes, rocks and stony masses of great size and weight would be but as the impalpable dust which a sunbeam renders visible . . .'. Moreover, 'it is a fact, established by the most indisputable evidence, that stony masses and lumps of iron do occasionally, and indeed by no means infrequently, fall upon the earth from the higher regions of our atmosphere (where it is obviously impossible they can have been generated), and that they have done so from the earliest times of history'. He continued:

> The heat which they possess when fallen, the igneous phaenomena which accompany them, their explosion on arriving within the denser regions of atmosphere, &c., are all sufficiently accounted for on physical principles, by the condensation of the air before them in consequence of their enormous velocity, and by the relations of air in a highly attenuated state to heat.[37]

Herschel also discussed 'shooting stars' in the same sections of his *Outlines*, remarking that these phenomena 'may not unreasonably be presumed to be bodies extraneous to our planet, which only became visible when in the act of grazing the confines of our atmosphere'. In particular, he believed that the 'circumstances in the history of shooting stars . . . very strongly corroborate the idea of their extraneous or *cosmical* origin, and their circulation round the sun in definite orbits'. Thus, 'on several occasions they have been observed to appear in . . . astonishing numbers, so as to convey the idea of rockets, or of snow-flakes falling, and brilliantly illuminating the whole heavens for hours together' on an annual period of recurrence. Such annual periodicity led directly 'to the conclusion that at that place [in its orbit] the earth incurs a liability to *frequent* encounters or concurrences with a stream of meteors in their progress of circulation round the sun'.[38]

Herschel's highly suggestive discussion of the zodiacal light made up of solid bodies which clearly had the same *cosmical* status as meteors and shooting stars thus provided an authoritative basis for William Thomson's employment of the zodiacal light as the probable source of the sun's heat. According to Thomson's version, which assumed a resisting medium throughout space, continuous with the atmospheres of sun and earth – the air–ether (ch. 12) – these meteoric bodies would be drawn into the sun in a very gradual spiral although much more rapidly than the planets in virtue of relatively minute inertia. As they reached denser, more resisting regions of the sun's atmosphere, they would ignite: 'after a few seconds of time all the dynamical energy the body had . . . is converted into heat and radiated off; and the mass itself settles incorporated in the Sun'.[39]

Thomson supported the general meteoric hypothesis with the claim that several cosmological phenomena could be accounted for on that view. Most important for his purposes was its adequacy for explaining the present rate of

[37] *Ibid.*, pp. 617–18. [38] *Ibid.*, pp. 618–19. [39] MPP, **2**, 8.

emission of energy from the sun. Again employing Pouillet's data, he estimated that on Waterston's view the quantity of meteoric matter required to strike the sun per square foot per second – or about a pound every five hours – would cover the sun's surface to a depth of thirty feet in a year, while on his own view, the whole surface would be covered to a depth of sixty feet in a year. Either way, however, the growth in apparent diameter over a historical period of 4000 years would 'scarcely be perceived by the most refined of modern observations'. Unlike the other hypotheses, then, the meteoric theory left 'for the speculations of geologists on ancient natural history a wide enough range of time with a sun not sensibly less [in size] than our present luminary'; and, by way of further dramatizing the adequacy of the meteoric theory, he stated that at this rate it would take two million years for the sun to grow in reality by as much as he appears to grow from June to December by the variation of the earth's distance – an amount 'quite imperceptible to ordinary observation'.[40]

Apart from accounting for the sun's heat, Thomson suggested that the meteoric theory could explain both past changes in the earth's climate and the appearance and disappearance of bright stars. Meteors striking the earth might have originally maintained the globe in a melted state. At a more recent time, the earth would have cooled enough for vegetation, but with a climate maintained at a much higher temperature than at present 'by the heat of meteors falling through its atmosphere' and by the sun itself having 'a more copious meteoric supply'.[41] It is likely that these remarks on climate derived from a discussion with Stokes. In a letter of 8th May, 1854, Stokes observed that Thomson's theory 'seems to account more simply than any other I know for the fluctuations of temperature of the Earth in past ages which geologists consider we have evidence of'. Stokes also noted that some geological changes which geologists assume 'seem almost too great to be produced by mere upheaval & sinking [of continents] and merely altering the state of the Earth as to configuration of land and water, current &c' – implying that Thomson's view of greater meteoric activity in the past would account more satisfactorily for major geological changes than any uniformitarian view (namely, that the present agents, acting at present rates, were sufficient).[42]

The appearance of bright stars could similarly be very simply explained by the meteoric theory. As Thomson told Stokes: 'A star suddenly appearing is merely a dark body perhaps as large as the sun moving into a cloud of [meteoric] dust in space. In a few seconds its surface might become incandescent by the impacts'.[43] On moving out of the cloud, the star might almost as suddenly become dark again.

In the course of this seminal paper, and his dialogue with Stokes, Thomson had convinced himself, not only of the adequacy of the meteoric theory, but of its near certainty. He summed up:

40 *Ibid.* 41 *Ibid.*, pp. 8–9.
42 G.G. Stokes to William Thomson, 8th May, 1854, S371, ULC.
43 William Thomson to G.G. Stokes, 20th April, 1854, K68, Stokes correspondence, ULC.

In conclusion, then, the source of energy from which solar heat is derived is undoubtedly meteoric . . . The principal source . . . is in bodies circulating round the Sun at present inside the earth's orbit, and probably seen in the sunlight by us and called 'the Zodiacal Light'. The store of energy for future sunlight is at present partly dynamical, that of the motions of these bodies round the Sun; and partly potential, that of their gravitation towards the Sun. This latter is gradually being spent, half against the resisting medium, and half in causing a continuous increase of the former. Each meteor thus goes on moving faster and faster, and getting nearer and nearer the centre, until some time, very suddenly, it gets too much entangled in the solar atmosphere, as to begin to lose velocity. In a few seconds more, it is at rest on the Sun's surface, and the energy given up is vibrated in a minute or two across the district where it was gathered during so many ages, ultimately to penetrate as light the remotest regions of space.[44]

The continuing dialogue with Stokes, however, and his own restless intellect, produced no less than four important additions to the 1854 paper between May and August – additions which further refined and developed his views on the nature of the sun's heat 'consistently with other astronomical phenomena'. First, he evaluated the effect on the earth's motion of matter reaching the sun from extra-planetary space rather than from within the earth's orbit. Second, he suggested that the sun's heat, on further analysis, was due not to solids striking the sun but to 'friction in an atmosphere of evaporated meteors'. Third, he linked those refined views on the meteoric theory to John Herschel's account of the sun's spots. And, fourth, he published his first estimate of the age of the sun.

In his earliest response to the gravitation theory, on 28th May, Stokes had raised the question of the effect of an increase of the sun's mass on the earth's motion. Thomson, anxious to anticipate such objections from physical astronomy, estimated in his original paper that in forty-seven and one-half years a mass of meteoric matter equal to the earth's mass would be drawn into the sun. To account for around 5000 years heat, then, a mass of meteoric matter one-hundred times that of the earth would be required either from the zodiacal light or from meteors revolving inside the orbit of Mercury. Large though this amount of matter was, he argued, it would nevertheless present no difficulties – even for physical astronomy. Thus, the matter 'is drawn from a [region of] space [inside the planetary orbits] where it acts on the planets with nearly the same forces as when incorporated in the Sun'. Thomson's form of the gravitation theory required 'not more [meteoric matter] than it is perfectly possible does fall into the Sun' and he concluded: 'Hence I think we may regard the adequacy of the meteoric matter theory to be fully established'.[45]

Discussing the same issue with Stokes in the letter of 26th April, Thomson remarked that he had 'not yet considered what amount of effect such an add^{n} to the Sun's mass if coming from extra-planetary space would produce on the planetary motions but I shall try directly'. So directly did he try

[44] MPP, **2**, 13. [45] *Ibid.*, pp. 9–10.

that in the very same letter he calculated that, given Herschel's estimate for the mass of the sun relative to the mass of the earth, and given the amount of meteoric matter required by the extra-planetary theory, the acceleration of the earth's motion about the sun would be 5/24 of a year in a 5000 year period. If so, Thomson concluded 'the extra-planetary meteoric theory must be false, & we may conclude that the sun's heat is generated by the falling in of meteors which for ages have been far inside the earth's orbit'.[46] He published this estimate, in an expanded form, as the first addition, dated 9th May, 1854, to his paper.

A second addition, entitled 'Friction between vortices of meteoric vapour and the sun's atmosphere the immediate cause of solar heat' also derived in large part from his dialogue with Stokes. Thomson explained that, since the meteors had been shown to be 'for thousands of years within the Earth's orbit before falling into the Sun', they must move in elliptical or circular orbits. He argued further that the atmosphere or resisting medium in space was 'carried in a vortex round the Sun by the meteors and other planets' or, as he remarked to Stokes on 20th April: 'I am much disposed to go back to the vortices, differing only from Descartes in [their] being dragged round by the planets instead of dragging them round'. Thomson then suggested that the great body of meteors circulating round the sun, and carrying the resisting medium along with them, may be moving through the medium with very small relative velocities, with the result that the effects of the resistance would be extremely gradual in bringing meteors into the sun. Small meteors revolving close to the sun would acquire the temperature of that part of space and would thus be converted into vapour. Consequently, 'each meteor, when volatilized, will contribute the actual energy it had before evaporation to a vortex of revolving vapours, approaching the sun spirally to supply the place of the inner parts which, from moving with enormously greater velocities than the parts of the Sun's surface near them, first lose motion by intense resistance, and then, from want of centrifugal force, fall in to the Sun . . .'. Thomson thus concluded that the sun's heat was caused 'not by solids striking him, or darting through his atmosphere, but by friction in an atmosphere of evaporated meteors . . .'.[47]

Though Thomson's conception of the nature of the sun's heat did not depend upon his identification of the sun's atmosphere with the luminiferous ether, nevertheless his model of the sun became here a central component in his grand scheme of etherial vortices. That scheme encompassed electromagnetism, heat, light, and vortex atoms. Expressed practically in James Thomson's vortex turbine, it transported the engineer's concepts of work and waste to macrocosmic and microcosmic domains (ch. 12). In fact, Thomson's meteoric theory of the sun's heat constituted nothing other than a vast reproduction of the vortex

[46] William Thomson to G.G. Stokes, 26th April and 23rd May, 1854, K69 and K70, Stokes correspondence, ULC; MPP, **2**, 16–19.

[47] MPP, **2**, 19–21; William Thomson to G.G. Stokes, 20th April, 1854, K68, Stokes correspondence, ULC.

turbine, in which the meteors delivered their gravitational energy to the sun through friction with its atmosphere rather than by impelling rigid blades. They therefore produced heat and light directly for distribution through interplanetary space via vibrations in the continuation of the solar atmosphere, or ether. He elaborated the requirements for such distribution in his 1854 'Note on the possible density of the luminiferous medium and on the mechanical value of a cubic mile of sunlight', basing his calculation on the assumption that 'there must be a medium forming continuous material communication throughout space to the remotest visible body'.[48]

In the same context, he also discussed the fate of the planet Mercury. The issue had arisen in his letter to Stokes of 26th April in which he had speculated:

Mercury falling in [to the sun] w[d] give 10 years heat. If he falls in as a whole, how will the luminiferous ether stand it? Of course the whole surface of the earth would be instantly scorched. Mercury will probably fall in before the earth has varied very much from its present orbit; but the solidity of his mass would not so far as I can see be a guarantee that he w[d] fall in as a whole. It is just possible that he might burst or evaporate into zodiacal light upon getting to the region of intense resistance (ignition) and not cause any increase in the sun's heat, but I think it more probable that he would make a great increase. I believe that if the solar system is permitted to go on long enough fulfilling the laws of matter which we know, the surface of the earth will in a few seconds or minutes be scorched & melted by Mercury falling in to the Sun. Till very lately I [was in doubt. *del*] saw no decided mechanical reason for anticipating that the Earth would first cease to be habitable for man by the Sun becoming cold or by itself becoming scorched by getting too near the Sun.[49]

Stokes, for his part, remarked that 'one is rather startled at first to give the Sun such an appetite as to devour 10 Mercurys in a century. Still, we do not know that he is more abstemious'. Thomson in fact reversed his very pessimistic speculation here for publication in the second addition to his 1854 paper, though the reasons for the changes are obscure. Thus, he asserted, in falling into the sun, Mercury 'will, in all probability, be dissipated in vapour long before it reaches the region of intense resistance, instead of . . . falling in solid, and in a very short time . . . generating three years' heat, to be radiated off in a flash which would certainly scorch one half of the earth's surface, or perhaps the whole, as we do not know that such an extensive disturbance of the luminiferous medium would be confined by the law of rectilineal propagation'.[50]

In the third addition to his 1854 paper, he argued that the vortex theory would account for the nearly uniform radiation from different parts of the sun's surface

48 William Thomson, 'Note on the possible density of the luminiferous medium and on the mechanical value of a cubic mile of sunlight' (read 1st May, 1854), *Trans. Royal Soc. Edinburgh*, **21** (1857), 57–61; MPP, **2**, 28–33, on p. 28.

49 William Thomson to G.G. Stokes, 26th April, 1854, K69, Stokes correspondence, ULC.

50 G.G. Stokes to William Thomson, 8th May, 1854, S371, ULC; William Thomson to G.G. Stokes, 31st May, 1854, K71, Stokes correspondence, ULC; MPP, **2**, 20.

– a uniformity not explained on the hypotheses by which solid bodies supplied the sun's heat. Solid bodies, revolving about the sun in planes nearly coinciding with its equator and striking there, would produce more radiation from that region than elsewhere. In general, the vortex theory implied that the temperature of the surface of the sun would be of uniform distribution – near to that of equilibrium between the vapours and the solid or liquid body of the sun into which they were distilling. Deviations from uniformity observed in the sun were 'probably due to eddies which must be continually produced throughout the atmosphere of intense resistance between his surface and the great vortex of meteoric vapour'.[51]

In order to add still more weight to his meteoric theory, Thomson then correlated his views with John Herschel's intricate interpretation of the spots and associated phenomena observable on the sun's disc.[52] While the details will not concern us, it is of interest that Thomson carried his meteoric theory into a very speculative mechanical theory of solar phenomena, indicating how much less cautious he had become compared to the 1840s and suggesting that at least since his father's death in 1849 the wilder tendencies of William's imaginative faculties, whether in his scientific or personal life, could no longer be restrained.

The fourth and final addition to his paper, entitled 'On the age of the sun', represented Thomson's earliest numerical estimate of the age of the sun. Conservation of angular momentum provided the basis of the calculation whereby the rotatory motion of meteors was transferred to the sun and its atmosphere. Given earlier estimates of the required quantity of meteoric matter falling in annually, and knowing the present moment of the sun's rotatory motion, he only needed to work out how long the present rate of fall of meteoric matter would take to produce the sun's present rotatory motion. Thus, if we suppose a planet orbiting the sun at its surface, and equate the angular momentum of the sun to that of the planet, we find (using Kepler's third law to calculate the planet's angular velocity) the planet's mass. Such an imaginary planet would be equivalent to the quantity of meteoric matter which would fall in to the sun in 25 000 years at the present rate, assuming that the meteors were revolving in the plane of the sun's equator. On the vortex hypothesis, however, he modified the estimate to 32 000 years.

Having concluded that it was 'not improbable that the Earth has been efficiently illuminated by the Sun alone for not many times more or less than 32 000 years', Thomson suggested that a future problem would be to determine the mass of the zodiacal light by examining the perturbations it produced in the motions of the visible planets. Since such perturbations had not been noticed, he estimated that at most the zodiacal light could amount to $\frac{1}{50}$ of the sun's mass,

[51] MPP, **2**, 21–2.

[52] William Thomson to G.G. Stokes, 31st May, 1854, K71, Stokes correspondence, ULC; MPP, **2**, 22; Herschel, *Outlines*, pp. 227–9.

and since in 3000 years meteors amounting to $\frac{1}{5000}$ of the sun's mass were falling in at the present rate, Thomson concluded 'that Sunlight cannot last as at present for 300 000 years'.[53]

In several important respects, William Thomson's account of the sun's heat in accordance with a gravitational theory bears a close resemblance to that of J.R. Mayer enunciated independently six years earlier. Mayer rejected the chemical hypotheses as inadequate for more than 4500 years of light and heat at its present rate and, like Thomson, took up John Herschel's considerations of the zodiacal light to locate the principal source of the sun's heat in meteoric matter brought to the sun by the effects of a resisting medium. Like Thomson, he also employed Pouillet's data to show the adequacy of a gravitational meteoric theory to account for the vast heat supplied by the sun but, unlike Thomson, he supposed the constancy of the sun's mass on the supposition that the fall of meteoric masses was balanced by a decrease of mass due to radiation. He also assumed a replenishment of the zodiacal light from outside the solar system, making any discussion of the finiteness of the sun and solar system meaningless. Mayer's version nevertheless came much closer than did Waterston's to Thomson's more refined gravitational theory, with the result that, in later years, when Thomson became acquainted with Mayer's work, he acknowledged Mayer's priority.[54]

William Thomson's vision of the solar system integrated the energy principles, the luminiferous ether, and a vortex conception into a unified cosmology. A cartesian-like vortex, but with the planets carrying the ether, meant that the ether offered very little resistance to the planets even though its high density at first sight posed a major conceptual problem. Behind the fragmentary insights lies a profound unity, a unity manifested at the level of the vortex atom (ch. 12). From the immensity of the solar system to the minuteness of the vortex atom, Thomson's search for a coherent synthesis is everywhere apparent.

The origins of the sun and solar system

During the summer of 1854, William Thomson presented a more general account of his cosmological system to the meeting of the British Association, held in Liverpool. 'On mechanical antecedents of motion, heat, and light' was essentially a cosmology of energy, which traced the various conversions and transformations of energy in the solar system, and illustrated them by specific examples. Animal heat and work, the explosion of gunpowder, the work and

53 MPP, **2**, 23–5; Burchfield, *Age of the earth*, pp. 24–5. David Wilson, *William Thomson. Lord Kelvin. His way of teaching natural philosophy* (Glasgow, 1910), pp. 15–16, suggests that Thomson's estimate of endings may have been part of an attack on contemporary religious fanatics who proclaimed that the end of the world was at hand. Such a suggestion is entirely consistent with Thomson's latitudinarian position.

54 Mayer, 'On celestial dynamics', pp. 241–8, 387–402; PL, **1**, 364; James, 'Thermodynamics and solar heat', pp. 157–61.

heat derived from wood or coal, and the growth of plants all derived from the potential energy of chemical affinity. Such a store of energy was, however, in turn due in general to the sun – the great source 'from which the mechanical energy of all the motions of heat of living creatures, and all the motion, heat, and light derived from fires and artificial flames is supplied' – and to the motions and forces among bodies of the solar system. From these major sources of energy derived both natural motions and those 'called into existence through man's directing powers' (ch. 18).[55]

Thomson then summarized his latest thinking on the nature of the sun's heat. This heat was 'probably due to friction in the atmosphere between his surface and a vortex of vapours, fed externally by the evaporation of small planets, in a region of very high temperature round the sun, which they reach by gradual spiral paths, and falling in torrents of meteoric rain, down from the luminous atmosphere of intense resistance, to the sun's surface'. Further inquiry, however, raised the question of the source from which all the planets, large and small, derived their motions. This question, he stated, was one 'to the answering of which mechanical reasoning may legitimately be applied'. In other words, the changing *arrangements* of matter and energy could be treated mechanically (or thermodynamically) between their origins and endings, although those end-points themselves, for example the *creation* of matter or energy *ex nihilo*, remained quite beyond mechanical or any other conceivable kind of scientific reasoning. He was thus careful to stress that all conclusions about origins and ends were 'subject to limitations, as we do not know at what moment a creation of matter or energy may have given a beginning, beyond which, mechanical speculations cannot lead us'. Having recognized this fundamental contingency upon the divine will we may, he argued, legitimately push our speculations 'into endless futurity, and we can be stopped by no barrier of past time' except by finding 'at some finite epoch a state of matter derivable from no antecedent by natural laws' – what would, for Thomson, constitute a *beginning* to the distribution or arrangement of energy and matter in the system (ch. 6).[56]

Thomson explained to his audience that, although such initial states were conceivable in abstract distributions of heat, 'yet we have no indications whatever of natural instances of it, and in the present state of science we may look for mechanical antecedents to every natural state of matter which we either know or can conceive at any past epoch however remote'. By tracing into the past the motions at present observed, using known laws of motion and heat, he suggested that all the bodies now constituting our solar system must have been 'at infinitely greater distances from one another in space than they are now'. An original system of bodies at great distances from one another 'may, by their mutual gravitations, and by the resistance their motions must experience in the

[55] William Thomson, 'On the mechanical antecedents of motion, heat and light', *BAAS Report*, **24** (1854), 59–63; *Comptes rendus*, **40** (1855), 1197–202; MPP, **2**, 34–40, on pp. 34–7. See also Burchfield, *Age of the earth*, 25–6, for an outline of this address. [56] MPP, **2**, 37–8.

gaseous atmosphere [which is] evaporated from them by the heat of their collisions after a vast period of time, come into a state of motion, heat, and light, analogous to the present condition of our solar system and the stars'. He explained, moreover, the origin of rotatory motion 'by showing that different systems starting from rest will influence one another so as to acquire contrary rotatory motions, without any aggregate of rotatory momentum being acquired by the whole'. Thomson's bold conclusion on origins, therefore, was that '*the potential energy of gravitation may be in reality the ultimate created antecedent of all the motion, heat, and light at present in the universe*' while on ends he reiterated his 1852 view that 'the end of this world as habitation for man, or for any living creature or plant at present existing in it, is mechanically inevitable'.[57]

Thomson's striking remark on gravity as the originally created form of all energy corresponds directly to his ether speculations four years later in 1858, on the 'universal plenum' with its primitive 'contractility' (ch. 12). He supposed that it had been 'created with such a distribution as to density (and possibly also with motion though the potential energy of contractility might be the created origin of all motion) that the present and all past phases of dead matter may have followed from it in accordance with constant mechanical laws'. In 1862 he published a similar speculation on gravity as the cause of cohesion, imagining continuous matter to be so distributed as to form a cellular structure, with high mass density only in its walls (ch. 11).[58]

The continuum version of Thomson's gravitational hypothesis evidently constituted a reformulation of J.P. Nichol's nebular hypothesis which had also been designed to account for progression 'in obedience to *known mechanical laws*', beginning from an originally nebular form of matter distributed throughout space and proceeding to the formation of the solar system with its fit habitation for man.[59] The two versions agreed also on a fiery end for this habitation when the solar system collapsed into the sun.

Thomson stressed, however, that the 'ordinarily stated' nebular theory (including Nichol's) assumed a primitive *gaseous* state, implying – to Thomson, but not to Nichol – high kinetic energy, the reverse of his own supposition of purely potential energy. His theory showed 'evaporation as a necessary consequence of heat generated by collisions and friction' and held the past and present tendency of matter to be 'the conglomeration of solids and liquids, accompanied by a gradual increase of the quantity of gaseous fluid occupying space'.[60] In this context he drew particular attention to Helmholtz's 1854 lecture 'On the interaction of natural forces' delivered at Königsberg on 7th February. A few

[57] *Ibid.*, p. 40.

[58] William Thomson, 'Journal and research notebook, 1858–1863', NB35, ULC; Ole Knudsen, 'From Lord Kelvin's notebook: ether speculations', *Centaurus*, **16** (1971), 41–53, on p. 47. William Thomson, 'Note on gravity and cohesion' (read 21st April, 1862), *Proc. Royal Soc. Edinburgh*, **4** (1866), 604–6; PL, **1**, pp. 59–63.

[59] Nichol, *Architecture of the heavens*, p. 146.

[60] MPP, **2**, 38.

months previously Helmholtz had attended the Hull meeting of the British Association, where he may have heard Waterston's paper on the gravitational meteoric theory, and now advanced a gravitational theory of his own. Thomson responded favourably to Helmholtz's estimate that 'if the particles at present constituting the sun's mass have been drawn together by mutual gravitation from a state of infinite diffusion as supposed in the nebular theory' – though from a static state rather than from a gaseous one – the whole heat generated would amount to twenty-eight million thermal units centigrade per pound of the sun's mass. Helmholtz's estimate, Thomson observed, would be essentially unaltered whether 'we assume, as the antecedent condition of the solar mass, a state of infinite diffusion or a state of aggregation in solid masses of any dimensions small compared with his present dimensions and separated from one another at comparatively great distances; provided always there has been no relative motion among them except what is generated by mutual gravitation'. This quantity represented about twenty million times as much heat as was at present radiated off in one year.[61]

Thomson argued, however, that, while Helmholtz's idea would account for the sun's origin in a manner consistent with his own theory, the heat of original conglomeration of the sun's mass could not be the source of the sun's *present* heat, for it would have been 'nearly all radiated off immediately on being generated'. Apparently he supposed that the process of conglomeration, once begun by collisions, heating, and evaporation of retarding atmospheres, would proceed very rapidly. He maintained his own meteoric theory with passionate conviction:

> That the present solar radiation is supplied chiefly from a store of heat contained in the mass, whether created there or generated mechanically by the impact of meteors which have fallen in during remote periods of past time, appears very improbable. On the contrary, there must in all probability be some agency continually supplying heat to compensate the loss constantly experienced by radiation from the sun; and that agency . . . can be no other than the mechanical action of masses coming from a state of very rapid motion round the sun to rest on his surface.[62]

By 1871, Thomson had sharpened the distinction between the 'old' nebular theory and his own (and Helmholtz's) theory. Thus, 'the old nebular hypothesis supposes the solar system . . . to have originated in the condensation of fiery nebulous matter. This hypothesis was invented before the discovery of thermodynamics, or the nebulae would not have been supposed to be fiery'. Helmholtz had shown in 1854 'that it was not necessary to suppose the nebulous matter to have been originally fiery, but that mutual gravitation between its parts may have generated the heat to which the present high temperature of the Sun is due'. Even as early as 1855, however, William Thomson's departure from an 'ortho-

[61] *Ibid.*, pp. 38–9. [62] *Ibid.*, pp. 39–40.

dox' nebular theory – in the tradition of Laplace – had been recognized in France.[63]

The Abbé François Moigno's appraisal of Thomson's 'cosmogony' in his journal *Cosmos* began with a comment on the 'popular' young British savant, the 'spoilt child of English science', who had recently addressed the *Académie des sciences* under the title 'Mechanical antecedents of motion, heat, and light'. Moigno's account, nevertheless, dramatized Thomson's new departure as 'entirely opposed to the traditions of the Laplacian school', and 'very much outside the reigning French doctrines'. He contrasted it with Jacques Babinet's cosmogony, essentially in the Laplacian tradition, which had been pubished in *Revue des deux mondes* at about the same time as Thomson had addressed the *Académie*.[64]

For Babinet, the earth had originated in the blazing fiery atmosphere of the sun. A belt or ring of fire had separated from the solar nebula and, condensing while cooling, had formed the earth and moon. For Nichol too the earth and other planets began as nebular rings, supposed to be separated by centrifugal action from the main nebula. In contrast, Thomson held that in the beginning the earth had been at an infinitely greater distance from the sun than it was at the present time. As Babinet also argued, Laplace had established – contrary to the implication drawn from Newton's non-conservative system, that the Creator had not been provident enough to give the universe a stable structure of its own – that the solar system was indeed subject to laws so wise that its permanence was not at risk. Moigno reminded his readers, however, that Babinet had not been affirming the indefinite existence of the world, contrary to revelation, but had rather suggested that there was no reason for the human race to fear, for a very long time, the end of the world.[65] Thomson, on the other hand, went further even than Newton, positively asserting that the end of the world was mechanically inevitable. The two views differed as 'day and night'.

The eventual end of the earth and solar system, Moigno argued, would be by cold in the Laplacian theory, as a result of continued radiation of its original heat, while for Thomson the end would be by fire, that is, by gravitational collapse into the sun. Moigno saw the return of the most advanced science (Thomson's) to the solution of a great inspiration, the revelation of Peter: 'the heavens shall pass away with a great noise, and the elements shall melt with fervent heat, the

[63] *Ibid.*, pp. 25–7. For a discussion of recent historical analyses of nebular hypotheses and the stabilist nature of Laplace's ideas see J.H. Brooke, 'Nebular contraction and the expansion of naturalism', an essay review of R.L. Numbers, *Creation by natural law: Laplace's nebular hypothesis in American thought* (Seattle and London, 1977), in *Brit. J. Hist. Sci.*, **12** (1979), 200–11. For the contrast with Nichol (discussed below) see Simon Schaffer, 'The nebular hypothesis and the science of progress', in J.R. Moore (ed.), *The humanity of evolution*, (Cambridge, forthcoming).

[64] F.N.M. Moigno, 'Cosmogenie', *Cosmos*, **6** (1855), 659–64; Jacques Babinet, 'Astronomie cosmogonique. La terre avant les époques géologique', *Revue des Deux Mondes*, **10** (1855), 702–26. For a whiggish discussion of Laplace, see S.L. Jaki, 'The five forms of Laplace's cosmogony', *Am. J. Phys.*, **44** (1976), 4–11.

[65] Moigno, 'Cosmogonie', pp. 661–2.

earth also and the works that are therein shall be burned up'. This pessimistic future was, however, happily tempered by the consoling assurance that 'we, according to His promise, look for new heavens and a new earth . . .'.[66]

In Thomson's theory, then, Moigno recognized a fundamental break with Laplacian tradition: the replacement of harmonious stability with progressive decay as the grand metaphor of the economy of nature. That break had of course occurred much earlier in Nichol's nebular hypothesis, which responded specifically to new conditions apparently set by Encke's comet (ch. 4). Generally, however, Nichol's hypothesis expressed an intellectual mood shared in the 1830s and after by a wide variety of writers on all aspects of the natural and political economy. Progression appeared on every side. Thomson's natural philosophy had begun in that context with the dilemma posed by directionality in physical processes, and with Nichol's teachings as a primary source of inspiration. In his own nebular theory he had now realized Nichol's visionary prospectus by transforming into a cosmogony the insights he had won originally for machines. Such, then, was the new world view, with implications far beyond the cosmological and geological, which William Thomson had presented to both the British Association and the *Académie des sciences* scarcely half a decade before Charles Darwin's *Origin of species*, which would inspire Thomson to develop further his views on the origins of the solar system, and in particular on the ages of the earth and sun.

[66] *Ibid.*, pp. 662–4.

15

The age of the sun controversies

> As to the sun, we can now go both backwards and forwards in his history, upon the principles of Newton and Joule. A large proportion of British popular geologists of the present day have been longer contented than other scientific men, to look upon the sun as Fontenelle's roses looked upon their gardener. 'Our gardener', say they, 'is a very old man: within the memory of roses he is the same as he has always been; it is impossible he can ever die, or be other than he is'. *Sir William Thomson, 1868*[1]

The *prima facie* reason for William Thomson's publication in 1862 of articles 'On the age of the sun's heat' and 'On the secular cooling of the earth' was to provide a critical response to the immense time scales apparently demanded by Charles Darwin's 1859 *The origin of species by means of natural selection.* As Thomson himself posed the question at the heart of his *Macmillan's Magazine* article on the sun's heat: 'What then are we to think of such geological estimates as 300 million years for the "denudation of the Weald?"'. Thomson was referring directly here to Darwin's discussion of the wide valley, known as the Weald of Kent, lying between the North and South Downs of southeast England, a valley which Darwin had supposed to result from the encroachment of the sea upon the line of chalk cliffs at a rate of one inch in a century. Darwin's estimate of geological time, then, provided a specific target for Thomson's attacks.[2]

Yet Thomson did not merely aim to correct false assumptions in Darwin's text. His critique ultimately sought to undermine both Darwinian natural selection and the Lyellian perspective on geological time upon which Darwin's theory had been founded. Underlying his attack, therefore, we shall find a continuing commitment to progression and design, shared with his fellow critics of Darwin, viz., Sedgwick, Hopkins, and John Phillips (1800–74). As Joule expressed Thomson's intentions in a letter of 13th May, 1861:

[1] Sir William Thomson, Discussion of Professor Grant's paper 'On the physical constitution of the sun', *Proc. Phil. Soc. Glasgow*, **7** (1869–71), 111–12.

[2] William Thomson, 'On the age of the sun's heat', *Macmillan's Mag.*, **5** (1862), 288–93; PL, **1**, 349–68, on p. 361; Charles Darwin, *The origin of species by means of natural selection* (Pelican reprint of first edition, 1968), pp. 296–7; J.D. Burchfield, *Lord Kelvin and the age of the earth* (London, 1975), pp. 70–2.

I am glad you feel disposed to expose some of the rubbish which has been thrust on the public lately. Not that Darwin is so much to blame because I believe he had no intention of publishing any finished theory but rather to indicate difficulties to be solved . . . It appears that nowadays the public care for nothing unless it be of a startling nature. Nothing pleases them more than parsons who preach against the efficacy of prayer and philosophers who find a link between mankind and the monkey or gorilla – certainly a most pleasing example of what *muscular* Christianity may lead to.[3]

Thomson's earliest opportunity 'to expose some of the rubbish' came at the September, 1861, meeting of the British Association held in Manchester. A serious leg injury in December, 1860, however, prevented him from attending the meeting in person, and his paper on the sun's heat was probably read by his colleague, H.D. Rogers (professor of natural philosophy), who had assisted with the natural philosophy class during Thomson's long absence. Significantly, the Manchester meeting followed the 1860 Oxford meeting at which Bishop Samuel Wilberforce and Thomas Henry Huxley (1825–95) had engaged in celebrated debate.[4] Joule's letter, therefore, almost certainly relates to this context, urging as it did an authoritative ruling from Thomson for the benefit of a misguided public.

Thomson's 1862 papers, then, are immediate responses, set within the framework of his own developing views on the nature and origin of the solar system, to the Darwinian debates. With regard to the sun's heat, however, his response takes place in the light of a fairly radical transition from commitment to the gravitational (condensation) form of meteoric theory of 1854 to a gravitational (contraction) form of primitive heat theory advocated by Helmholtz in the same year. Throughout the period from 1861 until his death in 1907, William Thomson would remain staunchly committed, on both religious and practical grounds, to this cosmological synthesis, which would set limits to human habitation of the earth and explain the great source of energy available to man for the production of mechanical effect. Speaking on the sun's heat to the Royal Institution in 1887, he gave implicit expression to his own underlying regrets that his harvesting of science in other fields had diverted him from such grand questions:

The sun, a mere piece of matter of the moderate dimensions which we know it to have, bounded all round by cold ether, has been doing work at the rate of 476×10^{21} horse-power for 3000 years, and possibly more [than this rate], certainly not much less, for a few million years. How is this to be explained? Natural philosophy cannot evade the question, and no physicist who is not engaged in trying to answer it can have any other justification than that his whole working time is occupied with work on some subject or subjects of his province by which he has more hope of being able to advance science.[5]

[3] J.P. Joule to William Thomson, 13th May, 1861, J269, ULC.

[4] On the Huxley–Wilberforce confrontation see J.R. Moore, *The post-Darwinian controversies* (Cambridge, 1979), pp. 60–2.

[5] PL, **1**, 370–1.

On the age of the sun's heat

In the previous chapter we drew attention to William Thomson's verdict in a footnote of 4th May, 1854, to his paper 'On the mechanical energies of the solar system' that, even if the cooling of the sun were not merely by conduction, the sun's thermal capacity (assumed to be as great as an equal mass of water) would not be large enough to account for the annual emission of heat at the present rate. Thus he rejected at that time two versions of a primitive heat theory: first, cooling by conduction only and second, cooling by a form of convection (suggested by Stokes a few days earlier) whereby the sun could be 'in a state of fusion, in which case if the surface cooled a little, and so became denser, the other portions would descend, and be renewed from beneath'.[6] In order to compensate for the inadequate thermal capacity of either version, therefore, Thomson had committed himself to meteoric agency within the earth's orbit.

From early 1860, however, Thomson began to entertain doubts about the meteoric theory and switched to another compensatory agency which had already been suggested by Helmholtz. Shrinkage or contraction of the sun would replace meteoric action. But Thomson's own position in 1861-2 was far from clear, and it is only in subsequent papers that the radical nature of the change of theory becomes fully explicit. Thus in 1871 he explained:

> [In 1854 Helmholtz] made the important observations that the potential energy of gravitation in the Sun is even now far from exhausted; but that with further and further shrinking more and more heat is to be generated, and that thus we can conceive the Sun even now to possess a sufficient store of energy to produce heat and light, almost as at present, for several million years of time future. It ought, however, to be added that this condensation [contraction] can only follow from cooling, and therefore that Helmholtz's gravitational explanation of future Sun-heat amounts really to showing that the Sun's thermal capacity is enormously greater, in virtue of the mutual gravitation between the parts of so enormous a mass, than the sum of the thermal capacities of separate and smaller bodies of the same material and same total mass.[7]

Thomson's explicit commitment to Helmholtz's model emerged only gradually, however, beginning with his 1854 opinion of Helmholtz's lecture.

In Thomson's 1854 British Association address he had inserted a footnote to his conclusion that Helmholtz's theory and his own meteoric theory might complement one another in so far as the heat generated by mutual gravitation 'must have been nearly all radiated off immediately' and that meteors supplied the compensating agency. The footnote explained that 'it is quite certain that it [the source of the sun's heat] cannot, as the nebular theory has led some to suppose it may, be the energy of gravitation effecting any continued condensation of the sun's present mass, since without increased pressure, it is only by

[6] MPP, **2**, 3*n*; G.G. Stokes to William Thomson, 24th April, 1854, S370, ULC.
[7] MPP, **2**, 26.

Hermann Helmholtz first met Thomson in 1855. His views on cosmology, as on vortex motion, frequently entered Thomson's natural philosophy. For his part, Helmholtz from the beginning claimed that Thomson 'exceeds all the great men of science with whom I have made personal acquaintance, in intelligence, and lucidity, and mobility of thought, so that I felt quite wooden beside him sometimes'. [From *Nature*, **15** (1876–7), facing p. 389.]

cooling that any condensation can be taking place; and the heat emitted in consequence of condensation by cooling, would depend merely on the specific heat of the whole mass in its actual circumstances of temperature and pressure, and might (for all we know of the properties of matter at such high temperatures) be greater than, equal to, or less than the thermal equivalent of the work done by gravity on the contracting body'.[8] At this time, then, the contraction

[8] William Thomson, 'On the mechanical antecedents of motion, heat, and light', *BAAS Report*, **24** (1854), 59–63, on p. 62*n*. This footnote is omitted from MPP, **2**, 34–40. We would like to thank Frank James for this information.

model provided no decisive answer to the demand for a compensatory agency, while the meteoric theory clearly did.

By 1859, however, the French astronomer U.J.J. Leverrier's 'great researches on the motion of the planet Mercury' provided grounds for Thomson's abandonment of his conviction that the zodiacal light (in meteoric form) provided the present source of the sun's heat. In August, 1854, Thomson had estimated that the zodiacal atmosphere must at present amount to at least $\frac{1}{5000}$ of the sun's mass in order to account for a future supply of 3000 years heat. He suggested that 'disturbance in the motions of visible planets' should be looked for in order to determine the possible amount of matter in the zodiacal light.[9] Speaking to the Glasgow Philosophical Society in December, 1859, he noted that Leverrier's researches afforded the 'kind of evidence of the existence of matter circulating round the Sun within the earth's orbit' which he had called for in 1854. In other words, the perturbation in Mercury's motion could be accounted for by the existence of planetary matter seen 'as the Zodiacal Light, long before conjectured to consist of a cloud of corpuscles circulating round the Sun, and, in the dynamical theory of the Sun's radiation, supposed to contain the reserve of force from which this Earth, as long as it continues a fit habitation for man as at present constituted, is to have its fresh supplies of heat and light'.[10]

A few weeks later, however, in early January, 1860, Thomson concluded that 'if matter has been really falling in at the rate supposed by my dynamical theory of the solar radiation, the place from which it has been falling must be either nearer the Sun or more diffused from the plane of Mercury's orbit' than he had supposed. Thus the survival of the meteoric theory would have depended on finding 'a place for a sufficient future dynamic supply without supposing a denser distribution of meteors or meteoric vapours than is consistent with what we know of the motion of comets before and after passing very close to the Sun'.[11]

During the remainder of 1860, two different but significant events occurred which set the scene for the following year's British Association paper, 'Physical considerations regarding the possible age of the sun's heat'. First, in June, Thomson attended the Oxford meeting of the British Association, at which Huxley and Wilberforce had their celebrated and (what later became) symbolic confrontation. We do not know of Thomson's immediate reaction, but the remarks of Joule quoted earlier suggest Thomson's desire to raise the post-Darwinian debate above the level of popular controversy. And second, in August, Helmholtz visited Thomson in Arran. This visit followed Thomson's

[9] MPP, **2**, 24–5; PL, **1**, 353.

[10] William Thomson, 'Recent investigations of M. Le Verrier on the motion of Mercury', *Proc. Phil. Soc. Glasgow*, **4** (1855–60), 263–6; MPP, **5**, 134–7.

[11] William Thomson, 'On the variation of the periodic times of the earth and inferior planets, produced by matter falling into the Sun', *Proc. Phil. Soc. Glasgow*, **4** (1855–60), 272–4; MPP, **5**, 138–40.

initial doubts about the consistency of the meteoric theory with the motion of Mercury and preceded Helmholtz's Royal Institution discourse 'On the application of the law of the conservation of force to organic nature' in the spring of 1861 where he summarized his earlier views on the sun's heat.[12]

In his 1854 paper Helmholtz had included an analysis of continued release of heat from the sun due to gradual contraction of its solid or liquid mass. He explained that 'the measure of the work performed by the condensation of the mass from a state of infinitely small density is the potential of the condensed mass upon itself' and set this potential equal to the quantity of heat produced in the sun's mass. The resulting equation gave an estimate of about twenty-eight million degrees centigrade for the elevation of temperature. He concluded further, however, that 'a diminution of 1/10 000 of the radius of the sun [assumed to be of uniform density] would generate work, in a water mass equal to the sun, equivalent to 2861 degrees Centigrade. And as, according to Pouillet, a quantity of heat corresponding to $1\frac{1}{4}$ degrees is lost annually in such a mass, the condensation referred to would cover the loss for 2289 years'. Helmholtz also added that, for greater density nearer the centre of the sun, 'the potential of its mass and the corresponding quantity of heat will be still greater'.[13] So great, however, had been Thomson's commitment to his own theory in 1854 that he had not then paid much attention to Helmholtz's discussion of contraction.

In his 1861 address to the physics section of the British Association, William Thomson claimed that 'the sun is probably an incandescent liquid mass, radiating away heat without any appreciable compensation by the influx of meteoric matter'.[14] He had thus publicly abandoned his meteoric theory. As to the nature of the replacement theory, however, his position was rather less decisive. In the draft of his address he struggled to find a form of words which would suggest a gentle transition, rather than a radical shift, from his own earlier condensation (meteoric) theory to Helmholtz's contraction theory. He therefore referred to the meteoric theory of solar heat 'in a form somewhat modified from that in which he [Thomson] had himself formerly held it' but deleted this version in favour of 'a form nearly agreeing with that in which it was stated by Helmholtz' in 1854.[15] The final published version, however, spoke of the meteoric theory 'in the form in which it was advocated by Helmholtz'.[16] Thus he abandoned at last his own explicit claims to parentage, but retained a clear link between 'the meteoric theory' and Helmholtz's theory, thereby blurring the transition from his former views to his new position and making no mention of contraction as a

[12] SPT, **1**, 409–17; Hermann Helmholtz, 'On the application of the law of the conservation of force to organic nature', *Proc. Royal Inst.*, **3** (1858–62), 347–57.

[13] Hermann Helmholtz, 'On the interaction of natural forces', *Phil. Mag.*, [series 4], **11** (1856), 489–518, esp. pp. 516–17.

[14] William Thomson, 'Physical considerations regarding the possible age of the Sun's heat', *BAAS Report*, **31** (1861), 27–8.

[15] William Thomson, Draft of 'Physical considerations regarding the possible age of the Sun's heat', October, 1861, NB46, ULC.

[16] Thomson, 'Physical considerations', p. 28.

compensating agency, which, unlike meteoric action, required no diffused meteoric vortex around the sun.

The immediate aim of Thomson's 1862 *Macmillan's Magazine* article on the sun's heat was an application of his energy principles 'to the discovery of probable limits to the periods of time, past and future, during which the sun can be reckoned as a source of heat and light'. He discussed the subject in three general sections: first, the secular cooling of the sun (in which he rejected both the meteoric and chemical theories as sources for the sun's *present* heat, concluding that the sun was merely an incandescent liquid mass cooling); second, the present temperature of the sun; and, third, the origin and total amount of the sun's heat. In this final section, he maintained a gravitational form of the meteoric theory as the cause of the sun's origin, and concluded that there was no difficulty in accounting for twenty million years' heat by continuing contraction. It seemed probable, however, that the sun had not illuminated the earth for a hundred million years, and that it was almost certain that the sun had not done so for 500 million years.[17] This latter figure, then, was Thomson's upper limit for the age of the sun.

Turning in the first section to the question of the secular cooling and age of the sun's heat, Thomson remarked that 'we have data on which we might plausibly found a probable estimate, and from which we might deduce, with at first sight seemingly well-founded confidence, limits, not very wide, within which the present true rate of the sun's cooling must lie'. Such estimates were based on the acceptance, first, of the 'independent but concordant' investigations of Herschel and Pouillet on the amount of heat radiated by the sun, and, second, that 'the sun's substance is very much like the earth's' – a conviction supported by Stokes's, Bunsen's, and Kirchhoff's researches in solar chemistry, showing the presence of sodium, iron, manganese and other metals in the sun. On this second assumption, Thomson stated, the mean specific heat of the sun might be taken *at first sight* to be less – certainly not much greater – than that of water. Taking the mean specific heat of the sun, then, to be unity, yielded a present annual rate of cooling of 1.4°C per annum.[18]

If, however, we supposed the sun's matter to be the same as the earth's, it would be plausible to assume not only a similarity in specific heats but also in expansibility. Thus, Thomson argued, if the sun contracted on cooling to the same extent as an average terrestrial body – such as solid glass – 'there would be in 860 years a contraction of one percent on the sun's diameter, which could scarcely have escaped detection by astronomical observation'. More especially, such a large contraction would, on Helmholtz's estimate of the amount of work performed by the mutual gravitation between the different parts of the sun's contracting mass, release unacceptably large amounts of heat. On Helmholtz's

[17] PL, **1**, 349–68.

[18] *Ibid.*, pp. 354–6. The draft for this part of Thomson's paper may be located in his Research notebook, July, 1861, NB45, ULC.

figures, a contraction of 0.1% of the sun's diameter – assuming uniform density through the sun – would be equal to 20 000 times the annual radiation of heat from the sun, or 200 000 times for a 1% contraction. Thus, 'even without historical evidence as to the constancy of his diameter, it seems safe to conclude that no such contraction [1% in 860 years] . . . can have taken place in reality'. Thomson suspected, therefore, 'that the physical circumstances of the sun's mass [especially the enormous pressure borne by the interior] render the condition of the substances of which it is composed, as to expansibility and specific heat, *very different from* that of the same substances when experimented on in our terrestrial laboratories'.[19]

In order to place limits on the period of time during which the sun had existed in something like its present state, Thomson argued first that it must have radiated away at least as much heat in cooling as there was mechanical energy generated in contraction, for it seemed to him 'in the highest degree improbable [presumably from the dissipation principle] that mechanical energy can in any case increase in a body contracting in virtue of cooling'. He therefore set the heat loss in cooling equal to that generated by contraction, which provided a relation between specific heat and expansibility cubed. Arguing further that a more rapid contraction of the sun's diameter than 0.1% in 20 000 years was improbable and anything faster than 0.1% in 8600 years was 'scarcely possible', and that an expansibility of its mass less than one-tenth that of glass 'improbable', he concluded that its specific heat must be at least ten times that of water. Similarly, because the expansibility was unlikely to be greater than that of glass, the specific heat could not exceed 10 000 times that of water.

Given these limits on specific heat between ten and 10 000 times that of water, the earth would decrease in temperature by 100°C in something between 700 and 700 000 years. Employing this reasoning to criticize Darwin's 300 million years, Thomson's upper limit of 100°C in 700 000 years would give a cooling of about 21 000°C in 300 million years, an unacceptably large figure for the secular cooling of the sun. He therefore concluded his discussion with a crucial rhetorical question:

> Whether is it more probable that the physical conditions of the sun's matter differ 1000 times more than dynamics compel us to suppose they differ from those of matter in our laboratories; or that a stormy sea, with possible channel tides of extreme violence, should encroach on a chalk cliff 1000 times more rapidly than Mr Darwin's estimate of one inch per century?[20]

In his elliptical style, Thomson here invited his reader to consider whether it was more difficult to accept channel tides of extreme violence than to accept that the

[19] PL, **1**, 356–7. Our emphasis. In his July, 1861, notebook he had concluded that it was 'not improbable that the enormous pressures to which the Sun's mass at considerable depths below his surface are subjected may greatly increase its specific heat and it would not be safe with our present knowledge to assume the mean specific heat of the Sun's mass to be anything less than thousands of times the specific heat of water as we have it on earth'. [20] PL, **1**, 357–61.

properties of the sun's matter differed in magnitude from those of the earth by much more (1000 times) than the above dynamical reasoning *already* demanded. By implication, the answer had to be that there could be no obstacle in the way of accepting the probability of violent tides, thus making Darwin's estimate of the time for denudation quite unacceptable.

In the short second part of his 1862 paper, Thomson argued for a distinction between the present temperatures at the surface and in the interior of the sun. Since the heat radiated from each square foot of the sun's surface was only about 7000 horse-power – a figure first mentioned in his 1854 correspondence with Stokes – and since coal burning in a locomotive furnace could produce an estimated $\frac{1}{15}$ to $\frac{1}{45}$ of this power per square foot, Thomson believed that the surface temperature was not 'incomparably higher than temperatures attainable artificially in our terrestrial laboratories'. On the other hand, the fluid interior of the sun would be at a far higher temperature since 'direct conduction can play no sensible part in the transference of heat between the inner and outer portions of his mass, and there must be an approximate *convective* equilibrium throughout the whole'.[21]

The final section of Thomson's article discussed the origin and total amount of the sun's heat. Having adopted the view of the sun as an incandescent but contracting liquid mass, he now considered the origin of that heated mass. As before, the fundamental dissipation principle constrained his range of options:

> It is certain that it [the sun's heat] cannot have existed in the sun through an infinity of past time, since, as long as it has so existed, it must have been suffering dissipation, and the finiteness of the sun precludes the supposition of an infinite primitive store of heat in his body. The sun must, therefore, either have been created as an active source of heat at some time of not immeasurable antiquity, by an over-ruling decree; or the heat which he has already radiated away, and that which he still possesses, must have been acquired by a natural process, following permanently established laws.[22]

The former supposition – direct divine intervention – he regarded not as incredible, but as improbable, especially if one could discover a natural process 'not contradictory to known physical laws'. That natural process was his theory of meteoric origins and subsequent contraction.

Employing Helmholtz's calculation, Thomson concluded that we may accept 'as a lowest estimate, 10 000 000 times a year's supply at present rate'. Allowing for dissipation of energy by impact and resistance before the final conglomeration, and for a much greater density towards the centre of the sun, however, we may regard '50 000 000 or 100 000 000 as possible'. Taking into account the implications of his earlier discussion of the possible specific heat of the sun, its rate of cooling, and its surface temperature, Thomson further stated that the sun 'must have been very sensibly warmer one million years ago than now', and so must have 'radiated away considerably more than the correspond-

[21] *Ibid.*, pp. 361–3. [22] *Ibid.*, pp. 363–4.

ing number of times the present yearly amount of loss'. He therefore concluded with the crucial claim (though without detailed calculation) that it seemed 'on the whole most probable that the sun has not illuminated the earth for 100 000 000 years, and most certain that he has not done so for 500 000 000 years'. As for the future, Thomson claimed, 'inhabitants of the earth cannot continue to enjoy the light and heat essential to their life, for many million years longer, unless sources now unknown to us are prepared in the great storehouse of creation'.[23] Consistent with his theology of nature, which emphasized the contingency of all nature on God's will, he left open the possibility of Providence preparing for the solar system other sources of energy which would ensure its continuation beyond the time that the present laws allowed.

Thomson's cosmological vision

Thomson had adopted in 1854 the view that 'the potential energy of gravitation may be in reality the ultimate created antecedent of all the motion, heat, and light at present in the universe'. In other words, it was 'the original form of all the energy in the universe'.[24] Such a speculation conformed to his theology of nature in which God had created energy *ex nihilo* in the beginning by His absolute power and had sustained its quantity by His ordained power. But while Thomson could remain content with potential energy as the original form of *energy*, the meteoric theory of the origin of the sun's heat raised the problem of the original or primitive forms of *matter*. Discussion of this important issue took place in a letter to his brother-in-law, Rev David King (1806–83), on 3rd February, 1862.

Thomson made clear that he regarded as 'a very improbable supposition' the idea of 'shapeless detached stones being a primitive form of matter' and rejected as 'mere hypothesis' the notion that the meteors had 'arisen from the disruption of more dignified masses'. Thus, while he readily accepted these 'shapeless fragments of matter' which we knew to exist in space as the probable cause of the original heat of the sun, his real objection concerned rather the assumption of meteors as the ultimate, primitive form of matter existing from eternity:

What is *large* and what is *small* even to our ideas, enlarged and enlightened by science? We are equally far from comprehending an act of creation out of nothing, whether it be of matter in a finished and approximately round globe like the earth, or in small solid fragments or in a general diffused medium; although perhaps the last may seem the most probable to us in our present state of feeble enlightenment. But without attempting anything so much beyond our powers as the discovery of *the* primitive condition of matter, we successfully investigate the present condition, and argue from analogies and

[23] *Ibid.*, pp. 365–8.

[24] MPP, **2**, 40; William Thomson and P.G. Tait, 'Energy', *Good Words*, **3** (1862), 601–7, on p. 606.

from strict dynamical reasoning, [to] what must have been the antecedent condition . . . back to more or less ancient times.[25]

These remarks made clear that neither a finished globe nor matter in meteoric form were *inconsistent* with his voluntarist theology of nature which placed the act of creation of matter (in whatever form) *ex nihilo* in the hands, not of human science, but of God. However, the concept of a 'general diffused medium' allowed much greater scope for the operation of secondary (mechanical) causes and so for the activity of human science in tracing antecedent conditions 'back to more or less ancient times'. Thomson's words in fact echo those of his Cambridge coach, William Hopkins, when in 1847 he explained to the British Association at Oxford his preference for a fluid, rather than a solid, earth at some former epoch as the basis for explaining volcanic action. According to the fluid hypothesis 'the terrestrial mass must have been more entirely free to receive all the modifications of constitution and form which physical and mechanical causes may have tended to impress upon it up to the present epoch, than if it had always existed in a state of solidity'.[26] We must also recall Thomson's strong preference, from at least 1854, for a continuum theory of matter, leading to his vortex atoms (ch. 12).

The same air–ether of his letters to Stokes in 1854 and to Tait in 1862 appeared also in a letter to David King dated 8th January, 1862, reaffirming Thomson's attachment to a continuum theory of the world: 'I think it possible that the sun's heat, which we know to be radiating away dissipated through space, may have the effect of evaporating matter in very distant regions, and preparing a suitable medium or *atmosphere* (for I think the word atmosphere, or interplanetary air, is quite as appropriate as ether, which is in fact aer, or a luminiferous medium, by which the fluid occupying the region in space through which the earth moves is more commonly called) for the propagation of light, and generally for the requisites of a "world" '.[27] Thomson's powerful commitment to an ether here suggests a strong belief in the unity, economy, and practical simplicity of God's creation. Thus the creation of a single, unifying ether from which all other forms of matter derived was preferable to either the multiple creation of different material entities or to the creation of vast numbers of separate atoms. Thomson's 'aer' performed the same role as J.P. Nichol's nebular matter; likewise his views on the unity of creation matched those of Nichol:

We are all too easily inclined to look on creation as made up of isolated parts – of independent or individual classes of beings . . . and most of what we divide and parcel out into isolated bundles, is nothing other than the parts of the same grand scheme. Philosophy has taught this for ages – it is in fact the secret of her life; for she aims to gather

[25] William Thomson to Rev David King, 3rd February, 1862, in SPT, **1**, 422–3.

[26] William Hopkins, 'Report on the geological theories of elevation and earthquakes', *BAAS Report*, **17** (1847), 33–92, on p. 37.

[27] William Thomson to Rev David King, 8th January, 1862, in SPT, **1**, 421–2; PL, **1**, 421–2.

up all fragments and to present the Universe united, compact, tending to one end – a type of its August CREATOR.[28]

By 1862, Thomson's convictions had become sufficiently developed for him to present, along with Tait, a popular account of his cosmos, directed principally against John Tyndall (1820–93), in the Presbyterian magazine *Good Words*. Tyndall's so-called naturalism, which assigned no role to God, and which placed all things (including mind) under the rule of deterministic, materialistic law, regularly provoked the wrath of his Scottish contemporaries (ch. 18), as did his failure to comprehend the Second Law of Thermodynamics. The authors outlined Thomson's new view of the sun as 'a heated body cooling' and referred to his 1854 view of 'the possible origin of energy at the creation' as 'excessively instructive'. Thus, 'created simply as difference of position of attracting masses, the potential energy of gravitation was the original form of all the energy in the universe'. Given the tendency of all energy 'ultimately to become heat which cannot be transformed without a new creative act into any other modification . . . the result will be an arrangement of matter possessing no realizable potential energy, but uniformly hot . . . chaos and darkness as "*in the beginning*" [Genesis 1:1–2]'. Prior to this ending of the universe, however, our solar system will have undergone 'tremendous throes and convulsions, destroying every now existing form'. And so 'as surely as the weights of a clock must run down to their lowest position, from which they can never rise again, unless fresh energy is communicated to them from some source not yet exhausted, so surely must planet after planet creep in, age by age, towards the sun' and a fiery end. Thomson and Tait emphasized the consistency between their cosmos and the biblical text:

we have the sober scientific certainty that heavens and earth shall 'wax old as doth a garment' [Psalm 102:26]; and that this slow progress must gradually, by natural agencies which we see going on under fixed laws, bring about circumstances in which 'the elements shall melt with fervent heat' [2 Peter 3:10]. With such views forced upon us by the contemplation of dynamical energy and its laws of transformation in dead matter, dark indeed would be the prospects of the human race if unillumined by that light which reveals 'new heavens and a new earth' [2 Peter 3:13].[29]

Although these biblical references appeared in a popular article aimed at a religious readership, the account was essentially the same as that provided by Thomson's 1851 draft in which the theological framework was an unambiguous expression of his own deepest convictions.

Within a few months, James Challis responded to Thomson and Tait's view of the sun's heat. Challis believed that the differing views advocated by Thomson in 1854 and 1862 'give evidence . . . that the problem of the sun's heat has not yet been solved'. Challis, however, was not content merely to criticize. From his assumptions that 'all visible and tangible substances consist of discrete

[28] J.P. Nichol, *Views of the architecture of the heavens: in a series of letters to a lady* (Edinburgh, 1838), pp. 166–8. [29] Thomson and Tait, 'Energy', pp. 606–7.

inert atoms of spherical form and constant magnitude' and that 'the force of heat is due to the dynamical action of the direct vibrations of the aether', he claimed that the theoretical explanation of the sun's internal heat was to be given in terms of 'stellar-undulations, of whatever kind, entering his vast body . . . [and having] their effects multiplied to an incalculable amount by the reaction of the immense number of atoms'. In this way 'the continual generation of new undulations both maintains the heat, and supplies the loss resulting from the continual propagation of undulations from the body of the sun . . . [and] the sun's heat will be constant so long as the calorific action of the universe is constant'. Similarly, 'on the hypothesis that the same operations are going on in the stars as in the sun, the theory at the same time accounts for the sensible light and heat, and for the internal heat, of every body of the universe'.[30]

In the same article, Challis referred to Sir John Herschel on 'The Sun', published in *Good Words* (1863), where Herschel had described certain willow leaf-like phenomena on the sun's surface which 'we cannot refuse to regard . . . as *organisms* of some peculiar and amazing kind . . . [and] we do know that vital action *is* competent to develop both heat, light, and electricity'. Challis interpreted these remarks as a sign of Herschel's dissatisfaction with Thomson's most recent views, and as an attempt to account for the *permanence* of the sun's light and heat by 'vitality'. Herschel, however, had only referred briefly, and without explicit comment, to the orbital version of the meteoric theory of the sun's heat (Thomson's 1854 views) as a development of his own frictional suggestions of 1833. He appeared, in fact, to regard the willow leaf phenomena as 'the *immediate sources of the solar light and heat*, by whatever mechanism or whatever processes, they may be enabled to develop . . .'.[31] Such responses from the astronomers, however, did not lead Thomson – or even Tait – into any major controversy over the question of the sun's heat. So obviously did it violate Thomson's fundamental conservation of dissipation principles, and so hypothetical and speculative were Challis's and Herschel's alternatives, that Thomson undoubtedly felt that his own views were rapidly consolidating into an almost unassailable position in the context of nineteenth-century cosmology.

If the consequences of Thomson's estimates for Darwin's theory had been in any doubt in 1862, they were no longer so by 1867. Fleeming Jenkin's well-known review of 'Darwin and the origin of species' brought the whole doctrine of energy to bear on the question of geological and cosmological time, and made it clear that Darwin's calculation of the time required for the denudation of the Weald 'savours a good deal of that known among engineers as "guess at the half and multiply by two"'. Apart from providing one of the most lucid explanations of the energy doctrines given at the time, Jenkin focussed on the

30 James Challis, 'On the source and maintenance of the sun's heat', *Phil. Mag.*, [series 4], **25** (1863), 460–7. See also Burchfield, *Age of the earth*, p. 123.

31 Challis, 'Sun's heat', p. 461; Sir John Herschel, 'The sun', *Good Words*, **4** (1863), 273–84, on p. 282.

implications of Thomson's irreversible cosmos. Emphasizing that 'the rate at which the planetary system is thus dying is perfectly measurable, if not yet perfectly measured', he concluded:

we are assured that the sun will be too cold for our or Darwin's purposes before many millions of years – a long time, but far enough from countless ages; quite similarly past countless ages are inconceivable, inasmuch as the heat required by the sun to have allowed him to cool from time immemorial, would be such as to turn him into mere vapour, which would extend over the whole planetary system, and evaporate us entirely . . . [This] reasoning concerning the sun's heat does not depend on any one special fact, or set of facts, about heat, but is the mere accidental form of decay, which in some shape is inevitable, and the very essential condition of action.[32]

A year later, William Thomson delivered his address 'On geological time' to the Geological Society of Glasgow. Taking as his text a passage from John Playfair's *Illustrations of the Huttonian theory* (1802), he attacked not only uniformitarian geology, but uniformitarian cosmology also. Because Playfair appeared to support the indefinite duration of both earth and solar system, it is likely that Thomson chose to criticize him rather than Lyell on this occasion. The central thrust of Thomson's attack on Playfair concerned his alleged confusion of natural philosophy with natural history, dynamical laws with present, past and future arrangements. Thomson believed that Playfair's views were 'pervaded by a confusion between "present order" or "present system" [natural history], and "laws now existing" [natural philosophy] – between destruction of the earth as a place habitable to beings such as now live on it, and a decline or failure of law and order in the universe'.[33] To have conflated the two conceptions was the error of the uniformitarian geologists when they claimed the indefinite duration of the present order and present laws. The present fundamental laws, of energy conservation and energy dissipation, were immutable except by divine will. The 'present order', however, according to the dissipation principle, was not eternal or indefinite in duration. The present order of the material world of Thomson, like the material world of J.P. Nichol, Thomas Chalmers, and William Whewell, was not co-eternal with God.

The important issue of tidal friction took up much of Thomson's 1868 address, providing a striking illustration of the universal dissipation of energy and further undermining Playfair's steady-state cosmos (ch. 17). But Thomson also deployed the sun's heat in this role:

We depend on the sun very much for the existing order of things . . . When Playfair spoke of the planetary bodies as being perpetual in their motion, did it not occur to him to ask, what about the sun's heat? . . . what an amount of mechanical energy is emitted from the sun every year! . . . And yet Playfair and his followers have totally disregarded this

[32] Fleeming Jenkin, 'Darwin and the origin of species', *North Brit. Rev.*, **46** (1867), 277–318; *Papers and memoir of Fleeming Jenkin* (2 vols., London, 1887), **1**, pp. 215–63, on pp. 241–2.

[33] William Thomson, 'On geological time', (read 27th February, 1868), *Trans. Glasgow Geol. Soc.*, **3** (1871), 1–28; PL, **2**, 10–64, on pp. 11–13; John Playfair, *Illustrations of the Huttonian theory* (Edinburgh, 1802), pp. 119–21.

prodigious dissipation of energy. He speaks of the existing state of things as if it must or could have been perennial.[34]

Thomson reiterated his view on the gravitational theory of the sun's origin and his conclusion that he could not see 'a *decided* reason against admitting that the sun may have had in it one hundred million years of heat, according to its present rate of emission, in the shape of energy'. While admitting again that his 1862 estimates were 'necessarily very vague', he emphasized that he did not think it possible, 'upon any reasonable estimate founded on known properties of matter, to say that we can believe the sun has really illuminated the earth for five hundred million years'.[35]

In 1869 T.H. Huxley launched his famous counterattack, though he did not dwell at length on the issue of the sun's heat. He did, however, try to exploit two admissions by Thomson – namely, that his 1862 estimates were 'necessarily very vague', and that 'only fifteen years ago he entertained a totally different view of the origin of the sun's heat and believed that the energy radiated from year to year was supplied from year to year – a doctrine which would have suited Hutton perfectly'.[36] Huxley had undoubtedly misinterpreted Thomson's 1854 view, as the Glasgow professor was quick to point out in his reply of 1869: 'if Professor Huxley will "hansardize" ['Hansard' being the official report of proceedings in Parliament] me by looking to my original paper . . . he will see that my contribution to the meteoric theory of solar heat was to prove the insufficiency of any chemical theory, and to point out that *meteoric supply cannot be perennial in even approximate uniformity with the existing order of things*'.[37] Whatever else had changed regarding Thomson's particular view of the nature of the sun's heat, and whatever might be admitted about the vagueness of the numerical estimates, the fundamental dissipation principle upon which the 1854 paper had been based ruled out any 'reconciliation' of Thomson with Hutton or Playfair. For his part, Huxley had been anxious to emphasize that a limited time scale – one hundred million years or more – 'may serve the needs of geologists perfectly well'. Thus, although eager to point out any possible defects in Thomson's case, he was in fact accepting the probability of a much restricted time scale.

The progress of solar dynamics

In *Good Words* of 1887, Sir William Thomson published his last major paper on the question of the sun's heat. The same issue also formed the subject of a Friday evening lecture delivered before the Royal Institution on 21st January of that year. Maintaining his earlier contraction theory, he stressed that 'the sun is *not* a

[34] PL, **2**, 45–9. [35] *Ibid.*, pp. 52–3.

[36] T.H. Huxley, 'Geological reform', *Quart. J. Geol. Soc. London*, **25** (1869), xxxviii–liii; *Discourses: biological and geological* (London, 1902), pp. 305–39, on pp. 335–6.

[37] PL, **2**, 96–8.

burning fire, [but] . . . *is* merely a white-hot fluid mass cooling, with some little accession of fresh energy by meteors occasionally falling in, but of very small account in comparison with the whole energy of heat which he gives out from year to year.[38] Given that premise, he aimed now to elucidate the process by which the sun's vast store of gravitational potential energy became gradually converted into heat. The discussion is of interest because it is wholly characteristic of his engineering mode of thinking, and provides a detailed example of the move from simple machines to physical reality. First, in order to illustrate the vast amount of heat 'continually carried up [by convection within the great mass of flaming fluid] to the sun's surface and radiated into space', Thomson estimated the mechanical value of the radiation per square metre to be 78 000 horse-power – equivalent to the engines of eight ironclad warships applied 'by ideal mechanisms . . . in perpetuity driving one small paddle in a fluid contained in a square-metre vat'. The same heat would be generated in the vat as from an equal area of the sun's surface.[39]

Second, Thomson invited his audience 'to pass from a practically impossible combination of engines and a physically impossible paddle and fluid and containing vessel, towards a more practical combination of matter for producing the same effect'. Shifting from a steam-engine illustration to a gravitational model, he retained the ideal vat, paddle, and fluid, but imagined the vat to be placed on the surface of a cool, solid homogeneous globe of the same size and density as the sun. To drive the paddle, a weight descended into a pit excavated below the vat. In making this transition, we may note the connection between the gravitational model and the central paradigm for thermodynamics, namely Joule's experiment employing a falling weight to heat water by turning a paddle wheel. The sun's heat was to be interpreted in terms of the conversion of mechanical effect into heat. As the simplest mechanism, Thomson suggested a long vertical shaft with the paddle mounted at the top, such that, as the shaft turned, the paddle rotated horizontally. The weight was simply a nut working its way down frictionless threads on the vertical shaft, but prevented from turning by frictionless guides fixed to the sides of the pit. He supposed the 'pit' to be a metre square at its upper end, but tapering uniformly to a point at the centre of the globe, while the nut was the excavated matter (about 326 million tons) of the sun's mass, 'with merely a little clearance between it and the four sides of the pit, and with a kilometre or so cut off the lower, pointed end to allow space for its descent'. In order to produce 78 000 horse-power, or '78 000 metre-tons solar surface-heaviness per hour', Thomson estimated that the 'nut' must descend at the rate of one metre in 313 hours, or about twenty-eight metres per year.[40]

Third, Thomson took another step 'still through impracticable mechanism,

[38] Sir William Thomson, 'On the sun's heat', *Proc. Royal Inst.*, **12** (1889), 1–21; *Good Words*, **28** (1887), 149–53, 262–9; PL, **1**, 369–422, on p. 371. [39] *Ibid.*, pp. 374–6.
[40] *Ibid.*, pp. 376–8.

towards the practical method, by which the sun's heat is produced'. He now suggested that we consider the thread of the screw of 'uniformly decreasing steepness from the surface downwards so that the velocity of the weight, as it is allowed to descend by the turning of the screw, shall be in simple proportion to distance from the sun's centre'. In other words, the weight's descent would be made faster at the outside and slower towards the centre of the sun's mass, yielding linear shrinkage throughout. This mechanism gave work at a rate of four-fifths that of the previous model – equivalent to the top end of the weight descending at thirty-five metres per year.

Fourth, Thomson developed this more-refined mechanism into what he called a 'model mechanical sun', capable of giving out heat and light to the same amount as the real sun. He supposed the whole surface of the mechanical sun to be divided into square metres, and the whole mass into long pyramidal rods each meeting at the centre and with screws, nuts, and paddles fitted as in the model. If the viscosity of the fluid and the size of each paddle were arranged so as to permit the top end of each rod to descend at the rate of thirty-five metres per year, and if the fluid were a few thousand metres deep over the paddles, 'it would be impossible, by any of the appliances of solar physics [the spectroscope, for example], to see the difference between our model mechanical sun and the true sun'.[41]

Finally, Thomson aimed 'to do away with the last vestige of impracticable mechanism in which the heavinesses of all parts of each long rod are supported on the thread of an ideal screw cut on a vertical shaft of ideal matter, absolutely hard and absolutely frictionless'. He considered first a single 'pit' excavated to the centre of the sun, and surrounded by rigid matter impervious to heat. Supposing the pit to be filled now by a pyramidal rod of incandescent fluid, consisting of the actual ingredients of the solar substance, the surface of the fluid would quickly cool by radiation while the fluid itself would remain incandescent. Convection currents would begin, bringing up hotter fluid from below, but from a depth insufficiently great to maintain the high surface temperature originally maintained by the (imaginary) paddle. This cooling would result in solidification at the surface, most probably followed by a falling in of the thin films at the surface until 'for several metres downwards, the whole mass of mixed solid and fluid becomes stiff enough . . . to prevent the frozen film from falling down from the surface'. Subsequently, the surface film would quickly thicken and become less than red-hot on its upper surface while 'the whole pit would go on cooling with extreme slowness until, after possibly about a million million million years or so, it would be all the same temperature as the space to which its upper end radiates'.[42]

Adapting this model to a complete globe, Thomson imagined every 'pit' in the globe to be so filled, but separated initially from the others by thin partitions

[41] *Ibid.*, pp. 378–80. [42] *Ibid.*, pp. 380–2.

impervious to heat. Furthermore, the globe was supposed to rotate at the same rate as the sun. If we then removed the partitions, 'so that there shall be perfect freedom for currents to flow unresisted in any direction', and if we left the globe – 'which we may now call the Sun' – to itself, it 'will immediately begin showing all the phenomena known in solar physics'. Sunspots, for instance, Thomson now regarded as 'due to the sun's own substance, and not to external influences of any kind' – although he did still leave open the possibility 'that some of the chief phenomena due to sunspots arise from influxes of meteoric matter circling round the sun'. While he did not elaborate the internal causes, he suggested that chemical combinations and dissociations *within* the sun might determine some of the features of non-uniformity of brightness 'in the grand phenomena of sunspots, hydrogen flames, and corona, which make the province of solar physics'.[43]

Having developed a realistic model of the sun from the imaginary dynamical models, Thomson summarized in two propositions the manner in which his model explained sunlight and sun-heat:

1. Gigantic currents throughout the sun's liquid mass are continually maintained by fluid, slightly cooled by radiation, falling down from the surface, and hotter fluid rushing up to take its place [i.e. a state of convective equilibrium].
2. The work done in any time by the mutual gravitation of all parts of the fluid, as it shrinks in virtue of the lowering of its temperature, is but little less than (so little less than that we may regard it as practically equal to) the dynamical equivalent of the heat that is radiated from the sun in the same time.[44]

From his earlier dynamical model, Thomson had estimated the rate of shrinkage, corresponding to the present rate of solar radiation, to be thirty-five metres on the radius per year – or 1% of the radius in 200 000 years. He then argued that if the sun's effective thermal capacity could be maintained by shrinkage until twenty million times the present year's amount of heat had been radiated away, the sun's radius would decrease by half and its density increase eightfold. If the present mean density of the sun were taken as 1.4 times that of water, its density would be 11.2 times that of water at that future time – equivalent to the density of lead. This figure, Thomson claimed, would probably be too high 'to allow the free shrinkage as of a cooling gas to be still continued without obstruction through over crowding of the molecules'.[45] Therefore:

It seems . . . most probable that we cannot for the future reckon on more of solar radiation than, if so much as, twenty million times the amount at present radiated out in a year. It is also to be remarked that the greatly diminished radiating surface, at a much lower temperature, would give out annually much less heat than the sun in his present condition gives.[46]

Support for Thomson's view came from the American astronomer Simon

[43] *Ibid.*, pp. 383–4. [44] *Ibid.*, p. 385. [45] *Ibid.*, pp. 385–7. [46] *Ibid.*, p. 387.

Newcomb (1835–1909), who had estimated a figure of ten million years for the future supply by the sun to support life on earth.[47]

As for the past age of the sun, Thomson for the first time modified Pouillet's (and Herschel's) estimate of solar radiation. Pointing to J.D. Forbes's 1844 method which corrected for atmospheric absorption, he noted Forbes's result as being 1.6 times Pouillet and Herschel's. More recent work by S.P. Langley yielded a factor of 1.7. Given a horse-power of 133 000 per square metre on Langley's result, the former twenty million years reduced to twelve. Thomson concluded that it would 'be exceedingly rash to assume as probable anything more than twenty million years of the sun's light in the past history of the earth'. Similarly, for the future, he thought it unwise 'to reckon on more than five or six million years of sunlight for time to come'.[48]

Thomson's unequivocal commitment to Helmholtz's theory in this address, and his reassessment of the sun's age, were essentially responses to an exchange of views with his friend George Darwin – Charles Darwin's son – in 1886. The younger Darwin had written to Thomson from Cambridge on 4th June, 1886:

> Your estimate of the Sun's heat is derived from the amount of energy lost in concentration from infinite dispersion & does not allow anything for energy residing in the primitive nebula. Now if any idea like [Norman] Lockyer's is true & if there is a possible dissociation of elements in the primitive nebula is it not at least possible (I do not say probable) that the energy of concentration may be largely underestimated by computing simply from gravitation. Then also is there anything intrinsically erroneous in [James] Croll's idea that the primitive nebula may have been hot? If either or both these ideas are *possible* is not a very large margin of uncertainty possible as to the total energy in the Sun?[49]

Thomson's immediate response came on a postcard one day later. For him, Lockyer's ideas of chemical dissociation were 'wildly nugatory'. He continued to believe that 'chemical energy is infinitely too meagre in proportion to gravitational energy, in respect to matter falling into the Sun, to be almost worth thinking of'.[50] It may also be noted that W.M. Williams in 1870 and C.W. Siemens in 1882 both put forward cyclical hypotheses of chemical dissociation and recombination to account for the sun's heat in a manner which violated the dissipation principle. Siemens sought to avoid 'the idea of prodigious waste through dissipation of energy into space' by suggesting a self-sustaining action, while Williams received support from A.R. Wallace, W.R. Grove, and Charles Lyell.[51] Such discussions would scarcely have commended chemical dissociation hypotheses to Sir William.

[47] Simon Newcomb, *Popular astronomy* (London, 1878), pp. 491–519, esp. pp. 505–11. See also Burchfield, *Age of the earth*, p. 111. [48] PL, **1**, 389–90.

[49] G.H. Darwin to William Thomson, 4th June, 1886, D35, Kelvin correspondence, ULC.

[50] William Thomson to G.H. Darwin, 5th June, 1886, D97, Kelvin correspondence, ULG; SPT, **2**, 860. Thompson mis-dates this postcard and erroneously regards it as a response to a meeting between Darwin and Thomson in Cambridge.

[51] See Burchfield, *Age of the earth*, pp. 123–5.

Croll's hypothesis of the origin of the sun's heat was, following Helmholtz and Thomson, gravitational in nature. Croll, however, assumed that the collision of two celestial bodies accounted for the existence of *hot* primitive nebula. As a consequence, he was able to lengthen the time scale by as much as he wanted.[52] Thomson found Croll's suggestion 'a wildly improbable hypothesis . . . and it can help us nothing'. In particular, he believed the 'exceedingly exact aim of the one body at the other' to be 'on the dry theory of probability, exceedingly improbable'.[53]

In his letter of 4th June, 1886, Darwin had also drawn Thomson's attention to Langley's estimate of the sun's radiation as greater than Pouillet's 'in about the proportion of 7 to 4'. In consequence, 'your results would have to be diminished in the proportion of 4 to 7'. Replying to Darwin, Thomson noted the greater radiation found by Langley and, as we have seen, reduced his estimates accordingly. Darwin for his part had made clear his own aims at the end of his June, 1886, letter to Thomson: 'I do not wish to combat the fundamental proposition at all, & only wish to speak against such dogmatism as I find in Tait's writings & not in yours. It appears to me that we know far too little as yet to be sure that we may not have overlooked some important point'. Thomson's reduced estimate (on Langley's data) was now in line with a confident assertion by Tait in 1869 of ten million years for the sun's age, and Darwin was most anxious that Thomson not commit himself to an exact figure in the light of so little certain knowledge. At the same time, as he stated in his September address as president of Section A of the British Association in 1886, he regarded this line of argument 'by which a superior limit is sought for the age of the solar system' as 'by far the strongest' of Thomson's three approaches – the secular cooling of the earth, the sun's heat, and tidal friction – to the question of geological time.[54]

William Thomson concluded his forty-year investigation of the sun's heat in his 1892 paper 'On the dissipation of energy'. Towards the end of his 1887 paper, he had spoken of the developments in Laplace's nebular theory of the evolution of the solar system now 'converted by thermodynamics into a necessary truth, if we make no other uncertain assumption than that the materials at present constituting the dead matter of the solar system have existed under the laws of dead matter for a hundred million years'. He inferred that 'there may in reality be nothing more of mystery or of difficulty in the automatic progress of the solar system from cold matter diffused through space, to its present manifest order and beauty, lighted and warmed by its brilliant sun, than there is in the winding up of a clock and letting it go till it stops'. In his 1892 paper, he likened the solar system to a clock running down, employing an illustration first seen in Thomson and Tait's 'Energy' (1862). Given a flywheel or clockwheel driven by a weight, he explained that when the weight had run down, and the initial

[52] James Croll, 'Age of the sun in relation to evolution', *Nature*, **17** (1877–8), 206–7; Burchfield, *Age of the earth*, pp. 64–5, 125–6. [53] SPT, **2**, 860; PL, **1**, 404–6.

[54] G.H. Darwin, 'Presidential address', *BAAS Report*, **56** (1886), 517.

potential energy had been 'all spent in heat', that heat was not available for raising the weight and giving the clockwork a renewed lease of 'motivity'. 'The solar system', he argued, 'according to the best of modern scientific belief, is dynamically analogous to the clockwork, in all the essentials of our consideration'. In particular, he emphasized that 'the running down of the weight in the clockwork has its perfect analogue, as Helmholtz was . . . the very first to point out, in the shrinkage of the sun from century to century under the influence of the mutual gravitational attractions between its parts'.[55]

The sun and solar system, like the clock, functioned not as perpetual motion machines, but as part of that 'physical law of irreversibility according to which there is a universal tendency in nature to the dissipation of mechanical energy'.[56] Sir William's public discussions of the sun's heat ended in 1892 effectively where they had begun in 1852 – with an unwavering commitment to the cosmic dissipation of energy. He offered a final glimpse into a grand cosmological vision:

> We do not know for certain whether the light which left the sun three thousand years ago is still travelling outwards with almost undiminished energy, or whether nearly all is dissipated in heat, warming the luminiferous ether, or ponderable bodies which have obstructed its course. But we may, I think, feel almost sure that it is partly still travelling outwards as radiant heat, and partly spent (or dissipated) in warming ponderable matter (or ponderable matter and the luminiferous ether).[57]

Lord Kelvin and the age of the sun

The first major critique of Sir William Thomson's estimate of the sun's age came in 1894–5 from a former pupil, John Perry (1850–1920), who had served as a member of Sir William's laboratory corps in the mid-1870s.[58] The central issue which concerned Perry was not the argument from the sun's heat but the secular cooling of the earth (ch. 17). Nevertheless, during the course of a lengthy debate in *Nature* and between Perry and Lord Kelvin (as he was by 1892), Perry critically examined many of Kelvin's assumptions relating to the sun's heat. In this section, we shall focus first on Perry's review of the sun's heat argument, and second on the quite different implications of radioactivity for that argument during the 1900s.

By contrast to George Darwin, Perry stated in a letter to Tait dated 26th November, 1894, that 'the argument from the sun's heat seems to me quite weak. Even a geologist without mathematics can see that the time given by Lord Kelvin will be increased if we assume that in past times the sun radiated energy at a smaller rate than at present, much of its mass being possibly cold and in the meteor form, and the rate may have varied greatly from time to time'. Such an alternative hypothesis was, Perry believed, 'not only possible but probable, and

[55] PL, **1**, 414–15; **2**, 470–3. [56] *Ibid.*, p. 469. [57] *Ibid.*, p. 472. [58] SPT, **2**, 742, 853.

it is for you and Lord Kelvin to prove a negative'.[59] In that belief, Perry cited the full support of his friend G.F. FitzGerald. Kelvin himself wrote to Perry on 13th December, 1894, and confidently asserted that 'Helmholtz, Newcomb, and another [probably a whimsical way of referring to himself] are inexorable in refusing sunlight for more than a score or a very few scores of million years of past time'.[60] Perry replied:

> Perhaps I had no right to avoid the sun's heat argument . . . I have only now received a copy of Newcomb's Astronomy. You will notice that he prefaces his remarks with 'If we take the doctrine of the Sun's contraction as furnishing the complete explanation of the solar heat during the whole period of the sun's existence, we can readily compute . . .'. Surely this is begging the whole question . . . Great meteor feeding is not now going on & the Helmholtz method is accepted as right. But what proof is there of no great feeding by streams of meteors for millions of years in the past. Surely Helmholtz was wiser than Fontenelle's roses.[61]

The uniformitarian assumption of Fontenelle's roses (see epigraph) that present activity was a guide to past activity – in kind rather than intensity – applied as much to Kelvin and Helmholtz as to the 'British popular geologists', and Perry reminded Kelvin that it was no more than an assumption.

Continuing his critique of Newcomb – and in effect of Kelvin – Perry referred to Newcomb's attempt 'to give a justification for your uniformitarian idea', a justification to which Kelvin had apparently already 'specially called' Perry's attention. Newcomb had remarked:

> If we reflect that a diminution of the solar heat by less than one-fourth its amount would probably mean an earth so cold that all the water on its surface would freeze, while an increase of much more than one-half would probably boil all the water away, it must be admitted that the balance of causes which would result in the sun radiating heat just fast enough to preserve the earth in its present state has probably not existed more than 10 million years.[62]

For Perry, this was 'a very off-hand statement, but who has proved it?' Yet it was 'the cornerstone of the Sun's Heat Argument'. On such a view, Perry reflected, 'if the earth were now 15½ percent further away from the sun there would be no water & no life, only ice; and if we were 18.4 percent nearer the sun, there again would be no water or life, only steam'. He went on facetiously, 'I often feel thankful to have lived in the reign of Queen Victoria, but I have not been sufficiently grateful for the average distance of the earth from the sun being so exactly what it is. Surely there is no other place in the universe with life!'

[59] John Perry to P.G. Tait, 26th November, 1894 (copy), P59c, ULC; *Nature*, **51** (1894–5), 226.
[60] Lord Kelvin to John Perry, 13th December, 1894 (copy), P60, ULC; *Nature*, **51** (1894–5), 227.
[61] John Perry to Lord Kelvin, 26th December, 1894, P62, ULC.
[62] Newcomb, *Popular astronomy*, p. 511. Clarence King, 'The age of the earth', *Am. J. Sci.*, **45** (1893), 1–20, on p. 19, quotes this passage from Newcomb in support of his own (and Kelvin's) view of a limited time scale.

Perry then raised two important questions deriving from his ironic criticism of Newcomb. First, was there no water on the planet Venus, and was its atmosphere mostly steam? Surely, he argued, 'it is well known that the atmos[e] of Venus is just like that of the earth'. Second, was there only ice on Mars and can there be no life? Yet Newcomb, while he had elsewhere ventured to say 'If there are any astronomers on Mars', seemed to Perry to be going too far when he made a statement proclaiming the utter sterility and iciness of Mars 'in all past times & for millions of years to come'.

Even supposing Newcomb's statement about an earth such as exists now to be true, Perry argued, 'great blanketings of the earth by vapour are possible in the past'. He asked, 'Is it not even probable that the earth surface may have been receiving much less heat & light from a smaller sun for hundreds of millions of years & yet being much warmer inside its blankets because of a very rapid conduction supply from the inside'. Perry confessed that he was 'beyond my depth in these speculations' – but asked rhetorically if Newcomb wasn't so also? Remarking that much in Newcomb's book was given at second hand, Perry turned to the source of Newcomb's views: Kelvin himself. Thus, as to the time estimate:

> surely this is not by Newcomb or Helmholtz, but another [Kelvin!]. Newcomb gives 18×10^6 years. But I very much prefer to take such a calculation at first rather than second hand. In your P. L. & A. [*Popular lectures and addresses*] you say 'which showed it to be possible that the sun may have already illuminated the earth for as many as one hundred million years, but at the same time also rendered it almost certain that he had not illuminated the earth for five hundred millions of years'. Let us give then to the geologists 400 million (of the present rate) & more than double it because the rate has fluctuated & add some hundreds of millions for a blanketed earth with a cold sun & you give Geikie and Huxley much more than all the time they want.

For Perry, 'a large planet like Jupiter, as it cools more slowly, exhibits very well the conditions of the earth long ago'. Jupiter was now in a blanketed state, and the *Brachiopoda* (a kind of bivalve mollusc) in Jupiter's seas 'don't care much whether the sun shines or not'. But 'without his blankets, if he has a solid surface, Jupiter would be cold indeed'.

On 8th January, 1895, Perry wrote again to Lord Kelvin. He once more confessed that with the sun's heat, unlike the secular cooling of the earth, he was quite beyond his depth – or, rather, 'my *assertions* can be of no value in comparison with your & Newcomb's'. Perry shrewdly side-stepped the temptation to offer rival speculations on the nature of the sun's heat. He merely reviewed Kelvin's assertions in the light of available evidence, evidence not compatible with those assertions. Thus, 'Mars has only a solar radiation of 40% of ours & yet its water is *not* all ice, for its polar caps melt in summer; and Venus has twice the solar radiation which we have & yet her atmosphere is like ours'. More importantly:

It is not a mere assertion to say that the actual temperature of our seas will not depend upon the solar radiation so much as upon the kind of atmosphere; the kind of blanketing. If the earth had no atmosphere what would be the average of temperature of its surface, say in equatorial Africa, even if solar radiation were twice as great as it is? Would it not be awfully cold[?][63]

In a letter addressed directly to *Nature*, and published on 7th March, 1895, Kelvin drew attention to the recent estimate of twenty-four million years for the earth's age by the American geologist Clarence King. Kelvin not only stated that he was 'not led to differ much' from King's estimate, but that the final section of King's paper carefully considered the estimates of the age of the sun's heat by Helmholtz, Newcomb and Kelvin. In particular, he quoted with evident pleasure King's conclusion:

The earth's age, about 24 million of years, accords with the 15 or 20 millions found for the sun. In so far as future investigation shall prove a secular augmentation of the sun's emission from early to present time in conformity with Lane's law [Homer Lane's determination of the internal density of the sun], his age may be lengthened, and further study of terrestrial conductivity will probably extend that of the earth. Yet the concordance of results between the ages of sun and earth certainly strengthens the physical case and throws the burden of proof upon those who hold to the vaguely vast age, derived from sedimentary geology.[64]

Perry, however, was not impressed by such *prima facie* concordance. In a long letter published in *Nature* on 18th April, 1895, he not only aimed to comment upon 'Lord Kelvin's friendly article of March 7 . . . but to deal more generally with the subject, in the hope of clearing away the misapprehensions which exist between modern geologists and palaeontologists, who are no longer uniformitarians, and physicists who are represented by Lord Kelvin'.[65]

First, Perry questioned Kelvin's 1887 justification of Helmholtz's hypothesis of mere mutual attraction and his criticism of Croll's idea of two moving bodies coming together to form the sun. In early times, Perry argued, we might postulate a sun of half the present mass, but many times the present diameter, with its radiant heat supplied by meteors. Excess feeding would lead to an increased diameter; reduced feeding would lead to contraction. Meteors from stellar space might possess great initial velocities, but if their paths were enormously out of line with one another and with the centre of the sun, 'we need not imagine them to alter much the moment of momentum of the sun about its axis'.[66]

Second, Perry published a revised version of his criticism of Helmholtz's, Newcomb's and Kelvin's uniformitarian assumption that the sun had been radiating always at this present rate. Departing from this assumption, 'we may

[63] John Perry to Lord Kelvin, 8th January, 1895, P63, ULC.
[64] Lord Kelvin, 'The age of the earth', *Nature*, **51** (1894–5), 438–40, on p. 440.
[65] John Perry, 'The age of the earth', *Nature*, **51** (1894–5), 582–5. [66] *Ibid.*, p. 584.

imagine that for long periods the sun radiated at a smaller rate, whether because his mass was smaller, or because of his atmosphere'. The consequence would again be an increase in the estimated age. Newcomb's defence of the uniformitarian assumption in terms of the effect of increased or reduced solar radiation on the earth required 'the careful consideration of men who know more about astronomical physics than I do'. Nevertheless, Perry drew attention to the implications – already discussed in his letter to Kelvin – of such an argument for the earth, Venus and Mars, implications which ran counter to accepted evidence. Furthermore, he turned one of Kelvin's remarks to his own account:

> On this question I venture to quote Lord Kelvin, who said, in 1887 . . . that 'the intensity of the solar radiation to the earth is 6 ½ per cent greater in January than in July; and neither at the equator nor in the northern or southern hemispheres has this difference been discovered by experience or general observation of any kind'. It is difficult to imagine that if the effect of 6 ½ per cent cannot be detected, 25 per cent should convert all the water to ice and destroy all life.[67]

His tentative answer, as we saw, centred rather on the possibility of a very much greater blanketing by the earth's atmosphere in the past.

Perry concluded that the age of the sun may have been very considerably under-estimated. The physicists gave maximum estimates of the age of life on earth as 1000 million years based on tidal retardation, 400 million years from the earth's heat, and 500 million years from the sun's heat. Therefore 'if we exclude everything but the arguments from mere physics, the probable age of life on the earth is much less . . . but if the palaeontologists have good reason for demanding much greater times, I see nothing from the physicists' point of view which denies them four times the greatest amount of these estimates'.

The possibility of 4000 million years as the age of the solar system may have been a far cry from Tait's ten million years, but it was not necessarily intended as a *refutation* of Kelvin's methods. Indeed, Kelvin himself had remarked in his letter to Perry of 13th December, 1894, that 'I thought by range from 20 millions to 400 millions was probably wide enough, but it is quite possible that I should have put the superior limit a good deal higher, perhaps 4000 instead of 400'. Nevertheless Kelvin continued to believe that the more precise physical data available in recent years pointed to a shorter time scale. Perry, recognizing that Kelvin was ignoring the geological evidence, had concluded that adjustments to Kelvin's assumptions could just as easily lead to a much longer time scale, which would then be consistent with those very demands of geology.

Kelvin did not respond further to Perry's critique. His 1899 paper 'The age of the earth as an abode fitted for life' focussed on terrestrial time scales without discussing Perry's arguments. He referred to the question of the age of the sun's heat only briefly in terms of 'the well-founded dynamical theory of the sun's

[67] *Ibid.*, p. 585.

heat carefully worked out and discussed by Helmholtz, Newcomb, and myself' – a theory which still limited the period of radiation at a rate sufficient to support some kind of animal and vegetable life on the earth to twenty to twenty-five million years.[68]

With the discovery of radioactivity by Henri Becquerel in 1896, and the revelation by Pierre Curie and Albert Laborde in March, 1903, that radium had the extraordinary property of continuously radiating heat without itself cooling down to the temperature of surrounding objects, W.E. Wilson was able to announce in a letter to *Nature* in July, 1903, that a clue had now been afforded to the source of energy in the sun and stars. On the basis of the Curies' measurement of the one-hundred calories per hour supplied by one gram of radium, and Langley's estimate of solar radiation, Wilson computed the amount of radium which would suffice to supply the sun's entire output of energy as a mere 3.6 grams per cubic metre of the sun's volume.[69]

In September, 1903, George Darwin wrote independently to *Nature* on the implications of radioactivity for the sun's heat. Quoting from Kelvin's 1862 paper, Darwin pointed to the phrase which Kelvin had there employed to qualify his estimates of a 500 million year upper limit: 'unless sources now unknown to us are prepared in the great storehouse of creation'. We had recently learnt the existence of another source of energy, Darwin explained, such that 'the amount of energy available is so great as to render it impossible to say how long the sun's heat has already existed, or how long it will last in the future'.[70]

On the basis of Langley's data, and on the assumption of a homogeneous sphere, the lost energy of concentration of the sun would suffice to give a supply for twelve million years. Again with Langley's data, but with a 'conjectural augmentation of the lost energy to allow for the concentration of the solar mass towards its central parts', Kelvin's sun would supply heat for sixty million years. Using Ernest Rutherford's (1871–1937) measurement that a gram of radium could emit 10^9 calories, Darwin estimated that, if the sun were made of such radioactive material, it would be capable of emitting 'nearly 40 times as much as the gravitational lost energy of the homogeneous sun, and 8 times as much as Lord Kelvin's conjecturally concentrated sun'. He concluded:

Knowing, as we now do, that an atom of matter is capable of containing an enormous store of energy in itself, I think we have no right to assume that the sun is incapable of liberating atomic energy to a degree at least comparable with that which it would do if made of radium. Accordingly, I see no reason for doubting the possibility of augmenting

[68] MPP, **5**, 229.

[69] W.E. Wilson, 'Radium and solar energy', *Nature*, **68** (1903), 222. For a full discussion of radioactivity and the age of the earth debates in the period 1896–1946, see Burchfield, *Age of the earth*, pp. 163–211.

[70] G.H. Darwin, 'Radio-activity and the age of the sun', *Nature*, **68** (1903), 496.

the estimate of solar heat as derived from the theory of gravitation by some such factor as ten or twenty.[71]

In contrast to the younger generation of physicists, Kelvin reacted unenthusiastically. On 19th January, 1906, a former pupil, James Orr (1844–1913) wrote to him to ask his opinion of the contention – urged by G.H. Darwin, W.C.D. Whetham and others – 'that if the sun or earth contains even a small amount of *radioactive* matter, all physical calculations of age & heat are overturned, & the old ratios of the geologists are restored (or may be)'. Orr, therefore, wanted to know if this contention 'really sets aside, or essentially modifies, your previous calculations'.[72]

In his reply, dated 29th January, 1906, Kelvin did not think there was 'any serious probability in the suggestions of Professor Darwin and Mr Whetham that either the heat of the Sun, or the underground heat of the Earth is practically due, in any considerable proportion, to radioactive matter'. Rather, he remained committed to the primacy of the gravitational theory:

> The gravitational theory is amply sufficient to account for the heat of both bodies [sun and earth], and of all the stars in the Universe, and it seems almost infinitely improbable that Radium adds practically to their energy for emission of heat and light. It may be indeed more probable that the energy of Radium may have come originally in connection with the excessively high temperatures, which we know to have been produced and to be at present being produced by gravitational action throughout the Universe.[73]

Kelvin, indeed, concluded that his own papers offered 'a strong body of evidence that the age of the Earth as an abode fitted for life cannot probably be vastly greater than twenty million years'. Most of that evidence, he claimed, 'would not be seriously affected even if radium concurred appreciably with gravitation in producing the heat and light which we have at present in the universe'. Again, in his letters to *The Times* during August of that year, Kelvin simply stated that there was no experimental foundation for the hypothesis that the heat of the sun was due to radium, ascribing the sun's heat instead to gravitation.[74]

71 *Ibid.* See also Burchfield, *Age of the earth*, pp. 166–7. Support for Darwin and Wilson came quickly from John Joly in *Nature*, **68** (1903), 526, and doubts expressed by W.B. Hardy in *Nature*, **68** (1903), 548, that beta and gamma rays from the sun had not been detected on earth were soon answered by R.J. Strutt and Joly in *Nature*, **68** (1903), 572. Essential to their reply was the argument that such radiation would be filtered out by the atmosphere. Ernest Rutherford in his 1904 address to the Royal Institution concluded with a brief reference to the conviction that 'the presence of radium in the sun, to the extent of about four parts in one million by weight, would of itself account for the present rate of emission of heat'. See *The collected papers of Lord Rutherford of Nelson* (3 vols., London, 1962–5), **1**, p. 657. Further early support appeared in W.C.D. Whetham, *The recent development of physical science* (London, 1904), pp. 320–2.

72 James Orr to Lord Kelvin, 19th January, 1906, O13, ULC.

73 Lord Kelvin to James Orr, 29th January, 1906 (copy), O14, ULC. On Orr, see Moore, *The post-Darwinian controversies*, p. 71.

74 See Frederick Soddy, 'The recent controversy on radium', *Nature*, **74** (1906), 516. Whetham, *Recent development*, p. 322, acknowledged that 'the spectrum of radium does not show in the sun's

The radium theory did not solve the problem of the sun's heat. It did, however, open up vast spans of time for geological and biological evolution, spans since increased to unthinkable proportions by the discovery of nuclear fusion. In retrospect, then, the geologists are vindicated and Kelvin appears a reactionary, holding tenaciously to doctrines outmoded even in his own lifetime. Yet the basic thrust of his half-century long campaign was also vindicated if it is seen as a struggle for recognition of the 'Universal tendency in nature to a dissipation of mechanical energy'. Rutherford's remarks in 1905 provide a fitting epitaph:

> Although such considerations [of the heating effects of radium] may increase our estimate of the probable duration of the sun's heat [possibly by one-hundred times Kelvin's estimate], science offers no escape from the conclusion of Kelvin and Helmholtz that the sun must ultimately grow cold and this earth become a dead planet moving through the intense cold of empty space.[75]

Put differently, Kelvin and the physicists of his generation had succeeded in imposing a profound restriction on any idea of progress and perfectibility through evolution in nature. All such progress had to be built upon decay. The resolution of one of the earliest conflicts in his personal life, which reconciled Chalmers with Nichol, providence with fixed natural law, and dissipation with conservation, had become an established tenet of all cosmology.

From this perspective, Thomson's opposition to the geologists' time and to the edifice it supported, of evolution by natural selection, is entirely comprehensible. The same commitment that had grounded his dissipation axiom grounded also his opposition. God had created a world with a built-in principle of decay, in which only the directing-power of will and mind could, for local systems, reverse the natural order of progression by applying externally obtained energy (ch. 18).

Kelvin's opposition to vast geological time, however, rested not merely on theological grounds. It involved also his profound attachment to measurement and machines. These criteria of reality in terms of practicality and concrete conceivability, which he constantly imposed throughout his researches, operated here in particularly evident ways through his frequent references to properties of matter consistent with those measurable on earth and in his machine models of the sun's internal action. An identical commitment underlies his parallel concern with the secular cooling of the earth.

light'. He believed that laboratory experiments indicated that 'a large proportion of radium would be necessary to make visible its characteristic lines' and that 'the prevalence of helium suggests the occurrence of radioactive processes, during which, as we know, helium may be formed'. Kelvin clearly found such evidence unacceptable.

[75] Ernest Rutherford, 'Radium – the cause of the earth's heat', *Harper's Mag.*, **49** (1904–5), 390–6; *Collected Papers*, **1**, 776–85.

16

The secular cooling of the earth

> Do you or geologists in general give adhesion to Darwin's prodigious durations for geological epochs? If so I think you must contrive your ancient plants and animals to have lived without a sun . . . The theory of central heat is about equally refractory against the assumption of great antiquity. The widest stretch I can make is some 200 to 1000 million years since the whole was melted if it ever was, or a correspondingly shorter history possible for organic life if the signs of central heat which we have are due to any ancient heat sufficient to produce them, but not intense enough to melt. I must say that Darwin's principle of estimating 340 million years for some of your ancient South of England headlands seems to me singularly defective; and that thousand might be substituted for million without violating probabilities in connection with any of the facts he mentions. *William Thomson to John Phillips, 1861.*[1]

William Thomson's 1846 inaugural dissertation 'On the distribution of the heat through the body of the earth' was an application of his 1844 work on heat conduction to problems of terrestrial temperatures and the age of thermal distributions (ch. 6). As such, the dissertation not only confirmed him as a disciple of Fourier, but also aligned him with the progressionist geological dynamics of Adam Sedgwick, William Whewell, and William Hopkins. Furthermore, as Thomson later explained, he had suggested in the dissertation 'that a perfectly complete geothermic survey would give us data for determining an initial epoch in the problem of terrestrial conduction'.[2]

Hopkins's researches in physical geology up to 1857 formed a prelude to Thomson's own studies of the earth in 1861–2. It was always Hopkins for whom Thomson retained the greatest respect and deepest admiration. In 1862, for instance, he expressed his verdict thus: 'good books on geology are mixed descriptive & dynamical. In the best of them the dynamical part is very bad,

[1] William Thomson to John Phillips, 7th June, 1861, Phillips papers, The University Museum, Oxford.

[2] William Thomson, 'On the secular cooling of the earth', *Phil. Mag.*, [series 4], **25** (1863), 1–14; MPP, **3**, 295–311, esp. pp. 296–7. On Sedgwick, Whewell, Hopkins, and their progressionist geological dynamics, see Crosbie Smith, 'Geologists and mathematicians: the rise of physical geology' in P.M. Harman (ed.), *Wranglers and physicists. Studies on Cambridge physics in the nineteenth century* (Manchester, 1985), pp. 49–83.

Hopkins alone excepted'.[3] Fundamental to the experimental researches of both Hopkins and Thomson in the 1850s was a concern with thermal properties of matter. With respect to his solar theorizing, Thomson attached much importance to investigating those properties which, under extremes of pressure and temperature, might differ markedly from the properties of the same substances measured in the laboratory. The same kinds of questions arose for the interior of the earth, and Hopkins devoted much effort to establishing, within his progressionist framework, that high pressures would account for a largely solid, rather than fluid, earth.

This work brought together Hopkins's interests in physical geology with those of Joule in measurement of physical quantities such as temperature, pressure, and volume, and of William Fairbairn in engineering and the strengths of materials. Although the investigations took place in Manchester, the guiding hand of the Glasgow professor of natural philosophy was evident at every stage. With his own visits to Manchester in connection with the Joule–Thomson experiments on the thermal properties of fluids in motion, with his regular correspondence with Joule, and with his own developing laboratory interests in the measurement of the properties of matter, the Hopkins–Joule–Fairbairn researches provide a vital part of the context for Thomson's studies of the secular cooling of the earth.

William Hopkins and the doctrine of central heat

Following the courses plotted by Sedgwick and Whewell, William Hopkins aimed to advance geological science to the kind of high and secure status long enjoyed by physical astronomy. To achieve these aims, he attempted to employ a rigorous 'inductive' method which would enable the physical geologist to avoid the fallacies of mere hypothetical or metaphysical speculation. Beginning with a phenomenal geology concerned with geometrical description, Hopkins had proceeded as early as 1836 to the level of a geological dynamics involving the notion of a general elevatory force as the indisputable cause of the phenomena of elevation (ch. 6). From this important but limited range of phenomena he had moved by 1847 to investigations of the nature of that general elevatory force by analogy with the action of volcanoes and earthquakes.

The expansive force of heat common to all these phenomena pointed towards the widely accepted and more general cause known as central or primitive heat, a cause which promised to achieve for physical geology what gravity had achieved for physical astronomy. The doctrine of central heat, therefore, not only received empirical support from increase-of-heat-with-depth measurements, but offered consilience (to use Whewell's term) between the diverse phenomena of, for example, the figure of the earth, volcanoes, earthquakes, and

[3] William Thomson, Notebook regarding Thomson and Tait's *Treatise* (1862), NB48, ULC.

elevation. It was a doctrine of great generality and simplicity, providing not only the fundamental agency in geological dynamics but the principal tenet of progressionist (directionalist) geological science.

Vigorous opposition to the central heat doctrine (and hence to progression) had come from Lyell as early as 1831–3. Because of the increase of temperature with depth, Lyell had argued against a solid earth (to which Fourier's techniques could be applied), and in favour of a liquid interior maintained at a uniform temperature of fusion in order that the crust could exist. He had further argued for compensating sources of heat in the form of a circle of perpetual motion with no discernible beginning or end. By contrast, Hopkins's ardent support for the central heat doctrine had led him in 1836 (and subsequently) to argue for an originally fluid earth which cooled, leaving cavities of still-molten matter in the solid mass as the source of that 'elevatory force' which produced observable geometrical phenomena as well as volcanoes. Thus a solid earth 'defeats' the uniformitarians by returning to Fourier's argument that the present distribution of temperature over the surface and with depth is consistent with the cooling of a solid by conduction. The full coherence of Hopkins's counterattack on Lyell, however, demanded a dramatic change in the properties of solid matter with increased temperature and pressure inside the earth.

In his presidential address to the Geological Society of London in February, 1852, Hopkins defined progression as a change by which the inorganic matter of our planet 'has passed, in the long process of time, from a primitive to its present condition, and may still pass to some ultimate and different state'. This state had rendered the surface of the earth 'more fit for the habitation of the higher orders of organized beings' (ch. 18) and may have come about by either paroxysmal or uniform action of geological forces. Paroxysmal here corresponded to Whewell's term 'catastrophist' (greater intensity of geological agencies than at present) and uniform to Whewell's 'uniformitarian' (equal intensity with the present). In contrast, Hopkins defined non-progression as the absence, not of periodic change, but of permanent change.[4]

Hopkins's address reflects a close acquaintance with Thomson's dynamical theory of heat, for Hopkins's completely general doctrine of cosmical progression derived here from a knowledge of the irreversible nature of heat. Hopkins therefore argued that 'the theories of uniformity and of non-progression appear to me incompatible with our most certain knowledge of the properties of heat – that ever-active agent in the work of terrestrial transformation'. He carefully refrained from discussing 'the evidence which might be deduced from recognized geological phenomena in favour of the theory of progression', basing his

4 William Hopkins, 'Anniversary address of the President', *Quart. J. Geol. Soc. London*, **8** (1852), xxiv–lxxx, esp. pp. lxxii–lxxiii, lvii. On Whewell's terms, see Smith, 'Geologists and mathematicians', p. 59; W.F. Cannon, 'The uniformitarian–catastrophist debate', *Isis*, **51** (1960), 38–55, esp. pp. 54–5.

conclusion on the 'incontrovertible' experimental truth of the simple proposition that 'if a mass of matter, such, for instance as the earth . . . be placed in space of which the temperature is lower than its own, it will necessarily lose a portion of its heat by radiation, until its temperature ultimately approximates to that of the circumambient space, unless this reduction be prevented by the continued generation of heat'.[5] The coincidence with Thomson's thinking is scarcely surprising since they probably discussed the subject at Manchester in December, 1851, or January, 1852, when engaged in joint experiments.

For the earth's interior temperature to be consistent with non-progression, Hopkins continued, it would have to be maintained by some cause internal to the earth. Chemical causes might act for a finite, but certainly not for an infinite, time, and no other internal cause seemed any more likely to avoid the violation of the impossibility of perpetual motion. External causes, such as solar or stellar radiation, could not produce an absolutely constant, stationary temperature which increased with depth. Hopkins concluded: 'I cannot conceive . . . the present state of terrestrial temperature to be a *permanent* state'. On the other hand, a perpetually recurring series of changes might be attributed – along the lines advocated by Poisson (ch. 6) – to the external causes. Hopkins regarded 'such hypotheses as extremely unsatisfactory, and utterly unfit to be made the foundation on which a great speculative theory may rest'. In the end, he argued, the celestial bodies themselves must be subject to permanent changes of temperature. Thus, 'reasoning from all we know of the properties of matter and heat, I am unable in any manner to recognize the seal and impress to those laws alone by which we conceive it at present to be governed'.[6]

As a further, closely related, criticism of the doctrine of non-progression, Hopkins argued that it was contrary to the very essence of geology as a physical science. Such a doctrine seemed 'to involve the rejection of the notion of a *beginning* of the actual physical condition of our planet' and to imply that the sequence of periodical changes had been of infinite duration. The alternative was to assert that 'the earth must have been created at once, at some finite distance of time, as fit a dwelling-place for organic beings, as it has been rendered, according to the theory of progression, only by a long series of superficial operations'. Such an immediate act of creation would precede the operation of antecedent physical causes – the study of which formed the essence of physical geology – and so would 'sap the foundations on which alone geology can rest as physical science'. As we saw in the previous chapter, William Thomson implicitly reiterated Hopkins's and Whewell's position here, namely, that while an immediate act of creation was within the power of an omnipotent God, God had chosen to work through certain laws, and it was this choice which made physical science possible as a study of 'antecedent conditions' and 'future evolutions'.

[5] Hopkins, 'Anniversary address', p. lxxiii. [6] *Ibid.*, p. lxxiv.

In the same paper, Hopkins stressed that *approximate* non-progression was perfectly consistent with the progressive refrigeration of the earth for enormous periods of time. In this context he referred to his own very recent and more specific studies which purported to show that 'climatal conditions . . . may, consistently with the earth's continual refrigeration, have remained sensibly unaffected by the internal heat . . . for millions of centuries; and the very theory which tells us that these conditions can never be sensibly altered in all future time (external circumstances remaining the same) essentially involves the hypothesis of progressive change towards an ultimate limit'.[7] These more detailed investigations of the effects of the earth's internal heat responded directly to a new challenge to the validity of the central heat doctrine.

The acceptance (albeit slowly and reluctantly) by British geologists of a recent and widespread glacial epoch in Western Europe posed a significant problem for the central heat doctrine. The notion of a gradually cooling earth seemed consistent with a warmer climate in times past, but inconsistent with a change of climate from comparative warmth to a recent glacial epoch. Louis Agassiz's attempt in 1840 to reconcile his ice age with central heat by suggesting a modified doctrine, that of oscillatory cooling of the globe (whereby the temperature would fall steeply and then rise slightly to a stable plateau before falling again), scarcely achieved general acceptance.[8] During the 1850s, on the other hand, Hopkins aimed to show that such climatal changes as the glacial epoch could scarcely be affected by a central heat, but were due rather to 'superficial causes' such as changes in the distributions of land and sea.[9]

Having thus satisfied himself regarding the consistency of the central heat doctrine with the occurrence of a recent glacial epoch, Hopkins turned to the problems which the doctrine raised concerning the internal character of the earth, and in particular the thermal properties of matter. As early as 1839–42, he had investigated the earth's precession and nutation and concluded that a largely solid rather than almost wholly fluid terrestrial sphere was required. The immediate problem posed by Lyell for the central heat doctrine, however, had been that the law of increase of heat with depth pointed to a central molten mass at depths of not much more than sixty or seventy miles beneath the earth's surface. As a solution, Hopkins had therefore suggested that the effects of pressure at great depths might increase the temperature at which rocks melted, and that the thermal conductivities of the lower rocks might be considerably greater than those of the superficial sedimentary beds. In either case, the thickness of the earth's solid crust would then be much enhanced, and consistency with the findings of physical astronomy obtained. Hopkins thus sought to institute empirical researches on the thermal properties of matter in order to

[7] *Ibid.*, p. lxxv.

[8] M.J.S. Rudwick, 'The glacial theory', *Hist. Sci.*, **8** (1969), 136–57, esp. pp. 148–50.

[9] William Hopkins, 'On the causes which may have produced changes in the earth's superficial temperature', *Proc. Geol. Soc. London*, **8** (1852), 56–92.

provide an answer to this important question of solidity, a question which would also prove important for Thomson.[10]

In his presidential address to the British Association in 1853, Hopkins drew attention to the purpose and progress of his experimental investigations with Joule and Fairbairn at Manchester on the effects of great pressure upon the thermal properties (notably the temperature of fusion) of matter. They aimed to establish that the temperature of fusion of common substances increased under pressure, suggesting for the earth's crust a much greater thickness than that indicated by increase-of-heat-with-depth arguments based on the melting of rocks under ordinary pressures. Hopkins reported promising first results:

At present our experiments have been restricted to a few substances, and those of easy fusibility; but I believe our apparatus to be now so complete for a considerable range of temperature, that we shall have no difficulty in obtaining further results. Those already obtained indicate *an increase in the temperature of fusion proportional to the pressure to which the fused mass is subjected.* In employing a pressure of about 13 000 lbs. to the square inch on bleached wax, the increase in the temperature of fusion was not less than 30° Fahr. . . .[11]

The experimental researches had begun with discussions between Hopkins and Fairbairn in April, 1851. Hopkins's friend Robert Willis had informed him that a lever employed by Eaton Hodgkinson of Manchester to produce enormous pressures on cylinders now resided with Fairbairn. Fairbairn thus supplied the initial apparatus, and, as Hopkins remarked with gratitude for Fairbairn's very considerable contributions, 'without the engineering resources . . . at Mr Fairbairn's command, success would have been hopeless'. The Royal Society supplied a grant of £250 under the superintendence of a committee which included Joule.[12] But the presence of Professor William Thomson soon became manifest in several respects.

At a time when Joule and Thomson had begun their own researches on the properties of fluids in motion, it is scarcely surprising to find Thomson being consulted by Hopkins and Joule from the early stages. Thomson's first explicit contribution came in the form of an ingenious and simple experimental technique for determining the temperature of fusion inside brass tubes or cylinders. He suggested placing a small magnetized steel wire within the substance such that, on the substance melting, the wire would fall to the bottom of the tube. A small compass outside the tube would enable the detection of the movement, which corresponded to the moment of fusion. Brass, rather than glass, could thus be safely employed under severe pressures and high temperatures. In August, 1853, Joule reported to Fairbairn that 'our experiment this morning was

[10] William Hopkins, 'Researches in physical geology – first series', *Phil. Trans.*, (1839), 381–423, esp. pp. 381–5; 'Report on the geological theories of elevation and earthquakes', *BAAS Report*, **17** (1847), 33–92, esp. pp. 51–3.

[11] William Hopkins, 'Presidential address', *BAAS Report*, **23** (1853), xli–lvii, on pp. li–lii.

[12] William Hopkins to William Fairbairn, 25th April, 1851, in William Pole (ed.), *The life of Sir William Fairbairn, Bart.* (London, 1877), pp. 289–90.

satisfactory, the needle having fallen at 373, the exact temperature expected by Mr Hopkins, and which shows that the temperature of fusion rises with the pressure in arithmetic progression'. This result, together with a drawing of the apparatus despatched to John Phillips, Hopkins intended to communicate to the 1853 Hull meeting of the British Association over which he presided.[13]

Thomson, however, contributed more than experimental technique. Much of the investigation of the temperature of fusion derived from Thomson's thermodynamic formulae, and the results provided generally promising agreement between theory and observation. Bismuth, for example, showed no increase in the temperature of fusion for, like ice, it contracted on melting. As Hopkins explained to Fairbairn on 8th October, 1855:

> According to our new theory of heat [Thomson's dynamical theory], this ought to be the case, provided there be no increase of volume while the substance passes from the solid to the fluid state. I had just time before I left to try the experiment, and assure myself that bismuth in a fluid state probably occupies less volume than in its solid state, while such substances as wax, spermaceti, &c., of which the temperature of fusion is so much increased by pressure, occupy much *larger* space in a fluid than in a solid state. This is all accordant with theory; but requires still to be worked out with accuracy, which there will be no difficulty now in doing.[14]

In other words, the investigations of Hopkins, Joule, and Fairbairn into the temperatures of fusion of various substances under high pressures accorded directly with the thermodynamics of William and James Thomson.

Meanwhile, Hopkins had embarked around 1855 on studies of the effects of pressure on the conductivities of various substances. Greater conductivities in the inferior rocks would, like increased temperature of fusion, also allow for a greater thickness of solid crust for, given the same amount of heat conducted outwards, a higher conductivity would not require as steep a thermal gradient. In other words, the temperature of fusion would be reached at a greater depth. Aided once more by Fairbairn and Joule, Hopkins had carried out another long series of investigations which yielded the conductive powers of a variety of mineral substances under different conditions of pressure, discontinuity, temperature, and moisture.[15] His findings revealed that igneous rocks possessed much greater conductivity than sedimentary rocks, indicating, in accordance with his earlier conclusion, that 'the thickness of the crust could not be so small as 200 or 300 miles'.[16]

[13] J.P. Joule to William Fairbairn, 12th August, 1853; William Hopkins to William Fairbairn, 22nd August, 1853, in W. Pole (ed.), *Fairbairn*, pp. 295–6. See also William Hopkins, 'An account of some experiments on the effect of pressure on the temperature of fusion of different substances', *BAAS Report*, **24** (1854), 57–8.

[14] William Hopkins to William Fairbairn, 8th October, 1855, in W. Pole (ed.), *Fairbairn*, pp. 303–4. See also PL, **2**, 306–8, for William Thomson's agreement.

[15] William Hopkins, 'Experimental researches on the conductive powers of various substances, with the application of the results to the problem of terrestrial temperature', *Phil. Trans.*, (1857), 805–49, esp. pp. 805–21; W. Pole (ed.), *Fairbairn*, pp. 301–8.

[16] William Hopkins, 'On the earth's internal temperature and the thickness of its solid crust', *Proc. Royal Inst.*, **3** (1858–62), 139–43, esp. p. 141.

Hopkins attempted to go further, however, along the road towards increasing the thickness of the earth's crust. He argued that the apparent uniformity of temperature gradient obtained from measurements in deep mines actually conflicted with theory. Theory showed a variation of temperature gradient with conductivity, which implied that the temperature gradient in very different localities, with different rock strata, ought not to be uniform. He therefore concluded his paper with a speculation that 'superficial causes' – which he did not specify but which in his previous discussions meant changes in the relative positions of land and sea – might play a greater role in increasing the temperature gradient near the surface, while at the same time reducing that part due to the primitive heat and so allowing for a still-thicker solid crust:

> After the preceding investigations, it appears to me extemely difficult, if not impossible, to avoid the conclusion that a part at least of the heat now existing in the superficial crust of our globe is due to superficial and not to central causes. It should be remarked, however, that the argument thus afforded is not directly against the theory of a *primitive* heat, but only against the manifestation of the remains of such heat as the sole cause of the existing terrestrial temperatures at depths beyond the direct influence of solar heat. The argument in favour of the earth's original fluidity (a state only conceivable as the effect of heat) founded on the spheroidal form of the earth, remains unaffected. Whatever cogency it may have been supposed to possess, it possesses still. At the same time, all those collateral arguments derived from the existing temperature of the earth's crust, or the climatal changes which we believe to have taken place on its surface, are deprived, I conceive, of a great part of their weight.[17]

Hopkins made no attempt here or elsewhere, however, to say how much of the heat could be ascribed to superficial causes. At the same time, his commitment to the central heat doctrine remained as strong as ever.

Thomson, Helmholtz, and the importance of terrestrial temperatures

By comparison with Hopkins's extensive experimental and theoretical investigation into problems of terrestrial heat, William Thomson's own researches in that field during the period 1846–62 appear of relatively minor character. In 1855, however, Thomson read to the British Association a paper 'On the use of observations of terrestrial temperature for the investigation of absolute dates in geology'. He began with an assessment of J.D. Forbes's recent investigation of thermal conductivities, his omission of any reference to Hopkins being probably due to the fact that Hopkins's researches into conductivities had only just begun:

[17] Hopkins, 'Experimental researches', pp. 836–7. See also S.G. Brush, 'Nineteenth-century debates about the inside of the earth: solid, liquid or gas?', *Ann. Sci.*, **36** (1979), 225–54, esp. pp. 233–4. Brush's remarks that 'it is not clear that Hopkins had serious doubts about the doctrine of central primitive heat in 1842, but his research in the following years did lead him to challenge its adequacy' are misleading. Hopkins maintained a fundamental commitment to central heat throughout his researches. That Hopkins and Thomson do not appear to have communicated much after 1855 seems entirely due to the failure of Hopkins's health late in that year. He died in October, 1866, aged 73. See W. Pole (ed.), *Fairbairn*, pp. 304–5.

The relative thermal conductivities of different substances have been investigated by many experimenters; but the only absolute determinations yet made in this most important subject are due to Professor James Forbes, who has deduced the absolute thermal conductivity of the trap rock of Calton Hill, of the sandstone of Craigleith quarry, and of the sand below the soil of the Experimental Gardens, from observations on terrestrial temperature, which were carried on for five years in these three localities (all in the immediate neighbourhood of Edinburgh), by means of thermometers constructed and laid, under his care, by the British Association.[18]

In his paper, Thomson drew attention to specific results which showed a greater conductivity of the trap rock at the greater depths – a variation not established for the sandstone or sand. Even allowing for inaccuracies in the calibration of the thermometers, Thomson suggested that these results indicated 'the comparatively modern time at which the trap rock of Calton Hill has burst up in an incandescent fluid state'. Furthermore, he felt that his conjecture had been confirmed by a similar result for porphyry in the locality of the Rhenish spa town of Kreuznach, which he had recently visited with his ailing wife and where he had met Helmholtz for the first time.[19]

Thomson concluded that 'the mathematical theory of heat, – with data as to absolute conductivities of rocks, such as those supplied by Professor Forbes, and with the assistance of observation on the actual cooling of historical lava streams . . . – may be applied to give estimates, within determined limits of accuracy, of the absolute dates of eruption of actual volcanic rocks of prehistoric periods of geology, from observations of temperature in bores made into the volcanic rocks themselves and the surrounding strata'.[20] So far, however, he had not attempted to approach again the question of the cooling of the earth as a whole – still less in relation to geological time. His 1855 paper simply held out the possibility of assigning dates to particular rock formations on the earth's surface.

The precise issues discussed by Thomson and Helmholtz at Kreuznach are not known. But Helmholtz's 1854 lecture 'On the interaction of natural forces' – translated for the *Philosophical Magazine* by John Tyndall in 1856 – contained an important discussion of terrestrial heat which began in the context of nebular hypotheses as to the origin of the solar system:

Our earth bears still the unmistakable traces of its old fiery fluid condition. The granite formations of her mountains exhibit a structure, which can only be produced by the crystallization of fused masses. Investigation still shows that the temperature in mines and borings increases as we descend . . . But the cooled crust of the earth has already become so thick, that, as may be shown by calculations of its conductive power, the heat coming to

[18] William Thomson, 'On the use of observations of terrestrial temperature for the investigation of absolute dates in geology', *BAAS Report*, **25** (1855), 18–9; MPP, **2**, 175–7; J.D. Forbes, 'Account of some experiments on the temperature of the earth at different depths and in different soils near Edinburgh', *Trans. Royal Soc. Edinburgh*, **16** (1846), 189–236; William Thomson, 'On the reduction of observations of underground temperature; with application to Professor Forbes' Edinburgh observations, and the continued Calton Hill series', *Trans. Royal Soc. Edinburgh*, **22** (1861), 405–27; MPP, **3**, 261–94. [19] SPT, **1**, 308–10. [20] MPP, **2**, 177.

the surface from within, in comparison with that reaching the earth from the sun, is exceedingly small, and increases the temperature of the surface only about $\frac{1}{30}$ of a degree Centigrade; so that the remnant of the old store of force which is enclosed as heat within the bowels of the earth, has a sensible influence upon the processes at the earth's surface only through the instrumentality of volcanic phenomena.[21]

For Helmholtz, the processes at the earth's surface at present owed their power 'almost wholly to the action of other heavenly bodies, particularly to the light and heat of the sun, and partly also, in the case of the tides, to the attraction of the sun and moon'. Likewise, as to the future of our planetary system in general and of our earth in particular, 'the heat of the sun is the only thing which essentially affects the question' since the earth's internal heat 'has but little influence on the temperature of the surface'. With regard to past history, 'the time during which the earth generated organic beings is . . . small when we compare it with the ages during which the world was a ball of fused rocks'. According to Helmholtz's estimate, the earth's cooling from 2000° to 200°C would, on the experimental data provided for basalt by Bischof, require 350 million years.[22]

This brief statement regarding terrestrial time was intended by Helmholtz to illustrate to his audience both the finiteness of the planetary system and the vastness of its time scale in relation to the estimated 6000 year duration of human history or the one to nine million year duration of successive plant and animal species on earth: 'the history of man . . . is but a short ripple in the ocean of time'. With Helmholtz, then, we see already the public reinterpretation of man's place in nature within the framework of the new energy principles of conservation and dissipation. With Hopkins we have seen the conviction that the rapid advance of geology as a science will be best aided by the attention of men 'whose primary studies have been those of mechanical and physical science'.[23] And, with Thomson, we shall now see the powerful development of these themes when, stimulated by the biological and geological debates of this time, he embarked upon his paper 'On the secular cooling of the earth'.

The secular cooling of the earth

During the spring and early summer of 1861, while recovering from his broken leg, Thomson contemplated a challenge to uniformitarian geology at the forthcoming meeting of the British Association. His disposition indicated in Joule's letter of May, 1861, 'to expose some of the rubbish which has been thrust on the public lately', shows that the 'incomprehensibly vast' time scales demanded by Darwin's *Origin of species* were the *immediate* cause of his response to the uniformitarian claims, even though his outlook had been for many years

[21] Hermann Helmholtz, 'On the interaction of natural forces', (lecture at Königsberg, 7th February, 1854), *Phil. Mag.* [series 4], **11** (1856), 489–518, on p. 508. [22] *Ibid.*, pp. 515–16.

[23] William Hopkins, 'Anniversary address of the President', *Quart. J. Geol. Soc. London*, **9** (1853), xcii.

fundamentally at variance with Lyell's. Similarly, his letter of June, 1861, to John Phillips, quoted at the beginning of the present chapter, focussed on the defective nature of Darwin's computation of the Weald of Kent.[24]

In his reply, Phillips, a principal organizer of the British Association since its foundation in 1831 and appointed professor of geology at Oxford in 1860, agreed with Thomson's remarks and drew attention to his own recent maximum of ninety-six million years for the period of stratification:

Darwin's computations are something *absurd* as to the wasting of the sea. *For* the calcul[ation] includes as a coeff[icient] the *height*! of the cliff. This astonishing error is not the only one. No one who ever does calculate (among our geologists) attaches any weight to the *result*! In my 'Life on Earth' I have given some calcul[ations] which for the period of stratification, rise to 96 millions only – & have added not unlikely estimate, as to the period of coal deposition & other things of the sort.[25]

Phillips had published his *Life on the earth* in 1860, a book which contained the substance of the Rede Lecture delivered at Cambridge in May of that year. Although Thomson was not present, and did not receive a copy of the book from Phillips until 1869,[26] two features of Phillips's remarks on the 'Antiquity of the earth' stand out as significant for Thomson's subsequent paper on secular cooling: Phillips's discussion of Leibniz's *consistentior status* (the consolidation of the earth from red- or white-hot molten matter), and his estimate of ninety-six million years for the period of stratification.

Phillips's concern was to bring the epoch of life on the earth into a scale of solar time – to estimate 'not only the *relative* antiquity of the several races of plants and animals, but the *absolute* antiquity of the earliest inhabitants of the earth'. The natural phenomenon upon which he based his estimate was stratification but the estimate also depended on whether one assumed a uniformitarian or Leibnizian theory. Assuming equality of present and past rates of sedimentation, Phillips explained:

Take any large surface of the land, which yields to the atmospheric agency unequally in different parts . . .; observe and measure what is carried away by one or more rivers from the surface to the sea in one year. Assume this to be a fair average for the whole surface of the land, and, to save trouble, suppose the whole of the sediment to be spread out on the

[24] Charles Darwin's phrase 'incomprehensibly vast' – altered to 'how vast' in the fifth edition of *The origin of species* published in August, 1869 – and his discussion of the denudation of the Weald – omitted from the third edition of April, 1861, onwards – are contained in chap. 9, 'On the imperfection of the geological record' under the sub-heading 'On the lapse of time'. See Morse Peckham (ed.), *The origin of species by Charles Darwin. A variorum text* (Philadelphia, 1959), pp. 478–87. See also J.D. Burchfield, 'Darwin and the dilemma of geological time', *Isis*, **65** (1974), 301–21, and *Lord Kelvin and the age of the earth* (London, 1975), pp. 70–2, for an account of Darwin's calculation and its early critics.

[25] John Phillips to William Thomson, 12th June, 1861, P72, ULC; SPT, **1**, 539*n*. On Phillips and the British Association, see Jack Morrell and Arnold Thackray, *Gentlemen of science. Early years of the British Association for the Advancement of Science* (Oxford, 1981), esp. pp. 439–44.

[26] John Phillips to William Thomson, 5th January, 1869, P73, ULC. Thomson did not visit Cambridge between 1852 and 1866. See William Thomson to G.G. Stokes, 7th July, 1871, K176, Stokes correspondence, ULC.

sea-bed, over an area equal to that from which it was derived. The thickness wasted from the land, in one year, is thus the same as that added to the sea-bed in one year. Divide by this thickness the measured thickness of the sedimentary strata, the result is the number of Uniformitarian years employed in depositing the strata . . .[27]

As an example, Phillips stated that the Ganges would deliver $\frac{1}{111}$ of an inch of sediment to the Bay of Bengal each year, and so, for a maximum thickness of strata of about 72 000 feet (a figure also cited by Darwin), the antiquity of the base of the stratified rocks would be ninety-six million years. Such an estimate, however, might be too large – given that the sediment did not in fact spread over such a wide area – or too small – given the outstanding ability of the Ganges, compared to other rivers, to deposit sediment.[28]

Turning from uniformitarian assumptions to the Leibnizian theory, Phillips noted that 'we first must settle the limits of the atmospheric power to waste the surface in the early geological periods'. At the origin of the oceans 'there would indeed be a mighty power in action – perpetually falling floods – the fountains of heaven – wasting a hot and friable [likely to crumble] surface'. But, leaving aside this epoch, he moved forward to the base of the Palaeozoic system:

adopting as high a limit for the temperature of the surface of the globe as might be possible to suit the races of Mollusca and Crustacea then coming into view, we may allow 20° [F] higher mean temperature than at present. From this point it must be supposed to have declined to the actual condition, and with the depression of temperature in a more rapid progression.[29]

Assuming that the quantity of moisture sustained in air varied in geometrical proportion to the temperature, a former temperature 20°F above the present mean surface temperature of 56°F would, Phillips argued, imply a quantity of moisture nearly double that now supported by the atmosphere. Assuming also that the causes of rain and vicissitudes of seasons remained the same in kind, 'we may admit the atmospheric power . . . to have been double what it is now, and from that time to the present it has sunk in geometrical progression'. He concluded, 'the time consumed . . . could not be so little as two-thirds of that computed on the hypothesis of uniform action', viz., about sixty-four million years. On the other hand, supposing the resisting powers of the earth's crust to have been half as great in the past, 'we shall have the earlier atmospheric waste effectively four times as great as at present. This allowed, we shall find the whole time . . . cannot be reduced to so little as $\frac{2}{5}$, or 38 million years'.

In concluding his discussion of geological time, however, Phillips insisted that a thirty-eight million year period for the deposition of fossiliferous strata was too short:

It is rather to be supposed very much longer, if reduction of temperature is to be taken as the mainly influential condition of deviation from uniform action. For if this reduction of

[27] John Phillips, *Life on the earth. Its origin and succession* (London and Cambridge, 1860), pp. 121–4. [28] *Ibid.*, pp. 125–6; *The origin of species by Charles Darwin*, p. 481.

[29] Phillips, *Life on the earth*, pp. 134–5.

temperature were taken at 20° it must have required ages of ages to be accomplished by the excess of radiation into space over the heat received from the sun; the period may not elude calculation, but it lies quite beyond the power of the mind to contemplate with steadiness.[30]

With such a statement, Phillips might well be said to have set up the problem to which Thomson was about to turn his attention. His alignments with progressionist thinking in cosmical physics made him a natural ally of Phillips and Hopkins against Lyell and Darwin. But at no time did he hold geologists generally in contempt. On the contrary, his insistence upon consistency among different theories and different branches of natural science demanded that there be no great gulf between the findings of the geological scale of time – based on sedimentation rates – and those of the thermodynamic scale of time – based on heat conduction rates. With Phillips's estimates, uncertain and approximate though they were, Thomson had that necessary consistency. Thus, even if later geological approaches might produce far longer time scales quite at variance with his own, in 1861–2 Thomson did not ignore the findings of a purely geological approach to terrestrial time.[31]

The correspondence between Thomson and Phillips in 1861, however, reveals a more specific interaction of Glasgow natural philosopher and Oxford geologist. In his letter of 7th June, Thomson explained that he had just received results on underground temperatures at Greenwich from a former pupil, J.D. Everett. These results 'are very striking and unless the thermometers are *far wrong* demonstrate cold rather than heat in going downwards'. On consideration of possible causes of these results, Thomson rejected as inadequate a cooling effect produced by 'a glacial era of a few thousand centuries, ending 40,000 years ago'. Such an effect would 'only diminish the rate of increase [of temperature] and not make an actual diminution of temperature in going down'. An estimated cooling effect of $\frac{1}{4}$ to $\frac{1}{2}$°F per one-hundred feet of descent had to be set against an 'increase of temperature due to a more ancient heat' of about 2°F per one-hundred feet of descent. He therefore remarked that 'it seems on the whole more probable that the thermometers have gone wrong', but concluded with a question to Phillips:

is it quite certain that in all parts of the world there is an increase of temperature in going downwards at some such rate as 1°Fahr. per 50 feet, except in limited localities such as Calton Hill, or Creuznach, where eruptions of trap rocks in comparatively recent times gives a greater increase, or others where some local source of cold exists or has existed?

Since the 1855 BAAS meeting, indeed, Thomson had been urging the need for special geothermic surveys for the purpose of estimating absolute dates in geology.

[30] *Ibid.*, pp. 135–8.

[31] Burchfield, *Age of the earth*, pp. 32, 59–60, discusses Phillips's estimates in relation to Thomson's, but does not explore the nature of any interaction between geologist and natural philosopher.

In his reply of 12th June, John Phillips regretted that 'your wishes as to the subterranean temperature have not been more attended to', and suggested local cold – including the downward flow of cold water – as a possible cause of the variations. Thomson, in his 1862 published paper 'On the secular cooling of the earth' pointed out that Professor Phillips had 'well shown in a recent inaugural address to the Geological Society' that 'the disturbing influences affecting underground temperature . . . are too great to allow us to expect any very precise or satisfactory results'.[32] In other words, the great variations near the earth's surface seemed attributable only to local causes rather than to any more general causes on a par with central heat. Thomson apparently felt able to ignore, therefore, Hopkins's remarks concerning superficial causes.

William Thomson's ideas on the age of the sun's heat in relation to what he saw as the excessive time demanded by geologists for life on earth had emerged in his notebook for July, 1861, under the heading – remarkably similar to that employed by Phillips – 'On the antiquity of the earth as a habitable world'. He began his draft with the issue of geological time:

> Geologists have discovered monuments of ancient life on the Earth affording irrefragable evidence of a vast antiquity to be measured in thousands of centuries. But speculation outruns evidence, and, not contented with demanding time enough for the generation of all existing fossiliferous rocks, assumes ancestral lives of indefinite length for the Silurian beings, and draws upon an unlimited fund of time, without a thought of the force required for their living, and for the destruction of all traces of their forms after death. In reality the material of the Earth and Sun cannot have afforded in all time more than a finite quantity of mechanical energy; and life, as at present regulated, essentially involves a continual dissipation of energy by solar radiation into space.[33]

In parallel with his remarks on the limitations of the sun's heat, he also noted that 'the present thermal condition of the earth's mass, presenting in every locality in which it has hitherto been investigated, an increase of temperature downwards from the surface, either must have originated in comparatively recent times in a general elevation of the external temperature, which appears very improbable, or must proceed from an antecedent condition of measurable antiquity when the earth was unfit for life, whether from being at too high a temperature or from being composed of unoxydized metal up to its surface'.

Thomson began his 1861 paper to the British Association with the assertion that it was 'quite certain that the solar system cannot have gone on, even as at present, for a few hundred thousand or a few million years without the irrevocable loss (by *dissipation*, not by *annihilation*) of a very considerable proportion of the entire energy initially in store for such heat, and for plutonic

[32] MPP, **3**, 297.

[33] William Thomson, Research notebook, July, 1861, NB45, ULC. For the significance of 'an unlimited fund of time' in the relation of geology to political economy see M.J.S. Rudwick, 'Poulett Scrope on the volcanoes of Auvergne: Lyellian time and political economy', *Brit. J. Hist. Sci.*, **8** (1974), 205–42.

[igneous] action'. Although the whole store of energy was greater in the past, it was '*conceivable* that . . . the rate at which it was drawn upon may have been nearly equable, or may even have been less rapid'. But he believed it 'far more *probable* that the rate of secular consumption had been in some direct (not *simple* . . .) proportion to the total amount of energy in store at any time after the commencement of the present order of things'.[34]

This view he had 'endeavoured to prove . . . for the case of the sun' in a separate paper at the same meeting. He had there 'shown that most probably the sun was sensibly hotter a million years ago than he is now'. Hence, 'geological speculation assuming somewhat greater extremes of heat, more violent storms and floods, more luxuriant vegetation, and hardier and coarser grained plants and animals in remote antiquity are more probable than those of the extreme quietist school'. For his part, 'a "middle path", not generally safest in scientific speculation, seems to be so in this case'. Thus, he concluded, it was 'probable that hypotheses of grand catastrophes destroying all life from the earth and ruining its whole surface at once are greatly in error; it is impossible that hypotheses assuming an equability of sun and storms for 1 000 000 years can be quite correct'.[35]

Now, however, it was the earth itself which provided further 'proof' of the errors of recent geological speculation. As he expressed the essence of the problem in his draft:

Fourier's mathematical theory of the conduction of heat is a beautiful working out of a particular case belonging to the general doctrine of the 'Dissipation of Energy'. A general character of the practical solutions it presents is that in each case a distribution of temperature becoming gradually equalized through an unlimited future, is expressed as a function of the time, which is infinitely divergent for all times longer past than a definite determinable epoch. The distribution of heat at such an epoch is essentially *initial*, that is to say, it cannot result from any previous condition of matter by natural processes. It is, then, well called an *arbitrary* initial distribution of heat, in Fourier's [grand del.] great mathematical poem, since it could only be realized by an action of a free will having power to interfere with the laws of dead matter.[36]

Thomson's clear distinction here (and in the published paper) between mind, will, or power on the one hand, and dead matter on the other will figure prominently in his views of life on earth in relation to the laws of thermodynamics (ch. 18).

His 1862 paper 'On the secular cooling of the earth' opened with the prefatory declaration that 'for eighteen years it has pressed on my mind, that essential principles of thermodynamics have been overlooked by those geologists who uncompromisingly oppose all paroxysmal hypotheses, and maintain not only

[34] NB45, ULC. Commencing at the back of the notebook, this draft (addressed to Professor Rogers who would read the paper on Thomson's behalf) was separate from 'On the antiquity of the earth as a habitable world'. These remarks, and those which followed, were all published in the 1862 paper. See MPP, **3**, 295–6. [35] NB45, ULC. [36] *Ibid.*

that we have examples now before us, on the earth, of all the different actions by which its crust has been modified in geological history, but that these actions have never, or have not on the whole, been more violent in past time than they are at present'.[37] Since at least 1844 Thomson had been concerned with Lyell's views on terrestrial heat. Now, in the wake of Darwin's *Origin of species* and a decade of his own work on thermodynamics – particularly in relation to questions of the sun's heat – he attacked frontally Lyell's neglect of those 'essential principles of thermodynamics', the most important being that of energy dissipation – of which Fourier's law of heat conduction gave a perfect illustration.

Referring to his paper of early 1844 on the 'age' of temperature distributions, Thomson stated that he had there given the mathematical criterion for 'an essentially initial distribution'. Now, in 1862, he was concerned to apply that criterion to estimate 'from the known general increase of temperature in the earth downwards, the date of the first establishment of that *consistentior status* which, according to Leibniz's theory is the initial date of all geological history'.[38]

In his approach to the problem of terrestrial heat, Thomson not only took up once more his earlier interest (1844–6), but implicitly followed the path trodden by his great French predecessors, Laplace and Fourier. Their aims, however, had been quite different from Thomson's. The Frenchmen had been primarily concerned to produce a temperature map over the earth's surface, and not with a determination of the earth's age. They, unlike Thomson, had treated the full spherical problem and took into account the heating effect of the sun, making their results much more complex than his.[39] Thomson aimed in 1862 simply to offer a rough estimate, between upper and lower limits, of the possible age of the earth in order to show 'the untenability of the enormous claims for TIME which, uncurbed by physical science, geologists and biologists had begun to make and to regard as unchallengeable'.[40]

Having set forth his primary aims and commitments, Thomson directed his attention to Lyell's *Principles of geology*. Thomson regarded the increase of temperature with depth as an empirical fact which implied 'a continual loss of heat from the interior, by conduction outwards through or into the upper crust' of the earth. Since, he argued, that crust does not become hotter from year to year, 'there must be a secular loss of heat from the whole earth'. While the bulk of the earth was not necessarily cooling – if the heat loss derived from the

[37] MPP, **3**, 295.

[38] *Ibid.*, p. 297. Thomson's explication of the term *consistentior status* is given in MPP, **5**, 215, but had already been employed by Phillips in his outline of Leibniz's views. See Phillips, *Life on the earth*, pp. 122–4.

[39] Fourier, 'Extrait d'un mémoire sur le refroidissement séculaire du globe terrestre', *Ann. Chim. Phys.*, **13** (1820), 418–38; P.S. Laplace, 'De la chaleur de la terre, et de la diminution de la durée du jour par son refroidissement', *Traité de mécanique céleste* (5 vols., Paris, 1798–1827), **5**, pp. 72–85.

[40] MPP, **1**, 39.

exhaustion of potential energy in the form of chemical affinity between substances in the earth's mass – it *was* certain that this loss of available energy was quite irreversible. His comments on Lyell were caustic:

To suppose, as Lyell, adopting the chemical hypothesis, has done, that the substances, combining together, may be again separated electrolytically by thermo-electric currents, due to the heat generated by their combination, and thus the chemical action and its heat continued in an endless cycle, violates the principles of natural philosophy in exactly the same manner, and to the same degree, as to believe that a clock constructed with a self-winding movement may fulfil the expectations of its ingenious inventor by going for ever.[41]

Widening his criticism of Lyell to 'geological writers of the "Uniformitarian" school' who had argued 'in a most fallacious manner against hypotheses of violent action in past ages', Thomson attacked, not only the blatant violation of irreversibility by the uniformitarians, but also what he saw as their naive assumption of an equal intensity of activity throughout geological history. With regard to volcanic activity, therefore, Thomson declared that 'it would be very wonderful, but not an absolutely incredible result, that volcanic action has never been more violent on the whole than during the last two or three centuries'. Either way, he continued, 'it is certain that there is now less volcanic energy in the whole earth than there was a thousand years ago, as it is [certain] that there is less gunpowder in a "Monitor" after she has been seen to discharge shot and shell, whether at a nearly equable rate or not, for five hours without receiving fresh supplies, than there was at the beginning of the action'.[42]

In order to justify the assumptions of his estimate for the 'age' of the earth, Thomson needed to eliminate the alternative hypotheses which accounted for terrestrial heat. First, he rejected the chemical hypothesis. If the increase of temperature with depth occurred merely in isolated localities, chemical action might be taken seriously. But it was 'extremely improbable' that chemical combinations were 'going on at some great unknown depth under the surface everywhere, and creeping inwards gradually as the chemical affinities in layer after layer are successively saturated'. Cautious as ever, Thomson remarked that this hypothesis nevertheless could not be pronounced 'to be absolutely impossible, or contrary to all analogies in nature'. In the present state of science, however, the less hypothetical view, that the earth was 'merely a warm chemically inert body cooling', was to be preferred.[43]

Second, he rejected Poisson's hypothesis that 'the present underground heat is due to a passage, at some former period, of the solar system throughout hotter stellar regions'. From Forbes's data on underground temperatures, Thomson

41 MPP, **3**, 298; cf. PL, **2**, 108–10.

42 MPP, **3**, 298–9. The United States Navy's prototype monitor (a shallow draft, very low freeboard warship of heavy gunpower and armour) named *Monitor* had a famous but inconclusive contest with the Confederate Navy's ironclad *Virginia* on 9th March, 1862. See *Conway's all the world's fighting ships. 1860–1905* (London, 1979), p. 119.

43 MPP, **3**, 299.

estimated that, to account for the present general rate of increase of temperature downwards, 'taken as 1°F in 50 feet', the temperature of that stellar region through which the earth was passing must have been from 25° to 50°F above the present mean temperature of the earth's surface between 1250 and 5000 years ago, or more than 100°F 20 000 years ago. In the first case, 'human history negatives this supposition; while in the second, animal and vegetable life would have been destroyed – contrary to the evidence of palaeontology'.[44]

For the geologists, Thomson claimed, the best theory was that of Leibniz, which 'simply supposes the earth to have been at one time an incandescent liquid, without explaining how it got into that state'. That theory, provided the *whole* earth solidifies, also happened to be the one best suited to an application of Fourier's methods! The estimated time scale which he could offer geologists – founded on Fourier's solution – ranged from twenty million to 400 million years. The problem of accurate data for the temperature of melting rock, and for the variation in conductivities, specific heats, and latent heats of fusion of rocks at high temperatures and in different localities, meant that 'we must therefore allow very wide limits in such an estimate as I have attempted to make'. As a more precise estimate of the time available to the geologists, Thomson found a figure of 200 million years based on 10 000°F as the temperature of melting rock; or, supposing the temperature to be 7000°F – 'more nearly what it is generally supposed to be' – 'we may suppose the consolidation to have taken place ninety-eight million years ago'. The very wide limits, twenty and 400 million years, would if exceeded imply, respectively, more underground heat than we actually have, or less underground heat than that entailed by the least observed increment of temperature.[45]

Thomson based his estimates on Fourier's mathematical treatment of heat conduction in an infinite solid, not however with spherical geometry but with the linearizing supposition that at an initial epoch the temperature had two different, but constant, values on the two sides of an arbitrary infinite plane in an infinite solid. Such a distribution is an 'essentially initial distribution', because of the discontinuity between the two sides of the central plane which is not derivable from a preceding distribution. The partial differential equation for the conduction of heat in the linear case is:

$$\frac{\partial v}{\partial t} = K\frac{\partial^2 v}{\partial x^2}\,.$$

The elementary solutions given by Fourier to the problem of finding at any time the rate of variation of temperature from point to point, and the actual temperature at any point in such a solid were:

$$\frac{\partial v}{\partial x} = \frac{V}{(\pi K t)^{\frac{1}{2}}}\mathrm{e}^{-x^2/4Kt}$$

[44] *Ibid.*, pp. 299–300. [45] *Ibid.*, p. 300.

and

$$v = v_0 + \frac{2V}{\pi^{\frac{1}{2}}} \int_0^{x/2(Kt)^{\frac{1}{2}}} dz\, e^{-z^2},$$

where

K = conductivity of the solid measured in terms of the thermal capacity of the unit of bulk,
V = half the difference of the two initial temperatures,
v_0 = their arithmetical mean,
t = time,
x = distance of any point from the middle plane,
v = temperature of point x at time t, and

$\frac{\partial v}{\partial x}$ = rate of variation of temperature with distance x.

In other words, if the initial constant temperatures on each side of the infinite plane were v_1 and v_2, then for
$t=0$ and x positive, $V=v_1=v_0+V$
and for
$t=0$ and x negative, $V=v_2=v_0-V$.
In an accompanying graphical representation (figure 16.1), Thomson traced out the respective curves for the expressions $\frac{\partial v}{\partial x}$ and v.[46]

Thomson argued that the solution would apply for a certain time, without sensible error, 'to the case of a solid sphere primitively heated to a uniform temperature, and suddenly exposed to any superficial action, which for ever after keeps the surface at some other constant temperature'. In a globe, for example, 8000 miles in diameter, he showed that the solution would apply 'with scarcely sensible error' for 1000 million years since the gradient of temperature $\frac{\partial v}{\partial x}$ was insensible at depths exceeding 568 miles.

Taking a figure of one-hundred million years, he referred to his curve for $\frac{\partial v}{\partial x}$ in order to provide an estimate for the variation in temperature which would now exist *in the earth* if 'its whole mass being first solid and at one temperature 100 million years ago, the temperature of its surface had been everywhere suddenly lowered by V degrees, and kept permanently at this lower temperature'. For $V=7000$°F, his estimated melting point of rock, Thomson found the rate of increase to be about $\frac{1}{51}$ of a degree per foot for the first 100 000

[46] *Ibid.*, pp. 300–4.

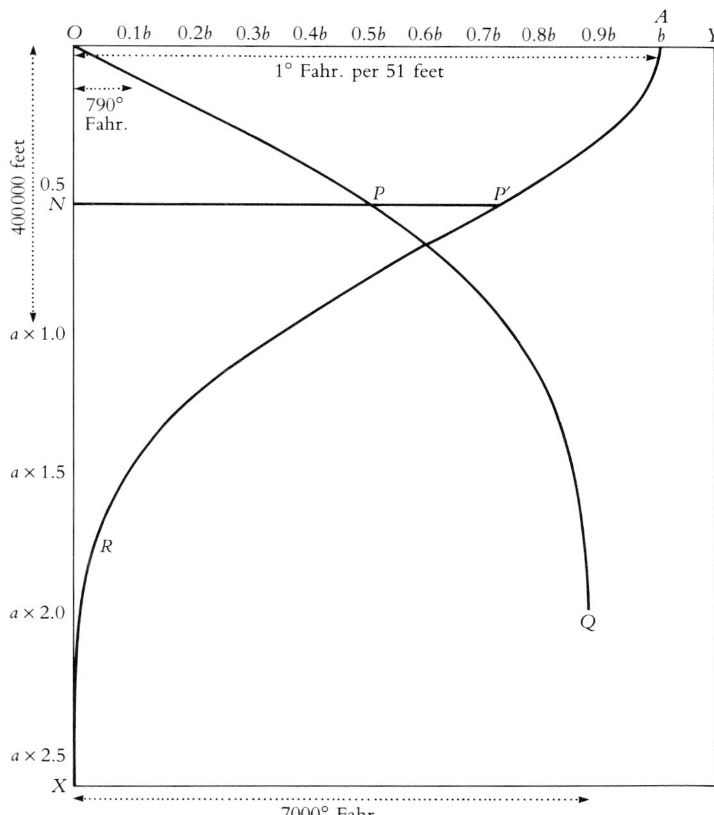

Figure 16.1 The curve *OPQ* shows excess of temperature above that of the surface. The curve *AP′R* shows rate of augmentation of temperature downwards.

feet or so, diminishing thereafter. He thus concluded that 'such is, on the whole, the most probable representation of the earth's present temperature at depths of from 100 feet, where the annual variations cease to be sensible, to 100 miles'.[47]

Turning to more specific geological issues, Thomson first considered the effect of terrestrial heat on climate over long periods. In the above instance, he argued, there would be an increment of 2°F per foot downwards near the surface 10 000 years after the beginning of the cooling. Since, he believed, the radiation from earth and atmosphere into space would probably be so rapid as to prevent the surface temperature being sensibly augmented by conduction from below through a gradient of 2°F per foot, he inferred 'that the general climate cannot be

[47] *Ibid.*, pp. 302–4.

sensibly affected by conducted heat at any time more than 10 000 years after the commencement of superficial solidification'[48] – a conclusion consistent with Hopkins's paper of 1852.

On the other hand, Thomson remarked, even at three or four million years – with a rate of increase of $\frac{1}{10}$°F per foot – vegetation would be 'influenced by the sensibly higher temperature met with by roots extending a foot or more below the surface'. On these estimates, then, the rate of increase of temperature over the ninety-six million years (from four to one-hundred million years after the initial epoch) had diminished by one-fifth – from $\frac{1}{10}$°F to $\frac{1}{50}$°F per foot – and 'the thickness of the crust through which any stated degree of cooling has been experienced has gradually increased in that period from $\frac{1}{5}$th of that thickness up to its present thickness'. He thus asked rhetorically:

> Is not this, on the whole, in harmony with geological evidence, rightly interpreted? Do not the vast masses of basalt, the general appearances of mountain-ranges, the violent distortions and fractures of strata, *the great prevalence of metamorphic action* (which must have taken place at depths of not many miles, if so much), all agree in demonstrating that the rate of increase of temperature downwards must have been much more rapid, and in rendering it probable that volcanic energy, earthquake shocks, and every kind of so-called plutonic action have been, on the whole, more abundantly and violently operative in geological antiquity than in the present age?[49]

In other words, Thomson defended the possibility of more violent geological effects in earlier times (catastrophism in Whewell's sense) against the uniformitarian view that terrestrial activity in the past was neither greater nor less than at present.

The remainder of the paper, devoted to answering possible objections to his arguments, included a discussion of the possible state of the globe prior to the formation of a solid crust on its surface. The geological evidence which had convinced geologists that the earth had a fiery beginning, Thomson facetiously argued, 'goes but a very small depth below the surface, and affords us absolutely no means of distinguishing between the actual phenomena, and those which would have resulted from either an entire globe of liquid rock, or a cool nucleus covered with liquid to any depth exceeding 50 or 100 miles'. Consequently, he tried to ensure that his conclusions as to the formation of the crust did not depend on a prior commitment to either a solid or a liquid state of the globe beyond the fifty or one-hundred mile-deep ocean of lava. Nevertheless, Thomson did appear to favour as 'a probable hypothesis regarding the earth's antecedents' from which 'the earth's initial fiery condition may have followed by natural causes', a form of meteoric theory similar to that adopted from 1854 to account for the sun's heat.[50] He concluded his paper with an evaluation, in the light of thermodynamic principles, of the possible ways by which solidification might have occurred. He rejected the possibility of a stable crust forming over a molten

[48] *Ibid.*, pp. 304–5 [49] *Ibid.*, p. 305. [50] *Ibid.*, pp. 306–7.

interior, and considered instead two models which conformed to Hopkins's ideas. In the first model, the earth's crust would not solidify until the whole globe was solid. This model was 'most consistent with what we know of the physical properties of the matter concerned' and 'agrees . . . with the view adopted as most probable by Mr Hopkins'.[51] In the second model solidification might occur by ingredients crystallizing at different temperatures such that 'when the whole globe, or some very thick superficial layer of it, still liquid or viscid [semifluid], has cooled down to near its temperature of perfect solidification, incrustation at the surface must commence'.

Essentially that advocated by Hopkins in 1847, this preferred model opened up the possibility of accounting for a number of geological phenomena, viz., earthquakes and volcanoes. A crust thus formed might be temporarily held up either by its own 'vesicular' (cellular) character deriving from the 'ebullition [boiling] of the liquid' in some places or by the viscidity of the liquid. At some point, the heavier crust would sink below the lighter liquid, which in turn would solidify. The process would continue, Thomson argued, 'until the sunk portions of crust build up from the bottom a sufficiently close-ribbed solid skeleton or frame to allow fresh incrustations to remain bridging across the now small area of lava pools or lakes'. The result, then, was the formation of a honeycombed solid and liquid mass in which the lighter liquid would tend to work its way up.[52]

Having argued for the consistency of this terrestrial model with thermodynamic laws – with 'a consideration of the necessary order of cooling and consolidation' – with experimental results, and with the occurrence of earthquakes and volcanoes, Thomson concluded his famous 1862 paper with the claim that his model was also 'in perfect accordance' with what he had recently demonstrated regarding the present condition of the earth's interior: 'that it is not, as commonly supposed, all liquid within a thin solid crust of from 30 to 100 miles thick, but that it is on the whole more rigid certainly than a continuous solid globe of glass of the same diameter, and probably than one of steel'.[53]

The rigidity of the earth

William Thomson's 1862–3 Royal Society paper 'On the rigidity of the earth' – complemented by a long mathematical paper on the deformation of elastic spheroids entitled 'Dynamical problems regarding elastic spheroidal shells and spheroids of incompressible liquid' and published in the same volume of the

[51] *Ibid.*, p. 310.

[52] *Ibid.*, p. 311. Once again Thomson here employed the same model for large scale structure as he used at the micro-level. See William Thomson, 'Note on gravity and cohesion', *Proc. Royal Soc. Edinburgh*, **4** (1862), 604–7; PL, **1**, 59–63. See also the associated correspondence with Tait (ch. 11) and PA146A, ULC, which contains notes on a paper by W.J. Russell and Augustus Matthiessen, 'On the cause of vesicular structure in copper', *Phil. Mag.*, [series 4], **23** (1862), 81–4, dating from February, 1862.

[53] PL, **3**, 311.

Philosophical Transactions – aimed to prove that 'unless the solid substance of the earth be on the whole of extremely rigid material, more rigid for instance than steel, it must yield under the tide-generating influence of sun and moon to such an extent as to very sensibly diminish the actual phenomena of the tides, and of precession and nutation'.[54]

Considerations of tide-generating influences quickly became of supreme importance for Thomson's arguments in favour of the earth's solidity. As early as October and November, 1861, he announced to Stokes that 'the earth as a whole must be far more rigid than glass otherwise the solid would yield so much to tidal influence of sun & moon as to leave no sensible tides of water relative to solid land', and that 'my theory of elastic tides is even more decisive than Hopkins' argument against the prevalent geological theory of internal fluidity'. Stokes for his part found the results 'very remarkable, and not at all what I should have expected', but suggested further the likelihood 'that the rigidity of a solid is greatly increased when the solid is subjected to an enormous hydrostatic pressure [rather] than that the materials of which the earth are composed are under like conditions vastly more rigid than glass'. Thomson agreed that the great rigidity in the interior could be explained by the great pressure, which altered the ordinary properties of matter: 'Indeed it seems almost quite certain that, if it is solid at all, the great pressure must make it much more rigid than the same matter under ordinary pressure. The coefficient of resistance to compression must be enormously increased, for certain'.[55]

Finding that the actual phenomena of the tides would, for an earth of the same rigidity as glass, be only two-ninths, and for an earth of the same rigidity as steel, be only three-fifths of the amount which a perfectly rigid spheroid of the same dimensions, figure, and homogeneous density would exhibit in the same circumstances, Thomson concluded that any such discrepancy as three-fifths seemed 'scarcely possible', and that it was 'therefore almost necessary to conclude that the earth is on the whole much more rigid than steel'. He recommended, however, a comparison of theory and observation by observing the lunar fortnightly and solar half-yearly tides, and suggested Iceland and Teneriffe as two stations 'well adapted for the differential observations that would be required'.[56] As he wrote to Stokes in July, 1862:

> What do you think of Iceland and Teneriffe for observations on the fortnightly tides? We propose, Tait and I, to go one to one place & the other to the other, and spend a month observing, before we write our chapter on the Tides for [T & T′] . . . I should choose Teneriffe for myself . . .[57]

[54] William Thomson, 'On the rigidity of the earth', *Proc. Royal Soc.*, **12** (1862–3), 103–4; *Trans. Royal Soc.*, (1863), 573–82.

[55] William Thomson to G.G. Stokes, 29th October and 7th November, 1861, K130, K131, Stokes correspondence, ULC; G.G. Stokes to William Thomson, 4th November, 1861, S397, ULC.

[56] Thomson, 'Rigidity of the earth', *Proc. Royal Soc.*, pp. 103–4.

[57] William Thomson to G.G. Stokes, 8th July, 1862, K138, Stokes correspondence, ULC.

Thomson's excuse for choosing Teneriffe was the opportunity afforded for experimental work on atmospheric electricity, though we may suppose that climatal reasons offered no small inducement.

The associated problem of estimating the effects on precession and nutation of assuming a fluid rather than a solid interior and of elastic yielding of the crust presents an interesting and complex historical story, but one which we cannot enter upon in detail here. Rather, we can only survey the stages of its development in terms of the series of conclusions which Thomson published.

In his first draft 'On the rigidity of the earth' dated October, 1861, he supported William Hopkins's arguments wholeheartedly, arguments for the solidity of the earth based on a mathematical investigation of precessional and nutational effects for a fluid and solid earth:

> In its movements of precession and nutation the Earth presents, so far as our astronomical tests have yet shown, the same phenomena as a body perfectly rigid throughout would do; which are excessively different from those that would be presented by a body externally like the Earth but consisting of a thin shell of solid matter with the whole interior fluid. On these grounds is founded Mr Hopkins' celebrated argument for the solidity of the Earth; an argument which has always seemed to me valid and which I believe he rightly maintains to be decisive against the supposition entertained by so many geologists that there is melted lava everywhere below a crust of something less than 100 miles thickness.[58]

At the same time, Thomson rejected the possibility that viscosity arguments might upset Hopkins's verdict against the fluid earth. Thus while 'such a degree of viscosity as we can conceive melted rock or metal to possess might cause the interior to be dragged round frictionally so that its axis of rotation could never be left very far behind that of the crust in the great precessional movement', it seemed 'highly improbable that any amount of internal friction consistent with the geological idea of fluidity could load the light crust . . . with sufficient lunar and solar influences and experiencing great monthly and annual nutations'. Consequently, as he wrote to Stokes at the end of November, 1861, 'these considerations incline me to give full weight to Hopkins' argument'.[59]

By April, 1862, however, Thomson had concluded 'on looking into Hopkins' papers that, so far as the mathematical problems he attacks are concerned, they are all wrong'. In particular, he believed that Hopkins's fundamental proposition concerning the same precessional and nutational effects in both a spheroidal homogeneous liquid mass enclosed in a rigid shell and a body rigid throughout was 'obviously wrong'. So too was Hopkins's proposition in his third memoir concerning isothermal surfaces.[60]

[58] William Thomson, Draft of 'On the rigidity of the earth', October, 1861, NB46, ULC.

[59] William Thomson to G.G. Stokes, 27th November, 1861, K132, Stokes correspondence, ULC.

[60] William Thomson to G.G. Stokes, 19th April, 1862, K137, Stokes correspondence, ULC.

As Thomson explained to Stokes, this verdict against his honoured and aging mentor caused him 'great difficulty':

I have therefore felt great difficulty in referring to his investigation. It was necessary I should do so because he was the first to propose the argument, & I think his conclusion valid. Yet, having experienced the greatest possible benefit from his teaching, I would certainly shrink from the task of finding fault with his investigation: and, as it was necessary I should not commit myself to assenting to it, I therefore introduced a statement in a foot note near the beginning . . . I fear Hopkins, if he notices it, may not be pleased with the foot note, & I should be very sorry to be called on to justify it to him.[61]

Thomson, probably following Stokes's advice, opened his *Philosophical Transactions* paper by referring tactfully to Mr Hopkins 'to whom is due the grand idea of thus learning the physical condition of the interior from phenomena of rotatory motion presented by the surface', and explaining that 'although the mathematical part of the investigation might be objected to, I have not been able to perceive any force in the arguments by which this conclusion has been controverted, and I am happy to find my opinion in this respect confirmed by so eminent an authority as Archdeacon Pratt'.[62]

Thomson had thus cleared the way for a discussion of the major new question of the earth's rigidity. He seems originally to have thought that he could merely extend Hopkins's results, arguing not only (with Hopkins) that the earth's crust must be not less than 800 or 1000 miles thick, but that it was 'extremely improbable that any crust thinner than 2000 or 2500 miles could maintain its figure with sufficient rigidity against the tide-generating forces of the sun and moon, to allow the phenomena of the ocean tides and of precession and nutation to be as they are'.[63] Rather than correct Hopkins's erroneous propositions, Thomson, however, now continued with his own calculations on these effects of imperfect rigidity.

With regard to the effects of elastic yielding on precession and nutation, he employed a qualitative, typically visualizable argument which avoided the mathematical complexities, and indicated that if the earth is liquid it will yield in such a way as to remain in equilibrium under the forces of the sun tending to produce precession, and so cancelling that effect. In order to obtain the observed precession, Thomson then concluded that the earth would have to be more rigid than steel.[64]

In an 1872 letter to *Nature*, however, Thomson felt compelled to counter an argument put forward by Delaunay in 1868 (but already considered by Thomson himself in 1861) that sufficient viscosity in a fluid interior would cause the interior to follow the crust just as though it were solid. Thomson showed the absurdity of Delaunay's case by estimating that a viscosity ten million million million times that of water would be required to produce the effect. At the same

61 *Ibid.* 62 Thomson, 'Rigidity of the earth', *Trans. Royal Soc.*, p. 573.
63 Thomson, 'Rigidity of the earth', *Proc. Royal Soc.*, p. 104.
64 Thomson, 'Rigidity of the earth', *Trans. Royal Soc.*, pp. 578–81.

time, Thomson repeated his earlier conclusion that far more dramatic limitations on fluidity derived from the notable absence of large deformations of figure in the crust.[65]

During his 1876 visit to the United States, a conversation with Professor Newcomb cast doubt on the 1862 argument, but inspired a new argument based on 'a rigorous application of the perfect hydrodynamical equations' which 'leads still more decidedly to the same conclusion' concerning the earth's rigidity. As Thomson reported the discussion to the British Association at Glasgow on his return:

> Admitting fully my evidence for the rigidity of the earth from the tides, he [Newcomb] doubted the argument from precession and nutation. Trying to recollect what I had written on it [in 1862] . . . my conscience smote me, and I could only stammer out that I had convinced myself that so-and-so and so-and-so, at which I had arrived by a non-mathematical short cut, were true. He hinted that viscosity might suffice to render precession and nutation the same as if the earth were rigid, and so vitiate the argument for rigidity. This I could not for a moment admit, any more than when it was first put forward by Delaunay. But doubt entered my mind regarding the so-and-so and so-and-so; and I had not completed the night journey to Philadelphia . . . before I had convinced myself that they were grievously wrong.[66]

Thomson now substituted a new argument from the effective rigidity of vortex motion in the interior of a fluid earth, an argument which showed that with a rigid shell the ellipticity of the earth would have to be considerably larger than it is in order to compel the liquid interior to follow the crust in precessional and nutational motion. Although not conclusive in all cases, the calculation was, in the case of the fortnightly lunar nutation, absolutely decisive against 'a thin rigid shell full of liquid'.

Hopkins's 'problem', assuming as it did a crust 'perfectly stiff and unyielding in its figure', could not be decisive against the earth's interior liquidity, however, because no material is infinitely rigid. While the full problem of precession and nutation in a continuous revolving liquid spheroid with imperfectly rigid crust had 'not yet been coherently worked out', Thomson felt that 'precession and nutation will be practically the same in it as in a solid globe'. Hence, he concluded, 'precession and nutation yield nothing to be said against such hypotheses as that of [Charles] Darwin, that the earth as a whole takes approximately the figure due to gravity and centrifugal force, because of the fluidity of the interior and the flexibility of the crust'.[67]

[65] William Thomson, 'The rigidity of the earth', *Nature*, **5** (1871–2), 223–4; 'On the internal fluidity of the earth', *Nature*, **5** (1871–2), 257–9; MPP, **5**, 163–70.

[66] William Thomson, 'Review of the evidence regarding the physical condition of the earth', *BAAS Report*, **46** (1876), 8–12, on p. 5. See also MPP, **3**, 320–4, and Brush, 'Nineteenth-century debates', pp. 238–42.

[67] Thomson, 'Physical condition of the earth', p. 7; MPP, **3**, 325. Thomson was alluding to Charles Darwin, 'Observations on the parallel roads of Glen Roy, and other parts of Lochaber in Scotland, with an attempt to prove that they are of marine origin', *Trans. Royal Soc.*, (1839), 39–82, on p. 81.

Tidal considerations, on the other hand, still weighed decisively against an imperfectly rigid crust which would thus 'yield so freely to the deforming influence of sun and moon that it would simply carry the waters of the ocean up and down with it, and there would be no sensible tidal rise and fall of water relatively to land'. He therefore summarized the state of his case in 1876:

The hypothesis of a perfectly rigid crust containing liquid violates physics by assuming preternaturally rigid matter, and violates dynamical astronomy in the solar semiannual and lunar fortnightly nutations; but tidal theory has nothing to say against it. On the other hand, the tides decide against any crust flexible enough to perform the nutations correctly with a liquid interior, or as flexible as the crust must be unless of preternaturally rigid matter. But now thrice to slay the slain: suppose the earth this moment to be a thin crust of rock or metal resting on liquid matter; its equilibrium would be unstable! And what of upheavals and subsidences? They would be strikingly analogous to those of a ship which has been rammed – one portion of crust up and another down, and then all down.[68]

Two years later, in an address to the Geological Society of Glasgow, William Thomson offered a definitive survey of his views of the internal condition of the earth as to temperature, fluidity, and rigidity. Much of the paper was taken up with an explicit demonstration, on the basis of the 1862 paper on secular cooling, that increase-of-temperature-with-depth arguments did not necessarily lead to internal fluidity. While still citing Hopkins's work on the influence of pressure on melting points and the implication of James Thomson's thermodynamic researches, he explained that 'the view which I wish to put before you just now . . . is of a very different nature' for 'I mean to deny altogether the intensely high temperature which Hopkins accepted'. Thomson argued that we 'too generally supposed that the rate of increase of underground temperature as we proceed inwards is uniform, or nearly so . . .'. He thus concluded 'that at the greatest depths the temperature is not high enough for fusion of the material' and showed from his 1862 graphs that below about 800 000 feet the 'underground temperature does not sensibly increase'.[69]

This conclusion then removed the old Lyellian obstacle to the positive arguments in favour of the earth's solidity. Thus Thomson reiterated his 1876 verdict that 'the arguments derived from precession and nutation present considerable difficulties, and indeed do not afford us at the present time a decisive answer', while, by contrast, 'the phenomena of the tides . . . lead us to no uncertain conclusion' that 'the earth is not a thin shell filled with fluid, but is on the whole or in great part solid'.[70]

[68] MPP, **3**, 325.

[69] William Thomson, 'The internal condition of the earth; as to temperature, fluidity, and rigidity', *Trans. Geol. Soc. Glasgow*, **6** (1882), 38–49; PL, **2**, 299–318, esp. pp. 308–15.

[70] *Ibid.*, pp. 316–18. For a survey of the scientific literature on the fate of the solid earth after Thomson, see Brush, 'Nineteenth-century debates', 252–4.

17

The age of the earth controversies

I think what my father said about time is quite justifiable from a biological point of view; 100 or 200 million years is 'incomprehensibly vast' – even a million is not conceivable in the way a hundred is. I have no doubt however that if my father had had to write down the period he assigned at that time, he w[oul]d have written a 1 at the beginning of the line & filled the rest up with 0's. Now I believe that he cannot quite bring himself down to the period assigned by you, but does not pretend to say how long may be required. I fail to see the justice of your remark that a few hundred million years would be insufficient to allow of transmutation of species by nat[ural] selection. What possible datum can one have for the rate at which it has or can work? *George Darwin to William Thomson, 1878.*[1]

By tracing for the terrestrial fabric the consequences of energy dissipation, William Thomson undermined any Lyellian beliefs in non-progression, in equal intensity of geological agents through time, and in an indefinitely large scale of geological time. He aligned himself with those whom he termed 'true geologists' – Sedgwick, Phillips, Hopkins, Forbes, Murchison – in opposition to those who would thrust rubbish upon the public by ignoring the fundamental principles of natural philosophy.[2] Far from an isolated figure, unaware of geological thinking and merely carrying the torch of physics to lighten the darkness of geology, Thomson saw himself in the mainstream of British geology. Thus he asked in 1869:

Is geology not a branch of physical science? Are investigations, experimental and mathematical, of underground temperature, not to be regarded as an integral part of geology? Are suggestions from astronomy and thermodynamics, when adverse to a tendency in geological speculation recently become extensively popular in England through the brilliancy and eloquence of its chief promoters, to be treated by geologists as an invitation to meddle with their foundations, which a 'wise discrimination' declines? For myself, I am anxious to be regarded by geologists, not as a mere passer-by, but as one constantly interested in their grand subject, and anxious, in any way, however slight, to assist them in their search for truth.[3]

[1] G.H. Darwin to William Thomson, 1st November, 1878, D8, ULC.
[2] PL, **2**, 111. [3] *Ibid.*, p. 113.

Again, when addressing the Glasgow Geological Society in 1878, Sir William claimed that 'we, the geologists, are at fault for not having demanded of the physicists experiments' on the properties of matter.[4]

Taking seriously Thomson's self-identification with geologists, we shall not characterize the controversies over the age of the earth in terms of a dichotomy, or breakdown of dialogue, between physics and geology. The protagonists were often personal acquaintances moving easily from section to section at the annual British Association meetings, where the constraints of a highly specialized and departmentalized scientific world scarcely existed. We shall emphasize, therefore, Sir William's social and scientific alignments with a relatively intimate circle of colleagues, and shall leave to others the assessment of his general influence on geology.[5]

On geological and biological time

Three years after his paper on secular cooling, Thomson again attacked the 'doctrine of uniformity' in geology 'as held by many of the most eminent of British geologists'. In particular, he was keen to undermine what he saw as one of their most basic assumptions – 'that the earth's surface and upper crust have been nearly as they are at present in temperature, and other physical qualities, during millions of millions of years'. If the present annual amount of heat conducted out of the earth 'had been going on with any approach to uniformity for 20 000 million years', Thomson calculated that the heat thus lost would be 'more than enough to melt a mass of surface rock equal in bulk to the *whole earth*'. He therefore concluded that no hypothesis – whether of chemical action, internal fluidity, effects of pressure at great depth, or possible character of substances in the interior of the earth – could justify the supposition 'that the earth's upper crust has remained nearly as it is, while from the whole, or from any part, of the earth, so great a quantity of heat has been lost'.[6] With the publication of this direct attack on the 'principle of uniformity', the stage was set for the famous age of the earth debates in the period between the late 1860s and the early 1900s.

Fleeming Jenkin's review of Darwin's theory of evolution by natural selection linked Thomson's arguments on the age of the sun and the secular cooling of the earth to a criticism of Darwin in a way that Thomson himself had not yet attempted. Jenkin explained that the distribution of heat in the earth afforded 'another method by which the rate of decay of our planetary system can be measured'. While admitting that the data for the calculation were still very

4 *Ibid.*, p. 304.

5 J.D. Burchfield, *Lord Kelvin and the age of the earth* (London, 1975) provides an excellent survey of Kelvin's wide-ranging influence. J.D. Burchfield, 'Darwin and the dilemma of geological time', *Isis*, **65** (1974), 301–21, focusses on Darwin's circle.

6 William Thomson, 'The "doctrine of uniformity" in geology briefly refuted', *Proc. Royal Soc. Edinburgh*, **5** (1866), 510–2; PL, **2**, 6–9.

imperfect, making the estimate a mere approximation, Jenkin claimed that 'unless our information be wholly erroneous as to the gradual increase of temperature as we descend towards the centre of the earth, the main result of the calculation, that the centre is gradually cooling, and if uninterfered with must, within a limited time, have been in a state of complete fusion, cannot be overthrown'.[7]

From this basis, Jenkin launched his criticism both of Darwin's theory of evolution by natural selection and of a uniformitarian theory of geological formations. Time, he argued, was not only limited, but was 'limited to periods utterly inadequate for the production of species according to Darwin's views'. In order to produce a species differing only slightly from the parent stock, Jenkin believed that Darwin would regard a million years as 'no long time to ask'. Yet, it was doubtful if a thousand times more change than had actually taken place in wild animals in historic times would 'produce a cat from a dog, or either from a common ancestor'. Thus, he concluded, 'how preposterously inadequate are a few hundred times this unit [a million years] for the action of the Darwinian theory!'[8]

With geology, unlike Darwinian theory, Jenkin attempted a reconciliation. By considering the implications of 'the general theory of the gradual dissipation of energy' for geological formations, he suggested a solution to the problem of reconciling the estimates of secular cooling with the results deduced from the denudation or deposition of geological strata: 'if there have been a gradual and continual dissipation of energy, there will on the whole have been a gradual decrease in the violence or rapidity of all physical changes'. Activity, then, on the whole diminished through time – an argument already put forward in William Thomson's 1862 paper. For Jenkin, however, the significance for geology was dramatic:

> Once this is granted, the calculations as to the length of geological periods, from the present rates of denudation and deposit, are blown to the winds. They are rough, very rough, at best . . . The rates of denudation and deposition have been gradually, on the whole, slower and slower, as the time of fusion has become more and more remote. There has been no age of cataclysm . . . no time, when the physical laws were other than they now are, but the results were as different as the rates of a steam engine driven with a boiler first heated to 1500 degrees Fahrenheit, and gradually cooling to 200.[9]

Jenkin's arguments were precise, forceful, and had employed Thomson's most basic principle – dissipation – to demonstrate that the uniformitarian assumption of a rate of denudation at times past equal to the present rate was unjustified.

'A great reform in geological speculation seems now to have become necessary'. Thus did William Thomson open his February, 1868, address 'On

[7] Fleeming Jenkin, 'Darwin and the origin of species', *North Brit. Rev.*, **46** (1867), 277–318; *Papers and memoir of Fleeming Jenkin* (2 vols., London, 1887), **1**, pp. 215–63, on p. 242–3. See also Burchfield, *Age of the earth*, pp. 73–4. [8] Jenkin, *Papers*, **1**, p. 243. [9] *Ibid.*, p. 245.

geological time' to the Glasgow Geological Society, an address which took as its text a passage from Playfair's *Illustrations of the Huttonian theory* for an assault on both non-progressionist geology and steady-state cosmology (ch. 15). Furthermore, the address initiated the important public debate with T.H. Huxley over the issue of geological reform. Thomson considered, first, the motions of the heavenly bodies (including the earth), and, second, phenomena presented by the earth's crust (notably underground heat).[10] The first division, concerning principally the new question of tidal retardation, dominated his arguments. James Thomson's interest in tidal retardation, closely related to his concern with loss of power or mechanical effect in machines, extended back to 1840 (ch. 9). Similarly, William's criticism of J.H. Pratt's view of compensatory agents, maintaining the earth's motion, found expression in the 1851 draft: 'the motions of the earth & other planets are losing vis viva w^{h} is converted into heat; and . . . although some vis viva may be restored . . . the loss cannot be *precisely* compensated & I think it probable that it is undercompensated'.[11]

Questions of the earth's rotation did not reappear in William Thomson's work until 1866. In the course of his Rede Lecture at Cambridge, 'On the dissipation of energy', he discussed publicly for the first time the observations and calculations required to find the tidal retardation of the earth's rotation. The estimate of the possible amount of such diminution of the earth's rotation through tidal friction was founded 'on calculating the moment round the earth's centre of the attraction of the moon on a regular spheroidal shell of water symmetrical about its longest axis' – the axis being kept, by fluid friction, 'in a position inclined backwards at an acute angle to the line from the earth's centre to the moon'. In order to calculate this moment, Thomson employed the techniques recently prepared for the *Treatise*. By the theorem that a homogeneous prolate spheroid of revolution attracted points outside it approximately as if its mass were collected in a uniform bar having its end in the foci of an equipotential spheroid, he calculated that the line of action of the attracting point (the moon) would pass 0.02 feet from the centre of the bar (the earth's centre). Finding the whole attraction of the moon on a globe of water equal in bulk to the earth to be 3.3×10^{15} tons force, Thomson had an estimate of the moment of the force acting through a line passing the centre at 0.02 feet distance – 'the same as a simple frictional resistance (as of a friction-brake) consisting of 3.3×10^{15} tons force acting tangentially against the motion of a pivot or axle of about $\frac{1}{2}$ inch diameter'.[12]

10 William Thomson, 'On geological time', *Trans. Glasgow Geol. Soc.*, **3**, (1871), 1–28; PL, **2**, 10–64. See also Burchfield, *Age of the earth*, pp. 80–6. Thomson had in mind such widely read works as David Page, *Advanced text-book of geology. Descriptive and industrial*, 4th ed. (Edinburgh and London, 1867). Page (p. 399) wrote that 'the student should never lose sight of the element TIME – an element to which we can set no bounds in the past, any more than we know of its limits in the future'. See also MPP, **5**, 206.

11 William Thomson, Preliminary draft for the 'Dynamical theory of heat', PA 128, ULC, p. 8.

12 William Thomson, 'On the observations and calculations required to find the tidal retardation of the earth's rotation', *Phil. Mag.*, [series 4], **31** (1866), 533–7; PL, **2**, 65–72, esp. pp. 65–7.

Assuming Laplace's 'probable law of increasing density inwards', Thomson estimated the loss of angular velocity of the earth per year as 3.6 seconds. Thus, 'the accumulation of effect of uniform retardation at that rate would throw the earth as a time-keeper behind a perfect chronometer (set to agree with it in rate and absolute indication at any time) by 180 seconds at the end of a century, 720 seconds at the end of two centuries, and so on'. Tidal friction, however, was only one of many disturbing influences which rendered the earth 'a very untrustworthy time-keeper'. Other possible disturbances were the melting of ice from the polar regions, the arrival of meteors on the earth's surface, and the shrinking of the earth by cooling. Thomson observed in a footnote:

> It seems hopeless, without waiting for some centuries, to arrive at any approach to an exact determination of the amount of the actual retardation of the earth's rotation by tidal friction, *except* by extensive and accurate observation of the amounts and times of the tides on the shores of continents and islands in all seas, and much assistance from *true* dynamical theory [Laplace's theory rather than Newton's equilibrium theory of the tides] to estimate these elements all over the sea. But supposing them known for every part of the sea, the retardation of the earth's rotation could be calculated by quadratures [i.e. integration].[13]

Thomson's subsequent part in the Tidal Committee of the British Association, his untiring enthusiasm for tidal data from all corners of the Empire, from India to the River Clyde, and his encouragement of G.H. Darwin in his extensive mathematical analyses of tides, largely derived from his strong feeling that the problem of tidal retardation – with its implications for the origins and future of the earth – demanded a much more complete understanding than was possible in 1866.[14]

In his 1868 address, Sir William argued that the celebrated theorem of the French physical astronomers on the stability of planetary motions was one 'of approximate application, and one which professedly neglected frictional resistance of every kind'. Still, British naturalists such as Playfair (and Herschel and Dr Thomson) had taken the theorem to imply a 'perpetuity of the "existing order", past and future' (chs. 4, 6, and 15). Even Laplace, who had been 'perfectly aware of the existence of resistance to fluid motion' in his theory of the tides, had not explicitly stated that 'tidal resistance influences the rotation of the earth, or, by reaction, the motions of the moon and sun'. But 'The modern theory of energy was imperfectly understood by Laplace and Lagrange . . . The theory of energy declares, in perfectly general terms, that as there is frictional resistance, there must be loss of energy somewhere . . . the modern theory must account for what becomes of that energy'. With regard to the tides, then, 'it becomes obvious that if there is resistance to the motion of the water that constitutes the tides, that resistance must directly affect the earth, and must react on those bodies, the moon and the sun, whose attractions cause the tides'. Fluid friction due to the

[13] *Ibid.*, 69*n*. [14] SPT, **1**, 581; **2**, 611, 619, 681, 730, 783.

movement of the sea and friction between the waters and the seabed generated heat, which would in the end be dissipated through space.[15]

With imperfect data as to the tides, Thomson stated that a calculation of the actual diminution of the earth's rotation was impossible. Yet because the diminution was undoubtedly 'something very sensible', Laplace could not be correct in calculating that the length of the day had not varied 'by one ten-millionth part of twenty-four hours from 721 years before the Christian era'. Even the data upon which the calculation was based, however, agreed 'in *demonstrating* . . . that the moon's mean angular motion has been accelerated somewhat relatively to the earth as time-keeper'.[16] Laplace accounted for this apparent secular acceleration by showing that the planets indirectly accelerate the moon's angular velocity through their influence in producing a secular diminution of the eccentricity of the earth's orbit, and his results seemed to square with the constancy of the earth's rotation. But in 1853 J.C. Adams had discovered an error in Laplace's calculation, which reduced roughly by half the amount of the theoretical acceleration. The other half of the observed acceleration had still to be accounted for, and as a consequence of communication with Adams, the French mathematician Delaunay had, early in 1866, suggested that the true explanation might be the retardation of the earth's rotation by tidal friction:

> Using the hypothesis that the cause of the discrepancy is retardation by tidal friction, and allowing for the consequent retardation of the moon's mean motion, Adams, in an estimate which he has recently worked out in conjunction with Professor Tait and myself, found, on a certain assumption as to the proportion of retardations due to the moon and the sun, that 22 seconds of time is the error by which the earth would in a century get behind a thoroughly perfect clock rated at the beginning of the century. Thus the probable result that physical astronomy gives us up to the present time is that the earth is not an accurate chronometer, but, on the contrary, is getting slower and slower.[17]

From these important arguments within physical astronomy, Thomson proceeded to a full-scale attack on excessive time scales. Given that the earth's great rigidity effectively prohibited any change in its figure since the *consistentior status* (the starting point for geological, and so for biological, history), a calculation of the centrifugal forces acting at any assumed past time could determine whether or not solidification at that time could have produced the present figure of the earth. From the present rate of loss of angular velocity, Thomson estimated the centrifugal forces acting 10 000 million years ago (when the earth rotated more than twice as fast as at present) and 100 million years ago, arguing that at the latter time a solidifying earth could have been in all respects like the present earth, but at the earlier time its figure would have been 'something totally different from what it is'. He concluded this 1868 discussion of tidal retardation with remarks sure to provoke Huxley:

[15] PL, **2**, 12–19. [16] *Ibid.*, pp. 36–8. [17] *Ibid.*, pp. 39–40.

Now, here is direct opposition between physical astronomy, and modern geology as represented by a very large, very influential, and I may also add, in many respects, philosophical and sound body of geological investigators, constituting perhaps a majority of British geologists. It is quite certain that a great mistake has been made – that British popular geology at the present time is in direct opposition to the principles of natural philosophy . . . [Whether] the earth's lost time is 22 seconds, or considerably more or less than 22 seconds, in a century, the principle is the same. There cannot be uniformity. The earth is filled with evidences that it has not been going on for ever in the present state, and that there is a progress of events towards a state infinitely different from the present.[18]

Responding in 1869, Huxley admitted the 'undoubted fact that the tides tend to retard the rate of the earth's rotation upon its axis'. What he doubted, however, was the 'estimation of the practical value of that tendency'. He emphasized, therefore, the availability of three hypotheses to account for the acceleration of the moon's mean motion. First, there was Delaunay's suggestion of tidal retardation, followed through by Adams, Thomson, and Tait. Second, there was Dufour's suggestion that the retardation 'may be due in part, or wholly, to the increase of the moment of inertia of the earth by meteors falling upon its surface' – a hypothesis discussed by Thomson in his 1866 Rede lecture and one which, for Huxley, met with 'the entire approval of Sir W. Thomson, who shows that meteor-dust, accumulating at the rate of one foot in 4000 years, would account for the remainder of retardation'.[19] Thomson had actually referred to Dufour's 'excellent suggestion, supported by calculations which show it to be not improbable', and had concluded that the only direct test of 'the probable truth of M Dufour's very interesting hypotheses' was to analyze chemically quantities of natural dust taken from suitable localities. Thus, 'should a considerable amount of iron with a large proportion of nickel be found or not found, strong evidence for or against the meteoric origin of a sensible part of the dust would be afforded'.[20]

The third hypothesis, the effect of the melting of ice from the polar regions, also originated from Thomson's 1866 Rede lecture. On Thomson's estimate, the melting of a foot thickness of ice would, over the whole globe, 'raise the sea-level by only some such undiscoverable difference as three-fourths of an inch or an inch'. Were this melting to occur in one year, the earth's rate as a time-keeper would slacken by one-tenth of a second. Huxley, however, supposed that the accumulation of polar ice since the Miocene epoch – when the ice was 'many feet thinner than it has been during, or since, the Glacial epoch' – was only sufficient 'to produce ten times the effect of a coat of ice one foot thick'. The earth's rotation would thus be one second faster. Tidal retardation, on the other hand, amounted to one-fifth of a second on Thomson's calculation of twenty-two

18 *Ibid.*, pp. 42–4.

19 T.H. Huxley, 'Geological reform', *Quart. J. Geol. Soc. London*, **25** (1869), xxxviii–liii; *Discourses: biological and geological* (London, 1902), pp. 305–39, on pp. 329–32.

20 PL, **2**, 70–1.

seconds per century. For Huxley, then, the ice accumulation 'covers all the loss from tidal action, and leaves a balance of $\frac{4}{5}$ second per annum in the way of acceleration'.[21]

Although Huxley was prepared to admit the correctness of each of these calculations, his attack rested partly on the claim that there was no agreed hypothesis to account for tidal retardation, but more especially upon the lack of adequate data: 'mathematics may be compared to a mill of exquisite workmanship, which grinds you stuff of any degree of fineness; but nevertheless, what you get out depends upon what you put in; and as the grandest mill in the world will not extract wheat-flour from peascods, so pages of formulae will not get a definite result out of loose data'. He concluded:

> If tidal retardation can thus be checked and overthrown by other temporary conditions, what becomes of the confident assertion, based upon the assumed uniformity of tidal retardation, that ten thousand million years ago the earth must have been rotating more than twice as fast as at present, and, therefore, that we geologists are 'in direct opposition to the principles of Natural Philosophy' if we spread geological history over that time?[22]

Thomson answered in his 1869 address 'Of geological dynamics' (a title used earlier by Whewell) to the Glasgow Geological Society, dealing first with the issue of polar ice. He simply added to a summary of Huxley's argument the remark that the *observed* result was retardation and that Huxley's hypothesis, if it were valid, would prove retardation by the tides 'six times as much as that which we have ventured to estimate'. Cleverly turning Huxley's argument against him, Thomson thanked his critic for offering an account of why the observed retardation might be considerably less than that due to the tides. It was therefore 'conceivable that something of this accumulation of ice suggested by Professor Huxley, or erosion of matter suggested by Mr Croll, may to a considerable extent, have temporarily counteracted the tidal retardation' since 'the dynamical theory of the tides, and known facts regarding the interval between "full and change of moon", and the times of the spring tides, render it difficult to see how tidal retardation of the earth's rotation can be so little as to make the integral of lost time in a century amount to only twenty-two seconds'.[23] He then rejected Huxley's conclusion that the effect of tidal retardation could be compensated – and so permanently overthrown – by temporary conditions such as the melting or accumulation of polar ice. Finally, Thomson stated that his expectations from tidal dynamics weighed with him very decidedly against Dufour's meteoric hypothesis, although he still looked forward to its test by chemical analysis of the dust accumulated over Egyptian or other monuments during two or three thousand years.[24]

By comparison with these new questions raised by tidal retardation,

[21] Huxley, *Discourses*, pp. 332–4. [22] *Ibid.*, p. 335.

[23] William Thomson, 'Of geological dynamics', *Trans. Glasgow Geol. Soc.*, **3** (1871), 215–40; PL, **2**, 73–131, esp. pp. 90–1. [24] *Ibid.*, pp. 91–6.

Thomson's consideration in his 1868 address of his second theme, that of phenomena presented by the earth's crust, mainly pointed up arguments already stated in his earlier discussions of secular cooling. For example, he told his audience that for Playfair to have asserted that 'the earth could have been for ever as it is now' was thus about as reasonable as taking 'a hot water jar, such as is used in carriages, and say[ing] that that bottle has been as it is for ever'.[25]

Huxley simply replied that Thomson's limit based on secular cooling was 'of the vaguest', ranging from fifty million to 300 million years. For the geologist, 'one or two hundred million years might serve the purpose, even of a thorough-going Huttonian uniformitarian, very well'. If for some reason it did not prove sufficient, then 'we must closely criticize the method by which the limit is reached'. 'The argument is simple enough. *Assuming* the earth to be nothing but a cooling mass, the quantity of heat lost per year, *supposing* the rate of cooling to have been uniform, multiplied by any given number of years, will be given the maximum temperature that number of years ago.'[26] Huxley therefore asked, first, whether the earth was really nothing but a cooling mass, and, second, if its cooling had actually been uniform. 'The validity of the calculations on which Sir W. Thomson lays so much stress' depended upon 'an affirmative answer to both these questions'. Even given affirmative answers – by their nature hypothetical – 'other suppositions have an equal right to consideration'. For instance, at the prodigious temperatures 100 miles down, metallic bases might behave as mercury at red heat and refuse to combine with oxygen, while nearer the surface such a combination would occur to produce heat 'totally distinct from that which they possess as cooling bodies. Again, the quality of the earth's atmosphere 'may immensely affect its permeability to heat' and so 'profoundly modify the rate of cooling [of] the globe as a whole'. Huxley concluded that it could not be denied that 'such conditions may exist, and may so greatly affect the supply, and the loss, of terrestrial heat as to destroy the value of any calculations which leave them out of sight'.[27]

The general part of Huxley's address set forth an evolutionary cosmology embracing both geological and biological phenomena. Assigning the term 'evolutionism' to this view, he discussed the rival systems of 'catastrophism' and 'uniformitarianism'. Catastrophism was for Huxley 'any form of geological speculation which . . . supposes the operation of forces different in their nature, or immeasurably different in power, from those we at present see in action in the universe'. He implicitly rejected the version that assumed 'the operation of extra-natural power', as in Mosaic cosmogony, but accepted the doctrine that the earth had developed from a state in which its form, and the forces which it exerted, were 'very different from those we now know'. By the use of the phrase 'immeasurably different in power', however, Huxley ingeniously altered the meaning of catastrophism employed by Whewell and Thomson (i.e., a greater,

[25] *Ibid.*, pp. 53–4. [26] Huxley, *Discourses*, p. 338. [27] *Ibid.*, pp. 338–9.

but in no way unlimited, intensity of geological activity in the past) to an insistence upon 'the existence of a practically unlimited bank of force', thereby placing catastrophism on a par with uniformitarianism.[28]

Uniformitarianism, as originally conceived, had introduced *arbitrary limitations* into geological science. Huxley cited the words of James Hutton and Charles Lyell – foremost of the uniformitarians – to show that both agreed 'in their indisposition to carry their speculations a step beyond the period recorded in the most ancient strata now open to observation in the crust of the earth'. These arbitrary limitations apart, uniformitarianism had, *with equal justice to* catastrophism's bank of force, 'insisted upon a practically unlimited bank of time', keeping 'before our eyes the power of the infinitely little' and compelling us 'to exhaust known causes before flying to the unknown'.[29]

'Evolutionism', Huxley claimed, avoided the errors of its two predecessors, and embraced 'all that is sound in both'. In geology, as in biology, we needed to go beyond the natural history stage to the causal or aetiological stage and thereby ultimately 'attempt to deduce the history of the world, as a whole, from the known properties of the matter of the earth, in the conditions in which the earth has been placed'.[30] At first sight, therefore, Huxley's evolutionism appeared quite similar to Hopkins's progressionism, but, very quickly, Huxley's refusal to accept a cosmological and universal doctrine of progression became apparent.

Huxley regarded Immanual Kant as the heroic creator of evolutionism, the doctrine which traced the conversion of chaos into cosmos. An infinite expanse of formless and diffused matter attracted towards a single centre would reclaim more and more of the molecular waste until, by the heat evolved, the systems were 'converted once more into molecular chaos' such 'as that in which it began'. Thus, 'the cosmos has a reproductive operation "by which a ruined constitution may be repaired"'. This essentially cyclical view 'forestalled' Hutton but inevitably aroused the wrath of Thomson![31]

In his 1869 reply, Thomson dealt with Huxley's points directly. On the question of the earth being nothing but a cooling mass he referred his listeners to the arguments of his 1862 paper in which rival hypotheses had been considered and rejected. On the issue of uniform cooling he gave a decisively negative answer. Huxley, he hinted, had not examined 'the analytical investigation [which] shows the law of the *greater* rate of conduction outwards in past times, and demonstrates a much closer limit for the whole time during which the earth has been solid and continuously cool enough at its surface to be habitable without break of continuity to life, than can be estimated without taking into account the *deviation from uniformity* which I assert'.[32] In other words, non-uniform cooling would provide a *shorter* time scale than would uniform cooling, contrary to what Huxley had implied in his criticism of Thomson.

[28] *Ibid.*, pp. 307, 324. [29] *Ibid.*, pp. 313–16, 324. [30] *Ibid.*, pp. 316–19, 325.
[31] *Ibid.*, pp. 320–3. [32] PL, **2**, 98–9.

Turning to Huxley's 'other suppositions', Thomson argued that the first possibility – deriving from the heat of combination of elements – constituted 'merely an addition to the sum of the thermal capacities of the several elements separately reckoned, to give the effective thermal capacity of the composite mass'. The value of calculations which neglected this possibility was not destroyed as Huxley had claimed, 'though an altered figure in the result might be necessitated by an altered estimate of specific heat'. Thomson observed, however, that 'in my calculations I have left a wide enough margin to give due weight on Professor Huxley's side to the smallness of our knowledge regarding specific heats, thermal conductivities, and temperatures of fusion, of the earth's material'. As to the second possibility, he answered that the cloudiness or clearness of the atmosphere did not affect the secular cooling of the earth, since 'my calculations depend only on the assumption that through geological history . . . [the resulting temperature of the upper surface of land and sea] has been suitable for such life as now exists on the earth'. Commenting also upon the alleged vagueness of his estimates, Thomson argued that if he had calculated from the data a limit for the past duration of life on the earth of one or ten million years, 'so ill drawn an inference could scarcely embarrass those who are still disposed to trust a practically unlimited bank of time'. But,

> it is obvious that they must be seriously embarrassed by even a superior limit of four hundred million years: especially when the declaration of it is coupled with the assertion of a *very strong probability* that all geological history showing continuity of life is in reality to be condensed into a period not exceeding *one* hundred million years.[33]

Thomson ended by pointing out that Lyell and Huxley were both guilty of that 'direct opposition to the principles of natural philosophy' which he had attributed to British popular geology in 1868. Quoting at length from the 1868 edition of Lyell's *Principles*, Thomson made it clear that the famous geologist was still committed to a cyclical view of the solar system.[34] Firing on Huxley at the same time, Thomson remarked that 'Kant's hypothesis of the restoration of a new chaos . . . with potential energy for a repetition of cosmogony, described by Professor Huxley, was not a more violent contravention of thermodynamic law'. Furthermore, Huxley had violated the energy conservation principle in allowing to 'evolutionism' an unlimited bank of force and time:

> In the Catastrophism of Leibniz, Newton, Sedgwick, Phillips, Hopkins, Forbes, Murchison, and many other true geologists, which is no respect different as a geological doctrine from that now described by Professor Huxley under the new name 'evolutionism', there has been no 'unlimited bank of force'. And it is because the whole amount of energy existing in the earth has always been essentially finite, that physical science supports their theory, and rejects, as radically opposed to the principles of natural philosophy, the uniformitarianism described by Professor Huxley [in terms of a practically unlimited bank of time].[35]

[33] *Ibid.*, pp. 99–103.

[34] *Ibid.*, pp. 103–7; Charles Lyell, *Principles of geology*, 10th edn. (2 vols., London, 1867–8), **2**, p. 213.

[35] PL, **2**, 108–11.

On the harmony of geology and physics

Sir William's recollection of a conversation with the geologist Sir Andrew Ramsay at the Dundee meeting of the British Association in 1867 vividly illustrated the distinctive assumptions upon which the two sciences of geology and natural philosophy had been proceeding with regard to geological time. Thomson and Ramsay heard 'a brilliant and suggestive lecture by Professor Geikie on the geological history of the actions by which the existing scenery of Scotland was produced'. Sir William asked Ramsay how long a time he would allow for that history. Ramsay replied that 'he could suggest no limit to it'. Sir William persisted:

> 'You don't suppose things have been going on always as they are now? You don't suppose geological history has run through 1000 million years,'
>
> 'Certainly I do'.
>
> '10 000 million years',
>
> 'Yes'.
>
> 'The sun is a finite body. You can tell how many tons it is. Do you think it has been shining on for a million million years?'
>
> 'I am as incapable of estimating and understanding the reasons which you physicists have for limiting geological time as you are incapable of understanding the geological reasons for our unlimited estimates'.
>
> 'You can understand physicists' reasoning perfectly if you give your mind to it'.[36]

And there, according to Sir William's recollection, the argument ended with 'a friendly agreement to temporarily differ'.

Ramsay's commitment to an indefinitely long time scale placed him firmly in the tradition of Lyell's non-progressionist uniformitarianism. Yet that tradition did not command the allegiance of all geologists. The running dispute between progressionist and non-progressionist geology had never been finally resolved. Lyell had remained an unrepentant uniformitarian in the tenth edition of his *Principles* (1867–8). Phillips, in his *Life on the earth* (1860), had carefully avoided committing himself to either a uniformitarian or a catastrophist approach, though he clearly opposed any notion of indefinite time scales (ch. 16). And Sir Roderick Murchison, in the fourth edition of his *Siluria* (1867), still remained a staunch catastrophist who could now claim for support Thomson's recent work on the secular cooling of the earth.[37]

As a much younger geologist, Archibald Geikie (1835–1924) was probably less constrained by the convictions of an earlier generation. In his 1868 review of

[36] William Thomson, 'The age of the earth as an abode fitted for life', *Phil. Mag.*, [series 5], **47** (1899), 66–90; MPP, **5**, 205–30, on pp. 209–10.

[37] Sir Roderick Murchison, *Siluria*, 4th edn. (London, 1867), pp. 499–500. Murchison had noted with satisfaction that 'the view I long ago adopted solely from an appeal to geological phenomena has since been supported on these independent [thermodynamic] grounds by the reasoning of one of our leaders in physical science'.

the fourth edition of Murchison's *Siluria* he attempted to settle some of the old issues between uniformitarians and catastrophists by an appeal to geological evidence rather than geological authority. Yet he acknowledged that geological evidence alone would be insufficient to settle major questions of geological principle; some further criteria were necessary. Geikie sought for those criteria in both physics and astronomy, making possible a prolonged period of comparative harmony between geology and physics. At the same time, however, he acknowledged no infallibility of physicists in pronouncing on matters geological, Sir William Thomson not excepted. The physicist might offer assistance, but not authority.[38]

Geikie claimed that the role of the physicist had emerged dramatically in Thomson's 'remarkable memoir on the "Secular cooling of the earth" '. Using the simile of a 'monitor' – having less ammunition at the end than at the beginning of a battle – Sir William had inferred that uniformitarian arguments against a more violent intensity in the past were fallacious, and that the catastrophist doctrines were in fact supported by independent evidence from modern physics. Furthermore, Sir William had himself appealed to purely geological data to support the view of greater violence and intensity in past epochs (ch. 16). But Geikie remained sceptical, without questioning the fundamental principle 'that there must be less potential energy in the solar system now than there was originally'.[39]

To begin with, Geikie noted that Sir William had assumed that such geological phenomena as upheaval, fracture, and metamorphism 'depend for their production directly upon the effects of underground heat, and naturally infers that when this heat near the surface was greater, phenomena of that kind must have been more abundant and more violent'. But, Geikie asserted, there was certainly no geological evidence for this assumption; on the contrary, 'fracture and contortion of the crust of the earth are more probably referable to contraction due to cooling, and, if so, ought to have been less severe in ancient than in more recent times' since the thicker the earth's crust and the smaller the earth's diameter, the more marked would be the effects of each successive shrinkage. Thus, even granting more subterranean heat in early times than today, 'we are not called upon to admit that this necessitates any former greater intensity of earthquakes or upheavals'.

Secondly, 'Sir William Thomson's simile of the monitor furnishes an excellent illustration against himself'. It was true that just as such a ship must have less

[38] [Archibald Geikie], 'Sir Roderick Murchison and modern schools of geology', *Quart. Rev.*, **125** (1868), 188–217. Geikie subsequently published his *Life of Sir Roderick I. Murchison* (2 vols., London, 1875). See also Archibald Geikie, 'Geological time', *Geol. Mag.*, **4** (1867), 171–2; Burchfield, *Age of the earth*, pp. 60–2. Burchfield remarks that Geikie's conviction concerning the role of physics and astronomy led him 'firmly into Kelvin's camp' in 1868. Geikie's conviction derived from James Croll rather than directly from Thomson, of whose 1862 paper Geikie was distinctly critical. [39] [Geikie], 'Sir Roderick Murchison', pp. 203–4.

ammunition on board at a later stage of battle than at the beginning, so the earth must have a lesser store of heat now than at first. But:

is it not as evident that the effects produced by each gun in the vessel in no way depend upon the amount of powder in the magazine? The last shot fired is as loud and may be as destructive as the first, and so it would be even though there did not remain powder enough to fire a single discharge more. If the simile is to hold, we must grant that the upheavals and fracture of the crust of the earth, like the discharges of the monitor, have not been growing weaker in proportion as the internal magazine whence they came has parted with its stores of energy, but that the latest are, at least, equal in violence to the earliest.[40]

Geikie's third criticism of Thomson's application of the dissipation principle to the earth concerned his misuse of geological evidence. Like Murchison, Thomson had asked rhetorically if all geological evidence did not agree in demonstrating the former greater intensity of subterranean forces. Geikie, however, claimed that it was 'absolutely certain that no such demonstration can be drawn from any geological data yet discovered'. To say that 'powerful fractures or great contortions must have been produced by sudden and violent agencies' was 'precisely the point to be proved'. Moreover, the fact that older rocks had suffered more fracture and contortion could imply a longer exposure to geological change rather than a greater intensity of disturbance in early times. Indeed, Geikie argued, the inversion of the miocene strata of the Alps, cited by Sir Roderick as proof of far more violent operations of former times, was quite recent in geological terms. The inversion offered no proof 'that the internal force is growing weaker in its effects upon the surface'. And, similarly, volcanic outbursts such as those of northwest Britain in miocene times, had been much greater compared to those among the Lower Silurian rocks.

Geikie concluded that 'a serious study of the earth's crust furnishes no evidence of a diminution in the intensity of those agents by which the rocks have been upheaved, depressed, or fractured'. In reality, 'such evidences as can be brought to bear upon the question rather goes to show that the intensity, so far from diminishing, has, on the whole, been augmenting'. Compared to these potential fallacies of catastrophism, then, the uniformitarian school had proceeded 'on a safer basis of inquiry', having 'taken the present economy of Nature as their guide'. Provided that such an approach was regarded, not as a proven fact, but as a provisional rule based upon assumption, it was 'unquestionably the best which a geologist can follow'. Since our experience was 'at the best but scanty', such that there may be sources of geological change of which we had not even dreamed, the principle of uniformity acted as a methodological guide, and not a principle true of nature itself.[41]

40 *Ibid.*, p. 204.

41 *Ibid.*, pp. 205–6. Burchfield, *Age of the earth*, p. 62, argues that Geikie accepted 'uniformity of *kind*'; but not 'uniformity of *degree*'. If, however, Geikie had adopted such a position, there would have been no conceptual conflict between 'uniformitarianism' and 'catastrophism' (in Whewell's

Having 'freed' geological science from the 'dogmas' of these older schools, Geikie gave an enthusiastic welcome to the recent work of his fellow Scot, James Croll (1821–90).[42] Croll's work, though largely derivative from Thomson and Helmholtz on the question of the age of the sun, was, unlike Thomson's, explicitly uniformitarian rather than catastrophist in its approach.[43] Although Croll acknowledged that it was 'the modern and philosophic doctrine of uniformity that has chiefly led geologists to overestimate the length of geological periods', he still advocated the doctrine itself:

This philosophic school teaches, and that truly, that the great changes undergone by the earth's crust must have been produced not by great convulsion and cataclysms of nature, but by those ordinary agencies that we see at work every day round us . . . Now, when we reflect that with such extreme slowness do these agents perform their work . . . we are necessitated to conclude that geological periods must be enormous. And the conclusion at which we thus arrive is undoubtedly correct. It is, in fact, impossible to form an adequate conception of the length of geological time. It is something too vast to be fully grasped by our conception.[44]

We may note in Croll's phrase 'great convulsion and cataclysms of nature' the tendency to render 'catastrophism' in some sense 'miraculous' or at least associated with anomalous events in nature. Rhetorical moves such as this one (or Huxley's discussed earlier) obscure the main feature of catastrophism, the decreasing intensity of geological activity. Croll, like Geikie, maintained at least approximate uniformity of kind *and* degree, in contrast to Thomson's commitment to a much greater activity in past time.

sense of those terms), since both doctrines were committed to the same *kind* of geological agents for past and present activity, differing as they did over the issue of *degree* or *intensity*. From the evidence of Geikie's essay, there can be little doubt that he wished to maintain uniformity of both kind *and* degree, not as a rigid dogma, but as a flexible methodological rule.

42 Croll had been janitor of the Andersonian Museum in Glasgow from 1859 until 1867. He owed this lowly position to Thomson's father-in-law, Walter Crum, who was chairman of the Andersonian. His intellectual energies he then devoted to physics. He was appointed in 1867 as Edinburgh office-keeper, with the initial rank of assistant geologist, for the Geological Survey of Scotland – largely through the efforts of the Director (Geikie), aided by Murchison and Thomson. See J. Campbell-Irons, *Autobiographical sketch of James Croll with memoir of his life and work* (London, 1896), pp. 91–174.

43 James Croll, 'On geological time, and the probable date of the glacial and the upper miocene period', *Phil. Mag.*, [series 4], **35** (1868), 363–84; **36** (1868), 141–54, 362–86. Croll explained the aim of his paper – 'to show that geology and physical science bear the same testimony as to time' – in a letter dated 25th April, 1868, published in Campbell-Irons, *James Croll*, p. 188. See Burchfield, *Age of the earth*, pp. 62–7, for a full discussion of Croll's glacial hypothesis. Burchfield does not interpret Geikie's work as intimately related to that of Croll. Furthermore, when arguing for Croll's 'rejection of uniformitarianism', Burchfield means Croll's rejection of an indefinitely long time scale rather than uniformitarianism as Whewell had defined the term.

44 Croll, 'On geological time', p. 377. Passages from Croll's 1868 paper such as this one were much-welcomed by Charles Darwin and incorporated into the fifth edition of *The Origin*, published in August, 1869. In that edition the phrase 'incomprehensibly vast' was changed to 'how vast', and Darwin added Croll's opinion that it was the numerical estimates of geologists which were erroneous, rather than their conception of the vastness of geological time. See Morse Peckham (ed.), *The origin of species by Charles Darwin. A variorum text* (Philadelphia, 1959), pp. 485–6; Burchfield, *Age of the earth*, pp. 75–6; Campbell-Irons, *James Croll*, pp. 199–203, 215–22.

The principal issue now was the removal of any conflict between geological and physical time scales. Croll therefore argued that 'If the geologist could find a method of ascertaining the actual rate at which these denuding agents do perform their work . . . then we should have a means of ascertaining whether or not the agents to which we refer were really capable of producing the required amount of change in the earth's surface in the allotted time'. There did indeed exist a method, he believed, for determining, 'with the most perfect accuracy', the rate at which the face of the earth was being denuded by subaerial agency, namely 'that of determining the amount of solid materials which is being carried down annually by our rivers into the sea'. Since the materials all derived from the surface of the country drained by these rivers, we could determine exactly the extent to which the drainage area was being lowered annually by subaerial denudation. In the case of the Mississippi, Croll estimated that the whole area of drainage (with an average height of 748 feet) would be reduced to sea-level in less than four and a half million years, provided no elevation of the land took place. Estimates for other rivers suggested that the mean level of the country would be lowered by 1000 feet in six million years, a rate sufficiently rapid for Croll to emphasize as one of his major conclusions: 'we have thus seen that geology, alike with physics, is opposed to the idea of an unlimited age to our globe. And it is perfectly plain that if there be physical reasons, as there certainly are, for limiting the age of the earth to something less than 100 millions of years, geological phenomena, when properly interpreted, do not offer any opposition'.[45]

By 1868, then, Geikie and Croll had developed very similar assumptions concerning the terrestrial fabric. Both were uniformitarians in their belief that the intensity of geological agencies had remained *more or less* the same over time. Both, therefore, rejected catastrophism. Both were sceptical about the ability of the geological record to afford data for computing the length of its period in years, and both were convinced that geology had much to gain from a closer relationship with physics and astronomy. And Geikie, as Croll had done, was about to accept the consistency of Thomson's one-hundred million year time scale with the results of denudation. Whatever Geikie's reservations about Thomson's support for catastrophism, the physics of Thomson and Croll, and the geology of Geikie, were now very much in harmony.

Nowhere was the new synthesis more apparent than in Geikie's 1871 paper 'On modern denudation'. Acknowledging that his attention had been first called to the subject of subaerial denudation by Croll's 1867 paper, which instanced the Mississippi 'as a measure of denudation and thereby of geological

[45] Croll, 'On geological time', pp. 141, 378–84. Croll drew attention to Geikie's 'valuable paper' on 'Denudation as a measure of geological time' read to the Geological Society of Glasgow on 26th March, 1868. This paper was published as 'On modern denudation', *Trans. Geol. Soc. Glasgow*, **3** (1871), 153–90. Croll's account of denudation was very similar – at times indistinguishable – from that of Geikie. And both Croll and Geikie argued that marine denudation was of trifling proportions alongside subaerial denudation.

time', Geikie engaged in a wide ranging discussion of several rivers, large and small, and the respective rates of denudation of their areas of drainage. He concluded, like Croll, that geologists had been over-extravagant in their demand on time. And in a footnote connecting the 'recent researches in physics' specifically to the work of Thomson and Croll, Geikie stated that 'about 100 millions of years is the time assigned within which all geological history must be comprised'.[46]

At the same time, however, Geikie felt the need to answer the objection that palaeontological evidence, in direct defiance of physical evidence, demanded an expansion of geological time: 'under any view of the origin of species, a long time must needs be demanded for the appearance and disappearance of successive tribes of plants and animals', such that the palaeontologist would not readily abandon an appeal to unlimited time. He therefore resorted to the argument, deriving from Croll, 'that such reluctance would mainly arise from the difficulty of adequately conceiving the length [beyond a few hundreds of thousands] of even comparatively brief periods'. The palaeontologist, then, would probably find 'that the periods, as defined by purely physical evidence, would still remain quite vast enough for the accomplishment of the long history of life'. Since changes in the organic world were largely regulated by those in the inorganic world, it would now be necessary to consider whether, as in geology, 'the periods demanded for the growth and extinction of species and genera might not likewise have been exaggerated' and whether the intervals of time which palaeontology postulated 'might not all be easily comprised within the limits required by physical data'.[47]

Geikie's 1868–71 papers marked the beginning of a fifteen-year period during which the physics of Sir William Thomson and the geology of the day seemed to co-exist in comparative harmony. In his 1869 address, replying to Huxley, Thomson was able to cite not only the support which his own estimates had received from 'Professor Phillips's careful analysis of "the geological scale of time"', but also from Geikie's recent address to the Geological Society of Glasgow 'declaring his secession from the prevailing orthodoxy' and maintaining 'that all the erosion of which we have monumental evidence in stratified

[46] Geikie, 'On modern denudation', pp. 158*n*, 189*n*. See James Croll, 'On the excentricity of the earth's orbit, and its physical relations to the glacial epoch', *Phil. Mag.*, [series 4], **33** (1867), 119–31, esp. pp. 130–1.

[47] Geikie, 'On modern denudation', pp. 189–90. Geikie summarized the methods – geological and physical – of measuring geological time in his famous *Text-book of geology* (London, 1882), pp. 54–6, which was an expanded version of his article 'Geology' for the ninth edition of the *Encyclopaedia Britannica* (1880). Geikie appeared to give the secular cooling argument pride of place, while emphasizing Thomson's admission that 'very wide limits were necessary'. He further noted that Thomson now inclined 'rather towards the lower [twenty million years] than the higher antiquity [400 million years], but concludes that the limit, from a consideration of all the evidence, must be placed within some such period of past time as 100 millions of years'. By contrast, the sun's heat argument led 'to results confessedly less emphatic'. In this context he referred to Tait's figure of fifteen to twenty million years for the age of the sun. See below (note 64).

rocks, and in the shapes of hills and valleys over the world, could have taken place several times over in the period of 100 million years'.[48]

The 'new orthodoxy' which now prevailed involved a geological time scale derived by Thomson's methods, and restricted to estimates within his limits. For example, the Cambridge geologist W.J. Sollas (later professor of geology at Oxford) wrote in 1877 that geologists' estimates of the time since the deposition of the earliest recorded strata (fifty to 500 million years) could be brought 'into very close proximity with that when the sun first began his existence as our luminary [100–500 million years ago according to Sir William Thomson], and when our earth was first permanently crusted over [100–200 million years ago, again according to Thomson]'. Sollas emphasized, however, that while all these estimates may be 'erroneous to a considerable degree', there 'will yet remain a very considerable excess of energy, both in the sun and in our planet, beyond what they at present possess'. The influence of this excess of energy, Sollas argued, 'must have been distinctly sensible on the rate of early geological change'. In other words, 'the decreasing energy of the sun and of the earth must have led to diminishing rapidity in the action of three of the main factors of geologic change, viz. on the *denudation, reproduction*, and the *elevation* and *depression* of strata'. For Sollas, this 'catastrophist' deduction from physical law rendered independent geological estimates of time based on constancy of action wholly unreliable.[49]

Other geologists were sometimes less ready to commit themselves wholeheartedly to Thomson's time scale. For example, the professor of geology at Trinity College, Dublin, Samuel Haughton (1821–97), at first considered Thomson's one-hundred million years as sufficient. Subsequently, his own estimates of geological time took him to over 2000 million years. By 1878, however, he had settled for 153 million years, apparently by compelling his results to yield a figure within Thomson's range.[50] Equally, the geologist and engineer T.M. Reade (1832–1909) criticized, in 1878, Thomson's estimated maximum age of the earth, proposing instead a minimum figure of about 526 million years, derived from a consideration of chemical denudation. By 1893, however, he had accepted one-hundred million years as adequate.[51]

48 PL, **2**, 86–7. Thomson had received from Phillips a copy of his *Life on the earth* in January, 1869. In his 1864 address as president of the geological section of the British Association, Phillips had referred to 'a careful computation by Professor W. Thomson, on selected data' which led to a period of 100 million years within which 'all our speculations regarding the solid earth must be limited'. On the other hand, he also drew attention to Samuel Haughton's much greater estimate, discussed below. See *BAAS Report*, **34** (1864), 45–9.

49 W.J. Sollas, 'On evolution in geology', *Geol. Mag.*, **4** (1877), 1–7.

50 Burchfield, *Age of the earth*, pp. 100–3, provides a thorough discussion of Haughton's rather eccentric and erratic approach.

51 *Ibid.*, pp. 98–100, 129–30. As Burchfield points out, Reade was especially critical in 1878 of Thomson's secular cooling approach as a 'tremendous super-structure of inference'. In particular Thomson's data for the earth's temperature gradient and internal conductivity were both local and unreliable. See T.M. Reade, 'The age of the world as viewed by the geologist and the mathematician', *Geol. Mag.*, **5** (1878), 145–54; 'Measurement of geological time', *Geol. Mag.*, **10** (1893), 97–100.

For their part, the evolutionary biologists tended to look towards Croll and Geikie, rather than to capitulate directly to Thomson. A.R. Wallace (1823–1913), for example, had accepted Croll and Geikie's arguments by 1870, and had begun to develop the notion that high orbital eccentricity of the earth was linked to rapid changes of species. By 1880, indeed, Wallace was even working within Thomson's reduced estimates of twenty to thirty million years.[52] Charles Darwin himself, however, while welcoming into the fifth edition of *The origin* (1869) Croll's remarks on the difficulty of humanly comprehending the meaning of millions of years, became more cautious about any quantitative estimates of geological time:

> Mr Croll, judging from the amount of heat-energy in the sun and from the date which he assigns to the last glacial epoch, estimates that only 60 million years have elapsed since the deposition of the first Cambrian formation. This appears a very short period for so many and such great mutations in the forms of life, as have certainly since occurred. It is admitted that many of the elements in the calculation are more or less doubtful, and Sir W. Thomson gives a wide margin to the possible age of the habitable world. But as we have seen, we cannot comprehend what the figures 60 million really imply; and during this, or perhaps a longer roll of years, the land and the waters have everywhere teemed with living creatures, all exposed to the struggle for life and undergoing change.[53]

In the sixth and final edition of *The origin* (1872), Darwin referred to Thomson's criticism that the lapse of time since the consolidation of our planet was not sufficient to permit the assumed amount of organic change as 'probably one of the gravest as yet advanced'. He now conceded for the first time the probability, as Thomson insisted, 'that the world at a very early period was subjected to more rapid and violent changes in its physical conditions than those now occurring; and such changes would have tended to induce changes at a corresponding rate in the organisms which then existed'. But Darwin tried to avoid giving ground on whether or not Thomson's twenty to 400 million (or, more probably, ninety-eight to 200 million) year range was long enough to permit the assumed amount of organic change: on the one hand, 'we do not know at what rate species change as measured by years', while on the other, 'many philosophers are not as yet willing to admit that we know enough of the constitution of the universe and of the interior of our globe to speculate with safety on its past duration'.[54]

The younger Darwin: a critique of tidal friction and secular cooling

George Howard Darwin (1845–1912) provides a fascinating bridge between two of the most famous and widely acclaimed scientific intellects of the

[52] Burchfield, *Age of the earth*, pp. 76–9, 103–5. St George Mivart (1827–1900), on the other hand, had used Thomson's conclusions to attack natural selection in 1871. *Ibid.*, pp. 74–5. See also Burchfield, 'Darwin and geological time', pp. 316–19.

[53] Peckham (ed.), *The origin of species*, p. 486. See especially Burchfield, 'Darwin and geological time', pp. 309–13.

[54] Peckham (ed.), *The origin of species*, pp. 513, 728. See Burchfield, *Age of the earth*, pp. 78–9.

Victorian age. Born at Down in Kent as second son of the distinguished naturalist, this younger Darwin subsequently became an adopted intellectual son of the eminent natural philosopher. George's academic activities, indeed, came much closer to Sir William's than to his father's. Emerging from Trinity College in 1868 as second wrangler and second Smith's prizeman, he was elected a fellow of his college later that year. By 1873, he had settled in Trinity, having abandoned attempts to read for the Bar as his health deteriorated. Two years later, he noted in his diary that a 'paper on equipotentials [was] much approved by Sir W. Thomson'. And soon afterwards Sir William reported favourably to the Council of the Royal Society on Darwin's first major memoir, 'On the influence of geological changes in the earth's axis of rotation', published in the 1877 volume of the *Philosophical Transactions*.[55]

Sir William's method of refereeing, however, was rather unconventional. According to Darwin, Thomson 'seemed to find that on these occasions the quickest way of coming to a decision was to talk over the subject with the author himself – at least this was frequently so as regards myself'. As a result of the initial meeting in 1877, Darwin regarded his work on oceanic tides and lunar disturbances of gravity, and on tidal friction and cosmogony – occupying a substantial proportion of his published memoirs – 'as the scientific outcome of our conversation of the year 1877'.[56] A year later, the Royal Society accepted his first major paper on the tides, again apparently with Sir William's support. In this context, Charles Darwin wrote to his son on 29th October, 1878:

> All of us are delighted, for considering what a man Sir William Thomson is, it is most grand that you should have staggered him so quickly, and that he should speak of your 'discovery, etc.' . . . Hurrah for the bowels of the earth and their viscosity and for the moon and for the heavenly bodies and for my son George (F. R. S. very soon) . . .[57]

The passage quoted from George Darwin's 1878 letter at the beginning of this chapter indicated his special interest in defending his father's work. Darwin added, however, that his father did not, as was often alleged, attribute all to natural selection: 'he undoubtedly thinks it a leading cause, but has written long chapters & even a whole book on other causes'.

George Darwin's 1886 address as President of Section A of the British Association dealt with specific approaches to the issue of geological time. In his opening remarks he noted that 'great as have been the advances of geology during the present century, we have no precise knowledge of one of its fundamental units', that of time. Referring to Phillips's estimate of geological time as offering 'some insight into the order of magnitude of the periods with

[55] Sir Francis Darwin, 'Memoir of Sir George Darwin', in G.H. Darwin, *Scientific papers* (5 vols., Cambridge, 1907–16), **5**, pp. ix–xxxii, esp. xv.

[56] Darwin, *Scientific papers*, **3**, p. v.

[57] *Ibid.*, p. xv. George Darwin was elected F.R.S. in 1879. He succeeded James Challis in 1883 as Plumian Professor of astronomy and experimental physics at Cambridge.

which we have to deal' and as giving us confidence 'that a million years is not an infinitesimal fraction of the whole of geological time', Darwin nevertheless thought it unlikely that 'we shall ever attain to any very accurate knowledge of the geological time scale from this kind of [denudation] argument'. Equally, he believed that the imprecision, doubt, and uncertainty surrounding Croll's astronomical theory of geological climate, which he, like his father, had admired since at least 1868, could give little satisfaction on the question of geological time.[58] Darwin then turned to Sir William's methods:

> Amongst the many transcendent services rendered to science by Sir William Thomson, it is not the least that he has turned the searching light of the theory of energy on to the science of geology. Geologists have thus been taught that the truth must lie between the cataclysms of the old geologists and the uniformitarianism of forty years ago. It is now greatly believed that we must look for a greater intensity of geologic action in the remote past, and that the duration of the geologic ages must bear about the same relation to the numbers which were written down in the older treatises on geology, as the life of the ordinary man does to the age of Methuselah.[59]

Evaluating the argument from tidal friction, Darwin first drew attention to the modifications introduced into successive editions of the *Treatise* with regard to Adams's estimate of the secular acceleration of the moon's mean motion. In the second edition (1879–83) – in the editing of which Darwin had taken a prominent part – the addition of the sentence 'It is proper to add that Adams lays but little stress on the actual numerical values which have been used in this computation, and is of opinion that the amount of tidal retardation of the earth's rotation is quite uncertain' had been prompted by a recent estimate by Simon Newcomb in which Adams's value was reduced by two-thirds. Darwin thus stressed that 'in the opinion of our great physical astronomer, a datum is still wanting for the determination of a limit to geological time according to Thomson's argument'.

Second, Darwin considered the validity of the conclusion, subject to the uncertainty already emphasized, that 1000 million years ago the earth was rotating twice as fast as at present. The most recent edition (1883) of the *Treatise* had refined the argument that, if the earth had consolidated then, or earlier, the figure of the earth would have been very different from what it is now, and concluded 'that the date of consolidation is considerably more recent than a thousand million years ago.'[60] For Darwin, this estimate did not yet allow a sufficiently large margin of uncertainty. While once again expressing his hope that he was not presumptuous in criticizing 'the views of my great master, at

[58] G.H. Darwin,'Presidential address', *BAAS Report*, **56** (1886), 511–18, esp. 511–14. Darwin discussed at some length Croll's theory as published in his *Climate and time in their geological relations. A theory of secular changes of the earth's climate* (London, 1875). Evidence of George's early interest in Croll's views is provided by the correspondence between Charles Darwin and Croll in 1868–9. See Campbell-Irons, *James Croll*, 203, 215.

[59] Darwin, 'Presidential address', p. 514.

[60] *Ibid.*, p. 515.

whose intuitive perception of truth in physical questions I have often marvelled', he proceeded to criticize Sir William on three points – one geological, another astronomical, and the third physical – which challenged the assumption that 'the earth possessed rigidity of such a kind as to prevent its accommodation to the figure and arrangement of density appropriate to its rotation'.

The geological criticism was founded on James Croll's 1885 conclusion that 'nearly one mile may have been worn off the equator during the past 12 million years, if the rate of denudation all along the equator be equal to that of the basin of the Ganges'. Since, Darwin argued, the equatorial protuberance when the ellipticity of the earth was $\frac{1}{230}$ exceeded that when the ellipticity was $\frac{1}{300}$ by fourteen miles, it followed that 170 million years would suffice to wear down the surface to the equilibrium figure. Even at a much reduced rate of denudation, Darwin suggested, external evidence of excess ellipticity would be obliterated.

Darwin then drew attention to the possibility of unobserved internal evidence for excess ellipticity. In the absence of visible external evidence, he argued, 'we must rely on the incompatibility of the known value of the precessional constant with an ellipticity of internal strata of equal density greater than that appropriate to the actual ellipticity of the surface'. Might there not, he asked rhetorically, be 'a considerable excess of internal ellipticity without our being cognizant of the fact astronomically?'

Finally, Darwin founded his physical criticism of the tidal friction argument on a suggestion first made in his 1878 letter concerning the rigidity of the earth. He had then told Sir William that he did not see 'the force of the consideration that if the earth was rotating twice or thrice as fast as at present when it was first solid, that it could not be such as it is'. In particular, Tresca's memoirs on the flow of solids had shown 'iron flowing like butter', so that 'even if the Earth was as rigid to resist breaking by shearing as iron, I conceive it could only stand a small departure from the figure of equilibrium without flowing – & if it could once flow, why not admit that it might continuously adjust itself indefinitely'. Have we any right, he now asked his British Association audience, to feel 'so confident of the internal structure of the earth as to be able to allege that the earth would not through its whole mass adjust itself almost completely to the equilibrium figure?'[61] Darwin thus concluded that he could 'neither feel the cogency of the argument from tidal friction itself, nor, accepting it, can I place any reliance on the limits which it assigns to geological history'.

Darwin next referred to Sir William's 'celebrated essay' on the subject of secular cooling. While entirely agreeing with Thomson's general conclusion as to 'more abundantly and violently operative' plutonic activity in geological antiquity than at present, Darwin offered reasons against the assumption that the earth was simply a cooling globe:

[61] *Ibid.*, pp. 515–16.

The solidification of the earth probably began from the middle and spread to the surface. Now is it not possible, if not probable, that after a firm crust had been formed, the upper portion still retained some degree of viscosity? If the interior be viscous, some tidal oscillations must take place in it, and, these being subject to friction, heat must be generated in the viscous portion; moreover, the diurnal rotation of the earth must be retarded.[62]

Referring back to his 1879 *Philosophical Transactions* paper on the tides of a spheroid, viscous throughout the whole mass, in which he had estimated the amount and distribution of the heat generated while the planet's rotation was being retarded and the satellite's distance was being increased, Darwin noted that it had then seemed that 'the distribution of heat must be such that it would only be possible to attribute a very small part of the observed temperature gradient to such a cause [as viscosity]'.

He therefore suggested a 'more probable' hypothesis for the internal constitution of the earth. By supposing that 'it is only those strata which are within some hundreds of miles of the surface which are viscous, whilst the central portion is rigid', he argued that 'when tidal friction does its work the same amount of heat is generated as on the hypothesis of the viscosity of the whole planet, but instead of being distributed throughout the whole mass . . . it is now to be found in the more superficial layers'. From his 1879 estimate of the amount of heat generated during the lengthening of the day by one hour – equivalent to twenty-three million years at the present rate of heat loss by the earth – Darwin concluded first, that if the new hypothesis were true, and the same amount of heat were generated near the earth's surface, 'the temperature gradient in the earth must be largely due to it, instead of to the primitive heat of the mass', and, second, that such a hypothesis 'would greatly prolong the possible extension of geological time'. Fundamental to this hypothesis, however, was the assumption, developed in his attack on the tidal friction issue, that the earth could adjust itself to the equilibrium figure adapted to its rotation.

In a footnote to this address, Darwin outlined Sir William's objection to the hypothesis: 'since the meeting of the Association, Sir William Thomson has expressed to me his absolute conviction that, with any reasonable hypothesis as to the degree of viscosity of the more superficial layers, and as to the activity of tidal friction, the disturbance of temperature gradient through internal generation of heat must be quite infinitesimal'.[63] In concluding his discussion of the secular cooling argument, however, Darwin remarked that 'our data for the average gradient of temperature may be somewhat fallacious'. Given immense variations of temperature at or near the earth's surface, 'it does not then seem impossible that the mean temperature gradient for the whole earth should differ sensibly from the mean gradient in the borings already made'. In particular, the

[62] *Ibid.*, p. 516. [63] *Ibid.*, p. 517*n*.

recent *Challenger* expedition had revealed the temperature of the lower stratum of the ocean to be nearly at the freezing temperature.

In summing up his 1886 address, Darwin made it clear that he was protesting more against the precision of Tait's ten million year time scale than against the achievements of Thomson in turning 'the searching light of the theory of energy on to the science of geology'.[64] While Darwin admitted to having 'adduced some reasons' against the validity of the tidal friction argument, and to having shown that there were 'elements of uncertainty' surrounding the secular cooling argument, uncertainties which were 'far too great to justify us in accepting such a narrowing of the conclusions' as Tait's limit, nevertheless both arguments 'undoubtedly constitute a contribution of the first importance to physical geology'. Indeed, Darwin stated that we were 'fully justified in following Sir William Thomson who says that "the existing state of things on the earth, life on the earth, all geological history showing continuity of life, must be limited within some such period of past time as 100 million years"'.

Although Darwin could agree with Thomson's 1862 estimate for geological time, he was anxious that his father's theory would not be rejected because of cosmical physics. He therefore emphasized that 'at present our knowledge of a definite limit to geological time has so little precision that we should do wrong to summarily reject any theories which appear to demand longer periods of time than those which now appear allowable'. While not specifying which theories he had in mind, his final paragraph, methodological in nature, implied that Sir William himself, and certainly his ardent supporters, had been too unyielding in their demands for *consistency* between their dynamical theories and theories in other, less exact, branches of knowledge:

> In each branch of science, hypothesis forms the nucleus for the aggregation of observation, and as long as facts are assimilated and co-ordinated we ought to follow our theory. Thus, even if there be some inconsistencies with a neighbouring science, we may be justified in still holding to a theory, in the hope that further knowledge may enable us to remove the difficulties. *There is no criterion as to what degree of inconsistency should compel us to give up a theory*, and it should be borne in mind that many views have been utterly condemned when later knowledge has only shown us that we were in them only seeing the truth from another side.[65]

Lord Kelvin and the age of the earth: 1892–1907

In his presidential address to the 1892 meeting of the British Association at Edinburgh, Geikie still accepted Kelvin's notions of a calculable age for the solar system based on 'a gradual dissipation of energy from some definite starting

[64] Tait's estimates were published in [P.G. Tait], 'Geological time', *North Brit. Rev.*, **50** (1869), 406–39, esp. pp. 430–3, 438, and his *Lectures on some recent advances in physical science* (London, 1867), pp. 166–7. For discussion of Tait and his critics, see Burchfield, *Age of the earth*, pp, 76, 92–4, 109–10.

[65] Darwin, 'Presidential address', p. 518.

point' and the need to trim the extravagant demands on time to the 'tolerably ample period' of one-hundred million years as possibly quite sufficient 'for the transaction of all the prolonged sequence of events recorded in the crust of the earth'. The subsequent sweeping reductions of the time allowed for the evolution of the planet, however, had placed the geologist 'in the plight of Lear when his bodyguard of one hundred knights was cut down'. The 'inexorable physicist' had 'remorselessly struck slice after slice for his allowance of geological time' with the demand of Goneril: 'What need you five and twenty, ten, or five?'. Geikie thus observed that by 1892 Lord Kelvin was willing 'to grant us some twenty millions of years', while Professor Tait 'would have us content with less than ten millions'.[66]

With Geikie's geology and Thomson and Tait's physics no longer in harmony, Geikie saw two options: either to revise rigorously the interpretation given to the record of the rocks, or to revise the physical considerations. He confidently emphasized that, with regard to the latter option, 'the geological record constitutes a voluminous body of evidence regarding the earth's history which cannot be ignored, and must be explained in accordance with ascertained natural laws'. Echoing Huxley, Geikie argued that 'the mathematical mill is an admirable piece of machinery, but . . . the value of what it yields depends on the quality of what is put into it':

> That there must be some flaw in the physical argument I can, for my own part, hardly doubt, though I do not pretend to be able to say where it is to be found. Some assumption, it seems to me, has been made, or some consideration has been left out of sight, which will eventually be seen to vitiate the conclusions, and which when duly taken into account will allow time enough for any reasonable interpretation of the geological record.[67]

By comparison with Geikie's measured appraisal of the troubled relations between geology and physics, Lord Salisbury's presidential address to the British Association two years later must have seemed – as it certainly did to John Perry – both partisan and provocative. The former prime minister did confess to feeling like a country gentleman who, as a volunteer colonel, had been appointed to review a crack army corps. Nevertheless, Salisbury showed little restraint in his pronouncements upon Darwin's theory of evolution. He asserted that the 'deepest obscurity still hangs over the origin of the infinite variety of life'. In particular, he claimed that Kelvin's limitation of terrestrial time was one of 'the strongest objections to the Darwinian explanation' and, as such, still appeared to retain all its force. For their part, geologists and biologists had 'lavished their millions of years with the open hand of a prodigal heir indemnifying himself by present extravagance for the enforced self-denial of his youth'. Yet the biologists especially were 'making no extravagant claim when they demand[ed] at least

[66] Sir Archibald Geikie, 'Presidential address', *BAAS Report*, **62** (1892), 3–26, esp. pp. 18–20. See also Burchfield, *Age of the earth*, pp. 130–1.

[67] Geikie, 'Presidential address', p. 20.

many hundred million years for the stupendous process' of evolutionary development. If the mathematicians were right, he argued, the biologists could not have what they demanded, since any organic life existing on the globe more than one-hundred million years ago must, by Kelvin's theory, 'have existed in a state of vapour'. The controversy, as shown by Geikie's 1892 address, was still very much alive: 'the mathematicians sturdily adhere to their figures, and the biologists are quite sure the mathematicians must have made a mistake'. While carefully denying any wish to intervene in the controversy itself, Salisbury nonetheless concluded his remarks by 'returning a verdict of "not proven" upon the wider issues the Darwinian school has raised'.[68]

In the autumn of 1894, Perry reacted directly to Salisbury's address, while at the same time taking up Geikie's remarks as a challenge. Writing at length to both G.F. FitzGerald and Joseph Larmor, Perry explained that when asked by friends to criticize Kelvin's calculation of the probable age of the earth, he would reply that it was 'hopeless to expect that Lord Kelvin should have made an error in calculation'. Furthermore, 'in every class in mathematical physics in the whole world since 1862 the problem has been put before students, and, as the subject is of enormous interest, if there had been any error it certainly would have been discovered before now'. At the same time, Perry stated that his own dislike of any quantitative problem set by a geologist was due to the given conditions being much too vague in nearly every case for the matter to be in any sense satisfactory since 'a geologist does not seem to mind a few millions of years in matters relating to time':

> I never till about three weeks ago seriously considered the problem of the cooling of the earth except as a mere mathematical problem, as to which definite conditions were given. But the best authorities in geology and palaeontology are satisfied with evidences in their sciences of a much greater age than the one hundred million years stated by Lord Kelvin; and if they are right, there must be something wrong in Lord Kelvin's conditions. On the other hand, his calculation is just now being used to discredit the direct evidence of geologists and biologists, and it is on this account that I have considered it my duty to question Lord Kelvin's conditions.[69]

Perry's questioning of Kelvin's conditions centred on replacing Kelvin's assumption of a homogeneous mass by a superficial layer of one conductivity and an internal mass of higher conductivity, leading to the conclusion that the present (high) temperature gradient just below the surface would, given the (lower) temperature gradient in the internal mass, demand a greater age for the earth. He also suggested that a larger heat capacity would increase the age. Yet

[68] The Marquis of Salisbury, 'Presidential address', *BAAS Report*, **64** (1894), 3–15, esp. pp. 12–13. See also Burchfield, *Age of the earth*, pp. 121–2, 134.

[69] John Perry to Joseph Larmor, 12th October, 1894, P56 (copy), ULC; John Perry, 'On the age of the earth', *Nature*, **51** (1894–5), 224–7, on pp. 224–5. See also Burchfield, *Age of the earth*, pp. 134–7.

his attempts to extend the age foundered not on the hectoring replies he received from Tait, but on Kelvin's measured consideration of the thermal conductivities of known rocks at high temperatures.

In his reply to Perry, Kelvin argued that 'for all we know at present . . . I feel that we cannot assume as in any way probable the enormous differences of conductivity and thermal capacity at different depths which you take for your calculations'. Indeed, in his 1862 'Secular cooling' paper, he had referred to this very question and had, in the light of so much uncertainty in the data, allowed what he believed were very wide limits in his own estimate of the time involved since the emergence of the *consistentior status*:

> I thought my range from 20 millions to 400 millions was probably wide enough, but it is quite possible that I should have put the superior limit a good deal higher, perhaps 4000 instead of 400. The subject is intensely interesting; in fact, I would rather know the date of the *Consistentior Status* than of the Norman Conquest; but it can bring no comfort in respect to demand for time in palaeontological geology. Helmholtz, Newcomb, and another, are inexorable in refusing sunlight for more than a score or a very few scores of million years of past time . . . So far as underground heat alone is concerned you are quite right that my estimate was 100 millions, and please remark . . . that that is all Geikie wants; but I should be exceedingly frightened to meet him now with only 20 million in my mouth.[70]

These remarks indicate a degree of flexibility not always recognized by commentators. Kelvin certainly had little appreciation for the independent force of geological evidence, as insisted on by Geikie. But too often a portrait of a man with increasingly dogmatic views has been offered at the expense of a man whose methodological criterion of consistency between different branches of knowledge, founded upon quantitative data on the properties of matter, led him to reject theories which for him had an isolated, *ad hoc* character.[71]

Kelvin was able to communicate further to *Nature* in March, 1895, the news that 'Prof Perry and I had not to wait long . . . to learn that there was no ground for the assumption of greater conductivity of rock at higher temperatures, on which his effort to find that the consolidation of the earth took place far earlier than 400 million years ago, is chiefly founded'. He quoted from Robert Weber's latest results which showed either no change, or a considerable diminution, of thermal conductivity with rise of temperature and which were clearly at variance with Perry's claims. These new results contradicted Weber's earlier one upon which Perry had built his case. Kelvin also corresponded with E.E.N. Mascart over the search for more definite data on the thermal properties of rock up to its melting point. His explicit aim was to find 'closer limits' for the time that had passed since the consolidation of the earth. Furthermore, preliminary

[70] Lord Kelvin to John Perry, 13th December, 1894 (copy), P60, ULC; Perry, 'Age of the earth', p. 227. [71] cf., Burchfield, *Age of the earth*, p. 123.

experiments on slate, sandstone, and granite by Kelvin himself indicated a diminution of conductivity with a rise of temperature.[72] On the question of specific heats, Kelvin believed that '*they* increase with the temperature up to the melting point of rock, but the rate of augmentation assumed by Prof Perry is about five times as much as that determined up to 1200° by the recent experiments of A.W. Rücker and W.C. Roberts-Austen on basalt and of Carl Barus on diabase.[73]

With the aim of revising his estimates of the earth's age, Kelvin discussed the implications of the recent work of Barus and Clarence King in connection with the United States Geological Survey. In the light of Barus's results on the melting of diabase, a typical basalt, at 1100–1200°C, Kelvin concluded that his own 1862 estimate of 7000°F (3871°C) for the temperature of melting rocks was too high. The correction, with other assumptions unchanged, would reduce his 1862 estimate to just under ten million years. Given that the effect of pressure had to be taken into account, King had concluded 'that without farther experimental data "we have no warrant for extending the earth's age beyond 24 millions of years"'. By taking the case for specific heat increasing up to the melting point according to the results of Rücker, Roberts-Austen, and Barus, and for conductivity constant, and by taking into account the augmentation of melting temperature with pressure 'in a somewhat more complete manner than that adopted by Mr Clarence King', Kelvin found that he was 'not led to differ much from his estimate of 24 million years'. Nevertheless, any closer estimate would be 'quite uninteresting' until we knew something more 'as to the probable diminution, or still conceivably possible augmentation, of thermal conductivity with increasing temperature'.[74]

Perry's last letter to *Nature* in April, 1895, implicitly admitted, by its apologetic tone, that his criticism of Kelvin's conditions had been but a partial success. Referring to his previous communications to *Nature*, Perry made clear that his aims had been to show, and illustrate by a number of mathematically workable examples, that, if we assumed greater conductivity in the interior than at the surface, 'we increase this limit of age'. He 'did not pretend that any one of these represented the actual state of the earth'; rather, 'they merely proved that there were *possible* internal conditions which might give enormously greater ages than physicists had been inclined to allow'. Perry emphasized, indeed, that, while some of his results 'may have seemed more probable than others', he had 'tried to show that it was impossible for a physicist to obtain such an estimate, as

[72] Lord Kelvin to John Perry, 15th January, 1895, P68 (copy), ULC; R.H. Weber to Lord Kelvin, 13th January, 1895, W52, ULC, enclosing copies of Weber's papers of 1881–2 published in *Bull. Soc. Sci. Nat. Neuchatel*, **12** (1881–2), 394–418, 687–95; Lord Kelvin to E.E.N. Mascart, 2nd February, 1895, in SPT, **2**, 942; Lord Kelvin, 'The age of the earth', *Nature*, **51** (1894–5), 438–40. The results of the experiments were published by Lord Kelvin and J.R.E. Murray, 'On the temperature variation of the thermal conductivity of rocks', *Proc. Glasgow Phil. Soc.*, **26** (1895), 227–32; MPP, **5**, 198–203.

[73] W.C. Roberts-Austen and A.W. Rücker, 'On the specific heat of basalt', *Phil. Mag.*, [series 5], **35** (1893), 296–307, esp. pp. 301–3.

[74] Kelvin, 'Age of the earth', p. 439.

there were all kinds of possible assumptions which led to many different answers': 'I was not looking for a *probable* age of the earth from the point of view of mere physics. I wished to show that the physics' [*sic*] higher limit was greater than a few hundred of millions of years'.

Perry's concern with possible, as distinct from Kelvin's concern with probable, internal conditions, meant that no ultimate reconciliations could, or indeed need, occur between pupil and master. The correctness of Weber's earlier results, or of the assumption of masses of liquids inside the earth, for example, supported Perry's argument 'if they are correct, but they do not in any way destroy it if they are wrong'. For Perry, unlike Kelvin, the paper by King appeared 'somewhat inconclusive', being limited to a solid earth of uniform conductivity and uniform temperature. The effect of pressure on the melting temperature of rock, having the properties of diabase, allowed estimation of the temperatures above which liquidity would occur at various depths. Yet, Perry concluded, 'it is evident that if we take any probable law of temperature of convective equilibrium at the beginning and assume that there may be greater conductivity inside than on the surface rocks, Mr King's ingenious test for liquidity will not bar us from almost any great age'.[75]

Overall, Perry's review of the leading arguments for the age of the earth and solar system directed him to the conclusion that the lowest *maximum* age yet published by the physicists was Kelvin's 400 million years derived from the secular cooling argument. The tidal retardation and sun's heat arguments gave maxima of 1000 and 500 million years, respectively. Thus, while 'mere physics' pointed to a *probable* age much less than 400 million years, nevertheless Perry could see nothing from the physicists' point of view which would deny the palaeontologists four times the greatest of these estimates should they 'have good reason for demanding much greater times'.

Just as the debates concerning the new phenomenon of radioactivity in the early 1900s impinged upon the whole issue of the nature of the sun's energy (ch. 15), the same investigators soon questioned nineteenth-century assumptions regarding terrestrial heat. As early as May, 1904, Frederick Soddy (1877–1956) summarized the impact of the new science of radioactivity:

> It has been recognized that there is a vast and hitherto almost unsuspected store of energy bound up in, and in some ways associated with, the unit of elementary matter, represented by the atom of Dalton . . . Since the relations between energy and matter constitute the ultimate groundwork of every philosophical science, the influence of these generalizations on allied branches of knowledge is a matter of extreme interest at the present time. It would seem that they must effect, sooner or later, little short of a revolution in astronomy and cosmology. They will certainly be eagerly received, for it is only fair to add that they have been long awaited, by the biologist and geologist.[76]

[75] John Perry, 'The age of the earth', *Nature*, **51** (1894–5), 582–5.

[76] Frederick Soddy, *Radio-activity: an elementary treatise from the standpoint of the disintegration theory* (London, 1904), p. iv.

The first public challenge to Kelvin's secular cooling assumptions came in a letter to *Nature* one week after George Darwin's communication on 'Radioactivity and the age of the sun'. Under the title 'Radium and the geological age of the earth', John Joly – professor of geology at University College, Dublin, and recently responsible for estimates of geological time based on the supply of salt to the oceans – put the case for a wholly new interpretation of terrestrial heat:

Quite equivalent to increased supplies from the interior would be a source of supply of heat in every element of the material. The establishment of the existing gradient of temperature inwards may, in fact, have been deferred indefinitely during the exhaustion of stores of radium and similar bodies at greater or shallower depths. In fact, we find these bodies here; the only question is as to how much of them exists, or at one time existed, in the earth's interior . . . [It] would appear that the estimates derived from physical speculations are now subject to modification in just the direction which geological data required. The hundred million years which the doctrine of uniformity requires may, in fact, yet be gladly accepted by the physicist.[77]

Addressing the Royal Institution early in 1904, Ernest Rutherford drew particular attention to Kelvin's remark of 1862 that 'as for the future, we may say, with equal certainty, that the inhabitants of the earth cannot continue to enjoy the light and heat essential to their life for many million years longer, unless sources of heat – now unknown to us – are prepared in the great storehouse of creation'. In view of the discovery of radioactive substances, said Rutherford, 'this remark seems almost prophetic'. For, taking both temperature gradient and thermal conductivity employed by Kelvin, 'it can be calculated that the amount of heat conducted by the earth's surface each year and lost by radiation could be supplied by the presence of radium (or an equivalent amount of other radioactive matter) to the minute extent of about five parts in ten thousand million by weight', this estimate being based on Elster and Geitel's observations of the amount of radioactive matter present in the soil.[78]

Rutherford's erstwhile collaborator, Soddy, took a very similar line in his 1904 elementary treatise on radioactivity, with the crucial difference that Soddy, unlike Rutherford, raised the question of a 'reversible' cosmos and offered a highly speculative answer in terms of the possible reconstruction of matter at a subatomic level: 'with regard to radio-activity, an independent limit of the past age of the earth is set by our present ignorance of any concomitant process of

77 John Joly, 'Radium and the geological age of the earth', *Nature*, **68** (1903), 526.

78 Ernest Rutherford, 'The radiation and emanation of radium', *Sci. Am. Suppl.*, **58** (1904), 24073–4, 24086–8; *The collected papers of Lord Rutherford of Nelson* (3 vols., London, 1962–5), **1**, pp. 650–7, on pp. 656–7. As early as 1902, Rutherford claimed to have 'made some calculations to determine how much radium (or other radioactive matter expressed in terms of radium) would be required to be uniformly distributed throughout the earth in order to maintain the present observed temperature gradient'. His conclusion then was that the amount of radium actually present in the earth was of the right order of magnitude to maintain the temperature gradient. *Ibid.*, p. 926. The 1904 estimates were set out in Rutherford's *Radio-activity* (Cambridge, 1904), pp. 342–6. On Rutherford and radioactivity, see also Burchfield, *Age of the earth*, pp. 163–71.

atomic reconstruction'. In the absence of such a process, Soddy explained that 'the past age of radioactive minerals is fixed simply by the period of the average life of the elements uranium and thorium', set at somewhere between 100 and 1000 million years. By ascertaining the diminishing proportion of uranium remaining at various intervals, Soddy argued that even if the whole earth had originally been uranium it could not have survived more than 1000 or 10 000 million years without 'reproduction'.[79]

In the light of this conclusion, Soddy considered the possibility of the reconstruction of the parent radio-elements. While recognizing the problems as to the source of the energy necessary for such reconstruction, he called into question the Second Law of Thermodynamics, the limitations of which had, thanks to James Clerk Maxwell, 'long been clearly recognized'. Soddy therefore posed the 'real question underlying the possibility of a reconstruction of the elements with the absorption of energy':

> Can the second law of thermodynamics be applied to sub-atomic change? . . . All processes of evolution at present revealed proceed in the one direction, the energy being 'degraded' into heat of uniform temperature, which is no longer available for any useful purpose . . . The end of evolution is definitely fixed as occurring when all the available energy shall have run its course to exhaustion. Correspondingly, a sudden beginning of the universe – the time when present laws began to operate – is also fixed. It is necessary to suppose that the universe, as a thing in being, had its origin in some initial creative act, in which a certain amount of energy was conferred upon it sufficient to keep it in being for some period of years . . . The alternative . . . is that the second law . . . does not universally apply, and that, in the infinitely varied processes of Nature, a cyclic scheme of evolution is possible.[80]

That cyclic scheme would depend, for its operation, upon the reversal of the dissipation of available energy. To achieve such a result, the lighter elements would be 'continuously growing, by the gradual accretion of masses, possibly of electronic dimensions, and at the same time storing up the "waste" energy produced in the opposite [dissipation] process'.

Soddy believed, unlike Kelvin, that such a cyclical system would be 'a much more complete and satisfactory one' than a corresponding directional system. If true, 'the universe would then appear as a conservative system, limited with reference neither to the future nor the past, and demanding neither an initial creative act to start it nor a final state of exhaustion as its termination'. Although 'at present a mere speculation', Soddy thought it not unreasonable to anticipate that ultimately the laws of nature 'will be recognized to operate, not only universally with regard to space . . . but also consistently to time'; that is, the laws of nature would be recognized as having been 'in continuous operation without external interference'.[81]

[79] Soddy, *Radio-activity*, pp. 185–9. See also Burchfield, *Age of the earth*, pp. 169–70.
[80] Soddy, *Radio-activity*, p. 188. Maxwell's interpretation of the Second Law of Thermodynamics will be discussed in the next chapter. [81] *Ibid.*, p. 189.

In 1905 Rutherford argued that if 'the whole world had originally been composed of pure radium, its activity 20 000 years later would not be greater than that observed in pitchblende today'. Clearly the earth was much older than 20 000 years and so, to account for the presence of radium 'it is necessary to suppose that radium is a transition element continuously produced by the breaking up of some other substance'. In Rutherford's view, the most likely parent of radium was the element uranium with which it was always found associated in pitchblende. He therefore concluded:

If uranium is the parent of radium, the amount of radium present in the earth will at all times be proportional to the amount of uranium, so that the duration of radium is dependent on the life of uranium. It is probable that uranium or thorium would be half transformed in 1000 million years, so that if the radioactive matter in the earth is mainly derived from the breaking up of one or both of these substances, it is to be expected that the heat generated in the earth would fall to half value in about 1000 million years. The earth would thus cool very slowly with the time and must have been hotter in past times, but no estimate of the age of the earth can be made from the amount of internal heat observed today.[82]

In a lecture delivered before the Royal Astronomical Society of Canada in April, 1907 – the year of Kelvin's death – Rutherford returned to the question of the age of radioactive minerals. Pointing to the capacity of radium to produce the stable element helium, first shown by Ramsay and Soddy, at a rate of about 0.1 cc of helium per year from one gram of radium, and given that helium had only been found in quantity in the radioactive minerals, Rutherford argued that the total amount of helium trapped in a dense radioactive mineral would be proportional to the age of the mineral and the amount of radium contained in it. As an example, he calculated that the mineral thorianite was at least 500 million years old, but this was a minimum estimate, as some helium had probably escaped from the mineral since its formation in the earth's crust. Most other minerals, he stated, were apparently in the range 500–1000 million years old. Thus, Rutherford believed, 'when the constants involved in these calculations are accurately determined, I feel great confidence that this method will prove of the utmost value in determining with accuracy the age of the radioactive minerals and indirectly of the geologic strata in which they are found'.[83]

Kelvin's unenthusiastic response to the new views of the sun's energy and the earth's internal heat had appeared already in his reply to a letter from James Orr in January, 1906, (ch. 15). Six months later, in response to a query concerning the same issue, Kelvin made clear that he would not readily change his opinions:

[82] Ernest Rutherford, 'Radium – the cause of the earth's heat', *Harper's Mag.*, **49** (1904–5), 390–6; *Collected Papers*, **1**, 776–85, on pp. 783–4. On subsequent attempts at radioactive dating, see Burchfield, *Age of the earth*, pp. 171–9.

[83] Ernest Rutherford, 'Some cosmical aspects of radioactivity', *J. Royal Astron. Soc. Canada*, **1** (1907), 145–65; *Collected papers*, **1**, 917–31. See Burchfield, *Age of the earth*, pp. 173–4.

I do not think it at all probable that radioactivity can have been practically influential in connection with the primitive heat of the earth, which I believe to have been due to gravitational energy spent in the coalition of smaller bodies. I scarcely think that radioactivity can possibly have had any considerable influence on the subsequent cooling and shrinkage of the earth in its present condition.[84]

[84] Lord Kelvin to T.E. Young, 9th June, 1906, Y6, ULC.

18

The habitation of earth

> My task has been rigorously confined to what, humanly speaking, we may call the fortuitous concourse of atoms, in the preparation of the earth as an abode fitted for life . . . Mathematics and dynamics fail us when we contemplate the earth, fitted for life but lifeless, and try to imagine the commencement of life upon it. This certainly did not take place by any action of chemistry, or electricity, or crystalline grouping of molecules under the influence of force, or by any possible kind of fortuitous concourse of atoms. We must pause, face to face with the mystery and miracle of the creation of living creatures. *Lord Kelvin, 1899.*[1]

Fundamental to William Thomson's conception of life on earth was the principle of energy dissipation, which affected his view of the origin and nature of living creatures in three significant ways. First, the secular cooling of the earth (a particular instance of the dissipation principle) 'gave a very decisive limitation to the possible age of the earth as a habitation for living creatures; and proved the untenability of the enormous claims for time which, uncurbed by physical science, geologists had begun to make and to regard as unchallengeable' (chs. 16 and 17).[2] His investigations of the nature of the sun's heat led him to the same conclusion (chs. 14 and 15). The physical conditions of earth and sun set limits for the earliest time at which the earth could be regarded as an abode fitted for life. Second, Thomson held that living creatures and dead matter were distinguished by the former's privileged possession of free will, a will which enabled them to direct the energies of nature in accordance with the dissipation principle. And third, from these two beliefs, he drew the implication that the origin and evolution of life on earth had not taken place by any fortuitous concourse of atoms but by 'a creating and directing Power'.

Mind over matter

A distinctively Scottish view of man's relation to the material world expressed in the works of Reid and Robison depended fundamentally upon mind's power,

[1] Lord Kelvin, 'The age of the earth as an abode fitted for life', *Phil. Mag.*, [series 5], **47** (1899), 66–90; MPP, **5**, 205–30, on pp. 229–30. [2] MPP, **1**, 39.

manifested through will, intention, and design, to employ material means to gain ends, although never to abrogate the laws of nature. Man, as 'artist', utilized skills and instruments to act on matter by analogy with the Supreme Mind or Artist. This form of dualism emphasized the interrelation, and not merely the separateness, of mind and matter (ch. 4).

A basic dualism also ran through William Thomson's student notes in logic, moral philosophy, and natural philosophy at Glasgow College. Buchanan, for example, explained that, while 'a physical cause [of a stone falling] is a cause in nature', a 'moral cause depends on the will'. Thomson's own 1846 'Introductory lecture' emphasized that 'Mind and Matter, the two great provinces of Nature, are very remarkably and wonderfully distinct subjects of investigation'.[3] The distinction between mind and matter, free will and determinism, extended right through his thinking on the dissipation principle and would have far-reaching implications for his (and Maxwell's) interpretation of that principle during the 1860s and beyond.

In the 1851 draft for his 'Dynamical theory of heat', Thomson again sharply distinguished material and moral worlds. The material world was subject to determinate laws of nature, which at this point he thought fully expressible in terms of information accessible to man. Knowledge of 'the position & motion of each atom of matter' at any instant would determine 'the position & motion of each at any time past or future'. Given this Laplacian determinism, man could in principle, excepting distinct cases which 'we are justified in calling miracles, foresee the future with certainty' in the material world; but the moral world, the realm of mind, was different: man 'never will be able to foresee even the simplest fact with certainty in the operations of mind'.

Here the operations of mind refer to free will, rather than to the acquisition of knowledge by sensory perception and reason, which he believed fell under laws of mind. We have therefore to deal with four categories: laws of matter, laws of mind, free will and miracles. Of course Thomson's sense-based methodology supposedly guaranteed that laws of mind, properly applied, would never yield results in conflict with the laws of matter (ch. 13). He included in a footnote a more elaborate discussion of miracles:

> If a stone should stand in the air and we should be able to assure ourselves that there is no thread supporting it, no *sufficient* magnetic electrical or other action bearing it, we should assert that a miracle was wrought before our eyes. But if a man should resist the strongest apparent motives to commit some action; if we could be certain that he was convinced that this action would if committed, give him immediate gratification, we should not say a miracle was wrought upon him but we might recognise the influence of the spirit working the will of God in a way which man cannot investigate & reduce to 'laws' like those of matter.[4]

[3] William Thomson, Notes on philosophy lectures (Robert Buchanan's logic class), NB16, ULC; Notes on lectures in moral philosophy (William Fleming's class), NB20, ULC; SPT, **1**, 239.

[4] William Thomson, Preliminary draft for the 'Dynamical theory of heat', PA128, ULC, footnote facing pp. 7 and 8. Hereafter 'Draft'.

Actions of will, then, stood between laws of matter and mind, on the one hand, and miracles; they could not be reduced to laws, but neither did they violate laws. This intermediate status is of interest because Thomson did believe in miracles, especially miraculous visions, which gave knowledge not normally accessible to human minds, and therefore in violation of laws of mind, if not of matter. But why did not the free will of individuals also violate laws of mind?

Because, he answered, it involves singular acts of individual minds, which do not fall under the laws of mind: 'there are no absolute laws manifested to man' of the operation of single minds. However, 'there are absolute laws regarding the mutual actions of different minds, or of one mind to future events or unknown past and present events, certainly regarding mutual consciousness'. He attempted to explain more precisely what 'miraculous deviation' from such laws would mean:

The vision of Peter [Acts 10] & some of the circumstances connected with it are I think satisfactory illustrations of such miracles. Any vision, however intense, of a single mind cannot be called miraculous. If the vision related to persons & those [persons are] dead, we cannot know in this world whether those other beings were conscious. I think such visions are generally confined to the consciousness of the person seeing them. If the person by means of one [a vision] gets knowledge which he could not have acquired by recognizable means. I think we should be right in calling the vision miraculous . . . The speculations or pretended prophesies or ravings of enthusiasts or mad men are not miraculous but the utterances of a true prophet are.

The significance of Peter's vision, as distinguished from speculations and ravings, lay in its shared character. Although his own vision of a strange vessel descending from heaven concerned a single mind, the centurion Cornelius had also had a vision, requesting that he send for Peter. On his arrival with Cornelius, Peter received the message that the truth of the gospel should be conveyed to Jews and Gentiles alike.[5] This knowledge of divine will, acquired by more than one mind but not by 'recognisable means', was properly miraculous. Knowledge of the will of another person, acquired through a dream, would also be miraculous according to Thomson.

Miracles violate laws, whether of mind or of matter. Free will, however, like the ravings of a madman, falls under the doctrine of unique singular events, not governed by laws and thus not in violation of them. Man obtains knowledge of laws, both of mind and matter, 'by recognisable means' common to all. Free will allows him as an individual to employ those laws to his own ends. Thomson's problem was to make room for such singular action within the set of all law-governed events. But the task seemed impossible if Laplacian determinism had to be maintained, in the form of the predictive certainty of atomistic

[5] We are grateful to Colin Russell, *Time, chance and thermodynamics* (Milton Keynes, 1981), pp. 31–2, for pointing out the most probable source of the 'vision of Peter' as Acts 10. The prophecy in 2 Peter 3:10–12 is discussed in Chapter 15 above in relation to Thomson's 1854 and 1862 'cosmology of energy', but is almost certainly not what Thomson had in mind in the 1851 draft.

mechanics enunciated above. Slowly he would escape from the impasse, to the detriment of particle mechanics; but he had not yet moved so far (ch. 12).

Within this confusing context of laws, free will, and miracles Thomson formulated his notion of dissipation in 1851. Unlike God, man could not 'miraculously' create energy; but like God, man could, by his *will*, direct energy in its course, and so employ available energy to his own account and to his material benefit through machines. Upon this power to control and direct rested much of nineteenth-century technical achievement, steam power in particular. Why then could man not redirect energy so as to reverse the dissipation process? To work out a response to this question, Thomson had set out his discussion of miracles, beginning from the assertion that 'The material world could not come back to any previous state without a violation of the laws which have been manifested to man, that is without a creative act or an act possessing similar power'.[6] Clearly he wished to ascribe the same status to the Second Law of Thermodynamics as the First, but nothing quite like a creation of energy stood in the way of reversing dissipation. He had only a law 'manifested to man' (mutually, by recognizable means), as an apparent limitation on man's directing power. Was the law actually a law of matter; or was it merely a limitation of free will; or might it in fact be violated by free will? Given his position on miracles, the latter violation seemed untenable to Thomson, but he only gradually settled his position.

In the 1851 published paper, he took care to exclude living beings from the second axiom: 'It is impossible, *by means of inanimate material agency*, to derive mechanical effect from any portion of matter by cooling it below the temperature of the coldest of the surrounding objects'. The 1852 dissipation paper at first similarly excluded vegetative action and even chemical action: 'When radiant heat or light is absorbed, *otherwise than in vegetation, or in chemical action*, there is a dissipation of mechanical energy, and perfect restoration is impossible'. His conclusions, however, drawn from the foundation axiom and from 'known facts with reference to the mechanics of animal and vegetable bodies' nearly eliminated the qualifications, claiming that any restoration of energy without more than an equivalent of dissipation was 'probably never effected by means of organized matter, either endowed with vegetable life or subjected to the will of an animated creature' (ch. 14).[7] In other words, although Thomson believed that any reversal by organized matter of the Second Law would be miraculous, he left open the possibility. And the mere possibility of a violation of a supposed law by an act of will left its status as a law in doubt.

The depth of Thomson's dilemma is apparent from another 1852 paper, 'On the power of animated creatures over matter'. Accepting that animated creatures could not set matter in motion 'in virtue of an inherent power of producing a mechanical effect', he nevertheless asserted the directing power of will in

[6] Draft, p. 6. [7] MPP, **1**, 179, 511–14.

unusually explicit and forceful terms. The means remained uncertain by which the animal body converted chemical forces into external mechanical effects, yet:

> Whatever be the nature of these means, consciousness teaches every individual that they are, to some extent, subject to the direction of his will. It appears, therefore, that animated creatures have the power of immediately applying, to certain moving particles of matter within their bodies, forces by which the motions of these particles are directed to produce desired mechanical effects.[8]

Will could apply forces immediately to the particles of matter to direct their motions. He did not say why this power did not extend to a violation of the Second Law. It certainly violated the determinism of Laplacian particle mechanics in the material world.

In his lectures for the session of 1852–3, Thomson clarified somewhat his views on directing power. Concerning the power of galvanism (battery and current) to produce heat and to raise weights, he asked: 'What is the source of power? No energy can be deployed by inanimate materials without there being a source to draw from'. Chemical energy provided the answer for galvanism, with no question of directing power. But he extended his question to animals and plants, which also produced mechanical effect from chemical action by means of electrical currents, some heat being produced in the process. Concerning animals he offered two alternatives, both beginning with electrical currents derived from chemical forces, but one of which violated the Second Law. Either the currents produced heat directly, which 'in the animal frame acts so as to produce mech^{l} effect, from some of that heat' – with no fall of temperature, thereby violating the Second Law – or 'the will of the animal can make these currents produce mech^{l} effect', directly, with heat as a mere side effect and no violation of the Second Law. 'There is a great degree of probability', he asserted, 'in favour of the 2nd of these 2. If there be one or other of these hypotheses, the 2nd is nearly established'.

Similarly, with respect to plants, 'It is highly probable [they] do not convert heat into mechanical effect . . . If this be true, heat is never converted into mech^{l} effect, except by taking from a high & giving to a low temperature', in conformity with the Second Law. Most likely, 'The plant by its life has given it the faculty of directing the mechanical energy [of] light to separate carbon from carbonic acid'.[9] In the cases of both animals and plants, therefore, Thomson distinguished a violation of the Second Law from effects producible by directing power. His usage clarifies the distinction between directing power, which would never involve an abrogation of natural law, and creative power, which would. It also suggests a distinction between the directing power of life (plants) and that of will (animals). But the latter distinction is not sharply drawn.

[8] *Ibid.*, pp. 507–9.

[9] William Thomson, in William Jack, Lecture notes of the Glasgow College natural philosophy class taken during the 1852–3 session, Ms Gen. 130, ULG.

Presumably will allowed a great deal more freedom in the design of directive action and the use of instruments. In any case, Thomson later discussed the issues in terms of free will alone.

Not until an exchange of letters with James in 1862 did William examine these questions further. At the centre of the discussion was James's desire to allow for the possibility that mind – or vitality or life – could actually reverse the course of nature, producing mechanical effect or potential energy from the *vis viva* of heat. Since this discussion both illuminates William's ideas *and* forms a prelude to the important interpretation of the dissipation principle at a molecular level afforded by Maxwell's 'demon', we shall examine the letters in full.

The debate between the Thomson brothers contributed to a major discourse among Scottish natural philosophers. During the 1860s and early 1870s, Maxwell, Jenkin, Tait, and Balfour Stewart (1828–87) all participated.[10] Their special concern with thermodynamics and molecular physics in relation to free will was at first largely a response to their German friend, Helmholtz, although the debate subsequently widened to include 'popular' forms of materialism and determinism, epitomized in Tyndall's 1874 Belfast address to the British Association. For Helmholtz, the subjection of biological phenomena to the energy conservation principle supported a reductionist and deterministic view of physiology. Addressing the Royal Institution in April, 1861, in his paper 'On the application of the law of conservation of force to organic nature', Helmholtz argued:

> Therefore this opinion [among older physiologists] that the chemical or mechanical power of the elements can be suspended, or changed, or removed in the interior of the living body [by the 'vital principle'], must be given up if there is complete conservation of force. There may be other agents acting in the living body, than those which act in the inorganic world; but those forces, as far as they cause chemical and mechanical influences in the body, must be quite of the same character as inorganic forces, in this at least, that their effects must be ruled by necessity, and must always be the same, when acting in the same conditions, and that there cannot exist any arbitrary choice in the direction of their actions.[11]

Neither William nor James was present, although William was in regular correspondence with Helmholtz and was ready to adopt Helmholtz's views on the nature of the sun's heat in the same year (ch. 15). On the other hand, Maxwell, then at King's College, London, almost certainly was present and in fact delivered his own first lecture to the Royal Institution a month after

[10] The best survey and analysis of this discourse, from Maxwell's perspective, appears in Theodore M. Porter, *The rise of statistical thinking, 1820–1900* (Princeton, 1986), pp. 193–208, and 'A statistical survey of gases: Maxwell's social physics', *Hist. Stud. Phys. Sci.*, **12** (1981), 77–116. We aim here at broadening and deepening his basic interpretation, supplementing it with the Thomson brothers, with vortex atoms, and especially with the role of infinities and instabilities in the continuum.

[11] Hermann Helmholtz, 'On the application of the law of conservation of force to organic nature', *Proc. Royal Inst.*, **3** (1858–62), 347–57, on p. 357.

Helmholtz.[12] Like the Thomsons, Maxwell had the highest regard for Helmholtz. All the more reason, then, to take the German physicist's views seriously, though not necessarily to accept them in full. Divergence came over the issue of free will. In April, 1862, after the publication of Helmholtz's address in the *Proceedings*, both James Clerk Maxwell and James Thomson penned their private thoughts.

Maxwell wrote to Rev Lewis Campbell on 21st April, 1862, with high praise for Helmholtz as one 'who prosecutes physics and physiology, and acquires therein not only skill in discovering any desideratum, but wisdom to know what are the desiderata, e.g., he was one of the first, and is one of the most active, preachers of the doctrine that since all kinds of energy are convertible, the first aim of science at this time should be to ascertain in what way particular forms of energy can be converted into each other, and what are the equivalent quantities of the two forms of energy'. With regard to living beings, Maxwell agreed with Helmholtz that 'the soul is not the direct moving force of the body' for if it were 'it would only last till it had done a certain amount of work, like the spring of a watch'. But, unlike Helmholtz, Maxwell went on to discuss the positive nature of the soul: 'food is the mover, and perishes in the using, which the soul does not'. The action and reaction between body and soul was not of a kind in which energy passed from the one to the other, but the soul nevertheless directed the action of material agents. Thus, 'when a man pulls a trigger it is the gunpowder that projects the bullet, or when a pointsman shunts a train it is the rails that bear the thrust'. These analogies suggest that will performs a switching action at a point, taking advantage of an instability in existing mechanical arrangements. Maxwell seems always to have favoured analogies to such discontinuous actions. His mention of the rails, however, suggests another action which would not require a transfer of energy, a continuous guiding force, exerted perpendicular to the direction of a natural motion. Aware of the merely analogical status of both alternatives, Maxwell concluded in a cautious manner: 'the constitution of our nature is not explained by finding out what it is not. It is well that it will go, and that we remain in possession, though we do not understand it'.[13]

Writing from his Belfast home to William on 7th April, James Thomson's letter was more radical. James began by stating succinctly the basic conclusions of William's thermodynamics:

I understand that you consider that when mechanical energy disappears, it is replaced by an equivalent amount of *vis viva* in the molecular or other motions, which are supposed to constitute heat; and that while man can apply *mechanical vis viva*, or *mechanical energy of motion* (by which terms I mean to express the energy of the motion of bodies of finite size or of size decidedly larger than what can be regarded as molecular; and to distinguish it from the energy of such motions as may be supposed to constitute heat), so as to produce

[12] James Clerk Maxwell to Rev Lewis Campbell, 21st April, 1862, in Lewis Campbell and William Garnett, *The life of James Clerk Maxwell* (London, 1882), pp. 314–15.

[13] *Ibid.*, p. 336.

potential energy, he cannot possibly turn to such account, by any actions of matter on matter induced by him nor in any way, the *vis viva* of heat, if the heat he has at command be all at one temperature: also that he can obtain potential energy from heat only through allowing other portions of heat at different temperatures to equalize themselves in temperature, and that so the continual tendency of man's obtaining potential from differences of temperature and expending it, must be towards bringing about a state of uniform temperature from which it would be impossible for him to obtain potential energy again.[14]

While agreeing with this analysis of man's capacity to produce potential energy from mechanical *vis viva*, and of his inability to recover the *vis viva* of heat at uniform temperature, James Thomson nevertheless felt that it might be necessary 'to qualify the theory . . . by excepting from its decided statements, and leaving as still unknown and uncertain, the mysterious influence of spirit, life, or the vital principle in animals and plants over the matter composing their living bodies'. This influence, he thought, must be looked upon 'as being contrary to, or predominant over, the ordinary laws of matter'. Thus, given that the molecules composing a living body possessed much potential energy in virtue of their relation to one another, these molecules would, 'if left to themselves', continually pass to states of lower potential energy.

He believed, however, that spirit or life could hinder these natural actions and even 'cause them to go through quite different sets of actions':

sets of actions which, if we believe in free will, as I do, we must consider as not following one another in a natural sequence of cause and effect according to the physical laws of matter; as not being fixed for the future by the present condition in which the matter exists; and as being absolutely at the command or under the predominant influence of something which is not matter . . . [It] cannot be disputed by anyone who believes the mind or life to be distinct from matter and to be no mere result of arrangement of particles of matter and of their energies, that the vital principle has the power of regulating and applying the potential energies of the matter composing the living body with which it is connected, so as to keep them from applying themselves as they would do if left to themselves.

James inferred that 'this influence of mind or life over matter is altogether mysterious', implying, he thought, 'a perfect suspension or contravention of the physical laws of matter; and may be regarded as a part of a great miracle which constantly goes on around us'. The brothers agreed, at least, on the meaning of 'miracle', but James would include free will under miracles while William would not.

James speculated on the consequences of allowing 'that mind or life really has the power of counteracting the physical laws of matter'. He accepted as generally admitted 'that the vitality can and does divert the course of action of energies', that 'mind has the power of controlling the courses of energies during

[14] James to William Thomson, 7th April, 1862, in James Thomson, *Papers*, lv–lvii, on p. lv.

their passage to the lower state, or state of dissipation'. But he wished to go further and 'say that perhaps the vitality of animals and plants may be able to reverse processes which would go on if not influenced by the vitality'; that vitality 'at least stops some of their courses of action or even reverses some'.

'It seems to me,' he argued, 'a perfectly admissible supposition that mind or vitality may have the power, in the living body, of collecting, and applying as potential energy, the energy of the heat motions, or of bringing out in the living body from the *vis viva* stored indefinitely everywhere as heat, the potential energy which you regard as a *true equivalent* of the *vis viva* of the heat'. This ability of living agency to collect and reapply the energy of heat will look strikingly similar to William's later interpretation of Maxwell's demon. James emphasized that he was not asserting that 'vitality really has that power', but rather 'what I chiefly want to say is that I do not think we are entitled to set down as a *proved impossibility* the conversion, through vital influence, of heat into mechanical energy'. He concluded his letter with a suggestion as to the manner in which such a reversal might take place:

> may not the impulse of volition communicated from the mind to the body consist in a reversal of an action or set of actions in the body, perhaps in the brain or nerves, which would go on under the ordinary physical laws; the antecedent and subsequent conditions, whether in the forward or reversed action, being . . . truly equivalent for one another. Then when one part of an action is reversed or stopped it is easy to conceive that the whole remaining course of the actions may be thereby turned into new channels.

James's reflections here may have been prompted by recollections of Roget's *Bridgewater treatise* on *Animal and vegetable physiology* which he had first read in 1846. Roget had argued:

> The animal machine, in common with every other mechanical contrivance, is subject to wear and deteriorate [*sic*] by constant use . . . Provision must accordingly be made for remedying these constant causes of decay [friction and evaporation, for example] by the supply of those peculiar materials which the organs require for recruiting their declining energies.[15]

Unlike non-living, mechanical systems, then, living animal systems possessed a capacity for restoring the system: 'It would appear that, during the continuance of life, the progress of decay is arrested at its very commencement; and that the particles, which first undergo changes unfitting them for the exercise of their functions, and which, if suffered to remain, would accelerate the destruction of the adjoining parts, are immediately removed, and their place supplied by particles that have been modified for that purpose, and which, when they afterwards lose these salutary properties, are in their turn discarded and replaced

[15] P.M. Roget, *Animal and vegetable physiology considered with reference to natural theology* (2 vols., London, 1834), **2**, pp. 1–2. Roget's was the fifth *Bridgewater treatise*. In James to William Thomson, 12th May, 1846, T418, ULC, James stated that 'I am reading Roget's Bridgewater treatise on animal physiology'.

by others'.[16] In his 1862 letter, James Thomson may well have been reinterpreting, in the light of William's energy principles, Roget's distinction between living and mechanical systems.

In his reply of 21st April, 1862, William agreed with James's account 'on the whole', and he reminded his brother that he had 'excepted plants and animals from any assumption (founded merely on what we know of inorganic matter) of either [!] the First or Second Law' such that 'neither can be held to be an *axiom* when life is concerned, in the same sense as it can for dead matter'. This exception with respect to the *First* Law appears to be new, though William had earlier, as he now reiterated, excepted action involving life 'from the axiom on which I found the Second Law'. Apparently he had become more convinced that the two laws had equivalent status. To his brother's suggestion 'that energy may be called into existence by free will', William now responded that, since life was not perpetually reproducible in a cycle, 'the sound objections to assumptions involving the "perpetual motion" are not applicable' to this case. Referring to his 1852 article 'On the power of animated creatures over matter', however, he stated that 'analogy from known facts renders it probable (almost certain) that energy is *not* so called into existence'. Actually, in 1852, he had expressed no uncertainty on this point and had said that it could be 'asserted with confidence'. Concerning the Second Law, the real focus of James's argument, he concluded his letter with the same views he had held in his 1852–3 lectures: although 'physiology as yet proves nothing as to whether the 2nd law is really violated or fulfilled, I think it probably is fulfilled in the vital processes'.[17] The symmetrical manner in which he now treated the two laws suggests that he wished to take a very general stand. Either vital processes could violate the laws of matter or they could not. He believed they could not. Given that vitality involved directing powers, the nature of this limitation remained in doubt.

The demon

James Thomson's concern in 1862 with the distinction between the energy of motion of sensible bodies and the energy of molecular motions; with man's capacity to control and divert energy; and with the possibility that mind or life might stop some courses of action or even reverse some, indicates the pressing nature of the problem that led to the introduction of Maxwell's famous 'demon' five years later. The actions of this demon, as William developed them, followed very much the lines of James's suggestions. Although we know little of the detailed discussions among natural philosophers in Scotland, especially after Maxwell's return there from London in 1865, we do know that William himself annotated a seminal letter from Maxwell to Tait in December, 1867. This letter

[16] Roget, *Physiology*, **2**, 8–9.
[17] William to James Thomson, 21st April, 1862, in James Thomson, *Papers*, lvii–lviii.

provides the first recorded discussion of the demon concept, which aimed to show how a purely imaginary living being might indeed reverse the natural tendency towards dissipation. Furthermore, Maxwell's correspondence shows that William assigned the name 'demon' to that remarkable being.

As Maxwell originally explained the notion to Tait, one imagines two vessels *A* and *B* separated by a diaphragm. Each vessel contains an equal [finite] number of molecules, but those in *A* have 'the greatest energy of motion', i.e. the sum of the squares of the velocities is greater in *A*, but there will be velocities of all magnitudes in *A* as in *B*. Now, says Maxwell, 'conceive a finite being who knows the paths and velocities of all the molecules by simple inspection but who can do no work except open and close a hole in the diaphragm by means of a slide without mass'. This finite being then permits a molecule in *A*, the square of whose velocity is less than the mean square velocity of molecules in *B*, to pass into *B*; and a molecule in *B*, the square of whose velocity is greater than the mean square velocity of molecules in *A*, to pass into *A*. Consequently:

> the number of molecules in *A* and *B* are the same as at first, but the energy in *A* is increased and that in *B* diminished, that is, the hot system has got hotter and the cold colder and yet no work has been done, only the intelligence of a very observant and neat-fingered being has been employed. Or in short if heat is the motion of finite portions of matter and if we can apply tools to such portions of matter so as to deal with them separately, then we can take advantage of the different motion of different portions to restore a uniformly hot system to unequal temperatures or to motions of large masses. Only we can't, not being clever enough.[18]

William Thomson added in pencil the remarks: 'Very good. Another way is to reverse the motion of every particle of the Universe and to preside over the unstable motion thus produced'. These remarks would provide the framework for his own 1874 paper on 'The kinetic theory of the dissipation of energy'.

Several interrelated points about the purpose, nature, and origin of the 'demon' need to be made. First, the purpose or 'chief end' of the demon was, in Maxwell's words, 'to show that the 2nd Law of Thermodynamics has only a statistical certainty', when regarded as expressing the behaviour of finite portions of matter regulated only by the dynamics of particles. Just as the demon could reverse the dissipation process by his will, so statistical violations of it would occur in any finite system of particles moving randomly. Furthermore, the Second Law expressed our inability to control motions. We have (at present) 'no power of handling the separate molecules' making up bodies. In other words, our faculties are not sharp enough to enable us to direct these molecules. As Maxwell expressed the issue in his article 'Diffusion' for the ninth edition of the *Encyclopaedia Britannica* (1878):

[18] J.C. Maxwell to P.G. Tait, 11th December, 1867, in C.G. Knott, *The life and scientific work of Peter Guthrie Tait* (Cambridge, 1911), pp. 213–14. On Maxwell's demon see, in addition to Porter, *Statistical thinking*, esp. M.J. Klein, 'Maxwell, his demon and the second law of thermodynamics', *Am. Sci.*, **58** (1970), 84–97; also the papers cited in M.N. Wise, 'The Maxwell literature and British dynamical theory', *Hist. Stud. Phys. Sci.*, **13** (1982), 175–205, esp. pp. 199–200.

> the idea of dissipation of energy depends on the extent of our knowledge. Available energy is energy which we can direct into any desired channel. Dissipated energy is energy which we cannot lay hold of and direct at pleasure, such as the energy of the confused agitation of molecules which we call heat . . . [The] notion of dissipated energy could not occur to a being who could not turn any of the energies of nature to his own account, or to one who could trace the motion of every molecule and seize it at the right moment. It is only to a being in the intermediate stage, who can lay hold of some forms of energy while others elude his grasp, that energy appears to be passing inevitably from the available to the dissipated state.[19]

Maxwell was concerned with two interrelated limitations of human faculties, with our inability to *know* molecular motions and our parallel inability to devise tools to *control* them. He made no claims about the ultimate nature of reality. His thinking conforms to the claims made by Thomson since 1851 concerning man's remarkable, and remarkably limited, ability to direct the energies of nature.

Second, Thomson and Maxwell agreed not only on the purpose of the demon, but in large degree on its nature. The demon was essentially a *finite* being who dealt with finite molecules in finite numbers. Such beings came with different intelligence quotients. Maxwell's 1867 demon was 'very observant and neat-fingered' and capable of calculating. Subsequently, he considered more primitive creatures: 'less intelligent demons can produce a difference in pressure as well as temperature by merely allowing all particles going in one direction while stopping all those going the other way. This reduces the demon to a valve. As such value him. Call him no more a demon but a valve like that of the hydraulic ram . . .'. Maxwell was not here objecting to Thomson's use of the term demon, but simply considering a dull being who played only the role of a valve or switch, without actually exerting forces on molecules.[20]

By contrast, Thomson in the 1870s discussed a range of tasks performed by superior types of demon who actually transferred energy from one molecule to another. In 1874, he publicly defined the demon 'according to the use of this word by Maxwell' as 'an intelligent being endowed with free will and fine enough tactile and perceptive organization to give him the faculty of observing and influencing individual molecules of matter'.[21] Such a finite being possessed the same faculties as Thomson ascribed to man, and to other animated beings with free will, who could direct, though probably not reverse, the energies of nature. Transferred down to the molecular level, however, the capacity to direct actually gave such demons the ability to reverse the motions of individual molecules. Wielding molecular cricket bats as they guarded an interface be-

[19] J.C. Maxwell, *The scientific papers of James Clerk Maxwell*, W.D. Niven (ed.) (2 vols., Cambridge, 1890), **2**, p. 646.

[20] Knott, *Life of Tait*, pp. 214–15. Cf. P.M. Harman, *Energy, force, and matter. The conceptual development of nineteenth-century physics* (Cambridge, 1982), p. 140, who claims that Maxwell 'objected to Thomson's use of the term "demon" . . . there were no supernatural resonances to the argument'.

[21] William Thomson, 'The kinetic theory of the dissipation of energy', *Proc. Royal Soc. Edinburgh*, **8** (1875), 325–34; MPP, **5**, 11–20, on p. 12*n*.

tween two regions, they would absorb energy and momentum from some molecules and deliver it to others. If this were done so that each demon absorbed as much total momentum from molecules moving in one direction as in the other, thus maintaining his position, but from a larger number of molecules each with a lower kinetic energy in the one direction and from a smaller number with larger energies in the other, then a net transfer of energy would occur, producing a disequalization of temperature across the interface. In this way, a process of diffusion of heat could be produced 'by an army of Maxwell's "intelligent demons" stationed at an interface separating the hot and cold portions of a bar or gas'.

Five years later, in his lecture to the Royal Institution on 'The sorting demon of Maxwell', Thomson made even more explicit the nature of the superior demon. The demon was 'a creature of imagination having certain perfectly well-defined powers of action, purely mechanical in their character'. He was 'a being with no preternatural qualities, [i.e., qualities outside the ordinary course of nature] and differs from real living animals only in extreme smallness and agility'. He could not create or destroy energy, but 'just as a living animal does, he can store up limited quantities of energy, and reproduce them at will' and so 'push or pull each atom *in any direction*'. Then, 'by operating selectively on individual atoms, he can reverse the natural dissipation of energy' which 'follows in nature from the fortuitous concourse of atoms'. The demon concept showed that the loss of available energy ('motivity'), at least when considered at the level of a finite number of discrete particles obeying the laws of dynamics, 'is essentially not restorable otherwise than by an agency dealing with individual atoms; and the mode of dealing with the atoms to restore motivity is essentially a process of assortment, sending this way all of one kind . . . that way all of another kind'.[22]

The difficulty with this understanding, however, is that it seems to make the Second Law not a fundamental law of matter but a law contingent on our particular incapacity. Another finite being, essentially like us, only smaller, could abrogate the law, thereby undermining the independence of laws of matter from the action of mind. Since Thomson wished to maintain the Second Law, like the first, as a fixed law of nature, this potential action of the demon suggests that he did not consider the statistical account to be an ultimate explanation, but only an account consistent on the average with the dynamics of a finite number of particles. Its true basis would have to lie at a deeper level, perhaps at the level of continuum mechanics, where only an infinite intelligence could control all the variables required for a complete reversal of the processes of nature. We return to this point below.

At the same time, Thomson made clear that the 'sorting demon' was not

[22] William Thomson, 'The sorting demon of Maxwell', *Proc. Royal Inst.*, **9** (1879–81), 113–14; MPP, **5**, 21–3.

invented 'to help us to deal with questions regarding the influence of life and of mind on the motions of matter, questions essentially beyond the range of mere dynamics'. The demon did not provide an answer to the problem of whether or not animate beings possessed free will. Nevertheless, and thirdly, the historical *origin* of the demon lies in Thomson's and Maxwell's emphasis on the capacity of will to direct the energies of nature. Free will was for them a basic assumption, and a direct personal perception, of an endowment bestowed on man by God. They shared a Scottish tradition in natural philosophy which stressed both the separation of mind from matter and the directing power of mind over matter. To maintain consistently these two positions required strict limitations on free will. The 'sorting demon', concomitantly, exhibited the distinctive nature of living creatures, while serving 'to help us to understand the "Dissipation of Energy" in nature'.[23]

It is convenient to begin a deeper discussion of the demon with Thomson's annotation to Maxwell's 1867 letter. Another way to pick a hole in the Second Law, he stated, 'is to reverse the motion of every particle of the Universe and to preside over the unstable motion thus produced'. Two issues arise here: the essentially *finite* character of particle dynamics; and the *instability* of mechanical systems. We begin with particle dynamics. In a letter to J.W. Strutt (Lord Rayleigh) dated 6th December, 1870, Maxwell developed Thomson's line of argument as follows:

> If this world is a purely dynamical system, and if you accurately reverse the motion of every particle of it at the same instant, then all things will happen backwards to the beginning of things, the raindrops will collect themselves from the ground and fly up to the clouds, etc. etc. and men will see their friends passing from the grave to the cradle till we ourselves become reverse of born, whatever that is. We shall then speak of the impossibility of knowing about the past except by analogies taken from the future and so on. The possibility of executing this experiment is doubtful, but I do not think it requires such a feat to upset the 2nd law of thermodynamics [as based on particle dynamics].[24]

The lesser feat, Maxwell then explained, could be carried out by his essentially finite being, or sorting demon, 'a doorkeeper very intelligent and exceedingly quick, with microscopic eyes'. The being's intelligent action gave it the role of a guiding agent 'like a pointsman on a railway with perfectly acting switches [consuming no energy] who sends the express along one line and the goods along another'. Maxwell indeed suggested here that 'even intelligence might . . . be dispensed with and the thing made self-acting'. By implication, man might some day devise clever devices which, by acting on individual atoms, could reverse the natural tendency towards equal temperatures in a finite system of atoms, a feat which was 'at present impossible for us'.[25] The reversal of the

[23] *Ibid.*, p. 21.
[24] R.J. Strutt, *John William Strutt. Third Baron Rayleigh* (London, 1924), p. 47.
[25] J.C. Maxwell, *Theory of heat* (London, 1871), pp. 308–9.

whole course of nature, however, could be carried out only by a being capable of reversing the motion of every particle in the universe. It was this latter case that Thomson took up in 1874.

His 'kinetic theory of the dissipation of energy' began by stressing the distinction (already employed in the 1867 *Treatise*) between *abstract dynamics*, where total reversal was conceivable in the idealized world of an isolated, finite system of discrete particles and conservative forces, and *physical dynamics*, where reversibility necessarily failed in the infinite real world of imperfectly isolated systems and of friction and diffusion phenomena. Explaining that 'the essence of Joule's discovery [of the all-pervading law of energy conservation] is the subjection of physical phenomena to dynamical law' (abstract dynamics), Thomson argued that, if 'the motion of every particle in the universe were precisely reversed at any instant, the course of nature would be simply reversed for ever after'. For example, 'the bursting bubble of foam at the foot of a waterfall would reunite and descend into the water; the thermal motions would reconcentrate their energy, and throw the mass up the fall in drops re-forming into a close column of ascending water' and boulders 'would recover from the mud materials required to rebuild them into their previous jagged forms, and would become reunited to the mountain peak from which they had formerly broken away'. Thus Thomson and Maxwell fully agreed on the ideal implications of reversibility for physical phenomena. With regard to animate nature, Thomson amplified:

> if also the materialistic hypothesis of life were true, living creatures would grow backwards, with conscious knowledge of the future, but no memory of the past, and would become again unborn. *But the real phenomena of life infinitely transcend human science; and speculation regarding consequences of their imagined reversal is utterly unprofitable.* Far otherwise, however, is it in respect to the reversal of the motions of matter uninfluenced by life, a very elementary consideration of which leads to the full explanation of the theory of dissipation of energy.[26]

He, like Maxwell, then turned to a discussion of 'demons', though on the grand scale of armies, for an elucidation of the Second Law.

Thomson's remarks here were subtle and double edged. By 'the materialistic hypothesis of life', he no doubt referred to the popularizations of Helmholtz's views by Huxley and Tyndall, who sought to subsume the processes of life under the law of energy conservation alone, i.e. under abstract dynamics, yielding complete determinism. To defeat their strategy, Thomson would employ abstract dynamics to explain the law of energy dissipation, but only by showing in the process how that law, as a principle of physical dynamics, exceeded the grasp of our ideal science. How much less, by implication, could the materialistic hypothesis encompass the processes of life. The actions of life-

[26] MPP, **5**, 11–12. Our emphasis.

like demons on an abstract dynamical system would reveal the nature of the Second Law by demonstrating the requirements for violating it.

Accordingly, Thomson considered an isolated vessel of gas, consisting of a relatively small number of molecules, and showed both that an army of demons wielding cricket bats could reverse its dissipation of energy, without violating conservation of energy, and that such reversals would occur naturally as a statistical matter. Both kinds of violations of the Second Law, however, depended on the system being finite. Real physical systems, since they could never be fully isolated, would involve an infinite number of molecules (ch. 12). Only an infinite intelligence, therefore, or an infinite army of demons, could carry out the reversal in reality: 'Do away with this impossible ideal [of isolation], and believe the number of molecules in the universe to be infinite; then we may say one-half of the bar will never become warmer than the other'.[27]

Even at the level at which particle dynamics could describe reality, therefore, the Second Law was a true law of matter, violated neither statistically nor by the free will of demons whose directing power was limited to a finite number of variables. But particle dynamics itself was, as Thomson saw it, a rather primitive creation of our finite minds. It reduced the properties of macroscopic matter to molecular properties, themselves unexplained (ch. 12). A more sophisticated natural philosophy would explain the molecular properties, not by once again reducing them to yet smaller particles, but by referring them to an irreducible and undifferentiated continuum, the proper medium for an artificer of *infinite* directing power.

Such arguments from the difference between finite and infinite directing power played a most important role in Thomson's understanding of the Second Law as a true law of matter, inviolate under any conceivable human action. The same arguments, however, tended to remove further from comprehension the directing power that life did possess. The problem is most obvious in the case of the cricket-bat army. In order to reverse the motion of molecules, while conserving energy, Thomson's demons had to absorb finite amounts of energy from molecules moving in one direction and deliver it to those moving in the opposite direction. One could not regard such action as a direct action of will without mixing the categories of mind and matter. Mind could not itself store material energy, but only direct it. As Thomson stated in 1879, therefore, the demon was not invented to explain the actions of mind on matter, 'questions essentially beyond the range of mere dynamics'.

The same problem existed for Maxwell's sorting demon. So long as the demon had to employ material tools – cricket bats, valves, arms, trap doors, or switches – in order to direct molecules, any description of his action within

[27] *Ibid.*, p. 15.

abstract dynamics would require the transfer of finite amounts of energy from mind to the finite molecules of the directing material. The demon's action, therefore, could serve at best as a mere analogy for free will, limited to an idealized action that did not require mind to store energy. A number of authors in Thomson's circle employed Maxwell's analogy in this sense.

Fleeming Jenkin published his essay on 'Lucretius and the atomic theory' in 1867 only three months after Maxwell's letter to Tait and with repeated reference to both Thomson's vortex atoms and Maxwell's kinetic theory of gases. Arguing that free will was quite consistent with energy conservation, he replaced the infamous Lucretian 'swerve', produced at 'quite uncertain times and places' by the free will of the atoms themselves, with the deflecting force of external wills. It was, said Jenkin, 'a principle of mechanics that a force acting at right angles to the direction in which a body is moving does no work, although it may continually and continuously alter the direction in which the body moves'. Thus 'the will, if it so acted, would add nothing sensible to, nor take anything sensible from, the energy of the universe'. This argument depended critically on the will acting 'continually and continuously', for only in so acting could the will exert its effect without having to deliver energy in finite amounts either immediately to the body being directed or to a body employed as an intermediate directing tool. Concomitantly, Jenkin offered only a plausibility argument, not a proof of free will. Like his associate, he preferred to trust 'to our direct perception of free will', to free will as a personal intuition and belief. Still, the longer the group worked with the analogy the more the analogy functioned as a realistic model. As Jenkin said, 'The modern believer in free will will probably adopt this view, which is certainly consistent with observation, although not proved by it'.[28]

In August, 1868, Balfour Stewart and Norman Lockyer published in *Macmillan's Magazine* an essay entitled 'The place of life in a universe of energy'. Alongside the 'materialist' view, they considered three other hypotheses, viz., (i) that life had the capacity for creating energy, (ii) that life could transmute a finite quantity from one form to another, and (iii) that the living being was 'an organization of infinite delicacy, by means of which a principle in its essence distinct from matter, by impressing upon it an infinitely small amount of directive energy, may bring about perceptible results'. They quickly rejected (i) and (ii), and, while stating that it was not their aim to decide between the materialist view and (iii), nevertheless clearly treated (iii) more sympathetically.[29]

[28] [Fleeming Jenkin], 'The atomic theory of Lucretius', *North Brit. Rev.*, [new series], **9** (1868), 211–42; *Papers and memoir of Fleeming Jenkin* (2 vols., London, 1887), **1**, pp. 177–214, esp. pp. 192–4. For the uses of Lucretius in the period, see Frank M. Turner, 'Lucretius among the Victorians', *Vict. Stud.*, **16** (1972–3), 329–48; and *The Greek heritage in Victorian Britain* (New Haven, 1981).

[29] Balfour Stewart and J. Norman Lockyer, 'The sun as a type of the material universe. Part II: The place of life in a universe of energy', *Macmillan's Mag.*, **18** (1868), 319–27; reprinted in Lockyer, *Contributions to solar physics* (London, 1874), pp. 85–103, esp. pp. 98–100. Alfred Tennyson's famous

Maxwell developed his own cautious remarks of 1862 concerning the trigger and pointsman analogies in his February, 1873, essay on science and free will. He likened the soul's function to the steersman of a vessel, 'not to produce, but to regulate and direct animal powers'. But whereas the latter analogy, like most earlier ones, suggested a continuous force acting always perpendicularly to the motion of a body, Maxwell actually placed his emphasis on switching actions at singular points. Believing apparently that considerations of instability at such singular points would throw more light on the question of free will, he took up suggestions that Stewart would develop in his forthcoming textbook *On the conservation of energy*, a popular account of the energy physics shared among the Scots. Maxwell's essay and Stewart's textbook make use of a distinction between stable arrangements of forces, with the characteristic of *calculability*, and unstable arrangements, characterized by *incalculability*. Stewart, as hinted in 1868, committed himself to the notion that human beings and animals were machines of practically infinite delicacy, analogous to an egg balanced in unstable equilibrium at the edge of a table, to a rifle at full cock with a delicate hair-trigger or to the commander of a great army: 'Life is not a bully, who swaggers out into the open universe, upsetting the laws of energy in all directions, but rather a consummate strategist, who, sitting in his secret chamber, before his [telegraph] wires, directs the movements of a great army'.

The same assumptions underlay Maxwell's remarks in his 1873 essay that 'in the course of this mortal life we more or less frequently find ourselves on a physical or moral watershed, where an imperceptible deviation is sufficient to determine into which of two valleys we shall descend. The doctrine of free will asserts that in some such cases the Ego alone is the determining cause'.[30]

These statements of Stewart and Maxwell should remind us that, in his correspondence with Stokes in December, 1872, and January, 1873, Thomson had made quasi-stability the very essence of his vortex atoms and of his understanding of the Second Law (ch. 12). Throughout the 1870s and 80s he continued to produce papers on problems of stability and instability. Recall too that, when Stokes would annihilate vortices by bringing their rotational parts up to a boundary, Sir William answered that the infinitesimal difference between approaching and actually contacting the boundary was quite sufficient to distinguish living from dead vortices, like the nearly drowned from the drowned man. The infinitesimal switch between radically different states came easily to his mind as a metaphor for the singular, unique processes of life.

Juxtaposing the instability problem in vortex motion with Thomson's views on the relation of free-will to the Second Law, it appears that he supposed the

poem 'Lucretius' was published in the May issue of the same volume of *Macmillan's*, i.e. just two months after Jenkin's article in the *North British*.

[30] Campbell and Garnett, *Life of Maxwell*, pp. 434–44, esp. pp. 438–41; Balfour Stewart, *The conservation of energy, being an elementary treatise on energy and its laws* (London, 1874), pp. 154–67. The preface was dated August, 1873.

true mechanism of free-will to exceed our understanding in much the way the Second Law does, through our inability actually to grasp – in our minds and in our hands – the infinitesimal and the infinite. Neither free will nor the Second Law had its ultimate foundation at the level of Maxwell's intelligent demon and his control of abstract dynamical systems. Yet man could approach an understanding of both by a limiting procedure which suggested what abstract dynamics would become with infinite numbers of infinitesimal particles. Thomson approached that limit through his vortex atoms in the universal plenum. He sought to locate the inviolability of the Second Law in the limited stability of vortex atoms and their consequent progress by infinitesimal degrees towards states of minimum energy.

Similarly, he supposed that free will acted on matter by acting on the infinitesimal parts of the continuum, causing infinitesimal changes in the direction of motion. Such changes could cause observable effects just because of instabilities in the motion. The effects would be incalculable in detail, however, because of this instability and because of the infinity of variables involved in describing the continuum. The situation mirrored the 'interstreaking' of vortex atoms in their progress towards the minimum energy state. One might give a qualitative account of how the sponge-like mixture of rotationally and irrotationally moving fluid developed, but an exhaustive calculation of the process was not possible for a finite mind.

Although this interpretation of Thomson's idea of free will goes beyond any evidence on the subject, it gains plausibility by association with the writings of his friends in Scotland. Fleeming Jenkin's reinterpretation of the Lucretian swerve, while it did not depend on vortex atoms, nevertheless presented those atoms as the most promising basis for natural philosophy. More explicit was Stewart's and Tait's *The unseen universe*, published anonymously in 1875. The authors employed vortex atoms of both mundane and ultra-mundane kinds (referring to Thomson's version of Le Sage) to show how the behaviour of our visible world of gravitating matter might be traced back to an 'unseen' world where intelligence would have its domicile and its sphere of action and where the teachings of Christianity would find their unification with science. The progression from seen to unseen followed the principle of continuity, the most basic of all philosophical principles in Stewart and Tait's view, requiring that every natural state be derived from a preceding natural state, with no breaks in either explanation or action.[31] If smoke rings were vortices of matter, and if the atoms of matter were vortices in an underlying fluid (the highest level of the invisible universe), then the fluid might consist in vortices of a yet subtler

[31] Balfour Stewart and P.G. Tait, *The unseen universe, or physical speculations on a future state*, 4th edn. (1876, reprint, London, 1889), pp. 79–98. See P.M. Heimann, 'The unseen universe: physics and the philosophy of nature in Victorian Britain', *Brit. J. Hist. Sci.*, **6** (1972), 73–9. For Thomson's (and Maxwell's) dislike of the work, see Crosbie Smith, 'From design to dissolution: Thomas Chalmers' debt to John Robison', *Brit. J. Hist. Sci.*, **12** (1979), 59–70, esp. pp. 69–70.

substance, and so on, leading by an infinite regress into the infinite unseen 'with no halting-place for the mind, except in the belief that the universe as a whole participates in every motion which takes place even in the smallest of atoms'. Assuming consciousness to accompany the motion of atoms, it too would extend throughout the universe, for it 'cannot logically be confined to the apparently moving body or atom, but must in some sense extend to the Unseen Universe in its various orders'. Here the mind would unite with higher intelligences and ultimately with God, with '*a Divine over-life in which we live and move and have our being*'.[32]

Mind exerted its influence on body through the latter's 'delicacy of construction', analogous to the instability of balanced eggs, but at the chemical level (like nitroglycerine), and these chemical arrangements in turn owed their existence to the directing power of a higher intelligence: 'The body then owes its delicacy to its chemically unstable nature; to a peculiar collocation of particles which certainly would not, in virtue of their own merely physical forces, have united themselves together as we find them in the body'.[33] We can consciously control these chemical instabilities, thus controlling our bodies, Stewart and Tait apparently believed, because of an unconscious action of our minds on the underlying substance of the molecules in our nervous systems. Intelligences at a level above us would have *conscious* control though whether they would exert this control through the control of instabilities in vortex motion, effected through a yet deeper, and again unconscious, level of mind and substance, the authors did not say.[34] They did ascribe a similar view to 'others'.[35]

Stewart and Tait made Thomson's vortex atoms into a realistic basis for mind–matter dualism, all the while insisting that they did so only for concrete illustration. Through the atoms, the principle of continuity, operating in the first instance as a methodological principle, acquired ontological status as a principle of the continuity of matter. This essential role for vortex atoms makes

[32] Stewart and Tait, *The unseen universe*, pp. 244–5. In following the principle of continuity to its logical conclusion, Stewart and Tait were driven to reject the *creation* of vortex atoms by non-natural means and thus to reject Thomson's perfect fluid, in which only a supernatural agent could create vortices. Instead the atoms were to be seen as 'developing' out of processes in the unseen as vortices in a not-so-perfect fluid. Concomitantly, Stewart and Tait demoted the law of conservation of energy in the material world to a secondary principle, dependent on our limited observations, thus allowing for energy to pass back and forth from the seen to the unseen universe, and ultimately for the material world to disappear completely in the process of dissipation. They regarded the Second Law as only a regularity of large numbers of material particles, constantly violated for small systems. When all energy had passed into the unseen, the Second Law would not necessarily obtain, so that structures in that realm, unlike vortex atoms of matter, might be eternal, and the intelligences resident there might control the finest motions of substance. See pp. 85, 122–5, 155–7, 210.

[33] *Ibid.*, p. 186.

[34] *Ibid.*, pp. 235–6. Stewart and Tait considered it unacceptable that the effect on molecules of our exertions of free will should throw higher intelligences into 'inextricable confusion'. Our free will had somehow to harmonize with the rationality of these superior beings, so that the problem of free will could not be left at the level of our arbitrary directing action on molecules, as Fleeming Jenkin had left it, and also Thomson.

[35] *Ibid.*, p. 190.

it surprising that they did not directly connect their notion of free will, depending on instabilities in material systems, with either the atoms or the continuum. Perhaps they had not yet grasped Thomson's instability arguments for vortex motion or did not accord them much plausibility. Maxwell, however, soon came to think that the continuum would provide a solution to free will. The French theorist of hydrodynamics, Joseph Boussinesq, showed in 1878 that solutions to the equations of fluid flow would be multiple valued at certain singular points, so that the motion of the fluid (described as the motion of a point in phase space) would be indeterminate at those points. Maxwell no doubt regarded this mathematical solution as an advance on Thomson's qualitative arguments for instability in vortex motion and certainly as superior to any of Stewart's or his own switching suggestions. In a letter to Francis Galton early in 1879 he drew the grand consequence for 'Fixt Fate, Free Will, &c':

> There are certain cases in which a material system, when it comes to a phase in which the particular path which it is describing coincides with the envelope of all such paths, may either continue in the particular path or take to the envelope (which in these cases is also a possible path) and which course it takes is not determined by the forces of the system (which are the same for both cases) but when the bifurcation of path occurs, the system, ipso facto, invokes some determining principle which is extra physical (but not extra natural) to determine which of the two paths it is to follow . . . In most of the former methods, Dr Balfour Stewart's, &c. there was a certain small but finite amount of travail decrochant or trigger-work for the will to do. Boussinesq has managed to reduce this to mathematical zero, but at the expense of having to restrict certain of the arbitrary constants of the motion to mathematically definite values, and this I think will be found in the long run very expensive. But I think Boussinesq's method is a very powerful one against metaphysical arguments about cause and effect and much better than the insinuation that there is something loose about the laws of nature, not of sensible magnitude but enough to bring her round in time.[36]

Maxwell's solution, like Thomson's and Stewart's and Tait's, for preserving the full force of the laws of nature while allowing room for free will, could succeed only by introducing a fundamental indeterminacy into the Laplacian scheme of abstract dynamics, an indeterminacy dependent on instability. Their instability arguments, however, in no way dissuaded Galton from his own enthusiastic determinism, nor did it dissuade others of the same persuasion.

Maxwell's letter, like *The unseen universe*, responded directly to the notorious claims of Tyndall in his 1874 address. Tyndall asserted that the doctrine of the conservation of energy '"binds nature fast in fate" to an extent not hitherto recognised, exacting from every antecedent its equivalent consequent, from every consequent its equivalent antecedent, and bringing vital as well as physical

[36] J.C. Maxwell to Francis Galton, 26th February, 1879, Francis Galton Papers, The Library, University College, London; quoted from Porter, *Statistical thinking*, p. 206, where the full letter and discussion of its context will be found.

phenomena under the dominion of that law of causal connexion which, so far as human understanding has yet pierced, asserts itself everywhere in nature'.[37] The Scottish circle of natural philosophers regarded such views as a total miscarriage of the legitimate conclusions of physical science. Tyndall, however, simply expressed in dramatic form views which had been current for a decade or more in Britain, views fashionable among so-called 'scientific naturalists', many of them, like Tyndall and Huxley, prominent members of the X Club (founded in 1864).[38]

Thomson was not present for Tyndall's address, but his disapproval could hardly have been mild. To expose the gulf between them, as manifest in their respective presidential addresses to the British Association in 1871 and 1874, we turn to Thomson's fierce opposition to Tyndall's view that we may discern in matter 'the promise and potency of all terrestrial Life'.[39]

The origin and evolution of life on earth

In a lengthy footnote to his 1836 article on discoveries concerning the nebulae, J.P. Nichol evaluated the case for the 'transpossibility' or transmutation of species in accordance with a progressionist system (analogous to the evolution of the universe in accordance with the nebular hypothesis):

> The 'intranspossibility' [fixity] of what are termed the 'limits of species' is by no means settled; and it seems that the holders of the dogmatic belief to this effect rest their chief authority on their power to ridicule Lamarck, who grasped at a philosophical conception before he knew of any of the facts by which it could well be illustrated. Zoology is too much in its infancy – too much a mere science of classification on the ground of observed differences – to permit of dogmatism on either side of this question; but unquestionably, when Lamarck asserted, in the face of much obloquy, that a 'transpossibility' and a progression *might* exist, he was far nearer the truth than his noisy opponents.[40]

Nichol urged inquirers in this department 'to investigate closely the *powers of life*, and to ask whether there, as with the nebulae, a plastic *misno* exists capable of solving the gradation' established not by the 'living creations' alone but by 'these in connexion with the creatures whose relics have been preserved in their strong coffins'. It was, he concluded, full time 'that such speculations cease to be confounded with "Atheism", – at least by our Learned!'

[37] John Tyndall, 'Presidential address', *BAAS Report*, **44** (1874), lxvi–xcvii, on p. lxxxviii.

[38] On scientific naturalism and its relation to theology, see Robert M. Young, *Darwin's metaphor: Nature's place in Victorian culture* (Cambridge, 1985), esp. ch. 5, 'Natural theology, Victorian periodicals, and the fragmentation of a common context'. Young argues that by the 1870s natural theology no longer served as a unifying context for British intellectuals, as it had before mid-century.

[39] SPT, **2**, 649.

[40] J.P. N[ichol], 'State of discovery and speculation concerning the nebulae', *London and Westminster Rev.*, **3** (1836), 390–409, on pp. 404–5*n*. For a more general perspective of such evolutionary speculation in the period, see P.J. Bowler, *Fossils and progress. Palaeontology and the idea of progressive evolution in the nineteenth century* (New York, 1976).

While, on the one hand, Nichol's remarks prepared the way for Chambers's *Vestiges*, on the other hand, they also prepared the ground for William Thomson's openness towards a law-like evolution of species. Nichol's typically radical, anti-dogmatic views, based nevertheless upon a strictly voluntarist conception of laws of nature (ch. 4), suggest once again why Thomson himself would strongly oppose Darwinian natural selection while never aligning himself with biblical literalists and anti-evolutionists.

He had by the mid-1850s developed a doctrine of *inorganic* evolution based on the principle of energy dissipation. That doctrine, as he expressed it in his 1854 British Association paper 'On the mechanical antecedents of motion, heat, and light', posed three increasingly evolutionary options:

(1) The creation of matter and energy, and their initial arrangements, are contingent upon God's will. We do not know the moment of such a creation which could, presumably, have been at any moment up to the present.
(2) We may find at some finite epoch in the past a state of matter derivable from no antecedent by natural laws. That state would represent a beginning of the universe and of the present order of things. We have, however, 'no indications whatever of natural instances of it'.
(3) Tracing back the arrangements of energy and matter would lead to the conclusion that the matter of the solar system had once been at infinitely greater distances apart than now. Thomson adopted this view, arguing that the potential energy of gravitation – probably in fact kinetic energy in the universal plenum – was in reality the ultimate created antecedent of all the energy at present in the universe (chs. 12 and 14).[41]

Whichever possibility held good, however, 'purely mechanical reasoning' (based on known laws of energy and matter) showed 'a time when the earth must have been tenantless', i.e. devoid of living beings. The same reasoning taught us 'that our own bodies, as well as all living plants and animals, and all fossil organic remains, are organized forms of matter to which *science can point no antecedent except the Will of a Creator*'. This truth was furthermore 'amply confirmed by the evidence of geological history'.[42] In 1854, therefore, Thomson strongly emphasized the creation of living beings as wholly contingent upon the Creator's will, a conviction strengthened by the geological record, and serving to remind his audience of the real *possibility* of the first option above in respect of the creation of the inorganic universe. In these remarks of 1854 we have also Thomson's first public reference to the origins of life.

During the early 1850s, Sedgwick, Lyell, Hopkins, and others had debated questions of life on earth in relation to the issue of progression and non-progression. Sedgwick eloquently presented his own view thus:

[41] MPP, **2**, 37–40. [42] *Ibid.*, p. 38.

[I think there are], among the old deposits of the Earth . . . traces of an organic progression among the successive forms of life . . . [This] historical development of the forms and functions of organic life during successive epochs . . . seems to mark a gradual evolution of Creative Power manifested by a gradual ascent towards a higher type of being . . . [But] the elevation of the *Fauna* of successive periods was not . . . made by transmutation, but by creative additions; and it is by watching these additions that we get some insight into Nature's true historical progress.[43]

Organic progression *without transmutation of species*, and a corresponding hostility to Chambers's *Vestiges*, was thus the leading characteristic of Sedgwick's pre-Darwinian interpretation of life on earth. In 1849, Hugh Miller's *Footprints of the Creator* had presented very similar views in which (in Lyell's words) 'the successive development of the living inhabitants of the globe kept pace with a corresponding improvement in its habitable condition'. Progression in geology (which Miller held also to be catastrophist in Whewell's sense of the term) went hand-in-hand with development in the province of organic beings.[44]

Not surprisingly, Lyell set out to criticize such views. In his 1851 presidential address to the Geological Society of London, he questioned the palaeontological evidence in support of the doctrine of successive development, and avoided any definite commitment concerning the creation of species or the origins of geological phenomena. With regard to the former, 'whether such commencements be brought about by the direct intervention of the First Cause, or by some unknown Second Cause or Law appointed by the Author of Nature, is a point upon which I will not venture to offer a conjecture'. With regard to the latter, Lyell preferred to cite Hutton's conclusion that in geology we neither see the beginning nor the end. Lyell did, however, admit the recent origin of man whose intellect and spiritual and moral nature 'are the highest works of creative power known to us in the universe'.[45]

Hopkins responded to Sedgwick's and Lyell's views in his own 1852 presidential address to the Geological Society. He acknowledged that both advocates and opponents of the doctrine of progression of organic forms during successive geological epochs repudiated any notion of transmutation of species, and he emphasized the lack of demonstrative evidence for or against the doctrine. Hopkins nevertheless hinted that Lyell might have gone too far towards asserting the truth of the opposite doctrine. He therefore insisted 'on that philosophic caution and reserve which may leave us unshackled in our future speculations, and free to modify our opinions so far as future evidence may call

[43] Adam Sedgwick, 'Preface to the fifth edition', *A discourse on the studies of the University of Cambridge* (London and Cambridge, 1850), pp. xliv, cliv, ccxvi. Quoted (with slight alteration) by Lyell, 'Anniversary address of the president', *Quart. J. Geol. Soc. London*, **7** (1851), xxxii–lxxv, on pp. xxxii–xxxiv. See also Michael Ruse, *The Darwinian revolution* (Chicago, 1979), esp. pp. 87–8.

[44] Hugh Miller, *Footprints of the Creator, or, the asterolepis of Stromness* (London, 1849). Lyell's summary appears in his 'Anniversary address', pp. xxxiv–xxxv.

[45] *Ibid.*, pp. xxxvi–lxxv; Ruse, *Darwinian revolution*, pp. 76–8.

upon us to do'.[46] Of the doctrine of progression with respect to inanimate matter, Hopkins had no such reservations (ch. 16).

William Thomson's 1854 views conform to Hopkins's cautious approach. A decisive progressionist with regard to the inorganic world, Thomson only ruled out the evolution of life from antecedent arrangements of energy and matter. Although he did not explicitly reject transmutation, his remarks did seem to imply the separate creation of human beings, animals, living plants, and fossil organic remains. Thomson did not consider at this stage whether or not those creations took place according to a doctrine of progression, but it is evident that he, like Hopkins, was more disposed to follow the progressionists Sedgwick and Miller than the non-progressionist Lyell with respect to animate beings as well as inanimate matter.

Five years later, Charles Darwin published his *Origin of species*. Reviewers such as Sedgwick, Hopkins, and Phillips, chose to attack Darwin's methodology on the grounds that his conclusions violated 'the true spirit of inductive philosophy'. Hopkins, for example, stressed the importance of trying the theories of the naturalists by the same standards of evidence as those of the physicists. What he demanded was demonstrative, rigorous proof rather than mere belief or assertion. Since there was no *a priori* evidence in favour of natural selection, the theory had to be established inductively. And here the evidence of continuity in palaeontology – of intermediate links from one species to another – was lacking. To claim, as Darwin had done, that the palaeontological evidence was by its nature incomplete and unsatisfactory was to fail to do justice to geological science. For Hopkins, then, it was not the evidence which was deficient, but Darwin's theory.[47] Hence we can see what Thomson had meant by his phrase of 1854, 'amply confirmed by the evidence of geological history'. The evidence of geology, for Sedgwick, Hopkins, Thomson, and others, pointed to the separate creation of each species, to discontinuity rather than continuity of life on earth.

William Thomson's own response to Darwin came in 1861–2. In his 1861 British Association address, 'Physical considerations regarding the possible age of the sun's heat', he asserted not only the universal dissipation of energy and the finite age of the sun and solar system (ch. 15), but also that it was 'impossible to conceive either the beginning or the continuance of life without a creating and overruling power'. In the 1862 *Macmillan's Magazine* version, 'On the age of the sun's heat', Thomson changed the phrase 'creating and overruling power' to 'overruling creative power', a shift which may well mark his acceptance of a

[46] William Hopkins, 'Anniversary address of the president', *Quart. J. Geol. Soc. London*, **8** (1852), xxiv–lxxx, on pp. lxxi–lxxii.

[47] William Hopkins, 'Physical theories of the phenomena of life', *Fraser's Mag.*, **61** (1860), 739–53; **62** (1860), 74–90; Ruse, *Darwinian revolution*, pp. 238–9. See also Adam Sedgwick to Charles Darwin, 24th December, 1859, in J.W. Clark and T. McK. Hughes, *The life and letters of the Reverend Adam Sedgwick* (2 vols., Cambridge, 1890), **2**, pp. 356–9. Sedgwick there believed Darwin to have deserted the true method of induction.

progressionist theory of organic life, though with a strong emphasis on the divinely controlled nature of the process.[48] In other words, while the earlier phrase suggests a traditional distinction between creation and providence, the later one evokes the notion of an on-going divinely guided creation of life forms. Cautious as ever, Thomson did not make any commitment here to either transmutation or successive creation. What was explicit, however, was his repudiation of a directionless, random process of evolution of life on earth. By adding that 'therefore no conclusions of dynamical science regarding the future condition of the earth can be held to give dispiriting views as to the destiny of the race of intelligent beings by which it is at present inhabited', Thomson emphasized his conviction that life's history and destiny were neither subject to chance nor to dynamical law alone, but to the 'overruling creative power' of God.

During his 1868–9 debate with Thomson, T.H. Huxley carefully omitted discussion of 'natural selection', an omission which Thomson seized upon. Observing that Huxley was prepared to modify or correct the geological time scale and to accept that the naturalist must 'modify his notions of the rapidity of change accordingly', Thomson concluded:

> The limitation of geological periods, imposed by physical science, cannot, of course, disprove the hypothesis of transmutation of species; but it does seem sufficient to disprove the doctrine that transmutation has taken place through 'descent with modification by natural selection'.[49]

Thomson was here making more explicit Jenkin's conclusion in the 1867 review of Darwin 'that countless ages cannot be granted to the expounder of any theory of living beings, . . . [and] that the age of the inhabited world is proved to have been limited to a period wholly inconsistent with Darwin's views'. Jenkin had also undermined Darwin's concept of natural selection itself by arguing that discontinuous variation would be immediately swamped by blending.[50] Thus, as Darwin retreated from his estimates of geological time, Thomson and his circle advanced on the theory of natural selection.

Although Huxley's 1869 address was principally a defence of geology rather than biology, his advocacy of what he termed 'evolutionism', replacing 'catastrophism' and 'uniformitarianism' as the framework for geological science, related closely to biology, not only through an analogous method, but more especially through continuity and interrelation of subject-matter:

> [The] value of the doctrine of Evolution to the philosophic thinker [is not] diminished by the fact that it applies the same method to the living and non-living world; and embraces,

[48] William Thomson, 'Physical considerations regarding the possible age of the sun's heat', *BAAS Report*, **31** (1861), 27–8; 'On the age of the sun's heat', *Macmillan's Mag.*, **5** (1862), 288–93; PL, **1**, 349–68, on p. 350. [49] PL, **2**, 89–90.

[50] Fleeming Jenkin, 'Darwin and the origin of species', *North Brit. Rev.*, **47** (1867), 277–318; *Papers and memoir*, **1**, pp. 215–63, on p. 248. See also J.R. Moore, *The post-Darwinian controversies* (Cambridge, 1979), pp. 128–31, for an analysis of Jenkin's critique of natural selection.

in one stupendous analogy, the growth of a solar system from molecular chaos, the shaping of the earth from the nebulous cubhood of its youth, through innumerable changes and immeasurable ages, to its present form; and the development of a living being from the shapeless mass of protoplasm we term a germ.[51]

Such a version of evolutionism was quite incompatible with both Thomson's approach to cosmical evolution *and* with his developing views on divinely guided evolution of living beings. Although he here avoided the controversial issue of evolution by natural selection, Huxley failed to avert confrontation with those staunch opponents of the doctrine, Thomson and Jenkin.

The closing pages of William Thomson's 1871 presidential address to the British Association (read on 2nd August) contain not only some of his most quoted passages, but also form, as a whole, some of his most illuminating ideas on the nature of life in relation both to the material world and to God. The address had cost him much time and effort. Following Huxley as president, he felt called upon to respond to Huxley's 1870 presidential address. And when Huxley introduced Sir William as both 'an intellectual giant' and a 'gentle knight', he was describing a protagonist with whom he had already met in private to discuss the forthcoming occasion.[52]

To begin with, Thomson asserted that 'the essence of science, as is well illustrated by astronomy and cosmical physics, consists in inferring antecedent conditions, and anticipating future evolutions, from phenomena which have actually come under observation'. Such an 'actualist' approach characterized Thomson's own work on the sun's heat, on the origin of the solar system, and on the secular cooling of the earth. With biology, he went on, the difficulties of successfully realizing this ideal were prodigious. Nevertheless, 'the earnest naturalists of the present day' were 'struggling boldly and laboriously to pass out of the mere "Natural History" stage of their study, and to bring Zoology within the range of Natural Philosophy'.[53]

One such attempt, Thomson claimed, was mistaken. 'A very ancient speculation, still clung to by many naturalists . . . supposes that, under meteorological conditions very different from the present, dead matter may have run together or crystallised or fermented into "germs of life", or "organic cells", or "protoplasm"'. Science, however, had brought 'a vast mass of inductive evidence against this hypothesis of spontaneous generation', as set out by Prof Huxley in his presidential capacity the previous year. For Thomson, then, careful scrutiny of the evidence had in every case 'discovered life as antecedent to life'. He was in fact employing Huxley's address for his own purposes. Huxley had phrased his conclusions quite tentatively:

[51] T.H. Huxley, 'Geological reform', *Quart. J. Geol. Soc. London*, **25** (1869), xxxviii–liii; *Discourses: biological and geological* (London, 1902), pp. 305–39, on p. 325.

[52] SPT, **1**, 550–1; **2**, 598.

[53] William Thomson, 'Presidential address', *BAAS Report*, **41** (1871), lxxxiv–cv; PL, **2**, 132–204, on p. 197.

[Looking] back through the prodigious vista of the past, I find no record of the commencement of life, and therefore I am devoid of any means of forming a definite conclusion as to the conditions of its appearance . . . To say . . . in the admitted absence of evidence, that I have any belief as to the mode in which the existing forms of life have originated, would be using words in the wrong sense. But expectation is permissible where belief is not; and if it were given to me to look beyond the abyss of geologically recorded time to the still more remote period when the earth was passing through physical and chemical conditions, which it can no more see again than a man can recall his infancy, I should expect to be a witness of the evolution of living protoplasm from not living matter.[54]

Thus, while Huxley admitted the absence of evidence in favour of the origin of life from matter [abiogenesis], he did affirm his 'expectation' of that hypothesis turning out to be the true one. Thomson, however, used the negative evidence in favour of 'abiogenesis' as positive evidence in favour of 'biogenesis', i.e. 'that living matter always arises by the agency of pre-existing living matter'.

Thomson therefore asserted that the belief that 'dead matter cannot become living without coming under the influence of matter previously alive' was 'as sure a teaching of science as the law of gravitation'. As a biological *law* – and so representing the natural philosophy stage of biological inquiry – Thomson was 'ready to adopt, as an article of scientific faith, true through all space and time, that life proceeds from life, and from nothing but life'. This 'axiom', then, was the true basis from which to proceed in biology, to infer antecedent conditions and to anticipate future evolutions from phenomena which have actually come under observation.[55]

In conformity to this method which he referred to as 'philosophical uniformitarianism', Thomson next raised the question of how life originated on earth. If we traced 'the physical history of the Earth backwards, on strict dynamical principles, we are brought to a red-hot melted globe on which no life could exist'. Thus 'when the earth was first fit for life', with 'rocks solid and disintegrated, water, air all round, warmed and illuminated by a brilliant Sun, ready to become a garden', 'there was no living thing on it'. Two possibilities were then open to Thomson. Either 'grass and trees and flowers' were brought into existence, 'in all the fullness of ripe beauty', by a fiat of Creative power, or 'vegetation, growing up from seed sown, spread and multipl[ied] over the whole Earth'. Since 'we must not invoke an abnormal act of Creative Power' if we could find 'a probable solution, consistent with the ordinary course of nature', the latter possibility was to be preferred and pursued.[56]

The path by which Thomson pursued the preferred solution to the question of life's origin on earth was that of analogy, rather than direct investigation by

[54] T.H. Huxley, 'Presidential address', *BAAS Report*, **40** (1870), lxxxii–lxxxix, on pp. lxxxiii–lxxxiv.

[55] PL, **2**, 198–9. See also Stewart and Tait, *The unseen universe*, pp. 168–79, for Stewart and Tait's adherence to the same principle *pace* Tyndall and others.

[56] PL, **2**, 199–200.

observation or experiment. Life, he noted, quickly became established upon the lava of volcanoes or volcanic islands through the transport of seed by living creatures, wind, or sea:

Is it not possible, and if possible, is it not probable, that the beginning of vegetable life on the Earth is to be similarly explained? Every year thousands, probably millions, of fragments of solid matter fall upon the Earth – whence came these fragments? . . . Should the time when this Earth comes into collision with another body, comparable in dimensions to itself, be when it is still clothed as at present with vegetation, many great and small fragments carrying seed and living plants and animals would undoubtedly be scattered through space. Hence and because we all confidently believe that there are at present, and have been from time immemorial, many worlds of life besides our own, we must regard it as probable in the highest degree that there are countless seed-bearing meteoric stones moving about through space.[57]

While Thomson did not attempt to discuss 'the many scientific objections' to this hypothesis, he believed them 'to be all answerable'. He admitted that 'the hypothesis that life originated on this Earth through moss-grown fragments from the ruins of another world may seem wild and visionary', but maintained that 'it is not unscientific'. As in his earlier use of a meteoric theory, and the zodiacal light, to account for the sun's heat, Thomson was employing a cause (meteors) which was known to exist *independently* of the phenomena (the arrival of life on earth) to be explained (ch. 14).

Having brought life to earth in conformity with the axiom that life proceeds from life, Thomson accepted that 'all creatures now living on earth have proceeded by orderly evolution from some such origin' as vegetation. The step, from such beginnings, to the earth 'as teeming with all the endless variety of plants and animals which now inhabit it' was a prodigious one, but in accordance with the doctrine of continuity 'most ably laid before the Association by a predecessor in this Chair', W.R. Grove.[58] Thomson then quoted verbatim from the concluding passages of Darwin's *Origin of species* concerning 'an entangled bank' with its immense variety of plant and animal life all 'produced by laws acting around us'. Thus, for Darwin, 'there is grandeur in this view of life with its several powers, having been originally breathed by the Creator into a few forms or into one; and that, whilst this planet has gone cycling on according to the fixed law of gravity, from so simple a beginning endless forms, most beautiful and most wonderful, have been and are being evolved'. With these feelings, Thomson 'most cordially' sympathized, for Darwin's expression of orderly, law-like evolution of life conformed to his own view of cosmic evolution.[59]

[57] *Ibid.*, pp. 200–2.

[58] W.R. Grove, 'Presidential address', *BAAS Report*, **36** (1866), liii–lxxxii. See Stewart and Tait, *The unseen universe*, on the principle of continuity.

[59] PL, **2**, 202–4; Morse Peckham (ed.), *The origin of species by Charles Darwin. A variorum text* (Pennsylvania, 1959), pp. 758–9. First published as Charles Darwin, *On the origin of species by means of natural selection, or the preservation of favoured races in the struggle for life* (London, 1859).

Significantly, however, Thomson excluded Darwin's concluding remarks on 'the hypothesis of "the origin of species by natural selection" '. 'This hypothesis', he felt, 'does not contain the true theory of evolution, if evolution there has been, in biology'. Sir John Herschel, he pointed out, had 'expressed a favourable judgment on the hypothesis of zoological evolution', though 'with some reservation in respect to the origin of man', but had objected to the doctrine of natural selection on the grounds 'that it was too like the Laputan method of making books, and that it did not sufficiently take into account a continually guiding and controlling intelligence'. This criticism, Thomson believed, was 'a most valuable and instructive' one.[60]

That William Thomson should have rejected a random, chance mechanism for the evolution of life in favour of an ordered, law-like process subject to divine guidance and control was perfectly in accordance with his theology of nature. For that reason, he concluded his address with well-known remarks on Paley's argument from design:

> I feel profoundly convinced that the argument of design has been greatly too much lost sight of in recent zoological speculations. Reaction against frivolities of teleology . . . has I believe had a temporary effect in turning attention from the solid and irrefragable argument so well put forward in that excellent old book [Paley's *Natural theology*]. But overpoweringly strong proofs of intelligent and benevolent design lie all round us, and if ever perplexities, whether metaphysical or scientific, turn us away from them for a time, they come back upon us with irresistible force, showing to us through nature the influence of a free will, and teaching us that all living beings depend on one ever-acting Creator and Ruler.[61]

Thomson's remarks were probably a direct reply to J.D. Hooker's (1817–1911) criticism of natural theology at the end of his 1868 presidential address to the British Association. For Hooker, natural theology was 'the most dangerous of all two-edged weapons . . . a science, falsely so called, when, not content with trustfully accepting truths hostile to any presumptuous standard it may set up, it seeks to weigh the infinite in the balance of the finite, and shifts its ground to meet the requirement of every new fact that science establishes, and every old error that science exposes. Thus pursued, Natural Theology is to the scientific man a delusion, and to the religious man a snare, leading too often to disordered intellects and to atheism'.[62] But Thomson believed that his cosmical physics had shown, not only that the present order of the world was not eternal in either past or future, but that the finite time scale demonstrated the inadequacy of natural selection to account for the evolution of life on earth. On the one hand, therefore, the cosmic evolution of the earth as an abode *fitted* for life seemed to have been the result of benevolent and wise design by a God who had chosen well the initial arrangements of matter and energy. On the other hand, the

[60] PL, **2**, 204. [61] *Ibid.*, pp. 204–5.

[62] J.D. Hooker, 'Presidential address', *BAAS Report*, **38** (1868), lviii–lxxv, on p. lxxiv.

evolution of life on earth had occurred, if at all, in a manner too rapid and too ordered to be other than the result of the working, through laws, of 'one ever-acting Creator and Ruler'.

Some years later, Darwin's son George, in his letter of November, 1878, objected to Thomson's remark 'that a few hundred million years would be insufficient to allow of transmutation of species by nat[ural] selection'. In particular, he asked, 'what possible datum can one have for the rate at which it has or can work?' And he added that it was 'very frequently supposed that my father attributes all to Nat[ural] Sel[ection]; he undoubtedly thinks it a leading cause, but has written long chapters & even a whole book on other causes'. Unfortunately we do not know Thomson's reply, verbal or written, although in response to a correspondent twenty years later he reiterated his view that the limitation of geological time 'is an argument against "natural selection", as having the great potency attributed to it by Darwin and some of his followers'.[63]

More immediate reactions to Thomson's address occurred during the month of August, 1871. Huxley did not engage in direct debate, but he gave his witty verdict in a letter dated 23rd August to J.D. Hooker, written from Arrochar on the shores of the fjord-like Loch Long:

> I like what I have seen of Thomson much. He is, mentally, like the scene which lies before my windows, grand and massive but much encumbered with mist – which adds to his picturesqueness but not to his intelligibility . . . I cannot say I greatly admire the address. It wants cohesion and resembles a flash of his own aerolite [meteorite] more than anything else – bright points in the midst of much nebulosity.[64]

More specifically, but no less facetiously, Huxley had asked Hooker on the 11th August: 'What do you think of Thomson's "creation by cockshy" – God Almighty sitting like an idle boy at the seaside and shying aerolites (with germs), mostly missing, but sometimes hitting a planet!'[65]

Later, Huxley stated his belief that Thomson's meteoric theory of the origin of life on earth had no bearing on the evolutionary process itself since 'the germs brought to us by meteorites, if any, were not ova of elephants or crocodiles . . . but only those of the lowest form of animal and vegetable life'.[66] By implication, Thomson's theory could not resolve the difficulty of geological time with respect to evolution. Such, indeed, was the thrust of E.B. Poulton's criticism in his 1896 address as president of the zoological section of the British Association.

[63] George Darwin to William Thomson, 1st November, 1878, D8, ULC; J.D. Burchfield, 'Darwin and the dilemma of geological time', *Isis*, **65** (1974), 300–21, on pp. 320–1*n*; Lord Kelvin to E. Davys, 5th October, 1898, LB5/159, ULG. In the letter to Davys, Kelvin also explained: '"Evolution" is a larger question. We cannot put any limit to the rapidity with which evolution may have taken place, so far as there has been evolution at all. "Protoplasm" seems a very mythical affair, though the word is largely used by modern speculative writers'.

[64] Leonard Huxley, *Life and letters of Sir Joseph Dalton Hooker* (2 vols., London, 1918), **2**, pp. 165–6. [65] *Ibid.*, p. 126*n*. [66] SPT, **2**, 607*n*.

Kelvin's suggestion could not reasonably be used to shorten the time required for evolution of life on earth.[67] Thomson's belief in orderly evolution, however, did not require the meteoric theory on the grounds of time. Rather, the theory was introduced in conformity to methodological principles of uniformity and continuity. Life's origin, for Thomson, might be carried back to the very creation of matter and energy rather than merely to the beginning of earth history. If ultra-mundane vortices flying about in space could be the created origin of gravitation, and thus of all motions of mundane matter, why not germs flying about as the carriers of life?

Meanwhile, Hooker had already written to Charles Darwin on 5th August, anxious to hear Darwin's opinion. Hooker was full of critical admiration for the address: 'What a belly-full it is, and how Scotchy! It seems very able indeed, and what a good notion it gives of the gigantic achievements of mathematicians and physicists – it really makes one giddy to read of them!' But with regard to the meteoric theory of the origin of life on earth, Hooker offered a cogent criticism:

> The notion of introducing life by Meteors is astounding and very unphilosophical, as being dragged in head and shoulders apropos of the speculations of the 'Origin' of life from or amongst existing matter – seeing that Meteorites are after all composed of the same matter as the Globe is. Does he suppose that God's breathing upon Meteors or their progenitors is more philosophical than breathing on the face of the earth? I thought too that Meteors arrived on the earth in a state of incandescence, – the condition under which T. assumes that the world itself could not have sustained life. For my part I would as soon believe in the Phoenix as in the Meteoric import of life. After all the worst objections are to be found in the distribution of life, and the total want of evidence of renewal by importation such as meteoric visitations would suggest the constant recurrence of.[68]

To Thomson, thinking on a cosmic, rather than simply terrestrial, scale, such accusations of 'unphilosophical speculation' would have carried little weight. Given progression, uniformity, and continuity in dynamical science and life science alike, what grander conception of God could there be than to suppose matter, energy, and life to have had a common beginning in time?

In the same letter, Hooker expressed his perplexity at the meaning of Thomson's concluding remarks regarding design: 'how the Deuce can "proofs of intelligent design" (in Nature) show us "through nature the influence of a free will"?'[69] Taken by themselves, Thomson's remarks were undoubtedly much encumbered with mist. Nevertheless, his linking of intelligent design with free will connects the earlier discussions in this chapter with the present concern with the origin and evolution of life on earth. That connection takes the form of a belief in the unique directing capacity of life and mind over matter and energy, a power which could not originate from dead matter.

In the wake of the new statistical interpretation of the Second Law of

[67] E.B. Poulton, 'A naturalist's contribution to the discussion upon the age of the earth', *BAAS Report*, **66** (1896), 808–28, esp. pp. 818–19. [68] Huxley, *Life of Hooker*, **2**, pp. 126–7.

[69] *Ibid.*, p. 127.

Thermodynamics, Thomson's separation of abstract from physical dynamics, and his belief that 'the fortuitous concourse of atoms' could be the only foundation in abstract dynamics for the Second Law, encouraged and deepened his separation of life and mind from matter. In stark contrast to the fortuitous concourse of atoms in particle dynamics was the directing power of living beings, as he made clear in 1892:

The influence of animal or vegetable life on matter is infinitely beyond the range of any scientific inquiry hitherto entered on. Its power of directing the motions of moving particles, in the demonstrated daily miracle of our human free will, and in the growth of generation after generation of plants from a single seed, are infinitely different from any possible result of the fortuitous concourse of atoms; and *the fortuitous concourse of atoms is the sole foundation in Philosophy on which can be founded* [the Second Law of Thermodynamics] . . . The considerations of ideal reversibility, by which Carnot was led to his theory, and the true reversibility of every motion in pure dynamics have no place in the world of life.[70]

During the final two decades of William Thomson's life, we find several reaffirmations of his beliefs in the essential separateness of mind and matter, and in the impossibility of either the indefinite, past or future, duration of life on earth or the origin of life from dead matter. In his 1887 paper 'On the sun's heat' he emphasized that 'the beginning and the maintenance of life on the earth is absolutely and infinitely beyond the range of all sound speculation in dynamical science', adding that 'the only contribution of dynamics to theoretical biology is absolute negation of automatic commencement or automatic maintenance of life'.[71] Again, in the passage quoted at the beginning of this chapter from his aptly entitled paper of 1899, 'The age of the earth as an abode fitted for life', he concluded that his task had been confined to what, 'humanly speaking' – indicating the limited perspective of finite human science – 'we may call the fortuitous concourse of atoms, in the preparation of the earth as an abode fitted for life'. Mathematics and dynamics could account in this way for the earth's readiness to receive life, but 'fail us when we contemplate the earth, fitted for life but lifeless, and try to imagine the commencement of life upon it'. Such an origin of life 'certainly did not take place by any action of chemistry, or electricity, or crystalline grouping of molecules under the influence of force, or by any possible kind of fortuitous concourse of atoms'. Looking for antecedent conditions – the essence of science for Thomson – ceased when we came 'face to face with the mystery and miracle of the creation of living creatures'.[72]

Four years later, in 1903, he again stressed that 'every action of free will is a miracle to physical and chemical and mathematical science',[73] and in a letter of 1906 to Professor Helder he put the point yet more forcefully:

70 Lord Kelvin, 'On the dissipation of energy', *Fortnightly Rev.*, **51** (1892), 313–21; PL, **2**, 451–74, on pp. 464–5. 71 PL, **1**, 415. 72 MPP, **5**, 229–30.

73 [Lord Kelvin], 'Lord Kelvin on science and theism', *The nineteenth century*, **53** (1903), 1068–9. The basis of the short article was a letter from Kelvin to James Knowles.

The perception of every one of the human race of his own individuality and free will seems to me to absolutely disprove all materialistic doctrines and to give us scientific ground for believing in the Creator of the Univ. in whom we live & move & have our being.[74]

Because of our personal belief in free will inexplicable by science (and here he strongly echoed Jenkin), 'science positively affirms Creative Power'.[75] By the study of 'the physics and dynamics of living and dead matter all around', we are compelled to accept *as an article of belief* the existence of 'a creating and directing Power' other than physical, or dynamical, or electrical forces. This unknown power was 'a vital principle' quite distinct from a fortuitous concourse of atoms, and was, Thomson implied, the essence of the design argument. Since 'we only know God in His Works', such a directing power in nature – clearly distinguished from the effects of chance – was to be taken as a manifestation of divine activity. Thomson therefore concluded that 'if you think strongly enough you will be forced by science to the belief in God, which is the foundation of all religion'.[76] Overpoweringly strong proofs of intelligent and benevolent design did, he believed, lie all around us. The design argument, then, had certainly not been lost sight of in Lord Kelvin's zoological speculations, even though the precise nature of the argument in his 1871 address had been much encumbered with mist.

By now it will be apparent how radically Thomson's views of the origin and evolution of life on earth differed from those of X Club contemporaries such as Huxley, Hooker, Tyndall, and Herbert Spencer. Tyndall's 1874 address combined many of the ingredients of Spencerian evolution from matter and energy to mind which Thomson so vigorously opposed throughout his writings.[77] He did not himself undertake any public challenge to Tyndall, preferring perhaps to leave popular controversies of this nature to others, and probably secure in the view that his own (and Maxwell's) conception of man's directing powers was more than an intellectual match for Tyndall, Huxley or Spencer. But Thomson's contempt for Spencer's writings, often admired by Tyndall and Hooker, though not by Darwin, appears in his later verdict that he had 'never been of opinion that the philosophical writings of the late Mr Herbert Spencer had the value or importance which has been attributed to them by many readers [of *Nature*] of high distinction. In my opinion a national memorial would be unsuitable'.[78] Spencerian evolution of matter from mind and Thomsonian dualism could not co-exist.

[74] Lord Kelvin to Professor J. Helder, 12th May, 1906, LB31.2, ULC. Quoted in D.B. Wilson, 'Kelvin's scientific realism: the theological context', *Phil. J.*, **11** (1974), 41–60, on p. 60.

[75] [Kelvin], 'Lord Kelvin', p. 1068. [76] *Ibid.*, p. 1069.

[77] Tyndall, 'Presidential address', pp. lxxxviii–xciv.

[78] Lord Kelvin, Letter to *Nature*, **74** (1906), 521; SPT, **2**, 1124. On Spencer and Darwin, see Moore, *The post-Darwinian controversies*, pp. 153–73.

IV

Energy, economy, and Empire: the relief of man's estate

19

The telegraphic art

A strong recommendation of the study of Natural Philosophy arises from the importance of its results in improving the physical condition of mankind. At no period of the world's history have the benefits of this kind conferred by science been more remarkable than during the present age . . . Who would have believed that we should at present consider twenty-five or thirty miles an hour a slow average rate of travelling? or that our messages should now be communicated for thousands of miles by sea or land, literally with the speed of lightning? These are only single instances of the vast resources which we derive from direct applications of modern science; . . . every one is convinced of the immense practical importance of the principles of Natural Philosophy at present known. *William Thomson, 'Introductory lecture'.*[1]

As a Glasgow College student, William Thomson had in 1839 taken note of Prof Meikleham's professed aims of natural philosophy: worship of the Creator, intellectual satisfaction, and practical benefits. Natural philosophy, in its third aim, extended 'our power over nature by unfolding the principles of the most useful arts'.[2] In his own introductory lecture (above), Thomson amplified the powerful conviction that natural philosophy improved the material condition of mankind. Such expressions of Baconian ideology and the fruits of knowledge continue the Scottish tradition of natural philosophy, epitomized in the wide-ranging *Encyclopaedia Britannica* articles of John Robison (ch. 4).

To late Victorian and early twentieth-century writers, submarine telegraphy was one of the sublime achievements of the age of British supremacy in political, economic, and naval power. Electric telegraphy had begun in 1837, the first year of Queen Victoria's reign, when the work of Cooke and Wheatstone provided the basis for a vast network of land telegraph lines. The first great international exhibition in the Crystal Palace, in 1851, coincided with the first successful submarine telegraph cable linking England and France. And, above all, the Atlantic telegraph of 1866 provided almost instant communication between the Old World of Europe and the boundless New World of America. Throughout

[1] SPT, **1**, 247–9. The reference to submarine telegraphy must have been added to the text of the original 1846 lecture at some date after the laying of the earliest cross-channel cables around 1851.

[2] William Thomson, 'Notebook of Natural Philosophy class, 1839–40', NB9, ULC.

the remainder of Queen Victoria's reign, not only the British Empire, but the world was being united by this entirely new physical means. Telegraphy presented none of the double-edged consequences of early industrialization, with its social evils as well as its material benefits. Telegraphy seemed an undeniable blessing, bringing only prosperity to mankind and aiding the development of civilization, commerce, and peace. Rudyard Kipling captured the spirit in his poem 'The deep-sea cables':

The wrecks dissolve above us; their dust drops down from afar –
Down to the dark, to the utter dark, where the blind white sea-snakes are.
There is no sound, no echo of sound, in the deserts of the deep,
Or the great grey level plains of ooze where the shell-burred cables creep.

Here in the womb of the world – here on the tie-ribs of earth
Words, and the words of men, flicker and flutter and beat –
Warning, sorrow, and gain, salutation and mirth –
For a Power troubles the Still that has neither voice nor feet.

They have wakened the timeless Things: they have killed their father Time:
Joining hands in the gloom, a league from the last of the sun.
Hush! Men talk to-day o'er the waste of the ultimate slime.
And a new Word runs between: whispering, 'Let us be one!'[3]

The Atlantic cable enterprise, whose ultimate success would bring Thomson his knighthood in 1866, reinforced confidence in human progress, especially British progress, and gave tangible form and support to all previous optimism about the ability of man to exploit and control the mighty powers of nature. As early as 1857, for example, that most eminently practical of journals, *The Engineer*, included a whole hymn to human progress among its otherwise prosaic accounts of cable dimensions and properties. The Atlantic telegraph project was adding yet another link 'to the chain of events which, from the earliest dawn of history, has, under the influence of some mysterious law, determined the whole course of civilization from the East towards the West':

The great tidal wave, whose seconds are centuries in the march of human progress, has been moving on slowly and deviously, breaking now on this shore, now on that, now retrograding, but ever advancing, from the rising to the setting sun, with the silent cumulative momentum of the earth's rotation from the beginning of time . . . The whole history of the past is but a record of successive conquering and dominant races for ever issuing from the teeming East, like a series of waves falling into each other, and driven onwards by the same resistless force.[4]

[3] [Rudyard Kipling], *Rudyard Kipling's verse. Inclusive edition. 1885–1932*, 4th edn. (London, 1933), p. 173. For an account of the early years of submarine telegraphy, see especially Charles Bright, *Submarine telegraphs. Their history, construction, and working* (London, 1898), pp. xv, 1–22.

[4] *The engineer*, **3** (1857), 82.

Loading the Atlantic cable at Plymouth for the 1858 attempt. The yacht-like lines of the United States' *Niagara* (right) contrasted with the solid bulk of the Royal Navy's *Agamemnon* (left). The laying was a success, but the cable itself soon ceased to function. [From the *Illustrated London News*.]

According to *The Engineer*, all kingdoms of Europe had linked themselves together by the electric telegraph, and America was 'in a high state of electric tension'. The conductor, which was 'to establish an equilibrium between the old and new worlds', would provide yet another link in the 'electric nerve' system of the world. Addressed to mid-Victorian entrepreneurs, such stirring words affirmed the reality of material progress and human advancement through investment and innovation.

With steam-engines, railways, iron bridges, Cunard's liners, and now the electric telegraph, mankind (notably the British nation) had come, it seemed, a long way from the superstition and barbarism of the Middle Ages and pre-Reformation times. Charles Kingsley dramatized this technological vision in 1851:

> Give me the political economist, the sanitary reformer, the engineer: and take your saints and virgins, relics and miracles. The spinning-jenny and the railroad, Cunard's liners and the electric telegraph, are to me, if not to you, signs that we are, on some points at least, in harmony with the universe; that there is a mighty spirit working among us, who cannot be your anarchic and destroying Devil, and therefore may be the Ordering and Creating God.[5]

[5] Charles Kingsley, *Yeast. A problem* (London, 1851), p. 96. Quoted in W.E. Houghton, *The Victorian frame of mind, 1830–1870* (New Haven and London, 1957), pp. 43–4.

Such remarks echo exactly the practical thrust, which contrasted with all things metaphysical, dogmatic, symbolic and sectarian, in the values of the Thomsons and their circle.

Following the Great Exhibition, further triumphs of British engineering in the 1850s and beyond became symbolic of the conviction that the old days of social unrest, of Chartism, of cholera, and of famine were gone forever. The fruits of previous labours could now be harvested with increasing prosperity to the nation and Empire; old troubles were but a memory, and constant religious and political strife were features no longer so evident in national life.[6] Addressing the citizens of Glasgow just a few months after the 1858 cable actually failed, Thomson reasserted his total confidence in human advancement:

> After the harassments and disappointments of a year, when wealth and labour, care and anxiety, skill and invention might appear to have been absolutely thrown away, and to have gone to swell the vast amount of profitless labour which is done under the sun, it is no small solace to meet with such sympathy as you now manifest . . . What has been done can be done again . . . improbable, impossible as it seemed only six months ago – chimerical and merely visionary as such a project seemed ten short years earlier – instantaneous communication between the Old and the New Worlds is now a fact. It has been attained. What has been done will be done again. *The loss of a position gained is an event unknown in the history of man's struggle with the forces of inanimate Nature.*[7]

On the harmony of theory and practice in engineering: principles of economy

Economic historians have long spoken of a second industrial revolution in the last third of the nineteenth century, referring to the emergence of science-based industries, notably those producing and utilizing chemical fertilizers, synthetic dyes, and electrical power.[8] From the economists' perspective, the consequences of this transformation (relative to coal, iron and steel, steam power, and machinery) only became apparent after about 1870, but its foundations had been laid considerably earlier. In their study of these foundations, historians of science

[6] Compare, for example, W.L. Burn, *The age of equipoise. A study of the mid-Victorian generation* (London, 1964); Asa Briggs, *The age of improvement, 1783–1867* (London, 1959). See also Chapters 2 and 5 above, particularly on the earlier Scottish social context.

[7] William Thomson, Speech reported in the *Glasgow Herald*, 21st January, 1859; reprinted in SPT, **1**, 389–96, on pp. 389–90.

[8] For example, D.S. Landes, *The unbound Prometheus* (Cambridge, 1969), pp. 4, 196, 235. Modern historians have modified the too-simple picture of the *first* industrial revolution as the child of practical inventors ignorant of all science. Nevertheless, the image of the 'practical man' clearly persisted in the mid-nineteenth century. For the traditional view, see Eric Ashby, *Technology and the academics. An essay on universities and the scientific revolution*, 2nd edn. (London, 1963), pp. 50–5; Peter Mathias, *The first industrial nation. An economic history of Britain, 1700–1914* (London, 1969), pp. 134–44. For a reassessment of the relation of science and technology in the industrial revolution, see Peter Mathias, 'Who unbound Prometheus? Science and technical change, 1600–1800', in Peter Mathias (ed.), *Science and society. 1600–1900* (Cambridge, 1972), pp. 54–80; A.E. Musson and Eric Robinson, *Science and technology in the industrial revolution* (Manchester, 1969).

Uniting the Empire: labour, as well as capital, engineering, and electrical science, was of necessity employed in the submarine telegraphic enterprises which linked Britain to her Empire. Here a stage in the vital Indo–European telegraph, which would employ patents such as Thomson's siphon recorder (ch. 20), involved landing the cable in the mud at Fao, Persian Gulf, in mid-1865. [From the *Illustrated London News.*]

and technology point especially to the birth of the research laboratory and the birth of engineering as an academic discipline.[9]

In Britain, although chemical laboratories had been established by the 1830s, laboratories for research in natural philosophy did not appear until after 1840, and then largely as private laboratories loosely attached to apparatus rooms for lecture demonstrations and open to students only on a selective and voluntary basis. Not until the late 1860s were the laboratories funded institutionally for research as well as teaching. But between 1866 and 1874 ten such establishments were founded, including the Cavendish Laboratory at Cambridge under the directorship of the new professor of 'experimental physics', James Clerk Maxwell. In the next decade, some fourteen new physics laboratories appeared.[10]

The story in academic engineering is similar. Prior to the 1840s, engineering barely existed in British universities, although a few technical colleges had been established. Engineers defined themselves largely as practical men rather than philosophers, took their education as apprentices rather than as students, and did

[9] For example, Romualdas Sviedrys, 'The rise of physics laboratories in Britain', *Hist. Stud. Phys. Sci.*, **7** (1976), 405–36; R.A. Buchanan, 'The rise of scientific engineering in Britain', *Brit. J. Hist. Sci.*, **18** (1985), 218–33. [10] Sviedrys, 'Rise of physics', pp. 407–9, 415–21, 430–3.

not engage in research as an essential feature of their practice. The proliferation of steam-engines and railways after about 1830 began to change dramatically the context of both engineering and natural philosophy. The Cambridge association of Willis and Whewell in the writing of engineering textbooks in 1840–1 is symptomatic. From the more practical side, Willis wrote the science of mechanism, while Whewell as natural philosopher treated of the science of force applied to and transmitted by mechanism. Simultaneously, Glasgow University received the first chair of engineering in Britain, with Lewis Gordon its first occupant. Within two years, chairs also appeared in London and Dublin. Even with such formal recognition of engineering as a subject requiring academic study, however, an engineer did not qualify by academic degree but only by apprenticeship. Practical knowledge continued to take priority over theory. In the 1840s, the scientific engineer, like the research physicist, had yet to be born into Britain.[11]

The British Association for the Advancement of Science provided one of the strongest institutional forces behind the gradual professionalization of both scientific engineering and research physics. With its explicit aim of bringing science and industry together, with its popular format, and with its peripatetic schedule of meetings over a circuit of manufacturing towns and academic centres, it gave form, visibility and prestige to the goal of practical science and brought even science for profit within the domain of the gentleman.[12]

On the more local and everyday level, however, groups such as the Glasgow Philosophical and the Manchester Literary and Philosophical Societies brought industrial and academic interests into regular interaction.[13] The Glasgow Society, in the period 1840–70 especially, facilitated the interaction and mutual stimulation of emergent *scientific engineering* on the one hand (represented in particular by Gordon, W.J.M. Rankine, J.R. Napier, and James Thomson) with emergent *research science* on the other (represented by Thomas Thomson and William Thomson).

In his 1874 presidential address to the Society of Telegraph Engineers, Sir William Thomson paid a warm tribute to the Society's first president, William Siemens, for his strenuous and effective labour 'to promote the harmony of theory and practice, not only in the department to which this Society is devoted, but in all branches of the grand profession of engineering'.[14] This emphasis on the harmony of theory and practice in engineering has not only provided us with a leading characteristic of the work of Thomson and his Glasgow circle, but will suggest further that their conception of the professional engineer fundamentally depended upon the promotion of such harmony.

[11] *Ibid.*, pp. 405–7; Buchanan, 'Scientific engineering', pp. 218–23; and Chapters 2 and 3 above.

[12] Jack Morrell and Arnold Thackray, *Gentlemen of science. Early years of the British Association for the Advancement of Science* (Oxford, 1981), esp. pp. 202–22, 256–66, 449–531. The Thomson brothers' involvement in the 1840 Glasgow meeting is discussed in Chapter 2 above.

[13] See especially R.H. Kargon, *Science in Victorian Manchester. Enterprise and expertise* (Manchester, 1977), esp. pp. 5–14, 41–85.

[14] PL, **2**, 207–8.

In April, 1858, a few months before the laying of the first complete, but short-lived, Atlantic telegraph, Rankine presented a 'Report on the progress and state of applied mechanics' to a meeting of the Glasgow Philosophical Society. The report, prepared by a committee comprising the Clydeside engineers J.R. Napier (iron shipbuilder), Walter Neilson (mechanical engineer) and Rankine himself as convener, articulated a view of the relation between theory and practice shared by Thomson himself. Essential to this view was an understanding of the goal and criterion of *economy*:

> In the perfecting of Applied Mechanics, whether as a science or as an art, the end aimed at, and the criterion by which true is to be distinguished from false progress, may be expressed by the word ECONOMY: that is, the production of every desired effect by those means which are exactly adequate to produce it, and no more . . . perfect economy never is, nor can be attained in human works and in them the economy realized is expressed by some fraction, falling short of unity by a quantity which expresses the *waste of means*.[15]

The function of theory, the authors then explained, is to determine 'by experiment and by reasoning, the exact amount of and causes of waste, and how it is to be reduced' while 'practice strives, by continually improving skill, to effect that reduction; and both [theory and practice] tend to bring the fraction that denotes actual economy, continually nearer and nearer to that UNIT, which expresses the unattainable, though not unapproachable, limit of the result of human efforts'.

Applied mechanics, for the Philosophical Society committee, included 'every application of the laws of force and motion to works of human art'. The two great classes of objects of applied mechanics were 'structures', whose parts remained fixed relative to each other and whose requisites were stability and strength, and 'machines', whose parts performed work and which required in addition efficiency. Accordingly, the authors divided applied mechanics, when considered as a science, into tectonics and energetics, and, when considered as an art, into construction and mechanism. Many artifices, such as iron steamships, clearly belonged to both classes.

A structure involved three considerations: materials, the mode of putting the materials together, and the purpose for which the structure was to be used. The report devoted considerable attention to advances in materials, discussing both organic and inorganic materials, and dividing the inorganic into stony and metallic sorts. Stony materials had undergone a variety of improvements: in the arts of blasting rock in order to minimize waste of both powder and stone, in the manufacture and use of artificial building materials (bricks) for the lofty furnace chimneys of Clydeside, and in the economic employment of concrete for piers and breakwaters. Similarly, concerning metallic materials, the report instanced progress in increasing the speed and diminishing the cost of iron production as well as improvements in its qualities and in the techniques of iron working, such

[15] J.R. Napier, Walter Neilson, and W.J.M. Rankine, 'Report on the progress and state of applied mechanics', *Proc. Phil. Soc. Glasgow*, **4** (1855–60), 207–30, on p. 208. Publication of the report took place *after* the laying of the 1858 cable.

as casting, forging, and riveting. Next to iron in order of abundance and utility came copper (along with its alloys of brass and bronze).

The principal organic material remained timber. Economy of time and money, the authors noted, had been effected by seasoning timber in a hot air oven at Robert Napier's shipbuilding yard, while new processes of preservation of timber had also made rapid progress with economic advantages. New 'organic' materials, however, notably india-rubber and gutta-percha, had contributed much to the advancement of practical mechanics. Gutta-percha in particular, 'though softened by a moderate degree of heat, possesses a strength and elasticity, at ordinary temperatures, which enable it to be employed as a substitute for leather belts in machinery' as well as for the coating and insulating of telegraph wires. The committee also commented upon a recent example of the substitution of artificially manufactured wire-ropes for natural fibrous substances, enabling the new arts of suspension bridges and telegraph cables to be developed on a grand scale.

This concern with materials is easily comprehended within the context of recent industrial advances in Glasgow and the British Empire. Correspondingly, we must view the research work of Thomson's laboratory, devoted to measuring the properties of materials, very much within this context (ch. 5). It is no accident that after uttering his famous words to the Institution of Civil Engineers in 1883 on how, if you cannot measure what you are speaking about, you have not advanced to the stage of science (ch. 20, epigraph), he cited the backwardness of the science of strength of materials, 'so all important in engineering' but 'little advanced, and the part of it relating to the so-called hardness of different solids [precious stones and metals] least of all; there being in it no step toward quantitative measurement or reckoning in terms of a definite unit'. By contrast, he cited recent progress in electric science and in measuring the electric properties of matter.[16]

The second feature of structures, the art of putting together the materials, required observance of principles of stability and strength. The Philosophical Society report cited with particular approval Fairbairn's experiments, 'which have recently contributed most to the advancement of our knowledge of the strength of iron', and which were mostly made at the instance of the British Association, and also Thomson's discovery 'in the course of the present year [1858], with the assistance of two students of his class . . . [of] a kind of resistance in elastic solids, analogous to friction, inasmuch as it retards, without finally preventing, both the strain produced by the application of a load, and the recovery from that strain when the load is removed'.[17] With regard to the processes of techniques of construction, the report drew attention to advances in metal-working, especially accurate workmanship, which was 'the most effec-

[16] PL, **1**, 73–5.

[17] Napier *et al.*, 'Report on applied mechanics', pp. 216–7; Morrell and Thackray, *Gentlemen of science*, pp. 497–8.

tual means of diminishing friction, wear, and breakage, or obtaining economy of time, money, and materials, and of insuring efficiency of action'.

The third feature of structures, their purposes, divided into mines, houses, lines of conveyance, harbours, and vehicles (including ships). The committee focussed on lines of conveyance which included roads, railways, canals (including water supply and drainage), and signalling (notably the electric telegraph). In their discussion of recent improvements in railway engineering, they emphasized the need to combine economy in construction and economy in working. Thus the former might demand steep gradients and sharp curves, while the latter demanded minimal gradients and broad curves. Though the solution depended on the amount of traffic (differing for main lines and branch lines) 'a weak and perishable style of construction is never truly economical'. Canals, on the other hand, 'continue to be the most economical lines of conveyance for all articles in whose transport speed is of little importance'. But it was in signalling that the most recent improvements in lines of conveyance had been made:

> that which has superseded all others is the Electric Telegraph. The construction of telegraph lines on land is simple and well known, and has not recently been marked by any great improvement. In long lines of Submarine Telegraph, a difficulty in making signals, arising from the electrostatic charge of the conductor, was predicted from a theoretical investigation by Professor William Thomson, and means of overcoming that difficulty were invented by Mr Whitehouse [cf. below]. For such lines, batteries of great power, and receiving instruments of extraordinary delicacy, are required; and in both respects, the latest step in the march of improvement has been made by Professor William Thomson, as is shown by his instruments having succeeded in transmitting intelligible messages through the damaged Atlantic Cable, when all other means had failed.[18]

Economy mattered most of all for Clydeside engineers, however, in regard to the iron shipbuilding industry. One cause of retarded progress had been 'the practice of imitating the structure of a wooden ship, with keel, ribs and planking; a construction which is the most suitable for timber, but quite unsuitable for iron'. The authors referred to Scott Russell's *Great Eastern* as 'an admirable example of the use in ship-building of the true principles of construction in iron', and they expressed the hope that soon 'all iron ships may be constructed so as to give the greatest strength and capacity with least weight, *and so to realize the great principle of economy*'.[19]

Consideration of iron ships, not just as structures requiring stability and strength, but as machines requiring efficient working, led Rankine, Napier, and Neilson to the subject of machines in general. The efficiency of a machine was its economy of power or energy, part of the available work being lost or wasted in overcoming resistance foreign to the purposes of the machine, and the remain-

[18] Napier *et al.*, 'Report on applied mechanics', p. 223.

[19] *Ibid.*, p. 224. 'Scientific shipbuilding' in the nineteenth century is discussed by Sidney Pollard and Paul Robertson, *The British shipbuilding industry, 1870–1914* (Cambridge, MA, and London, 1979), pp. 130–50.

der being the useful work. Consequently 'the great end of improvement in machines is to diminish the lost work'. This analysis depended directly on William and James Thomson's, as well as Rankine's, formulation of thermodynamics a few years earlier (chs. 9 and 10). At the same time, widespread acceptance of the labour theory of value (seen, for example, in Whewell's writings both on political economy and engineering) provided an important link between efficiency of machinery and economy of expenditure.[20]

The steam-engine took pride of place in the committee's analysis on account of its central role both in Clydeside's growing heavy industries and in the science developed at Glasgow University by Thomson and Rankine. The report took up steam-engine efficiency in three ways. First, the furnace and boiler had to maximize the heat transferred from fuel to steam. Second, the steam had to be utilized in the most efficient way (such as super-heating). Here, the authors noted that while the most economical single-acting pumping engines yielded a duty from one pound of coal of one million foot-pounds, locomotive engines might only yield 0.2 or 0.5 million foot-pounds, while a recent experiment with a marine engine (presumably a compound engine) yielded almost two million foot-pounds. And, third, the 'trains of mechanism' had to diminish lost work by reducing friction and avoiding shocks.[21]

The ubiquitous interest of these Glasgow engineers in economy, with 'the production of every desired effect by those means which are exactly adequate to produce it, and no more' found its most thorough expression in their concern to improve machines, especially marine steam-engines. Their concern belongs very much to the continuing interaction of theory and practice which had led to the creation of thermodynamics at Glasgow. William Thomson himself developed a wider perspective for the York meeting of the British Association in 1881 (its half-century) by presenting a comparative analysis of different kinds of motive power (tides, wind, and rain as well as coal) in terms of their economy. For example:

> The subterranean coal-stores of the world are becoming exhausted surely, and the price of coal is upward bound – upward bound on the whole, though no doubt it will have its ups and downs in the future as it has had in the past, and as must be the case in respect to every marketable commodity . . . [Therefore] it is most probable that windmills or wind-motors in some form will again be in the ascendant.[22]

Thomson weighed, among other cases, tidal power in quantitative costs against steam power and asked whether forty acres producing 100 horsepower from tidal action would be more economical than the value of the land, with the same power produced by a steam-engine occupying an insignificant fraction of that land.

[20] M. Norton Wise (with the collaboration of Crosbie Smith), 'Work and waste: political economy and natural philosophy in nineteenth century Britain', *Hist. Sci.*, (forthcoming).

[21] Napier *et al.*, 'Report on applied mechanics', pp. 225–9.

[22] William Thomson, 'On the sources of energy in nature available to man for the production of mechanical effect', *BAAS Report*, **51** (1881), 513–18; PL, **2**, 433–50, on pp. 441–2.

Improvement of machine economy, however, was only the most explicit exemplification of the drive for economy. The basic programme, articulated by Rankine, and epitomized by William Thomson, sought to fulfil the criterion of economy through the harmony of theory and practice. James Thomson's earlier experience with the unfortunate Horseley Iron Works, which based its engineering upon 'rule of thumb' practice, exemplified what Fleeming Jenkin referred to in another context as 'guess at the half and multiply by two'.[23] The company's excessive waste and rapid liquidation supported the Scottish view of the need to harmonize theory and practice. Significantly, Rankine presented the most direct attack on the separation of theory and practice in his inaugural dissertation read on his appointment to the Glasgow chair of engineering in 1855, part of which address also formed his opening remarks on the objects of the mechanical science section at the 1855 Glasgow meeting of the British Association. In his attack, he explicitly aligned himself with Newton's preface to the *Principia* – 'the errors are not in the art, but in the artificers. He that works with less accuracy is an imperfect mechanic' – and with John Robison's emphasis upon the importance of applying scientific principles to useful arts such as carpentry. In particular, Rankine condemned Thomas Babington Macaulay for promoting a false dichotomy between idealized, rational, dynamical theory and real, empirical practice, leading to a double set of laws.[24]

Rankine went on to systematize his views on the new science of engineering, particularly as it was to be taught at Glasgow University. He distinguished three kinds of mechanical knowledge – purely scientific, purely practical, and that 'intermediate knowledge which relates the application of scientific principles to practical purposes, and which arises from understanding the harmony of theory and practice'. Each form of knowledge was distinguished by its purposes rather than by its conflict with the others. Purely scientific knowledge had as its object, first, 'to improve the mind of the cultivator intellectually and morally; and, secondly, to qualify him, if possible, for assisting in the advancement and diffusion of knowledge; and with this view each subject requires to be treated so as to investigate how the laws of particular phenomena are connected with the general economy of nature and the structure of the universe'. Here, machines are 'looked upon merely as natural bodies are: – namely, as furnishing experimental data for the ascertaining of principles, and examples for their illustration'. Purely practical knowledge, on the other hand, was acquired by experience and

[23] [Fleeming Jenkin], *Papers and memoir of Fleeming Jenkin* (2 vols., London, 1887), **1**, pp. 241–2. The remarks were directed against Darwin's calculations of geological time discussed in Chapter 15 above.

[24] W.J.M. Rankine, 'Preliminary dissertation on the harmony of theory and practice in mechanics', *A manual of applied mechanics*, 17th edn. (London, 1904), pp. 1–11; reprinted in C.A. Russell and D.C. Goodman (eds.), *Science and the rise of technology since 1800* (Bristol, 1972), pp. 266–71; 'Opening remarks on the objects of the [mechanical science] section', *BAAS Report*, **25** (1855), 201–3. See also D.F. Channel, 'The harmony of theory and practice: the engineering science of W.J.M. Rankine', *Technology and Culture*, **23** (1982), 39–52; Buchanan, 'Scientific engineering', pp. 225–7.

observation in everyday affairs of business and industry for the purpose of judging quality of materials and workmanship, and of assessing commercial profit. The procedure was that of following established practical rules – approximation – especially in matters connected with practical mechanical pursuits.

Rankine, however, aimed to promote the third kind of knowledge: 'the advancement of science as applied to practice in the Mechanical Arts' for which both the 'Mechanical Science' section of the British Association and the Glasgow chair of engineering had been established:

> It enables its possessor to plan a structure or machine for a given purpose without the necessity of copying some existing example – to compute the theoretical limit of the strength and stability of a structure, or the efficiency of a machine of a particular kind – to ascertain how far an actual structure or machine fails to attain that limit, and to discover the cause and remedy of such shortcoming – to determine to what extent, in laying down principles for practical use, it is advantageous, for the sake of simplicity, to deviate from the exactness required by pure science; and to judge how far an existing practical rule is founded on reason, how far on mere custom, and how far on error.[25]

Apart from the obvious advantages for 'designers and constructors of great works of mechanical art', Rankine emphasized that the 'mutual dependence and harmony between sound theory and good practice' not only benefited 'the diffusion and appreciation of theoretic knowledge' by impressing the public with the importance of scientific principles through practical application, but also advanced science itself by suggesting problems for scientific investigation, affording data for their solution, or leading to the improvement of the instruments for scientific experiment.

In these addresses and reports, Rankine articulated a programme for professional engineering characteristic of William Thomson's associates, particularly within the Glasgow Philosophical Society. As early as 1852, for example, Rankine and John Thomson (son of Dr William Thomson, professor of medicine) had read a short paper 'On telegraphic communication between Great Britain and Ireland'.[26] The art of telegraphy would soon emerge as the most striking branch of engineering with which to demonstrate the need for the harmony of theory and practice. We should recall that William Thomson, immediately on finding the theoretical equation of signal transmission in submarine cables in 1854, had interpreted the problem of signalling in economic terms through the 'law of the squares' (ch. 13). Addressing the 1855 British Association meeting he insisted that economic benefits in short cables, and indeed the entire success of long-distance submarine telegraphy, depended on attention to theory:

[25] Rankine, 'Opening remarks', p. 202.

[26] W.J.M. Rankine and John Thomson, 'On telegraphic communication between Great Britain and Ireland', *Proc. Phil. Soc. Glasgow*, **3** (1848–55), 265–6.

The theory shows how, from careful observations on such a wire as that between Varna and Balaklava, an exact estimate of the lateral dimensions required for greater distances, or sufficient for smaller distances, may be made. Immense economy may be practised in attending to these indications of theory in all submarine cables constructed in future for short distances; and the non-failure of great undertakings can alone be *ensured* by using them in a preliminary estimate.[27]

On the discord of theory and practice in telegraph engineering

By emphasizing theory in his 1855 paper, Thomson ran counter to the empirical approach of Wildman Whitehouse (*b.* 1815; by profession not an engineer but a medical man) who, in the following year at the British Association's Cheltenham meeting, challenged Thomson's law of the squares. Indeed, Whitehouse's remarks to the 1855 meeting at Glasgow under the title 'Experimental observations on an electric cable' demonstrated vividly his essentially non-mathematical and non-theoretical approach, an approach which emphasized 'facts', observations, and practical experience. They also showed that he regarded the problem of long-distance submarine telegraphy as merely a matter of scaling up the tried and true methods for shorter distances: 'Mr Whitehouse said that he regarded it as an established fact, that the nautical and engineering difficulties which at first existed had already been overcome, and that the experience gained in submerging the shorter lengths had enabled the projectors to provide for all contingencies affecting the greater'. 'And may we not', he added, 'fairly conclude also, that India, Australia, and America, are accessible by telegraph without the use of wires larger than those commonly employed in submarine cables?'[28]

Whitehouse entitled his 1856 address 'The law of the squares – is it applicable or not to the transmission of signals in submarine circuits?', responding directly to Thomson's 1855 remark that 'a wire of six times the length of the Varna and Balaklava wire, if of the same lateral dimensions, would give thirty-six times the retardation'. Whitehouse then related his own experimental researches which he claimed showed 'most convincingly that the law of the squares is not the law which governs the transmission of signals in submarine circuits', but rather that the retardation increased 'very little beyond the simplest arithmetical ratio'. Whitehouse also argued that, if the law of the squares had valid application to submarine circuits, the result in the case of the transatlantic line would:

necessitate the use, for a single conductor only, of a cable so large and ponderous, as that probably no ship except Mr Scott Russell's leviathan could carry it, – so unwieldy in the manufacture, that its perfect insulation would be a matter almost of practical impossibil-

[27] Willliam Thomson, 'On peristaltic induction of electric currents in submarine telegraph wires', *BAAS Report*, **25** (1855), 22; MPP, **2**, 77–8. The full paper appeared in *Proc. Royal Soc.*, **8** (1856), 121–32; MPP, **2**, 79–91.

[28] Wildman Whitehouse, 'Experimental observations on an electric cable', *BAAS Report*, **25** (1855), 23–4. See also SPT, **1**, 330–2.

ity, – and so expensive, from the amount of materials employed, and the very laborious and critical nature of the processes required in making and paying it out, that the thing would be abandoned as being practically and commercially impossible. If, on the other hand, the law of the squares be proved to be inapplicable to the transmission of signals by submarine wires . . . then we may shortly expect to see a cable not much exceeding one ton per mile containing three, four or five conductors, stretched from shore to shore, and uniting us to our Transatlantic brethren, at an expense of less than one-fourth that of the large one above mentioned, able to carry four or five times the number of messages, and therefore yielding about twenty times as much return in proportion to the outlay.[29]

Whitehouse's eloquent address was clearly intended to convince the commercial interests involved in the formation of the new Atlantic Telegraph Company. He ended his emphatic denial of the law of the squares with a stress upon the desirable divorce of theory and practice. Like Macaulay, Whitehouse stressed the distinction between ideal, rational principles of theory and real, practical rules of working which alone were of value for commercial and technical enterprises. He sounded a note which would have both amused and angered Rankine and Thomson:

And what, I may be asked, is the general conclusion to be drawn as a result of this investigation of the laws of the squares applied to submarine circuits? In all honesty, I am bound to answer, that I believe nature knows no such application of that law, and I *can only regard it as a fiction of the schools, a forced and violent adaptation of a principle in physics,* good and true under other circumstances but misapplied here.[30]

Whitehouse's attack prompted a debate with Thomson in the pages of *The Athenaeum*. Writing from Invercloy, Isle of Arran, on 24th September, 1856, Thomson made it clear that the confidence he placed in his own conclusions was a confidence in theory:

Mr Whitehouse's communication not only professes to overturn my theoretical conclusions, but it gives what might at first sight appear to be sufficient experimental evidence of the validity of an ordinary submarine cable for telegraphic communication between this country and America, in opposition to my warning that more than ordinary lateral dimensions of wire or insulating coat might be necessary to allow sufficient rapidity in the communication of intelligence through a conductor so much larger than hitherto used in practical operations. I therefore think it right to say, that all Mr Whitehouse's experimental results are perfectly consistent with my theory; but at the same time I wish it to be understood that my ground for saying so, is . . . knowledge of the theory itself which, like every theory, is merely a combination of established truths.[31]

In this debate, as in many later episodes concerned with his promotion of scientific engineering, Thomson had to defend himself against what seemed a *prima facie* failure of theory. Since practical tests apparently showed him wrong,

[29] Wildman Whitehouse, 'The law of the squares – is it applicable or not to the transmission of signals in submarine circuits', *BAAS Report*, **26** (1856), 21–3.

[30] *Ibid.*, p. 23. Our emphasis.

[31] William Thomson, *The Athenaeum*, 4th October, 1856; MPP, **2**, 92–3.

he could only attack the tests, a task which in the present case presented no difficulty. Whitehouse had not actually studied Thomson's theory nor had he carefully interpreted Thomson's published graphs of the variation of signal strength with time at the remote end of a cable, but had merely treated the law of squares as a simple result which represented the entire theory and which was subject to a crude test, viz., apply a strong signal to one end of a cable and see how long it takes to reach the other end. Thomson's graphs showed that the result would depend sensitively on how the signal was sent (that is, on how long the wire was connected to a battery of constant strength before sudden grounding) and at what fraction of its maximum strength the signal could be counted as received. In addition, any electromagnetic induction in the receiving indicator or meter would retard the indication. Taking account of none of these subtleties, Whitehouse simply connected a powerful battery (of decidedly non-constant voltage) and measured, for his particular indicator (with large self-induction), the time taken to respond.

With more than a hint of condescension, Thomson calmly but firmly replied to the attack, pointing out the sources of discrepancy, the necessity for carefully controlled experiments, and the fact 'that in the one case of comparison in which a manifestation of the law of squares could be expected, that law is manifested by Mr Whitehouse's results'.[32] In response to Whitehouse's attempt to discredit his claim that a much larger copper conductor or a number of smaller wires twisted together would allow a faster rate of signalling, Thomson pointedly cast doubt on Whitehouse's integrity:

Mr Whitehouse tests this proposal by three wires connected at their ends, so as to afford a triple conducting channel, *but separated throughout their lengths by their gutta-percha coats.* Now it is perfectly clear, that an electrical impulse through three wires so arranged, cannot differ from the sum of three separate electrical impulses of one-third strength each through any one of the wires, except in virtue of peristaltic induction [ch. 13]. With these very wires, Mr Whitehouse found the effects of mutual peristaltic induction to be extremely slight, requiring some of his finest tests to be shown at all. How then he found 'the retardation nearly twice as great with the triple wire as with one of the wires alone', is an anomaly which is not my part to explain.[33]

Having defended himself on all counts, Thomson ended with an implied challenge to 'those engaged in projecting the Ocean Telegraph'. He announced that he would publish a table relating different dimensions of cables to their estimated signalling rates which the planners might use 'if they have any confidence in scientific deductions from established principles'.[34] The organizers of the project apparently decided not to pit Whitehouse's practical knowledge against theoretical design requirements, for a few weeks later Thomson found that the dimensions he had recommended coincided with those

[32] William Thomson, *The Athenaeum*, 1st November, 1856; MPP, **2**, 94–102, on p. 98.
[33] *Ibid.*, p. 99. [34] *Ibid.*, p. 102.

adopted in a specimen of cable submitted to him for approval. Soon afterwards he himself became a director of the Atlantic Telegraph Company.[35]

Yet the controversy with Whitehouse was not an isolated debate, attributable merely to the differing personalities of the protagonists. Rather, the controversy had its roots at a deeper level. The discord of theory and practice was not confined to the academic discussions of a Rankine or a Thomson, but appears to have existed generally in the minds of the practical telegraph engineers. Nowhere is this characteristic better illustrated than in a series of letters exchanged between William Thomson and his former colleague, Lewis Gordon, now engaged in telegraphic enterprises through the cable-manufacturing firm of R.S. Newall and Company of Birkenhead. The correspondence took place just prior to the first attempt to lay an Atlantic cable in the summer of 1857 and concerned the 'return metallic circuit', in which two wires, insulated from one another, were placed within the same iron sheath and operated one from the negative and the other from the positive side of a battery, the far ends being grounded. Werner Siemens, the famous telegraph engineer, and others thought that they could by this means reduce or eliminate inductive retardation.

In the first letter of the series, Thomson expressed the hope that Gordon, at least, was

> free from the most vain fallacy of the 'return metallic circuit' as diminishing induction in a two wire submarine cable. I heard this morning with astonishment Siemens is not. If you care I can in two minutes at any time show you exactly how much (it is not *very* much) the capacity of the positive wire is increased by working the other negatively at the same time and *vice versa*.
>
> I should be very glad however that the class of practical men whom no reasoning affects should have a demonstration on a large scale: because as soon as they had worked their return metallic circuit for a week, they might next give the two wires a turn of working, both positively or both negatively at the same time. They would get on *rather* faster (not much) and then by the end of another week they might begin to work 'metallic circuit', and parallel currents simultaneously, giving one message by the one method and another independently by the other either in the same or in the opposite direction. Thus two wires under one sheath could be used to convey two independent messages simultaneously.[36]

Lewis Gordon replied from the Submarine Telegraph Works at Birkenhead with some irony: 'you have to purge me of a fallacy which I have accepted chiefly on a series of large scale experiments conducted by Werner Siemens and which prove that when two conductors insulated or isolated from each other are placed as near together [as possible] in the centre of the same insulating cover of gutta percha the charge or retardation is diminished by about $\frac{3}{4}$'.[37]

[35] 'Minutes of evidence taken before the Submarine Telegraph committee', *Parliamentary Papers*, **62** (1860), 111. See also SPT, **1**, 337–40.

[36] William Thomson to Lewis Gordon, 23rd May, 1857, G131 (copy), ULC.

[37] Lewis Gordon to William Thomson, 24th May, 1857, G132, ULC.

Once again, Thomson's theoretical views seemed obviously false in the face of an actual trial. Gordon graciously conceded that on Siemens's theory of the metallic circuit, there should have been 'no induction at all and that the retardation would be no more than that due to the resistance of the copper wire'. So the reduction by only three-quarters disproved their theory as well.

Before entering on the experimental result, let us first examine the idea of electrostatic action that led the telegraph engineers to suppose there would be no induction if the wires were worked oppositely. It seems that they regarded induction in a single-wire cable not as an action *between* the wire and the sheath but as an action of the charged wire *on* the originally neutral sheath. Two wires charged oppositely and simultaneously running down the centre of the sheath would thus produce no net inductive action on it. Furthermore, no action of either wire on the other would occur because each would already possess the charge that the other would have induced on it. Conduction along each wire would therefore occur without any retardation corresponding to the time required for induction.

Thomson registered astonishment: 'I do not wonder that the experiments proved the theory of no induction . . . to be erroneous. I am only surprised at its not being seen to be erroneous before the experiments were made'.[38] From his point of view, since the capacity for induction in a single-wire cable was a relation between the wire and the sheath (depending on geometry and the dielectric material), and since the double-wire arrangement did not alter that relation, the second wire could not alter the inductive capacity of the first with respect to the sheath. The metallic circuit split the original inductive action into two single-wire actions of half intensity while adding a peristaltic inductive action between the two wires tending to increase the charge of both, and thus increase the retardation.

If, instead of working the two wires oppositely, Thomson explained, one operated them together, this peristaltic induction would tend to decrease their respective charges and thus decrease the retardation. Siemens's finding of decreased retardation on working oppositely could not derive from electrostatic effects. Electromagnetic action, Thomson suggested, might have been the cause. A week of reflection and calculation left him 'not a doubt, or scarcely a doubt' of the truth of this conjecture.[39] Over the two or three miles of sample single-wire cable tested by Siemens, electromagnetic self-induction would seriously retard the signal current since the pulse would have a very steep leading edge and since the effect depended on the rate of change of the current. After a few miles, however, the leading edge would flatten out, eliminating further retardation.

In a short, two-wire cable worked in 'metallic circuit', mutual electromagnetic induction between the oppositely directed currents would largely offset

[38] William Thomson to Lewis Gordon, 25th May, 1857, G133 (copy), ULC.
[39] William Thomson to Lewis Gordon, 2nd June, 1857, G130, ULC.

the self-induction, yielding Siemens's result: 'The effects . . . are undoubtedly, as is clearly shown by Faraday's original experiments on the subject, very much diminished if not reduced quite to zero, by the arrangement experimented on by Werner Siemens which is in fact *precisely one of Faraday's* arrangements'.[40]

Faraday, indeed, would have had as little trouble as Thomson analyzing, qualitatively, the entire problem. And anyone thoroughly at home with potential theory, or with Thomson's heat analogy, would soon have come to his conclusion for the electrostatic effects. Yet philosophic electricity remained a subject alien even to the best practical electrical engineers. Gordon and Siemens, unlike the blundering Whitehouse, belonged with the very best, but even they did not approach their work from a thoroughly articulated theoretical perspective. Thomson saw telegraphy as an integral part of mathematical physics. There could be no divorce between theory and practice, no 'double set of natural laws', no idealized rational science to set against real, practical engineering, in which approximate working rules, far more valuable for practical purposes than rigorous deductions from theory, would be the order of the day.

The telegraph project added to Thomson's verdict an economic imperative that could not be ignored by commercial interests. Gordon had written to Thomson: 'If you can show me that [Siemens's metallic circuit] is . . . fallacious . . . you will greatly disappoint me but also do me a great kindness for we are now projecting experiments that will cost 4 or 5000 pounds to decide the *real* practical value of the system'.[41] Contrite when Thomson's analysis promised to save his company such a large and useless expense, Gordon wrote that he could 'only thank you sincerely for your letters on the subject of the Metallic Circuit . . . I am very much inclined to adopt your opinion simply on your Authority'.[42]

Such incidents demonstrated forcefully to Gordon and his colleagues that, for the engineer, economics and natural philosophy were not separate subjects. He wrote again to Thomson within a month to inform him:

> We are sadly at a loss for a Philosophic assistant who is at [the] same time a practical man; or who could readily become one. We could afford to pay him well and to give him large scope for experiments. Can you help us[?] If you would like us to supply *you* with means for experiments we should be very happy to supply them on the natural principle of the industrial results being for our advantage. But Newall would like to have a skilful experimenter & suggester beside him and if you know an aspirant philosopher and experimentalist let us hear of him.[43]

Science-based industry had here made an explicit appearance, complete with the emphatic priority of scientific education over practical experience in the

[40] *Ibid.* [41] Gordon to Thomson, *op. cit.* (note 37).
[42] Lewis Gordon to William Thomson, 3rd June, 1857, G134, ULC.
[43] Lewis Gordon to William Thomson, 3rd July, 1857, G136, ULC.

training of technical personnel, well-paid employment as the reward for such training, and finance from industry to fund the research laboratories of university professors. This nascent scientific–industrial complex would extend in Thomson's case to a personal share in the rewards of telegraphic industry through numerous telegraphic patents. Gordon in his letter agreed on behalf of Newall and Company to 'the important proposition you make as to our patenting your Telegraph inventions for you, under an arrangement that you should have a share of any profits that may accrue or have the right to give the use of them to the Atlantic Telegraph Company gratis'. The great venture of attempting to lay the first Atlantic telegraph was about to begin, as Gordon reported on 3rd July, 1857:

> The whole of our part of the Atlantic Co. cable will be shipped from our premises this week. They are getting it very well on board the *Niagara*, and by the end of the month or sooner I suppose the expedition will set sail. Are you going in either vessel?[44]

The making of Sir William Thomson

The technical problems challenging the Atlantic telegraph engineers were both enormous in scale and manifold in number. A 2000 mile stretch of stormy and unpredictable North Atlantic Ocean, in parts some three miles in depth, had to be crossed by the cable laying ships. The vessels, necessarily steamships of a breed still relatively primitive, had to be large enough to carry the cable, and had to be fitted with paying-out machinery adequate for the immense strains involved. When one reflects that the first Cunarder, an 1100 ton paddle steamer smaller than many coasters of the mid-twentieth century, had crossed the Atlantic as recently as 1840, one gains some perspective on the courage and confidence of these Victorian telegraph entrepreneurs.[45]

Apart from the retardation effects peculiar to the cable itself, and the technical problems of actually manufacturing the cable, the weather posed the greatest challenge. The late spring and summer would be the only possible period, but even then the Atlantic was notorious for its frequent eastward moving depressions, bringing gale-force winds in the best of summer seasons. Furthermore, these storms often produced great, heaving swells, affecting calmer seas many miles away, swells which could easily bring a cable ship to a sudden stop and then send her forward in a violent surge. The chance of a slender cable about half an inch in diameter surviving the strain of such erratic movements was poor indeed. However, as Thomson noted to Helmholtz at the end of December, 1856: 'the practical men have all the experience of previous

[44] *Ibid.*

[45] Among many accounts of the Atlantic telegraph, three are particularly useful: W.H. Russell, *The Atlantic telegraph (1865)* (reprinted Newton Abbott, 1972); Bright, *Submarine telegraphs*, pp. 23–56, 78–105; Bern Dibner, *The Atlantic cable* (New York, 1959). See also SPT, **1**, 325–96, 481–508.

failures, and it is to be hoped have learned some of the causes and will know how to avoid them. Altogether, I think there is a good chance of success'.[46]

Preparations for the Atlantic cable had been under way since 1854 when Cyrus W. Field's syndicate of the New York, Newfoundland, and London Telegraph Company had acquired rights to land cables in Newfoundland. Ocean surveys had revealed a 'telegraph plateau' extending across the ocean as a bed of soft mud made up mainly of microscopic 'infusoria' and gently undulating in depths of 1700 to 2400 fathoms. The engineers decided upon a course from Valencia Island in South West Ireland to Trinity Bay in Newfoundland, and by 1856 both island countries had been linked into their respective British and American networks of telegraph lines. The Atlantic Telegraph Company was registered in the same year with an initial capital of £350 000, the merchants and shipowners of Liverpool being the most enthusiastic subscribers to the 350 ordinary shares of £1000 each. The chief engineer to the new company was Charles Bright, the electrician was Whitehouse, and the secretary was George Saward. The British Government, whose blessing may have derived from the fact that the main cable would link the British Isles with British North America rather than the United States, guaranteed £14 000 per annum to the company, and promised ships to assist in the laying of the cable. The US government, rather reluctantly, gave a similar guarantee.[47]

In the short space of four months, 2500 miles of cable, weighing about one ton per nautical mile, were manufactured, the copper conductor and gutta-percha insulation being produced by the Gutta Percha Company of London, and the sheathing by Glass, Elliot and Company of Greenwich and Newall and Company of Birkenhead at a total contract price of £225 000. Meanwhile, HMS *Agamemnon*, a 3200 ton, twin-screw, ninety-one gun 'battleship', flagship of Admiral Lyons at the bombardment of Sebastopol during the recent Crimean War of 1854–5, prepared to load half of the cable at Greenwich, while the 5000 ton United States naval 'frigate' *Niagara*, her yacht-like lines contrasting with the solid bulk of the *Agamemnon*, picked up the other half at Birkenhead.[48] The great expedition made ready to weigh anchor in the summer of 1857, with William Thomson aboard the British ship.

Bright and Whitehouse differed on the most effective strategy for laying the cable. Because of the time and weather factors, Bright favoured a mid-ocean starting point. Under such conditions, the time of laying would be halved, and the initial splicing could be carried out when the weather proved suitable. Whitehouse, on the other hand, who avoided going to sea, regarded the advantage of continuous communication with land as most important, and his views prevailed. On 6th August, 1857, the *Niagara* began laying from the Irish coast at a speed of two knots, later increased to five. After only minor hitches, the

[46] William Thomson to Hermann Helmholtz, 30th December, 1856, in SPT, **1**, 335–7.
[47] Bright, *Submarine telegraphs*, pp. 28–33. [48] *Ibid.*, pp. 33–7.

cable snapped on 11th August when sudden braking occurred as the *Niagara* plunged into a heavy head sea. The distance run had been 274 nautical miles. With the summer already well advanced, the ships returned to unload the remaining cable for winter storage at Plymouth.[49]

By the spring of 1858, several developments had occurred. Bright and his team had entirely changed the paying-out machinery. An extra length of heavier cable had been manufactured to replace that lost in 1857, and to provide for additional losses. A decision had been taken to commence laying in mid-ocean, with a period of preliminary trials to be undertaken in the Bay of Biscay. And, of particular practical value, William Thomson provided a new testing and receiving instrument, his first marine galvanometer, whose extremely sensitive reflecting mechanism was capable of detecting very weak signals.[50]

After an extremely violent storm on passage, during which the *Agamemnon* came close to foundering and taking William Thomson with her, the ships rendezvoused in mid-ocean during June, 1858. Some 500 miles or more of cable was then lost as a result of three partings, and the ships returned to Queenstown (now Cobh) in Ireland for supplies. But then, in only one week, between 29th July and 5th August, the two vessels finally laid the first transatlantic cable, even though the unfortunate *Agamemnon* encountered head winds throughout most of her eastbound passage. With the initial communications between the Old World and the New, the celebrations were spontaneous and brilliant, not least the setting on fire of New York's Town Hall during a torchlight procession. At the age of twenty-six, Charles Bright received his knighthood, and *The Times* observed that 'since the discovery of Colombus, nothing has been done in any degree comparable to the vast enlargement which has thus been given to the sphere of human activity'.[51] The President of the United States, James Buchanan, in his reply by telegraph to Queen Victoria's congratulatory message (which had taken sixteen hours to transmit) also captured the emotional appeal of the technical achievement by two nations whose relations since the War of Independence had not always been so warm:

> The President cordially reciprocates the congratulations of Her Majesty the Queen on the success of the great international enterprise accomplished by the science, skill, and indomitable energy of the two countries. It is a triumph more glorious because far more useful to Mankind than was ever won by conqueror on the field of battle. May the Atlantic Telegraph under the blessing of Heaven prove to be a bond of perpetual peace and friendship between the kindred nations and an instrument destined by Divine Providence to diffuse religion, civilization, liberty, and laws throughout the world.[52]

New York's mayor, in spite of the accident to his Town Hall, regarded the success of the cable as a triumph of science over time and space which would

[49] *Ibid.*, pp. 38–41. [50] *Ibid.*, pp. 41–4. [51] *Ibid.*, pp. 44–50.

[52] President James Buchanan to Queen Victoria, 16th August, 1858, A26, ULG; Russell, *The Atlantic telegraph*, pp. 26–7.

unite 'more closely the bonds of peace and commercial prosperity, introducing an era in the world's history pregnant with results beyond the conception of a finite mind'. And the humble folk of Sackville, New Brunswick, no longer felt 'as distant colonists, but that we actually form a part of the glorious British Empire – God save the Queen!'[53]

The triumph was short-lived. The signals rapidly became confused and often unintelligible, communication being maintained only by Thomson's mirror instruments. Having transmitted a total of some 732 messages, the cable finally ceased to function in October, 1858. Congratulatory messages could no longer be exchanged. As Captain Moriarty of the *Agamemnon* expressed the sad news in a letter to William Thomson on 4th October: 'I am really very sorry for the poor Atlantic Cable but fear it will never speak again'. And it never did. The first transatlantic cable was dead. The direct cause of death was almost certainly faulty insulation, due in turn partly to manufacturing weaknesses, partly to inexperience in handling and storing the cable, partly to the abuse it had received during the *Agamemnon*'s June ordeal, and perhaps above all to Whitehouse's over-zealous attempts to send messages through it with five foot long, 2000 volt induction coils. The Atlantic Telegraph Company made Whitehouse the scapegoat for the cumulative failures, and gave him the sack as company electrician when he violated orders not to attempt any drastic remedies on his own initiative.[54]

Although Whitehouse and Thomson had generally been on friendly terms since their debate in the *Athenaeum*, Whitehouse's claim to have employed his own relay patents to receive signals from America, whereas all the while he had been using Thomson's instruments, finally ruptured their relations.[55] Even apart from this direct clash, the two men were of such differing outlooks that divergence of opinion seemed always inevitable. Whitehouse, as we have seen, emphasized the virtue of practice over theory, tending to regard practical difficulties as obstacles to be overcome by sheer perseverance and relentless attack which at times seemed little more than blind brute force. Rather than admit the advantages of Thomson's delicate instruments, Whitehouse persisted in trying to force the cable to operate with his own heavy patents. Again, Whitehouse had promised that he could get seven words per minute through the cable with his apparatus. He achieved one word a minute when the cable was under test at Devonport, and none at all when the cable was laid.[56] Whitehouse's rejection of theory – or at least his failure to grasp theory – thus led him to adopt an empirical approach which, though successful enough with small-scale

[53] D.H. Tieman (Mayor of New York) to Sir Walter Carden (Lord Mayor of London), 21st August, 1858, A39, ULG; Sackville, New Brunswick to Sir W. Carden, 4th September, 1858, A56, ULG.

[54] Bright, *Submarine telegraphs*, pp. 50–4; SPT, **1**, 367–86; Captain Moriarty to William Thomson, 4th October, 1858, M167, ULC.

[55] SPT, **1**, 374–7.

[56] William Thomson to J.P. Nichol, 18th March, 1859, N30, ULC.

Receiving messages from the *Great Eastern* in the instrument room of the telegraph house on Valencia Island, south-west Ireland. Here Whitehouse's use of induction coils had probably shortened the life of the troubled 1858 cable, while in the same setting Thomson's marine mirror-galvanometer had begun its distinguished career. [From the *Illustrated London News*.]

projects, proved disastrously expensive when used on the Atlantic cable. To ignore theory on such a project was a luxury which simply could no longer be afforded. The 2000 mile cable was not a cheap, expendable piece of apparatus, but an extremely valuable commercial property.

Thomson, for his part, continued to maintain his stand on the harmony of theory and practice. Writing to J.P. Nichol a few months later, he claimed that 'as to the Whitehouse controversy I never yielded a single point but on the contrary affirmed on every occasion, in the most uncompromising terms, the truth of each prop^{n} w^{h} had been shown to be a consequence of the theory. That I was right on each point is now proved in point of fact'. Thomson added that, as he became acquainted with Whitehouse, he found 'how to explain his experiments & to show that his conclusions did not follow from them: but I assented too readily to his own sanguine expectations of what his instruments could do'. Indeed, he confessed that Whitehouse's confident and undoubting answers to some of Thomson's fundamental questions had made even Thomson think that 'he might possibly work with advantage by his instruments, although I never believed he could get the seven words a minute w^{h} he promised'. For this reason Thomson had been hesitant about openly condemning Whitehouse prior to the direct clash brought about by Whitehouse's failure to admit using Thomson's instruments. Yet Thomson had been privately doubtful of Whitehouse's promises for some time:

> Even till the end of Feb[ruary] last year [1858] he [Whitehouse] was confidently promising in printed report 4 words a minute. Here we were reading in the newspapers of 5 words a minute till suddenly in a letter to the Directors he congratulated us that not less than one word a minute was secure! Then, & not till then, I began most reluctantly on my reserve: because I was much better pleased to let the practical telegraphic work be carried out by others.[57]

Thomson's hesitation did not find favour with the other anxious directors of the Atlantic Telegraph Company. They saw no virtue in sparing Whitehouse's feelings. C.M. Lampson wrote to Thomson, then resting at Invercloy before the College session of 1858–9, on 22nd October, 1858: 'I must not hide from you that the course you took in relation to our recent difficulties with Mr Whitehouse added greatly to our troubles at a most critical period of the Company's affairs & I am therefore much pleased to find that you are at length convinced that we acted wisely in dismissing Mr Whitehouse'. Lampson's condemnation of Whitehouse was wholehearted, as he added: 'You will perhaps remember that I expressed my views very firmly when at Plymouth; subsequent experience has shown most clearly that this great undertaking has been jeopardised & perhaps ruined by at first placing the electrical department in the hands of a man so inefficient, selfish & unscrupulous'.[58] After a costly failure, such recriminations and *post mortems* were inevitable. The autumn of 1858 was a time of great disappointment for all concerned in the project. Thomson's delight turned to boredom, as he wrote to Joule in late September:

> Instead of telegraphic work which, when it has to be done through 2400 miles of submarine wire, and when its effects are instantaneous exchange of ideas between the old and new worlds, possesses a combination of physical and (in the original sense of the word) *metaphysical* [i.e. mathematical] interest, which I have never found in any other scientific pursuit – instead of this, to which I looked forward with so much pleasure, I have had, almost ever since I accepted a temporary charge of this station [Valencia], only the dull and heartless business of investigating the pathology of faults in submerged conductors.[59]

The Company dispensed with the services of most of its employees by 30th November, since it was at present in no financial state to undertake a fresh attempt. The directors recognized, however, that a future attempt would rely increasingly on the advice of William Thomson, and they rejected totally the authority of Whitehouse as a discredited force:

[57] *Ibid.* [58] C.M. Lampson to William Thomson, 22nd October, 1858, L9, ULC.
[59] William Thomson to J.P. Joule, 25th September, 1858, in SPT, **1**, 378–9.

William Thomson's mirror-galvanometer as employed at Newfoundland on the 1858 Atlantic cable. [From The Science Museum Library, London.]

They [the Committee of the Company] say that they think the directors *as a body* had better not have anything further to do with Whitehouse as he will not be convinced by anything they may say; and that their collective statement upon a scientific subject would not be regarded. They however think that it would be perfectly correct on your part to disabuse the public upon the scientific errors into which Whitehouse has fallen if you do not object to the trouble. Your exalted position in science would give force and effect to what you thought proper to say – which would not be the case if the observations emanated from any other Person in the Company.[60]

At the same time, William Bottomley expressed the hopes that the Government would guarantee the interest on the capital of the Company and that the public could be convinced of the importance of the project both politically and commercially. He added that 'even as a speculative question it is interesting to consider the *effect* of the Telegraph between Gt Britain & America'. For instance: 'the effect certainly would be the equalizing of prices and the restraining of speculation. The stock of cotton for example in America & this country would come to be regarded as one general quantity, the Manchester manufacturers would cease to depend so much as they do on the Liverpool merchants, and the latter would be unable to produce those frequent fluctuations in prices according to the accidental demand which affect the Lancashire spinners so injuriously'.[61]

[60] George Saward to William Thomson, 9th October, 1858, A109, ULC.
[61] William Bottomley to William Thomson, 19th October, 1858, B275, ULC.

Thomson himself contributed to maintaining public interest during the banquet given in his honour by the citizens of Glasgow on 20th January, 1859. In his speech to the assembled dignitaries he spoke of the success and failures of the past year's telegraphic endeavours, and of the 'astonishing result of science, from which no degree of familiarity can remove wonder', namely the transmission of messages over long distances. Thomson's speech was published in the *Glasgow Herald*, a paper then, as now, with a wide readership, particularly among the commercial parts of Scottish society.[62] Some attempts were therefore being made to rebuild the ruins of the 1858 venture.

The ruins, however, were widespread. Most of the Directors of the Atlantic Telegraph Company resigned, apparently leaving C.M. Lampson to pay out of his own pocket the secretary, clerks, and office expenses until 1864, when the revival of the enterprise became more certain.[63] Meanwhile, the old Company was in general unable to repay in full all claims upon it. Some of the assets were or had been sold, although of course the largest asset of all was irrecoverable. Cyrus Field, for instance, had sold to the New York jewellers, Messrs Tiffany & Company, twenty miles of the cable upon the arrival of the *Niagara*. Field reported to Thomson that, on examination, the copper wire in parts of the cable had either separated from or been forced through the gutta-percha insulation, and he asked for Thomson's opinion regarding the failure of the 1858 cable.[64]

Thomson replied to Field's queries in a lengthy letter dated 2nd July, 1859. His remarks subsequently formed the basis of his evidence on 17th December to the Submarine Telegraph Committee consisting of a Committee appointed by the Board of Trade (Douglas Galton, Charles Wheatstone, William Fairbairn and G.P. Bidder) and a Committee of the Atlantic Telegraph Company (Edwin Clark, C.F. Varley, Latimer Clark and George Saward). In his letter to Field, Thomson attributed the defect in cable insulation to three possible causes – 'bending the core in the gutta percha works too soon after leaving the machines, or afterwards while heated by workmen engaged either in repairing abrasions or in making joints'; allowing portions of the cable to remain too long in the hot pitch bath; or exposing parts of the cable to the sun, as was the case at Greenwich. In Thomson's opinion, the penetration of the gutta-percha by the copper core had led to the failure of the cable. Portions of the copper conductor had come into contact with the moist tarred yarn outside the gutta-percha, or been exposed to seawater percolating through the yarn. Thomson then outlined the fateful course of events:

> Defective insulation, which must have been occasioned by flaws of this kind admitting water, began to appear to an alarming extent before the ends of the cable were landed. During the first 4 days we received at Valencia scarcely the slightest indication of a signal

[62] Thomson, *op. cit.* (note 7).
[63] Sir George Lampson to William Thomson, 12th July, 1897, L10, ULC.
[64] C.W. Field to William Thomson, 29th June, 1859, A111, ULC.

. . . On Monday the 9th August, the insulation proved so bad to tests at the Valencia end, that even if it should not become worse, it seemed doubtful whether it would be possible to work through the cable. About midnight following that day messages began to be received at Valencia, which altho' they were read with care on the mirror instrument, could not be received by relay in consequence of the weakness of the signal currents which allowed them to be continually overborne by irregular earth currents. After a few days more, it came to be only at times that messages could be read and there began to be longer and longer intervals during which no sensible effect of the signal currents could be discovered by the most delicate instruments. The gradual deterioration of the insulation which thus gave rise to the ultimate failure of the insulation could not but be accelerated by the effect of the signal currents escaping at the leak or leaks, and in doing so melting away the ragged edges of the gutta percha at the places of contact betw[een] water, copper & g.p. so as to expose a larger area of copper to the water.[65]

With regard to the improvement of signalling in a future Atlantic cable, Thomson urged Field above all that 'every possible means should be taken in the manufacture to ensure as high a standard of conductivity (of the copper conductor) as possible, as the speed ultimately obtained will depend very much on this'. Again, a good speed of working could be achieved with sensitive receiving and recording instruments, in which business Thomson was rapidly becoming an expert. In due course his display of technical competence in instrument design was to impress his contemporaries no less than his ability as a mathematical physicist, yet for him harmony would always obtain between the theories of mathematical physics and the design of instruments (chs. 20 and 22).

In April, 1861, the Submarine Telegraph Committee reported on 'the best form for the composition and outer covering of submarine telegraph cables'. The first part of the Report, 'an account of the principal telegraph lines which have been laid', presented detailed analyses of the stark fact that of over 11 000 miles of submarine telegraph lines laid since 1851, little more than 3000 miles were actually working. The length of failed lines had been greatly enhanced by the 2200 miles of the Atlantic cable and the 3500 miles of the Red Sea and India telegraph of 1859–60. Most of the casualties had occurred indeed with deep-sea cables, defined in the Report as lines laid at depths greater than about 100 fathoms.[66]

Judging from its membership and from the manner and content of its questioning, the Committee had been established with the conscious intent of formally discrediting Whitehouse and, to a considerable degree, of making him the scapegoat for the failure. The purpose of such an exercise would not only have been to allay the fears of potential public investors that any future Atlantic cable would be doomed, but more importantly to secure the technical basis for future success by discovering where past inadequacies could be remedied. The

[65] William Thomson to C.W. Field, 2nd July, 1859, A112 (draft), ULC.

[66] 'Report of the Submarine Telegraph Committee', *Parliamentary Papers*, **62** (1860), v–xxxvi, esp. pp. v–vii.

testimony thus ranged over all aspects, mechanical and electrical, of the design, manufacture, testing and laying of submarine cables. The Committee succeeded unequivocally in exposing the weaknesses of purely empirical telegraph engineering and in consolidating, even institutionalizing, the new status of theoretical design, laboratory testing, and precision measurements.

Of particular interest are the testimonies of Whitehouse and Thomson. The examiners elicited from Whitehouse a thorough statement of his trial-and-error techniques which, while intended as a self-justification, served as a condemnation both of himself and his methods when juxtaposed with Thomson's testimony on basic theory and sensitive measurements.

In its very first question the Committee required Whitehouse to identify himself as an amateur: 'What is your profession? – I am a member of the College of Surgeons, but not now practising; lately I have devoted myself to electro-telegraphy, and that must be called my profession'. By contrast, the Committee allowed Thomson to agree to his expert status: 'You are professor of natural philosophy in the University of Glasgow – Yes'. With identities established, credentials followed. Whitehouse had conducted practical experiments on 'electrical questions' for eight or nine years; Thomson had communicated his work on the 'science of electricity' to the Royal Society.[67]

The most telling difference between Whitehouse and Thomson emerged, however, in descriptions of their actual work. Concerning factors affecting velocity of transmission, for example, Whitehouse stated:

> I experimented upon different forms of current, and upon different modes of using the same form. I mean the voltaic current as continuously employed by interruptions or repetitions of similar currents, or otherwise, alternately, a positive and a negative. Then I tried the magneto-electric current from the induction process, and from permanent magnets also. My object was to work out, as far as I could practically, the best form of instrument for use in submarine lines, the best form of current and the best mode of using it.[68]

In comparison to this see-if-it-works approach, Thomson described in precise language the results of his mathematical theory: 'The capacity of the conductor for charge, or the electrostatic capacity as it is technically called, which influences most seriously the rate of signalling through it, I shewed to depend really on the ratio of the diameter of the gutta-percha to the diameter of the copper . . . The mathematical expression which I have found for the electrostatic capacity per unit of length is the specific inductive capacity of the gutta-percha divided by twice the Naperian logarithm of the ratio of the diameters. I concluded also that the rate of charging and discharging the cable must be proportioned to the square of the length.'[69]

From this point onwards, while the examiners solicited a technical education from the professor, they often drove the practical man into corners from which

[67] 'Minutes of evidence', *op.cit.* (note 35), pp. 69, 110 [68] *Ibid.*, p. 70. [69] *Ibid.*, p. 110.

he could do little but embarrass himself in weak defence of his opinions. Cromwell Varley (1828–83), engineer of the Atlantic Telegraph Company (for whom Thomson had high praise and whom we shall meet again as his business partner), forced Whitehouse to admit that his five foot long induction coils would produce a spark of a quarter of an inch in the air lasting for a considerable portion of a second and producing so severe a shock in anyone unfortunate enough to suffer it that 'they have been almost ready to faint'. Whitehouse claimed that this discharge would do little damage to the cable in comparison to a Daniell's battery (Thomson's preferred signal source). Having established that a Daniell's battery of 400 cells could not produce a spark over $\frac{1}{200}$th of an inch, Varley asked Whitehouse to explain why the battery, acting for the same time as the coils, would produce more damage. 'I believe the difference to be due to the incomparably greater heating power of the Daniell', responded the miserable witness.[70]

Less entertaining, but of great importance for the developing relation between the telegraph industry and laboratory research, was Thomson's evidence concerning testing procedures. Whitehouse's test of a cable had amounted to little more than attempts to get a signal through by brute force. Thomson wanted to know instead the conductivity of the copper, the inductive capacity of gutta-percha, the precise resistance of any part of the cable, and so on. He had found in his laboratory, to his own and everyone else's horror, that different samples of commercial copper differed in conductivity by 2:1, implying from his theory a corresponding message (and *profit*) ratio.

For the diagnosis of faults in the cable during both manufacture and laying, Thomson advocated in the strongest terms that standardized resistance measurements be continuously made:

> I consider that there was a very great omission in the apparatus on board in the want of standard resistance coils. I had urged on the electrician of the company . . . the very high importance of having a set of resistance coils properly made . . . admitting of variation to the smallest measurable quantity . . . but [he] had his own system of testing which he considered satisfactory.[71]

Now, however, Thomson was talking to the right people. Among the examiners, Varley had constructed the standards and Professor Wheatstone had developed the instrument for comparison, the famous Wheatstone Bridge. Asked to state his preferred system of testing a submarine line, Thomson answered decisively: 'Testing entirely by comparison with absolute standards of resistance. I am not aware that this system of testing by absolute standards of resistance was ever brought into practice by any practical electrician, except Mr Varley, at that time, but the principles were shown long before by Professor Wheatstone'.[72]

A higher degree of complicity between examiners and witness could hardly

[70] *Ibid.*, pp. 79–80. [71] *Ibid.*, p. 115. [72] *Ibid.*, p. 118.

be imagined, but neither could a closer relationship between industry and science. Pulled by economic necessity and pushed by entrepreneurial scientists such as Thomson, the twin professions of scientific engineering and research science were rapidly emerging.

In its verdict on the 1858 failure of the Atlantic cable, the Committee thus fully set their seal of approval on Thomson's evidence. They drew particular attention to the alleged objection of Field to extensive experimental tests on grounds of time prior to the cable's manufacture, to the absence of protection from the sun at the Glass, Elliot works, and to the considerable variations in conductivity, a standard only being adopted after most of the cable had been manufactured. Further damage had occurred before and during storage in Plymouth. On the other hand, the Report made special mention of the use of Professor Thomson's very delicate marine galvanometer after a serious fault in insulation appeared during laying and to Varley's claim that the working speed of one word per minute by relay had been doubled with Thomson's instrument. The Committee concluded:

> We attribute the failure of this enterprise to the original design of the cable having been faulty owing to the absence of experimental data, to the manufacture having been conducted without proper supervision, and to the cable not having been handled, after manufacture, with sufficient care. We have had before us samples of the bad joints which existed in the cable before it was laid; and we cannot but observe that practical men ought to have known that the cable was defective, and to have been aware of the locality of the defects, before it was laid.[73]

The second part of the Report considered the construction (divided into conducting wire, insulating covering, and external protection) and laying (divided into preliminary survey, apparatus, and contracts) of submarine cables. The Committee noted that 'the whole subject of submarine telegraphy may yet be said to be in its infancy, and all that has been done has been rather the result of bold though successful tentative processes than of the application of any well ascertained data to the ends to be obtained'. They also noted that the success of the early submarine lines set unfortunate precedents, such that later cables were laid down 'with no variation as regards the principles of construction' under circumstances and conditions entirely different from the original.[74] Taken together, these remarks pointed in the direction of telegraphic *engineering*, with its growing stock of precise quantitative data and its harmony of theory and practice enabling the design of cables for different purposes and conditions (notably length) rather than the mere copying of successful precedents.

Following submission of the Report, the Atlantic Telegraph Company appointed a scientific consultative committee (consisting of Galton, Fairbairn, Thomson, Wheatstone, and Joseph Whitworth) to examine the tenders and samples of cable from various manufacturers. The Committee rejected those

[73] 'Report', *op. cit.* (note 66), p. ix. [74] *Ibid.*, pp. xiii–xxxii.

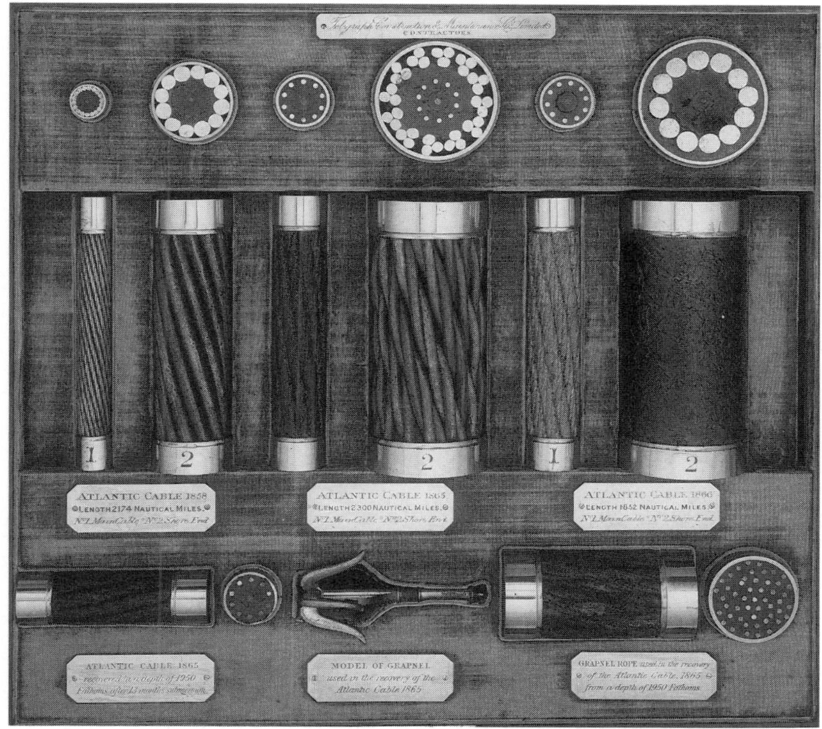

Specimens of Atlantic cable: 1858, 1865, and 1866. [From The Science Museum Library, London.]

cables employing a new material, india-rubber, as insulation on the grounds that 'it would be most unwise to adopt any untried material or design in so great an undertaking as the Atlantic telegraph'. By August, 1863, the Committee had recommended accepting the tender of Messrs Glass, Elliot and Company because of that firm's 'successful and varied experience in the manufacture and submergence of cables in different parts of the world'. Furthermore, the Company had offered to subscribe £65 000 to the ordinary stock of the Telegraph Company and 'to manufacture and submerge the cable at simple cost without charging any profit of any kind except in case of success'. In that event they were to receive a profit of 20% in the Company's shares, 'deliverable to them in instalments during a period of twelve months provided the cable continue in perfect working order'.[75]

In April, 1864, the Gutta Percha Company merged with Glass, Elliot to

[75] C.W. Field to William Thomson, 29th August, 1863, F5, ULG (report of the Committee enclosed). Russell, *The Atlantic telegraph*, p. 30, states the sum to be £25 000.

The *Great Eastern* under weigh from Valencia, 23rd July, 1865. [The Science Museum.]

become the Telegraph Construction and Maintenance Company, with the Scotsman John Pender as chairman. Pender had made his fortune as a merchant in textile fabrics in Glasgow and Manchester, and his personal security apparently did much to see the enterprise through an initially uncertain financial period when Field could raise a mere £70 000 in the United States towards a project now requiring capital of at least £600 000. The Telegraphic Construction and Maintenance Company, with an authorized capital of one million pounds, absorbed over half of the Atlantic Telegraph Company's stock and offered £100 000 of bonds. The scientific Committee, meanwhile, guided the cable specifications.[76]

Since America was engaged in its painful Civil War, no United States naval contribution was to be expected. However, the best-known and most controversial ship of the age, child of I.K. Brunel and John Scott Russell, was available for the Atlantic telegraph project. The *Great Eastern*, or *Leviathan* as she had been popularly known, was indeed a sea monster of over 20 000 tons. Although Rankine's remarks on the scientific nature of her construction had been appropriate, her career as a passenger vessel had been an unqualified failure, plagued as she was by accidents and other problems magnified by her unique size. She had been launched over a period of weeks – surely the longest launch in shipping history – in 1857–8, and she had first put to sea rather uncertainly in

[76] *Ibid.*, pp. 29–31; Dibner, *The Atlantic cable*, pp. 84–6; 'Sir John Pender', *DNB*.

The breaking of the 1865 Atlantic telegraph cable aboard the *Great Eastern*. Over 1000 miles of cable had been paid out when it parted and sank two miles to the ocean floor. [From the *Illustrated London News*.]

1859. Her early failures coincided, therefore, with the early failures of the Atlantic telegraph. Whitehouse had somewhat sarcastically remarked to the 1856 British Association that, were the cable dimensions advocated by Thomson to be adopted, Scott Russell's *Leviathan* would be the only ship afloat capable of carrying the cable. His words were wonderfully prophetic. The enormous vessel was destined to carry the very cable that Whitehouse had scorned as unnecessary. In its appraisal of the laying of submarine cables, the 1859–61 Committee expressed the view that, for the purpose of laying a new Atlantic cable, 6000 tons in weight, 'a vessel would have to be specially constructed, as no vessel existing, except the *Great Eastern*, would be adapted to the work'. As a suitable expedient, therefore, the *Leviathan* emerged from lay-up.[77]

[77] 'Report', *op. cit.* (note 66), p. xxx. See also G.S. Emmerson, *John Scott Russell. A great Victorian engineer and naval architect* (London, 1977), pp. 65–157.

Preparing for the final attempt in 1865 to grapple the lost cable. Apparent waste of capital turned to success in 1866 when both the 1865 cable and a new cable were completed by the *Great Eastern*. [From the *Illustrated London News*.]

With the ship refitted in time for a cable laying attempt in the summer of 1865, and with 500 people aboard, including William Thomson as a consulting expert, the *Great Eastern* steamed west from Valencia with the new design of cable paying out. Snapping around the half-way point, the cable sank two miles to the ocean bed. In vain the epic ship grappled for the thread. But the following year a cable was successfully laid in a period of two weeks, and the 1865 cable was picked up and completed also. The major step in the unification of the British speaking world had at last been achieved, and in recognition of his contribution to the prestige of the British Empire, Queen Victoria knighted Thomson at Windsor Castle on 10th November, 1866.[78]

To mathematical physics, and especially to Fourier, Sir William Thomson owed his strength in commercial telegraphy. He had written to his old teacher Nichol in 1859, the year of Nichol's death:

> The analysis [of the electric telegraph] is you will see all Fourier's – that w^h you set me to read & which I took up with so much delight after my session of Natural Philosophy under you. This – the first piece of physical mathematics I ever took up, has been since Fourier's time ready & *quite complete* for the telegraphic problems, including every practical detail – resistance in receiving instrument (radiating power of the end of a bar), imperfect insulation (loss of heat from the sides of a bar) &c.[79]

[78] Bright, *Submarine telegraphs*, pp. 78–105; SPT, **1**, 481–508.
[79] Thomson to Nichol, *op. cit.* (note 56).

The *Great Eastern*'s return in 1865. An armada of small craft greet the leviathan off Brighton, while ship's officers and Company men take stock from the giant ship's starboard paddle box. [From the *Illustrated London News*.]

Such, then, were the first fruits of mathematical physics. Thomson could truly regard the Atlantic telegraph enterprise as one which provided both intellectual satisfaction to himself and material benefits to mankind through the harmony of theory and practice. With this British achievement of 1866 the era of the second industrial revolution had truly dawned:

> the progress of science is sure and strong, and is not dependent on the weak powers of individuals. A few years sooner or a few years later such results as are now spoken of must have been achieved; there can be no doubt of that. Abstract science has tended very much to accelerate the results, and to give the world the benefit of those results earlier than it could have had them, if left to struggle for them and try for them by repeated efforts and repeated failures, unguided by such principles as can be evolved from the abstract investigations of science.[80]

[80] William Thomson, Reply to toast 'Sir William Thomson, our youngest Burgess' celebrating conferment of the freedom of the city of Glasgow, 1st November, 1866, in SPT, **1**, 504.

20

Measurement and marketing: the economics of electricity

In physical science a first essential step in the direction of learning any subject is to find principles of numerical reckoning and methods for practicably measuring some quality connected with it. I often say that when you can measure what you are speaking about, and express it in numbers, you know something about it; but when you cannot measure it, when you cannot express it in numbers, your knowledge is of a meagre and unsatisfactory kind: it may be the begining of knowledge, but you have scarcely, in your thoughts, advanced to the stage of *science* . . . *Sir William Thomson, 'Electrical units of measurement', 1883*[1]

Sir William's public success with the Atlantic cable, and the close relation thereby established between laboratory testing and electrical industry, opened the way for his ambitious marketing of scientific knowledge. In particular, through a carefully developed and cleverly exploited system of patents and partnerships, his financial returns on scientific capital reached significant heights from 1869, in which year the Atlantic telegraph companies agreed to a settlement of £7000, with a future annual payment of £2500, to Sir William and his partners, Fleeming Jenkin and Cromwell Varley.

In all large-scale electrical projects, such as the development of electric lighting, the introduction of electric traction for trams and trains, and the use of hydroelectric power for the new British aluminium industry, Sir William regarded the question of economy as paramount. Again and again he sought both to minimize waste of useful work or available energy, and to compare the economy of different systems of power transmission or of different forms of motive power. With his enduring commitment to the concept of work, therefore, and to the theme of economy, his specific concerns with the economics of electricity provide an explicit exemplification or embodiment of the labour theory of value.

Electrical units of measurement: the absolute system

In his obituary notice of Lord Kelvin, Joseph Larmor perceptively noted the close relation between Thomson's energy doctrines and his concern with units of measurement:

[1] PL, **1**, 73.

If one had to specify a single department of activity to justify Lord Kelvin's fame, it would probably be his work in connexion with the establishment of the science of Energy, in the widest sense in which it is the most far-reaching construction of the last century in physical science. This doctrine has not only furnished a standard of industrial values which has enabled mechanical power in all its ramifications, however recondite its sources may be, to be measured with scientific precision as a commercial asset; it has also, in its other aspect of the continual dissipation of available energy, created the doctrine of inorganic evolution and changed our conceptions of the material universe.[2]

'Not only', Larmor went on, 'did he enlarge and enforce the advantages of a universal correlated system of units, such as had been developed in the narrower field of the distribution of terrestrial gravity and terrestrial magnetism by Gauss and Weber because in fact they were indispensable to international cooperation in these subjects: he was also the prime mover in starting those determinations of absolute constants of nature and of numerical relations between the various natural standards, which, repeated and refined by a long line of eminent successors, are now the special care of governments, as affording the universal data on which modern exact engineering is ultimately based.'[3]

Larmor's comments here remind us of the wider context for Thomson's commitment to precision laboratory measurement as fundamental to the all-pervasive quest for economy. Writing, for example, to Stokes in April, 1866, Thomson praised Maxwell's latest report of his experimental researches on the viscosity of gases, particularly for the very satisfactory evidence 'it contains as to the accuracy of the results in absolute measure'. He went on:

I hope Maxwell will *soon* be able to settle the very important question of slip for liquids. Investigations of this kind are of national importance, and any thing that money can do to promote them, whether by supplying a convenient experimental laboratory, and a sufficient number of thoroughly qualified operators to carry out the work, or in any other way ought to be done by the government. If the government knew its own interest even on a [*sic*] strictly and simply economical grounds, it would do everything that money can do to promote the *execution* of good experiments . . . Could not the Royal Society move government for the establishment of laboratories for investigation, in which teaching would be thoroughly subordinate to the search for new knowledge of properties of matter.[4]

Thomson concluded his plea with a conviction that were such a laboratory to be established in London with Stokes or Maxwell, or both, in charge, 'the results would be worth hundreds of fold the annual cost, even in material economies w^h^ would arise from the knowledge gained'. Furthermore, 'the directors of such an institution . . . would naturally form an advisory council for the gov^t^ (admiralty, army, customs &c) which would save hundreds of thousands wasted in useless experiments on a large scale, for every thousand spent in keeping up the proposed laboratory'.

[2] JL, xxix. [3] *Ibid.*, p. li.

[4] William Thomson to G.G. Stokes, 11th April, 1866, RR.6.178, Royal Society of London. See also PL, **2**, 144, for Sir William's public promotion of the idea in 1871.

Thirty years later, Thomson's hopes had yet to be fulfilled. In 1896 the name of Kelvin headed a list of sixty-two scientific signatories on behalf of the British Association to a memorandum petitioning the government for the establishment of a National Physical Laboratory. The petitioners argued that 'Experience has shewn that there are certain operations and measurements many of which have been proved to be of great practical utility which can be best undertaken by highly trained officials attached to an Institution specially devoted to scientific investigations'. These investigations had three principal divisions:

(1) The observation of natural phenomena, the study of which must be prolonged through periods of time longer than the average duration of life;
(2) The testing and verification of instruments for physical investigation, and the preservation of standards for reference; and
(3) The systematic accurate determination of physical constants and of numerical data which may be useful either for scientific or industrial purposes.[5]

These divisions reflected very precisely the nature of the numerous committees of the British Association in which Thomson had played such a leading role since the 1860s. Whether devoted to the subject of underground temperatures, tides, or electrical standards, all had been committed to the goal of numerical observation and accurate measurement.

The ultimate outcome of the British Association campaign was the opening of the government-funded National Physical Laboratory near Hampton Court by the Prince of Wales (later King George V) in March, 1902. Seconding the vote of thanks, Kelvin emphasized again 'the scientific importance of exceedingly minute and accurate measurements, which, though they might not strike the popular imagination, were the foundation of the most brilliant discoveries'.[6] We now turn to the earlier history of those measurements, measurements which, in the absence of a National Physical Laboratory, were largely undertaken in university physical laboratories on behalf of the specialist committees of the British Association.

[5] British Association to Lord Kelvin enclosing copy of printed memorandum 'National Physical Laboratory', 30th January, 1897, B303–3a, ULC. The memorandum was dated 2nd December, 1896. It pointed out that, as early as 1891, Oliver Lodge had urged the British Association to promote the creation of such an institution. The Royal Society and other scientific bodies joined in support of the scheme. The signatories included industrialists such as Lord Armstrong as well as physicists, chemists and engineers. The supporters appealed particularly to the precedent of the *Physikalisch-Technische Reichsanstalt* established near Berlin in the 1880s through the inspiration of Helmholtz and Werner von Siemens (who had contributed £12 500) to promote both pure research and 'delicate operations of standardising and testing to meet the wants of investigators and to facilitate the application of science to industry'. See also David Cahan, 'Werner Siemens and the origin of the Physikalisch-Technische Reichsanstalt, 1872–1887', *Hist. Stud. Phys. Sci.*, **12** (1982), 253–83.

[6] SPT, **2**, 1165. See also Edward Pyatt, *The National Physical Laboratory. A history* (Bristol, 1983), esp. pp. 12–35, for a general background.

As is well known, it was at the 1861 Manchester meeting of the British Association, following a paper by Latimer Clark and Sir Charles Bright, that William Thomson (who was not actually present) proposed the formation of a British Association committee on standards of electrical resistence.[7] Latimer Clark later explained the significance of the paper:

It is interesting not only as being the earliest document on the subject, but also because it is a proposal for a system of perfectly inter-dependent units founded on metrical measure, & on an electro-static base. As a system of *practical* units it is in fact almost identical with that which has been finally adopted and is now in use, the Volts, Ohms, Farads &c. having the same unitary correlation. I knew nothing of Weber's work, and though I knew what Joule was doing and knew of the numerical relationship of heat, power & electricity, I was not mathematician enough to see the enormous value of an absolute system, founded on mass, time, & space. It is this which has gained for the British System of Electrical Measurement its universal acceptance by mankind.[8]

Clark's concluding sentence here underlines once again the universal nature of the new language of absolute units, crucially identified with the British system of electrical measurement, which we first noted in Chapter 11 with Thomson's facetious remarks to Tait concerning the displacement of Latin as the universal language of mankind. The age of the new imperialism, founded upon the language of physical science, had truly dawned.

The specific aim of the BAAS Committee, fulfilled in six principal reports published between 1862 and 1869, was to determine 'what would be the most convenient *unit* of resistance' and 'what would be the best form and material for the *standard* representing that unit'. As early as the provisional report of 1862, the Committee laid down as one of several desirable qualities that the unit of resistance, in common with the other units of the system, should, as far as possible, bear a definite relation to the unit of work, 'the great connecting link between all physical measurements'.[9] In the adoption of mechanical work as the measure of all things we immediately perceive the commanding presence of William Thomson himself, who had asserted this role for work as early as 1845 and, with Tait, had made it the foundation of all natural philosophy in the *Treatise*.

The second report (1863) offers the most concise account of the ingredients in

[7] SPT, **1**, 417–18. See Latimer Clark and Sir Charles Bright, 'On the formation of standards of electrical quantity and resistance', *BAAS Report*, **31** (1861), 37–8, pointing out the need for standards or units of electrical potential, absolute quantity, current, and resistance, asking 'the aid and authority of the British Association in introducing such standards into practical use', and indicating the necessity for a nomenclature 'in order to adapt the system to the wants of practical telegraphists'. The same *BAAS Report* (pp. xxxix–xl) announced the appointment of the electrical resistance standards Committee consisting initially of Profs Williamson, Wheatstone, Thomson, and Miller, Dr Matthiessen (all FRSs) and Mr Fleeming Jenkin.

[8] Latimer Clark to Sir William Thomson, 3rd May, 1883, C91, ULC. Clark added that, although the paper was a joint one, 'the *original ideas* emanated from me'.

[9] 'Provisional report of the Committee appointed by the British Association on standards of electrical resistance', *BAAS Report*, **32** (1862), 125–63, on p. 126.

a satisfactory absolute system of electrical measurement. From the beginning, the Committee made clear that it viewed 'this comparatively limited question [of resistance] as one part only of the much larger subject of general electrical measurement'. Not surprisingly, given Thomson's central role, the Committee 'after mature consideration' held the view that 'the system of so-called absolute electrical units, based on purely mechanical measurements, is not only the best system yet proposed, but is the only one consistent with our present knowledge both of the relations existing between the various electrical phenomena and of the connexion between these and the fundamental measurements of time, space, and mass'. Implementation of this desirable system, the Committee noted, depended upon the accurate measurement of electrical resistance in absolute units, now made possible by the success of Maxwell's, Stewart's, and Jenkin's employment of a method devised by Thomson.[10]

The Committee began by making clear the meaning of such a system:

The word 'absolute' in the present sense is used as opposed to the word 'relative'; . . . in other words, it does not mean that the measurements or units are absolutely correct, but only that the measurement, instead of being a simple comparison with an arbitrary quantity of the same kind as that measured, is made by reference to certain fundamental units of another kind treated as postulates.[11]

The report illustrated the issue by reference to the 'relative' power of an engine expressed 'as equal to the power of so many horses' without reference to units of space, mass, or time ('although all these ideas are necessarily involved in any idea of work'). On the other hand, the power of an engine 'when expressed in foot-pounds is measured in a kind of absolute measurement, i.e. not by reference to another source of power, such as a horse or a man, but by reference to the units of weight and length simply – units which have been long in general use, and may be treated as fundamental'.

The unit of force assumed here, however, weight, was itself arbitrarily chosen and inconstant, being dependent upon latitude. The Committee therefore adopted Gauss's dynamical unit, just as Thomson had urged since his early lectures at Glasgow (ch. 11). In Gauss's 'true absolute measurement the unit of force is defined as the force capable of producing the unit velocity in the unit of mass when it has acted on it for the unit of time. Hence this force acting through the unit of space performs the absolute unit of work. In these two definitions, time, mass, and space are alone involved and the units in which these are measured, i.e. the second, gramme, and metre, will alone, in what follows, be considered as fundamental units'. Consequently, they emphasized:

the word absolute is intended to convey the idea that the natural connexion between one kind of magnitude and another has been attended to, and that all the units form part of a

[10] 'Report of the Committee appointed by the British Association on standards of electrical resistance', *BAAS Report*, **33** (1863), 111–76, esp. p. 111. SPT, **1**, 418–19, states that the main part of the report (pp. 111–24) was drafted by Thomson. [11] 'Report', *op. cit.* (note 10), p. 112.

coherent system. It appears probable that the name of 'derived units' would more readily convey the required idea than the word 'absolute', or the name of mechanical units might have been adopted; but when a word has once been generally accepted, it is undesirable to introduce a new word to express the same idea. The object or use of the absolute system of units may be expressed by saying that it avoids useless coefficients in passing from one kind of measurement to another . . . [the introduction of numerous factors] is a very serious annoyance, and moreover, where the relations between various kinds of measurement are not immediately apparent, the use of the coherent or absolute system will lead much more rapidly to a general knowledge of these relations than the mere publication of formulae.[12]

In harmony with Thomson's enduring commitment to simplifying techniques and the saving of brains, the report stated that the absolute system was thus 'not only the best practical system, but it is the only rational system' since 'Every one will readily perceive the absurdity of attempting to teach geometry with a unit of capacity so defined that the contents of a cube should be $6\frac{1}{2}$ times the arithmetical cube of one side . . . but geometry so taught would not be one whit more absurd than the science of electricity would become unless the absolute system of units were adopted'.[13]

Having thereby set out persuasively the rationale for the absolute system, the Committee explained that in 'determining the unit of electrical resistance and the other electrical units, we must simply follow the natural relation existing between the various electrical quantities, and between these and the fundamental units of time, mass, and space'. The relations to one another of the four 'electrical phenomena susceptible of measurement' – current, electromotive force, resistance, and quantity – were expressed by two simple equations determined experimentally: Ohm's law from which it followed that 'the unit electromotive force must produce the unit current in a circuit of unit resistance'; and Faraday's relation of quantity of electricity to current and time, from which it followed that 'the unit of quantity must be the quantity conveyed by the unit current in the unit of time'. Any other relations would yield 'useless and absurd' factors or coefficients in the two equations.[14]

The report next moved to a crucial phase in the argument. From Ohm's law and Faraday's relation alone, 'it follows that only two of the electrical units could be arbitrarily chosen, even if the natural relation between electrical and mechanical measurements were disregarded . . . Such a system would be coherent; and if all mechanical, chemical, and thermal effects produced by electricity could be neglected, such a system might perhaps be called absolute'. But such effects could not be neglected:

all our knowledge of electricity is derived from the mechanical, chemical, and thermal effects which it produces, and these effects cannot be ignored in a true absolute system. Chemical and thermal effects are, however, now all measured by reference to the

[12] *Ibid.*, pp. 112–13. [13] *Ibid.*, p. 113. [14] *Ibid.*, pp. 113–14.

mechanical unit of work; and therefore, in forming a coherent electrical system, the chemical and thermal effects may be neglected, and it is only necessary to attend to the connexion between electrical magnitudes and the mechanical units. What, then, are the mechanical effects observed in connexion with electricity?[15]

First, as shown by Joule, the work (or heat or chemical action equivalent to work) performed by a current in a circuit is directly proportional to the square of the current, to the time during which it acts, and to the resistance of the circuit. This third fundamental equation affecting the four electrical quantities 'represents the most important connexion between them and the mechanical units' whereby 'the unit current flowing for a unit of time through a circuit of unit resistance will perform a unit of work or its equivalent'. The three equations still left the series of units undefined, however, since one unit might be arbitrarily chosen from which the three other units would be deduced by the three equations. A second relation between mechanical and electrical measurements existed in the form of the inverse square law from which it followed that 'the unit quantity [of electricity] should be that which at a unit distance repels a similar and equal quantity with unit force'. Thus the four equations were:

$$I = E/R \quad \text{(Ohm's law)}, \qquad (1)$$
$$Q = It \quad \text{(Faraday's relation)}, \qquad (2)$$
$$W = I^2Rt \quad \text{(Joule's law)}, \qquad (3)$$
$$F = Q/d^2 \quad \text{(Coulomb's law)}, \qquad (4)$$

where I = current, E = electromotive force, R = resistance, Q = quantity, t = time, d = distance, F = force, W = work. These four equations 'are sufficient to measure all electrical phenomena by reference to time, mass, and space only, or, in other words, to determine the four electrical units by reference to mechanical units. Equation (4) at once determines the [electrostatic] unit of quantity, which, by equation (2), determines the unit current; the unit of resistance is then determined by equation (3), and the unit electromotive force by equation (1). Here, then, is one absolute or coherent system, starting from an effect produced by electricity when at rest'. This 'electrostatical' system of units had been so-called by Weber, though without reference to equation (3), i.e. 'without reference to the idea of work, introduced into the system by Thomson and Helmholtz'.[16]

As this system was based on a statical phenomenon, 'whereas at present the chief applications of electricity are dynamic, depending on electricity in motion, or on voltaic currents with their accompanying electromagnetic effects', a more useful system would involve the retention of the first three equations and the replacement of equation (4) by a different relation between electrical and mechanical magnitudes. Thus 'the force exerted on the pole of a magnet by a current in its neighbourhood is a purely mechanical phenomenon' such that

[15] *Ibid.*, p. 114. [16] *Ibid.*, pp. 114–15.

$$f = ILm/k^2, \qquad (5)$$

where f = force, m = magnetic strength of the pole (measured in mechanical units), k = radius of the conductor placed in a circle round the pole, and L = length of conductor. It followed that 'the unit length of the unit current must produce the unit force on a unit pole at the unit distance'. Equations (1), (2), (3), and (5) thus yielded a distinct absolute system of units, called by Weber the 'electromagnetic' units.[17]

The Committee next explained how electrical measurements can be practically made in such electromagnetic units. Provided the horizontal force (H) of the earth's magnetism be known, currents were easily measured by means of a tangent galvanometer and an equation derived from (5):

$$I = Hk^2/L \tan d, \qquad (6)$$

where d = the deflection produced by the current. Measurement of quantity of electricity could be simply obtained, for example, by measuring the swing of a galvanometer needle when a single instantaneous discharge is allowed to pass through it. Then

> [if] we could measure resistance in absolute measure, the whole system of practical absolute measurement would be complete, since, when the current and resistance are known, equation (1) (Ohm's law) directly gives the electromotive force producing the current. The object of the experiments of the Sub-Committee (made at King's College . . .) was therefore to determine the resistance of a certain piece of wire in the absolute system, in order from this one careful determination to construct the material representative of the absolute unit with which all other resistances would be compared by well-known methods.[18]

The Committee summarized the methods by which the absolute resistance of a wire could be measured. First, starting from equation (3) in 1851, Professor Thomson 'had determined absolute resistance by means of Dr Joule's experimental measurement of the heat developed in the wire by a current; and by this method he obtained a result which agrees within about 5 per cent. with our latest experiments. This method is the simplest of all, so far as the mental conception is concerned, and is probably susceptible of very considerable accuracy'.[19]

Second, 'Indirect methods depending on the electromotive force induced in a wire moving across a magnetic field have . . . now been more accurately applied'. In the simplest (though barely practicable) case, a straight conductor with its two ends resting on two conducting rails of large section in connexion with the earth would move perpendicularly to the magnetic lines of force and to

[17] *Ibid.*, p. 115. [18] *Ibid.*, pp. 116, 143–4.

[19] *Ibid.*, p. 116. See William Thomson, 'Applications of the principle of mechanical effect to the measurement of electromotive forces and of galvanic resistance in absolute units', *Phil. Mag.*, [series 4], **2** (1851), 551–62; MPP, **1**, 490–502.

its own length, thereby developing a current in the circuit. The action of the magnetic force on this current would also resist the motion. Given uniform motion, the work done by the current would be equivalent to that done in moving the conductor against the resistance.

From the equation for the resisting force,

$$f = SLI, \tag{7}$$

where S = pole strength or intensity. With V the velocity of the conductor, the work was

$$W = VSLIt.$$

By equation (3), $W = I^2Rt$, and hence the resistance was

$$R = VSL/I. \tag{8}$$

Since both I and S could be obtained in absolute measure this relation gave absolute resistance. As a consequence, 'the resistance of a conductor in absolute measure is really expressed by a velocity . . . that is . . . the resistance of a circuit is the velocity with which a conductor of unit length must move across a magnetic field of unit intensity in order to generate a unit current in the circuit'. Similarly, from equations (1) and (8) we have

$$E = VSL, \tag{9}$$

so that 'the unit length of a conductor moving with unit velocity perpendicularly across the lines of force of a magnetic field will produce a unit electromotive force between its two ends'. This relation is the one Weber had made fundamental in advance of equation (3), Joule's work–heat relation.[20]

A more practicable method of determining absolute resistance involved 'a circular coil of known dimensions revolving with known velocity about an axis in a magnetic field of known intensity. Although Weber had determined absolute resistance of many wires by this method, the laborious determination of the intensity of the magnetic field, and its inconstant value for the earth, rendered the method less useful than that due to Thomson and adopted by the sub-Committee at King's College. In Thomson's method, a knowledge of the intensity of the magnetic field was unnecessary, the equation for resistance deriving from (1), (3), and (5) above:

$$R = L^2V/4k^2\tan d. \tag{10}$$

The resistance was thus expressed in electromagnetic absolute units: 'The essence of Professor Thomson's method consists in substituting, by aid of the laws of electromagnetic induction, the measurements of a velocity and a deflection for the more complex and therefore less accurate measurements of work and force

[20] 'Report', *op. cit.* (note 10), pp. 116–18.

required in the simple fundamental equations'. The economy, simplicity and practicality of the method used and apparatus employed had all the hallmarks of Thomson's scientific style:

The apparatus consisted of two circular coils of copper wire, about one foot in diameter, placed side by side, and connected in series; these coils revolved round a vertical axis, and were driven by a belt from a hand-winch, fitted with Huyghen's gear to produce a sensibly constant driving-power. A small magnet, with a mirror attached, was hung in the centre of the two coils, and the deflections of this magnet were read by a telescope from the reflection of a scale in the mirror. A frictional governor controlled the speed of the revolving coil.[21]

So sensitive was the apparatus, indeed, that the oscillations in deflection 'produced by the passage of steamers on the Thames at no great distance from the place of experiments were of very sensible magnitude'.[22]

The sub-Committee's results yielded an absolute unit of resistance about 8% larger than the unit as derived from a German-silver coil recently measured by Weber; 6.5% larger than that published by Weber of Siemens's mercury units; 5% smaller than that derived from coils issued by Thomson in 1858 based on Jacobi's standard and a previous determination from Joule's silver wire. The values agreed most closely with an old determination of a copper standard made by Weber for Thomson. At the same time, the sub-Committee's experiments agreed with one another to within 1–2%.[23] The Committee concluded its discussion of the subject of absolute units thus:

The sub-Committee especially urge the repetition of the experiments, as with the improvements already enumerated, and other minor alterations, they confidently expect a considerably closer approximation to the absolute unit than they have hitherto obtained. It will be well here to remark that, according to the resolution of the Committee of 1861, the coils, when issued, will not be called absolute units, but the units of the British Association; so that any subsequent improvement in experimental absolute measurement will not entail a change in the standard, but only a trifling correction in those calculations which involve the correlation of the physical forces.[24]

Subsequent reports of the Committee told of the progress made in the practical development of the conclusions of 1863 which in effect had announced the adoption of the absolute electromagnetic system of measurement based on the metre, gramme and second. Thus the 1864 Report dealt with the rigorous measurements of resistance made since the 1863 experiments and set out a detailed table of the approximate relative values of various units of electrical resistance against the BA unit or Ohmad (equal to 10^7 m/s according to experiments of the Standards Committee).[25] In 1865 the Committee reported 'that the object for which they were first appointed has now been accomplished':

[21] *Ibid.*, pp. 118–19, 163–76. [22] *Ibid.*, p. 120. [23] *Ibid.*, p. 121. [24] *Ibid.*, p. 122.

[25] 'Report of the Committee on standards of electrical resistance', *BAAS Report*, **34** (1864), 345–67, esp. pp. 345–9.

The unit of electrical resistance has been chosen and determined by fresh experiments; the standards have been prepared, and copies of these standards have been made with the same care as was employed in adjusting the standards themselves; seventeen of these copies have been given away, and sixteen have been sold.[26]

The recipients of the free copies included the Directors of Public Telegraphs in nine continental states as well as India and Australia. Copies sold at £2 10s each. The purchasers included telegraphic companies in Switzerland, while in Britain the leading commercial operators promised to use the unit exclusively. Overall the Committee aimed the standard coils not at institutions 'where they would probably have laid [*sic*] on a shelf useless and unknown, but rather to distribute them widely, where they might become available to practical electricians'.[27]

In the fifth Report (1867) Thomson provided a comprehensive analysis of electrometers. Here the familiar concept of work formed the basic principle for the construction of his instruments. He had been assigned the task of making experimental determination of the difference of potentials or electromotive force in absolute measure. The task of determining a unit of capacity had meanwhile occupied Matthiessen, Jenkin, and others for two years, but no progress had been made with current measurements.[28] The sixth Report (1869) set out the experiments of Thomson and Maxwell on the determination of v, the ratio between the electrostatic and electromagnetic units.[29] A final notice from the Committee in 1870 recommended its replacement by three smaller committees to determine and issue, first, a condenser representing the unit of capacity; second, a gauge for showing the unit difference of potential; and, third, an electrodynamometer adapted to measure the intensity of currents in a decimal multiple of the absolute measure. The Committee noted that its principal achievement had been to make the absolute measure of resistance 'a tangible and practical operation', a choice subsequently 'ratified by men of science over a great portion of the globe'.[30]

A decade later, that ratification took the form of the first International Congress on electrical standards, held at Paris in September, 1881. Helmholtz and Thomson were foreign vice-presidents, and publicly disputed whether or not to accept Siemens's unit of resistance (a column of mercury one metre in length) or the BA unit. The eventual compromise accepted the BA ohm

[26] 'Report of the Committee on standards of electrical resistance', *BAAS Report*, **35** (1865), 308–13, on p. 308. [27] *Ibid.*, pp. 310–11.

[28] 'Report of the Committee on standards of electrical resistance', *BAAS Report*, **37** (1867), 474–522, esp. pp. 476–9. The Committee also reported Joule's determination of the mechanical equivalent of heat by electrical methods using the Committee's standard of resistance (pp. 474–5). For a detailed and illustrated description of Thomson's electrical instruments in general, and his electrometers in particular, see George Green and J.T. Lloyd, *Kelvin's instruments and the Kelvin museum* (Glasgow, 1970), pp. 18–44.

[29] 'Report of the Committee on standards of electrical resistance', *BAAS Report*, **39** (1869), 434–8.

[30] 'Report of the Committee on standards of electrical resistance', *BAAS Report*, **40** (1870), 14–15.

William Thomson's original divided ring electrometer (*c*.1857–60) typified his enduring concerns with sensitive, delicate measuring instruments. In his 1867 *BAAS Report* Thomson classified this type and its more advanced descendant, the quadrant electrometer, as symmetrical electrometers, in accordance with 'the shape and kinematical relations of their parts'. This instrument was adapted for measuring potentials (of batteries, cables, or atmosphere, for example) in the range *c*.100–*c*.500 V. [From The Science Museum Library; further details of construction in George Green and J.T. Lloyd, *Kelvin's instruments and the Kelvin museum*, pp. 21–2.]

represented by a mercury column of appropriate length to be decided after further research. The Congress agreed to the names of the units: ohm, volt, farad, coulomb, and ampere (rather than weber). In 1884 the fourth meeting of the Paris Congress agreed to a standard of resistance to be called the 'legal ohm', the resistance of a column of mercury 106 centimetres long and one square millimetre in section.[31]

Meanwhile, the BAAS Committee on standards of electrical measurement

[31] See SPT, **2**, 773–5. The names 'ohm', 'volt' and 'farad', as well as the dynamical units of force and work in the CGS system, the 'dyne' and 'erg', had been adopted in 'The first report of the Committee for the selection and nomenclature of dynamical and electrical units', *BAAS Report*, **43** (1873), 222–5. Thomson, Maxwell and Jenkin were members. J.D. Everett, on behalf of the Committee, published his *Illustrations of the centimetre-gramme-second . . . system of units* (London, 1875). See also 'Report of the Committee . . . appointed for the purpose of constructing and issuing practical standards for use in electrical measurements', *BAAS Report*, **54** (1884), 29–32, for the adoption of the 'legal ohm' and its relation to the BA ohm.

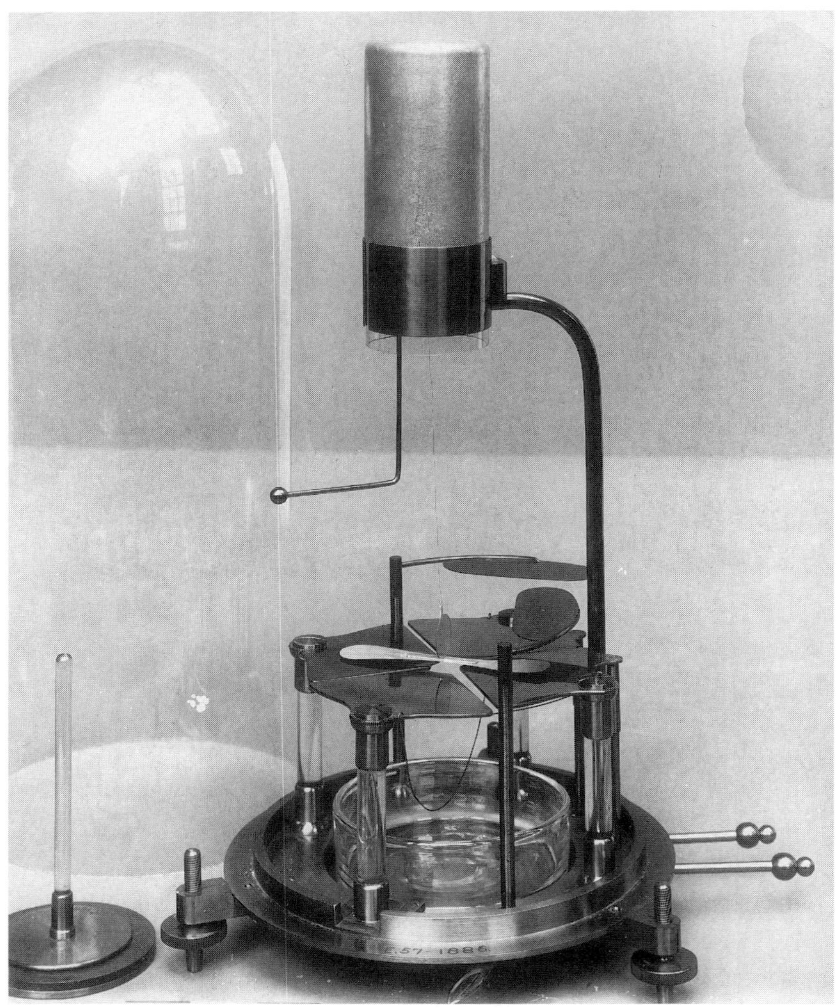

William Thomson's quadrant electrometer (1872). A highly sensitive improved form of the divided ring electrometer, it had four quadrants instead of two half rings as the symmetrical fixed conductors at different electric potentials. The indication of the force was produced by means of an electrified body moveable symmetrically in either direction from a middle position in this field. The instrument measured potential difference to 0.01 V. [From The Science Museum Library; further details in George Green and J.T. Lloyd, *Kelvin's instruments and the Kelvin museum*, pp. 22–4; *BAAS Report*, **37** (1867), 490–7.]

William Thomson's electrostatic voltmeter (1887), based on electrometer principles, but developed for use in the electrical industry (lighting and power). This type of instrument measured potentials in the range 400–10 000 V. [From The Science Museum Library; further details in George Green and J. T. Lloyd, *Kelvin's instruments and the Kelvin museum*, pp. 25–7.]

had been reactivated in 1880. It continued to report annually and Sir William remained a member until his death. Much of its activity centred on Lord Rayleigh's redeterminations of the BA ohm at Cambridge with the work of testing resistance coils taking place at the Cavendish Laboratory.[32] The nature of the work convinced Thomson of the pressing need for a government laboratory. Thus, as President of the Institution of Electrical Engineers in 1889, he used his authority to promote 'the establishment of an electrical standardizing laboratory, which we hope is to be taken in hand by our Government'.[33] Yet it had been the British Association, and not the government, which, in Thomson's view, had done most to serve the needs of commerce in the nineteenth century:

> Those who perilled and lost their money in the original Atlantic Telegraph were impelled and supported by a sense of the grandeur of their enterprise, and of the world-wide benefits which must flow from its success . . . but they little thought . . . when the assistance of the British Association was invoked to supply their electricians with methods for absolute measurement (which they found necessary to secure the best economical return for their expenditure, and to obviate and detect those faults in their electric material which had led to disaster), they were laying the foundation for accurate electric measurement in every scientific laboratory in the world, and initiating a train of investigation which now sends up branches into the loftiest regions and subtlest ether of natural philosophy. Long may the British Association continue a bond of union, and a medium for the interchange of good offices between science and the world![34]

Partnerships and patents

'Moreover I do like this bloodless, painless combat with wood and iron, forcing the stubborn rascals to do my will, licking the clumsy cubs into an active shape, seeing the child of today's thought working tomorrow in full vigour at his appointed task.' Penned from Birkenhead in April, 1858, to his fiancée at the beginning of a career largely devoted to telegraph engineering, these remarks of Fleeming Jenkin (1833–85) serve to illuminate something of the *raison d'etre* for the subsequent business partnership and personal friendship between Jenkin and Thomson from early 1859.[35]

Son of a Royal Navy officer who had earlier served as a midshipman on the guardship watching over Napoleon's St Helena exile and of a mother whose wild, partly Scottish family had settled in Jamaica, Jenkin was born in Kent and

[32] For example, 'Report of the Committee . . . appointed for the purpose of constructing and issuing practical standards for use in electrical measurements', *BAAS Report*, **52** (1882), 70–1; **53** (1883), 41–5. See also Sir William Thomson to Lord Rayleigh, 19th October, 1882, in SPT, **2**, 789–90, referring to the 1882 Paris Congress.

[33] Sir William Thomson to E.E.N. Mascart, 4th June, 1889, in SPT, **2**, 886–7. The Committee had expressed the desirability of a National Standardizing Laboratory for Electrical Instruments in their 1886 Report. See *BAAS Report*, **56** (1886), 146.

[34] PL, **2**, 161–2.

[35] Fleeming Jenkin to Anne Austin, 18th April, 1858, in R.L. Stevenson (ed.), *Papers and memoir of Fleeming Jenkin* (2 vols., London, 1887), **1**, pp. lxxv–lxxvi.

named after an Admiral Fleeming, one of his father's protectors in the navy. Fleeming's early years proved no less exotic than his ancestry. After attending Edinburgh Academy (where Maxwell was his senior and Tait his class-mate), he moved with his family to Frankfurt, Paris and Genoa in turn. In large part impelled by difficult financial circumstances at home, the family would have found attractive both the favourable living costs in continental Europe and the opportunities for a rich and varied education for Fleeming. Staunchly liberal in politics, Jenkin found himself not only in sympathy with the revolutions of 1848, but actually present in the heart of the Paris and Genoa disturbances.[36]

Impressed with the 'progress of liberal ideas' in Italy, Fleeming enrolled as the first protestant student at the University of Genoa, where he encountered electromagnetism in the physical laboratory of Professor Bancalari. Obtaining Master of Arts with first class honours, he returned to England in 1851 and became an apprentice to Fairbairns of Manchester. As though treading the earlier track of James Thomson, Jenkin moved five years later to work as a draughtsman for Penn's marine engine works at Greenwich building steam-engines for ships of war. By then wholly absorbed in the 'poetry of engineering' he moved to Lewis Gordon's firm of Liddell and Gordon in 1857 and became deeply involved in marine telegraph engineering thereafter, especially through Gordon's association with the cable-manufacturing firm of R.S. Newall of Birkenhead, one of the Atlantic cable contractors.[37]

Given Jenkin's interests, education, and political commitments, it is not surprising to find Thomson describing their first meeting thus:

> In the beginning of the year 1859 my former colleague, Lewis Gordon . . . came to Glasgow to see apparatus for testing submarine cables and signalling through them . . . As soon as he had seen something of what I had in hand, he said to me, 'I would like to show this to a young man of remarkable ability, at present engaged in our works at Birkenhead'. Fleeming Jenkin was accordingly telegraphed for, and appeared next morning in Glasgow. He remained for a week, spending the whole day in my class room and laboratory, and thus pleasantly began our lifelong acquaintance. I was much struck, not only with his brightness and ability, but with his resolution to understand everything spoken of, to see if possible thoroughly through every difficult question, and (no *if* about this!) to slur over nothing. I soon found that thoroughness of honesty was as strongly engrained in the scientific as in the moral side of his character.[38]

Thomson further explained how in that week 'the electric telegraph and, particularly, submarine cables, and the methods, machines, and instruments for laying, testing, and using them, formed naturally the chief subject of our conversations' as it did much of the subsequent 'well-sustained fire of letters on each side about the physical qualities of submarine cables, and the practical results attainable in the way of rapid signalling through them'. Yet Thomson

[36] *Ibid.*, pp. xi–xlvi. [37] *Ibid.*, pp. xlvii–lxii.

[38] Sir William Thomson, 'Note on the contributions of Fleeming Jenkin to electrical and engineering science', in *Papers and memoir of Fleeming Jenkin*, **1**, pp. clv–clix, on p. clv.

soon found Jenkin 'remarkably interested in science generally, and full of intelligent eagerness on many particular questions of dynamics and physics'.

With regard to the Birkenhead works, Jenkin evidently fulfilled the role of philosophical assistant, the position which Gordon and Newall had so anxiously sought to fill in 1857. Availing himself of the opportunities and facilities for experiment, Jenkin there 'began definite scientific investigation of the copper resistance of the conductor, and the insulating resistance and specific inductive capacity of its gutta-percha coating, in the factory, in various stages of manufacture; and he was the very first to introduce systematically into practice the grand system of absolute measurement founded in Germany by Gauss and Weber'. It can scarcely be known generally, Thomson noted, how much of this step, with its immense value to telegraphy in particular, was due to Jenkin. In 1859, for example, Jenkin's measurements of the specific resistance of gutta-percha were apparently the only results 'in the way of absolute measurements of the electric resistance of an insulating material which had then been made'.[39]

Jenkin's work from the beginning exemplified to perfection Thomson's own deep commitment to scientific engineering, precision measurement and laboratory research and testing. More than that, Jenkin's researches and their publication owed a very great deal to Thomson's guidance. As Jenkin wrote to the professor in August, 1859:

I am very happy to say that Messrs R.S. N[ewall] & Co. have consented to my communicating my experiments through you either to the R.S. or the B^{h} Ass^{n} . . . Should both these papers be sent to the Royal Society or one of them to the British Association? Should they be written fully or concisely? How soon should they be ready? Pray give me some advice on these points in order that I may do as much credit as I can both to yourself as my introducer and to Messrs Newall. Where can I buy or procure the various papers on electricity you have published? I am working at the higher calculus and do not despair of managing it.[40]

A year after the first meeting of Thomson and Jenkin, William wrote to James in February, 1860, with important news:

Did I tell you that a Mr F. Jenkin, whose experiments on cables were communicated to the British Association, has joined me in my patent, I having assigned him a charge, and that we have made joint proposals to the Red Sea Directors which I think have a good chance of being accepted? I am still working hard at the electrometer, and I hope at last have something convenient for general use.[41]

[39] *Ibid.*, pp. clvi–clvii. Sir. William drew particular attention to Jenkin's determination at Birkenhead of the specific inductive capacity of gutta-percha: 'This was the very first true measurement of the specific inductive capacity of a dielectric which had been made after the discovery by Faraday of the existence of the property, and his primitive measurement of it for the three substances, glass, shellac, and sulphur; and at the time when Jenkin made his measurements, the existence of specific inductive capacity was either unknown, or ignored, or denied, by almost all the scientific authorities of the day'. See also MPP, **6**, 336–8.

[40] Fleeming Jenkin to William Thomson, 4th August, 1859, J13, ULG.

[41] William to James Thomson, 14th February, 1860, in SPT, **1**, 408.

Thomson's fourth patent, a joint one with Jenkin under the general heading of 'Improvements in the means of telegraphic communication' (1860), marked the beginning of his famous business partnership. Yet it was to be some nine years before substantial economic rewards accrued to the partners from the Atlantic telegraph itself.

In the meantime, the initial aim of the 1860 patent partnership seems to have been a bid to exploit Jenkin's close involvement with the important but technically troubled Red Sea cable which would form a major part of telegraphic communication between Britain and her Indian empire. The final agreement between the new partners had three main clauses. First, the 'first three thousand pounds of clear profits (from either patent) shall be divided equally between us'. Second, the 'second three thousand pounds of clear profits shall be divided into three shares of which you [Thomson] shall receive two and I [Jenkin] shall receive one'. And third, thereafter 'our interest in the patents shall be equal'. Under this arrangement, then, there would be complete pooling of patents, a sharing of patent property, with 'no difference in our interest in the two patents' and 'no special clauses having reference to particular transactions'. In short, the arrangement aimed at efficient marketing of the patents.[42]

The correspondence between the partners quickly reveals Jenkin to be as shrewd a businessman as Thomson's father. Thus in July, 1863, he delivered a stern lecture to their electrical instrument supplier, James White of Glasgow:

> You must try to keep your accounts against Prof Thomson & myself in a little more intelligible fashion . . . In order to avoid confusion Prof Thomson & myself have agreed that all instruments made under our patents shall be charged to 'Thomson & Jenkin' . . . You will also please to make out your bill regularly at Xmas & Midsummer of each year and these accounts should be made out in the following form [Here Jenkin gave detailed instructions on how the accounts should be kept] . . . As an example of how I wish this done I will have your almost unintelligible account of last year abstracted in the way I propose and sent you as a model.[43]

Again, in late 1865, Jenkin advised against Thomson's patenting of his improvement to electrometers on the grounds that firms would not pay royalties on testing instruments but would simply seek 'to devise some dodge which will do without infringing patents'. By contrast 'The signalling instruments are a wholly different case – if they really are better than other peoples they will earn for the Atlantic *so* many pounds for every pound they earn for us – but no testing instrument will do this, at least not obviously'.[44]

In 1865 Jenkin and Thomson added C.F. Varley (1828–83) to their marketing partnership. First involved in practical telegraphy as early as 1846–7, Varley was an engineer with the Electric & International Telegraph Company, and had

[42] Fleeming Jenkin to William Thomson, 3rd January, 1860, J25, ULG.

[43] Fleeming Jenkin to James White, 23rd July, 1863, J48, ULG.

[44] Fleeming Jenkin to William Thomson, 17th November, 1865, J54, ULG.

cleverly estimated the position of a fault on the Company's Dutch cable in 1853.[45] Varley had been prominent also in the Atlantic telegraph project, and once again employed his expertise to locate faults. As early as 1858, Thomson wrote that 'Varley's report [to the Company] is, in my opinion, evidence of high scientific and practical talent', and a few months later Varley asked Thomson 'where I may see copies of your papers on Electricity'.[46] Varley also played a key role at the Board of Trade Inquiry, indicating a complete convergence of Varley and Thomson in opposition to Whitehouse (ch. 19). Not surprisingly, therefore, the three telegraphic experts (Thomson, Jenkin and Varley) came together just prior to the *Great Eastern*'s first cable-laying expedition.

Writing to Thomson in February, 1865, Varley reminded him that seven years of his first patent had expired. 'It seems to me', he explained, 'most important for all, not only that the term should be elongated as much as possible, but that foreign patents should be secured. It would be very hard indeed after so much labour, to see others making the apparatus, and using them, abroad'. But Varley already had an answer:

> I have consulted Fleming [*sic*] Jenkin as to what would be the best course to adopt, and he is of opinion that I should be invested with power on behalf of all to negotiate with the Tel[egraph] Construction & M[aintenan]ce Co. for the sale to them of our invention for it is clear that if they were made to understand its value they might make tremendous profits by it, while paying in a comparatively moderate sum, which would still be a large profit for us . . . As soon as what we have done becomes a 'fait accompli', all the world will be striving to rival us, & it is very desirable to avoid litigation. Could we get the Te. Construction & Mnce. Co. to take it up, no one unless possessed of great means, would dare to fight it. What is more important their influence would be secured to protect it at home & abroad.[47]

Varley added in a postscript that it 'is most desirable to keep our operations as dark as much as possible as there are others in the field trying to see what they can do'. He therefore aimed to get the necessary agreement drawn up and submitted to Thomson for approval.

That agreement between the partners and the Telegraph Construction and Maintenance Company was signed in June, 1865. Varley and Thomson meanwhile took out a patent for 'Improvements in electric telegraphs' which included Varley's patent for the interposition of condensers to accelerate signalling.[48] The partners thus seemed in a commanding position to benefit from the double success of the Atlantic telegraph project in 1866 and from the great expansion of ocean telegraphy which rapidly followed.

Not all of the business arrangements proceeded smoothly, however. Jenkin

[45] C.F. Varley to William Thomson, 10th October, 1859, V2, ULG.

[46] William Thomson to George Saward, 24th September, 1858, in SPT, **1**, 377–8; C.F. Varley to William Thomson, 2nd April, 1859, V1, ULG.

[47] C.F. Varley to William Thomson, 20th February, 1865, V9, ULG.

[48] C.F. Varley to William Thomson, 16th May, 1865, V10, ULG; SPT, **1**, 552–3.

noted in 1867 that he had been asked for assistance concerning a new (possibly the French) Atlantic cable and had answered 'that I could give every assistance except that of expediting the transmission of messages'. He believed he could 'report on other systems' but felt that 'that Atlantic agreement places one in a most awkward position and may be the means of not only fettering one directly but of preventing the obtaining an appointment as Engineer to an independent company'.[49]

At the same time, the agreement itself did not yield the rewards expected by the partners. The root of the trouble appeared to be the 1866 financial reconstruction whereby the Anglo American Telegraph Company had been formed to finance and operate the still-troubled enterprise. Varley in July, 1867, feared 'we shall have trouble because it appears that the Atlantic Tel. Co. have not raised the £100 000 to pay off the construction co. & therefore they are not likely to pay a dividend this year . . . If the 8% [Stock] be paid a *deferred* dividend we should at least be paid a deferred percentage'.[50] A few months later, while attempting to secure patents in the United States, Varley urged Sir William to 'remember if the A. A. T. Co. disown our claim on them under the Agreement with the A. T. Co. we can charge them what we like'.[51]

Meanwhile, some form of litigation seemed inevitable. In October, 1867, the Atlantic Telegraph Company's solicitors advised the partners' solicitors that the Company, acting 'fully in accordance with the terms of that agreement', proposed paying 3% upon the net earnings of the Company, earnings which amounted to just over £35 000, equivalent to about £1000 in royalties. After consultation with Archibald Smith, the partners' solicitors replied that the claim was for 3% 'before the deductions to the Anglo American Co. & that unless the Company agreed to that proposition we were instructed to file a Bill'.[52]

Following sustained wrangling but eventual compromise without litigation, a new agreement favourable to the partners appeared in draft form in mid-1868, though not in final form until 1869. This time the agreement was drawn up between the partners and both the Atlantic and the Anglo American Telegraph Companies. First, the Atlantic Company agreed to pay £7000 to Thomson, Varley and Jenkin in satisfaction of their claims under the old agreement. In return, all legal proceedings by the partners against the Company would henceforth cease. Second, both Companies 'shall pay to Thomson, Varley & Jenkin £3000 [reduced to £2500 in the final agreement] a year, quarterly for ten years from the 13th of May 1868 subject to the provisions for suspension and determination of the payment hereinafter contained'. In return, the Companies 'shall be entitled to use in the present Atlantic Cables and in any other Cables

[49] Fleeming Jenkin to Sir William Thomson, 3rd July, 1867, J71, ULG.

[50] C.F. Varley to Sir William Thomson, 16th February, 1867, V16, ULG.

[51] C.F. Varley to Sir William Thomson, *c.* November, 1867, C18, ULG.

[52] Freshfields to Sharpe, Parkers & Pritchard, 31st October, 1867; Sharpe, Parkers & Pritchard to Sir William Thomson, 2nd November, 1867, S25, ULG.

having the same Termini the inventions described in the Specifications mentioned in the annexed list'.[53]

The patents included in the 1868–9 Agreement consisted of three by Varley (1856, 1860, and 1862), one by Thomson (the 1858 'Improvements in testing and working electric telegraphs' which centred on the mirror galvanometer), one by Thomson and Jenkin (1860), and one by Varley and Thomson (the 1865 'curb key' for improved rates of signalling).[54] Once again, the pooling of patent property improved profitability and marketing to the commercial telegraph concerns.

Third, the Agreement of 1868–9 laid down that 'If there shall be a total interruption of communication across the Atlantic caused by the failure of all the Companies Atlantic Cables, the payment shall be suspended during the period of total interruption'. Fourth, 'Thomson, Varley and Jenkin shall not give to any other Company the right to use their inventions in any cable laid across the Atlantic on more favourable terms than those of this agreement'.

A crucial fifth clause gave the partners a specially privileged relationship with the Companies compared to any future rival patentees:

> If any inventions shall be made and furnished to the Companies by any other persons whereby the Companies shall be enabled without using any of the means described in the Said Specification as furnished to them by Thomson, Varley & Jenkin to signal through the Cables better and faster than they can do by using such means then the Companies may give a twelve months notice to determine this Agreement but in that case they will give Thomson, Varley & Jenkin every opportunity and facility for studying and endeavouring to supersede the rival invention, and if before the end of the twelve months Thomson, Varley & Jenkin shall furnish the Companies with the means of signalling through the Cables as well and as fast as they can do by the rival inventions the notice shall drop. And in consideration of the services rendered to the Companies throughout by Sir William Thomson, Mr Varley and Mr Jenkin the Directors will construe this clause in the most liberal manner.

This important clause effectively ruled out successful rivals for at least a decade. Not surprisingly did Sir William claim in 1883 that *all* signalling on ocean telegraphs from 1866 to 1883 was carried out with his instruments.[55]

Final agreement, reached only in mid-1869, had been delayed by much

[53] Draft 'Articles of agreement . . . 1868 between Sir William Thomson, Cromwell Fleetwood Varley and Fleeming Jenkin of the first part, the Atlantic Telegraph Company of the second part and The Anglo American Telegraph Company (Limited) of the third part', J92, ULG. See especially C.F. Varley to Sir William Thomson, 6th July, 1868, V22, ULG, where Varley stated that he (on behalf of the partners) had negotiated the financial settlement with Sir Richard Glass (on behalf of the telegraph companies). Archibald Smith drew up the draft agreement. The £7000 was not paid to the partners until July, 1869. £1500 was credited to Sir William, £500 retained for Varley's use, and the balance of £5000 was placed on deposit to the credit of the three patentees. See Sharpe, Parker & Pritchard to Sir William Thomson, 19th July, 1869, S86, ULG. The subsequent annual payment seems to have been £2500 in the final agreement. See Bircham & Co. to Sharpe, Parkers & Pritchard, 2nd March, 1869, S30 (copy), ULG.

[54] Draft 'Articles of agreement', *op. cit.* (note 53).

[55] *MPP*, **2**, 650*n*.

wrangling over relatively minor clauses insisted upon by the Companies. The legal problems prompted an exasperated Varley to exclaim that 'English Law & its mode of administration is the most unholy institution ever devised by man'.[56] Yet the partners had evidently struck a bargain very much to their advantage. From this success, Sir William would have the resources not only to fund his laboratory and industrial interests, but to enhance dramatically his life-style on Clydeside.

In 1881, for example, Sir William reported to Jenkin that the quadrant electrometer, the mirror galvanometer, and the last recorder patent were bringing the partners £3000 from the Eastern Telegraph Company, £2100 from the Eastern Extension, and £1500 from the Anglo.[57] Two years earlier, Sir William had derived an income of £5500 from telegraph patents, while the Thomson–Jenkin partnership as consulting engineers to different telegraph undertakings apparently brought them several thousand pounds each year.[58] Sir William had purchased his large schooner yacht in 1870 and had begun construction of his Largs home, Netherhall, in the mid-1870s, a modest baronial-style mansion which cost some £12 000 up to 1890. At the same time, he had offered substantial sums (£1000 in 1869 and £2000 in 1874) to Glasgow College, sums which, by funding 'Thomson Experimental Scholarships' to assist deserving students and the 'Neil Arnott Demonstratorship' in the physical laboratory, in effect represented reinvestment of capital in a highly lucrative enterprise.[59]

Among the large-scale telegraphic projects which followed the success of the 1866 Atlantic cable, the French Atlantic cable of 1869 figured prominently. Though French financiers apparently negotiated the required concessions, the Company was essentially British. Varley and Thomson served as consulting electricians, Jenkin as one of the three engineers, the Telegraph Construction Company of Blackwall provided the cable, and the *Great Eastern* laid the line from Brest (where Sir William was stationed) to St Pierre off Newfoundland.[60] Once again, the project demonstrated British industrial and technological domination of the most advanced engineering of the period. Sir William and his partners were especially active in the early 1870s promoting further ocean lines (notably to Brazil) and assisting in the design of purpose-built cable-laying

[56] C.F. Varley to Sir William Thomson, 5th June, 1869, V5, ULC. The appointment of an independent barrister to draw up the agreement in its final form seems to have brought the protracted dispute to an end. [57] SPT, **2**, 650*n*.

[58] See H.I. Sharlin, *Lord Kelvin: the dynamic Victorian* (Pittsburgh, 1979), p. 199; SPT, **2**, 650*n*.

[59] SPT, **1**, 554–7; **2**, 585, 649–50.

[60] *Ibid.*, 552–3. C.F. Varley to Sir William Thomson 17th July, 1868, V26, ULG, where Varley had arranged for Thomson and himself to receive £3000 for one year from the French Company in their capacity as consultants. See also C.F. Varley to Sir William Thomson, 21st June, 1869, V6, ULC. Varley confidently reported from the *Great Eastern* that 'Everything works like success'. Clark, Jenkin and Varley reported in July that 'we find that for all commercial purposes the line is in perfect working order'. See *Société du Cable Transatlantique Français* to Sir William Thomson, 15–16th July, 1869, S276, ULC.

vessels which would supersede for all time the limitations of adapted ships such as the *Great Eastern* (ch. 21).

At precisely the time when Sir William and his partners began to prosper from the patent system, public dissatisfaction with British patent laws reached new levels. The patent system had remained largely unreformed up to 1852 when a new Patent Law Amendment Act introduced a single United Kingdom patent whereby patentees no longer had to apply for separate patent rights in England and Scotland. Commissioners were appointed to administer the system and, most importantly, the initial cost of a patent fell from about £300 to only £25, leading to a sharp increase in applications. Widespread dissatisfaction continued, however, and in the period 1852–83 commissions of enquiry, parliamentary committees, and public debates subjected the system to severe criticism with demands for abolition. The rejection of patent protection by North German and Dutch governments in the late 1860s, and the consequent threat to British competitiveness, may have added fuel to the repeal movement. In general, however, British industrialists seem to have been divided over the issue, while engineers and inventors naturally preferred reform to abolition.[61] One such debate, to which Sir William contributed, took place in late 1869 during three successive meetings of the Glasgow Philosophical Society. It illustrates some of the interests underlying Sir William's and his partners' support of reform rather than abolition.

Introducing the debate, St John Vincent Day set the benefits of the patent system in the context of Glasgow's economic prosperity and the wealth of the nation:

while recognising every man's right to the *materially embodied results* (not the mere ideas on paper, bear in mind) of mental origination, experiment, expenditure of time and money, the State's chief object, under the patent system, has been to secure the invention to the nation. And I suppose it is hardly necessary for me to ask you to look around this great city – with its weaving and spinning mills, its iron works, forges and rolling mills, its ship yards and engine works, its alkali works, its oil works, its sugar refineries, its calico-printing and Turkey-red dye works, its potteries and glass works, its paper mills and its corn mills . . . and to inquire if all this teeming industry, all this wondrous productive ability, forcibly tells us whether we have profited or not by securing to the nation the thought, the foresight, the skilled plans, the cunningly-wrought devices of such men as [Siemens, Bessemer, Whitworth, Napier, Watt and others associated with these industries]; and does the work, the offspring of their labours, prove to us that we should have

[61] Fritz Machlup and Edith Penrose, 'The patent controversy in the nineteenth century', *J. Econ. Hist.*, **10** (1950), 1–29. See also Klaus Boehm, *The British patent system. 1. Administration* (Cambridge, 1967), esp. pp. 14–30. A more recent historical analysis of patents up to 1852 is provided by H.I. Dutton, *The patent system and inventive activity during the industrial revolution, 1750–1852* (Manchester, 1984). Arguing for a much greater economic role for patents in the industrial revolution, Dutton puts the case for interpreting inventors not as men motivated by simple love of invention or by philanthropic instincts but by the expectation of profit (pp. 103–21). He furthermore discussed invention itself in terms of an industry or trade whereby inventors employed various means to market their products (pp. 122–49).

been better off and as far on in the race towards the acmé of economy and productive capability, but for the value imparted to inventive talent by a patent law?[62]

Day's analysis of the British patent system draws attention therefore to the perceived interdependence of the patent law and the economic and social system, the political economy of the nation. Several of his further remarks underline this point, and serve to show just how intimately the whig values of William Thomson cohered with the commercial and industrial economy of his adopted city.

First, the patent system maintained a spirit of competition. 'Why is not the man of greatest enterprise – he who is foremost in the race – to reap the greatest benefit', asked Day rhetorically, 'which, as in every other transaction or pursuit, is what all who work at it labour after?' Such views reflected perfectly Thomson's own intense competitiveness, whether in his earliest scientific papers for the *Cambridge Mathematical Journal* and his keen anxiety over priority, his approach to the Mathematics Tripos, or his attitude towards the inter-collegiate boat races on the Cam.[63]

Second, Day recognized the patent system as part of the accepted legal right to personal property, in this case the 'materially embodied results' of a man's intellectual and physical labours rather than mere intellectual property:

If property in the results of matured thought, of experiment after the expenditure of time and means, is essentially wrong and inflictive of injury on others, why, then, it is equally wrong to possess property of any kind, because, if the possessor is better off than his neighbours in one case, he must also be in the other. The abolitionist [of patents], in order to be consistent, must equally urge that yonder owner of landed property, or house property, of capital realized out of speculation, or a careful watching of commercial trading and enterprise, must at once give it all up, after his years of labour . . .[64]

Thomson too was deeply attached to the notion of scientific knowledge as wealth or capital which generated compound interest available for reinvestment in intellectual capital or for exploitation in industrial application (ch. 5). Patented inventions represented the latter component in the capitalism of intellectual property, that is, the marketing of the 'materially embodied' products of scientific research to commercial interests.

Third, Day pointed to the defects of an alternative system of state rewards which would, by its nature, lack any but an arbitrary standard of payment. By contrast, the patent system was 'the only one by which the real value of an invention can be ascertained, and, therefore, the only one which can secure not merely a reward, but a due reward – precisely its exact worth – to the inventor'.[65] In other words, the patent system rewarded the inventor according

[62] St J.V. Day in 'Discussion on patents for inventions', *Proc. Glasgow Phil. Soc.*, **7** (1869–71), 158–220, on p. 166. [63] *Ibid.*, pp. 166–7. See Chapters 3 and 7.

[64] *Ibid.*, p. 167. See also Dutton, *The patent system*, pp. 69–85, for a discussion of patent property rights and the courts.

[65] Day, 'Discussion on patents', p. 168; Dutton, *The patent system*, pp. 25–6.

to the market value of his patent, and not by the decision of any professional body or any commission of experts. Here the skills of Thomson and his partners as entrepreneurs, and not merely inventors, came into full play.

Fourth, Day appealed to his audience's sense of imperial mission through the diffusion of useful knowledge and the fruits of industry for the advancement of all mankind:

> If the gentlemen who so urgently pursue the cry of abolition would help us to amend the discrepancies in the present law, we should then have no fault to find with them. As a nation which lives so largely upon the returns secured to us by the protection granted to the embodied results of inventive genius, – a nation which not only lives, but influences the whole world thereby in so many ways, which scatters the direct products of its inventive talent to feed and clothe, to teach, and emancipate from the thraldom of ignorance the sons of toil in every clime, – a nation which, by the very essence of invention, influences so effectively the march of civilisation, and, in return, brings home to her own door such gains; – I say, then we could tolerate these gentlemen.[66]

Few words from a contemporary source could express so emphatically the role of the inventor in the ideology of 'advancement' and in the cultural and economic imperialism centred on British science and engineering at whose heart stood Glasgow and its industry.

Although Sir William would certainly have agreed with Day's remarks, his only direct contribution to the debate consisted of a letter supporting the Society's resolution against total abolition of the Patent Law but in favour of remedying its defects by 'any well-advised scheme of amendment'. He drew particular attention to the need for international agreement, the issue in which he and his partners had personal interest:

> One of the worst features in the present system is the heterogeneousness of the patent laws within the British Empire and in the nations of Europe and the United States of America . . . I hope the grand object of obtaining a common patent law among all civilized nations, which would give a great stimulus to useful invention, and do away with much of the confusion and inconvenience both to inventors and users of inventions . . . will be kept in view by the committee of the Society appointed to consider the subject.[67]

Sir William's belief in the four values just outlined – the competitive spirit, the legal right to property and profits, the system of reward by marketing, and the diffusion of benefits to the world – was embodied in his invention and marketing of the famous siphon telegraph recorder. He had recognized the potential value of a self-recording receiving instrument in his earliest telegraph papers but it was not until 1867 that he took out a patent for 'Improvements in receiving or

[66] Day, 'Discussion on patents', p. 169.

[67] Sir William Thomson to Dr Bryce, 14th December, 1869, in 'Discussion on patents', p. 181. Sir William became a prominent member of a British Association committee on patents. See 'Report of the Committee . . . appointed for the purpose of watching and reporting to the Council on patent legislation', *BAAS Report*, **48** (1878), 157. Other members included St J.V. Day, C.W. Siemens, Neilson Hancock, and J.R. Napier. The Committee was most active at the time of the 1883 legislation.

Siphon recorder. With its automatic recording of telegraph signals on moving paper tape, the instrument's designer aimed to improve economy of working and minimize waste of time. Its commercial success generated still more wealth for Sir William. [From The Science Museum Library; further details in George Green and J.T. Lloyd, *Kelvin's instruments and the Kelvin museum*, pp. 34–5.]

recording instruments for electric telegraphs' which included the siphon recorder.[68] Jenkin announced it to the secretary of the Atlantic Telegraph Company on 1st July, 1867:

We have much pleasure in informing you that Sir Wm Thomson has invented, and is ready to bring into practical operation, a recording instrument, adapted to receive & record messages through long submarine lines. This instrument combines the advantages

[68] MPP, **2**, 108–9; SPT, **1**, 334, 570–6.

of the Mirror Galvanometer, with those of the Morse instrument. Every message which could be read on the Mirror Galvanometer, by an expert clerk, would be as accurately recorded by the new instrument. We are ready to communicate this invention in confidence, to you, and we shall be glad to know, at your earliest convenience whether you wish to exercise the option of taking out a patent or patents for the new instrument, given you by the existing agreement.[69]

The secretary, George Saward, reported the next day that he had lost no time in sending Jenkin's news to the directors of the Anglo American Telegraph Company 'to whom belong the entire control and responsibility of everything relating to the working of the cables'.[70] This move, however, raised sensitive issues of competition for patents and commercial secrecy. Saward's communication prompted the partners to seek legal advice from Archibald Smith: 'I do not see', wrote Jenkin to a concerned Varley, 'that we are bound to tell the Anglos anything but will ask Smith'. 'I must see Smith at once about the effect or drift of Saward's answer. The Anglos will almost certainly decline to say anything', he told Thomson.[71]

Archibald Smith, as a chancery barrister, advised that if there were the least danger that 'some one may be on the track of Thomson's discovery and that the communication to the Anglo American may lead to the method being protected before you can protect it' then 'you had better lose no time, & not wait for an answer, but take out the provisional protection at once'. He added that he thought 'Saward had no right to communicate your letter to the other Company & if you like to tell him so, you might send him such a note as I enclose'.[72] Jenkin's version of Smith's note to Saward gave no grounds:

It did not at first occur to me on reading your letter of the 2nd that your communication to the Anglo American Company of my letter of the 1st is in fact a breach of Article 5 of our agreement, with the Atlantic Company, which may lead to very injurious results both to us and to the Atlantic Company. It is possible that other Electricians may be on the track of the same discovery with Sir Wm Thomson and it may become a race for a patent, and we may be distanced by reason of your communication. Pray believe that in writing . . . I am actuated by no spirit of capriciousness but solely by a sincere regard for your as well as our own interests.[73]

When we realize that Sir William after 1870 could expect from each of the Falmouth–Lisbon, Anglo–Mediterranean and British–Indian Telegraph companies royalties of £1000 per annum on the siphon recorder alone, we may well understand the motives behind the partners' sensitivity.[74]

[69] Fleeming Jenkin to George Saward, 1st July, 1867, J71 (copy), ULG.
[70] George Saward to Fleeming Jenkin, 2nd July, 1867, J71 (copy), ULG.
[71] Fleeming Jenkin to C.F. Varley, 3rd July, 1867, J71 (copy), ULG; to Sir William Thomson, 3rd July, 1867, J71, ULG.
[72] Archibald Smith to Fleeming Jenkin, 5th July, 1867, J71 (copy), ULG.
[73] Fleeming Jenkin to George Saward, 5th July, 1867, J71 (copy), ULG.
[74] 'Copy memorandum of arrangement with Professor Jenkin', Archibald Smith to Sir William Thomson, 22nd August, 1870, S200, ULC; SPT, **1**, 575.

Thomson's usual enthusiasm for precision measurement and delicate instruments found expression yet again in the design and construction of the siphon recorder. Adapting the system of signalling which employed currents of different polarities to cause movements to left or right (the Steinheil plan of signals) rather than the system of dots and dashes (Morse's method) on the grounds of economy of time and reduction of cable retardations, he needed a practical means of automatically recording the signals formerly registered by delicate deflections of the galvanometer mirror.[75]

Since an ordinary pen marking a moving tape would create an unworkable degree of friction, Thomson devised a delicate ink capillary siphon in its place and as a substitute for the mirror in the galvanometer. A coil of fine wire suspended between the poles of an electromagnet had an ink-filled vaccine tube attached. A small electrifying machine charged the ink, making it spurt in a fine jet on to a moving paper ribbon, so that as the signals moved the coil to left or right the ink traced a permanent record. By 1869, Sir William felt that his new instrument was ready for competitive trial against the mirror instrument on commercial telegraph systems.[76] The recorder was first publicly exhibited at John Pender's London residence in June, 1870, on the completion of the British–Indian line from Falmouth to Bombay.[77] As Sir William reported to Helmholtz in July:

> The completion of the cables between England and India two months ago has led to an urgent demand for my recording instrument . . . Even in the specimen [of ribbon] I send you you may see an advantage of the new instrument. The repetition of the word 'Anglais' would, had the practical operators been using it, have been unnecessary; and half a minute of time would have been saved. The cable is full of work throughout the twenty-four hours.[78]

Three years later an official of the Eastern Telegraph Company reported from the Malta station that the siphon recorder 'when worked by careful and experienced clerks, proves a great source of general economy to the Company and cannot fail to diminish the number of errors'.[79] Minimizing waste of time and labour, maximizing useful work: these aims lay behind Sir William's improvements to telegraphic instruments.

Prospects for the recorder were immense. Thomson noted that his assistant, Leitch, would soon 'have to commence progress eastwards with seven recorders to be installed at Lisbon, Gibraltar, Malta, Alexandria, Suez, Aden, and Bombay',[80] that is at all the principal coaling stations for steamers en route to India via the so-called Overland Route, which from 1869 had become a direct voyage

[75] *Ibid.*, pp. 572–3.
[76] Sir William Thomson to Sir James Anderson, 9th March, 1869, in SPT, **1**, 574.
[77] *Ibid.*, p. 575.
[78] Sir William Thomson to Hermann von Helmholtz, 29th July, 1870, in SPT, **1**, 577–9.
[79] Robert Portelli to W.T. Ansell, 21st November, 1873, E10, ULC.
[80] Sir William Thomson to Jessie Crum, 1st August, 1870, in SPT, **1**, 579.

through the new Suez Canal. Now for the first time the Empire was becoming a unified entity held together not merely by vessels of trade and war but by the wires of an ocean telegraph system largely immune from enemy sabotage. The nation that ruled the seas also commanded the world's communications.

Power and light: the economics of electricity

As early as 1857, when the Institution of Civil Engineers debated the impracticality of economic electromagnetic engines, William Thomson concluded that 'until some mode is found of producing electricity as many times cheaper than that of an ordinary galvanic battery as coal is cheaper than zinc, electromagnetic engines cannot supersede the steam-engine'. Thomson's remarks echo those of Joule who, in his earliest experimental investigations, had sought improvements to electromagnetic engines but had found that such engines compared unfavourably with the economy of steam-engines. He had concluded in 1841:

> This comparison [between Joule's electromagnetic engine and the best Cornish steam engines] is so very unfavourable that I confess I almost despair of the success of electro-magnetic attractions as an economical source of power; for although my machine is by no means perfect, I do not see how the arrangement of its parts could be improved so far as to make the duty per lb. of zinc superior to the duty of the best steam-engines per lb. of coal. And even if this were attained, the expense of the zinc and exciting fluids of the battery is so great, when compared with the price of coal, as to prevent the ordinary electro-magnetic engine from being useful for any but very peculiar purposes.[81]

By 1857, however, the problem lay not with any intrinsic defect of the electromagnetic engine, but with the economic generation of electric power, a quantity which could now be measured in terms of its work equivalent and compared with the mechanical effect derived from other sources.

Six years later in a letter to Thomas Andrews, Thomson explained Nollet's electric light apparatus for lighthouses. Persuaded to do so by Tait, he reluctantly communicated clinical details of dimensions, horse-power required to drive the apparatus, and total cost. As soon as the question of economy emerged, however, his initial lack of enthusiasm changed to lively interest:

> There, I have made a clean breast of it. I know nothing more of the matter *except that Nollet does not reverse his connections, and therefore does have alternately reversed current in his flame*; whereas Holmes [designer of the British apparatus] does reverse and does not leave reversals in the flame. Thus Nollet escapes the commutator, a *great evil*, and gets a flame which does not burn one of the points faster than the other – a small but sensible benefit. The reverse of each proposition applies to Holmes. *The commutator is a frightful thing*. I don't mean that Holmes' is bad, because I do not know it, and it may be very good; but

[81] J.P. Joule, 'On a new class of magnetic forces', *Ann. Elec.*, **8** (1842), 219–24; *The scientific papers of James Prescott Joule* (2 vols., London, 1887), **1**, pp. 46–53, on p. 48; SPT, **1**, 397–9. For a general historical account of electrical engineering, see Percy Dunsheath, *A history of electrical engineering* (London, 1962).

the thing to be done at the requisite speed is appalling . . . I believe it cannot be done except theoretically without great waste of energy and consequent burning of contact surfaces.[82]

Confessing that he had now told Andrews 'rather more than all I know about it', Thomson concluded with a suggestion that 'a large voltaic battery will be more economical than any electromagnetic machine'. He aimed to discover by experiment 'how expensive its habits are, and multiply by the number required for a lighthouse'. Thus in relation to his earliest interest in electric lighting, the all-pervasive themes of economy and waste again predominated.

During the 1870s Sir William Thomson's inventive capacities were largely directed towards the improvement of navigational instruments (chs. 21 and 22). When electrical power first showed signs of developing into a new industry, he manifested but minimal interest. At the Philadelphia Exhibition of 1876, for example, he reported on Gramme's commercial electric machines, but not until another meeting of the Institution of Civil Engineers early in 1878 did a remark by C.W. Siemens on the possibility of distant transmission raise the question of economy and thus fire Thomson's enthusiasm:

[Thomson] believed that with an exceedingly moderate amount of copper it would be possible to carry the electric energy for one hundred, or two hundred, or one thousand electric lights to a distance of several hundred miles. The economical and engineering moral of the theory appeared to be that towns henceforth would be lighted by coal burned at the pit's mouth, where it was cheapest. The carriage expense of electricity was nothing, while that of coal was sometimes the greater part of its cost. The dross at the pit's mouth (which formerly was wasted) could be used for working dynamo-engines of the most economical kind, and in that way . . . the illumination of great towns would be reduced to a small fraction of the present expense.[83]

Electric power 'was not simply sufficient for sewing-machines and turning-lathes, but by putting together a sufficient number [of electric machines], any amount of horsepower might be developed', while for lighthouses 'the great adaptability of the electric light to furnish increase of power when wanted gave it a value which no other source of light possessed'. Furthermore, the reduction of heat generated by electric light and the complete absence of fumes 'meant not only increased economy, but a great advantage in regard to health'. Thus 'the use of such a [water]fall as that of Niagara, or the employment of waste coal at the pit's mouth' offered that vast economy which had been lacking in 1857 when he and other members of the Institution had doubted the practicality of electromagnetic engines.

These remarks formed the basis for Sir William's evidence in 1879 to the

82 William Thomson to Thomas Andrews, 4th March, 1863, A41B, ULC; SPT, **1**, 426–7. For Nollet's and Holmes's contributions to lighthouse illumination, see C.MacK. Jarvis, 'The generation of electricity', in Charles Singer, E.J. Holmyard, A.R. Hall and T.I. Williams (eds.), *A History of technology* (8 vols., Oxford, 1954–84), **5**, pp. 177–207, esp. pp. 181–3.

83 SPT, **2**, 683–5. On Gramme's machines, see Jarvis, 'The generation of electricity', pp. 188–92.

House of Commons Select Committee on Electric Light. True to his liberal values, he sought the removal by legislation of all restrictions and the promotion of invention and innovation as the best means of advancement. At the same time he estimated the quantity of copper required for the economical transmission of electric power to any distance. Applying the calculation to Niagara, he concluded that a copper wire of half-inch diameter would transmit 21 000 horse-power from the Falls to a distance of 300 miles, the capital cost of the copper being £60 000 or under £3 per horse-power at the destination.[84] He subsequently developed this preliminary estimate into various well-known forms for the economic long-distance transmission of electric power, communicating the results of his calculations to the British Association in 1881.

It was in the spring of that year that Sir William began his association with Joseph Swan's new company set up to manufacture the Swan glow-lamp in Newcastle,[85] and also became acquainted with Camille Faure's accumulator for the storage of electrical energy. This 'box of electricity', as Sir William referred to it, soon underwent rigorous tests and measurements in the Glasgow laboratory. 'It is splendidly powerful, but I have yet to find whether it does the whole amount of work specified by Mr Reynier (and Mr Faure) and how much actual work must be spent on it each time to renew its charge', he reported in mid-May from Netherhall.[86] Two weeks later he informed Jenkin that his staff was 'as hard at work as possible in the laboratory on the Faure storage cells, which are doing *splendidly*. It is going to be a most valuable practical affair – as valuable as water cisterns to people whether they had or had not systems of water-pipes and water-supply'.[87] Fascinated by this simple solution to the great problem of storing up available energy, Sir William saw in the Faure cell a method of preventing failure of light or power should the generating machinery break down.

So captivated had he become by the 'potted-energy' that he aimed to light both his yacht and his university house by its aid. He also sent his nephew, James Thomson Bottomley (Anna's son, 1845–1926) to Paris 'to help them with their scientific work there', and asked Jenkin to provide a good man 'who knows something of electricity and engineering' to leave in Paris to 'take charge of testing and inspecting in the manufacture of Faure accumulators'. Meanwhile, James White had been appointed, under Sir William's direction, 'sole agent for the manufacture and sale of Faure accumulators in the United Kingdom'. During the summer, however, Sir William took a leading role in an attempt to form a new company to manufacture and market the patent. The French proprietors of Faure's patents, however, appeared to have more liking for the 'formations of monster companies' rather than for 'the real work of manufactur-

[84] *Nature*, **20** (1879), 110–11; SPT, **2**, 685, 690–1. [85] *Ibid.*, p. 765.

[86] Sir William Thomson to J.H. Gladstone, 17th May, 1881, *ibid.*, p. 766.

[87] Sir William Thomson to Fleeming Jenkin, 3rd June, 1881, *ibid.*, pp. 766–7.

ing and supplying accumulators' with the result that the size of the proposed capital of £400 000 (of which Sir William was to receive 10% in return for his agreement to act as technical adviser) deterred the investors, and he withdrew completely from the costly venture.[88]

In his 1881 presidential address to Section A of the British Association, Sir William discussed the economy of Faure cells for the storage of electrical energy obtained from non-constant sources of power such as windmills:

> The charging may be done uninjuriously, and with good dynamical economy, in any time from six hours to twelve or more. The drawing off of the charge for use may be done safely, but somewhat wastefully, in two hours, and very economically in any time of from five hours to a week or more. Calms do not last often longer than three or four days at a time . . . One of the twenty kilogramme cells charged when the windmill works for five or six hours at any time, and left with its sixty-candle hours' capacity to be used six hours a day for five days, gives a two-candle light.[89]

He concluded, however, that 'windmills as hitherto made are very costly machines; and it does not seem probable that, without inventions not yet made, wind can be economically used to give light in any considerable class of cases, or to put energy into store for work of other kinds'.

Altogether, Sir William read to Section A in 1881 six papers (including a joint paper with J.T. Bottomley) on electricity in relation to power and light. 'On the economy of metal in conductors of electricity' provided a full technical derivation of the results given to the 1879 Parliamentary Committee. He explained that 'The most economical size of the copper conductor for the electric transmission of energy, whether for the electric light or for the performance of mechanical work, would be found by comparing [equating] the annual interest of the money value of the copper with the money value of the energy lost annually in the heat generated in it by the electric current'. This is 'Kelvin's law' in electrical engineering.[90]

A parallel paper set out theoretically based proportions for the design of dynamos. Providing formulae for the work wasted in heating the coils and the useful work in the external circuit, Sir William explained that the key question was 'how ought R [the resistance of the wire] and R' [the resistance of the working coil] be proportioned to make the ratio of waste to work a minimum, with any given speed? Or, which comes to the same thing, to make the speed required for a given ratio of work to waste a minimum?' Given that for good economy the ratio of whole work to waste r 'must be but little greater than

[88] Lady Thomson to G.H. Darwin, 17th June, 1881, *ibid.*, p. 768. Sir William Thomson to Fleeming Jenkin, 22nd and 26th June, 1881, *ibid.*, pp. 768–9, 769–70.

[89] Sir William Thomson, 'On the sources of energy in nature available to man for the production of mechanical effect', *BAAS Report*, **51** (1881), 513–18; PL, **2**, 433–50, esp. pp. 442–4.

[90] Sir William Thomson, 'On the economy of metal in conductors of electricity', *BAAS Report*, **51** (1881), 526–8; MPP, **5**, 432–4. For Kelvin's law see SPT, **2**, 773.

unity', he concluded that the ratio R'/R must be made very small in the derived formula $r=1+2\ (R'/R)^{\frac{1}{2}}$.[91]

In 1882 Sir William became acquainted with Sebastian de Ferranti (1864–1930), whose name remains one of the most famous among Britain's electrical and defence industries. Not yet twenty years old, Ferranti had recently given up a one pound per week job with Siemens to enter into partnership with Francis Ince, a City lawyer and amateur electrician, and Alfred Thompson, an engineer. Thompson and Ince provided financial backing for Ferranti's alternator (patented in July, 1882) but already Sir William had become involved after recognizing that Ferranti's machine effectively coincided with his own design of 1881. Under an agreement of October, 1882, therefore, the Company of Ferranti, Thompson, and Ince would market the patent and act as consulting engineers. Ferranti would receive half the profits, and his partners a quarter each, while Sir William would receive a minimum royalty of £500 per annum. As early as December, 1882, Sir William had begun to promote the Ferranti machine for lighting a new Atlantic liner ordered by the Cunard Line.[92]

The so-called Thomson–Ferranti alternator had several new and significant features including the absence of the commutator which Sir William had condemned in 1863 as a great evil. *The Times* reported in September, 1882:

> Electrical scientists have been diligently at work trying to improve upon the bulky and expensive dynamo machines now in use, and we understood that Sir William Thomson patented a new invention for a simpler and more efficient dynamo machine only a short time before an electrician [Ferranti] in Messrs Siemens establishment hit upon much the same thing. The great feature in the new machine is the absence of iron in the revolving armature, very greatly decreasing its weight, and, by enabling the field magnets to be brought very close together, greatly increasing its efficiency. In fact it is stated that a Ferranti machine to produce 10 000 incandescent lights can be manufactured for less than one-fifth of the cost of the cheapest dynamos at present before the public. The increased efficiency of the new machine is aided by the abolition of the commutator.[93]

The machine's 'exceptionally small size as compared to the work it is capable of doing' particularly impressed the journal *Engineering* in December, 1882, which carried a full description. The prototype illuminated a circuit of over 300 'twenty-candle' Swan lamps, while it weighed little over half a ton, measuring some twenty-four inches by twenty inches by eighteen inches overall.[94]

[91] Sir William Thomson, 'On the proper proportions of resistance in the working coils, the electro-magnets, and the external circuits of dynamos', *BAAS Report*, **51** (1881), 528–31; MPP, **5**, 435–9.

[92] Noel Currer-Briggs, *Doctor Ferranti. The life and work of Sebastian Ziani de Ferranti, F.R.S. 1864–1930* (London, 1970), pp. 120–5; G.Z. de Ferranti and Richard Ince, *The life and letters of Sebastian de Ferranti* (London, 1934), pp. 50–2; SPT, **2**, 754. On the Cunard promotion see John Rennie to Francis Ince, 20th December, 1882, LB4.109, ULC.

[93] *The Times*, 22nd September, 1882. Quoted in Ferranti and Ince, *Sebastian de Ferranti*, p. 53.

[94] 'The Ferranti–Thomson electrical machine', *Engineering*, **34** (1882), 526–7. Quoted in Ferranti and Ince, *Sebastian de Ferranti*, p. 53.

Meanwhile, Robert Hammond of the Hammond Electric Light and Power Supply Company had become the largest single shareholder in Ferranti, Thompson and Ince, and promoted a new company, the Ferranti, Hammond Electric Light Company of Hampstead to purchase the Ferranti patents. Hammond also applied to light Hampstead under the recent Electric Lighting Act (1882). Ferranti's products, however, suffered delays, as Ince complained:

> We have a large number of machines there [in the works] in a more or less skeletal form, and they appear from day to day . . . only to increase in number . . . Pray let us finish the machines in hand . . . and let improvements come as they can later on. It is of great importance that *no* machine should ever *stand* for improvements and if you insist on this we shall be improved off the face of the earth and others with a less perfect machine will do the work and laugh at us while we are theorising . . . I am beginning to feel very dissatisfied with the practical progress we are making with our work . . .[95]

In a similar vein, Hammond reported to Sir William that 'If a new lease of life is [to be] given to the works we shall have to restrain Ferranti from experimenting as at the present time. £20 000 a year would easily be absorbed by him in this direction'.[96] As yet lacking the business experience and entrepreneurial skill of a Jenkin or a Varley, Ferranti's brilliant technical qualities thus actually weighed against the initial success of the enterprise. Financial difficulties mounted with the collapse of Hammond's interests early in 1884. One of Ferranti's achievements, however, had been the installation of a Ferranti alternator and associated equipment in St Peter's College (Peterhouse), Cambridge, an innovation presented to the College by Sir William in 1884 and partly paid for out of his royalties from the Ferranti–Thomson patent.[97]

By late 1885, liquidation of the Hammond Electric Light and Power Supply Company made possible Ferranti's announcement to Sir William: 'You will be pleased to hear that Mr Ince and I have repurchased the Ferranti patents from the Hammond C^oy^ and consequently in dealing with the Alternating Current Machine in the future we shall only have to deal with you as to the best course to be adopted'.[98] Once free of these early difficulties, Ferranti's reputation as an electrical engineer quickly developed. The cause of this change of fortune had been his involvement with Sir Coutts Lindsay, whom Sir William had apparently advised to consult Ferranti over the troubled electric lighting of his Grosvenor Art Gallery in fashionable Bond Street. Ferranti explained the problems to Sir William in late December, 1885:

[95] Francis Ince to Sebastian de Ferranti, *c.*1883, in Currer-Briggs, *Doctor Ferranti*, pp. 133–4.

[96] Robert Hammond to Sir William Thomson, 1st October, 1883, H33, ULC.

[97] Sebastian de Ferranti to Sir William Thomson, 3rd December, 1883, F49, ULC; Robert Hammond to Sir William Thomson, 7th December, 1883, H34, ULC; William Theobald to Sir William Thomson, 24th October, 1885, T150b, ULC. SPT, **2**, 798–9, notes that the installation probably cost over £2000. Sir William accepted a reduced royalty of £300 in 1884. In total his royalties under the 1882 agreement came to £950 including a final payment from the liquidator. See William Theobald to Sharpe, Parkers & Pritchard, 10th and 23rd December, 1885, T153–4, ULC.

[98] Sebastian de Ferranti to Sir William Thomson, 21st December, 1885, F52, ULC. See also Francis Ince to Sir William Thomson, 28th October and 7th December, 1885, I11–12, ULC.

Mr Ince has already sent you a photo of my Alternating Current Motor . . . It became necessary for me to perfect this Motor as Sir Coutts Lindsay & Co. Limited who are putting up the Grosvenor Gallery Installation of 6000 lights wanted an Alternating Current Motor & you will be pleased to hear that I have entered into a contract with them to supply them with all motors which they may require. . . . the installation is being run with Siemens Dynamos & Goullard [*sic*] & Gibbs Convertors [transformers]; these Convertors are being run in series with a high tension current the result of which is that a constant quantity of current is always flowing thro' each Convertor in the circuit. The E. M. F. at the terminals of each Convertor varies therefore as the number of Lamps which are put on instead of remaining constant; the Convertors therefore cannot have their Lamps turned on & off. You can understand how serious this difficulty is & how it really puts an end to a system of electric lighting. The current has to be kept constant in the main circuit by varying the exciting current mechanically instead of having as they should a Dynamo which keeps a constant e.m.f. at its terminals as ours does.[99]

Ferranti added that he did not know whether Sir Coutts Lindsay & Co. were 'fully alive to the difficulty they are in altho' they have already found out that it is almost impossible for them to carry on their installation satisfactorily'. Indeed in two months running 'they have stopped the lighting during the evening at least 12 times which means that people are becoming very dissatisfied'.

Ferranti's solution to the problems included the patenting of a new convertor or transformer (connected in parallel rather than in series) 'made strictly in regard to electrical principles & also so as to be cheap for manufacture', and with very small internal resistance such that 'In consequence of this we can turn off 49 lights out of 50 on one circuit without in the slightest way affecting the remaining lamp; this no one else has accomplished'. He concluded his five-page letter:

I shall be so very grieved to see the Grosvenor Gallery Installation fail thro' incompetence as it certainly will do as matters now stand & I am perfectly sure that I can show them how to put everything right & that I can turn the installation into a success and that to[o] with the least possible expenditure. After all the speculation that has taken place in the electric lighting world I am anxious to shew that electric lighting can be done & that well & not that it is a mere something for Stockbrokers to speculate in. Mr Whiteley tells me that our 1200 light Machine . . . has been working regularly now for 12 months & during this period it has worked to his entire satisfaction & he has never had a single stoppage of the lights.

With two, 750 horse-power Ferranti alternators replacing the Siemens generators, the supply from the Grosvenor Gallery installation covered some one hundred miles of West London streets by 1889. Many of the aristocratic town houses and fashionable public buildings changed from gas to electricity. After years as a novelty rather than as a serious rival, electricity now seemed ready to displace gas altogether. To this end, the London Electric Supply Company (into

[99] Ferranti to Thomson, *op. cit.* (note 98). See also Currer-Briggs, *Doctor Ferranti*, pp. 132–40; Ferranti and Ince, *Sebastian de Ferranti*, pp. 54–6.

which merged the Grosvenor Gallery concern) was formed early in 1888 with a capital of one million pounds. Ferranti became its contracting engineer, responsible for the first large-scale, central electricity generating station at Deptford (which began work in 1891) designed to supply up to a quarter of London's needs but in pratice inhibited by the fragmented nature of local government which favoured small-scale non-centralized installations.[100]

Though Sir William did not involve himself directly in Ferranti's later projects, his opinions in 1888 reflected a definite shift away from small-scale generation with the use of storage cells to large-scale economic generation of electricity:

I quite expect that it [electric light] will altogether supersede gas lighting in cities, although it is impossible to say how soon. If it were at present supplied to houses at twice the price of gas, the light bill would be less than in most private houses than at present, because the electric light can be extinguished in an instant in any place where it is not wanted, and relighted in an instant, with the greatest possible ease. The storage of electricity is too costly for very general use,but it is quite unnecessary in any very large-scale electric lighting.[101]

Two years later, Sir William chaired the so-called Niagara Commission formed to act as a body of consulting experts on another very large-scale generating project of which he had long been an advocate. He favoured a high pressure, direct-current system over the alternating-current system actually adopted for long-distance transmission, arguing characteristically in 1892 that a fascination with 'the mathematical problems and experimental illustrations presented by the alternating current system . . . have . . . tended to lead astray even engineers, who ought to be insensible to everything except estimates of economy and utility'.[102] His well-known preference for direct-current systems rested heavily on this criterion of economy as he made even more explicit in a letter of 1896:

For electric transmission of power over very long distances, *very* high pressure is necessary for economy. I object to the use of alternate currents for this practical purpose, primarily because 41 per cent higher pressure can be transmitted to a great distance by direct current than by alternate current, with the same conductors and insulation in the two cases. For distance exceeding 50 miles it would probably be advisable to use two wires, one of them at +20 000, and the other at −20 000 volts' difference of potentials from the earth. I believe it would not be possible to obtain, for such distances, as good economy by alternate current as by direct current. The prime cost of direct-current dynamos for such purposes has never been gone into practically.[103]

[100] Currer-Briggs, *Doctor Ferranti*, pp. 138–41, 144–63. For a recent study see I.C.R. Byatt, *The British electrical industry 1875–1914. The economic returns to a new technology* (Oxford, 1979).

[101] Sir William Thomson to ?, 28th February, 1888, in SPT, **2**, 875. The identity of the correspondent is unknown.

[102] SPT, **2**, 894–6*n*.

[103] Lord Kelvin to Campbell Swinton, 25th February, 1896, in SPT, **2**, 960–1. See also p. 1197, for Kelvin's final words on direct *versus* alternating current in March, 1907.

Closely allied to Sir William's association with the generation of electricity, his interest in the development of electric trams and trains also began around 1882 when he subscribed to one hundred £10 shares in the Giant's Causeway, Portrush and Bush Valley Railway and Tramway (Electric). This very early electric line opened in the summer of 1883 to convey tourists from the mainline station of Portrush along the scenic North of Ireland coastline to the famous geological spectacle of regular basaltic columns known as the Giant's Causeway. Sir William Siemens had provided much of the expertise and equipment, but after his death in the opening year his place on the Company's Board of Directors was filled by Thomson. Throughout 1884 he advised the Company on technical matters which largely focussed on the need to devise a means of automatic power regulation for the water turbines and electrical system. As W.A. Traill explained to Sir William in February: 'When the current in the car is cut off there is an instantaneous increase of speed and rise of potential before any turbine governor can possibly act'.[104]

Early in his years as a member of the House of Lords, Kelvin was appointed in May, 1892, to a Joint Select Committee of the Lords and Commons on the schemes for underground railways in London. He particularly favoured an overhead arrangement for supply to electric traction at a maximum of 500 volts.[105] In January, 1893, he acted as a consultant to the nascent Central London Railway and wrote at length in answer to six questions put to him by the Company. His first five answers advanced the case of electric over steam power in respect of safety, reliability, health and comfort, speed and frequency for underground railways. These answers merely confirmed the wisdom of the City & South London Railway which had operated with electric traction since 1890. But the answer to the sixth question offered a characteristically Kelvinese comparison of the *economy* of electric with steam traction:

I have carefully considered the estimated cost of working electrically the Inner Circle of the Metropolitan Railway put before me by Mr Alexander Siemens, and I am satisfied that it allows an ample margin for expenditure under the various heads. It shows a cost of 9.2d per train mile, each train being adapted to carry 250 or 300 passengers. This compares favourably with the actual cost of working on that line by steam which, I am informed, is about 10d per train mile. From the considerable experience already obtained in the successful working of electric railways in different parts of the world, and from all the

[104] John Rennie to W.A. Traill, 4th May, 1882, LB4.36, ULC; W.A. Traill to Sir William Thomson, 30th November, 1883; 7th February, 1884, T604 and T607, ULC; Sir William Thomson to Anthony Traill, 30th January, 1884, T591 (draft), ULC. There are many other letters between the Traills and Sir William concerning the railway. See also SPT, **2**, 797. Following Siemens's death, Thomson seems to have negotiated with the Siemens Company on behalf of the railway in order to maintain favourable links with the contractors on matters of technical improvement and repair. Oliver Lodge also offered technical advice. See Oliver Lodge to Sir William Thomson, 11th April, 1884, L74, ULC.

[105] Electric and cable railways (Metropolis): Report from Joint Select Committee, 23rd May, 1892, especially Appendix A.

means of judging which I have, I believe that the economy of electric traction, in the manner proposed . . . will be at least as good as, and probably better than, working steam could be.[106]

Kelvin's analysis here most probably offered a precise answer to an anxious company during a period of economic slump. Unlikely as it was that the railway would in fact have employed steam, the company derived from Kelvin the strong assurance that electric traction was not only more desirable than steam for underground use, but that it was indeed no less economic. The Central London Railway (now part of the Central Line) finally opened with electric locomotives in 1900.[107]

Illustrative of the same undeviating quest for economy and for man's best use of the resources of nature, Lord Kelvin's role as scientific adviser to the British Aluminium Company in the 1890s demonstrates that his commitment to industrial advancement did not diminish with age and status. He made a number of reports to the Company which was then actively engaged in developing a major new aluminium smelter at the Falls of Foyers in the Scottish Highlands. The beautiful waterfall would no longer waste its power into Loch Ness, but would become both the source of the large quantity of electricity required in the electrolytic production of aluminium from its ore and of a reversal of the century-old evil of progressive Highland depopulation:

That magnificent piece of work of the Aluminium Company [at Foyers] was the beginning of something that would yet transform the whole social economy of countries such as the Highlands, where water abounded. He [Kelvin] looked forward to the time when the Highlands would be re-peopled to some degree with cultivators of the soil, but re-peopled also with industrious artizans doing the work which that utilization of the water would provide for them. The British Aluminium works were very popular in the locality. It was only at a distance that the sentimental question 'What is to become of the beautiful Falls of Foyers?' was asked . . . He thought when the time came that every drop of water that now fell over the Falls of Foyers was used for the benefit of mankind, no wise man, no man who ever thought of the good of the people, would regret that the power in the waterfall was developed for the benefit of mankind.[108]

Shortly afterwards, he emphasized his vision of a revival of Highland life and prosperity, 'to the present crofters being succeeded by a happy industrial population occupied largely in manufactories rendered possible by the utilization of all the water power of the country. It seems to me a happy thought

[106] Lord Kelvin to F.A. Lucas, 6th January, 1893, in Michael Robbins, 'Lord Kelvin on electric railways', *J. Transport Hist.*, **3** (1958), 235–8, on pp. 235–7; F.A. Lucas to Lord Kelvin, 24th January, 1893, C73, ULC. Robbins (p. 237) seems to suggest that the subject was rather outside the range of Kelvin's own special interests, which, as we have seen, is hardly the case. Kelvin was not only familiar with electric traction since the Portrush line in 1883, but in 1895 considered the possibility of mainline electric trains being built at the famous Crewe locomotive works. See SPT, **2**, 944.

[107] Robbins, 'Lord Kelvin', pp. 237–8.

[108] SPT, **2**, 1001. Kelvin was opening a carbon factory at Greenock for the British Aluminium Company on 4th August, 1897.

that the poor people of the country will be industrious artizans, rather than mere guides to tourists.'[109]

In promoting this idyllic image of a subsistence economy transformed into industrial prosperity, Lord Kelvin did not attempt to distinguish the profits of capital from the well-being of mankind, to appreciate the intrinsic beauty of an unexploited Falls of Foyers, and to consider seriously other possibilities for the Highland economy. His perspective could scarcely fail to provoke the anger of late-twentieth-century critics of industrialization. Yet his remarks did have a powerful logic of their own. The pestilence, poverty and over-population accompanying the industrial development of his beloved Glasgow had as their counterpart the depopulation and decline of the Highland economy. Now the advancement of science, and especially science-based industry, which was transforming Glasgow into a healthier, more spacious and very prestigious Second City of Empire, would equally bring economic and human salvation to the vast Highland regions, for so long, like Ireland, the mere reservoirs for Glasgow's labour.

[109] *Ibid.*, pp. 1002–3. Kelvin expressed these views at Niagara in 1897 en route to the Toronto meeting of the British Association.

21

Rule, Britannia: the art of navigation

Navigation, in the technical sense of the word, means the art of finding a ship's place at sea, and of directing her course for the purpose of reaching any desired place . . . To find a ship's place at sea is a practical application of Pure Geometry and Astronomy. It is on this piece of practical mathematics that I am now to speak to you. *Sir William Thomson, Lecture on navigation delivered in Glasgow's City Hall, 1875.*[1]

In the spring of 1878, a young Polish seaman joined a British steamer at Marseilles. The future master mariner and celebrated English novelist Joseph Conrad (1857–1924) was to serve some fifteen years – and as many ships – under the Red Ensign which, in the period 1870–1914, flew over almost half of the world's shipping tonnage. No one ever dramatized the significance of the interdependence of practical geometry, the merchant marine, and the economic wealth of the British Empire more effectively than Conrad. One famous novel, *The secret agent*, published in 1907, approached this theme through a tale of anarchists in Victorian London. Deeply angered by England's toleration of all kinds of revolutionaries plotting the overthrow of more autocratic regimes, the foreign diplomat Vladimir planned a terrorist outrage which would provoke the complacent British middle classes into taking repressive action against their dangerous guests. The target was to be a no less significant one than Airy's Greenwich Observatory. Thus Vladimir explained to his agent:

The fetish of today is neither royalty nor religion. Therefore the palace and the church should be left alone . . . The sacrosanct fetish of today is science . . . [The middle class] believe that in some mysterious way science is at the source of their material prosperity . . . The attack must have all the shocking senselessness of gratuitous blasphemy . . . What do you think of having a go at astronomy? . . . The whole civilised world has heard of Greenwich . . . the blowing up of the first meridian is bound to raise a howl of execration . . . Go for the first meridian.[2]

In the late Victorian and Edwardian eras, Britannia's rule depended less on the military power of earlier empires than on geometrical conventions imposed

[1] PL, **3**, 1–138, on pp. 1–2.

[2] Joseph Conrad, *The secret agent. A simple tale* (London, 1907), pp. 41–50. Jerry Allen, *The sea years of Joseph Conrad* (London, 1967), traces Conrad's life under the Red Ensign.

upon the material world in order to make possible the maritime trade and communication, without which the widely scattered and largely island Empire could scarcely have existed. Without those geometrical divisions of space and time – the lines of latitude and of longitude, the British Admiralty charts, the lighthouses and the day-marks – the foundation of Britain's wealth and security would disintegrate into the chaos of anarchy. And at the symbolic centre stood the Greenwich Observatory – marking the first meridian and enabling the calculation of longitude anywhere at sea from an accurate knowledge of Greenwich Mean Time, which in 1884 became the internationally accepted basis of the world's time zones.[3]

Sir William Thomson's 1875 lecture on navigation, delivered in Glasgow's City Hall under the auspices of the Glasgow Science Lectures Association, might well have been cited by Conrad's Vladimir as evidence for the close links between geometrical and astronomical science, practical navigation, and Victorian commercial prosperity. At the same time, Thomson's lecture epitomizes his own conviction that mathematics was the only true metaphysics, and that mathematics should be of a geometrical and eminently practical kind. His lecture began with a discussion of four modes, employed separately or jointly, for finding the position of a ship at sea: pilotage, astronomical navigation, 'dead reckoning', and deep-sea sounding.

First, he defined pilotage as 'navigation in the neighbourhood of land', usually by means of terrestrial objects such as headlands, lighthouses, landmarks, or hills. He made clear the overwhelming importance of geometrical reasoning for the competent pilot:

> Mere acquaintance with the general appearance of the visible objects no longer suffices, and the pilot, however unscholarly may have been his training, *becomes of necessity a practical mathematician*. The principle of clearing marks for [outlying] dangers is of the purest geometry. A certain line is described or specified by the aid of two objects seen in line or nearly so . . . An outlying danger is completely circumscribed by three lines.[4]

In the interests of pilotage, Sir William campaigned vigorously between 1872 and 1881 to improve the distinguishing characteristics of lights at sea. With others, he advocated a system of group flashes in rotating lights and group occultations in fixed lights to make them more easily visible and also recognizable.[5] *The Times* enthusiastically endorsed recommendations that Sir William made in 1879 as having the 'authority of a distinguished man of science and of a

[3] See Derek Howse, *Greenwich time and the discovery of the longitude* (Oxford, 1980), pp. 45–171; David Landes, *Revolution in time. Clocks and the making of the modern world* (Cambridge, MA and London, 1983), pp. 285–7. [4] PL, **3**, 3, 55–6. Our emphasis.

[5] Sir William Thomson, 'On the identification of lights at sea', *BAAS Report*, **42** (1872), 251; 'Lighthouses of the future', *Good Words*, **14** (1873), 217–24; 'On lighthouse characteristics', read at the Naval and Marine Exhibition, Glasgow, 11th February, 1881, PL, **3**, 389–421. See also SPT, **2**, 659–60, 724–9, for relevant correspondence.

practical yachtsman' as well as bearing 'one of the most distinctive marks of genius – simplicity'.[6]

Astronomical navigation, the second mode of finding a ship's position, required the fixing of the ship's latitude and longitude from sights of celestial objects. To begin with, Sir William noted especially that elementary texts on navigation, geography, and astronomy often proceeded on the supposition that the earth is an exact sphere. Upon this supposition, they defined key terms such as terrestrial latitude and longitude, meridians, horizontal planes, verticals, and altitudes. Characteristically, however, Sir William preferred 'definitions of a more practical kind, which, be the figure of the earth what it may, shall designate in each case the thing found when the element in question is determined in practice by actual observation'. Thus, he defined a vertical at any position as simply 'the direction of the plumb line there, when the plummet hangs at rest'. A vertical plane is 'any plane through a vertical' and a horizontal plane is a plane perpendicular to the plumb-line or vertical. The altitude is 'the inclination to the horizontal plane of a line directed to the object'. The meridian of a place is the vertical plane passing through the pivot point of the sky defined as the celestial pole. Consequently, the latitude of a place is 'the altitude there of the celestial pole'; and the longitude is 'the angle between its meridian and that of Greenwich'.[7] Proceeding on the basis of observation, measurement, and geometry, Sir William thereby explicitly avoided the hypothetical model of the earth as an exact sphere and reduced the problem to one of local geometry in much the same way that he (and Fourier) had long before reduced the problem of heat flow to observable geometric relations.

Equally characteristic of his scientific style was Sir William's preference for a particular method of determining latitude and longitude from astronomical observation, known as Sumner's method (after Captain Thomas Sumner of Boston). This method had been used by Captain Moriarty, RN, during the transatlantic cable expeditions of 1858, 1865, and especially 1866, when it greatly facilitated recovery of the lost cable of 1865. In Sir William's view:

> [Sumner's method] is not only valuable as giving us a clear view of *the geometrical process underlying the piece of calculation by logarithmic tables* which is performed morning and evening by the practical navigator at sea, but it actually gives him a much more useful practical way of working out the results of his observations than that which is ordinarily taught in schools and books of navigation, and ordinarily practised on board ship.[8]

[6] Sir William Thomson, 'Distinguishing lights for lighthouses', *Nature*, **21** (1879), 109–10.

[7] PL, **3**, 68–72.

[8] *Ibid.*, pp. 87–98, on p. 91. Our emphasis. See also Captain S.T.S. Lecky, '*Wrinkles' in practical navigation* (London, 1881), pp. 196–217; W.E. May, *A history of marine navigation* (Henley-on-Thames, 1973), pp. 38–9, 172–3; S.T.S. Lecky to Sir William Thomson, 16th February, 1880, CS 612, ULC. At this stage, Lecky was not in favour of Sumner's method. In his '*Wrinkles*' he spoke highly of the method, a change which almost certainly resulted from his dialogue with Sir William whose opinion Lecky greatly valued.

For its application, Sumner's method assumed the availability on board of both an accurate chronometer (for correct Greenwich Mean Time) and the *Nautical Almanac*, which together would enable the navigator to locate the sun's position directly above the earth. Making an altitude measurement of the sun, he could then in principle draw a circle corresponding to that altitude about the sun's position as centre. If he were to draw at a later time a second circle coupled with a simultaneous translation of the first circle along the line of running, the intersection of the circles would in principle give the ship's position. With only a local chart, however, the navigator would find this exercise difficult in practice since he could not simply swing a circle about the sun's position. Instead, he needed to draw segments of the circles (ovals on Mercator's projection) on his local charts. As the calculations needed to obtain these arcs or 'Sumner lines' proved quite laborious and time-consuming, Sir William provided simplifying tables which made the graphical method wholly practicable.

Sir William's warm support for Sumner's method in the *Proceedings of the Royal Society* in 1871 prompted Airy to express alarm to Stokes (as the Society's secretary) lest the method 'should lead to disasters in consequence of people forgetting that the place so determined is liable to a shift east or west depending on the unknown error of the chronometer'. Stokes published the Astronomer Royal's criticisms, but added a devastating reply of his own:

> From a general recollection of a conversation I had with Sir W. Thomson before the presentation of his paper, I do not imagine his object to have been exactly what the Astronomer Royal here describes . . . Of course the place so determined is liable to an error east or west corresponding to the unknown error of the chronometer; and doubtless, under ordinary circumstances, this forms the principal error to which the determination of a ship's place is liable. This remains precisely as it did before; and it is hard to suppose that the mere substitution of a graphical for a purely numerical process could lead a navigator to forget that he is dependent upon his chronometer . . .[9]

'Dead reckoning', the third mode of finding a ship's position, concerned navigation in cloud covered skies and restricted visibility, with the magnetic compass as the only guide, other than soundings in comparatively shallow waters. Given especially our limited knowledge of ocean currents, and the inaccuracies of the compass in iron ships, Sir William argued forcefully that 'undue trust in dead reckoning has produced more disastrous shipwrecks of seaworthy ships . . . than all other causes put together'. Even allowing for currents, he believed that it would be unsafe to trust a ship's place within fifteen or twenty miles per twenty-four hours of dead reckoning. And even with half-quarter point (one or two degrees) compass errors in wooden vessels, and with very careful steering, he estimated that a total error of a quarter of a point

[9] Sir William Thomson. 'On the determination of a ship's place from observations of altitude', *Proc. Royal Soc. London*, **19** (1871), 259–66, 524–6; G.B. Airy to G.G. Stokes, 5th April, 1871, *Proc. Royal Soc. London*, **19** (1871), 448–50; Sir William Thomson, *Tables for facilitating Sumner's method at sea*' (London, 1876).

would lead to the ship being ten miles out of her course in a run of 200 miles. The only solution for a competent navigator approaching land in cloudy weather was to feel his way by deep-sea soundings until 'he makes the land and makes sure by lighthouse and landmark of where he really is'. Deep-sea soundings not only complemented dead reckoning, but constituted Sir William's fourth mode of finding a ship's position at sea.[10]

In this and the next chapter we shall be principally concerned with the historical origin and commercial context of Sir William's famous navigational innovations during the 1870s. His deep-sea sounding machines and his dry-card magnetic compasses facilitated the third and fourth modes of navigation just outlined. His enthusiasm for navigation had special sources of inspiration during this period. There was, for example, his long-standing friendship with Archibald Smith and their mutual work on magnetic problems. Again, his purchase of the yacht *Lalla Rookh* in 1870, shortly after the death of Margaret and following on more than a decade of submarine telegraphic work, gave him the opportunity to resolve the practical problems of compass and sounding machine. During the 1860s and 1870s also, the coming-of-age of the large iron steamship, centred on Clydeside, provided the opportunity for Sir William to exercise his shrewd entrepreneurial skills. Above all, however, it was his commitment to 'practical mathematics' which yielded the fruits of navigational improvement. With good reason, therefore, Sir William referred his audience in 1875 to Lieutenant Raper's 'excellent book on navigation' which opened with the telling prefatory remarks:

> Those who have been brought up to the sea, and who have experienced the distaste for long calculations which that kind of life inspires, will not hesitate to admit that the only means of inducing seamen generally to profit by the numerous occasions which offer themselves for finding the place of the ship, is extreme brevity of solution. It is not, however, merely as a concession to indolence, that rules should be made as easy and simple as possible; the nature of a sea life demands that every exertion should be made to abridge computation, which has often to be conducted in circumstances of danger, anxiety, or fatigue.[11]

The shipyard of Empire

When, in the summer of 1907, the 31 550 gross-ton Cunard liner *Lusitania* ran her speed trials on the Firth of Clyde within close view of Lord Kelvin's home near Largs, the eighty-three-year-old peer must have reflected that here was indeed more than a reassuring symbol that Britain's shipbuilders could create the largest, most prestigious, and most elegant ships in the world. Constructed at Clydebank, near Glasgow, the four-funnelled, turbine-driven giant for the Liverpool to New York passenger and mail service might well have reminded

[10] PL, **3**, 105–16.

[11] Henry Raper, *The practice of navigation and nautical astronomy* (London, 1840), p. vi.

Lord Kelvin that less than a century had passed since his father had spent four days on an uncertain passage in a lime smack from Ireland to Scotland, en route for Glasgow College. Now, the mighty *Lusitania*, navigated by Lord Kelvin's patent compasses, would regularly cross the North Atlantic in a mere five days.[12]

During the first half of the nineteenth century, ocean trade had been wholly dependent on sail, and the best sailing ships belonged not to Britain, but to the United States. American sailing packets – such as the fast and reliable ships of the Black Ball Line founded in 1816 – ruled North Atlantic passenger routes. The first of the tea clippers, the *Helena*, came from a New York builder in 1841, and, in the period up to 1860, American built China clippers excelled on the long-distance passages from the East to London. British builders, located mainly in Aberdeen and Greenock, eventually replied with more durable hardwood clippers which could equal and even surpass the performance of their American softwood rivals. In 1866, for example, three clippers, *Taeping*, *Ariel*, and *Serica*, all built by Robert Steele of Greenock, arrived in London from China on the very same day to win the most famous tea-clipper race in maritime history.[13]

Britain's pre-eminence on ocean routes had been by no means assured prior to 1860. The fierce American competition only dwindled with the Civil War of the 1860s and, more importantly with the progress of the iron and screw steamship. Samuel Cunard's Liverpool-based and Glasgow-financed British and North American Royal Mail Steam Packet Company was already in existence by the early 1840s, as were two major new Southampton-based mail services. The Peninsular and Oriental Steam Navigation Company, founded in 1837 by the Scotsman Arthur Anderson (1792–1868), and better known as P & O, quickly opened up the England to India route with an overland link via Egypt. The Royal Mail Steam Packet Company, founded in 1839 by another Scotsman, James MacQueen (1778–1870), served the West Indies. All three companies employed wooden paddle steamers assisted by a traditional arrangement of masts and sails.[14] Indeed, the British Admiralty, whose agent travelled with the mails, insisted on wooden hulls for mail ships until the early 1850s on safety grounds, mail ships being regarded as reserve warships. Iron hulls, the Admiralty believed, would incur greater damage and inflict more injury in a conflict than timber, while also suffering from serious deviations of their

[12] 'The Cunard liner *Lusitania*', *The shipbuilder*, **2** (1907–8), 83–8; Thomson's patent compass books: record of ships supplied with compass (1876–1918), Glasgow University archives. Kelvin's firm supplied five compasses to the *Lusitania* in April, 1907.

[13] Sidney Pollard and Paul Robertson, *The British shipbuilding industry, 1870–1914* (Cambridge, MA and London, 1979), pp. 9–13; R.G. Albion, *Square-riggers on schedule. The New York sailing packets to England, France and the cotton ports* (Princeton, 1938); John Shields, *Clyde built* (Glasgow, 1949), pp. 121–4.

[14] F.E. Hyde, *Cunard and the North Atlantic, 1870–1973* (London, 1975); Boyd Cable, *A hundred year history of the P & O* (London, 1937); T.A. Bushell, *'Royal Mail'. A centenary history of the Royal Mail Line. 1839–1939* (London, 1939).

magnetic compasses. Only after the burning at sea in 1852 of the wooden Royal Mail steamer *Amazon*, with the loss of 105 lives, did the change to iron become a reality for these high-class ships.[15]

Between 1840 and 1870, Britain became for iron ships the shipyard and shipowner of the world and the Clyde the shipyard of Empire. To the Napiers of Clydeside must go the lion's share of the credit for the swift change in Britain's maritime fortune. The River Clyde had been made navigable for sea-going vessels only at the end of the eighteenth century (ch. 2). Greenock, at the head of the deep Firth and mouth of the river, was indeed to remain the centre of wooden shipbuilding on the Clyde well into the nineteenth century. But the effect of the change to iron on Greenock proved devastating: 'Gone was the industrious hum of saw-milling and the pleasing sound of caulking mallet striking caulking iron that had been heard . . . from shipyards that had specialized in wooden shipbuilding only. Hundreds of craftsmen skilled in the art of wooden shipbuilding, many of whom had never had a single day of idleness, walked the streets of Greenock or stood in small groups, with little to talk about except their idleness and misery'.[16]

By the 1840s, then, the marine engineering activities of the Napier cousins, Robert and David, had altered the balance firmly in favour of an iron steamship era centred on the rapidly growing industrial city more and more easily accessible from the sea. Robert's engineering of the first Cunard liners, and his success in obtaining for Cunard the financial backing needed around 1840, formed the prelude to a Napier era of iron shipbuilding in Glasgow. Although David Napier had already left for London, two of his enterprising managers, David Tod and John MacGregor, had opened their own yard in 1836.[17] In 1850, they launched to their own account the 1600 ton *City of Glasgow* for a Glasgow to New York service. Unlike the usual Atlantic mail steamer of the 1840s, the ship boasted iron hull and screw propulsion. Within the same year, the *City of Glasgow* was bought by the new Liverpool and Philadelphia Steam Ship Company (best known subsequently as the Inman Line), and, with other Tod and MacGregor ships soon in service, the line prospered, even without a mail subsidy and even after the loss of two ships in 1854. By contrast, the rival Cunard and Collins Lines looked outdated. The American owned Collins Line, with its wooden hulls and paddle wheels, ceased trading in 1858 after a series of disastrous losses, while Cunard switched to iron with the Robert Napier-built *Persia* of 1856, though not to screw propulsion until the early 1860s.[18]

The business network established by Robert Napier provided a solid founda-

[15] Bushell, *'Royal Mail'*, pp. 65–79.

[16] Shields, *Clyde built*, pp. 121–2; Pollard and Robertson, *British shipbuilding industry*, pp. 13–69.

[17] James Napier, *Life of Robert Napier of West Shandon* (Edinburgh and London, 1904), pp. 1–165; Anthony Slaven, *The development of the west of Scotland: 1750–1960* (London, 1975), pp. 125–33, 178–82.

[18] N.R.P. Bonsor, *North Atlantic seaway* (5 vols., Newton Abbot and Jersey, 1975–9), **1**, pp. 218–46, 201–8, 72–84.

tion for the development of the Clyde's shipbuilding reputation in the second half of the century. The phrase 'Clyde-built' became synonymous with economy, quality and reliability, and with an enormous variety of high-class ships ranging from ocean liner and battleship to tug and dredger. Some insight into why the Clyde deserved such a reputation may be gleaned from a comparison made by Joule in 1858 between Brunel's *Great Eastern* (or *Leviathan*) and Napier's *Persia*:

> I saw in London 3 weeks ago the blundering arrangements for launching the *Leviathan* and could not help contrasting them with those for the launch of the *Persia* at Glasgow, a vessel of at least one quarter the weight . . . I made a calculation what her speed would be when fully laden and find it somewhat less than that of *Persia*. I think some limit should be placed on the enormous consumption of coal by steam ships which is rapidly exhausting our coalmines.[19]

Increasingly, mailship owners such as P & O turned from the long-established English shipbuilders of the Thames or Bristol to the comparatively new iron shipbuilders of the Clyde for their first-class tonnage. Cunard's English friends could no longer advise him, as they did when he first approached Napier, that on the Clyde he would have 'neither substantial work nor completed in time'.[20]

Almost every member of the new generation of Clyde shipbuilders and marine engineers had received early experience and training in Robert Napier's works. William Denny, John Elder, J. and G. Thomson, A.C. Kirk, and William Pearce – best known among the leaders of Clyde shipbuilding in the period 1850–80 – had all been through Napier's stable at some stage of their early careers. So too had J.R. Napier, Robert's eldest son and important associate of William Thomson in the period 1855–75.[21] The new generation differed from Napier in one respect – their even greater enthusiasm for technical innovation. It was therefore on Clydeside that the marine compound, triple expansion, and turbine engines were first developed in commercial form, that steel replaced iron in hull construction, and that new departures such as electric lighting of ships were first introduced. It is within this commercial context of technical innovation that we must subsequently view Sir William's instruments of navigation.

The most innovative of the firms after mid-century was that of John Elder (1824–69). Elder, one-time pupil of Lewis Gordon at Glasgow College and chief draughtsman for Robert Napier, entered into partnership with a firm of millwrights in 1852. As Randolph, Elder and Company, the firm began marine engine building and, in 1854, fitted a compound steam-engine, designed and patented by Elder, to the paddle steamer *Brandon*. It is highly probable, given

[19] J.P. Joule to William Thomson, 1st January, 1858, J248, ULC. The broadside launch of the *Great Eastern* took several weeks. See also Napier, *Life of Robert Napier*, pp. 92–3, for the Napiers' verdict on the 'blundering arrangements', and see pp. 192–7 for an account of the *Persia* contract.

[20] Samuel Cunard to Robert Napier, 21st March, 1839, in Napier, *Life of Robert Napier*, p. 135; Hyde, *Cunard*, p. 6.

[21] Napier, *Life of Robert Napier*, pp. 85–8, 150–1, 182–207.

Elder's friendship with Rankine, that a knowledge of the new science of thermodynamics guided Elder in the design. A dramatic reduction in fuel consumption due to the use of high-pressure steam in two manageable stages led to orders for engines for two ships for the Pacific Steam Navigation Company of Liverpool. The consequent economy made possible steam navigation along the western seaboard of South America where coal supplies were scarce or non-existent. Pacific Steam returned again and again to Elder for new ships, and grew into one of the world's largest shipping companies by the 1880s.[22]

The marine compound engine also made possible the economic development of long-distance ocean navigation by steam rather than sail. Although steamship owners were generally slower than Pacific Steam to invest in the more complex – and thus more difficult to service – compound engine, the P & O Company had adopted a version in the early 1860s, and Alfred Holt's new Ocean Steamship Company of Liverpool ordered three compound-engined ships from Scott's of Greenock in 1865–6. These three vessels opened up the Far East to long-distance steam, just in time for the completion of the Suez Canal in 1869, and thus brought to a close the golden age of the tea clipper.[23] Henceforth, the high-quality imports and the manufactured exports of Britain would be loaded into iron or steel steamships designed to run on scheduled routes across the world.

Although Elder's untimely death came in 1869, the firm continued to develop its reputation for innovation. Under the chairmanship of a former manager at Napier's, Sir William Pearce, the firm built for Liverpool owners in 1874 the world's first triple expansion steamer, the *Propontis* (well known in every maritime history book). Her designer, A.C. Kirk (1830–92) had been an apprentice in Napier's foundry, and was to return to that firm as senior partner after Napier's death in 1876. Kirk had attended J.D. Forbes's classes in natural philosophy, but was indirectly indebted to William Thomson's teaching of the new theory of heat. He wrote to Thomson in 1883:

> About the year 1852 or 1853 while I was an apprentice at the Vulcan foundry (R. Napier) one of your students . . . shewed me a printed class exercise . . . you had given regarding the calculation based on the mechanical theory of heat of a refrigerating machine of compressed air, & that was my first introduction to the mechanical theory of heat, for Professor Forbes in Edinburgh had not touched on it.[24]

With its relatively high-pressure operation in the first of three successive stages, the triple expansion engine would remain the most popular, most reliable, and most economical form of driving unit for the workhorses of the sea

[22] W.J.M. Rankine, *A memoir of John Elder, engineer and shipbuilder* (Edinburgh and London, 1871).

[23] F.E. Hyde, *Blue Funnel. A history of Alfred Holt and Company of Liverpool from 1865 to 1914* (Liverpool, 1957), pp. 15–39.

[24] A.C. Kirk to Sir William Thomson, 14th December, 1883, K101, ULC. See also Napier, *Life of Robert Napier*, pp. 188–9.

until the mid-twentieth-century's substitution of highly developed diesel engines. Though the steam turbine found favour with twentieth-century thoroughbreds of liner and capital ship, the reciprocating steam-engine continued as the symbol of the Clyde shipbuilder's art.

If one had to point to a single year of particular significance for the convergence of Clyde shipbuilding and Glasgow's University, that year would probably be 1870. The Suez Canal had opened a year earlier. The College had just vacated its old High Street site near the heart of the city, and removed itself west to Gilmorehill within sight of the busy shipbuilding yards of the upper Clyde. William Thomson had just purchased the *Lalla Rookh*, W.J.M. Rankine had reached the height of his fame for work on steam-engines, naval architecture, and shipbuilding, and the shipbuilders themselves were already well enough established along the banks of the river (since 1841, two-thirds of Britain's steamship output had been Clyde-built) to take advantage of an unprecedented demand for iron ships during the early 1870s.

A depression in the industry during the late 1860s had virtually finished London and Liverpool as major shipbuilding centres. Important wooden shipbuilding towns such as Sunderland had only just begun the transition to iron. The River Tyne was still at a comparatively early phase of iron shipbuilding. Thus, when in 1870–1 the new boom was fully under way, the Clyde flourished. At the end of 1871, over 300 000 gross tons of new ships were on order or under construction there. And, in 1876, the Clyde achieved the remarkable feat of building more iron ships than the rest of the world put together.[25]

Although depression and boom alternated with one another thereafter in what is, after all, a notoriously cyclical industry, the years 1870–1914 were ones of British pre-eminence in world shipping and shipbuilding. Around half of the world's seaborne trade by both value and volume was carried in British ships. In 1874, the Suez Canal was used by almost 900 British ships, over ten times the number of Britain's nearest rival, and builder of the Canal, France. Between 1900 and 1913, 60% of the world's shipping tonnage was built in Britain with almost a quarter of the new tonnage for export. Altogether, there were over fifty shipbuilding yards with an annual output exceeding 10 000 gross tons each, over twenty of which were located on Clydeside. About one-third of Britain's huge output was produced on the Clyde between 1870 and 1914.[26] Yet, at mid-century, Britain's place as the shipyard of the world had been far from guaranteed. The pace of historical change, created by a close partnership between technical innovation and economic demand, was indeed dramatic in its consequences for Britain, the Clyde, and for Thomson himself.

[25] Pollard and Robertson, *British shipbuilding industry*, pp. 46–69; Sidney Pollard, *The economic history of British shipbuilding, 1870–1914* (University of London PhD thesis, 1951).

[26] W.S. Lindsay, *History of merchant shipping and ancient commerce* (4 vols., London, 1874), **4**, p. 643; *The shipbuilder*, **2** (1907–8), 123–38; Pollard and Robertson, *British shipbuilding industry*, pp. 25–6, 61–2.

The Lalla Rookh

From the industrial and commercial world of the River Clyde, Sir William moved quite naturally into a relationship with the sea itself, which, as Conrad expressed it, 'plays with men till their hearts are broken, and wears stout ships to death'.[27] Life at sea even in the second half of the nineteenth century was often a grim struggle for the survival of ships and men. Casualties were high, often – in winter – very high. In the five years from 1877 to 1881, for example, fifty-two ships foundered on the coast of County Down, many of them on passage to the Clyde, Liverpool, or Belfast. In November, 1880, as many as 130 ship losses were posted at Lloyd's in a single day, while altogether in that year over 1200 British seamen alone lost their lives. And, even as late as 1894, a west-north-west gale wrecked forty-seven vessels around the British Isles and caused the disappearance of another forty-nine British-flag vessels at sea, all but fourteen of the known total being sailing craft.[28]

Given such alarming casualty figures, it is perhaps surprising that a well-established Glasgow professor should have purchased, in September of 1870, a large schooner-yacht from the wealth accumulated through his recent telegraph patents. Following the death of his wife Margaret on 17th June, 1870, Sir William had spent two weeks at the end of August on a yacht cruise among the Western Islands of Scotland. Very soon after his return he began a quest for a yacht of his own. Constructed of oak in 1853, the *Lalla Rookh* measured 126 tons and was, by any standards, a sizeable pleasure craft, requiring a professional master and crew. Even Archibald Smith, familiar as he was with all things maritime, expressed his shock at the news of Thomson's new passion: 'You quite take my breath by your plan for a schooner of 120 tons. I take for granted you have considered the question of size well, for such a vessel would be too large for ordinary West of Scotland sailing'. Yet Sir William's impulsive purchase at the end of the 1870 season was altogether characteristic, reflecting in fact, on a considerably larger scale, his controversial investment of 1842 in a rowing boat 'built of oak, and as good as new'. And so now he justified the *Lalla Rookh* to his brother with the explanation that she too was 'of oak & very strong and in perfect condition, also a very good model & said to be a fast sailer'.[29]

Towards the end of September, Smith accompanied Thomson to Cowes to inspect the yacht at her layup berth and to help make a final decision on purchase. Within a few days Thomson had made ready for the voyage north to the Clyde. A brief call in Belfast Lough on 16th October enabled his brother, James, and

[27] Joseph Conrad, *The mirror of the sea* (London, 1906).

[28] Ian Wilson, *Shipwrecks of the Ulster coast* (Coleraine, 1979), p. 1; Allen, *The sea years of Joseph Conrad*, p. 113; Ian Wilson, 'The great gale of 1894', *Sea breezes*, **58** (1984), 41–53.

[29] Sir William Thomson to Hermann von Helmholtz, 8th September, 1870, in SPT, **1**, 581–2; Archibald Smith to Sir William Thomson, 19th September, 1870, S201, ULC; Sir William to James Thomson, 21st September, 1870 (copy), T480, ULC.

nephews, David Thomson King (Elizabeth's son) and James Thomson and William Bottomley (Anna's sons), to partake in a short Scottish cruise before the *Lalla Rookh*'s winter layup in the Gareloch, opposite Greenock. This annual layup conveniently coincided each year with the six month Glasgow College session from November to April.[30]

After a spring refit at a Greenock shipyard – from which she emerged with the varnish of her upper works gleaming and her bottom freshly copper-sheeted to reduce fouling – the *Lalla Rookh* awaited her first full season under Sir William Thomson's ownership. Although his plans varied from year to year, the general pattern was a short, working-up cruise in April (limited to the Firth of Clyde), followed by a passage to the South Coast of England, the departure point for some of the longer passages to Lisbon (1871), Gibraltar (1872), and Madeira (1874 and 1877). Cruising among the islands of the West Coast of Scotland would often take place in late August or early September, with more Clyde sailing to complete the season in October, prior to a layup and refit, for which in 1871–2 the famous shipbuilder, Robert Steele of Greenock, charged no less than £325.[31]

The *Lalla Rookh*'s first season set the pattern for the next two decades of Sir William's life. Following an opening cruise to Arran in April, 1871, the yacht departed for the South Coast with calls at Penzance, Plymouth and Dartmouth. Also aboard were James Thomson Bottomley and Elizabeth's husband, Rev David King, the latter, in Sir William's words, 'faintly denying that he would have enjoyed it still more if she had been on the slip at Greenock all the time', while the former 'was more reticent, but I believe felt as deeply'. Having replaced Rev King, William Bottomley (Thomson's brother-in-law) remarked to Sir William that 'the best thing about yachting was going on shore'.[32] His companions did not thus always share his special delight in ploughing the swelling ocean waves. S.P. Thompson has described Sir William's distinctive relationship with the sea:

> He was a daring navigator, and would sail far into the season when other yachts were laid up, sometimes in darkness and in severe weather. Once when he was sailing in the teeth of a gale his assistant John Tatlock, who often was with him as amanuensis, heard Captain Flarty saying half-aloud in Sir William's presence: 'You will not rest till you have your boat at the bottom'. He took no notice. He never seemed to tire. With all the sailors he was extremely popular; their only grievance was that he would sometimes pop up on deck in the small hours of the morning to make sure the watch was at his post and awake. In all the operations of sailing he took his keenest interest, and became a most expert navigator.[33]

[30] Sir William Thomson to Jessie Crum, 20th September, 1870, in SPT, **1**, 582; James Thomson, *Papers*, p. lxiii; SPT, **2**, 585.

[31] Sir William Thomson to Mrs P.G. Tait, 29th March, 1871, in SPT, **2**, 586–7; Robert Steele & Co. to Sir William Thomson, 4th April, 1872, S75, ULG.

[32] Sir William Thomson to Mrs P.G. Tait, 8th May, 1871, in SPT, **2**, 592–3.

[33] *Ibid.*, pp. 594–5.

Sir William Thomson's schooner-yacht *Lalla Rookh* of 126 tons which he purchased in 1870. The grand scale of the yacht mirrored Sir William's character, and for many years served as his summer residence, as floating laboratory, and as a means of satisfying his sense of adventure and love of the sea. [From SPT.]

Sir William's membership of the Admiralty Committee upon the Designs of Ships of War, set up in the wake of the tragic and controversial overturning of the new turret ship HMS *Captain* with the loss of 472 hands in September, 1870, at first restricted the 1871 season. Gun turrets in place of traditional broadside arrangements were new to the Royal Navy. HMS *Monarch*, designed and built by the Admiralty, was commissioned in 1869 as the first sea-going turret ship. HMS *Captain*, designed and built by Lairds of Birkenhead to the specification of Captain Cowper Coles, followed in January, 1870. Coles had been the strongest advocate of the turret system and, following his criticism of the *Monarch*'s limited arc of fire, mobilized public opinion in favour of building the *Captain*. The Court of Inquiry attributed her subsequent capsize, one of the most notorious maritime events of the Victorian era, to 'pressure of sail assisted by a heave of the sea'. Scott Russell even hinted that the Admiralty had all along wanted to prove her inferiority to the tiresome political lobby which favoured turret ships. A major Committee, chaired by the Ulster peer, Lord Dufferin, and comprised not only of Admiralty officials and senior naval officers but of Sir William Thomson, W.J.M. Rankine, and William Froude, pronounced on the *Captain*'s design faults, notably the fatal combination of a very low freeboard and top-heavy superstructure, and reported on the stability and structural strength of other large modern fighting ships such as HMS *Monarch*. As a result, new rules were laid down governing the construction of future vessels. The Committee met every two weeks, necessitating much travelling on the part of Sir William from Glasgow to London during the College session, and the delegation of duties to J.T. Bottomley.[34]

With the *Lalla Rookh* on the South Coast from May, however, Sir William's Admiralty duties eased and the adjournment of the Committee for several weeks permitted longer voyages to Cherbourg and Lisbon. As the Committee prepared to submit its report in July, Sir William undertook short South Coast cruises with Froude and his old Cambridge friend Shedden. Lord and Lady Dufferin also came to Cowes for the yachting season. Suddenly the knight of some five years standing had begun to participate in a fashionable British pastime, the recreation of peers and royalty alike.[35]

Yet the *Lalla Rookh* was for Sir William less a tool for social advancement, than a place of retreat from the demands of university duty, telegraphic business, and government or British Association committees. He 'really found the L.R. the quietest and best place attainable for work'; not least, in 1871, for the arduous preparation for his presidential address to the British Association. A year later he told Helmholtz likewise that he could 'only get mathematical work done in the yacht, as elsewhere there are too many interruptions'. In particular, the *Lalla Rookh* had become his vacation residence, a veritable maritime home unfettered by the limits of stone and mortar, ready to move closer to or more distant from the centres of scientific and commercial activity as Sir William required, and always in readiness to receive his close relations or distinguished colleagues during those six months away from Glasgow. Stokes, who seemed not to relish the seafaring life, occasionally felt the frustration of Thomson's elusive wanderings: 'It is not easy to say where to find a man who owns a yacht'. For his part, Thomson would tease his friend: 'Will you not come and have a sail with us and see & *feel* waves. We would take you away out to the west of Scilly for a day or two if that would suit best'.[36]

After the 1871 British Association meeting at Edinburgh, the *Lalla Rookh* received her most distinguished guest of all: Helmholtz. Sir William had aimed to have aboard not only his Berlin colleague, but also Tait, Maxwell, Huxley, and Tyndall. Tait, however, according to Thomson, had 'a great aversion to being afloat'. While denying this accusation, Tait himself expressed a strong preference for spending 'my few holidays in active physical work, such as the

[34] *Ibid.*, pp. 731–3; SPT, **1**, 582–4; 'The story of the *Captain*', *Gentleman's Mag.*, **227** (1870), 701–14; John Scott Russell, 'The loss of the *Captain*', *MacMillan's Mag.*, **22** (1870), 473–80; Pollard, *Economic history of British shipbuilding*; Robert Gardiner (ed.), *Conway's all the world's fighting ships. 1860–1905* (London, 1979), pp. 20–1; Sir William Thomson to G.G. Stokes, 7th January and 3rd March, 1871, K171 and K173, Stokes correspondence, ULC. [35] SPT, **2**, 595–8.

[36] Sir William Thomson to Jessie Crum, 1st July, 1871; to Alexander Crum, 11th July, 1871; to Hermann von Helmholtz, *c.* 1st June, 1872, in SPT, **2**, 597–8, 625; G.G. Stokes to William Thomson, 9th September, 1880, S434, ULC; William Thomson to G.G. Stokes, 14th July, 1880, K244, Stokes correspondence, ULC.

game of golf'. Helmholtz, indeed, confirmed this preference when he stayed with Tait at St Andrews, en route to join the *Lalla Rookh*: 'Tait is a peculiar sort of savage; lives here, as he says, only for his muscles, and it was not till today, Sunday, when he dared not play, and did not go to church either, that he could be brought to talk of rational matters . . . From Sir William we had yesterday two telegrams and two letters, today two telegrams with changing directions. W. Thomson must be now just as much absorbed in yachting as Mr Tait in golfing'.[37]

Helmholtz, in fact, was the only one of the five famous names actually to join the *Lalla Rookh*, lying at anchor off the Duke of Argyll's castle at Inverary near the head of Loch Fyne, a northern finger of the Firth of Clyde. The eighth Duke (1823–1900), chieftain of the celebrated Campbell clan, was about to mark the marriage in March of that year of his son, the Liberal MP for Argyllshire (1868–78), later Governor-General of Canada (1887–8) and ninth Duke of Argyll (1900–14), to Princess Louise, Queen Victoria's fourth daughter, by a gathering of his clans-folk on land and some forty yachts off shore.[38] The eighth Duke became a prominent Liberal Unionist with whom Sir William would make common cause in the opposition to Irish Home Rule (ch. 23), as well as a frequent host to Queen Victoria herself.

Helmholtz's first impression of the yacht's accommodation and of life on board is revealing:

> My cabin is just so large that I can stand upright in it beside the narrow bed: the rest of the space is less lofty, yet it contains wash-table, dressing-table, and three drawers, so that I can arrange my things well. For washing the space is rather small, particularly when the ship rolls and one cannot stand firm. Today we began the morning by running on deck wrapped in a plaid and sprang straight from bed into the water. After that an abundant breakfast was very pleasant. Then came visits to the other yachts, and so the day has up to now passed very pleasantly in spite of the rain.[39]

From rain-swept Inverary, the yacht sailed for Greenock, and thence to Belfast Lough for visits to Thomas Andrews's laboratory in Queen's College and Lord Dufferin's estate at Clandeboye, County Down. The fifth Baron Dufferin (1826–1902) was himself an adventurous yachtsman and skilled navigator, having made a celebrated voyage in 1856 in his schooner-yacht *Foam* to Iceland, Jan Mayen, and Spitzbergen, as vividly described in his *Letters from high latitudes*.[40] He served as Under Secretary of State for India (1864–6) and for War (1866) under Liberal governments. For his public services, Lord Dufferin became Earl of Dufferin and Viscount Clandeboye in November, 1871. He was appointed Governor-General of the Dominion of Canada (1872–8), and Viceroy of India (1884–8). Although his sympathies were with liberal politics, he, like the

[37] Sir William Thomson and P.G. Tait to Hermann von Helmholtz, 30th March, 1871; von Helmholtz to Frau von Helmholtz, 20th August, 1871, in SPT, **2**, 587–8, 612–13.

[38] Hermann von Helmholtz to Frau von Helmholtz, 24th August, 1871, in SPT, **2**, 613–14.

[39] *Ibid.*, 614.

[40] Lord Dufferin, *Letters from high latitudes* (London, 1857).

Duke of Argyll and Sir William himself, would split with Gladstone over Home Rule for Ireland. Sir William's subsequent social and political alignments were here in the making.

Taking aboard his brother James and William Bottomley and his two sons, the *Lalla Rookh* set a northerly course for the Western Islands of Scotland, encountering foul enough weather from the open Atlantic to prostrate 'even our Admiral', according to Helmholtz. The weather improved, however, for a passage to Loch Ailort, where Professor Hugh Blackburn resided between sessions, and northwards to the Gair Loch, before returning through the Sound of Mull.[41] James meanwhile recorded some of the professional activity aboard:

> William during the almost perfect calm [in the Sound of Mull], noticed the very slight speed of the yacht through the water as being fit to enable him to make experiments on ripples and waves regarding which he had been making out mathematical theories and discussing them with Professor Helmholtz. He did it by having a nearly vertical fishing line hanging in the water; and observing by himself with aid of W.B. and of me and of J.T. junr. the conditions as to the mode of spreading of the waves, and as to the speed of the yacht that he wanted. He found the results very satisfactory to him.[42]

These wave discussions evidently formed a major part of the scientific dialogue between the two giants of nineteenth-century European physics. As in other spheres of his life (such as his undergraduate academic and rowing activities), Thomson approached the subject in a highly competitive spirit, facetiously warning his colleague from the outset not to work at waves while he went ashore at Inverary.[43]

Helmholtz's views, from a wholly different cultural perspective, throw much light on both his own, comparatively serious and formal manner, and Thomson's extraordinarily relaxed and unconstrained character, the very personality which made him equally at home with students, engineers, seafarers, and aristocrats. Helmholtz wrote to his wife:

> It was all very friendly and unconstrained [at the Blackburn's house]. W. Thomson presumed so far on the freedom of his surroundings that he always carried his mathematical notebook about with him, and as soon as anything occurred to him, in the midst of company, he would begin to calculate, which was treated with a certain awe by the party. How would it be if I accustomed the Berliners to the same proceedings? But the greatest naiveté of all was when . . . as soon as the ship was on her way, and everyone was settled on deck as securely as might be in view of the rolling, he vanished into the cabin to make

[41] SPT, **2**, 615–16; James Thomson, *Papers*, lxiii–lxv; Archibald Smith to Sir William Thomson, 21st August, 1871, S204, ULC; C.S. Smith to Sir William Thomson, 27th December, 1872, S222, ULC.

[42] James to his wife Elizabeth Thomson, 14th September, 1871, in James Thomson, *Papers*, lxiv–lxv.

[43] SPT, **2**, 614–18; Sir William Thomson, 'Ripples and waves', *Nature*, **5** (1871), 1–3; MPP, **4**, 86–92; Sir William Thomson to G.G. Stokes, 7th October, 31st October, 20th November, and 29th November, 1871, K178, K179, K368, and K180, respectively, Stokes correspondence, ULC.

calculations, while the company were left to entertain each other so long as they were in the vein; naturally they were not exactly very lively.[44]

For the remainder of the season, Sir William lived aboard the *Lalla Rookh*, moored either at Largs or Greenock for the most part, but making a couple of brief runs to Arran and Loch Fyne. This arrangement enabled him to move towards the proof-stage of his *Electricity and magnetism* as well as to think more about waves.[45]

At the same time, however, the *Lalla Rookh* became the testing laboratory for a 'pressure log' designed by Robert Napier's son, J.R. Napier, for the measurement of a vessel's speed through the water. J.R. Napier's concerns were primarily with ship performance (ch. 2). He instituted, for example, elaborate measured-mile trials for new ships on the Clyde in order to provide owners and builders with accurate data on performance. Like his Glasgow academic friends, Rankine and Thomson, Napier participated actively in the Glasgow Philosophical Society, and his several papers on practical maritime subjects often referred to Sir William's scientific work. A frequent visitor on board the *Lalla Rookh*, he had a particular interest in compass improvement (ch. 22) and rapid soundings. No doubt J.R. Napier was one of the key inspirations for Sir William's instruments of navigation.[46]

With trials on Napier's log occupying even the final passage of the 1871 season from Largs to Gareloch, Sir William recorded the last day thus:

I am now on the point of 'flitting', as we say in Scotland, from my summer Quarters on board the *Lalla Rookh* to the College. I am alone with one man on board waiting for my train, the others having just sailed away in the 'cutter' and 'gig' for Greenock to leave the boats there for the winter, and to find places, chiefly no doubt in foreign-going ships, for themselves . . .[47]

The *Lalla Rookh*'s first full season, then, brought together a remarkable number of diverse strands. On a social level, the yacht carried Sir William firmly into the sphere of Lord Dufferin and the Duke of Argyll. Similarly, cruising enabled a more relaxed dialogue with distinguished colleagues, providing always that they could adapt as well as Helmholtz to life afloat. And again, Sir William could entertain at sea scientific men with strong practical and technical interests, particularly in maritime affairs: James Napier and William Froude, for example. On a more personal level, the *Lalla Rookh* provided a unique place of work for Sir William – free from the distractions of Glasgow, and yet open to the wide family circle of Thomsons, Crums, Bottomleys, and Kings. And, on an intellec-

[44] Hermann von Helmholtz to Frau von Helmholtz, 9th September, 1871, in SPT, **2**, 616.

[45] Sir William Thomson to Hermann von Helmholtz, 29th October, 1871, in SPT, **2**, 616–17.

[46] J.R. Napier, 'Description of an instrument for measuring the velocity of ships, currents, &c.,' *Proc. Phil. Soc. Glasgow*, **8** (1871–3), 146–59; Napier, *Life of Robert Napier*, pp. 199–201; Sir William Thomson to Thomas Andrews, 2nd November, 1871 (copy), A42C, ULC; SPT, **2**, 618, 636.

[47] Sir William Thomson to J.H. Gladstone, 4th November, 1871, in SPT, **2**, 619.

tual level, the intimate links between practical navigation, geometrical mathematics, and cosmical physics, gave full expression to Thomson's enduring scientific perspective.

Sounding the deep sea

The purpose of sounding at sea is to ascertain the depth of water at any position, and perhaps also the character of the seabed.[48] In thick weather, with astronomical navigation impossible and landmarks invisible, a ship's position might be found with a series of soundings on an accurate local chart. Such sounding constituted Sir William's fourth mode of establishing position, and was the function of his navigational sounding machine patented for commercial use in 1876. The instrument originally derived, however, from his involvement with submarine telegraphy.

In their 1859–61 report, the Submarine Telegraph Committee emphasized the importance of a 'careful and detailed survey of the nature and inequalities of the bottom of the sea' prior to the laying of cables, remarking that 'it would be of great advantage for this purpose if some instrument could be devised which would enable the actual outline of the bottom of the sea to be traced'.[49] Thomson's trials with sounding apparatus in the period 1872–4 aimed to fulfil this specialized requirement.

He probably gained his first experience of deep-sea sounding for navigational purposes during the Atlantic cable expeditions. Both Sir James Anderson (a Cunard shipmaster seconded as Captain of the *Great Eastern*) and Captain Moriarty, RN, employed a navigational technique using soundings which, by its practical simplicity and effectiveness, greatly appealed to Thomson. On the edge of a slip of card they simply marked points corresponding, on the scale of the relevant chart, to the actual distance estimated as having been run by the ship in the intervals between successive soundings. Assuming the ship had run a straight course, and writing on the card beside each point the depth and character of the bottom found by the lead, they merely placed the card on the chart and moved it about until they found agreement between soundings on the chart and the series on the card. The result yielded, in Sir William's words, 'an admirably satisfactory certainty as to the course over which the ship has passed':

> Sir James Anderson tells me that he has run [his Cunard liners] from the Banks of Newfoundland for two days through a thick fog at twelve knots, never reducing speed for soundings, but sounding every hour by the deep sea lead and Massey fly [depth-gauge], has brought up his last sounding [showing] black mud opposite to the mouth of Halifax Harbour, and has gone in without ever once having got a sight of sun or stars all the way from England . . . In moderate weather, with her engines in working order, and

[48] Raper, *Practice of navigation*, pp. 91–2.

[49] 'Report of the Submarine Telegraph Committee', *Parliamentary papers*, **62** (1860), xxxv.

coal enough on board to keep up steam, no steamer making land from the ocean, in a well explored sea, need ever, however thick the fog, be lost by running on the rocks. Nothing but neglect of the oldest of sailors' maxims, 'lead, log, and look-out', can possibly ever, in such circumstances, lead to such a disaster.[50]

In 1872, during a second deep-sea voyage with the *Lalla Rookh*, Sir William's direct involvement with sounding began. After spending the early part of the season on the South Coast and Thames Estuary while engaging in telegraphic tests in London, he departed during late June from Gravesend on passage for Torquay and Gibraltar. With him went a new recorder instrument for the Gibraltar telegraph station, and apparatus for both deep-sea sounding and for speed measurement.[51] Recalling his collaborative experimental work with J.R. Napier at the end of the yacht's 1871 season, and given Napier's major interest in the development of accurate speed, distance, and depth measurements aboard ship, we may suppose that the son of the famous Clyde shipbuilder inspired Thomson's trials of deep-sea sounding and speed measurement.

On the 29th June, Sir William conducted the first real tests of the sounding apparatus while in the Bay of Biscay. As he described it in his 1874 paper 'On deep-sea sounding by pianoforte wire', communicated, significantly, to the Society of Telegraph Engineers, the simple apparatus comprised a 30 lb lead weight, fitted with brass tube and valve for seabed samples, 'hung by 19 fathoms of cod-line from another lead weight of 4 lb attached to one end of a three-mile coil made up of lengths of pianoforte wire spliced together, and wound on a light wheel about a fathom in circumference, made of tinned iron plate'. A groove in the circumference permitted the application of a braking resistance via the friction of a cord kept tight by weights. The lead weights descended smoothly and uniformly at a position north of Spain determined by Sir William's use of Sumner lines. The predicted depth from the chart was less than 2600 fathoms. Sir William described what happened:

> the wheel suddenly stopped revolving as I had expected it to do a good deal sooner [at 2000–2500 fathoms]. The impression on the men engaged was that something had broken; and nobody on board except myself had, I believe, the slightest faith that the bottom had been reached. The wire was then hauled up by four or five men pulling on an endless rope round a groove on one side of the wheel's circumference. After about 1000 fathoms of wire had been got in, the wheel began to show signs of distress. I then perceived, for the first time (and I felt much ashamed that I had not perceived it sooner), that every turn of wire under a pull of 50 lb must press the wheel on the two sides of any diameter with opposing forces of 100 lb, and that therefore 2240 turns . . . must press the wheel together with a force of 100 tons, or else something must give way.[52]

According to Thomson's own account, the wheel did give way to the extent of preventing further use of the endless cord, and necessitating a slow turning of the wheel by hand. All the time Sir William 'was in the greatest anxiety,

[50] PL, **3**, 113–16. [51] SPT, **2**, 623–7. [52] PL, **3**, 337–9.

expecting at any moment to see the wheel get so badly out of shape that it would be impossible to carry it round in its frame'. Indeed, he 'half expected to see it collapse altogether and cause a break of the wire'. In the end, the 30 lb sinker came aboard, bringing with it the brass tube and valve. 'An abundant specimen of soft grey ooze' in the tube demonstrated that, with the wire 'so nearly vertical that the whole length of line out cannot have exceeded the true depth by more than a few fathoms', bottom had been reached at 2700 fathoms. The recorded chart depth was therefore misleading: the yacht had found a greater actual depth much nearer the north coast of Spain. Thomson, however, regarded his initial trial as a triumph: 'That one trial was quite enough to show that the difficulties which had seemed to make the idea of sounding by wire a mere impracticable piece of theory have been altogether got over'.[53]

Pianoforte wire had a striking advantage over rope of comparable strength in its small cross-section and smooth surface, and in the fact that 22 gauge wire (with a weight of about 14 ½ lb per nautical mile and a breaking strain of about 230 lb) was 'less cumbersome and heavy and acts with less friction than the hempen line now used'. Furthermore, as Sir William reported to the 1872 meeting of the British Association, 'it needs not the heavy mass of iron, weighing from two to four hundred weight, hitherto employed to sink it, 30 lb being amply sufficient for sounding in 3000 fathoms'.[54] In his 1874 address he emphasized that his simple design for deep-sea soundings avoided the American Navy's wasteful method of detaching the sinker when it reached the bottom in order to overcome the difficulty of hauling in: 'I never throw away a pound of lead if I can help it'. Only in very great depths, 4000 fathoms or more, would a heavy sinker have to be detached. Sir William summed up in 1874 the merits of his deep-sea method compared to the older methods:

> You see the simplicity of the apparatus, and the comparative inexpensiveness of it; no donkey-engine required, no three or four hundred pounds of iron cast away every time, as in the ordinary method of deep-sea soundings . . . The apparatus at present in use in our navy, which is better than that of any other navy in the world at this moment, except the American, is, as I know by actual experience of it, more difficult and tedious, and less sure at 500 fathoms, than sounding by the pianoforte wire at 2000 fathoms.[55]

In 1872, however, there were practical defects to be overcome, problems not only of safe splicing and prevention of rust, but of the destructive stress on the wheel.

On his return to England via Lisbon by 1st August, Sir William wasted no time in communicating his conclusions, not only to the British Association, at Brighton, but to the Admiralty, whose screw corvette HMS *Challenger* had been selected as a survey vessel and scientific research ship for a round-the-world

[53] *Ibid.*, pp. 339–40.

[54] Sir William Thomson, 'On the use of steel wire for deep-sea soundings', *BAAS Report*, **42** (1872), 251.

[55] PL, **3**, 358–9.

voyage to investigate the physical, chemical and biological constitution of the ocean. It is probable that W.B. Carpenter, whom Sir William introduced at Brighton as his presidential successor, discussed the sounding apparatus with Thomson in view of Carpenter's close association with the *Challenger* project. Carpenter was already well known for his association with deep-sea researches such as those carried out by the surveying ship *Porcupine* in 1870 off the Portuguese coast and in the Mediterranean. The realization that 'the vast ocean lay scientifically unexplored' would have been as much a challenge to Sir William as to the Circumnavigation Committee of the Royal Society.[56] In October, 1872, less than two months prior to the *Challenger*'s departure from Sheerness dockyard where she had been refitting as Britain's first ship exclusively commissioned for oceanography, he informed Stokes that the Hydrographer of the Royal Navy, Admiral Richards, 'was disposed to give it [the sounding apparatus] a trial in the circumnavigation expedition and thought it might be useful as a ready means of making soundings whether from boats or from the ship when a mere sounding, with a small specimen of the bottom, is required'. Sir William admitted, however, that 'the heavy apparatus for dredging which the expedition requires cannot of course be superseded by anything such as I have used, the chief value of which must be to allow deep-sea soundings to be made at small expense, and by ships not specifically fitted for the purpose'.[57]

During his discussions with the Admiralty, Sir William had telegraphic projects very much in mind:

> It will be of great importance for telegraphic enterprise that a line of soundings from the Azores to Bermuda, and soundings in the neighbourhood of Bermuda, and between Bermuda and St Thomas [Virgin Islands] should be taken as early as possible. I understand that such soundings can conveniently be taken in the commencement of the work, and would properly belong to the programme of the expedition. For the sake of the Great Western Telegraph Company, of which Professor Jenkin and I are engineers, I hope this may be done, and I understand that favourable replies have been given officially to an application to that effect made to the Admiralty by the Directors of the Company.[58]

Given the *Challenger*'s very different goals, it is not surprising that she went to sea without taking advantage of Sir William's offer. The ship had a variety of complex measurements to carry out using specially constructed apparatus and could scarcely have welcomed an additional device of little value for their purposes. Sir William, however, subsequently indicted the Admiralty's conser-

[56] W.B. Carpenter and J.G. Jeffreys, 'Report on deep-sea researches carried on during the months of July, August, and September 1870, in H.M. Survey-ship *Porcupine*', *Proc. Royal Soc. London*, **19** (1870–1), 146–221; G.S. Ritchie, *Challenger. The life of a survey ship* (London, 1957), pp. xiv–xxi; Anita McConnell, *No sea too deep. The history of oceanographic instruments* (Bristol, 1982), pp. 59–61, 106–16.

[57] Sir William Thomson to G.G. Stokes, 14th October, 1872, K185, Stokes correspondence, ULC.

[58] *Ibid.*

vatism. He had tried, he said, for four years to bring his apparatus to the attention of Admiralty and general public alike, but was ultimately 'obliged to take a patent [in late 1876] and work the thing out myself if the public was to get any benefit of it at all'. The American Navy, on the other hand, 'found my apparatus full of defects. They never asked me to perfect it, but they perfected it in their own way, and obtained excellent results'.[59]

Telegraphy occupied a great deal of Sir William's time over the winter of 1872–3. The centre of attention was the new, purpose-built, cable-laying vessel, the *Hooper*, under construction at Charles Mitchell's shipyard, Walker-on-Tyne, a yard which, five years earlier, had begun building naval hulls for fitting out with ordnance from W.G. Armstrong's Newcastle armaments firm.[60] The *Hooper*, a large and specialized vessel of almost 5000 tons, was fitted with hydraulic side thrusters at the stern to provide maximum manoeuvrability on the powerful insistence of Thomson and Jenkin. After consultation with his brother, James, Sir William appears to have got his way with owners and builders alike.[61] He was, as his sister had once said, 'a most excellent logician'.

On the *Hooper*'s completion in the spring of 1873, Sir William accompanied her on the delivery voyage from Newcastle to the Thames. As the *Hooper* prepared for transatlantic cable laying, Sir William was busy making ready sounding apparatus for despatch on a steamer from Liverpool to Para (Belem) in Brazil, along with one of his laboratory students 'indoctrinated in the use of it' for the purpose of taking preparatory soundings along the Brazilian coast from Para to Pernambuco (Recife) in advance of the *Hooper*'s arrival with the South American cable. Leaving the *Lalla Rookh* behind for most of the summer of 1873, Sir William joined the *Hooper* in late June. Arriving in Lisbon, she began laying the 2500 miles of cable with a first call at Madeira.[62]

Jenkin wrote to his wife that the *Hooper* had arrived off Madeira at seven o'clock in the morning: 'Thomson has been sounding with his special toy ever since half-past three (1087 fathoms of water)'.[63] Sir William had apparently been trying out a method of avoiding the destructive stress on the wheel. But 'stopping the hauling every twenty turns, taking the strain off the wire by aid of a clamp, and easing it round the wheel', proved insufficient to relieve the problem. Even the tedious process of stopping every ten turns accomplished little. As a result of this experience he introduced an auxiliary hauling-in pulley 'by which the pull on the wire is very much reduced before it is coiled on the main sounding wheel'. This method relieved two-thirds or more of the strain

[59] Sir William Thomson to Captain W.J.L. Wharton, 26th November, 1888, in SPT, **2**, 722–3; PL, **3**, 359–60; Captain Maclear's remarks, in *RUSI*, **22** (1879), 116–17.

[60] Pollard and Robertson, *British shipbuilding industry*, p. 98.

[61] Sir William to James Thomson, 30th October, 5th and 18th November, 1872 (copies), T483, T484 and T486, ULC; SPT, **2**, 624, 628–30; Charles Mitchell & Co. to Sir William Thomson, 30th April, 1873, M149, ULC.

[62] Sir William Thomson to Thomas Andrews, 25th May, 1873, A42E (copy), ULC; SPT, **2**, 635–7.

[63] Fleeming Jenkin to Mrs Jenkin, 29th June, 1873, in SPT, **2**, 637.

Figure 21.1. Perspective view of Sir William Thomson's sounding apparatus.

from the wire before it reached the main sounding wheel.[64] The modified sounding apparatus shown in figure 21.1 was that described to the Glasgow Philosophical Society early in 1874 and designed for use aboard Siemens's new cable ship *Faraday* launched at Newcastle in February, 1874.[65]

One essential feature of the apparatus was lightness of the large wheel (a fathom in circumference) so that the inertia of the wheel would not 'shoot the wire out so far as to let it coil on the bottom' and so lead to those kinks which were almost always fatal to the wire. Even so, about five fathoms of hemp line between wire and sounding weight was needed to prevent the wire reaching the bottom. According to Sir William, 'the art of deep-sea sounding is to put such a resistance on the wheel as shall secure that the moment [within one second] the weight reaches the bottom the wheel will stop'. To achieve this result, Sir

[64] *Ibid.*, pp. 345–9.

[65] Sir William Thomson, 'On deep-sea sounding by pianoforte wire', *Proc. Phil. Soc. Glasgow*, **9** (1873–5), 111–17; PL, **3**, 352; Sir William Thomson to G.G. Stokes, 18th February, 1874, RR.7.294, Royal Society of London.

William used a form of brake patented by him for cable laying in 1858: 'the rule I have adopted in practice is to apply resistance always exceeding by 10 lb the weight of the wire out'. A 34 lb sinker would then provide a moving force of 24 lb. Consequently, when the 34 lb weight reached the bottom, instead of there being a pull of 24 lb downwards on the wire, 'there will suddenly come to be a resistance of 10 lb against its motion'. Such a technique gave virtually instantaneous perception of the bottom', a result of inestimable value for so-called 'flying-soundings'.[66]

A particular advantage of the new device was that the ship need not be kept hove to during the hauling-in period, especially when the length out did not exceed 2500 fathoms. As the wire came aboard, the vessel's speed could be steadily increased, thereby effecting 'a great saving of time' since 'in the ordinary process the hemp rope must be kept as nearly as possible up and down'. Sir William pointed out, in fact, that a 34 lb sinker would take not more than thirty minutes to run out to a depth of 2000 fathoms, and would require, using a dozen men hauling an endless rope, only fifteen minutes to haul in. Further acceleration might be achieved using a donkey engine instead of the men. The ship moreover had only to be kept stopped while the lead descended, thereafter 'going ahead full speed'. Indeed, the apparatus lent itself to the making of flying-soundings, that is, to the taking of soundings without stopping the ship at all during the cable laying. Flying-soundings from the *Hooper* were necessitated by the fact that, over the first section of the Western and Brazilian Company's cable from Pernambuco to Para, the Brazilian Government's gunboat *Paraense*, ordered to take soundings for the *Hooper*, had insufficient coal for the whole passage. Although the route had been previously sounded, it was vital that soundings be taken during the actual laying:

> Accordingly, Captain Edington arranged that my sounding-wheel should be set up over the stern of the *Hooper*, and soundings were taken every two hours without stopping the ship. A 30 lb weight was hung by a couple of fathoms of cord from the ring at the top of the wire. Then the wheel was simply let go, with a resistance of about 6 lb on its circumference, the ship running at the rate of . . . 6 knots relatively to the bottom; and after, perhaps, 150 fathoms had run out . . . suddenly the wheel would stop revolving . . . Thus we achieved flying-soundings . . . and obtained information of the greatest possible value with reference to the depth of the water and the course to be followed by the cable . . . I never would like to go to lay a cable without an apparatus for flying-soundings.[67]

Flying-soundings by this method, Thomson added to his 1874 address to the Society of Telegraph Engineers, would without doubt 'be found useful in ordinary navigation'. In answer to a question about the allowance needed for the non-verticality, Sir William replied that, although the soundings were only

[66] PL, **3**, 352–6, 365; Thomson, 'On deep-sea sounding', p. 113.
[67] PL, **3**, 366–7; Thomson, 'On deep-sea sounding', pp. 115–16.

approximate, 'they are sufficiently near for many practical purposes, and a little experience gives data for making allowances with considerable accuracy'. The *Hooper*'s soundings in the depth range from 40 to 170 fathoms, he believed, were within 3–10% of the actual depth, the estimate of the actual depth being simply based on the horizontal distance separating ship and sinker at the time the sinker touched the bottom, and the length of wire out.[68]

Here, then, was a device well suited both to the practical needs of cable laying and to the practical needs of ordinary navigation in depths of up to 200 fathoms. The contrast, Sir William emphasized, 'between the ease with which the wire and sinker are got on board from a depth of 200 fathoms by a single man . . . in this process, and the labour of hauling in the ordinary deep-sea lead and line, by four or five men, from a ship going through the water at four or five knots in depths of from 30 to 60 fathoms, is remarkable'. He concluded in 1874 that he was now having constructed 'for the purpose of navigation, a small wire wheel of 12 inches diameter, to have 400 fathoms of pianoforte wire coiled on it, for flying-soundings in depths of from 5 to 200 fathoms, without any reduction of the speed of the ship, or at all events, without reducing it below five or six knots'.[69] With Sir William Thomson's astonishing capacity to bring together the diverse elements of Victorian drawing rooms and nineteenth-century steam navigation, a commercial sounding machine seemed, by 1874, very much in prospect.

Coincidentally, a fault in the cable, 400 miles down one of the coils stowed below decks, detained the *Hooper* for some sixteen days in Funchal Bay. Not all that time was spent in uncoiling, splicing, and recoiling, as Sir William reported home:

> We had some admirable lamp signalling several evenings at Funchal between the *Hooper* and Mr Blandy's house, about 1½ miles distant. The Miss Blandys learned 'Morse' very well and quickly, and both sent and read long telegrams the first evening they tried it, to the admiration of France and other old telegraphers on board.[70]

One of the Miss Blandys, Frances Anna (*c.*1838–1916), was to marry Sir William on his fiftieth birthday, 24th June, 1874. Having returned to England in October, 1873, Sir William, in characteristically enterprising fashion, sailed next May from Falmouth in the *Lalla Rookh* on the 1200 mile voyage to Funchal Bay in under seven days, there to propose to Fanny. As he wrote to Helmholtz a day before the wedding, Lady Thomson would share one important quality with her husband: 'The future mistress of the *Lalla Rookh* promises to be a very good sailor, having already been out a good many times for a day's sail . . . and always hitherto escaped sea-sickness'. And he concluded facetiously: 'My present

[68] PL, **3**, 368–71; Thomson, 'On deep-sea sounding', p. 116.
[69] PL, **3**, 371; Thomson, 'On deep-sea sounding', p. 117.
[70] Sir William Thomson to Jessie Crum, 8th August, 1873, in SPT, **2**, 638.

happiness is due to a fault in the cable which kept the *Hooper* for sixteen days in Funchal Bay last summer'.[71]

Sir William and Lady Thomson returned to England in late July. While on passage they received a reminder of the power of wind and sea even in midsummer. Off Finisterre the *Lalla Rookh* broke her main gaff, and had to complete the voyage under top-sail.[72] They lived on board during the remainder of the season, but encountered another unpleasant taste of the ferocity of the elements, this time in the very Gareloch itself:

> We said good-bye with much regret to the *Lalla Rookh* on Friday last, and left her to be laid up for the winter in the Gareloch. After a succession of severe gales, in one of which . . . she dragged both anchors, and went ashore in the Gareloch, we had a week of fine weather and a little beautiful sailing to Arran and in the Firth of Clyde . . . We were fortunately not on board the night of her shipwreck, by a mere chance of an unexpected meeting of the 'University Court' keeping us in Glasgow. She was got off without damage and by ourselves without assistance or expense, after being 36 hours on soft sand and in a very good position.[73]

Later that same year, the 2400 ton wooden paddle steamer *La Plata* (built for the Royal Mail Steam Packet Company's South American service in 1852 but sold in 1871 as one of that firm's last wooden ships), carrying cable to South America for Siemens Brothers, foundered off Ushant with the loss of sixty lives including David Thomson King, Sir William's nephew.[74] Although navigational problems did not contribute to the disaster, the *La Plata*'s sinking served both to emphasize the general need for much greater safety at sea and to confirm the specific conclusion that large wooden vessels had proved inferior to iron ships in the face of an ocean's fury. The waters of the Bay of Biscay which had almost taken the *Agamemnon* in 1858 could indeed play with men till their hearts were broken, and wear stout ships to death.

The navigational sounding machine

An even greater maritime disaster than the loss of the *La Plata* occurred on 5th May, 1875. The 3400 ton steamer *Schiller*, completed by Robert Napier only a year earlier as one of eight iron screw liners ordered on the Clyde by the newly established German Transatlantic Steamship Company, was homeward bound from New York when, in dense fog, she grounded on the Isles of Scilly in the Western Approaches. In heavy seas, the *Schiller* rapidly broke up with the loss of 312 passengers and crew. The nature of the disaster led to Sir William's letter to *The Times* of 12th May, in which he stated that a method of taking rapid soundings would have enabled the *Schiller* to remain outside the fifty fathom

71 Sir William Thomson to Hermann von Helmholtz, 23rd June, 1874, in SPT, **2**, 646; Sir William Thomson to Elizabeth King, 12th May, 1874, in SPT, **2**, 645. 72 *Ibid.*, p. 647.

73 Sir William Thomson to Thomas Andrews, 5th November, 1874, A42F, ULC; SPT, **2**, 654.

74 *Ibid.*, pp. 654–5; Bushell, '*Royal Mail*', pp. 72–5, 100, 122, 257.

line, while, at the official inquiry shortly after the disaster, the *Schiller*'s chief mate stated that during three days of fog the lead had not once been cast.[75]

Sir William had, in his own sounding apparatus, just such a method for safe navigation. The opportunity to test that apparatus and 'to realise the thing for practical use' on large merchant ships came during his first transatlantic voyage as a passenger in May, 1876. Thomson travelled to New York on the Cunard liner *Russia*, considered one of the most graceful steamships ever built. He returned aboard the *Scythia* at the end of July. Both Cunarders were compound-engined iron-screw vessels, high class products of the Clydebank yard of J. and G. Thomson in 1867 and 1875, respectively. Yet such was the progress of the Atlantic liner in this period that the *Scythia* exceeded the *Russia*'s 3000 gross tons by half as much again in a bid to meet stiff competition from the new White Star Line.[76] Cunard ships had, however, an unrivalled record of regularity and safety of service. Though two ships had been wrecked, neither life nor letter entrusted to the firm's care had been lost in thirty-five years of transatlantic operation. From the 1840s, one of the Scottish partners, Charles MacIver, had demanded of his captains the greatest attention to detail while at sea and in port. A prime example of such a demand for efficiency and discipline in the name of safety was the frequent use of soundings (evidently neglected by the unfortunate rival *Schiller* in 1875):

> It is to be borne in mind that every part of the coast board of England and Ireland can be read off by the lead and ships from abroad making their landfall should never omit to verify their position by soundings. But masters eager to obtain the credit of making a short passage rather than lose a few minutes in heaving the ship to, will run the risk of losing the vessel and all the lives on board.[77]

By the 1870s, the owners of large, fast, and prestigious liners faced more acutely than ever the old dilemma: they could neither afford to have their ships meet with accidents, nor could they afford to reduce speed in poor visibility. Safety and regularity, upon which their reputation among customers was built, often presented conflicting requirements to a shipmaster. Writing to John Perry in August, 1876, Sir William announced that his sounding machine would meet these very demands:

> I made soundings from the *Russia* and *Scythia* going at 14 knots without reducing speed. I found it perfectly easy to haul in the wire, of which I sometimes had as much as 300 fathoms with a 22 lb iron sinker and a pressure-gauge for measuring the depth. I found bottom in 68 fathoms quite unexpectedly in a place where 1900 fathoms was marked on the chart.[78]

[75] Bonsor, *North Atlantic seaway*, **3**, pp. 949–52; Lindsay, *Merchant shipping*, **4**, p. 241*n*; SPT, **2**, 662. [76] Bonsor, *North Atlantic seaway*, **1**, 90–3, 149, 152; SPT, **2**, 668–72.

[77] 'Captains' memoranda', McIver papers, 25th March, 1848, quoted in Hyde, *Cunard*, pp. 45–9. See also Lindsay, *Merchant shipping*, **4**, 239–50, esp. p. 242*n*, where Cunard safety is discussed and the instructions to captains are fully printed.

[78] Sir William Thomson to John Perry, 2nd August, 1876, in SPT, **2**, 672.

Sir William's friend, Captain S.T.S. Lecky, engaged in a fruitless search for 'Thomson's bank' while on passage to North America in June, 1879. Just what 'shoal' Thomson had found at sixty-eight fathoms, we may never know.[79]

In the same year that Sir William tested his sounding apparatus on board the Cunard liners, he took out his first navigational sounding machine patent. The purpose of the machine was to enable soundings to be obtained from a ship running at full speed in water of any depth not exceeding 100 or 150 fathoms. The first steamship provided with a machine for navigational purposes was the *Palm*, an iron-screw vessel of under 1300 tons built in Greenock in 1869 for Messrs Horsfall of Liverpool. Her master, Captain Leighton, reported to Sir William in April, 1877:

> I feel a great honour was conferred upon me by having the *first* of your sounding machines issued for *actual* service. During the voyage in the *Palm* . . . I took frequent opportunities of testing it when I had a chance of cross bearings as shewn by chart & always found it most accurate . . . The first real use I made of the machine was in the Black Sea during a fog which obscured everything. Wishing to make sure of my position I put the ship's head for the land, & kept the machine at work. After running up to 30 fath at full speed I slowed down & went in to 12 fath, then hauled out to a convenient depth & put her on the course up the coast. When it became clear I found myself in a proper position, and no time had been lost by stopping to sound.[80]

Leighton concluded with a powerful commendation of the sounding machine. Many shipmasters, he believed, 'let hours go by without obtaining soundings either because of the delay, or on account of the danger of rounding-to in heavy weather when, if they were provided with your sounding machine they could have their minds set at ease by having timely warning of danger, or by knowing that they were in a good position'.

To obtain sure evidence of the depth the lead had descended, Thomson included a self-recording depth gauge. The entire apparatus was designed as follows:

> [It] consists of a drum, about a foot in diameter and four inches wide, upon which 300 fathoms of steel pianoforte wire are tightly wound. To the wire is attached 9 feet of log-line, and to this is fastened an iron sinker, about twice the length of the ordinary lead, but not so thick. On the log-line, between the wire and the sinker, a small copper tube is securely seized. The lower end of this tube is perforated; the upper end being opened or shut at pleasure by means of a close-fitting cap. When ready for sounding, the copper tube contains a smaller sized glass one. This latter is open at the bottom end, and hermetically sealed at the other. The interior surface is coated with a chemical preparation of a light salmon colour (chromate of silver). The drum is fitted with a brake cord, which,

[79] S.T.S. Lecky to William Bottomley, 2nd June, 1879, CS506, ULC. Navigational charts of the North Atlantic name the Kelvin Seamount among the New England Seamounts, but the depth recorded is about 800 fathoms. The surrounding seabed is about 2700 fathoms.

[80] Lecky, '*Wrinkles*', p. 94. See also Sir William Thomson, 'On compass adjustment in iron ships, and on a new sounding apparatus', *RUSI*, **22** (1879), 91–119, esp. pp. 107–14; PL, **3**, 372–4, 377–88.

on a cast being taken, controls its speed, and ultimately arrests it when the lead touches the bottom. A pair of small winch handles wind up the wire again . . . The sinker is 'armed' in the usual way.[81]

Captain Leighton's one criticism of the early sounding machine concerned the self-registering depth gauge. Like Ericsson's lead, Sir William's depth gauge depended on the compression of air, with the extent of compression marked directly on the inside of the tube by the chemical action of seawater on the preparation of salmon-coloured chromate of silver lining the tube internally. Applying the glass tube to a graduated boxwood scale yielded the depth descended.[82] Leighton's complaint, however, was that at great depths the liquid would be found to have gone higher up the tube on one side than on the other. Sir William's initial response to the problem was to provide tubes of smaller bore. But the difficulty of designing a satisfactory depth gauge proved to be an enduring one. As James Thomson explained to George Darwin in 1880, successive changes had been rigorously tested on board the *Lalla Rookh* 'in their latest forms of development by *natural selection*'.[83]

Abandoning the chemical method in 1880, Sir William devised a modified version of the 'Ericcson lead' which he called the 'triple depth gauge'. This patent device comprised three independent gauges measuring *successive* ranges of depths from eleven to over 120 fathoms. By 1885, however, Sir William had patented a mechanical depth gauge with a piston which pushed a marker into a cylinder as the air was compressed and which, being restored by a spring, left the marker at the maximum depth attained as indicated on a graduated scale. This mechanical depth gauge, like the triple depth gauge, could be used again and again without the need for a continuous supply of chemically prepared glass tubes. In the mechanical arrangement, the depth gauge fitted neatly inside the sinker weight. Sir William could at last write in 1885 that the new depth recorder 'after 9 years hitherto unavailing attempts now promised success'.[84]

After Captain Leighton's report, Sir William's second major 'satisfactory experience' with his navigational sounding machine was a personal one while navigating the *Lalla Rookh* up Channel, running before a southwesterly gale in characteristically thick weather while on passage home from Madeira early in August, 1877 (see figure 21.2). From the one-hundred fathom line he took a sounding every hour in order to apply the Moriarty–Anderson technique of navigation by a series of soundings marked on the edge of a card and transferred to the chart. More than twenty-four hours went by:

[81] Captain Leighton to Sir William Thomson, 4th April, 1877, CS2, ULC; PL, **3**, 382–3; *Lloyd's register of British and foreign shipping from 1st July, 1877, to the 30th June, 1878* (London, 1878).

[82] PL, **3**, 381–2.

[83] James Thomson to George Darwin, 25th June, 1880, in SPT, **2**, 758–9. See also pp. 760, 766, 847.

[84] Sir William Thomson, 'A new navigational sounding machine and depth-gauge', *RUSI*, **25** (1882), 374–86; Lecky, '*Wrinkles*', pp. 364–5; PL, **3**, 375–6; SPT, **2**, 723; Sir William Thomson to Elizabeth King, 30th September, 1885, in SPT, **2**, 850–1; McConnell, *No sea too deep*, pp. 73–5.

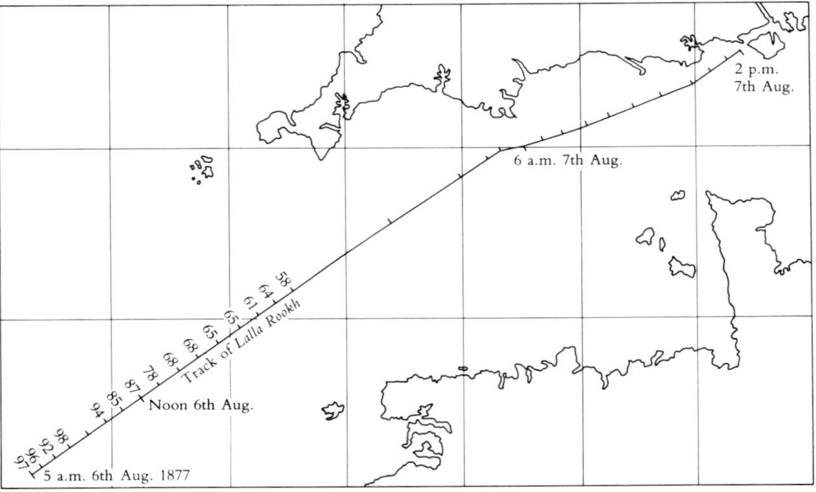

Figure 21.2. Track of the *Lalla Rookh*, August, 1877.

nothing of the land was to be seen through the haze and rain; . . . with the assistance of about ten more casts of the lead (by which I was saved from passing south of St Catherine's) I made the Needles Lighthouse right ahead, at a distance of about three miles, at 2 p.m., having had just a glimpse of the high cliffs east of Portland, but no other sight of land since leaving Madeira and Porto Santo. In the course of the 280 miles from the point where I struck the 100 fathom line, to the Needles, I took about thirty casts in depths of from 100 fathoms to 19 fathoms without once rounding-to or reducing speed; during some of the casts the speed was ten knots.[85]

Sir William soon afterwards received another highly favourable and exceptionally vivid report, this time from Captain J.J. Walter, master of the 1000 ton screw steamer *Iberia* owned in Ardrossan on the Clyde:

[On the 22nd November, 1877] while scudding before a heavy westerly gale in a mineral loaded steamboat about 40 miles S.W. of Scilly I obtained accurate soundings at noon with your machine, which indicated 72 fathoms fine sand, the ship going about 8½ knots; this I verified by sighting Scilly at 5 p.m. The soundings were taken by myself with the assistance of the steward & carpenter – the sinker & tube being hove in by the two last named without trouble, although the spray was flying heavily over the taffrail. Now, Sir, I venture to say with no other apparatus could the soundings have been obtained with so few hands & so little risk of being washed away from the winding gear.[86]

This kind of report must have brought immense personal satisfaction to Sir William, not only because of the commercial value of a series of favourable

[85] Thomson, 'On compass adjustment', p. 110; PL, **3**, 383–6; SPT, **2**, 723–4.

[86] Captain J.J. Walter to Sir William Thomson, 27th November, 1877, CS46, ULC; *Lloyd's register of shipping, (1877–8).*

reports within the tightly knit communities of shipowners and masters, but also because anyone with actual experience of a westerly gale in the Western approaches to the British Isles would be unable to put a value on such an instrument in those conditions. As Conrad wrote of just such an approach under sail, 'under their feet the ship rushes at some twelve knots in the direction of the lee shore; and only a couple of miles in front of her swinging and dripping jib-boom . . . a grey horizon closes the view with a multitude of waves surging upwards violently as if to strike at the stooping clouds . . . To see! to see! – this is the craving of the sailor . . . I have heard a reserved, silent man, with no nerves to speak of, after three days of hard running in thick south-westerly weather, burst out passionately: "I wish to God we could get sight of something!"'[87] Sir William Thomson could indeed provide the seafarer with a precious new kind of sight. His tool of the simplest but most practical kind occupied a symbolic role in one of the two supporters of the coat of arms assigned in 1892 to the barony of Kelvin of Largs. Like the Glasgow University student with marine voltmeter on the right, and the sailor with sounding line on the left, Sir William was both student and sailor at heart.[88]

[87] Conrad, *Mirror of the sea*, pp. 134–42. The 'blind' approach by Conrad's captain invites comparison with the precision of Thomson's same approach to the Isle of Wight.

[88] SPT, **2**, 696, 914.

22
The magnetic compass

> The compass causes a quite unprecedented addition to my occupations, but it is very interesting, and as it takes me a good deal about shipping it is not like plodding at writing or 'book work'. Willy Bottomley is most helpful ... He was down with me ... in a new ship, Balmoral Castle, belonging to Messrs Donald Currie & Co., which has been fitted out with three of my compasses. We went to Gareloch head on Friday evening and 'swung' early next morning to adjust compasses. *Sir William Thomson to Elizabeth King, 1877.*[1]

In the thirty-year period from Sir William's first commercial compass patent in 1876 until his death in 1907, his firm supplied no fewer than 10 000 compasses to the world's merchant ships and fighting navies. Production in the Glasgow workshops of James White rose from little more than a dozen in 1876 to around 150 in 1880, 350 in 1885, and an all-time peak of over 500 in each of the years 1892 and 1893. These sales figures coincided to a certain extent with the boom years of the early 1880s in shipbuilding (especially on the Clyde), with a contraction in the later 1880s, and with a generally prosperous though fluctuating trend in the 1890s. Up to 1907, indeed, the annual production of the Kelvin compass never fell much below 300, and exceeded 400 in the years 1898–1901 and 1906–7. While the figures show an approximate correlation with shipbuilding output, many of the compasses were for delivery to new, high-class liner tonnage building on the Clyde, in Belfast, and on the Tyne as well as to ships of the Royal Navy. Such tonnage was not subject to the same large fluctuations of output during trade cycles of boom and slump as was ordinary merchant ship tonnage, and it is therefore not surprising that production of Kelvin compasses manifested relative stability over these decades.[2]

According to S.P. Thompson, it was 'about the year 1871' that Sir William's friend, the Rev Norman Macleod, asked the professor to contribute to his magazine, *Good Words*. Sir William chose the mariner's compass as his subject for two articles which appeared in 1874 and 1879, and in doing so became aware

[1] Sir William Thomson to Elizabeth King, 18th February, 1877, in SPT, **2**, 679–80.

[2] Thomson's patent compass books: record of ships supplied with compass (1876–1918), Glasgow University archives; Sidney Pollard and Paul Robertson, *The British shipbuilding industry, 1870–1914* (Cambridge, MA and London, 1979), p. 26, giving graph of fluctuating shipbuilding output for the United Kingdom and its regions (1870–1913).

of the very serious defects in existing compass construction, an awareness which led to his world famous compass patents.[3] Yet the choice of subject did not depend upon mere accidental circumstance 'about 1871'. That year marked the beginning of a decade of unparalleled expansion in iron shipbuilding centred on Glasgow (ch 21). That year, too, the *Lalla Rookh*, with all her eminent qualities as Sir William's laboratory afloat, had her first full season under his ownership. Above all, in that year Archibald Smith, a former fellow-student of Macleod at Glasgow College in the period 1828–32, had almost reached the end of a thirty-year study of the magnetism of ships. Following Smith's sudden death in December, 1872, his old friend Thomson not only became the author (at Stokes's request early in 1873) of his Royal Society obituary notice, but the heir to his researches on ships' magnetism. In fact, Thomson himself had played a prominent role in those researches since the 1840s.[4]

Archibald Smith and the magnetism of ships

Archibald Smith, as one of the 'whig mathematicians', had instigated the *Cambridge Mathematical Journal*. Throughout his life he staunchly supported liberal causes, and in 1869 (after characteristic hesitation) he stood unsuccessfully as liberal candidate for the parliamentary seat of Glasgow and Aberdeen Universities. Although by profession a Chancery barrister, Smith's greatest passion was sailing and navigation, especially on the west coast of Scotland, for which he compiled his own local charts. While weighing up the prospects of becoming Meikleham's successor, he had written to his sister Isabella in 1846: 'Then there are six months holidays in the year instead of about *two* and I should . . . get a yacht and make philosophical cruises all summer and live an easy pleasant respectable dull stupid life not toiling and moiling all day long and much of the night as I now do'.[5]

Unable then or later to escape the treadmill of his chosen profession, Smith employed for compass investigation what for others constituted leisure time and sleeping hours. The consequent strain contributed to the breakdown of his health in 1870 and the abandonment of his profession. Combining his expertise in mathematics and his passion for the sea with his concern to advance the prosperity and well-being of mankind, he had (in Sir William's words) 'conferred never-to-be-forgotten benefits on the marine service of the world, and made contributions to nautical science which have earned credit for England

[3] SPT, **2**, 627; Sir William Thomson, 'The mariner's compass', *Good Words*, **15** (1874), 69–72; PL, **3**, 228–45; 'Terrestrial magnetism and the mariner's compass', *Good Words*, **20** (1879), 383–90, 445–53; PL, **3**, 242–322.

[4] Sir William Thomson to G.G. Stokes, 8th and 11th January, 1873, K190, Stokes correspondence, ULC; G.G. Stokes to Sir William Thomson, 18th January, 1873, NB 21.57, Stokes correspondence, ULC; Sir William Thomson to G.G. Stokes, 21st November, 1873, K192A, Stokes correspondence, ULC.

[5] Archibald to Isabella Smith, 14th May, 1846, TD1/676/5, Smith papers, Strathclyde Regional Archives.

among maritime nations'. His contributions were five-fold: (1) harmonic reduction of observations; (2) practical expression of the full mathematical theory; (3) investigation of heeling error; (4) dygograms (ch. 11); and (5) rules for the positions of needles on the compass card, with dynamical and magnetic reasons.[6]

The first two aspects of Smith's work relate to his association with Major Edward Sabine, best known as the driving force behind the so-called 'magnetic crusade', inaugurated by Captain James Clark Ross's Antarctic expedition aboard the *Erebus* and *Terror* in 1840–1 and culminating by 1845 in the establishment of a world-wide (and especially Empire-wide) network of magnetic observatories.[7] In 1844 Smith introduced Thomson to Sabine, who subsequently communicated Thomson's 1849–50 paper on magnetism to the Royal Society (ch. 8), the paper which led to Thomson becoming an FRS in 1851.[8]

The earlier Arctic expedition of 1818 by the *Isabella* and *Alexander* had drawn Poisson's attention in 1824 to the problem of compass deviation, and so the last of his three famous papers on the mathematical theory of magnetic induction was concerned with the application to ships' magnetism. Poisson's applications, however, involved simplifying assumptions no longer adequate to achieve the accuracy demanded for reduction of the Antarctic observations. Thomson summed up in 1871 the special achievement of Archibald Smith:

A vast mass of precious observations, made chiefly on board ship, were brought home from this [Antarctic] expedition. To deduce the desired results from them, it was necessary to eliminate the disturbance produced by the ships' magnetism; and Sabine asked his friend Archibald Smith to work out from Poisson's mathematical theory, then the only available guide, the formulae required for the purpose . . . It was the beginning of a series of labours carried on with most remarkable practical tact, with thorough analytical skill, and with a rare extreme of disinterestedness, in the intervals of an arduous profession, for the purpose of perfecting and simplifying the correction of the mariner's compass – a problem which had become one of vital importance for navigation, on account of the introduction of iron ships.[9]

The first step in Smith's work involved reduction of observations by Fourier's harmonic analysis. In Sir William's words, 'we may . . . say, in Fourier's

[6] Sir William Thomson, 'Archibald Smith and the magnetism of ships', *Proc. Royal Soc. London*, **22** (1874), i–xxiv; MPP, **6**, 306–34, esp. pp. 306–12.

[7] MPP, **6**, 310–12; Lieut-Colonel Edward Sabine, 'Contributions to terrestrial magnetism', *Phil. Trans. Royal Soc. London*, (1843), 145–231, esp. pp. 145–50; J.H. Lefroy, in G.F.C. Stanley (ed.), *In search of the magnetic north. A soldier-surveyor's letters from the North-West. 1843–44* (Toronto, 1955); John Cawood, 'The magnetic crusade: science and politics in early Victorian Britain', *Isis*, **70** (1979), 493–518; Jack Morrell and Arnold Thackray, *Gentlemen of science. Early years of the British Association for the Advancement of Science* (Oxford, 1981), pp. 353–70.

[8] Archibald Smith to William Thomson, *c.* April, 1844, S143. ULC; William to Dr Thomson, 4th April, 1844, T255, ULC; SPT, **1**, 67, 212, 226.

[9] Sir William Thomson, 'Presidential address', *BAAS Report*, **41** (1871), lxxxiv–cv; PL, **2**, 132–205, on pp. 150–2; Archibald Smith, Memoranda, included in Sabine, 'Terrestrial magnetism', (1843), pp. 145–50; (1844), pp. 116–19; (1846), pp. 248–50.

language, that the disturbance of the compass is a periodic function of the angle between the vertical plane of any line fixed relatively to the ship [say the horizontal line from the stern towards the bow] and any fixed vertical plane [say the magnetic meridian] when the ship . . . is turned into different azimuths – the period of this function being four right angles'. One of Smith's 'earliest contributions to the compass problem was the application of Fourier's grand and fertile theory of the expansion of a periodic function in series of sines and cosines of the argument and its multiples, now commonly called the harmonic analysis of a periodic function'.[10]

For practical application of this analysis, Smith provided tables and calculational rules, which were soon employed in the Admiralty's Compass Department. To this Department every ship in the Navy sent her table of observed deviations at least once a year for harmonic analysis in order to obtain a full history of the ship's magnetic condition. As a result, Sir William believed in 1871 'that it is to the thoroughly scientific method thus adopted by the Admiralty, that no iron ship of Her Majesty's Navy has ever been lost through errors of the compass'. He also reminded his readers in 1874 that Smith's analysis had 'proved exceedingly valuable in many other departments of practical physics besides ships' magnetism'. For example, Thomson had employed Smith's tables in reducing Forbes's observations of underground temperature in 1849 and, through the Tidal Committee of the British Association, in the harmonic analysis of tidal observations for various parts of the world in the period 1868–72 (ch. 11).[11]

The second step in Smith's work on ships' magnetism was his practical expression of Poisson's mathematical theory, i.e. his reduction of the theory to a few simple and easily applied formulae. Poisson, in 1824, and the Astronomer Royal, G.B. Airy, in 1838, had attempted such a simplification. But 'Poisson himself, in making practical application of his theory, had simplified it by assuming particular conditions as to symmetry of the iron in the ship, and even with these restrictions had left it in a form which seemed to require further simplification before it could be rendered available for general use'. Poisson had also omitted as irrelevant the disturbing effects of permanent magnetism upon a properly placed compass in a wooden ship with only isolated iron masses. But even with the earliest iron steamers and sailing ships these simplifications could no longer be made. In 1838, therefore, Airy carried out his detailed investigations of compass deviations and their correction on the iron steamer *Rainbow* at Deptford on behalf of the Admiralty and on the first ocean-going iron sailing ship, *Ironsides*, in Liverpool's Brunswick Dock on behalf of her owners.[12]

Airy's calculations were founded, not on Poisson's full mathematical theory

[10] MPP, **6**, 312–13. [11] PL, **2**, 151–2; MPP, **6**, 313.

[12] *Ibid.*, pp. 313–15; G.B. Airy, 'Account of experiments on iron-built ships, instituted for the purpose of discovering a correction for the deviation of the compass produced by the iron of the ships', *Phil. Trans. Royal Soc. London*, (1839), 167–213; C.H. Cotter, 'George Biddell Airy and his mechanical correction of the magnetic compass', *Ann. Sci.* **33** (1976), 263–74.

of induction, but on the supposition that 'by the action of terrestrial magnetism every particle [i.e. element $dxdydz$] of iron is converted into a magnet whose direction is parallel to that of the dipping needle, and whose intensity is proportional to the intensity of terrestrial magnetism'. He thereby eliminated mutual influence between magnetized parts. As Thomson diplomatically put it, Airy believed that his theory would produce results 'sufficiently accurate and complete for practical purposes'. Airy admitted that 'it would have been desirable to make the calculations on Poisson's [complete] theory, which undoubtedly possesses greater claims on our attention (as a theory representing accurately the facts of some very peculiar cases) than any other. The difficulties, however, in the application of this theory to complicated cases are great, perhaps insuperable'.[13] To have overcome these difficulties was Smith's supreme achievement.

The 'achievement', however, depended upon a dialogue between Thomson and Smith in the summer of 1848. Poisson's theory of magnetic induction in non-isotropic media required nine constants of inductive capacity to connect three components of magnetizing force with three components of magnetization. On 22nd July, 1848, Thomson wrote in his notebook: 'I have today found that the nine const^s which, as it has appeared to me for more than a year (since spring, 1847) enter into the expressions for the magnetic moment of a [non-isotropic] mass of soft iron influenced by the terrestrial force, are reducible to six (as I always thought must be the case)'. The result followed from requiring that the magnetized mass should not experience a continuous rotational acceleration, which would constitute a perpetual source of mechanical effect. It suggested an immense simplification of the problem of ship's magnetism. The ship, like every magnetized body, could be treated as an ellipsoid, with three principal axes and only three corresponding constants determining its rotational behaviour (the other three determining translational motion).[14] To find how to correct the compass, one had only to construct an analogous argument, similarly simplified to three constants, for the effect of the ship on the needle.

Thomson immediately wrote to Smith, but omitted the conclusion for compass corrections. He noted with apparent dismay a month later: 'I had actually put in at the end of my letter a remark such as the following: "analogous considerations would probably lead us to the establishment of a relation among the 9 constants in Poisson's formulae. In fact if we express that [for] a ship, allowed to turn round the vertical axis of its compass, the needle being held at

[13] Airy, 'Experiments on iron-built ships', pp. 177–88; MPP, **6**, 314.

[14] William Thomson, 22nd July, 1848, 'Journal and research notebook, 1845–56', NB34, ULC, pp. 95–7. Thomson also remarks on 14th August, 1848, that he has sent an abstract to the British Association, 'On the equilibrium of magnetic or diamagnetic bodies of any form, under the influence of the terrestrial magnetic force', *BAAS Report*, **18** (1848), 8–9; MPP, **1**, 88–90. The ellipsoid analysis derives from work in March, 1847, NB34, ULC, p. 51. See also William Thomson, 'On the theory of magnetic induction in crystalline and non-crystalline substances', *Phil. Mag.*, [series 4], **1** (1851), 177–86; E&M, 465–81; abstract in *BAAS Report*, **20** (1850), 23.

rest, there can be no accelerated rotation, we shall have probably the 9 constants reduced to 6". I [erased this and burned it, *del.*] removed this & concluded the letter without it when I found that the conclusion was erroneous; but I did not see the flaw'. The conclusion was correct; but he had not included all of the interactions between earth, ship, and needle.[15]

Having corrected his analysis shortly afterwards, Thomson wrote again to Smith, 'explaining my views of the possibility of correcting the deviations due to soft iron by three soft iron bars fixed in the ship vertical, horiz^{l} across, & along the ship'.[16] But the letter crossed one from Smith that had been delayed ten days, 'in which [Smith] said he intended to incite Cap. Johnstone to try vertical soft iron bars abaft & below, or before and above, the needle'.[17] Thus Smith was the first to announce the basic idea of compass correction for induced magnetism that Thomson had contemplated.

In May, 1850, Thomson reiterated his more complete views to Smith:

> I have reconsidered the question of the possibility of correcting by bars of soft iron the effect of the magnetization of the ship's iron (if all this be perfectly soft) and after our conversation of yesterday I am now convinced that, with any three given bars of soft iron the entire effect, both on the direction and the directive force, upon the compass . . . may be corrected by placing the bars in positions which are determinate when the directions of the bars are specified (provided the bars are infinitely thin).[18]

Two months later, Thomson wrote to Stokes from Greenhithe, downriver from London, that 'I am here to see "the Retribution" *swing*, for det^{g} the dev^{n} of his compass'.[19] This dry humour alludes to the long tradition of execution by hanging on the banks of the Thames. Thomson's presence at the 'swinging' highlights the importance he attached to the pressing problems of ships' magnetism long before his *Good Words* articles. Indeed his correspondence with Smith and involvement with Sabine indicate how intimately related were his early theoretical papers on magnetism, especially magnetic induction, to Britain's rule of the seas.

Archibald Smith's investigations actually culminated, not in corrections for compasses themselves, but in corrections for the readings they gave. The methods entered the famous *Admiralty manual for ascertaining and applying the deviations of the compass* (1862) which was translated into French, German, Russian, and Spanish. This version replaced earlier publications which began with 'practical rules', the aim being 'to enable the seaman, by the process of swinging his ship, to obtain a table of the deviations of his compass on each point, and then to apply the tabular corrections to the courses steered'. The rules had been drawn up in 1842 by the Admiralty Compass Committee (formed in 1837)

[15] William Thomson, 24th August, 1848, NB34, ULC, pp. 103–5. [16] *Ibid.*

[17] *Ibid.* See also Archibald Smith to William Thomson, 7th and 22nd August, 1848, S160–161, ULC.

[18] William Thomson to Archibald Smith (rough draft), 17th May, 1850, NB34, ULC.

[19] William Thomson to G.G. Stokes, 19th July, 1850, K42, Stokes correspondence, ULC.

consisting of Admiral Beaufort, Admiral Sir J.C. Ross, Captain Edward Johnson, Mr S.H. Christie and General Sabine. The Committee also designed the Admiralty standard compass, defined as that from which bearings were to be taken and from which all courses were to be shaped. Once introduced, it remained unchanged until 1889 when the Admiralty adopted Thomson's compass.[20] The Committee's 'practical rules' now constituted Part I of the *Manual*. Part II consisted of a graphic method ('Napier's method' after J.R. Napier) for representing deviations.[21]

Part III of the *Manual* constituted Smith's special contribution. First published separately in 1851, and then as a supplement to the 'practical rules' in 1855, Smith rewrote it with exact formulae instead of the approximate ones employed in these earlier versions, the revision being desirable on account of the very large deviations found in the new iron-plated ships of war. It aimed to provide accurate practical formulae derived from Poisson's fundamental equations. Poisson's equations, Smith explained, 'derived from the hypothesis that the magnetism of the ship, except so far as it is permanent, is *transient induced* magnetism, the intensity of which is proportional to the intensity of the inducing force, and that the length of the compass-needle is infinitesimal compared to the distance of the nearest iron'. On this hypothesis the following formula gave an exact representation of compass deviation:

$$\sin\,\delta = A\cos\delta + B\sin\theta' + C\cos\theta' + D\sin(2\theta' + \delta) + E\cos(2\theta' + \delta),$$

where

δ = deviation,
θ = correct magnetic course,
θ' = compass course,
A, D, E are coefficients depending solely on the soft iron of the ship,
B, C are coefficients each depending on the soft iron and hard iron and on the intensity of the earth's magnetic force at any given position.

If the coefficients were so small that their squares and products could be neglected, the formula became:

$$\delta = A + B\sin\theta' + C\cos\theta' + D\sin 2\theta' + E\cos 2\theta',$$

which was sufficiently exact for deviations of less than 20°. The terms were interpreted as follows:

(1) A is the 'constant part of the deviation' caused by soft iron arranged unsymmetrically around the compass.

[20] MPP, **6**, 320.

[21] F.J. Evans and Archibald Smith, *The Admiralty manual for ascertaining and applying the deviations of the compass* (London, 1862); 'Report on the three reports of the Liverpool Compass Committee and other recent publications on the same subject', *BAAS Report*, **32** (1862), 87–101, esp. pp. 87–8.

(2) $B\sin\theta' + C\cos\theta'$ make up the *semicircular deviation*, so-called because the harmonic effects reached a maximum every 180°, which was the most difficult to handle on account not only of each coefficient being made up of two parts which could not be distinguished by observations made in one latitude, but also from the fact that part of the ship's permanent magnetism was in fact only *subpermanent*, a term introduced by Airy.

(3) $D\sin2\theta' + E\cos2\theta'$ make up the *quadrantal deviation*, so-called because the harmonic effects reached a maximum every 90°, caused only by horizontal magnetic induction in soft iron. *E* was caused by horizontal induction in soft iron unsymmetrically distributed but, with compasses placed in the midship line, unsymmetrical arrangements of soft iron had rarely much effect, so that both *A* and *E* were usually very small. *D* was caused by symmetrical arrangements of soft iron both transverse and fore-and-aft.[22]

The principal aim of Part III of the *Manual* was to find the means of computing *A*, *B*, *C*, *D*, and *E* from the observed deviations or from Napier's curve for a certain number of equidistant compass points. This computation could be carried out easily by formulae founded on the method of least squares, simplified in its application by the use of tables. Further formulae enabled computation of the exact coefficients to a reasonable approximation.[23] Smith's concern here, not with correcting the compass itself by magnetic arrangements of hard and soft iron, but only with the computation of deviations in iron ships, reflects Admiralty practice. The navigator made the correction for himself from tables or curves, and not by automatic adjustments within the instrument. This characteristic Admiralty preference for direct control through the mathematical skill of the navigator would contrast strikingly with the mercantile marine's willingness to introduce Sir William's system of compass correction and management.

Investigation of heeling error constituted the third key feature of Smith's work. Iron sailing ships and steamers with auxiliary sail (as was the case with virtually all steamers) were pressed by sail to very considerable degrees of constant heel. When the ship heeled over to one side, there was more iron to the high or weather side of the compass than to the lee. Although Airy had investigated the heeling error in his earliest papers, he had confined his practical correction of compass error to a ship on an even keel. In Thomson's words, Smith 'took up the question with characteristic mathematical tact and practical ability, and gave the method of correcting the heeling error – which is now, I believe, universally adopted in the Navy, and too frequently omitted (without the substitution of any other method) in the mercantile marine'.[24]

As early as 31st July, 1848, Smith mentioned in a letter to Thomson the deviation of a compass made by Dent, chronometer maker, for the Queen's

[22] Evans and Smith, 'Report on Liverpool Compass Committee, pp. 88–9.
[23] *Ibid.*, p. 90. [24] *Ibid.*, pp. 90–1; MPP, **6**, 314–15.

yacht. If its axis were not perfectly vertical, he argued, the compass would show a serious deviation unless the card (the rotating horizontal disc divided according to the points of the compass and carrying the magnetic needles) had been accurately weighted so as to counterbalance the vertical part of the magnetic force.[25] He concluded: 'I daresay her majesty little suspects that if the inclination of her compass to the vertical be i there will be a deviation =

$$\{[f\sin\theta - \omega)/f\cos\theta]\sin\alpha - [\omega'/f\cos\omega]\cos\alpha)i\text{'}.$$

Smith's full treatment of the heeling error came with Part III of the 1862 *Manual*. He explained that it was 'desirable to deduce [in the *Manual*] from Poisson's formulae, expressions for the alteration of the coefficients introduced by the inclination of the ship'. The heeling error left unaltered coefficients depending on fore-and-aft action, B and D; gave a small value to A and E; but significantly altered C.[26]

Arrangements of iron corresponding to a transverse horizontal bar of soft iron together with a vertical bar of hard or soft iron magnetized by induction, or of hard iron magnetized permanently, caused the two-part alteration of C. Such an arrangement would cause a deviation of the north point of the compass needle to the weather side. To correct this deviation, 'the vertical magnetism must either act upwards, or the transverse magnetism must be such as would be caused by a horizontal transverse rod on each side of the compass, the formula indicating the relation which must exist between the vertical and transverse horizontal magnetism in order that the heeling error may be zero'.[27]

The final aspect of Smith's contribution to the magnetism of ships was his rule for the positions of the needles mounted on the card. In his letter to Thomson in July, 1848, Smith enunciated his rule for placing the needles:

Dent the chronometer maker . . . has been also making compasses. He makes them with an axis instead of a cap and point. He showed me one with four bars which he's had by trial arrayed so as to make the moments of inertia as nearly as possible equal but wanted a rule. I had formerly worked at this and given Col Sabine numbers by which all Admiralty compasses are made but when I worked it again I found it comes to the very simple construction

$AB' = 60°$
$A'B = 60°$

This gives a very simple rule for placing any even number of bars at equal distances so that the moments are equal. The dist. $= 60/n$ [where n is the number of bars on each side of the north–south axis].[28]

[25] Archibald Smith to William Thomson, 31st July, 1848, S159, ULC.

[26] Evans and Smith, 'Report on Liverpool Compass Committee', p. 90. See also G.B. Airy, 'On the connexion between the mode of building iron ships, and the ultimate correction of their compass', *Trans. Inst. Naval Arch.*, **1** (1860), 105–9, esp. p. 107.

[27] Evans and Smith, 'Report on Liverpool Compass Committee', pp. 90–1.

[28] Smith. *op. cit.* (note 25).

For example, two needles should be placed with their ends on the card 60° apart and four needles (as in the Admiralty standard compass) with their ends 30° apart.[29] As Sir William explained in 1874, the initial reason was a dynamical one – to give equal moments of inertia round all horizontal axes and so prevent the 'wobbling' motion of the compass card when balanced on its pivot – but the same arrangement was found later to have important magnetic benefits in overcoming so-called 'sextantal' and 'octantal' terms in the deviations of the compass.[30]

F.J. Evans (1815–85; Johnson's successor at the Admiralty's Compass Department and Smith's collaborator on the 1862 *Manual*) observed deviations of up to 6° during the *Great Eastern*'s experimental voyage from the Thames to Portland, yet the compass was nearly correct on cardinal and quadrantal points using Airy's methods. The error was thus neither semicircular nor quadrantal, and was attributed to the ship's two 11 $\frac{1}{2}$ inch needles in her standard compass. Smith and Evans's joint paper to the Royal Society in 1861 subjected the observations to harmonic analysis and showed sextantal and octantal terms (i.e. terms of coefficients multiplying the sines or cosines of six and eight times the ship's magnetic azimuth) very large for the *Great Eastern*'s compass but comparatively small for the Admiralty standard compass. Smaller needles gave smaller sextantal and octantal terms, and double needles of the same magnitude as single needles also reduced the terms. But, above all, later mathematical investigation showed that the arrangement of needles giving equal moments of inertia also prevented sextantal deviation in the case of correcting magnets, and octantal deviation in the case of soft iron correctors.[31]

The practical problems of compass correction

Airy had recognized the existence of two kinds of deviation which combine to disturb the compass: semicircular deviation due to the permanent [polar] magnetism of the ship's hard iron; and quadrantal deviation due to the horizontal component of the transient [induced] magnetism of the ship's soft iron. Two small bar magnets, one placed parallel to the fore-and-aft line but athwart (i.e. to one or other side of) the compass position, and the other transverse to the fore-and-aft line but ahead of or abaft (i.e. behind) the compass, corrected the semicircular deviation. Adjusting these magnets with the ship's head on the cardinal points neutralized the ship's permanent magnetism. A mass of wrought

[29] See also H.L. Hitchins and W.E. May, *From lodestone to gyro-compass* (London, 1952), pp. 75–8.

[30] MPP, **6**, 320–2.

[31] *Ibid.*, 320–2; F.J. Evans, 'Reduction and discussion of the deviations of the compass observed on board of all the iron-built ships, and a selection of the wood-built steam-ships in Her Majesty's navy, and the iron steam-ship *Great Eastern*; being a report to the Hydrographer of the Admiralty', *Phil. Trans. Royal Soc. London*, (1860), 337–78; Archibald Smith and F.J. Evans, 'On the effect produced on the deviation of the compass by the length and arrangement of the compass needles; and on a new mode of correcting the quadrantal deviation', *Proc. Royal Soc. London*, **11** (1860–2), 179–81.

Archibald Smith earned a high reputation as a Chancery barrister but laboured through the night hours on the mathematical problems of ship's magnetism. As a reward for services to the Empire, the Admiralty presented him with a watch in 1862. After Smith's early death in 1872, Thomson pursued practical solutions to these problems, and his resulting patents generated considerable wealth. [From Smith papers, Strathclyde Regional Archives.]

iron (soft) placed on one or other side of the compass corrected the quadrantal deviation.[32]

Airy, however, chose to neglect the change in the vertical component of the earth's magnetism with magnetic latitude and made no practical allowance for heeling error. His conclusions, moreover, did not allow for a major change in the ship's permanent magnetism over time (especially when the ship was newly launched). As a result of these shortcomings, the Admiralty and many shipowners for long remained distrustful of such attempts at compass correction.[33] But the Astronomer Royal's methods received their severest criticism in the mid-1850s following a particularly melancholy shipwreck.

In January, 1854, the new 2000 ton *Tayleur* sailed from the Mersey bound for Australia with over 500 passengers and crew. When just two days out and when encountering very heavy weather, the compasses (previously adjusted by permanent magnets) were observed to differ materially from one another. Trusting to the steering compass, the master believed that he was on course between West Wales and the Irish coast. Very soon, however, the sight of Lambay Island, north of Dublin Bay, provided evidence to the contrary. All attempts to clear the island failed, and the ship was driven broadside on to the rocks with the subsequent loss of nearly 300 lives. The Liverpool Marine Board inquiry concluded that the wreck occurred through the deviation of the compass, the cause of which they had been unable to determine. This tragedy provided the case for the Rev Dr William Scoresby's attack at the Liverpool meeting of the British Association in 1854 on Airy's principles of compass correction.[34]

The *Tayleur*'s compasses had been adjusted by permanent magnets, most probably in her fitting-out dock (as Airy had done with the *Ironsides*) in Liverpool. Unlike large steamers, the sailing ship did not require extensive sea trials and would almost certainly, on completion, have proceeded directly on her maiden voyage. Scoresby, son of a celebrated whaling captain, and himself a navigator and man of science, argued that the severe weather encountered by the new ship had been sufficient to alter the magnetic condition acquired when building. He therefore emphasized that 'the plan of correcting, or adjusting, the compass on board such [iron] ships by the antagonistic action of steel magnets, must be delusive and might be extremely dangerous'. In the case of the unfortunate *Tayleur*, he argued, the authorized reports suggested that after the change in ship's magnetism the adjusting magnets had 'seriously augmented the new errors'. Had there been no adjusting magnets, Scoresby concluded, 'the captain would have been guarded against the delusion that he was making a fair course down the channel'. To support his case by direct evidence, Scoresby

[32] Airy, 'Experiments on iron-built ships', pp. 210–13; Cotter, 'George Biddell Airy', pp. 265–6.

[33] MPP, **6**, 315; C.H. Cotter, 'The early history of ship magnetism: the Airy-Scoresby controversy', *Ann. Sci.*, **34** (1977), 589–99, esp. p. 593.

[34] William Scoresby, 'On the loss of the *Tayleur*, and the changes in the action of compasses in iron ships', *BAAS Report*, **24** (1854), 49–53; Cotter, 'George Biddell Airy', pp. 271–4; 'Airy–Scoresby controversy', pp. 591–7.

demonstrated the ease with which the magnetic conditions of iron bars and plates may be dramatically altered by stress or vibration.[35]

Replying in the pages of *The Athenaeum*, Airy argued that iron ships did not, as Scoresby had suggested, alter their magnetic character suddenly enough to render mechanical correctors (rather than numerical correction) dangerous. He also considered that the *Tayleur*'s compass error might have been due to heeling or to a trifling disturbance in the position of the compass rather than to a sudden change in her magnetic condition.[36]

The ensuing controversy provoked considerable discussion among William Thomson's close colleagues and friends. Joule, for example, wrote to him in January, 1855:

> I recollect well when with you at Glasgow College that Mr Napier showed results of the extraordinary alteration of the action of a vessel on its compass by careening. I have not seen Scoresby or Airy's remarks. There seems to be so much ignorance and stupidity among practical men on this subject that I certainly would rather decline sailing in an iron ship.[37]

Mr J.R. Napier himself read a paper entitled 'On ships' compasses' to the Glasgow Philosophical Society in January, 1855. William Thomson was present, and the paper included letters to Napier from both Scoresby and Smith. Smith and Napier had a common interest in the iron steamship's mastery of the oceans. Thus, in October, 1854, Smith had hoped that Napier 'will come out tonight to stay all night, & then we will have a regular magnetic conference'. Two days later he reported to his wife that 'James Napier came to dinner – we had a great discussion about ship's compasses & their scientific mysteries'.[38]

Napier's paper of 1855 treated Scoresby very favourably while speaking cautiously about Airy's methods of compass correction, the distrust of which had clearly become an emotional issue among seafarers. Indeed, what emerges from Napier's discussion is that all three men – Napier, Smith, and Scoresby – were in 1855 quite sceptical about the reliability of compass correction, preferring instead a neutral solution. Napier concluded his paper thus: 'Though many captains of vessels still adhere to the principle of having their compass errors corrected by magnets, the custom of placing a compass on a mast or pole high above the vessel to be free from the influence of iron, as urged by Dr Scoresby and implied by Mr Smith, is becoming more frequent. It is to be hoped this custom will soon become universal'.[39]

[35] Scoresby, 'Loss of the *Tayleur*', pp. 50–3.

[36] G.B. Airy, 'Correction of the compass in iron ships', *The Athenaeum*, no. 1409 (1854), 1303–5; William Scoresby, 'Correction of the compass in iron ships by magnets', *The Athenaeum*, no. 1415 (1854), 1494–5; no. 1416 (1854), 1526–8; G.B. Airy, 'Correction of the compass in iron ships', *The Athenaeum*, no. 1423 (1855), 145–8; Cotter, 'Airy–Scoresby controversy', pp. 595–8.

[37] J.P. Joule to William Thomson, 30th January, 1855, J191, ULC.

[38] Archibald to Susan Smith, 14th and 16th October, 1854, TD1/887, Smith papers, Strathclyde Regional Archives.

[39] J.R. Napier, 'Remarks on ships' compasses', *Proc. Phil. Soc. Glasgow*, **3** (1848–55), 365–75, on p. 375.

Napier also explained that he was 'startled by the remarks [of Scoresby in a letter to the Liverpool Underwriters' Association] regarding the loss of the *Tayleur* when, from the evidence, he [Scoresby] says it appeared that the compasses were all correct on leaving Liverpool, and all wrong after getting to sea'. Napier then recalled a similar change in the compass deviations of the Glasgow-built iron brig *Haiti* swung, on completion, first in Glasgow and subsequently downriver at Gourock. The two sets of results differed markedly. The causes, Napier argued, were those assigned by Scoresby: the soft iron, hammered in a given direction when on the building slip, became 'highly penetrated with retentive magnetism' which was retained so long as the ship moved quietly but which was liable to change when vibrated in new positions:

> On the Clyde [unlike Liverpool's dock system] . . . our want of conveniences for swinging ships for compass deviations may have been the means of preserving our iron ships from those very sudden changes, as a voyage to Gareloch behind a tug steamer may have wrought the necessary change in the retained magnetism. With our steamers . . . I presume that the working of our engines for three hours or more on their way to Gareloch, and making more thumping and knocking perhaps than they ought to do, would be considered quite sufficient to redistribute the retained magnetism, so that when the ship was swung, there would be less risk of any sudden change when first proceeding to sea.[40]

Napier's paper then treated of two major features of ships' compasses of even more immediate interest to Thomson's circle than the comparatively emotional distrust of compass correctors. First, he considered the rival merits of single- and compound-needled compasses. And, second, he considered the causes and remedies of unsteadiness of the compass during rolling of the ship.

Most persons, Napier argued, would admit that, all other things being equal, 'the best compass is that which has the strongest magnetism in its needles, or the most directive power in proportion to the whole card and needles; for, by having this superior directive power, it is enabled to overcome pivot friction and other causes, which render useless less powerful compasses'. Scoresby's researches had aimed to show 'that a compass with a number of thin needles would be much more powerful than if the same weight of steel formed one solid bar needle'. Yet, Napier remarked, Snow Harris had apparently suggested that a single needle, simpler and cheaper than compound needles, might be shown by experience to be 'more than adequate to any practical purpose'. Consequently:

> To satisfy myself I had a compass made, which I conceived would combine the good qualities of most of the compasses I had read about. I adopted Scoresby's compound needles made of clock springs, placed them for economy flat on the card in two sets, four [placed together] in each set . . . I placed these two sets of needles [i.e. two compound needles] at the distances calculated by Mr Archibald Smith, viz. 60°, shown by him to be the position (of the needles) in which the compass card might pitch or roll in all positions without affecting its direction. I tested the power of the card, with its needles complete on

[40] *Ibid.*, pp. 370–3.

it, by the *torsion* balance . . . This method, when the *torsion* of wire is divided by the weight of card, gives I conceive a true measure of the powers which different cards would have of indicating correctly by overcoming friction, &c., when placing on their proper pivots.[41]

Napier compared this torsion–weight ratio for seven designs of compass including Snow Harris's azimuth and steering cards with single-bar needles made by Lilley of London, the Admiralty standard card with four compound needles, and cards by the Liverpool makers Gray's and Keen's. He found that his own Scoresby experimental card with two compound needles had three times the power of Keen's patent card, and a considerably greater power than the Harris and Admiralty cards. He therefore concluded that with cards of equal weight 'the compound needle has about one half more power; or if, as Sir Snow Harris says, the extra power is not needed, then a much lighter card with less pivot friction, and consequently more durability, will be as powerful and sensitive as Sir Snow Harris's'. Furthermore, he criticized the Liverpool makers' view that weight was 'a necessary element of a *steady* card', arguing that weight without power to overcome friction might produce steadiness, but would not combine steadiness with accuracy.[42]

Steadiness was the second major feature of ships' compasses which Napier discussed. Many, if not all, compasses were, he believed, very unsteady at sea in heavy weather. The particular cause he focussed upon arose 'from the influence of the induced magnetism of the iron . . . especially when the vessel sails in an easterly or westerly direction and rolls heavily, for then iron (the sides of an iron ship perhaps . . .), which at one roll or lurch become nearly parallel with the dip, and consequently powerfully magnetic, attracts or repels the north end of the compass needle, by the opposite roll, may become nearly at right angles to the dip, and therefore less magnetic, thereby causing the oscillations'. These effects he illustrated with results obtained from experiments on an eight foot iron boat.[43]

Following these discussions among his friends, William Thomson delivered his own verdict to Forbes in May, 1856. He admitted not having read Airy's original papers on ships' magnetism, but he believed the first paper (1839) contained a fundamental error 'in neglecting the effect of mutual influence of the parts of a piece of iron in influencing one anothers magnetism, according to which [error] the magnetization of a bar would be the same in intensity whether the bar is held along or across the lines of force instead of being as it really is many times greater in the former than in the latter case'. With regard to Airy's methods of correction:

I am quite convinced that his method of compensating the effect on the compass was founded on incorrect reasoning, and has proved very injurious in practice. I believe every attempt at correction if introducing either steel magnets or soft iron into closer proximity

[41] *Ibid.*, pp. 365–6; William Scoresby, *Magnetical investigations* (2 vols., London, 1839–52), **1**, pp. 282–329. [42] Napier, 'Remarks on ship's compasses', pp. 366–9. [43] *Ibid.*, p. 369.

than several yards of the needle *must* no matter how well devised & how much in accordance with any theory as yet applied, be *most pernicious*.[44]

He concluded his indictment of Airy's system by stating that 'every system which makes the captain think he can rely on his compass without constant observation is liable to produce great evils', whereas by contrast 'Smith's system can do no harm unless it is trusted to when direct observation can be had, and is I believe the best system when observation cannot be had'.

Three years later the controversy erupted again, this time between Airy and Smith over Smith's introduction to Scoresby's posthumous *Journal of a voyage to Australia and round the world for magnetical research*. Smith's lengthy introduction (written after consultations with Thomson) aimed 'to ascertain what was correct or mistaken on each side of the [Airy–Scoresby] controversy'. While pointing out that Scoresby had probably misunderstood Airy's meaning with regard to the effects of different magnetic latitudes on the semicircular deviation (Scoresby attributing the effect to retentive or subpermanent magnetism rather than to transient induced magnetism from the vertical component of the earth's field), Smith criticized the imperfections of Airy's theory which purported to bypass the complexities of Poisson's by ignoring mutual influence.[45] Airy's caustic reply in *The Athenaeum* defended his approach on the grounds that it had been adequate for the kind of ship then in service: 'Mr Smith entered late into these investigations, when the general laws of magnetic disturbances in iron ships were perfectly established; and he can form little idea of the obscurity which oppressed the subject in 1838'.[46]

Thomson, writing to Smith soon after, again took sides against the Astronomer Royal:

I am surprised to see Airy in the light of an injured but forgiving person ... I am also much struck with the want of power shown in his section (I). How can he make so transcendent a difficulty of the mathematics of Poisson's investigation which I suppose would be repulsive to a 2nd or 3rd year man only from a want of neatness in his geometry of three dimensions? How can he think riveting not a good enough connection &c? Does he not see by pure mathematics that if a piece of iron were cut into any number of parts & put together with varnish or silk paper, or anything non magnetic of lineal dimensions small in comparison with the smallest diameter of the parts, that the magnetism induced in the whole in any field would not be sensibly altered? ... His justification would have been fair if it had been confined to explaining how he fell into a great mistake & if the mistake had been distinctly admitted. As it is it is plausible, & if uncorrected will lead even tolerably competent readers either temporarily or permanently into the same error.[47]

[44] William Thomson to J.D. Forbes, 16th May, 1856, Forbes papers, St Andrews University Library.

[45] William Scoresby, in Archibald Smith, ed., *Journal of a voyage to Australia and round the world for magnetical research* (London, 1859), pp. vii–xlviii, esp. pp. viii–ix, xxiii–xxxvi; Archibald Smith to William Thomson, 24th January, 1859, S51, ULG.

[46] G.B. Airy, 'Iron ships – the Royal Charter', *The Athenaeum*, no. 1672 (1859), 632–6, on p. 632.

[47] William Thomson to Archibald Smith, 15th November, 1859, S172, ULC; Airy, 'Iron ships', pp. 632–3.

In the same vigorous letter to Smith, Thomson also criticized Airy for misusing quotations (correct in themselves) from his 1839 paper to demonstrate that he had all along been aware of subpermanent magnetism while ignoring completely 'his controversy with Scoresby about the *Tayleur* in which I believe he denied the real substance of what he is now claiming priority for himself', namely, that the ship's magnetism did not vary much over time.

When, some twelve years later, Sir William began to develop his own patent compass, he used all the very considerable resources of his personal diplomacy to steer a difficult course between the various hazards left behind by these earlier battles. He aimed to employ the Astronomer Royal's methods of mechanical correction in accordance, not with any approximate theory, but with Poisson and Smith's theory of magnetism in iron ships and his own general theory of magnetic induction. In aiming once more for the harmony of theory and practice, Sir William had now to devise a compass and its correctors which would permit maximum flexibility for regular adjustment while at the same time retaining a practical simplicity which no numerical method of correction could rival.

Sir William Thomson's patent compass

Sir William Thomson attributed the five-year delay in completing his *Good Words* articles on the mariner's compass to his realization that 'when there seemed a possibility of finding a compass which should fulfill the conditions of the problem, I felt it impossible to complacently describe compasses which perform their duty ill, or less well than might be, through not fulfilling these conditions'. Sir William's famous compass patents were essentially his attempt to 'fulfill the conditions of the problem', conditions which previous compasses of the mercantile marine and Admiralty had not in general satisfied.[48] The compass requirements, as he appreciated them in the period 1871–6, were four in number.

First, the card must have as little friction as possible while turning on its pivot. As Napier had already pointed out in his 1855 paper, steadiness without accuracy was useless: the needle should rest in its true equilibrium position on a steady course. In his 1879 article, Sir William himself argued that the methods of procuring steadiness in common use nearly all relied upon friction on the bearing-point, whereas 'in reality friction introduces a peculiar unsteadiness of a very serious kind':

> In some [compass cards] (as in the Admiralty 'J' card, provided for use in stormy weather) there is a swelling in the middle of each of the steel needles to make them heavier; in others heavy brass weights are attached to the compass cards as near the centre as may be, being

[48] Thomson, 'Terrestrial magnetism', p. 449; PL, **3**, 296. See also W.E. May, 'Lord Kelvin and his compass', *J. Navigation*, **32** (1979), 122–34.

> sometimes, for instance, in the form of a small brass ring of about an inch and a half diameter. Another method, scarcely less scientific, is to blunt the bearing-point by grinding it or striking it with a hammer, as has not unfrequently been done to render the compass 'less lively'; or to fill the cup with brickdust . . . once done at sea by a captain who was surprised to find afterwards that his compass could not be trusted within a couple of points.[49]

Minimizing friction, on the other hand, also reduces wear on the pivot and so improved the durability and economy of the compass.

Second, the compass card should have a long vibrational period to eliminate oscillations due to so-called kinetic error. Sir William made this issue the subject of his own study in the period 1874–9. He drew attention to the general belief that 'the greater the magnetic moment [the proper expression for which in common language is often called 'power' or 'strength'] of the needles the better the compass'. Indeed, Napier himself had opened his 1855 paper with a statement of that very belief. Thomson, however, claimed that it was '*not* generally known that the greater the magnetic moment, other things being the same, the more unsteady will the compass be when the ship is rolling on ocean wave slopes'. On this account, as well as for another reason which will be apparent from his fourth condition (correction of quadrantal error), small magnetic moment was desirable.[50]

The complete rolling error in a ship's compass is made up of the heeling error (analyzed by Archibald Smith) and the kinetic error. The heeling error, Thomson noted, was in many cases 'a more potent cause of unsteadiness than the merely dynamical influence of the ship's rolling', the deflecting influence sometimes amounting to as much as two degrees for every degree of heel. The two influences often conspired together, both drawing the north point of the card towards the upper side of the ship in the northern hemisphere and the reverse in the southern, the effect being greatest when the ship's head was north or south and least when east or west.[51] Smith (and Thomson) had recognized the means of correcting heeling error by horizontal and vertical arrangements of soft iron and by vertical arrangements of hard iron. In his patent compass, Sir William employed permanent magnets in the binnacle below the card together with (in an improved version of the compass around 1879) a vertical bar ahead or abaft the binnacle. This vertical bar, known as a Flinders' bar, also compensated for the change with magnetic latitude of the vertical component of the earth's field.[52]

The kinetic or dynamical error consisted of two factors. First, Sir William

[49] Thomson, 'Terrestrial magnetism', p. 447; PL, **3**, 287–9; Napier, 'Remarks on ship's compasses', pp. 368–9.

[50] Thomson, 'Terrestrial magnetism', p. 447; PL, **3**, 289; Napier, 'Remarks on ship's compasses', p. 365.

[51] Thomson, 'Terrestrial magnetism', pp. 447–8; PL, **3**, 289–93.

[52] Thomson, 'Terrestrial magnetism', pp. 452–3; PL, **3**, 311–22; Evans and Smith, 'Report on Liverpool Compass Committee', pp. 90–1.

recognized in his 1874 paper, 'On the perturbations of the compass produced by the rolling of the ship', that due to the difference of apparent gravity (resulting from accelerations in rolling) from true gravity the compass card does not sit level.[53] The error thus introduced behaved like the heeling error, though depending on acceleration in the heeling motion rather than on the heel itself. Second, as he recognized by 1879, the centre of gravity of the compass card rests just slightly north of the support pivot (in the northern hemisphere) to compensate for the dip in the earth's magnetic force drawing the card downwards towards the north. Thus, when sailing north or south, the ship's rolling will throw the centre of gravity east and west, producing a rotational oscillation in the compass. In this analysis Sir William expressed his indebtedness to Froude's theory of the rolling of ships according to which 'the longer the vibrational period of the ship when set a-rolling in still water by men running from side to side, the steadier she will be in a seaway'.[54]

In order to correct both factors making up the kinetic part of the rolling error, Sir William concluded that he needed a compass with a period of oscillation several times that of the waves, to which end a small magnetic moment would contribute substantially. In reality, he noted that the longest period of a ship's actual rolling in a sea-way varied from fourteen to eighteen seconds. The vibrational periods of the 'A' card of the Admiralty standard (seven and one-half inch) compass was about nineteen seconds; and that of the ten inch compass of merchant steamers was about twenty-six seconds. For satisfactory steadiness, a much longer period still, achievable by a reduction of the magnetic moment, would be necessary.[55]

Third, he required short needles in order that the (corrected) magnetic field over the space of rotation of the needles be effectively constant. Then, for all compass courses *at one position on the earth*, a single correction would be valid.

And fourth, a small magnetic moment was not only important for the long vibrational period, but 'unless the magnetic moment be vastly smaller than that of any of the compasses ordinarily in use hitherto, the accuracy for all parts of the world, of the correction of what is called the quadrantal error in an iron ship, by the Astronomer Royal's method, is vitiated by the inductive influence of the compass upon the iron correctors'. In order for the soft iron spheres to respond proportionally to changes in the earth's magnetic field, and so provide the proper correction at all points of the earth, the effect upon them from the needles had to be minimized.[56]

A couple of years after his first patent in 1876, Sir William described the actual instrument designed by him to satisfy these key conditions:

[53] Sir William Thomson, 'On the perturbations of the compass produced by the rolling of the ship', *Nature*, **10** (1874), 388–9; *Phil. Mag.*, [series 4], **48** (1874), 363–9.

[54] Thomson, 'Terrestrial magnetism', p. 447; PL, **3**, 289. See also William Froude to Sir William Thomson, December, 1877 to October, 1878, F281–F289, ULC.

[55] Thomson, 'Terrestrial magnetism', pp. 448–9; PL, **3**, 293–5.

[56] Thomson, 'Terrestrial magnetism', p. 449; PL, **3**, 294–5.

Eight small needles of thin steel wire 3 ¼ inches to 2 inches long . . . are fixed (like the steps of a ropeladder) on two parallel silk threads, and slung from a light aluminium circular rim of 10 inches diameter by four silk threads through eyes in the four ends of the outer pair of needles. The aluminium rim is connected by thirty-two stout silk threads, the spokes as it were of the wheel, with an aluminium disk about the size of a fourpenny-piece forming the nave. A small inverted cup, with sapphire crown and aluminium sides and projecting lip, fits through a hole in this disk and supports it by the lip; the cup is borne by its sapphire crown on a fine iridium point soldered to the top of a thin brass wire supported in a socket attached to the bottom of the compass bowl. The aluminium rim and thirty-two silk-thread spokes form a circular platform which bears a light circle of paper constituting the compass card proper.[57]

The circle of paper, reduced to a two inch broad band to minimize weight and securely bound to the outer aluminium rim with paste and silk ribbon, was cut across in thirty-two places mid-way between the spoke attachments to prevent it from warping the aluminium rim by the shrinkage it underwent from the sun's heat. Sir William's compass card weighed only about 170 grains compared to the 1600 grains of the Admiralty standard compass and about 2900 grains or six ounces of the ordinary ten inch used in the best-found merchant steamers. The vibrational period of his ten inch compass was about forty-two seconds, while the frictional error did not exceed a quarter of a degree.

Having designed a card which fulfilled the requirements for a good compass, Sir William aimed to incorporate correctors within the binnacle according to Smith's (and his own) improved version of Airy's principles. He emphasized that 'the magnetism of a ship's iron is a very variable property, and it is almost as difficult to classify and describe it in words as it is to correct its effect on the compass'. As we saw in the previous section, its permanent constituent was not perfectly permanent (part, at least, being subpermanent), and its transient constitution not perfectly transient: 'the mariner must be constantly on his guard to determine and allow for unpredictable irregularities in his compass due to variations of the permanent magnetism, and to retention of some of the transient magnetism when the inducing influence is past'. The compass 'is not to be corrected perfectly once and for all by any possible operations or observations, however accurately performed . . . and readjustment becomes necessary; sooner generally in a new ship, but sooner or later in every ship . . . in the course of their service'.[58]

The familiar mariner's expression of 'swinging a ship' meant turning her round with her head successively on all points of the compass and determining 'the error of the compass for a sufficient number of different courses to allow it to

[57] Thomson, 'Terrestrial magnetism', pp. 449–50; PL, **3**, 296–8; 'On compass adjustment in iron ships, and on a new sounding apparatus', *RUSI*, **22** (1879), 91–119, on pp. 93–4; 'On compass adjustment in iron ships', *Nature*, **17**, (1878), 331–4, 352–4, 387–8, on pp. 331–2. A fourpenny piece, or groat, minted in 1838–55, was a small silver coin ⅝ of an inch, or about 1.5 cm, in diameter.

[58] Thomson, 'Terrestrial magnetism', pp. 451–2; PL, **3**, 303–8.

be estimated with sufficient accuracy for every course'. The procedure, though best carried out under way, could also be done at anchor with the aid of a tug or mooring buoys. Subsequent calculation of *A*, *B*, *C*, *D*, and *E* in Smith's theory was regularly performed at frequent intervals for every ship of the Navy. But, Sir William explained, Airy's method took advantage of swinging the ship 'not merely to determine the errors of the compass, but to annul them'.[59]

First, steel magnets placed close to the compass corrected the semicircular error on the north–south and east–west courses. In Sir William's binnacles these permanent magnets were adjustable; whereas in Airy's arrangement they had been fixed, corresponding to his belief that the permanent magnetism of ships varied but little. Second, soft iron corrected the residual error. With small needles and small magnetic moments, Sir William's compass enabled this correction to be made even when the quadrantal error was as much as the 15°–20° sometimes found in ironclads. His binnacle thus had 'appliances for placing and fixing once for all a pair of iron globes in proper positions on the two sides of the compass to correct the quadrantal error'. No further change was ever necessary except in the case of a change in the ship's iron or iron cargo. Third, a steel magnet placed below the centre of the compass corrected the heeling error, 'a subordinate but still very important part' of Airy's complete method of correction. Sir William's binnacle incorporated a further appliance for readjustment of this corrector. He claimed that by these means 'a complete realization of Airy's method is thus now for the first time rendered practically possible for all classes of ships'.[60]

To facilitate even further the correction of compasses in practice, Sir William patented an adjustable deflector, based on a method put forward by Sabine in the 1840s, by which errors could be completely compensated without sights of stars or landmarks. He described the system to Stokes in 1876:

> I am working out a plan by magnets used as deflectors to test the directive force on ship's compass, and then, by applying corrections to make it equal for all azimuths of the ship's length, to 'correct' the compass on all points without sights of landmarks or celestial objects. I think this plan promises to be readier than & as easy as the ordinary by sights or marks. I use the 'sinus-galvanometer' method: that is to say I turn the deflector (whose centre is a few inches above the centre of the compass) until its magnetic axis is perpendicular to the direction in which the magnetic axis of the deflected compass rests.[61]

The deflector made possible above all a great economy of time, being independent of visibility. As Kelvin explained to Admiral Fisher in 1892: 'A ship with us is never detained on account of weather for the adjustment of her compasses'.[62]

59 Thomson, 'Terrestrial magnetism', pp. 451–2; PL, **3**, 306–9.

60 Thomson, 'Compass adjustment in ships', pp. 95–100; 'Terrestrial magnetism', p. 452; PL, **3**, 311.

61 Sir William Thomson to G.G. Stokes, 9th October, 1876, K208, Stokes correspondence, ULC; Thomson, 'Compass adjustment in ships', pp. 103–6; PL, **3**, 322–9.

62 Lord Kelvin to J.A. Fisher, 3rd March, 1892, in SPT, **2**, 716–17.

Sir William Thomson's ten-inch compass and binnacle. In 1884 Sir William provided the capital for a new factory sited in Cambridge Street, Glasgow, and employing some 200 hands and a staff of trained electrical engineers in the testing department. Here were manufactured the famous navigational and electrical instruments. After 1900, the firm became a limited company, Kelvin and James White, Ltd. [From The Science Museum Library, London.]

The commercial compass and sounding machine

After many early experimental trials aboard the *Lalla Rookh*, Sir William, writing from the Cunard liner *Russia*, reported to Helmholtz in May, 1876, that a very fine transatlantic passage had given 'just enough of rough weather to test thoroughly a new compass, which I shall show you when you come to Glasgow [for the British Association]'. The compass had indeed 'behaved perfectly well throughout, notwithstanding a great shaking from the screw (which almost prevents me from being able to write legibly)'. That same year Sir William took out his earliest compass patent and fitted his first commercial compass to the cross-channel steamer *Buffalo* (built 1865), one of six vessels employed by the Glasgow firm of G. and J. Burns on the Glasgow to Belfast service.[63]

From the beginning, G. and J. Burns had been one of the three leading partners in Samuel Cunard's British and North American Royal Mail Steam Packet Company, soon to be reorganized into the Cunard Steamship Company in 1878, with John Burns (a son of the founders and later the first Lord Inverclyde) as chairman and dominant partner. Given its strong Glasgow connections, not least with the Napiers, it is scarcely surprising to find Sir William supplying compasses to each new generation of front-rank Cunarders, shown in table 22.1. In 1884, the Cunard Company provided Sir William with a free passage to New York by the *Servia* and a return on board the *Aurania* following delivery of his Baltimore Lectures. As Lord Kelvin he travelled to and from New York on the *Campania* in the late summer of 1897 on a visit which took in the Niagara Falls, the British Association meeting in Toronto, and the Canadian Pacific Railway to British Columbia. On his return, he reported to Lord Inverclyde on the *Campania*'s tendency to vibrate and roll rather heavily.[64]

The North Atlantic route, primarily to New York, did not have a complete monopoly on competition for size, speed, comfort, and safety, but these factors certainly received very great publicity on that most prestigious of steamer tracks. Even before the arrival of Cunard's greatest rival, T.H. Ismay's White Star Line, in the early 1870s, the Inman, National, and Guion Lines competed out of Liverpool for a share of the passenger (especially emigrant) traffic to North America. Together, indeed, these four Liverpool lines carried 96% of all transatlantic passenger traffic in 1870. In the 1870s, the coming of White Star, of expanding German lines, of the Dominion Line, Leyland Line, Beaver Line, Donaldson Line and others meant a dramatic intensification of competition.

[63] Sir William Thomson to Hermann von Helmholtz, 30th May, 1876, in SPT, **2**, 668–9; to Professor Peirce, 22nd April, 1875, in SPT, **2**, 659–61; Thomson's patent compass books, *op. cit.* (note 2); W.S. Lindsay, *History of merchant shipping and ancient commerce* (4 vols., London, 1874), **1**, pp. 606–8.

[64] F.E. Hyde, *Cunard and the North Atlantic, 1870–1973* (London, 1975), pp. 24–6, 119–31. N.R.P. Bonsor, *North Atlantic seaway* (5 vols., Newton Abbott and Jersey, 1975–9), **1**, pp. 93–102, 152–7; Thomson's patent compass books, *op. cit.* (note 2); SPT, **2**, 802–4, 1001–4; Lord Kelvin to Lord Inverclyde, 11th November, 1897, LB5/29, ULG.

Table 22.1. *Cunard Steamship Company Ltd, leading transatlantic vessels 1881–1907*

ship	year built	yard built	gross tons	remarks
Servia	1881	Clydebank	7392	First steel Cunarder
Aurania	1883	Clydebank	7269	
Umbria	1884	Elder	7718	
Etruria	1885	Elder	7718	
Campania	1893	Fairfield (formerly Elder)	12 950	First twin-screw, triple-expansion Cunarder
Lucania	1893	Fairfield	12 952	
Lusitania	1907	Clydebank	31 550	
Mauretania	1907	Swan, Hunter (Tyne)	31 938	Held Atlantic speed record until 1929

Altogether nearly thirty new Atlantic lines appeared in Europe and America during that decade, with more to come in the 1880s.[65] Consequently, public reputations were all-important for survival, especially during the depressed market of the late 1870s, and the fitting of the most modern and reliable navigational instruments became an issue inseparable from the maintenance of passenger confidence, especially after the *Schiller*'s loss in 1875.

With its forceful policy of technical innovation, White Star (founded as the Oceanic Steam Navigation Company in 1869) did not hesitate to adopt the Thomson compass and sounding machine. With the 5000 ton sister ships *Britannic* and *Germanic* of 1874–5, White Star had not only the largest, but (at over fifteen and one-half knots) the fastest liners in service anywhere in the world. Sir William supplied a ten inch compass to the *Britannic* at the end of 1876, while the *Germanic* received a compass early in 1877, with the older *Baltic* and *Adriatic* (built by Harland and Wolff of Belfast in 1871–2 as part of the original sextet of 3700 ton compound-engined steamers) following shortly. In the same year, Sir William also achieved a major breakthrough for his sounding machine:

> the White Star liner *Britannic* . . . now [early 1878] takes soundings regularly, running at 16 knots over the Banks of Newfoundland and in the English and Irish Channels in depths sometimes as much as 130 fathoms. In this ship, perhaps the fastest ocean-going steamer in existence, the sounding machine was carefully tried for several voyages in the hands of Captain Thompson, who succeeded perfectly in using it to advantage; and under him it was finally introduced into the service of the White Star Line.[66]

[65] Hyde, *Cunard*, p. 94; Bonsor, *North Atlantic seaway*, **1**, pp. 11–14.
[66] Thomson, 'Compass adjustment in ships', p. 107; PL, **3**, 378; Thomson's patent compass books, *op. cit.* (note 2); Bonsor, *North Atlantic seaway*, **2**, pp. 732–6.

The 17 000-ton White Star liner *Oceanic*, built at Belfast in 1899, and the first ship to exceed the *Great Eastern*'s length, exemplified the progress of nineteenth-century British shipbuilding and, like most of the prestigious merchant ships of the period, carried Kelvin patent compasses. [From The National Maritime Museum, Greenwich.]

When White Star built the 10 000 ton sisters *Teutonic* and *Majestic* in Belfast in 1889–90, Sir William supplied the compasses. These twin-screw, triple expansion-engined ships were the first Atlantic liners to abandon all auxiliary sail, and held for a time the speed record for westbound passages averaging over twenty knots. Then, in 1899, the giant 17 000 ton *Oceanic* became the first liner to exceed the *Great Eastern* in length. Liner size increased rapidly thereafter, with White Star seldom far from the leading position, and continuing to fit Sir William's compasses. The finest, and most tragic, hour came with the completion of the 45 300 ton *Olympic* and 46 300 ton *Titanic* at Belfast in 1911 and 1912. Both great ships had several Kelvin compasses on board, and both, like all new White Star liners, had been built in Kelvin's own birthplace.[67]

The case of the Guion Line, and its involvement with Sir William, makes a revealing contrast with White Star's spectacular successes in the late nineteenth century. In November, 1877, the Guion Line removed Sir William's compass, recently supplied to their 4300 ton steamer *Montana* (built 1874). The owners, having lost the *Montana*'s sistership, *Dakota*, on the Anglesey coast in May of that

[67] Thomson's patent compass books, *op. cit.* (note 2); Bonsor, *North Atlantic seaway*, **2**, pp. 737–45.

year, seemed especially nervous about compasses, particularly since the inquiry had blamed the *Dakota*'s master for laying his ship on the wrong course. Attempts to persuade Guion to reinstate Sir William's compass failed after the *Montana*'s master had condemned its unsteadiness. The owner brusquely requested Sir William to 'return to us the cost of your compass and order it to be removed from our quay'. Having diagnosed the problem as deriving from engine vibration, Sir William offered to provide a bowl and card with his improved mode of suspension in time for the ship's next voyage and with the option of a return of Guion's payment if the new arrangement still proved unsatisfactory. But Guion was adamant. Sir William's offer did not lead to a permanent installation and he had to fight another fruitless battle over a compass for Guion's new 5000 ton Elder-built steamer *Arizona* of 1879.[68]

Meanwhile, the loss of the 3100 ton *Idaho* (built 1869) on Ireland's southeast coast in June, 1878, compounded Guion's misfortunes. This time an inquiry attributed the loss to a failure to use the lead. Like the *Dakota*, the *Idaho* did not possess Sir William's compasses. Worse was to come, however, as the *Montana* herself ran ashore on 14th March, 1880, a few miles from where her sister ship had foundered on Anglesey three years before. Again, an inquiry blamed the master for navigational errors.[69] Guion at last agreed to place a Thomson compass on board the *Arizona* at the end of the same month. The *Arizona*'s master, Captain Murray, also complained, but by December, 1880, admitted to being perfectly satisfied. A favourable outcome was vital, as Sir William's Liverpool agent remarked to William Bottomley:

> I am extremely anxious that a compass should be made correct on this vessel & trust you will do all in your power to assist me, otherwise there is no saying what injury it may do us with other firms, as officials & captains are very fond of gossiping & it is the easiest thing in the world to damage the reputation of an instrument.[70]

Though the Line lost no more ships through navigational error, mounting financial difficulties in the 1880s brought the Guion Line to its knees.[71]

With a compass aboard the 3100 ton P & O mail steamer *Poonah* in 1876, Sir William had gone some way to winning over one of the world's most powerful shipping lines. Since its foundation as the Peninsular Steam Navigation Company in 1837 with a mail contract to the Iberian peninsula, the line had received a Royal Charter in 1840 to carry the mails to Egypt and, by 1842, to establish a

[68] J.H. McClure to Guion & Co., 6th November, 1877, CS20, ULC; Guion & Co., to Sir William Thomson, 19th January, 1878, CS114, ULC; Sir William Thomson to Guion & Co., 21st January, 1878, CS117, ULC; Guion & Co., to Sir William Thomson, 19th May, 1879, CS504, ULC; J. Sewill to William Bottomley, 6th and 15th September, 1879, CS538, ULC; Bonsor, *North Atlantic seaway*, **2**, pp. 701–11; *The nautical magazine*, **46** (1877), 715–16.

[69] Bonsor, *North Atlantic seaway*, **2**, pp. 703–5; *The nautical magazine*, **47** (1878), 769; **49** (1880), 530.

[70] J. Sewill to William Bottomley, 31st March, 25th and 27th October, and 7th December, 1880, CS634, CS691, CS694, and CS707, ULC; Guion & Co., to William Bottomley, 26th October, 1880, CS693, ULC. [71] Bonsor, *North Atlantic seaway*, **2**, pp. 705–8, 710.

mail steamship route between Egypt and India with extensions three years later to Penang, Singapore, and China. In 1852, P & O captured the Bombay to Suez mail service, last bastion of the old East India Company, and began a link with Australia. As a result, P & O controlled virtually all stages of the ancient 'Overland Route' to the East. The opening of the Suez Canal in 1869, however, not only brought a revolution in communication with the Orient, but threatened the very survival of P & O's fleet of liners, some of which had been designed for European and the rest for Indian Ocean service. The Company's response was to rebuild its fleet with much larger liners suitable for a through passage. From 1870 to 1879 it took delivery of over two dozen new vessels in the range 3000–4000 tons compared to a pattern of 1500–2000 tons in the 1860s. P & O had for long built its ships on the Thames, but most of the new vessels, all iron-hulled, screw-driven, compound-engined ships, came from the Clyde.[72]

At the end of 1877, William Bottomley reported to his uncle that the *Poonah* had arrived home and that her master was 'quite as well pleased with his compass now as he was when he was at home last. He has applied to the P & O Company for a sounding machine to take out with him on his next voyage and he seems determined to have it'. By April, 1878, another compass had been supplied to P & O.[73] At the same time, the Greenock shipbuilders, Caird and Company, asked Sir William to oblige them 'with the particulars of his standard compass and by stating the price at which he would supply one for the Peninsular & Oriental Company's steamship *Kaiser-i-Hind*'. This veritable Emperor of India was P & O's newest and, at over 4000 tons, easily their largest mail liner, with a speed of fifteen knots on trials. Within two days Sir William had offered to fit and adjust one of his ten inch compasses on board the *Kaiser-i-Hind* for the sum of £50. Swan (Sir William's London agent) supplied a sounding machine to the *Kaiser-i-Hind* in September, 1878, and reported to Bottomley that her master was 'the scientific captain of their fleet – I do hope there is no other . . . Altogether whatever science does for other people it certainly does not improve captains'. Nevertheless, Captain Methuen reported in January, 1879, that 'he was very well pleased with the sounding machine', although displeased with the standard of workmanship.[74]

In September, 1878, Swan reported to Bottomley on the determination of P & O to introduce Thomson's instruments and that as a result he hoped 'bye and bye to see every ship that sails from London with at least one of Sir William's instruments'. By the end of 1879, at least another eight principal P & O liners had

[72] Boyd Cable, *A hundred year history of the P & O* (London, 1937), pp. 30–168, 245–6.

[73] William Bottomley to Sir William Thomson, 28th December, 1877, CS77, ULC; P.M. Swan to Sir William Thomson, 9th April, 1878, CS190, ULC; Bonsor, *North Atlantic seaway*, **3**, pp. 951–2.

[74] Caird & Co., to Sir William Thomson, 23rd March and 1st April, 1878, CS168 and CS178, ULC; P.M. Swan to William Bottomley, 20th and 21st September, 1878, CS310 and CS312, ULC; Cable, *P & O*, pp. 269–70.

been supplied with the compass.[75] In 1880, the 3400 ton *Ravenna*, built by Denny, introduced steel hulls to the P & O, and in 1885 the 4700 ton *Coromandel* ushered in the triple-expansion engine. At the end of the 1880s, 6000 ton liners entered P & O service, with 10 000 tonners, powered by quadruple-expansion engines, following in the 1900s. These P & O ships, then, served the Raj in the later part of Queen Victoria's reign, ships which could be relied upon, with their Clyde-built hulls and engines and their Kelvin compasses and sounding machines, to convey the precious mails, Viceroys, administrators, army officers and trading entrepreneurs from Britain to the vast Indian Empire, the glittering jewel in Imperial Britannia's earth-circling crown.

Since its foundation by the Scotsman William MacKinnon (1823–93) in 1862, the British India Steam Navigation Company commanded services from Calcutta to Rangoon and Singapore, from Bombay to Karachi and the Persian Gulf, from Aden to Zanzibar, and the entire Indian coast from Calcutta to Bombay and beyond. By 1873, over thirty ships served the company, and, by 1893, the year of MacKinnon's death, the number had increased to almost ninety. After the cutting of the Suez Canal, British India opened direct links from London to the Gulf and to Colombo, Madras and Calcutta in direct competition with P & O. New vessels were ordered on the Clyde (principally from Denny of Dumbarton) and in 1874 alone seven new vessels of around 2000 tons each were delivered. There could be few greater prizes for Sir William Thomson's instruments than to win approval from this firm whose very name exemplified Britain's Imperial sway over the coasts and islands of the Indian Ocean.[76]

Sir William supplied his first British India compass to the new, Denny-built steamer *Chanda* of 2000 tons in April, 1877. The ship's maiden voyage took her via Suez to Calcutta, and it was from there that her master despatched a critical report to Sir William on 31st July:

> I regret to inform you that I am not able to report so favourably on your compass as I had hoped to do. I navigated the vessel out to Port Said by it, but it never proved itself a true compass, & required as much alteration to find the error of it, as any ordinary compass . . . I must mention that I never altered the arrangements of the magnets & thought it useless to do so, because the error was constantly altering. Pray understand that the Standard Compass was not true either . . . My limited practical experience of compasses in new vessels is that the error is constantly altering for the first six months, & then becomes pretty settled but still it decreases for some time after.[77]

Captain Hutcheson acknowledged, however, that Sir William's compass card

[75] P.M. Swan to William Bottomley, 11th September, 1878, CS301, ULC; Captain Shallard to William Bottomley, 24th February, 1879, CS457, ULC; P.M. Swan to William Bottomley, 22nd March, 1879, CS468, ULC; Thomson's patent compass books, *op. cit.* (note 2); Cable, *P & O*, p. 269.

[76] George Blake, *B.I. Centenary, 1856–1956* (London, 1956).

[77] Captain C.C.D. Hutcheson to Sir William Thomson, 31st July, 1877, CS7, ULC; Blake, *B.I. Centenary*, pp. 221, 255.

had been 'wonderfully steady & at the same time very sensitive showing at once when the ship's head swings to one side or the other'. But to have to constantly alter the magnets was 'too complicated for practical purposes', a deviation card being 'quite as simple'. He had no doubt that the compass 'is a most beautiful & probably correct instrument when used by a scientific man, but it is too complicated & delicate for rough sea use'. Whatever the cause of the *Chanda*'s problem, Captain Hutcheson had been replaced by early 1878. British India seemed happy to fit Sir William's instrument to four new steamers under construction by Denny on the Clyde in July of that year. MacKinnon, indeed, intimated that he would have Sir William's compass on board all his new vessels.

Denny, however, insisted that MacKinnon pay for the compass as an extra, the Scottish builder's margin of profit being so small that he claimed to be unable to afford to supply them within the contract.[78] As a price list for 1879 shows (table 22.2), a shipowner could pay up to £50 for each complete ten inch compass with binnacle and adjustment, costs which only the most flourishing firms could really afford. Similarly, the complete sounding machine, with 300 fathoms of wire, four sinkers, four brass tubes, and a box of 100 prepared glass tubes and 'other requisites' sold in 1879 for £20. Some notion of comparative costs may be gained by noting that a P & O captain had been receiving pay of £20 per month, and a seaman £2.10s in 1839, while in the 1880s a skilled Clydeside riveter might earn £1.10s per week.[79]

Early in 1878, J.H. McClure (Thomson's Liverpool agent) reported to Bottomley that he had extracted a promise from MacKinnon to have the sounding machine introduced to some of the Calcutta boats, the smaller vessels trading to and from the treacherous waters of the Hooghly River and among the poorly charted shallows of the Bay of Bengal. Given that in the period 1856–66, MacKinnon had lost eleven ships out of a fleet of thirty-one on coasts not only uncharted and unlighted, but subject to violent storms and sudden cyclones, he evidently considered £20 a small price to pay for a convenient sounding machine. Thus, in 1879, British India ordered from Lilley & Son, Swan's successor as London agent, sounding machines for four of their modern ships.[80]

The case of the *Chanda*'s compass shows that compass sales were not altogether free from complaints from anxious shipmasters and owners. Masters unfamiliar with compass correction sometimes found Sir William's patent rather complex, and unfavourable reports by masters to owners would carry much weight against the adoption of the new instrument. With British India, the issue had been quickly resolved in Sir William's favour, but not all leading

[78] P.M. Swan to Sir William Thomson, 13th July, 1878, CS273, ULC; Sir William Thomson to William Bottomley, 16th July, 1878, CS276, ULC.

[79] Prices quoted by F.M. Moore (Chronometer, Watch Maker, Optician &c., 102 High Street, Belfast and 23 Eden Quay, Dublin) to William Bottomley, 12th February, 1879, CS439, ULC; Cable, *P & O*, p. 71; Pollard and Robertson, *British shipbuilding industry*, p. 185.

[80] J.H. McClure to William Bottomley, 29th January, 1878, CS125, ULC; John Lilley & Son to William Bottomley, 25th February, 1880, CS617, ULC; Blake, *B.I. Cenentenary*, p. 45.

Table 22.2. *Price list*

SIR WM. THOMSON'S IMPROVED MARINER'S COMPASS (PATENT)

7 inch	Compass Card & Bowl with knife-edge Gimbals, suitable for Wooden and Iron vessels, when no adjt. required	£9
8	d[itt]o	£10
10	do	£12
10	suitable for Pole or Masthead	£15
10	Wheel-house Compass for steering by, with Lamp, Correctors, &c., and including adjt.	£30
	Teakwood Pillar Binnacle, 10-inch Compass Card, Bowl with knife-edge Gimbals, Brass Helmet and Side Lamps	£20
	do , all necessary Correctors, and Clinometer showing the Heel of the vessel	£40
	do do with Instrument for taking Shore Bearings or Azimuths of Heavenly Bodies	£45
	do do do including adjt.	£50

Estimates supplied for other sizes of Masthead, Pole, Steering or other special kinds of Compasses.

The above prices do not include carriage from Glasgow, or travelling expenses to place of adjt.

Terms – Nett Cash on delivery.

SIR WM. THOMSON'S IMPROVED SOUNDING MACHINE

Sounding machine, with 300 fathoms of Wire, four Sinkers, and four brass tubes, a box containing 100 prepared glass tubes and other requisites	£20

owners were as easily impressed as MacKinnon. One such man was Donald Currie (1823–1909).

Greenock-born Currie had been educated briefly at the Royal Belfast Academical Institution, and from 1842 until 1862 he had served Cunard through the Liverpool partners D. and C. MacIver. Thereafter his own firm of Donald Currie and Company quickly developed from a small but modern fleet of four 1200 ton Napier-built iron sailing ships engaged in trade with India to a powerful line of mail steamers between England and Cape Colony during the 1870s. Since 1857, the mail contract to the Cape had been in the hands of the Union Steamship Company, originally founded by P & O's Arthur Anderson to ship South Wales's coal to Southampton's mail ships but soon to become the established link with Britain's colonies of the Cape and Natal in southern Africa. Between 1866 and 1877, however, the Union company had introduced only second-hand tonnage to its services; while, by contrast, Currie had four new Clyde-built steamers on order for his Indian service in 1872. That same year he opened his rival, private mail service from Dartmouth to the Cape using some of his new ships to make record breaking passages. By late 1876 he had won a mail contract shared equally with his older rival, a contract which provided for a weekly mail service to Cape Town with alternate sailings by Union and Castle Line ships. At the same time he had five new 2800 ton mail steamers on order from Napier.[81]

As the extract quoted at the beginning of the present chapter shows, Sir William supplied three compasses to the Napier-built *Balmoral Castle* in February, 1877, these being adjusted under Sir William's personal supervision. On 8th March, 1877, Sir William wrote to Lord Rayleigh with an urgent request:

> I see by this morning's paper that you and Lady Rayleigh are to be passengers in the *Balmoral Castle* tomorrow from Dartmouth for the Cape. There are three of my compasses on board – an azimuth compass, and two steering compasses in the wheel house. I should be very glad if you would look at them a little and see how they behave . . . The enclosed printed papers about the compass . . . will tell you all you may care to know about my system of adjustment.[82]

By June, 1877, two more of Currie's brand new steamers, the *Taymouth* and *Dublin Castles*, had been supplied. Part of Currie's initial enthusiasm for Sir William's compass may have arisen from the total loss by stranding on Dassen Island, Cape Colony, in October, 1876, of the pride of the early Currie mail fleet, the 2700 ton *Windsor Castle* (built by Napier in 1872), fortunately without loss of life.[83]

In late 1877, however, Currie's Captain Small had formed very unfavourable opinions on both compass and sounding machine, making it unlikely that the last of the new steamers, *Conway Castle*, would receive either instrument. Currie, it seemed, had become convinced that Sir William had been 'experimenting with his ships'. Swan, however, skilfully used his success with the rival Union Steamship Company and P & O's favourable testimonial to persuade Currie to adopt a Thomson compass for the new ship in January, 1878. Currie also fitted the instrument to his new coastal steamer *Dunkeld*, designed for the Natal–Cape link, in 1878.[84]

Of great significance to Currie's change of heart was the supply of a sounding machine and compass to one of the Union Company's first new steamers since 1866, the 2900 ton *Durban*, in November, 1877. The *Durban's* master, Captain Warleigh, subsequently reported to Swan that it was impossible for him 'to

[81] Marischal Murray, *Union-Castle chronicle, 1853–1953* (London, 1953), pp. 1–108; James Napier, *Life of Robert Napier of West Shandon* (Edinburgh and London, 1904), pp. 261–2.

[82] Sir William Thomson to Lord Rayleigh, 8th March, 1877, in SPT, **2**, 680.

[83] Murray, *Union-Castle*, pp. 79–81.

[84] P.M. Swan to Sir William Thomson, 3rd November and 12th December, 1877, CS18 and CS69, ULC; 18th January and 26th February, 1878, CS110 and CS148, ULC; P.M. Swan to William Bottomley, 4th October, 1878, CS324, ULC; Thomson's patent compass books, *op. cit.* (note 2).

express how much pleased I am with the [sounding] apparatus or what a comfort it is'. He was, however, not satisfied with the chemical depth recorder, finding that the line marking the colour change was irregular.[85] As to the compass, in late 1878 the *Durban* became the first vessel to be fitted with a Flinders bar in connection with Sir William's compass, as Sir William wrote in 1879:

> an error of 34° growing up in the voyage from England to Algoa Bay [Port Elizabeth], and disappearing on her return to England, has been corrected by a Flinders bar attached to the front side of the binnacle, and the ship now goes and comes through that long voyage with no greater changes of compass error than might be experienced in the same time in a ship plying across the Irish Channel.[86]

Captain Warleigh, meanwhile, had reported very favourably to the directors of the company.[87] As a result the company ordered several more compasses for their steamers in 1878–9, including instruments for the new 3200 ton mail ships *Pretoria* and *Arab* which represented the Union Company's determination to match Currie's Castle Packets Company. By 1880, indeed, Sir William's instruments seem to have become standard equipment for both Union and Castle ships, even though Currie demanded a reduction in price 'before we decide to put them on board each ship of our fleet'.[88]

The final seal of approval placed on Sir William's instruments by these half-dozen major British lines (Cunard, White Star, P & O, British India, Union Steamship and Castle Line) effectively guaranteed massive sales not only to their own fleets but to smaller companies. Few, if any, of British and foreign liner companies could afford not to adopt the best in navigational instruments – and Sir William's compass and sounding machine were the best in the eyes of leading shipbuilders and shipowners. Less well-known, but often expanding, firms such as D.J. Jenkins's Shire Line and MacGregor Gow's Glen Line of Steamers, both trading to the Far East, Cayzer Irvine's Clan Line of Steamers, the Dominion Line, the Inman Line, the African Steamship Company, David McIver and Company, and many others all placed orders.

The widespread adoption of Sir William's instruments also owed much to the persuasive powers of Captain S.T.S. Lecky (1838–1902), master mariner, fellow Ulsterman and author of the highly successful textbook on practical navigation known popularly as Lecky's '*Wrinkles*' (1881), which reached its twelfth (greatly enlarged) edition by 1900 and its twenty-third by 1956. He made accurate tests of Sir William's compass using star azimuths during transatlantic voyages with the 3400 ton steamer *British Empire* in 1879, and reported to Sir William that he considered the results 'very wonderful': 'they certainly bear

[85] Captain H.S. Warleigh to P.M. Swan, *c.*25th April, 1878 (copy), CS213, ULC.

[86] Thomson, 'Terrestrial magnetism', p. 453.

[87] Captain Warleigh to Sir William Thomson, 12th October, 1878 and 22nd January, 1879, CS333 and CS422, ULC.

[88] Donald Currie to John Lilley and Son, 7th October, 1880, CS686a, ULC.

testimony to the extreme precision both of the compass & mirror'. Lecky aimed 'to get shipowners to benefit *themselves*' by adopting Thomson's navigational instruments, and he carried out a dramatic demonstration of the effectiveness of the compass and sounding apparatus in the same year. While in command of the 3600 ton steamer *British Crown* on her delivery voyage from Harland and Wolff's Belfast shipyard to Liverpool, Lecky encountered 'so thick a fog that we could get no point of departure'. Sir William's sounding machine and standard compass and patent taffrail log for measuring distance were his sole guides. He reported to Thomson:

> I kept your sounding machine going every half hour & felt my way round the Copelands [at the southern entrance to Belfast Lough] & Isle of Man with the greatest ease. Next morning . . . we made the [Mersey] Bar Lightship 4 points on the port bow, distant about 400 yards. We saw *nothing whatever* during the passage, though we several times *conversed* with the people in sailing vessels which we passed. Now I do not hesitate to say that I could not have done this without your navigational instruments & I cannot tell you how thankful I am to have them on board. We sail for Phil[a] [Philadelphia] on the 15th . . .[89]

With such practical experience of the compass and sounding machine behind him, Lecky could write with conviction in his '*Wrinkles*' that in his view the sounder which 'immeasurably excels all others' was that invented by Sir William Thomson, and that 'when the owner's pocket can afford it . . . there is no standard compass which in any way can rival the one invented and patented by Sir Wm. Thomson . . . Its mechanical construction is as near perfection as may be'.[90]

Sir William Thomson and the British Admiralty

Traditionally, the Royal Navy's wooden fighting ships had been constructed in the shipyards of England, notably on the Thames or at the Royal Dockyards. With the coming of steam, however, Robert Napier persuaded the Admiralty to place orders for marine engines with his Scottish firm, and subsequently demonstrated that, for similar engines, his products not only cost less to build, but were cheaper to maintain. He thereafter constructed the Navy's first iron vessels in 1844 and, through the efforts of his son, J.R. Napier, built the armour plated *Erebus* in just three and a half months during the Crimean War. But the greatest event of all at Napier's occurred in 1861 with the launch of the 9250 displacement-ton *Black Prince* which, with her Thames-built sister ship *Warrior*, became the world's first ocean-going, iron-armoured fighting ships or 'ironclads'. To mark the occasion, even Glasgow College's professor of Greek,

[89] S.T.S. Lecky to Sir William Thomson, 12th April and 12th October, 1879, CS479 and CS551, ULC; 23rd March, 1880, CS632, ULC.

[90] S.T.S. Lecky, '*Wrinkles' in practical navigation* (London, 1881), pp. 11, 94.

Lushington, adjourned his class with the comment that 'this was a sight the Athenians would have loved to see'.[91]

This powerful association between Clydeside contractor and Imperial guardian greatly enhanced Sir William's existing Admiralty connections through Archibald Smith and Captain Moriarty, RN. His appointment in 1871 to the Admiralty Committee on the Design of Ships of War, a Committee which not only met in Whitehall but actually examined the performance of fighting ships at sea, further strengthened these links. In March of that year, Sir William wrote to his sister-in-law that he had 'tried the Hotspur (Ram) [an ironclad ramming ship launched by Napier a year earlier and not yet commissioned] outside the [Plymouth] breakwater on Tuesday, to see the green water breaking over her bow and washing up to the turret'.[92]

With the causes of the *Captain*'s loss very much in mind, the Committee advised the Admiralty to increase the freeboard amidships of the first mastless turret ships, *Devastation* and *Thunderer*, under construction at the time. Both ships retained low freeboard forward – a considerable disadvantage in heavy weather – and so the Admiralty redesigned their immediate successor, *Dreadnought*, on the basis of the Committee's recommendations. Although not completed for over eight years, the 10 900 displacement-ton *Dreadnought* ushered in the age of the steam battleship, characterized by armour plate, gun turrets, generous freeboard, and a complete absence of auxiliary sail.[93]

At first, Sir William's compass met with no enthusiasm from the Hydrographic Office, as he explained in April, 1875:

> I have had two years' struggle with the compass department of our Hydrographic Office to induce them to take up some suggestions I have brought before them for the correction of the compass in iron ships. They are most obliging in giving me information, but are utterly immovable for anything like co-operation for anything new, as they were in the matter of sounding by pianoforte wire. I now see that if anything is to be done in the matter the whole business of it will fall upon myself, and I have therefore resolved to take out a patent for a new form of compensator and appliances which I have now made. I doubt whether with all the obligingness that has been shown to me I should be more successful in getting a trial of it than I have been in getting a trial of the pianoforte wire by our people, until I can offer it to them as a patented invention.[94]

The aging Astronomer Royal himself bestowed an equally cool verdict upon the compass; on being sent a prototype he tersely remarked: 'It won't do'. But Sir

[91] Napier, *Life of Robert Napier*, pp. 69–81, 153–65, 197–8, 208–15; Robert Gardiner (ed.), *Conway's all the world's fighting ships, 1860–1905* (London, 1979), p. 7. 'Displacement', employed in relation to warships, provides a measure of the actual weight of vessels, unlike 'gross tonnage' for merchant ships which measures volume of enclosed space.

[92] Sir William Thomson to Jessie Crum, 4th March, 1871, in SPT, **1**, 583–4; Gardiner (ed.), *Conway's all the world's fighting ships*, p. 22.

[93] *Ibid.*, pp. 23–4.

[94] Sir William Thomson to Professor Peirce, 22nd April, 1875, in SPT, **2**, 659–61; to F.J. Evans, 25th November, 1873, in SPT, **2**, 702–4.

William dismissed Airy with the words: 'So much for the Astronomer Royal's opinion!'[95]

By November, 1877, P.M. Swan, Sir William's London agent, could report to William Bottomley that he was 'in correspondence with the Swedish, French, Austro–Hungarian, [and] Russian Governments about the compass & sounding machine and expect the German–Danish, Italian & Chilean Naval Attaches to call on me in the course of a few days'. Before many months had passed, orders materialized from the German Admiralty at Kiel (who ordered three compasses during 1878), the Imperial Austrian Government, the Russian Navy, the Royal Dutch Navy, and the Italian and Brazilian Governments.[96] By contrast, only two large British warships had been fitted with the compass in 1877, both being supplied under unofficial, trial arrangements.

One of these ships, HMS *Minotaur*, belonged to a class which at the time of building in the 1860s comprised the world's three largest fighting ships. A huge ironclad of 10 700 displacement-tons and thus containing an enormous mass of iron, the *Minotaur* proved eminently suitable for rigorous compass trials, and through the enthusiasm of her captain, Lord Walter Kerr (1839–1927) (brother of the Marquess of Lothian and later the immediate predecessor of John Arbuthnot Fisher (1841–1920) as First Sea Lord of the Admiralty) both instruments underwent extensive tests in the autumn of 1877. Lord Walter wrote to Sir William explaining that he had just reported favourably to the Lords of the Admiralty concerning the instruments and 'suggesting that the Government should become the purchasers of them to admit of their being more fully tried on board *Minotaur* with the ulterior view of their being generally adopted in H.M. Service'.[97] With the sounding machine, *Minotaur*'s Captain had been especially impressed:

> We have been using it constantly when running up Channel, from the time of crossing the line of soundings to the time of reaching Plymouth; and, though running before a gale of wind, with a heavy sea, at the rate of ten knots, we were able to get soundings as if the ship had been at anchor. We were able to signal to the squadron each sounding as it was obtained; thus, in thick weather, verifying our position by soundings without having to round the ships to.[98]

The Admiralty did purchase the single compass, but continued to show little interest concerning the general introduction of Sir William's instruments into the Navy.

95 Sir William Thomson to G.B. Airy, 3rd March, 1876, in SPT, **2**, 708–9, 710.

96 P.M. Swan to William Bottomley, 22nd November, 1877, CS38, ULC; to Sir William Thomson, 9th April, 1878, CS190, ULC; Thomson's patent compass books, *op. cit.* (note 2).

97 Lord Walter Kerr to Sir William Thomson, 15th September, 1877, CS7B, ULC; Gardiner (ed.), *Conway's all the world's fighting ships*, pp. 10–11; J.H. McClure to Sir William Thomson, 26th November, 1877, CS42, ULC.

98 Lord Walter Kerr to Sir William Thomson, *c.* September, 1877, in Thomson, 'Compass adjustment in ships', pp. 110–11; PL, **3**, 386; J.H. McClure to Lord Walter Kerr, 13th November, 1877, CS26a, ULC.

The 10 700 displacement-ton ironclad HMS *Minotaur* symbolized British Imperial sea power in the 1860s and 1870s. Her captain, Lord Walter Kerr, undertook rigorous trials of Sir William's compass in 1877 and reported enthusiastically on the effectiveness of the sounding machine. But not until 1889 did the Admiralty adopt Sir William's as its standard compass. [From the *Illustrated London News*.]

The commanding officers of the other fighting ship fitted with a trial compass, HMS *Thunderer*, carried out tests at Portland on the effects of the huge revolving turret on the compass as well as the effects of heeling. Given the large quadrantal deviation for such a massive ironclad, ten inch globes had to be specially fitted. Before long, her commanding officers expressed the hope that Sir William's compass would be purchased by the Admiralty as a standard compass, but believed that 'the Hydrographic Office will be against it'. Thus the Hydrographic Office continued to be the principal source of Admiralty resistance to the widespread introduction of the Thomson compass.[99]

Early in February, 1878, Sir William delivered his lecture 'On compass adjustment in iron ships, and on a new sounding apparatus' to the Royal United Service Institution. One member of the audience commented soon afterwards that 'after hearing the discussion which followed the lecture, in which the Hydrographer of the Admiralty, Captain Evans, took a prominent part, I consider that Sir William Thomson fully maintained his position, that, for all practical purposes of navigation, his improved compass was the best that had yet

[99] Robert Jackson to Sir William Thomson, 12th and 18th September, 21st October, and 16th November, 1877, CS7A, CS7C, CS9A, and CS32, ULC; Robert Jackson to J.H. McClure, 28th October, 1877 (copy), CS13A, ULC.

been invented'.[100] F.J. Evans had collaborated with Archibald Smith on the *Admiralty manual*. An understanding of his attitude is important, for upon his verdict depended the official adoption of Sir William's compass by the Royal Navy.

Evans based his judgement on a recognition of two different systems of compass management aboard ship. The practice followed in the Navy was to determine experimentally the magnetic character of the ship and only then to fix correcting magnets permanently to the deck. The next important step was to 'insist upon the principle that the seaman is not to consider that his compass is perfect in its pointing; he must always consider it imperfect'. Under this system there must be no movable magnets to adjust, only the constant observation and recording of compass error from which course corrections should be made. In Sir William's system, on the other hand, one did not need to know the magnetic constitution of the ship. One did, however, require the technical skill and knowledge to make the delicate adjustments. Evans fully acknowledged that 'there is no doubt that any instrument, especially a compass, brought forward by Sir William Thomson, will be thoroughly perfect in all its theoretical details', and he confessed that he had followed him 'in the progress of this instrument to its present state' of theoretical perfection. But he believed that seamen as a rule did not have 'the intelligence to handle those delicate magnets which he applies'. Thus, 'theoretically, Sir William Thomson gives you something approaching to perfection; but practically, you are brought up by the performance of details, which demand far more intelligence and knowledge than we know to exist among seamen'.[101]

In reply, Sir William argued characteristically that he did not recognize such a division between theory and practice. To begin with, he rejected the practicability of Evans's rule not to admit movable correctors: 'as a matter of fact it is not, and cannot be, carried out in practice' since the magnets used to correct any compass could be moved by irregular, if not regular, means. For example, if on a voyage to the southern hemisphere the magnets were found to be increasing the error, the magnets nailed to the deck would have to be torn up. Rather, the system of having adjustable magnets within a locked-up binnacle, the key of which was placed in the hands of a responsible officer, was much to be preferred. This officer would act in accordance with definite general instructions 'to change the correctors gradually according to observations made at sea or in harbour from time to time' or else by handing over to professional adjusters who would occasionally come on board. Sir William skilfully summed up:

> Captain Evans has done ample justice to the instruments and methods before you in saying that they are theoretically perfect; but if they are not also suitable for practical work they have no right to be here, not to be admitted on board ship. The *Admiralty*

[100] Edward Smith to Vogel, 5th February, 1878 (copy), CS130A, ULC.
[101] F.J. Evans in Thomson, 'Compass adjustment in ships', pp. 114–15.

manual, by Captain Evans and Mr Archibald Smith, is perfect in theory; it gives the theory of the effect of ship's magnetism upon the compass most admirably, I believe as well as it can be given. I have only endeavoured very humbly to put in practice some of the theoretical conclusions to be found in the *Admiralty manual*.[102]

Sir William's instruments stood as the embodiment of magnetic theory.

Quadrantal error and the need for its correction provided a specific illustration of the different approaches to compass management by Evans and Sir William. In Evans's opinion, 'the quadrantal error . . . is not a very formidable error under ordinary circumstances, either in merchant ships or at the standard compass position in ships of war'. According to the Admiralty system, 'the quadrantal deviation is not often corrected mechanically, but is generally left for tabular corrections'. Sir William, however, argued that this rule was not easily defended as 'a good practical rule', whereas correction of the quadrantal error by the Astronomer Royal's method not only vastly simplified the compass problem for the navigator and reduced the risk of mistake, but diminished the labour of the analysis performed in the Compass Department by which the accuracy of observations made at sea was checked. Furthermore, Sir William believed that the quadrantal error was scarcely ever less than 6° and often 9° or 10° in merchant steamers and warships. The 'practical value of a complete correction of this error is very clear', he argued, since the quadrantal error changed twice as rapidly as the semicircular error and led to large compass errors even with small alterations of compass courses.[103]

The fundamental divergence of Evans and Sir William is particularly revealing. Evans combined a cautious approach to technical innovation with an unashamed expression of his and the Admiralty's superiority over the inferior intelligence of practical seamen in general and over the achievements of the merchant service as a whole. For his part, Sir William never underestimated the intelligence of either seamen or students. On the other hand, Admiralty caution towards technical innovation must have been promoted by the *Captain* disaster in 1870 as well as by an endless line of nineteenth-century inventors queuing to persuade the Navy to take on board their patent devices. After all, Queen Victoria's fleets totalled more than the combined fleets of the next two largest navies, making a huge market for any entrepreneur.[104]

On 11th April, 1878, Sir William wrote to the First Lord of the Admiralty, W.H. Smith of newsagent fame, asking him directly 'to consider the question of introducing them [compass and sounding machine] into the Royal Navy for practical use'. He enclosed a list, with printed testimonials for compass and sounding machine, of upwards of sixty large iron steamers and sailing ships in which the compass had been 'now amply tested at sea'. He also emphasized that 'after six months' trial on board the German ironclad *Deutschland* [launched on

[102] *Ibid.*, p. 117. [103] *Ibid.*, pp. 115, 117–19.

[104] Gardiner (ed.), *Conway's all the world's fighting ships*, p. 1; Oscar Parkes, *British battleships. Warrior 1860 to Vanguard 1950. A history of design, construction, and armament* (London, 1957).

the Thames in 1874 as Germany's last foreign-built capital ship], a second [compass] had been ordered by the German Imperial Admiralty, who have recently ordered two more sounding machines after a short trial of the first one'.[105]

In the absence of a decisive reply from the political First Lord, Sir William tried again in 1879, this time offering to supply one of his improved compasses at his own expense for trial 'in any ship or ships in which their Lordships may give orders to have it tried, and to allow it to remain on trial without charge as long as they may desire'. The Admiralty accepted in May, although it was November, 1879, before the improved compass was placed in the conning tower of HMS *Northampton*, a Napier-built, partly armoured cruiser (or second-class ironclad) launched in 1876 and employed for various experiments including the fitting of the first protective torpedo nets in 1880. Sir William supplied HMS *Minotaur* with a new ten inch compass a few weeks later. Ironically, the British-built Brazilian ironclad *Independenzia* had been requisitioned by the Admiralty as HMS *Neptune* during the Balkan crisis of 1878, with the result that the Navy had to write to Sir William requesting despatch of a pamphlet explaining the management of compass and sounding machine![106]

HMS *Northampton* underwent a series of rigorous trials in November, 1879, and *The Times* in a graphic report noted that both compass and sounding machine had given the utmost satisfaction. Sir William's agents in London and Liverpool requested 500 copies of the report for circulation among merchant shipowners.[107] Thomson himself had been on board the *Northampton*, where his task of compass adjustment was frustrated by faulty workmanship:

Sir William Thomson came to sea in the ship to try his compass; he was at it from daylight to dark. A lieutenant was told off to assist him, and a very cold job it was. For three days he failed to adjust it; there was always some error he could not eliminate, and his temper and language went from bad to worse. On the fourth day the Lieutenant observed that the compass card was only marked 359 degrees instead of 360 degrees, the missing degree having been cut off when the two ends of the paper were joined together. Then Sir William fairly exploded. He was landed at Portland, raced up to Glasgow, slaughtered some one and came back again with the mistake corrected. The compass was then adjusted and gave no further trouble.[108]

[105] Sir William Thomson to W.H. Smith, 11th April, 1878, CS198, ULC; SPT, **2**, 711–12.

[106] Admiralty to Sir William Thomson, 23rd May, 1879, CS505, ULC; J. Sewill to William Bottomley, 20th November, 1879, CS564, ULC; Alfred Thomas to Sir William Thomson, 22nd April, 1878, CS211, ULC; Gardiner (ed.), *Conway's all the world's fighting ships*, pp. 25, 64; Thomson's patent compass books, *op. cit.* (note 2); Colin White, *Victoria's navy. The heyday of steam* (Emsworth, 1983), pp. 52, 62.

[107] *The Times*, 19th November, 1879; J. Sewill to William Bottomley, 20th November, 1879, CS564, ULC; John Lilley and Son to Sir William Thomson, 24th November, 1879, CS568, ULC.

[108] Lieutenant (later Admiral Sir) George Egerton in Admiral Sir R.H. Bacon, *The Life of Lord Fisher of Kilverstone* (2 vols., London, 1929), **1**, pp. 64–5; Sir William Thomson, 'Distinguishing lights for lighthouses', *Nature*, **21** (1879), 109–10.

Significantly for the future, the *Northampton*'s flag-captain at this time was 'Jacky' (J.A.) Fisher, who in 1904 succeeded Lord Walter Kerr as First Sea Lord. As such, he presided over the *Dreadnought* era of the all-big-gun battleship and remained master of Britain's naval destiny until conflict with Winston Churchill led to his resignation over the Dardanelles campaign in 1915.[109] Even in 1879–80, however, Fisher's tireless campaigns for technical innovation and naval reform were very much in evidence. And it was to his enthusiasm above all that Sir William would be indebted for the final introduction of his compass into the Royal Navy. Although Fisher had first become familiar with the compass while flag-captain of HMS *Hercules* during the Balkan crisis, his experience aboard the *Northampton* early in 1880 convinced him of its superiority. He wrote to Sir William from the warship on 18th May:

> I have now arrived at that stage with your compass that I am able to write a 'sledge-hammer' report because we have been a good deal at sea and in all latitudes from 50°N to 10°N. We have swung several times and 5 days of heavy firing and to my mind the Admiralty compass is doomed. I shall not send you a copy of my report but I shall beg in my official letter that you should be furnished with a copy because that will enable you to go publicly to the 1st Lord on the subject . . . The sounding machine has proved equally satisfactory and I am going to report on that also at the same time.[110]

Fisher also enclosed a letter taken from *The Times* and written 'by the most experienced Admiral in the Navy', Admiral of the Fleet Thomas Symonds, a letter which in Fisher's opinion was 'so true it might be an extract from the Bible'. Symonds argued that the British ironclads were 'generally bad and obsolete, built under the advice of old sailors bigoted to masts and sails'. They had proved themselves both 'unmanageable under sail' and 'bad steamers, stopped by a moderate gale because they are over-masted', a feature which also 'ruins them as fighting ships'. Worse still, the sailing-ship stores made them ready for a bonfire. The Admiral therefore urged the abandonment of large masts, the introduction of fore-and-aft bulkheads, the adoption of twin screws, and the appointment of a parliamentary committee to investigate fully the state of the Navy 'on which the safety and welfare of England mainly depend'.[111]

For his part, Fisher exhorted Sir William to use his influence with eminent figures such as the Duke of Argyll. Increasingly, Fisher was to spearhead a campaign to revolutionize the Royal Navy from an era of sail-and-steam hybrids to the era of steam, armour plate and, ultimately, the big, long-range gun. Integral to this campaign would be the introduction of hitherto unwel-

[109] Bacon, *The life of Lord Fisher*, **1**, **2**; R.F. Mackay, *Fisher of Kilverstone* (Oxford, 1973); Richard Hough, *First Sea Lord. An authorized biography of Admiral Lord Fisher* (London, 1969).

[110] J.A. Fisher to Sir William Thomson, 18th May, 1880, F104, ULC; Captain Cole to Sir William Thomson, 6th January, 1880, CS589, ULC; F.J. Evans to Sir William Thomson, 30th January, 1880, CS605, ULC; Bacon, *The life of Lord Fisher*, **1**, pp. 60, 64.

[111] Admiral Thomas Symonds to *The Times*, 14th April, 1880, F104, ULC.

come innovations to permit navigational techniques to keep pace with the increasing size, speed, and firepower of nineteenth-century ships of war.

Following the *Northampton*'s trials, Sir William supplied a ten inch compass to HMS *Inflexible* in June, 1881, and personally supervized the adjustment at Portsmouth in July and trials off the south coast of England in October. *Inflexible*, now commanded by Fisher, had been launched at Portsmouth Dockyard in 1876 as the most prestigious warship of her time, mounting a pair of sixteen inch muzzle-loading guns in two midship turrets and protected by the thickest armour ever put afloat. Displacing about 11 900 tons, *Inflexible* was powered by Elder compound engines giving a speed of almost fifteen knots.[112]

Among many technical innovations, the *Inflexible* had been fitted with a.c. generators and Faure accumulators. Fisher's biographer has recorded Sir William's near-fatal involvement with this 600 volt supply, which Sir William nevertheless handled with characteristic flair:

> One day the Captain's Coxswain received a nasty shock through touching an arc lamp; and, as Sir William was on board, Fisher called his attention to what had happened. He visited the place and saw that a small arc had formed between one of the cables and earth. He diagnosed the matter as 'a nasty little leak, but not likely to be dangerous to life'. Just then the cable slipped through his hand and the bare wire touched his finger. He leapt into the air, and his immediate second diagnosis was 'Dangerous, very dangerous to life. I will mention this to the British Association'.[113]

The subsequent death of a stoker brought about by a similar leak led to the general adoption of eighty volts aboard ships of the Royal Navy.

In July, 1882, the *Inflexible* took part in a famous bombardment of Alexandria to quell an Egyptian mutiny and potential threat to the Suez Canal. Fisher later recorded that 'the firing of the eighty ton guns of the *Inflexible* with maximum charges, which blew my cap off my head and nearly deafened me, had no effect on his [Sir William's] compasses, and enabled us with supreme advantage to keep the ship steaming about rapidly and so get less often hit whilst at the same time steering the ship with accuracy amongst the shoals'. A similar report from the captain of HMS *Alexandra* confirmed his opinion.[114]

Fisher was appointed captain of the Portsmouth shore-training establishment HMS *Excellent* in April, 1883, and began a series of manoeuvres which he explained confidentially to Sir William. The battle for the adoption of the compass and sounding machine opened with the case of HMS *Wye*'s stranding. Fisher reported to Sir William:

> HMS *Wye* got aground in the Red Sea and the Court-Martial consisting of nine of our best Port Captains have this day written a most strongly worded letter to the Admiralty

[112] Thomson's patent compass books, *op. cit.* (note 2); SPT, **2**, 769, 776–7; Gardiner (ed.), *Conway's all the world's fighting ships*, p. 26.

[113] Bacon, *The life of Lord Fisher*, **1**, pp. 76–7; Mackay, *Fisher*, p. 152.

[114] [J.A. Fisher], *Records by Admiral of the Fleet Lord Fisher* (London, 1919), p. 62; Bacon, *The life of Lord Fisher*, **1**, pp. 73–92; SPT, **2**, 714.

to say that had she been supplied with your compass and sounding machine the disaster would not have occurred: I think this will wake up the Hydrographer & Co.... I have no doubt we shall eventually win the battle. I would suggest your remaining quite silent... As Capt. Lyon [who reported to the Admiralty on the subject of Sir William's compass and sounding machine] observed to me yesterday we must approach the subject like a burglar approaching a safe, getting in the thin end of the wedge and never cease driving it in and burst it must.[115]

Fisher later reported further to Sir William that 'our agitation was progressing favourably and thanks to our friend Lyon. The subject of your compasses and sounding machines is to be brought formally before the Board and the Hydrographer is to be had up and asked what he has to say against their adoption as Service Fittings – of course we have not won the victory yet but success I think is certain...' Success may well have become certain with the death of F.J. Evans in 1885.[116]

The sales figures of Sir William's compasses confirm a relentless conquest of the Royal Navy during this period of Fisher's campaign. He supplied a mere five in 1883, eighteen in 1884, fifty-eight in 1885–6, and very large numbers from 1887. The final triumph came in 1889 as his sister Elizabeth King described to her daughters on 23rd November:

Uncle William came back this morning quite fresh after his busy day in London. It was to meet with the Lords of the Admiralty that he went . . . He goes so often by night to London the railway attendant knows exactly how to make his bed, and all the little arrangements he likes, and attends most carefully to his comfort. He drove at once to Admiral Fisher's, where he had his bath before 8.30 breakfast, and then set off about his various business. The meeting with the Admiralty was most satisfactory to him, for it is now ordained that his be the standard compass, and be used throughout the Navy.[117]

But Sir William's appearance had been little more than a formality. As Fisher had put it, your 'strong point is to stand entirely aloof from the whole business and let your disciples do the fighting'. Fisher later noted that ridicule of the old Admiralty standard compass had won the day. When asked by the Judge at the inquiry whether the Admiralty compass was sensitive, Fisher, as Sir William's witness, replied 'No, you had to kick it to get a move on'. Such was the brash, unadorned style of the future First Sea Lord which so endeared him to the lower decks but which made him a controversial figure in the history of the Royal Navy. At the same time, prior to Sir William's arrival in London, Fisher informed him that no more Admiralty compasses would be made. Instead, 'twenty of yours to be kept in stock and all ships to be fitted with two, one for steering and one for taking bearings, but please do not mention anything about this'.[118]

115 J.A. Fisher to Sir William Thomson, *c.*1883–5, F107, ULC; SPT, **2**, 713.

116 J.A. Fisher to Sir William Thomson, *c.*1883–5, F109, ULC.

117 Elizabeth King to her daughters, 23rd November, 1889, in SPT, **2**, 889–90; Thomson's patent compass books, *op. cit.* (note 2).

118 J.A. Fisher to Sir William Thomson, *c.*1889, F111, ULC; [Fisher], *Records*, pp. 62–3.

One reason for the discreet nature of the campaign had been Fisher's apparent uncovering of, in Elizabeth's words, 'much mean and underhand work' at the Admiralty. Of sixty letters from Captains, one 'spoke of some slight objection to the compass, eight said they had not had sufficient experience, and the remainder spoke of it in terms of unbounded admiration and appreciation'. Yet these fifty-one 'were hidden away in pigeon-holes in the Hydrographic Office, and the disapproving ones made a great deal of... Admiral Fisher has been instrumental in exposing the abuse'.[119]

The 1880s marked the beginning of the naval revolution and with it the vast expansion in armaments manufacture which culminated in Fisher's 'armageddon' – the 1914–18 war – after which the British Empire, the Royal Navy, and the first great industrial nation went into rapid decline following a century of world dominance. The naval revolution coincided with the rise of W.G. Armstrong's vast Tyneside armaments empire which, in 1882, took over the Mitchell yard at Walker (builders of the cable ship *Hooper*) and, in 1884, established a shipyard at Elswick specifically for naval (ironclad) construction. By 1900, the firm of Sir W.G. Armstrong, Whitworth employed some 25 000 men in heavy engineering, armaments and armour plate manufacture, and shipbuilding. The Elswick yard not only built for the Royal Navy, but for many foreign navies. For example, in the Russo–Japanese War of 1904–5 when fourteen Russian battleships were lost, all six principal Japanese battleships had been British built, three at Elswick, two by the Thames Ironworks, and one by John Brown, the Sheffield steel and armour plate makers who had bought out the famous Clydebank yard of J. and G. Thomson in 1899. By 1912, Armstrong had supplied 128 warships totalling nearly half-a-million displacement tons for sixteen different nations.[120]

Most of these formidable fighting ships of the 1890s and early 1900s had Kelvin compasses. Japan's first armoured cruiser, the *Chiyoda*, launched at Clydebank in 1890, her twenty-three knot protected cruiser *Yoshino*, launched at Elswick in 1892, and her massive 15 200 displacement-ton battleship *Asahi*, launched at Clydebank in 1899, all had Kelvin compasses. So too had the Argentinian cruiser *Veinticinco de Mayo*, the United States cruiser *Albany*, and the Norwegian coast-defence battleship *Norge*, all launched by Armstrong in 1890, 1899, and 1900, respectively. These are only a cross-section of the large number of British-built ships of war thus supplied during a period when Britain was not only the shipyard of the world for merchant ships but for the world's navies as well.[121]

The eventual displacement of the Kelvin compass from fighting ships oc-

[119] Elizabeth King to her daughters, 23rd November, 1889, in SPT, **2**, 890.

[120] Pollard and Robertson, *British shipbuilding industry*, pp. 211–22, esp. pp. 219–21; Gardiner (ed.), *Conway's all the world's fighting ships*, pp. 170–2, 216, 29.

[121] Thomson's patent compass books, *op. cit.* (note 2).

curred with the changeover to liquid compasses. As early as 1890–1, Lord Walter Kerr, writing from the new battleship HMS *Trafalgar*, had warned Sir William of the potential threat to his dry-card patent. Lord Walter reported that the new Admiralty liquid compass was not sluggish and had no more oscillation than Sir William's dry card either when firing or when steaming at high speed. Sir William's on the other hand was 'much easier to correct' and had the advantage of a larger card, though the azimuth mirror arrangement did not find approval for taking rapid bearings or in dull weather. He subsequently concluded in favour of Sir William's compass for big ships, but speculated that the smaller, liquid compass might be well suited to the confined spaces of torpedo boats subject to more vibration.[122]

The change to liquid compasses occurred very rapidly around 1904–6 as the following remarks from Kelvin to J.T. Bottomley demonstrate:

> This Admiralty business is very serious. It seems they have quite resolved to have Captain Chetwynd's liquid compass all through the Navy and to displace mine everywhere. We have had no Admiralty orders for compasses during the last two or three years. They have about 3000 of my binnacles lying unused in dockyards, some of them not taken out of the cases in which they came from us . . . We shall probably be invited to offer to take them all . . . The Dreadnoughts it seems are to be fitted throughout with the liquid. F.C. [Frank Clark, one of the works managers of Kelvin and James White, Ltd] sees a prospect of our having the making of the binnacles for the liquid compasses; and possibly the compasses and bowls themselves . . . It is very important for us now to keep *in* with the Admiralty all we can. Help that F.C. or I can give to Cap. Chetwynd will be good for us as well as for him.[123]

By 1906, Chetwynd's version of the Admiralty liquid compass had decisively won the day for the big ships also, and Lord Kelvin's position had been reduced to negotiating to supply binnacles for the liquid compass and to manufacture Chetwynd's liquid compass bowl. Up to July, 1907, six binnacles had been supplied – three to HMS *Agamemnon* and three to HMS *Lord Nelson*, last of the pre-Dreadnought battleships.[124]

Though the effective loss of the Admiralty contracts had been a severe blow, the factory of Kelvin and James White, Ltd (formed into a limited company in 1900) continued to prosper through the supply of compasses to merchant vessels, as well as through the manufacture of sounding machines and electrical instruments. In the decade after Kelvin's death, for example, the firm sold over 3000 compasses, while even today the name of Kelvin Hughes as an Admiralty chart

[122] Lord Walter Kerr to Sir William Thomson, 7th October, 1890 and 18th February, 1891, K47 and K48, ULC.

[123] Lord Kelvin to J.T. Bottomley, *c*.14th December, 1906, B26, ULG.

[124] Lord Kelvin to Frank Clark, 14th December, 1906, and 27th July, 1907, C23 and C27, ULG; Lord Kelvin to Captain Chetwynd, 22nd March, 6th November, 13th and 19th December, 1905, LB25.51, LB28.106, LB29.66 and LB29.75, respectively, ULC.

agent and supplier of high quality navigational instruments such as radar, electronic echo sounders, and compasses is familiar to everyone who goes down to the sea in ships.[125]

[125] Thomson's patent compass books, *op. cit.* (note 2). Following various name changes, the firm of Kelvin, Bottomley and Baird amalgamated with the London instrument makers Henry Hughes & Son in 1947 to become Kelvin Hughes. With a takeover by Smiths Industries in the mid-1960s, the firm's manufacturing connections with Glasgow ceased. Under the trade name 'Kelvite', dry-card compasses and electrically driven sounding machines were supplied in 1936 to the most famous Clyde-built Cunarder of all, the *Queen Mary*.

23

Baron Kelvin of Largs

> The end of my electioneering was at Oban on Tuesday night, where and when Craig Sellar and I held a meeting of Liberal Unionists . . . There was much joy on board the *Lalla Rookh* . . . to hear 'Majority 613 Malcolm' . . . Unionist organization must be kept up rigorously and Unionists must act well together in Parliament; keeping AS FAR AS POSSIBLE all subjects on which, as conservatives and liberals and radicals, they may differ among themselves . . . until the two imps of mischief, Parnell and Gladstone, are finally deprived of all power for evil. What a blessing it would be if we could have Lord Salisbury, Lord Hartington, [Joseph] Chamberlain and Jesse Collings all in one government. *Sir William Thomson to Lord Rayleigh, 1886*[1]

Sir William Thomson's professional life coincided with much of Queen Victoria's long reign, during which the Island Empire, expanding across every ocean and sea around the globe, depended for its very existence upon the improvements to industry, commerce and navigation in which the Clyde had played so dominant a role. And each decade after 1850 contained at least one landmark symbolizing Sir William's advancement in the context of Imperial Britain. The 1850s saw his establishment of a physical laboratory at Glasgow College, dedicated to those precision measurements of the properties of matter on which so much of his science depended. The 1860s witnessed the successful completion of the Atlantic cable and his receipt of a knighthood for services to communication between English-speaking peoples. His purchase of the schooner-yacht *Lalla Rookh* following Margaret's death in 1870 initiated a dramatic new decade of personal and professional progress, ranging from his second marriage in 1874 and construction of his country house, Netherhall, outside Largs from 1875, to his world-wide fame for contribution to safety at sea.

Each of these strands, and others, we have investigated separately in the preceding chapters. Such a steady social advancement provided the solid foundation for the social summit of William Thomson's career, the conferment of a peerage in January, 1892, following a decade not merely of further professional achievements which included succeeding Stokes as Royal Society president in

[1] Sir William Thomson to Lord Rayleigh, 18th July, 1886, in SPT, **2**, 860–1.

Britain's first scientific peer, Baron Kelvin of Largs, discussing a problem with his sister, Elizabeth, eldest of the Thomson family. She died in 1896, four years after James. [From a photograph by A.G. King in A.G. King, *Kelvin the man*.]

1890, but of a very active involvement in Imperial politics from 1886. This involvement, together with his scientific and technical reputation, effectively guaranteed his elevation to the House of Lords as the first scientist to be so honoured.

Sir William Thomson's political campaigns, centred on his deep personal opposition to Gladstone's policy of Home Rule for Ireland, serve both to take us back to our starting point of James Thomson's home town of Ballynahinch (ch. 1), and to highlight the enormous distance traversed between the obscure beginnings in an Ulster market town and the years of glittering pomp and circumstance amid Royalty and aristocracy, culminating in funeral and burial alongside Sir Isaac Newton in Westminster Abbey.

The liberal, commercial, and protestant values which had originated in the Ulster context of his father not only remained with William throughout his life, but provided the principal motivation for his rigorous opposition to Home Rule. Home Rule as he saw it (along with many of his fellow countrymen) threatened the unity of an Empire standing as guardian over those values and offered only a return to inevitable sectarian strife and narrow factional nationalism. At the same time, the immensely successful prosecution of a professional life guided by the same values gained him influential aristocratic friends at court.

While Sir William's enthusiasm for the British Association and for technical advancement had brought him into the circles of eminent entrepreneurs, including Sir William Fairbairn, the telegraph financier Sir John Pender, and the armaments industrialist Sir William Armstrong, all of these entrepreneurs, like Sir William himself, had inherited no automatic right to a place in the higher ranks of the British establishment. They might, as whigs intent on the expansion of their personal property, material and social, aspire to such a place, but they needed friends among the elect to gain entry to a very exclusive club. Sir William had been especially fortunate and especially careful in his friendships with at least four powerful patrons: the 15th Earl of Derby (1826–93), the 8th Duke of Argyll (1823–1900), the Marquis of Hartington (1833–1908), and the Marquis of Dufferin and Ava (1826–1902). Of these, Derby and Hartington had been Lord Rectors of Glasgow University between 1868–71 and 1877–80, respectively, while Argyll, whose family had long connections with the College, had a strong amateur interest in geology. Dufferin, on the other hand, had chaired the 1871 Admiralty Committee of which Sir William had been a prominent member and had shared with him an enthusiasm for practical navigation and seamanship. All of them, however, shared a progressive (if aristocratic) liberalism which, after 1885–6, became a liberal unionism. It was Hartington, leader of the Liberal Unionists and heir to the famous Cavendish family properties, who provided the required support for Sir William's elevation to the peerage.[2]

[2] See 'Derby', 'Argyll', 'Devonshire', and 'Dufferin' entries in *Burke's genealogical and heraldic history of the peerage, baronetage and knightage*, 56th edn. (London, 1892).

The first 'scientific' peer

'Are you trying to re-collect the scattered supporters of Irish Nationality and make another effort for independence? Or do you fraternize with the Saxon, the enemy of your country?'[3] William Thomson's scornful but perceptive questions – posed in a letter of August, 1848, to his friend G.G. Stokes who was at the time visiting their native North of Ireland – referred facetiously to the Young Irelanders' ill-conceived attempt in the year of European revolutions to liberate Ireland at last from the tyranny of Saxon rule. Thomson implied that the conservative Stokes would enjoy better the geological spectacle of the Giant's Causeway or a bathe in the Atlantic Ocean with a good northwest wind sending in its mighty waves. Yet his comments offer a fascinating glimpse into the issues which eventually compelled him to assume an active political role against the advocates of Home Rule.

The Young Irelanders of the 1840s presented an ideology quite at variance with that of the United Irishmen and more particularly with that of the elder and younger Thomson. Thomas Davis, their chief theorist and himself a middle-class protestant, launched bitter attacks on all forms of 'utilitarianism' (here the application of science and reason to government). While these attacks could be represented as a polemic against English instruments of repression, they were in fact directed at the very basis of social improvement through state interference, improvement (supported by the whig alliance of the Westminster government and O'Connell) which threatened to destroy peasant (Irish) culture and promote the assimilation of Irish to British values. The Young Irelanders promoted instead a vision of Irish cultural nationalism which glorified the military virtues of the Irish Race and the agrarian values of the peasantry. Most significantly, Davis's romantic distaste for industrial society and his indictment of progressive, whig values set him in stark contrast to the practical aims of Dr Thomson during the very period when his mathematical textbooks sold thousands of copies for Ireland's National Schools. As Davis wrote in 1842:

> Utilitarianism, the creed of Russell and Peel . . . which measures prosperity by exchangeable value, measures duty by gain, and limits desire to clothes, food, and respectability – this damned thing has come into Ireland under the Whigs, and is equally the favourite of the 'Peel' Tories. It . . . threatens to corrupt the lower classes, who are still faithful and romantic . . . The 'Useful Knowledge Society' period arrived in Britain, and flooded that island with cheap tracts on algebra and geometry . . . Unluckily for us, there was no great popular passion in Ireland at the time, and our communication with England had been greatly increased by steamers and railways, by the Whig alliance, by democratic sympathy, and by the transference of our political capital to Westminster . . . the National Schools were spreading the elements of science and the means of study through the poorer classes, and their books were merely intellectual.[4]

[3] William Thomson to G.G. Stokes, 23rd August, 1848, K27, Stokes correspondence, ULC.
[4] Thomas Davis, quoted in Edward Norman, *A history of modern Ireland* (Harmondsworth, 1971), pp. 126–8.

Davis's remarks here exemplify perfectly the wide gulf between the 'modern' values of the Thomsons and the Young Irelanders' mystical desire for return to a rural ideal.

The significance of the Home Rule issue for our understanding of William Thomson's life is two-fold. First, a careful analysis of the grounds of his opposition will suggest a continuation of, rather than a reaction to, the liberal political and social outlook of his father. And, second, it will become clear that his peerage of 1892 was not achieved on scientific prestige alone, but on his powerful role in co-ordinating West-of-Scotland political opposition to a separate Irish parliament. Without such an active political interest, Sir William Thomson would not have become Lord Kelvin.

Since Peel's 1845 manoeuvres to set up the new colleges, Ireland had been undergoing further radical social transformations. On the one hand, famine had devastated the rural communities. On the other hand, Belfast had seen the foundation of Harland and Wolff's shipyard around 1860, an industry which brought new and unprecedented economic growth to the town. Linen manufacturing, too, was a booming industry. Belfast, by 1900, had a population of 350 000. The gap between the industrial, Protestant, North of Ireland and the rural, Catholic, South was widening irreversibly. O'Connell, too, was dead, and his movement for repeal of the Act of Union had given way to a new party agitating for Home Rule.[5]

Only from 1885, however, did the Home Rule issue become a crucial element in British politics. Until then, disestablishment of the Irish Church in 1868, and land reform in 1870 and 1881, set against a background of violence, had provided some of the main problems of the governments of Gladstone and Disraeli. The November, 1885, general election gave Charles Stewart Parnell's Irish party eighty-six seats – including that of a Liverpool MP – and the balance of power between Gladstone's Liberal party and Lord Salisbury's Conservatives. Salisbury's party did not long keep the support of Parnell. Already privately converted to Home Rule, Gladstone took over by February, 1886, and led a government now openly pledged to pass a Home Rule Bill. The effect was disastrous for the already divided Liberals. The party split, losing in the process famous radical parliamentarians like John Bright and Joseph Chamberlain. Most of the Liberal peers, too, left the party. When, therefore, the first Home Rule Bill came before the Commons in April–June 1886, proposing to exclude Ireland from representation at Westminster, over ninety Liberals voted against it, making its defeat certain.[6]

During the early 1880s, Sir William Thomson had been a faithful supporter of the Liberals, heirs to the Whig tradition. His brother-in-law, Alexander Crum of Thornliebank, had been returned as Liberal MP for Renfrewshire at a

[5] See especially J.C. Beckett, *The making of modern Ireland, 1603–1923* (London, 1966), pp. 336–88; Norman, *A history of modern Ireland*, pp. 109–95; J.C. Beckett and R.E. Glassock (eds.), *Belfast. The origin and growth of an industrial city* (London, 1967), pp. 88–119.

[6] Beckett, *Modern Ireland*, pp. 394–8; Norman, *A history of modern Ireland*, pp. 212–15.

by-election in 1881, a constituency which he represented until a redistribution of seats took place for the 1885 general election. Crum communicated to Sir William in 1884 the wishes of the Liberal whips that the Glasgow professor become the Liberal candidate for the joint university seat of Glasgow and Aberdeen, then occupied by a Conservative member. The other Scottish university seat, Edinburgh and St Andrews, was held for the Liberals by the former chemistry professor at Edinburgh, Lyon Playfair, from 1868 until 1885. Thomson, however, declined – as he had apparently also done in 1869 and 1880 – to stand against 'so moderate, useful, and popular a member' as the Conservative James Campbell, and so the natural philosophy professor never entered the House of Commons alongside his Edinburgh science colleague.[7]

With the Home Rule crisis of 1885–6 bringing to the fore the divisions within the Liberals, Sir William Thomson and many erstwhile West-of-Scotland Liberals left the party. During the spring of 1886 he argued with sarcasm and optimism against Home Rule as a solution to Ireland's ills:

> I wonder if it has never occurred to the great mind of Gladstone himself that some moderate encouragement and help toward business habits and arrangements between Landlords and Tenants, coupled with defence of honest dealing on each side by all the power of the empire, is the only remedy possible, and all the remedy that is needed, to make the best of bad times in Ireland as everywhere else.[8]

Prior to the election of July, 1886, in which Gladstone's Liberals were heavily defeated by the Conservatives, and in which some seventy-eight anti-Home Rule Liberals – Liberal Unionists as they became known – were returned, Sir William had taken an active and successful part in campaigning against Gladstone's Irish policy in the West-of-Scotland constituencies of Partick and Argyllshire. His friend and colleague, John Nichol – professor of English literature at Glasgow and son of J.P. Nichol, the late professor of astronomy – wrote to congratulate him on the success of the campaigns. Nichol enthusiastically noted that 'the Ides of July' had brought at least 'a temporary deliverance' from the tyranny of a 'despot half mad by failure of his traitorous schemes'.[9] Lord Salisbury took office for the next six years, his nephew Arthur Balfour being appointed Chief Secretary for Ireland in 1887 to carry out a policy of 'resolute government' as the alternative to Home Rule.[10]

At first sight, William Thomson's opposition to Home Rule seems quite at

[7] Alexander Crum to William Thomson, 7th and 10th March, 1884, C197E–F, ULC. On Playfair, see 'Baron Playfair of St. Andrews', *Burke's peerage*. In a draft letter from Sir William Thomson to George Douglas (later 8th Duke of Argyll), Research notebook 1864–8, NB54, ULC, written about 1868, it is apparent that Sir William was no mere passive supporter of the Liberals but the very active organizer of a political campaign to promote Archibald Smith as candidate for Glasgow and Aberdeen. Sir William felt 'confident that he [Smith] will be on the liberal side in the great questions regarding churches, and education, now in view, or such as may become before the new parliament'.

[8] Sir William Thomson to John Dodds, 22nd March, 1886, in SPT, **2**, 856.

[9] John Nichol to Sir William Thomson, 21st July, 1886, N27A, ULC.

[10] Beckett, *Modern Ireland*, pp. 400–10; Norman, *A history of modern Ireland*, pp. 215–20.

variance with his father's liberal and tolerant attitude towards his fellow countrymen. Unionism has been subsequently identified in the popular imagination with a narrow, if sincere, Ulster protestantism; Home Rule with the national aspirations of a down-trodden native people seeking to free themselves from British rule. Sir William Thomson's sympathy with a unionist cause, therefore, appears to align him with all the bigotry and dogmatism of a Henry Cooke, against whom Dr Thomson and his circle had once so boldly taken their stand. Yet such a view of British and Irish history distorts the issues of the late nineteenth-century. In particular, it is a view which drastically narrows the scope of Liberal Unionist opposition to Home Rule – the Liberal Unionism of men such as Joseph Chamberlain, the eighth Duke of Argyll, the eighth Duke of Devonshire, and Sir William Thomson.

Writing in 1893, for example, to express his aristocratic dislike of Irish nationalism, the Duke of Argyll quoted from Edmund Burke:

> For, in the name of God, what grievance has Ireland, to complain of with regard to Great Britain, unless the protection of the most powerful country on earth – giving all her privileges, without exception, in common to Ireland . . . – be a matter of complaint. The subject . . . is as free in Ireland, as he is in England. As a member of the empire, an Irishman has every privilege of a natural-born Englishman, in every part of it, in every occupation, and in every branch of commerce.[11]

The Duke commented that 'what Ireland wants above all things is the rule of a Government which *is above all her factions*, and which will maintain the authority of just and equal laws'. Such convictions, expressing the righteousness of imperialism, to be sure, but also a deep faith in the rule of law, constitute the essence of the cause to which Sir William became so passionately committed in the mid-1880s.

To capture Thomson's own conception of Liberal Unionism, and that of other leading members of the cause, requires that one de-emphasize the air of condescending superiority evident in their remarks and focus instead on their conviction that Britain and the British Empire were now the best guarantee of liberty for all the citizens of Ireland. When he spoke of the need for defence of honest dealing between landlords and tenants by all the power of empire, Sir William had in mind the British notion of equality before the law. When he expressed again and again the necessity of continued Irish representation at Westminster, he conceived of Ireland not as a down-trodden colony, but as an equal member of the United Kingdom. When he thought of Ulster's repudiation of Home Rule, he had in view not the bigotry of a fanatical protestantism, but the growing economic prosperity of a province – founded on industry, technology, and ingenuity – which he believed could prove the means of salvation for the rest of rural Ireland. At the July, 1887, jubilee of the electric

[11] Edmund Burke quoted in George Douglas, 8th Duke of Argyll, *Irish nationalism: an appeal to history* (London, 1893), pp. 265–6.

telegraph, Sir William emphasized, in a manner which implicitly mocked the cause of any Young Irelanders, the contribution of science to the unity of Empire and to benevolent rational rule:

I must say there is some little political importance in the fact that Dublin can now communicate its requests, its complaints, and its gratitudes to London at the rate of 500 words per minute. It seems to me an ample demonstration of the utter scientific absurdity of any sentimental need for a separate Parliament in Ireland. I should have failed in my duty in speaking for science if I had omitted to point out this, which seems to me a great contribution of science to the political welfare of the world.[12]

For the Liberal Unionist such as Thomson, therefore, Home Rule did not mean liberation and the establishment of nationhood but rather a return to internal strife, to rural barbarism, to a system which in the end could only emphasize sectarian differences and party rivalries by destroying the entire British constitution.

The alternatives were clear. The liberal ideals of Empire – based on universal law, freedom of the individual, freedom of worship, freedom of trade – were set in opposition to the particularist nationalism implicit in Home Rule. Imperialism, properly conceived, implied a bond operating above party and religious interests; nationalism meant increasing protectionism and a continuing struggle by the dominant political or religious faction to maintain its rule. Thus, in a speech of 1889, the Earl of Derby (at whose Knowsley Park home Sir William often stayed around Christmas) argued that an Irish Parliament would inevitably oppose liberal causes by advocating 'in commerce, protection; in regard to education and social questions, clericalism pushed to the utmost; in foreign affairs, an Ultramontane policy' whereby the absolute temporal and spiritual power of the Pope would be upheld.[13] Home Rule, Derby felt, would mean Rome Rule. Liberal Unionism thus typically implied anti-Catholicism, long a feature of Thomson's latitudinarian perspective.

More positively, Sir William's enthusiasm for the Salisbury government of 1886–92 derived in large measure from the way in which it appeared to unite disparate party factions – Conservatives and Liberal Unionists, Tories, Whigs and Radicals – into a coalition of British interests. He wrote to Stokes, who had been returned as Conservative MP for Cambridge University at a by-election in 1887:

I hope you have been enjoying Parliament. It must be very satisfactory, and pleasant, to see all going so well. We were pretty pleased with Doncaster and Deptford [two by-elections]. Everything seems to promise well for a long and useful tenure of the present ministry and *I am now feeling very hopeful that I may myself live to see the last of government by party.*[14]

12 SPT, **2**, 869–70.

13 Edward Henry, 15th Earl of Derby, *Speeches and addresses* (2 vols., London, 1894), **2**, p. 213. On Kelvin's visits to Knowsley Park, see SPT, **2**, 641, 761, 881.

14 William Thomson to G.G. Stokes, 2nd March, 1888, K283, Stokes correspondence, ULC. Our emphasis.

Had Sir William Thomson, between 1886 and 1892, been a mere passive supporter of Liberal Unionism, had he not become one of the most active campaigners for the cause alongside the Duke of Argyll and Lord Hartington, had he not shown himself a friend of Conservative candidates in constituencies where a Liberal Unionist did not stand lest the vote be split, then it is highly improbable that Lord Salisbury should have offered him the first peerage ever given to a man of science. Sir William's knighthood had been the reward for his technical contributions to the Atlantic cable. Between 1866 and 1892 he had served many useful, if unspectacular, hours on British Association Committees. In 1890 he had succeeded Stokes as President of the Royal Society. Yet, taken together, these activities scarcely supply a convincing reason for the peerage being offered at the time it was. To believe, as several newspapers of the day did, and as biographers of Kelvin subsequently have, that 'a peerage has at length been conferred upon a scientific man because he is a scientific man' is too naive by far.[15]

By 1891, Sir William Thomson had been appointed president of the West of Scotland Liberal Unionist Association, whose members included powerful aristocrats such as the Duke of Argyll and the Earl of Stair, wealthy shipowners such as Sir Donald Currie – founder of the Castle line of steamships – well known shipbuilders such as Peter Denny of Dumbarton, and a number of Glasgow College academics such as John Caird – the principal – and William Thomson's brother James.[16] The honorary president was Lord Hartington, son of the famous seventh Duke of Devonshire (patron of Cambridge University's Cavendish Laboratory) and himself Chancellor of Cambridge from 1892. Hartington's experience of Irish affairs had been considerable. He had been Chief Secretary for Ireland in the years 1870–4 during Gladstone's first ministry, and his brother, Lord Frederick Cavendish, had, as Chief Secretary for Ireland, been murdered in Dublin's Phoenix Park in 1882 during Gladstone's second ministry. He became one of the leading Unionists in the post-1886 era. On the death of his father, Hartington became the eighth Duke of Devonshire late in 1891.[17] The eighth Duke was indeed to prove a key influence with regard to Sir William Thomson's peerage.

On the 28th December, 1891, Salisbury wrote to Sir William Thomson,

[15] SPT, **2**, 905–6; H.I. Sharlin, *Lord Kelvin. The dynamic Victorian* (Pennsylvania, 1979), pp. 216–17. M.B. Hall, *All scientists now. The Royal Society in the nineteenth century* (Cambridge, 1984), p. 145, remarks that Sir William was not elected President of the Royal Society without opposition. His predecessor, T.H. Huxley, had apparently given as one of his reasons for accepting the presidency the desire to avoid the office falling to a 'commercial gent'. Certainly Sir William's entrepreneurial activities, combined with his passionate political campaigning, places him apart from the lofty, gentlemanly image of science often promoted by Britain's most prestigious scientific society.

[16] Robert Bird to William Thomson, 5th January, 1892, Pr 36, ULC; West of Scotland Liberal Unionist Association: list of Office Bearers and General Executive, January, 1892, M177, ULC.

[17] 'Devonshire', *Burke's peerage*; Bernard Holland, *The life of Spencer Compton, eighth Duke of Devonshire* (2 vols., London, 1911). Hartington had been in communication with Thomson about Liberal Unionist affairs during the late 1880s. See, for example, Hartington to William Thomson, 25th October, 1887, H41A, ULC.

stating that 'I have Her Majesty's authority for informing you that it is her design, if agreeable to yourself, to confer on you a Peerage of the United Kingdom in recognition of your varied and most valuable service to science & progress in this country'.[18] Salisbury's words avoided any exact specification of the reasons behind the peerage. The natural assumption, shared by nearly all the 438 letters of congratulation which poured in, was that Thomson had been afforded due recognition for his scientific and technical achievements. A few correspondents, however, felt that 'in honouring you, the local head of our [Liberal Unionist] Association, we have all been honoured' and that 'every Unionist will feel prouder today of your distinguished position'.[19] But one letter in particular, from the Duke of Devonshire, made clear the underlying story of the New Year honour:

> I can assure you that it has been a great pleasure to me if I have been in any way instrumental *in suggesting the well deserved recognition of your eminent services not only to science but to the Unionist cause in Scotland*; and I am sure that all our friends will join me in offering to you and Lady Thomson our most hearty congratulations.[20]

As leader of the Liberal Unionists in the Commons during Salisbury's coalition, Hartington wrote frequently to the Prime Minister to make recommendations for political honours. Thus in May, 1891, following a discussion of the merits or otherwise of Sir John Pender – chairman of the Eastern Telegraph Company and unsuccessful Liberal Unionist candidate in a Glasgow Govan by-election – for a peerage, Hartington expressed his opinion to Salisbury thus:

> If you should ever wish to make another scientific Peer, I believe that Sir William Thomson of the Glasgow University is one of the first electricians in the world. He takes an active interest in politics & is an ardent Unionist. He has no sons and is I believe fairly well off. He has never asked for anything, and the suggestion has not been made to me on purely political grounds, but if you thought it desirable as a recognition of his scientific merits, it would no doubt give much satisfaction.[21]

In effect, Hartington stated how eminently suitable Thomson was for a peerage and, in his reply, Salisbury heartily agreed. The honour would clearly be *seen* as a recognition of service to the nation, and not as a mere reward for party loyalty. Nevertheless, Sir William was a sound, active Unionist, whose elevation would certainly be acceptable to the West of Scotland Liberal Unionists. Furthermore,

[18] Lord Salisbury to William Thomson, 28th December, 1891, Pr 342, ULC. The whole subject of honours is discussed in H.J. Hanham, 'The sale of honours in late Victorian England', *Victorian Stud.*, **3** (1959–60), 277–89. Salisbury, it seems, was exceptionally skilled at rewarding his supporters within the coalition of Conservative and Liberal Unionists.

[19] Bird, *op. cit.* (note 16); M.S. Grady to William Thomson, 1st January, 1892, PR 185, ULC.

[20] Duke of Devonshire to William Thomson, 2nd January, 1892, Pr 125, ULC. Our emphasis.

[21] Lord Hartington to Lord Salisbury, 10th May, 1891, Salisbury collections, Hatfield House. Hartington noted that Pender had 'done a great deal for the telegraphic communications of the Empire'. By implication, Sir William Thomson had done even more to forge the unity of Empire. For Salisbury's favourable reply, see Salisbury to Hartington, 10th May, 1891, Devonshire collections, Chatsworth.

the conventional £5000 a year estate with which the first of the lineage would endow the peerage would be irrelevant in Sir William's case as he had no heirs, while at the same time his income – from the business interests in scientific instrument manufacture rather than from his professional salary – and his country house at Largs meant that he was indeed 'fairly well off'. Finally, unlike persons such as the invalid's wife who had once approached Hartington's secretary with the accusation that her husband's illness was the result of His Lordship's refusal to give him a baronetcy, Sir William had never been so vulgar as to ask for anything.[22]

In accepting the honour in December, 1891, Sir William requested that he be permitted a title connected with the University and City of Glasgow 'in which my home has been for sixty years, and where I have done nearly all my work'. Thus he chose to become Baron Kelvin of Largs, after the River Kelvin, a tributary of the Clyde that flows round two sides of the University before reaching the famous shipbuilding river. Lord Sandford (Sir Daniel Sandford's son and boyhood friend of William at Glasgow College) and Lord Rayleigh introduced the new Baron Kelvin of Largs to the House of Lords to take the oath of allegiance on 25th February, 1892.[23]

There can be little doubt that Devonshire had played a major part in the elevation of Sir William to the peerage as a recognition of his recent services to Unionism in Scotland and of the value his active support would have in the struggles ahead. The new Lord Kelvin would indeed prove a loyal and weighty force in a crucial election year. The Duke of Devonshire and Lord Kelvin now 'superseded' the Marquis of Hartington and Sir William Thomson as honorary president and president, respectively, of the West of Scotland Liberal Unionist Association. Throughout the months leading up to the general election of July, 1892, Kelvin chaired meetings of the West of Scotland Liberal Unionist Association, and addressed other Unionist rallies. His verdict of Balfour's achievements had nothing in common with the vitriolic comments from Home Rulers concerning 'Bloody Balfour's' coercive policies. Balfour, said Kelvin, had shown that what was needed to cure the ills of Ireland was 'good, honest, resolute, kindly government'.[24] On May 28th Lord Kelvin introduced the Duke of Devonshire at one such meeting, and, in doing so, again emphasized that, as in 1886, he as an Irishman felt 'that a frightful damage threatened against his native land: to take away Ireland from its grand position as a constituent equal member of the British Empire, to make it [merely] a naval and military station of the neighbouring island of England and Scotland'.[25]

Gladstone took office once more after the election of 1892 and led a minority Liberal Party in a Commons where the Irish members held the balance of

[22] Hanham, 'The sale of honours', p. 280. [23] SPT, **2**, 913–14. [24] *Ibid.*, pp. 911–12.

[25] *Ibid.*, pp. 920–1. Home Rule, of course, implied a very different conception of Ireland from that of the United Irishmen of 1798. The former was tending towards Catholic nationalism; the latter had been avowedly republican after the French and American 'models'.

power. Lyon Playfair, chemist turned politician, was rewarded for his faithfulness to the Liberal cause with a peerage in September, 1892. The second Home Rule Bill passed successfully through the House in 1893 – with the significant provision that Irish members would continue at Westminster – but the Lords overwhelmingly defeated the bill. Gladstone resigned for the last time, and the Conservatives (by now the Conservative and Unionist Party) returned to rule the country from 1895 until 1905. Home Rule ceased to be the dominant issue until the early twentieth century.[26]

With Chamberlain as colonial secretary and with Salisbury's policy of 'splendid isolation', Imperialism overshadowed domestic issues. To Chamberlain's aims of building the Empire into a more coherent unit, Lord Kelvin gave his undiluted support, and became a close friend of the famous statesman. Chamberlain was Rector of Glasgow University in 1897, when he delivered a speech on 'patriotism' which Lord Kelvin – with whom Chamberlain stayed – declared was the finest he had ever heard.[27] Yet the very tensions within the notion of Imperialism were destined to split asunder the unity of Tory, Whig, and Radical elements which Kelvin had so admired in the governments of Lord Salisbury. From 1903, Chamberlain openly advocated a system of Tariff Reform which would include preferential tariffs for the colonies by way of promoting Imperial ideals. In so doing, Chamberlain ran counter to the cherished Free Trade principles of the old Whig tradition still represented among the Liberal Unionists by the Duke of Devonshire. No longer united by a common opposition to Home Rule, the Unionists became more and more divided over the issue of protection, particularly at a time when the British economy seemed threatened by foreign competition.[28] As a result, the Liberals were returned to power with a massive majority in 1906, just a year before Kelvin's death. Within a little more than a decade the disintegration of the British Empire and of the old order was well and truly advancing. The post-war era of the 1920s, with Ireland finally partitioned and British industrial progress at an end, bore little resemblance to the promise and optimism which characterized the late Victorian age.

For an understanding of William Thomson, Lord Kelvin, his Irish context is essential. No other single theme in the scientist's life linked him so inseparably with the major British political developments of the day. No other single factor contributed so much to his elevation to a peerage. From his father, Thomson had inherited more than academic skills. From Dr Thomson, Lord Kelvin had adopted the liberal tradition which owed its origins to the Ulster of the late eighteenth century and which sought the establishment of a non-sectarian framework of government and institution, free from party rivalry and based on

[26] Beckett, *Modern Ireland*, pp. 410–11.

[27] Julian Amery and J.L. Garvin, *The life of Joseph Chamberlain* (6 vols., London, 1932–69), **3**, p. 200; SPT, **2**, 1004, 1128–30.

[28] R.A. Rempel, *Unionists divided. Arthur Balfour, Joseph Chamberlain, and the Unionist Free Traders* (Newton Abbot, 1972). Kelvin himself favoured free trade. See SPT, **2**, 1129–30.

Lord Kelvin's last lecture before his retirement from the chair of natural philosophy at Glasgow University in 1899. [From SPT.]

merit alone. With the progress and prosperity of the British Empire so conspicuously apparent after the mid-century – nowhere more so than in Glasgow and in Belfast – the crown of liberty passed to a new conception, that of Imperialism, in which the principles of equality, toleration, and freedom were embodied as prerequisites of economic and social progress in the unwritten British constitution. A line from a class essay of 1838 had thus come to have for William Thomson a more than symbolic significance: ''tis liberty that crowns Britannia's isle'.[29]

'No time spent in idleness'

Addressing university students at Rochester during his last visit to the United States in the spring of 1902, Lord Kelvin urged his audience to use their college days such that when they looked back 'you may see no day wasted, no time spent in idleness, no duty neglected'.[30] Waste, idleness, and neglect of duty – the deadly sins of the progressive Victorian man – certainly occupied no part of Kelvin's own life, even in the last fifteen years following his elevation to the peerage.

[29] William Thomson, 'Oration' [written for Professor Buchanan], 2nd April, 1838, NB6, ULC.
[30] SPT, **2**, 1167.

The period 1892–1907 was characterized by a series of ceremonial occasions and by the inevitable winter of old age and death of close friends and relatives. Glasgow University commemorated in 1896 his fifty years in the natural philosophy chair. Eminent representatives from all countries of the academic world came to honour the Lord of British physics. The celebrations ended with a steamer cruise around the islands of the Firth of Clyde. Three years later, Kelvin retired from the chair to the role of 'research student'. Then, in 1904, at the age of eighty, he became Chancellor of the University of Glasgow.[31]

Following retirement, Lord and Lady Kelvin acquired, in addition to Netherhall, a London home at 15 Eaton Place, in the heart of the fashionable West End. During his years as Lord Kelvin, he made some fourteen speeches in the House, six of which related to maritime affairs such as mail steamer contracts and load lines. As his last speech on 24th June, 1907, two days before his eighty-third birthday, he addressed the House on deck loads, a subject which combined in characteristic mode a dynamical problem of heavy moveable objects with the concerns of practical seamanship.[32] Other speeches ranged from labour questions to weights and measures. During the same period he published almost 130 papers, including six in the year of his death and two in the year after.

In the decade following his peerage, Lord Kelvin lost a large number of his closest friends and relatives. James died in May, 1892, aged seventy, having lived just long enough to see his younger brother crowned with the ultimate accolade of recognition and success in Imperial Britain. Elizabeth, the eldest of the family and last surviving sister, passed away four years later. Alexander Crum, Francis Sandford, and the Earl of Derby died in 1893. Of his closest scientific allies, Helmholtz went in 1894, Tait in 1901, and Stokes in 1903.[33] All of them were of Lord Kelvin's own generation and, like the death of Queen Victoria in 1901, symbolized the end of a remarkable era.

Through his last fifteen years, Kelvin's own health remained robust, although problems with his leg in 1895–6 and some severe attacks of facial neuralgia (to which he alluded as 'No. 5 demon') caused much discomfort.[34] Margaret Smith, daughter of Archibald, offered a fascinating glimpse into Lord Kelvin's character at this time when she wrote to her mother from Jordanhill in 1901:

> Then the dear Kelvins arrived. He is a source of great anxiety. He insists on doing everything but is not well and once or twice in the last week he has turned faint at dinner. He did so last night but gulped down some champagne & came round & went to a party in the evening. His troubles are connected with digestion & Maimie is distracted in hoping that she is providing the right food. We love having them but shall be thankful when they are safely away.[35]

[31] *Ibid.*, pp. 964–1011, gives a detailed account of Kelvin's jubilee celebrations and his retirement.

[32] *Hansard*, [4th series], **176**, pp. 811–12.

[33] SPT, **2**, 918–20; 932, 934, 938, 1163, 1173–4.

[34] *Ibid.*, pp. 953–8, 990, 1149, 1186; Lord Kelvin to George Chrystal, 16th January, 1905, C20, ULG.

[35] Margaret Smith to Susan Smith, 13th September, 1901, TD1/967, Smith papers, Strathclyde Regional Archives.

Lord and Lady Kelvin dressed for the coronation of King Edward VII. [From A.G. King, *Kelvin the man*.]

The new University of Glasgow (1870), a veritable cathedral of knowledge, which dominated the River Kelvin after which Sir William Thomson took his title. [From A.G. King, *Kelvin the man*.]

Margaret went on to say, however, that Lord Kelvin's Imperialist and Unionist convictions – which led him to announce in conversation that pro-Boers were lunatics – remained as vigorous as ever. At the same time, his talent for discussing physics on apparently equal terms with Smith's grandson, 'Archie', left a deep impression on Margaret's mind, for Archie had not yet entered 'the University of his ancestors'.

In September, 1907, Lady Kelvin suffered a serious stroke. Her illness rendered the aged Lord Kelvin anxious yet optimistic, but at the end of November he too became gravely ill. The doctor diagnosed the complaint as 'a severe chill of the liver (duodenal catarrh)'.[36] The fever which accompanied his illness did not diminish, and he died at Netherhall on 17th December. The funeral took place in Westminster Abbey, at the heart of the Empire, two days before Christmas. A very great name in British science had passed into history.

36 George Green to P.J. Freyer, 12th December, 1907 (draft), F41, ULG; SPT, **2**, 1202–13.

Bibliography

PRIMARY SOURCES

(a) William Thomson

A virtually complete list of Thomson's published papers is given in SPT, **2**, 1225–74, and we have avoided reproducing the list here. Full references to the relevant published memoirs, however, have been given in the footnotes wherever a specific reference has been involved. Similarly, full references to Thomson's manuscripts, notebooks, and correspondence given in the footnotes relate to the two major collections of Thomson's papers catalogued in *Catalogue of the manuscript collections of Sir George Gabriel Stokes and Sir William Thomson, Baron Kelvin of Largs, in Cambridge University Library* (Cambridge, 1976), and *Index to the manuscript collection of William Thomson, Baron Kelvin, in Glasgow University Library* (Glasgow, 1977). Other important sources for Thomson manuscripts have been the James Thomson papers at Queen's University Library, Belfast, and the J.D. Forbes papers at St Andrews University Library. Again specific references will be found in the footnotes.

The principal reprints etc. of William Thomson's papers are:

Reprint of papers on electrostatics and magnetism. London, 1872. (E&M).

Mathematical and physical papers. 6 vols. Cambridge, 1882–1911. (MPP).

Popular lectures and addresses. 3 vols. London, 1891–4. (PL).

Notes of lectures on molecular dynamics and the wave theory of light. Baltimore, 1884. (BL). Delivered at the Johns Hopkins University, Baltimore, stenographically reported by A.S. Hathaway, and published in papyrograph form. Very extensively reworked, with reprints of some of Thomson's relevant papers, as *Baltimore lectures on molecular dynamics and the wave theory of light*. London, 1904. BL is republished in Robert Kargon and Peter Achinstein (eds.), *Kelvin's Baltimore Lectures and modern theoretical physics*. Cambridge, MA and London, 1987, pp. 7–263.

(b) Reports

(i) *Parliamentary papers*. Relevant Scottish university reports are: 'Report to His Majesty by a Royal Commission of Inquiry into the state of the Universities in Scotland' (1831); 'Evidence taken before the Commissioners of the Universities of I. Edinburgh II. Glasgow III. St. Andrews IV. Aberdeen' (1837); 'Report of the Commissioners for visiting the University of Glasgow' (1839), and 'Second Report' (1839); 'General Report of Commissioners under Universities, Scotland, Act, 1858; with appendix, containing ordinances, minutes, reports on special subjects and other documents' (1863). The important telegraph report is: 'Report of Joint Committee appointed by Board of Trade and Atlantic Telegraph Company to inquire into construction of submarine telegraph cables; with evidence and appendix' (1860).

(ii) *Statistical accounts.* The most relevant are: *The new statistical account of Scotland, 6: Lanark*; 7: *Renfrew-Argyle* (London and Edinburgh, 1845) and *The third statistical account of Scotland: Glasgow* (Glasgow, 1958).

(iii) *British Association for the Advancement of Science.* Apart from papers and reports by named authors (listed in (c) below) we refer extensively to reports by BAAS committees, especially in relation to electrical units and standards. Details will be found in the footnotes.

(c) Other authors

Airy, G.B. *Mathematical tracts* (Cambridge, 1826); 2nd edn. (Cambridge, 1831).

Airy, G.B. 'Account of experiments on iron-built ships, instituted for the purpose of discovering a correction for the deviation of the compass produced by the iron of the ships', *Phil. Trans.*, (1839), 167–213.

Airy, G.B. 'Correction of the compass in iron ships', *The Athenaeum*, no. 1409 (1854), 1303–5; no. 1423 (1855), 145–8.

Airy, G.B. 'Iron ships – the Royal Charter', *The Athenaeum*, no. 1672 (1859), 632–6.

Airy, G.B. 'On the connexion between the mode of building iron ships, and the ultimate correction of their compass', *Trans. Inst. Naval Architects*, **1** (1860), 105–9.

Aitken, John. 'Experiments illustrating rigidity produced by centrifugal force', *Proc. Royal Soc. Edinburgh*, **9** (1878), 73–8.

Ampère, André-Marie. *Essai sur la philosophie des sciences, ou exposition analytique d'une classification naturelle de toutes les connaissances humaines* (Paris, 1834).

Argyll, 8th Duke of. *Irish nationalism: an appeal to history* (London, 1893).

Babbage, Charles. *Passages from the life of a philosopher* (London, 1864).

Babinet, Jacques. 'Astronomique cosmogonique. La terre avant les époques géologique', *Revue des deux mondes*, **10** (1855), 702–26.

Boole, George. 'On integration of linear differential equations with constant coefficients', *Cam. Math. J.*, **2** (1841), 114–19.

Boole, George. 'Symmetrical solutions of problems respecting the straight line and plane', *Cam. Math. J.*, **2** (1841), 179–88.

Boole, George. 'On the transformation of multiple integrals', *Cam. Math. J.*, **4** (1845), 20–8.

Boole, George. 'On the attraction of a solid of revolution on an external point', *Cam. and Dublin Math. J.*, **2** (1847), 1–7.

Boole, George. 'On a certain symbolical equation', *Cam. and Dublin Math. J.*, **2** (1847), 7–12.

Boole, George. *The mathematical analysis of logic: being an essay towards a calculus of deductive reasoning* (Cambridge, 1847).

Boole, George. *An investigation of the laws of thought, on which are founded the mathematical theories of logic and probabilities* (London, 1854).

Boole, George. *A treatise on differential equations* (Cambridge, 1859).

Boscovich, Roger. *A theory of natural philosophy*, translated by J.M. Child from the 2nd edn. (1763). (London, 1922).

Brett, J.W. 'On the origin of the submarine telegraph and its extension to India and America', *BAAS Report*, **24** (1854), 7–8.

Bristed, A.R. *Five years in an English University* (2 vols., New York, 1852).

Carnot, Sadi. 'Reflections on the motive power of fire and on machines fitted to develop that power' (1824), in E. Mendoza (ed.), *Reflections on the motive power of fire by Sadi Carnot and other papers on the second law of thermodynamics by E. Clapeyron and R. Clausius* (New York, 1960), pp. 1–59.

Carpenter, W.B. and Jeffreys, J.G. 'Report on deep-sea researches carried on during the months of July, August, and September 1870, in H.M. survey ship *Porcupine*', *Proc. Royal Soc.*, **19** (1870–1), 146–221.

Cayley, Arthur. 'On geometrical representation of the motion of a solid body', *Cam. and Dublin Math. J.*, **1** (1846), 164–7.

Cayley, Arthur. 'On the differential equations which occur in dynamical problems', *Cam. and Dublin Math. J.*, **2** (1847), 210–19.

Cayley, Arthur. 'Report on the recent progress of theoretical dynamics', *BAAS Report*, **27** (1857), 1–42.

Challis, James. 'On the source and maintenance of the sun's heat', *Phil. Mag.*, [series 4], **25** (1863), 460–7.

Chalmers, Thomas. *On political economy, in connexion with the moral state and moral prospects of society* (Glasgow, 1832).

Chalmers, Thomas. *The adaptation of external nature to the moral and intellectual constitution of man* (London, 1834).

Chalmers, Thomas. *The works of Thomas Chalmers* (25 vols., Glasgow, 1836–42).

Clapeyron, Emile. 'Memoir on the motive power of heat', in Richard Taylor (ed.), *Scientific memoirs*, vol. 1 (London, 1837), pp. 347–76.

Clark, Latimer, and Bright, Charles. 'On the formation of standards of electrical quantity and resistance', *BAAS Report*, **31** (1861), 37–8.

Clausius, Rudolf. 'On the motive power of heat, and on the laws which can be deduced from it for the theory of heat' (1850), in E. Mendoza (ed.), *Reflections on the motive power of fire by Sadi Carnot and other papers on the second law of thermodynamics by E.Clapeyron and R. Clausius* (New York, 1960), pp. 109–52.

Conrad, Joseph. *The mirror of the sea* (London, 1906).

Conrad, Joseph. *The secret agent. A simple tale* (London, 1907).

Croll, James. 'On the excentricity of the earth's orbit, and its physical relations to the glacial epoch', *Phil. Mag.*, [series 4], **33** (1867), 119–31.

Croll, James. 'On geological time, and the probable date of the glacial and the upper miocene period', *Phil. Mag.*, [series 4], **35** (1868), 363–84; **36** (1868), 141–54, 362–86.

Croll, James. *Climate and time in their geological relations. A theory of secular changes of the earth's climate* (London, 1875).

Croll, James. 'Age of the sun in relation to evolution', *Nature*, **17** (1877–8), 206–7.

Darwin, Charles. 'Observations on the parallel roads of Glen Roy, and other parts of Lochaber in Scotland, with an attempt to prove that they are of marine origin', *Phil. Trans.*, (1839), 39–82.

Darwin, Charles, in Morse Peckham (ed.), *The origin of species by Charles Darwin. A variorum text* (Philadelphia, 1959).

Darwin, G.H. 'Presidential address to Section A', *BAAS Report*, **56** (1886), 511–18.

Darwin, G.H. 'Radio-activity and the age of the sun', *Nature*, **68** (1903), 496.

Darwin, G.H., in F.J.M. Stratton and J. Jackson (eds.), *Scientific papers* (5 vols., Cambridge, 1907–16).

Day, St. J.V. 'Discussion on patents for inventions', *Proc. Phil. Soc. Glasgow*, **7** (1869–71), 158–220.

Delaunay, C.E. *Traité de mécanique rationelle*, 1st edn. (Paris, 1856); 2nd edn. (Paris, 1857); 3rd edn. (Paris, 1862).

De Morgan, Augustus. *Elements of algebra* (London, 1837).

De Morgan, Augustus. 'On the foundations of algebra', *Trans. Cam. Phil. Soc.*, **7** (1842), 173–87.

De Morgan, Augustus. *The differential and integral calculus* (London, 1842).

De Morgan, Augustus. *Formal logic, or, the calculus of inference, necessary and probable* (London, 1847).

Derby, 15th Earl of. *Speeches and addresses* (2 vols., London, 1894).

Drennan, William, in D.A. Chart (ed.), *The Drennan letters being a selection from the correspondence which passed between William Drennan M.D. and his brother-in-law and sister Samuel and Martha McTier during the years 1776–1819* (Belfast, 1931).

Dufferin, Lord. *Letters from high latitudes* (London, 1857).

Duhamel, J.M.C. *Coeurs de mécanique de l'école polytechnique*, 1st. edn. (Paris, 1845); 2nd edn. (2 vols., Paris, 1853–4); 3rd edn. (2 vols., Paris, 1862).

Duhem, Pierre, in P.P. Wiener (ed.), *The aim and structure of physical theory* (New York, 1962).

Earnshaw, Samuel. *Dynamics, or a treatise on motion; to which is added a short treatise on attractions*, 3rd edn. (Cambridge, 1844).

Ellis, R.L. 'On the lines of curvature of an ellipsoid', *Cam. Math. J.*, **2** (1841), 133–8.

Ellis, R.L. 'On the integration of certain differential equations', *Cam. Math. J.*, **2** (1841), 169–77, 193–201.

Ellis, R.L. 'Memoirs of the late D.F. Gregory', *Cam. Math. J.*, **4** (1845), 145–52.

Ellis, R.L. 'Some remarks on the theory of matter', *Trans. Cam. Phil. Soc.*, **8** (1849), 600–5.

Ellis, R.L. 'General preface to Bacon's philosophical works', in J. Spedding, R.L. Ellis and D.D. Heath (eds.) *The works of Francis Bacon* (15 vols., Boston, 1861), **1**, pp. 61–127.

Ellis, R.L., in William Walton (ed.), *The mathematical and other writings of Robert Leslie Ellis* (Cambridge, 1863).

Evans, F.J. 'Reduction and discussion of the deviations of the compass observed on board of all the iron-built ships, and a selection of the wood-built steam ships in Her Majesty's Navy, and the iron steam-ship *Great Eastern*; being a report to the Hydrographer of the Admiralty', *Phil. Trans.*, (1860), 337–78.

Evans, F.J., and Smith, Archibald. 'Report on the three reports of the Liverpool Compass Committee and other recent publications on the same subject', *BAAS Report*, **32** (1862), 87–101.

Evans, F.J., and Smith, Archibald. *The Admiralty manual for ascertaining and applying the deviations of the compass* (London, 1862).

Everett, J.D. *Illustrations of the centimetre-gramme-second . . . system of units* (London, 1875).

Faraday, Michael. *Experimental researches in electricity* (3 vols., London, 1839–55).

Faraday, Michael. 'On electric induction – associated cases of current and static effects', *Phil. Mag.*, [series 4], **7** (1854), 197–208.

[Fisher, J.A.] *Records by Admiral of the Fleet Lord Fisher* (London, 1919).

FitzGerald, G.F. 'Molecular dynamics', *Nature*, **31** (1855), 503.

FitzGerald, G.F. 'Sir Wm. Thomson and Maxwell's electro-magnetic theory of light', *Nature*, **32** (1885), 4–5.

FitzGerald, G.F. 'On a model illustrating some properties of the ether', *Sci. Proc. Royal Dublin Soc.*, **4** (1885), 407–19.

FitzGerald, G.F. 'Presidential address to Section A', *BAAS Report*, **58** (1888), 557–62.

FitzGerald, G.F., in Joseph Larmor (ed.), *Scientific writings of the late George Francis FitzGerald* (Dublin, 1902).

Fleming, William. *A manual of moral philosophy* (London, 1867).

Forbes, J.D. 'Account of some experiments on the temperature of the earth at different depths and in different soils near Edinburgh', *Trans. Royal Soc. Edinburgh*, **16** (1846), 189–236.

Fourier, Joseph. 'Extrait d'un mémoire sur le refroidissement séculaire du globe terrestre', *Ann. Chim. Phys.*, **13** (1820), 418–38.

Fourier, Joseph. *Théorie analytique de la chaleur* (Paris, 1822). Translated by A. Freeman (Cambridge, 1878).

Fourier, Joseph. Remarques générales sur les températures du globe terrestre et des espaces planétaires', *Ann. Chim. Phys.*, **27** (1824), 136–67.

Fresnel, Augustin. 'Elementary view of the undulatory theory of light', *Quart. J. Sci.*, **23** (1827), 127–40, 441–54; **24** (1827), 113–35, 431–48; **25** (1828), 198–215; **26** (1829), 168–91, 389–407; **27** (1829), 159–65.

Gauss, C.F. 'Allgemeine Lehrsätze in Beziehung auf die im verkehrten Verhältnis des Quadrats der Entfernung wirkenden Anziehungs – und Abstossungs-Kräfte', in C.F. Gauss and Wilhelm Weber (eds.), *Resultate aus den Beobachtungen des magnetischen Vereins im Jahr 1839* (Leipzig, 1840), pp. 1–51.

Geikie, Archibald. 'Geological time', *Geol. Mag.*, **4** (1867), 171–2.

[Geikie, Archibald.] 'Sir Roderick Murchison and modern schools of geology', *Quart. Rev.*, **125** (1868), 188–217.

Geikie, Archibald. 'On modern denudation', *Trans. Geol. Soc. Glasgow*, **3** (1871), 153–90.

Geikie, Archibald. *Text-book of geology* (London, 1882).

Geikie, Archibald. 'Presidential address', *BAAS Report*, **62** (1892), 3–26.

Gordon, Lewis. 'On dynamometrical apparatus; or, the measurement of the mechanical effect of moving powers', *Proc. Phil. Soc. Glasgow*, **1** (1841–4), 41–2.

Gordon, Lewis. *A synopsis of lectures to be delivered. Session 1847–8* (Glasgow, 1847).

Graham, Robert. *Practical observations on continued fever, especially that form at present existing as an epidemic* (Glasgow, 1818).

Greatheed, S.S. 'On general differentiation', *Cam. Math. J.*, **1** (1839), 11–21, 109–17.

Greatheed, S.S. 'Application of the symmetrical equations of a straight line [and plane] to various problems in analytical geometry of three dimensions', *Cam. Math. J.*, **1** (1839), 37–42, 135–42.

Green, George. *An essay on the application of mathematical analysis to the theories of electricity and magnetism* (Nottingham, 1828).

Green, George. 'On the determination of the exterior and interior attractions of ellipsoids of variable densities', *Trans. Cam. Phil. Soc.*, **5** (1835), 395–429.

Green, George. 'On the laws of the reflexion and refraction of light at the common surface of two non-crystallized media', *Trans. Cam. Phil. Soc.*, **7** (1842), 113–20.

Green, George. 'On the propagation of light in crystallized media' (1839). *Trans. Cam. Phil. Soc.*, **7** (1842), 121–40.

Green, George, in N.M. Ferrers (ed.), *Mathematical papers of the late George Green* (London, 1871).

Gregory, D.F. 'Preface', *Cam. Math. J.*, **1** (1839), 1–2.

Gregory, D.F. 'On the solution of linear differential equations with constant coefficients', *Cam. Math. J.*, **1** (1839), 22–32.

Gregory, D.F. 'On the solution of linear equations of finite and mixed differences', *Cam. Math. J.*, **1** (1839), 54–61.

Gregory, D.F. 'Notes on Fourier's heat', *Cam. Math. J.*, **1** (1839), 104–7.

Gregory, D.F. 'On the solution of certain partial differential equations', *Cam. Math. J.*, **1** (1839), 123–31.

Gregory, D.F. 'On the impossible logarithms of quantities', *Cam. Math. J.*, **1** (1839), 226–34.

Gregory, D.F. 'On existence of branches and curves in several planes', *Cam. Math. J.*, **1** (1839), 259–66.

Gregory, D.F. 'On the elementary principles of the application of algebraical symbols to geometry', *Cam. Math. J.*, **2** (1841), 1–9.

Gregory, D.F. 'On the theory of maxima and minima of functions of two variables', *Cam. Math. J.*, **2** (1841), 138–40.

Grove, W.R. 'Presidential address', *BAAS Report*, **36** (1866), liii–lxxxii.

Guthrie, Frederick. 'On approach caused by vibration', *Proc. Royal Soc.*, **18** (1870), 93–4.

Guthrie, Frederick. 'On approach caused by vibration', *Phil. Mag.*, [series 4], **39** (1870), 309; **40** (1870), 345–54.

[Hamilton, Sir William.] 'Patronage of universities', *Edinburgh Rev.*, **59** (1834), 196–227.

Hamilton, W.R. 'Address by the President' (1837), *Proc. Royal Irish Acad.*, **1** (1841), 212–21.

Hamilton, W.R. 'On symbolical geometry', *Cam. and Dublin Math. J.*, **1** (1846), 45–57, 137–54, 256–63; **2** (1847), 47–52, 130–3, 204–9; **3** (1848), 68–84, 220–5; **4** (1849), 84–9, 105–18.

[Harris, Walter]. *The antient and present state of the County of Down* (Dublin 1744).

Helmholtz, Hermann. 'On the interaction of natural forces' (lecture at Königsberg, 7th February, 1854), *Phil. Mag.*, [series 4], **11** (1856), 489–518.

Helmholtz, Hermann. 'Ueber Integrale der hydrodynamischen Gleichungen welche den Wirbelbewegungen entsprechen', *Journal für die reine und angewandte Mathematik*, **55** (1858), 22–55.

Helmholtz, Hermann. 'On the application of the law of the conservation of force to organic nature', *Proc. Royal Inst.*, **3** (1858–62), 347–57.

Herschel, J.F.W. 'Light', in *Encyclopaedia metropolitana* (London, 1830), **4**, pp. 341–586.

Herschel, J.F.W. *Preliminary discourse on the study of natural philosophy* (London, 1830).

[Herschel, J.F.W.] 'Mechanism of the heavens', *Quart. Rev.*, **47** (1832), 537–59.

Herschel, J.F.W. *A treatise on astronomy* (London, 1833).

[Herschel, J.F.W.] 'Whewell on inductive sciences', *Quart. Rev.*, **68** (1841), 177–238.

Herschel, J.F.W. 'Presidential address', *BAAS Report*, **15** (1845), xxvii–xliv.

Herschel, J.F.W. *Outlines of astronomy* (London, 1849).

Herschel, Sir John. 'The sun', *Good Words*, **4** (1863), 273–84.

Hind, John. *The elements of algebra: designed for the use of students in the University* (Cambridge, 1829).

Hooker, J.D. 'Presidential address', *BAAS Report*, **38** (1868), lviii–lxxv.

Hopkins, William. 'Researches in physical geology', *Trans. Cam. Phil. Soc.*, **6** (1835), 1–84.

Hopkins, William. 'An abstract of a memoir on physical geology, with a further exposition of certain points connected with the subject', *Phil. Mag.*, [series 3], **8** (1836), 227–36, 272–81, 357–66.

Hopkins, William. 'Researches in physical geology – first series', *Phil. Trans.*, (1839), 381–423.

Hopkins, William. *Remarks on certain proposed regulations respecting the studies of the University* (Cambridge, 1841).

Hopkins, William. 'Report on the geological theories of elevation and earthquakes', *BAAS Report*, **17** (1847), 33–92.

Hopkins, William. 'Anniversary address of the President', *Quart. J. Geol. Soc. London*, **8** (1852), xxiv–lxxx.

Hopkins, William. 'On the causes which may have produced changes in the earth's superficial temperature', *Proc. Geol. Soc. London*, **8** (1852), 56–92.

Hopkins, William. 'Presidential address', *BAAS Report*, **23** (1853), xli–lvii.

Hopkins, William. 'An account of some experiments on the effect of pressure on the temperature of fusion of different substances', *BAAS Report*, **24** (1854), 57–8.

Hopkins, William. 'Experimental researches on the conductive powers of various substances, with the application of the results to the problem of terrestrial temperature', *Phil. Trans.*, (1857), 805–49.

Hopkins, William. 'On the earth's internal temperature and the thickness of its solid crust', *Proc. Royal Inst.*, **3** (1858–62), 139–43.

Hopkins, William. 'Physical theories of the phenomena of life', *Frazer's Mag.*, **61** (1860), 739–53; **62** (1860), 74–90.

Hume, David, in H.D. Aitken (ed.), *Dialogues concerning natural religion* (1779) (New York, 1948).

Huxley, T.H. 'Geological reform', *Quart. J. Geol. Soc. London*, **25** (1869), xxxviii–liii.

Huxley, T.H. 'Presidential address', *BAAS Report*, **40** (1870), lxxxii–lxxxix.

Huxley, T.H. *Discourses: biological and geological* (London, 1902).

Hymers, John. *A treatise on plane trigonometry, and on trigonometrical tables and logarithms* (Cambridge, 1837).

Hymers, John. *A treatise on spherical trigonometry, together with a selection of problems* (Cambridge, 1841).

[Jenkin, Fleeming.] 'Darwin and the origin of species', *North Brit. Rev.*, **46** (1867), 277–318.

[Jenkin, Fleeming.] 'The atomic theory of Lucretius', *North Brit. Rev.*, [new series], **9** (1868), 211–42.

Jenkin, Fleeming, in R.L. Stevenson (ed.), *Papers and memoir of Fleeming Jenkin* (2 vols., London, 1887).

Joly, John. 'Radium and the geological age of the earth', *Nature*, **68** (1903), 526.

Joule, J.P. 'On a new class of magnetic forces', *Ann. Elec.*, **8** (1842), 219–24.

Joule, J.P. 'On the calorific effects of magneto-electricity, and on the mechanical value of heat', *Phil. Mag.*, [series 3], **23** (1843), 263–76, 347–55, 435–43.

Joule, J.P. 'On the changes of temperature produced by the rarefaction and condensation of air', *Phil. Mag.*, [series 3], **26** (1845), 369–83.

Joule, J.P. 'On matter, living force, and heat' (1847), *The scientific papers of J.P. Joule* (2 vols., London, 1887).

Joule, J.P. 'On shooting-stars', *Phil. Mag.*, [series 3], **32** (1848), 349–51.

Joule, J.P. 'On the œconomical production of mechanical effects from chemical forces', *Phil. Mag.*, [series 4], **5** (1853), 1–9.

Joule, J.P. *The scientific papers of J.P. Joule* (2 vols., London, 1887).

Kelland, Philip. *Theory of Heat* (Cambridge, 1837).

Kelland, Philip. 'On the dispersion of light, as explained by the hypothesis of finite intervals', *Trans. Cam. Phil. Soc.*, **6** (1838), 153–84.

King, Clarence. 'The age of the earth', *Am. J. Sci.*, **45** (1893), 1–20.

Kingsley, Charles. *Yeast. A problem* (London, 1851).

Kipling, Rudyard. *Rudyard Kipling's verse. Inclusive edition. 1885–1932*, 4th edn. (London, 1933).

Lagrange, J.-L. *Mécanique analytique* (Paris, 1788); 4th edn. (2 vols., Paris, 1888–9).

Lamé, Gabriel. *Coeurs de physique* (3 vols., Paris, 1836).

Laplace, P.S. de. *Exposition du système du monde* (Paris, 1796); 2nd edn. (2 vols., Paris, 1798–9); translated by J. Pond (2 vols., London, 1809); 4th edn. (2 vols., Paris, 1813).

Laplace, P.S. de. *Traité de mécanique céleste* (5 vols., Paris, 1798–1827).

Laplace, P.S. de. 'Mémoire sur les mouvements de la lumière dans les milieux diaphanes', *Mémoir de l'Institut*, **10** (1809), 300–42.

Laplace, P.S. de. *Oeuvres completes de Laplace* (14 vols., Paris, 1878–1912).

Larmor, Joseph (ed.) 'The origins of Clerk Maxwell's electric ideas, as described in familiar letters to W. Thomson', *Proc. Cam. Phil. Soc.*, **32** (1936), 695–750.

Lecky, S.T.S. *'Wrinkles' in practical navigation* (London, 1881).

Lefroy, J.H., in G.F.C. Stanley (ed.), *In search of the magnetic north. A soldier-surveyor's letters from the North-West, 1843–44* (Toronto,1955).

Le Sage, G.-L. 'Lucrèce Newtonien', *Nouveaux mémoires de l'Academie Royal des Sciences et Belles-lettres, année 1782* (Berlin, 1784), 404–32.

Le Sage, G.-L., in P. Prevost (ed.), *Traité de physique mécanique* (Geneva and Paris, 1818).

Lindsay, W.S. *History of merchant shipping and ancient commerce* (4 vols., London, 1874).

Lloyd, Humphrey. 'On the rise and progress of Mechanical Philosophy' (1834), in *Miscellaneous papers connected with physical science* (London, 1877), pp. 414–36.

Lockyer, J.N. *Contributions to solar physics* (London, 1874).

Lodge, O.J. 'Sketch of electrical papers in Section A at the recent Bath meeting of the British Association', *The Electrician*, **21** (1888), 622–5.

Lyell, Charles. *Principles of geology, being an attempt to explain the former changes of the earth's surface by reference to causes now in operation* (3 vols., London, 1830–3); 3rd edn. (3 vols., London, 1835); 10th edn. (2 vols., London, 1867–8).

Lyell, Charles. 'Anniversary address of the president', *Quart. J. Geol. Soc. London*, **7** (1851), xxxii–lxxv.

MacCullagh, James. 'On laws of crystalline reflection and refraction', *Trans. Royal Irish Acad.*, **18** (1837), 31–74.

MacCullagh, James. 'An essay towards a dynamical theory of crystalline reflexion and refraction', *Trans. Royal Irish Acad.*, **21** (1848), 17–50.

MacCullagh, James, in J.H. Jellet and Samuel Haughton (eds.), *The collected works of James MacCullagh* (Dublin, 1880).

Matthiessen, Augustus. 'On the cause of vesicular structure in copper', *Phil. Mag.*, [series 4], **23** (1862), 81–4.

Maxwell, J.C. 'On the equilibrium of elastic solids', *Trans. Royal Soc. Edinburgh*, **20** (1853), 87–120.

Maxwell, J.C. 'On Faraday's lines of force', *Trans. Cam. Phil. Soc.*, **10** (1856), 27–83.

Maxwell, J.C. *Theory of heat* (London, 1871).

Maxwell, J.C. *A treatise on electricity and magnetism* (2 vols., Oxford, 1873).

Maxwell, J.C., in W.D. Niven (ed.), *The scientific papers of James Clerk Maxwell* (2 vols., Cambridge, 1890).

Mayer, J.R. 'On celestial dynamics', *Phil. Mag.*, [series 4], **25** (1863), 241–8, 387–409, 417–28.

Mill, J.S. *A system of logic*, 9th edn. (2 vols., London, 1875).

Mill, J.S. 'Unpublished letters from John Stuart Mill to Professor Nichol', *Fortnightly Rev.*, **61** (1897), 660–78.

Miller, Hugh. *Footprints of the Creator, or, the asterolepsi of Stromness* (London, 1849).

Moigno, F.N.M. 'Cosmogenie', *Cosmos*, **6** (1855), 659–64.

Murchison, Sir Roderick. *Siluria*, 4th edn. (London, 1867).

Murphy, Robert. *Elementary principles of the theories of electricity, heat, and molecular actions: Part 1, On electricity* (Cambridge, 1833).

Napier, J.R. 'Remarks on ships' compasses', *Proc. Phil, Soc. Glasgow*, **3** (1848–55), 365–75.

Napier, J.R. 'Description of an instrument for measuring the velocity of ships, currents, &c.', *Proc. Phil. Soc. Glasgow*, **8** (1871–3), 146–59.

Napier, J.R., Nielson, Walter, and Rankine, W.J.M. 'Report on the progress and state of applied mechanics', *Proc. Phil. Soc. Glasgow*, **4** (1855–60), 207–30.

Napier, J.R., and Rankine, W.J.M. *Shipbuilding. Theoretical and practical* (London, 1866).

Newcomb, Simon. *Popular astronomy* (London, 1878).

Newton, Isaac. *Philosophiae naturalis principia mathematica* (London, 1687).

Newton, Isaac. *Opticks*, 4th edn. (London, 1730).

Newton, Isaac, in F. Cajori (ed.), *Mathematical principles of natural philosophy*, translated by A. Motte (1729) (Berkeley, 1934).

Newton, Isaac, in Alexander Koyré and I.B. Cohen (eds.), *Isaac Newton's Philosophiae naturalis principia mathematica: the third edition (1726) with variant readings* (2 vols., Cambridge, 1972).

[Nichol], J.P. 'State of discovery and speculation concerning the nebulae', *London and Westminster Rev.*, **3** (1836), 390–409.

Nichol, J.P. *Views of the architecture of the heavens. In a series of letters to a lady* (Edinburgh and London, 1837).

Nichol, J.P. 'Preliminary dissertation' to J. Willm, *The education of the people* (Glasgow, 1847), pp. xi–lxxx.

Nichol, J.P. *Thoughts on some important points relating to the system of the world*, 2nd edn. (Edinburgh, 1848).

O'Brien, Matthew. *Mathematical tracts* (Cambridge, 1840).

Page, David. *Advanced text-book of geology. Descriptive and industrial*, 4th edn. (Edinburgh and London, 1867).

Paley, William. *The works of William Paley, D.D.* (5 vols., London, 1823).

Peacock, George. *A treatise on algebra* (Cambridge, 1830).

Peacock, George. 'Report on the recent progress and present state of certain branches of analysis', *BAAS Report*, **3** (1833), 185–352.

Peel, Robert, in C.S. Parker (ed.), *Sir Robert Peel from his private correspondence* (3 vols., London, 1891–9).

Perry, John. 'On the age of the earth', *Nature*, **51** (1894–5), 224–7, 582–5.

Phillips, John. *Life on Earth. Its origin and succession* (London and Cambridge, 1860).

Phillips, John. 'Presidential address to the Geological Section', *BAAS Report*, **34** (1864), 45–9.

Playfair, John. *Illustrations of the Huttonian theory* (Edinburgh, 1802).

Playfair, John. 'La Place, Traité de mécanique céleste', *Edinburgh Rev.*, **22** (1808), 249–84.

Playfair, John. *Outlines of natural philosophy, being the heads of lectures delivered in the University of Edinburgh*, 3rd edn. (Edinburgh, 1819).

Poisson, S.D. *Traité de mécanique* (Paris 1811); translated by H.H. Harte (2 vols., London and Dublin, 1842).

Poisson, S.D. 'Mémoire(s) sur la distribution de l'électricité à la surface des corps conducteurs', *Mémoire de l'Institut* (1811), 1–92, 163–274.

Poisson, S.D. 'Mémoire(s) sur la théorie du magnétisme', *Mémoire de l'Académie des Sciences*, **5** (1821–2), 247–338, 488–533.

Poisson, S.D. *Théorie mathématique de la chaleur* (Paris 1835).

Poulton, E.B. 'A naturalist's contribution to the discussion upon the age of the earth', *BAAS Report*, **66** (1896), 808–28.

Pratt, J.H. *The mathematical principles of mechanical philosophy and their application to the theory of universal gravitation* (Cambridge, 1836).

Rankine, W.J.M. 'Laws of the elasticity of solids', *Cam. and Dublin Math. J.*, **10** (1851), 47–80.

Rankine, W.J.M. 'On the velocity of sound in liquid and solid bodies of limited dimensions, especially along prismatic masses of liquid', *Cam. and Dublin Math. J.* **10** (1851), 238–67.

Rankine, W.J.M. 'On the reconcentration of the mechanical energy of the universe', *Phil. Mag.*, [series 4], **4** (1852), 358–60.

Rankine, W.J.M. 'Opening remarks on the objects of the [mechanical science] section', *BAAS Report*, **25** (1855), 201–3.

Rankine, W.J.M., in W.J. Millar (ed.), *Miscellaneous scientific papers* (London, 1881).

Rankine, W.J.M. 'Preliminary dissertation on the harmony of theory and practice in mechanics', in *A manual of applied mechanics*, 17th edn. (London, 1904), pp. 1–11.

Rankine, W.J.M. and Thomson, John. 'On telegraphic communication between Great Britain and Ireland', *Proc. Phil. Soc. Glasgow*, **3** (1848–55), 265–6.

Raper, Henry. *The practice of navigation and nautical astronomy* (London, 1840).

Reade, T.M. 'The age of the world as viewed by the geologist and the mathematician', *Geol. Mag.*, **5** (1878), 145–54.

Reade, T.M. 'Measurement of geological time', *Geol. Mag.*, **10** (1893), 97–100.

Reid, Thomas, in Sir William Hamilton (ed.), *The works of Thomas Reid, D.D.* (2 vols., Edinburgh, 1846–63).

Roberts-Austen, W.C. and Rucker, A.W. 'On the specific heat of basalt', *Phil. Mag.*, [series 5], **35** (1893), 196–307.

[Robison, John.] 'Philosophy', in *Encyclopaedia Britannica* (Edinburgh, 1797–1801), **14**, pp. 573–600.

[Robison, John.] 'Physics', in *Encyclopaedia Britannica* (Edinburgh, 1797–1801), **16**, pp. 637–59.

Roget, P.M. *Animal and vegetable physiology considered with reference to natural theology* (2 vols., London, 1834).

Russell, John Scott. 'The loss of the *Captain*', *Macmillan's Mag.*, **22** (1870), 473–80.

Russell, W.H. *The Atlantic telegraph (1865)* (reprinted Newton Abbot, 1972).

Rutherford, Ernest. *Radio-activity* (Cambridge, 1904).

Rutherford, Ernest. 'The radiation and emanation of radium', *Sci. Am. Suppl.*, **58** (1904), 24073–4, 24086–8.

Rutherford, Ernest. 'Radium – the cause of the earth's heat', *Harper's Mag.*, **49** (1904–5), 390–6.

Rutherford, Ernest. 'Some cosmical aspects of radioactivity', *J. Royal Astron. Soc. Canada*, **1** (1907), 145–65.

Rutherford, Ernest. *The collected papers of Lord Rutherford of Nelson* (3 vols., London, 1962–5).

Sabine, Edward. 'Contributions to terrestrial magnetism', *Phil. Trans.*, (1834), 145–231; (1844), 87–224; (1846), 237–432.

Salisbury, Marquis of. 'Presidential address', *BAAS Report*, **64** (1894), 3–15.

Scoresby, William. *Magnetical investigations* (2 vols., London, 1839–52).

Scoresby, William. 'Correction of the compass in iron ships by magnets', *The Athenaeum*, no. 1415 (1854), 1494–5; no. 1416 (1854), 1526–8.

Scoresby, William. 'On the loss of the *Tayleur*, and the changes in the action of compasses in iron ships', *BAAS Report*, **24** (1854), 49–53.

Scoresby, William, in Archibald Smith (ed.), *Journal of a voyage to Australia and round the world for magnetical research* (London, 1859).

Sedgwick, Adam. *A discourse on the studies of the University*, 2nd edn. (Cambridge, 1834); 5th edn. (London and Cambridge, 1850).

Smith, Adam, in W.P.D. Wightman and J.C. Bryce (eds.), *Adam Smith. Essays on philosophical subjects* (Oxford, 1980).

Smith, Archibald. 'Investigation of the equation of Fresnel's wave surface', *Trans. Cam. Phil. Soc.*, **6** (1838), 85–9.

Smith, Archibald. 'Notes on the undulatory theory of light', *Cam. Math. J.*, **1** (1839), 3–9

Smith, Archibald. 'Elimination by means of cross multiplication', *Cam. Math. J.*, **1** (1839), 46.

Smith, Archibald and Evans, F.J. 'On the effect produced on the deviation of the compass by the length and arrangement of the compass needles; and on the new mode of correcting the quadrantal deviation', *Proc. Royal Soc.*, **11** (1860–2), 179–81.

Snowball, J.C. *The elements of plane and spherical trigonometry, with the construction and use of mathematical tables* (Cambridge, 1834).

Soddy, Frederick. *Radio-activity: an elementary treatise from the standpoint of the disintegration theory* (London, 1904).

Soddy, Frederick. 'The recent controversy on radium', *Nature*, **74** (1906), 516.

Sollas, W.J. 'On evolution in geology', *Geol. Mag.*, **4** (1877), 1–7.

Somerville, Mary. *Mechanism of the heavens* (London, 1831).

Stewart, Balfour. *The conservation of energy, being an elementary treatise on energy and its laws* (London, 1874).

Stewart, Balfour and Lockyer, J.N. 'The sun as a type of the material universe. Part II: the place of life in a universe of energy', *Macmillan's Mag.*, **18** (1868), 319–27.

[Stewart, Balfour and Tait, P.G.] *The unseen universe, or physical speculations on a future state* (London, 1875); 4th edn. (London, 1876).

Stokes, G.G. 'On some cases of fluid motion', *Trans. Cam. Phil. Soc.*, **8** (1849), 105–37, 409–14.

Stokes, G.G. 'On the theories of the internal friction of fluids in motion of elastic solids', *Trans. Cam. Phil. Soc.*, **8** (1849), 287–319.

Stokes, G.G. 'On the conduction of heat in crystals', *Cam. and Dublin Math. J.*, **10** (1851), 215–38.

Stokes, G.G. *Mathematical and physical papers* (5 vols., Cambridge, 1880–1905).

Tait, P.G. *The position and prospects of physical science: a public inaugural lecture* (Edinburgh, 1860).

Tait, P.G. 'Note on a quaternion transformation', *Proc. Royal Soc. Edinburgh*, **5** (1863), 115–19.

Tait, P.G. *Lectures on some recent advances in physical science* (London, 1867).

[Tait, P.G.] 'Geological time', *North Brit. Rev.*, **50** (1869), 406–39.

Tait, P.G. 'Force', *Nature*, **14** (1876), 459–63.

Tait, P.G. *Scientific papers* (2 vols., Cambridge, 1898).

Thomson, Allen. 'Prefatory notice of the new College buildings', in *Introductory addresses delivered at the opening of the University of Glasgow session 1870–71* (Glasgow, 1870), pp. vi–xxvi.

Thomson, F.H. 'Opening address by the president', *Proc. Phil. Soc. Glasgow*, **6** (1865–8), 233–6.

[Thomson, James.] 'Recollections of the Battle of Ballynahinch', *Belfast Mag.*, **1** (1825), 56–64.

[Thomson, James.] 'State of science in Scotland', *Belfast Mag.*, **1** (1825), 269–78.

[Thomson, James.] 'State of science in Ireland', *Belfast Mag.*, **1** (1825), 459–69.

Thomson, James. *A treatise on arithmetic in theory and practice*, 2nd edn. (Belfast, 1825).

Thomson, James. *An introduction to modern geography: with an appendix containing an outline of astronomy, and the use of the globes* (Belfast, 1827).

Thomson, James. *Elements of plane and spherical trigonometry, with the first principles of analytical geometry*, 2nd edn. (Belfast, 1830).

Thomson, James. *An introduction to the differential and integral calculus* (Belfast, 1831); 2nd edn. (London, 1848).

Thomson, James. *An elementary treatise on algebra, theoretical and practical* (London, 1844).

Thomson, James. 'Theoretical considerations on the effects of pressure in lowering the freezing point of water', *Trans. Royal Soc. Edinburgh*, **16** (1849), 575–80.

Thomson, James. 'On the vortex water-wheel', *BAAS Report*, **22** (1852), 130.

Thomson, James. 'On a centrifugal pump with exterior whirlpool, constructed for draining land', *Trans. Inst. Eng. Scotland*, **2** (1858), 20–6.

Thomson, James, in Sir Joseph Larmor and James Thomson (eds.), *Collected papers in physics and engineering* (Cambridge, 1912).

Thomson, Thomas. *History of the Royal Society from its institution to the end of the eighteenth century* (London, 1812).

Thomson, Thomas. *An outline of the sciences of heat and electricity* (London and Edinburgh, 1830).

Thomson, William, and Tait, P.G. 'Energy', *Good Words*, **3** (1862), 601–7.

Thomson, William, and Tait, P.G. *Elementary dynamics* (Oxford, 1867).

Thomson, William, and Tait, P.G. *Treatise on natural philosophy* (Oxford, 1867); new edn. (2 vols., Cambridge, 1879–83).

Thomson, William, and Tait, P.G. *Elements of natural philosophy* (Oxford, 1873); 2nd edn. (Cambridge, 1879).

Tyndall, John. 'Presidential address', *BAAS Report*, **44** (1874), lxvii–xcvii.

Walker, J. *A new system of arithmetic . . . A new edition, with an appendix by W. Russell* (London, 1823).

Walton, William. 'Symmetrical investigation of points of inflection', *Cam. Math. J.*, **4** (1845), 13–17.

Weisbach, Julius. *Principles of the mechanics of machinery and engineering* (2 vols., London, 1848).

Whethem, W.C.D. *The recent development of physical science* (London, 1904).

Whewell, William. *An elementary treatise on mechanics* (Cambridge, 1819).

Whewell, William. 'On the principles of dynamics, particularly as stated by French writers', *Edinburgh J. Sci.*, **8** (1829), 27–38.

Whewell, William. 'Observations on some passages of Dr. Lardner's *Treatise on mechanics*', *Edinburgh J. Sci.*, [new series], **3** (1830), 148–55.

Whewell, William. 'Theory of electricity', in *Encyclopaedia metropolitana* (London, 1830), **4**, pp. 140–70.

Whewell, William. 'Mathematical exposition of some doctrines of political economy', *Trans. Cam. Phil. Soc.*, **3** (1830), 191–230.

Whewell, William. 'Mathematical exposition of some of the leading doctrines in Mr. Ricardo's "Principles of political economy and taxation"', *Trans. Cam. Phil. Soc.*, **4** (1833), 155–98.

Whewell, William. *Astronomy and general physics considered with reference to natural theology* (London, 1833).

Whewell, William. 'On the nature of the truth of the laws of motion', *Trans. Cam. Phil. Soc.*, **5** (1834), 149–72.

Whewell, William. 'Report on the recent progress and present condition of the mathematical theories of electricity, magnetism and heat', *BAAS Report*, **5** (1835), 1–34.

Whewell, William. *An elementary treatise on mechanics: designed for the use of students in the University*, 5th edn. (Cambridge, 1836).

Whewell, William. *The mechanical Euclid* (Cambridge, 1837).

Whewell, William. *History of the inductive sciences from the earliest to the present times* (3 vols., London, 1837).

Whewell, William. *The philosophy of the inductive sciences* (2 vols., London, 1840).

Whewell, William. *The mechanics of engineering, intended for use in universities, and in colleges of engineers* (Cambridge, 1841).

Whewell, William. *Of a liberal education* (Cambridge, 1845).

Whitehouse, Wildman. 'Experimental observations on an electric cable', *BAAS Report*, **25** (1855), 23–4.

Whitehouse, Wildman, 'The law of the squares – is it applicable or not to the transmission of signals in submarine circuits?', *BAAS Report*, **26** (1856), 21–3.

Willis, Robert. *Principles of mechanism, designed for the use of students in the universities, and engineering students generally* (London, 1841).

Wilson, W.E. 'Radium and solar energy', *Nature*, **68** (1903), 222.

Wood, James. *The principles of mechanics: designed for the use of students in the University*, 8th edn. (Cambridge, 1830).

Young, Thomas. *A course of lectures on Natural Philosophy and the mechanical arts* (2 vols., London, 1807).

SECONDARY SOURCES

Addison, I.W. (ed.) *The matriculation album of the University of Glasgow from 1728 to 1858* (Glasgow, 1913).

Albion, G. *Square-riggers on schedule. The New York sailing packets to England, France and the cotton ports* (Princeton, 1938).

Allen, Jerry. *The sea years of Joseph Conrad* (London, 1967).

Alter, Peter. *The reluctant patron: science and the state in Britain, 1850–1920*, translated by A. Davis (Oxford, 1987).

Amery, Julian, and Garvin, J.L. *The life of Joseph Chamberlain* (6 vols., London, 1932–69).

Arnold, D.H. 'The *mécanique physique* of Siméon Denis Poisson: the evolution and isolation in France of his approach to physical theory (1800–1840). II. The Laplacian programme', *Arch. Hist. Exact Sci.*, **28** (1983), 267–87.

'IV. Disquiet with respect to Fourier's treatment of heat', *Arch. Hist. Exact Sci.*, **28** (1983), 299–320.

'VII. *Mécanique physique*', *Arch. Hist. Exact Sci.*, **29** (1983), 37–52.

'VIII. Applications of the *mécanique physique*', *Arch. Hist. Exact Sci.*, **29** (1983), 53–72.

'IX. Poisson's closing synthesis: *Traité de physique mathématique*', *Arch. Hist. Exact Sci.*, **29** (1983), 73–94.

Ashby, Eric. *Technology and the academics. An essay on universities and the scientific revolution*, 2nd edn. (London, 1963).

Bacon, R.H. *The life of Lord Fisher of Kilverstone* (2 vols., London, 1929).

Banbury, Philip. *Shipbuilders of the Thames and Medway* (Newton Abbot, 1971).

Beales, Derek. *From Castlereagh to Gladstone* (London, 1969).

Becher, Harvey. 'William Whewell and Cambridge mathematics', *Hist. Stud. Phys. Sci.*, **11** (1980), 1–48.

Beckett, J.C. *Protestant dissent in Ireland, 1687–1780* (London, 1948).

Beckett, J.C. *The making of modern Ireland, 1603–1923* (London, 1966).

Beckett, J.C., and Glasscock, R.E. (eds.) *Belfast. The origin and growth of an industrial city* (London, 1967).

Ben-David, Joseph. 'The profession of science and its powers', *Minerva*, **10** (1972), 363–83.

Best, Geoffrey. 'The Scottish Victorian city', *Vict. Stud.*, **11** (1967–8), 329–58.

Blake, George. *B.I. Centenary, 1856–1956* (London, 1956).

Boehm, Klaus. *The British patent system. 1. Administration* (Cambridge, 1967).

Bonsor, N.R.P. *North Atlantic seaway* (5 vols., Newton Abbot and Jersey, 1975–9).

Bottomley, J.T. 'Physical science in Glasgow University', *Nature*, **6** (1872), 29–32.

Bowler, P.J. *Fossils and progress. Palaeontology and the idea of progressive evolution in the nineteenth century* (New York, 1976).

Briggs, Asa. *The age of improvement. 1783–1867* (London, 1959).

Bright, Charles. *Submarine telegraphs. Their history, construction, and working* (London, 1898).

Brooke, J.H. 'Nebular contraction and the expansion of naturalism' (an essay review of R.L. Numbers, *Creation by natural law: Laplace's nebular hypothesis in American thought*), *Brit. J. Hist. Sci.*, **12** (1979), 200–11.

Brown, S.J. *Thomas Chalmers and the Godly Commonwealth in Scotland* (Oxford, 1982).

Brush, S.G. *The kind of motion we call heat: a history of the kinetic theory of gases in the 19th century* (2 vols., Amsterdam, 1976).

Brush, S.G. 'Nineteenth-century debates about the inside of the earth: solid, liquid, or gas?', *Ann. Sci.*, **36** (1979), 225–54.

Brush, S.G. *Statistical physics and the atomic theory of matter from Boyle and Newton to Landau and Onsager* (Princeton, 1983).

Buchanan, R.A. 'The rise of scientific engineering in Britain', *Brit. J Hist. Sci.*, **18** (1985), 218–33.

Buchwald, J.Z. 'William Thomson and the mathematization of Faraday's electrostatics', *Hist. Stud. Phys. Sci.*, **8** (1977), 101–36.

Buchwald, J.Z. 'Optics and the theory of the punctiform ether', *Arch. Hist. Exact Sci.*, **21** (1980), 245–78.

Buchwald, J.Z. 'The quantitative ether in the first half of the nineteenth century', in G.N. Cantor and M.J.S. Hodge (eds.), *Conceptions of ether: studies in the history of ether theories, 1740–1900* (Cambridge, 1981), pp. 215–37.

Buchwald, J.Z. *From Maxwell to microphysics: aspects of electromagnetic theory in the last quarter of the nineteenth century* (Chicago, 1985).

Burchfield, J.D. 'Darwin and the dilemma of geological time', *Isis*, **65** (1974), 301–21.

Burchfield, J.D. *Lord Kelvin and the age of the earth* (London, 1975).

Burn, W.L. *The age of equipoise. A study of the mid-Victorian generation* (London, 1964).

Bushell, T.A. *'Royal Mail'. A centenary history of the Royal Mail Line. 1839–1939* (London, 1939).

Byatt, I.C.R. *The British electrical industry 1875–1914. The economic returns to a new technology* (Oxford, 1979).

Cable, Boyd. *A hundred year history of the P & O* (London, 1937).

Cahan, David. 'Werner Siemens and the origin of the Physikalisch–Technische Reichsanstalt, 1872–1887', *Hist. Stud. Phys. Sci.*, **12** (1982), 253–83.

Campbell, Lewis, and Garnett, William. *The life of James Clerk Maxwell* (London, 1882).

Campbell-Irons, J. *Autobiographical sketch of James Croll with memoir of his life and work* (London, 1896).

Cannon, W.F. 'The uniformitarian–catastrophist debate', *Isis*, **51** (1960), 38–55.

Cantor, G.N. 'Henry Brougham and the Scottish methodological tradition', *Stud. Hist. Phil. Sci.*, **2** (1971), 69–89.

Cantor, G.N. 'The reception of the wave theory of light in Britain: a case study illustrating the role of methodology in scientific debate', *Hist. Stud. Phys. Sci.*, **6** (1975), 109–32.

Cantor, G.N. *Optics after Newton: theories of light in Britain and Ireland, 1704–1840* (Manchester, 1983).

Cardwell, D.S.L. *The organization of science in England* (London, 1957); revised edn., 1972.

Cardwell, D.S.L. *From Watt to Clausius. The rise of thermodynamics in the early industrial age* (London, 1971).

Cawood, John. 'The magnetic crusade: science and politics in early Victorian Britain', *Isis*, **70** (1979), 493–518.

Channell, D.F. 'The harmony of theory and practice: the engineering science of W.J.M. Rankine', *Technology and Culture*, **23** (1982), 39–52.

Clark, G. Kitson. *The making of Victorian England* (Edinburgh, 1962).

Clark, J.W. and Hughes, T. McK. *The life and letters of the Reverend Adam Sedgwick* (2 vols., Cambridge, 1890).

Constable, Thomas. *Memoir of Lewis D.B. Gordon* (Edinburgh, 1877).

Cotter, C.H. 'George Biddell Airy and his mechanical correction of the magnetic compass', *Ann. Sci.*, **33** (1976), 263–74.

Cotter, C.H. 'The early history of ship magnetism: the Airy–Scoresby controversy', *Ann. Sci.*, **34** (1977), 589–99.

Coutts, James. *A history of the University of Glasgow from its foundation in 1451 to 1901* (Glasgow, 1909).

Crosland, Maurice, and Smith, Crosbie. 'The transmission of physics from France to Britain: 1800–1840', *Hist. Stud. Phys. Sci.*, **9** (1978), 1–61.

Cross, J.J. 'Integral theorems in Cambridge mathematical physics, 1830–55', in P.M. Harman (ed.), *Wranglers and physicists. Studies on Cambridge physics in the nineteenth century* (Manchester, 1985), pp. 112–48.

Crowther, J.G. *British scientists of the nineteenth century* (London, 1935).

Currer-Briggs, Noel. *Doctor Ferranti. The life and work of Sebastian Ziani de Ferranti, F.R.S. 1864–1930* (London, 1970).

Daiches, David. *Glasgow* (London, 1977).

Daston, L.J. 'The physicalist tradition in early nineteenth century French geometry', *Stud. Hist. Phil. Sci.*, **17** (1986), 269–95.

Daub, E.E. 'The regenerator principle in the Stirling and Ericsson hot air engines', *Brit. J. Hist. Sci.*, **7** (1974), 259–77.

Davie, G.E. *The democratic intellect. Scotland and her universities in the late nineteenth century* (Edinburgh, 1961).

De Ferranti, G.Z., and Ince, Richard. *The life and letters of Sebastian de Ferranti* (London, 1934).

De Morgan, Sophie. *Memoir of Augustus de Morgan* (London, 1882).

Devine, T.M. *The tobacco lords. A study of the tobacco merchants of Glasgow and their trading activities c. 1740–1790* (Edinburgh, 1975).

Dibner, Bern. *The Atlantic cable* (New York, 1959).

Dickens, R.J. *Ulster emigration to Colonial America, 1718–1775* (London, 1966).

Dunsheath, Percy. *A history of electrical engineering* (London, 1962).

Dutton, H.I. *The patent system and inventive activity during the industrial revolution, 1750–1852* (Manchester, 1984).

Emmerson, G.S. *John Scott Russell. A great Victorian engineer and naval architect* (London, 1977).

Everitt, C.W.F. 'Maxwell's scientific creativity', in Rutherford Aris *et al.* (eds.), *Springs of scientific creativity* (Minneapolis, 1983), pp. 71–141.

Finn, B.S. 'Developments in thermoelectricity, 1850–1920'. PhD dissertation, University of Wisconsin, 1963.

Fisch, Menachem. 'Necessary and contingent truth in William Whewell's antithetical theory of knowledge', *Stud. Hist. Phil. Sci.*, **16** (1985), 275–314.

Fisch, Menachem. *William Whewell, philosopher of science* (Oxford, forthcoming).

Fisch, Menachem, and Schaffer, Simon. (eds.). *William Whewell. A composite portrait. Studies of his life, work, and influence* (Oxford, forthcoming).

Forrester, John. 'Chemistry and the conservation of energy: the work of James Prescott Joule', *Stud. Hist. Phil. Sci.*, **6** (1975), 273–313.

Fox, Robert. *The caloric theory of gases from Lavoisier to Regnault* (Oxford, 1971).

Fox, Robert. 'The rise and fall of Laplacian physics', *Hist. Stud. Phys. Sci.*, **4** (1974), 89–136.

Garber, Elizabeth. 'Molecular science in late-nineteenth-century Britain', *Hist. Stud. Phys. Sci.*, **9** (1978), 265–97.

Gardiner, Robert (ed.) *Conway's all the world's fighting ships. 1860–1905* (London, 1979).

Garland, M.M. *Cambridge before Darwin. The ideal of a liberal education 1800–1860* (Cambridge, 1980).

Gash, Norman. *Sir Robert Peel. The life of Sir Robert Peel after 1930* (London, 1972).

Geikie, Archibald. *Life of Sir Roderick I. Murchison* (2 vols., London, 1875).

Goldstein, Herbert. *Classical mechanics* (Reading, MA and London, 1950).

Gooding, David. 'Metaphysics versus measurement: the conversion and conservation of force in Faraday's physics', *Ann. Sci.*, **37** (1980), 1–29.

Gooding, David. 'Faraday, Thomson, and the magnetic field', *Brit. J. Hist. Sci.*, **13** (1980), 91–120.

Gooding, David. 'Final steps to the field theory: Faraday's investigation of magnetic phenomena: 1845–1850', *Hist. Stud. Phys. Sci.*, **11** (1981), 231–75.

Gooding, David. 'A convergence of opinion on the divergence of lines: Faraday and Thomson's discussion of diamagnetism', *Notes and Records Royal Soc. London*, **36** (1982), 243–59.

Grabiner, J.V. 'Changing attitudes toward mathematical rigor: Lagrange and analysis in the eighteenth and nineteenth centuries', in H.N. Jahnke and M. Otte (eds.), *Epistemological and social problems of the sciences in the early nineteenth century* (Dordrecht, 1981), pp. 311–30.

Graham, H.G. 'Glasgow University life in olden times', in *The book of the Jubilee. 1451–1901* (Glasgow, 1901), pp. 12–25.

Grattan-Guinness, Ivor. 'Mathematical physics, 1800–1835: genesis in France, and development in Germany', in H.N. Jahnke and M. Otte (eds.), *Epistemological and social problems of the sciences in the early nineteenth century* (Dordrecht, 1981), pp. 349–70.

Grattan-Guinness, Ivor. 'Psychology in the foundations of logic and mathematics: the cases of Boole, Cantor and Brouwer', *Hist. Phil. Logic*, **3** (1982), 33–53.

Grattan-Guinness, Ivor. 'Work for the workers: advances in engineering mechanics and instruction in France, 1800–1830', *Ann. Sci.*, **41** (1984), 1–33.

Grattan-Guinness, Ivor, and Ravetz, J.R. *Joseph Fourier, 1768–1830: a survey of his life and work based on a critical edition of his monograph on the propagation of heat, presented to the Institut de France in 1807* (Cambridge, MA and London, 1972).

Gray, Andrew. 'Lord Kelvin's laboratory in the University of Glasgow', *Nature*, **55** (1896–7), 486–92.

Gray, Andrew. *Lord Kelvin. An account of his scientific life and work* (London, 1908).

Green, E.R.R. *The Lagan valley. 1800–1850. A local history of the industrial revolution* (London, 1949).

Green, George, and Lloyd, J.T. *Kelvin's instruments and the Kelvin museum* (Glasgow, 1970).

Greenhill, Basil, and Gifford, Ann. *The British assault on Finland 1854–55: a forgotten naval war* (Teignmouth, 1988).

Hahn, Roger. *Laplace as a Newtonian scientist* (Los Angeles, 1967).

Hall, M.B. *All scientists now. The Royal Society in the nineteenth century* (Cambridge, 1984).

Hanham, H.J. 'The sale of honours in late Victorian England', *Vict. Stud.*, **3** (1959–60), 277–89.

Hankins, T.L. 'Eighteenth-century attempts to resolve the *vis viva* controversy', *Isis*, **56** (1965), 281–97.

Hankins, T.L. *Sir William Rowan Hamilton* (Baltimore and London, 1980).

Hanna, William. *Memoirs of the life and writings of Thomas Chalmers* (4 vols., Edinburgh, 1849–52).

Harman, P.M. *Energy, force, and matter. The conceptual development of nineteenth-century physics* (Cambridge, 1982).

Heimann, P.M. 'The unseen universe: physics and the philosophy of nature in Victorian Britain', *Brit. J. Hist. Sci.*, **6** (1972), 73–9.

Heimann, P.M. 'Voluntarism and immanence: conceptions of nature in eighteenth century thought', *J. Hist. Ideas*, **39** (1978), 271–83.

Herivel, John. *Joseph Fourier: the man and the physicist* (Oxford, 1975).

Hilken, T.J.N. *Engineering at Cambridge University, 1783–1965* (Cambridge, 1967).

Hitchens, H.L., and May, W.E. *From lodestone to gyro-compass* (London, 1952).

Holland, Bernard. *The life of Spencer Compton, eighth Duke of Devonshire* (2 vols., London, 1911).

Hough, Richard. *First Sea Lord. An authorized biography of Admiral Lord Fisher* (London, 1969).

Houghton, W.E. *The Victorian frame of mind. 1830–1870* (New Haven and London, 1957).

Howse, Derek. *Greenwich time and the discovery of the longitude* (Oxford, 1980).

Hunt, Bruce. '"Practice vs. theory": the British electrical debate, 1888–1891', *Isis*, **74** (1983), 341–55.

Hunt, Bruce. 'Experimenting on the ether: Oliver J. Lodge, and the great whirling machine', *Hist. Stud. Phys. Sci.*, **16** (1986), 111–34.

Hunt, Bruce. '"How my model was right": G.F. FitzGerald and the reform of Maxwell's theory' (unpublished paper).

Hunt, Bruce. 'Imperial science: telegraphy and physics in Victorian Britain' (unpublished paper).

Huxley, Leonard. *Life and letters of Sir Joseph Dalton Hooker* (2 vols., London, 1918).

Hyde, F.E. *Blue Funnel. A history of Alfred Holt and Company of Liverpool from 1865 to 1914* (Liverpool, 1957).

Hyde, F.E. *Cunard and the North Atlantic, 1870–1973* (London, 1975).

Hyman, Anthony. *Charles Babbage; pioneer of the computer* (Princeton, 1982).

Jaki, S.L. 'The five forms of Laplace's cosmogony', *Am. J. Phys.* **44** (1976), 4–11.

James, F.A.J.L. 'Thermodynamics and sources of solar heat, 1846–1862', *Brit. J. Hist. Sci.*, **15** (1982), 155–81.

Jamieson, John. *The history of the Royal Belfast Academical Institution, 1810–1960* (Belfast, 1959).

Jarvis, C. MacK. 'The generation of electricity', in Charles Singer *et al.* (ed.), *A history of technology* (8 vols., Oxford, 1954–84), **5**, pp. 177–207.

Kargon, R.H. *Science in Victorian Manchester. Enterprise and expertise* (Manchester, 1977).

Kidwell, P.A. 'Prelude to solar energy: Pouillet, Herschel, Forbes and the solar constant', *Ann. Sci.*, **38** (1981), 457–76.

King, A.G. *Kelvin the man. A biographical sketch by his niece* (London, 1925).

King, Elizabeth. *Lord Kelvin's early home* (London, 1909).

Klein, M.J. 'Maxwell, his demon and the second law of thermodynamics', *Am. Sci.*, **58** (1970), 84–97.

Klein, M.J. 'Closing the Carnot cycle', in *Sadi Carnot et l'essor de la thermodynamique* (Paris, 1974), pp. 213–19.

Knight, W. *Memoir of John Nichol* (Glasgow, 1896).

Knott, C.G. *Life and scientific work of Peter Guthrie Tait* (Cambridge, 1911).

Knudsen, Ole. 'From Lord Kelvin's notebook: ether speculations', *Centaurus*, **16** (1971), 41–53.

Knudsen, Ole. 'The Faraday effect and physical theory, 1845–1873', *Arch. Hist. Exact Sci.*, **15** (1976), 235–81.

Knudsen, Ole. 'Mathematics and physical reality in William Thomson's electromagnetic theory', in P.M. Harman (ed.), *Wranglers and physicists. Studies on Cambridge physics in the nineteenth century* (Mancheser, 1985), pp. 149–79.

Koppelman, Elaine. 'The calculus of operations and the rise of abstract algebra', *Arch. Hist. Exact Sci.*, **8** (1971), 155–242.

Kuhn, T.S. 'Energy conservation as an example of simultaneous discovery', in M. Clagett (ed.), *Critical problems in the history of science* (Madison, 1959), pp. 321–56.

Kuhn, T.S. 'Mathematical versus experimental traditions in the development of physical science', in *The essential tension. Selected studies in scientific tradition and change* (Chicago, 1977), pp. 31–65.

Kuklick, Henrika. 'Professionalization', in W.F. Bynum *et al.* (ed.), *Dictionary of the history of science* (London, 1981), pp. 341–2.

Laita, L.M. 'Influences on Boole's logic: the controversy between William Hamilton and Augustus de Morgan', *Ann. Sci.*, **36** (1979), 45–65.

Laita, L.M. 'Boolean logic and its extra-logical sources: the testimony of Mary Everest Boole', *Hist. Phil. Logic*, **1** (1980), 37–60.

Landes, David. *The unbound Prometheus* (Cambridge, 1969).

Landes, David. *Revolution in time. Clocks and the making of the modern world* (Cambridge, MA and London, 1983).

Larmor, Joseph. 'Lord Kelvin on optical and molecular dynamics', *Nature* (supplement), **70** (1904), iii–v.

Larmor, Joseph. 'William Thomson, Baron Kelvin of Largs. 1824–1907' (Obituary notice), *Proc. Royal Soc.*, [series A], **81** (1908), i–lxxvi.

Laudan, L.L. 'Thomas Reid and the Newtonian turn of British methodological thought', in R.E. Butts and J.W. Davis (eds.), *The methodological heritage of Newton* (Oxford, 1970), pp. 103–31.

Low, W.L. *David Thomson, M.A. Professor of natural philosophy in the University of Aberdeen. A sketch of his life and character* (Aberdeen, 1894).

McCaffrey, J.F. (ed.). *Glasgow, 1858. Shadow's midnight scenes and social photographs* (Glasgow, 1976).

McConnell, Anita. *No sea too deep. The history of oceanographic instruments* (Bristol, 1982).

McCullough, S. *Ballynahinch: centre of Down* (Belfast, 1968).

Machlup, Fritz, and Penrose, Edith. 'The patent controversy in the nineteenth century', *J. Econ. Hist.*, **10** (1950), 1–29.

Mackay, R.F. *Fisher of Kilverstone* (Oxford, 1973).

Mackie, J.D. *The University of Glasgow, 1451–1951* (Glasgow, 1954).

Mathew, W.M. 'The origins and occupations of Glasgow students, 1740–1839', *Past and present*, no. 33 (1966), 74–94.

Mathias, Peter. *The first industrial nation. An economic history of Britain, 1700–1914* (London, 1969).

Mathias, Peter. 'Who unbound Prometheus? Science and technical change, 1600–1800', in Peter Mathias (ed.), *Science and society. 1600–1900* (Cambridge, 1972), pp. 54–80.

May, W.E. *A history of marine navigation* (Henley on Thames, 1973).

May, W.E. 'Lord Kelvin and his compass', *J. Navigation*, **32** (1979), 122–34.

Mendelsohn, Everett. 'The emergence of science as a profession in nineteenth-century Europe', in Karl Hill (ed.), *The management of scientists* (Boston, 1964), pp. 3–48.

Moody, T.W., and Beckett, J.C. *Queen's Belfast 1845–1949* (2 vols., London, 1959).

Moore, J.R. *The post-Darwinian controversies* (Cambridge, 1979).

Morrell, J.B. 'Thomas Thomson: Professor of Chemistry and University reformer', *Brit. J. Hist. Sci.*, **4** (1969), 245–65.

Morrell, J.B. 'Practical chemistry in the University of Edinburgh, 1799–1843', *Ambix*, **16** (1969), 66–80.

Morrell, J.B. 'Individualism and the structure of British science in 1830', *Hist. Stud. Phys. Sci.*, **3** (1971), 183–204.

Morrell, J.B. 'Professors Robison and Playfair, and the *theophobia gallica*: natural philosophy, religion and politics in Edinburgh, 1789–1815', *Notes and Records Royal Soc. London*, **26** (1971), 43–63.

Morrell, J.B. 'The chemist breeders: the research schools of Liebig and Thomas Thomson', *Ambix*, **19** (1972), 1–46.

Morrell, Jack, and Thackray, Arnold. *Gentlemen of science. Early years of the British Association for the Advancement of Science* (Oxford, 1981).

Moyer, D.F. 'James MacCullagh', in *DSB*, **8**, pp. 591–3.

Moyer, D.F. 'Energy, dynamics, hidden machinery: Rankine, Thomson and Tait, Maxwell', *Stud. Hist. Phil. Sci.*, **8** (1977), 251–68.

Murray, David. *Lord Kelvin as professor in the Old College of Glasgow* (Glasgow, 1924).

Murray, Marischal. *Union Castle chronicle, 1853–1953* (London, 1953).

Musson, A.E., and Robinson, Eric. *Science and technology in the industrial revolution* (Manchester, 1969).

Napier, James. *Life of Robert Napier of West Shandon* (Edinburgh, 1904).

Norman, Edward. *A History of modern Ireland* (Harmondsworth, 1971).

Oakley, C.A. *The second city* (London and Glasgow, 1946).

Oakley, C.A. *A history of a faculty. Engineering at Glasgow University* (Glasgow, 1973).

Oakley, Francis. 'Christian theology and the Newtonian science', *Church Hist.*, **30** (1961), 433–57.

Olson, R.G. 'Scottish philosophy and mathematics', *J. Hist. Ideas*, **32** (1971), 29–44.

Olson, R.G. *Scottish philosophy and British physics 1750–1880. A study in the foundations of the Victorian scientific style* (Princeton, 1975).

Owen, D.J. *History of Belfast* (Belfast, 1921).

Pakenham, Thomas. *The year of liberty. The story of the great Irish rebellion of 1798* (London, 1969).

Parker, C.S. *Life and letters of Sir James Graham, 1792–1861* (2 vols., London, 1907).

Parkes, Oscar. *British battleships. Warrior 1860 to Vanguard 1950. A history of design, construction, and armament* (London, 1957).

Pole, William (ed.). *The life of Sir William Fairbairn, Bart* (London, 1877).

Pollard, Sidney. 'The economic history of British shipbuilding, 1870–1914', PhD dissertation, University of London, 1951.

Pollard, Sidney, and Robertson, Paul. *The British shipbuilding industry, 1870–1914* (Cambridge, MA and London, 1979).

Porter, Theodore. 'A statistical survey of gases: Maxwell's social physics', *Hist. Stud. Phys. Sci.*, **12** (1981), 77–116.

Porter, Theodore. *The rise of statistical thinking, 1820–1900* (Princeton, 1986).

Pyatt, Edward. *The National Physical Laboratory. A history* (Bristol, 1983).

Pycior, Helena. 'George Peacock and the British origins of symbolical algebra', *Historia Mathematica*, **8** (1981), 23–45.

Pycior, Helena. 'Early criticisms of the symbolic approach to algebra', *Historia Mathematica*, **9** (1982), 413–40.

Rankine, W.J.M. *A memoir of John Elder, engineer and shipbuilder* (Edinburgh and London, 1871).

Reader, W.J. *Professional men. The rise of the professional classes in nineteenth-century England* (London, 1966).

Rempel, R.A. *Unionists divided. Arthur Balfour, Joseph Chamberlain, and the Unionist Free Traders* (Newton Abbot, 1972).

Reynolds Osborne. 'Memoir of James Prescott Joule', *Proc. Manchester Lit. Phil. Soc.*, **6** (1892), 1–206.

Rice, D.F. 'Natural theology and the Scottish philosophy in the thought of Thomas Chalmers', *Scot. J. Theol.*, **24** (1971), 23–46.

Richards, J.L. 'The art and science of British algebra: a study in the perception of mathematical truth', *Historia Mathematica*, **7** (1980), 343–65.

Richards, J.L. 'Projective geometry and mathematical progress in mid-Victorian Britain', *Stud. Hist. Phil. Sci.*, **17** (1986), 297–325.

Richards, J.L. 'Augustus de Morgan, the history of mathematics and the foundations of algebra', *Isis*, **78** (1987), 7–30.

Richards, John. 'Boole and Mill: differing perspectives on logical psychologism', *Hist. Phil. Logic*, **1** (1980), 19–36.

Ritchie, G.S. *Challenger. The life of a survey ship* (London, 1957).

Robbins, Michael. 'Lord Kelvin on electric railways', *J. Transport Hist.*, **3** (1958), 235–8.

Rothblatt, Sheldon. *The revolution of the dons. Cambridge and society in Victorian England* (Cambridge, 1968).

Rudwick, M.J.S. 'The glacial theory', *Hist. Sci.*, **8** (1969), 136–57.

Rudwick, M.J.S. 'Poulett Scrope on the volcanoes of Auvergne: Lyellian time and political economy', *Brit. J. Hist. Sci.*, **8** (1974), 205–42.

Rudwick, M.J.S. *The great Devonian controversy. The shaping of scientific knowledge among gentlemanly specialists* (Chicago, 1985).

Ruse, Michael. *The Darwinian revolution* (Chicago, 1979).

Russell, C.A., and Goodman, D.C. (eds.). *Science and the rise of technology since 1800* (Bristol, 1972).

Russell, Colin. *Time, chance and thermodynamics* (Milton Keynes, 1981).

Saunders, L.J. *Scottish democracy: 1800–1840* (Edinburgh, 1950).

Schaffer, Simon. 'The nebular hypothesis and the science of progress', in J.R. Moore (ed.), *The humanity of evolution* (Cambridge, forthcoming).

Schweber, S.S. (ed.) *Aspects of the life and thought of Sir John Frederick Herschel* (2 vols., New York, 1981).

Shapin, Steven, and Barnes, Barry. 'Head and hand; rhetorical resources in British pedagogical writing, 1770–1850', *Ox. Rev. Ed.*, **2** (1976), 231–54.

Sharlin, H.I. *Lord Kelvin. The dynamic Victorian* (Pennsylvania, 1979).

Shields, John. *Clyde built* (Glasgow, 1949).

Siegel, Daniel. 'The origin of the displacement current', *Hist. Stud. Phys. Sci.*, **17** (1986), 99–146.

Silliman, R.H. 'Fresnel and the emergence of physics as a discipline', *Hist. Stud. Phys. Sci.*, **4** (1974), 137–62.

Slaven, Anthony. *The development of the west of Scotland: 1750–1960* (London, 1975).

Smith, Crosbie. '"Mechanical Philosophy" and the emergence of physics in Britain: 1800–1850', *Ann. Sci.*, **33** (1976), 3–29.

Smith, Crosbie. 'Faraday as referee of Joule's Royal Society paper "On the mechanical equivalent of heat"', *Isis*, **67** (1976), 444–9.

Smith, Crosbie. 'Natural philosophy and thermodynamics: William Thomson and "the dynamical theory of heat"', *Brit. J. Hist. Sci.*, **9** (1976), 298–304.

Smith, Crosbie. 'William Thomson and the creation of thermodynamics: 1840–1855', *Arch. Hist. Exact Sci.*, **16** (1976), 231–88.

Smith, Crosbie. 'From design to dissolution: Thomas Chalmers' debt to John Robison', *Brit. J. Hist. Sci.*, **12** (1979), 59–70.

Smith, Crosbie. 'Geologists and mathematicians: the rise of physical geology', in P.M. Harman (ed.), *Wranglers and physicists. Studies on Cambridge mathematical physics in the nineteenth century* (Manchester, 1985), pp. 49–83.

Stephens, M.D., and Roderick, G.W. 'Science, the working class and mechanics' institutes', *Ann. Sci.*, **29** (1972), 353–59.

Strutt, R.J. *John William Strutt. Third Baron Rayleigh* (London, 1924).

Sviedrys, Romualdas. 'The rise of physics laboratories in Britain', *Hist. Stud. Phys. Sci.*, **7** (1976), 405–36.

Terrall, Mary. 'Maupertuis and eighteenth-century scientific culture', PhD dissertation, University of California, Los Angeles, 1987.

Thompson, D'Arcy W. 'Obituary of Joseph Larmor', in *Yearbook of the Royal Society of Edinburgh* (1941–2), pp. 11–13.

Thompson, S.P. *The life of William Thomson, Baron Kelvin of Largs* (2 vols., London, 1910).

Turner, Frank M. 'Lucretius among the Victorians', *Vict. Stud.*, **16** (1972–3), 329–48.

Westfall, R.S. *Force in Newton's physics: the science of dynamics in the seventeenth century* (London and New York, 1971).

White, Colin. *Victoria's navy. The heyday of steam* (Emsworth, 1983).

Wiener, Martin. *English culture and the decline of the industrial spirit, 1850–1980* (Cambridge, 1981).

Williams, L.P. *Michael Faraday* (London, 1965).

Wilson, David. *William Thomson. Lord Kelvin. His way of teaching natural philosophy* (Glasgow, 1910).

Wilson, D.B. 'Kelvin's scientific realism: the theological context', *Phil. J.*, **11** (1974), 41–60.

Wilson, D.B. 'The educational matrix: physics education at early-Victorian Cambridge, Edinburgh and Glasgow Universities', in P.M. Harman (ed.), *Wranglers and physicists. Studies on Cambridge mathematical physics in the nineteenth century* (Manchester, 1985), pp. 12–48.

Wilson, Ian. *Shipwrecks of the Ulster coast* (Coleraîne, 1979).

Wilson, Ian. 'The great gale of 1894', *Sea breezes*, **58** (1984), 41–53.

Winstanley, D.A. *Early Victorian Cambridge* (Cambridge, 1955).

Wise, M. Norton. 'William Thomson's mathematical route to energy conservation: a case study of the role of mathematics in concept formation', *Hist. Stud. Phys. Sci.*, **10** (1979), 49–83.

Wise, M. Norton. 'The flow analogy to electricity and magnetism – Part I: William Thomson's reformulation of action at a distance', *Arch. Hist. Exact Sci.*, **25** (1981), 19–70.

Wise, M. Norton. 'German concepts of force, energy, and the electromagnetic ether: 1845–1880', in G.N. Cantor and M.J.S. Hodge (eds.), *Conceptions of ether: studies in the history of ether theories, 1740–1900* (Cambridge, 1981), pp. 269–307.

Wise, M. Norton. 'The Maxwell literature and British dynamical theory', *Hist. Stud. Phys. Sci.*, **13** (1982), 175–205.

Wise, M. Norton. 'Mediating machines', *Science in Context*, **2** (1988), 81–117.

Wise, M. Norton and Smith, Crosbie. 'Measurement, work, and industry in Lord Kelvin's Britain', *Hist. Stud. Phys. Sci.*, **17** (1986), 147–73.

Wise, M. Norton and Smith, Crosbie. 'The practical imperative: Kelvin challenges the Maxwellians', in Robert Kargon and Peter Achinstein (eds.), *Kelvin's Baltimore Lectures and modern theoretical physics* (Cambridge, MA, and London, 1987), pp. 323–48.

Wise, M. Norton (with the collaboration of Crosbie Smith). 'Work and waste: political economy and natural philosophy in nineteenth-century Britain', *Hist. Sci.* (forthcoming).

Woodham-Smith, Cecil. *The great hunger. Ireland 1845–1849* (London, 1962).

Young, R.M. 'Malthus and the evolutionists: the common context of biological and social theory', *Past and Present*, no. 43 (1969), 109–45.

Young, R.M. *Darwin's metaphor: nature's place in Victorian culture* (Cambridge, 1985).

Young, Thomas. 'Life of Robison', in George Peacock (ed.), *Miscellaneous works of the late Thomas Young* (3 vols., London, 1855), **2**, pp. 505–17.

Index

Numbers in **bold** type indicate pages on which illustrations appear